REVIEWS in MINERALOGY
and GEOCHEMISTRY

Volume 90   2024

Exoplanets: Compositions, Mineralogy, Evolution

EDITORS

Natalie R. Hinkel
Louisiana State University, Baton Rouge, LA, USA

Keith Putirka
California State University, Fresno, CA, USA

Siyi Xu (许偲艺)
Gemini Observatory/NOIRLab, HI, USA,

Series Editor: **Ian Swainson**

MINERALOGICAL SOCIETY of AMERICA
GEOCHEMICAL SOCIETY

COVER ILLUSTRATION

There is a wealth of variation and diversity in planets outside of our Solar system, as illustrated in this artistic representation, offering an exciting opportunity to understand their stellar interactions, formation, interior processes, and atmosphere in terms of habitability and the future of holistic characterization. Credit: NASA/JPL-Caltech

Original link: https://exoplanets.nasa.gov/resources/2319/exoplanet-types-illustration/

Reviews in Mineralogy and Geochemistry, Volume 90
Exoplanets: Compositions, Mineralogy, Evolution

ISSN 1529-6466 (print)
ISSN 1943-2666 (online)
ISBN 978-1-946850-12-6
COPYRIGHT 2024
THE MINERALOGICAL SOCIETY OF AMERICA
3635 CONCORDE PARKWAY, SUITE 500
CHANTILLY, VIRGINIA, 20151-1125, U.S.A.
WWW.MINSOCAM.ORG

The appearance of the code at the bottom of the first page of each chapter in this volume indicates the copyright owner's consent that copies of the article can be made for personal use or internal use or for the personal use or internal use of specific clients, provided the original publication is cited. The consent is given on the condition, however, that the copier pay the stated per-copy fee through the Copyright Clearance Center, Inc. for copying beyond that permitted by Sections 107 or 108 of the U.S. Copyright Law. This consent does not extend to other types of copying for general distribution, for advertising or promotional purposes, for creating new collective works, or for resale. For permission to reprint entire articles in these cases and the like, consult the Administrator of the Mineralogical Society of America as to the royalty due to the Society.

Exoplanets: Compositions, Mineralogy, Evolution

90   *Reviews in Mineralogy and Geochemistry*   **90**

DEDICATION

An Yin, of UCLA, was invited to write a chapter on planetary tectonics. Dr. Yin unfortunately passed away during a field trip to the eastern Sierra Nevada just months before chapters were due—a devastating loss for the scientific community and an immeasurable loss to our RiMG workshop and volume. An was invited to contribute to our volume because his approach to understanding tectonic processes was informed by decades of near peerless field studies of some of Earth's most perplexing structures, including the growth of the Tibetan Plateau, and before that, the nature of low-angle normal faults of the Basin and Range province (see Chapter 9 for references). These field studies were matched by a remarkably complete understanding of experimental and theoretical rock mechanics. These field and theoretical faculties combined in An to yield a view of tectonics unique from contributors who are rooted more heavily in the theory of convection, heat flow and deep mantle processes, with or without the nuances of computer simulations. This is not at all to say that the latter are of lesser value—they assuredly are not. But our hope was, if ever so slightly, to rebalance the scale with regard to the kinds of observations that are utilized to speculate on a planet's tectonic behavior. An's group has developed some fascinating insights and their work at Enceladus provides an enlightened example for testing our sundry models of tectonic processes. An's colleague, Mark Harrison, has an excellent discussion of An's work, which can be found here: https://epss.ucla.edu/news/in-memoriam-professor-an-yin/. For a sampling of what we have lost, An's outline follows:

1. Introduction: fundamental concepts and methods in tectonic studies; significance of planetary tectonic studies for understanding the solar-system evolution (resurfacing and recycling of thermal boundary layers). Importance of Earth-based research when placed in a sound physical contexts (e.g., Lithosphere/thermal boundary layer; tectonics; plate tectonics; plate-boundary processes; discrete vs. distributed deformation; oceanic vs. continental deformation).

2. Tectonics of rocky planets and rocky moons
a. Earth (early history, primitive plate tectonics, modern plate tectonics)
b. Mercury
c. Venus
d. Mars
e. Earth's moon
f. Jupiter's moon Io

3. Tectonics of icy satellites and dwarf planets
a. Europa
b. Enceladus
c. Titan
d. Miranda
e. Charon
f. Pluto

4. Discussion
a. Kinematic classification of tectonic deformation in the Solar System
b. Dynamic regimes of tectonic deformation in the Solar System
c. Scaling rheology, geometry, and gravity.

It is no understatement that An's passing is a devastating loss—a setback in planetary tectonics that may be impossible to quantify but is undoubtedly immense. Nevertheless, while there is no easy replacement for An's combination of field and theoretical knowledge, this loss may inspire others to take up the challenge of integrating field and theory, and using observations of our planetary neighbors to validate numerical and theoretical models. There could be no better tribute to An's memory than for a reader of this volume to take on this immense but invaluable task.

Keith D. Putirka (California State University, Fresno, CA)

PREFACE

We initiated this volume with the conviction that a better union of geologic insights with astronomical ideas and data would provide the clearest prospect for increasing our understanding of exoplanets and their evolution. We realized that close communication is key. Geologists, for example, can help refine testable hypotheses that would usefully influence the astronomical targets and measurements being made by current and upcoming telescopes. The focus for this RiMG volume is on rocky exoplanets because the search for truly Earth-like planets is of special interest. Our goal is to motivate communication between the disciplines so as to make the best use possible of existing data and data yet to be collected by the James Webb and the Nancy Grace Roman Space Telescopes, since the astronomy community is gathering data on stars and exoplanets at an accelerating rate. Such data now include exoplanet size and mass (i.e., density) as well as their atmospheric compositions, which are collectively telltale of mineralogy and evolution. Much of what is published may still fall in the realm of educated speculation, but our conjectures are metamorphosing into testable hypotheses. Another intriguing turn in the Astro-Geology connection is how the telescope must sometimes be turned back towards Earth, as our speculation on the exoplanets necessarily contends with incomplete knowledge of the origin and evolution of Earth and its planetary neighbors. We hope to share expertise that will our understanding of exoplanets to develop in pace with data collection.

We begin with chapters on exoplanet host stars (Hinkel et al.), and the compositions of proto-planetary disks (Zhang et al.). The properties and composition of host stars have been crucial to examining the range of possible exoplanet compositions that might exist in our part of the Milky Way; the following chapter on protoplanetary disks gives further constraints on the compositions of the materials from which rocky planets ultimately precipitate. This begins a conversation about how and why planets can differ in composition from the stars they orbit. The next chapters also have important implications for exoplanet compositions as they cover the physics of planetary formation (Mordasini and Burn), where location and timing of planetary accretion can affect the final product, and meteorites formed in our Solar system (Jones), which provide the clearest record we have of the earliest of planetary accretion processes. The next two chapters are on so-called "polluted white dwarfs"—their compositions (Xu et al.) and the physics of how white dwarf atmospheres become polluted (Veras et al.). The "polluted" varieties of white dwarfs normally have their otherwise pure H or He atmospheres polluted by the ingress of rocky planetary materials and so these very special stars may give us our most accurate estimates of exoplanet bulk compositions.

Our next three chapters use Earth and terrestrial processes as analogs to understand exoplanet mineralogy (Putirka), bulk rock type and fluid compositions (Guimond et al.) and tectonic processes (Putirka). For a planet to be Earth-like an as yet unknown range of minerals and rock types are needed so as to either allow or facilitate plate tectonics. Exoplanet mineralogy and rock types may still be uncertain, but these chapters explore some avenues of speculation. The next chapter reviews the cycling of "life-essential volatile elements" or LEVE (Dasgupta et al.) for rocky object within the inner Solar system, and then uses such, again as analogs, to interpret various observations of exoplanets. This review of LEVE is

1529-6466/24/0090-0000$00.00 (print)
1943-2666/24/0090-0000$00.00 (online)
http://dx.doi.org/10.2138/rmg.2024.90.00

followed by a chapter on planetary magnetic fields (Brain et al.), including how such fields are generated, and how variations in magnetic properties can influence planetary evolution.

Of special interest is whether any exoplanet might harbor life and our next three chapters approach the issue. We start with measurements of the compositions of exoplanetary atmospheres (Kempton et al.), and the possible biosignatures (Schwieterman and Leung) that such atmospheres might contain. These chapters are followed by an Earth-based perspective on biogeochemistry, especially focused on the early Earth, which at present provides our only testing ground for how life is initiated. We then end the volume with a summary of future directions in exoplanet studies (Foley) so as to maximize the rate at which this new scientific discipline might evolve and advance.

There is now a remarkably large amount of astronomical data (with even more on the way) that geochemists and petrologists can make much use of. But just as astronomers may benefit from geologic insights, geologists need our colleagues in astronomy to help interpret their data and their underlying implications to better understand its astronomical context. Our hopes for this volume will be fulfilled if readers initiate their own analyses of what at present may seem like novel or unusual data, and if new collaborations between academic departments and subfields are forged.

Natalie R. Hinkel (Louisiana State University, Baton Rouge, LA)

Keith D. Putirka (California State University, Fresno, CA)

Siyi Xu (许偲艺) (Gemini Observatory/NOIRLab, HI)

Exoplanets: Compositions, Mineralogy, Evolution

90   *Reviews in Mineralogy and Geochemistry*   **90**

TABLE OF CONTENTS

1   **Host Stars and How Their Compositions Influence Exoplanets**

Natalie R. Hinkel, Allison Youngblood, Melinda Soares-Furtado

2   **Chemistry in Protoplanetary Disks**

Ke Zhang

3   Planet Formation—Observational Constraints, Physical Processes, and Compositional Patterns

Christoph Mordasini, Remo Burn

4   Meteorites and Planet Formation

Rhian H. Jones

5   The Evolution and Delivery of
Rocky Extra-Solar Materials to White Dwarfs

Dimitri Veras, Alexander J. Mustill, Amy Bonsor

6   **The Chemistry of Extra-solar Materials**
from White Dwarf Planetary Systems

Siyi Xu (许偲艺), Laura K. Rogers, Simon Blouin

7   **Exoplanet Mineralogy**

Keith D. Putirka

8   From Stars to Diverse Mantles, Melts, Crusts, and Atmospheres of Rocky Exoplanets

Claire Marie Guimond, Haiyang Wang, Fabian Seidler,
Paolo Sossi, Aprajit Mahajan, Oliver Shorttle

9   Some Tectonic Concepts Relevant to the Study of Rocky Exoplanets

Keith D. Putirka

10   A Framework for the Origin and Deep Cycles of Volatiles in Rocky Exoplanets

Rajdeep Dasgupta, Debjeet Pathak, Maxime Maurice

13   An Overview of Exoplanet Biosignatures

Edward W. Schwieterman, Michaela Leung

14   The Early Earth as an Analogue for Exoplanetary Biogeochemistry

Eva E. Stüeken, Stephanie L. Olson, Eli Moore, Bradford J. Foley

15   Exoplanet Geology: What Can We Learn from Current and Future Observations?

Bradford J. Foley

Reviews in Mineralogy & Geochemistry
Vol. 90 pp. 1–26, 2024
Copyright © Mineralogical Society of America

1

Host Stars and How Their Compositions Influence Exoplanets

Natalie R. Hinkel,[1,2] Allison Youngblood,[3] Melinda Soares-Furtado[4,5,*]

*[1]Physics and Astronomy Department, Louisiana State University
Baton Rouge, LA 70803, USA*
[2]Southwest Research Institute, 6220 Culebra Rd, San Antonio, TX 78238, USA
*[3]Exoplanets and Stellar Astrophysics Laboratory, NASA Goddard Space Flight Center
Greenbelt, MD 20771, USA*
*[4]Department of Astronomy, University of Wisconsin-Madison, 475 N. Charter St.
Madison, WI 53703, USA*
*[5]Department of Physics and Kavli Institute for Astrophysics and Space Research
Massachusetts Institute of Technology, Cambridge, MA 02139, USA*

Corresponding author: Natalie R. Hinkel: natalie.hinkel@gmail.com

[*]NASA Hubble Postdoctoral Fellow

INTRODUCTION

Though distant and seemingly unreachable, planets outside the Solar System, or exoplanets, have captivated the imagination of scientists and stargazers alike. With more than 5,000 confirmed exoplanet detections to date, it has become apparent that the Solar System—with multiple small, rocky planets interior to the larger gaseous planets—is not the only possible architecture for planetary systems. For example, some systems have "hot-Jupiters" where Jupiter-sized planets orbit very close to their host star (at distances comparable to the Sun–Mercury separation, e.g., Dawson and Johnson 2018). There are also planets that orbit two stars at the same time—much like Luke Skywalker's home planet Tatootine (e.g., Bromley and Kenyon 2015). Planets have been detected that are unlike anything in the Solar System, with sizes that are in between Earth and Neptune (known as super-Earths), meaning it is unclear whether they are giant rocky planets or small gaseous planets. In addition, because not all stars are like our yellow Sun, some planets orbit much redder stars (e.g., Dressing and Charbonneau 2013) or stars that barely shine at all because they are the remaining cores leftover from a stellar explosion (Wolszczan and Frail 1992). These diverse worlds serve as invaluable test cases to investigate their origin and the possibilities for planetary habitability.

Stars are the building blocks of the Universe: they are the fundamental components of galaxies and the foundation of planetary systems. The size (mass and radius) of a star dictates its most basic properties such as temperature, evolution, and surface processes (like flares). All of these properties have a significant impact on an orbiting planet (Meadows and Barnes 2018). The composition of the star—the raw basic elements (like those found in the Periodic Table of the Elements)—is linked to prior generations of stars that have seeded the cosmos with elements heavier than hydrogen and helium through their various nucleosynthesis[1] processes (e.g., Burbidge et al. 1957; Nomoto et al. 2013). Stars and their planets form at the same time and are made up of the same elements, gas, and dust. This chemical link between stars and planets is vital to determining the interior composition of small rocky planets. A planet's interior composition

---

[1] Nucleosynthesis refers to the creation, or synthesis, of atomic nuclei.

1529-6466/24/0090-0001$05.00 (print)
1943-2666/24/0090-0001$05.00 (online)

http://dx.doi.org/10.2138/rmg.2024.90.01

dictates whether properties that are essential to habitability are present, such as a clement climate, tectonics, and the presence of magnetic fields (Foley and Driscoll 2016). Unfortunately, these properties often cannot be directly measured with current technological capabilities. Therefore, the composition of the host star is used as a proxy for the make-up of the planet's interior.

It has become a common practice within the exoplanet field to say that "to know the star is to know the planet." The properties of the host star have a strong, direct influence on the interior and surface conditions of the orbiting planet and oftentimes measurements of planetary properties are made relative to the star's properties. Not only are observational measurements of the star necessary to determine even the most basic aspects of the planet (such as mass and radius), but the stellar environment influences how the planet evolves. Therefore, in this chapter, we begin by discussing the basics of stars, providing an overview of stellar formation, structure, photon and particle emissions, and evolution. Next, we go over the possible ways to determine the age of a star. We then outline how different kinds of stars are distributed within the Milky Way galaxy. Afterwards, we explain how to measure the composition of stars and the underlying math inherent to those observations, including caveats that are important when using the data for research applications. Finally, we explain the underlying physics and observations that enable stellar composition to be used as a proxy for planetary composition. In addition, given that this chapter focuses more on astronomy/astrophysics and uses a variety of important terms that may not be familiar to all readers, we have defined many terms either within the text or as a footnote for better interdisciplinary comprehension.

BASICS OF STARS

Stars are bright, shining balls of gas that are balanced under hydrostatic equilibrium, meaning the outward pressure created from their internal nuclear fusion is balanced by the inward pull of gravity. Note, however, that there are multiple phases throughout a star's lifetime where they are not fusing hydrogen into helium (see the *Evolution* section) but are still called stars. Astronomers use a variety of parameters and classification schemes to describe stars, including spectral type, luminosity class, and the magnitude system[2]. Briefly, stars are grouped into spectral types (Capital letters: O, B, A, F, G, K, M) and luminosity classes (Roman numerals: I, II, III, IV, V, VI) based on the appearance of their optical or infrared spectra[3] (see the *Spectroscopy and stellar abundances* section). O-type stars are the largest, most massive, hottest, and shortest-lived stars, while M-type stars are the smallest, least massive, coolest, and longest-lived stars. The luminosity classes correspond to different evolutionary stages of a star. For instance, "V" corresponds to dwarf stars like our Sun that are in the longest-duration phase of their nuclear-burning lifetimes, akin to adulthood. Evolved stars are often called giants and have smaller Roman numerals. The Sun is a G V type dwarf star and the vast majority of known exoplanet host stars are F, G, K, and M-type stars of luminosity class V. In this section, we briefly describe the formation, structure, emissions and winds, and evolution of stars.

Formation

Stars are born in star-forming regions, where high-density molecular clouds[4] fragment into smaller regions of even higher density, ultimately forming many protostellar cores (e.g., Shu et al. 1987; McKee and Ostriker 2007; Luhman 2012). As these protostellar cores collapse

---

[2] The magnitude system is a logarithmic measurement of stellar brightness where larger numbers correspond to fainter stars. Apparent magnitude describes how bright an object appears to us here on Earth and depends on the star's absolute magnitude (related to its intrinsic brightness), distance from the Earth, and the amount of interstellar dust between Earth and the object.

[3] A spectrum is a way to visualize a star's brightness as a function of wavelength and contains a wealth of information about a star, including its gravity, temperature, and elemental composition.

[4] Molecular clouds are cold, dense regions of gas and dust that provide the raw materials for star formation.

under their own gravity, the conservation of angular momentum (or spin) leads to an increase in rotation rate and a protostellar disk forms with a protostar at the center. Gas from the protostellar nebula[5] is accreted[6] onto the protostar via the protostellar disk; however, the amount of material that can be accreted is oftentimes limited by powerful outflows from the protostar that emerge under conservation of angular momentum (e.g., Bally 2016; Hartmann et al. 2016).

As the protostar accretes more and more material, its internal temperature and pressure build until it reaches ~10 million K, which is the approximate threshold for the nuclear fusion process that turns hydrogen into helium (see, e.g., Burbidge et al. 1957). The protostar is now a fully formed star, termed a main sequence star for its position on the Hertzsprung–Russell (HR) diagram, which is a scatter plot of stellar luminosity and effective temperature. Figure 1 shows a color–magnitude diagram (CMD), which is similar to an observational HR diagram as color[7] is an observational proxy for effective temperature and absolute magnitude is directly related to luminosity. The timescale for stars to reach the main sequence ranges from less than 1 million

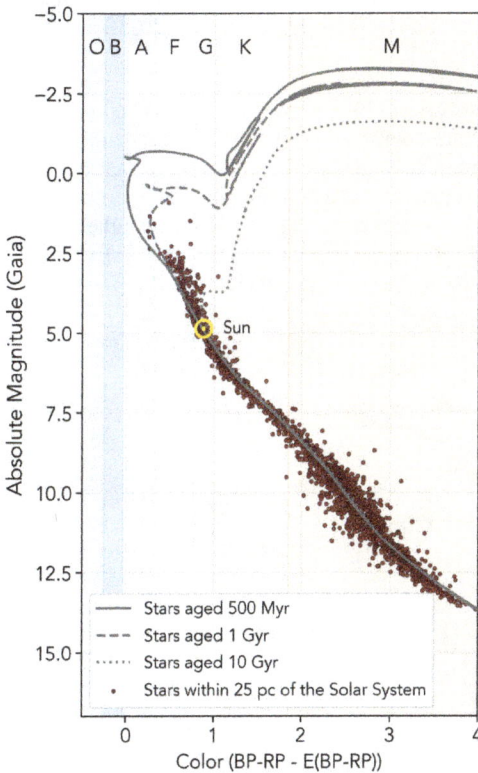

Figure 1. Color–magnitude diagram of stars within 25 pc (1 parsec = 3.086×10^{13} km) of our Solar System (Gaia Collaboration et al. 2016). The variable "BP-RP-E(BP-RP)" refers to the reddening-corrected color of a star, as measured by the Gaia space observatory (Gaia Collaboration et al. 2021). The location of the Sun is shown as a **gold circle with a central dot** (the symbol for the Sun). Like the Sun, the vast majority of stars in the solar neighborhood are on the main sequence. Also shown are the theoretically-determined locations of stars in a stellar system that is aged 500 Myr (**solid grey line**), 1 Gyr (**dashed grey line**), and 10 Gyr (**dotted grey line**) (Bressan et al. 2012). The optical color of each of the spectral types is illustrated with visually representative colors (Harre and Heller 2021).

---

[5] A nebula is a region of densely clumped gas and dust often present before and after star formation.

[6] Accretion is the accumulation of matter, often under the influence of gravity, leading to growth of an object.

[7] An object's color is defined as the difference in apparent magnitude between two distinct bandpasses.

years for the most massive stars to 1 billion years for the least massive stars. Note that young stars can deplete the gas in their protoplanetary disks (via accretion, winds, radiation pressure, and/or by planet formation) within only a few million years, while the planet formation process is thought to be complete within of order 1–10 million years (see, e.g., Lissauer 1993). The lifetime of protoplanetary disks may depend on stellar type, with more massive stars exhibiting shorter disk lifetimes (Ribas et al. 2015). Note also that more massive stars likely have more massive disks and therefore a greater reservoir of raw materials for planet formation (Andrews et al. 2013). This may explain why gas giants are rare around M-dwarfs, the lowest mass stars (e.g., Laughlin et al. 2004; Pass et al. 2023). However, lower mass stars are more likely to have at least one planet than higher mass stars (e.g., Mulders et al. 2015; Yang et al. 2020). Host star metallicity (or the composition of those elements heavier than H or He) also plays a role in the planet formation process, because stars with higher metallicity are more likely to have giant planets, presumably because their protoplanetary disks have more condensable solids (e.g., Gonzalez 1997; Fischer and Valenti 2005; Mulders et al. 2016). For an overview of the connection between planet formation theory and observed exoplanet demographic trends, see Gaudi et al. (e.g., 2021), Rodríguez Martínez et al. (e.g., 2023), and references within.

Structure

Stars are roughly spherical objects comprised of ionized gas and are bound together by their own gravity. They are composed mostly of hydrogen and helium (99.9% by number or 98% by mass for the Sun), with a small but important percentage of metals. Astronomers refer to most elements heavier than helium as a metal because, aside from trace amounts of lithium and beryllium, they were not produced during the Big Bang but rather arise from nucleosynthesis in stellar cores, supernovae (powerful explosion of massive stars), or other processes (Burbidge et al. 1957; Nomoto et al. 2013; Arcones and Thielemann 2023; also see Fig. 2).

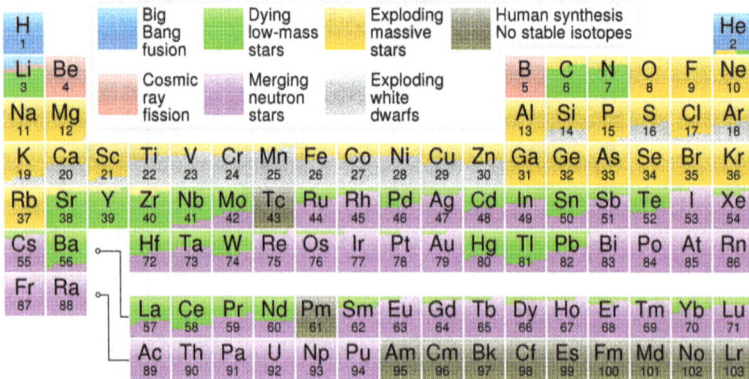

Figure 2. Periodic Table of the Elements color-coded to depict the primary formation channels that give rise to a given element. Note that this is a first-order simplification of a research field with many open questions, as well as a rough approximation of the data collected by Jennifer A. Johnson. While beryllium (Be) was produced in trace amounts during the Big Bang, the vast majority of Be is produced via cosmic ray fission (the splitting of an atomic nucleus by high-energy cosmic rays). Image credit: Cmglee, used under CC BY-SA 3.0 via Wikimedia Commons (unmodified). Data from Jennifer Johnson, OSU.

Stars inherit the heavy elemental composition (or metallicity) and angular momentum properties of their star-forming region (although pockets of inhomogeneous or stochastic processes can alter these stellar properties). The structure of stars are separated into interiors and atmosphere, with a thin "surface" layer called a photosphere dividing the two (Fig. 3). A star's radius is defined from the core to the photosphere because it is the layer of the star that can be "seen" in visible light because photons can freely travel into space from the photosphere.

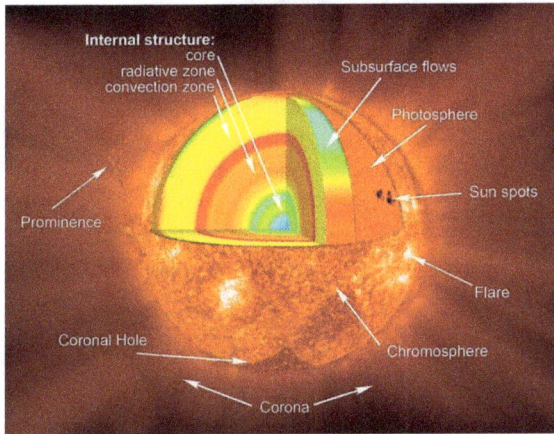

Figure 3. Schematic diagram of the Sun showing the interior and atmosphere. Image Credit: NASA.

At the center of a star's interior is its core, where temperatures are ~10 million K or more and hydrogen fuses into helium; fusion is the star's energy source. Nuclear fusion reactions are highly dependent on temperature and vary depending on the stellar type (e.g., Nomoto et al. 2013; Choi et al. 2016). More massive stars with hotter cores quickly burn through their vast hydrogen supplies. The smallest stars (M-dwarfs) have a total mass between approximately 0.08 and $0.50\,M_\odot$ (in units of solar masses) but will take longer than the current age of the Universe to burn through their hydrogen supply and evolve past the main sequence. In other words, no M-dwarf has died! On the other hand, a B-type star with $10\,M_\odot$ will burn through its hydrogen in only about 20 Myr, which is very fast, astronomically-speaking. For reference, Table 1 lists the properties corresponding to the various spectral type classes[8] (effective temperatures, masses, main sequence lifetimes), as well as the fraction of stars in the Universe represented by a given spectral type, and the eventual end product after the star has ceased all nuclear-burning processes (e.g., Heger et al. 2003).

Exterior to the stellar core is an envelope that carries the energy released from nuclear fusion to the photosphere via radiation (photons), convection (bulk motion of matter or plasma where material from deeper layers are moved vertically upwards into cooler layers and vice versa), or a mix of both (i.e., a radiation layer and/or a convection layer). The exact structure depends on the stellar mass (see Chabrier and Baraffe 1997; Paxton et al. 2011, 2018). The stellar temperature and density decline from the core to the photosphere. The temperature of the photosphere is given a special term called the effective temperature and is calculated through the Stefan-Boltzmann law, which describes the relationship between a star's luminosity (L; units: erg·s^{-1}), radius (R; units: cm), and effective temperature (T_{eff}; units: K): $L = \sigma_{\text{SB}} T_{\text{eff}}^4 4\pi R^2$, where σ_{SB} is the Stefan–Boltzmann constant ($\sigma_{\text{SB}} = 5.67 \times 10^{-5}\,\text{erg·cm}^{-2}\cdot\text{s}^{-1}\cdot\text{K}^{-4}$). Between M-type and O-type stars, effective temperatures range from approximately 2000 K to > 30,000 K (Table 1), respectively. For reference, the Sun's effective temperature is ~5780 K.

For the remainder of this section, we focus on the structure of the F, G, K, and M-type stars that comprise most of the known exoplanet host population. Above the photosphere is the stellar atmosphere, where the temperature continues to decline to a minimum before, surprisingly, rising again. The upper layers that comprise the stellar atmosphere in order of increasing altitude are called the chromosphere, transition region, and corona. The chromosphere contains the temperature minimum (Linsky 1980, 2017), and the corona contains the temperature maximum

---

[8] https://www.pas.rochester.edu/~emamajek/EEM_dwarf_UBVIJHK_colors_Teff.txt

Table 1. Properties corresponding to main sequence (MS) stars of a given spectral type (Pecaut and Mamajek 2013; Choi et al. 2016). Note that the most massive stars are thought to be limited to 150–300 $M_\odot$ (Figer 2005; Crowther et al. 2010) based primarily on observations of the most massive known stellar clusters, and the least massive stars are limited to ~0.08 $M_\odot$ (Chabrier et al. 2000) because of the core temperature threshold requirement for fusion of hydrogen into helium. Objects below 0.08 $M_\odot$ are called brown dwarfs, and their energy source is the fusion of deuterium into helium. Brown dwarfs are formed slightly less frequently than M-dwarfs, still making them extremely numerous throughout the galaxy.

Spectral type	Effective Temp. (K)	Mass ($M_\odot$)	Main Sequence (MS) Lifetime	Fraction of MS stars	End Product
O	>31,500 K	> 18	< 10 Myr	0.00003%	neutron star or black hole
B	10,000–31,500	2.7–18	10–500 Myr	0.13%	neutron star, black hole, or white dwarf
A	7,500–10,000	1.7–2.7	500 Myr–2 Gyr	0.6%	white dwarf
F	6,000–7,500	1.1–1.7	2–7 Gyr	3%	white dwarf
G	5,300–6,000	0.9–1.1	7–15 Gyr	7.6%	white dwarf
K	3,900–5,300	0.6–0.9	15–60 Gyr	12%	white dwarf
M	2,300–3,700	0.08–0.6	> 60 Gyr	76.5%	white dwarf

(Güdel and Nazé 2009); the transition region is aptly named as it is a narrow range of altitudes in the atmosphere where temperatures transition from chromospheric temperatures (~10,000 K) to coronal (~1–10 million K). Note that stars more massive than ~1.8 $M_\odot$ (O, B, and some A types) are thought not to have chromospheres or coronae. The large temperatures of the upper atmosphere, despite the low density and great distance from the star's energy source, cannot be explained by radiative or convective heat transport from the core, but appear to be related to magnetism. F, G, K, and M-type stars generate large-scale magnetic fields; the magnetic field is generated in the hot, rotating stellar core, and the convective layers in the envelope transport and magnify these fields to the surface (Ossendrijver 2003). The energy carried by these magnetic fields is dissipated in the chromosphere through the corona; the exact nature of this dissipation and energy propagation is an active area of research (e.g., Zweibel and Yamada 2009).

Photon and particle emissions

The vast majority of a star's light emanates from its hot photosphere, which emits roughly as a blackbody (i.e., according to Planck's radiation law that defines the behavior of an object that absorbs all radiation and re-radiates that energy) with different atomic and molecular absorption features[9] (see the *Spectroscopy and stellar abundances* section) superimposed. Broadly speaking, the stellar metallicity does not dictate the most prominent absorption features, but rather the temperature and density do by controlling the dominant ionization states[10] or molecular species of the gas (see the *Stellar abundance caveats* section). As described by Wien's displacement law, a star's photospheric emission peaks in different parts of the electromagnetic spectrum depending on the stellar effective temperature, with the

---

[9] Absorption features show up as dark lines or dips in the spectrum when viewed using a spectrograph, which is a tool used to separate light into its different wavelengths.

[10] The process of gaining or losing electrons is known as ionization. There are different levels of ionization (ionization states) that an atom can undergo, resulting in the formation of ions with varying numbers of electrons.

hottest stars emitting most of their photospheric radiation in the ultraviolet and the coolest stars emitting most of their energy in the infrared (see Fig. 1 for the colors of different spectral types as they would appear to the human eye). Cooler stellar photospheres are not warm enough to emit light in the ultraviolet, and no photospheres are hot enough to generate X-ray emission. Yet, many stars are UV and X-ray bright thanks to their magnetically-heated upper atmosphere (the chromosphere, transition region, and corona). Emission lines from atoms and ions tend to dominate the energy output in the UV and X-ray (Linsky 2017; Güdel and Nazé 2009). Sun-like stars (astronomers often refer to F, G, and K-type dwarfs as "Sun-like") emit ~10% of their total luminosity in the X-ray and UV, and M-dwarfs emit ~1% in the X-ray and UV. Although these emissions are small compared to the stellar luminosity, these UV and X-ray photons have a significant impact on the atmospheres and surface conditions of orbiting planets.

Stars also emit charged particles via winds and explosive eruptions. Massive stars' winds are driven by their intense radiation pressure (Castor et al. 1975), while low-mass stars like the Sun and M-dwarfs have hot, low-density outer layers (the corona) that expand, creating an outflow (Parker 1965). These stellar winds are streams of charged particles flowing away from the star through interplanetary space. Low-mass stars like the Sun also experience eruptions called coronal mass ejections, which are pockets of coronal material (ions) with their own magnetic fields that stream out into interplanetary space where they can interact with planets and their magnetic fields (Green et al. 2018). For an overview of winds and CMEs from exoplanet host stars, see Wood et al. (2021) and references therein.

Evolution

Stars are not static, homogeneous objects. They exhibit brightness variations over all time scales and have bright and dark regions spotting their surfaces (e.g., Penza et al. 2022). With timescales that range from seconds-to-hours, they exhibit brightness variations due to stellar flares (Benz and Güdel 2010) and acoustic standing waves excited by convection just beneath the stellar photosphere (see the *Age-dating stars* section and Kurtz 2022 for an overview of asteroseismology[11]). On days-to-months timescales, their brightness variations are due to the emergence, fading, or rotation (in and out of the line of sight of the observer) of starspots (dark spots) or faculae (bright spots). On multi-year timescales, stellar luminosity varies due to a more systematic change in the emergence of spots and faculae. For example, the Sun exhibits an 11-year magnetic activity cycle (Hathaway 2015); some active K and M-dwarfs have been found to have much shorter cycles, on the order of a few years (e.g., Boro Saikia et al. 2018).

Over the course of a star's long main-sequence lifetime (see Table 1 for typical values), stars will continue to rotate and fuse hydrogen into helium in their cores. As more of the material in the stellar core is converted into helium, fusion rates increase and stars become brighter over their lifetimes. For example, the Sun is 30% brighter today than it was 4.6 Gyr ago (Güdel 2007). Stellar winds can remove significant angular momentum thanks to the coupling between the star's magnetic field and the ionized material of the stellar wind and coronal mass ejections. The interaction of these magnetic fields generates a torque that slows down the rotation of the star over time; this well studied phenomenon is known as magnetic braking (Skumanich 1972). Eventually, when no more hydrogen remains for fusion, the star leaves the main sequence phase of its life and enters into stages marked by relatively rapid changes, which are highly dependent on the star's mass. In short, after passing through a red giant phase, stars that are less massive than ~8 $M_\odot$ will become compact stellar remnants, known as white dwarfs. Stars more massive than this approximate threshold will explode as supernovae and become neutron stars[12] or black holes. Note that white dwarfs in close binaries

---

[11] Asteroseismology is the study of the internal structure and properties of stars through the observation and analysis of their natural oscillations or vibrations.

[12] A neutron star is the dense remaining core leftover from a stellar supernova, see Table 1.

with red giants can accrete mass directly from the giant, exceeding a stability threshold and resulting in a supernova. Nucleosynthesis during these late stages of a star's life are important sources of many metals in the Universe. Remarkably, the first exoplanet system ever discovered was found orbiting a type of neutron star known as a pulsar (Wolszczan and Frail 1992). Exoplanets have also been observed around other late-stage hosts, such as red giants (Huber et al. 2013) and white dwarfs (Vanderburg et al. 2020). However, these exotic worlds represent a small minority of the known exoplanet population. There are active searches to detect more planets orbiting these late-stage hosts in hopes of building census demographics to help constrain theories of their formation and evolution.

AGE-DATING STARS

Determining both an accurate and precise age of a given star or stellar group is of critical significance in astronomy. These measurements have far-reaching implications for a wide range of phenomena in the Universe, setting constraints on the formation and evolution of planets, stars, clusters, and galaxies. There are a number of age-dating techniques available to the modern-day astronomer. These age estimates are generally determined using model-dependent or empirical methods; however, not all techniques are applicable to a given star. For an in-depth review of stellar age-dating, we refer the interested reader to Soderblom (2010, 2015).

Stars in controlled environments, such as stellar associations, co-moving groups, and clusters are among the most well-suited to age-dating analyses. Since these stars form together from a shared molecular cloud, they share a roughly common metallicity and are coeval (or of the same age, give or take a few million years). Stellar evolutionary models can be leveraged to age-date stars in such environments. This method relies on fitting stellar parameters within a CMD using model isochrones[13] (see Fig. 1 for tracks corresponding to three distinct isochronal ages; Soderblom 2010). This method is most applicable to coeval stellar populations of a known metallicity that exhibit a clearly discernible main-sequence turnoff[14] point. Knowing the metallicity of these environments is critical to accurate age-dating estimates using stellar evolutionary models, as a star's location on the CMD is metallicity-dependent. More specifically, stars that are metal-poor tend to appear brighter and bluer than their metal-rich counterparts, as light emanates from their photospheres more efficiently (due to reduced absorption and scattering).

CMD-determined age estimates are not well-suited for age-dating the most common star within our Milky Way galaxy: low-mass (K- and M-type stars) main sequence stars. This is because low-mass stars have long main-sequence lifetimes where their CMD positions do not significantly change over billions of years, making it hard to determine their specific age. So the CMD-position of many K- and M-type stars are consistent with stars (of a common mass) that span an age range of several billion years. As shown in Figure 1, there is an indistinguishable overlap in the CMD positions of low-mass main sequence stars corresponding to systems of considerably different ages (500 Myr, 1 Gyr, and 10 Gyr).

While a star's CMD position does not change appreciably when it's on the main sequence, its rotation rate slows in a fairly predictable way, resulting in a gyrochronological[15] relation (color–rotation sequence) that provides a useful means of age-dating young ($\lesssim 1$ Gyr) low-mass (F, G, K-type) stars (Barnes 2003). This age-dating technique, which was first empirically identified by Skumanich (1972), leverages the relationship between a main sequence star's

---

[13] Isochrones illustrate the CMD position of stars of a particular age, born from the same initial composition of elements.

[14] The main sequence turnoff point is a key feature of a CMD, denoting the location where stars in a coeval stellar population begin to evolve away from the main sequence phase.

[15] Gyrochronology literally translates from Greek to mean "rotation age study."

rotation period, its color (a useful proxy for mass, as shown in Fig. 1), and its age (the slope of the color–rotation relation changes with time). The gyrochronological relation corresponding to three clusters of differing ages is illustrated in Figure 4. Note, how the bluer, massive stars more readily converge onto a color–rotation relation—a result of more efficient angular momentum loss. The stellar rotation rates are often inferred from the photometric variability observed as starspots pass in and out of the line of sight of the observer, tracing out a regular rotation sequence. Such starspots are only present on stars with convective envelopes and reasonably rapid rotation rates (e.g., Berdyugina 2005). Therefore, this method is constrained to early-to-mid main sequence stars that have not lost considerable angular momentum ($\lesssim 1\,\text{Gyr}$). While rotation age-dating is the most precise tool available among low-mass stars (K and M-type), there are also challenges in using this tool to age-date young ($\lesssim 100\,\text{Myr}$), low-mass stars, as these stars take more time to shed angular momentum (via magnetic braking) and converge onto a well-behaved gyrochronological sequence (see Fig. 4).

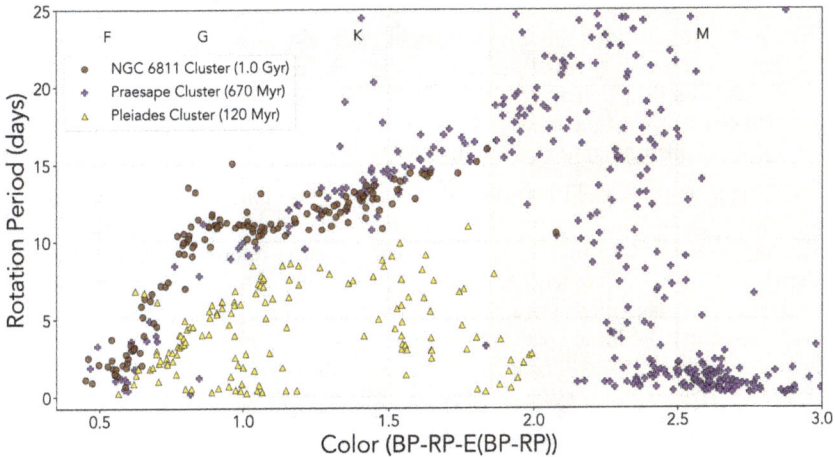

Figure 4. The gyrochronological rotation sequence of three known stellar clusters of differing ages: NGC 6811: 1 Gyr (Curtis et al. 2019), Praesape: 670 Myr (Douglas et al. 2017), and the Pleiades: 120 Myr (Rebull et al. 2016). Clusters older than a few hundred Myr trace out a tighter rotation–color relation, as they have shed sufficient angular momentum for this pattern to emerge. Stars more massive than those depicted here generally do not exhibit the starspots responsible for photometric rotation measurements. Low-mass M-type stars take many hundreds of Myr to shed angular momentum via magnetic braking and, therefore, do not trace out the rotation sequence. The optical color of each of the spectral types is illustrated with visually representative hexadecimal colors (Harre and Heller 2021).

Another useful age-dating technique relies on asteroseismic stellar data. More specifically, this method relies on variations in stellar brightness generated by oscillations on the surface of a star. The frequency spectrum of such photometric oscillations can be used to determine the mass, size, and internal properties (such as the depth of a convective zone) of a given target (Aerts et al. 2010). A star's observed oscillation frequencies are then compared to predicted frequencies (which vary with age-dependent changes in the star's size and internal structure), providing age estimates with a relatively high degree of accuracy (e.g., Soderblom 2010). Accurate asteroseismic age-dating requires high-quality observational data of a star's oscillation frequencies, which is generally not available for all stars of interest.

A star's spectroscopic element abundance signature can also offer a useful age indicator (see the *Spectroscopy and stellar abundances* section). More specifically, a main sequence star's lithium absorption strength is known to decrease with stellar age (e.g., Skumanich 1972; Carlos et al. 2016), likely due to extra mixing processes that remove lithium from the convective envelope of the star and/or depletion of this fragile trace element (e.g., Baraffe et al. 2017). Lithium is depleted via proton capture, but this process requires high temperatures (2.5 million Kelvin). As described in the *Structure* section, stellar core temperatures increase with stellar age, along with the temperature at the base of the convective envelope, as this region deepens with time (Iben 1967). The strength of a star's lithium feature is compared to another variable, such as stellar color, effective temperature, or rotation period, which is known to produce an age-dependent relation (e.g., Stanford-Moore et al. 2020). This method is optimal for low-mass ($< 1.5 \, M_\odot$), isolated, main sequence stars that do not exhibit anomalous rotation signatures. This is because the internal temperature, evolutionary state, rotational histories, and accretion histories impact the abundance signatures of lithium.

POPULATION DISTRIBUTION OF STARS IN THE GALAXY

There are approximately 100 billion stars in the Milky Way galaxy based on observed stellar densities and the structure of the galaxy. The vast majority of these stars formed in the Milky Way based on their kinematics (or the study of the position, velocity, and acceleration of objects over time) and/or metallicity, but interlopers have been observed and identified that have anomalous properties compared to other Milky Way stars.

Star-forming regions yield a regular distribution of stellar masses that appears to be independent of region-specific properties (e.g., size, metallicity), at least in the Milky Way (e.g., Bastian et al. 2010; Smith 2020). This distribution is referred to as the Initial Mass Function (Kroupa 2001; Chabrier 2003) and it accounts for the fact that a molecular cloud forms a small percentage of massive stars and an overwhelmingly large percentage of low-mass stars (see Table 1). More specifically, less than 4% of stars will be more massive than $1 \, M_\odot$, while more than 75% will constitute stars lower than $0.5 \, M_\odot$.

Many stars also have companion stars. The same gravitational fragmentation process that results in multiple protostars within a molecular cloud can result in gravitationally-bound protostars (Duchêne and Kraus 2013). Or, massive protoplanetary disks can gravitationally fragment, producing companion stars or brown dwarfs. The stellar multiplicity[16] rate is not constant for all stars; it varies depending on stellar characteristics and environmental conditions (e.g., Duchêne and Kraus 2013; Winters et al. 2019; Niu et al. 2021). The typical multiplicity rate for sun-like stars (in terms of mass) in our galaxy is ~50% (e.g., Raghavan et al. 2010). Empirically it is observed that more massive stars tend to have companions at wider separations and less massive stars are more likely to have companions at closer separations (Duchêne and Kraus 2013; Winters et al. 2019). Stars with companions are known to sometimes undergo mass exchange, mergers, and collisions, especially pre- or post-main sequence when their radii are substantially larger than their main sequence radii or when they reside in dense stellar environments where the frequency of a collision is higher. Mergers can result in nucleosynthesis, enriching future generations of stars with heavy elements. The existence of a companion star can have diverse effects on an exoplanet, influencing factors such as its dynamical stability, tidal heating, atmospheric conditions, and even the truncation of protoplanetary disks (Howell et al. 2022).

Abundance gradient in the Milky Way

The first stars in the Universe formed out of the primordial gas leftover after the Big Bang (hydrogen, helium, and trace amounts of lithium and beryllium) and thus were extremely

---

[16] Stellar multiplicity refers to the phenomenon where two or more stars are gravitationally bound and orbit each other in a system.

metal-poor (e.g., Bromm 2013). Without metals, stellar models predict that these first stars were ~1000 times more massive than the Sun, had very short lives, and quickly became massive black holes (see the *Evolution* section above). These black holes gravitationally attracted nearby gas, which helped trigger the formation of new stars near the black hole—likely forming a central galactic bulge[17] as shown in Figure 5. Accreted gas persistently fed bulge star formation as well as the central black hole, increasing its gravitational influence. For the Milky Way, it is generally believed that large, independent clouds of matter and debris were caught up by the black hole's gravity over time and eventually created the extended galactic halo[18] (Fig. 5), triggering a short burst of star formation, producing low metallicity halo stars and globular clusters[19] (Freeman and Bland-Hawthorn 2002, and references therein).

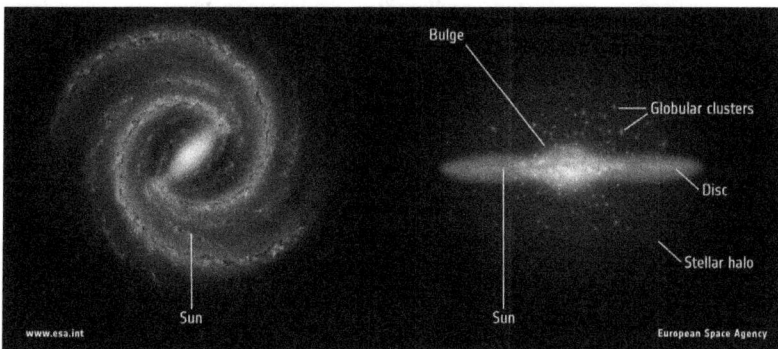

Figure 5. Diagram of the Milky Way spiral galaxy showing the key components such as the bulge, disk, and halo stars. Image credit left: NASA/JPL-Caltech; right: ESA; layout: ESA/ATG medialab—CC BY-SA IGO 3.0 (cropped).

At the same point in the Milky Way's history when the halo stars were formed, there was a large variation in the chemical composition (see *Spectroscopy and stellar abundances* section) of the halo stars, as compared to the bulge stars. For example, the various accretion events, gas dynamics, and overall star formation within the bulge resulted in rapid stellar formation and eventually post-main sequence evolution (death). The stellar death, often in the form of supernovae (Table 1), released new elements into the surrounding regions, thereby increasing the overall metallicity within the bulge.

The stars within the halo underwent a single, short (~1–2 Gyr, Freeman and Bland-Hawthorn 2002) burst of star formation. Without the cycling of enriched material, the metallicity of the halo stars and globular clusters are more likely reflective of the smaller debris clouds that were pulled into the Galaxy's gravity well. In addition, they are notably the most metal-poor stars compared to all other stars within the galaxy—creating a metallicity or abundance gradient from the younger stellar bulge stars to the much older halo stars (Hawkins et al. 2015, and references therein).

After some time, and with the help of the rotation of the Milky Way, matter began to spread out into a plane to form the galactic disk (Fig. 5). Galactic disks in general are divided into two components, the thick disk and the thin disk—the latter is where the galactic bar and/or spiral arms[20] reside. In our galaxy, it is currently thought that the thick disk formed first, shortly after the settling of the disk (e.g., Masseron and Gilmore 2015). The thick disk

---

[17] The bulge is a central, densely-populated region at the core of many spiral galaxies.

[18] The galactic halo is a large, spherical region surrounding spiral galaxies like the Milky Way. The halo contains older stars and globular clusters.

[19] A globular cluster is a gravitationally bound group of many thousands to millions of very old stars.

[20] Galactic bars and arms are the build-up of dust and gas into thin regions in the galactic disk.

is believed to have been formed from heating (e.g., causing an increase in temperature to) the early stellar disk via accretion events or minor mergers, for example with a low-mass galaxy (e.g., Walker et al. 1996; Rix and Bovy 2013), which created a vertical structure within the thick disk. Providing additional evidence that the thick disk stars are older, they have been empirically observed to have low iron-content (or [Fe/H], see the *Math formalism of stellar abundances* section) and also being enriched in α-elements[21], such as C, O, Mg, Si, etc. (e.g., Rix and Bovy 2013; Bensby et al. 2014; Recio-Blanco et al. 2014; Hawkins et al. 2015; Ness et al. 2019; Bland-Hawthorn et al. 2019, and references therein).

In comparison, the thin disk is much flatter due to the higher rotational velocity around the Galactic Bulge. The majority of gas within the Milky Way resides in the thin disk, which resulted in many generations of star formation (i.e., the youngest stars) and stellar death, often via supernovae. Due to the many cycles of frequent star formation within the thin disk, the overall content of metals within the galaxy was increased over time (e.g., Bromm 2013). Another event crucial to the production of new elements within the thin disk is the merger of neutron stars. While rare, these merger events are impactful, leading to the ejection of neutron-rich material, which is subsequently mixed into the surrounding interstellar medium. In particular, supernovae and neutron star mergers led to the generation of rapid-neutron capture elements—elements that are formed by the rapid assembly of neutrons before the onset of radioactive decay. These elements are denoted in purple and yellow in the Periodic Table of the Elements illustrated in Figure 2. Supernovae and neutron star merger events are responsible for producing most (stable) elements that are more massive than iron. As a result, thin disk stars (including the Sun) have the highest metallicity content compared to all other stars within the Galaxy.

Especially from our vantage point on Earth, orbiting the Sun at 8.5 kpc (1 parsec or pc = 3.086×10^{13} km) from the center of the Milky Way (Fig. 5), we often rely on the observed kinematics of gas, dust, and stars to infer the overall make-up of our galaxy. And while there are many kinematic and chemical differences between the thin and thick disk stars, identifying the original birthplace of a star is not always obvious (e.g., Hawkins et al. 2015). For example, the vertical structure within the thick disk is at a maximum near the bulge and tapers with radial distance (e.g., Bensby et al. 2012; Bovy et al. 2012, 2016; Rix and Bovy 2013), meaning that there are very few [Fe/H]-poor and α-rich stars in the outer disk of the Milky Way (>2 kpc from the central bulge). By taking into account a stellar age gradient within the thick disk, Martig et al. (2016) were able to help reconcile the discrepancies within the definitions of the disk components. However, the Martig et al. (2016) result provides a strong implication that there are geometric, chemical, and age gradients throughout both disk components, making their distinction a sometimes difficult determination.

USING SPECTROSCOPY TO MEASURE THE COMPOSITION OF STARS

We have generally discussed the metallicity content of stars or their elemental composition, however, we have not yet gotten into the specifics. Therefore, in this section, we will go over how astronomers measure element abundances within stars via spectroscopy, the math underlying stellar abundances (e.g., the abundance of elements within stars)—including how to convert them to molar fractions, and caveats to fully understanding stellar abundance measurements.

Spectroscopy and stellar abundances

As mentioned above, stars are hot, dense balls of gas whose internal fusion reactions cause them to emit light as radiation. Measuring the intensity of the stellar light across a wavelength range, usually using an instrument called a spectrograph, results in a stellar spectrum.

---

[21] Many of the elements with an even atomic number were formed by fusing multiple He atoms, known as the α-particle.

An object that is a perfect emitter of light or radiation, i.e., blackbody radiation, produces a continuous spectrum. However, stellar spectra are not continuous—they have dark spectral absorption lines due to the atoms (and sometimes molecules) that exist in the stellar photosphere. These absorption lines occur when the stellar radiation is absorbed by an atom, causing an electron to make the transition from a lower energy orbit to a higher orbit. Each atom absorbs light at unique wavelengths, depending on its atomic structure (electron shell configuration, quantum numbers, etc.), meaning that it's possible to identify which atoms are within the stellar photosphere via the absorption features. Because the presence of more atoms results in wider and deeper absorption features, it is possible to use spectroscopy to measure the area (or equivalent width) of the absorption features and determine the amount— or abundance—of an element within a star's photosphere.

In order to measure the spectroscopic abundances of elements within a star, first the fundamental properties of the star need to be understood—such as the effective temperature (T_{eff}), surface gravity (log g), and metallicity ([M/H], or [Fe/H] is often used as a proxy). These properties determine how radiation propagates and reacts within a star, ultimately defining the overall stellar atmospheric model. Jofré et al. (2019)—as part of a very thorough breakdown of current spectroscopic abundance techniques—provided a list of the current, regularly updated stellar models that are publicly available in their Table 1. In addition, the specifics of the atomic (and/or molecular) lines that are going to be measured need to be collected from theoretical calculations, published journal articles, or laboratory experiments. The actual stellar abundances can then be measured by using either 1) the "curve-of-growth" technique— which employs measuring the width of individual absorption lines, 2) a modeled synthetic spectrum that is compared to the measured spectrum, or 3) differential analysis to directly compare the differences in spectra between one star and another star on a line-by-line basis. Along with Jofré et al. (2019), we also refer the reader to Allende Prieto (2016) who provides an excellent explanation of the nuances in determining stellar abundances.

In addition to the differences between stellar abundance techniques, there are also variations between telescopes and their spectrographic instruments, which could have an impact on the measured abundances. For example, spectrograph resolution determines the extent to which spectral lines can be resolved, as well as the presence of lines that are blended together or lines that have asymmetries. This is of the utmost importance for precise abundance measurement (typically 0.05–0.2 dex, see the *Math formalism of stellar abundances* section), meaning that most stellar abundances are measured using high-resolution instruments, which can have varying resolutions from R ($\Delta\lambda/\lambda$) = 50,000–100,000 in the optical band (while high resolution in the infrared is $R \gtrsim 20,000$). Signal-to-noise (S/N) of a spectrograph defines the overall abundance precision, where a $S/N > 100$ is considered to be high.

Astronomers have been able to measure the spectroscopic (as opposed to photometric[22]) abundances of stars for decades. Over that time, a variety of stellar models, abundance techniques, spectrographs, and telescopes have been built. While these multitudes of methodologies provide an interesting way to observe, analyze, and compare stellar spectra, they also result in discrepancies. Fortunately, the community has come together in a variety of ways to better understand the differences and create important solutions. To begin, Smiljanic et al. (2014), Hinkel et al. (2016), and Jofré et al. (2017) studied a variety of abundance techniques within the community and analytically compared them to one another to better understand their strengths, weaknesses, inconsistencies, and overall relationships. At the same time, large spectroscopic surveys have been designed to obtain large, homogeneous datasets of thousands (if not millions) of stellar abundances. Some of the largest current or upcoming surveys are:

---

[22] Spectroscopy measures the flux or intensity of light from a star with respect to wavelength, while photometry measures flux with respect to an area/region of space at a very narrow wavelength range, similar to a photograph.

RAdial Velocity Experiment (RAVE, $R \sim 7500$) measuring 7 elements in $\sim$300,000 stars (Steinmetz et al. 2006), WHT Enhanced Area Velocity Explorer (WEAVE, $R \sim 20,000$) will obtain abundances (the exact elements have not yet been specified) for $\sim$1.5 million bright stars in the galactic field and in open clusters (Jin et al. 2022), the GALactic Archeology with HERMES (GALAH) survey will observe $\sim$30 element abundances in an estimated $\sim$1 million stars (currently at $\sim$350,000 observed stars Buder et al. 2021), Apache Point Observatory Galaxy Evolution Experiment (APOGEE, $R \sim 22,500$) has currently determined $\sim$30 elements in $\sim$350,000 stars (Buder et al. 2021), and the Gaia-ESO survey had a resolution of $R \sim 47,000$ to observe $\sim$30 elements in $\sim$115,000 stars (Gilmore et al. 2012).

While large surveys succeed in creating consistent abundance data from a single source telescope+instrument combination, they often sacrifice element diversity and abundance precision—measuring a smaller number of elements while incurring larger measurement uncertainties—compared to smaller, more specific ground-based studies that often have higher resolution and S/N. To this end, the Hypatia Catalog (hypatiacatalog.com) is an amalgamate database compiled from $\sim$300 literature sources that measured high-resolution abundances for Fe in addition to one other element for main sequence (F-, G-, K-, M-type) stars within 500pc, or any exoplanet host regardless of distance (Hinkel et al. 2014). The Hypatia Catalog is the largest element abundance database for nearby stars, currently containing abundance measurements for $\sim$90 elements and species within $\sim$11,100 stars, $\sim$1450 of which are confirmed exoplanet host stars. There is a plethora of data within the Hypatia Catalog, which is uniquely multidimensional because it incorporates multiple measurements of the same element within the same star from different literature sources. Therefore, a concerted effort was made to ensure the abundance data is more homogeneous by standardizing the solar normalization scale (see the *Math formalism of stellar abundances* section and Hinkel et al. 2014, 2022). Regardless of the heterogeneity, the Hypatia Catalog is a powerful tool because of the unique, unrivaled combination of the breadth (number of stars) and depth (number of elements).

Math formalism of stellar abundances

Stellar abundances provide key insight into the chemical history and make-up of a stellar system, while also creating a compositional connection between stars and planets (see the *Composition of stars and their planets* section). Therefore, it makes sense to understand how stellar abundances are mathematically defined, so that the measurements may be converted across disciplines. Here we provide a summary of the math formalism defined in Hinkel et al. (2022), which provides more detail as well as a walk-through example.

Stellar abundances are defined as a ratio between a generic element Q with respect to hydrogen (H), where the number of hydrogen atoms are considered to be a constant 10^{12} (Payne 1925a,b; Claas 1951), in a similar way that meteoritic abundances are often compared to 10^6 silicon atoms. In this way, Q represents the number of Q atoms for every hydrogen atom—which is another way of showing that stellar abundances are ultimately the amount or number of an element (e.g., abundances are not to be confused with mass ratios). In terms of respective number fractions q and h—where a number fraction of 0 indicates that this element is not present and 1 means that an object is composed entirely of that element, we see that

$$Q' = q/h \times 10^{12} \tag{1}$$

Now, we are able to define the absolute stellar abundance of Q, or $A(Q)$, as

$$A(Q) \equiv \log_{10}(Q') \tag{2}$$

where $\equiv$ indicates that the two values are mathematically equivalent to one another. Looking at hydrogen in particular, we see that the absolute value $A(H) = 12$ and the number fraction

$h \approx 1$, since it is the most ubiquitous element in the Universe. This means that Equation (2) is able to be defined as

$$
\begin{aligned}
A(Q) &= \log_{10}(q \times 10^{12}) \\
&= \log_{10}(q) + 12 \\
&= \log_{10}(q) + A(H) \\
&\equiv Q/H
\end{aligned}
\tag{3}
$$

Stellar abundances tend to be either defined with respect to H or Fe. Recognizing that Q/H = A(Q) and Fe/H = A(Fe), it is possible to convert between the two ratios using:

$$
\begin{aligned}
Q/Fe &\equiv Q/H - Fe/H \\
&= A(Q) - A(Fe)
\end{aligned}
\tag{4}
$$

Because the Sun is the closest star, and therefore the most highly observed star, it is customary in astronomy to normalize stellar abundances with respect to the Sun. In this way, [Q/H] = 0.0 dex means that a star has the same elemental abundance for Q as the Sun. Fortunately, normalizing to the Sun is a fairly intuitive subtraction in log-space:

$$
\begin{aligned}
[Q/H] &\equiv Q_*/H_* - Q_\odot/H_\odot \\
&= A(Q)_* - A(Q)_\odot
\end{aligned}
\tag{5}
$$

where the observed stellar abundances are indicated with an $*$, and the solar abundances are demarcated using $\odot$. We note here that logged abundances that are normalized to the Sun are defined with the "dex" unit, meaning "decadic logarithmic unit" (Lodders 2019). The "decadic" aspect of dex means that the log base is 10, similar to the decibel (dB), which is able to span a large dynamic range more easily defined using logarithmic (as opposed to linear) scales. See Figure 6 for an example plot showing stellar abundances for two ratios ([Si/Fe] vs [Fe/H]) in dex notation.

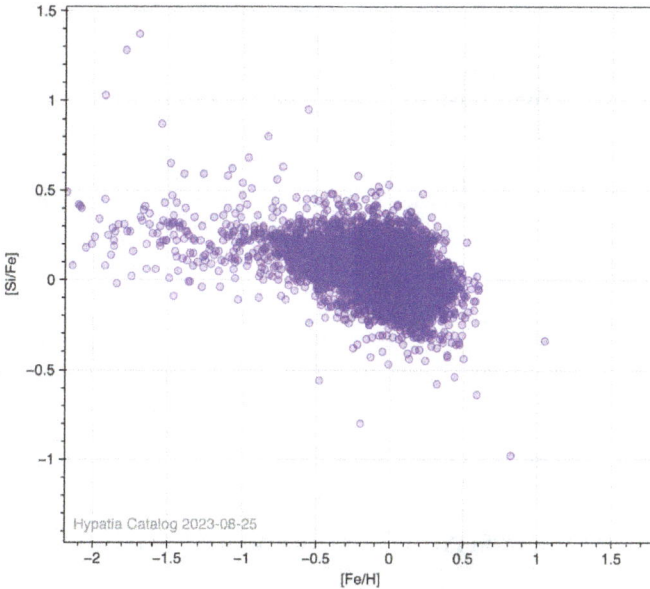

Figure 6. Classic stellar abundance plot (from the Hypatia Catalog, Hinkel et al. 2014) showing the ratio of [Si/Fe] with respect to [Fe/H], both with units in dex, where the Sun is located at (0, 0). Given the log-scale, a star with [Fe/H] = 0.5 dex has $10^{0.5} \approx 3.16$ times more Fe than the Sun.

While working in the dex-notation for stellar abundances is useful for astronomers, it is often more appropriate for other fields to work with mole or molar ratios. It is, therefore, necessary to convert from log- to linear-space while also removing the implicit solar normalization. We convert from the original stellar abundance, $[Q/H]_*$, to moles, Q_*, using:

$$Q_* = 10^{([Q/H]_* + Q_\odot)} \tag{6}$$

where $Q_\odot$ is the solar value used to normalize Q. We refer the reader to Hinkel et al. (2022) for an analysis of error propagation when converting from dex to moles (their Section 4.1) as well as a more in-depth discussion on the variety and comparison of solar normalizations within the stellar abundance field (their Section 6).

Stellar abundance caveats

When working with stellar abundance data, there are a number of technical considerations related to the underlying methods and models that need to be taken into account. To begin, most stellar models assume that small regions of the star's interior are in a thermally isolated, consistent (or steady-state) condition that can be well defined with equations, known as local thermodynamic equilibrium (LTE). However, a lot of physical processes happen within the stellar interior, creating varying spectral line strengths and unusual line shapes (Lodders 2019), indicating that there has been a deviation from LTE; these are called non-LTE (NLTE) effects. Introducing additional physics into the stellar models or investigating NLTE impacts on atomic and line data are time intensive, but also very important for resolving discrepancies between the data and models. Given the different underlying physical models between the calculations, we do not recommend combining (such as averaging) measurements that come from separate LTE and NLTE determinations. Their results, while comparable, are not equivalent.

In general, stellar abundances are defined as "elemental" abundances, meaning that they represent the overall amount of a specific element. While atomic lines are often used to determine abundances (mostly out of convenience), the term "elemental" doesn't preclude using molecular lines, such as FeH and H_2O, where the elements are disassociated to produce individual abundances, e.g., for Fe and O, respectively. The underlying implication is that, unless otherwise specified, the most common stable isotopes are usually measured and may be combined with other isotopes (Lodders 2019). Work is currently being done, for example by Lugaro et al. (2023), on separating stellar abundances into their distinct isotopic representations for certain elements.

Measuring different ionization states of an element—or when it's either gained an electron (anion) or lost an electron (cation)—can provide useful information for stellar abundance analysis. For example, many stellar abundance models iterate over the main stellar parameters (T_{eff}, log g, and [Fe/H]) until Fe I and Fe II are balanced (Jofré et al. 2015; Hinkel et al. 2016; Jofré et al. 2019). We note that astronomy ionization notation differs from chemistry such that a neutral element is indicated with a Roman numeral "I" (e.g., [Q I/H]) and a singly ionized element is denoted with "II" (e.g, [Q II/H]), an unnecessarily confusing alteration that is rooted in historical precedent and is therefore unlikely to be changed (Millikan and Bowen 1924). While some spectroscopists measure the ionization states separately, many combine them in the final measurement, often shown as [Q/H].

THE COMPOSITION OF STARS AND THEIR PLANETS

Nearly all elements in the Periodic Table were created in stars. Elements like Th and U are produced through rapid-process neutron capture and have been critical to the Earth's evolution. More specifically, the radioactive decay of Th and U (as well as ^{26}Al and ^{60}Fe) play a major role in heating the Earth's interior which is critical to mantle convection, the generation of the Earth's magnetic field, and the movement of the tectonic plates. Some rapid-process elements are vital to biological life, for example K is critical in maintaining proper cell function, nerve transmission, and muscle contraction. Stars within the Milky Way, even stars that are (currently) neighbors within the thin disk[23], have varying compositions that are the result of differing events that occurred prior to or during formation. As a result, the violent, seemingly rare astronomical events like supernovae and neutron star mergers seeded the Earth with elements that give rise to the everyday phenomena that make our lives possible.

Stellar abundances as a proxy for planetary interiors

It is not currently possible to directly measure the bulk mineralogical surface composition of an exoplanet, especially one with an atmosphere (Kreidberg et al. 2019; Zieba et al. 2023). While it's the goal for current and upcoming exoplanet missions like the James Webb Space Telescope (or JWST, launched 2021) and the Nancy Grace Roman Space Telescope (or Roman, expected launch 2026) to directly image planets and glean some sense of the planet's surface via color maps or measuring light reflected off the planet's surface (Lisse et al. 2020), it will be very difficult to obtain specific high-resolution data of the surface composition. To this end, no missions are currently expected to directly observe and measure a planet's interior composition and mineralogy.

Therefore, one of the most important tools currently available to understand the composition of a small planet is the relationship between its mass and radius, which helps define the planet's density and overall classification (e.g., super-Mercury, rocky, or mini-Neptune). While there are a variety of ways to detect an exoplanet, there are two that are the most common/popular: the radial-velocity technique that determines the gravitational wobble that the planet imparts on the star or the transit technique that measures how much stellar light is blocked when the planet passes in front of the star (from the Earth's perspective). Unfortunately, most current observing techniques for exoplanets make it very difficult to measure both planetary mass (radial-velocity) and radius (transit); for example, to date, only ~20% of discovered planets within the NASA Exoplanet Archive[24] have both planet and mass measurements. In addition, planetary interior models based only on mass and radius are fraught with degeneracies. A study by Schlichting and Young (2022) illustrated the impact of elements that can be evaporated at low temperatures (or volatiles) on bulk densities, where hydrogen and other light elements within iron-rich cores significantly changed the mass–radius relationship for small planets. One way to break the mass–radius degeneracy is to employ the use of stellar abundances, especially those important to forming rocky material, as proxies for the composition of small planets (e.g., Lodders 2003; Sotin et al. 2007; Santos et al. 2015, 2017; Dorn et al. 2015, 2017a,b; Hinkel and Unterborn 2018; Putirka and Rarick 2019; Plotnykov and Valencia 2020; Putirka et al. 2021; Putirka and Xu 2021; Schulze et al. 2021).

The interiors of stars have high enough temperatures and pressures to support atomic fusion. And while some time-dependent processes slowly change the abundances of old stars, the composition of the stellar photosphere is largely reflective of the original molecular cloud from which it formed (Lodders et al. 2009; Lodders 2019). The stellar birth cloud also gave rise to other objects within a stellar system, such as orbiting planets, moons, asteroids, etc. Because they originate from the same place, it is possible to use the abundance of the host star as a proxy for the composition of their planets. This is especially true for elements that can only be evaporated at high temperatures (or refractory elements), such as Mg, Si, and Fe, which are not strongly impacted by chemical and physical processes within the stellar disk. For example, the molar ratios of these three elements within the Sun, Earth, and Mars are all

---

[23] See abundances within the Hypatia Catalog (hypatiacatalog.com) to visualize the abundances differences between nearby stars.
[24] https://exoplanetarchive.ipac.caltech.edu/

the same to within 10% (McDonough 2003; Unterborn and Panero 2019). Importantly, Mg, Si, and Fe—in combination with O—are also especially important to the formation of small planets, making up 95 mol % of the Earth (McDonough 2003).

From a theoretical perspective, the one-to-one relationship between stellar abundances and planetary interiors was studied by Bond et al. (2010a,b) who simulated the dynamics and chemistry when small planets are formed. Taking into account the changes within the stellar disk as the solar nebula evolved, they found that the bulk refractory abundances in small planets did not significantly change. In fact, their simulations lead them to believe that the differences in the interior composition of small exoplanets were more likely the result of variations between host star abundances than from processing within the disk. Thiabaud et al. (2015) specifically analyzed where and how the direct star–planet compositional link breaks down, modeling the Fe/Si, Mg/Si, and C/O ratios within the protoplanetary disk to see how it impacted the formation of rocky planets, ice planets, and giant gaseous planets. Comparing the composition of the three planet types to the abundances of the host star, they found that the Fe/Si and Mg/Si ratios matched between the star and planets.

More observationally, Bonsor et al. (2021) utilized an extremely interesting testbed involving a binary system, or two stars orbiting each other. One star is a normal main-sequence star while the other is a white dwarf star, which is the end-of-life result for stars that were originally $\sim$1–8 M$_\odot$ (see Table 1); they are extremely dense, compact objects whose stellar spectra are fairly devoid of elemental or molecular lines. This particular white dwarf's spectrum has more metal absorption lines than a white dwarf should; the "pollution" indicates that a planetary object was accreted onto the surface of the white dwarf. Measuring the elemental abundances of polluted white dwarfs is the only direct way to determine the interior composition of the accreted object (e.g., Jura and Young 2014; Xu et al. 2017). However, chemical processing occurs during object accretion, which makes it difficult to identify the true composition of the original planetary body. Fortunately, the white dwarf (and planetary body) is in a binary orbit with a main sequence star, which all formed at the same time and should therefore exhibit similar abundances (e.g., Hawkins et al. 2020). This makes it possible to directly compare the composition of the planetary object to the original stellar composition. Through their analysis, Bonsor et al. (2021) found that the refractory abundances of the main-sequence K dwarf star and the make-up of a comet-like body that had accreted onto the polluted white dwarf matched within measurement uncertainty. Guimond et al. (2023) expanded on these results by calculating the interior structures and mantle mineralogies of hypothetical planets using the host's stellar abundances. They found that not only did the bulk refractory compositions of stars and planets agree, but that the compositional similarity extended also to the mantle mineralogies of the exoplanets.

Additional considerations for the star–planet chemical link

While stellar abundances are useful for a more holistic characterization and classification of an orbiting exoplanet, there are additional processes that occur during planet formation and evolution that alter the composition of a planet that must be taken into account. For example, Bonomo et al. (2019) analyzed two twin planets, Kepler-107 b and c, and found that the density of the c-planet was twice that of the b-planet. After considering multiple possible reasons, it was found that the only likely mechanism to explain the difference was a giant impact that must have stripped much or all of the c-planet's silicate mantle. This is similar to the current understanding of why Mercury has a dissimilar Si, Mg, and Fe abundance compared to other Solar System bodies and is markedly iron-rich compared with the Earth and Mars (Benz et al. 2008).

In addition to the possibility of giant impacts, the temperature of the stellar disk controls the condensation of elements, which creates radial variations in different locations of the disk that change over time (e.g., Lodders 2003). While these radial variations are most likely to impact volatile elements, as opposed to refractory, Dorn et al. (2019) found that a possible exception occurs when refractory elements partially condense at temperatures > 1200 K within the inner part of the stellar disk. They found that it is within this regime that super-Earths, such as HD 219134 b, 55 Cnc e, and WASP-47 e, could have potentially formed at high condensation temperatures ($T > 1200$ K) which results in Fe-depleted cores and enrichments in Al and Ca. Especially for super-Earth-type planets, which have no Solar System analog, it may be that the interior composition models are not better constrained by the inclusion of stellar abundances. This finding was supported by Otegi et al. (2020) who specifically studied stellar Fe/Si and Mg/Si molar ratios for planets with masses < 25 $M_\oplus$ and radii < 3.5 $R_\oplus$, although their results were dependent on uncertainties and the specific stellar abundance values.

Similarly, Adibekyan et al. (2021a,b) found that a radial gradient that is oxidizing within the disk, or possible Fe-enrichment coupled with a giant impact, may create differences during small planet formation. They used stoichiometry to balance mass fraction equations for those species dominant in a small planet, e.g., H, He, C, O, Mg, Si, and Fe (Santos et al. 2015). By analyzing the composition of 38 small planets, they tested the iron mass fraction and found that relationship between stellar abundances and planet composition was 4-to-1 instead of 1-to-1.

Free-floating planets. There is also the case of free-floating planets (FFPs): planetary-mass objects that do not orbit a star. Several hundred FFPs have been detected in young star-forming regions, such as ρ Ophiuchus (Bouy et al. 2022; Miret-Roig et al. 2022), Upper Scorpius (David et al. 2019; Lodieu et al. 2021; Bouy et al. 2022; Miret-Roig et al. 2022), Orion Trapezium Cluster (Lucas and Roche 2000), σ Orionis (Peña Ramírez et al. 2012), the Carina-Near moving group (Gagné et al. 2017), and NGC 1333 (Scholz et al. 2012). The age constraints provided by extremely young star-forming environments in which FFPs are found (commonly ranging between 1–10 Myr), indicate that the FFP formation process must occur rapidly. In fact, a recent investigation of the Trapezium Cluster (1–3 Myr) using the James Webb Space Telescope Near Infrared Camera Pearson and McCaughrean 2023 reported a population of 540 FFPs down to 0.6 M_J within an $11' \times 7.5'$ (1.2 pc $\times$ 0.8 pc) field of view.

Like all planets, the internal structure and composition of FFPs are tightly linked to the origin and evolution of these bodies. Some FFPs are expected to form via typical stellar disk accretion processes, only to be ejected from their host stars due to dynamical processes (e.g., stellar fly-bys, disk instabilities, or planet scattering events, Rasio and Ford 1996; Veras and Raymond 2012). Like their bound counterparts, gas giants formed in this manner can widely range in metallicity, depending on the initial disk location and the properties of the host star (e.g., Öberg et al. 2011; Helling et al. 2014; Mordasini et al. 2016). Some FFPs may have never been bound to a central star, forming via the disk fragmentation of a molecular cloud (similar to stars) and would contribute to the low-mass end of the Initial Mass Function. Like ejected planets, FFPs formed via instabilities in the disk are predicted to exhibit a wide range in composition. However, unlike ejected planets, they would share similar kinematic properties with co-moving neighbors fashioned from the same molecular cloud.

FFPs are valuable in ways that extend beyond their origin story. For example, they can serve as critical testbeds to probe the atmospheric properties of planets spared from the radiative effects of a central star, which are well known to be a dominant environmental influence. Contemporary and forthcoming surveys including JWST and Roman are expected to provide the FFP census data required to better investigate the origin of starless worlds.

In addition, the number of FFPs are certainly plentiful. Extrapolating their detections, an estimated several billion FFPs are expected to roam throughout the Milky Way galaxy (Miret-Roig et al. 2022). In fact, tens of thousands of FFPs have been predicted to occupy a spherical volume centered on Earth and extending to our nearest stellar neighbor, Proxima Centauri ($d = 1.3$ pc; Lingam et al. 2023). JWST is the first instrument capable of directly detecting and characterizing FFPs, probing masses in the 1–15 M_J mass regime. As such, several JWST observing programs have focused on an FFP search within nearby star-forming regions. Similarly, forthcoming microlensing observations conducted by the Roman are expected to significantly increase the number of detected FFPs, probing masses down to 0.01 $M_\oplus$ (e.g., Dai and Guerras 2018; Limbach et al. 2023).

Being able to characterize and classify small planets accurately—ones that orbit their host star and those that are free-floating—is of the utmost importance to upcoming NASA and ESA missions. It currently appears that the general, first-order assumption that stellar abundances can be used as a proxy for planet composition is valid until it (obviously) is not, e.g., for super-Earths and gas giants. However, it is vital to the community that we better understand crucial second- and third-order processes that occur within the disk and during planet formation itself (see Mordasini and Burn 2024: Zhang, both this volume), since they have an influential impact on the planet's make-up.

ACKNOWLEDGMENTS

The research shown here acknowledges the use of the Hypatia Catalog Database, an online compilation of stellar abundance data as described in Hinkel et al. (2014), which was supported by NASA's Nexus for Exoplanet System Science (NExSS) research coordination network and the Vanderbilt Initiative in Data-Intensive Astrophysics (VIDA). MSF gratefully acknowledges the generous support provided by NASA through Hubble Fellowship grant HST-HF2-51493.001-A awarded by the Space Telescope Science Institute, which is operated by the Association of Universities for Research in Astronomy, In., for NASA, under the contract NAS 5-26555.

REFERENCES

Adibekyan V, Santos NC, Dorn C, Sousa SG, Hakobyan AA, Bitsch B, Mordasini C, Barros SC, Mena ED, Demangeon OD, Faria JP (2021a) Composition of super-Earths, super-Mercuries, and their host stars. Commun Byurakan Astrophys Obs 68:447–453, doi:10.52526/25792776-2021.68.2-44710.48550/arXiv.2112.14512

Adibekyan V, Dorn C, Sousa SG, Santos NC, Bitsch B, Israelian G, Mordasini C, Barros SC, Delgado Mena E, Demangeon OD, Faria JP (2021b) A compositional link between rocky exoplanets and their host stars. Science 374:330–332, doi:10.1126/science.abg8794 10.48550/arXiv.2102.12444

Aerts C, Christensen-Dalsgaard J, Kurtz DW, Aerts C, Christensen-Dalsgaard J, Kurtz DW (2010) The future. *In:* Asteroseismology, Springer, p 669–677

Allende Prieto C (2016) Solar and stellar photospheric abundances. Living Rev Sol Phys 13:1, doi:10.1007/s41116-016-0001-6

Andrews SM, Rosenfeld KA, Kraus AL, Wilner DJ (2013) The mass dependence between protoplanetary disks and their stellar hosts. Astrophys J 771:129, doi:10.1088/0004-637X/771/2/129

Arcones A, Thielemann F-K (2023) Origin of the elements. Astron Astrophys Rev 31:1, doi:10.1007/s00159-022-00146-x

Bally J (2016) Protostellar outflows. Annu Rev Astron Astrophys 54:491–528, doi:10.1146/annurev-astro-081915-023341

Baraffe I, Pratt J, Goffrey T, Constantino T, Folini D, Popov MV, Walder R, Viallet M (2017) Lithium depletion in solar-like stars: Effect of overshooting based on realistic multi-dimensional simulations. Astrophys J Letters 845:L6, doi:10.3847/2041-8213/aa82ff

Barnes SA (2003) On the rotational evolution of solar- and late-type stars, its magnetic origins, and the possibility of stellar gyrochronology. Astrophys J 586:464–479, doi:10.1086/367639

Bastian N, Covey KR, Meyer MR (2010) A universal stellar initial mass function? a critical look at variations. Annu Rev Astron Astrophys 48:339–389, doi:10.1146/annurev-astro-082708-101642

Bensby T, Alves-Brito A, Oey MS, Yong D, Meléndez J (2012) Abundance trends in the inner and outer galactic disk. *In:* Galactic Archaeology: Near-Field Cosmology and the Formation of the Milky Way. ASP Conf Ser Vol 458. Vol 458. W Aoki, M Ishigaki, T Suda, T Tsujimoto, N Arimoto (eds), p 171–174

Bensby T, Feltzing S, Oey MS (2014) Exploring the Milky Way stellar disk. A detailed elemental abundance study of 714 F and G dwarf stars in the solar neighbourhood. Astron Astrophys 562:A71, doi:10.1051/0004-6361/201322631

Benz AO, Güdel M (2010) Physical processes in magnetically driven flares on the sun, stars, and young stellar objects. Annu Rev Astron Astrophys 48:241-287, doi:10.1146/annurev-astro-082708-101757

Benz W, Anic A, Horner J, Whitby JA (2008) The origin of Mercury. *In*: Mercury. Vol 26. Balogh A, Ksanfomality L, von Steiger R (eds). Space Science Reviews, p 7–20

Berdyugina SV (2005) Starspots: A key to the stellar dynamo. Living Rev Sol Phys 2:8, doi:10.12942/lrsp-2005-8

Bland-Hawthorn J, Sharma S, Tepper-Garcia T, Binney J, Freeman KC, Hayden MR, Kos J, De Silva GM, Ellis S, Lewis GF, Asplund M (2019) The GALAH survey and Gaia DR2: dissecting the stellar disc's phase space by age, action, chemistry, and location. Mon Not R Astron Soc 486:1167–1191, doi:10.1093/mnras/stz217

Bond JC, Lauretta DS, O'Brien DP (2010a) Making the Earth: Combining dynamics and chemistry in the Solar System. Icarus 205:321–337, doi:10.1016/j.icarus.2009.07.037

Bond JC, O'Brien DP, Lauretta DS (2010b) The compositional diversity of extrasolar terrestrial planets. I. In situ simulations. Astrophys J 715:1050–1070, doi:10.1088/0004-637X/715/2/1050

Bonomo AS, Zeng L, Damasso M, Leinhardt ZM, Justesen AB, Lopez E, Lund MN, Malavolta L, Silva Aguirre V, Buchhave LA, Corsaro E (2019) A giant impact as the likely origin of different twins in the Kepler-107 exoplanet system. Nat Astron 3:416–423, doi:10.1038/s41550-018-0684-9

Bonsor A, Jofré P, Shorttle O, Rogers LK, Xu S, Melis C (2021) Host-star and exoplanet compositions: a pilot study using a wide binary with a polluted white dwarf. Mon Not R Astron Soc 503:1877–1883, doi:10.1093/mnras/stab370

Boro Saikia S, Marvin CJ, Jeffers SV, Reiners A, Cameron R, Marsden SC, Petit P, Warnecke J, Yadav AP (2018) Chromospheric activity catalogue of 4454 cool stars. Questioning the active branch of stellar activity cycles. Astron Astrophys 616:A108, doi:10.1051/0004-6361/201629518

Bouy H, Tamura M, Barrado D, Motohara K, Rodríguez NC, Miret-Roig N, Konishi M, Koyama S, Takahashi H, Huélamo N, Bertin E (2022) Infrared spectroscopy of free-floating planet candidates in Upper Scorpius and Ophiuchus. Astron Astrophys 664:A111, doi:10.1051/0004-6361/202243850

Bovy J, Rix H-W, Hogg DW (2012) The Milky Way has no distinct thick disk. Astrophys J 751:131, doi:10.1088/0004-637X/751/2/131

Bovy J, Rix H-W, Schlafly EF, Nidever DL, Holtzman JA, Shetrone M, Beers TC (2016) The stellar population structure of the galactic disk. Astrophys J 823:30, doi:10.3847/0004-637X/823/1/30

Bressan A, Marigo P, Girardi L, Salasnich B, Dal Cero C, Rubele S, Nanni A (2012) PARSEC: stellar tracks and isochrones with the PAdova and TRieste stellar evolution code. Mon Not R Astron Soc 427:127–145, doi:10.1111/j.1365-2966.2012.21948.x

Bromley BC, Kenyon SJ (2015) Planet formation around binary stars: Tatooine made easy. Astrophys J 806:98, doi:10.1088/0004-637X/806/1/98

Bromm V (2013) Formation of the first stars. Rep Prog Phys 76:112901, doi:10.1088/0034-4885/76/11/112901

Buder S, Sharma S, Kos J, Amarsi AM, Nordlander T, Lind K, Martell SL, Asplund M, Bland-Hawthorn J, Casey AR, De Silva GM (2021) The GALAH+ survey: Third data release. Mon Not R Astron Soc 506:150–201, doi:10.1093/mnras/stab1242

Burbidge EM, Burbidge GR, Fowler WA, Hoyle F (1957) Synthesis of the elements in stars. Rev Mod Phys 29:547–650, doi:10.1103/RevModPhys.29.547

Carlos M, Nissen PE, Meléndez J (2016) Correlation between lithium abundances and ages of solar twin stars. Astron Astrophys 587:A100, doi:10.1051/0004-6361/201527478

Castor JI, Abbott DC, Klein RI (1975) Radiation-driven winds in Of stars. Astrophys J 195:157-174, doi:10.1086/153315

Chabrier G (2003) Galactic stellar and substellar initial mass function. Publ Astron Soc Pac 115:763–795, doi:10.1086/376392

Chabrier G, Baraffe I (1997) Structure and evolution of low-mass stars. Astron Astrophys 327:1039–1053, doi:10.48550/arXiv.astro-ph/9704118

Chabrier G, Baraffe I, Allard F, Hauschildt P (2000) Evolutionary models for very low-mass stars and brown dwarfs with dusty atmospheres. Astrophys J 542:464–472, doi:10.1086/309513

Choi J, Dotter A, Conroy C, Cantiello M, Paxton B, Johnson BD (2016) Mesa Isochrones and Stellar Tracks (MIST). I. Solar-scaled models. Astrophys J 823:102, doi:10.3847/0004-637X/823/2/102

Claas WJ (1951) The Composition of the Solar Atmosphere. University of Utrecht, Netherlands

Crowther PA, Schnurr O, Hirschi R, Yusof N, Parker RJ, Goodwin SP, Kassim HA (2010) The R136 star cluster hosts several stars whose individual masses greatly exceed the accepted 150M_solar stellar mass limit. Mon Not R Astron Soc 408:731–751, doi:10.1111/j.1365-2966.2010.17167.x

Curtis JL, Agüeros MA, Douglas ST, Meibom S (2019) A temporary epoch of stalled spin-down for low-mass stars: Insights from NGC 6811 with Gaia and Kepler. Astrophys J 879:49, doi:10.3847/1538-4357/ab2393

Dai X, Guerras E (2018) Probing extragalactic planets using quasar microlensing. Astrophys J Lett 853:L27, doi:10.3847/2041-8213/aaa5fb

David TJ, Hillenbrand LA, Gillen E, Cody AM, Howell SB, Isaacson HT, Livingston JH (2019) Age determination in Upper Scorpius with eclipsing binaries. Astrophys J 872:161, doi:10.3847/1538-4357/aafe09

Dawson RI, Johnson JA (2018) Origins of hot Jupiters. Annu Rev Astron Astrophys 56:175–221, doi:10.1146/annurev-astro-081817-051853

Dorn C, Khan A, Heng K, Connolly JAD, Alibert Y, Benz W, Tackley P (2015) Can we constrain the interior structure of rocky exoplanets from mass and radius measurements? Astron Astrophys 577:A83, doi:10.1051/0004-6361/201424915

Dorn C, Hinkel NR, Venturini J (2017a) Bayesian analysis of interiors of HD 219134b, Kepler-10b, Kepler-93b, CoRoT-7b, 55Cnc e, and HD 97658b using stellar abundance proxies. Astron Astrophys 597:A38, doi:10.1051/0004-6361/20162874910.48550/arXiv.1609.03909

Dorn C, Venturini J, Khan A, Heng K, Alibert Y, Helled R, Rivoldini A, Benz W (2017b) A generalized Bayesian inference method for constraining the interiors of super Earths and sub-Neptunes. Astron Astrophys 597:A37, doi:10.1051/0004-6361/20162870810.48550/arXiv.1609.03908

Dorn C, Harrison JHD, Bonsor A, Hands TO (2019) A new class of Super-Earths formed from high-temperature condensates: HD219134 b, 55 Cnc e, WASP-47 e. Mon Not R Astron Soc 484:712–727, doi:10.1093/mnras/sty343510.48550/arXiv.1812.07222

Douglas ST, Agüeros MA, Covey KR, Kraus A (2017) Poking the beehive from space: K2 rotation periods for Praesepe. Astrophys J 842:83, doi:10.3847/1538-4357/aa6e52

Dressing CD, Charbonneau D (2013) The occurrence rate of small planets around small stars. Astrophys J 767:95, doi:10.1088/0004-637X/767/1/95

Duchêne G, Kraus A (2013) Stellar multiplicity. Annu Rev Astron Astrophys 51:269-310, doi:10.1146/annurev-astro-081710-102602

Figer DF (2005) An upper limit to the masses of stars. Nature 434:192-194, doi:10.1038/nature03293

Fischer DA, Valenti J (2005) The planet–metallicity correlation. Astrophys J 622:1102–1117

Foley BJ, Driscoll PE (2016) Whole planet coupling between climate, mantle, and core: Implications for rocky planet evolution. Geochem Geophys Geosystems 17:1885–1914

Freeman K, Bland-Hawthorn J (2002) The new galaxy: Signatures of its formation. Annu Rev Astron Astrophys 40:487–537, doi:10.1146/annurev.astro.40.060401.093840

Gagné J, Faherty JK, Burgasser AJ, Artigau É, Bouchard S, Albert L, Lafrenière D, Doyon R, Bardalez Gagliuffi DC (2017) SIMP J013656.5+093347 Is likely a planetary-mass object in the Carina-Near Moving Group. Astrophys J Letters 841:L1, doi:10.3847/2041-8213/aa70e2

Gaia Collaboration, Brown AG, Vallenari A, Prusti T, De Bruijne JH, Mignard F, Drimmel R, Babusiaux C, Bailer-Jones CA, Bastian U, Biermann M, Evans DW (2016) Gaia data release 1. Summary of the astrometric, photometric, and survey properties. Astron Astrophys 595:A2, doi:10.1051/0004-6361/201629512

Gaia Collaboration, Brown AG, Vallenari A, Prusti T, De Bruijne JH, Babusiaux C, Biermann M, Creevey OL, Evans DW, Eyer L, Hutton A, Jansen F (2021) Gaia early data release 3. Summary of the contents and survey properties. Astron Astrophys 649:A1, doi:10.1051/0004-6361/202039657

Gaudi BS, Meyer M, Christiansen J (2021) The demographics of exoplanets. *In*: ExoFrontiers; Big Questions in Exoplanetary Science. Madhusudhan N (ed), p 2-1–2-21

Gilmore G, Randich S, Asplund M, Binney J, Bonifacio P, Drew J, Feltzing S, Ferguson A, Jeffries R, Micela G, Negueruela I (2012) The Gaia-ESO public spectroscopic survey. The Messenger 147:25–31

Gonzalez G (1997) The stellar metallicity–giant planet connection. Mon Not R Astron Soc 285:403–412, doi:10.1093/mnras/285.2.403

Green LM, Török T, Vrvsnak B, Manchester W, Veronig A (2018) The origin, early evolution and predictability of solar eruptions. Space Sci Rev 214:46, doi:10.1007/s11214-017-0462-5

Güdel M (2007) The Sun in time: Activity and environment. Living Rev Sol Phys 4:3, doi:10.12942/lrsp-2007-3

Güdel M, Nazé Y (2009) X-ray spectroscopy of stars. Astron Astrophys 17:309-408, doi:10.1007/s00159-009-0022-4

Guimond CM, Shorttle O, Rudge JF (2023) Mantle mineralogy limits to rocky planet water inventories. Mon Not R Astron Soc 521:2535–2552, doi:10.1093/mnras/stad148

Harre J-V, Heller R (2021) Digital color codes of stars. Astron Nachr 342:578-587, doi:10.1002/asna.202113868

Hartmann L, Herczeg G, Calvet N (2016) Accretion onto pre-main-sequence stars. Annu Rev Astron Astrophys 54:135–180, doi:10.1146/annurev-astro-081915-023347

Hathaway DH (2015) The solar cycle. Living Rev Sol Phys 12:4, doi:10.1007/lrsp-2015-4

Hawkins K, Jofré P, Masseron T, Gilmore G (2015) Using chemical tagging to redefine the interface of the galactic disc and halo. Mon Not R Astron Soc 453:758–774, doi:10.1093/mnras/stv1586

Hawkins K, Lucey M, Ting Y-S, Ji A, Katzberg D, Thompson M, El-Badry K, Teske J, Nelson T, Carrillo A (2020) Identical or fraternal twins? The chemical homogeneity of wide binaries from Gaia DR2. Mon Not R Astron Soc 492:1164–1179, doi:10.1093/mnras/stz3132

Heger A, Fryer CL, Woosley SE, Langer N, Hartmann DH (2003) How massive single stars end their life. Astrophys J 591:288–300, doi:10.1086/375341

Helling C, Woitke P, Rimmer PB, Kamp I, Thi W-F, Meijerink R (2014) Disk evolution, element abundances and cloud properties of young gas giant planets. Life 4:142–173, doi:10.3390/life4020142

Hinkel NR, Unterborn CT (2018) The star–planet connection. I. Using stellar composition to observationally constrain planetary mineralogy for the 10 closest stars. Astrophys J 853:83, doi:10.3847/1538-4357/aaa5b4

Hinkel NR, Timmes FX, Young PA, Pagano MD, Turnbull MC (2014) Stellar abundances in the solar neighborhood: The Hypatia catalog. Astron J 148:54, doi:10.1088/0004-6256/148/3/54

Hinkel NR, Young PA, Pagano MD, Desch SJ, Anbar AD, Adibekyan V, Blanco-Cuaresma S, Carlberg JK, Mena ED, Liu F, Nordlander T (2016) A comparison of stellar elemental abundance techniques and measurements. Astrophys J Suppl 226:4, doi:10.3847/0067-0049/226/1/4

Hinkel NR, Young PA, Wheeler CH, III (2022) A concise treatise on converting stellar mass fractions to abundances to molar ratios. Astron J 164:256, doi:10.3847/1538-3881/ac9bfa

Howell SB, Matson RA, Marzari F (2022) Editorial: The effect of stellar multiplicity on exoplanetary systems. Front Astron Space Sci 8:830980, doi:10.3389/fspas.2021.830980

Huber D, Carter JA, Barbieri M, Miglio A, Deck KM, Fabrycky DC, Montet BT, Buchhave LA, Chaplin WJ, Hekker S, Montalbán J (2013) Stellar spin–orbit misalignment in a multiplanet system. Science 342:331–334, doi:10.1126/science.1242066

Iben I, Jr. (1967) Stellar evolution.VI. Evolution from the main sequence to the red-giant branch for stars of mass 1 M_Sun, 1.25 M_Sun, and 1.5 M_Sun. Astrophys J 147:624, doi:10.1086/149040

Jin S, Trager SC, Dalton GB, Aguerri JA, Drew JE, Falcón-Barroso J, Gänsicke BT, Hill V, Iovino A, Pieri MM, Poggianti BM (2022) The wide-field, multiplexed, spectroscopic facility WEAVE: Survey design, overview, and simulated implementation. arXiv e-prints:arXiv:2212.03981

Jofré P, Heiter U, Soubiran C (2019) Accuracy and precision of industrial stellar abundances. Annu Rev Astron Astrophys 57:571–616, doi:10.1146/annurev-astro-091918-104509

Jofré P, Heiter U, Soubiran C, Blanco-Cuaresma S, Masseron T, Nordlander T, Chemin L, Worley CC, Van Eck S, Hourihane A, Gilmore G (2015) Gaia FGK benchmark stars: abundances of alpha and iron-peak elements. Astron Astrophys 582:A81, doi:10.1051/0004-6361/201526604

Jofré P, Heiter U, Worley CC, Blanco-Cuaresma S, Soubiran C, Masseron T, Hawkins K, Adibekyan V, Buder S, Casamiquela L, Gilmore G (2017) Gaia FGK benchmark stars: opening the black box of stellar element abundance determination. Astron Astrophys 601:A38, doi:10.1051/0004-6361/201629833

Jura M, Young ED (2014) Extrasolar cosmochemistry. Annu Rev Earth Planet Sci 42:45–67, doi:10.1146/annurev-earth-060313-054740

Kreidberg L, Koll DD, Morley C, Hu R, Schaefer L, Deming D, Stevenson KB, Dittmann J, Vanderburg A, Berardo D, Guo X (2019) Absence of a thick atmosphere on the terrestrial exoplanet LHS 3844b. Nature 573:87–90, doi:10.1038/s41586-019-1497-4

Kroupa P (2001) On the variation of the initial mass function. Mon Not R Astron Soc 322:231–246, doi:10.1046/j.1365-8711.2001.04022.x

Kurtz DW (2022) Asteroseismology across the Hertzsprung–Russell diagram. Annu Rev Astron Astrophys 60:31–71, doi:10.1146/annurev-astro-052920-094232

Laughlin G, Bodenheimer P, Adams FC (2004) The core accretion model predicts few jovian-mass planets orbiting red dwarfs. Astrophys J Letters 612:L73–L76, doi:10.1086/424384

Limbach MA, Soares-Furtado M, Vanderburg A, Best WM, Cody AM, D'Onghia E, Heller R, Hensley BS, Kounkel M, Kraus A, Mann AW (2023) The TEMPO survey. I. Predicting yields of transiting exosatellites, moons, and planets from a 30 days survey of Orion with the Roman Space Telescope. Publ Astron Soc Pac 135:014401, doi:10.1088/1538-3873/acafa4

Lingam M, Hein AM, Eubanks TM (2023) Chasing nomadic worlds: A new class of deep space missions. arXiv e-prints:arXiv:2307.12411, doi:10.48550/arXiv.2307.12411

Linsky JL (1980) Stellar chromospheres. Annu Rev Astron Astrophys 18:439–488, doi:10.1146/annurev.aa.18.090180.002255

Linsky JL (2017) Stellar model chromospheres and spectroscopic diagnostics. Annu Rev Astron Astrophys 55:159–211, doi:10.1146/annurev-astro-091916-055327

Lissauer JJ (1993) Planet formation. Annu Rev Astron Astrophys 31:129–174, doi:10.1146/annurev.aa.31.090193.001021

Lisse CM, Desch SJ, Unterborn CT, Kane SR, Young PR, Hartnett HE, Hinkel NR, Shim S-H, Mamajek EE, Izenberg NR (2020) A geologically robust procedure for observing rocky exoplanets to ensure that detection of atmospheric oxygen is a modern Earth-like biosignature. Astrophys J Letters 898:L17, doi:10.3847/2041-8213/ab9b91

Lodders K (2003) Solar System abundances and condensation temperatures of the elements. Astrophys J 591:1220–1247, doi:10.1086/375492

Lodders K (2019) Solar Elemental Abundances. The Oxford Research Encyclopedia of Planetary Science, Oxford University Press:arXiv:1912.00844

Lodders K, Palme H, Gail H-P (2009) Abundances of the elements in the Solar System. Landolt Boumlrnstein 4B:712, doi:10.1007/978-3-540-88055-4_34

Lodieu N, Hambly NC, Cross NJG (2021) Exploring the planetary-mass population in the Upper Scorpius association. Mon Not R Astron Soc 503:2265–2279, doi:10.1093/mnras/stab401

Lucas PW, Roche PF (2000) A population of very young brown dwarfs and free-floating planets in Orion. Mon Not R Astron Soc 314:858–864, doi:10.1046/j.1365-8711.2000.03515.x

Lugaro M, Ek M, Pet Ho M, Pignatari M, Makhatadze GV, Onyett IJ, Schönbächler M (2023) Representation of s-process abundances for comparison to data from bulk meteorites. Eur Phys J A 59:53, doi:10.1140/epja/s10050-023-00968-y

Luhman KL (2012) The formation and early evolution of low-mass stars and brown dwarfs. Annu Rev Astron Astrophys 50:65–106, doi:10.1146/annurev-astro-081811-125528

Martig M, Minchev I, Ness M, Fouesneau M, Rix H-W (2016) A radial age gradient in the geometrically thick disk of the Milky Way. Astrophys J 831:139, doi:10.3847/0004-637X/831/2/139

Masseron T, Gilmore G (2015) Carbon, nitrogen and alpha-element abundances determine the formation sequence of the Galactic thick and thin discs. Mon Not R Astron Soc 453:1855–1866, doi:10.1093/mnras/stv1731

McDonough WF (2003) 2.15 - Compositional model for the Earth's core. In: Treatise on Geochemistry. Pergamon, Oxford, p 547–568

McKee CF, Ostriker EC (2007) Theory of star formation. Annu Rev Astron Astrophys 45:565-687, doi:10.1146/annurev.astro.45.051806.110602

Meadows VS, Barnes RK (2018) Factors affecting exoplanet habitability. In: Handbook of Exoplanets. Deeg HJ, Belmonte JA (eds). SpringerLink, p 2771–2794

Millikan RA, Bowen IS (1924) Extreme ultra-violet spectra. Phys Rev 23:1–34, doi:10.1103/PhysRev.23.1

Miret-Roig N, Bouy H, Raymond SN, Tamura M, Bertin E, Barrado D, Olivares J, Galli PA, Cuillandre JC, Sarro LM, Berihuete A (2022) A rich population of free-floating planets in the Upper Scorpius young stellar association. Nat Astron 6:89–97, doi:10.1038/s41550-021-01513-x

Mordasini C, Burn R (2024) Planet formation—Observational constraints, physical processes, and compositional patterns. Rev Mineral Geochem 90:55–112

Mordasini C, van Boekel R, Mollière P, Henning T, Benneke B (2016) The imprint of exoplanet formation history on observable present-day spectra of hot Jupiters. Astrophys J 832:41, doi:10.3847/0004-637X/832/1/41

Mulders GD, Pascucci I, Apai D (2015) A stellar-mass-dependent drop in planet occurrence rates. Astrophys J 798:112, doi:10.1088/0004-637X/798/2/112

Mulders GD, Pascucci I, Apai D, Frasca A, Molenda-Żakowicz J (2016) A super-solar metallicity for stars with hot rocky exoplanets. Astron J 152:187, doi:10.3847/0004-6256/152/6/187

Ness MK, Johnston KV, Blancato K, Rix H-W, Beane A, Bird JC, Hawkins K (2019) In the galactic disk, stellar [Fe/H] and age predict orbits and precise [X/Fe]. Astrophys J 883:177, doi:10.3847/1538-4357/ab3e3c

Niu Z, Yuan H, Wang S, Liu J (2021) Binary fractions of G and K dwarf stars based on Gaia EDR3 and LAMOST DR5: Impacts of the chemical abundances. Astrophys J 922:211, doi:10.3847/1538-4357/ac2573

Nomoto Ki, Kobayashi C, Tominaga N (2013) Nucleosynthesis in stars and the chemical enrichment of galaxies. Annu Rev Astron Astrophys 51:457–509, doi:10.1146/annurev-astro-082812-140956

Öberg KI, Murray-Clay R, Bergin EA (2011) The effects of snowlines on C/O in planetary atmospheres. Astrophys J Letters 743:L16, doi:10.1088/2041-8205/743/1/L16

Ossendrijver M (2003) The solar dynamo. Astron Astrophysr 11:287–367, doi:10.1007/s00159-003-0019-3

Otegi JF, Dorn C, Helled R, Bouchy F, Haldemann J, Alibert Y (2020) Impact of the measured parameters of exoplanets on the inferred internal structure. Astron Astrophys 640:A135, doi:10.1051/0004-6361/20203800610.48550/arXiv.2006.12353

Parker EN (1965) The passage of energetic charged particles through interplanetary space. Planet Space Sci 13:9–49, doi:10.1016/0032-0633(65)90131-5

Pass EK, Winters JG, Charbonneau D, Irwin JM, Latham DW, Berlind P, Calkins ML, Esquerdo GA, Mink J (2023) Mid-to-late M dwarfs lack Jupiter analogs. Astron J 166:11, doi:10.3847/1538-3881/acd349

Paxton B, Bildsten L, Dotter A, Herwig F, Lesaffre P, Timmes F (2011) Modules for Experiments in Stellar Astrophysics (MESA). Astrophys J Suppl 192:3, doi:10.1088/0067-0049/192/1/3

Paxton B, Schwab J, Bauer EB, Bildsten L, Blinnikov S, Duffell P, Farmer R, Goldberg JA, Marchant P, Sorokina E, Thoul A (2018) Modules for Experiments in Stellar Astrophysics (MESA): Convective boundaries, element diffusion, and massive star explosions. Astrophys J Suppl 234:34, doi:10.3847/1538-4365/aaa5a8

Payne CH (1925a) Stellar Atmospheres; a Contribution to the Observational Study of High Temperature in the Reversing Layers of Stars. Radcliffe College.

Payne CH (1925b) Astrophysical data bearing on the relative abundance of the elements. PNAS 11:192–198, doi:10.1073/pnas.11.3.192

Pearson SG, McCaughrean MJ (2023) Jupiter mass binary objects in the Trapezium cluster. arXiv e-prints:arXiv:2310.01231, doi:10.48550/arXiv.2310.01231

Pecaut MJ, Mamajek EE (2013) Intrinsic colors, temperatures, and bolometric corrections of pre-main-sequence stars. Astrophys J Suppl 208:9, doi:10.1088/0067-0049/208/1/9

Peña Ramírez K, Béjar VJS, Zapatero Osorio MR, Petr-Gotzens MG, Martín EL (2012) New isolated planetary-mass objects and the stellar and substellar mass function of the sigma Orionis cluster. Astrophys J 754:30, doi:10.1088/0004-637X/754/1/30

Penza V, Berrilli F, Bertello L, Cantoresi M, Criscuoli S, Giobbi P (2022) Total solar irradiance during the last five centuries. Astrophys J 937:84, doi:10.3847/1538-4357/ac8a4b

Plotnykov M, Valencia D (2020) Chemical fingerprints of formation in rocky super-Earths' data. Mon Not R Astron Soc 499:932–947, doi:10.1093/mnras/staa2615 10.48550/arXiv.2010.06480

Putirka KD, Rarick JC (2019) The composition and mineralogy of rocky exoplanets: A survey of >4000 stars from the Hypatia Catalog. Am Mineral 104:817–829, doi:10.2138/am-2019-6787

Putirka K, Xu S (2021) On the lithology and mineralogy of polluted white dwarf materials. Bull Am Astron Soc 53:1044

Putirka KD, Dorn C, Hinkel NR, Unterborn CT (2021) Compositional diversity of rocky exoplanets. Elements 17:235, doi:10.48550/arXiv.2108.08383

Raghavan D, McAlister HA, Henry TJ, Latham DW, Marcy GW, Mason BD, Gies DR, White RJ, ten Brummelaar TA (2010) A survey of stellar families: Multiplicity of solar-type stars. Astrophys J Suppl 190:1–42, doi:10.1088/0067-0049/190/1/1

Rasio FA, Ford EB (1996) Dynamical instabilities and the formation of extrasolar planetary systems. Science 274:954–956, doi:10.1126/science.274.5289.954

Rebull LM, Stauffer JR, Bouvier J, Cody AM, Hillenbrand LA, Soderblom DR, Valenti J, Barrado D, Bouy H, Ciardi D, Pinsonneault M (2016) Rotation in the Pleiades with K2. I. Data and first results. Astron J 152:113, doi:10.3847/0004-6256/152/5/113

Recio-Blanco A, De Laverny P, Kordopatis G, Helmi A, Hill V, Gilmore G, Wyse R, Adibekyan V, Randich S, Asplund M, Feltzing S (2014) The Gaia-ESO Survey: the galactic thick to thin disc transition. Astron Astrophys 567:A5, doi:10.1051/0004-6361/201322944

Ribas Á, Bouy H, Merín B (2015) Protoplanetary disk lifetimes vs. stellar mass and possible implications for giant planet populations. Astron Astrophys 576:A52, doi:10.1051/0004-6361/201424846

Rix H-W, Bovy J (2013) The Milky Way's stellar disk. Mapping and modeling the galactic disk. Astron Astrophys 21:61, doi:10.1007/s00159-013-0061-8

Rodríguez Martínez R, Martin DV, Gaudi BS, Schulze JG, Asnodkar AP, Boley KM, Ballard S (2023) A comparison of the composition of planets in single-planet and multiplanet systems orbiting M dwarfs. Astron J 166:137, doi:10.3847/1538-3881/aced9a

Santos NC, Adibekyan V, Mordasini C, Benz W, Delgado-Mena E, Dorn C, Buchhave L, Figueira P, Mortier A, Pepe F, Santerne A (2015) Constraining planet structure from stellar chemistry: the cases of CoRoT-7, Kepler-10, and Kepler-93. Astron Astrophys 580:L13, doi:10.1051/0004-6361/201526850

Santos NC, Adibekyan V, Dorn C, Mordasini C, Noack L, Barros SC, Delgado-Mena E, Demangeon O, Faria JP, Israelian G, Sousa SG (2017) Constraining planet structure and composition from stellar chemistry: trends in different stellar populations. Astron Astrophys 608:A94, doi:10.1051/0004-6361/20173135910.48550/arXiv.1711.00777

Schlichting HE, Young ED (2022) Chemical equilibrium between cores, mantles, and atmospheres of super-Earths and sub-Neptunes, and implications for their compositions, interiors and evolution. Planet Sci J 3:127, doi:10.48550/arxiv.2107.10405

Scholz A, Jayawardhana R, Muzic K, Geers V, Tamura M, Tanaka I (2012) Substellar Objects in Nearby Young Clusters (SONYC). VI. The planetary-mass domain of NGC 1333. Astrophys J 756:24, doi:10.1088/0004-637X/756/1/24

Schulze JG, Wang J, Johnson JA, Gaudi BS, Unterborn CT, Panero WR (2021) On the probability that a rocky planet's composition reflects its host star. Planet Sci J 2:113, doi:10.3847/PSJ/abcaa8

Shu FH, Adams FC, Lizano S (1987) Star formation in molecular clouds: observation and theory. Annu Rev Astron Astrophys 25:23-81, doi:10.1146/annurev.aa.25.090187.000323

Skumanich A (1972) Time scales for Ca II emission decay, rotational braking, and lithium depletion. Astrophys J 171:565, doi:10.1086/151310

Smiljanic R, Korn AJ, Bergemann M, Frasca A, Magrini L, Masseron T, Pancino EL, Ruchti G, San Roman I, Sbordone L, Sousa SG (2014) The Gaia-ESO Survey: The analysis of high-resolution UVES spectra of FGK-type stars. Astron Astrophys 570:A122, doi:10.1051/0004-6361/201423937

Smith RJ (2020) Evidence for initial mass function variation in massive early-type galaxies. Annu Rev Astron Astrophys 58:577–615, doi:10.1146/annurev-astro-032620-020217

Soderblom DR (2010) The ages of stars. Annu Rev Astron Astrophys 48:581-629, doi:10.1146/annurev-astro-081309-130806

Soderblom DR (2015) Ages of stars: Methods and uncertainties. *In*: Asteroseismology of Stellar Populations in the Milky Way. Astrophysics and Space Science Proceedings, vol 39. Miglio A, Eggenberger P, Girardi L, Montalbán J (eds) Springer, Cham, p 3–9

Sotin C, Grasset O, Mocquet A (2007) Mass radius curve for extrasolar Earth-like planets and ocean planets. Icarus 191:337–351, doi:10.1016/j.icarus.2007.04.006

Stanford-Moore SA, Nielsen EL, De Rosa RJ, Macintosh B, Czekala I (2020) BAFFLES: Bayesian ages for field lower-mass stars. Astrophys J 898:27, doi:10.3847/1538-4357/ab9a35

Steinmetz M, Zwitter T, Siebert A, Watson FG, Freeman KC, Munari U, Campbell R, Williams M, Seabroke GM, Wyse RF, Parker QA (2006) The Radial Velocity Experiment (RAVE): First data release. Astron J 132:1645–1668, doi:10.1086/506564

Thiabaud A, Marboeuf U, Alibert Y, Leya I, Mezger K (2015) Elemental ratios in stars vs planets. Astron Astrophys 580:A30, doi:10.1051/0004-6361/201525963

Unterborn CT, Panero WR (2019) The pressure and temperature limits of likely rocky exoplanets. J Geophys Res (Planets) 124:1704–1716, doi:10.1029/2018JE005844

Vanderburg A, Rappaport SA, Xu S, Crossfield IJ, Becker JC, Gary B, Murgas F, Blouin S, Kaye TG, Palle E, Melis C (2020) A giant planet candidate transiting a white dwarf. Nature 585:363–367, doi:10.1038/s41586-020-2713-y

Veras D, Raymond SN (2012) Planet–planet scattering alone cannot explain the free-floating planet population. Mon Not R Astron Soc 421:L117–L121, doi:10.1111/j.1745-3933.2012.01218.x

Walker IR, Mihos JC, Hernquist L (1996) Quantifying the fragility of galactic disks in minor mergers. Astrophys J 460:121, doi:10.1086/176956

Winters JG, Henry TJ, Jao W-C, Subasavage JP, Chatelain JP, Slatten K, Riedel AR, Silverstein ML, Payne MJ (2019) The solar neighborhood. XLV. The stellar multiplicity rate of M dwarfs within 25 pc. Astron J 157:216, doi:10.3847/1538-3881/ab05dc

Wolszczan A, Frail DA (1992) A planetary system around the millisecond pulsar PSR1257 + 12. Nature 355:145–147, doi:10.1038/355145a0

Wood BE, Müller HR, Redfield S, Konow F, Vannier H, Linsky JL, Youngblood A, Vidotto AA, Jardine M, Alvarado-Gómez JD, Drake JJ (2021) New observational constraints on the winds of M dwarf stars. Astrophys J 915:37, doi:10.3847/1538-4357/abfda5

Xu S, Zuckerman B, Dufour P, Young ED, Klein B, Jura M (2017) The chemical composition of an extrasolar Kuiper-belt-object. Astrophys J Lett 836:L7, doi:10.3847/2041-8213/836/1/L7

Yang J-Y, Xie J-W, Zhou J-L (2020) Occurrence and architecture of Kepler planetary systems as functions of stellar mass and effective temperature. Astron J 159:164, doi:10.3847/1538-3881/ab7373

Zhang K (2024) Chemistry in protoplanetary disks. Rev Mineral Geochem 90:27-54

Zieba S, Kreidberg L, Ducrot E, Gillon M, Morley C, Schaefer L, Tamburo P, Koll DD, Lyu X, Acuña L, Agol E (2023) No thick carbon dioxide atmosphere on the rocky exoplanet TRAPPIST-1 c. Nature 620:746–749, doi:10.1038/s41586-023-06232-z

Zweibel EG, Yamada M (2009) Magnetic reconnection in astrophysical and laboratory plasmas. Annu Rev Astron Astrophys 47:291–332, doi:10.1146/annurev-astro-082708-1017

Reviews in Mineralogy & Geochemistry
Vol. 90 pp. 27–54, 2024
Copyright © Mineralogical Society of America

2

Chemistry in Protoplanetary Disks

Ke Zhang

University of Wisconsin-Madison
475 N Charter St, Madison, WI 53706 USA

ke.zhang@wisc.edu

INTRODUCTION

Planets are formed inside disks around young stars. The gas, dust, and ice in these natal disks are the building materials of planets, and therefore, their compositions fundamentally shape the final chemical compositions of planets. In this review, we summarize current observations of molecular lines in protoplanetary disks, from near-infrareffd to millimeter wavelengths. We discuss the basic types of chemical reactions in disks and the current development of chemical modeling. In particular, we highlight the progress made in understanding snowline locations, abundances of main carriers of carbon, oxygen, and nitrogen, and complex organic molecules in disks. Finally, we discuss efforts to trace planet formation history by combining the understanding of disk chemistry and planet formation processes.

OVERVIEW: CHEMISTRY IN PLANET-FORMING DISKS

The habitability of a planet depends not only on its physical properties such as mass, radius, and orbital location, but also on its chemical properties, such as the water and organic contents on its surface and atmosphere. A fundamental question of the planet formation field revolves around how planet formation processes determine the chemical compositions of nascent planets. Planets are formed inside disks around young stars (see reviews by Williams and Cieza 2011; Andrews 2020), the so-called protoplanetary disks. In these natal disks, the gas, dust, and ice serve as the raw building materials of planets. The compositions of these raw materials are first inherited from the compositions of the host stars themselves (see more details in the Hinkel et al. 2024, this volume). The materials are then further altered and redistributed at the stage of protoplanetary disks (e.g., Krijt et al. 2023; Öberg et al. 2023). Therefore, a thorough understanding of chemistry in these natal disks is an essential part of our understanding of planet formation and the potential habitability of extrasolar planetary systems.

The field of chemistry in protoplanetary disks has two goals. The first one is to understand the compositions of materials in the natal disks, and how the compositions of disk materials change over location and time in disks during planet formation. The second one, which goes beyond the pure chemistry aspect, aims to understand how we can use molecular and atomic lines as probes to study physical conditions and processes during planet formation, which are otherwise hard or even impossible to constrain. The two goals are intertwined as physical and chemical processes are often coupled with each other.

In this review, we discuss recent observations of chemistry in protoplanetary disks, theoretical expectations of chemical reactions in protoplanetary disks, the latest important insights into chemistry in disks, and connections between disk chemistry to compositions of exoplanets and solar system objects. Finally, we highlight a few promising frontiers in the field for the upcoming decade. There have been excellent recent reviews on chemistry in

1529-6466/24/0090-0002$05.00 (print)
1943-2666/24/0090-0002$05.00 (online)

http://dx.doi.org/10.2138/rmg.2024.90.02

protoplanetary disks (e.g., Henning and Semenov 2013; Öberg and Bergin 2021; Öberg et al. 2023). We are not trying to replicate these efforts; instead, this review aims to provide an easily accessible introduction material for interdisciplinary readers.

BASICS OF PROTOPLANETARY DISKS

Planets are born in gas- and dust-rich disks around young stars, the so-called protoplanetary disks (see reviews by Williams and Cieza 2011; Andrews 2020). These disks are analogs of our Solar system at 4.6 billion years ago, providing precious windows for us to witness the formation of planets. The gas, dust, and ice in these disks are raw materials to build planets. As a disk evolves, the amount of disk materials and their compositions may change dramatically. As a result, the final interior structure/atmosphere of a planet depends on how, when, and what types of materials are accreted.

The chemical compositions and structures of protoplanetary disks are largely determined by the physical properties of the disk, in particular, the structures of density, temperature, and radiation field (e.g., Bergin et al. 2007). The characterization of disk physical properties, such as the gas and dust mass distribution, has made revolutionary progress over the past decade, thanks to the Atacama Large Millimeter/submillimeter Array (ALMA) (see review by Miotello et al. 2023). So far, ~1500 disks in the nearby star-forming regions within 500 pc have been observed by ALMA at millimeter continuum, and around a few hundred disks with at least CO molecular line observations (e.g., Ansdell et al. 2016; Barenfeld et al. 2016; Long et al. 2017; Cieza et al. 2019; Tobin et al. 2020). A large diversity of physical properties has been seen among protoplanetary disks. For example, even in the same stellar mass, there are two orders of magnitude differences in the dust disk masses. This diversity in physical properties would naturally lead to diversity in chemical structures, which subsequently contributes to the large diversity seen in extrasolar planets.

The gas in protoplanetary disks can last for a few Myr (e.g., Fedele et al. 2010; Ercolano and Pascucci 2017). Gas can be accreted onto the central star, used to form gaseous envelops of planets, and blown away by disk winds. The characteristic lifetime scale of protoplanetary disks is estimated to be ~3 Myr, based on the decreasing fraction of young stars with near-infrared excess or stars showing accretions. This lifetime scale sets a fundamental limit for the timescale of the giant planet formation. The amount of dust masses (grains with sizes up to 1 cm) is found to generally decrease over time, with a median dust mass of $1 M_\oplus$ for 1–3 Myr disks to a median of $0.1 M_\oplus$ for 5–10 Myr disks (Manara et al. 2023). The gas masses of protoplanetary disks are still highly uncertain, because the dominant mass constitute H_2 does not have a strong emission line at the bulk temperature of disks (e.g., Bergin and Williams 2017; Trapman et al. 2022). Most of the gas disk estimations are based on dust mass and the assumption of a gas-to-dust mass of 100, which are probably upper limits. The majority of disks are likely to have a lower gas mass compared to the Minimum Solar Nebular Mass (~$0.01 M_\odot$, the minimum mass required to form solar system plants, Hayashi 1981). Besides the large intrinsic spread in global disk properties, there are other complications in the study of chemistry in protoplanetary disks. One complication is that these disks have large gradients in the temperature, density, and radiation fields, in the radial as well as the vertical direction (e.g., Calahan et al. 2021; Law et al. 2021). For example, the inner 1 au region can be quite hot, reaching 500–1500 K (e.g., Carr and Najita 2008; Salyk et al. 2008; Dullemond and Monnier 2010), while the bulk region outside 100 au is below 20 K (e.g., Dartois et al. 2003; Pinte et al. 2018; Law et al. 2021). As a result, the chemical compositions of a given disk vary with radial and vertical locations. Another complication is that the physical conditions in the disk evolve over time at a comparable timescale as that of many chemical reactions. As a result, our predictive power on chemistry in disks also depends on our understanding of the physical processes in disks.

OBSERVATIONS OF MOLECULES IN PROTOPLANETARY DISKS

Unlike studies of many objects in the Solar System objects, where in-situ samples can be collected and measured, the observations of protoplanetary disks are solely based on remote observations of ground-based and space telescopes. Molecules in protoplanetary disks have been detected at a wide range of wavelengths, from UV to cm wavelengths (see review by Henning and Semenov 2013). However, lines within a given wavelength range only trace a limited range of radial and vertical locations in the disk, due to large temperature and density gradients in protoplanetary disks. For example, the mid-IR wavelengths trace lines mostly arise from warm gas ($T > 500$ K), which is found in the inner few au around solar-like stars, while the pure rotational lines of heavy molecules are found from the colder and outer regions of disks. As a result, our observations of protoplanetary disks are similar to blind men's observations of an elephant, and therefore multi-wavelength observations are needed to characterize the chemical structures at different regions of protoplanetary disks.

Infrared and millimeter wavelengths are the two primary windows to probe chemistry in protoplanetary disks. The vibrational bands of molecular lines are mainly at infrared wavelengths and pure rotational lines at millimeter wavelengths. Below we describe the main molecules detected within these wavelength ranges.

INFRARED OBSERVATIONS OF PROTOPLANETARY DISKS

Near-IR wavelength range (1–5 μm)

At the near-IR wavelength range (1–5 μm), CO is the most commonly detected molecular line, seen in many protoplanetary disks around 0.5–3 $M_\odot$ stars (e.g., Carr et al. 1993; Blake and Boogert 2004; Brittain et al. 2007; Salyk et al. 2011; Brown et al. 2013; Banzatti and Pontoppidan 2015). The fundamental band of rovibrational lines ($v = 1$–0) at 4.7 μm is the most frequently detected feature, and the overtone band ($v = 2$–0) at 2.3 μm is also detected in a small number of protoplanetary disks. The CO line profiles can be spectrally resolved by high-resolution ground-based observations (R~25,000–100,000), which provides information on the velocity field of the CO-emitting areas. The line shapes of CO rovibrational lines often show two distinctive components, a narrow component with FWHM of 10–50 km/s, consistent with Keplerian rotation between 0.1—a few au, and a broad velocity component with FWHM of 50–200 km/s, arising from ~0.05 au (e.g., Banzatti et al. 2022).

H_2O has three vibrational modes, and the symmetric and asymmetric stretching bands between 2.5–3.5 μm have been detected in many disks from ground-based observations (e.g., Salyk et al. 2008, 2019; Banzatti et al. 2022). These detections suggest that hot water vapor (~1000 K) commonly exists inside 0.5 au region of protoplanetary disks. The detections of organic molecules at near-IR wavelengths have been challenging from ground-based observations. HCN and C_2H_2 were detected only in a few disks (Mandell et al. 2012). CH_4 has only been detected in absorption features in one disk (Gibb and Horne 2013).

Mid-IR wavelength (5–30 μm)

At mid-infrared wavelength range (5–30 μm), simple molecules like H_2O, CO, CO_2, HCN, C_2H_2, OH, and H_2, have been observed in ~100 protoplanetary disks by the Spitzer space telescope with a spectral resolution of R~600 (e.g., Carr and Najita 2008; Salyk et al. 2008, 2011; Pontoppidan et al. 2010; Carr and Najita 2011). These molecules are detected via their rovibrational or rotational lines at mid-IR. The upper energy levels of these lines are generally very high, spanning over 500–10,000 K (Pontoppidan et al. 2010). Therefore, these lines are expected to be emitted from the hot region inside a few aus from the central star. Simple slab models (assuming constant temperature and density) showed that typical excitation temperatures between 400–1600 K and emitting areas with a radius of ~1 au (e.g., Carr and Najita 2011; Salyk et al. 2011). In a few cases, a small number of mid-IR water lines are also observed by the ground-based highresolution spectrograph, which provided direct constraints on the line profiles and therefore the line emitting regions (Pontoppidan et al. 2011; Salyk et al. 2019). These lines showed broad line profiles with FWHM~ 20–35 km/s, consistent with the velocity range of Keplerian rotation from the inner 1 au region. In short, the mid-IR lines provide an important window to trace chemical compositions in the terrestrial planet-forming regions.

Interestingly, Spitzer's observations showed that detection rates of molecular lines in disks are correlated with the masses of the host stars. The detection rates of H_2O and small organic are generally high (~30–60%) in T Tauri disks (stellar mass below 1 solar mass), but drop to nearly zero for Herbig disks (stellar mass between 2–3 solar mass) (Pontoppidan et al. 2010). It is still under debate whether the dearth of water detection in Herbig disks is due to intrinsic difference in the water abundance between the two types of disks or due to the high IR continuum in the more luminous Herbig disks blanketing the weak molecular lines (e.g., Antonellini et al. 2016). Another interesting dependence of stellar masses was the HCN/H_2O line ratios seen in disks around late M-dwarf stars (0.3 $M_\odot$ or smaller) on average are significantly higher than that of the solar-like type stars, suggesting a higher C/O elemental ratio in the gas of inner disk regions in late M-dwarf sources (Pascucci et al. 2013). However, the late M-dwarf sample size was small (~10) and more observations are needed to confirm the trend.

The inventory of the molecules detected in mid-IR is expected to significantly expand over the next few years, thanks to the newly launched James Webb Space Telescope (JWST). Compared to Spitzer, JWST offers an order of magnitude higher sensitivity and a factor of few higher spectral resolution. See Figure 1 for an example of the JWST/MIRI spectrum of a protoplanetary disk.

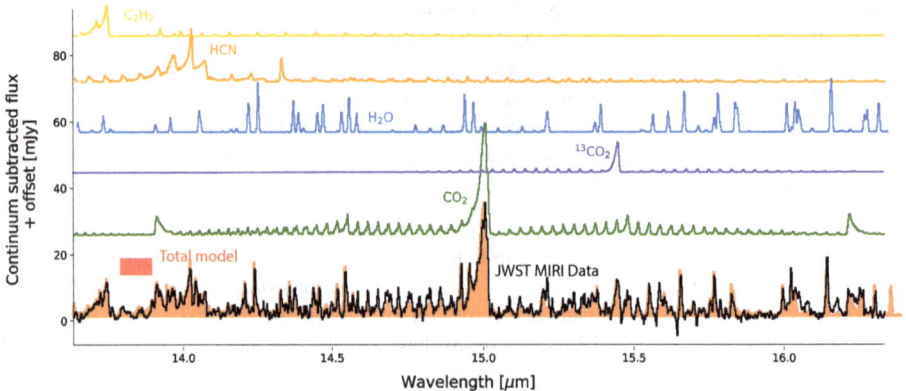

Figure 1. JWST MIRI/MRS continuum-subtracted spectrum of the GW Lup disk between 13.5–16.5 μm (Grant et al. 2023). The GW Lup protoplanetary is located in the Lupus star-forming region (1–3 Myr-old). The JWST-MIRI data (**black**) is compared to a slab model (**red**) composed of different molecules.

Recent JWST discoveries include the first detection of $^{13}CO_2$ in a disk around a solar-like star (Grant et al. 2023), and a suite of organics were detected in a disk around a very low mass young star ($M_* = 0.1 M_\odot$), including C_4H_2, C_6H_6, $^{13}C^{12}CH_2$, and a tentative detection of CH_4 (Tabone et al. 2023). Also, JWST recently detected CH_3^+ in a protoplanetary disk in the Orion star-forming region, supporting that gas-phase organic chemistry is activated by ultraviolet radiation (Berné et al. 2023).

Far-IR wavelength range (30–550 μm)

In the far-IR wavelength range (30–550 μm), most of the disk observations were provided by the Herschel Space Telescope. The HD $J = 1$–0 line was detected in three protoplanetary disks, which provided important constraints on the gas disk masses (Bergin et al. 2013; McClure et al. 2016). The far-IR wavelength range covers a wide range of water lines ($E_{up} = 53$–1100 K) that trace cold water vapors (150 K or lower) beyond the surface water snowline (Zhang et al. 2013; Blevins et al. 2016). However, there were few far-IR water line detections in protoplanetary disks, compared to the theoretical predictions. The cold water vapor was only detected in the TW Hya protoplanetary disk (a T Tauri disk), the HD 100546 disk (a Herbig disk) and tentatively in the HD 163296 disk (Herbgi disk) (Bergin et al. 2010; Hogerheijde et al. 2011; Fedele et al. 2012; Meeus et al. 2012; Riviere-Marichalar et al. 2012; Du et al. 2017). High J rotational lines of CO are commonly detected in Herbig disks and in some T Tauri disks (e.g., Bruderer et al. 2012; Fedele et al. 2012; Meeus et al. 2012; Bergin et al. 2013). NH_3 and N_2H^+ were detected in the TW Hya disk (Salinas et al. 2016). The dearth of cold water vapor lines was the first evidence that volatile may be largely depleted in the atmosphere of the outer disk region.

MILLIMETER WAVELENGTH OBSERVATIONS OF PROTOPLANETARY DISKS

Another important window to probe chemistry in protoplanetary disks is the millimeter wavelength range (~0.3–3 mm). A great advantage of the mm-wavelength observations is that they provide spatially resolved images (down to 30milliarcsec) and high spectral resolution (down to 0.01 km/s). This observational window has made revolutionary progress over the past decade, thanks to the Atacama Large Millimeter/submillimeter Array (ALMA). ALMA can observe molecular line emissions of nearby protoplanetary disks with spatial resolution down to 10–20 au (e.g., Öberg et al. 2021). The spatially resolved observations provide important constraints to the chemical abundances, temperature structures, and gas mass distributions in the 20–400 au regions of protoplanetary disks (e.g., Dartois et al. 2003; Schwarz et al. 2016; Pinte et al. 2018; Law et al. 2021; Öberg et al. 2021; Zhang et al. 2021). Figure 2 shows examples of ALMA images of various of molecular line emissions from the HD 163296 protoplanetary disk.

In mm wavelength observations, ^{12}CO is still the most wildly detected molecule in protoplanetary disks, ubiquitously seen in hundreds of nearby protoplanetary disks (e.g., Ansdell et al. 2016; Barenfeld et al. 2016; Long et al. 2017). Its main isotopologues, ^{13}CO and $C^{18}O$ are also commonly detected. The even more rare isotopologues, $C^{17}O$, $^{13}C^{18}O$, $^{13}C^{17}O$, have been detected in several bright protoplanetary disks (e.g., Booth et al. 2019; Booth and Ilee 2020; Zhang et al. 2020a, 2021). Due to its high abundance and stable chemical nature, CO and its isotopologue lines are usually used as a probe to study the distribution of gas mass and temperature in protoplanetary disks (e.g., Dartois et al. 2003; Schwarz et al. 2016; Zhang et al. 2017, 2019, 2021; Pinte et al. 2018; Law et al. 2021).

Simple polyatomic molecules are also commonly detected in protoplanetary disks, including H_2CO, HCN, DCN, HNC, CS, C_2H, c-C_3H_2, CN, DCN (e.g., Dutrey et al. 1997; Öberg et al. 2010; Bergin et al. 2016; Cleeves et al. 2016; van Terwisga et al. 2019; Pegues et al. 2020; Guzmán et al. 2021). These molecules are often studied with CO isotopologue lines together, to measure their relative abundances to CO, and to constrain the general elemental C/O, C/H, and N/H ratios in the disk gas (e.g., Bergin et al. 2016; Cleeves et al. 2018; Miotello et al. 2019; Bosman et al. 2021).

For molecular ions, N_2H^+, HCO^+, DCO^+, and $H^{13}CO^+$, are the most commonly detected species in protoplanetary disks (e.g., Öberg et al. 2010; Qi et al. 2013, 2015, 2019; Huang et al.

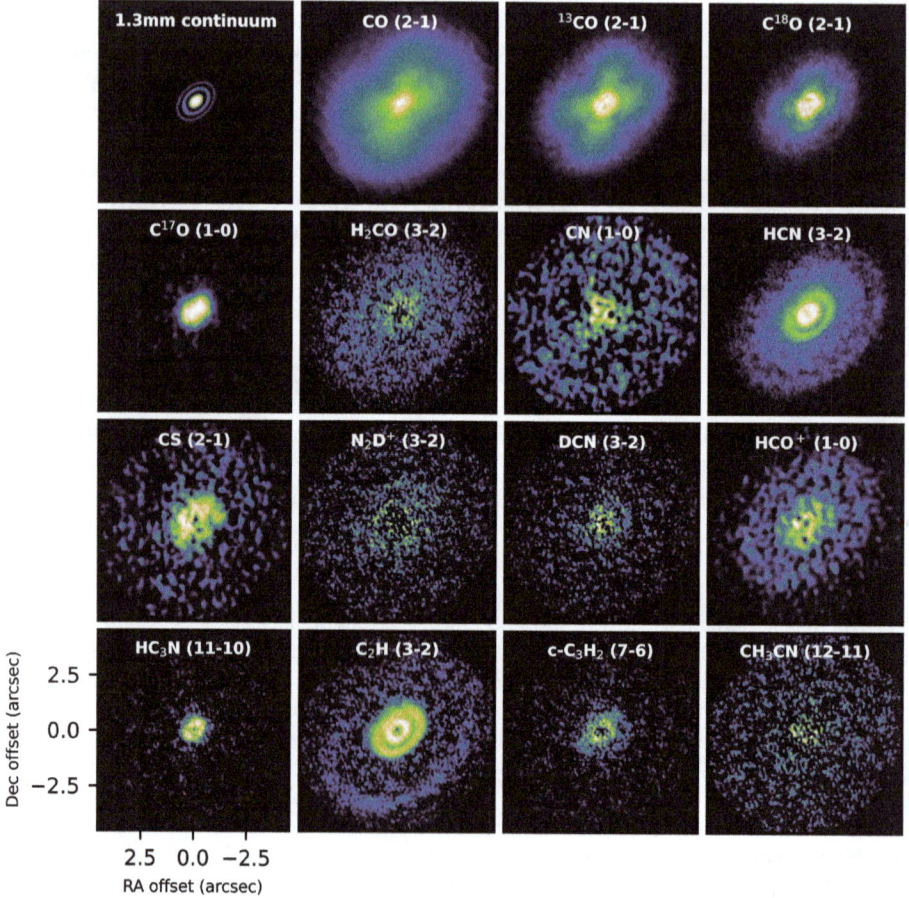

Figure 2. ALMA images of 15 molecular faces of the HD 163296 disk. Data are adopted from Öberg et al. 2021. These comprise a representative, but non-exhaustive, set of zeroth moment maps towards the HD 163296 disk. The diversity of molecular emission is due to the differences in the temperature of the emission region and abundance distributions of molecules.

2017; Anderson et al. 2019, 2022). These ions provide important constraints on the ionization rate in protoplanetary disks (e.g., Cleeves et al. 2015, 2017; Seifert et al. 2021; Trapman et al. 2022). Besides as an ionization tracer, N_2H^+ is also used as a tracer to probe the mid-plane CO snowline, because its formation and destruction are closely related to gas phase CO (e.g., Qi et al. 2013, 2015, 2019; van't Hoff et al. 2017; Anderson et al. 2019, 2021; Trapman et al. 2022).

The detections of H_2O in disks via ground-based millimeter observations are shown to be really challenging, especially in older than 1 Myr disks (Notsu et al. 2016, 2019; Carr et al. 2018). HDO and $H_2^{18}O$ lines have been detected in a young disk (< 1 Myr-old) which is particularly warm due to its ongoing accretion outburst (Tobin et al. 2023).

This is of great interest to study the abundances of complex organic molecules (COMs), defined as unsaturated carbon-bearing molecules with five or more atoms (Herbst and van Dishoeck 2009). However, COMs appear to be difficult to detect in disks, likely because most of the COMs freeze out on grain surfaces and therefore only have very small abundances in the gas phase. CH_3CN and HC_3N have only detected in ~10 disks (Öberg et al. 2015; Bergner et al. 2018; Loomis et al. 2018; Ilee et al. 2021). CH_3OH has only been detected in three disks (Walsh et al. 2016; van't Hoff et al. 2018a; Booth et al. 2021), and CH_3OCH_3 was only detected in the Oph IRS 48 disk at the location of the ice/dust trap (Brunken et al. 2022). The simplest organic acid (HCOOH) was detected in the TW Hya protoplanetary disk (Favre et al. 2018). In spatially resolved observations, the COMs seem to be highly concentrated in the inner 50–100 au region of the disks (Ilee et al. 2021). In contrast, smaller hydrocarbons and nitriles are more widely distributed.

THEORIES AND MODELS OF CHEMISTRY IN PROTOPLANETARY DISKS

Types of chemical reactions in protoplanetary disks

Even in the dense region of protoplanetary disks, the number density is around 10^{12-13} cm^{-3}, still orders of magnitude lower than that of planetary atmospheres. Furthermore, most regions of the disk are cold (< 100 K). These low-density and low-temperature environments lead to five important characteristics of chemistry in the bulk regions of protoplanetary disks (Öberg and Bergin 2021): (1) *Dominance of two-body reactions.* The average rates of two molecules colliding together are low, and therefore, the reactions are almost exclusively two-body reactions, three-body reactions are extremely rare except for the innermost region (e.g., Kamp et al. 2017). (2) *Time-dependent chemistry.* Many chemical reactions take a much longer than the disk lifetime to reach equilibrium. Therefore, chemical compositions depend on the initial conditions as well as the disk age, deviating significantly from thermochemical equilibrium. (3) *The importance of ions.* Ion reactions are generally much faster than neutral-neutral reactions. As a result, the ionization sources (UV, X-ray, and cosmic-ray ionization) are driving forces of chemistry in protoplanetary disks. (4) *Difficulty of exothermic reactions.* Due to the low temperature, atoms and molecules generally lack sufficient kinetic energy to overcome substantial reaction barriers. Exothermic reactions (reactions that generate energy) are more unlikely. (5) *Grain surface chemistry.* Molecules freeze out onto grain at low temperatures, and ice on grain surface can remain chemically active. Surface chemistry is an important channel for making molecules. These new molecules formed on the grain surface can go back into the gas phase through processes such as sublimation.

The gas-phase chemistry is classified into different types, based on the formation, destruction, and rearrangement of chemical bonds (e.g., Herbst and Leung 1989; Herbst and van Dishoeck 2009).

Formation of bonds. In two-body reactions ($A + B \rightarrow C$), new bonds can form when photons or electrons carry away the bond-formation energy, which is called *radiative association* and *associative detachments*, respectively.

Destruction of bonds. To break a bond, extra energy needs to be inserted into the molecule first. This extra energy can be obtained by absorption of photons (*photodissociation*), collision with electrons (*dissociative recombination*), and collision with other molecules in shocks. Typical covalent bonds have strengths above 5 ev, and therefore UV or X-ray are required to destroy bonds. Besides photon energy, cosmic-ray can also be an important source of energy to destroy bonds.

Rearrangement of bonds. There are reactions that rearrange chemical bonds to form new molecules. The process can be generalized as $AB + C \rightarrow A + BC$. The reactants can be ions, radicals, neutral molecules, and atoms. In most of the disk region, ion-molecule reactions are much faster than neutral-neutral reactions, as ions can cause induced dipole on neutral molecules. Therefore, UV, X-ray, and cosmic-ray are the main producers of ions in disks and therefore are important drivers of chemistry in planet-forming disks. Ion-molecule reactions are central in the chemistry of protoplanetary disks. Most of neutral–neutral reactions only become important in the high-temperature region ($T > 400$ K).

Nevertheless, some neutral–neutral reactions may be rapid even at low temperatures, especially reactions involve small radicals (e.g., C_2H) (e.g., Herbst and Woon 1997; Smith et al. 2004).

Grain surface reactions. Besides gas-phase reactions, reactions on grain surfaces can play an important role in the formation of molecules, especially in the formation of more complex molecules (e.g., Hasegawa et al. 1992; Garrod et al. 2008). Molecules freeze out onto grain surfaces, building up icy mantles. Light atoms, like H, can migrate on grain surfaces until they find a reactant on the grain surface to form a new molecule. The grains act as a third body to absorb bond formation energy and thus facilitate the formation of new bonds. This type of H addition process is called hydrogenation reaction, which is believed to play an important role in the formation of H_2 (e.g., Cazaux and Tielens 2004) and how water is formed in the cold ISM (e.g., van Dishoeck et al. 2013). Grain surface reactions are also important ways to make complex molecules, while the detailed pathways are still highly uncertain. The molecules formed on the grain surface can be released back into the gas phase, due to thermal desorption, photodesorption, electron-stimulated desorption, and other energetic events.

Chemical modeling of protoplanetary disks

To model the chemistry in protoplanetary disks, we need to know what types of chemical reactions happen (the choice of chemical networks), the local physical conditions (e.g., density, temperature, radiation field, available grain surface area), and the initial compositions of the materials as the chemistry is time-dependent. Furthermore, if the timescales of dynamical processes, such as diffusion, turbulence mixing, and transport of materials, are shorter or comparable to the chemical timescales, these processes should be added as additional source and sink terms.

The chemical reaction rates for chemistry in protoplanetary disks can be found in two main databases: the UMIST Database and the KIDA (McElroy et al. 2013; Wakelam et al. 2015). These databases collect comprehensive reaction rates for gas-phase chemical reactions, including two-body reactions, photodissociation and photoionization, cosmic-ray ionization, and cosmic-ray induced photodissociation and ionization. Some rates are from lab experiments, and many are from theoretical calculations. The rates are temperature dependent but rates at low temperature regions are often missing. Grain-surface reaction rates are still highly incomplete, and the networks often stop at the simplest complex organics (e.g., CH_3OH). Commonly used grain-surface reaction networks include the Hasegawa et al. (1992) and the Ohio State University (OSU) network (Garrod et al. 2008). Depending on the environmental conditions, the chemical networks are sometimes simplified to accelerate the computational speed (e.g., van't Hoff et al. 2017; Krijt et al. 2020) or sometimes additional networks are added to account for three-body and high-temperature reactions (e.g., Kamp et al. 2017; Anderson et al. 2021; Kanwar et al. 2023).

Current chemical models often adopt static density structures for gas and dust in protoplanetary disks. The stellar radiation is the dominant energy input for the thermal structure and radiation field in disks. The stellar photospheric spectra are often adopted from theoretical models, and UV and X-ray spectra from observations or theoretical models (e.g., Getman et al. 2005; Yang et al. 2012; Dionatos et al. 2019). Once a physical disk structure is set up, stellar spectra over the whole wavelength range (from X-ray to mm wavelength) are used as inputs to compute photon propagation and then determine the radiation field inside the disk (e.g., Woitke et al. 2009; Walsh et al. 2010; Bruderer et al. 2012; Du and Bergin 2014; Cleeves et al. 2015). Cosmic-ray ionization rate inside the disk usually uses a simple e-fold column density of $96 \, \mathrm{g \cdot cm^{-2}}$ from the ISM value of $10^{-17} \, \mathrm{s^{-1}}$ or some fixed values (Umebayashi and Nakano 1981). The dust temperature structures are usually based on radiative transfer calculation of heating and cooling balance between stellar light and thermal emission of dust grains (Bjorkman and Wood 2001; Dullemond and Dominik 2004; Pinte et al. 2006; Bruderer et al. 2012; Du and Bergin 2014). In the disk atmospheres, the dust and gas no longer have enough collisions to be thermally coupled, and the gas temperature can become much higher than the dust temperature (e.g., Kamp and Dullemond 2004; Najita et al. 2011; Bruderer et al. 2012). Some models calculate the gas temperature self-consistently based on the heating and cooling processes of gas, while others adopt a semi-analytical formula to correct the temperature differences between gas and dust. The initial conditions are usually set as atomic abundances based on molecular clouds, or atomic ratio and simple CO and H_2O abundances. Readers can find a more comprehensive review of chemical modelings of protoplanetary disks in Henning and Semenov (2013).

Most chemical models run time-dependent chemistry for 1 to a few Myrs and then use the abundance structure to compare with observations. The abundance and thermal structures are usually used as inputs for radiative transfer models to generate simulated observations. The line excitation is assumed to be at local-thermal-equilibrium (LTE) or is calculated to achieve collisional balances at the local density and temperature conditions (e.g., Brinch and Hogerheijde 2010). The line model images from radiative transfer calculations are then convolved with the proper spatial and spectral resolutions to compare with observations.

CURRENT INSIGHT INTO CHEMISTRY IN PROTOPLANETARY DISKS

Snowlines

Snowlines are where molecules freeze out in the disk. Due to radial and vertical temperature gradients in disks, different molecules have snowlines at different distances from the central star. It is believed that the snowlines of the abundant volatile molecules, such as water, CO_2, and CO, play an important role in planet formation (e.g., Hayashi 1981). In particular, the water snowline has long been used to explain the dichotomy of the terrestrial and giant planets in the solar system, i.e., the giant planets formed outside the water snowline because water ices provide extra solid masses and subsequently accelerated the growth of planetary cores (e.g., Stevenson and Lunine 1988; Pollack et al. 1996; Kennedy and Kenyon 2008).

There are several arguments for the importance of snowlines in planet formation: (1) *Solid surface density is enhanced beyond snowlines.* Beyond a particular snowline, gas phase molecules freeze out and become ices, which can enhance the surface density by a factor of 1–4 (depending on the rock–ice ratio). (2) *Snowlines facilitate the formation of planetesimals.* One line of argument is the particles' stickiness and fragmentation thresholds change across different snowlines as their icy composition changes (e.g., Gundlach and Blum 2015). As a result, the maximum grain sizes increase when the thresholds become higher, or grains become smaller when the thresholds become lower. For example, the fragmentation threshold of bare silicates is expected to be lower than that of water ice grains. When grains cross the water snowline, they become smaller due to the loss of icy mantle and the change of fragmentation threshold. Smaller grains drift inwards more slowly compared to the larger ones, making a local "traffic jam" just inside the snowline (e.g., Banzatti et al. 2015; Pinilla et al. 2017). Also, the ice evaporation and outward diffusion of water increases the abundance of icy pebbles outside the water snowline (Stevenson and Lunine 1988; Cuzzi and Zahnle 2004; Ciesla and Cuzzi 2006; Ros and Johansen 2013). The enhanced local dust-to-gas ratio around the water snowline may trigger planetesimal formation via streaming instability (e.g., Drazkowska and Alibert 2017). (3) *Snowlines set the boundary of disk regions with different elemental ratios in gas or solids* As different molecules have different snowlines, the partition of C and O elements in the solid and gas phases depends on the distance from the central star. Therefore, a gas giant that accretes its envelop materials between two snowlines would carry a birthmark in their C/O, C/H, and O/H ratios in their atmospheres (Öberg et al. 2011). This is discussed further in the section *Elemental ratios*.

Despite their great importance in planet formation theories, snowlines are challenging to be directly observed. Most of the snowlines like H_2O and CO_2 are inside 10 au for typical disks, where the dust emission becomes highly optically thick and therefore the line emission is only from the surface layer. The surface layer is expected to be hotter than the mid-plane and therefore the surface snowline is further out compared to the mid-plane snowline. Several studies measured the water snowline at the disk surface layer, by constraining where the column density of water vapor drops promptly (e.g., Zhang et al. 2013; Blevins et al. 2016). These results suggested the surface water snowlines are between 3–11 au. Still, we do not have direct constraints on their mid-plane water snowline. If a disk is ongoing an accretion outburst that dramatically increases its luminosity and subsequently warm up the disk, the water snowline can be moved outwards to 10–100 au, and becomes much easier to detect. Recently, HDO and $H_2^{18}O$ lines have been imaged in the V883 Ori outburst disk, suggesting a water snowline at 80 au (Tobin et al. 2023). There are also ideas of indirect tracers of the mid-plane water snowlines. One possibility is to use a molecule that is chemically anti-correlated with water vapor abundance. HCO^+ has been proposed to be such as a tracer (van't Hoff et al. 2018b; Leemker et al. 2021). Another idea is to look for a sharp change in the spectral slope of continuum emission, as a result of dust size changes around the water snowline (Banzatti et al. 2015; Cieza et al. 2016). But more tests are needed to evaluate the robustness of these methods.

Compared to water and CO_2, the mid-plane CO snowline is expected to be at a much larger distance, ~20–40 au for disks around solar-like stars and 60–100 au for disks around 2–3 $M_\odot$ stars. The abundant ^{12}CO, ^{13}CO, or sometimes even $C^{18}O$ become optically thick inside the mid-plane CO snowline, and therefore more rare CO isotopologues are needed to probe the mid-plane CO snowlines. Spatially resolved $^{13}C^{18}O$ line emission has been used to constrain the mid-plane CO snowlines in three disks, founding CO snowline between 20–100 au (Zhang et al. 2017, 2021; Loomis et al. 2020). However, observing the faint CO isotopologue lines is very timeconsuming, and therefore the number of disks with direct CO snowline constraints is still small.

Alternatively, N_2H^+ can be used as an indirect tracer of the mid-plane CO snowline, because this ion only becomes abundant when the gas-phase CO abundance is low. Spatially resolved N_2H^+ line emission often shows an inner cavity and the radius of the inner cavity is inferred as the mid-plane CO snowline (Qi et al. 2013, 2015, 2019). However, thermo-chemical models show that the cavity radius of N_2H^+ is probably an upper limit of the mid-plane CO snowline, and the offset between the two depends on the CO/N_2 abundance ratio and the detailed temperature structure in the disk (e.g., van't Hoff et al. 2017).

Abundances of carbon, oxygen, and nitrogen carriers

Carbon, oxygen, and nitrogen are the three most abundant elements beyond H and He. The molecules and ions with these elements are the backbones of chemistry in planet-forming disks, and therefore understanding the abundances and distributions of these carriers is an essential part of our understanding of chemistry in planet formation (e.g., Pontoppidan et al. 2014; Krijt et al. 2023; Öberg et al. 2023).

The initial elemental abundances of a planetary forming disk are assumed to be the same as that of its central star, considering the star and its disk are formed from the same molecular cloud. Measuring the CNO elemental abundances of individual young stars is challenging because the active accretion fills out the absorption lines in stellar photospheres. Due to this challenge, the total CNO elemental abundances are often assumed to be the solar abundances. The fractions of different CNO carriers at the birth time of protoplanetary disks are estimated based on ice compositions seen in protostellar envelopes and in solar system comets (e.g., Mumma and Charnley 2011; Öberg et al. 2011; Boogert et al. 2015). Below we discuss current understanding of the main CNO carriers in protoplanetary disks.

Carbon. Nearly half of the carbon atoms are thought to be in a refractory format, based on observations of carbon depletion in diffuse interstellar medium (Mishra and Li 2015). The rest of the carbon atoms are in volatile formats (in gas or ice), such as CO, CO_2, and CH_4 (e.g., Pontoppidan et al. 2008; Öberg et al. 2011; Boogert et al. 2015). CO is the main carbon carrier in a volatile format. Indeed, CO is the most widely detected molecule in protoplanetary disks.

For many years, the CO abundance in disks has been assumed to be the canonical interstellar medium (ISM) ratio of CO-to-H_2 ratio of 10^{-4}. However, recent ALMA surveys of 1–3 Myr-old protoplanetary disks showed that the ^{13}CO and $C^{18}O$ line fluxes are much lower than expected (e.g., Favre et al. 2013; Ansdell et al. 2016; Long et al. 2017), even after the correction of photodissociation and freeze-out (Miotello et al. 2017). One possibility is that the gas in disks dissipates much faster than previously assumed and therefore gas masses in these disks are low. However, observations showed that many of these disks still have relatively high accretion rates onto their central stars ($\sim 10^{-9} M_\odot$/year). A low gas mass disk would run out of gas extremely fast (0.1 Myr), which is in conflict with the age of 1–3 Myr of these disks (Manara et al. 2016, 2020). More direct evidence is three protoplanetary disks have independent gas mass constraints from HD (1–0) line fluxes (Bergin et al. 2013; McClure et al. 2016). Their $C^{18}O$ line flux suggests that the CO-to-H_2 ratios in these three disks are 1–2 orders of magnitude lower than the canonic ISM ratio (Favre et al. 2013; Schwarz et al. 2016; Calahan et al. 2021; Trapman et al. 2022). Zhang et al. (2020); Bergner et al. (2020) studied CO abundances of several youngest disks (< 1 Myr), showing that the CO-to-H_2 ratio starts with the ISM ratio at the youngest disks, but rapidly decrease with ages, with a timescale of ~1 Myr.

To explain the low CO gas abundances in 1–3 Myr-old disks, two broad types of mechanisms have been proposed. The first type is chemical processes. The idea is that once CO gas freezes out onto grain surfaces, surface reactions can process CO into other carbon-carriers like CO_2, CH_4, and CH_3OH (e.g., Bergin et al. 2015; Bosman et al. 2018; Schwarz et al. 2018). This chemical process require a cosmic ray ionization rate of 10^{-17} s^{-1} or higher to reduce the CO abundance by one order of magnitude within 1 Myr. However, it is still unclear if such a high cosmic ionization rate exists in disks (e.g., Cleeves et al. 2013, 2015; Aikawa et al. 2021; Seifert et al. 2021). The second type of mechanism proposed is dust growth. As dust grains grow into larger sizes, they settle onto the mid-plane of the disk. As a result, CO ices are carried with icy grains onto the mid-plane and gradually reduces the CO abundance in the disk atmosphere. The problem is still the efficiency, as current dust evolution models only predict a factor of few depletion in 1 Myr (e.g., Xu et al. 2017; Krijt et al. 2018). However, the two types of mechanisms may work together. Recent simulations including both chemical processes and dust growth showed the CO abundance can be depleted by 2 orders of magnitude in 1 Myr timescale (Krijt et al. 2020).

The CO_2 is another main carbon-carrier beyond CO. CO_2 gas has been detected via its rovibrational Q-branch around 15μm in ~10 protoplanetary disks by Spitzer (Pontoppidan et al. 2010; Carr and Najita 2011). In Spitzer observations, the detection rate of CO_2 was much lower than H_2O, HCN, and C_2H_2 in protoplanetary disks. The CO_2 gas abundances in disks are still highly uncertain, as the 15 μm feature is dominated by optically thick emission (Bosman et al. 2017). The inferred CO_2-to-H_2 abundances are on the order of 10^{-9}–10^{-7}, much lower than the expected ISM value of 10^{-5} (Bosman et al. 2017). Recently, $^{13}CO_2$ was detected in the JWST/MIRI spectra of the GW Lup protoplanetary disk, which suggested that the CO_2 abundance is orders of magnitude higher than previously derived from the Spitzer data (Grant et al. 2023). In the next few years, JWST observations will provide more accurate constraints on the CO_2 abundances in protoplanetary disks.

CH_4 is expected to be another main carriers of carbon in protoplanetary disks. However, it has only been detected in near-IR absorption features in the GV Tau N protoplanetary disk (Gibb and Horne 2013). CH_4 has vibrational bands at 7.5 μm, where the spectral resolution of Spitzer/ IRS was only 100 and did not provide sensitive constraints. A tentative detection of CH_4 was reported in JWST/MIRI spectra by Tabone et al. 2023. JWST/MIRI (with R ~ 3000), will provide one order of magnitude better constraints on the CH_4 abundances in many protoplanetary disks.

Oxygen. Based on extinction and absorption feature studies in the ISM, about 25% of the atomic oxygen is locked in silicates, 40% is in an unknown format, and the rest of 35% in volatiles, mainly H_2O, CO_2, and CO (Whittet 2010; Boogert et al. 2015). H_2O, CO, and CO_2 take ~ 10%, 25% and 10% of the oxygen budget, respectively.

Similar to the CO depletion seen in disks, the observations of far-IR water in disks suggested a significant depletion of H_2O in the atmosphere of the outer disk region (50au and beyond) (Hogerheijde et al. 2011; Du et al. 2017). Beyond the mid-plane water snowline, a low abundance of water vapor exists as UV radiation can release water from dust grain surfaces (e.g., Walsh et al. 2010). However, deep searches of cold water vapor in protoplanetary disks by the Herschel space telescope have very low detection rates (e.g., Bergin et al. 2010; Hogerheijde et al. 2011). Compared with predictions from thermo-chemical models, the results suggested that the water vapor abundance is 1–2 orders of magnitude lower than the models with ISM level O/H elemental ratio (Du et al. 2017). The depletion of water vapor is also consistent with the expectations of dust growth and sequestering water ice onto the mid-plane (Krijt et al. 2016).

In the inner disk region (< 5 au), the abundances of carbon and oxygen carriers are still highly uncertain. The main tracers of the inner warm disk regions are mid-IR molecular lines, but these lines are emitted from the surface disk layer, probably only represent a few percent of the total column density (Woitke et al. 2018; Bosman et al. 2022). Current dust evolution models suggest that large amount of icy grains drift into the inner disk region, and the inner disk can be enriched in carbon and oxygen after ices evaporate (e.g., Booth et al. 2017; Krijt et al. 2018, 2020; Booth and Ilee 2019). Trends of higher water column density and lower C/O ratios have been seen in more compact disks and less massive disks (Banzatti et al. 2020). This is consistent with simulations of dust evolution in disks. Upcoming JWST observations can test this scenario in a large number of disks.

Nitrogen. In protoplanetary disks, the majority of nitrogen is expected to be in the format of molecular nitrogen (N_2), based on N_2/NH_3 ratio measured in star-forming molecular clouds (Womack et al. 1992). Unfortunately, N_2 is hard to be directly detected as it does not have a permanent dipole moment and therefore lacks of strong lines. Therefore, tracer species are needed to understand the nitrogen abundance, with the assistance of chemical models. NH_3 and HCN are the most abundant N-carriers detected in comets (Mumma and Charnley 2011) and therefore natural candidates of tracer species. NH_3 has only been detected in one protoplanetary disk (Salinas et al. 2016), and upper limits of NH_3/H ratio were reported for a few disks (Pontoppidan et al. 2019). In contrast, HCN is widely detected in nearby disks, both in its vibrational branch at mid-IR and pure rotational lines at millimeter wavelengths (e.g., Salyk et al. 2011; Huang et al. 2017; Guzmán et al. 2021). Therefore, HCN and its isotopologues have been used to constrain the N/H elemental ratio in protoplanetary disks (e.g., Cleeves et al. 2018). CN is already readily detectable in disks (e.g., Dutrey et al. 1997; Öberg et al. 2010; Cazzoletti et al. 2018). N_2H^+ is another commonly detected nitrogen carrier in protoplanetary disks (e.g., Qi et al. 2013, 2019; Anderson et al. 2019), but its abundance depends on both the N_2 abundance and the cosmic-ray ionization rates in disks. The usage of different N-carriers to constrain nitrogen abundance is still relatively new and more tests are needed.

In general, N_2 is expected to stay in gas phase in the bulk region of the disk and less subjected to depletion in gas phase, due to its low condensation temperature (~19 K). Indeed, the elemental ratios seen in solar system bodies showed orders of magnitude depletion of nitrogen, suggesting that in the solar nebula most of nitrogen likely stayed in gas phase (e.g., Bergin et al. 2015). Chemical models of CO isotopologues and HCN observations of the IM Lup protoplanetary disk shows that its C/H ratio in gas is depleted by a factor of 10–20, but N/H ratio is still close to the ISM ratio of 10^{-5} (Cleeves et al. 2018). Combined N_2H^+ and CO isotopologue studies in two old Upper Sco disks also suggested that their CO gas abundances are heavily depleted but N/H is consistent with the ISM ratio (Anderson et al. 2019).

Complex organic molecules

Understanding the abundances and distributions of complex organic molecules (COMs) in protoplanetary disks is of great importance because the COMs in the Solar Nebular were potential seeds of life on Earth. The early Earth may have obtained most of its volatile materials at its surface from meteorites and comets (e.g., Anders 1989; Altwegg et al. 1999). Both in-situ and remote observations have revealed that comets and carbonaceous meteorites are rich in COMs. About a quarter of comet mass is in the format of complex organic refractory materials (dominated by carbon), 9% in small carbonaceous molecules, and a small contribution from simplest organics in water ice (e.g., Greenberg 1998). Most of the organics found in carbonaceous meteorites are macro-molecules (Cronin and Chang 1993). COMs have been widely seen in star-forming regions, both in cold gas in large scales, and in the inner envelop region close to forming protoplanetary disks (see recent reviews by Jørgensen et al. 2020). However, it is still unclear what the COMs abundances and distributions are in protoplanetary disks, and how they may depend on the stellar masses and/or the stellar environments.

The observations of COMs in protoplanetary disks have been challenging, because of three reasons. The first reason is that the temperature of the bulk region of the disk is below 100 K and therefore most of the COMs exist as ices and cannot be probed by gas-phase molecular emission line observations. The second reason is that large molecules have more degrees of freedom and more possible transitions, which dilutes the strength of individual transitions and makes their detections more challenging than the simple molecules. Finally, the sizes of protoplanetary disks are small, up to a few hundred of au, which is much smaller compared to the sizes of protostellar envelopes (a few thousand au) and the dark clouds in the ISM.

Given these challenges, the number of COM detections in protoplanetary disks is still small and mostly limited to the most luminous and massive disks. In the outer disk region, the most commonly detected COMs are HC_3N, CH_3CN, and c-C_3H_2 (e.g., Chapillon et al. 2012; Öberg et al. 2015; Bergin et al. 2016; Bergner et al. 2018; Ilee et al. 2021), which have been detected in ~20 disks. Spatially resolved observations showed that line emissions of these organics tend to concentrate at the inner 50–100 au region, more compact compared to the emitting areas of simple molecules, such as CO, CS, and H_2CO. In contrast, CH_3OH has only been detected two disks between (1–10 Myr), one in a T Tauri disk (Walsh et al. 2016) and the other one in a Herbig disk (Booth et al. 2021). Compared to the older disks, the young embedded disks (<1 Myr) are warmer and therefore there are better chances to detect COMs (Podio et al. 2020). In particular, multiple COMs lines, including CH_3OH, CH_3CHO, CH_3OCHO, and CH_3COCH_3, were detected in the V883 Ori disk, an FU Ori type source that has an on-going accretion outburst with an enhanced luminosity of 200 solar luminosity (van't Hoff et al. 2018a; Lee et al. 2019). So far, most of the COMs detections of protoplanetary disks have been at millimeter wavelengths, tracing line emission outside of 10 au. However, the recent JWST/MIRI detections of multiple COMs in a M-dwarf disk, including C_4H_2, C_6H_6, and $^{13}C^{12}CH_2$ (Tabone et al. 2023). This may open a new window to study COMs in the inner few au regions of the disk region.

The origin of COMs in protoplanetary disks is still under debate. One fundamental debate centered on how much the COM compositions are inherited from the interstellar and prestellar materials and how much the compositions are altered by chemical reactions inside the disk. It is generally thought a significant amount of materials may be inherited from the ISM and prestellar, as suggested by the similarity of relative compositions in comets and protostellar inventories (e.g., Mumma and Charnley 2011; Drozdovskaya et al. 2019). The COM ratios found in the outburst source V883 Ori disk also broadly agree with cometary values, consistent with the picture of the observed COMs are sublimated from inherited organic ices from the protostellar stage. On the other hand, chemical models suggest that there may be significant CO freeze-out onto grain surfaces, and these carbons can be processed into CH_3OH or more complex organics. Besides grain surface reactions, photochemical-dominated gas-phase chemistry may develop at the later stage of disk evolution, as dust grains grow, settle, and drift. This high gas-dust-ratio and C-rich environment allows for deeper UV penetration and therefore gas-phase formation of organic molecules (Calahan et al. 2023).

ELEMENTAL RATIOS:
LINK FROM DISK COMPOSITIONS TO PLANETARY COMPOSITIONS

One of the ultimate goals of planet formation studies is to understand how the initial conditions and planet formation processes determine planetary compositions. There are significant challenges to this mighty goal. Exoplanetary systems are known to be incredibly diverse, even within a narrow stellar mass range. This diversity could be the result of different initial conditions, formation history, and/or stochastic processes in planet formation. The hope is that some observable features of a planet are unique results of its initial conditions or formation history, and therefore we can use the observed characteristics to inform its formation history and deduct other important but unobservable properties of the planet, such as its interior structure and its surface conditions.

Link between disk compositions and exoplanetary atmospheric compositions

So far, the elemental ratios of carbon and oxygen elements (C/O, C/H, and O/H) have been the most widely explored tracers to connect the atmospheric compositions of hot-Jupiters and their formation/migration history (e.g., Öberg et al. 2011; Madhusudhan 2012; Mordasini et al. 2016). Interested readers can find more details about current studies of exoplanet atmospheres in the Kempton and Knutson (2024, this volume). Volatile molecules (e.g., H_2O, CO_2, CO, CH_4) are the most detectable molecules in exoplanets' atmospheres (see review by Madhusudhan 2019). The C/O elemental ratio can be measured in hot hydrogen-dominated atmospheres, as the C/O ratio critically influences the relative abundances of dominant species, like H_2O, CO_2, CO, CH_4 (Madhusudhan 2012). On the other hand, molecules of carbon, nitrogen, and oxygen (CNO) elements account for the majority of the mass of ices and gaseous species (beyond H and He) in protoplanetary disks (e.g., Salyk et al. 2013; Pontoppidan et al. 2014). Given their chemical importance and detectability of carbon and oxygen, C/O elemental ratio has been proposed as a key connective tissue between atmospheric compositions of planets and their formation/migration history (e.g., Öberg et al. 2011; Madhusudhan 2012; Mordasini et al. 2016; Cridland et al. 2020).

The key idea is that the C/O elemental ratio in the atmosphere of a gas giant is a birthmark from the planet's forming region in the natal disk. The simplest argument is that the main ice species H_2O, CO_2, CO, have different condensation temperatures, and therefore their snowlines are located at different distances from the central star. As a result, the elemental ratio of C/O in the gas-phase (solid-phase) of disk materials increases (decreases) like a multi-step function between the snowlines (see Fig. 3). If a gas giant planet accretes the bulk of its atmosphere masses within a region between two snowlines and there is no significant mixing between the planetary core and atmospheric materials, the C/O ratio of its atmosphere can tell its forming region (Öberg et al. 2011). For example, a giant planet with C/O ≥1 can only form in the region outside the CO snowline. However, this simple scenario did not consider the effects of time-dependent chemistry in disks, the transport of solid materials, and the alteration of volatile compositions during planet formation processes.

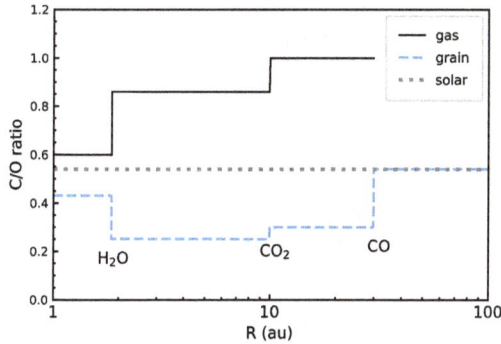

Figure 3. How the C/O elemental ratio in the gas and grains change at different radii in a typical proto-planetary disk around a solar-like star. The snowlines of H_2O, CO_2, and CO are boundaries of different elemental ratios. Figure is adapted from Öberg et al. (2011).

Since then, many works have explored the impacts of chemistry and formation processes on the C/O ratio in the planetary atmosphere, using different levels of complexity and considering various mechanisms of planet formation processes (e.g., Madhusudhan et al. 2014; Cridland et al. 2016, 2020; Mordasini et al. 2016; Öberg and Bergin 2016; Booth et al. 2017; Booth and Ilee 2019; Notsu et al. 2020). However, when different assumptions and processes are considered, the resulting C/O are C/H ratios are not always the same. For example, whether planetary atmospheric composition is dominated by the composition of the accreted gas can lead to distinctive predictions on the C/O ratio in planetary atmospheres. For example, Öberg et al. (2011) and Madhusudhan et al. (2014) predicted that C/O ≥ 1 can be achieved if a planet forms beyond the CO snowline, while other studies showed that if planetesimal accretion dominates the volatile and metal enrichment in planetary atmospheres, the models predict subsolar C/O ratio (Thiabaud et al. 2015; Morbidelli et al. 2016). Another question is whether the disk materials move across the disk. In the static disk models, the absolute ratios of C/H and O/H in the disk gas are always below the stellar ratio across the whole disk, because some C and O are locked in refractory materials and the rest of the C and O are further locked in ices at larger distance from the central star. In contrast, in models with pebble drift, a large amount of icy pebbles drift into the inner disk region. When icy pebbles cross a snowline, the corresponding ice mantle can evaporate and enrich the volatile abundances in the gas phase. As a result, the C/H or O/H ratios in the disk gas can become super-stellar (Öberg and Bergin 2016; Booth et al. 2017; Booth and Ilee 2019; Estrada and Cuzzi 2022; Bitsch and Mah 2023). Indeed, this super-stellar C/H ratio has been seen in the gas inside the CO snowline of the HD 163296 protoplanetary disk (Zhang et al. 2020b). An enhanced cold water components have been seen JWST/MIRI spectra of two compact disks compared with two extend disks, which is consistent with the expectation that more pebble drift occurs in the compact disks (Banzatti et al. 2023). These preliminary evidence require a more robust understanding of planet formation processes.

The assumption of the disk compositions is another main source of uncertainty of the C/O ratio as a tracer of planet birth region. The simplest models adopt the same chemical partition of C and O elements as that of the interstellar ices, assuming full heritage (e.g., Öberg et al. 2011), or assume chemical equilibrium based on the solar abundance, partitioning elements into refractory and volatiles at the mid-plane (e.g., Madhusudhan et al. 2014). The next level of complexity adopts gas and ice compositions from kinetic disk chemistry models after some evolution time (typically 1 Myr) and assumes no further evolution (e.g., Cridland et al. 2019; Turrini et al. 2021). More complex models start to consider the coupling between dust growth and chemistry, but so far have only considered simplified chemical networks (e.g., Krijt et al. 2020).

Besides the C/O and C/H ratios, recent studies have started to explore constraints from other abundant elements. Öberg and Wordsworth (2019) and Bosman et al. (2019) argued that the enrichment of nitrogen in Jupiter's atmosphere requires that the core of Jupiter formed exterior to the N_2 snowline beyond 30 au, because the main carrier nitrogen carrier N_2 only condenses at very low temperature. Crossfield (2023) proposed that C/S and O/S ratios can be constrained in exoplanet atmospheres via SO_2 and provide additional constraints to the formation history. Another idea is to add the Si/H ratio in addition to the C/O and C/H ratios to trace the planet formation history, as Si/H provides a baseline estimation of refractory materials accreted during planet formation (Chachan et al. 2023).

Isotopic ratios: Link between disk compositions and solar system objects

Many clues of the solar system formation can be obtained from isotopic ratios and the elemental ratios between radiative decay parent and daughter elements (e.g., Marty 2012; Füri and Marty 2015; Kruijer et al. 2017). Compared to high precision measurements of elemental and isotopic ratio for the solar system objects, the isotopic ratio measurements of protoplanetary disks are limited to the abundant volatiles (H, C, O, N, and S) and have much higher uncertainties.

D/H ratio. The D/H ratio in water has long been used to trace the origin of water on the Earth, by comparing the D/H ratio in Earth's oceans with that in a wide range of solar system bodies, including comets, ancient Martian water, and hydrated minerals in meteorites. The ocean water on Earth on average has a D/H ratio of 1.56×10^{-4}, much higher than the D/H elemental ratio of 2.5×10^{-5} measured in the Sun and the D/H ratio of 1.5×10^{-5} within 100 pc of the solar neighborhood (Linsky 1998). In contrast, the D/H ratio in water at prestellar cores and protostar envelopes, is one to two orders of magnitude higher than the ratio in Earth ocean water, generally between 10^{-3}–10^{-2} (see review by Ceccarelli et al. 2014).

This enhanced D/H ratio, so-called deuterium fractionation, is widely observed in prestellar and protostar environments. This enhancement is attributed to the result of cold-temperature and ion-driven chemistry (see review by Ceccarelli et al. 2014). As mentioned above, in cold molecular gas, chemical reactions with ions are much faster than neutral–neutral reactions. H_3^+ is the most important ion, which is produced by H_2 and H ionized by cosmic ray. It reacts with HD to form H_2D^+, i.e., $H_3^+ + HD \rightarrow H_2D^+ H_2 + \Delta E$. In cold gas ($\leq 30$ K), the ratio of H_2D^+/H_3^+ becomes higher than HD/H_2, because the ions do not have sufficient energy to overcome the energy barrier ($\Delta E \sim 124$ K) in the reverse direction of the reaction. When H_3^+ and H_2D^+ have dissociative recombination with electrons, the H and D atoms are then liberated. The liberated D atoms can then react with O on the grain surface to form HDO. Therefore, cold prestellar cores and protostar envelopes readily produce a high D/H ratio in water ice formed in these environments.

The cold environment can also be provided in the outer region of the protoplanetary disks (e.g., Drouart et al. 1999; Hersant et al. 2001; Yang et al. 2013). However, the cosmic ray ionization rate inside protoplanetary disks may be reduced by one to several orders of magnitude due to stellar and disk winds (Cleeves et al. 2013, 2015). If this is the case, current models predict the ion-drive chemistry in protoplanetary disks cannot produce an enhanced D/H ratio like that of the Earth's ocean water and therefore much of the enhanced D/H ratio must be inherited from the prestellar and protostar stage (Cleeves et al. 2014). However, recent chemical studies of several ion lines in the IM Lup protoplanetary disk suggested that the cosmic ray ionization rate reaches the ISM level between 100–300 au region of the disk (Seifert et al. 2021). In general, the cosmic-ray ionization rate in protoplanetary disks is still highly uncertain.

So far, there is only one direct HDO detection in a protoplanetary disk (V883 Ori disk). This disk is undergoing an accretion outburst and therefore has a water snowline at 80 au instead of a few au for typical quiescent disks. The measured D/H ratio of H_2O in the V883 Ori disk is consistent with that of protostars (see Fig. 4, suggesting significant H_2O in this disk is inherited from the protostar stage (Tobin et al. 2023).

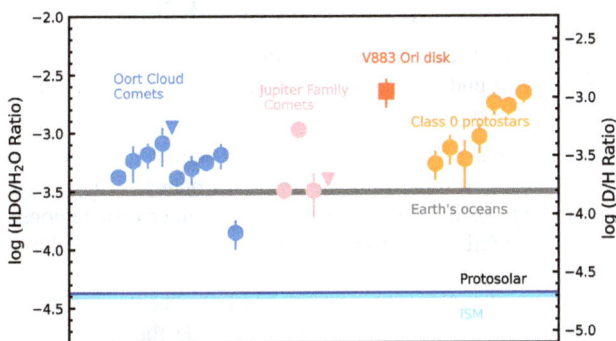

Figure 4. HDO:H$_2$O ratio in the V883 Ori protoplanetary disk, Class 0 protostars, and different objects in the solar system. The data are adapted from Tobin et al. (2023).

^{15}N/^{14}N ratio. The ^{15}N/^{14}N ratio of the Earth's atmosphere clearly differs from those of the protosolar nebula and most of the cometary values, and therefore provide important clues on the formation history of the Earth (Marty 2012). The ^{15}N/^{14}N ratio of the Earth's atmosphere is about 3.678×10^{-3}, 1.6 times higher than the protosolar value, but about 1.8 times lower than typical ratios measured in comets. Not only the Earth, but all inner solar system objects show enrichment of ^{15}N/^{14}N ratio compared to the protosolar nebula value.

In protoplanetary disks, the ^{15}N/^{14}N ratio has been measured in several disks, based on HC^{15}N/HCN or C^{15}N/CN (Guzmán et al. 2017; Hily-Blant et al. 2017, 2019; Booth et al. 2019). The ^{15}N/^{14}N ratios measured from HCN are broadly consistent with the ratios in comets, and always are enriched compared to the ratio of the Earth. In one disk where ^{15}N/^{14}N ratios were measured from both HCN and CN, the ^{15}N/^{14}N from CN is much less enriched in ^{15}N compared to HCN (Hily-Blant et al. 2017).

In two protoplanetary disks, the radial distribution of ^{15}N/^{14}N ratio has been measured from spatially resolved HC^{15}N images. Both cases showed that ^{15}N/^{14}N ratio shows a strong radial gradient, with the inner disk region being 2–3 times more enriched in ^{15}N (Guzmán et al. 2017; Hily-Blant et al. 2019).

^{18}O/^{16}O, ^{17}O/^{16}O ratios. Similar to the isotopes of hydrogen and nitrogen, oxygen isotope ratios (^{18}O/^{16}O, ^{17}O/^{16}O) also show variations among bodies in the solar system, providing critical insights into the formation, evolution, and distribution of materials in the solar nebula (e.g., Thiemens 2006). The Sun is known to be ^{16}O enriched, compared to the terrestrial planets, asteroids, and chondrules (e.g., McKeegan et al. 2011). A possible mechanism to account for the ^{18}O and ^{17}O enrichment in solar system bodies is self-shielding of CO (e.g., Thiemens and Heidenreich 1983; Clayton 2002). CO can be only be photodissociated by UV radiation at specific wavelengths, and different CO isotopologues are photodissociated by UV radiation at lightly different wavelength (van Dishoeck and Black 1988). Thus, when gas is irradiated by UV radiation, the optically depth of the ^{12}CO quickly becomes optically thick while less abundant CO isotopologues, such as ^{13}CO, C^{18}O, C^{17}O in the interior of the gas still experience photodissociation. This process frees more ^{18}O and ^{17}O that go into water ice or other condensates. It is still underdebate whether the process mainly happen in molecular cloud or protoplanetary disk stages. Compared to the D/H and ^{15}N/^{14}N ratios, the difference of ^{18}O/^{16}O or ^{17}O/^{16}O ratios relative to the Sun is only upto a few percent level. As discussed above, due the large temperature and abundance gradients in protoplanetary disks, the current abundance retrieval methods of protoplanetary disks have not yet had precision like that of solar system bodies.

PREDICTIONS OF FUTURE DIRECTIONS

The next decade represents tremendous opportunities to advance the study of chemistry in planetforming disks and our understanding of the chemical origins of extrasolar systems. We highlight a few promising directions here.

- *Transport of volatiles in protoplanetary disks.* One exciting opportunity is to test whether the volatile reservoirs in the terrestrial planet-forming region are enriched by large amounts of icy pebbles drift into the inner disk regions, as required by current pebble accretion models of planet formation. It is still unknown whether such significant pebble drift into the inner disk region can widely occur in protoplanetary disks. Evidence of volatile enrichment has been seen in a few cases: in the HD 163296 protoplanetary disk, the C/H ratio inside the CO snowline seems to be enhanced to superstellar level (Zhang et al. 2020b, 2021); recent JWST/MIRI spectra of four protoplanetary disks show that an extra excess of cold water emissions are seen in two compact disks compared to two large disks (Banzatti et al. 2023). In the outer disk region (>20 au), ALMA can measure chemical structures and constrain the spatial distributions of pebble trapping. In the terrestrial region (inner 5 au), JWST can trace the emission of water and simple organics from the surface layer. By combining the power of JWST and ALMA, we can trace the volatile budgets across the whole disk, and test the transport of volatiles in a large number of disks.

- *Chemical evolution over the disk lifetime.* The gas in protoplanetary disks can last for 1–10 Myr. The chemical composition of disk materials may change dramatically over the disk lifetime (e.g., Eistrup et al. 2016), driven by changes in physical conditions (e.g., gas-to-dust mass ratio, UV intensity, temperature), as well as dynamical processes inside the disk (e.g., turbulence and radial/vertical transport of materials). The lack of empirical constraints on volatile evolution over the disk lifetime poses a great challenge to link atmospheric compositions of transiting planets with their formation history. ALMA already showed some evidence of the CO gas abundance decreases at the timescale of 1 Myr (Bergner et al. 2020; Zhang et al. 2020a). However, the sample is still small, particularly needing more of the youngest (< 1 Myr) and the oldest disks (5–10 Myr). Except for CO, the existing ALMA observations of molecular lines are predominantly towards the large massive disks as well as the intermediate age (1–3 Myr) disks. A similar problem also existed in the Spitzer mid-IR spectroscopic studies of protoplanetary disks, largely due to the sensitivity limit. JWST/MIRI has orders of magnitude better sensitivity than Spitzer/IRS and therefore can probe fainter old disks. For young embedded disks, JWST provides better spatial resolution to separate the contributions from the disk and envelope regions. Therefore, in the next few years, not only the number of disks with chemistry will increase significantly, but also the age range as well as the stellar type range will dramatically expand. With a more diverse sample, there will be significant progress in the understanding of chemical evolution over the disk lifetime.

- *Coupling of physical and chemical models for protoplanetary disks.* So far, the modeling of physical and chemical processes in disks has been largely two separate camps. In the physical modeling camp, sophisticated models, such as dust growth, planetesimal formation, magnetohydrodynamic simulations of turbulence and disk wind, and planet-disk interactions, include detailed treatment of physical processes, but often consider minimum or no chemistry in the simulations. On the other side, thermo-chemical models of protoplanetary disks compute time-dependent chemistry using large chemical networks and self-consistently calculate heating-cooling processes. However, most of the chemical models are based on static physical structures and fixed dust populations. Growing observations demonstrate that the physical and chemical processes are probably tightly coupled in planet formation. Recent chemical studies have started to include turbulent mixing, disk wind, and dust growth models, but often used a simplified chemical network (e.g., Heinzeller et al. 2011; Semenov and Wiebe 2011; Furuya and Aikawa 2014; Krijt et al. 2020; Price et al. 2020; Eistrup et al. 2022; Van Clepper et al. 2022). In the upcoming decade, the coupling the physical and chemical modeling for planet formation is likely to become a new frontier.

- *Linking atmospheric compositions of nascent planets and the chemistry in the local disk areas.* To trace the chemical origin of planets, it is of great import to compare the atmospheric compositions of nascent planets to the compositions of disk materials around their forming location. Although the number of nascent planets detected in protoplanetary disks is still small, the number may quickly expand in the next decade. Embedded planets can now also be searched by tracing kinematic perturbation in the disk velocity field. With the new JWST imaging power and the upcoming 30 m-class ground-based telescope, we expect to discover more nascent planets while their natal disks are still around and follow up with studies of their atmospheric compositions. Since their natal gaseous disk is still around, we can also probe the chemical compositions around the location of the nascent planet. This type of study has been started for planets inside the PDS 70 protoplanetary disk (e.g., Cridland et al. 2023).

REFERENCES

Aikawa Y, Cataldi G, Yamato Y, et al. (2021) Molecules with ALMA at planet-forming scales (MAPS). XIII. HCO$^+$ and disk ionization structure. Astrophys J Suppl Ser 257:13, https://doi.org/10.3847/1538-4365/ac143c

Altwegg K, Balsiger H, Geiss J (1999) Composition of the volatile material in Halley's coma from in situ measurements. Space Sci Rev 90:3–18, https://doi.org/10.1023/A:1005256607402

Anders E (1989) Pre-biotic organic matter from comets and asteroids. Nature 342:255–257, https://doi.org/10.1038/342255a0

Anderson DE, Blake GA, Bergin EA, Zhang K, Carpenter JM, Schwarz KR, Huang J, Öberg KI (2019) Probing the gas content of late-stage protoplanetary disks with N$_2$H$^+$. Astrophys J 881:127, https://doi.org/10.3847/1538-4357/ab2cb5

Anderson DE, Blake GA, Cleeves LI, Bergin EA, Zhang K, Schwarz KR, Salyk C, Bosman AD (2021) Observing carbon and oxygen carriers in protoplanetary disks at mid-infrared wavelengths. Astrophys J 909:55, https://doi.org/10.3847/1538-4357/abd9c1

Anderson DE, Cleeves LI, Blake GA, Bergin EA, Zhang K, Carpenter JM, Schwarz KR (2022) New constraints on protoplanetary disk gas masses in Lupus. Astrophys J 927:229, https://doi.org/10.3847/1538-4357/ac517e

Andrews SM (2020) Observations of protoplanetary disk structures. Annu Rev Astron Astrophys 58:483–528, https://doi.org/10.1146/annurev-astro-031220-010302

Ansdell M, Williams JP, van der Marel N, et al. (2016) ALMA survey of Lupus protoplanetary disks. I. Dust and gas masses. Astrophys J 828:46, https://doi.org/10.3847/0004-637X/828/1/46

Antonellini S, Kamp I, Lahuis F, Woitke P, Thi W-F, Meijerink R, Aresu G, Spaans M, Güdel M, Liebhart A (2016) Mid-IR spectra of pre-main sequence Herbig stars: An explanation for the non-detections of water lines. Astron Astrophys 585:A61, https://doi.org/10.1051/0004-6361/201526787

Banzatti A, Pontoppidan KM (2015) An empirical sequence of disk gap opening revealed by rovibrational CO. Astrophys J 809:167, https://doi.org/10.1088/0004-637X/809/2/167

Banzatti A, Pinilla P, Ricci L, Pontoppidan KM, Birnstiel T, Ciesla F (2015) Direct imaging of the water snow line at the time of planet formation using two ALMA continuum bands. Astrophys J 815:L15, https://doi.org/10.1088/2041-8205/815/1/L15

Banzatti A, Pascucci I, Bosman AD, et al. (2020) Hints for icy pebble migration feeding an oxygen-rich chemistry in the inner planet-forming region of disks. Astrophys J 903:124, https://doi.org/10.3847/1538-4357/abbc1a

Banzatti A, Abernathy KM, Brittain S, et al. (2022) Scanning disk rings and winds in CO at 0.01–10 au: A high-resolution M-band spectroscopy survey with IRTF-iSHELL. Astron J 163:174, https://doi.org/10.3847/1538-3881/ac52f0

Banzatti A, Pontoppidan KM, Pére Chávez J, et al. (2023) The kinematics and excitation of infrared water vapor emission from planet-forming disks: Results from spectrally resolved surveys and guidelines for JWST Spectra. Astron J 165:72, https://doi.org/10.3847/1538-3881/aca80b

Barenfeld SA, Carpenter JM, Ricci L, Isella A (2016) ALMA observations of circumstellar disks in the upper Scorpius OB association. Astrophys J 827:142, https://doi.org/10.3847/0004-637X/827/2/142

Bergin EA, Williams JP (2017) The determination of protoplanetary disk masses. *In*: Formation, Evolution, and Dynamics of Young Solar Systems. Vol 445. Astrophysics and Space Science Library, p 1

Bergin EA, Aikawa Y, Blake GA, van Dishoeck EF (2007) The chemical evolution of protoplanetary disks. *In*: Physical Processes in Circumstellar Disks Around Young Stars. Garcia (ed) University of Chicago Press, Chicago, p 751

Bergin EA, Hogerheijde MR, Brinch C, et al. (2010) Sensitive limits on the abundance of cold water vapor in the DM Tauri protoplanetary disk. Astron Astrophys 521:L33, https://doi.org/10.1051/0004-6361/201015104

Bergin EA, Cleeves LI, Gorti U, et al. (2013) An old disk still capable of forming a planetary system. Nature 493:644–646, https://doi.org/10.1038/nature11805

Bergin EA, Blake GA, Ciesla F, Hirschmann MM, Li J (2015) Tracing the ingredients for a habitable earth from interstellar space through planet formation. Proc Nat Acad Sci 112:8965–8970, https://doi.org/10.1073/pnas.1500954112

Bergin EA, Du F, Cleeves LI, Blake GA, Schwarz K, Visser R, Zhang K (2016) Hydrocarbon emission rings in protoplanetary disks induced by dust evolution. Astrophys J 831:101, https://doi.org/10.3847/0004-637X/831/1/101

Bergner JB, Guzmán VG, Öberg KI, Loomis RA, Pegues J (2018) A survey of CH_3CN and HC_3N in protoplanetary disks. Astrophys J 857:69, https://doi.org/10.3847/1538-4357/aab664

Bergner JB, Öberg KI, Bergin EA, et al. (2020) An evolutionary study of volatile chemistry in protoplanetary disks. Astrophys J 898:97, https://doi.org/10.3847/1538-4357/ab9e71

Berné O, Martin-Drumel M-A, Schroetter I, et al. (2023) Formation of the methyl cation by photochemistry in a protoplanetary disk. Nature 621:56–59, https://doi.org/10.1038/s41586-023-06307-x

Bitsch B, Mah J (2023) Enriching inner discs and giant planets with heavy elements. Astron Astrophys 679:A11, https://doi.org/10.1051/0004-6361/202347419

Bjorkman JE, Wood K (2001) Radiative equilibrium and temperature correction in Monte Carlo radiation transfer. Astrophys J 554:615–623, https://doi.org/10.1086/321336

Blake GA, Boogert ACA (2004) High-resolution 4.7 Micron Keck/NIRSPEC spectroscopy of the CO emission from the disks surrounding Herbig Ae stars. Astrophys J Lett 606:L73–L76, https://doi.org/10.1086/421082

Blevins SM, Pontoppidan KM, Banzatti A, Zhang K, Najita JR, Carr JS, Salyk C, Blake GA (2016) Measurements of water surface snow lines in classical protoplanetary disks. Astrophys J 818:22, https://doi.org/10.3847/0004-637X/818/1/22

Boogert ACA, Gerakines PA, Whittet DCB (2015) Observations of the icy universe. Annu Rev Astron Astrophys 53:541–581, https://doi.org/10.1146/annurev-astro-082214-122348

Booth RA, Ilee JD (2019) Planet-forming material in a protoplanetary disc: the interplay between chemical evolution and pebble drift. Mon Not R Astron Soc 487:3998–4011, https://doi.org/10.1093/Mon Not R Astron Soc/stz1488

Booth AS, Ilee JD (2020) $^{13}C^{17}O$ suggests gravitational instability in the HL Tau disc. Mon Not R Astron Soc 493:L108–L113, https://doi.org/10.1093/Mon Not R Astron Socl/slaa014

Booth AS, Walsh C, Ilee JD (2019a) First detections of $H^{13}CO^+$ and $HC^{15}N$ in the disk around HD 97048. Evidence for a cold gas reservoir in the outer disk. Astron Astrophys 629:A75, https://doi.org/10.1051/0004-6361/201834388

Booth AS, Walsh C, Ilee JD, Notsu S, Qi C, Nomura H, Akiyama E (2019b) The first detection of $^{13}C^{17}O$ in a protoplanetary disk: A robust tracer of disk gas mass. Astrophys J Lett 882:L31, https://doi.org/10.3847/2041-8213/ab3645

Booth AS, Walsh C, Terwisscha van Scheltinga J, van Dishoeck EF, Ilee JD, Hogerheijde MR, Kama M, Nomura H (2021) An inherited complex organic molecule reservoir in a warm planet-hosting disk. Nat Astron 5:684–690, https://doi.org/10.1038/s41550-021-01352-w

Bosman AD, Bruderer S, van Dishoeck EF (2017) CO_2 infrared emission as a diagnostic of planet-forming regions of disks. Astron Astrophys 601:A36, https://doi.org/10.1051/0004-6361/201629946

Bosman AD, Walsh C, van Dishoeck EF (2018) CO destruction in protoplanetary disk midplanes: Inside versus outside the CO snow surface. Astron Astrophys 618:A182, https://doi.org/10.1051/0004-6361/201833497

Bosman AD, Cridland AJ, Miguel Y (2019) Jupiter formed as a pebble pile around the N_2 ice line. Astron Astrophys 632:L11, https://doi.org/10.1051/0004-6361/201936827

Bosman AD, Alarcón F, Bergin EA, et al. (2021) Molecules with ALMA at Planet-forming Scales (MAPS). VII. Substellar O/H and C/H and superstellar C/O in planet-feeding gas. Astrophys J Suppl Ser 257:7, https://doi.org/10.3847/1538-4365/ac1435

Bosman AD, Bergin EA, Calahan JK, Duval SE (2022) Water UV-shielding in the terrestrial planet-forming zone: Implications for carbon dioxide emission. Astrophys J Lett 933:L40, https://doi.org/10.3847/2041-8213/ac7d9f

Brinch C, Hogerheijde MR (2010) LIME - a flexible, non-LTE line excitation and radiation transfer method for millimeter and far-infrared wavelengths. Astron Astrophys 523:A25, https://doi.org/10.1051/0004-6361/201015333

Brittain SD, Simon T, Najita JR, Rettig TW (2007) Warm gas in the inner disks around young intermediate-mass stars. Astrophys J 659:685–704, https://doi.org/10.1086/511255

Brown JM, Pontoppidan KM, van Dishoeck EFv, Herczeg GJ, Blake GA, Smette A (2013) VLT-CRIRES survey of rovibrational CO emission from protoplanetary disks. Astrophys J 770:94

Bruderer S, van Dishoeck EF, Doty SD, Herczeg GJ (2012) The warm gas atmosphere of the HD 100546 disk seen by Herschel. Evidence of a gas-rich, carbon-poor atmosphere? Astron Astrophys 541:A91, https://doi.org/10.1051/0004-6361/201118218

Brunken NGC, Booth AS, Leemker M, Nazari P, van der Marel N, van Dishoeck EF (2022) A major asymmetric ice trap in a planet-forming disk. III. First detection of dimethyl ether. Astron Astrophys 659:A29, https://doi.org/10.1051/0004-6361/202142981

Calahan JK, Bergin E, Zhang K, et al. (2021) The TW Hya Rosetta Stone project. III. Resolving the gaseous thermal profile of the disk. Astrophys J 908:8, https://doi.org/10.3847/1538-4357/abd255

Calahan JK, Bergin EA, Bosman AD, et al. (2023) UV-driven chemistry as a signpost of late-stage planet formation. Nat Astron 7:49–56, https://doi.org/10.1038/s41550-022-01831-8

Carr JS, Najita JR (2008) Organic molecules and water in the planet formation region of young circumstellar disks. Science 319:1504–1506, https://doi.org/10.1126/science.1153807

Carr JS, Najita JR (2011) Organic molecules and water in the inner disks of T Tauri stars. Astrophys J 733:102, https://doi.org/10.1088/0004-637X/733/2/102

Carr JS, Tokunaga AT, Najita J, Shu FH, Glassgold AE (1993) The inner-disk and stellar properties of the young stellar object WL 16. Astrophys J Lett 411:L37, https://doi.org/10.1086/186906

Carr JS, Najita JR, Salyk C (2018) Measuring the water snow line in a protoplanetary disk. Res Not Am Astron Soc 2:169, https://doi.org/10.3847/2515-5172/aadfe7

Cazaux S, Tielens AGGM (2004) H_2 formation on grain surfaces. Astrophys J 604:222–237, https://doi.org/10.1086/381775

Cazzoletti P, van Dishoeck EF, Visser R, Facchini S, Bruderer S (2018) CN rings in full protoplanetary disks around young stars as probes of disk structure. Astron Astrophys 609:A93, https://doi.org/10.1051/0004-6361/201731457

Ceccarelli C, Caselli P, Bockelée-Morvan D, Mousis O, Pizzarello S, Robert F, Semenov D (2014) Deuterium fractionation: The Ariadne's Thread from the precollapse phase to meteorites and comets today. In: Protostars and Planets VI. H Beuther, RS Klessen, CP Dullemond, T Henning (eds) University of Arizona Press, p 859–882

Chapillon E, Dutrey A, Guilloteau S, et al. (2012) Chemistry in disks. VII. First detection of HC_3N in protoplanetary disks. Astrophys J 756:58, https://doi.org/10.1088/0004-637X/756/1/58

Ciesla FJ, Cuzzi JN (2006) The evolution of the water distribution in a viscous protoplanetary disk. Icarus 181:178–204, https://doi.org/10.1016/Jicarus.2005.11.009

Cieza LA, Casassus S, Tobin J, et al. (2016) Imaging the water snow-line during a protostellar outburst. Nature 535:258–261, https://doi.org/10.1038/nature18612

Cieza LA, Ruíz-Rodríguez D, Hales A, et al. (2019) The Ophiuchus DIsc Survey Employing ALMA (ODISEA)—I: Project description and continuum images at 28 au resolution. Mon Not R Astron Soc 482:698–714, https://doi.org/10.1093/Mon Not R Astron Soc/sty2653

Clayton RN (2002) Solar System: Self-shielding in the solar nebula. Nature 415:860–861, https://doi.org/10.1038/415860b

Cleeves LI, Adams FC, Bergin EA (2013) Exclusion of cosmic rays in protoplanetary disks: Stellar and magnetic effects. Astrophys J 772:5, https://doi.org/10.1088/0004-637X/772/1/5

Cleeves LI, Bergin EA, Alexander CMO, Du F, Graninger D, Öberg KI, Harries TJ (2014) The ancient heritage of water ice in the solar system. Science 345:1590–1593, https://doi.org/10.1126/science.1258055

Cleeves LI, Bergin EA, Qi C, Adams FC, Öberg KI (2015) Constraining the X-ray and cosmic-ray ionization chemistry of the TW Hya protoplanetary disk: Evidence for a sub-interstellar cosmic-ray rate. Astrophys J 799:204, https://doi.org/10.1088/0004-637X/799/2/204

Cleeves LI, Öberg KI, Wilner DJ, Huang J, Loomis RA, Andrews SM, Czekala I (2016) The coupled physical structure of gas and dust in the IM Lup protoplanetary disk. Astrophys J 832:110, https://doi.org/10.3847/0004-637X/832/2/110

Cleeves LI, Bergin EA, Öberg KI, Andrews S, Wilner D, Loomis R (2017) Variable $H^{13}CO^+$ emission in the IM Lup disk: X-ray driven time-dependent chemistry? Astrophys J Lett 843:L3, https://doi.org/10.3847/2041-8213/aa76e2

Cleeves LI, Öberg KI, Wilner DJ, Huang J, Loomis RA, Andrews SM, Guzmán VV (2018) Constraining gas-phase carbon, oxygen, and nitrogen in the IM Lup protoplanetary disk. Astrophys J 865:155, https://doi.org/10.3847/1538-4357/aade96

Cridland AJ, Pudritz RE, Alessi M (2016) Composition of early planetary atmospheres—I. Connecting disc astrochemistry to the formation of planetary atmospheres. Mon Not R Astron Soc 461:3274–3295, https://doi.org/10.1093/Mon Not R Astron Soc/stw1511

Cridland AJ, van Dishoeck EF, Alessi M, Pudritz RE (2019) Connecting planet formation and astrochemistry. A main sequence for C/O in hot exoplanetary atmospheres. Astron Astrophys 632:A63, https://doi.org/10.1051/0004-6361/201936105

Cridland AJ, van Dishoeck EF, Alessi M, Pudritz RE (2020) Connecting planet formation and astrochemistry. C/Os and N/Os of warm giant planets and Jupiter analogues. Astron Astrophys 642:A229, https://doi.org/10.1051/0004-6361/202038767

Cridland AJ, Facchini S, van Dishoeck EF, Benisty M (2023) Planet formation in the PDS 70 system. Constraining the atmospheric chemistry of PDS 70b and c. Astron Astrophys 674:A211, https://doi.org/10.1051/0004-6361/202245619

Cronin JR, Chang S (1993) Organic matter in meteorites: Molecular and isotopic analyses of the Murchison meteorite. *In:* The Chemistry of Life's Origins. NATO ASI Series, vol 416. JM Greenberg, CX Mendoza-Gómez, V Pirronello (eds) Springer, Dordrecht, p 209–258

Cuzzi JN, Zahnle KJ (2004) Material enhancement in protoplanetary nebulae by particle drift through evaporation fronts. Astrophys J 614:490–496, https://doi.org/10.1086/423611

Dartois E, Dutrey A, Guilloteau S (2003) Structure of the DM Tau outer disk: Probing the vertical kinetic temperature gradient. Astron Astrophys 399:773-787, https://doi.org/10.1051/0004-6361:20021638

Dionatos O, Woitke P, Güdel M, et al. (2019) Consistent dust and gas models for protoplanetary disks. IV. A panchromatic view of protoplanetary disks. Astron Astrophys 625:A66, https://doi.org/10.1051/0004-6361/201832860

Drazkowska J, Alibert Y (2017) Planetesimal formation starts at the snow line. Astron Astrophys 608:A92, https://doi.org/10.1051/0004-6361/201731491

Drouart A, Dubrulle B, Gautier D, Robert F (1999) Structure and transport in the solar nebula from constraints on deuterium enrichment and giant planets formation. Icarus 140:129–155, https://doi.org/10.1006/icar.1999.6137

Drozdovskaya MN, van Dishoeck EF, Rubin M, Jørgensen JK, Altwegg K (2019) Ingredients for solar-like systems: protostar IRAS 16293-2422 B versus comet 67P/Churyumov-Gerasimenko. Mon Not R Astron Soc 490:50–79, https://doi.org/10.1093/Mon Not R Astron Soc/stz2430

Du F, Bergin EA (2014) Water vapor distribution in protoplanetary disks. Astrophys J 792:2, https://doi.org/10.1088/0004-637X/792/1/2

Du F, Bergin EA, Hogerheijde M, et al. (2017) Survey of cold water lines in protoplanetary disks: Indications of systematic volatile depletion. Astrophys J 842:98, https://doi.org/10.3847/1538-4357/aa70ee

Dullemond CP, Dominik C (2004) Flaring vs. self-shadowed disks: The SEDs of Herbig Ae/Be stars. Astron Astrophys 417:159–168, https://doi.org/10.1051/0004-6361:20031768

Dullemond CP, Monnier JD (2010) The inner regions of protoplanetary disks. Annu Rev Astron Astrophys 48:205–239, https://doi.org/10.1146/annurev-astro-081309-130932

Dutrey A, Guilloteau S, Guelin M (1997) Chemistry of protosolar-like nebulae: The molecular content of the DM Tau and GG Tau disks. Astron Astrophys 317:L55

Eistrup C, Walsh C, van Dishoeck EF (2016) Setting the volatile composition of (exo)planet-building material. Does chemical evolution in disk midplanes matter? Astron Astrophys 595:A83, https://doi.org/10.1051/0004-6361/201628509

Eistrup C, Cleeves LI, Krijt S (2022) Chemical evolution in planet-forming regions with growing grains. Astron Astrophys 667:A121, https://doi.org/10.1051/0004-6361/202243981

Estrada PR, Cuzzi JN (2022) Global modeling of nebulae with particle growth, drift, and evaporation fronts. III. Redistribution of refractories and volatiles. Astrophys J 936:40, https://doi.org/10.3847/1538-4357/ac81c6

Ercolano B, Pascucci I (2017) The dispersal of planet-forming discs: theory confronts observations. R Soc Open Sci 4:170114, https://doi.org/10.1098/rsos.170114

Favre C, Cleeves LI, Bergin EA, Qi C, Blake GA (2013) A significantly low CO abundance toward the TW Hya protoplanetary disk: A path to active carbon chemistry? Astrophys J Lett 776:L38, https://doi.org/10.1088/2041-8205/776/2/L38

Favre C, Fedele D, Semenov D, et al. (2018) First detection of the simplest organic acid in a protoplanetary disk. Astrophys J Lett 862:L2, https://doi.org/10.3847/2041-8213/aad046

Fedele D, van den Ancker ME, Henning T, Jayawardhana R, Oliveira JM (2010) Timescale of mass accretion in pre-main-sequence stars. Astron Astrophys 510:A72, https://doi.org/10.1051/0004-6361/200912810

Fedele D, Bruderer S, van Dishoeck EFv, Herczeg GJ, Evans NJ, Bouwman J, Henning T, Green J (2012) Warm H_2O and OH in the disk around the Herbig star HD 163296. Astron Astrophys 544:L9

Füri E, Marty B (2015) Nitrogen isotope variations in the Solar System. Nat Geosci 8:515–522, https://doi.org/10.1038/ngeo2451

Furuya K, Aikawa Y (2014) Reprocessing of ices in turbulent protoplanetary disks: Carbon and nitrogen chemistry. Astrophys J 790:97, https://doi.org/10.1088/0004-637X/790/2/97

Garrod RT, Widicus Weaver SL, Herbst E (2008) Complex chemistry in star-forming regions: An expanded gas–grain warm-up chemical model. Astrophys J 682:283–302, https://doi.org/10.1086/588035

Getman KV, Flaccomio E, Broos PS, et al. (2005) Chandra Orion Ultradeep Project: Observations and source lists. Astrophys J Suppl Ser 160:319–352, https://doi.org/10.1086/432092

Gibb EL, Horne D (2013) Detection of CH_4 in the GV Tau N protoplanetary disk. Astrophys J Lett 776:L28, https://doi.org/10.1088/2041-8205/776/2/L28

Grant SL, van Dishoeck EF, Tabone B, Gasman D, Henning T, Kamp I, Güdel M, Lagage EA (2023) MINDS. The detection of $^{13}CO_2$ with JWST-MIRI indicates abundant CO_2 in a protoplanetary disk. Astrophys J Lett 947:L6, https://doi.org/10.3847/2041-8213/acc44b

Greenberg JM (1998) Making a comet nucleus. Astron Astrophys 330:375–380

Gundlach B, Blum J (2015) The stickiness of micrometer-sized water-ice particles. Astrophys J 798:34, https://doi.org/10.1088/0004-637X/798/1/34

Guzmán VV, Öberg KI, Huang J, Loomis R, Qi C (2017) Nitrogen fractionation in protoplanetary disks from the $H^{13}CN/HC^{15}N$ ratio. Astrophys J 836:30, https://doi.org/10.3847/1538-4357/836/1/30

Guzmán VV, Bergner JB, Law CJ, et al. (2021) Molecules with ALMA at Planet-forming Scales (MAPS). VI. Distribution of the small organics HCN, C_2H, and H_2CO. Astrophys J Suppl Ser 257:6, https://doi.org/10.3847/1538-4365/ac1440

Hasegawa TI, Herbst E, Leung CM (1992) Models of gas–grain chemistry in dense interstellar clouds with complex organic molecules. Astrophys J Suppl Ser 82:167, https://doi.org/10.1086/191713

Hayashi C (1981) Structure of the solar nebula, growth and decay of magnetic fields and effects of magnetic and turbulent viscosities on the nebula. Prog Theor Phys Suppl 70:35–53, https://doi.org/10.1143/PTPS.70.35

Heinzeller D, Nomura H, Walsh C, Millar TJ (2011) Chemical evolution of protoplanetary disks—The effects of viscous accretion, turbulent mixing, and disk winds. Astrophys J 731:115, https://doi.org/10.1088/0004-637X/731/2/115

Henning T, Semenov D (2013) Chemistry in protoplanetary disks. Chem Rev 113:9016–9042, https://doi.org/10.1021/cr400128p

Herbst E, Leung CM (1989) Gas phase production of complex hydrocarbons, cyanopolyynes, and related compounds in dense interstellar clouds. Astrophys J Suppl Ser 69:271, https://doi.org/10.1086/191314

Herbst E, van Dishoeck EF (2009) Complex organic interstellar molecules. Annu Rev Astron Astrophys 47:427–480, https://doi.org/10.1146/annurev-astro-082708-101654

Herbst E, Woon DE (1997) The rate of the reaction between C_2H and C_2H_2 at interstellar temperatures. Astrophys J 489:109–112, https://doi.org/10.1086/304786

Hersant F, Gautier D, Huré J-M (2001) A Two-dimensional model for the primordial nebula constrained by D/H measurements in the Solar System: Implications for the formation of giant planets. Astrophys J 554:391–407, https://doi.org/10.1086/321355

Hily-Blant P, Magalhaes V, Kastner J, Faure A, Forveille T, Qi C (2017) Direct evidence of multiple reservoirs of volatile nitrogen in a protosolar nebula analogue. Astron Astrophys 603:L6, https://doi.org/10.1051/0004-6361/201730524

Hily-Blant P, Magalhaes de Souza V, Kastner J, Forveille T (2019) Multiple nitrogen reservoirs in a protoplanetary disk at the epoch of comet and giant planet formation. Astron Astrophys 632:L12, https://doi.org/10.1051/0004-6361/201936750

Hinkel NR, Youngblood A, Soares-Furtado M (2024) Host stars and how their compositions influence exoplanets. Rev Mineral Geochem 90:1–26

Hogerheijde MR, Bergin EA, Brinch C, et al. (2011) Detection of the water reservoir in a forming planetary system. Science 334:338, https://doi.org/10.1126/science.1208931

Huang J, Öberg KI, Qi C, Aikawa Y, Andrews SM, Furuya K, Guzmán VV, Loomis RA, van Dishoeck EF, Wilner DJ (2017) An ALMA survey of DCN/H^{13}CN and DCO$^+$/H^{13}CO$^+$ in protoplanetary disks. Astrophys J 835:231, https://doi.org/10.3847/1538-4357/835/2/231

Ilee JD, Walsh C, Booth AS, et al. (2021) Molecules with ALMA at Planet-forming Scales (MAPS). IX. Distribution and properties of the large organic molecules HC_3N, CH_3CN, and c-C_3H_2. Astrophys J Suppl Ser 257:9, https://doi.org/10.3847/1538-4365/ac1441

Jørgensen JK, Belloche A, Garrod RT (2020) Astrochemistry during the formation of stars. Annu Rev Astron Astrophys 58:727–778, https://doi.org/10.1146/annurev-astro-032620-021927

Kamp I, Dullemond CP (2004) The gas temperature in the surface layers of protoplanetary disks. Astrophys J 615:991–999, https://doi.org/10.1086/424703

Kamp I, Thi W-F, Woitke P, Rab C, Bouma S, Ménard F (2017) Consistent dust and gas models for protoplanetary disks. II. Chemical networks and rates. Astron Astrophys 607:A41, https://doi.org/10.1051/0004-6361/201730388

Kanwar J, Kamp I, Woitke P, Rab C, Thi W-F, Min M (2023) Hydrocarbon chemistry in inner regions of planet forming disks. arXiv e-prints:arXiv:2310.04505, https://doi.org/10.48550/arXiv.2310.04505

Kempton EM-R, Knutson HA (2024) Transiting exoplanet atmospheres in the era of JWST. Rev Mineral Geochem 90:411–464

Kennedy GM, Kenyon SJ (2008) Planet formation around stars of various masses: The snow line and the frequency of giant planets. Astrophys J 673:502–512, https://doi.org/10.1086/524130

Krijt S, Ciesla FJ, Bergin EA (2016) Tracing water vapor and ice during dust growth. Astrophys J 833:285, https://doi.org/10.3847/1538-4357/833/2/285

Krijt S, Schwarz KR, Bergin EA, Ciesla FJ (2018) Transport of CO in protoplanetary disks: consequences of pebble formation, settling, and radial drift. Astrophys J 864:78, https://doi.org/10.3847/1538-4357/aad69b

Krijt S, Bosman AD, Zhang K, Schwarz KR, Ciesla FJ, Bergin EA (2020) CO depletion in protoplanetary disks: A unified picture combining physical sequestration and chemical processing. Astrophys J 899:134, https://doi.org/10.3847/1538-4357/aba75d

Krijt S, Kama M, McClure M, Teske J, Bergin EA, Shorttle O, Walsh KJ, Raymond SN (2023) Chemical habitability: Supply and retention of life's essential elements during planet formation. *In:* Protostars and Planets VII. S Inutsuka, Y Aikawa, T Muto, K Tomida, M Tamura (eds). Astron Soc Pac Conf Ser Vol 534, p 1031

Kruijer TS, Burkhardt C, Budde G, Kleine T (2017) Age of Jupiter inferred from the distinct genetics and formation times of meteorites. Proc Natl Acad Sci 114:6712–6716, https://doi.org/10.1073/pnas.1704461114

Law CJ, Teague R, Loomis RA, et al. (2021) Molecules with ALMA at Planet-forming Scales (MAPS). IV. Emission surfaces and vertical distribution of molecules. Astrophys J Suppl Ser 257:4, https://doi.org/10.3847/1538-4365/ac1439

Lee J-E, Lee S, Baek G, Aikawa Y, Cieza L, Yoon S-Y, Herczeg G, Johnstone D, Casassus S (2019) The ice composition in the disk around V883 Ori revealed by its stellar outburst. Nat Astron 3:314–319, https://doi.org/10.1038/s41550-018-0680-0

Leemker M, van't Hoff MLR, Trapman L, van Gelder ML, Hogerheijde MR, Ruíz-Rodríguez D, van Dishoeck EF (2021) Chemically tracing the water snowline in protoplanetary disks with HCO^+. Astron Astrophys 646:A3, https://doi.org/10.1051/0004-6361/202039387

Linsky JL (1998) Deuterium abundance in the local ISM and possible spatial variations. Space Sci. Rev 84:285-296

Long F, Herczeg GJ, Pascucci I, et al. (2017) An ALMA Survey of CO isotopologue emission from protoplanetary disks in chamaeleon I. Astrophys J 844:99, https://doi.org/10.3847/1538-4357/aa78fc

Loomis RA, Cleeves LI, Öberg KI, Aikawa Y, Bergner J, Furuya K, Guzmán VV, Walsh C (2018) The distribution and excitation of CH_3CN in a solar nebula analog. Astrophys J 859:131, https://doi.org/10.3847/1538-4357/aac169

Loomis RA, Öberg KI, Andrews SM, et al. (2020) An unbiased ALMA spectral survey of the LkCa 15 and MWC 480 Protoplanetary Disks. Astrophys J 893:101, https://doi.org/10.3847/1538-4357/ab7cc8

Manara CF, Rosotti G, Testi L, et al. (2016) Evidence for a correlation between mass accretion rates onto young stars and the mass of their protoplanetary disks. Astron Astrophys 591:L3, https://doi.org/10.1051/0004-6361/201628549

Manara CF, Natta A, Rosotti GP, et al. (2020) X-shooter survey of disk accretion in Upper Scorpius. I. Very high accretion rates at age > 5 Myr. Astron Astrophys 639:A58, https://doi.org/10.1051/0004-6361/202037949

Manara CF, Ansdell M, Rosotti GP, Hughes AM, Armitage PJ, Lodato G, Williams JP (2023) Demographics of young stars and their protoplanetary disks: Lessons learned on disk evolution and its connection to planet formation. *In:* Protostars and Planets VII. S Inutsuka, Y Aikawa, T Muto, K Tomida, M Tamura (eds). Astron Soc Pac Conf Ser Vol 534, p 539

Mandell AM, Bast J, van Dishoeck EF, Blake GA, Salyk C, Mumma MJ, Villanueva G (2012) First detection of near-infrared line emission from organics in young circumstellar disks. Astrophys J 747:92, https://doi.org/10.1088/0004-637X/747/2/92

Marty B (2012) The origins and concentrations of water, carbon, nitrogen and noble gases on Earth. Earth and Planetary Science Letters 313:56–66, https://doi.org/10.1016/Jepsl.2011.10.040

McClure MK, Bergin EA, Cleeves LI, et al. (2016) Mass measurements in protoplanetary disks from hydrogen deuteride. Astrophys J 831:167, https://doi.org/10.3847/0004-637X/831/2/167

McElroy D, Walsh C, Markwick AJ, Cordiner MA, Smith K, Millar TJ (2013) The UMIST database for astrochemistry 2012. Astron Astrophys 550:A36, https://doi.org/10.1051/0004-6361/201220465

McKeegan KD, Kallio APA, Heber VS, et al. (2011) The oxygen isotopic composition of the sun inferred from captured solar wind. Science 332:1528, https://doi.org/10.1126/science.1204636

Meeus G, Montesinos B, Mendigutía I, et al. (2012) Observations of Herbig Ae/Be stars with Herschel/PACS. The atomic and molecular contents of their protoplanetary discs. Astron Astrophys 544:A78, https://doi.org/10.1051/0004-6361/201219225

Miotello A, van Dishoeck EF, Williams JP, et al. (2017) Lupus disks with faint CO isotopologues: low gas/dust or high carbon depletion? Astron Astrophys 599:A113, https://doi.org/10.1051/0004-6361/201629556

Miotello A, Facchini S, van Dishoeck EF, Cazzoletti P, Testi L, Williams JP, Ansdell M, van Terwisga S, van der Marel N (2019) Bright C_2H emission in protoplanetary discs in Lupus: High volatile C/O > 1 ratios. Astron Astrophys 631:A69, https://doi.org/10.1051/0004-6361/201935441

Miotello A, Kamp I, Birnstiel T, Cleeves LC, Kataoka A (2023) Setting the stage for planet formation: Measurements and implications of the fundamental disk properties. *In:* Protostars and Planets VII. S Inutsuka, Y Aikawa, T Muto, K Tomida, M Tamura (eds). Astron Soc Pac Conf Ser Vol 534, p 501

Mishra A, Li A (2015) Probing the role of carbon in the interstellar ultraviolet extinction. Astrophys J 809:120, https://doi.org/10.1088/0004-637X/809/2/120

Morbidelli A, Bitsch B, Crida A, Gounelle M, Guillot T, Jacobson S, Johansen A, Lambrechts M, Lega E (2016) Fossilized condensation lines in the Solar System protoplanetary disk. Icarus 267:368–376, https://doi.org/10.1016/Jicarus.2015.11.027

Mordasini C, van Boekel R, Mollière P, Henning T, Benneke B (2016) The imprint of exoplanet formation history on observable present-day spectra of hot Jupiters. Astrophys J 832:41, https://doi.org/10.3847/0004-637X/832/1/41

Mumma MJ, Charnley SB (2011) The chemical composition of comets–Emerging taxonomies and natal heritage. Annu Rev Astron Astrophys 49:471–524

Najita JR, Ádámkovics M, Glassgold AE (2011) Formation of organic molecules and water in warm disk atmospheres. Astrophys J 743:147

Notsu S, Nomura H, Ishimoto D, Walsh C, Honda M, Hirota T, Millar TJ (2016) Candidate water vapor lines to locate the H_2O snowline through high-dispersion spectroscopic observations. I. The case of a T Tauri star. Astrophys J 827:113, https://doi.org/10.3847/0004-637X/827/2/113

Notsu S, Akiyama E, Booth A, Nomura H, Walsh C, Hirota T, Honda M, Tsukagoshi T, Millar TJ (2019) Dust continuum emission and the upper limit fluxes of submillimeter water lines of the protoplanetary disk around HD 163296 observed by ALMA. Astrophys J 875:96, https://doi.org/10.3847/1538-4357/ab0ae9

Notsu S, Eistrup C, Walsh C, Nomura H (2020) The composition of hot Jupiter atmospheres assembled within chemically evolved protoplanetary discs. Mon Not R Astron Soc 499:2229–2244, https://doi.org/10.1093/Mon Not R Astron Soc/staa2944

Öberg KI, Bergin EA (2016) Excess C/O and C/H in outer protoplanetary disk gas. Astrophys J 831:L19, https://doi.org/10.3847/2041-8205/831/2/L19

Öberg KI, Bergin EA (2021) Astrochemistry and compositions of planetary systems. Phys Rep 893:1–48, https://doi.org/10.1016/Jphysrep.2020.09.004

Öberg KI, Wordsworth R (2019) Jupiter´s composition suggests its core assembled exterior to the N_2 snowline. Astron J 158:194, https://doi.org/10.3847/1538-3881/ab46a8

Öberg KI, Qi C, Fogel JKJ, Bergin EA, Andrews SM, Espaillat C, van Kempen TA, Wilner DJ, Pascucci I (2010) The disk imaging survey of chemistry with SMA. I. Taurus protoplanetary disk data. Astrophys J 720:480–493, https://doi.org/10.1088/0004-637X/720/1/480

Öberg KI, Boogert ACA, Pontoppidan KM, van den Broek S, van Dishoeck EF, Bottinelli S, Blake GA, Evans NJ, II (2011) The Spitzer ice legacy: Ice evolution from cores to protostars. Astrophys J 740:109, https://doi.org/10.1088/0004-637X/740/2/109

Öberg KI, Guzmán VV, Furuya K, Qi C, Aikawa Y, Andrews SM, Loomis R, Wilner DJ (2015) The comet-like composition of a protoplanetary disk as revealed by complex cyanides. Nature 520:198–201, https://doi.org/10.1038/nature14276

Öberg KI, Guzmán VV, Walsh C, et al. (2021) Molecules with ALMA at Planet-forming Scales (MAPS). I. Program overview and highlights. Astrophys J Suppl Ser 257:1, https://doi.org/10.3847/1538-4365/ac1432

Öberg KI, Facchini S, Anderson DE (2023) Protoplanetary disk chemistry. Annu Rev Astron Astrophys 61:287–328, https://doi.org/10.1146/annurev-astro-022823-040820

Pascucci I, Herczeg G, Carr JS, Bruderer S (2013) The atomic and molecular content of disks around very low-mass stars and brown dwarfs. Astrophys J 779:178, https://doi.org/10.1088/0004-637X/779/2/178

Pegues J, Öberg KI, Bergner JB, et al. (2020) An ALMA Survey of H_2CO in protoplanetary disks. Astrophys J 890:142, https://doi.org/10.3847/1538-4357/ab64d9

Pinilla P, Pohl A, Stammler SM, Birnstiel T (2017) Dust density distribution and imaging analysis of different ice lines in protoplanetary disks. Astrophys J 845:68, https://doi.org/10.3847/1538-4357/aa7edb

Pinte C, Ménard F, Duchêne G, Bastien P (2006) Monte Carlo radiative transfer in protoplanetary disks. Astron Astrophys 459:797–804, https://doi.org/10.1051/0004-6361:20053275

Pinte C, Ménard F, Duchêne G, et al. (2018) Direct mapping of the temperature and velocity gradients in discs. Imaging the vertical CO snow line around IM Lupi. Astron Astrophys 609:A47, https://doi.org/10.1051/0004-6361/201731377

Podio L, Garufi A, Codella C, et al. (2020) ALMA chemical survey of disk-outflow sources in Taurus (ALMA-DOT). II. Vertical stratification of CO, CS, CN, H_2CO, and CH_3OH in a Class I disk. Astron Astrophys 642:L7, https://doi.org/10.1051/0004-6361/202038952

Pollack JB, Hubickyj O, Bodenheimer P, Lissauer JJ, Podolak M, Greenzweig Y (1996) Formation of the giant planets by concurrent accretion of solids and gas. Icarus 124:62–85, https://doi.org/10.1006/icar.1996.0190

Pontoppidan KM, Blake GA, van Dishoeck EF, Smette A, Ireland MJ, Brown J (2008) Spectroastrometric imaging of molecular gas within protoplanetary disk gaps. Astrophys J 684:1323–1329, https://doi.org/10.1086/590400

Pontoppidan KM, Salyk C, Blake GA, Käufl HU (2010a) Spectrally resolved pure rotational lines of water in protoplanetary disks. Astrophys J Lett 722:L173–L177, https://doi.org/10.1088/2041-8205/722/2/L173

Pontoppidan KM, Salyk C, Blake GA, Meijerink R, Carr JS, Najita J (2010b) A Spitzer survey of mid-infrared molecular emission from protoplanetary disks. I. Detection rates. Astrophys J 720:887–903, https://doi.org/10.1088/0004-637X/720/1/887

Pontoppidan KM, van Dishoeck E, Blake GA, et al. (2011) Planet-forming regions at the highest spectral and spatial resolution with VLT-CRIRES. The Messenger 143:32–36

Pontoppidan KM, Salyk C, Bergin EA, et al. (2014) Volatiles in protoplanetary disks. *In:* Protostars and Planets VI. H Beuther, RS Klessen, CP Dullemond, T Henning (eds). University of Arizona Press, p 363

Pontoppidan KM, Salyk C, Banzatti A, Blake GA, Walsh C, Lacy JH, Richter MJ (2019) The nitrogen carrier in inner protoplanetary disks. Astrophys J 874:92, https://doi.org/10.3847/1538-4357/ab05d8

Price EM, Cleeves LI, Öberg KI (2020) Chemistry along accretion streams in a viscously evolving protoplanetary disk. Astrophys J 890:154, https://doi.org/10.3847/1538-4357/ab5fd4

Qi C, Öberg KI, Wilner DJ, D'Alessio P, Bergin E, Andrews SM, Blake GA, Hogerheijde MR, van Dishoeck EF (2013) Imaging of the CO snow line in a solar nebula analog. Science 341:630–632, https://doi.org/10.1126/science.1239560

Qi C, Öberg KI, Andrews SM, Wilner DJ, Bergin EA, Hughes AM, Hogherheijde M, D'Alessio P (2015) Chemical imaging of the CO snow line in the HD 163296 disk. Astrophys J 813:128, https://doi.org/10.1088/0004-637X/813/2/128

Qi C, Öberg KI, Espaillat CC, Robinson CE, Andrews SM, Wilner DJ, Blake GA, Bergin EA, Cleeves LI (2019) Probing CO and N_2 snow surfaces in protoplanetary disks with N_2H^+ emission. Astrophys J 882:160, https://doi.org/10.3847/1538-4357/ab35d3

Riviere-Marichalar P, Ménard F, Thi WF, et al. (2012) Detection of warm water vapour in Taurus protoplanetary discs by Herschel. Astron Astrophys 538:L3, https://doi.org/10.1051/0004-6361/201118448

Ros K, Johansen A (2013) Ice condensation as a planet formation mechanism. Astron Astrophys 552:A137, https://doi.org/10.1051/0004-6361/201220536

Salinas VN, Hogerheijde MR, Bergin EA, et al. (2016) First detection of gas-phase ammonia in a planet-forming disk. NH_3, N_2H^+, and H_2O in the disk around TW Hydrae. Astron Astrophys 591:A122, https://doi.org/10.1051/0004-6361/201628172

Salyk C, Pontoppidan KM, Blake GA, Lahuis F, van Dishoeck EF, Evans I, N.~J (2008) H_2O and OH gas in the terrestrial planet-forming zones of protoplanetary disks. Astrophys J Lett 676:L49–L52, https://doi.org/10.1086/586894

Salyk C, Blake GA, Boogert ACA, Brown JM (2011a) CO rovibrational emission as a probe of inner disk structure. Astrophys J 743:112, https://doi.org/10.1088/0004-637X/743/2/112

Salyk C, Pontoppidan KM, Blake GA, Najita JR, Carr JS (2011b) A Spitzer survey of mid-infrared molecular emission from protoplanetary disks. II. Correlations and local thermal equilibrium models. Astrophys J 731:130

Salyk C, Herczeg GJ, Brown JM, Blake GA, Pontoppidan KM, van Dishoeck EF (2013) Measuring protoplanetary disk accretion with H I Pfund β. Astrophys J 769:21, https://doi.org/10.1088/0004-637X/769/1/21

Salyk C, Lacy J, Richter M, Zhang K, Pontoppidan K, Carr JS, Najita JR, Blake GA (2019) A high-resolution mid-infrared survey of water emission from protoplanetary disks. Astrophys J 874:24, https://doi.org/10.3847/1538-4357/ab05c3

Schwarz KR, Bergin EA, Cleeves LI, Blake GA, Zhang K, Öberg KI, van Dishoeck EF, Qi C (2016) The radial distribution of H_2 and CO in TW Hya as revealed by resolved ALMA observations of CO isotopologues. Astrophys J 823:91, https://doi.org/10.3847/0004-637X/823/2/91

Schwarz KR, Bergin EA, Cleeves LI, Zhang K, Öberg KI, Blake GA, Anderson D (2018) Unlocking CO depletion in protoplanetary disks. I. The warm molecular layer. Astrophys J 856:85, https://doi.org/10.3847/1538-4357/aaae08

Seifert RA, Cleeves LI, Adams FC, Li Z-Y (2021) Evidence for a cosmic-ray gradient in the IM Lup protoplanetary disk. Astrophys J 912:136, https://doi.org/10.3847/1538-4357/abf09a

Semenov D, Wiebe D (2011) Chemical evolution of turbulent protoplanetary disks and the solar nebula. Astrophys J Suppl Ser 196:25, https://doi.org/10.1088/0067-0049/196/2/25

Smith IWM, Herbst E, Chang Q (2004) Rapid neutral–neutral reactions at low temperatures: a new network and first results for TMC-1. Mon Not R Astron Soc 350:323–330, https://doi.org/10.1111/J1365-2966.2004.07656.x

Stevenson DJ, Lunine JI (1988) Rapid formation of Jupiter by diffuse redistribution of water vapor in the solar nebula. Icarus 75:146–155, https://doi.org/10.1016/0019-1035(88)90133-9

Tabone B, Bettoni G, van Dishoeck EF, et al. (2023) A rich hydrocarbon chemistry and high C to O ratio in the inner disk around a very low-mass star. Nat Astron 7:805–814, https://doi.org/10.1038/s41550-023-01965-3

Thiabaud A, Marboeuf U, Alibert Y, Leya I, Mezger K (2015) Gas composition of the main volatile elements in protoplanetary discs and its implication for planet formation. Astron Astrophys 574:A138, https://doi.org/10.1051/0004-6361/201424868

Thiemens MH (2006) History and applications of mass-independent isotope effects. Annu Rev Earth Planet Sci 34:217–262, https://doi.org/10.1146/annuRevearth.34.031405.125026

Thiemens MH, Heidenreich JE, III (1983) The mass-independent fractionation of oxygen—A novel isotope effect and its possible cosmochemical implications. Science 219:1073–1075, https://doi.org/10.1126/science.219.4588.1073

Tobin JJ, Sheehan PD, Megeath ST, et al. (2020) The VLA/ALMA Nascent Disk and Multiplicity (VANDAM) survey of Orion protostars. II. a statistical characterization of Class 0 and Class I protostellar disks. Astrophys J 890:130, https://doi.org/10.3847/1538-4357/ab6f64

Tobin JJ, van't Hoff MLR, Leemker M, et al. (2023) Deuterium-enriched water ties planet-forming disks to comets and protostars. Nature 615:227–230, https://doi.org/10.1038/s41586-022-05676-z

Trapman L, Zhang K, van't Hoff MLR, Hogerheijde MR, Bergin EA (2022) A novel way of measuring the gas disk mass of protoplanetary disks using N_2H^+ and $C^{18}O$. Astrophys J Lett 926:L2, https://doi.org/10.3847/2041-8213/ac4f47

Turrini D, Schisano E, Fonte S, Molinari S, Politi R, Fedele D, Panić O, Kama M, Changeat Q, Tinetti G (2021) Tracing the formation history of giant planets in protoplanetary disks with carbon, oxygen, nitrogen, and sulfur. Astrophys J 909:40, https://doi.org/10.3847/1538-4357/abd6e5

Umebayashi T, Nakano T (1981) Fluxes of energetic particles and the ionization rate in very dense interstellar clouds. Publ Astron Soc Jpn 33:617

van't Hoff MLR, Walsh C, Kama M, Facchini S, van Dishoeck EF (2017) Robustness of N_2^+ as tracer of the CO snowline. Astron Astrophys 599:A101, https://doi.org/10.1051/0004-6361/201629452

van't Hoff MLR, Tobin JJ, Trapman L, Harsono D, Sheehan PD, Fischer WJ, Megeath ST, van Dishoeck EF (2018a) Methanol and its relation to the water snowline in the disk around the young outbursting star V883 Ori. Astrophys J Lett 864:L23, https://doi.org/10.3847/2041-8213/aadb8a

van't Hoff MLR, Persson MV, Harsono D, Taquet V, Jørgensen JK, Visser R, Bergin EA, van Dishoeck EF (2018b) Imaging the water snowline in a protostellar envelope with $H^{13}CO^+$. Astron Astrophys 613:A29, https://doi.org/10.1051/0004-6361/201731656

Van Clepper E, Bergner JB, Bosman AD, Bergin E, Ciesla FJ (2022) Chemical feedback of pebble growth: Impacts on CO depletion and C/O ratios. Astrophys J 927:206, https://doi.org/10.3847/1538-4357/ac511b

van Dishoeck EF, Black JH (1988) The photodissociation and chemistry of interstellar CO. Astrophys J 334:771, https://doi.org/10.1086/166877

van Dishoeck EF, Herbst E, Neufeld DA (2013) Interstellar water chemistry: From laboratory to observations. Chem Rev 113:9043-9085, https://doi.org/10.1021/cr4003177

van Terwisga SE, van Dishoeck EF, Cazzoletti P, et al. (2019) The ALMA Lupus protoplanetary disk survey: Evidence for compact gas disks and molecular rings from CN. Astron Astrophys 623:A150, https://doi.org/10.1051/0004-6361/201834257

Wakelam V, Loison J-C, Herbst E, et al. (2015) The 2014 KIDA network for interstellar chemistry. Astrophys J Suppl Ser 217:20, https://doi.org/10.1088/0067-0049/217/2/20

Walsh C, Millar TJ, Nomura H (2010) Chemical processes in protoplanetary disks. Astrophys J 722:1607–1623, https://doi.org/10.1088/0004-637X/722/2/1607

Walsh C, Loomis RA, Öberg KI, Kama M, van't Hoff MLR, Millar TJ, Aikawa Y, Herbst E, Widicus Weaver SL, Nomura H (2016) First detection of gas-phase methanol in a protoplanetary disk. Astrophys J Lett 823:L10, https://doi.org/10.3847/2041-8205/823/1/L10

Whittet DCB (2010) Oxygen depletion in the interstellar medium: implications for grain models and the distribution of elemental oxygen. Astrophys J 710:1009–1016, https://doi.org/10.1088/0004-637X/710/2/1009

Williams JP, Cieza LA (2011) Protoplanetary disks and their evolution. Annu Rev Astron Astrophys 49:67–117, https://doi.org/10.1146/annurev-astro-081710-102548

Woitke P, Kamp I, Thi W-F (2009) Radiation thermo-chemical models of protoplanetary disks. I. Hydrostatic disk structure and inner rim. Astron Astrophys 501:383–406, https://doi.org/10.1051/0004-6361/200911821

Woitke P, Min M, Thi W-F, Roberts C, Carmona A, Kamp I, Ménard F, Pinte C (2018) Modelling mid-infrared molecular emission lines from T Tauri stars. Astron Astrophys 618:A57, https://doi.org/10.1051/0004-6361/201731460

Womack M, Wyckoff S, Ziurys LM (1992) Observational constraints on solar nebula nitrogen chemistry: N_2/NH_3. Astrophys J 401:728, https://doi.org/10.1086/172100

Xu R, Bai X-N, Öberg K (2017) Turbulent-diffusion mediated CO depletion in weakly turbulent protoplanetary disks. Astrophys J 835:162, https://doi.org/10.3847/1538-4357/835/2/162

Yang H, Herczeg GJ, Linsky JL, Brown A, Johns-Krull CM, Ingleby L, Calvet N, Bergin E, Valenti JA (2012) A far-ultraviolet atlas of low-resolution Hubble Space Telescope spectra of T Tauri stars. Astrophys J 744:121, https://doi.org/10.1088/0004-637X/744/2/121

Yang L, Ciesla FJ, Alexander CMOD (2013) The D/H ratio of water in the solar nebula during its formation and evolution. Icarus 226:256–267, https://doi.org/10.1016/j.icarus.2013.05.027

Zhang K, Pontoppidan KM, Salyk C, Blake GA (2013) Evidence for a snow line beyond the transitional radius in the TW Hya protoplanetary disk. Astrophys J 766:82, https://doi.org/10.1088/0004-637X/766/2/82

Zhang K, Bergin EA, Blake GA, Cleeves LI, Schwarz KR (2017) Mass inventory of the giant-planet formation zone in a solar nebula analogue. Nat Astron 1:0130, https://doi.org/10.1038/s41550-017-0130

Zhang K, Bergin EA, Schwarz K, Krijt S, Ciesla F (2019) Systematic variations of CO gas abundance with radius in gas-rich protoplanetary disks. Astrophys J 883:98, https://doi.org/10.3847/1538-4357/ab38b9

Zhang K, Schwarz KR, Bergin EA (2020a) Rapid evolution of volatile CO from the protostellar disk stage to the protoplanetary disk stage. Astrophys J Lett 891:L17, https://doi.org/10.3847/2041-8213/ab7823

Zhang K, Bosman AD, Bergin EA (2020b) Excess C/H in protoplanetary disk gas from icy pebble drift across the CO snowline. Astrophys J Lett 891:L16, https://doi.org/10.3847/2041-8213/ab77ca

Zhang K, Booth AS, Law CJ, et al. (2021) Molecules with ALMA at Planet-forming Scales (MAPS). V. CO gas distributions. Astrophys J Suppl Ser 257:5, https://doi.org/10.3847/1538-4365/ac1580

Reviews in Mineralogy & Geochemistry
Vol. 90 pp. 55–112, 2024
Copyright © Mineralogical Society of America

3

Planet Formation—Observational Constraints, Physical Processes, and Compositional Patterns

Christoph Mordasini and Remo Burn

Physikalisches Institut,
Universität Bern
3012 Bern, Switzerland
and
Max-Planck-Institut für Astronomie
69117 Heidelberg, Germany

christoph.mordasini@unibe.ch; burn@mpia.de

INTRODUCTION

Motivation and chapter content

A general theory of planet formation has been a topic of intense study over many years. The interest in such a theory emerges naturally from asking the question of where our planet came from. However, a general picture is also required which explains not only the planet Earth but the whole Solar System with its diverse planets. Moreover, after the first discovery of an exoplanet around a Sun-like star (Mayor and Queloz 1995) the exoplanet revolution added numerous additional constraints from thousands of systems of planets orbiting stars different than the Sun.

The goal of planet formation as a field of study is not only to provide the understanding of how planets come into existence. It is also an interdisciplinary bridge which links astronomy to geology and mineralogy. Recent observations of young stars accompanied by their protoplanetary disks provide direct insights into the conditions at which planets are forming (Manara et al. 2023). These astronomical observations can be taken as initial conditions for the models of planet formation. In particular, we are currently gaining insights into the composition of the inner, planet-forming region within the disks thanks to observations from the James Webb Space Telescope (van Dishoeck et al. 2023).

The task for planet formation modelling is then to link these observational properties and compositional content to the physical properties, and also the elementary inventory of meteorites, the Moon, Earth, and the other planets. If successful, this global approach can provide useful constraints for geological studies.

Here, we will review some of the recent steps that have been undertaken in this task. In particular, we will provide a short overview of observational constraints from the Solar System and from exoplanets. This is followed by a simple, educational review of a number of relevant physical processes before we put the pieces together to show some state-of-the art model results regarding the physical and compositional properties of model planetary systems.

Challenges in planet formation theory

In the classical planet formation theory based on the Solar System alone, it was expected that there should be no massive planets inside of about 3 astronomical units (AU) for physical reasons that we will discuss below, and that the overall architecture of the Solar System with lower mass inner rocky and massive outer gaseous and icy planets should be universal.

1529-6466/24/0090-0003$10.00 (print)
1943-2666/24/0090-0003$10.00 (online)

http://dx.doi.org/10.2138/rmg.2024.90.03

However, in the last nearly three decades, a very large population of exoplanets has been found exactly where the Solar System-based formation theory did not predict them. This pointed towards a serious gap in the understanding of planet formation derived from our planetary system alone and pointed at the necessity of developing a general theory that is not only based on a handful, or even a single planetary system, but can address both the Solar System and the exoplanets in a general way.

While considerable progress has been made including the shift of several paradigms (see the *Predictions from original Solar System theory and the shift of paradigms* section), it has proven very challenging to come up with such a general theory based on the first principles of physics alone (conservation laws of mass, energy, and momentum), and the field is currently still far from a complete understanding of how planets from and evolve. This implies a current absence of a highly predictive theory of planet formation—the field is rather driven by observational discoveries, often made possible thanks to new observational facilities on the ground or in space. Observational facilities that have played a key role are depicted in Figure 1.

Developing a theory of planet formation is challenging for a number of reasons. First, planet formation is a process covering a huge range in spatial scales from dust grains with a size on the order of 10^{-4} cm to giant planets with a radius on the order of 10^{10} cm. At each size scale, there are different governing physical processes involving multiple input physics like gravity, hydrodynamics, thermodynamics, radiative transport, magnetic fields, high pressure physics, and so on.

Second, there are many strong nonlinear mechanisms and feedbacks. For example, we now understand that in the classical picture from the Solar System, where giant planets were predicted to exist only at large orbital distances, the effect of orbital migration (the radial motion of protoplanets through the natal gas disk) was missing. Orbital migration is caused

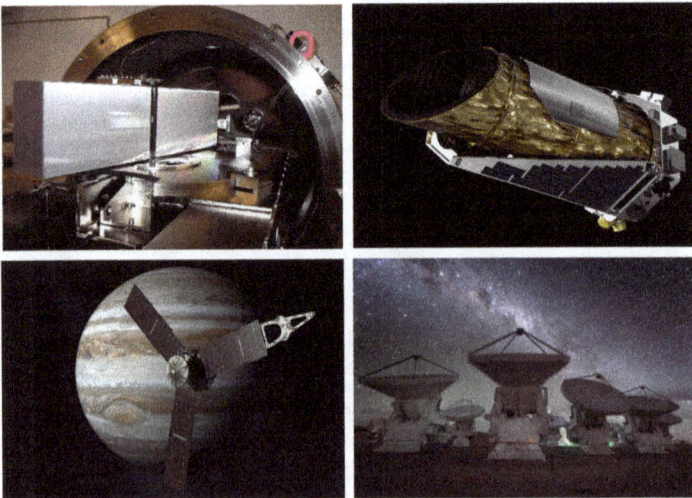

Figure 1. Observational facilities yielding key constraints for planet formation theory. **Top left:** the HARPS spectrograph (Mayor et al. 2003) installed at ESO's 3.6 Meter telescope in Chile. Through its unprecedented radial velocity precision of 1 m/s, it allowed for the first time the discovery of low-mass extrasolar planets, and the study of exoplanetary system architectures in the mass–distance diagram. [Photo: ©ESO]. **Top right:** the NASA Kepler satellite (Borucki et al. 2010). With ultra-precise transit photometry, it has unveiled the demographics of the transiting exoplanet population [Image credit: NASA]. **Bottom left:** the NASA Juno satellite in orbit around Jupiter has deeply changed our picture of the interior of giant planets [Image credit: NASA/JPL-Caltech]. **Bottom right:** the joint American–European–Japanese Atacama Large Millimeter/Submillimeter Array (ALMA) has started a new era in the study of protostellar and protoplanetary disks, the cradles of planets [CC BY 4.0 ESO/Y. Beletsky].

by an exchange of angular momentum between the embedded protoplanet and the disk. This process does, however, not only change the planet's orbit and thus its angular momentum, but (because overall, angular moment needs to be conserved) also changes the structure of the gas disk, leading in particular to the formation of gaps in the disk. This implies that the different processes acting during planet formation (here the co-evolution of the planets' orbit and of the disks' structure) must be described in a self-consistently coupled way.

Third, in contrast to other fields where laboratory experiments can be used as a source of ground truth, for planet formation, laboratory experiments are only possible for special regimes, namely for the early growth regimes from dust to pebbles. But the approach via numerical simulations also has its limits: global three-dimensional high-resolution radiation-magneto-hydrodynamic simulations covering the entire growth of a planetary system would be much too computationally expensive. Taken together, these challenges demonstrate the importance of observational guidance to develop a conclusive theory of planet formation and the necessity of quantitatively confronting theoretical and observational results, which is by itself not trivial. It is also important to take into account various observational constraints as they concern and constrain different aspects of the theory.

OVERVIEW OF OBSERVATIONAL CONSTRAINTS

Our current understanding of the origin and evolution of planets is mainly based on three different data sets: the Solar System, the extrasolar planets, and especially since the advent of ALMA observations, the protoplanetary disks surrounding young stellar objects. The latter yield crucial observational constraints on the initial and boundary conditions of planet formation. In a few instances, ongoing planet formation can even be observed directly (see Keppler et al. 2018, for the discovery of the first bona fide extrasolar gas giant planet unergoing active gas accretion). For an overview of the constraints coming from disk observations, see Andrews (2020), Manara et al. (2023), and Miotello et al. (2023). In the following, we first discuss selected constraints from the Solar System and then from the extrasolar planet population.

Selected constraints from the Solar System

Before the detection of the exoplanets and protoplanetary disks, the classical theory of planet formation (Safronov 1969) was based on centuries of Solar System studies going back to Kant (1755) who first proposed that planets form in a flattened disk revolving about the Sun. These studies include remote and in-situ observations of major and minor bodies (planets, minor planets, asteroids, comets), laboratory measurements of meteorites, theoretical studies, as well as sample return missions. Today, the Solar System remains the benchmark system for planet formation theory, because only here we have access to a very high number of unique detailed constraints not accessible for exoplanets: the interior structures of the planets, the properties and dynamics of small bodies not directly observable around other stars, and cosmo-chemical constraints obtained from the meteoritic record and the interior and atmospheric composition and structure of the various bodies. The latter yields the abundances of elements and isotopes providing information on composition and origin of planetary building blocks, accretion processes, the timing, and early planetary-wide chemical differentiation.

A non-exhaustive list of central constraints for planet formation theory including in particular some elements not available for exoplanets (or contrasting them) is given by the following points. There are, of course, many more constraints, see reviews of, e.g., Cameron (1988), Lissauer (1993), Encrenaz et al. (2013), Spohn (2015), or Encrenaz and Lequeux (2021):

- The chronology of important events during the formation of the Solar System like the formation of calcium–aluminium-rich inclusions (CAI) as first solids in the Solar system representing the temporal zero point (Anand and Mezger 2023), the formation of chondrites (Kleine et al. 2009), the accretion of the terrestrial planets (Yin et al. 2002), core formation (Kleine et al. 2002), the moon forming impact (Benz et al. 1989; Canup and Asphaug 2001), the formation timescale of Jupiter (Kruijer et al. 2017), the lifetime of the Solar gas nebula (Wang et al. 2017) or the timing of a potential late dynamical instability leading to a re-rise of impacts (the late heavy bombardment, Bottke and Norman 2017).

- The orbits of the terrestrial planets, in particular their small eccentricities (varying over Milankovich cycles, see Berger and Loutre (1991), with present-day values of 0.017, 0.007, 0.094, and 0.2 for Earth, Venus, Mars, and Mercury) and the absence of planets inside of 0.39 AU (the orbit of Mercury). The latter is in stark contrast to very frequent extrasolar systems with close-in (inside of 0.1 AU) compact systems of multiple super-Earth and sub-Neptune planets (Petigura et al. 2018; Weiss et al. 2023).

- The masses of the terrestrial planets, in particular Mars' small mass and its short formation timescale (Dauphas and Pourmand 2011). Based on classical formation models (Lissauer 1993) adopting a Minimum-Mass-Solar-Nebula (MMSN)-like distribution of solid building block (definition see below), one would rather expect that planet mass increases with semimajor distance. Likely, the small mass is an imprint of the important role of Jupiter in shaping the entire Solar System, as supposed for example in the Grand tack model (Walsh et al. 2011). In this model, it is assumed, based on the Masset–Snellgrove mechanism of outward migrating pairs of resonant giant planets (Masset and Snellgrove 2001), that Jupiter migrated first to 1.5 AU before tacking to migrate again outwards after Saturn had caught up, thereby truncating the planetesimal disk and depriving Mars from further building blocks. Mars does thus probably represent the protoplanet stage rather than a fully grown terrestrial planet emerging from the final giant impact growth phase.

- The structure of the asteroid belt including the absence of big bodies and the presence of a radial compositional gradient (DeMeo and Carry 2014) from relatively dry ordinary chondrites (less than 0.1% water by mass) to carbonaceous chondrites with 5–20% water by mass (Raymond and Izidoro 2017), as well as the size distribution of the asteroids which gives clues about the likely typical primordial size of planetesimals (probably ~100 km, Morbidelli et al. 2009).

- Earth's total water content, usually thought to have been delivered in the form of water-rich primitive asteroidal material (Morbidelli et al. 2000). Alternatively, it might have formed to some extent intrinsically from magma ocean–primordial atmosphere interactions (Young et al. 2023) or was delivered in parts by Theia, the impactor of the moon forming impact (Budde et al. 2019; Mezger et al. 2021).

- The masses and orbits (semimajor axes, eccentricities, inclinations) of the gas and ice giant planets, characterized and contrasting the extrasolar planets by the absence of mean motion resonances and low eccentricities and inclinations. It has been proposed that the orbital distances of Jupiter and Saturn (5.2 and 9.5 AU) are larger than what is typical for extrasolar giant planet systems (confined to rather 1–3 AU) which would mean that other systems are typically more compact than the Solar system (Fernandes et al. 2019; Fulton et al. 2021). However, this question is not yet settled because of the detection bias against distant exoplanets, and other studies suggest that current data on exoplanets does not indicate a decline in frequency outside of 3 to 8 AU (Lagrange et al. 2023).

- The giant planets' internal structure and their bulk and atmospheric composition as found from measurement of the gravitational moments (Wahl et al. 2017) or by entry probes and remote observations as well as their intrinsic luminosity (Guillot and Gautier 2015; Helled et al. 2022). This includes the over-luminosity of Saturn and the under-luminosity of Uranus relative to predictions of fully convective interior models (Nettelmann et al. 2013; Linder et al. 2019). Particularly important is Jupiter's complex structure as informed by the JUNO probe (Wahl et al. 2017) with a diluted core and non-convecting regions with entropy and compositional gradients (Debras and Chabrier 2019; Miguel et al. 2022; Howard and Guillot 2023). This contrasts the traditional view of a single fully convective adiabatic interior of giant planets (see Guillot et al. 2023, for a review).

- The structure, dynamics, and evolution of the Kuiper belt which constrains the size of the nascent planetesimals disk and which might have played a crucial role triggering a dynamical instability among the giant planets as supposed in the Nice Model for the formation of the Solar System (Levison et al. 2008).
- Further compositional and isotopic constrains (e.g., for deuterium and noble gases) for various bodies (Earth's atmosphere, asteroids, comets like 67P), which are of key interest for example for the origin of Earth's water and its atmospheres (Altwegg et al. 2015; Bekaert et al. 2020).

Statistical constraints: the demographics of extrasolar planets.

On the other hand, there is the continuously growing population of extrasolar planets. The situation for exoplanets is quite different from the Solar System. We have typically little knowledge about an individual exoplanetary system, although there are of course, also notable exceptions. But there is a large number of exoplanets known (currently about 5000), meaning that the exoplanet population yields statistical constraints on, for example, the frequencies of certain planet types, the distributions of fundamental properties, or on numerous correlations between different planetary properties, between stellar properties, and between planetary system properties. For quantitative statistical constraints, large surveys play a primary role, as for them it is possible to account for the observational bias and to correct for it. Important examples are the HARPS survey (Mayor et al. 2011) and the California Legacy survey (Rosenthal et al. 2021) for the radial velocity method; the Kepler satellite for photometric transits (Borucki et al. 2010; Petigura et al. 2018); direct imaging surveys like GEMINI (Nielsen et al. 2019) or SPHERE (Vigan et al. 2021); and finally microlensing surveys (Suzuki et al. 2016a). These different techniques probe planets with different properties (like close-in versus distant planets). This is important, as different sub-populations are most important in constraining different aspects of planet formation theory. This also implies that for a general theory, the constraints of all different methods should be considered together self-consistently. In the following, we present a selection of important results obtained from these observations. We start with an overview, and then address the frequency of different types of exoplanet, followed by the distribution of planetary characteristics, and finally a number of correlations, for example with the properties of the host star.

More detailed and comprehensive information can be found in the reviews of Winn and Fabrycky (2015), Zhu and Dong (2021), Lissauer et al. (2023), and Weiss et al. (2023).

The mass–distance diagram. As an overview, Figure 2 shows one of the most important diagrams when it comes to extrasolar planets, which is the observed mass–distance diagram. Its importance is comparable to the Hertzsprung–Russel diagram for stars. The coloured points show the exoplanets discovered by the radial velocity (red), transit photometry (cyan), direct imaging (magenta), and microlensing (green) techniques.

While giving an impression of the diversity of the outcome of the planet formation process that theoretical models must reproduce, this observed diagram also gives a highly biased view, making for example hot Jupiters appear frequent while when correcting for the detection bias, they are a rare planet type. The absence of planets in the bottom right of the diagram is also simply caused by the current observational bias, that is, the current inability to detect distant low-mass planets. Nevertheless, besides the diversity, one can identify in the diagram some structures in the form of over- and under-densities as well as the aforementioned complementarity of the different detection methods.

Frequency of different types of planets. Despite their prominence in the observed mass–distance diagram, close-in giant planets (hot Jupiters, also known as pegasids after the prototypical first exoplanet 51 Pegasi b (Mayor and Queloz 1995) of this kind) have a bias-corrected frequency of occurrence of only about 0.5–1% for Solar-like stars (Howard et al. 2010; Mayor et al. 2011; Dawson and Johnson 2018). This is much less than the frequency of giant planets at larger orbital distances, where an upturn in frequency is seen at about 1 AU (Udry and Santos 2007). Within 5–10 AU, about 10 to 20% of FGK stars have giant planets (Cumming et al. 2008; Mayor et al. 2011; Fulton et al. 2021). This means that the clear majority of Solar-like stars do not host a giant planet companion, which might have had profound consequences for the Solar System (Morbidelli et al. 2015). For lower stellar masses, there is a clear trend towards less frequent occurrence of giant planets (e.g. Gaidos et al. (2013) see also the *Correlations with host star properties, age, and intra-system architecture* section).

About half of the extrasolar giant planets are members of multi-giant systems (Bryan et al. 2016) like our Jupiter–Saturn system. As shown initially by high precision radial velocity surveys (Lovis et al. 2006) and then the Kepler transit survey, low-mass respectively small close-in planets are a very frequent type of planet. Depending on the exact criteria such super-Earth and sub-Neptunian planets at orbital distances of fractions of an AU with masses of about 1 to 30 Earth masses ($M_\oplus$), or radii of $R \lesssim 4$ Earth Radii ($R_\oplus$) have a high frequency (20–50%) for Solar-like stars (Mayor et al. 2011; Fressin et al. 2013; Petigura et al. 2013; Zhu et al. 2018). These planets are frequently members of multiple systems with compact architectures (Weiss et al. 2023). This implies again that many Solar-like stars have a planetary system that differs clearly from the Solar System. Moving to (very) large orbital distances of tens to hundreds of AU, there is a lower frequency on a level of about 1–5% of stars with detectable, i.e., sufficiently luminous distant giant planets as probed by direct imaging surveys (Vigan et al. 2021). An implication of this is that the frequency of giant planets must decrease with distance by a factor of several as we move outwards. This puts constraints on the efficiency of planet formation by gravitational instability which would be the prime mechanism to populate this region (Forgan et al. 2018). The occurrence rate of such planets probably scales with the mass of the host star (Bowler 2016). As visible in Figure 2, microlensing surveys mainly probe an intermediate orbital distance range of cold, roughly Neptunian-mass planets around M-dwarfs, which are found to be an abundant population (Cassan et al. 2012; Suzuki et al. 2016a). Finally, if we consider the overall fraction of stars with planets that can be detected by a high-precision radial velocity survey at about 1 m/s precision, a value of about 75% is found (Mayor et al. 2011). A similarly high number applies to M-dwarfs (Bonfils et al. 2013).

Distributions of important planetary properties. There are several essential distributions like the mass, radius, eccentricity, and orbital distance distributions that characterize the exoplanet population (Fig. 3). The planetary mass function, or PMF, is known from radial velocity surveys mainly for Solar-like stars and from microlensing for M dwarfs. It is an important distribution that has been addressed and predicted by various theoretical works (e.g. Ida and Lin 2005; Mordasini et al. 2009; Emsenhuber et al. 2021a). Focusing on Solar-like stars, the observed distribution as found by radial velocity surveys scales with mass M approximately as $1/M$ (Marcy et al. 2005) in the mass range of about $30 M_\oplus$ to roughly $4 M_J$ (with M_J being the mass of Jupiter). The distribution is thus bottom heavy in the sense that there are more low-mass giants than there are high-mass giant planets. Above this value, the frequency decreases even stronger (Santos et al. 2017). Whether there is a clear upper end of the PMF is unknown, but there are only very few brown dwarfs around Solar-like stars inside of about 5 AU (Grether and Lineweaver 2006). The driest part, and thus a potential upper end of the PMF might be at around $30 M_J$

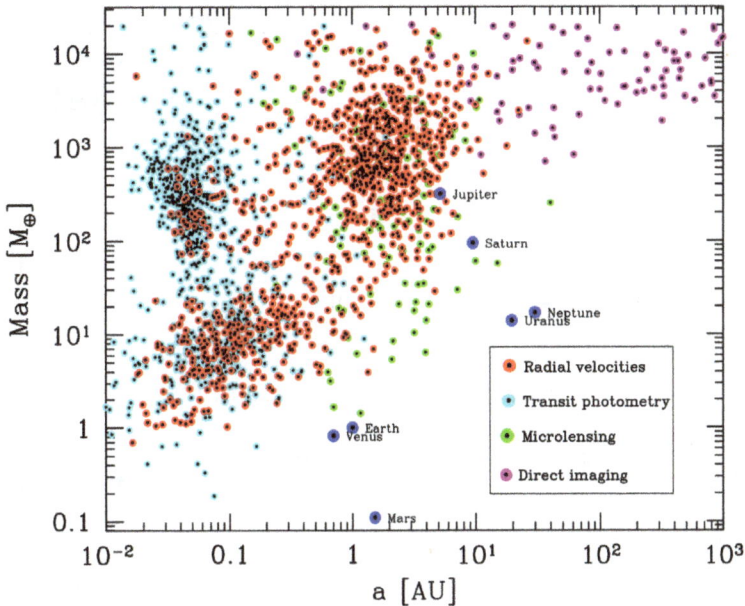

Figure 2. The observed mass–distance diagram of extrasolar planets. The **colors** indicate the detection method. The planets of the Solar System are also included for comparison. While illustrating the diversity of extrasolar planets and pointing at the existence of some structure, the diagram also gives a highly biased view of the exoplanet population.

(Sahlmann et al. 2011). Moving to small masses, there is a break in the PMF at around 30 $M_\oplus$ below which the frequency increases strongly (Howard et al. 2010; Mayor et al. 2011) as had been predicted by models based on the core accretion theory, where at a similar total mass (core and envelope), gas accretion becomes important (Mordasini et al. 2009). Thus, the break would correspond to the transition from a slope that is governed by solid accretion to one governed by gas accretion. Microlensing observations, however, challenge this picture (Suzuki et al. 2018).

Regarding the semimajor axes of giant planets (Fig. 3, bottom right), there is a local maximum at a period of about 3–4 days that is formed by the hot Jupiters. It is followed by a less populated region known as the period valley (Udry et al. 2003). At about 1 AU, the frequency increases by about a factor 4 (Fulton et al. 2021). Already outside of about 3–10 AU, the frequency could be decreasing (Bryan et al. 2016; Fernandes et al. 2019; Fulton et al. 2021) but this might be an effect of the observational bias (Lagrange et al. 2023).

While the eccentricities of the Solar System planets are very low, there is a broad distribution of eccentricities for extrasolar giant planets with about half of them having an eccentricity in excess of 0.3 (Ford and Rasio 2008). Some exoplanets even have eccentricities exceeding 0.9 (Naef et al. 2001). The distribution resembles for the higher values a Rayleigh distribution which is the distribution expected from planet–planet scattering (Jurić and Tremaine 2008). This is an indication that in some extrasolar systems, strong dynamical interactions have happened (Winn and Fabrycky 2015), an effect that is also seen in numerical simulations (Rasio and Ford 1996; Emsenhuber et al. 2023a). When taking into account that radial velocity measurements are biased towards overestimating eccentricities (Lucy and Sweeney 1971; Hara et al. 2019), then a significant number of orbits are also consistent with being circular. Planets detected via radial velocity with lower masses (less than about 30 $M_\oplus$) tend to have lower eccentricities of less than about 0.5 (Mayor et al. 2011). Transiting planets detected by the Kepler mission in multiple systems planets have orbits that are almost circular

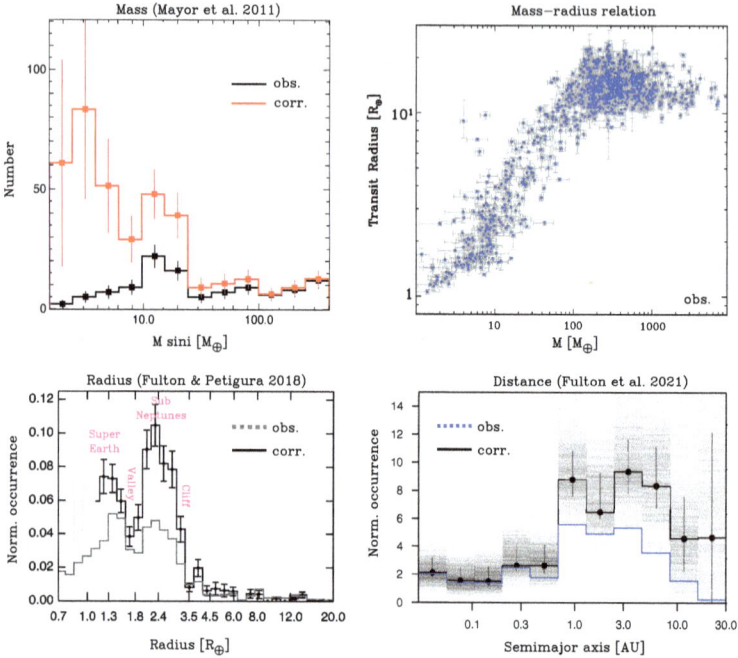

Figure 3. Statistical observational constraints from the exoplanet population. **Top left** (modified from Mayor et al. 2011): mass distribution. **Top right**: observed (NASA Exoplanet Archive, biased) mass-radius relation. **Bottom left** (modified from Fulton and Petigura 2018): radius distribution. Some important features are labelled. **Bottom right** (modified from Fulton et al. 2021): semimajor axis distribution for planets more massive than $30\,M_\oplus$. Raw observed and bias-corrected results are given.

(average eccentricity of about 0.02–0.05). They also have low mutual inclinations of about 1.4 degrees or less, scaling inversely with the number of planets in the system (Weiss et al. 2023). Eccentricity and inclination are correlated with each other (Xie et al. 2016).

The distribution of radii of planets detected by the Kepler satellite (Fig. 3, bottom left) shows a peak at about the same size as Jupiter, which is consistent with the theoretical relation between mass and radius (Mordasini et al. 2012b). The radius distribution is roughly constant in $\log(R)$ for planets between 4 and $10\,R_\oplus$. For smaller planets, the number of planets increases sharply (Fressin et al. 2013), which is sometimes referred to as the "radius cliff" (Kite et al. 2019). There is a local minimum in the distribution at around $1.7\,R_\oplus$ known as the radius gap or radius valley (Fulton et al. 2017; Van Eylen et al. 2018). It separates smaller, super-Earth planets, which are mostly rocky, from larger sub-Neptunes, which have thick H/He atmospheres or water layers (Mousis et al. 2020). This could be because of the loss of the primordial H/He atmospheres due to stellar radiation (Owen and Wu 2013; Jin and Mordasini 2018) or different formation pathways for these two types of planets (Venturini et al. 2020; Burn et al. 2024).

Correlations with host star properties, age, and intra-system architecture. The most well-known correlation with stellar properties is the higher frequency of giant planets around stars with higher metallicity which was noted already early on (Gonzalez 1997; Santos et al. 2004; Fischer and Valenti 2005; Dong et al. 2018). In the domain of metallicities above the Solar value, the frequency of giant planets increases by about an order of magnitude from [Fe/H] = 0 to [Fe/H] = 0.5 (the brackets indicate logarithmic abundance ratios relative to the Solar value). This is often seen as evidence that core accretion is the main mode of giant planet formation (Ida and Lin 2004; Mordasini et al. 2012a). For lower-mass or smaller planets, the dependency on metallicity becomes weaker and weaker, and for planets with the smallest radii currently known ($\lesssim 1.7 R_{\oplus}$), there seems even to be a preference for slightly sub-Solar values (Mayor et al. 2011; Narang et al. 2018; Petigura et al. 2018).

The occurrence of giant planets around host stars with lower masses (M-dwarfs) is about 2–6% which is lower than for Solar-like stars (Bonfils et al. 2013; Ghezzi et al. 2018; Sabotta et al. 2021; Ribas et al. 2023). The discovery of giant planets around stars of even very-low mass (~0.1 Solar masses, $M_{\odot}$) is seen as a challenge to both core accretion and gravitational instability (Morales et al. 2019; Burn et al. 2021; Schlecker et al. 2022), but recent simulations combining growth by pebble accretion and giant impacts find for special conditions (high disk masses, high turbulence level) giant planet formation also around such stars (Pan et al. 2023). In contrast to giant planets, low-mass stars have 2–3 times more low-mass planets (around 10 Earth masses) than G-type stars as found via the transit method using Kepler (Mulders et al. 2015) and later supported using radial velocity (Sabotta et al. 2021). However, a recent re-analysis of the Kepler data challenges this trend (Bergsten et al. 2023). For stars with masses above $1 M_{\odot}$, the occurrence of giant planets detected by radial velocity first rises to a peak at about 1.7 to $2 M_{\odot}$, and then declines for even more massive stars (Reffert et al. 2015; Wolthoff et al. 2022).

The relationship between stellar age and planetary properties and their statistics is now being explored with various surveys (e.g., Mann et al. 2016; Grandjean et al. 2021; Capistrant et al. 2024). For these studies, it is important that Gaia has allowed to greatly expand our knowledge of young stellar groups (Bouma et al. 2022). Several close-in planets have been found around young (T-Tauri, pre-main-sequence and young main sequence) stars (David et al. 2016; Donati et al. 2016; Plavchan et al. 2020). They reveal that some planets close to their stars can form within a few million years, probably due to orbital migration caused by planet–disk interactions but that dynamical effects (high eccentricity migration) also plays a role (Dai et al. 2023). T-Tauri stars may have more hot Jupiters than main sequence stars (Yu et al. 2017). At larger separations, direct imaging can also inform about the properties of young planets with ages of a few tens of million years (Lagrange et al. 2010; Macintosh et al. 2015; Wagner et al. 2019; Vigan et al. 2021). First direct detections of planets that are still in the formation process have occurred revealing emission not only from the planets' photosphere but also from the gas accretion shock and from the surrounding circumplanetary disk (Keppler et al. 2018; Haffert et al. 2019; Benisty et al. 2021; Wang et al. 2021). Such observations have the potential to put much more direct constraints on physical processes like gas accretion (Hashimoto et al. 2020; Marleau et al. 2022). Furthermore, there are indirect indications of the presence of forming planets via their impact on their parent protoplanetary disk (gaps, kinks in the rotational velocities) (Huang et al. 2018; Teague et al. 2018; Pinte et al. 2020). In the coming years, the PLATO transit survey (Rauer et al. 2014) will provide much more statistical information on how the population of transiting planets changes over time.

As illustrated by the mass–distance diagram, the exoplanet population shows a lot of variation and diversity among different systems, but interestingly, there is in contrast intra-system uniformity within individual systems of small close-in planets, called the peas-in-a-pod pattern (Millholland et al. 2017; Weiss et al. 2018, 2023): planets in the same system have comparable sizes, masses, and relative distances from each other. The study of extrasolar planetary systems and their architectural patterns, as opposed to individual planets, is currently a field of active research (e.g., Lissauer et al. 2011; Alibert 2019; Adams et al. 2020; Bashi and Zucker 2021; Emsenhuber et al. 2023a; Mishra et al. 2023a,b).

Predictions from original Solar System theory and the shift of paradigms

The classical theory of planetary formation based on the Solar System alone had predicted that the fundamental architecture of the Solar System (refractory, low-mass terrestrial planets inside, then (two) massive gas giant planets outside of the iceline, and finally ice giants of intermediate mass, all on nearly circular orbits) should be universal (Dole 1970). Assuming that planets form in situ (i.e., grow at the position where they are found now), it was thought that giant planets are always found outside of the iceline, i.e., the distance outside of which it is cold enough (<170 K) for water to condense in the protoplanetary nebula surrounding the young Solar nebula (which is at about 2.7 AU in optically thin, irradiated disk models, Ida and Lin 2004). The amount of planetary building blocks (assumed to be only kilometer-sized planetesimals) thus increases there, providing the material to form massive cores (about 10 Earth masses) necessary to start efficient gas accretion (Pollack et al. 1996). At even larger distances r, the collisional growth timescale is long since it is proportional to the local Keplerian frequency

$$\Omega_K\left(r\right) = \sqrt{\frac{GM_*}{r^3}} \tag{1}$$

where M_* and G are the stellar mass and Newtonian gravitational constant, respectively. Thus, growth is reduced such that the outwardly decreasing surface density of planetesimals would only allow the formation of lower-mass, core-dominated ice giants during the finite lifetime of the Solar nebula of a few million years (Pollack et al. 1996). Finally, inside of the water iceline, due to limited availability of building blocks, only low-mass refractory terrestrial planets should form.

An important concept in this in-situ theory is the Minimum Mass Solar Nebula (MMSN) (Weidenschilling 1977b; Hayashi 1981). In this approach, the masses in solids (refractory rocky material and volatile ices) contained in the Solar System planets (quantities which can only be inferred indirectly for the giant planets) are radially spread over touching annuli, and the gas is complemented in a Solar-composition ratio. The result (Fig. 4) is that the MMSN should have contained about 80 $M_\oplus$ of solids, and 0.013 $M_\odot$ of gas, and that the surface density of planetesimals and gas should have decreased with orbital distance as a smooth power law proportional to distance $\propto r^{-1.5}$ with a jump (increase) by a factor 4.2 at the water condensation front (iceline) at 2.7 AU for the planetesimals (Hayashi 1981). This indirect inference of the

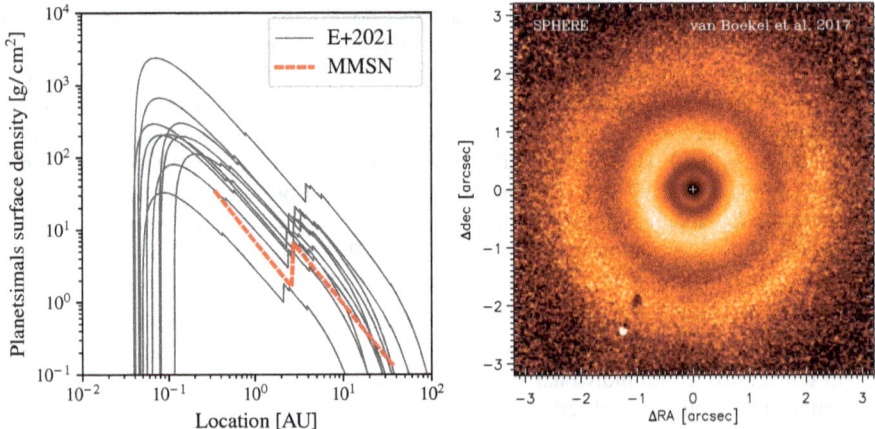

Figure 4. Left panel: Theoretical surface densities of planetesimals as a function of orbital distance. The **red dashed line** is the Minimum Mass Solar Nebula (Hayashi 1981). The jump at 2.7 AU represents the condensation front of water (the water iceline), leading to an increase of the surface density. Besides that, the MMSN follows a smooth unstructured power-law. The **gray lines** show surface densities informed by observations of protoplanetary disks (Tychoniec et al. 2018). Jumps correspond to icelines of different species. While informed by observed disks, these surface density profiles still assume an unstructured power-law profile (Figure adapted from Emsenhuber et al. 2021b). This contrasts the observations of numerous disk structures as illustrated in the **right panel** (Figure adapted from van Boekel et al. 2017). It shows the scattered light image of the TW Hya disk, tracing sub-micron dust grains. The structures indicate gas surface density variations of 50–80% across the three radial gaps.

nebula properties contrasts today's approach of obtaining the properties of protoplanetary disks from direct astronomical observations in star-forming regions (Andrews et al. 2009; Miotello et al. 2023). While the MMSN still remains a point of reference for modern models, the importance of structured disk instead of smooth power-law disks (Andrews 2020), and high diversity in the sense of a large spread in inferred masses and sizes are stressed today (Manara et al. 2023).

However, the discovery of the diversity of extrasolar planets has shattered this static picture, profoundly shifting the paradigm of planet formation. Already the first extrasolar planet 51 Peg b (Mayor and Queloz 1995) as a giant planet orbiting at a few days period was diametrically opposed to the predictions of the in-situ theory. The subsequently revealed proprieties of exoplanets has in their sum drawn a much more dynamic picture of planet formation, i.e., a process where the (radial) mobility and redistribution, but also trapping, as well as dynamical interactions of building blocks at different growth scales are key. This paradigm of mobility and dynamics regards both roughly cm-sized drifting pebbles and large migrating protoplanets. More recently, also the view of the planet's formation environment, the protoplanetary disks, has shifted to a complex structured picture (Andrews 2020; Bae et al. 2023), away from the smooth power-laws inspired by the Solar System. The strong impact of disk structures on planet formation has already been stressed by several early works (Matsumura et al. 2007; Hasegawa and Pudritz 2013; Coleman and Nelson 2016).

The assumed fundamental mechanism driving accretion in protoplanetary disks is also shifting away from the traditional picture of viscous disks (Lynden-Bell and Pringle 1974) where a turbulent viscosity redistributes angular momentum (Shakura and Sunyaev 1973) to MHD-wind-driven disk evolution where magnetic fields extract angular momentum (Lesur et al. 2023). Finally, as an additional shift of paradigm, it is now understood that planetesimals are not the only potential agents of planetary solid growth, but also smaller pebbles (Ormel and Klahr 2010), in particular for the growth of the cores of (distant) giant planets (Lambrechts and Johansen 2012). It should be noted that these paradigm shifts are ongoing processes and no convergence of the relative importance of the different concepts has been reached at this time. It could well be that eventually, a more complex picture will arise that synthesizes aspects of both the classical and new paradigms (Voelkel et al. 2022).

In the context of the paradigm of mobility of the building blocks and protoplanets, it is interesting to note that both processes, the early radial drift of pebbles (Weidenschilling 1977a) as well as gas-driven orbital migration of protoplanets (Goldreich and Tremaine 1980; Lin and Papaloizou 1986) were theoretical concepts derived well before the discovery of 51 Peg b. Orbital migration, because of its seeming absence or inefficiency as judged from the Solar System alone, was however not seen as of high import or was even dismissed completely until the exoplanets were discovered.

PHYSICAL PROCESSES GOVERNING PLANET FORMATION

While the emergence of planets from a protoplanetary disk is undisputed, several competing or complementary paradigms to form the bodies exist. Similar to stars, giant gaseous planets may form directly in a top-down fashion from the self-gravitational collapse of a large unstable part of a protoplanetary disk (Boss 1997)—the so-called disk instability or gravitational instability mechanism (Kratter and Lodato 2016). Since the most unstable wavelength is relatively large (Toomre 1964), gravitational instability will lead typically to giant planets at large orbital distances (Schib et al. 2023). The mechanism needs massive gas disks and likely occurs during, or just at the end, of the infall phase when the protoplanetary disk is forming (Schib et al. 2021). Its efficiency is still unknown, but it might be the formation mechanism of some distant massive planets (Marois et al. 2008, 2010).

On the other hand, the bottom-up formation of planets starting out small and growing over time is seen as the standard pathway to form planets, both gas giants and small rocky planets. In the following, we will review some of the key processes of planet formation via this core accretion paradigm which needs to connect micrometer-sized dust inherited from the interstellar medium to the final planets with sizes of several ten thousand kilometers.

While doing so, we follow the simplified picture of sequential growth shown in Figure 5, addressing physical processes that are relevant along the way. In the figure, the different stages are shown in the upper part, while in the lower part, physical processes relevant at these sizes are shown. Characteristic timescales are also given, but it should be noted that they are only very rough indicative values and depend for example on the orbital distance that is considered.

We begin on the left of Figure 5 with the protoplanetary gas disk in the following *Gas disk* section that sets the background structure in which planets form. The next section discusses the dynamics of the small building blocks, micrometer-sized dust and roughly cm-sized pebbles, where the latter form rapidly on a timescale of 10^4 years through coagulation. At this stage, the radial drift through the disk is a key process.

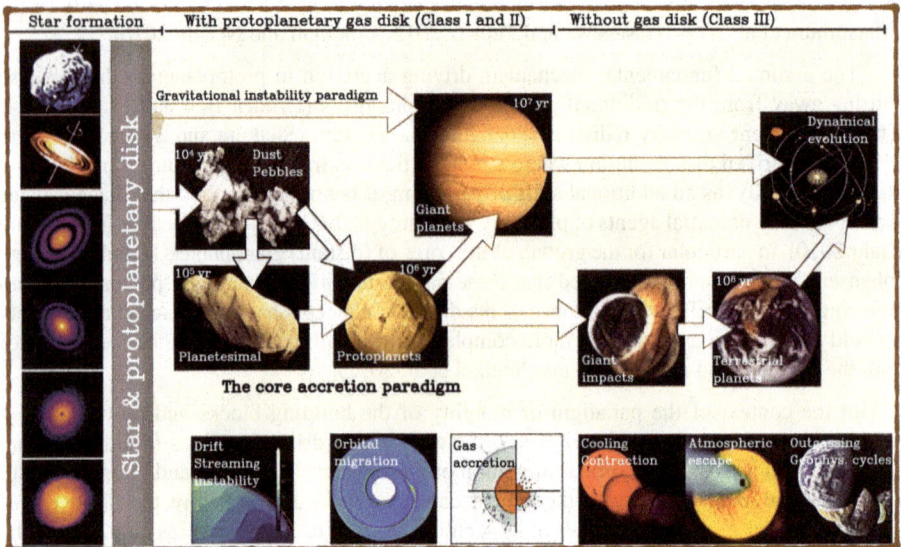

Figure 5. A cartoon view of the different stages of planet formation and evolution and the processes involved. At the top, the main classes of disk evolution are indicated by **black horizontal bars**. In the **middle**, the panels connected by **white arrows** show the main growth stages and involved processes of growth in the core accretion paradigm which is a bottom-up process. The gravitational instability paradigm for giant planet formation, a top-down process, is also shown. The **bottom panels** show various additional physical processes governing at the different stages. A number of typical timescales are also given. [Image Credits: W. Benz/C. Mordasini (composite). Subfigures from Andrews et al. (2018), Baruteau et al. (2014), Birnstiel et al. (2012), ESA, A. Vidal-Madjar (IAP, CNRS), NASA]

The next step, that happens after ∼10^5 years is the growth from *pebbles to planetesimals* as the first gravitationally bound objects. It likely involves the streaming instability to concentrate pebbles to a point that a pebble cloud becomes self-gravitating and collapses to directly form planetesimals.

For the next step, the way in which protoplanets (objects of a size of about 1000 km) grow further by the accretion of solid building blocks, several possibilities exist (*Solid accretion mechanisms* section), namely *pebble accretion* and *planetesimal accretion*. This is a process that occurs on a ~10^6 year timescale. These protoplanets exchange angular momentum with the gas disk, leading to orbital migration, which is discussed in the following *Orbital migration* section.

Some protoplanets grow massive enough (~$10 M_\oplus$) to be able to accrete nebular gas efficiently (on a timescale shorter than 10 Ma), leading to the formation of gas giants. The process of gas accretion is discussed in the *Gas accretion* section.

After the dissipation of the gas disk, terrestrial planets form via giant impacts on a timescale of about 100 Ma, although some more recent models propose that substantial growth may already happen earlier (Ogihara et al. 2018; Brož et al. 2021; Batygin and Morbidelli 2023; Woo et al. 2023).

Finally, in principle during the entire lifetime of a planetary system, the different components interact gravitationally and settle into a state that is stable over Gyr-timescales. On the individual planets, evolutionary processes like cooling and contraction, atmospheric escape, and various mechanisms like interior–atmosphere interactions (in/outgassing) and geophysical cycles take place (Lichtenberg et al. 2023).

Gas disk

On the modelling side, the structure of protoplanetary disks is simulated with 2D and 3D hydrodynamic numerical simulations that can include the effects of radiative energy transport and magnetic fields, where it is important for the latter to include non-ideal effects (see Turner et al. 2014; Lesur et al. 2023). While yielding a detailed understanding of the structure of the disks, these simulations can only simulate short temporal intervals because of their computational costs.

To simulate the full temporal evolution on a million-year timescale, one dimensional (radial) axisymmetric models of gas disk evolution were developed by Lynden-Bell and Pringle (1974) and Pringle (1981) and, for example, reviewed in Armitage and Kley (2019). For our purposes, it is useful to recall the basic concept which will help to introduce the used quantities.

Protoplanetary disks are often assumed to rotate at the Keplerian orbital frequency Equation (1), although the gas is actually rotating slightly slower because of the radial pressure support as discussed in *Radial drift* section. Furthermore, if the disk is in vertical direction isothermal and in hydrostatic equilibrium, the vertical profile of the gas density is

$$\rho_g(z) = \rho_{g,0} \, \exp\left(-\frac{z^2}{2H_g^2}\right) \tag{2}$$

where z is the elevation above the midplane, $\rho_{g,0}$ is the gas density at the midplane,

$$H_g = \sqrt{\frac{k_B T}{\Omega_K^2 \mu m_H}} \tag{3}$$

is called the gas scale height, k_B is the Boltzmann constant, T is the temperature, and μm_H is the mean molecular weight conventionally expressed in hydrogen atomic mass units m_H.

Protoplanetary disks are observed to dissipate in about 1 to 10 Ma (Haisch et al. 2001; Pfalzner et al. 2022) (with a very large spread in lifetimes) and relatively high accretion rates onto the host star ~$10^{-8} M_\odot \cdot yr^{-1}$ are observed (e.g., Alcalá et al. 2017; Manara et al. 2023). The initial mass of the disks is usually assumed to be a few percents of the stellar mass but could also be an order of magnitude higher, in broad agreement with dust continuum measurements of young disks (Tobin et al. 2020). Even under this optimistic assumption, the high accretion rates imply that most of the disk mass needs to be transported toward the star.

Therefore, its angular momentum needs to be removed on the same timescale. The microscopic viscosity of the disk gas cannot account for this, which motivated the introduction of a turbulent α viscosity $\nu = \alpha c_{sT} H_g$ with the Shakura–Sunyaev α parameter (Shakura and Sunyaev 1973). Here, $c_{sT} = \sqrt{k_B T / \mu m_H}$ is the speed of sound in an isothermal medium. The assumption of viscous shear transporting angular momentum and Keplerian orbital motion allows for deriving a single disk evolution equation which controls the radial motion of the gas disk and can be easily implemented in one dimensional (rotationally symmetric) models (Lüst 1952; Lynden-Bell and Pringle 1974)

$$\frac{d\Sigma}{dt} = \frac{3}{r}\frac{d}{dr}\left(r^{1/2}\frac{d}{dr}\left(r^{1/2}\nu\Sigma\right)\right) \tag{4}$$

This yields the evolution of the gas surface density Σ as a function of time t and distance from the star r.

In addition, a significant effect is expected from energetic radiation heating the molecules in the upper- or outermost regions of the disk. In addition to heating, this leads to ionization and dissociation of the molecules. From the absorption of the high-energy radiation the thermal energy of the lighter monoatomic hydrogen, or hydrogen ions, can exceed the local gravitational potential energy (Hollenbach et al. 1994). Thus, they become unbound and leave the disk as a wind. This photoevaporative process can be driven by radiation of the host star (Ercolano et al. 2009; Gorti and Hollenbach 2009) or from external stars (Facchini et al. 2016), typically more massive O or B stars whose spectrum includes orders of magnitudes more high-energy photons. Depending on the source of the radiation and the wavelength, different prescriptions (Haworth et al. 2018; Picogna et al. 2019, 2021; Ercolano et al. 2021) for an additional sink term in Equation (4) can be used. Recently, disk photoevaporation was reviewed in detail by Winter and Haworth (2022).

Another sink or source term can be added in Equation (4) for those minor constituent gases other than H/He that can condense, respectively evaporate or sublimate, from solid grains under nebular conditions. This happens in a narrow region at the icelines of the various volatile species, such as H_2O, CO, and CO_2. The transport of such gaseous species should be modelled individually using the transport equation for tracer fluids (e.g., Ciesla 2011). While in terms of mass, these species other than H/He are of negligible importance, they are of key interest for the resulting chemical composition of planets. More details on the chemical composition of protoplanetary disks are reviewed in Zhang (2024, this volume).

More recently, an alternative paradigm to the classic viscous evolution of the disk has become the focus of the research (see review by Lesur et al. 2023) building upon the results of non-ideal magnetohydrodynamic simulations (e.g., Bai 2016; Bai et al. 2016). The magnetohydrodynamic instability which can drive significant turbulent viscosities as assumed in the classic picture is only active in regions where gas is sufficiently ionized (Gammie 1996). This is the case at large temperatures where thermal ionization can take place, or at low surface densities where cosmic rays can ionize the full vertical extent of the disk. Between these two regions, a viscously 'dead' zone is expected where turbulence levels are low and viscous angular momentum transport is inhibited. In this region, angular momentum could be primarily removed by magnetically-driven winds from the surface layer of the disk. In this case, the evolution equation needs to be changed to include an advective term (Suzuki et al. 2016b), such as used in recent global disk evolution studies (Tabone et al. 2022; Weder et al. 2023).

Despite these developments, it is still helpful to study as a reference model the classical viscous picture with a constant α value, as shown in Figure 6. The evolution of the overall gas surface density, the disk temperature and scale height, as well as the radial gas flow is shown. The viscosity is assumed to heat the disk midplane due to viscous dissipation.

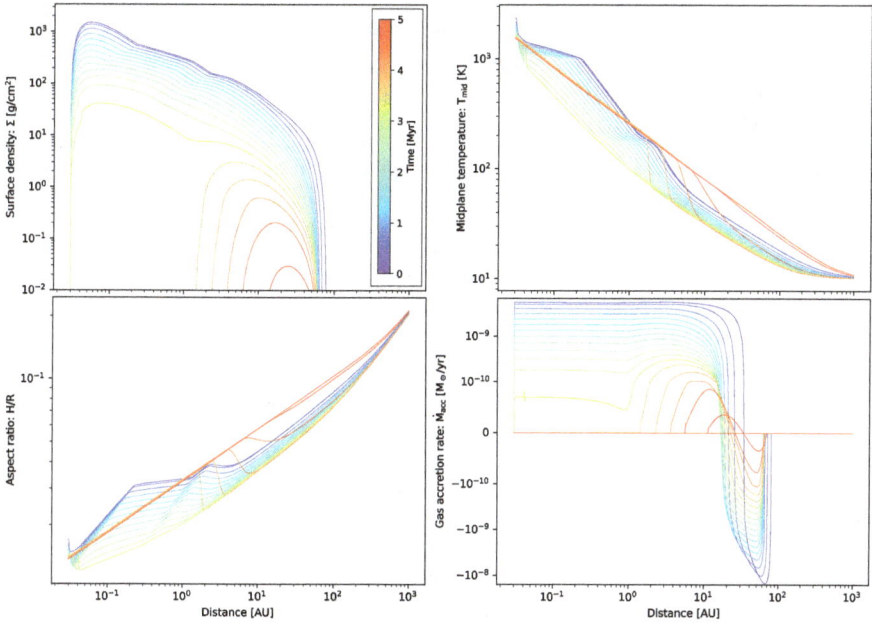

Figure 6. Evolution of a protoplanetary gas disk as a function of distance from the star and time in the classical turbulent α-viscosity paradigm. Internal and external photoevaporation is also included. The panels show the gas surface density (**top left**), midplane temperature (**top right**), aspect ratio H/r (**bottom left**), and accretion rate (**bottom right**). In the latter plot, positive values correspond to a gas flow towards the star (accretion) while negative values are outward spreading parts (decretion).

This effect can be prominently seen in the temperature distribution interior of approximately 5 AU where the disk temperature slope is modified. In the viscously heated part, the temperature is sensitive to the opacity of the dust grains which in turn depend on the composition, here modelled following Bell and Lin (1994). In their opacity parameterization, a transition region at the water iceline is introduced smoothly from 150 to 170 K. It manifests as a region with a shallow temperature slope which moves inwards as time evolves. Interior to the water opacity transition, the radial temperature gradient is large up to the silicate evaporation front at 1000 K. These temperature regimes leave an imprint visible in the bottom left panel on the aspect ratio since the scale height H_g depends on temperature. In the same panel, we can see that the disk is flaring, that is, the aspect ratio increases with distance. In principle, this has an effect on the absorbed radiation via shadowing. Here, instead of a fully coupled treatment, a constant, but roughly consistent, flaring angle $d\ln H/d\ln r = 9/7$ (Chiang and Goldreich 1997) is assumed to calculate the absorbed irradiation from the star. This was used in several works (Hueso and Guillot 2005; Emsenhuber et al. 2021b) to obtain numerically stable solutions.

We note that although the temperatures in the disk can be sufficient to evaporate also refractory species, the region in which this occurs is limited to an annulus close to the star. Despite the increased surface density, little mass resides in this ring, which is unlikely to leave an imprint in the mass budget of planets under standard assumptions. Therefore, the composition in refractory elements in planets is generally expected to be similar to that of the star (Thiabaud et al. 2014).

As the viscosity depends on temperature, the steady-state surface density will also evolve to a profile which shows corresponding features. Since the viscosity is significant, such a steady-state accretion is reached over relatively short timescales. Therefore, the gas accretion rate (bottom right) is constant up to the transition at around 20 AU where the accretion switches to *decretion*, i.e., to an outward flow of gas. Such a transition is fundamental in solutions to any diffusion equation. The transition occurs where most of the mass reservoir is located from which the rest of the disk can draw.

The outer edge of the protoplanetary disk is given by the equilibrium between viscous spreading against external photoevaporation (e.g., Coleman et al. 2024). In Figure 6, the initial profile is more extended than this equilibrium radius, thus the disk is initially shrinking.

The later stages of the disk evolution are dominated by photoevaporation since those sink terms are here assumed to be constant and independent on the local surface density. In the surface density evolution plot, it can be seen that an inner hole opens. This is a typical effect of internal photoevaporation, here following the prescription of Clarke et al. (2001). As the inner disk is cleared, radiation from the star can heat the disk through the midplane which leads to an increase in temperature to the equilibrium temperature in the absence of a disk. An increasing temperature front which propagates outwards can be seen in the top right panel for the latest timesteps (orange to red colors).

Dynamics of dust and pebbles

Like the interstellar medium, the protoplanetary disks which form together with the stars due to (partial) angular momentum conservation during infall do not only contain gas but also small particles. From multiwavelength spectroscopic observations, we know that they range in size from $0.01\,\mu m$ to $1\,\mu m$ (Mathis et al. 1977). As for the gas, we refer to Zhang (2024, this volume) or the literature (Pollack et al. 1994; Thiabaud et al. 2014) for a discussion of their composition.

How closely the particle dynamics is linked to the gas dynamics is expressed with the Stokes number St which corresponds to the dimensionless friction time. It relates the stopping time t_{stop} of a particle due to gas drag to a typical timescale for the gas motion, that is, in practice the orbital timescale. Thus, $St = t_{stop}\,\Omega_K$, where we assume the gas orbits at a Keplerian frequency Ω_K. For small particles, which are in most of the disk in the Epstein drag regime (see Weidenschilling 1977a, for different drag regimes) the Stokes number at the midplane is

$$St = \frac{\pi a \rho_s}{2\Sigma} \tag{5}$$

for a given gas surface density Σ and particle with size a and bulk density ρ_s, which is on the order of $1\,g\cdot cm^{-3}$ but is expected to be higher if the local temperature led to evaporation of volatile species and lower if grains are porous or fractal which is currently only weakly constrained.

Growth from dust to pebbles. For the initially micrometer-sized particles inherited from the interstellar medium, sometimes referred to as monomers, to grow, they need to collide with other particles. The collision rate between a spherical particle of size a_t with particles of size a_p is given by $\Gamma = \Delta v \sigma n$, where Δv is the relative velocity, $\sigma = \pi(a_t + a_p)^2$ is the cross sectional area of the two spherical particles, and n is the number density of particles in the gas.

To estimate the local number density it is important to account for particles settling towards the midplane which locally increases their number. This process is relatively fast and will reach an equilibrium where, at regions close to the midplane, the particles of a given Stokes number can be described with a reduced scale height H_d compared to the gas (Youdin and Lithwick 2007)

$$H_d \approx H_g \sqrt{\frac{\alpha}{St}} \tag{6}$$

At larger elevations ($z > H_g$) a simple scale height approach is imprecise (Fromang and Nelson 2009). It is further noteworthy that there exists a lower limit to the dust scale height as the shear between the settled layer of solids and the low-viscosity gas above it will trigger the Kelvin–Helmholtz instability which stirs-up the solids. Thus, the effect increases the local turbulent α (Youdin and Shu 2002; Johansen et al. 2006).

While the number density and size of the particles is usually constrained from disk models, the relative velocity among grains or pebbles depends on which process dominates the particle velocity (see Birnstiel et al. 2016, for a review). For a large portion of a turbulent disk and for small particles (low St), the velocity due to turbulence is dominating the approach speed of two particles and is approximately given by Ormel and Cuzzi (2007)

$$\Delta v \approx \sqrt{\frac{3\alpha}{St + St^{-1}}} c_s \tag{7}$$

where α is the Shakura and Sunyaev (1973) parameter for turbulent strength. We note that depending on the nature of the turbulence, the induced relative velocities might be lower because particles move in concert with each other. This is the case when the turnover times of the large eddies is longer than the stopping time of the particles. Such a situation can be expected if turbulence is driven by the vertical shear instability (Nelson et al. 2013; Lin and Youdin 2015).

The relative velocities are not only used to estimate collision rates but are also key to determine the outcome of a collisional encounter. When the relative velocities are larger than a threshold velocity v_{frag}, particles can break apart and prevent further growth. Laboratory experiments place v_{frag} in the range from $1\,\mathrm{m \cdot s^{-1}}$ to $10\,\mathrm{m \cdot s^{-1}}$ (Blum and Wurm 2008; Gundlach et al. 2018; Steinpilz et al. 2019) depending on the composition. However, this is part of ongoing research and numerical simulation of porous grains hint at larger speeds giving rise to uninhibited growth to meter- or kilometer-sized bodies, that is, planetesimals (Okuzumi et al. 2012; Kobayashi and Tanaka 2021). However, on the way to planetesimals, other potential barriers need to be overcome. Instead of fragmenting, the collision of two grains could also lead to the impacting particle bouncing-off the target particle which also prevents sticking and growth (Seizinger and Kley 2013). Recent work indicates that discrepancies (Schräpler et al. 2022) between limits derived from laboratory and numerical approaches could be resolved by taking into account the size of the particle (Arakawa et al. 2023) which should motivate the adoption of those limits in modern dust evolution models Stammler and Birnstiel (2022).

Radial drift. A third barrier for growth is not related to collisions between particles but to their aerodynamic interaction with the gaseous material (Weidenschilling 1977a). It is a key element of today's dynamic picture of planet formation because of the mobility of building blocks discussed in the *Predictions from original Solar System theory and shift of paradigms* section.

Considering a parcel of gas, the pressure support of the gas interior to it reduces the orbital velocity relative to the Keplerian value to (Adachi et al. 1976)

$$v_g = v_K \sqrt{1 - 2\eta} \tag{8}$$

where

$$\eta \equiv -\frac{r}{2 v_K^2 \rho} \frac{dP}{dr} \tag{9}$$

v_K is the Keplerian orbital velocity ($\Omega_K \rho$), and ρ and P are the density and pressure of the gas. Although η is small for most disk profiles, the reduced gas speed introduces a headwind for dust particles orbiting at v_K. At 1 AU, it is on the order of 100 m/s. While this is small relative to v_K, this is non-negligible in absolute terms.

Therefore, gas drag is acting and braking the dust particle, removing its angular momentum and leading to a radial velocity component. Assuming the radial velocity to be in steady-state and small compared to the orbital velocity, a low dust-to-gas ratio, and a small value of η, it follows (Nakagawa et al. 1986)

$$v_{d,r} = \frac{-2\eta v_K + St^{-1} v_{g,r}}{St^{-1} + St} \tag{10}$$

We note that η is typically positive due to the minus sign and pressure decreasing with orbital distance, thus the particles indeed move—or drift—towards the star.

From Equation (10), we see that particles with Stokes number close to unity will experience the strongest radial drift. As can be seen in Figure 7, this has the consequence that such objects are removed extremely quickly from the inner parts of the Solar nebula (less than 100 years for roughly meter-sized bodies). For typical disk profiles, the radial drift timescale becomes shorter than dust growth timescales below a Stokes number of unity. Therefore, radial drift can act as a barrier which removes particles instead of allowing for growth. This fact was termed the meter-sized barrier and is a common issue discussed since the early works of Whipple (1972), Weidenschilling (1977a), and Adachi et al. (1976).

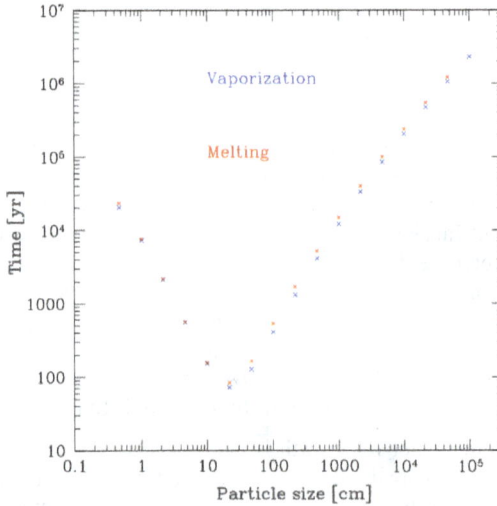

Figure 7. Radial drift resulting from the different rotation velocities of solid particles (pebbles, planetesimals) and nebular gas around the star. The figure shows the time for rocky (silicate) bodies of different sizes to drift from 1 AU to their thermal destruction (either via vaporisation or melting) in the inner hot part of the disk. Destruction happens at the rockline, which corresponds to temperatures of about 1800 K usually found at orbital distances of a few 0.01 AU to 0.1 AU. In the size domain between 10 and 100 cm, bodies spiral to destruction extremely quickly (60–70 yr). Smaller and larger bodies spiral in much slower.

However, if a local pressure maximum exists in the radial gas distribution, η and the drift velocity can change sign. Thus, the solids accumulate at the pressure maximum and it will act as a dust trap. In these locations, the drift limit can be overcome. Symmetric dust features observed with the Atacama Large Millimeter/submm Array (ALMA) show observational evidence of this scenario (Partnership et al. 2015; Andrews et al. 2018).

However, today, radial drift is no longer perceived as a process that is only detrimental to growth, but that on the contrary can critically help to form planetesimals and grow planets because of two key mechanisms: first, when considering the collective behavior of drifting particles (instead of individual ones as in the previous considerations) and when including their back-reaction onto the gas disk, the following instability occurs: spontaneously forming over-densities of drifting particles make the surrounding gas orbit slightly super-Keplerian because of the back-reaction. This reduces the inward drift of this clump of particles. They are then catching other individual particles that drift faster, which further amplifies the process, making it run away. As discussed further in the following *Pebbles to planetesimals* section, this *streaming instability* (Youdin and Goodman 2005) probably plays a key role for the formation of planetesimals.

Second, if the drifting particles can be accreted by planets (Ormel and Klahr 2010), they become a large reservoir for solid growth. At Stokes number of 0.01 to 0.1, radial drift is efficient and since particles with these Stokes numbers correspond roughly to a few centimeters in size, the objects were termed pebbles and the accretion of these particles is known as *pebble accretion* which will be discussed in its own section.

Entrainment of grains in photoevaporative winds. In contrast to heating of the nebular gas, high energy radiation will not lead to heating of dust particles to thermal energies sufficient become unbound. However, small dust can be entrained in the gaseous photoevaporative wind which can affect the mass budget and composition and might provide direct observational clues (Facchini et al. 2016; Hutchison et al. 2016; Franz et al. 2020; Sellek et al. 2020). For externally driven winds, it can be assumed that all grains are entrained with a size below a critical size (Sellek et al. 2020)

$$a_{ent,ext} = \frac{v_{th} \dot{\Sigma} r^2}{\rho_s \mathcal{F} G M_*} \tag{11}$$

where v_{th} is the thermal velocity of the heated gas ($T \sim 1000\,\mathrm{K}$), $\dot{\Sigma}$ is the gas surface density change due to photoevaporation, and $\mathcal{F} \approx H_g / r$ is a geometrical factor. Equation (11) is obtained by equating gravity to the gas drag from the escaping wind. For internally driven winds, the limiting size is most likely given by the size of the particles which can be swept to higher elevations above the midplane by a vertical flow in the disk. For this case, Booth and Clarke (2021) derived another limiting particles size which lies for typical disks close to the monomer size of the grains on the order of 10^{-5} cm implying that the entrainment can be stalled completely for disks with weak photoevaporative winds. Burn et al. (2022) discuss the effects of entrainment on the solid mass budget for a range of parameters.

Pebbles to planetesimals

To proceed on the way of converting the solids in disks from pebble size to eventually planet size, the bouncing, fragmentation, or drift limits discussed in the last section need to be overcome. This can be the case if particles spend little time as aggregates where they would drift or fragment and instead a process exists which directly bridges these critical size regimes to at least several kilometer-sized objects.

Direct formation of large planetesimals overcomes several other threats which would affect intermediate-sized planetesimals. First, at tens of kilometer sizes, they would be sturdy enough to not fragment in mutual collisions (Benz and Asphaug 1999). Second, erosion of the planetesimal by interaction with the gas becomes inefficient and limited to a smaller region close to the star for larger planetesimals (Demirci et al. 2020; Schönau et al. 2023). Similarly, erosion due to collisions with pebbles are negligible at larger sizes (Burn et al. 2019). Lastly, as visible in Figure 7, large planetesimals are not affected by radial drift as their mass out-scales the surface-dependent drag force in the turbulent drag regime leading to negligible specific drag (Weidenschilling 1977a; Birnstiel et al. 2016; Burn et al. 2019).

For these reasons, a scenario where planetesimals form relatively large is preferred. This can be achieved if dust is concentrated enough to trigger the gravitational collapse of a dust cloud. Goldreich and Ward (1973) proposed this mechanism as a pathway for planetesimal formation where the required large dust-to-gas ratio is achieved by settling of dust to the midplane. However, for global dust-to-gas ratios on the order of a percent, including the aforementioned (see the *Growth from dust to pebbles* section) lower limit due to self-induced turbulence of the dust layer, it was shown that this is impossible to realize (Sekiya 1998; Youdin and Shu 2002). Thus, the dust-to-gas ratio needs to be locally enhanced with a different process. To achieve this, Youdin and Shu (2002) found factor ~10 enhancements from radial drift only in a smooth disk which is a fringe case to trigger the collapse. Later, it was understood that the collective behavior of radially drifting dust can trigger what is now called the streaming instability via its back-reaction on the gas (Youdin and Goodman 2005; Squire and Hopkins 2018). It has an automatic dust-concentration effect (Johansen and Youdin 2007; Simon et al. 2016) aiding the potential gravitational collapse but it is also a source of turbulence which could act against the gravitational collapse (Klahr and Schreiber 2020). The growth of the streaming instability is only efficient at enhanced super-Solar pebble-to-gas ratios and particles with Stokes numbers $St > 0.01$ (Bai and Stone 2010; Carrera et al. 2015). Furthermore, if a distribution of particle sizes (Krapp et al. 2019) or a background turbulence (Umurhan et al. 2020) is present, the range of conditions suitable for streaming instability shrinks. Thus, it might be a process which does not trigger planetesimal formation on its own. However, it is of great aid in the final clumping process. These considerations are discussed in the context of formulating a criterion for the formation of planetesimals as a function of disk conditions (Gerbig et al. 2020).

Therefore, to trigger planetesimal formation aided by the streaming instability concentrating pebbles in clouds eventually undergoing gravitational collapse, some other process could be required to reach the necessary dust concentration. It could be that planetesimals only form at preferential locations in the protoplanetary disk, such as the inner edge of the unionized and thus magnetically dead-zone (Gammie 1996; Drążkowska et al. 2016) or the iceline where water sublimates (Drążkowska and Alibert 2017; Schoonenberg and Ormel 2017). Alternatively, local concentrations of pebbles were assumed as put forward by Lenz et al. (2019) to construct a flux-regulated prescription for planetesimal formation useful for global planet formation models. Such an approach is motivated by a number of theoretically expected processes which can create zonal flows and therefore local over-densities. The zonal flow can for example be caused by magnetic instabilities (Dittrich et al. 2013; Bai and Stone 2014; Béthune et al. 2016). Particle rings could originate also from a dust-driven instability in the outer magnetically active regions (Dullemond and Penzlin 2018). Alternatively, particles could accumulate in forming vortices (Raettig et al. 2015; Manger and Klahr 2018) caused by the vertical shear instability (Nelson et al. 2013; Lin and Youdin 2015; Pfeil and Klahr 2019). Most importantly in an observationally driven field such as planet formation, such rings are observed with ALMA in many protoplanetary disks (e.g., ALMA-Partnership et al. 2015; Dullemond et al. 2018).

While there is certainly a need for more research in determining the mechanism(s) of planetesimal formation, another important topic is the resulting size distribution. For the streaming instability, mass distributions of clumps of dust have been determined (Simon et al. 2016, 2017; Schäfer et al. 2017; Abod et al. 2019; Li and Youdin 2021). However, these works are often using already enhanced dust concentrations and missing the resolution to resolve the latest final gravitational collapse phase which might reduce the final planetesimal mass further (Klahr and Schreiber 2020; Nesvorný et al. 2021; Polak and Klahr 2023). Nevertheless, the order of magnitude in size of these first primordial planetesimals is likely similar to or larger than 100 km. This size regime might be problematic for both pebble and planetesimal accretion presented below without further fragmenting the primordial bodies or making use of the concentration of solids in rings (Lorek and Johansen 2022).

Solid accretion mechanisms

In this section, we discuss how protoplanets grow by accreting solids (as opposed to gas accretion). There are three different fundamental types, involving increasingly large bodies: pebble accretion, planetesimal accretion, and giant impacts. We first give a short overview describing how the understanding of solid growth has changed over time and then address the physics of the mechanism separately.

Development of the concepts of solid accretion. In the past fifty years, the concept of how protoplanets grow by accreting solids has evolved. At the beginning of modern theories still considered valid in a general sense stands the planetesimal accretion paradigm, proposed by Viktor Safronov in the 1960s (Safronov 1969). According to this concept, protoplanets (also called *planetary embryos*) grow through the gradual collisional accretion of small solid bodies, the planetesimals, which have sizes of about 1 to 100 km. This idea stems from the observations of the ubiquity of such bodies in the Solar System (asteroid and comets).

Within the planetesimal paradigm, an important process is the one of planetesimal runaway growth (Wetherill and Stewart 1993). During planetesimal runaway accretion, the more massive bodies have a large gravitational focusing factor (the Safronov factor, see Eqn. 17). This means that their effective collisional cross section is much larger than their physical size. In turn, the more massive bodies grow faster than less massive ones. This represents an unstable (runaway) situation that splits an originally monodispersed (same sized) planetesimal population into massive rapidly growing runaway bodies (protoplanets) and small slowly growing background planetesimals, i.e., a bimodal distribution.

However, Ida and Makino (1993) showed that the fast runaway growth regime cannot be sustained ad infinitum. Instead, once the large bodies are massive enough to dynamically heat the surrounding planetesimals (i.e., increase their eccentricities and inclination), the growth mode changes to the slower oligarchic mode (Ormel et al. 2010). In this regime, several neighboring protoplanets (now called *oligarchs*) grow in lockstep, mutually separated by about 5 to 10 Hill spheres. The oligarchs still grow faster than the background planetesimals. In the oligarchic regime, the growth timescale becomes an increasing function of planet mass, opposite to the runaway regime.

However, in oligarchic growth there could be a timescale problem when it comes to the formation of giant planet cores in the outer regions of a planetary system, especially outside of about 10 AU. Given the high amounts of nebular gas that gas giants contain, the massive cores required to trigger gas runaway accretion (about 10 Earth masses) obviously need to form before the dissipation of the gas nebula i.e., within 3–10 Ma. Thommes et al. (2003) highlighted this issue, indicating that the growth of these cores may take too long to occur within the observed lifetimes of protoplanetary disks. The problem becomes more acute the further one moves away from the star since the growth timescale scales with the Keplerian frequency. While the in-situ formation of Jupiter from 100 km planetesimals at 5.2 AU is still feasible in a nebula with about 2–4 times the surface density of the MMSN and modern models (Podolak et al. 2020), it becomes very difficult to form in-situ the ice giants or the giant extrasolar planets observed at distances of several tens of AUs. In this context it should be noted that the ice giants might have formed closer in (in the Nice model, they rather start forming at 6–8 AU, Walsh et al. 2011), and the gravitational instability mechanism could have formed at least some of the massive distant extrasolar planets (Boss 2011). An alternative explanation could be scattering events (Marleau et al. 2019).

To address the timescale problem, Ida and Lin (2004) suggested small planetesimals, around 1 km in size, as a solution. Planetesimal growth is the faster the smaller the planetesimals are, for two reasons: First, there is a stronger damping of eccentricities and inclinations by the gas drag of the surrounding protoplanetary disk, leading to smaller random velocities of the planetesimals and thus a larger gravitational focusing factor. Second, the enhancement of the capture radius of the protoplanet because of drag in its gaseous envelope of the protoplanet is also increased for small planetesimals[1].

However, observations of the asteroid belt (Morbidelli et al. 2009) and various streaming instability and planetesimal formation models (e.g., Polak and Klahr 2023) rather indicated that planetesimals are born relatively large, around 100 km in size. This would pose a challenge to the idea of small planetesimals being the primary building blocks of planets, but it should be noted that contrasting evidence also exists (see Emsenhuber et al. 2021a, for an extended discussion).

To reconcile these different perspectives, a new paradigm was introduced in a dedicated way by Ormel and Klahr (2010) which proposes that bodies much smaller than km-sized planetesimals are actually important planetary building blocks. This *pebble accretion* paradigm suggests that small pebbles (usually mm- to dm-sized), can more efficiently accrete onto larger bodies, especially at larger orbital distances (Lambrechts and Johansen 2012). The fundamental difference to planetesimal accretion lies in the fact that during the encounter, not only gravity is important but also dissipative drag forces (Ormel 2017). This process is fast and can occur even in the outer regions of a planetary system, as the capture radius scales with the Hill sphere which in turn scales with the semi-major axis. In contrast, the physical size of the core, which is more relevant for planetesimal accretion, does not increase with distance.

However, pebble accretion in the outer system requires a large enough starting seed because pebbles cannot be accreted by another pebble. This can be problematic if the streaming instability does not lead to the formation of such bodies (Lorek and Johansen 2022). This would bring us back to the need for planetesimals during the intermediate growth stage (from the largest mass formed directly out of the planetesimal formation mechanism to the lowest mass where efficient pebble accretion starts), which would in turn bring back the aforementioned timescale problem in the outer disk.

One potential solution to this new challenge is the formation of protoplanets in a structured disk that contains rings (Lau et al. 2022; Jiang and Ormel 2023), or more accurately pressure maxima where drifting pebbles accumulate. The high concentration of pebbles and the low headwind velocity inside the ring renders pebble accretion very efficient. These studies suggest that the presence of pressure maxima can facilitate the formation of massive protoplanets even at large orbital distances (50–100 AU). Disk structures can furthermore act as traps for the orbital migration of protoplanets, further enhancing the efficiency of planet formation (e.g., Coleman and Nelson 2016).

In summary, solid accretion involves various paradigms and models, including the planetesimal paradigm with runaway and oligarchic growth stages, pebble accretion and giant impacts. The latter is thought to be the final stage in the growth of terrestrial planets. The role of pebbles, planetesimal, and impacts in different regions and phases of planetary system growth are central subjects of ongoing research and hybrid models start to be developed (Coleman 2021; Voelkel et al. 2022).

Basic concepts of planetesimal accretion. A key concept in collisional growth via planetesimals is the one of the gravitational focusing. In this section, we address this effect in a more educative fashion than processes addressed in the other parts. More realistic state-of-the-art calculations are considerably more complex (Inaba et al. 2001), but the simple picture presented here illustrates the basic concepts.

---

[1] When the effect of aerodynamic fragmentation of planetesimals in protoplanetary atmosphere is included besides drag, then the capture radius effect is also efficient for large planetesimals, see Podolak et al. (2020).

Consider two bodies of masses m_1 and m_2 with radii r_1 and r_2 approach each other at an initial relative velocity v_∞ and with an impact parameter b (Fig. 8). The velocity is formally the one at infinity, or in practice when the planets are still very far from each other.

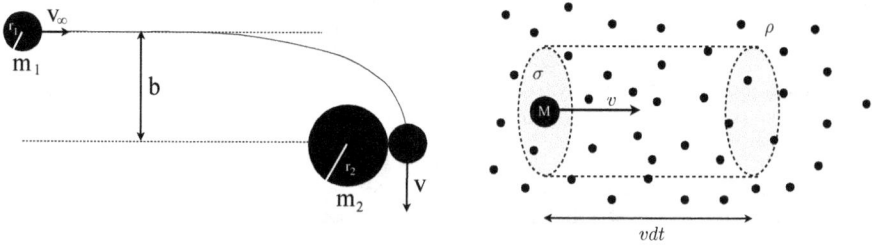

Figure 8. Illustration of the approach of two gravitating bodies (**left**) and the cylinder of accreted mass onto a moving body (**right**). The cylinder (**dotted**) is visualized for a particle with mass M moving at velocity v through a medium with density ρ for a time dt at a certain interaction cross-section σ. In the left schematic, particle 1 with mass m_1 and radius r_1 is drawn at two different times on its path (**arrow**), once at infinite separation from particle 2, where it has a velocity v_∞, and at its closest approach to particle 2 where it has a differing velocity v due to the gravitational attraction.

In a billiard game, the collisional cross sections of two bodies is simply given by the geometrical cross sections

$$\sigma = \sigma_{geo} = \pi\left(r_1 + r_2\right)^2 \tag{12}$$

However, the attracting nature of gravity leads for planetary growth to an increase of the collisional cross section over the geometrical one. This is called gravitational focusing. Energy and angular momentum conservation allows one to calculate the collisional cross section for two arbitrary-sized, gravitating (spherical) bodies, neglecting the influence of the Sun (two body approximation).

Energy conservation yields (with values at infinity on the left and values at the closest approach on the right):

$$\frac{1}{2}\mu_m v_\infty^2 = \frac{1}{2}\mu_m v^2 - G\frac{m_1 m_2}{r_1 + r_2} \tag{13}$$

with μ_m the reduced mass $m_1 m_2/(m_1 + m_2)$. The conservation of angular momentum yields

$$\mu_m b v_\infty = \mu_m\left(r_1 + r_2\right)v \tag{14}$$

which can be solved to yield the velocity at the closest approach, $v = v_\infty/(r_1 + r_2)$. Together with the definition of the escape velocity

$$v_{esc} = \sqrt{\frac{2G\left(m_1 + m_2\right)}{r_1 + r_2}} \tag{15}$$

Gives

$$b^2 = \left(r_1 + r_2\right)^2\left(1 + \frac{v_{esc}^2}{v_\infty}\right) \tag{16}$$

This yields

$$\sigma = \pi b^2 = \pi (r_1 + r_2)^2 \left(1 + \frac{v_{esc}^2}{v_\infty^2}\right) = \sigma_{geo}(1 + \Theta) \tag{17}$$

where we have used the so-called Safronov number, $\Theta = v_{esc}^2 / v_\infty^2$, in honour of Viktor Safronov who was the first to develop this collisional accretion scenario. We thus find that the collisional cross section is increased over the geometrical cross section by a factor which is proportional to the square of the escape velocity to the random velocity. Clearly, this ratio can become quite big thus strongly increasing the collision probability.

The accretion of mass per time is now a simple geometrical task. As visualized in the right panel of Figure 8, a particle with an interaction cross-section of σ moving through a medium of density ρ will accrete over a time dt the mass in the visualized cylinder

$$\dot{M} dt = \sigma \rho v \, dt \tag{18}$$

where we introduced the mass accretion rate $\dot{M}$. Thus, for our derived cross-section, we get

$$\dot{M} = \rho v_\infty \pi (r_1 + r_2)^2 \left(1 + \frac{v_{esc}^2}{v_\infty^2}\right) \tag{19}$$

where the velocity with which the particle under consideration (with mass M) can be set to v_∞ which should be interpreted as the average relative velocities between the considered particle and all other particles at large distance.

For applications, we need to describe the density ρ used in the above formula. Assuming the medium to be composed of planetesimals, that is, particles well decoupled from the gas, ρ can be estimated using the eccentricity e and inclination i distribution of the particles. Under these assumptions, the particles move on Keplerian orbits. Therefore, their velocity is on the order of $v \approx \sqrt{e^2 + i^2} v_K$ (Lissauer 1993).

This velocity can also be used to describe the planetesimals as a fluid and derive a planetesimal scale height $H_{pla} = v / \Omega_K = a\sqrt{e^2 + i^2}$ with a corresponding approximate planetesimal midplane density for a given planetesimal surface density Σ_{pla} of

$$\rho_{pla} \approx \frac{\Sigma_{pla}}{\sqrt{2\pi(e^2 + i^2)}} \tag{20}$$

(Safronov 1969). Combined with Equation (19), this yields an easy-to-use estimate of the planetesimal accretion rate

$$\dot{M}_{pla} \approx \Sigma_{pla} v_K (r_1 + r_2)^2 \left(1 + \frac{v_{esc}^2}{(e^2 + i^2) v_K^2}\right) \tag{21}$$

where we omitted factors of order unity to not overstate the precision of the approach. We see that the more massive a growing planet, the larger the escape velocity and the more rapid its growth. Therefore, if the other factors are kept constant this accretion formula describes relatively well the *runaway planetesimal accretion*.

As seen in this derivation, the complexity emerges when trying to estimate the exact approach speed of two bodies. For non-circular orbits, it depends on the location on the orbit of both the target and the impactor. The integration then becomes non-trivial and when additionally accounting for approaches which do not neglect the gravity of the central star, the problem is no longer solvable analytically. Therefore, numerical studies were conducted

(Greenzweig and Lissauer 1990, 1992; Inaba et al. 2001) who obtained much more precise prescriptions for the accretion rate of planetesimals.

As the aforementioned insights were gained, it also became clear that the eccentricities and inclinations of the planetesimals are key parameters which are not independent of the planetary mass. In the vicinity of a relatively massive growing protoplanet, the eccentricities and inclinations of the planetesimals are affected due to gravitational stirring (Ida 1990; Ohtsuki 1999) and thus the velocities and accretion probabilities are affected. This reduces the fast, runaway accretion via the denominator in Equation (21) and leads to a transition the slower so-called *oligarchic* accretion. Prescriptions to model the gravitational stirring were formulated and applied in a number of works (Ohtsuki 1999; Chambers 2006; Fortier et al. 2013).

The particles in the gas disk are not only subject to gravitational stirring of the planet but also experience a similar effect from gas turbulence (Ormel and Kobayashi 2012; Kobayashi et al. 2016) and from each other (Ohtsuki et al. 2002). The stirring effects are compensated by aerodynamic gas drag which slows down the planetesimals as long as the gaseous disk is present (Adachi et al. 1976). Since the drag force scales with the surface of the planetesimals while the other processes mainly depend on the mass, gas drag is more efficient for small planetesimals. Thus, planetesimal accretion becomes more efficient if planetesimals are born small (of order km in radius) or fragments are produced frequently.

In addition, another aerodynamic effect emerges from the gaseous envelope of the planet. As the gas density is increased in the vicinity of the planet over the density in the disk, the aerodynamic drag increases as particles get close to the planet. Instead of passing the planet, it can then be accreted (Inaba and Ikoma 2003; Mordasini et al. 2006).

While one might be tempted to assume that this is an effect which takes only place close to the planet when the particles are anyways on collision-course, one needs to consider that during the gas disk stage, the planetary envelope can be very extended and have a size comparable to the Hill sphere (see the *Gas accretion* section).

Pebble accretion. The pebble accretion process builds on the aforementioned gas-drag enhanced capture radius of the planet. This effect was studied with focus on larger planetesimals for decades (Podolak et al. 1988; Inaba and Ikoma 2003). For even smaller particles, which are experiencing the strongest drag effects (i.e. Stokes numbers close to unity), the cross section of the planet is even more significantly increased over the physical cross-section due to aerodynamic drag.

The idea to accrete the small, millimeter- to centimeter-sized particles was brought forward by Klahr and Bodenheimer (2006), although it was applied to giant planets forming in a vortex instead of classical core accretion, which was only done in Ormel and Klahr (2010). The advantages over planetesimal accretion are faster growth and a larger reservoir of solids, in principle all solids exterior to the growing planet's location. This motivated the full development of an alternative paradigm of solid accretion termed pebble accretion (see Johansen and Lambrechts 2017; Ormel 2017, for in-depth reviews).

Here, we briefly review the fundamentals of the approach. As for planetesimals, an accretion rate of solids can be expressed as the flux of particles through the cross sectional area of the planet

$$\dot{M}_{peb} = \pi R_{acc,peb}^2 \rho_{peb} v_{rel,peb} \tag{22}$$

Since the particles drift through the protoplanetary disk, the relative velocity $v_{rel,peb}$ contains a significant contribution from the radial motion. Approximately, the sum of the difference between Keplerian orbits and the radial velocity can be used for the relative velocity (Ormel 2017).

$$v_{\text{rel, peb}} = v_{\text{d,}r} + \frac{3}{2}\Omega_{\text{K}} R_{\text{acc, peb}} \tag{23}$$

This expression neglects flow patterns of the gas due to the planet or turbulence which can become relevant for low Stokes numbers.

On the other hand, eccentricities and inclinations of pebbles are damped due to efficient gas drag to a degree where they can be neglected. As for planetesimals, the relevant mass density of pebbles distributed in space (not to be confused with the bulk density of an individual pebble) is a key quantity to determine. In steady state, it can be approximated from the reduced scale height of pebbles or dust (Eqn. 6) using the definition of a surface density $\rho_{\text{peb}} \approx \Sigma_{\text{peb}} / \left(\sqrt{2\pi} H_{\text{peb}} \right)$. This expression is accurate if the planet's radius of influence $R_{\text{acc,peb}}$ (see below), is small with respect to the scale height of the pebbles. However, once the planet's envelope becomes vertically more extended than the scale height of the pebble disk, it transitions from this so-called 3D regime to a 2D case, where the full vertical extent of the pebble distribution can be accreted resulting in $\dot{M}_{\text{peb}} = 2R_{\text{acc,peb}} v_{\text{rel,peb}}$ instead of Equation (22).

The final quantity to be determined is the accretion radius $R_{\text{acc,peb}}$. It can be estimated by comparing the timescale that particles require to settle toward the planetary core against the timescale they reside within the planet's sphere of influence. The latter time is either determined from the pebble radially drifting toward the star (headwind regime) or passing azimuthally due to the difference in Keplerian orbital velocities (shear regime). Following Ormel (2017), this leads to an expression for the headwind regime

$$R_{\text{acc,peb,hw}} \sim \sqrt{\frac{2GM \text{St}\Omega_{\text{K}}}{v_{\text{d,}r}}} \tag{24}$$

while in the shear regime it evaluates to $R_{\text{acc,peb,hw}} \sim \text{St}^{1/3} R_{\text{H}}$, where the Hill radius

$$R_{\text{H}} = \left(\frac{M}{3M_{\star}} \right)^{\frac{1}{3}} a \tag{25}$$

is used with a being the semimajor axis of the planet.

This approximate treatment can be used for pebbles with Stokes numbers of order 0.01 to 1. For lower Stokes number, that is, for dust, particles follow the gas streamlines around the planet without accretion (Ormel 2017). For larger Stokes number, that is, entering the regime of planetesimals, the assumptions of the Epstein drag regime and zero inclinations and eccentricities break.

At larger planetary masses, pebble accretion terminates once the planet significantly perturbs the gas disk (see the *Orbital migration* section below) to create a local pressure maximum where pebble drift is stopped (Lambrechts et al. 2014; Ataiee et al. 2018). While this effect depends on the gas disk properties, such as viscosity and scale height, the typical pebble isolation mass is on the order of 10 to 40 $M_{\oplus}$.

Recently, hybrid models combining planetesimal and pebble accretion have started to be constructed (e.g., Alibert et al. 2018; Voelkel et al. 2022; Kessler and Alibert 2023). They find a highly complex interplay with phases where one or the other mechanism is dominant resulting in multiple generations of protoplanets, where the additional planet luminosity caused by planetesimal accretion can delay gas accretion relative to the pure pebble case. They also find that giant planet formation alone from 100 km planetesimals is efficient in forming giant planets with final orbital distances out of about 1 AU, but not further away (Voelkel et al. 2020). This could indicate a distance dependency of the importance of pebbles (larger distances) and planetesimals (smaller distances) (Brügger et al. 2020). Together with the effect of planetesimal fragmentation (Kaufmann and Alibert 2023), that can link the two scenarios, more work is warranted to study the relative roles of planetesimal and pebble accretion and their interplay.

Gas accretion

The process of gas accretion can be separated into two stages, an initial cooling-limited stage at early times and lower planet masses. During this first stage, the planet's envelope is smoothly attached to the surrounding protoplanetary disk. This phase can be followed by a later disk-limited stage for sufficiently high planet masses where the planet's surface is detached from the disk.

Cooling-limited, attached stage. In the initial stage, as a protoplanet grows by solid accretion, its gravitational pull will increasingly attract H/He gas from the protoplanetary disk until it is balanced by the pressure support of already attracted gas. The pressure support is sustained by the luminosity in the planet's envelope, which is in turn generated by the envelope gas itself or by the accretion of solids. Before the advent of 3D hydrodynamic simulations, this situation was treated as a fully static 1D spherically symmetric problem where the cooling and contraction of the already accreted gas allows new nebular gas to flow into the planet's sphere of influence, regulating thereby the gas accretion rate (Bodenheimer and Pollack 1986; Ikoma et al. 2000). Originally, this sphere of influence was taken to be simply the smaller of the Bondi R_B and the Hill sphere radius R_H (Bodenheimer and Pollack 1986). However, hydrodynamic simulations show that only about the inner $0.25\,R_H$ are actually bound to the planet, while the outer layers participate in the disk's surrounding shear flow (Lissauer et al. 2009).

In this inner part, the relevant equations are similar to the stellar structure equations. Spherical symmetry is assumed. Then, mass conservation, momentum conservation in the form of hydrostatic equilibrium, energy conservation, and energy transport can be written using the radius from the planet's center r as coordinate (Bodenheimer and Pollack 1986):

$$\frac{\partial m}{\partial r} = 4\pi r^2 \rho \tag{26}$$

$$\frac{\partial P}{\partial r} = -\frac{Gm\rho}{r^2} \tag{27}$$

$$\frac{\partial l}{\partial r} = 4\pi r^2 \rho \left(\varepsilon - P\frac{\partial V}{\partial t} - \frac{\partial u}{\partial t} \right) \tag{28}$$

$$\frac{\partial T}{\partial r} = \frac{T}{P}\frac{\partial P}{\partial r}\nabla(T,P) \tag{29}$$

where G is the gravitational constant, m the enclosed mass at radius r, and P, T, and ρ are the pressure, temperature, and density in a layer. The gradient ∇ depends on the process by which energy is more efficiently transported (radiation or convection) as judged by the Schwarzschild criterion. In the case of convection, tabulated adiabatic gradients from specialized non-ideal equations of state are used which also yield the density as a function of pressure and temperature. Assuming radiative diffusion, the radiative gradient is given by

$$\nabla_{\text{rad}} = \frac{3\kappa l}{64\pi\sigma_{\text{SB}}GmT^3} \tag{30}$$

where κ is the Rosseland mean opacity and σ_{SB} is the Stefan–Boltzmann constant.

We notice that the energy equation is the only equation which explicitly depends on time t. This implies that it drives the temporal evolution. In this equation, l is the luminosity (energy flux), ε an extra energy source such as impacts or radiogenic heating (for stars it would be nuclear fusion), $V = 1/\rho$ is the specific volume, and u is the specific internal energy. Using the first law of thermodynamics, one can also consider the temporal change of the gas' entropy instead of the change of the volume and internal energy separately. This is particularly useful if (most of) the envelope is adiabatic and thus characterized by one entropy.

These equations can be solved numerically (Bodenheimer and Pollack 1986; Ikoma et al. 2000; Mordasini et al. 2014b; Piso and Youdin 2014; Kimura and Ikoma 2020; Emsenhuber et al. 2021b) given boundary conditions, opacities, and an equation of state to close the system of equations. In the initial stage, the outer boundary conditions are given by the disk background conditions modified by the effects of the circulating flow (Ali-Dib et al. 2020). Specialized equations of state (EOS) that cover the required regime of pressures and temperatures in giant planet interiors deviate strongly from the ideal gas case but must include also the degenerate limit (Saumon et al. 1995; Chabrier and Debras 2021). Such EOS are specially developed for the interior of planets and brown dwarfs (see Helled et al. 2020, for a review). Modern equations of state can to some extent account for varying helium fractions and metallicities Z, but many questions still remain (Howard and Guillot 2023).

For the early stage in which solid accretion is the main source of luminosity for the planet, a useful simplification is to assume a constant luminosity throughout the structure of the planet and that for sufficiently large impactors (planetesimals), energy is deposited at the envelope to core boundary (Mordasini et al. 2012a). This is a good approximation when the gaseous envelope is thin enough to allow for planetesimals to reach the solid core (Podolak et al. 1988; Mordasini et al. 2006). For small initial relative velocities between the accreted material and the planet, the luminosity of accreted material is given by the released potential energy in the gravitational field of the planet,

$$L_{\text{solid}} = G \frac{\dot{M}_{\text{acc,solid}} M_{\text{c}}}{R_{\text{c}}} \tag{31}$$

where $\dot{M}_{\text{acc,solid}}$ is the accretion rate of solids, and M_{c} and R_{c} are the core mass and core radius. This establishes a link between gas and solid accretion (*Solid accretion mechanisms* section). Once the envelope becomes "opaque" to the incoming bodies, the liberated energy depends on the radius at which the particles are mainly stopped and the corresponding encompassed mass at this radius (Podolak et al. 1988) instead of the core radius and mass.

In the initial cooling-limited stage, the temporal evolution is controlled by the timescale of the Kelvin–Helmholtz (KH) cooling of the envelope τ_{KH}, therefore the characteristic timescale for the accretion of gas for a planet of mass M can be estimated as

$$\dot{M}_{\text{gas}} = \frac{M}{\tau_{\text{KH}}} \tag{32}$$

The solution of the structure equations shows that under simplifying assumptions (for example, no solid accretion) the Kelvin–Helmholtz timescale τ_{KH} can be parameterized as (Ikoma et al. 2000; Mordasini et al. 2014)

$$\tau_{\text{KH}} = 10^{p_{\text{KH}}} \, yr \left(\frac{M}{M_{\oplus}} \right)^{q_{\text{KH}}} \left(\frac{\kappa}{1 \, \text{gcm}^{-2}} \right) \tag{33}$$

In this equation, the parameters p_{KH} and q_{KH} are found by fitting the accretion rate found by solving the internal structure equations (e.g., Ida and Lin 2004; Miguel et al. 2011; Mordasini et al. 2014). Mordasini et al. (2014), for example, derive $p_{\text{KH}} = 10.4$, $q_{\text{KH}} = -1.5$, and $\kappa = 3 \times 10^{-3} \, \text{cm}^2/\text{g}$. One can see that the gas accretion rate is thus an increasing function of planet mass. Once the planet mass is sufficient high (about 5–10 Earth masses), τ_{KH} becomes comparable to the disk lifetime, implying that gas accretion becomes important.

Disk-limited, detached stage. As the planet's mass increases further (where its envelope mass becomes increasingly important relative to the core mass), the gas accretion process speeds up further, meaning that runaway gas accretion sets in (Perri and Cameron 1974; Mizuno 1980; Stevenson 1982). The exact critical mass at which this occurs depends on several factors including the opacity in the envelope (given mainly by solid grain opacity, Mordasini 2014; Ormel 2014; Kimura and Ikoma 2020) but also the solid accretion rate. At some point, core and envelope mass become equal, which is known as the crossover mass (Pollack et al. 1996). In addition to the solid accretion luminosity, contributions from the gas' cooling and contraction become relevant and eventually dominate.

At some point in the runaway phase, the outer radius of the planet contracts so rapidly (but still quasi-statically, Bodenheimer and Pollack 1986) that the gas disk can no longer supply nebular gas at a rate sufficient to fill the rapidly emptying shell—the planet's surface thus detaches from the protoplanetary disk (Bodenheimer et al. 2000; Mordasini et al. 2014b) and the growth enters the disk-limited accretion stage. The planet's radius now shrinks to a value that is much smaller than the Hills sphere of about 1.5–5 Jovian radii depending on the entropy of the gas in the interior (Mordasini et al. 2012b; Marleau et al. 2017). This is also the moment when the structure changes from an approximately spherically symmetric shape during the attached stage to a flattened one (Ayliffe and Bate 2012; Szulágyi and Mordasini 2017) with a circumplanetary disk surrounding the growing gas giant (Adams and Batygin 2022). The outer boundary conditions of the structure equations are also modified as gas now falls with high velocity on the surface of the planet where it shocks (Mordasini et al. 2012a; Marleau et al. 2023). The resulting emission in the H-α line has been observed recently for forming extrasolar gas giants (Haffert et al. 2019).

In this later stage of massive planets ($M \gtrsim 100 M_\oplus$), the gas accretion rate is no longer limited by the cooling of the envelope but by the supply rate of gas from the protoplanetary disk. The exact limit is influenced by the three-dimensional structure of the gas around the planet. A first estimate yields the classical Bondi/Hill accretion rate (D'Angelo and Lubow 2008; Mordasini et al. 2012a), i.e., the rate at which the planet sweeps nebular gas given relative velocities caused by the Keplerian shear

$$\dot{M}_{\text{e,Bondi}} \approx \frac{\Sigma}{H}\left(\frac{R_{\text{H}}}{3}\right)^3 \Omega_{\text{K}} \tag{34}$$

In this equation, Σ, H, R_{H}, and Ω_{K} are the gas surface density averaged over the planet's feeding zone, the vertical scale height of the disk, the Hill sphere radius, and the Keplerian orbital frequency at the planet's position. More accurate rates can be derived from 2D and 3D hydrodynamic simulations (e.g., D'Angelo and Lubow 2008; Machida et al. 2010; Bodenheimer et al. 2013; Choksi et al. 2023) that can be used to calibrate the 1D approaches (Schib et al. 2022).

Three-dimensional hydrodynamic simulations have in recent years also revealed that envelope gas can be recycled back to the protoplanetary disk with potential influence on the energy budget of the planet (Ormel et al. 2015; Moldenhauer et al. 2021, 2022). Especially at low planet masses and small orbital distances, protoplanetary envelopes are not closed hydrostatic 1D systems, but dynamically exchange gas with the surrounding disk. The resulting advection of high entropy material can delay gas accretion (Cimerman et al. 2017). The basic picture that the KH-contraction of the inner bound region ultimately regulates growth remains however valid, but the outer layers need to take into account the multi-dimensional hydrodynamical effects (Bailey and Zhu 2023). Regarding the composition, further high-resolution investigations are required to assess to what extent this effect could potentially reset the atmospheric composition of a migrated gas-rich planet to a more local composition.

Figure 9 shows the mass of H/He as a function of the heavy element (core) mass predicted by the Generation III Bern global planet formation model (Emsenhuber et al. 2021b) that is based on the core accretion paradigm. The initial conditions of the model were varied according to observed properties of protoplanetary disks (Emsenhuber et al. 2021a) to synthesize a population of model planets. The host star mass is $1\,M_\odot$ and the age of the synthetic planets is 5 Ga.

The envelope masses were derived by solving the internal structure equations in 1D. The opacity caused by grains in the protoplanetary atmospheres was assumed to be a factor 0.003 reduced (Mordasini 2014) relative to ISM grain opacities (Bell and Lin 1994), meaning that the KH-contraction of the envelopes is relatively efficient. This has the consequence that even relatively low-mass planets can accrete some H/He which is then reflected in the planetary mass–radius relation (Mordasini et al. 2014b). The disk-limited gas accretion rate was found in a way similar to Equation (34).

At low masses, the envelope mass increases with increasing heavy element mass. This is expected from the general scaling of the KH-timescale with mass (Eqn. 33). There is, however, a large spread. This is on one hand caused by different formation pathways which result in different efficiencies of gas accretion during the nebular phase. This diversity cannot be fully captured by simpler semi-analytical expressions for gas accretion (Alibert and Venturini 2019). On the other hand, additional effects modify (reduce) the envelope mass post-formation: during the long-term evolutionary phase, i.e., from the end of the lifetime of the disk to 5 Ga, the planets can lose their H/He via impact stripping (Denman et al. 2020) or atmospheric escape (e.g., Jin et al. 2014; Owen 2019; Affolter et al. 2023). Planets without H/He are shown by red points.

Then, at higher heavy element / core masses of about $10\text{--}20\,M_\oplus$, runaway gas accretion sets in. In the plot, this shifts the planets almost vertically upwards in the figure to high H/He masses, whereby they become gas giants.

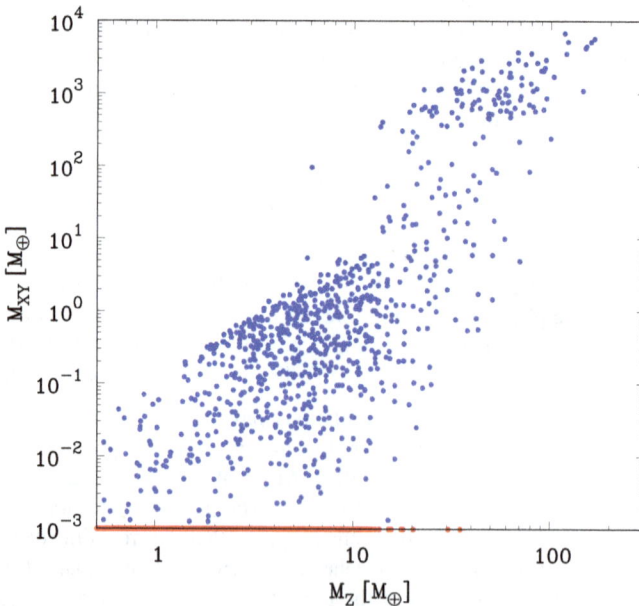

Figure 9. Illustration of the effect of envelope accretion (and loss). The **blue points** show the mass of hydrogen and helium M_{XY} as a function of the mass of heavy elements M_Z (iron, silicate, potentially water) of synthetic planets predicted by the global planet formation model of Burn et al. (2024). This model also includes the processes of envelope loss via impact stripping and hydrodynamic atmospheric escape driven by high-energy photons. **Red points** are planets without H/He. The age of the planets is 5 Ga.

At intermediate masses of about 20–$200 M_\oplus$, there are fewer planets, known as the "planetary desert" (Ida and Lin 2004). It is caused by the relatively high gas accretion rates in the runaway and initial disk-limited phases of about a few 10^{-4} to $10^{-3} M_\oplus/\mathrm{yr}$ (Mordasini et al. 2017). This makes it less likely that the gas disk dissipates at just the moment of intermediate masses. Whether this "smoking gun" of core accretion is actually visible in the observed planetary mass function is a subject of debate. While some observational studies support the existence of the desert (Mayor et al. 2011; Bertaux and Ivanova 2022), others do not (Suzuki et al. 2018; Bennett et al. 2021).

Orbital migration

The mentioned orbital migration of planets within the gaseous disk is likely affecting the overall population of exoplanets which leave observable imprints on the planetary population level (Burn et al. 2024). Thus, we briefly review the process. More detailed reviews were published in Baruteau et al. (2016); Paardekooper et al. (2023).

The physical reason for orbital migration lies in the perturbation of the disk gas in the vicinity of the planet which in turn exerts a torque on the planet. The leading order contribution is due to the mass accumulating at the Lindblad resonance, which is both trailing behind the planet exterior to the planetary orbit as well as leading in front of the planet just interior to the planetary orbit. Due to the geometry of the problem, the outer over-density typically exerts the larger torque on the planet and therefore decelerates it. This leads to a radial motion towards the star (Goldreich and Tremaine 1979; Ward 1997; Tanaka et al. 2002). In detailed studies, the process is usually discussed with reference to a contribution from the Lindblad torque

$$\Gamma_0 = \left(\frac{q}{h}\right)^2 \Sigma_g a_p^4 \Omega_K^2 \tag{35}$$

where the Keplerian orbital frequency Ω_K needs to be evaluated at the planet location a_p, q is the planet-to-star mass ratio and h is the local aspect ratio in the disk H/a_p. A torque Γ_{tot} exerted on the planet on a circular orbit will lead to a change in semi-major axis of

$$\frac{da}{dt} = 2a_p \frac{\Gamma_{tot}}{J} \tag{36}$$

where $J = M\sqrt{GM_\star a_p}$. In addition to the Lindblad torque, there are several other contributions to Γ_{tot}, which can be categorized into four regimes based on planetary mass and disk viscosity.

In the high viscosity, low planetary mass regime, the contributions are well-studied and analytical expressions are available. This regime is the classical viscous type I regime (Ward 1997; Tanaka et al. 2002). In addition to the Lindblad torque, the co-rotation torque can contribute significantly and even invert the typically inwards direction of the orbital motion. The co-rotation torque originates from material orbiting the star in close vicinity to the planet. It is sensitive to the gradient in thermal properties over the region where parcels of gas describe a U-turn motion in proximity of the planet (Paardekooper et al. 2010, 2011; Kley and Nelson 2012; Jiménez and Masset 2017), that is, over a region of width

$$r_{co} \approx \frac{a_p}{\gamma^{1/4}} \sqrt{\frac{Ma_p}{M_\star h}} \tag{37}$$

where γ is the adiabatic index. For typical disk properties, the co-rotation torque alone would lead to an outward motion of the planet. However, it can be weaker than the Lindblad torque and it further saturates at a certain planetary mass. This is the case when gradients in the gas properties are not re-established by material flowing into the relevant region due to viscous diffusion. The timescale of viscous diffusion over the corotation region is $\tau_v = r_{co}^2 / \nu$.

By comparing this timescale to the time it takes the gas to complete a full libration (Hellary and Nelson 2012)

$$t_{\text{lib}} = \frac{8\pi a_{\text{p}}}{3\Omega_{\text{K}} r_{\text{co}}} \tag{38}$$

and with inclusion of more precise factors of order unity, Emsenhuber et al. (2023a) derived a useful critical planetary mass for saturation of the co-rotation torque called the saturation mass for planetesimal accretion

$$M_{\text{sat}} = \left(\frac{8\pi\alpha}{3}\right)^{2/3} \sqrt{\frac{\gamma}{C_{\text{HS}}^4}} M_* \left(\frac{H}{a_{\text{p}}}\right)^{7/3} \tag{39}$$

where $C_{\text{HS}} \approx 1.1$ is based on numerical experiments (Paardekooper et al. 2010). The direct dependence on α is visible in the upper left panels of Figure 10 where saturation is reached at the upper mass end of the red regions caused by the corotation torque.

In the high viscosity, high planetary mass regime, the torques on the planet decrease because the region around the planet is emptied and a gap forms. Formulas for gap emergence and width were obtained by Crida et al. (2006)

$$\frac{3H_{\text{g}}}{4R_{\text{H}}} + \frac{50\nu M_*}{Ma_{\text{p}}^2\Omega_{\text{K}}} \leq 1 \tag{40}$$

and Kanagawa et al. (2017, 2018)

$$M_{\text{gap}} = 8 \times 10^{-5} M_* \sqrt{\frac{\alpha}{10^{-3}}} \left(\frac{H_{\text{g}}/a_{\text{p}}}{0.05}\right)^{2.5} \tag{41}$$

In Figure 10, the transitions are shown and smoothed (following Emsenhuber et al. 2021b). Dürmann and Kley (2015) and Kanagawa et al. (2018) suggest to use the type I torque (including the co-rotation torque) and reduce it linearly with the reduction of the surface density at the bottom of the gap. Prior works suggested a more pronounced transition of regimes where the planet migration would be linked to the radial velocity of the gas (Dittkrist et al. 2014) but with a reduction once the planet mass dominates over the local disk mass (Alexander and Armitage 2009). This latter approach is used in Figure 10, which causes an outward directed type II migration at the locations where the gas flow is also directed outwards.

For low-viscosity disks ($\alpha \lesssim 10^{-4}$), the field is currently determining accurate migration rates (McNally et al. 2019; Lega et al. 2022). Of large importance could be the dynamical co-rotation torque (Paardekooper 2014; Pierens 2015). It emerges in the same fashion as the co-rotation torque in a viscous disk described above but for cases where the viscous co-rotation torque would saturate. When a planet grows to significant mass in such a disk, due to the lack of viscous diffusion, gas is trapped in the horseshoe region and migrates together with the planet, effectively slowing down its Lindblad-torque-driven migration. To quantify it, tracking the history of the planet is required.

At larger planetary mass, in low-viscosity disks, there is still research to be done to find a conclusive migration rate. Vortices and magnetic field lines penetrating into the planetary gap (Aoyama and Bai 2023; Wafflard-Fernandez and Lesur 2023) make this regime particularly challenging to explore (Paardekooper et al. 2023).

Finally, under certain circumstances and not limited to low-viscosity disks, further processes can exert a considerable torque, such as the torque from planetesimals (Levison et al. 2010; Ormel et al. 2012) or dust (Benítez-Llambay and Pessah 2018) in the vicinity of the planet as well as heating effects (Lega et al. 2014; Benítez-Llambay et al. 2015; Masset 2017).

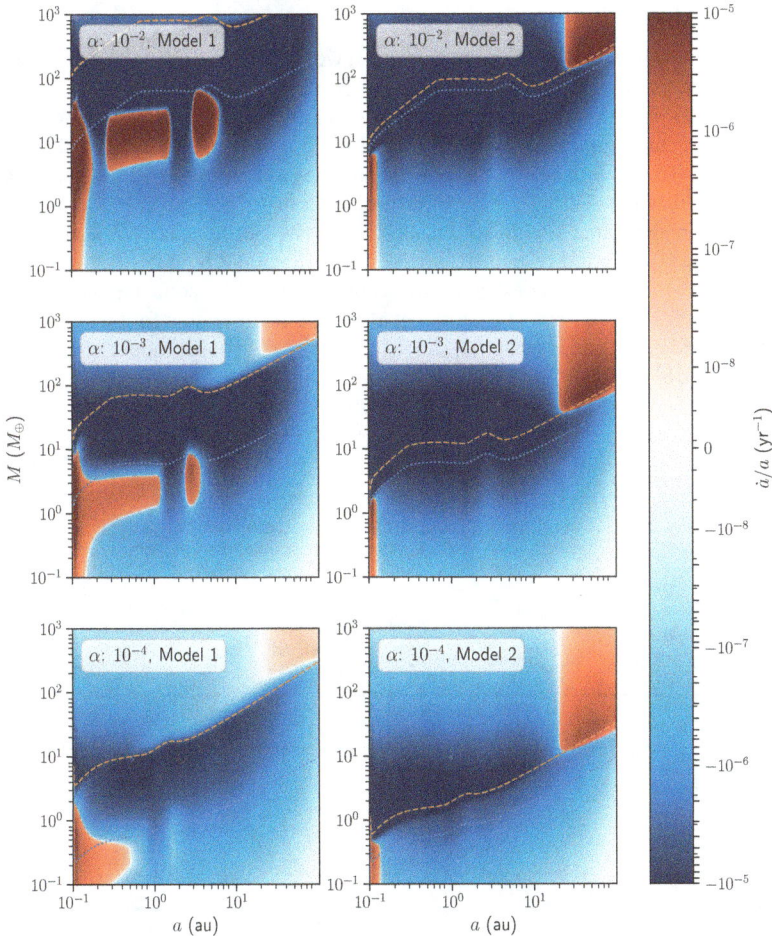

Figure 10. Migration rate of planetary embryos as a function of distance and planetary mass in viscous protoplanetary disks after 0.1 Ma of evolution. The rate is normalized by the distance (thus showing the inverse timescale). The same α is used here for migration, disk evolution, and viscous heating. Model 1 uses a distinct transition (Eqn. 40, **dashed line**) and classical migration rates (Crida et al. 2006; Paardekooper et al. 2011; Dittkrist et al. 2014), while Model 2 uses more recently derived type I rates (Jiménez and Masset 2017), which are also used in the type II regime but suppressed (Kanagawa et al. 2018) at masses exceeding Equation (41). The saturation mass (Eqn. 39, **dotted line**) provides a useful estimate of the upper mass limit of outward migration for Model 1.

PUTTING THE PIECES TOGETHER: GLOBAL MODELS AND PLANETARY POPULATION SYNTHESIS

In this section, we aim to demonstrate how global models of planet formation, which combine several of the physical processes discussed in the last section, bridge the gap from protoplanetary disks to the final planetary systems. As an example, we will focus on the results obtained with the Bern Model of planet formation (Alibert et al. 2005, 2013; Mordasini et al. 2012a; Emsenhuber et al. 2021b). Other global models were developed by several authors (Ida and Lin 2004; Coleman and Nelson 2014; Bitsch et al. 2015; Chambers 2018; Lambrechts et al. 2019; Kimura and Ikoma 2020; Alessi and Pudritz 2022) and are reviewed in detail in Mordasini (2018) and Drążkowska et al. (2023).

Coupled planet formation models

To highlight the interplay of different processes, we show in Figure 11 the time evolution of a planet modelled using different assumptions. The protoplanetary disk evolution is calculated and follows a gas disk with identical parameters but twice the mass of the one shown in Figure 10 (initial mass of $0.043\,M_\odot$). The initial solid-to-gas mass ratio is 0.05 ($715\,M_\oplus$), which is a massive and solid-rich, but not unrealistic case compared to the population of observed disks (Tobin et al. 2020). The solids are either distributed as dust and pebbles or as already formed planetesimals with 100 km diameter. The protoplanet is injected at 5.2 AU and grows by solid accretion of planetesimals in the oligarchic regime (Fortier et al. 2013) or pebbles (Ormel 2017) and consistent gas accretion given the solid luminosity and later limited by the gas disk supply (Bodenheimer et al. 2013). Its migration follows either Jiménez and Masset (2017) and Kanagawa et al. (2018) (Model 2) or Paardekooper et al. (2011) and Dittkrist et al. (2014) (Model 1).

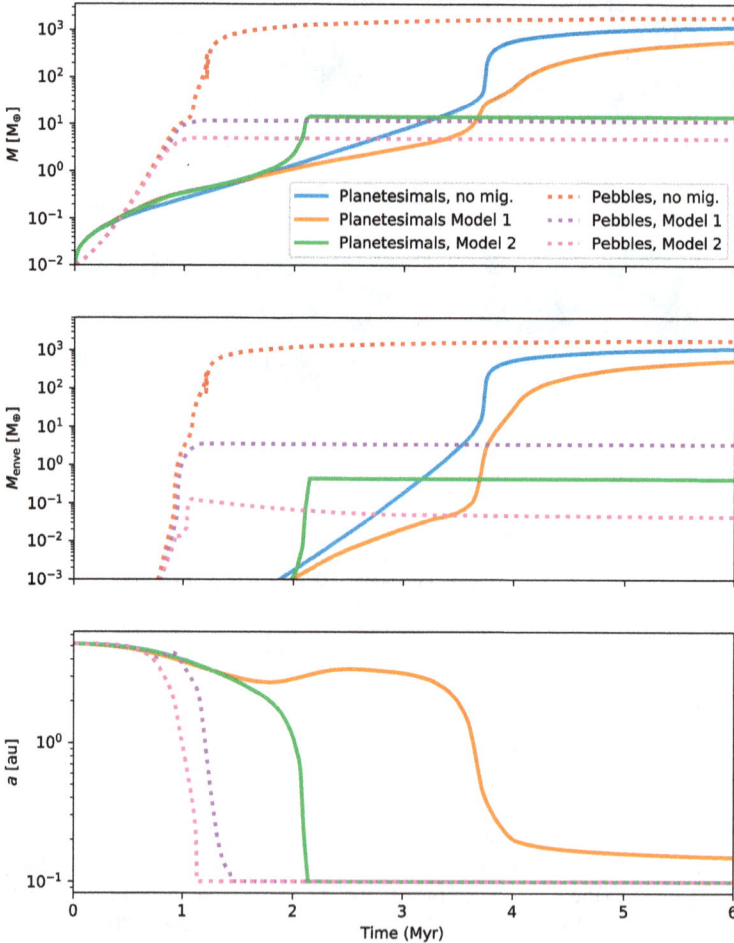

Figure 11. Time evolution of the formation of planets for different models. A protoplanetary embryo with a mass of $0.01\,M_\oplus$ is injected into the same gas disk and its migration is calculated following Model 1 or Model 2 (Orbital migration section). Accretion of solids follows the presented pebble- or planetesimal-accretion recipes.

From the results, we can first see that migration often limits growth to larger masses. To form a giant planet, the critical timescale to overcome is typically the type I migration timescale. This is particularly true if no outward migration region forms. Such an outward-migration zone keeps the planet growing with planetesimal accretion and Model 1 migration at a larger distance until it reaches several Earth masses after 3.5 Myr. It then proceeds to become a hot Jupiter aided by accretion of planetesimals as the planet migrates, a scenario which is, however, not likely when multiple embryos would be considered since larger bodies are expected to emerge on shorter timescales in the inner system (Emsenhuber et al. 2021b).

If migration is turned off, cold giant planets can form under a wide range of parameters. Although this case is chosen for illustrative purposes here, it might be a realistic scenario if the disk is structured in a configuration resulting in migration traps even before the first planets emerge (Matsumura et al. 2007; Lyra et al. 2010; Hasegawa and Pudritz 2013; Coleman and Nelson 2016).

The rest of the model runs with migration produce hot sub-Neptunes which is a typical outcome of planet formation models and might explain the large number of observed sub-Neptunes and potentially super Earths (Burn et al. 2024).

Solid accretion in the pebble scenario is proceeding faster than planetesimal accretion at the chosen initial distance of 5.2 AU. At larger distances, the differences grow even larger with pebble accretion as a likely scenario for giant planet emergence from regions exterior to 10 AU while planetesimal accretion becomes inefficient at these separations, unless the planetesimals are km-sized. These differences were also highlighted by Brügger et al. (2020) and Voelkel et al. (2020). Here, we note that the high efficiency of pebble accretion can also be detrimental to planets growing to become giants because there is a large disk mass at early times which leads to shorter migration timescales. After pebble accretion terminates at the pebble isolation mass, gas accretion proceeds at these early stages more slowly than type I migration. However, this depends on the chosen viscosity (here, α: 10^{-3}).

We note that the timing of the emergence of an embryo at the chosen location is in principle not unconstrained. Here we chose time zero, but in reality, it would take considerable time to assemble an object with a mass of $0.01 M_\oplus$. Voelkel et al. (2021, 2022) create an analytic model to realistically inject these objects.

Planetary population synthesis

To gain a more general understanding of planet formation, the example given above is far from sufficient. As hinted, the chosen disk conditions are tuned towards giant planet formation and the chosen starting location was further aiding in promoting growth.

To address the influence of these initial conditions, it is required to sample from realistic distributions and run the global planet formation model many times. This is exactly the idea of planetary population synthesis. Disk initial conditions are nowadays observed for large statistical samples. This allows for drawing initial conditions as Monte Carlo variables from the distribution of observed disks. The key distributions are disk gas mass, size, solid content, and lifetime (e.g., Benz et al. 2014; Mordasini 2018; Emsenhuber et al. 2023a, for reviews).

The population synthesis approach was recently used to generate a set of new populations yielding many quantities that can be directly compared with observations (Emsenhuber et al. 2021a; Schlecker et al. 2021a). The underlying Generation III Bern Model uses the aforementioned Model 1 for migration and planetesimal accretion as detailed in Emsenhuber et al. (2021b). An important change was to use the distribution of younger Class I objects to start the simulations (Tychoniec et al. 2018). The initial conditions were further tested against observations of more evolved disks (Emsenhuber et al. 2023b) with some further suggestions for future works.

The resulting population of planets shares several important features with the observed exoplanetary population (Emsenhuber et al. 2021a). As an illustration, Figure 12 shows the synthetic mass–distance diagram and the mass–radius diagram at an age of the population of 5 Ga (compare with the observed versions in Figs. 2 and 3). One can identify a number of features that are also seen in the observed population, such as the observed increase of the frequency of giant planets outside of about 1 AU and the large population of close-in lower mass planets. Furthermore, thanks to including *N*-body interactions, the regime of low-mass planets down to Earth mass can be realistically explored.

Collisions among protoplanets erase to a large degree the imprint of the previously discussed solid accretion timescale and limits on the final population. While the inner system consists at an early stage of low-mass planets with properties similar to those predicted by analytical theory, the system gets subsequently dynamically excited and polluted by migrated planets. These complications also limit the precision with which modern statistical methods can classify the resulting synthetic planetary population (Schlecker et al. 2021b).

Despite the challenges in this low-planet mass regime, the radius valley (Burn et al. 2024) and the peas-in-a pod trend (Weiss et al. 2018) are reproduced by the model (Mishra et al. 2021). Furthermore, the occurrence of the Solar System-like configuration with an inner super Earth and an outer giant can be discussed (Schlecker et al. 2021a). An extension toward lower stellar masses (Burn et al. 2021) was also tested against available radial velocity surveys (Schlecker et al. 2022), where important observed features can be reproduced.

However, some rare massive planets around low-mass stars (e.g., Morales et al. 2019) are not explained by this population synthesis model. Furthermore, the observed trend of increasing planet occurrence with decreasing stellar mass from transit surveys (Mulders et al. 2015) is not reproduced at the same quantitative level (Burn et al. 2021). Further challenges remain the eccentricity distribution of giant planets (potentially requiring a revision of the planet-disk interplay, Bitsch et al. 2020) and the production of hot Jupiters with the same model that can also create cold Jupiters. This is the case because, typically, a change in migration and maximum gas accretion rate tilts the outcome to favor only one of the two categories. For a more comprehensive review of recent population synthesis calculations, we refer to Mordasini (2018); Emsenhuber et al. (2023a).

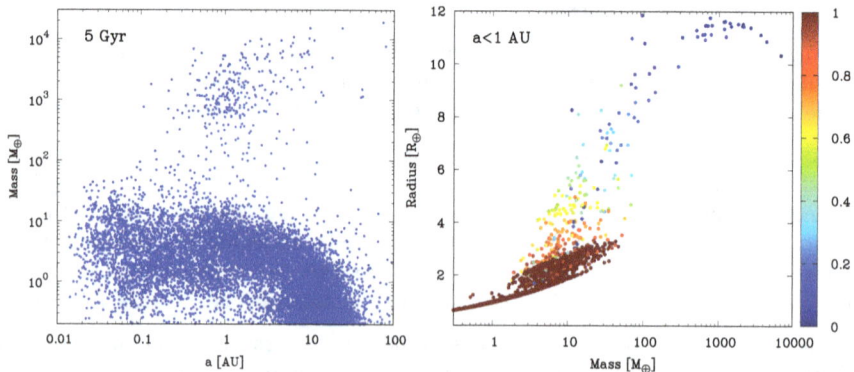

Figure 12. Mass–distance diagram (**left**) and mass-radius relation in the nominal planetary population of $1 M_\odot$ stars obtained from planetary population synthesis with the Generation III Bern Model (Emsenhuber et al. 2021b; Burn et al. 2024). The age is 5 Ga. The population consists of 1000 planetary system which each is initially seeded with 100 lunar-mass planetary embryos. On the **right**, only planets with $a < 1$ AU are included as a zero-order representation of the observational bias of the transit method. The **color code** gives the envelope metallicity Z, where zero would correspond to a pure H–He envelope.

THE COMPOSITIONAL OPPORTUNITY

Chemical and compositional clues to the origin and evolution of planets

As planet formation occurs in a protoplanetary disk, the chemical inventory thereof is inherited as a starting point for a planet's composition. When trying to model the process (see Öberg et al. 2023, for a recent review), the initial conditions are key. However, the initial disk composition is still disputed with cold and low-density regions not allowing for chemical evolution to reach an equilibrium state. Instead, the composition in the outer disk is likely inherited from the star-forming region (Eistrup et al. 2016). More recently, the static picture, which was for simplicity assumed in the seminal work of Öberg et al. (2011), was substantially revised by works that consider the time evolution of the protoplanetary disk composition (Eistrup et al. 2018; Booth and Ilee 2019; Krijt et al. 2020).

To illustrate this, we present in Figure 13 the time-evolution of the composition of a protoplanetary disk using the Bern Model of planet formation (Emsenhuber et al. 2021b) extended with solid evolution and entrainment in photoevaporative winds (Voelkel et al. 2020; Burn et al. 2022) as well as compositional tracking of the pebbles and gas as outlined in the *Gas disk* section. The evaporation and recondensation of the volatile species use the saturation pressure curves collected by Fray and Schmitt (2009). For numerical consistency, a physical diffusion flux limit is enforced, and the process of evaporation is slowed down such that the concentration of gaseous species decreases in radial direction, at most, by a factor of e over a gas scale height H_g. A complete description of this model is currently in preparation.

Figure 13 shows the time evolution of a standard disk with initial mass of $0.038\,M_\odot$ and an exponential cut-off radius of 30 AU. Several stages of evolution take place, before the composition evolves significantly, dust grows to pebble mass, then pebbles drift and transport material to the inner system. At the same stage, the disk cools down and the snowlines move closer to the star. Pebble drift becomes inefficient once all dust has grown and drifted towards the star. As the supply of fresh material subsides, the inner disk gas becomes depleted of volatiles except CO, which has a snowline exterior of where the bulk of the disk mass resides.

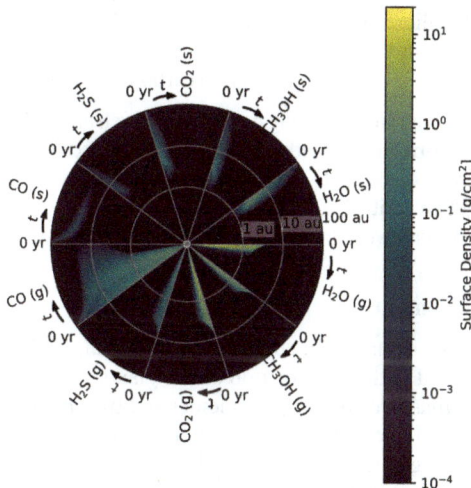

Figure 13. Evolution over 15 Ma of the surface density of solid and gaseous species in a protoplanetary disk around a Solar-mass star (Öberg et al. 2011). In each sector, the linear time evolution is shown in clockwise orientation and the surface density is **color-coded**. Distance to the star is shown in logarithmic spacing. The upper half of the plot shows the volatile species in the solid phase where an inner cut-off of the surface density corresponds to the species iceline. The lower half shows the same species in the gaseous phase.

Finally, internal photoevaporation opens a gap (visible in the CO (g) sector in Fig. 13) and an outer ring gradually disperses.

For CO, we see a ring of solids which is made of condensed CO ice. This ice exists because CO gas is moving radially outwards in this decreting part of the disk and subsequently condenses. Rings can also exist in the accreting part of the gas disk, as long as pebbles are supplied and volatile gas, which is evaporated from the pebbles, diffuses outwards (cold-finger effect, Drążkowska and Alibert 2017). Here, the effect is not as prominent as the one caused by CO condensation which persists until the end of the gas disk lifetime. Other species accrete on the star in a timeframe dominated by pebble drift which is efficient under reasonable assumptions of moderate turbulence (α: 10^{-3}) and sturdy grains (v_{frag}: 2 m/s) at all locations of the disk. A change in fragmentation velocity or turbulence with distance to the star can cause further features (see Drążkowska et al. 2023, and references therein for an overview). We note that the exact feedback and amount of condensation depends on the viscosity of the gas but also on the numerical and physical smoothing as icelines are in principle a 2D surface in radial-vertical extent instead of a sharp line.

Furthermore, it was shown by Booth and Ilee (2019) that chemical evolution plays a role over Ma timescales in the inner disk, which was not included in the simulations here. A simple reaction network was used in that work to keep the computation time short. More extensive chemical modelling is required to resolve the lower-density regions and for comparison to observations (Woitke et al. 2009; Semenov and Wiebe 2011; Walsh et al. 2015). On the other hand, by excluding radial transport of ices, Cridland et al. (2016, 2017) could use a more extended chemical network when modelling planet formation and obtaining planetary compositions. Another effect which was so far not coupled to other dynamical processes affecting the disk composition is that of radioactive heating leading to outgassing of water from 100 km-sized planetesimals (Lichtenberg et al. 2019). These shortcomings show that there is still need of theoretical work to fully do justice to all relevant effects within one model.

With the advent of the James Webb Space Telescope (JWST), we have the opportunity to test models of protoplanetary disk composition using infrared spectroscopy. The technique can probe molecular lines in the warm ($T \gtrsim 300\,\mathrm{K}$) gas as well as the continuum emission which is typically dominated by the silicate grain features. First results (see van Dishoeck et al. 2023, for an overview) highlight a surprising diversity with some water-rich (Banzatti et al. 2023; Perotti et al. 2023), over CO_2-rich (Grant et al. 2023), as well as a carbon-rich (Tabone et al. 2023) inner disk.

Several larger JWST programs are being conducted which aim to resolve the compositional variability and potential correlations with stellar type, large-scale disk structure, or age. In particular, the synergy with constraints from ALMA that can probe species in the outer disk (Öberg et al. 2021) might be fruitful. The ALMA data on its own are more challenging to link to planetary compositions as fewer planets form in the outer disk (at least in the conventional core accretion scenario).

Disk carbon to oxygen content

Despite the challenging nature of the subject (Mollière et al. 2022), the protoplanetary disk composition has been linked to the composition of gas giants (e.g., Madhusudhan et al. 2014; Thiabaud et al. 2014; Mordasini et al. 2016; Schneider and Bitsch 2021; Bitsch et al. 2022). To illustrate the link, it is useful to consider here the pioneering work of Öberg et al. (2011) who discussed the carbon to oxygen ratio in the protoplanetary disk gas and solids. The C/O ratio changes with distance to the star at the respective snowlines. This is shown in Figure 14 for the same disk for which the evolution was discussed above. For example, exterior to the CO_2 snowline, the gas phase contains only CH_4 and CO as carbon or oxygen-bearing species. This means moving interior to the CO_2 snowline, a jump towards a C/O of 0.5 (from CO_2) occurs. Even more oxygen is present interior to the water snowline, further reducing C/O.

The basic idea is that once the C/O in the atmosphere of a giant planet can be measured, its formation location can be identified. The formation location is a key information for planet formation. Originally, it was thought that the disk gas phase is the determining factor (solid line in Fig. 14). However, Mordasini et al. (2016) showed that for typical enrichment levels of Hot Jupiters, rather the accreted solids (the dashed line in Fig. 14) determine the atmospheric C/O if atmosphere and interior mix. The reason is that nebular gas usually contains only few C and O molecules compared to the solids in which these elements are major components (for example oxygen in silicate rocks or C in carbon-bearing ices). As discussed in this section, the picture has become even more complex with effects of the disk's temporal evolution, pebble drift and evaporation, inhomogeneous planetary interiors, and other factors.

Figure 14 also shows the time evolution of the modelled protoplanetary disk. In this case, we see a change of the inner protoplanetary gas disk C/O to a quasi-static value of approximately 0.3. This value is sensitive to a number of assumptions on the molecular inventory of the disk, such as the water to CO and CO_2 ratio (Marboeuf et al. 2014). More recently, it was also speculated whether hydrocarbons could be present and evaporate at the soot line which would be located in the hot inner disk (approx. 500 K, Li et al. 2021). This is one of the hypotheses to obtain a carbon-rich gas composition, similar to what was now measured in a disk (Tabone et al. 2023). Alternatively, Mah et al. (2023) proposed that the carbon-rich disk could be in a later evolutionary stage in the disk. The process proposed in that work is the same as what develops in Figure 14: At later stages, all the material with high condensation temperatures, such as water, has accreted onto the star due to, first, radial drift of solids followed by viscous accretion of gas. Instead, what remains in the inner disk is carbon-rich material from the outer disk (2 Ma line in Fig. 14). It is the species which exist primarily in the gas phase (CO and CH_4) which are not as quickly depleted as the volatiles in the rapidly drifting solid phase. Qualitatively this scenario is a common outcome for disks with fast pebble drift (i.e., for fragmentation velocities larger than 1 m/s, Mah et al. 2023). The prediction of this hypothesis is that the disk is strongly depleted in overall solids. These novel results motivate more detailed studies of the microphysics at the snowlines and their efficiency to trap and cycle volatiles from the gas phase back to the solid phase.

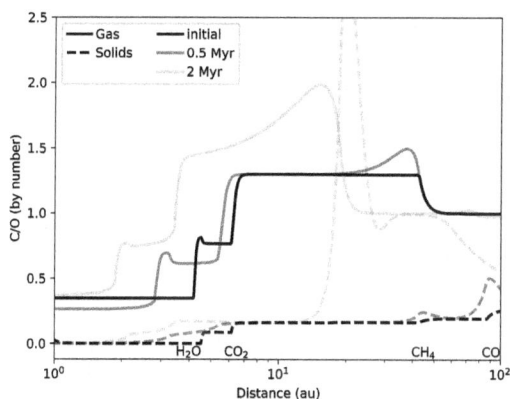

Figure 14. Carbon-to-oxygen number ratio of a protoplanetary disk around a Solar mass star as a function of distance. In contrast to Öberg et al. (2011), we used the initialization of Marboeuf et al. (2014) and did not add refractory carbon to the solids (C/O of refractories is zero); thus, the solid C/O drops to zero interior to the methanol and water lines in the inner disk. The same disk evolution with the species shown in Figure 13 is used to calculate C/O ratios. The CH_4 and CO snowlines significantly alter the local composition at later times. In the inner disk, the short viscous timescale (for α: 10^{-3}) leads to a quick equilibration instead of a local accumulation of material. We note that there is a minor impact by the methanol line which is not labelled.

Bulk and atmospheric composition

While the disk composition is key to determine the bulk composition of planets, since planets are built from this material, the partitioning of elements to different reservoirs in the planet as well as the chemical form in which they present themselves to observers is not trivial. Figure 15 shows various pathways to different atmospheric compositions for hot Jupiters in terms of their C/H, O/H, and C/O, which are most easily observed.

The plot is based on Mordasini et al. (2016) but updated to include several new aspects like pebble evaporation (Booth et al. 2017). In addition to the disk-related aspects (formation location with respect to icelines), the efficiency of solid accretion and the mixing of species within the atmosphere are key aspects. In particular, if mixing is not efficient (which seems to be the case, at least for our Jupiter, see the *Selected constraints from the Solar System* section), the observable composition is sensitive to the latest formation stage where the planet has already obtained its bulk mass content but might not stop solid accretion completely (Shibata et al. 2023).

Going from left to right and from top to bottom in Figure 15, we first consider that solid accretion is efficient. The Hot Jupiter will then have a comparatively high mean density and solids will determine the bulk enrichment of the atmosphere, which will be at a superstellar level. If interior and atmosphere are well mixed, the composition of the accreted solids will also dominate the atmospheric enrichment. Accreted solids will be the prime source of heavy elements in the planetary envelope and atmosphere, dominating over the heavy elements accreted together with the H/He gas. If the planet formed outside of the water iceline or the CO_2 line, the accreted solids would contain oxygen (and carbon), but the composition of icy solids containing water ice is overall oxygen dominated. This can be seen with the dashed line in Figure 14 with C/O values of about 0.1 to 0.3. This finally results in an atmosphere with a superstellar O/H, C/H and a low substellar C/O.

Next, we consider still the case that the bulk and atmospheric enrichment is dominated by the accreted solids, but for an accretion of the solids inside of the water iceline. Here it depends on the composition of the refractory solids that are accreted. If these solids are carbon

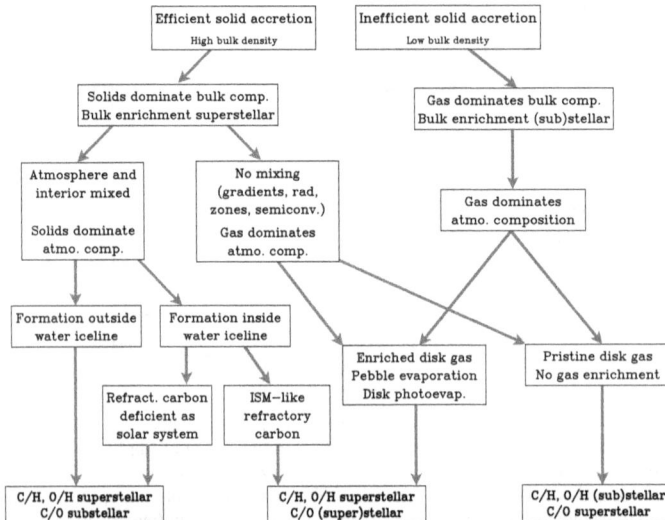

Figure 15. Formation pathways and physical processes leading to different atmospheric composition in terms of C/H, O/H, and C/O for Hot Jupiters. The bulk enrichment is also noted as another observational constraint accessible via the planet's mean density.

depleted relative to the ISM as it is the case in the inner Solar System (Bergin et al. 2015) and as might be generally the case as suggested by most polluted white dwarfs (Xu and Bonsor 2021), then again a superstellar O/H, C/H and substellar C/O will result, because of the oxygen in the accreted silicates—at least if the silicates do not condense out again in the deeper layers (Kitzmann et al. 2024). On the other hand, if the solids inside of the iceline contain large quantities of refractory carbon similar to the ISM, a planet with a superstellar C/H, O/H and a stellar or superstellar C/O will result.

The right part of Figure 15 discusses the outcome if solid accretion was comparatively inefficient and thus the atmospheric C and O is dominated by the one of the accreted gas. This applies also to the case that atmosphere and interior are poorly mixed because of compositional gradients and inefficient semi-convection (Vazan et al. 2013), even if solid accretion was efficient overall. This case further branches into sub-scenarios depending on whether the gas has a pristine background composition or whether effects like disk photoevaporation (Guillot and Hueso 2006) and pebble drift and evaporation (Booth et al. 2017) enrich the nebular gas at the place of the planet. In the former case, a planetary atmosphere with a (sub)stellar O/H and C/H and a superstellar C/O would result, while the latter would lead to enrichment levels of O/H, C/H and a C/O ratio which are (super)stellar. Despite being strongly simplified, these different pathways already give a hint of the complexity of linking formation and planetary atmospheres. To address this, other elements like N, S or refractories can be used to add further complementary constraints (Crossfield 2023; Ohno and Fortney 2023; Polman et al. 2023).

Additionally, for lower planetary mass, i.e., for sub-Neptunes and even more so for super-Earths, the solid core of a planet can be a substantial reservoir for volatile elements. The core can then control the atmospheric composition over long timescales. But whenever a primordial atmosphere of significant mass is present, both reservoirs need to be included and modelled consistently (Bower et al. 2019, 2022; Lichtenberg et al. 2023). The self-consistent coupling between outer atmosphere and the deeper envelope (which can, for example, be water-dominated) is also important (Guzmán-Mesa et al. 2022) and allows to combine spectroscopic constraints on the atmospheric composition with observational constraints on the planet's mean density and thus bulk composition. Sub-Neptunes fall in the regime where large quantities of volatile elements different from hydrogen and helium might make up most of the envelope of the planet (Burn et al. 2024) which could even mix with rocky material for hot planets in a supercritical phase (Vazan et al. 2022). More interdisciplinary research connecting geophysics with planet formation and astronomical constraints are required to determine the exact partitioning of elements and the interior evolution (e.g., Kite et al. 2019; Dasgupta et al. 2024; Guimond et al. 2024, both this volume).

SUMMARY AND CONCLUSIONS

In this chapter, we have reviewed the theory of planet formation. In the first part, key observational constraints from the Solar System and from the exoplanet population were discussed. Then, some of the physical mechanisms governing planetary system formation were introduced. We then addressed how these mechanisms can be put together to form global planet formation models that can be used in planetary population synthesis. Population synthesis makes quantitative statistical comparisons possible between formation theory and exoplanet observations. Finally, we showed how the compositional links from protoplanetary disks to planetary atmospheres put novel constraints on planet formation theory.

Triggered mainly by observational progress, the understanding how planets form has evolved from a static picture inspired solely by the Solar System to a highly dynamic one, which is reflected in the diversity of the population of extrasolar planets. This evolution is accompanied by three major ongoing shifts of paradigms: first, from in situ formation to mobility of the building blocks both on the dust/pebble scale (drift) and on the scale of protoplanets (orbital migration). Second, from solid accretion only via planetesimals to a multifaceted process were pebbles, planetesimals, and giant impacts play a role. Third, from a formation in smooth viscous disks to structured and potentially MHD wind-driven disks. At the moment, it is not yet possible to say to what extent these new paradigms will replace the older ones or if in the end a synthesis of older and newer views will result.

In order to make progress, it will be important to address a number of key questions and frontiers:

1. Are the structures in protoplanetary disks cause or consequence of planet formation? What is the formation mechanism of the protoplanets that seem to emerge at large orbital distances already at early times as suggested by observed disk structures?

2. Is the evolution of protoplanetary disks driven by viscous accretion or by MHD-winds and what are the consequences for planet formation, like potentially reduced orbital migration?

3. What is the importance of pebbles, planetesimals, and giant impacts as mechanisms of solid growth? Is there a distance dependency? How does this influence the planetary composition and structure?

4. How large are planetary gas accretion rates in the runaway and disk-limited phase and which mechanisms control it? Is there a planetary desert in the mass function?

5. How did the Solar System form, capitalizing on the special constraints we only have in our system like the meteoritic record and the minor bodies? Was there a grand tack? How much material passed Jupiter? How special is its architecture?

6. How can the complexity of the interiors of the giant planets in our Solar System (especially Jupiter) be translated to exoplanets for which typically simple fully adiabatic interiors are assumed?

7. What is the origin of the very abundant population of close-in low-mass/small exoplanets? Are some of these planets ocean worlds that have migrated from beyond the iceline to their current position? Are there hints from architectural patterns?

8. What is the role of gravitational instability as a giant planet formation mechanism and/or potentially as a seed for core accretion?

9. What is the impact of the cluster/stellar environment on planet formation via late infall, external photoevaporation of disks, and stellar encounters?

10. What can be learned from observations of ongoing planet formation? Is it possible to derive much more direct constraints on processes like formation timescales, formation location, gas accretion, migration, and circumplanetary disks? Can we identify an evolutionary sequence like for stars?

11. Do planets keep an atmospheric memory about their origin? Can it be accessed remotely via spectroscopic observations? How are bulk and observable atmospheric composition linked?

12. What is the diversity of low-mass planets in the habitable zone in terms of water content, bulk and atmospheric composition, or remaining primordial H/He envelope? How does this depend on host star properties like stellar mass and on the architecture of the host planetary system? Which fraction is really Earth-like and habitable?

Fortunately, chances are high that significant advancements will be possible in the coming years for this dozen points because the number of observational constraints will continue to grow rapidly. The JWST is a first already active example. Despite its recent launch, it already leads to important insights and triggers the development of more precise models for the composition of the inner disk, including effects like a soot line or late carbon-rich disk gas. This showcases the potential of linking the outer reservoir of the disk to the inner disk composition and finally the planetary atmospheres, at least if they can be related to the bulk of the material accreted during formation.

In the Solar System, space missions like JUICE, BepiColombo, or the Uranus mission but also ultra-precise laboratory measurements will help to better understand our own planetary system, while ALMA, ANDES, or RISTRETTO will probe the formation environment, the atmospheric composition, and ongoing planet formation from the ground. In space, GAIA, PLATO, and the Roman Space Telescope will critically extend the parameter space of the known exoplanet demographics in the high- and low-mass regime to areas currently unknown. ARIEL will add, via atmospheric spectroscopy, statistical constraints on planetary composition for a larger sample of planets, which is important given the complexity of linking formation to atmospheres. Finally, over longer timescales, the LIFE mission and the Habitable Worlds Observatory should be able to observationally address the last question listed above about the diversity of potentially habitable planets and maybe even life.

Taken together, these observational efforts will yield very rich data to build a set of many multifaceted constraints for planetary formation theory. Theoretical results can be confronted to them, improving eventually our understanding of the origin of planets.

ACKNOWLEDGEMENTS

We thank Willy Benz, Alexandre Emsenhuber, Benjamin Fulton, Erik Petigura, Roy van Boekel, and Jesse Weder for important input. C.M. acknowledges the support from the Swiss National Science Foundation under grant 200021_204847 "PlanetsInTime". R.B. acknowledges financial support from the German Excellence Strategy via the Heidelberg Cluster of Excellence (EXC 2181 – 390900948) "STRUCTURES" under Exploratory Project 8.4. Parts of this work has been carried out within the framework of the NCCR PlanetS supported by the Swiss National Science Foundation under grants 51NF40_182901 and 51NF40_205606.

REFERENCES

Abod CP, Simon JB, Li R, Armitage PJ, Youdin AN, Kretke KA (2019) The mass and size distribution of planetesimals formed by the streaming instability. II. The effect of the radial gas pressure gradient. Astrophys J 883:192, https://doi.org/10.3847/1538-4357/ab40a3

Adachi I, Hayashi C, Nakazawa K (1976) The gas drag effect on the elliptical motion of a solid body in the primordial solar nebula. Prog Theor Phys 56:1756–1771, https://doi.org/10.1143/PTP.56.1756

Adams FC, Batygin K (2022) Analytic approach to the late stages of giant planet formation. Astrophys J 934:111, https://doi.org/10.3847/1538-4357/ac7a3e

Adams FC, Batygin K, Bloch AM, Laughlin G (2020) Energy optimization in extrasolar planetary systems: the transition from peas-in-a-pod to runaway growth. Mon Not R Astron Soc 493:5520–5531, https://doi.org/10.1093/mnras/staa624

Affolter L, Mordasini C, Oza AV, Kubyshkina D, Fossati L (2023) Planetary evolution with atmospheric photoevaporation. II. Fitting the slope of the radius valley by combining boil-off and XUV-driven escape. Astron Astrophys 676:A119, https://doi.org/10.1051/0004-6361/202142205

Alcalá JM, Manara CF, Natta A, et al. (2017) X-shooter spectroscopy of young stellar objects in Lupus. Accretion properties of Class II and transitional objects. Astron Astrophys 600:A20, https://doi.org/10.1051/0004-6361/201629929

Alessi M, Pudritz RE (2022) Combined effects of disc winds and turbulence-driven accretion on planet populations. Mon Not R Astron Soc 515:2548–2577, https://doi.org/10.1093/mnras/stac1782

Alexander RD, Armitage PJ (2009) Giant planet migration, disk evolution, and the origin of transitional disks. Astrophys J 704:989–1001, https://doi.org/10.1088/0004-637X/704/2/989

Ali-Dib M, Cumming A, Lin DNC (2020) The imprint of the protoplanetary disc in the accretion of super-Earth envelopes. Mon Not R Astron Soc 494:2440–2448, https://doi.org/10.1093/mnras/staa914

Alibert Y (2019) New metric to quantify the similarity between planetary systems: application to dimensionality reduction using T-SNE. Astron Astrophys 624:A45, https://doi.org/10.1051/0004-6361/201834592

Alibert Y, Venturini J (2019) Using deep neural networks to compute the mass of forming planets. Astron Astrophys 626:A21, https://doi.org/10.1051/0004-6361/201834592

Alibert Y, Mordasini C, Benz W, Winisdoerffer C (2005) Models of giant planet formation with migration and disc evolution. Astron Astrophys 434:343–353, https://doi.org/10.1051/0004-6361:20042032

Alibert Y, Carron F, Fortier A, Pfyffer S, Benz W, Mordasini C, Swoboda D (2013) Theoretical models of planetary system formation: mass vs. semi-major axis. Astron Astrophys 558:A109, https://doi.org/10.1051/0004-6361/201321690

Alibert Y, Venturini J, Helled R, et al. (2018) The formation of Jupiter by hybrid pebble–planetesimal accretion. Nat Astron 2:873–877, https://doi.org/10.1038/s41550-018-0557-2

Altwegg K, Balsiger H, Bar-Nun A, et al. (2015) 67P/Churyumov–Gerasimenko, a Jupiter family comet with a high D/H ratio. Science 347:1261952–1261952, https://doi.org/10.1126/science.1261952

Anand A, Mezger K (2023) Early Solar System chronology from short-lived chronometers. Chem Erde / Geochem 83:126004, https://doi.org/10.1016/j.chemer.2023.126004

Andrews SM (2020) Observations of protoplanetary disk structures. Annu Rev Astron Astrophys 58:483–528, https://doi.org/10.1146/annurev-astro-031220-010302

Andrews SM, Wilner DJ, Hughes AM, Qi C, Dullemond CP (2009) Protoplanetary disk structures in Ophiuchus. Astrophys J 700:1502–1523, https://doi.org/10.1088/0004-637X/700/2/1502

Andrews SM, Terrell M, Tripathi A, Ansdell M, Williams JP, Wilner DJ (2018) Scaling relations associated with millimeter continuum sizes in protoplanetary disks. Astrophys J 865:157, https://doi.org/10.3847/1538-4357/aadd9f

Aoyama Y, Bai X-N (2023) Three-dimensional global simulations of Type-II planet–disk interaction with a magnetized disk wind. I. Magnetic flux concentration and gap properties. Astrophys J 946:5, https://doi.org/10.3847/1538-4357/acb81f

Arakawa S, Okuzumi S, Tatsuuma M, Tanaka H, Kokubo E, Nishiura D, Furuichi M, Nakamoto T (2023) Size dependence of the bouncing barrier in protoplanetary dust growth. Astrophys J Lett 951:L16, https://doi.org/10.3847/2041-8213/acdb5f

Armitage PJ, Kley W (2019) From Protoplanetary Disks to Planet Formation. Springer, Berlin

Ataiee S, Baruteau C, Alibert Y, Benz W (2018) How much does turbulence change the pebble isolation mass for planet formation? Astron Astrophys 615:A110, https://doi.org/10.1051/0004-6361/201732026

Ayliffe BA, Bate MR (2012) The growth and hydrodynamic collapse of a protoplanet envelope. Mon Not R Astron Soc 427:2597–2612, https://doi.org/10.1111/j.1365-2966.2012.21979.x

Bae J, Isella A, Zhu Z, Martin R, Okuzumi S, Suriano S (2023) Structured distributions of gas and solids in protoplanetary disks. Protostars and Planets VII, ASP Conf Ser 534:423, https://doi.org/10.48550/arXiv.2210.13314

Bai X-N (2016) Toward a global evolutionary model of protoplanetary disks. Astrophys J 821:80, https://doi.org/10.3847/0004-637X/821/2/80

Bai X-N, Stone JM (2010) Dynamics of solids in the midplane of protoplanetary disks: Implications for planetesimal formation. Astrophys J 722:1437–1459, https://doi.org/10.1088/0004-637X/722/2/1437

Bai X-N, Stone JM (2014) Magnetic flux concentration and zonal flows in magnetorotational instability turbulence. Astrophys J 796:31, https://doi.org/10.1088/0004-637X/796/1/31

Bai X-N, Ye J, Goodman J, Yuan F (2016) Magneto-thermal disk winds from protoplanetary disks. Astrophys J 818:152, https://doi.org/10.3847/0004-637x/818/2/152

Bailey A, Zhu Z (2023) Growing planet envelopes in spite of recycling flows. arXiv:2310.03117, https://doi.org/10.48550/arXiv.2310.03117

Banzatti A, Pontoppidan KM, Carr JS, et al. (2023) JWST reveals excess cool water near the snow line in compact disks, consistent with pebble drift. Astrophys J Lett 957:L22, https://doi.org/10.3847/2041-8213/acf5ec

Baruteau C, Bai X, Mordasini C, Mollière P (2016) Formation, Orbital and Internal Evolutions of Young Planetary Systems. Space Sci Rev 205:77–124, https://doi.org/10.1007/s11214-016-0258-z

Bashi D, Zucker S (2021) Quantifying the similarity of planetary system architectures. Astron Astrophys 651:A61, https://doi.org/10.1051/0004-6361/202140699

Batygin K, Morbidelli A (2023) Formation of rocky super-Earths from a narrow ring of planetesimals. Nat Astron 7:330–338, https://doi.org/10.1038/s41550-022-01850-5

Bekaert DV, Broadley MW, Marty B (2020) The origin and fate of volatile elements on Earth revisited in light of noble gas data obtained from comet 67P/Churyumov–Gerasimenko. Sci Rep 10:5796, https://doi.org/10.1038/s41598-020-62650-3

Bell KR, Lin DNC (1994) Using FU Orionis outbursts to constrain self-regulated protostellar disk models. Astrophys J 427:987, https://doi.org/10.1086/174206

Benisty M, Bae J, Facchini S, et al. (2021) A circumplanetary disk around PDS70c. Astrophys J Lett 916:L2, https://doi.org/10.3847/2041-8213/ac0f83

Benítez-Llambay P, Pessah ME (2018) Torques induced by scattered pebble-flow in protoplanetary disks. Astrophys J Lett 855:L28, https://doi.org/10.3847/2041-8213/aab2ae

Benítez-Llambay P, Masset F, Koenigsberger G, Szulágyi J (2015) Planet heating prevents inward migration of planetary cores. Nature 520:63–65, https://doi.org/10.1038/nature14277

Bennett DP, Ranc C, Fernandes RB (2021) No sub-Saturn-mass planet desert in the CORALIE/HARPS radial-velocity sample. Astron J 162:243, https://doi.org/10.3847/1538-3881/ac2a2b

Benz W, Asphaug E (1999) Catastrophic disruptions revisited. Icarus 142:5–20, https://doi.org/10.1006/icar.1999.6204

Benz W, Cameron AGW, Melosh HJ (1989) The origin of the Moon and the single-impact hypothesis III. Icarus 81:113–131, https://doi.org/10.1016/0019-1035(89)90129-2

Benz W, Ida S, Alibert Y, Lin D, Mordasini C (2014) Planet population synthesis. Protostars and Planets VI:691–713, https://doi.org/10.2458/azu_uapress_9780816531240-ch030

Berger A, Loutre MF (1991) Insolation values for the climate of the last 10 million years. Quat Sci Rev 10:297–317, https://doi.org/10.1016/0277-3791(91)90033-Q

Bergin EA, Blake GA, Ciesla F, Hirschmann MM, Li J (2015) Tracing the ingredients for a habitable earth from interstellar space through planet formation. Proc Natl Acad Sci 112:8965–8970, https://doi.org/10.1073/pnas.1500954112

Bergsten GJ, Pascucci I, Hardegree-Ullman KK, Fernandes RB, Christiansen JL, Mulders GD (2023) No evidence for more Earth-sized planets in the habitable zone of Kepler's M versus FGK stars. Astron J 166:234, https://doi.org/10.3847/1538-3881/ad03ea

Bertaux J-L, Ivanova A (2022) A numerical inversion of $m \sin i$ exoplanet distribution: the sub-Saturn desert is more depleted than observed and hint of a Uranus mass gap. Mon Not R Astron Soc 512:5552–5571, https://doi.org/10.1093/mnras/stac777

Béthune W, Lesur G, Ferreira J (2016) Self-organisation in protoplanetary discs. Global, non-stratified Hall–MHD simulations. Astron Astrophys 589:A87, https://doi.org/10.1051/0004-6361/201527874

Birnstiel T, Fang M, Johansen A (2016) Dust evolution and the formation of planetesimals. Space Sci Rev 205:41–75, https://doi.org/10.1007/s11214-016-0256-1

Bitsch B, Lambrechts M, Johansen A (2015) The growth of planets by pebble accretion in evolving protoplanetary discs. Astron Astrophys 582:A112, https://doi.org/10.1051/0004-6361/201526463

Bitsch B, Trifonov T, Izidoro A (2020) The eccentricity distribution of giant planets and their relation to super-Earths in the pebble accretion scenario. Astron Astrophys 643:A66, https://doi.org/10.1051/0004-6361/202038856

Bitsch B, Schneider AD, Kreidberg L (2022) How drifting and evaporating pebbles shape giant planets. III. The Formation of WASP-77A b and τ Boötis b. Astron Astrophys 665:A138, https://doi.org/10.1051/0004-6361/202243345

Blum J, Wurm G (2008) The growth mechanisms of macroscopic bodies in protoplanetary disks. Annu Rev Astron Astrophys 46:21–56, https://doi.org/10.1146/annurev.astro.46.060407.145152

Bodenheimer P, Pollack JB (1986) Calculations of the accretion and evolution of giant planets: The effects of solid cores. Icarus 67:391–408, https://doi.org/10.1016/0019-1035(86)90122-3

Bodenheimer P, Hubickyj O, Lissauer JJ (2000) Models of the in situ formation of detected extrasolar giant planets. Icarus 143:2–14, https://doi.org/10.1006/icar.1999.6246

Bodenheimer P, D'Angelo G, Lissauer JJ, Fortney JJ, Saumon D (2013) Deuterium burning in massive giant planets and low-mass brown dwarfs formed by core-nucleated accretion. Astrophys J 770:120, https://doi.org/10.1088/0004-637X/770/2/120

Bonfils X, Delfosse X, Udry S, et al. (2013) The HARPS search for southern extra-solar planets. XXXI. The M-dwarf sample. Astron Astrophys 549:A109, https://doi.org/10.1051/0004-6361/201014704

Booth RA, Clarke CJ (2021) Modelling the delivery of dust from discs to ionized winds. Mon Not R Astron Soc 502:1569–1578, https://doi.org/10.1093/mnras/stab090

Booth RA, Ilee JD (2019) Planet-forming material in a protoplanetary disc: the interplay between chemical evolution and pebble drift. Mon Not R Astron Soc 487:3998–4011, https://doi.org/10.1093/mnras/stz1488

Booth RA, Clarke CJ, Madhusudhan N, Ilee JD (2017) Chemical enrichment of giant planets and discs due to pebble drift. Mon Not R Astron Soc 469:3994–4011, https://doi.org/10.1093/mnras/stx1103

Borucki WJ, Koch D, Basri G, et al. (2010) Kepler planet-detection mission: Introduction and first results. Science 327:977–980, https://doi.org/10.1126/science.1185402

Boss AP (1997) Giant planet formation by gravitational instability. Science 276:1836–1839, https://doi.org/10.1126/science.276.5320.1836

Boss AP (2011) Formation of giant planets by disk instability on wide orbits around protostars with varied masses. Astrophys J 731:74, https://doi.org/10.1088/0004-637X/731/1/74

Bottke WF, Norman MD (2017) The late heavy bombardment. Annu Rev Earth Planet Sci 45:619–647, https://doi.org/10.1146/annurev-earth-063016-020131

Bouma LG, Kerr R, Curtis JL, et al. (2022) Kepler and the behemoth: Three mini-Neptunes in a 40 million year old association. Astron J 164:215, https://doi.org/10.3847/1538-3881/ac93ff

Bower DJ, Kitzmann D, Wolf AS, Sanan P, Dorn C, Oza AV (2019) Linking the evolution of terrestrial interiors and an early outgassed atmosphere to astrophysical observations. Astron Astrophys 631:A103, https://doi.org/10.1051/0004-6361/201935710

Bower DJ, Hakim K, Sossi PA, Sanan P (2022) Retention of water in terrestrial magma oceans and carbon-rich early atmospheres. Planet Sci J 3:93, https://doi.org/10.3847/PSJ/ac5fb1

Bowler BP (2016) Imaging extrasolar giant planets. Publ Astron Soc Pac 128:102001–102001, https://doi.org/10.1088/1538-3873/128/968/102001

Brož M, Chrenko O, Nesvorný D, Dauphas N (2021) Early terrestrial planet formation by torque-driven convergent migration of planetary embryos. Nat Astron 5:898–902, https://doi.org/10.1038/s41550-021-01383-3

Brügger N, Burn R, Coleman GAL, Alibert Y, Benz W (2020) Pebbles versus planetesimals. The outcomes of population synthesis models. Astron Astrophys 640:A21, https://doi.org/10.1051/0004-6361/202038042

Bryan ML, Knutson HA, Howard AW, et al. (2016) Statistics of long period gas giant planets in known planetary systems. Astrophys J 821:89, https://doi.org/10.3847/0004-637X/821/2/89

Budde G, Burkhardt C, Kleine T (2019) Molybdenum isotopic evidence for the late accretion of outer Solar System material to Earth. Nat Astron 3:736–741, https://doi.org/10.1038/s41550-019-0779-y

Burn R, Marboeuf U, Alibert Y, Benz W (2019) Radial drift and concurrent ablation of boulder-sized objects. Astron Astrophys 629:A64, https://doi.org/10.1051/0004-6361/201935780

Burn R, Schlecker M, Mordasini C, Emsenhuber A, Alibert Y, Henning T, Klahr H, Benz W (2021) The New Generation Planetary Population Synthesis (NGPPS). IV. Planetary systems around low-mass stars. Astron Astrophys 656:A72, https://doi.org/10.1051/0004-6361/202140390

Burn R, Emsenhuber A, Weder J, Völkel O, Klahr H, Birnstiel T, Ercolano B, Mordasini C (2022) Toward a population synthesis of disks and planets. I. Evolution of dust with entrainment in winds and radiation pressure. Astron Astrophys 666:A73, https://doi.org/10.1051/0004-6361/202243262

Burn R, Mordasini C, Mishra L, Haldemann J, Venturini J, Emsenhuber A, Henning T (2024) A radius valley between migrated steam worlds and evaporated rocky cores. Nat Astron, Advanced Online Publication, https://doi.org/10.1038/s41550-023-02183-7

Cameron AGW (1988) Origin of the solar system. Annu Rev Astron Astrophys 26:441–472, https://doi.org/10.1146/annurev.aa.26.090188.002301

Canup RM, Asphaug E (2001) Origin of the Moon in a giant impact near the end of the Earth's formation. Nature 412:708–712, https://doi.org/10.1038/35089010

Capistrant BK, Soares-Furtado M, Vanderburg A, et al. (2024) TESS Hunt for Young and Maturing Exoplanets (THYME). XI. An Earth-sized planet orbiting a nearby, Solar-like host in the 400 Myr Ursa Major moving group. Astron J 167:54, https://doi.org/10.3847/1538-3881/ad1039

Carrera D, Johansen A, Davies MB (2015) How to form planetesimals from mm-sized chondrules and chondrule aggregates. Astron Astrophys 579:A43, https://doi.org/10.1051/0004-6361/201425120

Cassan A, Kubas D, Beaulieu JP, et al. (2012) One or more bound planets per Milky Way star from microlensing observations. Nature 481:167–169, https://doi.org/10.1038/nature10684

Chabrier G, Debras F (2021) A new equation of state for dense hydrogen–helium mixtures. II. Taking into account hydrogen–helium interactions. Astrophys J 917:4, https://doi.org/10.3847/1538-4357/abfc48

Chambers J (2018) Planet formation: An optimized population-synthesis approach. Astrophys J 865:30, https://doi.org/10.3847/1538-4357/aada09

Chambers JE (2006) A Semi-analytic model for oligarchic growth. Icarus 180:496–513, https://doi.org/10.1016/j.icarus.2005.10.017

Chiang EI, Goldreich P (1997) Spectral energy distributions of T Tauri stars with passive circumstellar disks. Astrophys J 490:368–376, https://doi.org/10.1086/304869

Choksi N, Chiang E, Fung J, Zhu Z (2023) The maximum accretion rate of a protoplanet: how fast can runaway be? Mon Not R Astron Soc 525:2806–2819, https://doi.org/10.1093/mnras/stad2269

Ciesla FJ (2011) Residence times of particles in diffusive protoplanetary disk environments. II. Radial motions and applications to dust annealing. Astrophys J 740:9, https://doi.org/10.1088/0004-637X/740/1/9

Cimerman NP, Kuiper R, Ormel CW (2017) Hydrodynamics of embedded planets' first atmospheres—III. The role of radiation transport for super-Earth planets. Mon Not R Astron Soc 471:4662–4676, https://doi.org/10.1093/mnras/stx1924

Clarke CJ, Gendrin A, Sotomayor M (2001) The dispersal of circumstellar discs: the role of the ultraviolet switch. Mon Not R Astron Soc 328:485–491, https://doi.org/10.1046/j.1365-8711.2001.04891.x

Coleman GAL (2021) From dust to planets—I. Planetesimal and embryo formation. Mon Not R Astron Soc 506:3596–3614, https://doi.org/10.1093/mnras/stab1904

Coleman GAL, Nelson RP (2014) On the formation of planetary systems via oligarchic growth in thermally evolving viscous discs. Mon Not R Astron Soc 445:479–499, https://doi.org/10.1093/mnras/stu1715

Coleman GAL, Nelson RP (2016) Giant planet formation in radially structured protoplanetary discs. Mon Not R Astron Soc 460:2779–2795, https://doi.org/10.1093/mnras/stw1177

Coleman GAL, Mroueh JK, Haworth TJ (2024) Photoevaporation obfuscates the distinction between wind and viscous angular momentum transport in protoplanetary discs. Mon Not R Astron Soc 527:7588–7602, https://doi.org/10.1093/mnras/stad3692

Crida A, Morbidelli A, Masset F (2006) On the width and shape of gaps in protoplanetary disks. Icarus 181:587–604, https://doi.org/10.1016/j.icarus.2005.10.007

Cridland AJ, Pudritz RE, Alessi M (2016) Composition of early planetary atmospheres—I. Connecting disc astrochemistry to the formation of planetary atmospheres. Mon Not R Astron Soc 461:3274–3295, https://doi.org/10.1093/mnras/stw1511

Cridland AJ, Pudritz RE, Birnstiel T, Cleeves LI, Bergin EA (2017) Composition of early planetary atmospheres—II. Coupled dust and chemical evolution in protoplanetary discs. Mon Not R Astron Soc 469:3910–3927, https://doi.org/10.1093/mnras/stx1069

Crossfield IJM (2023) Volatile-to-sulfur ratios can recover a gas giant's accretion history. Astrophys J 952:L18, https://doi.org/10.3847/2041-8213/ace35f

Cumming A, Butler RP, Marcy GW, Vogt SS, Wright JT, Fischer DA (2008) The Keck planet search: Detectability and the minimum mass and orbital period distribution of extrasolar planets. Publ Astron Soc Pac 120:531–531, https://doi.org/10.1086/588487

D'Angelo G, Lubow SH (2008) Evolution of migrating planets undergoing gas accretion. Astrophys J 685:560–583, https://doi.org/10.1086/590904

Dai Y-Z, Liu H-G, Yang J-Y, Zhou J-L (2023) Understanding the Planetary formation and evolution in star Clusters (UPiC). I. Evidence of hot giant exoplanets formation timescales. Astron J 166:219, https://doi.org/10.3847/1538-3881/acff67

Dasgupta R, Pathak D, Maurice M (2024) A framework for the origin and deep cycles of volatiles in rocky exoplanets. Rev Mineral Geochem 90:323–374

Dauphas N, Pourmand A (2011) Hf–W–Th evidence for rapid growth of Mars and its status as a planetary embryo. Nature 473:489–492, https://doi.org/10.1038/nature10077

David TJ, Hillenbrand LA, Petigura EA, et al. (2016) A Neptune-sized transiting planet closely orbiting a 5–10-million-year-old star. Nature 534:658–661, https://doi.org/10.1038/nature18293

Dawson RI, Johnson JA (2018) Origins of hot Jupiters. Annu Rev Astron Astrophys 56:175–221, https://doi.org/10.1146/annurev-astro-081817-051853

Debras F, Chabrier G (2019) New models of Jupiter in the context of Juno and Galileo. Astrophys J 872:100, https://doi.org/10.3847/1538-4357/aaff65

DeMeo FE, Carry B (2014) Solar System evolution from compositional mapping of the asteroid belt. Nature 505:629–634, https://doi.org/10.1038/nature12908

Demirci T, Schneider N, Teiser J, Wurm G (2020) Destruction of eccentric planetesimals by ram pressure and erosion. Astron Astrophys 644:A20, https://doi.org/10.1051/0004-6361/202039312

Denman TR, Leinhardt ZM, Carter PJ, Mordasini C (2020) Atmosphere loss in planet–planet collisions. Mon Not R Astron Soc 496:1166–1181, https://doi.org/10.1093/mnras/staa1623

Dittkrist KM, Mordasini C, Klahr H, Alibert Y, Henning T (2014) Impacts of planet migration models on planetary populations. Effects of saturation, cooling and stellar irradiation. Astron Astrophys 567:A121, https://doi.org/10.1051/0004-6361/201322506

Dittrich K, Klahr H, Johansen A (2013) Gravoturbulent planetesimal formation: The positive effect of long-lived zonal flows. Astrophys J 763:117, https://doi.org/10.1088/0004-637X/763/2/117

Dole SH (1970) Computer simulation of the formation of planetary systems. Icarus 13:494–508, https://doi.org/10.1016/0019-1035(70)90095-3

Donati JF, Moutou C, Malo L, et al. (2016) A hot Jupiter orbiting a 2-million-year-old solar-mass T Tauri star. Nature 534:662–666, https://doi.org/10.1038/nature18305

Dong S, Xie J-W, Zhou J-L, Zheng Z, Luo A (2018) LAMOST telescope reveals that Neptunian cousins of hot Jupiters are mostly single offspring of stars that are rich in heavy elements. Proc Nat Acad Sci 115:266–271, https://doi.org/10.1073/pnas.1711406115

Drążkowska J, Alibert Y (2017) Planetesimal formation starts at the snow line. Astron Astrophys 608:A92, https://doi.org/10.1051/0004-6361/201731491

Drążkowska J, Alibert Y, Moore B (2016) Close-in planetesimal formation by pile-up of drifting pebbles. Astron Astrophys 594:A105, https://doi.org/10.1051/0004-6361/201628983

Drążkowska J, Bitsch B, Lambrechts M, Mulders GD, Harsono D, Vazan A, Liu B, Ormel CW, Kretke K, Morbidelli A (2023) Planet formation theory in the era of ALMA and Kepler: from pebbles to exoplanets. Protostars and Planets VII, ASP Conf Ser 534:717, https://doi.org/10.48550/arXiv.2203.09759

Dullemond CP, Penzlin ABT (2018) Dust-driven viscous ring-instability in protoplanetary disks. Astron Astrophys 609:A50, https://doi.org/10.1051/0004-6361/201731878

Dullemond CP, Birnstiel T, Huang J, et al. (2018) The Disk Substructures at High Angular Resolution Project (DSHARP). VI. Dust Trapping in thin-ringed protoplanetary disks. Astrophys J Lett 869:L46, https://doi.org/10.3847/2041-8213/aaf742

Dürmann C, Kley W (2015) Migration of massive planets in accreting disks. Astron Astrophys 574:A52, https://doi.org/10.1051/0004-6361/201424837

Eistrup C, Walsh C, van Dishoeck EF (2016) Setting the volatile composition of (exo)planet-building material. Astron Astrophys 595:A83, https://doi.org/10.1051/0004-6361/201628509

Eistrup C, Walsh C, van Dishoeck EF (2018) Molecular abundances and C/O ratios in chemically evolving planet-forming disk midplanes. Astron Astrophys 613:A14, https://doi.org/10.1051/0004-6361/201731302

Emsenhuber A, Mordasini C, Burn R, Alibert Y, Benz W, Asphaug E (2021a) The New Generation Planetary Population Synthesis (NGPPS). II. Planetary population of solar-like stars and overview of statistical results. Astron Astrophys 656:A70, https://doi.org/10.1051/0004-6361/202038863

Emsenhuber A, Mordasini C, Burn R, Alibert Y, Benz W, Asphaug E (2021b) The New Generation Planetary Population Synthesis (NGPPS). I. Bern global model of planet formation and evolution, model tests, and emerging planetary systems. Astron Astrophys 656:A69, https://doi.org/10.1051/0004-6361/202038553

Emsenhuber A, Mordasini C, Burn R (2023a) Planetary population synthesis and the emergence of four classes of planetary system architectures. Eur Phys J Plus 138:181, https://doi.org/10.1140/epjp/s13360-023-03784-x

Emsenhuber A, Burn R, Weder J, Monsch K, Picogna G, Ercolano B, Preibisch T (2023b) Toward a population synthesis of disks and planets. II. Confronting disk models and observations at the population level. Astron Astrophys 673:A78, https://doi.org/10.1051/0004-6361/202244767

Encrenaz T, Lequeux J (2021) The Solar System 1: Telluric and Giant Planets, Interplanetary Medium and Exoplanets. Wiley

Encrenaz T, Bibring JP, Blanc M, Barucci MA, Roques F, Zarka P (2013) The Solar System, Translated by Dunlop S. Springer, Berlin Heidelberg

Ercolano B, Clarke CJ, Drake JJ (2009) X-ray irradiated protoplanetary disk atmospheres. II. Predictions from models in hydrostatic equilibrium. Astrophys J 699:1639–1649, https://doi.org/10.1088/0004-637X/699/2/1639

Ercolano B, Picogna G, Monsch K, Drake JJ, Preibisch T (2021) The dispersal of protoplanetary discs—II: photoevaporation models with observationally derived irradiating spectra. Mon Not R Astron Soc 508:1675–1685, https://doi.org/10.1093/mnras/stab2590

Facchini S, Clarke CJ, Bisbas TG (2016) External photoevaporation of protoplanetary discs in sparse stellar groups: the impact of dust growth. Mon Not R Astron Soc 457:3593–3610, https://doi.org/10.1093/mnras/stw240

Fernandes RB, Mulders GD, Pascucci I, Mordasini C, Emsenhuber A (2019) Hints for a turnover at the snow line in the giant planet occurrence rate. Astrophys J 874:81, https://doi.org/10.3847/1538-4357/ab0300

Fischer DA, Valenti J (2005) The planet–metallicity correlation. Astrophys J 622:1102–1117, https://doi.org/10.1086/428383

Ford EB, Rasio FA (2008) Origins of eccentric extrasolar planets: Testing the planet–planet scattering model. Astrophys J 686:621–636, https://doi.org/10.1086/590926

Forgan DH, Hall C, Meru F, Rice WKM (2018) Towards a population synthesis model of self-gravitating disc fragmentation and tidal downsizing II: the effect of fragment–fragment interactions. Mon Not R Astron Soc 474:5036–5048, https://doi.org/10.1093/mnras/stx2870

Fortier A, Alibert Y, Carron F, Benz W, Dittkrist KM (2013) Planet formation models: the interplay with the planetesimal disc. Astron Astrophys 549:A44, https://doi.org/10.1051/0004-6361/201220241

Franz R, Picogna G, Ercolano B, Birnstiel T (2020) Dust entrainment in photoevaporative winds: The impact of X-rays. Astron Astrophys 635:A53, https://doi.org/10.1051/0004-6361/201936615

Fray N, Schmitt B (2009) Sublimation of ices of astrophysical interest: A bibliographic review. Planet Space Sci 57:2053–2080, https://doi.org/10.1016/J.PSS.2009.09.011

Fressin F, Torres G, Charbonneau D, Bryson ST, Christiansen J, Dressing CD, Jenkins JM, Walkowicz LM, Batalha NM (2013) The false positive rate of Kepler and the occurrence of planets. Astrophys J 766:81, https://doi.org/10.1088/0004-637X/766/2/81

Fromang S, Nelson RP (2009) Global MHD simulations of stratified and turbulent protoplanetary discs. Astron Astrophys 496:597–608, https://doi.org/10.1051/0004-6361/200811220

Fulton BJ, Petigura EA (2018) The California-Kepler Survey. VII. Precise planet radii leveraging Gaia DR2 reveal the stellar mass dependence of the planet radius gap. Astron J 156:264, https://doi.org/10.3847/1538-3881/aae828

Fulton BJ, Petigura EA, Howard AW, et al. (2017) The California-Kepler Survey. III. A gap in the radius distribution of small planets. Astron J 154:109, https://doi.org/10.3847/1538-3881/aa80eb

Fulton BJ, Rosenthal LJ, Hirsch LA, et al. (2021) California legacy survey. II. Occurrence of giant planets beyond the ice line. Astrophys J Suppl Ser 255:14, https://doi.org/10.3847/1538-4365/abfcc1

Gaidos E, Fischer DA, Mann AW, Howard AW (2013) An understanding of the shoulder of giants: Jovian planets around late K dwarf stars and the trend with stellar mass. Astrophys J 771:18, https://doi.org/10.1088/0004-637X/771/1/18

Gammie CF (1996) Layered Accretion in T Tauri Disks. Astrophys J 457:355, https://doi.org/10.1086/176735

Gerbig K, Murray-Clay RA, Klahr H, Baehr H (2020) Requirements for gravitational collapse in planetesimal formation—The impact of scales set by Kelvin–Helmholtz and nonlinear streaming instability. Astrophys J 895:91, https://doi.org/10.3847/1538-4357/ab8d37

Ghezzi L, Montet BT, Johnson JA (2018) Retired A stars revisited: An updated giant planet occurrence rate as a function of stellar metallicity and mass. Astrophys J 860:109, https://doi.org/10.3847/1538-4357/aac37c

Goldreich P, Tremaine S (1979) The excitation of density waves at the Lindblad and corotation resonances by an external potential. Astrophys J 233:857–871, https://doi.org/10.1086/157448

Goldreich P, Tremaine S (1980) Disk–satellite interactions. Astrophys J 241:425–441, https://doi.org/10.1086/158356

Goldreich P, Ward WR (1973) The formation of planetesimals. Astrophys J 183:1051–1062, https://doi.org/10.1086/152291

Gonzalez G (1997) The stellar metallicity–giant planet connection. Mon Not R Astron Soc 285:403–412, https://doi.org/10.1093/mnras/285.2.403

Gorti U, Hollenbach D (2009) Photoevaporation of circumstellar disks by far-ultraviolet, extreme-ultraviolet and X-ray radiation from the central Star. Astrophys J 690:1539–1552, https://doi.org/10.1088/0004-637X/690/2/1539

Grandjean A, Lagrange AM, Meunier N, et al. (2021) A SOPHIE RV search for giant planets around young nearby stars (YNS). A combination with the HARPS YNS survey. Astron Astrophys 650:A39, https://doi.org/10.1051/0004-6361/202039672

Grant SL, van Dishoeck EF, Tabone B, et al. (2023) MINDS. The detection of $^{13}CO_2$ with JWST-MIRI indicates abundant CO_2 in a protoplanetary disk. Astrophys J Lett 947:L6, https://doi.org/10.3847/2041-8213/acc44b

Greenzweig Y, Lissauer JJ (1990) Accretion rates of protoplanets. Icarus 87:40–77, https://doi.org/10.1016/0019-1035(90)90021-Z

Greenzweig Y, Lissauer JJ (1992) Accretion rates of protoplanets II. Gaussian distributions of planetesimal velocities. Icarus 100:440–463, https://doi.org/10.1016/0019-1035(92)90110-S

Grether D, Lineweaver CH (2006) How dry is the brown dwarf desert? Quantifying the relative number of planets, brown dwarfs, and stellar companions around nearby Sun-like stars. Astrophys J 640:1051–1062, https://doi.org/10.1086/500161

Guillot T, Gautier D (2015) Giant planets. *In*: Treatise on Geophysics. Schubert G, (ed), p 529–557

Guillot T, Hueso R (2006) The composition of Jupiter: sign of a (relatively) late formation in a chemically evolved protosolar disc. Mon Not R Astron Soc 367:L47–L51, https://doi.org/10.1111/j.1745-3933.2006.00137.x

Guillot T, Fletcher LN, Helled R, Ikoma M, Line MR, Parmentier V (2023) Giant planets from the inside-out. Protostars and Planets VII, ASP Conf Ser 534:947, https://doi.org/10.48550/arXiv.2205.04100

Guimond CM, Wang H, Seidler F, Sossi P, Mahajan A, Shorttle O (2024) From stars to diverse mantles, melts, crusts, and atmospheres of rocky exoplanets. Rev Mineral Geochem 90:259–300

Gundlach B, Schmidt KP, Kreuzig C, Bischoff D, Rezaei F, Kothe S, Blum J, Grzesik B, Stoll E (2018) The tensile strength of ice and dust aggregates and its dependence on particle properties. Mon Not R Astron Soc 479:1273–1277, https://doi.org/10.1093/mnras/sty1550

Guzmán-Mesa A, Kitzmann D, Mordasini C, Heng K (2022) Chemical diversity of the atmospheres and interiors of sub-Neptunes: a case study of GJ 436 b. Mon Not R Astron Soc 513:4015–4036, https://doi.org/10.1093/mnras/stac1066

Haffert SY, Bohn AJ, de Boer J, Snellen IAG, Brinchmann J, Girard JH, Keller CU, Bacon R (2019) Two accreting protoplanets around the young star PDS 70. Nat Astron 3:749–754, https://doi.org/10.1038/s41550-019-0780-5

Haisch KE, Jr., Lada EA, Lada CJ (2001) Disk frequencies and lifetimes in young clusters. Astrophys J 553:L153–L156, https://doi.org/10.1086/320685

Hara NC, Boué G, Laskar J, Delisle JB, Unger N (2019) Bias and robustness of eccentricity estimates from radial velocity data. Mon Not R Astron Soc 489:738–762, https://doi.org/10.1093/mnras/stz1849

Hasegawa Y, Pudritz RE (2013) Planetary populations in the mass–period diagram: A statistical treatment of exoplanet formation and the role of planet traps. Astrophys J 778:78, https://doi.org/10.1088/0004-637X/778/1/78

Hashimoto J, Aoyama Y, Konishi M, Uyama T, Takasao S, Ikoma M, Tanigawa T (2020) Accretion properties of PDS 70b with MUSE. Astron J 159:222, https://doi.org/10.3847/1538-3881/ab811e

Haworth TJ, Clarke CJ, Rahman W, Winter AJ, Facchini S (2018) The FRIED grid of mass-loss rates for externally irradiated protoplanetary discs. Mon Not R Astron Soc 481:452–466, https://doi.org/10.1093/mnras/sty2323

Hayashi C (1981) Structure of the solar nebula, growth and decay of magnetic fields and effects of magnetic and turbulent viscosities on the nebula. Prog Theor Phys Suppl 70:35–53, https://doi.org/10.1143/PTPS.70.35

Hellary P, Nelson RP (2012) Global models of planetary system formation in radiatively-inefficient protoplanetary discs. Mon Not R Astron Soc 419:2737–2757, https://doi.org/10.1111/j.1365-2966.2011.19815.x

Helled R, Mazzola G, Redmer R (2020) Understanding dense hydrogen at planetary conditions. Nat Rev Phys 2:562–574, https://doi.org/10.1038/s42254-020-0223-3

Helled R, Stevenson DJ, Lunine JI, Bolton SJ, Nettelmann N, Atreya S, Guillot T, Militzer B, Miguel Y, Hubbard WB (2022) Revelations on Jupiter's formation, evolution and interior: Challenges from Juno results. Icarus 378:114937–114937, https://doi.org/10.1016/j.icarus.2022.114937

Hollenbach D, Johnstone D, Lizano S, Shu F (1994) Photoevaporation of disks around massive stars and application to ultracompact H II regions. Astrophys J 428:654, https://doi.org/10.1086/174276

Howard S, Guillot T (2023) Accounting for non-ideal mixing effects in the hydrogen–helium equation of state. Astron Astrophys 672:L1, https://doi.org/10.1051/0004-6361/202244851

Howard AW, Marcy GW, Johnson JA, Fischer DA, Wright JT, Isaacson H, Valenti JA, Anderson J, Lin DNC, Ida S (2010) The occurrence and mass distribution of close-in super-Earths, Neptunes, and Jupiters. Science 330:653–655, https://doi.org/10.1126/science.1194854

Huang J, Andrews SM, Dullemond CP, et al. (2018) The Disk Substructures at High Angular Resolution Project (DSHARP). II. Characteristics of annular substructures. Astrophys J Lett 869:L42, https://doi.org/10.3847/2041-8213/aaf740

Hueso R, Guillot T (2005) Evolution of protoplanetary disks: constraints from DM Tauri and GM Aurigae. Astron Astrophys 442:703–725, https://doi.org/10.1051/0004-6361:20041905

Hutchison MA, Laibe G, Maddison ST (2016) On the maximum grain size entrained by photoevaporative winds. Mon Not R Astron Soc 463:2725–2734, https://doi.org/10.1093/mnras/stw2191

Ida S (1990) Stirring and dynamical friction rates of planetesimals in the solar gravitational field. Icarus 88:129–145, https://doi.org/10.1016/0019-1035(90)90182-9

Ida S, Lin DNC (2004) Toward a deterministic model of planetary formation. I. A desert in the mass and semimajor axis distributions of extrasolar planets. Astrophys J 604:388–413, https://doi.org/10.1086/381724

Ida S, Lin DNC (2005) Toward a deterministic model of planetary formation. III. Mass distribution of short-period planets around stars of various masses. Astrophys J 626:10451060, https://doi.org/10.1086/429953

Ida S, Makino J (1993) Scattering of planetesimals by a protoplanet: Slowing down of runaway growth. Icarus 106:210–227, https://doi.org/10.1006/icar.1993.1167

Ikoma M, Nakazawa K, Emori H (2000) Formation of giant planets: Dependences on core accretion rate and grain opacity. Astrophys J 537:1013–1025, https://doi.org/10.1086/309050

Inaba S, Ikoma M (2003) Enhanced collisional growth of a protoplanet that has an atmosphere. Astron Astrophys 410:711–723, https://doi.org/10.1051/0004-6361:20031248

Inaba S, Tanaka H, Nakazawa K, Wetherill GW, Kokubo E (2001) High-accuracy statistical simulation of planetary accretion: II. Comparison with N-body simulation. Icarus 149:235–250, https://doi.org/10.1006/icar.2000.6533

Jiang H, Ormel CW (2023) Efficient planet formation by pebble accretion in ALMA rings. Mon Not R Astron Soc 518:3877–3900, https://doi.org/10.1093/mnras/stac3275

Jiménez MaA, Masset FS (2017) Improved torque formula for low- and intermediate-mass planetary migration. Mon Not R Astron Soc 471:4917–4929, https://doi.org/10.1093/mnras/stx1946

Jin S, Mordasini C (2018) Compositional imprints in density–distance–time: A rocky composition for close-in low-mass exoplanets from the location of the valley of evaporation. Astrophys J 853:163, https://doi.org/10.3847/1538-4357/aa9f1e

Jin S, Mordasini C, Parmentier V, van Boekel R, Henning T, Ji J (2014) Planetary population synthesis coupled with atmospheric escape: a statistical view of evaporation. Astrophys J 795:65, https://doi.org/10.1088/0004-637X/795/1/65

Johansen A, Youdin A (2007) Protoplanetary disk turbulence driven by the streaming instability: Nonlinear saturation and particle concentration. Astrophys J 662:627–641, https://doi.org/10.1086/516730

Johansen A, Lambrechts M (2017) Forming planets via pebble accretion. Annu Rev Earth Planet Sci 45:359–387, https://doi.org/10.1146/annurev-earth-063016-020226

Johansen A, Henning T, Klahr H (2006) Dust sedimentation and self-sustained Kelvin–Helmholtz turbulence in protoplanetary disk midplanes. Astrophys J 643:1219–1232, https://doi.org/10.1086/502968

Jurić M, Tremaine S (2008) Dynamical origin of extrasolar planet eccentricity distribution. Astrophys J 686:603–620, https://doi.org/10.1086/590347

Kanagawa KD, Tanaka H, Szuszkiewicz E (2018) Radial migration of gap-opening planets in protoplanetary disks. I. The case of a single planet. Astrophys J 861:140, https://doi.org/10.3847/1538-4357/aac8d9

Kanagawa KD, Tanaka H, Muto T, Tanigawa T (2017) Modelling of deep gaps created by giant planets in protoplanetary disks. Publ Astron Soc Jpn 69:97–97, https://doi.org/10.1093/pasj/psx114

Kant I (1755) Allgemeine Naturgeschichte und Theorie des Himmels

Kaufmann N, Alibert Y (2023) The influence of planetesimal fragmentation on planet formation. Astron Astrophys 676:A46, https://doi.org/10.1051/0004-6361/202345901

Keppler M, Benisty M, Müller A, et al. (2018) Discovery of a planetary-mass companion within the gap of the transition disk around PDS 70. Astron Astrophys 617:A44, https://doi.org/10.1051/0004-6361/201832957

Kessler A, Alibert Y (2023) The interplay between pebble and planetesimal accretion in population synthesis models and its role in giant planet formation. Astron Astrophys 674:A144, https://doi.org/10.1051/0004-6361/202245641

Kimura T, Ikoma M (2020) Formation of aqua planets with water of nebular origin: Effects of water enrichment on the structure and mass of captured atmospheres of terrestrial planets. Mon Not R Astron Soc 496:3755–3766, https://doi.org/10.1093/mnras/staa1778

Kite ES, Fegley B, Jr., Schaefer L, Ford EB (2019) Superabundance of exoplanet sub-Neptunes explained by fugacity crisis. Astrophys J Lett 887:L33, https://doi.org/10.3847/2041-8213/ab59d9

Kitzmann D, Stock JW, Patzer ABC (2024) FASTCHEM COND: equilibrium chemistry with condensation and rainout for cool planetary and stellar environments. Mon Not R Astron Soc 527:7263–7283, https://doi.org/10.1093/mnras/stad3515

Klahr H, Bodenheimer P (2006) Formation of giant planets by concurrent accretion of solids and gas inside an anticyclonic vortex. Astrophys J 639:432–440, https://doi.org/10.1086/498928

Klahr H, Schreiber A (2020) Turbulence sets the length scale for planetesimal formation: Local 2D simulations of streaming instability and planetesimal formation. Astrophys J 901:54, https://doi.org/10.3847/1538-4357/abac58

Kleine T, Münker C, Mezger K, Palme H (2002) Rapid accretion and early core formation on asteroids and the terrestrial planets from Hf-W chronometry. Nature 418:952–955, https://doi.org/10.1038/nature00982

Kleine T, Touboul M, Bourdon B, Nimmo F, Mezger K, Palme H, Jacobsen SB, Yin Q-Z, Halliday AN (2009) Hf–W chronology of the accretion and early evolution of asteroids and terrestrial planets. Geochim Cosmochim Acta 73:5150–5188, https://doi.org/10.1016/j.gca.2008.11.047

Kley W, Nelson RP (2012) Planet–disk interaction and orbital evolution. Annu Rev Astron Astrophys 50:211–249, https://doi.org/10.1146/annurev-astro-081811-125523

Kobayashi H, Tanaka H (2021) Rapid formation of gas-giant planets via collisional coagulation from dust grains to planetary cores. Astrophys J 922:16, https://doi.org/10.3847/1538-4357/ac289c

Kobayashi H, Tanaka H, Okuzumi S (2016) From planetesimals to planets in turbulent protoplanetary disks. I. Onset of runaway growth. Astrophys J 817:105, https://doi.org/10.3847/0004-637X/817/2/105

Krapp L, Benítez-Llambay P, Gressel O, Pessah ME (2019) Streaming instability for particle-size distributions. Astrophys J Lett 878:L30, https://doi.org/10.3847/2041-8213/ab2596

Kratter K, Lodato G (2016) Gravitational instabilities in circumstellar disks. Annu Rev Astron Astrophys 54:271–311, https://doi.org/10.1146/annurev-astro-081915-023307

Krijt S, Bosman AD, Zhang K, Schwarz KR, Ciesla FJ, Bergin EA (2020) CO depletion in protoplanetary disks: A unified picture combining physical sequestration and chemical processing. Astrophys J 899:134, https://doi.org/10.3847/1538-4357/aba75d

Kruijer TS, Burkhardt C, Budde G, Kleine T (2017) Age of Jupiter inferred from the distinct genetics and formation times of meteorites. Proc Natl Acad Sci 114:6712–6716, https://doi.org/10.1073/pnas.1704461114

Lagrange AM, Bonnefoy M, Chauvin G, et al. (2010) A giant planet imaged in the disk of the young star beta Pictoris. Science 329:57–59, https://doi.org/10.1126/science.1187187

Lagrange AM, Philipot F, Rubini P, Meunier N, Kiefer F, Kervella P, Delorme P, Beust H (2023) Radial distribution of giant exoplanets at Solar System scales. Astron Astrophys 677:A71, https://doi.org/10.1051/0004-6361/202346165

Lambrechts M, Johansen A (2012) Rapid growth of gas-giant cores by pebble accretion. Astron Astrophys 544:A32, https://doi.org/10.1051/0004-6361/201219127

Lambrechts M, Johansen A, Morbidelli A (2014) Separating gas-giant and ice-giant planets by halting pebble accretion. Astron Astrophys 572:A35, https://doi.org/10.1051/0004-6361/201423814

Lambrechts M, Morbidelli A, Jacobson SA, Johansen A, Bitsch B, Izidoro A, Raymond SN (2019) Formation of planetary systems by pebble accretion and migration. How the radial pebble flux determines a terrestrial-planet or super-Earth growth mode. Astron Astrophys 627:A83, https://doi.org/10.1051/0004-6361/201834229

Lau TCH, Drążkowska J, Stammler SM, Birnstiel T, Dullemond CP (2022) Rapid formation of massive planetary cores in a pressure bump. Astron Astrophys 668:A170, https://doi.org/10.1051/0004-6361/202244864

Lega E, Crida A, Bitsch B, Morbidelli A (2014) Migration of Earth-sized planets in 3D radiative discs. Mon Not R Astron Soc 440:683–695, https://doi.org/10.1093/mnras/stu304

Lega E, Morbidelli A, Nelson RP, Ramos XS, Crida A, Béthune W, Batygin K (2022) Migration of Jupiter mass planets in discs with laminar accretion flows. Astron Astrophys 658:A32, https://doi.org/10.1051/0004-6361/202141675

Lenz CT, Klahr H, Birnstiel T (2019) Planetesimal population synthesis: Pebble flux-regulated planetesimal formation. Astrophys J 874:36, https://doi.org/10.3847/1538-4357/ab05d9

Lesur G, Ercolano B, Flock M, et al. (2023) Hydro-, magnetohydro-, and dust–gas dynamics of protoplanetary disks. Protostars and Planets VII, ASP Conf Ser 534:465, https://doi.org/10.48550/arXiv.2203.09821

Levison HF, Morbidelli A, Van Laerhoven C, Gomes R, Tsiganis K (2008) Origin of the structure of the Kuiper belt during a dynamical instability in the orbits of Uranus and Neptune. Icarus 196:258–273, https://doi.org/10.1016/j.icarus.2007.11.035

Levison HF, Thommes EW, Duncan MJ (2010) Modeling the formation of giant planet cores. I. Evaluating key processes. Astrophys J 139:1297, https://doi.org/10.1088/0004-6256/139/4/1297

Li R, Youdin A (2021) Thresholds for particle clumping by the streaming instability. Astrophys J 919:107, https://doi.org/10.3847/1538-4357/ac0e9f

Li J, Bergin EA, Blake GA, Ciesla FJ, Hirschmann MM (2021) Earth's carbon deficit caused by early loss through irreversible sublimation. Sci Adv 7:eabd3632, https://doi.org/10.1126/sciadv.abd3632

Lichtenberg T, Golabek GJ, Burn R, Meyer MR, Alibert Y, Gerya TV, Mordasini C (2019) A water budget dichotomy of rocky protoplanets from ^{26}Al-heating. Nat Astron 3:307–313, https://doi.org/10.1038/s41550-018-0688-5

Lichtenberg T, Schaefer LK, Nakajima M, Fischer RA (2023) Geophysical evolution during rocky planet formation. Protostars and Planets VII, ASP Conf Ser 534:907, https://doi.org/10.48550/arXiv.2203.10023

Lin DNC, Papaloizou J (1986) On the tidal interaction between protoplanets and the primordial solar nebula. II. Self-consistent nonlinear interaction. Astrophys J 307:395, https://doi.org/10.1086/164426

Lin M-K, Youdin AN (2015) Cooling requirements for the vertical shear instability in protoplanetary disks. Astrophys J 811:17, https://doi.org/10.1088/0004-637X/811/1/17

Linder EF, Mordasini C, Mollière P, Marleau G-D, Malik M, Quanz SP, Meyer MR (2019) Evolutionary models of cold and low-mass planets: cooling curves, magnitudes, and detectability. Astron Astrophys 623:A85, https://doi.org/10.1051/0004-6361/201833873

Lissauer JJ (1993) Planet formation. Annu Rev Astron Astrophys 31:129–174, https://doi.org/10.1146/annurev.aa.31.090193.001021

Lissauer JJ, Hubickyj O, D'Angelo G, Bodenheimer P (2009) Models of Jupiter's growth incorporating thermal and hydrodynamic constraints. Icarus 199:338–350, https://doi.org/10.1016/j.icarus.2008.10.004

Lissauer JJ, Ragozzine D, Fabrycky DC, et al. (2011) Architecture and dynamics of Kepler's candidate multiple transiting planet systems. Astrophys J Suppl Ser 197:8, https://doi.org/10.1088/0067-0049/197/1/8

Lissauer JJ, Batalha NM, Borucki WJ (2023) Exoplanet science from Kepler. Protostars and Planets VII, ASP Conf Ser 534:839, https://doi.org/10.48550/arXiv.2311.04981

Lorek S, Johansen A (2022) Growing the seeds of pebble accretion through planetesimal accretion. Astron Astrophys 666:A108, https://doi.org/10.1051/0004-6361/202244333

Lovis C, Mayor M, Pepe F, et al. (2006) An extrasolar planetary system with three Neptune-mass planets Nature 441:305–309, https://doi.org/10.1038/nature04828

Lucy LB, Sweeney MA (1971) Spectroscopic binaries with circular orbits. Astron J 76:544–556, https://doi.org/10.1086/111159

Lüst R (1952) Die Entwicklung einer um einen Zentralkörper rotierenden Gasmasse. I. Lösungen der hydrodynamischen Gleichungen mit turbulenter Reibung. Z Naturforsch Teil A 7:87–98, https://doi.org/10.1515/zna-1952-0118

Lynden-Bell D, Pringle JE (1974) The evolution of viscous discs and the origin of the nebular variables. Mon Not R Astron Soc 168:603–637, https://doi.org/10.1093/mnras/168.3.603

Lyra W, Paardekooper S-J, Mac Low M-M (2010) Orbital migration of low-mass planets in evolutionary radiative models: Avoiding catastrophic infall. Astrophys J Lett 715:L68–L73, https://doi.org/10.1088/2041-8205/715/2/L68

Machida MN, Kokubo E, Inutsuka S-i, Matsumoto T (2010) Gas accretion onto a protoplanet and formation of a gas giant planet. Mon Not R Astron Soc 405:1227–1243, https://doi.org/10.1111/j.1365-2966.2010.16527.x

Macintosh B, Graham JR, Barman T, et al. (2015) Discovery and spectroscopy of the young jovian planet 51 Eri b with the Gemini planet imager. Science 350:64–67, https://doi.org/10.1126/science.aac5891

Madhusudhan N, Amin MA, Kennedy GM (2014) Toward chemical constraints on hot Jupiter migration. Astrophys J Lett 794:L12, https://doi.org/10.1088/2041-8205/794/1/L12

Mah J, Bitsch B, Pascucci I, Henning T (2023) Close-in ice lines and the super-stellar C/O ratio in discs around very low-mass stars. Astron Astrophys 677:L7, https://doi.org/10.1051/0004-6361/202347169

Manara CF, Ansdell M, Rosotti GP, Hughes AM, Armitage PJ, Lodato G, Williams JP (2023) Demographics of young stars and their protoplanetary disks: lessons learned on disk evolution and its connection to planet formation. Protostars and Planets VII, ASP Conf Ser 534:539, https://doi.org/10.48550/arXiv.2203.09930

Manger N, Klahr H (2018) Vortex formation and survival in protoplanetary discs subject to vertical shear instability. Mon Not R Astron Soc 480:2125–2136, https://doi.org/10.1093/mnras/sty1909

Mann AW, Gaidos E, Mace GN, et al. (2016) Zodiacal exoplanets in time (Zeit). I. A Neptune-sized planet orbiting an M4.5 dwarf in the Hyades star cluster. Astrophys J 818:46, https://doi.org/10.3847/0004-637X/818/1/46

Marboeuf U, Thiabaud A, Alibert Y, Cabral N, Benz W (2014) From stellar nebula to planetesimals. Astron Astrophys 570:A35, https://doi.org/10.1051/0004-6361/201322207

Marcy G, Butler RP, Fischer D, Vogt S, Wright JT, Tinney CG, Jones HRA (2005) Observed properties of exoplanets: masses, orbits, and metallicities. Prog Theor Phys Suppl 158:24–42, https://doi.org/10.1143/PTPS.158.24

Marleau G-D, Klahr H, Kuiper R, Mordasini C (2017) The planetary accretion shock. I. Framework for radiation-hydrodynamical simulations and first results. Astrophys J 836:221, https://doi.org/10.3847/1538-4357/836/2/221

Marleau G-D, Coleman GAL, Leleu A, Mordasini C (2019) Exploring the formation by core accretion and the luminosity evolution of directly imaged planets. The case of HIP 65426 b. Astron Astrophys 624:A20, https://doi.org/10.1051/0004-6361/201833597

Marleau GD, Aoyama Y, Kuiper R, et al. (2022) Accreting protoplanets: Spectral signatures and magnitude of gas and dust extinction at H alpha. Astron Astrophys 657:A38, https://doi.org/10.1051/0004-6361/202037494

Marleau G-D, Kuiper R, Béthune W, Mordasini C (2023) The Planetary accretion shock. III. Smoothing-free 2.5D simulations and calculation of H alpha emission. Astrophys J 952:89, https://doi.org/10.3847/1538-4357/accf12

Marois C, Macintosh B, Barman T, Zuckerman B, Song I, Patience J, Lafrenière D, Doyon R (2008) Direct imaging of multiple planets orbiting the star HR 8799. Science 322:1348–1352, https://doi.org/10.1126/science.1166585

Marois C, Zuckerman B, Konopacky QM, Macintosh B, Barman T (2010) Images of a fourth planet orbiting HR 8799. Nature 468:1080–1083, https://doi.org/10.1038/nature09684

Masset FS (2017) Coorbital thermal torques on low-mass protoplanets. Mon Not R Astron Soc 472:4204–4219, https://doi.org/10.1093/mnras/stx2271

Masset F, Snellgrove M (2001) Reversing type II migration: resonance trapping of a lighter giant protoplanet. Mon Not R Astron Soc 320:L55–L59, https://doi.org/10.1046/j.1365-8711.2001.04159.x

Mathis JS, Rumpl W, Nordsieck KH (1977) The size distribution of interstellar grains. Astrophys J 217:425, https://doi.org/10.1086/155591

Matsumura S, Pudritz RE, Thommes EW (2007) Saving planetary systems: Dead zones and planetary migration. Astrophys J 660:1609–1623, https://doi.org/10.1086/513175

Mayor M, Queloz D (1995) A Jupiter-mass companion to a solar-type star. Nature 378:355–359, https://doi.org/10.1038/378355a0

Mayor M, Pepe F, Queloz D, et al. (2003) Setting new standards with HARPS. The Messenger 114:20–24

Mayor M, Marmier M, Lovis C, et al. (2011) The HARPS search for southern extra-solar planets XXXIV. Occurrence, mass distribution and orbital properties of super-Earths and Neptune-mass planets. arXiv:1109.2497, https://doi.org/10.48550/arXiv.1109.2497

McNally CP, Nelson RP, Paardekooper S-J, Benítez-Llambay P (2019) Migrating super-Earths in low-viscosity discs: unveiling the roles of feedback, vortices, and laminar accretion flows. Mon Not R Astron Soc 484:728–748, https://doi.org/10.1093/mnras/stz023

Mezger K, Maltese A, Vollstaedt H (2021) Accretion and differentiation of early planetary bodies as recorded in the composition of the silicate Earth. Icarus 365:114497–114497, https://doi.org/10.1016/j.icarus.2021.114497

Miguel Y, Guilera OM, Brunini A (2011) The diversity of planetary system architectures: contrasting theory with observations. Mon Not R Astron Soc 417:314–332, https://doi.org/10.1111/j.1365-2966.2011.19264.x

Miguel Y, Bazot M, Guillot T, et al. (2022) Jupiter's inhomogeneous envelope. Astron Astrophys 662:A18, https://doi.org/10.1051/0004-6361/202243207

Millholland S, Wang S, Laughlin G (2017) Kepler multi-planet systems exhibit unexpected intra-system uniformity in mass and radius. Astrophys J Lett 849:L33, https://doi.org/10.3847/2041-8213/aa9714

Miotello A, Kamp I, Birnstiel T, Cleeves LI, Kataoka A (2023) Setting the stage for planet formation: measurements and implications of the fundamental disk properties. Protostars and Planets VII, ASP Conf Ser 534:501, https://doi.org/10.48550/arXiv.2203.09818

Mishra L, Alibert Y, Leleu A, Emsenhuber A, Mordasini C, Burn R, Udry S, Benz W (2021) The New Generation Planetary Population Synthesis (NGPPS) VI. Introducing KOBE: Kepler Observes Bern Exoplanets. Theoretical perspectives on the architecture of planetary systems: Peas in a pod. Astron Astrophys 656:A74, https://doi.org/10.1051/0004-6361/202140761

Mishra L, Alibert Y, Udry S, Mordasini C (2023a) Framework for the architecture of exoplanetary systems. II. Nature versus nurture: Emergent formation pathways of architecture classes. Astron Astrophys 670:A69, https://doi.org/10.1051/0004-6361/202244705

Mishra L, Alibert Y, Udry S, Mordasini C (2023b) Framework for the architecture of exoplanetary systems. I. Four classes of planetary system architecture. Astron Astrophys 670:A68, https://doi.org/10.1051/0004-6361/202243751

Mizuno H (1980) Formation of the Giant Planets. Prog Theor Phys 64:544–557, https://doi.org/10.1143/PTP.64.544

Moldenhauer TW, Kuiper R, Kley W, Ormel CW (2021) Steady state by recycling prevents premature collapse of protoplanetary atmospheres. Astron Astrophys 646:L11, https://doi.org/10.1051/0004-6361/202040220

Moldenhauer TW, Kuiper R, Kley W, Ormel CW (2022) Recycling of the first atmospheres of embedded planets: Dependence on core mass and optical depth. Astron Astrophys 661:A142, https://doi.org/10.1051/0004-6361/202141955

Mollière P, Molyarova T, Bitsch B, et al. (2022) Interpreting the atmospheric composition of exoplanets: Sensitivity to planet formation assumptions. Astrophys J 934:74, https://doi.org/10.3847/1538-4357/ac6a56

Morales JC, Mustill AJ, Ribas I, et al. (2019) A giant exoplanet orbiting a very-low-mass star challenges planet formation models. Science 365:1441–1445, https://doi.org/10.1126/science.aax3198

Morbidelli A, Chambers J, Lunine JI, Petit JM, Robert F, Valsecchi GB, Cyr KE (2000) Source regions and time scales for the delivery of water to Earth. Meteorit Planet Sci 35:1309–1320, https://doi.org/10.1111/j.1945-5100.2000.tb01518.x

Morbidelli A, Bottke WF, Nesvorný D, Levison HF (2009) Asteroids were born big. Icarus 204:558–573, https://doi.org/10.1016/j.icarus.2009.07.011

Morbidelli A, Lambrechts M, Jacobson S, Bitsch B (2015) The great dichotomy of the Solar System: Small terrestrial embryos and massive giant planet cores. Icarus 258:418–429, https://doi.org/10.1016/j.icarus.2015.06.003

Mordasini C (2014) Grain opacity and the bulk composition of extrasolar planets. II. An analytical model for grain opacity in protoplanetary atmospheres. Astron Astrophys 572:A118, https://doi.org/10.1051/0004-6361/201423702

Mordasini C (2018) Planetary population synthesis. *In*: Handbook of Exoplanets. Deeg HJ, Belmonte JA (eds) Springer Living Reference Work, p 143

Mordasini C, Alibert Y, Benz W (2006) Destruction of planetesimals in protoplanetry atmospheres. *In*: Tenth Anniversary of 51 Peg-b: Status of and prospects for hot Jupiter studies. Arnold L, Bouchy F, Moutou C (eds), p 84–86

Mordasini C, Alibert Y, Benz W, Naef D (2009) Extrasolar planet population synthesis. II. Statistical comparison with observations. Astron Astrophys 501:1161–1184, https://doi.org/10.1051/0004-6361/200810697

Mordasini C, Alibert Y, Klahr H, Henning T (2012a) Characterization of exoplanets from their formation. I. Models of combined planet formation and evolution. Astron Astrophys 547:A111, https://doi.org/10.1051/0004-6361/201118457

Mordasini C, Alibert Y, Georgy C, Dittkrist KM, Klahr H, Henning T (2012b) Characterization of exoplanets from their formation. II. The planetary mass–radius relationship. Astron Astrophys 547:A112, https://doi.org/10.1051/0004-6361/201118464

Mordasini C, Klahr H, Alibert Y, Miller N, Henning T (2014) Grain opacity and the bulk composition of extrasolar planets. I. Results from scaling the ISM opacity. Astron Astrophys 566:A141, https://doi.org/10.1051/0004-6361/201321479

Mordasini C, van Boeckel R, Mollière P, Henning T, Benneke B (2016) The Imprint of exoplanet formation history on observable present-day spectra of hot Jupiters. Astrophys J 832:41, https://doi.org/10.3847/0004-637X/832/1/41

Mordasini C, Marleau GD, Mollière P (2017) Characterization of exoplanets from their formation. III. The statistics of planetary luminosities. Astron Astrophys 608:A72, https://doi.org/10.1051/0004-6361/201630077

Mousis O, Deleuil M, Aguichine A, Marcq E, Naar J, Aguirre LA, Brugger B, Gonçalves T (2020) Irradiated ocean planets bridge super-Earth and sub-Neptune populations. Astrophys J Lett 896:L22, https://doi.org/10.3847/2041-8213/ab9530

Mulders GD, Pascucci I, Apai D (2015) An increase in the mass of planetary systems around lower-mass stars. Astrophys J 814:130, https://doi.org/10.1088/0004-637X/814/2/130

Naef D, Mayor M, Pepe F, Queloz D, Santos NC, Udry S, Burnet M (2001) The CORALIE survey for southern extrasolar planets. V. 3 new extrasolar planets. Astron Astrophys 375:205–218, https://doi.org/10.1051/0004-6361:20010841

Nakagawa Y, Sekiya M, Hayashi C (1986) Settling and growth of dust particles in a laminar phase of a low-mass solar nebula. Icarus 67:375–390, https://doi.org/10.1016/0019-1035(86)90121-1

Narang M, Manoj P, Furlan E, Mordasini C, Henning T, Mathew B, Banyal RK, Sivarani T (2018) Properties and occurrence rates for Kepler exoplanet candidates as a function of host star metallicity from the DR25 Catalog. Astron J 156:221, https://doi.org/10.3847/1538-3881/aae391

Nelson RP, Gressel O, Umurhan OM (2013) Linear and non-linear evolution of the vertical shear instability in accretion discs. Mon Not R Astron Soc 435:2610–2632, https://doi.org/10.1093/mnras/stt1475

Nesvorný D, Li R, Simon JB, Youdin AN, Richardson DC, Marschall R, Grundy WM (2021) Binary planetesimal formation from gravitationally collapsing pebble clouds. Planet Sci J 2:27, https://doi.org/10.3847/PSJ/abd858

Nettelmann N, Helled R, Fortney JJ, Redmer R (2013) New indication for a dichotomy in the interior structure of Uranus and Neptune from the application of modified shape and rotation data. Planet Space Sci 77:143--151, https://doi.org/10.1016/j.pss.2012.06.019

Nielsen EL, De Rosa RJ, Macintosh B, et al. (2019) The Gemini planet imager exoplanet survey: Giant planet and brown dwarf demographics from 10 to 100 AU. Astron J 158:13, https://doi.org/10.3847/1538-3881/ab16e9

Öberg KI, Murray-Clay R, Bergin EA (2011) The effects of snowlines on C/O in planetary atmospheres. Astrophys J Lett 743:L16, https://doi.org/10.1088/2041-8205/743/1/L16

Öberg KI, Guzman VV, Walsh C, et al. (2021) Molecules with ALMA at Planet-forming Scales (MAPS) I: Program overview and highlights. Astrophys J Suppl Ser 257:1, https://doi.org/10.3847/1538-4365/ac1432

Öberg KI, Facchini S, Anderson DE (2023) Protoplanetary disk chemistry. Annu Rev Astron Astrophys 61:287–328, https://doi.org/10.1146/annurev-astro-022823-040820

Ogihara M, Kokubo E, Suzuki TK, Morbidelli A (2018) Formation of close-in super-Earths in evolving protoplanetary disks due to disk winds. Astron Astrophys 615:A63, https://doi.org/10.1051/0004-6361/201832720

Ohno K, Fortney JJ (2023) Nitrogen as a tracer of giant planet formation. I. A universal deep adiabatic profile and semianalytical predictions of disequilibrium ammonia abundances in warm exoplanetary atmospheres. Astrophys J 946:18, https://doi.org/10.3847/1538-4357/acafed

Ohtsuki K (1999) Evolution of particle velocity dispersion in a circumplanetary disk due to inelastic collisions and gravitational interactions. Icarus 137:152–177, https://doi.org/10.1006/icar.1998.6041

Ohtsuki K, Stewart GR, Ida S (2002) Evolution of planetesimal velocities based on three-body orbital integrations and growth of protoplanets. Icarus 155:436–453, https://doi.org/10.1006/icar.2001.6741

Okuzumi S, Tanaka H, Kobayashi H, Wada K (2012) Rapid coagulation of porous dust aggregates outside the snow line: A pathway to successful icy planetesimal formation. Astrophys J 752:106, https://doi.org/10.1088/0004-637X/752/2/106

Ormel CW (2014) An atmospheric structure equation for grain growth. Astrophys J Lett 789:L18, https://doi.org/10.1088/2041-8205/789/1/L18

Ormel CW (2017) The emerging paradigm of pebble accretion. In: Formation, Evolution, and Dynamics of Young Solar Systems. Vol 445. Pessah M, Gressel O, (eds). p 197–228

Ormel CW, Cuzzi JN (2007) Closed-form expressions for particle relative velocities induced by turbulence. Astron Astrophys 466:413–420, https://doi.org/10.1051/0004-6361:20066899

Ormel CW, Klahr HH (2010) The effect of gas drag on the growth of protoplanets. Analytical expressions for the accretion of small bodies in laminar disks. Astron Astrophys 520:A43, https://doi.org/10.1051/0004-6361/201014903

Ormel CW, Kobayashi H (2012) Understanding how planets become massive. I. Description and validation of a new toy model. Astrophys J 747:115, https://doi.org/10.1088/0004-637X/747/2/115

Ormel CW, Dullemond CP, Spaans M (2010) A new condition for the transition from runaway to oligarchic growth. Astrophys J Lett 714:L103–L107, https://doi.org/10.1088/2041-8205/714/1/L103

Ormel CW, Ida S, Tanaka H (2012) Migration rates of planets due to scattering of planetesimals. Astrophys J 758:80, https://doi.org/10.1088/0004-637X/758/2/80

Ormel CW, Shi J-M, Kuiper R (2015) Hydrodynamics of embedded planets' first atmospheres—II. A rapid recycling of atmospheric gas. Mon Not R Astron Soc 447:3512–3525, https://doi.org/10.1093/mnras/stu2704

Owen JE (2019) Atmospheric escape and the evolution of close-in exoplanets. Annu Rev Earth Planet Sci 47:67–90, https://doi.org/10.1146/annurev-earth-053018-060246

Owen JE, Wu Y (2013) Kepler planets: A tale of evaporation. Astrophys J 775:105, https://doi.org/10.1088/0004-637X/775/2/105

Paardekooper SJ (2014) Dynamical corotation torques on low-mass planets. Mon Not R Astron Soc 444:2031–2042, https://doi.org/10.1093/mnras/stu1542

Paardekooper SJ, Baruteau C, Crida A, Kley W (2010) A torque formula for non-isothermal type I planetary migration—I. Unsaturated horseshoe drag. Mon Not R Astron Soc 401:1950–1964, https://doi.org/10.1111/j.1365-2966.2009.15782.x

Paardekooper SJ, Baruteau C, Kley W (2011) A torque formula for non-isothermal Type I planetary migration—II. Effects of diffusion. Mon Not R Astron Soc 410:293–303, https://doi.org/10.1111/j.1365-2966.2010.17442.x

Paardekooper S, Dong R, Duffell P, Fung J, Masset FS, Ogilvie G, Tanaka H (2023) Planet–disk interactions and orbital evolution. Protostars and Planets VII, ASP Conf Ser 534:685, https://doi.org/10.48550/arXiv.2203.09595

Pan M, Liu B, Johansen A, Ogihara M, Wang S, Ji J, Wang SX, Feng F, Riba I (2023) Forming giant planets around late–M dwarfs: Pebble accretion and planet–planet collision. arXiv:2311.10317, https://doi.org/10.48550/arXiv.2311.10317

Partnership A, Brogan CL, Pérez LM, et al. (2015) The 2014 ALMA long baseline campaign: First results from high angular resolution observations toward the HL Tau region. Astrophys J Lett 808:L3, https://doi.org/10.1088/2041-8205/808/1/L3

Perotti G, Christiaens V, Henning T, et al. (2023) Water in the terrestrial planet-forming zone of the PDS 70 disk. Nature 620:516–520, https://doi.org/10.1038/s41586-023-06317-9

Perri F, Cameron AGW (1974) Hydrodynamic instability of the solar nebula in the presence of a planetary core. Icarus 22:416–425, https://doi.org/10.1016/0019-1035(74)90074-8

Petigura EA, Howard AW, Marcy GW (2013) Prevalence of Earth-size planets orbiting Sun-like stars. Proc Nat Acad Sci 110:19273–19278, https://doi.org/10.1073/pnas.1319909110

Petigura EA, Marcy GW, Winn JN, Weiss LM, Fulton BJ, Howard AW, Sinukoff E, Isaacson H, Morton TD, Johnson JA (2018) The California-Kepler survey. IV. Metal-rich stars host a greater diversity of planets. Astron J 155:89, https://doi.org/10.3847/1538-3881/aaa54c

Pfalzner S, Dehghani S, Michel A (2022) Most planets might have more than 5 myr of time to form. Astrophys J Lett 939:L10, https://doi.org/10.3847/2041-8213/ac9839

Pfeil T, Klahr H (2019) Mapping the conditions for hydrodynamic instability on steady-state accretion models of protoplanetary disks. Astrophys J 871:150, https://doi.org/10.3847/1538-4357/aaf962

Picogna G, Ercolano B, Owen JE, Weber ML (2019) The dispersal of protoplanetary discs—I. A new generation of X-ray photoevaporation models. Mon Not R Astron Soc 487:691–701, https://doi.org/10.1093/mnras/stz1166

Picogna G, Ercolano B, Espaillat CC (2021) The dispersal of protoplanetary discs—III. Influence of stellar mass on disc photoevaporation. Mon Not R Astron Soc 508:3611–3619, https://doi.org/10.1093/mnras/stab2883

Pierens A (2015) Fast migration of low-mass planets in radiative discs. Mon Not R Astron Soc 454:2003–2014, https://doi.org/10.1093/mnras/stv2024

Pinte C, Price DJ, Ménard F, et al. (2020) Nine localized deviations from keplerian rotation in the DSHARP circumstellar disks: Kinematic evidence for protoplanets carving the gaps. Astrophys J Lett 890:L9, https://doi.org/10.3847/2041-8213/ab6dda

Piso A-MA, Youdin AN (2014) On the minimum core mass for giant planet formation at wide separations. Astrophys J 786:21, https://doi.org/10.1088/0004-637X/786/1/21

Plavchan P, Barclay T, Gagné J, et al. (2020) A planet within the debris disk around the pre-main-sequence star AU Microscopii. Nature 582:497–500, https://doi.org/10.1038/s41586-020-2400-z

Podolak M, Pollack JB, Reynolds RT (1988) Interactions of planetesimals with protoplanetary atmospheres. Icarus 73:163–179, https://doi.org/10.1016/0019-1035(88)90090-5

Podolak M, Haghighipour N, Bodenheimer P, Helled R, Podolak E (2020) Detailed calculations of the efficiency of planetesimal accretion in the core-accretion model. Astrophys J 899:45, https://doi.org/10.3847/1538-4357/ab9ec1

Polak B, Klahr H (2023) High-resolution study of planetesimal formation by gravitational collapse of pebble clouds. Astrophys J 943:125, https://doi.org/10.3847/1538-4357/aca58f

Pollack JB, Hollenbach D, Beckwith S, Simonelli DP, Roush T, Fong W (1994) Composition and radiative properties of grains in molecular clouds and accretion disks. Astrophys J 421:615, https://doi.org/10.1086/173677

Pollack JB, Hubickyj O, Bodenheimer P, Lissauer JJ, Podolak M, Greenzweig Y (1996) Formation of the giant planets by concurrent accretion of solids and gas. Icarus 124:62–85, https://doi.org/10.1006/icar.1996.0190

Polman J, Waters LBFM, Min M, Miguel Y, Khorshid N (2023) H_2S and SO_2 detectability in hot Jupiters. Sulphur species as indicators of metallicity and C/O ratio. Astron Astrophys 670:A161, https://doi.org/10.1051/0004-6361/202244647

Pringle JE (1981) Accretion discs in astrophysics. Annu Rev Astron Astrophys 19:137–162, https://doi.org/10.1146/annurev.aa.19.090181.001033

Raettig N, Klahr H, Lyra W (2015) Particle trapping and streaming instability in vortices in protoplanetary disks. Astrophys J 804:35, https://doi.org/10.1088/0004-637X/804/1/35

Rasio FA, Ford EB (1996) Dynamical instabilities and the formation of extrasolar planetary systems. Science 274:954–956, https://doi.org/10.1126/science.274.5289.954

Rauer H, Catala C, Aerts C, et al. (2014) The PLATO 2.0 mission. Exp Astro 38:249–330, https://doi.org/10.1007/s10686-014-9383-4

Raymond SN, Izidoro A (2017) Origin of water in the inner Solar System: Planetesimals scattered inward during Jupiter and Saturn's rapid gas accretion. Icarus 297:134–148, https://doi.org/10.1016/j.icarus.2017.06.030

Reffert S, Bergmann C, Quirrenbach A, Trifonov T, Künstler A (2015) Precise radial velocities of giant stars. VII. Occurrence rate of giant extrasolar planets as a function of mass and metallicity. Astron Astrophys 574:A116, https://doi.org/10.1051/0004-6361/201322360

Ribas I, Reiners A, Zechmeister M, et al. (2023) The CARMENES search for exoplanets around M dwarfs. Guaranteed time observations data release 1 (2016–2020). Astron Astrophys 670:A139, https://doi.org/10.1051/0004-6361/202244879

Rosenthal LJ, Fulton BJ, Hirsch LA, et al. (2021) The California Legacy Survey. I. A catalog of 178 planets from precision radial velocity monitoring of 719 nearby stars over three decades. Astrophys J Suppl Ser 255:8, https://doi.org/10.3847/1538-4365/abe23c

Sabotta S, Schlecker M, Chaturvedi P, et al. (2021) The CARMENES search for exoplanets around M dwarfs. Planet occurrence rates from a subsample of 71 stars. Astron Astrophys 653:A114, https://doi.org/10.1051/0004-6361/202140968

Safronov VS (1969) Evolution of the Protoplanetary Cloud and Formation of the Earth and the Planets. Moscow: Nauka

Sahlmann J, Ségransan D, Queloz D, Udry S, Santos NC, Marmier M, Mayor M, Naef D, Pepe F, Zucker S (2011) Search for brown-dwarf companions of stars. Astron Astrophys 525:A95, https://doi.org/10.1051/0004-6361/201015427

Santos NC, Israelian G, Mayor M (2004) Spectroscopic [Fe/H] for 98 extra-solar planet-host stars. Exploring the probability of planet formation. Astron Astrophys 415:1153–1166, https://doi.org/10.1051/0004-6361:20034469

Santos NC, Adibekyan V, Figueira P, et al. (2017) Observational evidence for two distinct giant planet populations. Astron Astrophys 603:A30, https://doi.org/10.1051/0004-6361/201730761

Saumon D, Chabrier G, van Horn HM (1995) An equation of state for low-mass stars and giant planets. Astrophys J Suppl Ser 99:713, https://doi.org/10.1086/192204

Schäfer U, Yang C-C, Johansen A (2017) Initial mass function of planetesimals formed by the streaming instability. Astron Astrophys 597:A69, https://doi.org/10.1051/0004-6361/201629561

Schib O, Mordasini C, Helled R (2022) Calibrated gas accretion and orbital migration of protoplanets in 1D disc models. Astron Astrophys 664:A138, https://doi.org/10.1051/0004-6361/202141904

Schib O, Mordasini C, Helled R (2023) The link between infall location, early disc size, and the fraction of self-gravitationally fragmenting discs. Astron Astrophys 669:A31, https://doi.org/10.1051/0004-6361/202244789

Schib O, Mordasini C, Wenger N, Marleau GD, Helled R (2021) The influence of infall on the properties of protoplanetary discs. Statistics of masses, sizes, lifetimes, and fragmentation. Astron Astrophys 645:A43, https://doi.org/10.1051/0004-6361/202039154

Schlecker M, Mordasini C, Emsenhuber A, Klahr H, Henning T, Burn R, Alibert Y, Benz W (2021a) The New Generation Planetary Population Synthesis (NGPPS). III. Warm super-Earths and cold Jupiters: a weak occurrence correlation, but with a strong architecture-composition link. Astron Astrophys 656:A71, https://doi.org/10.1051/0004-6361/202038554

Schlecker M, Pham D, Burn R, Alibert Y, Mordasini C, Emsenhuber A, Klahr H, Henning T, Mishra L (2021b) The New Generation Planetary Population Synthesis (NGPPS). V. Predetermination of planet types in global core accretion models. Astron Astrophys 656:A73, https://doi.org/10.1051/0004-6361/202140551

Schlecker M, Burn R, Sabotta S, Seifert A, Henning T, Emsenhuber A, Mordasini C, Reffert S, Shan Y, Klahr H (2022) RV-detected planets around M dwarfs: Challenges for core accretion models. Astron Astrophys 664:A180, https://doi.org/10.1051/0004-6361/202142543

Schneider AD, Bitsch B (2021) How drifting and evaporating pebbles shape giant planets I: Heavy element content and atmospheric C/O. Astron Astrophys 654:A71, https://doi.org/10.1051/0004-6361/202039640

Schönau L, Teiser J, Demirci T, Joeris K, Bila T, Onyeagusi FC, Fritscher M, Wurm G (2023) Forbidden planetesimals. Astron Astrophys 672:A169, https://doi.org/10.1051/0004-6361/202245499

Schoonenberg D, Ormel CW (2017) Planetesimal formation near the snowline: in or out? Astron Astrophys 602:A21, https://doi.org/10.1051/0004-6361/201630013

Schräpler RR, Landeck WA, Blum J (2022) Collisional properties of cm-sized high-porosity ice and dust aggregates and their applications to early planet formation. Mon Not R Astron Soc 509:5641–5656, https://doi.org/10.1093/mnras/stab3348

Seizinger A, Kley W (2013) Bouncing behavior of microscopic dust aggregates. Astron Astrophys 551:A65, https://doi.org/10.1051/0004-6361/201220946

Sekiya M (1998) Quasi-equilibrium density distributions of small dust aggregations in the Solar nebula. Icarus 133:298–309, https://doi.org/10.1006/icar.1998.5933

Sellek AD, Booth RA, Clarke CJ (2020) The evolution of dust in discs influenced by external photoevaporation. Mon Not R Astron Soc 492:1279–1294, https://doi.org/10.1093/mnras/stz3528

Semenov D, Wiebe D (2011) Chemical evolution of turbulent protoplanetary disks and the Solar nebula. Astrophys J Suppl Ser 196:25, https://doi.org/10.1088/0067-0049/196/2/25

Shakura NI, Sunyaev RA (1973) Black holes in binary systems. Observational appearance. Astron Astrophys 24:337–335

Shibata S, Helled R, Kobayashi H (2023) Heavy-element accretion by proto-Jupiter in a massive planetesimal disc, revisited. Mon Not R Astron Soc 519:1713–1731, https://doi.org/10.1093/mnras/stac3568

Simon JB, Armitage PJ, Li R, Youdin AN (2016) The mass and size distribution of planetesimals formed by the streaming instability. I. The role of self-gravity. Astrophys J 822:55, https://doi.org/10.3847/0004-637X/822/1/55

Simon JB, Armitage PJ, Youdin AN, Li R (2017) Evidence for universality in the initial planetesimal mass function. Astrophys J Lett 847:L12, https://doi.org/10.3847/2041-8213/aa8c79

Spohn T (2015) Physics of terrestrial planets and moons: An introduction and overview. *In*: Treatise on Geophysics. Schubert G, (ed), p 1–22

Squire J, Hopkins PF (2018) Resonant drag instabilities in protoplanetary discs: the streaming instability and new, faster growing instabilities. Mon Not R Astron Soc 477:5011–5040, https://doi.org/10.1093/mnras/sty854

Stammler SM, Birnstiel T (2022) DustPy: A Python package for dust evolution in protoplanetary disks. Astrophys J 935:35, https://doi.org/10.3847/1538-4357/ac7d58

Steinpilz T, Teiser J, Wurm G (2019) Sticking properties of silicates in planetesimal formation revisited. Astrophys J 874:60, https://doi.org/10.3847/1538-4357/ab07bb

Stevenson DJ (1982) Formation of the giant planets. Planet Space Sci 30:755–764, https://doi.org/10.1016/0032-0633(82)90108-8

Suzuki D, Bennett DP, Sumi T, et al. (2016a) The exoplanet mass-ratio function from the MOA-II Survey: Discovery of a break and likely peak at a Neptune mass. Astrophys J 833:145, https://doi.org/10.3847/1538-4357/833/2/145

Suzuki TK, Ogihara M, Morbidelli A, Crida A, Guillot T (2016b) Evolution of protoplanetary discs with magnetically driven disc winds. Astron Astrophys 596:A74, https://doi.org/10.1051/0004-6361/201628955

Suzuki D, Bennett DP, Ida S, et al. (2018) Microlensing results challenge the core accretion runaway growth scenario for gas giants. Astrophys J Lett 869:L34, https://doi.org/10.3847/2041-8213/aaf577

Szulágyi J, Mordasini C (2017) Thermodynamics of giant planet formation: shocking hot surfaces on circumplanetary discs. Mon Not R Astron Soc 465:L64–L68, https://doi.org/10.1093/mnrasl/slw212

Tabone B, Rosotti GP, Lodato G, Armitage PJ, Cridland AJ, van Dishoeck EF (2022) MHD disc winds can reproduce fast disc dispersal and the correlation between accretion rate and disc mass in Lupus. Mon Not R Astron Soc 512:L74–L79, https://doi.org/10.1093/mnrasl/slab124

Tabone B, Bettoni G, van Dishoeck EF, et al. (2023) A rich hydrocarbon chemistry and high C to O ratio in the inner disk around a very low-mass star. Nat Astron 7:805–814, https://doi.org/10.1038/s41550-023-01965-3

Tanaka H, Takeuchi T, Ward WR (2002) Three-dimensional interaction between a planet and an isothermal gaseous disk. I. Corotation and Lindblad torques and planet migration. Astrophys J 565:1257–1274, https://doi.org/10.1086/324713

Teague R, Bae J, Bergin EA, Birnstiel T, Foreman-Mackey D (2018) A kinematical detection of two embedded Jupiter-mass planets in HD 163296. Astrophys J Lett 860:L12, https://doi.org/10.3847/2041-8213/aac6d7

Thiabaud A, Marboeuf U, Alibert Y, Cabral N, Leya I, Mezger K (2014) From stellar nebula to planets: The refractory components. Astron Astrophys 562:A27, https://doi.org/10.1051/0004-6361/201322208

Thommes EW, Duncan MJ, Levison HF (2003) Oligarchic growth of giant planets. Icarus 161:431–455, https://doi.org/10.1016/S0019-1035(02)00043-X

Tobin JJ, Sheehan PD, Megeath ST, et al. (2020) The VLA/ALMA Nascent Disk and Multiplicity (VANDAM) survey of Orion protostars. II. A statistical characterization of Class 0 and Class I protostellar disks. Astrophys J 890:130, https://doi.org/10.3847/1538-4357/ab6f64

Toomre A (1964) On the gravitational stability of a disk of stars. Astrophys J 139:1217–1238, https://doi.org/10.1086/147861

Turner NJ, Fromang S, Gammie C, Klahr H, Lesur G, Wardle M, Bai XN (2014) Transport and accretion in planet-forming disks. Protostars and Planets VI:411, https://doi.org/10.48550/arXiv.1401.7306

Tychoniec Ł, Tobin JJ, Karska A, Chandler C, Dunham MM, Harris RJ, Kratter KM, Li Z-Y, Looney LW, Melis C (2018) The VLA Nascent Disk and Multiplicity survey of perseus protostars (VANDAM). IV. Free–free emission from protostars: Links to infrared properties, outflow tracers, and protostellar disk masses. Astrophys J Suppl Ser 238:19, https://doi.org/10.3847/1538-4365/aaceae

Udry S, Santos NC (2007) Statistical properties of exoplanets. Annu Rev Astron Astrophys 45:397–439, https://doi.org/10.1146/annurev.astro.45.051806.110529

Udry S, Mayor M, Santos NC (2003) Statistical properties of exoplanets. I. The period distribution: Constraints for the migration scenario. Astron Astrophys 407:369–376, https://doi.org/10.1051/0004-6361:20030843

Umurhan OM, Estrada PR, Cuzzi JN (2020) Streaming instability in turbulent protoplanetary disks. Astrophys J 895:4, https://doi.org/10.3847/1538-4357/ab899d

van Boekel R, Henning T, Menu J, et al. (2017) Three radial gaps in the disk of TW Hydrae imaged with SPHERE. Astrophys J 837:132, https://doi.org/10.3847/1538-4357/aa5d68

van Dishoeck EF, Grant S, Tabone B, et al. (2023) The diverse chemistry of protoplanetary disks as revealed by JWST. Faraday Discuss 245:52–79, https://doi.org/10.1039/D3FD00010A

Van Eylen V, Agentoft C, Lundkvist MS, Kjeldsen H, Owen JE, Fulton BJ, Petigura E, Snellen I (2018) An asteroseismic view of the radius valley: stripped cores, not born rocky. Mon Not R Astron Soc 479:4786–4795, https://doi.org/10.1093/mnras/sty1783

Vazan A, Kovetz A, Podolak M, Helled R (2013) The effect of composition on the evolution of giant and intermediate-mass planets. Mon Not R Astron Soc 434:3283–3292, https://doi.org/10.1093/mnras/stt1248

Vazan A, Sari Re, Kessel R (2022) A new perspective on the interiors of ice-rich planets: Ice–rock mixture instead of ice on top of rock. Astrophys J 926:150, https://doi.org/10.3847/1538-4357/ac458c

Venturini J, Guilera OM, Haldemann J, Ronco MP, Mordasini C (2020) The nature of the radius valley. Hints from formation and evolution models. Astron Astrophys 643:L1, https://doi.org/10.1051/0004-6361/202039141

Vigan A, Fontanive C, Meyer M, et al. (2021) The SPHERE infrared survey for exoplanets (SHINE). III. The demographics of young giant exoplanets below 300 au with SPHERE. Astron Astrophys 651:A72, https://doi.org/10.1051/0004-6361/202038107

Voelkel O, Klahr H, Mordasini C, Emsenhuber A, Lenz C (2020) Effect of pebble flux-regulated planetesimal formation on giant planet formation. Astron Astrophys 642:A75, https://doi.org/10.1051/0004-6361/202038085

Voelkel O, Deienno R, Kretke K, Klahr H (2021) Linking planetary embryo formation to planetesimal formation. II. The effect of pebble accretion in the terrestrial planet zone. Astron Astrophys 645:A132, https://doi.org/10.1051/0004-6361/202039245

Voelkel O, Klahr H, Mordasini C, Emsenhuber A (2022) Exploring multiple generations of planetary embryos. Astron Astrophys 666:A90, https://doi.org/10.1051/0004-6361/202141830

Wafflard-Fernandez G, Lesur G (2023) Planet–disk–wind interaction: The magnetized fate of protoplanets. Astron Astrophys 677:A70, https://doi.org/10.1051/0004-6361/202245305

Wagner K, Apai D, Kratter KM (2019) On the mass function, multiplicity, and origins of wide-orbit giant planets. Astrophys J 877:46, https://doi.org/10.3847/1538-4357/ab1904

Wahl SM, Hubbard WB, Militzer B, et al. (2017) Comparing Jupiter interior structure models to Juno gravity measurements and the role of a dilute core. Geophys Res Lett 44:4649–4659, https://doi.org/10.1002/2017GL073160

Walsh KJ, Morbidelli A, Raymond SN, O'Brien DP, Mandell AM (2011) A low mass for Mars from Jupiter's early gas-driven migration. Nature 475:206–209, https://doi.org/10.1038/nature10201

Walsh C, Nomura H, van Dishoeck E (2015) The molecular composition of the planet-forming regions of protoplanetary disks across the luminosity regime. Astron Astrophys 582:A88, https://doi.org/10.1051/0004-6361/201526751

Wang H, Weiss BP, Bai X-N, Downey BG, Wang J, Wang J, Suavet C, Fu RR, Zucolotto ME (2017) Lifetime of the solar nebula constrained by meteorite paleomagnetism. Science 355:623–627, https://doi.org/10.1126/science.aaf5043

Wang JJ, Vigan A, Lacour S, et al. (2021) Constraining the nature of the PDS 70 protoplanets with VLTI/GRAVITY. Astron J 161:148, https://doi.org/10.3847/1538-3881/abdb2d

Ward WR (1997) Protoplanet migration by nebula tides. Icarus 126:261–281, https://doi.org/10.1006/icar.1996.5647

Weder J, Mordasini C, Emsenhuber A (2023) Population study on MHD wind-driven disc evolution. Confronting theory and observation. Astron Astrophys 674:A165, https://doi.org/10.1051/0004-6361/202243453

Weidenschilling SJ (1977a) Aerodynamics of solid bodies in the solar nebula. Mon Not R Astron Soc 180:57–70, https://doi.org/10.1093/mnras/180.2.57

Weidenschilling SJ (1977b) The distribution of mass in the planetary system and solar nebula. Astrophys Space Sci 51:153–158, https://doi.org/10.1007/BF00642464

Weiss LM, Marcy GW, Petigura EA, et al. (2018) The California-Kepler survey. V. Peas in a pod: Planets in a Kepler multi-planet system are similar in size and regularly spaced. Astron J 155:48, https://doi.org/10.3847/1538-3881/aa9ff6

Weiss LM, Millholland SC, Petigura EA, Adams FC, Batygin K, Bloch AM, Mordasini C (2023) Architectures of compact multi-planet systems: diversity and uniformity. Protostars and Planets VII, ASP Conf Ser 534:863, https://doi.org/10.48550/arXiv.2203.10076

Wetherill GW, Stewart GR (1993) Formation of planetary embryos: Effects of fragmentation, low relative velocity, and independent variation of eccentricity and inclination. Icarus 106:190–209, https://doi.org/10.1006/icar.1993.1166

Whipple FL (1972) On certain aerodynamic processes for asteroids and comets. *In*: From Plasma to Planet, Proceedings of the Twenty-First Nobel Symposium. Evlius A (ed) Wiley Interscience Division, p 211–232

Winn JN, Fabrycky DC (2015) The occurrence and architecture of exoplanetary systems. Annu Rev Astron Astrophys 53:409–447, https://doi.org/10.1146/annurev-astro-082214-122246

Winter AJ, Haworth TJ (2022) The external photoevaporation of planet-forming discs. Eur Phys J Plus 137:1132, https://doi.org/10.1140/epjp/s13360-022-03314-1

Woitke P, Kamp I, Thi WF (2009) Radiation thermo-chemical models of protoplanetary disks. I. Hydrostatic disk structure and inner rim. Astron Astrophys 501:383–406, https://doi.org/10.1051/0004-6361/200911821

Wolthoff V, Reffert S, Quirrenbach A, Jones MaI, Wittenmyer RA, Jenkins JS (2022) Precise radial velocities of giant stars. XVI. Planet occurrence rates from the combined analysis of the Lick, EXPRESS, and PPPS giant star surveys. Astron Astrophys 661:A63, https://doi.org/10.1051/0004-6361/202142501

Woo JMY, Morbidelli A, Grimm SL, Stadel J, Brasser R (2023) Terrestrial planet formation from a ring. Icarus 396:115497–115497, https://doi.org/10.1016/j.icarus.2023.115497

Xie J-W, Dong S, Zhu Z, et al. (2016) Exoplanet orbital eccentricities derived from LAMOST-Kepler analysis. Proc Nat Acad Sci 113:11431–11435, https://doi.org/10.1073/pnas.1604692113

Xu S, Bonsor A (2021) Exogeology from polluted white dwarfs. Elements 17:241, https://doi.org/10.2138/gselements.17.4.241

Yin Q, Jacobsen SB, Yamashita K, Blichert-Toft J, Télouk P, Albarède F (2002) A short timescale for terrestrial planet formation from Hf–W chronometry of meteorites. Nature 418:949–952, https://doi.org/10.1038/nature00995

Youdin AN, Goodman J (2005) Streaming instabilities in protoplanetary disks. Astrophys J 620:459–469, https://doi.org/10.1086/426895

Youdin AN, Lithwick Y (2007) Particle stirring in turbulent gas disks: Including orbital oscillations. Icarus 192:588–604, https://doi.org/10.1016/j.icarus.2007.07.012

Youdin AN, Shu FH (2002) Planetesimal formation by gravitational instability. Astrophys J 580:494–505, https://doi.org/10.1086/343109

Young ED, Shahar A, Schlichting HE (2023) Earth shaped by primordial H_2 atmospheres. Nature 616:306–311, https://doi.org/10.1038/s41586-023-05823-0

Yu L, Donati JF, Hébrard EM, et al. (2017) A hot Jupiter around the very active weak-line T Tauri star TAP 26. Mon Not R Astron Soc 467:1342–1359, https://doi.org/10.1093/mnras/stx009

Zhang K (2024) Chemistry in protoplanetary disks. Rev Mineral Geochem 90:27–54

Zhu W, Dong S (2021) Exoplanet statistics and theoretical implications. Annu Rev Astron Astrophys 59:291–336, https://doi.org/10.1146/annurev-astro-112420-020055

Zhu W, Petrovich C, Wu Y, Dong S, Xie J (2018) About 30% of Sun-like stars have Kepler-like planetary systems: A study of their intrinsic architecture. Astrophys J 860:101, https://doi.org/10.3847/1538-4357/aac6d5

Reviews in Mineralogy & Geochemistry
Vol. 90 pp. 113–140, 2024
Copyright © Mineralogical Society of America

4

Meteorites and Planet Formation

Rhian H. Jones

Department of Earth and Environmental Sciences
The University of Manchester
Manchester, M13 9PL, U.K.

Rhian.jones-2@manchester.ac.uk

ROLE OF METEORITES IN UNDERSTANDING PLANET FORMATION

Meteorites are a remarkable resource. They capture the imagination of people worldwide with their spectacular entry through Earth's atmosphere as fireballs, and their exotic character of being pieces of other worlds. Scientifically, they are critical to interpreting the early stages of formation of the Solar System, as well as the geological evolution of asteroids, the Moon, and Mars, and they are vital to understanding planetary formation processes. With the burgeoning exploration of extrasolar planetary systems, knowledge of the fundamental process of planetary growth from protoplanetary disks has taken on a new significance. Meteorites provide essential and detailed insight into the formation of planetary systems, although we must bear in mind that they only represent one reference point (our own Solar System) in what is clearly a wide spectrum of possible chemical and physical parameters governing the diverse realm of extrasolar planets. This chapter summarises the nature of our meteorite collections, and the ways in which meteorites contribute to our understanding of the formation and evolution of our own Solar System, with broader implications for planetary systems in general.

Within our own Solar System, each planet and minor body (such as the Moon, Pluto, Ceres) has a unique composition and history. The asteroid population is also extremely diverse in terms of chemistry and geologic history, within the main asteroid belt as well as in other populations such as Jupiter's Trojan asteroids. There are many fundamental unknowns relating to these observations, and aspects of our own Solar System that are not well understood. For example: How were the different chemical and isotopic compositions of individual rocky planets established? To what extent are planetary compositions related to their initial formation locations in the protoplanetary disk? How do current planet compositions compare with initial compositions shortly after planet formation? What is the volatile element inventory of planetary mantles? When and how was water incorporated into the terrestrial planets? What role did Jupiter play in controlling the compositions of inner and outer Solar System material? Meteorites play an important role in addressing such questions. This chapter mainly focuses on the meteorites that derive from asteroids, and their importance for understanding the early stages of formation of our planetary system. The mineralogy and isotope geochemistry of these meteorites provide a tangible record of physical and chemical environments in the protoplanetary disk, and the nature of potential planet-building materials. In addition to asteroidal meteorites, lunar and martian meteorites provide essential geochemical information about larger bodies, enabling us to compare the chemistry of three planetary bodies, Earth, Moon and Mars. These planetary samples also inform us about topics such as the importance of magma oceans in the early stages of planet formation, the effect of giant impacts on planetary evolution, and crustal and surface processes on airless bodies.

1529-6466/24/0090-0004$05.00 (print)
1943-2666/24/0090-0004$05.00 (online)   http://dx.doi.org/10.2138/rmg.2024.90.04

METEORITES FROM ASTEROIDS

In order to discuss the relationship between meteorites and the protoplanetary disk environment, it is necessary to explain the classification of meteorites and the essential characteristics of the different meteorite groups. A glossary of meteorite terms is summarised in Table 1. There are two important divisions of asteroid-derived meteorites: undifferentiated and differentiated meteorites. The term differentiated refers to the separation of a body into distinct chemical layers. In the case of meteorite parent bodies, differentiation is the result of heating that has produced extensive melting, allowing for density-driven separation of a metallic core and rocky mantle. In contrast, the parent body of an undifferentiated meteorite has not melted. As a result, undifferentiated meteorites, known as chondrites, preserve protoplanetary disk materials. An important heat source in meteorite parent bodies was decay of the short-lived isotope ^{26}Al (half-life 0.717 million years), which was present in sufficient abundance in early-formed planetesimals to drive extensive melting (McSween et al. 2002). The potency of this energy source diminished towards the later stages of disk evolution, leading to milder heating of chondrite parent bodies (e.g., Desch et al. 2018). Impacts are an alternative heat source (e.g., Rubin 1995), but they are unlikely to account for the global-scale heating of asteroids recorded by many meteorites because they produce only limited amounts of impact melt (Keil et al. 1997; McSween et al. 2002) However, even though the collision rate has decreased throughout Solar System history, there is no a priori time limitation to the effects of impact-related heating (e.g., Wakita et al. 2022).

Undifferentiated meteorites: Chondrites

Chondrites are the most abundant type of meteorites in our collections, 90% of all meteorites (which total around 73,000 to date: Meteoritical Bulletin Database). There are three major classes of chondrites: ordinary, carbonaceous and enstatite classes, and each class is divided into multiple groups and subgroups (Krot et al. 2014; Scott and Krot 2014). There are additional groups which do not fit into any of these classes (Kakangari-like (K) chondrites, Rumuruti-like (R) chondrites), as well as individual anomalous chondrites, but we will not concern ourselves with the details here. For the purposes of this discussion, it is important to understand divisions at the level of the major classes and their comprised groups. The essential characteristics of these groups are described below, following a brief summary of the basic chondrite components. Differences in oxygen isotope compositions among chondrite groups are discussed in the section on oxygen isotopes later in this chapter.

Chondrites are made up of a set of components which have very different compositions and origins (e.g., Scott and Krot 2014). Chondrites are named after their most abundant component, chondrules (Fig. 1). These are (sub-) millimeter-sized beads that formed from individual melt droplets (Jones 2012; Connolly and Jones 2016). Chondrules typically consist of the (iron–magnesium–calcium) silicate minerals olivine and pyroxene, as well as either glassy or crystalline interstitial "mesostasis" material that has a feldspathic composition (feldspar is a calcium–sodium aluminosilicate mineral). (See Table 2 for mineral formulae and details of the major mineral groups.) Many chondrules also include beads of iron–nickel metal (Fe,Ni) and iron sulfide (FeS). Individual chondrules have a wide range of chemical compositions and oxidation states, with the Mg/(Mg+Fe) atomic ratio in olivine and pyroxene varying from 0.99 to 0.70 or lower. The heating mechanism that produced chondrule melt droplets is a matter of debate. One of the most favoured current models proposes heating of dust in shock fronts, such as in the bow shock of planetesimals as they migrated through the disk (Morris and Boley 2018). Planetesimal collision models are also considered (Johnson et al. 2018). Heating models for chondrule formation are necessitated by the fact that chondrules exist: if we did not observe them in chondrites, they would not be predicted in our own or any other planetary system, based on current observations and disk formation models.

Table 1. Glossary of meteorite terms.

Term	Description
Achondrites	Rocky (stony) meteorites derived from mantles and crusts of differentiated asteroids. Examples are eucrite, howardite, diogenite, angrite, ureilite, aubrite (= enstatite achondrite).
Basalt	A volcanic igneous rock, typically comprised of pyroxene and plagioclase feldspar.
Breccia	Rock comprised of an aggregate of broken rock fragments, resulting from impact processes.
Chondrites	Meteorites derived from asteroid parent bodies that have not undergone melting. Major classes of chondrites include ordinary chondrites (OC), carbonaceous chondrites (CC) and enstatite chondrites (EC).
Chondrules	(Sub-)millimeter-sized rocky beads found in chondrites. Chondrules formed from individual melt droplets in short heating events in the protoplanetary disk.
Condensation temperature	The 50% condensation temperature, T_c, of an element is the temperature at which 50% of that element is present in the condensed (solid) phase (for a solar composition) Elements are classed as refractory ($T_c > 1350$ K), main component (Fe, Mg, Si).
Differentiated meteorites	Meteorites derived from asteroid parent bodies that have melted significantly, resulting in formation of an iron-nickel core and rocky mantle. Includes iron meteorites (from cores of differentiated bodies), achondrites (from mantles) and stony-irons (mixtures of iron and rocky material).
Geochemical classification	Classification of elements based on affinities for different materials
HED meteorite group	Suite of related achondrite meteorites including howardite, eucrite and diogenite groups. Thought to be derived from asteroid 4 Vesta.
Iron meteorites	Meteorites rich in metallic iron-nickel (Fe,Ni). Includes magmatic irons (from cores of differentiated asteroids) and non-magmatic irons (derived from impact processes).
Nucleosynthetic isotope reservoirs	Two groups of planetary materials defined by characteristic ratios of nucleosynthetic isotopes
Oxidation state	The degree of oxidation of a rock, roughly equating to the availability of oxygen. Iron is present as Fe^{3+} in a highly oxidized rock, Fe^{2+} in most silicate minerals such as olivine and pyroxene, and Fe^0 (metallic iron) in reduced systems.
Planetary meteorites	Meteorites derived from large, planet-scale bodies, including lunar and martian meteorites.
Presolar grains	Individual grains, inherited from the molecular cloud, found in chondrite matrix. Identified as presolar by their non-solar isotope compositions.
Refractory inclusions	(Sub-)millimeter- to centimeter-sized inclusions found in chondrites, including calcium-, aluminum-rich inclusions (CAIs). Formed by gas-solid condensation at high temperatures.
Stony-iron meteorites	Meteorites consisting of Fe,Ni metal and silicate (rocky) material, derived from differentiated parent bodies. Includes mesosiderites and pallasites.
Undifferentiated meteorites	Meteorites derived from parent bodies that have not undergone melting, also known as chondrites

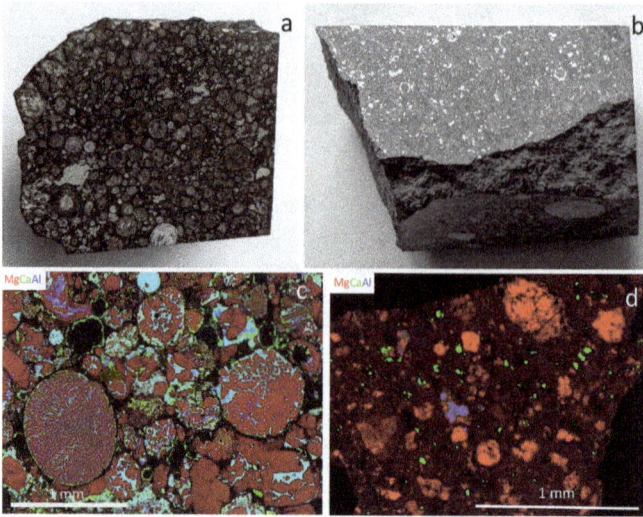

Figure 1. Chondrite meteorites. **a)** CV3 carbonaceous chondrite, Northwest Africa (NWA) 3118, 4 cm across. Circular objects are chondrules, white irregular objects are CAIs, dark material between chondrules and CAIs is the fine-grained matrix. **b)** Aba Panu L3 ordinary chondrite, 5 cm across. Circular objects are chondrules. Highly reflective material, white on the upper surface, is metal and sulfide which commonly surrounds chondrules. Striations on the upper surface are saw marks. **c)** False-color combined X-ray map of L3 ordinary chondrite, OUT 18016: Mg **red**, Ca **green**, Al **blue**. Chondrules (circular objects) show different textures including barred olivine (large chondrule, lower left) and porphyritic (e.g., large chondrule centre right). Chondrule mesostasis is **blue**. Metal and sulfide are **black**; matrix is **dark**, interstitial to chondrules. X-ray maps courtesy of Jane MacArthur, Lost Meteorites of Antarctica project, The University of Manchester. **d)** False-color combined X-ray map of CM2 carbonaceous chondrite, Murchison: Mg **red**, Ca **green**, Al **blue**. Chondrules appear **red**, CAIs are **blue**, carbonates are **green**. X-ray maps courtesy of Catherine Harrison, Natural History Museum / The University of Manchester.

The second dominant component of chondrites is matrix (Fig. 1). This is a fine-grained, sub-micrometer, mixture of different materials which includes low-temperature components that must have escaped chondrule heating events (Scott and Krot 2014). Matrix surrounds and supports chondrules, and also occurs as well-defined fine-grained rims adhering to chondrules. In chondrites that represent the most primitive materials, i.e., those that have not been affected by the secondary processes discussed below, matrix consists predominantly of amorphous silicates, as well as crystalline silicates (olivine, pyroxene) with abundant nano-scale sulfides and carbides (Abreu and Brearley 2010; Dobrică and Brearley 2020). Because of its fine-grained nature, matrix is highly susceptible to alteration, and this primitive mineralogy is only retained in a few rare chondrites. Carbon-rich materials are also an important matrix component, including organic material that consists of insoluble macromolecular carbon and soluble organics such as amino acids (Grady and Wright 2003). Matrix is also the host of presolar grains, which formed prior to formation of the Solar System in various stellar environments such as outflow from supernovae and AGB stars (Nittler and Ciesla 2016). Presolar grains are identified by their non-solar isotopic compositions (e.g., noble gases, Si, Ti, C, O, N isotopes), and include graphite, diamond, silicon carbide, oxides, and presolar silicates (Nittler and Ciesla 2016). Primitive matrix therefore is an accumulation of dust particles that represents a low-temperature protoplanetary disk environment. Because we are unable to recognise presolar material with solar isotopic compositions, the proportion of matrix material that is inherited directly from the presolar molecular cloud is unknown.

Refractory inclusions, ranging from sub-millimetre up to centimetres in size, are less abundant than chondrules (up to 3 vol%, but more typically less than 1 vol% in most chondrite

Table 2. Common minerals in meteorite.

Mineral group	Mineral names and formulae
Olivine	$(Mg,Fe)_2SiO_4$ Forsterite (Fo): Mg_2SiO_4 Fayalite (Fa): Fe_2SiO_4
Pyroxene	$(Mg,Fe,Ca)_2Si_2O_6$ Enstatite (En): $Mg_2Si_2O_6$ Ferrosilite (Fs): $Fe_2Si_2O_6$ Diopside (Di): $MgCaSi_2O_6$
Feldspar	Plagioclase: $CaAl_2Si_2O_8$–$NaAlSi_3O_8$ Anorthite (An): $CaAl_2Si_2O_8$ Albite (Ab): $NaAlSi_3O_8$
Phyllosilicate	Serpentine: $(Mg,Fe)_3Si_2O_5(OH)_4$
Oxide	Magnetite: Fe_3O_4
Carbonate	Calcite: $CaCO_3$
Fe,Ni metal	Fe,Ni Kamacite: Ni < 5% Taenite: Ni > 10%
Sulfide	Troilite: FeS

groups: Scott and Krot 2014; Figs. 1a,d). Despite their low abundance, they are highly significant for interpreting the earliest stages of formation of solids in the protoplanetary disk: they are thought to represent the first solids that were produced within the Solar System (MacPherson 2014; Krot 2019). Two main types of refractory inclusions are calcium-, aluminum-rich inclusions (CAIs) and amoeboid olivine aggregates (AOAs). CAIs consist of highly refractory minerals, in the sense that these minerals are the first to condense (and last to vaporize) from a gas of solar composition. Their constituent minerals, rich in Ca, Al, Si, Ti and O, include perovskite, hibonite, spinel, melilite, pyroxene, anorthite and a multitude of other minerals that are highly exotic to a planetary geologist. Refractory inclusions have distinct oxygen isotope compositions, which range to ^{16}O-rich values that approach the oxygen isotope composition of the Sun (see discussion of oxygen isotopes below). The oldest absolute ages measured in Solar System materials are Pb–Pb ages of CAIs, 4.567 Ga (billion years) (Connelly et al. 2012; see discussion of chronometry below). The above lines of evidence indicate that CAIs formed in a hot, reducing environment (Righter et al. 2016), possibly at the inner edge of the protoplanetary disk. However, their subsequent distribution throughout the disk must be accounted for in disk models (Cuzzi et al. 2003; Desch et al. 2018). CAI-like inclusions, as well as chondrule-like inclusions, have been identified in samples from comet Wild 2, which implies a disk-wide transport mechanism (Joswiak et al. 2017). In turn, if disk-wide transport is required, the timing and mechanism of transport must be reconciled with the origin and separation of the two (nucleosynthetic) isotopic reservoirs of Solar System materials discussed below.

In order to understand the distinctions between chondrite groups it is important to understand the framework of alteration that individual meteorites have undergone. This is described in a scheme of "petrologic types" (Brearley and Jones 1998; Scott and Krot 2014). The most primitive chondrites, i.e., those that have undergone little alteration since accretion of their parent bodies, are classified as petrologic type 3, and the least altered of these are subtype 3.00, on a scale of 3.00 to 3.9. Most chondrites have been affected to varying degrees by secondary processes since accreting to their parent bodies. Progressive effects of parent-body heating are denoted as petrologic types 3, 4, 5 and 6 (and sometimes 7). Progressive effects of aqueous alteration on chondrite parent bodies are described on a scale from petrologic type 3 (unaltered) to 2 then 1 (completely altered). Effects of shock are described in a scheme of shock stages from S1 (unshocked) to S6 (highly shocked), as defined by Stöffler et al. (1991) and revised by Fritz et al. (2017).

Ordinary chondrites (OCs) (Fig. 1b,c) are the most abundant type of meteorites to fall (94% of all chondrites). They have a high chondrule / matrix ratio, typically consisting of around 80% chondrules. There are three main OC groups, H, L and LL, which have high (H) or low (L) amounts of metallic iron, or low metallic iron as well as low total iron (LL). Abundances of metallic iron vary from 8 to 3 to 1.5 vol% in H, L and LL chondrites respectively (Scott and Krot 2014), which reflects differences in bulk chemical composition as well as oxidation state on their parent asteroids. It is likely that the H, L, and LL chondrites are derived from separate parent asteroids (Kallemeyn et al. 1989; Krot et al. 2014). Most ordinary chondrites were heated on their parent asteroids, in the first few million years after accretion (McSween et al. 2002). This resulted in thermal metamorphism (petrologic types 3 to 6), characterised by recrystallisation of the initial chondrule / matrix texture, chemical equilibration of minerals such as olivine and pyroxene that were initially very heterogeneous in composition, and formation of new minerals including feldspar, oxides, and phosphates (Huss et al. 2006). The early stages of heating occurred in the presence of a fluid (a process known as metasomatism), which included formation of water-bearing minerals such as phyllosilicates, and mobilisation of water-soluble elements (e.g., Lewis et al. 2022). Bulk water contents of OCs are typically less than 1 wt% (Fig. 2), but it is important to recognise that most OCs have lost their initial water during heating (Grant et al. 2023). Average water and carbon contents of the most primitive OCs (petrologic subtypes 3.00 to 3.2) are both around 0.5 wt% (Fig. 2). Peak metamorphic temperatures in OC parent bodies were around 950 °C, the lowest temperature at which melting occurs in material comprised of Fe, Ni, and S (otherwise described as the eutectic temperature in the Fe–Ni–S system; Buono and Walker 2011). However, by definition a chondrite has not experienced any melting, so this is a somewhat self-fulfilling limitation. The extent to which the OC parent bodies might have melted in their interiors, producing material that we would not classify as chondritic, is unresolved. Paleomagnetic records show that some partially molten planetesimals might have had a core and mantle, overlain by a chondritic crust (e.g., Elkins-Tanton et al. 2011; Weiss and Elkins-Tanton 2013; Dodds et al. 2021).

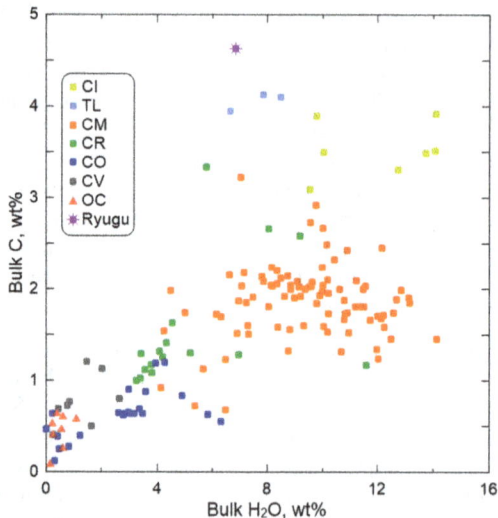

Figure 2. Bulk chondrite carbon and water contents for ordinary chondrites (OC), carbonaceous chondrites (individual groups are prefixed with C), the unique carbonaceous chondrite, Tagish Lake (TL), and for asteroid Ryugu which is comparable to CI1 chondrites. Data are from Alexander et al. (2012), Vacher et al. (2020), and a compilation of literature data in Piani et al. (2021), with additional data for Winchcombe (CM2) from Bates et al. (2023) and Verchovsky et al. (2023). Data for Ryugu and one data point for Ivuna (CI1) are from Yokoyama et al. (2022).

Carbonaceous chondrites (CCs) (Fig. 1a,d) comprise 5% of all chondrites. They have lower chondrule / matrix ratios than OCs, typically consisting of around 50% chondrules. CC groups are named after a typical meteorite in that group, preferably an observed fall, for example the CI group is named after the Ivuna meteorite which fell in Tanzania in 1938. Several CC groups, notably the CI, CM and CR groups, have been altered, some extensively, by the action of water (aqueous alteration: petrologic types 1 and 2) as demonstrated by high abundances of phyllosilicates, carbonate minerals, and the iron oxide, magnetite (Fig. 1d). Phyllosilicate is a hydrated (water-bearing) mineral, the product of reaction of anhydrous (water-free) primary minerals (olivine, pyroxene) and water. Carbonate minerals and magnetite precipitated from fluids. The most likely model for alteration is that water ice accreted initially to CC parent bodies, and then mild heating to <200 °C melted the ice, resulting in fluid flow (e.g., Vacher and Fujiya 2022). In heavily altered CCs, chondrules have been erased by alteration reactions, resulting in high apparent matrix abundances that are not necessarily reflective of the primary abundance. For example, CI chondrites (petrologic type 1) consist of 100% matrix, but they, like similar material from asteroid Ryugu, might have initially contained chondrules (Liu et al. 2022). It is important to recognise that not all CCs are aqueously altered in a similar way. CO, CV and CK chondrites have considerably lower abundances of phyllosilicates and carbonate minerals than CI, CM and CR groups. The CO and CV groups, mostly petrologic type 3, experienced higher-temperature metasomatism in the presence of fluids (up to 300 °C), and the CK group shows a range of metamorphic effects comparable to those observed in OCs (i.e., peak temperatures up to 950 °C, petrologic types 3 to 6). The CB and CH groups are distinct from other CCs, in that they are very metal-rich and their chondrules are younger than other chondrite groups. They are thought to be the products of a giant impact plume resulting from a collision between planetesimals (e.g., Krot et al. 2005, 2014).

Overall, the carbonaceous chondrite class is very diverse (Braukmüller et al. 2018). The bulk chemical composition of the CI group is a very close match to the solar photosphere for most condensable elements (Lodders 2003). However, other CC groups are relatively depleted in moderately volatile elements such as Mn, Na and K; for example, the bulk Mn/Mg ratio in CM chondrites is around 0.7 times the ratio in CI chondrite (Lodders and Fegley 1998; Scott and Krot 2014). Measured bulk water contents of CCs also vary widely, from <1 wt% H_2O in most CO chondrites to more than 10 wt% H_2O, in aqueously altered CI, CM and CR chondrites (Alexander et al. 2012; Vacher et al. 2020; Piani et al. 2021; Yokoyama et al. 2022; Bates et al. 2023; Verchovsky et al. 2023: Fig. 2). Calculations of the initial accreted water abundances for carbonaceous chondrites range between 3% to 28% H_2O for CO and CI groups respectively (Alexander 2019). The wide range of measured values reflects different degrees of secondary alteration, and whether a chondrite retains the products of aqueous alteration reactions. For example, CM chondrites record a range of degrees of alteration, from essentially unaltered to those that are almost completely altered, in which very few primary anhydrous (water-free) silicate mineral grains remain. In addition, a number of CM chondrites, as well as the CY chondrites, have undergone heating that resulted in dehydration following aqueous alteration, resulting in lower H_2O abundances (King et al. 2019, 2021; Suttle et al. 2023). Terrestrial contamination is a potential problem when H_2O abundances are measured (Vacher et al. 2020). The H_2O content of pristine CI-like material returned from asteroid Ryugu, 6.8 wt%, is much less than that measured in the same study in the Ivuna CI chondrite (12.7 wt%). This difference is interpreted as terrestrial adsorption of H_2O in CI chondrites, most of which are old falls (Yokoyama et al. 2022). H_2O contents of CI and CM chondrites determined by Vacher et al. (2020) and Piani et al. (2021), in which adsorbed water is degassed prior to analysis, are mostly less than 10 wt%.

Since carbon is initially concentrated in matrix, and CCs have high matrix abundances, many CCs have high bulk carbon contents as their name implies. Carbon was accreted initially into chondrite parent bodies as organic material and / or CO/CO_2 ices (Alexander et al. 2012, 2015).

CI chondrites and the ungrouped, highly altered carbonaceous chondrite, Tagish Lake, have C contents up to 4 wt% (Fig. 2), and CI-like material from asteroid Ryugu has 4.6 wt% C (Yokoyama et al. 2022). However, not all carbonaceous chondrites have high bulk carbon content, for example carbon contents of thermally altered CO and CV groups overlap with carbon contents of OCs, around 0.5 wt% C (Fig. 2).

Enstatite chondrites (ECs) are a rarer class of chondrites, 1% of all chondrites. Their chondrule / matrix ratios are comparable to OCs, up to 80% chondrules. The EH and EL groups have high and low amounts of metal (plus sulfide), respectively, and high abundances of the Mg-rich pyroxene, enstatite (hence their name). They are rich in sulfur (3–6 wt% in bulks: Lodders and Fegley 1998) and EH chondrites in particular have higher abundances of moderately volatile elements such as Mn, Na, K, and Cl than most carbonaceous chondrite groups (Scott and Krot 2014; Brearley and Jones 2018). The ECs are extremely reduced rocks: the Mg/(Fe+Mg) atomic ratio of pyroxene, and less abundant olivine, is very high, >0.99. They also contain an exceptionally diverse array of sulfide minerals, including Mn, Ca and K sulfides that are otherwise very rare in planetary environments because these are lithophile (rock-seeking) elements that are normally found in silicate and oxide minerals. Carbon contents of ECs are 0.2–0.7 wt%, and much of the carbon is present as graphite (Grady and Wright 2003; Alexander et al. 2007; Storz et al. 2021). Measured hydrogen abundances of ECs give equivalent water contents of <1 wt% (Piani et al. 2020), but it is not clear that H is present as H_2O. The ECs show the same range of metamorphism as OCs (petrologic types 3 to 6), which resulted in chemical and textural equilibration to the point where the initial chondrule / matrix texture was erased. Because they are highly reduced, ECs are thought to originate in the inner regions of the protoplanetary disk where little H_2O was available to oxidise the environment.

Chondritic meteorites are derived from asteroids, and matches between meteorites and asteroid types can be made from reflectance spectra. In general, OCs are derived from S-type, CCs from C-type, and ECs from Xc-type spectral classes (DeMeo et al. 2022), but the details are quite complex. Some meteorite-asteroid connections have been confirmed by recent return missions of asteroid samples: the JAXA Hayabusa mission to S-type asteroid Itokawa returned samples similar to LL ordinary chondrites (Nakamura et al. 2011), and the JAXA Hayabusa2 mission to Cb-type asteroid Ryugu returned samples similar to CI carbonaceous chondrites (Yokoyama et al. 2022). Preliminary results from samples returned by the NASA OSIRIS-REx mission to B-type asteroid Bennu in September 2023 show that it also has affinities to CI carbonaceous chondrites. The asteroid population in the main asteroid belt is zoned, with S-type asteroids concentrated towards the inner part of the belt and C-type asteroids concentrated at greater distances (DeMeo and Carry 2014). The presence in the main asteroid belt of chondritic asteroid bodies with very different compositions, matrix abundances, and oxidation states, which are thought to have originated across a wide span of heliocentric distances, is thought to be attributable to perturbance of the initial planetesimal distribution, caused by migration of the giant planets (Walsh et al. 2011; Morbidelli et al. 2015), and / or ring structures in the disk (Izidoro et al. 2022; Morbidelli et al. 2022). An important aspect of such a model is that it could have brought material that formed in the outer regions of the disk into the inner Solar System, a point that is discussed further below. It is important to recognise that the distribution of meteorites in our collections is not representative of the asteroid population, which is dominated by CC material. Several factors contribute to this sampling bias, including recent breakup of specific asteroids, proximity of asteroids to the Kirkwood gap orbital resonances that enhance delivery to Earth-crossing orbits, and the susceptibility of friable materials such as CI and CM chondrites to fragmentation, during asteroid collisions and atmospheric entry.

Differentiated meteorites: Achondrites and irons

The term differentiated meteorites encompasses a wide range of meteorite types, all sourced from parent bodies (planetesimals) on which melting and core formation processes have taken place. At a basic level, the main types are iron meteorites, derived from planetesimal cores, and achondrite meteorites, which are rocky (stony) materials derived from mantles and crusts. (The word achondrite literally indicates simply that a meteorite is not a chondrite, but it is commonly used to refer to only the rocky material of a differentiated planetesimal.) There are also various groups of "primitive achondrites", which derive from bodies that have seen less extensive melting, and which we will not discuss here. An additional division of differentiated meteorites, named stony-irons, consists of meteorites that contain approximately equal abundances of stony material (i.e., silicate minerals) and Fe,Ni metal. This includes the pallasite and mesosiderite groups. However, the stony-iron descriptor includes materials with very different origins, and does not uniquely identify a formation process. Pallasites consist of approximately equal abundances of iron metal and olivine (and/or more rarely, pyroxene). Various origins have been proposed for pallasites, including mixing at the core / mantle boundary of a differentiated planetesimal, disruption of a partly solidified planetesimal, and ferrovolcanism arising from a partially molten core (Yang et al. 2010; Boesenberg et al. 2012; Johnson et al. 2020). Disruptive impacts have played a prominent role in the evolution of differentiated bodies: several groups of irons (classed as non-magmatic irons), as well as mesosiderites, are the products of significant mixing of material from different depths in their parent bodies, and their impactors, during the period when parent bodies were partially molten (Scott et al. 2011; Goldstein et al. 2009; Haba et al. 2019). Iron meteorites that represent the relatively undisturbed products of core formation, named magmatic irons, have of course subsequently been excavated from the depths at which they originally formed, by collisional erosion over an extended time period. The following discussion focuses on achondrites and irons, which are the essential groups needed to further our understanding of planet-formation processes in the early Solar System.

Achondrites. Achondrites are igneous rocks that crystallised from silicate melts, either at the surface (volcanic or extrusive rocks) or at depth (intrusive rocks) in their parent planetesimals. Silicate melts originate from partial melting of a solidified mantle, and may undergo processes such as fractionation (i.e., partial crystallisation, with removal of the solid fraction) or assimilation (i.e., incorporation of surrounding rock) as they rise to the surface of the body. In addition to the range of chemical compositions that can be attributed to such processes, achondrites in our meteorite collections have diverse chemical compositions that attest to a wide range of mantle compositions and oxidation states on different parent bodies. These chemical groupings, coupled with distinct oxygen isotope compositions (see below), define different achondrite groups that likely each originate from different parent bodies (Mittlefehldt 2014). The origin of this primary diversity among differentiated planetesimals is likely related to location and timing of accretion within the protoplanetary disk. The requirement for localised accretion of diverse planetesimals that all currently reside within the main asteroid belt must be taken into account in disk models. Here we will focus on a few selected achondrite groups to illustrate the diversity, avoiding considerable complexity and detail that can be found elsewhere.

The best understood achondrites are the HED suite (e.g., Mittlefehldt 2014). The acronym stands for howardite, eucrite and diogenite, three achondrite groups that are closely related chemically, petrologically, and isotopically. They are the most abundant achondrites (75% of all achondrites, out of a total of around 3,800 known achondrites). Eucrites are basaltic rocks (Fig. 3a,b), and the group includes volcanic rocks that erupted as surface flows as well as intrusive cumulates. The dominant silicate minerals are pyroxene and plagioclase feldspar. Diogenites consist predominantly of pyroxene, specifically low-calcium orthopyroxene, and they represent cumulates that crystallised at depth. Many eucrites and diogenites are highly brecciated (i.e., they consist of an aggregate of broken rock fragments), as a result of extended impact bombardment at the surface of their parent asteroid (Fig. 3b). The howardites are mechanical mixtures of eucrite and diogenite material, including rock clasts and individual mineral grains, that were formed as surface breccias. The various compositions of the HED

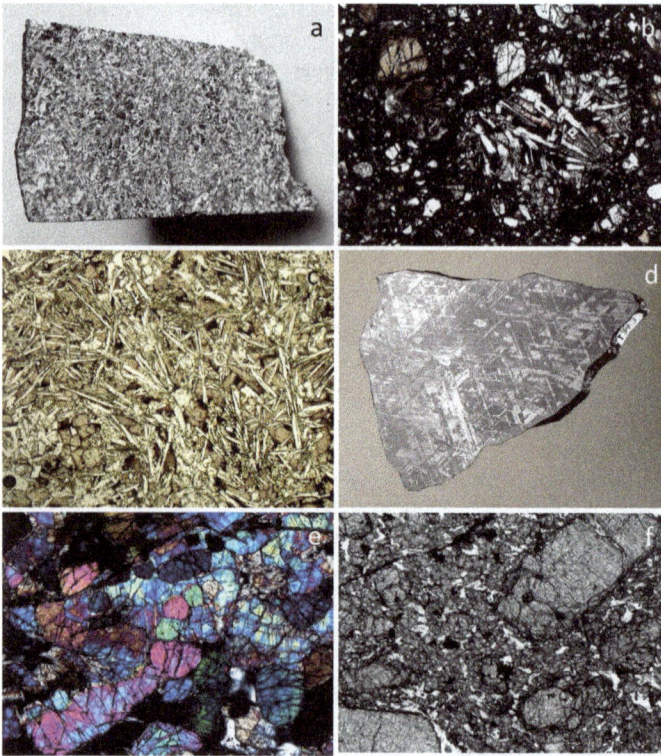

Figure 3. Differentiated and planetary meteorites. **a)** Eucrite NWA 11245, consisting predominantly of white grains of plagioclase feldspar and grey grains of pyroxene. 5 cm across. **b)** Eucrite NWA 1109. Plane-polarised transmitted light, field of view is 5 mm. NWA 1109 is brecciated, and is composed of comminuted (broken) fragments of individual mineral grains of plagioclase and pyroxene, as well as rock fragments (clasts). The larger coarse-grained rock clast on the right of the image consists predominantly of plagioclase feldspar (white) and pyroxene (brown) **c)** Angrite, Sahara (SAH) 99555. Plane-polarised transmitted light, field of view is 4 mm. Elongate white grains are feldspar, and pink / brown grains are pyroxene. **d)** Polished and etched surface of the iron meteorite, Gibeon, showing the Widmanstätten pattern and rounded inclusions of iron sulfide. Photo courtesy of Institute of Meteoritics, University of New Mexico. **e)** Lunar gabbro, NWA 8127. Cross-polarised light, field of view 1.25 mm. Igneous texture includes pyroxene (colored grains with colored lineations showing exsolution), olivine (rounded, colored grains), oxides (black, angular) and maskelynite (shock-transformed plagioclase feldspar: black / grey). Fracturing in olivine and pyroxene results from shock. **f)** Martian meteorite, Tissint, an olivine-phyric shergottite (basalt). Plane polarised light, field of view 2.5 mm. Large grains of olivine are surrounded by a fine-grained groundmass consisting of pyroxene (grey), maskelynite (white) and oxides (black / opaque). Olivine and pyroxene are highly fractured from shock. The olivine grain in the upper right has two circular melt inclusions. The linear black feature in the upper left is an impact-melt vein.

achondrites are consistent with derivation from the V-type asteroid 4 Vesta, which was the subject of intense remote study by NASA's Dawn mission (McCoy et al. 2015). This connection is the strongest link between a group of achondrites and a known asteroid source.

The ureilites are also an abundant group of meteorites (18% of achondrites). These achondrites are composed predominantly of pyroxene and olivine, and they represent restites (the residues after melts have been extracted) and cumulates derived from the mantle of a carbon-rich parent body (e.g., Goodrich et al. 2004; Mittlefehldt 2014). They contain carbon in the form of graphite and diamond, and they have diverse oxygen isotope ratios related to carbonaceous chondrites (see below).

The angrites are a rather rare group of achondrites (1% of achondrites). They are basaltic, predominantly consisting of pyroxene, olivine, and plagioclase feldspar (Fig. 3c), and include volcanic as well as intrusive (cumulate) rocks; vesicles (gas bubbles) are common in volcanic angrites (e.g., Mittlefehldt 2014). Angrites are enriched in refractory elements (i.e., elements with high condensation temperatures) and have unusual Al,Ti-rich pyroxene and Ca-rich olivine compositions. They are extremely depleted in volatile elements, including moderately volatile Na and K, which appears to be controlled by partial condensation of materials accreted to the angrite parent body (e.g., Tissot et al. 2022). Unlike most other achondrites, and notably the eucrites, many angrites are remarkably unshocked and unbrecciated: this could be attributed to early catastrophic impact and breakup of the original parent body (Scott and Bottke 2011).

The enstatite achondrites (2% of achondrites), also known as aubrites, are the most highly reduced igneous rocks in the Solar System. Like their chondrite counterparts, they consist predominantly of enstatite, a pyroxene mineral with a very high $Mg/(Mg+Fe)$ atomic ratio >0.99 (e.g., Mittlefehldt 2014). Aubrites and enstatite chondrites have similar oxygen isotope ratios (see below) and similar chemical compositions, but several aspects of these rocks rule out formation of aubrites directly from an EC parent body (Keil 2010). Most aubrites are highly brecciated.

Iron meteorites. As mentioned above, iron meteorites are described as either magmatic or non-magmatic irons (e.g., Goldstein et al. 2009). Iron meteorites consist predominantly of iron–nickel metal (Fe,Ni) that occurs as two main minerals, kamacite (Ni $<$ 5 wt%) and taenite (Ni $>$ 10 wt%). Iron sulfide, FeS (mineral name troilite), is also common, as are minor phosphides, carbides, and graphite. Iron meteorites are classified according to two different criteria. One criterion, the structural classification, is based on the (micro)structure of the Widmanstätten pattern, an intergrowth of kamacite and taenite that is produced during slow cooling (over millions of years) in the solid state, from solid metal of uniform composition (Fig. 3d). This gives rise to classifications such as octahedrite and hexahedrite, dependent on the presence of the Widmanstätten pattern and the widths of kamacite lamellae. Cooling rates of iron meteorites, determined from stranded chemical diffusion profiles in kamacite and taenite, indicate that their parent bodies were hundreds of kilometers in diameter (Goldstein et al. 2009; Benedix et al. 2014). The second criterion is a chemical classification, based on the bulk composition of the sample. Discriminatory classification diagrams of Ir vs. Ni and Ge vs. Ni content (Fig. 4) are used to identify different chemical groups that have names such as IAB: the four original groups, I to IV, were defined in order of decreasing abundances of volatile elements Ga and Ge (e.g., see Goldstein et al. 2009). There are also numerous ungrouped iron meteorites that do not fit into well-defined groups. Based on these classifications, it appears that the (currently) 1,375 known iron meteorites represent the cores of numerous differentiated planetesimals: more than 50 distinct bodies are likely sampled (Benedix et al. 2014). The asteroid parent bodies of iron meteorites are thought to be the M-type asteroids. M-type asteroid 16 Psyche is the target of NASA's current Psyche mission, although silicate minerals detected at the surface by infrared spectroscopy, and the relatively low bulk density of Psyche (~4 kg·m^{-3} vs. ~8 kg·m^{-3} for iron meteorites), indicates that this asteroid does not simply represent a metallic core comparable to magmatic iron meteorites (Elkins-Tanton et al. 2020, 2022). Since reflectance spectra of metals are very featureless, and all iron meteorites consist of the same basic mineralogy, there is little possibility of matching the different iron groups to individual asteroids.

Examples of magmatic irons include the IIAB, IIC, IIIAB, IVA and IVB groups. Within each magmatic iron group, the chemical compositions of individual meteorites show a well-defined, negative and often steep trend on the Ir vs. Ni classification diagram (Fig. 4a), consistent with fractional crystallisation of solid metal from liquid metal as the planetesimal cools and the core solidifies (Goldstein et al. 2009). Each magmatic iron group defines a different trend, with a range of Ni abundances from about 5 to 20 wt%, indicating different initial chemical compositions of the individual planetesimal cores. The Ge vs. Ni classification

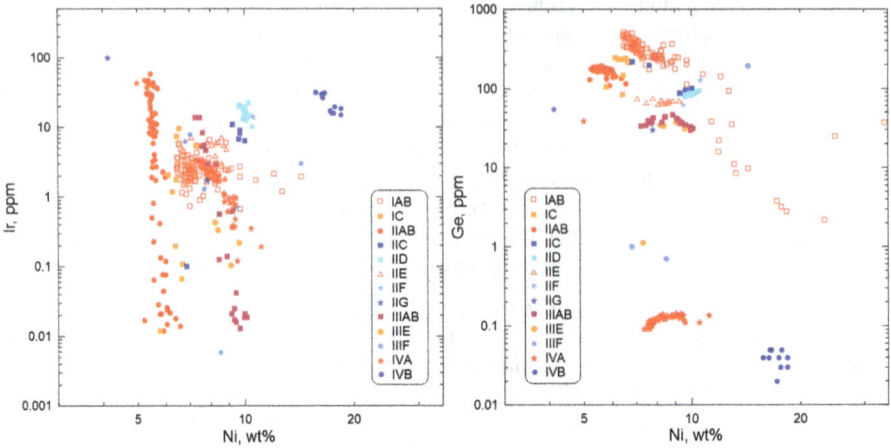

Figure 4. Bulk compositions of iron meteorites. (Iron metal composition only for non-magmatic irons.) Magmatic irons are **solid symbols**; non-magmatic irons are **open symbols**. The CC isotopic group is in shades of **blue**; NC isotopic group is in shades of **red and orange**. Data sources are those listed in Table 1 of Goldstein et al. (2009).

diagram (Fig. 4b) illustrates differences in volatile element compositions, since Ge is a volatile element (50% condensation temperature, T_c, is 883 K: Lodders 2003). Ge contents of the magmatic irons vary over four orders of magnitude, which is attributed to significant differences in the bulk compositions of differentiated planetesimals.

The non-magmatic irons include the IAB complex and the IIE group: together these groups represent around 30% of all iron meteorites. These meteorites have high abundances of iron-nickel metal, but they are brecciated and contain silicate clasts. As mentioned above, disruptive impacts played a significant role in forming these rocks. Non-magmatic irons show compositional clusters on the Ir vs. Ni and Ge vs. Ni classification diagrams that are not related to solidification processes. Ir contents are in the range 1–10 ppm, but Ge contents are more variable. The IAB complex shows a wide range of Ge vs. Ni, and silicate compositions are broadly chondritic, suggesting formation on a partially differentiated parent body (Wasson and Kallemeyn 2002; Benedix et al. 2014).

PLANETARY METEORITES FROM THE MOON AND MARS

Lunar meteorites

There are over 600 known lunar meteorites, although many of these are paired and the number of unique meteorite falls is much lower, less than 150 (Joy et al. 2023). Lunar meteorites include mare basalts, highland anorthosites, breccias that contain multiple lithologies, and impact melt breccias (e.g., Korotev 2005; Joy et al. 2023). These correspond to the most abundant lunar rock types known from remote measurements of the lunar surface and from returned samples. They are typically highly shocked, as a result of the impacts that lofted them from the Moon's surface: for example, feldspar is commonly transformed to the amorphous shocked phase maskelynite in lunar meteorites, indicating shock pressures 20–30 GPa (Rubin 2015; Chen et al. 2019: Fig. 3e). Lunar meteorites are very important to our understanding of lunar evolution (Korotev 2005; Joy and Arai 2013; Tartèse et al. 2019; Joy et al. 2023). In contrast to returned samples from the Apollo, Luna and more recent (2020) Chang'e 5 missions, their original locations on the Moon's surface are unknown. However, the random nature of their source locations means that they provide a more accurate overall picture of the Moon's bulk chemistry, very likely including high-latitude regions and the far side of the Moon that have not been sampled by missions to date. Lunar meteorites play an important role in interpreting the impact history of the Moon, including discussion of the "late heavy bombardment", a proposed period of intense impact events at around 3.9 Ga that would have affected all of the terrestrial planets, not only the Moon, and which has implications for the early development of life (e.g., Bottke and Norman 2017).

Martian meteorites

There are around 360 known martian meteorites, although as for lunar meteorites, many of these are paired and the number of unique meteorite falls is lower. Most martian meteorites belong to a suite of geochemically related mafic and ultramafic igneous rocks, including volcanic basalts, and intrusive gabbros, olivine cumulates, and pyroxene cumulates (e.g., McSween 2015: Fig. 3f). Their relatively young geologic ages (as young as 200 Ma), similarity to rocks on the martian surface observed by remote sensing and lander missions, and the match of their gas compositions to the martian atmosphere, provide convincing evidence that they originate from Mars (McSween 2015). Like lunar meteorites, most martian meteorites were highly shocked during the impact that lifted them from Mars, including transformation of plagioclase to maskelynite. More recently, an important group of martian regolith breccias has been recognised that probably originate from the ancient martian highland region (Agee et al. 2013; McCubbin et al. 2016; Cassata et al. 2018). These rocks contain clasts (fragments) of numerous rock types, including more geochemically evolved rocks than the main groups of martian meteorites which expand the available geological record significantly. Martian meteorites contain important evidence for volatile element abundances and hydrothermal activity on Mars (e.g., McCubbin and Jones 2015; Filiberto et al. 2016).

Since martian meteorites are currently our only samples of Mars, they are a unique resource for investigating the isotopic composition, mantle chemistry, geologic evolution, volatile content, and surface processes of a different planet in the Solar System, giving us two data points rather than just one (the Earth) for a detailed comparison of the geochemical properties of planets. The details of these topics are mostly beyond the scope of this article. Isotopic compositions of martian meteorites are included in the discussion below.

CHEMICAL AND ISOTOPIC HETEROGENEITY IN THE SOLAR SYSTEM

Chemical heterogeneity in the inner Solar System

From the above discussion, it is apparent that there is considerable chemical variability in the compositions of rocky bodies in the inner Solar System, including planets, the Moon, and asteroids. A detailed discussion of the bulk chemistry of individual meteorite parent bodies and planets is beyond the scope of this review (see Putirka 2024, this volume). Some of the complexity in understanding the chemical differences among the mantles of differentiated bodies can be illustrated with the following example. The most abundant rock type on rocky bodies is basalt, a rock that is typically comprised of pyroxene and plagioclase feldspar. Basalt arises from partial melting of a planetary mantle, followed by migration of the melt to the surface where it crystallizes. Thus, the chemistry of the minerals in a basalt gives (indirect) insight into the chemistry of the mantle where it originated. Basaltic rocks from different meteorite parent bodies, including the planetary bodies Earth, Moon and Mars, have clearly defined Mn/Fe ratios in their silicate minerals, olivine and pyroxene (Fig. 5). The different Mn/Fe ratios are well enough constrained that they are a useful parameter to classify the parentage of basaltic meteorites. Manganese and Fe behave similarly in igneous processes, because they both form divalent ions (charge of 2+) that have similar ionic radii. Therefore, differences in

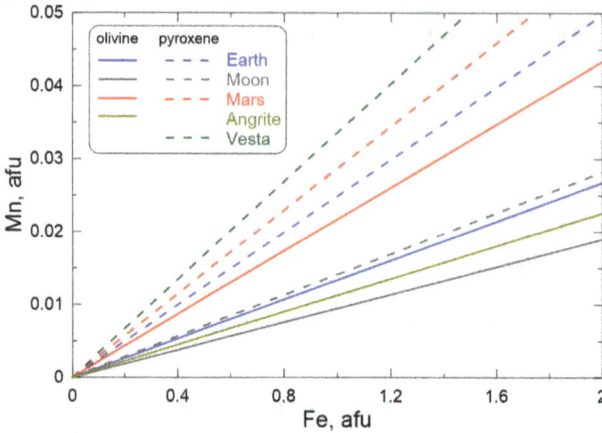

Figure 5. Mn vs. Fe atoms per 4-oxygen formula unit (afu) for olivine, and per 6-oxygen formula unit for pyroxene, in basaltic igneous rocks from various bodies: HED achondrites (presumed from asteroid 4 Vesta), angrites, Mars, Earth, Moon. Slopes of lines are taken from Papike et al. (2003, 2017).

Mn/Fe ratios in minerals such as olivine and pyroxene must reflect distinctly different bulk Mn/Fe ratios in planetesimal or planetary mantles. Such differences may partly be attributable to the fundamental chemistry of the body: since Mn is more volatile than Fe (50% condensation temperatures, T_c, for Mn and Fe are 1,158 K and 1,334 K respectively: Lodders 2003), the higher Mn/Fe ratio of Mars than Earth could be attributed to accretion of Mars at a greater radial distance from the Sun, where disk temperatures were lower, and more Mn had condensed (e.g., Papike et al. 2003). If volatility is the main control on Mn/Fe ratios, asteroid 4 Vesta would have formed closer to Mars, and the angrite parent body closer to Earth (Fig. 5: Papike et al. 2017). However, an additional factor is the oxidation state of the planet: a higher degree of reduction will lead to greater sequestering of metallic Fe into a planetary core which will raise the initial Mn/Fe ratio of the derivative mantle (Papike et al. 2017). As a result, we would expect a higher Mn/Fe ratio closer to the Sun which counteracts the volatility trend. Although it is not clear which factor exerts more control, the observation that each body has a unique mantle Mn/Fe ratio clearly points to fundamental chemical heterogeneity among planetary bodies.

One of the key questions related to the composition and chemical evolution of a planet is the abundance of water, as well as other volatiles, including carbon, nitrogen, and the halogen elements, fluorine and chlorine (e.g., Dasgupta and Grewal 2019; Broadley et al. 2022; Halliday and Canup 2023; Lodders and Fegley 2023). Water plays a major role in mantle and crust evolution, and it is of course essential for life. Water is also critical in determining the oxidation state of a planet, which in turn determines the relative size of mantle and core, and mantle chemistry. As discussed above, most primitive chondritic materials, including ordinary as well as carbonaceous chondrites, contain evidence for initial presence of water, which probably accreted to planetesimals as water ice. In addition, achondrite parent bodies, including the eucrite, angrite and ureilite parent bodies, initially contained water at abundances of tens to hundreds of parts per million (e.g., McCubbin and Barnes 2019; Peterson et al. 2023). Isotopic compositions of hydrogen (D/H ratios), as well as C and N isotopes, contribute to debate about whether achondrite parent bodies accreted water during planetesimal growth, or whether water was added later in the form of carbonaceous chondrite material, as well as the extent of planetary degassing and the contribution of interstellar water (e.g., McCubbin and Barnes 2019; Newcombe et al. 2023). Similarly, the source of water on the Earth and other terrestrial planets has been proposed to be either from late accretion of a cometary or carbonaceous chondrite-like component, or from primary accretion of inner Solar System material (Alexander et al. 2012, 2018; Marty 2012; Piani et al. 2020; Broadley et al. 2022; Izidoro and Piani 2022; Tissot et al. 2022; Halliday and Canup 2023). Although these discussions continue to evolve, overall it is clear that water ice was a ubiquitous primary component of most of the planetary materials from rocky parent bodies that have been sampled to date, and that water and other volatiles can be considered as inevitable components of planets, at least in our own planetary system.

Oxygen isotope compositions of planetary materials

Oxygen isotope compositions of meteorites show significant diversity, and they are an essential aspect of understanding the evolution of the protoplanetary disk and our planetary system (e.g., Ireland et al. 2020). Oxygen isotope compositions are typically illustrated on an oxygen three-isotope diagram (Fig. 6a–c), with compositions given in the delta notation such that $^{17}O/^{16}O$ and $^{18}O/^{16}O$ ratios are referenced to a terrestrial standard, standard mean ocean water (SMOW) by the relationship: $\delta^{17,18}O = [(^{17,18}O/^{16}O)_{sample} / {}^{17,18}O/^{16}O_{SMOW}) - 1]$ × 1000 ‰. On this diagram, the terrestrial fractionation (TF) line, which defines the oxygen isotope compositions of Earth materials, is used as a reference line. The TF line has a slope of 0.52, derived from geochemical reactions and processes that are controlled by mass-dependent fractionation effects. A useful construct is the definition of $\Delta^{17}O$, given as $\Delta^{17}O = \delta^{17}O - (0.52 \times \delta^{18}O)$, which defines the vertical offset of either a point or a mass-dependent fractionation line from the TFL on this diagram, as illustrated in Figure 6a. Two other reference lines are shown in Figures 6a–c: the carbonaceous chondrite anhydrous minerals (CCAM) line, which has a slope of 0.94, and the primitive chondrule mixing (PCM) line which has a slope of 0.99, both of which are mass-independent fractionation or mixing lines (Dauphas and Schauble 2016), and which are discussed further below.

Chondrites show a range of oxygen isotope compositions among the different classes and groups (Fig. 6a). Ordinary, enstatite and carbonaceous chondrites have bulk oxygen isotope compositions with decreasing values of $\Delta^{17}O$: enstatite chondrite values are similar to terrestrial compositions and lie on the TF line. For most carbonaceous chondrite groups, arrays of chondrite bulk compositions have slopes around one, parallel to the CCAM / PCM lines. Within carbonaceous chondrite groups that have been affected extensively by aqueous alteration (CI, CM, CR, and CY groups), oxygen isotope compositions lie along arrays with shallower slopes as a result of reactions between primitive chondrule and matrix materials, and water. The bulk oxygen isotope composition of asteroid Ryugu, sampled by the Hayabusa2 mission, is similar to CI chondrites (Greenwood et al. 2023; Tang et al. 2023).

Individual chondrules and refractory inclusions show remarkable heterogeneity in the oxygen 3-isotope diagram (Fig. 6b). Most bulk analyses of chondrules have similar compositional ranges to their respective bulk chondrite compositions. However, bulk analyses of refractory inclusions (CAIs and AOAs) lie on an extended array with a slope close to one: this observation originally defined the CCAM line (Clayton and Mayeda 1999). The ^{16}O-rich end-member of this array has $\delta^{17,18}O$ values around −45 ‰ (Fig. 6b). Individual in situ point analyses of mineral grains in refractory inclusions, made using secondary ion mass spectrometry (SIMS), also spread along the entire length of the CCAM line. For chondrules, in situ SIMS analyses of constituent mineral grains lie on a slightly different slope-1 array, the PCM line (Ushikubo et al. 2012; Tenner et al. 2018). These observations are true for chondrules and refractory inclusions in ordinary as well as carbonaceous chondrites (e.g., Ushikubo et al. 2012; Williams et al. 2020; Piralla et al. 2021; Marrocchi et al. 2022). Interpretation of these oxygen isotope distributions in primitive chondrite components is currently a matter of debate (Dauphas and Schauble 2016; Ireland et al. 2020). The ^{16}O-rich end of the line is well understood, because it is similar to and consistent with the oxygen isotope composition of the Sun (McKeegan et al. 2011: Fig. 6b). However, the ^{16}O-poor endmember is less well defined. The most accepted interpretation of the slope-1 distribution is that solar system solids

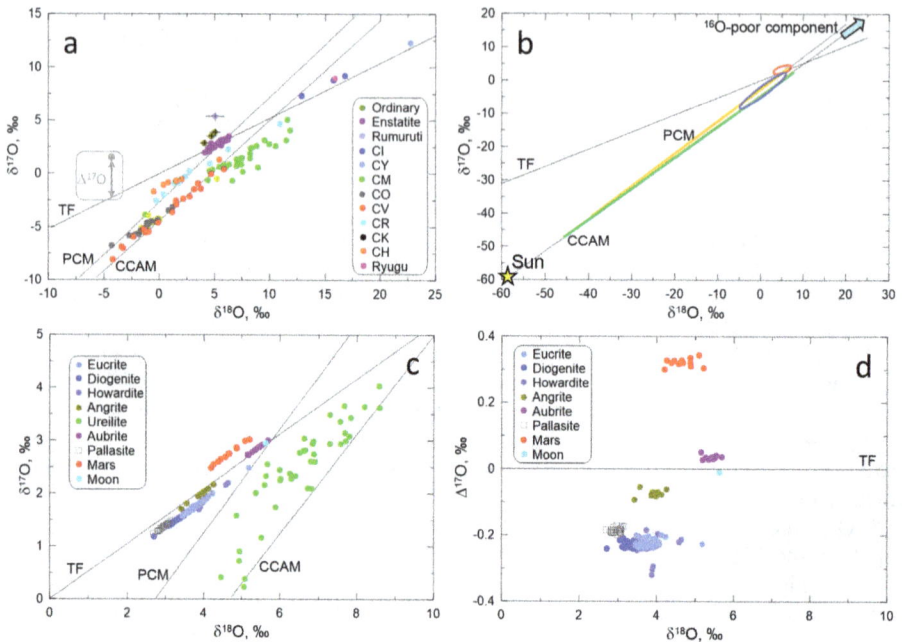

Figure 6. Oxygen isotope compositions of meteorites and their components. Reference lines are: TF, terrestrial fractionation line (slope = 0.52 on $\delta^{17}O$ vs. $\delta^{18}O$, slope = 0 on $\Delta^{17}O$ vs. $\delta^{18}O$); CCAM, carbonaceous chondrite anhydrous mineral line (slope = 0.94 on $\delta^{17}O$ vs. $\delta^{18}O$); PCM, primitive chondrule mixing line (slope = 1 on $\delta^{17}O$ vs. $\delta^{18}O$). **a)** Oxygen isotope compositions for bulk chondrites and asteroid Ryugu. Data points for carbonaceous chondrites are individual analyses; data points for the H, L and LL groups of ordinary chondrites, and Rumuruti (R) group chondrites, are means for individual groups with 2σ standard deviation indicated. **b)** Range of oxygen isotope compositions in chondrite components and the Sun. Solar value is from McKeegan et al. (2011). Open ellipses show fields for bulk compositions of chondrules in ordinary chondrites (**red**) and selected groups of carbonaceous chondrites that have not undergone extensive aqueous alteration (**blue**). These fields are similar to fields for bulk chondrites in (a): note difference in axis scales. The orange line shows the array of oxygen isotope compositions from individual in situ measurements of chondrule mineral grains, which includes OC and CC chondrules. The green line shows the array of oxygen isotope compositions from individual in situ measurements of grains in CAIs. The arrow points towards a "heavy", ^{16}O-poor oxygen isotope component on the slope-1 line. **c)** Bulk meteorite oxygen isotope compositions for selected groups of achondrites, and martian meteorites. Data points are individual analyses apart from lunar basalts which is an average from Hallis et al. (2010). Most achondrite groups and martian meteorites lie on distinct linear arrays with different values of $\Delta^{17}O$. In contrast, ureilites show a wide scatter. **d)** Achondrite, lunar and martian meteorite data from (c), excluding ureilites. Data are from: Clayton and Mayeda (1999), Franchi et al. (1999), Hallis et al. (2010), Bischoff et al. (2011), McKeegan et al. (2011), Greenwood et al. (2015, 2023), Greenwood et al. (2017) and a compilation therein.

exchanged oxygen with a "heavy", ^{16}O-poor, component. This isotopic component could have been introduced into the disk from the molecular cloud, in the form of H_2O ice that originated from a region where CO self-shielding produced H_2O enhanced in ^{17}O and ^{18}O (e.g., Dauphas and Schauble 2016; Ireland et al. 2020). A possible identification of the ^{16}O-poor component is a "cosmic symplectite" (COS: an intergrowth of the iron oxide, magnetite and the iron-sulphide mineral, pentlandite) observed in a primitive chondrite, which has $\delta^{18}O$ values of +180‰, lying on an extension of the slope-one lines (Sakamoto et al. 2007; Seto et al. 2008: Fig. 6b). However, this is a unique and rare occurrence. Further evidence for a similar ^{16}O-poor component comes from hydrated interplanetary dust particles (IDPs) which likely record aqueous alteration in Kuiper belt bodies (Snead et al. 2017; Keller and Flynn 2022).

An alternative interpretation of the ^{16}O-poor endmember is that it had an oxygen isotope composition of Δ^{17}O ~0 ‰; this composition was established very early, during thermal processing of ^{16}O-rich solar materials and water-dominated ice (Δ^{17}O ~ +24‰), in regions of the disk with elevated (dust + ice)/gas ratios (Alexander et al. 2017).

In contrast to chondrites, most achondrites, as well as mesosiderites and (main group) pallasites, show well-defined mass-dependent fractionation trends along arrays of slope 0.52, parallel to the TF line, with separate fractionation lines for each group (Fig. 6c,d). For example, the HEDs, angrites, and aubrites have mean Δ^{17}O values of –0.24, –0.07 and +0.03 ‰ respectively (Greenwood et al. 2017). (Oxygen isotope compositions of mesosiderites overlap with those of the HEDs, and they may be derived from the same parent body, Vesta: Haba et al. 2019.) Martian meteorites also define a unique fractionation line, with a Δ^{17}O value of +0.32 ‰ (Franchi et al. 1999; Fig. 6c,d). The Moon has a near-identical Δ^{17}O value to the Earth (Hallis et al. 2010; Cano et al. 2020; Fig. 6c,d), which can be attributable to either similar oxygen isotope compositions of the Earth and the impactor, Theia, or mixing during the Moon-forming giant impact (Cano et al. 2020). These observations illustrate that Earth, Mars, possibly Theia, and the large asteroid 4 Vesta (thought to be the source of HEDs) each has a unique bulk oxygen isotope composition, and that oxygen isotope heterogeneity also occurs at a scale of individual differentiated planetesimals.

One group of achondrites, the ureilites, shows considerable heterogeneity in oxygen isotope compositions (Fig. 6c). The range of individual oxygen isotope analyses is comparable to the heterogeneity in carbonaceous chondrites, and individual analyses do not lie on a mass-dependent fractionation line. This indicates that the ureilite parent body did not undergo global-scale melting (e.g., Greenwood et al. 2017).

Oxygen isotope compositions of essentially all solid bodies that we have currently sampled lie within a limited range, with d^{18}O values of most bulk meteorites lying between –5 and +10 ‰ (Fig. 6). This suggests an overall similar degree of exchange between solar values and the proposed ^{16}O-poor component. It also requires that a large mass of material is affected by the exchange process, which in turn requires a large mass of isotopically heavy oxygen. Whatever the end-members of isotope exchange, refractory inclusions and chondrules capture the processes and environment in which exchange was taking place. The process that controls planet-scale and asteroid-scale heterogeneity in oxygen isotopes must relate to the timing and location of formation of primary accretionary building blocks, and the spatial extent of the accretionary zone from which each body is constructed. Although there is general acceptance of a model for two-component oxygen isotope exchange, it is important to understand the disk dynamics necessary to account for the preservation of planet-scale and planetesimal-scale oxygen isotope heterogeneity in an environment where large-scale isotopic exchange processes are taking place.

Nucleosynthetic isotope heterogeneity in planetary materials

Much recent work on meteorites has focussed on observations of mass-independent, nucleosynthetic isotopic anomalies that show an isotopic division, or dichotomy, among planetary materials (Trinquier et al. 2009; Warren 2011; Dauphas and Schauble 2016; Bermingham et al. 2020; Kleine et al. 2020; Kruijer et al. 2020). Nucleosynthetic isotope variations reflect the fact that the products of stellar nucleosynthesis were not fully homogenised in the protoplanetary disk. Isotopic diagrams that illustrate these effects (such as Fig. 7) are typically shown in epsilon units, ε, which reference the ratio of the isotope in question to another isotope of the same element, and to a standard, in parts per 10,000. As an example, ε^{50}Ti = [(^{50}Ti/^{47}Ti)$_{sample}$ / (^{50}Ti/^{47}Ti)$_{standard}$ – 1] × 10,000. One ε unit therefore represents a 0.01% deviation in the isotopic ratio of the sample relative to the standard.

Two distinct isotopic reservoirs have been defined, on the basis of isotope ratios for multiple elements. These are named the carbonaceous chondrite (CC) and non-carbonaceous chondrite

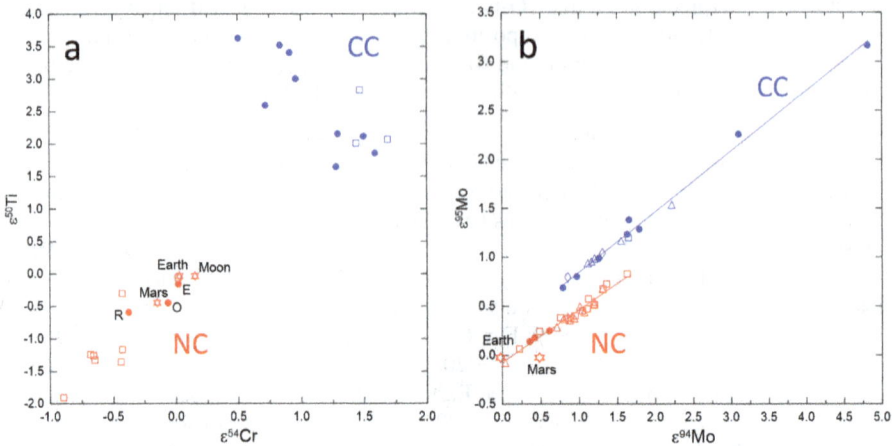

Figure 7. Nucleosynthetic isotope variations in meteorites: carbonaceous chondrite (CC) isotopic group in **blue**, non-carbonaceous (NC) isotopic group in **red**. Chondrite groups are **filled circles**, achondrites **squares**, pallasites **diamonds**, iron meteorites **triangles**, planets **star symbols**. **a)** ε^{50}Ti vs. ε^{54}Cr. Data are from a compilation by Rüfenacht et al. (2023), also Trinquier et al. (2009) and Sanborn et al. (2013, 2019). **b)** ε^{95}Mo vs. ε^{94}Mo. Data are from a compilation by Budde et al. (2019), also Burkhardt et al. (2011).

(NC) reservoirs, since they are largely defined by carbonaceous vs. other chondrite groups (Fig. 7). However, the division is not limited to chondrites alone: although most achondrites fall in the NC group, some ungrouped achondrites lie in the CC group (Sanborn et al. 2013, 2019). In Figure 7, CC-group achondrite data points are for individual meteorites whereas the NC-group data points are averages for different large achondrite groups including HEDs, aubrites, angrites and ureilites. The two isotopic groups are clearly distinguished on plots of ε^{50}Ti (reference isotope ^{47}Ti) vs. ε^{54}Cr (reference isotope ^{52}Cr) (Fig. 7a): ^{50}Ti and ^{54}Cr are neutron-rich isotopes thought to be produced in similar astrophysical environments, but the site and method of their production are still somewhat uncertain (Qin and Carlson 2016). The two isotopic groups are also commonly shown on plots of Δ^{17}O vs. ε^{54}Cr (e.g., Scott et al. 2018; Kruijer et al. 2020), although oxygen isotope variations are not nucleosynthetic in origin (see above).

Molybdenum isotopes are particularly useful for defining the two nucleosynthetic isotope reservoirs because isotopic heterogeneities in Mo can be attributed to carriers of Mo nuclides produced in different nucleosynthetic processes. Also, Mo is measurable in a wider range of materials, including iron-metal-rich meteorites (Fig. 7b). On a plot of ε^{95}Mo vs. ε^{94}Mo (both referenced to ^{96}Mo), CC-group and NC-group meteorites define two separate sub-parallel arrays (Fig. 7b). Each separate array is produced as a result of variable mixing between an *s*-process component (^{96}Mo) and a component with a fixed ratio of *p*-process (^{94}Mo) to *r*-process nuclides (^{95}Mo is produced in both the *r*-process and *s*-process). The offset between the CC and NC arrays reflects different ratios of *r*-process to *p*-process nuclides, and could be interpreted as a uniform *r*-process excess in the CC reservoir (Kruijer et al. 2020). A key observation is that there are iron meteorite and pallasite members of both the CC and NC Isotopic groups (Kruijer et al. 2017a; Budde et al. 2019: Fig. 7b). The CC-group pallasites are the rare Eagle Station and Milton groups which represent less than 10 meteorites out of a total of 170 known pallasites. The CC-group iron meteorites are also some of the less abundant groups (see Fig. 4): the relevant groups are represented in total by < 100 meteorites out of 1375 known iron meteorites. Given that carbonaceous chondrites are themselves a small proportion of all chondrites (5%, see above), our current sampling of extraterrestrial material is strongly weighted to the NC isotopic group. To date, we are only able to place two planets on the nucleosynthetic isotope diagrams, Mars and the Earth / Moon system (Earth and the Moon are essentially isotopically indistinguishable). Both Mars and Earth belong to the NC group (Fig. 7), consistent with their formation predominantly from material similar to enstatite and ordinary chondrites, and/or associated differentiated planetesimals. Enstatite chondrites are a match to the Earth for multiple (but not all) isotopic systems, which leads to the suggestion that Earth and enstatite chondrites formed from the same isotopic reservoir, even though they are chemically distinct (Javoy et al. 2010; Dauphas and Schauble 2016; Dauphas 2017).

The prevailing explanation for the two isotopic reservoirs is that the protoplanetary disk was divided into two regions because growth of Jupiter created a gap in the disk, preventing mixing of inner disk (NC) and outer disk (CC) materials (Kruijer et al. 2017a, 2020; Desch et al. 2018; Bermingham et al. 2020; Kleine et al. 2020). Models with additional components such as CAI-like dust have also been proposed (e.g., Yap and Tissot 2023). This model leads to a common assumption that the CC isotopic reservoir consists of outer disk material that has higher volatile element contents than those of the OC reservoir, and an inference that the two reservoirs are "wet" (CC) and "dry" (NC), forming outside and inside the water snow line. However, in detail the division is not so clear. As discussed above, there is ample evidence that ordinary chondrites and NC-group achondrites accreted water ice, and NC-group chondrites, including ordinary and enstatite chondrites, as well as (NC) ureilites accreted carbon, including organic material. Ureilites also have oxygen isotope compositions similar to carbonaceous chondrites (Fig. 6), which would not be expected for a simple divided disk. The nitrogen inventory of Earth could also be accounted for entirely by accretion from inner Solar System reservoirs (Grewal et al. 2021; Steller et al. 2022). Among the iron meteorites, CC-group irons and NC-group irons have variable and overlapping Ge contents (Fig. 4). Moderately volatile elements such as Mn, Na and K are depleted in most groups of carbonaceous chondrites relative to solar abundances, and to a greater extent than in ordinary chondrites. The enstatite chondrites also have relatively high abundances of moderately volatile elements, even though they are commonly assumed to have accreted close to the Sun. Hence, the inner Solar System was not completely devoid of volatile elements during planetesimal accretion and volatility trends are not systematic.

Recognition of the nucleosynthetic isotope dichotomy has sparked much discussion of disk dynamics that are relevant to observations of other young planetary systems undergoing planet formation. There is also debate about the origin of nucleosynthetic isotope anomalies, for example whether they represent addition or subtraction of presolar carrier components, such as supernova dust, to the inner vs. outer Solar System, and when such processes might have taken place in relation to the timing of planet formation (e.g., Qin and Carlson 2016; Nagashima et al. 2018; Ek et al. 2020; Lichtenberg et al. 2021; Hopp et al. 2022). There is a need to reconcile such models with variation in oxygen isotope compositions, bulk chemical compositions, and the relationship of planetesimal or planet accretion zones to the snow line. If CC materials formed beyond the orbit of Jupiter, later perturbations of the system must account for the current presence of CC-type asteroids (both chondritic and differentiated) in the main asteroid belt. Also, disk models must account for the fact that physical processes, such as the heating events that formed chondrules, and differentiation of early-formed planetesimals, took place in both regions of the disk.

TIMELINE OF EVENTS INFERRED FROM METEORITES

Meteorites and returned samples allow us to measure the chronology (timing) of the formation and evolution of Solar System bodies with high accuracy and precision. Two types of chronometers are used, both based on radiometric methods. Firstly, absolute chronometers provide a direct measurement of the date of a particular process, using radioactive decay schemes with long (billions of years) half-lives. For example, using the decay of uranium to lead, a precise formation age for calcium-, aluminum-rich inclusions (CAIs) has been determined, 4567.30 ± 0.16 Ma (Connelly et al. 2012). This age defines the timing of formation of the first solid particles within the protoplanetary disk, and it is taken to be time $t = 0$, as a reference for subsequent events. (Two recent evaluations of early Solar System chronology by Desch et al. (2023a) and Piralla et al. (2023) suggest that CAI ages ($t = 0$) of 4,568.4 or 4,568.7 Ma should be adopted.) The second dating method uses relative chronometers, which are based on decay of short-lived radioisotopes that were present in the early Solar System. Since these isotopes have now decayed completely, it is not possible to use a conventional chronometer approach. Instead, they must be referenced to a measurement that anchors a precise absolute age to a known abundance of the radioactive isotope, relative to a stable isotope of the same element. As an example, decay of ^{26}Al to ^{26}Mg has a half-life of 0.7 m.y.. The initial $^{26}Al/^{27}Al$ ratio, determined from CAIs, is 5.25×10^{-5}, and this is taken to be the reference ratio at $t=0$ (Kita et al. 2013). A chondrule might have an initial $^{26}Al/^{27}Al$ ratio of around 1×10^{-5}, indicating that it formed around 2 million years after $t=0$. The absolute age of the chondrule has not been measured, but the relative chronology is established. A second important short-lived chronometer, decay of ^{182}Hf to ^{182}W (half-life 8.9 m.y.) is used to determine the timing of core formation (Kleine et al. 2009). In a molten planetesimal or planet, Hf, a lithophile (rock-seeking) element, partitions into the mantle and W, a siderophile (metal-seeking) element, partitions into the core. If core formation is completed before ^{182}Hf has completely decayed, an excess of ^{182}W will be observed in mantle materials. Use of relative chronometers relies on the assumption that the radioactive isotope is distributed homogeneously throughout the protoplanetary disk, and such assumptions are questioned repeatedly (e.g., Connelly et al. 2017; Nagashima et al. 2018; Bollard et al. 2019; Desch et al. 2023a,b; Piralla et al. 2023).

Detailed dating of planetary processes establishes the timeline of events in the protoplanetary disk. The older view that chondrites are the oldest materials, and that differentiated planetesimals formed later, has been overturned. There is now abundant evidence for early core formation, and it is clear from achondrites and iron meteorites that accretion and melting of many differentiated parent bodies took place within two million years after CAI formation, and that the parent bodies of most iron meteorites accreted in the first million years (Kleine et al. 2009; Desch et al. 2018; Bermingham et al. 2020; Kruijer et al. 2020; Lichtenberg et al. 2021). This is consistent with early-formed planetesimals having a higher abundance of ^{26}Al, an important heat source, leading to extensive melting. Planetesimals that accreted later (e.g., 2–4 m.y. after CAIs), when ^{26}Al had substantially decayed, did not reach such high peak temperatures: these are the parent bodies of chondritic meteorites in which primitive disk materials are preserved. This model opens questions such as whether the materials that accreted to form differentiated planetesimals were chondritic, i.e., whether they included chondrules and matrix, and whether the chondrule formation period extended to a time prior to formation of differentiated planetesimals. (Some chondrule-formation models, such as formation in the bow shock of a planetesimal migrating through the disk, require planetesimals to have formed before chondrules: Morris and Boley 2018.) Most ^{26}Al chondrule ages are around 1–4 m.y. after CAI formation (Kita and Ushikubo 2012; Nagashima et al. 2018), but Pb–Pb ages extend chondrule formation to older dates overlapping with CAI formation (Connelly et al. 2017). A further perennial question is how CAIs were stored for 1–2 m.y. until they mixed with other chondrite components to form chondritic planetesimals, and how they were distributed throughout the disk from their presumed initial locations close to the Sun (Desch et al. 2018; Jongejan et al. 2023).

Heating of chondritic planetesimals as a result of ^{26}Al decay began soon after accretion. Using the chronometer based on decay of short-lived ^{53}Mn to ^{53}Cr (half-life 3.7 m.y.), it has been shown that aqueous alteration resulted in carbonate formation, at times of around 3–5 m.y. after CAI formation (e.g., Fujiya et al. 2012). Carbonate ages from asteroid Ryugu are significantly younger, 1.8 m.y. after CAI formation, indicating very early accretion and heating of the initial

Ryugu parent body (McCain et al. 2023). More extensive heating, and slow cooling from the high peak temperatures of metamorphism, such as in ordinary chondrites, continued for tens of millions of years (e.g., Blackburn et al. 2017). On differentiated planetesimals, achondrites such as angrites and eucrites crystallised within a few million years of CAI formation (e.g., Hublet et al. 2017). Since this early period of geologic activity, essentially for the last 4.5 billion years, the dominant geologic process on asteroid-sized bodies has been impacts. The record of impacts includes brecciation (i.e., solid-state fragmentation), heating to varying degrees above and beyond the point of melting, and high-pressure shock effects.

Accretion of the terrestrial planets occurred very early, likely within the first few tens of million years after CAI formation (e.g., Kruijer et al. 2017b; Halliday and Canup 2023), and the Moon-forming giant impact could have taken place as early as 60 m.y. after CAI formation (Barboni et al. 2017; Thiemens et al. 2019). The oldest mineral grains, zircons, from the Earth and Moon record ages of 4.38 Ga and 4.4 Ga, respectively (Harrison 2009; Nemchin et al. 2009), constraining the timing of crust formation.

METEORITES AND PLANET FORMATION: OVERVIEW

The above discussion illustrates how meteorites contribute significantly to our fundamental understanding of formation of planetary systems. They enable us to address the chemical evolution of disk materials, the chemical and isotopic heterogeneity of planets and planetesimals, and the timing of accretion and early geologic processes on rocky bodies. While fundamental major differences among the planets such as the relative sizes of planetary cores have been understood for a long time, it is only through laboratory studies of meteorites and returned samples that the details of mantle geochemistry can be surveyed.

Timing and location are fundamental to establishing the bulk chemical and isotopic properties of individual planets. Factors related to location of the planet-forming region include abundances of moderately volatile elements (such as Mn, Na, S, Ge), as well as volatile compounds such as ices (H_2O, CO/CO_2) and organic material, the availability of which would be expected to vary initially as a function of heliocentric distance, hence temperature, in the protoplanetary disk. However, any systematic, volatility-controlled chemical gradation in the disk may be overprinted by planet migration, and the consequent perturbation of the distribution of small bodies (Walsh et al. 2011; Morbidelli et al. 2015; Nesvorný 2018). Identifying the feeder zones for the terrestrial planets is therefore a complex problem, as exemplified by continuing discussion of the chemical and isotopic nature of meteorite components that could have contributed to the bulk composition of the Earth and Mars (e.g., Dauphas and Schauble 2016; Budde et al. 2019; Liebske and Kahn 2019; Mezger et al. 2020; Sossi et al. 2022; Halliday and Canup 2023; Paquet et al. 2023). Factors related to timing include the changing nature of available material as the snow line evolves, and the timing of planet formation in relation to formation of a Jupiter-related gap in the disk. Also, early-formed planetesimals are more likely to have differentiated, and could potentially have been degassed, prior to accretion to a planet (Dhaliwal et al. 2018; Hirschmann et al. 2021), while later-formed chondritic planetesimals are more likely to preserve low-temperature components. Giant impacts between planet-sized bodies, such as the Moon-forming impact (Asphaug 2014; Lock et al. 2018), and impacts that could have removed a large part of Mercury's mantle (Asphaug and Reufer 2014; Chau et al. 2018), also contribute to more randomised planetary chemistry. This is particularly the case in the early stages of planet formation when major collisions are more likely. Therefore, interpretations of planet compositions, including exoplanets, from remote observations, need to bear in mind that the current composition could have been modified substantially from material that originally accreted.

To sum up, it is serendipitous that small bodies are preserved as asteroids in the inner regions of our Solar System. The asteroid population preserves protoplanetary disk material with a wide variety of compositions, as well as a detailed chronology, and delivers the record of it to Earth in the form of meteorites. If we were to attempt to understand the Solar System without the benefit of insights from meteorites, we would be ignorant of much of our current understanding, and we would likely have a very different perspective of the formation and evolution of the terrestrial planets. Even though we are only able to build this detailed picture for a single planetary system, the lessons learned can be applied to interpretations of other planetary systems, and the nature of exoplanets that they contain.

ACKNOWLEDGEMENTS

This work was partially supported by STFC grant ST/V000675/1. The author is grateful to Jamie Gilmour, Rajdeep Dasgupta, and an anonymous reviewer for discussions and reviews that improved the manuscript.

REFERENCES

Abreu NM, Brearley AJ (2010) Early solar system processes recorded in the matrices of two highly pristine CR3 carbonaceous chondrites, MET 00426 and QUE 99177. Geochim Cosmochim Acta 74:1146–1171

Agee CB, Wilson NV, McCubbin FM, Ziegler K, Polyak VJ, Sharp ZD, Asmerom Y, Nunn MH, Shaheen R, Thiemens MH, Steele A (2013) Unique meteorite from early Amazonian Mars: Water-rich basaltic breccia Northwest Africa 7034. Science 339:780–785

Alexander CMO'D (2019) Quantitative models for the elemental and isotopic fractionations in chondrites: The carbonaceous chondrites. Geochim Cosmochim Acta 254:277–309

Alexander CMO'D, Fogel M, Yabuta H, Cody G (2007) The origin and evolution of chondrites recorded in the elemental and isotopic compositions of their macromolecular organic matter. Geochim Cosmochim Acta 71:4380–4403

Alexander CMO'D, Bowden R, Fogel ML, Howard KT, Herd CD, Nittler LR (2012) The provenances of asteroids, and their contributions to the volatile inventories of the terrestrial planets. Science. 337:721–723

Alexander CMO'D, Bowden R, Fogel ML, Howard KT (2015) Carbonate abundances and isotopic compositions in chondrites. Meteorit Planet Sci 50:810–833

Alexander CMO'D, Nittler LR, Davidson J, Ciesla FJ (2017) Measuring the level of interstellar inheritance in the solar protoplanetary disk. Meteorit Planet Sci 52:1797–821

Alexander CMO'D, McKeegan KD, Altwegg K (2018) Water reservoirs in small planetary bodies: Meteorites, asteroids, and comets. Space Sci Rev 214:36

Asphaug E (2014) Impact origin of the Moon? Annu Rev Earth Planet Sci 42:551–578

Asphaug E, Reufer A (2014) Mercury and other iron-rich planetary bodies as relics of inefficient accretion. Nat Geosci 7:564–568

Barboni M, Boehnke P, Keller B, Kohl IE, Schoene B, Young ED, McKeegan KD (2017) Early formation of the Moon 4.51 billion years ago. Sci Adv 3:e1602365

Bates HC, King AJ, Shirley KS, Bonsall E, Schroeder C, Wombacher F, Fockenberg T, Curtis RJ, Bowles NE (2023) The bulk mineralogy, elemental composition, and water content of the Winchcombe CM chondrite fall. Meteorit Planet Sci, https://doi.org/10.1111/maps.14043

Benedix GK, Haack H, McCoy TJ (2014) Iron and stony-iron meteorites. *In*: Meteorites and Cosmochemical Processes, Treatise on Geochemistry vol 1, Davis AM (ed) Oxford: Elsevier–Pergamon p 267–285

Bermingham KR, Füri E, Lodders K, Marty B (2020) The NC–CC isotope dichotomy: Implications for the chemical and isotopic evolution of the early Solar System. Space Sci Rev 216:1–29

Bischoff A, Vogel N, Roszjar J (2011) The Rumuruti chondrite group. Chem Erde 71:101–133

Blackburn T, Alexander CM, Carlson R, Elkins-Tanton LT (2017) The accretion and impact history of the ordinary chondrite parent bodies. Geochim Cosmochim Acta 200:201–217

Boesenberg JS, Delaney JS, Hewins RH (2012) A petrological and chemical re-examination of Main Group pallasite formation. Geochim Cosmochim Acta 89:134–158

Bollard J, Kawasaki N, Sakamoto N, Olsen M, Itoh S, Larsen K, Wielandt D, Schiller M, Connelly JN, Yurimoto H, Bizzarro M (2019) Combined U-corrected Pb–Pb dating and ^{26}Al–^{26}Mg systematics of individual chondrules— Evidence for a reduced initial abundance of ^{26}Al amongst inner Solar System chondrules. Geochim Cosmochim Acta 260:62–83

Bottke WF, Norman MD (2017) The late heavy bombardment. Annu Rev Earth Planet Sci 45:619–47

Braukmüller N, Wombacher F, Hezel DC, Escoube R, Münker C (2018) The chemical composition of carbonaceous chondrites: Implications for volatile element depletion, complementarity and alteration. Geochim Cosmochim Acta 239:17–48

Brearley AJ, Jones RH (1998) Chondritic meteorites. Rev Mineral Geochem 36:3-01–3-398

Brearley AJ, Jones RH (2018) Halogens in chondritic meteorites. *In:* The Role of Halogens in Terrestrial and Extraterrestrial Geochemical Processes. Harlov DE, Aranovich L (eds) Springer Cham, p 871–958

Broadley MW, Bekaert DV, Piani L, Füri E, Marty B (2022) Origin of life-forming volatile elements in the inner Solar System. Nature 611:245–255

Budde G, Burkhardt C, Kleine T (2019) Molybdenum isotopic evidence for the late accretion of outer Solar System material to Earth. Nat Astron 3:736–741

Buono AS, Walker D (2011) The Fe-rich liquidus in the Fe–FeS system from 1 bar to 10 GPa. Geochim Cosmochim Acta 75:2072–2087

Burkhardt C, Kleine T, Oberli F, Pack A, Bourdon B, Wieler R (2011) Molybdenum isotope anomalies in meteorites: Constraints on solar nebula evolution and origin of the Earth. Earth Planet Sci Lett 312:390–400

Cano EJ, Sharp ZD, Shearer CK (2020) Distinct oxygen isotope compositions of the Earth and Moon. Nat Geosci 13:270–274

Cassata WS, Cohen BE, Mark DF, Trappitsch R, Crow CA, Wimpenny J, Lee MR, Smith CL (2018) Chronology of martian breccia NWA 7034 and the formation of the martian crustal dichotomy. Sci Adv 4:eaap8306

Chau A, Reinhardt C, Helled R, Stadel J (2018) Forming Mercury by giant impacts. Astrophys J 865:835

Chen J, Jolliff BL, Wang A, Korotev RL, Wang K, Carpenter PK, Chen H, Ling Z, Fu X, Ni Y, Cao H (2019) Petrogenesis and shock metamorphism of basaltic lunar meteorites Northwest Africa 4734 and 10597. J Geophys Res: Planets 124:2583–2598

Clayton RN, Mayeda TK (1999) Oxygen isotope studies of carbonaceous chondrites. Geochim Cosmochim Acta 63:2089–2104

Connolly Jr HC, Jones RH (2016) Chondrules: The canonical and noncanonical views. J Geophys Res: Planets 121:1885–1899

Connelly JN, Bizzarro M, Krot AN, Nordlund Å, Wielandt D, Ivanova MA (2012) The absolute chronology and thermal processing of solids in the solar protoplanetary disk. Science 338:651–655

Connelly JN, Bollard J, Bizzarro M (2017) Pb–Pb chronometry and the early solar system. Geochim Cosmochim Acta 201:345–363

Cuzzi JN, Davis SS, Dobrovolskis AR (2003) Blowing in the wind. II Creation and redistribution of refractory inclusions in a turbulent protoplanetary nebula. Icarus 166:385–402

Dasgupta R, Grewal DS (2019) Origin and early differentiation of carbon and associated life-essential volatile elements on Earth. *In:* Deep Carbon: Past to Present. Orcutt B, Daniel I, Dasgupta R (eds) Cambridge University Press, Cambridge, p 4–39

Dauphas N (2017) The isotopic nature of the Earth's accreting material through time. Nature 541:521–524

Dauphas N, Schauble EA (2016) Mass fractionation laws, mass-independent effects, and isotopic anomalies. Annu Rev Earth Planet Sci 44:709–783

DeMeo FE, Carry B (2014) Solar System evolution from compositional mapping of the asteroid belt. Nature 505:629–634

DeMeo FE, Burt BJ, Marsset M, Polishook D, Burbine TH, Carry B, Binzel RP, Vernazza P, Reddy V, Tang M, Thomas CA (2022) Connecting asteroids and meteorites with visible and near-infrared spectroscopy. Icarus 380:114971

Desch SJ, Morris MA, Connolly Jr HC, Boss A (2012) The importance of experiments: Constraints on chondrule formation models. Meteorit Planet Sci 47:1139–1156

Desch SJ, Kalyaan A, Alexander CM (2018) The effect of Jupiter's formation on the distribution of refractory elements and inclusions in meteorites. Astrophys J Supp 238:11

Desch SJ, Dunlap DR, Dunham ET, Williams CD, Mane P (2023a) Statistical chronometry of meteorites. I A Test of ^{26}Al homogeneity and the Pb–Pb age of the solar system's $t = 0$. Icarus 402:115607

Desch SJ, Dunlap DR, Williams CD, Mane P, Dunham ET (2023b) Statistical chronometry of meteorites: II. Initial abundances and homogeneity of short-lived radionuclides. Icarus 402:115611

Dhaliwal JK, Day JM, Moynier F (2018) Volatile element loss during planetary magma ocean phases. Icarus 300:249–260

Dobrică E, Brearley AJ (2020) Amorphous silicates in the matrix of Semarkona: The first evidence for the localized preservation of pristine matrix materials in the most unequilibrated ordinary chondrites. Meteorit Planet Sci 55:649–668

Dodds KH, Bryson JF, Neufeld JA, Harrison RJ (2021) The thermal evolution of planetesimals during accretion and differentiation: Consequences for dynamo generation by thermally-driven convection. J Geophys Res: Planets 126:e2020JE006704

Ek M, Hunt AC, Lugaro M, Schönbächler M (2020) The origin of *s*-process isotope heterogeneity in the solar protoplanetary disk. Nat Astron 4:273–281

Elkins-Tanton LT, Weiss BP, Zuber MT (2011) Chondrites as samples of differentiated planetesimals. Earth Planet Sci Lett 305:1–10

Elkins-Tanton LT, Asphaug E, Bell III JF, Bercovici H, Bills B, Binzel R, Bottke WF, Dibb S, Lawrence DJ, Marchi S, McCoy TJ (2020) Observations, meteorites, and models: a preflight assessment of the composition and formation of (16) Psyche. J Geophys Res: Planets 125:e2019JE006296

Elkins-Tanton LT, Asphaug E, Bell III JF, Bierson CJ, Bills BG, Bottke WF, Courville SW, Dibb SD, Jun I, Lawrence DJ, Marchi S (2022) Distinguishing the origin of asteroid (16) Psyche. Space Sci Rev 218:17

Filiberto J, Baratoux D, Beaty D, Breuer D, Farcy BJ, Grott M, Jones JH, Kiefer WS, Mane P, McCubbin FM, Schwenzer SP (2016) A review of volatiles in the Martian interior. Meteorit Planet Sci 51:1935–1958

Franchi IA, Wright IP, Sexton AS, Pillinger CT (1999) The oxygen-isotopic composition of Earth and Mars. Meteorit Planet Sci 34:657–661

Fritz J, Greshake A, Fernandes VA (2017) Revising the shock classification of meteorites. Meteorit Planet Sci 52:1216–1232

Fujiya W, Sugiura N, Hotta H, Ichimura K, Sano Y (2012) Evidence for the late formation of hydrous asteroids from young meteoritic carbonates. Nat Commun 3:627

Goldstein JI, Scott ER, Chabot NL (2009) Iron meteorites: Crystallization, thermal history, parent bodies, and origin. Chem Erde 69:293–325

Goodrich CA, Scott ER, Fioretti AM (2004) Ureilitic breccias: Clues to the petrologic structure and impact disruption of the ureilite parent asteroid. Chem Erde 64:283–327

Grady MM, Wright IP (2003) Elemental and isotopic abundances of carbon and nitrogen in meteorites. Space Sci Rev 106:231–248

Grant H, Tartèse R, Jones R, Piani L, Marrocchi Y, King A, Rigaudier T (2023) Bulk mineralogy, water abundance, and hydrogen isotope composition of unequilibrated ordinary chondrites. Meteorit Planet Sci 58:1365–1381

Greenwood RC, Barrat JA, Scott ERD, Haack H, Buchanan PC, Franchi IA, Yamaguchi A, Johnson D, Bevan AW, Burbine TH (2015) Geochemistry and oxygen isotope composition of main-group pallasites and olivine-rich clasts in mesosiderites: Implications for the "Great Dunite Shortage" and HED-mesosiderite connection. Geochim Cosmochim Acta 169:115–136

Greenwood RC, Burbine TH, Miller MF, Franchi IA (2017) Melting and differentiation of early-formed asteroids: The perspective from high precision oxygen isotope studies. Chem Erde 77:1–43

Greenwood RC, Franchi IA, Findlay R, Malley JA, Ito M, Yamaguchi A, Kimura M, Tomioka N, Uesugi M, Imae N, Shirai N (2023) Oxygen isotope evidence from Ryugu samples for early water delivery to Earth by CI chondrites. Nat Astron 7:29–38

Grewal DS, Dasgupta R, Hough T, Farnell A (2021) Rates of protoplanetary accretion and differentiation set nitrogen budget of rocky planets. Nat Geosci 14:369–76

Haba MK, Wotzlaw JF, Lai YJ, Yamaguchi A, Schönbächler M (2019) Mesosiderite formation on asteroid 4 Vesta by a hit-and-run collision. Nat Geosci 12:510–515

Halliday AN, Canup RM (2023) The accretion of planet Earth. Nature Rev Earth Environ 4:19–35

Hallis LJ, Anand M, Greenwood RC, Miller MF, Franchi IA, Russell SS (2010) The oxygen isotope composition, petrology and geochemistry of mare basalts: Evidence for large-scale compositional variation in the lunar mantle. Geochim Cosmochim Acta 74:6885–6899

Harrison TM (2009) The Hadean crust: evidence from > 4 Ga zircons. Annu Rev Earth Planet Sci 37:479–505

Hirschmann MM, Bergin EA, Blake GA, Ciesla FJ, Li J (2021) Early volatile depletion on planetesimals inferred from C–S systematics of iron meteorite parent bodies. Proc Nat Acad Sci 118:e2026779118

Hopp T, Dauphas N, Spitzer F, Burkhardt C, Kleine T (2022) Earth's accretion inferred from iron isotopic anomalies of supernova nuclear statistical equilibrium origin. Earth Planet Sci Lett 577:117245

Hublet G, Debaille V, Wimpenny J, Yin QZ (2017) Differentiation and magmatic activity in Vesta evidenced by ^{26}Al–^{26}Mg dating in eucrites and diogenites. Geochim Cosmochim Acta 218:73–97

Huss GR, Rubin AE, Grossman JN (2006) Thermal metamorphism in chondrites. *In*: Meteorites and the Early Solar System II Lauretta DS, McSween Jr HY (eds), The University of Arizona Press, Tucson AZ, p 567–586

Ireland TR, Avila J, Greenwood RC, Hicks LJ, Bridges JC (2020) Oxygen isotopes and sampling of the solar system. Space Sci Rev 216:225

Izidoro A, Piani L (2022) Origin of water in the terrestrial planets: insights from meteorite data and planet formation models. Elements 18:181–186

Izidoro A, Dasgupta R, Raymond SN, Deienno R, Bitsch B, Isella A (2022) Planetesimal rings as the cause of the Solar System's planetary architecture. Nat Astron 6:357–366

Javoy M, Kaminski E, Guyot F, Andrault D, Sanloup C, Moreira M, Labrosse S, Jambon A, Agrinier P, Davaille A, Jaupart C (2010) The chemical composition of the Earth: Enstatite chondrite models. Earth Planet Sci Lett 293:259–268

Johnson BC, Ciesla FJ, Dullemond CP, Melosh HJ (2018) Formation of chondrules by planetesimal collisions. *In*: Chondrules: Records of Protoplanetary Disk Processes. Russell SS, Krot AN, Connolly Jr HC (eds), Cambridge University Press, p 343–360

Johnson BC, Sori MM, Evans AJ (2020) Ferrovolcanism on metal worlds and the origin of pallasites. Nat Astron 4:41–44

Jones RH (2012) Petrographic constraints on the diversity of chondrule reservoirs in the protoplanetary disk. Meteorit Planet Sci 47:1176–1190

Jones RH, Villeneuve J, Libourel G (2018) Thermal histories of chondrules: Petrologic observations and experimental constraints. *In:* Chondrules: Records of Protoplanetary Disk Processes. Russell SS, Krot AN, Connolly Jr HC (eds), Cambridge University Press, p 57–90

Jongejan S, Dominik C, Dullemond C (2023) The effect of Jupiter on the CAI storage problem Astron Astrophys 679:A45

Joy KH, Arai T (2013) Lunar meteorites: New insights into the geological history of the Moon. Astron Geophys 54:4–28

Joy KH, Gross J, Korotev RL, Zeigler RA, McCubbin FM, Snape JF, Curran NM, Pernet-Fisher JF, Arai T (2023) Lunar meteorites. Rev Mineral Geochem 89:509–562

Joswiak DJ, Brownlee DE, Nguyen AN, Messenger S (2017) Refractory materials in comet samples. Meteorit Planet Sci 52:1612–1648

Kallemeyn GW, Rubin AE, Wang D, Wasson JT (1989) Ordinary chondrites: Bulk compositions, classification, lithophile-element fractionations and composition-petrographic type relationships. Geochim Cosmochim Acta 53:2747–2767

Keil K (2010) Enstatite achondrite meteorites (aubrites) and the histories of their asteroidal parent bodies. Chem Erde 70:295–317

Keil K, Stöffler D, Love SG, Scott ERD (1997) Constraints on the role of impact heating and melting in asteroids. Meteorit Planet Sci 32:349–363

Keller LP, Flynn GJ (2022) Evidence for a significant Kuiper belt dust contribution to the zodiacal cloud. Nat Astron 6:731–735

King AJ, Bates HC, Krietsch D, Busemann H, Clay PL, Schofield PF, Russell SS (2019) The Yamato-type (CY) carbonaceous chondrite group: Analogues for the surface of asteroid Ryugu? Chem Erde 79:125531

King AJ, Schofield PF, Russell SS (2021) Thermal alteration of CM carbonaceous chondrites: Mineralogical changes and metamorphic temperatures. Geochim Cosmochim Acta 298:167–190

Kita NT, Ushikubo T (2012) Evolution of protoplanetary disk inferred from ^{26}Al chronology of individual chondrules. Meteorit Planet Sci 47:1108–1119

Kita NT, Yin QZ, MacPherson GJ, Ushikubo T, Jacobsen B, Nagashima K, Kurahashi E, Krot AN, Jacobsen SB (2013) ^{26}Al–^{26}Mg isotope systematics of the first solids in the early solar system. Meteorit Planet Sci 48:1383–1400

Kleine T, Touboul M, Bourdon B, Nimmo F, Mezger K, Palme H, Jacobsen SB, Yin QZ, Halliday AN (2009) Hf–W chronology of the accretion and early evolution of asteroids and terrestrial planets. Geochim Cosmochim Acta 73:5150–5188

Kleine T, Budde G, Burkhardt C, Kruijer TS, Worsham EA, Morbidelli A, Nimmo F (2020) The non-carbonaceous–carbonaceous meteorite dichotomy. Space Sci Rev 216:1–27

Korotev RL (2005) Lunar geochemistry as told by lunar meteorites. Chem Erde 65:297–346

Krot AN (2019) Refractory inclusions in carbonaceous chondrites: Records of early solar system processes. Meteorit Planet Sci 54:1647–1691

Krot AN, Amelin Y, Cassen P, Meibom A (2005) Young chondrules in CB chondrites from a giant impact in the early Solar System. Nature 436:989–992

Krot AN, Keil K, Scott ER, Goodrich CA, Weisberg MK (2014) Classification of meteorites and their genetic relationships. *In*: Meteorites and Cosmochemical Processes, Treatise on Geochemistry vol 1, Davis AM (ed) Oxford: Elsevier–Pergamon, p 1–63

Kruijer TS, Burkhardt C, Budde G, Kleine T (2017a) Age of Jupiter inferred from the distinct genetics and formation times of meteorites. Proc Nat Acad Sci 114:6712–6716

Kruijer TS, Kleine T, Borg LE, Brennecka GA, Irving AJ, Bischoff A, Agee CB (2017b) The early differentiation of Mars inferred from Hf–W chronometry. Earth Planet Sci Lett 474:345–354

Kruijer TS, Kleine T, Borg LE (2020) The great isotopic dichotomy of the early Solar System. Nat Astron 4:32–40

Lewis JA, Jones RH, Brearley AJ (2022) Plagioclase alteration and equilibration in ordinary chondrites: Metasomatism during thermal metamorphism. Geochim Cosmochim Acta 316:201–229

Lichtenberg T, Drążkowska J, Schönbächler M, Golabek GJ, Hands TO (2021) Bifurcation of planetary building blocks during Solar System formation. Science 371:365–70

Liebske C, Khan A (2019) On the principal building blocks of Mars and Earth. Icarus 322:121–34

Liu MC, McCain KA, Matsuda N, Yamaguchi A, Kimura M, Tomioka N, Ito M, Uesugi M, Imae N, Shirai N, Ohigashi T (2022) Incorporation of ^{16}O-rich anhydrous silicates in the protolith of highly hydrated asteroid Ryugu. Nat Astron 6:1172–1177

Lock SJ, Stewart ST, Petaev MI, Leinhardt Z, Mace MT, Jacobsen SB, Cuk M (2018) The origin of the Moon within a terrestrial synestia. J Geophys Res: Planets 123:910–51

Lodders K (2003) Solar system abundances and condensation temperatures of the elements. Astrophys J 591:1220

Lodders K, Fegley B (1998) The Planetary Scientist's Companion. Oxford University Press, USA

Lodders K, Fegley Jr B (2023) Solar system abundances and condensation temperatures of the halogens fluorine, chlorine, bromine, and iodine. Chem Erde:125957

MacPherson GJ (2014) Calcium-aluminum-rich inclusions in chondritic meteorites. *In*: Meteorites and Cosmochemical Processes, Treatise on Geochemistry vol 1, Davis AM (ed) Oxford: Elsevier-Pergamon p 139–179

Marrocchi Y, Piralla M, Regnault M, Batanova V, Villeneuve J, Jacquet E (2022) Isotopic evidence for two chondrule generations in CR chondrites and their relationships to other carbonaceous chondrites. Earth Planet Sci Lett 593C:117683

Marty B (2012) The origins and concentrations of water, carbon, nitrogen and noble gases on Earth. Earth Planet Sci Lett 313:56–66

McCain KA, Matsuda N, Liu MC, McKeegan KD, Yamaguchi A, Kimura M, Tomioka N, Ito M, Imae N, Uesugi M, Shirai N (2023) Early fluid activity on Ryugu inferred by isotopic analyses of carbonates and magnetite. Nat Astron 7:309–317

McCoy TJ, Beck AW, Prettyman TH, Mittlefehldt DW (2015) Asteroid (4) Vesta II: Exploring a geologically and geochemically complex world with the Dawn Mission. Chem Erde 75:273–285

McCubbin FM, Barnes JJ (2019) Origin and abundances of H_2O in the terrestrial planets, Moon, and asteroids. Earth Planet Sci Lett 526:115771

McCubbin FM, Jones RH (2015) Extraterrestrial apatite: Planetary geochemistry to astrobiology. Elements 11:183–188

McCubbin FM, Boyce JW, Novák-Szabó T, Santos AR, Tartèse R, Muttik N, Domokos G, Vazquez J, Keller LP, Moser DE, Jerolmack DJ (2016) Geologic history of Martian regolith breccia Northwest Africa 7034: Evidence for hydrothermal activity and lithologic diversity in the Martian crust. J Geophys Res: Planets 121:2120–2149

McKeegan KD, Kallio AP, Heber VS, Jarzebinski G, Mao PH, Coath CD, Kunihiro T, Wiens RC, Nordholt JE, Moses Jr RW, Reisenfeld DB (2011) The oxygen isotopic composition of the Sun inferred from captured solar wind. Science 332:1528–1532

McSween Jr HY (2015) Petrology on Mars. Am Mineral 100:2380–2395

McSween Jr HY, Ghosh A, Grimm RE, Wilson L, Young ED (2002) Thermal evolution models of asteroids. *In:* Asteroids III Bottke WF, Cellino A, Paolicchi P, Binzel RP (eds), Univ Arizona Press, Tucson, p 559–571, Meteoritical Bulletin Database: https://www.lpi.usra.edu/meteor/metbull.php

Mezger K, Schönbächler M, Bouvier A (2020) Accretion of the earth—missing components? Space Sci Rev 216:27

Mittlefehldt DW (2014) Achondrites. *In*: Meteorites and Cosmochemical Processes, Treatise on Geochemistry vol 1, Davis AM (ed) Oxford: Elsevier-Pergamon p 235–266

Morbidelli A, Walsh KJ, O'Brien DP, Minton DA, Bottke WF (2015) The dynamical evolution of the asteroid belt. *In*: Asteroids IV, Michel P, DeMeo FE, Bottke WF (eds) Univ Arizona Press, Tucson, p 493–507

Morbidelli A, Baillie K, Batygin K, Charnoz S, Guillot T, Rubie DC, Kleine T (2022) Contemporary formation of early Solar System planetesimals at two distinct radial locations. Nat Astron 6:72–79

Morris MA, Boley AC (2018) Formation of chondrules by shock waves. *In:* Chondrules: Records of Protoplanetary Disk Processes. Russell SS, Krot AN, Connolly Jr HC (eds) Cambridge University Press, p 375–399

Nagashima K, Kita NT, Luu T-H (2018) ^{26}Al–^{26}Mg systematics of chondrules. *In:* Chondrules: Records of Protoplanetary Disk Processes. Russell SS, Krot AN, Connolly Jr HC (eds) Cambridge University Press, p 247–275

Nakamura T, Noguchi T, Tanaka M, Zolensky ME, Kimura M, Tsuchiyama A, Nakato A, Ogami T, Ishida H, Uesugi M, Yada T (2011) Itokawa dust particles: a direct link between S-type asteroids and ordinary chondrites. Science 333:1113–1116

Nemchin A, Timms N, Pidgeon R, Geisler T, Reddy S, Meyer C (2009) Timing of crystallization of the lunar magma ocean constrained by the oldest zircon. Nat Geosci 2:133–136

Nesvorný D (2018) Dynamical evolution of the early Solar System. Annu Rev Astron Astrophys 56:137–174

Newcombe ME, Nielsen SG, Peterson LD, Wang J, Alexander CO, Sarafian AR, Shimizu K, Nittler LR, Irving AJ (2023) Degassing of early-formed planetesimals restricted water delivery to Earth. Nature 615:854–857

Nittler LR, Ciesla F (2016) Astrophysics with Extraterrestrial Materials. Annu Rev Astron Astrophys 54:53–93

Papike JJ, Karner JM, Shearer CK (2003) Determination of planetary basalt parentage: A simple technique using the electron microprobe. Am Mineral 88:469–472

Papike JJ, Burger PV, Bell AS, Shearer CK (2017) Mn–Fe systematics in major planetary body reservoirs in the solar system and the positioning of the Angrite Parent Body: A crystal-chemical perspective. Am Mineral 102:1759–1762

Paquet M, Moynier F, Yokoyama T, Dai W, Hu Y, Abe Y, Aléon J, O'D. Alexander CM, Amari S, Amelin Y, Bajo KI (2023) Contribution of Ryugu-like material to Earth's volatile inventory by Cu and Zn isotopic analysis. Nat Astron 7:182–189

Peterson LD, Newcombe ME, Alexander CM, Wang J, Sarafian AR, Bischoff A, Nielsen SG (2023) The H_2O content of the ureilite parent body. Geochim Cosmochim Acta 340:141–157

Piani L, Marrocchi Y, Vacher LG, Yurimoto H, Bizzarro M (2021) Origin of hydrogen isotopic variations in chondritic water and organics. Earth Planet Sci Lett 567:117008

Piani L, Marrocchi Y, Rigaudier T, Vacher LG, Thomassin D, Marty B (2020) Earth's water may have been inherited from material similar to enstatite chondrite meteorites. Science 369:1110–1113

Piralla M, Villeneuve J, Batanova V, Jacquet E, Marrocchi Y (2021) Conditions of chondrule formation in ordinary chondrites. Geochim Cosmochim Acta 313:295–312

Piralla M, Villeneuve J, Schnuriger N, Bekaert DV, Marrocchi Y (2023) A unified chronology of dust formation in the early solar system. Icarus 394:115427

Putirka KD (2024) Exoplanet mineralogy. Rev Mineral Geochem 90:199–258

Qin L, Carlson RW (2016) Nucleosynthetic isotope anomalies and their cosmochemical significance. Geochem J 50:43–65

Righter K, Sutton SR, Danielson L, Pando K, Newville M (2016) Redox variations in the inner solar system with new constraints from vanadium XANES in spinels. Am Mineral 101:1928–1942

Rubin AE (1995) Petrologic evidence for collisional heating of chondritic asteroids. Icarus 113:156–167

Rubin AE (2015) Maskelynite in asteroidal, lunar and planetary basaltic meteorites: An indicator of shock pressure during impact ejection from their parent bodies. Icarus 257:221–229

Rüfenacht M, Morino P, Lai YJ, Fehr MA, Haba MK, Schönbächler M (2023) Genetic relationships of solar system bodies based on their nucleosynthetic Ti isotope compositions and sub-structures of the solar protoplanetary disk. Geochim Cosmochim Acta 355:110–125

Sakamoto N, Seto Y, Itoh S, Kuramoto K, Fujino K, Nagashima K, Krot AN, Yurimoto H (2007) Remnants of the early solar system water enriched in heavy oxygen isotopes. Science 317:231–233

Sanborn ME, Yamakawa A, Yin QZ, Irving AJ, Amelin Y (2013) Chromium isotopic studies of ungrouped achondrites NWA 7325, NWA 2976, and NWA 6704. Meteorit Planet Sci Supp 76:5220

Sanborn ME, Wimpenny J, Williams CD, Yamakawa A, Amelin Y, Irving AJ, Yin QZ (2019) Carbonaceous achondrites Northwest Africa 6704/6693: Milestones for early Solar System chronology and genealogy. Geochim Cosmochim Acta 245:577–596

Scott ERD, Bottke WF (2011) Impact histories of angrites, eucrites, and their parent bodies. Meteorit Planet Sci 46:1878–1887

Scott ERD, Krot AN (2014) Chondrites and their components. *In*: Meteorites and Cosmochemical Processes, Treatise on Geochemistry vol 1, Davis AM (ed) Oxford: Elsevier-Pergamon p 65–137

Scott ERD, Haack H, Love SG (2001) Formation of mesosiderites by fragmentation and reaccretion of a large differentiated asteroid. Meteorit Planet Sci 36:869–881

Scott ERD, Krot AN, Sanders IS (2018) Isotopic dichotomy among meteorites and its bearing on the protoplanetary disk. Astrophys J 854:164

Seto Y, Sakamoto N, Fujino K, Kaito T, Oikawa T, Yurimoto H (2008) Mineralogical characterization of a unique material having heavy oxygen isotope anomaly in matrix of the primitive carbonaceous chondrite Acfer 094. Geochim Cosmochim Acta 72:2723–2734

Snead CJ, McKeegan KD, Keller LP, Messenger S (2017) Ion microprobe measurements of comet dust and implications for models of oxygen isotope heterogeneity in the Solar System. Lunar Planet Sci Conf Abstract 2623

Sossi PA, Stotz IL, Jacobson SA, Morbidelli A, O'Neill HS (2022) Stochastic accretion of the Earth. Nat Astron 6:951–960

Steller T, Burkhardt C, Yang C, Kleine T (2022) Nucleosynthetic zinc isotope anomalies reveal a dual origin of terrestrial volatiles. Icarus 386:115171

Stöffler D, Keil K, Scott ERD (1991) Shock metamorphism of ordinary chondrites. Geochim Cosmochim Acta 55:3845–3867

Storz J, Ludwig T, Bischoff A, Schwarz WH, Trieloff M (2021) Graphite in ureilites, enstatite chondrites, and unique clasts in ordinary chondrites – Insights from the carbon-isotope composition. Geochim Cosmochim Acta 307:86–104

Suttle MD, King AJ, Harrison CS, Chan QH, Greshake A, Bartoschewitz R, Tomkins AG, Salge T, Schofield PF, Russell SS (2023) The mineralogy and alteration history of the Yamato-type (CY) carbonaceous chondrites. Geochim Cosmochim Acta 361:245–64

Tang H, Young ED, Tafla L, Pack A, Di Rocco T, Abe Y, Aléon J, Alexander CM, Amari S, Amelin Y, Bajo KI (2023) The oxygen isotopic composition of samples returned from asteroid Ryugu with implications for the nature of the parent planetesimal. Planet Sci J 4:144

Tartèse R, Anand M, Gattacceca J, Joy KH, Mortimer JI, Pernet-Fisher JF, Russell S, Snape JF, Weiss BP (2019) Constraining the evolutionary history of the Moon and the inner solar system: A case for new returned lunar samples. Space Sci Rev 215:54

Tenner TJ, Ushikubo T, Nakashima D, Schrader DL, Weisberg MK, Kimura M, Kita NT (2018) Oxygen isotope characteristics of chondrules from recent studies by secondary ion mass spectrometry. *In:* Chondrules: Records of Protoplanetary Disk Processes. Russell SS, Krot AN, Connolly Jr HC (eds), Cambridge University Press, p 196–246

Thiemens MM, Sprung P, Fonseca RO, Leitzke FP, Münker C (2019) Early Moon formation inferred from hafnium–tungsten systematics. Nat Geosci 12:696–700

Tissot FL, Collinet M, Namur O, Grove TL (2022) The case for the angrite parent body as the archetypal first-generation planetesimal: Large, reduced and Mg-enriched. Geochim Cosmochim Acta 338:278–301

Trinquier A, Elliott T, Ulfbeck D, Coath C, Krot AN, Bizzarro M (2009) Origin of nucleosynthetic isotope heterogeneity in the solar protoplanetary disk. Science 324:374–376

Ushikubo T, Kimura M, Kita NT, Valley JW (2012) Primordial oxygen isotope reservoirs of the solar nebula recorded in chondrules in Acfer 094 carbonaceous chondrite. Geochim Cosmochim Acta 242–264

Vacher LG, Fujiya W (2022) Recent advances in our understanding of water and aqueous activity in chondrites. Elements 18:175–180

Vacher LG, Piani L, Rigaudier T, Thomassin D, Florin G, Piralla M, Marrocchi Y (2020) Hydrogen in chondrites: Influence of parent body alteration and atmospheric contamination on primordial components. Geochim Cosmochim Acta 281:53–66

Verchovsky AB, Abernethy FA, Anand M, Barber SJ, Findlay R, Franchi IA, Greenwood RC, Grady MM (2023) Quantitative evolved gas analysis: Winchcombe in comparison with other CM2 meteorites. Meteorit Planet Sci, https://doi.org/10.1111/maps.13983

Wakita S, Genda H, Kurosawa K, Davison TM, Johnson BC (2022) Effect of impact velocity and angle on deformational heating and postimpact temperature. J Geophys Res Planets 127:e2022JE007266

Walsh KJ, Morbidelli A, Raymond SN, O'Brien DP, Mandell AM (2011) A low mass for Mars from Jupiter's early gas-driven migration. Nature 475:206–209

Warren PH (2011) Stable-isotopic anomalies and the accretionary assemblage of the Earth and Mars: A subordinate role for carbonaceous chondrites. Earth Planet Sci Lett 311:93–100

Wasson JT, Kallemeyn GW (2002) The IAB iron-meteorite complex: A group, five subgroups, numerous grouplets, closely related, mainly formed by crystal segregation in rapidly cooling melts. Geochim Cosmochim Acta 66:2445–2473

Weiss BP, Elkins-Tanton LT (2013) Differentiated planetesimals and the parent bodies of chondrites. Annu Rev Earth Planet Sci 41:529–560

Williams CD, Sanborn ME, Defouilloy C, Yin QZ, Kita NT, Ebel DS, Yamakawa A, Yamashita K (2020) Chondrules reveal large-scale outward transport of inner Solar System materials in the protoplanetary disk. Proc Nat Acad Sci 117:23426–23435

Yang J, Goldstein JI, Scott ER (2010) Main-group pallasites: Thermal history, relationship to IIIAB irons, and origin. Geochim Cosmochim Acta 74:4471–92

Yap TE, Tissot FL (2023) The NC–CC dichotomy explained by significant addition of CAI-like dust to the Bulk Molecular Cloud (BMC) composition. Icarus 405:115680

Yokoyama T, Nagashima K, Nakai I, Young ED, Abe Y, Aléon J, Alexander CM, Amari S, Amelin Y, Bajo KI, Bizzarro M (2022) Samples returned from the asteroid Ryugu are similar to Ivuna-type carbonaceous meteorites. Science 379:eabn7850

Reviews in Mineralogy & Geochemistry
Vol. 90 pp. 141–170, 2024
Copyright © Mineralogical Society of America

5

The Evolution and Delivery of
Rocky Extra-Solar Materials to White Dwarfs

Dimitri Veras[1,2,3], Alexander J. Mustill[4], Amy Bonsor[5]

[1]*Centre for Exoplanets and Habitability, University of Warwick, Coventry CV4 7AL, UK*
[2]*Centre for Space Domain Awareness, University of Warwick, Coventry CV4 7AL, UK*
[3]*Department of Physics, University of Warwick, Coventry CV4 7AL, UK*
[4]*Lund Observatory, Department of Astronomy and Theoretical Physics, Lund University,
Box 43, 221 00, Lund, Sweden*
[5]*Institute of Astronomy, University of Cambridge, Madingley Road,
Cambridge CB3 0HA, UK*

INTRODUCTION

The most direct and substantial way to measure the geochemistry of extrasolar objects is to perform chemical autopsies on their broken-up remains. However, traditional ways of probing extrasolar planetary systems around Sun-like stars cannot achieve this goal, instead yielding just a planet's bulk density or chemical details of its atmosphere. Fortunately, new techniques have emerged which allow us to perform the desired geochemical postmortems. These techniques work only on *old* planetary systems: specifically, systems which are old enough so that their parent stars have already exhausted their fuel and have evolved into burnt cores known as *white dwarfs*.

Hence, understanding stellar evolution (how the star transforms itself over time) and its effect on planetary systems is crucial for correctly interpreting the chemical constraints of exo-planetary material that can be given to us by white dwarfs. The previous two chapters (Jones 2024; Mordasini and Burns 2024, both this volume) have focussed on the formation of planetary systems and their evolution when they are young and in middle-age. The content in those chapters provide the initial conditions for the late-stage evolution of those systems. The subsequent chapter (Xu et al. 2024, this volume) will then outline the compositional measurements obtained in white dwarf planetary systems.

This chapter will detail the transition from young to old, and describe how asteroids, moons, and comets, as well as boulders, pebbles and dust, evolve into eventual targets for chemical spectroscopy, and how planets and companion stars play a vital role in reshaping system architectures for this purpose. Related reviews from the astrophysical literature which cover these themes include Veras (2016a), Bonsor and Xu (2017), Jackson and Carlberg (2018), van Lieshout and Rappaport (2018), Vanderburg and Rappaport (2018) and Veras (2021); existing reviews which are focussed more on the content of the subsequent chapter (Xu et al. 2024, this volume) are Jura and Young (2014), Farihi (2016), Zuckerman and Young (2018) and Xu and Bonsor (2021).

For pedagogical purposes, we divide the future evolution of planetary systems from their middle-aged states into three stages, each of which is covered in a separate section in this chapter. The first section ("Stage 1") addresses the changes to a planetary system when a Sun-like star undergoes significant variations in mass, size and luminosity as it convulses

1529-6466/24/0090-0005$05.00 (print)
1943-2666/24/0090-0005$05.00 (online)

http://dx.doi.org/10.2138/rmg.2024.90.05

into a white dwarf[1]. This violent transition is known in the astronomy community as the *giant branch phases* of evolution. The second section ("Stage 2") then outlines how planetary system components move around in the system after the parent star has become a white dwarf and eventually reach the white dwarf's *Roche sphere*. The Roche sphere is where objects break up easily and become observable. The final section of the chapter ("Stage 3") details our knowledge of the compact debris environment within the white dwarf Roche sphere. This environment is the immediate precursor to accretion onto the white dwarf, where chemical autopsies can be performed (Xu et al. 2024, this volume).

STAGE 1: TRAVERSING THE GIANT BRANCH PHASES

Introduction

By the time a planetary system is a few tens of Myr old, its constituent planets, moons, asteroids and comets have become fully formed. Later one-off events, hundreds of Myr or several Gyr in the future, may significantly reshape these primordial configurations or bodies. Potential examples of these one-off events in the solar system include the creation of Earth's moon due to a giant impact (Canup 2004) or the reordering of the ice and gas giant planets due to a crossing of a gravitational resonance (Thommes et al. 1999; Tsiganis et al. 2005).

Nevertheless, the older a Sun-like star is, the more likely that its planets have settled into a steady state. Further, although smaller reservoirs of bodies, such as collections of asteroids or planetary rings, are consistently being ground down and replenished, the regions in which they are concentrated would not significantly change unless the planets' orbits do.

However, this relative post-formation quiescence does not last forever. Each star's store of fuel is finite. As the different layers of this fuel, in the form of different elements, are depleted, then the star is transitioning between different giant branch phases. During the giant branch phases the quiescence is broken and the planetary system undergoes significant transformations.

The first layer of the stellar fuel is hydrogen, which the Sun has been fusing into helium for the last 4.6 Gyr. The Sun contains enough hydrogen to continue in its current state, known as the *main sequence*, for approximately the next 6.5 Gyr (Veras 2016b). The vast majority of known exoplanet host stars also have many Gyr remaining on the main sequence before transitioning to the giant branches[2].

Consequently, understanding if and how planetary systems would change significantly during this substantial remaining time on the main sequence is important (Davies et al. 2014). Within the solar system, the four giant planets will remain in their current stable orbits until the end of the Sun's main sequence (Duncan and Lissauer 1998; Veras 2016b; Zink et al. 2020) except in the highly unlikely case that a different star flies close enough to these planets to create a significant perturbation (Brown and Rein 2022).

The evolution of Mercury, Venus, Earth and Mars, is, however, not as straightforward. Primarily due to Mercury's large eccentricity relative to the other planets, Laskar and Gastineau (2009) found in about 1% of their simulations, Mercury will collide with Venus or the Sun, destabilizing the rest of the inner solar system. This 1% value has been subsequently scrutinized. However, because the inner solar system is mathematically chaotic, predicting its future is impossible, even with much more accurate ephemerides (Zeebe 2015; Abbot et al. 2021, 2023; Mogavero and Laskar 2021; Hoang et al. 2022; Brown and Rein 2023; Mogavero

---

[1]For the purposes of this chapter, "Sun-like" stars are defined to be any stars which will become white dwarfs; significantly more massive stars have different fates.
[2]See the NASA Exoplanet Archive at https://exoplanetarchive.ipac.caltech edu/.

et al. 2023). Nevertheless, the community consensus is that the solar system's four inner planets will very likely survive until the end of the Sun's main sequence lifetime.

This consensus is not surprising, especially given all of the exoplanets that are currently known to orbit stars that have already left the main sequence. Observational limitations restrict our planet discoveries around giant branch stars to giant planets only (as opposed to terrestrial planets). Nevertheless, well over 100 of these giant planets have been discovered (Reffert et al. 2015; Grunblatt et al. 2019, 2023; Huber et al. 2019; Luhn et al. 2019; Niedzielski et al. 2021), and some giant branch stars exhibit evidence of having recently accreted planets (Adamów et al. 2012; Stephan et al. 2020; Sevilla et al. 2022). Further, the discovery of planetary debris disks around these types of stars (Bonsor et al. 2013, 2014; Lovell et al. 2022) indicates the survival of exo-Kuiper belts, and in particular exo-asteroids and the dust they produce from mutual collisions.

This enticing observational evidence motivates exploration of how planetary systems are transformed when their host stars leave the main sequence, which is the subject of the remainder of this section.

Orbital shifts

When stars leave the main sequence, they begin shedding their mass through winds. These winds carry the mass completely out of the system, beyond the star's gravitational influence. As a result, the gravitational potential in the system changes, affecting the orbit of every planet, asteroid, comet, boulder and speck of dust. The effect on the planetary system is significant because the star ends up losing at least 50% of its mass, and up to 80%, through these winds (Hurley et al. 2000).

Further, the loss of mass is not constant because the giant branch phases consist of multiple phases. The two most important of these phases are the *red giant phase* and the *asymptotic giant branch phase*. The timescales of and mass lost during each phase are a strong function of the stellar mass.

Figure 1 provides an explicit example. For a star with the same mass and metallicity as the Sun, the red giant and asymptotic giant branch phases last for, respectively, about 700 Myr and 5 Myr, with equal mass loss fractions of approximately 25%, although this percentage can vary significantly depenending on the model parameters chosen. Instead, for a star whose mass is three times greater, the duration of these phases are about 2.4 Myr and 4.2 Myr, with respective mass loss fractions of approximately 0.019% and 75%.

The net result of a star losing mass is the expansion of a body's orbit. A rule-of-thumb is that an orbit typically expands by a factor equal to the ratio of the initial to final mass of the star. If the mass loss rate is particularly large, or a body such as a comet is particularly far away, or the star loses mass very asymmetrically, then the orbit can also stretch, perhaps to a breaking point when the body escapes the system (Omarov 1962; Hadjidemetriou 1963; Veras et al. 2011, 2013a, 2014d; Veras and Wyatt 2012; Adams et al. 2013; Dosopoulou and Kalogera 2016a,b; Regály et al. 2022).

However, the orbital changes to a single body become much more interesting in the context of the consequences for other bodies in the system. Although the orbit expansion factor is the same for multiple bodies, the change in gravitational potential alters the conditions in which they can remain stable: because the planet-to-star mass ratio increases, both planet–planet and planet–asteroid interactions become stronger (Debes and Sigurdsson 2002; Veras et al. 2013b; Voyatzis et al. 2013). As a result, multiple planets whose orbits were far enough apart on the main sequence to remain stable during that stellar phase might not remain stable after their parent star starts traversing the giant branches. This phenomenon is not limited to just planets: a planet and an asteroid, for example, can experience this same "late" instability.

Stellar evolution:

(constant mass & radius) → | Main sequence | → | Giant branches | → | White dwarf | (constant mass & radius)

Giant branch evolution of a 1.0 Solar mass star

Giant branch evolution of a 3.0 Solar mass star

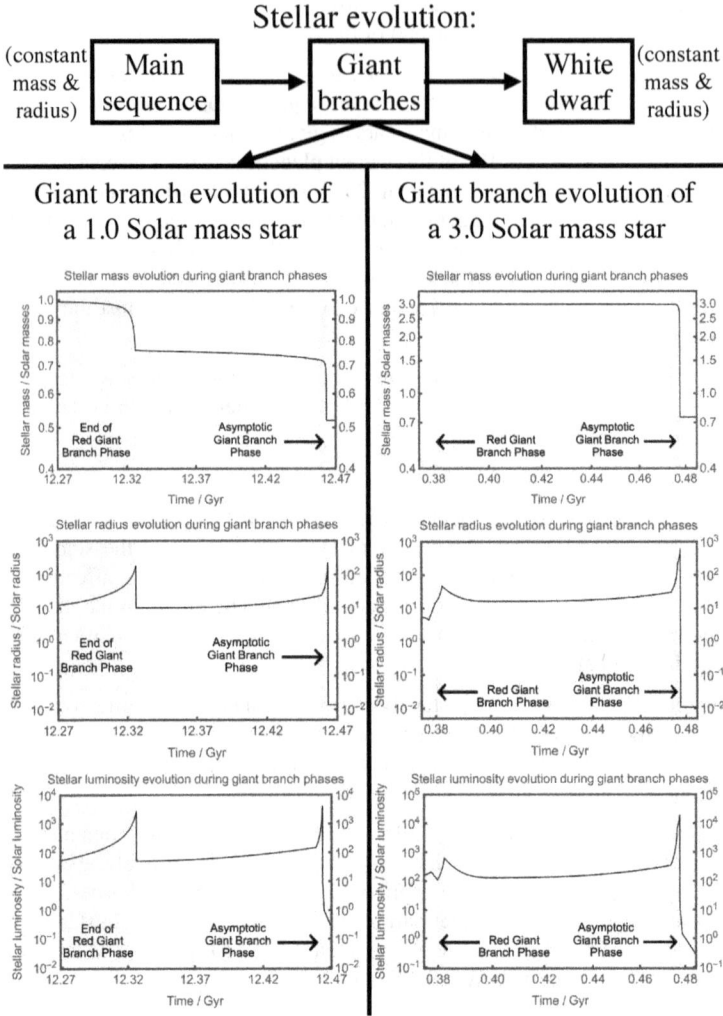

Figure 1. Representative mass, radius and luminosity evolution profiles of both Sun-like stars and stars which are three times as massive during the giant branch phases. The stellar mass and radius only change appreciably during the red giant branch and asymptotic giant branch phases, and massive stars reach the giant branch phases more quickly than Sun-like stars. Data for these plots was computed from Hurley et al. (2000).

This situation becomes more complex when a companion star is involved, in a so-called *binary star system*. Planets, asteroids and comets can orbit one or both of the stars; the former case is referred to as *circumstellar* and the latter as *circumbinary*. In these binary star planetary systems, both stars may or may not traverse the giant branches concurrently. Mass loss from the system forces the mutual orbit of the stars and any planetary material to all expand their orbits appropriately (Kratter and Perets 2012; Portegies Zwart 2013; Kostov et al. 2016), except if the stars are close to each other (5–10 au). In this case, the stars may form a *common envelope* or trigger one or two supernovae, depending on the mass of the stars, leading to much greater mass loss rates than in the single star case and potential ejection of circumbinary material (Veras and Tout 2012).

How planetary orbits shift due to stellar mass loss determines their final configurations and is hence a crucial consideration in late-stage population synthesis studies of the currently known exoplanet population (Andryushin and Popov 2021; Maldonado et al. 2020a,b, 2021, 2022). Part of this synthesis includes understanding which systems remain stable and which ones do not. The fates of notable individual systems such as ones with multiple giant planets like HR 8799 (Veras and Hinkley 2021) or multiple stars like the circumbinary systems found with the *Kepler* space telescope (Kostov et al. 2016) or the triple star system HD 131399 (Veras et al. 2017c) have received dedicated studies because of their complexity.

Finally, orbital shifts are not generated solely by stellar mass loss. Another feature of stars which ascend the giant branch phases is that their luminosity increases by up to a factor of about 10^4. This drastic luminosity increase has two important consequences. The first is that dust is drawn into the star quickly due to an effect known as *Poynting–Robertson drag* (Bonsor and Wyatt 2010; Dong et al. 2010; Zotos and Veras 2020). The second is that pebbles, cobbles, boulders and asteroids may be propelled either inwards or outwards due to an effect known as the *Yarkovsky effect* (Veras et al. 2015c, 2019b). Overall, orbital shifts from both radiation and gravity would need to be incorporated concurrently in self-consistent treatments of these evolved planetary systems, a prospect which has become more feasible with advances in *N*-body numerical codes (Baronett et al. 2022; Ferich et al. 2022).

Physical survival

We have so far described how a star's mass and luminosity changes during the giant branch phases of stellar evolution, and how these changes affect orbital evolution. Another, potentially more destructive aspect of giant branch stars, is their great size. A star's radius can increase by a factor of hundreds while traversing the giant branches, reaching out to a distance of one or more au (one au is equivalent to 215 Solar radii, or about 6.96×10^5 km).

The consequences for orbiting planets are significant. Consider first the solar system. Its predicted fate is illustrated in Figure 2. The Sun's outer envelope will extend far enough to engulf Mercury and Venus, and probably the Earth (Goldstein 1987; Sackmann et al. 1993; Rybicki and Denis 2001; Schröder and Smith 2008; Iorio 2010). Mars and the outer planets, however, will survive, and approximately double their current semi-major axes due to the Sun losing about half of its mass[3].

This increase in the star's size occurs while it is losing mass partly because envelope material is more weakly bound at such large radii. Hence, there is a competition between a planetary orbit expanding and the star's radius expanding. In a sense, the planet is trying to outrun the star. Complicating this picture further is the important concept of tidal interactions between the star and planet, and specifically the tidal distortion of the star induced by the planet. This distortion draws the planet closer to the star as the surfaces of the star and planet approach each other. The strength of the tidal interactions is strongly dependent on the mass and radius of the planet and the star, which is why the dashed curve in Figure 2 bends to the right for higher planetary masses.

Understanding the interplay between tidal interactions and mass loss has led many researchers to compute critical engulfment distances for planetary bodies orbiting stars which traverse the giant branches (Villaver and Livio 2009; Kunitomo et al. 2011; Mustill and Villaver 2012; Adams and Bloch 2013; Nordhaus and Spiegel 2013; Villaver et al. 2014; Madappatt et al. 2016; Ronco et al. 2020). The need for so many studies arose because tidal theory is poorly constrained and dependent on a large number of parameters. One of these parameters is stellar mass. Overall, a rough rule-of-thumb is that for each extra solar mass, the critical engulfment distance increases by another au (Mustill and Villaver 2012).

---

[3]The orbital eccentricity of the surviving planets will also increase by at least 10^{-5}–10^{-4} due to tidal convective motions in the star (Lanza et al. 2023).

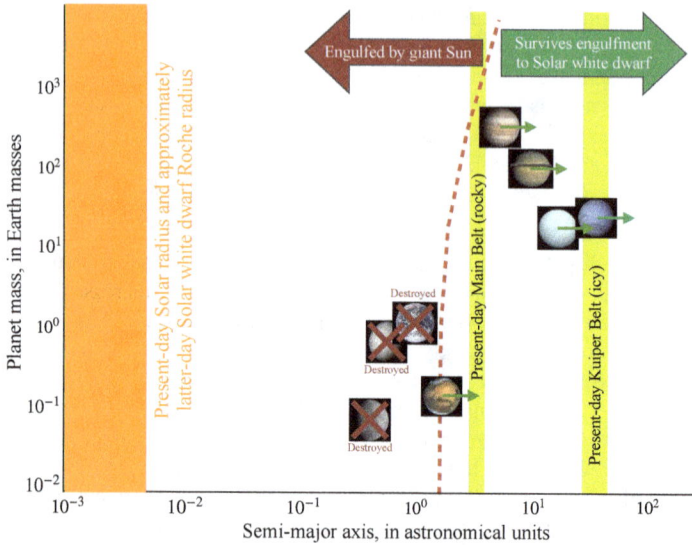

Figure 2. The fate of the Solar System's eight planets after the Sun leaves the main sequence in about 6.5 Gyr from now and traverses the giant branch phases. The Sun's radius will increase by a factor of hundreds, enveloping Mercury, Venus and probably Earth. However, Mars, Jupiter, Saturn, Uranus and Neptune will all survive, and be pushed away from the Sun by the amount indicated by the **dark green arrows** due to the Sun shedding about half of its mass through winds. Although both the asteroid Main Belt and Kuiper Belt (**vertical light green strips**) will survive engulfment, they will be affected in other ways; due to the Sun's increased luminosity, the Main Belt will be largely ground down and the Kuiper belt will disperse and spread apart. After the Sun becomes a white dwarf, its Roche radius will coincidentally be at a similar distance to its present-day radius ($R_\odot = 5 \times 10^{-3}$ au). The solar white dwarf's actual photospheric radius will be approximately 5×10^{-5} au.

The majority of all currently known exoplanets reside within the critical engulfment distance, even when accounting for variations in this distance depending on the tidal theory used. This fraction is not necessarily representative of the true demographics of exoplanets, but rather may reflect our observational bias for more easily finding planets on compact orbits. Many studies about the discovery or dynamics of these compact systems now include estimates for when these planets would be engulfed (e.g., Li et al. 2014; Jiang et al. 2020), and can highlight peculiar evolutionary histories of the parent star (Campante et al. 2019; Hon et al. 2023).

When a star's outer envelope engulfs a planet, the destruction is not instantaneous (Staff et al. 2016; Jia and Spruit 2018; MacLeod et al. 2018; O'Connor et al. 2023a). In fact, if the planet is large enough, it may survive and expel the envelope before being destroyed (Nelemans and Tauris 1998; Bear et al. 2011, 2021; Passy et al. 2012; Chamandy et al. 2021; Lagos et al. 2021; Merlov et al. 2021; Yarza et al. 2023), or expel the envelope while fragmenting into smaller planets (Bear and Soker 2012). The critical planet mass above which survival is possible is roughly one Jupiter mass, although survival is likely only for those planets much larger than a Jupiter mass. Planets which do survive this so-called *common envelope* evolution will have a net inwards motion due to drag from the envelope, overpowering any orbital expansion from stellar mass loss. Common envelope evolution remains highly uncertain and represents an active area of research (Röpke and De Marco 2023).

Because of the high luminosity of a giant branch star, the atmospheres of surviving planets which are sufficiently close to the star are in danger of partial evaporation (Villaver and Livio 2007; Bear and Soker 2011; Gänsicke et al. 2019; Schreiber et al. 2019). Surviving terrestrial planets might have their entire atmospheres evaporated, potentially with implications for the prospects of habitability on those planets (Ramirez and Kaltenegger 2016; Kozakis and Kaltenegger 2019) and for determining if a new atmosphere is generated on detected planets orbiting stars which have left the main sequence (Lin et al. 2022). An eroding atmosphere, coupled with heating of the surface from increased irradiation, and the loss of a planetary magnetosphere due to giant branch stellar winds (Veras and Vidotto 2021), poses challenges for life surviving during these phases of stellar evolution.

The combination of high stellar luminosity and strong winds also has several other physically significant effects in planetary systems. One is the creation of *planetary nebulae*, a term which does not necessarily actually refer to planets, but instead describes a structure of glowing gas around a giant branch star. The shape of these planetary nebulae may be partially due to the presence of planets interacting with the stellar wind (Soker 1996; De Marco and Soker 2011; Sabach and Soker 2018; Hegazi et al. 2020), but the origin of the gas itself may be due to sublimation of comets (Stern et al. 1990; Jura 2004; Su et al. 2007; Marshall et al. 2022).

Dust belts and disks are subject to dust destruction, blowout and collisional evolution, while coupled with orbital expansion from stellar mass loss (Bonsor and Wyatt 2010; Dong et al. 2010; Veras et al. 2015c; Zotos and Veras 2020). Even large asteroids may be broken apart from spinning too quickly (named the YORP effect), all entirely due to high stellar luminosity (Veras et al. 2014c; Veras and Scheeres 2020), perhaps after changing shape (Katz 2018), and maybe followed by inward or outward drift from the Yarkovsky effect (Veras et al. 2015c, 2019b).

The resulting picture is one where extant minor planets or dust may be significantly redistributed within the system during the giant branch phases. Planets which survive stellar engulfment may then exist in a sea of smaller particles, rather than being located at a specific distance from more concentrated reservoirs of smaller particles, as on the main sequence. Figure 3 provides a summary of the different forces involved in determining where and how planetary constituents can survive the giant branch phases.

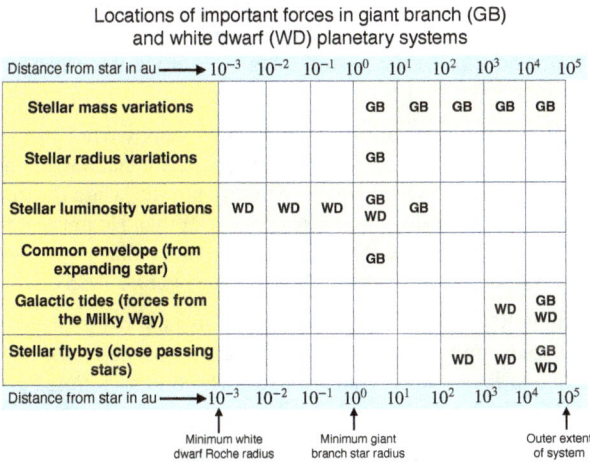

Locations of important forces in giant branch (GB) and white dwarf (WD) planetary systems

Distance from star in au $\rightarrow$	10^{-3}	10^{-2}	10^{-1}	10^{0}	10^{1}	10^{2}	10^{3}	10^{4}	10^{5}
Stellar mass variations					GB	GB	GB	GB	GB
Stellar radius variations				GB					
Stellar luminosity variations	WD	WD	WD	GB WD	GB				
Common envelope (from expanding star)				GB					
Galactic tides (forces from the Milky Way)								WD	GB WD
Stellar flybys (close passing stars)							WD	WD	GB WD
Distance from star in au $\rightarrow$	10^{-3}	10^{-2}	10^{-1}	10^{0}	10^{1}	10^{2}	10^{3}	10^{4}	10^{5}

Minimum white dwarf Roche radius   Minimum giant branch star radius   Outer extent of system

Figure 3. Order-of-magnitude distance estimates of which forces act upon planetary systems after the parent star has left the main sequence. The first four forces are internal to the planetary systems, whereas the last two, from the cumulative effect of all stars in the Milky Way and from individual passing stars, are external. Unlike for giant branch stars, white dwarf stars have constant masses and radii. Also, giant branch lifetimes are relatively short, allowing Galactic tides and stellar flybys to penetrate more deeply in white dwarf planetary systems.

Compositional alteration

Planetary bodies which do survive the giant branch phases of stellar evolution may be compositionally altered in two important ways. The first is through depletion of volatiles due to the star's high luminosity (Jura and Xu 2010; Jura et al. 2012; Malamud and Perets 2016, 2017a,b; Katz 2018), leading to largely "dry" asteroids except perhaps in the deep interiors of the larger members. Observations of chemical species after the giant branch phases, combined with a quantification of volatile depletion, can also help one infer formation properties of particular systems (Harrison et al. 2021).

The second way is through the compositional effect of the stellar wind on planetary atmospheres. Giant planet atmospheres can accrete enough of the stellar wind to change their C/O, O/H, H_2O, and CH_4 abundances by a potentially observable amount (Spiegel and Madhusudhan 2012). For terrestrial planets, the change in mixing ratios must be treated in concert with the extent of the evaporation of the atmospheres (Kozakis and Kaltenegger 2019; Lin et al. 2022).

The consequences of these compositional alterations, as well as the general survival of planetary bodies during the giant branch phases of evolution, are observed in older systems, which is the subject of the remainder of this chapter.

STAGE 2: DELIVERING MATERIAL TO THE WHITE DWARF'S ROCHE SPHERE

Introduction

At the conclusion of a star's giant branch phases, a critical point is reached with the star's remaining fuel, which would be composed of primarily C and O. The star is no longer hot enough to fuse such heavy elements. At this point, these elements build up into an inert core, and all of the outer layers of the star are shed. What remains is a burnt core known as a *white dwarf*.

A white dwarf's radius is relatively tiny; when the Sun becomes a white dwarf, its radius will shrink to about 5×10^{-5} au (1.2 Earth radii, or 7500 km). However, throughout this phase, the Sun will lose only about half of its mass. Therefore, its surface gravity will be six orders of magnitude higher than the current surface gravity of the Sun. This sudden increase in density will drag down any elements in the star's photosphere that are heavier than H or He (Paquette et al. 1986; Koester 2009).

Hence, white dwarfs have almost pristine atmospheres composed of either H and/or He. This concept becomes very important for planetary science because it enables chemical autopsies to be performed on any planetary material which accretes onto the white dwarf atmosphere: we know that any elements in a white dwarf atmosphere heavier than He must come from accreted planetary material, with few exceptions (for instance, the star's radiation can sometimes levitate certain atomic species for the youngest white dwarfs; Chayer et al. 1995).

Further, and crucially, we know that accreted planetary material is a very common phenomenon. In fact, dedicated surveys of different patches of the Galactic neighborhood of the Sun reveal that one in every two to four white dwarfs contain accreted planetary material (Zuckerman et al. 2003, 2010; Koester et al. 2014).

The subsequent chapter (Xu et al. 2024, this volume) will cover the exciting chemical trends and links with composition, differentiation and formation that we observe. The remainder of this chapter will instead focus on how all of the surviving planetary material from the giant branch phases of evolution reaches the tiny white dwarf. In fact, reaching the white dwarf's atmosphere is not requisite; planetary material needs only to reach the white dwarf's *Roche sphere radius* (or *Roche radius*), where material will fragment due to strong tidal forces, subsequently evolve and then eventually accrete onto the photosphere. The approximate location of a white dwarf's Roche radius is, coincidentally, at about $1R_\odot$, or hundreds of white dwarf radii (see Fig. 2).

This section will describe how planetary material might be delivered down to $1R_\odot$ from au-scale distances after the star has become a white dwarf, and the next section will describe the complex debris environment that we observe around and within $1R_\odot$ of white dwarfs. First we describe our knowledge from observations.

Known planetary material exterior to the white dwarf Roche sphere

Around giant branch stars, planets and disks are primarily observed soon after the star has left the main sequence, on the red giant branch. Observations of planets or disks around asymptotic giant branch stars are rarer, and only one unconfirmed planet candidate actually exists (Kervella et al. 2016) in this final phase before the star technically becomes a white dwarf. Unfortunately, due to observational biases, we have little information on the architectures and frequencies of wide-orbit planets that can survive stellar evolution.

The transition to a white dwarf is not necessarily sharp (Soker 2008), particularly when the star is still very hot and luminous, even after its outer layers have mostly dissipated. In this transition phase, observations remain rare but important. The Helix planetary nebula was discovered in this transition phase, with dust detected at tens of au (Su et al. 2007). Although the planetary origin of this dust is still debated (Clayton et al. 2014; Marshall et al. 2022), its presence provides observational evidence of planetary material that might just be starting to make its journey to the parent white dwarf's Roche radius (Veras and Heng 2020).

Newly formed white dwarfs have temperatures of 10^5 K, but cool quickly. This temperature is halved after just 10 Myr, and reduced to 10^4 K after 1 Gyr. The age of a star after it has become a white dwarf is known as its *cooling age*. Young white dwarfs are luminous enough to easily evaporate atmospheres of nearby planets (Schreiber et al. 2019): white dwarf WD J0914+1914, with a cooling age of just 13.3 Myr, is evaporating an ice giant planet at a separation of just 0.07 au (Gänsicke et al. 2019; Veras 2020; Veras and Fuller 2020; Zotos et al. 2020).

The other known planets orbiting white dwarfs are all gas giants and orbit much older white dwarfs, all with cooling ages over 1 Gyr old. These planets include WD 0806-661 b, at a planet–star separation of about 2500 au (Luhman et al. 2011; Rodriguez et al. 2011), PSR B1620-26 (AB) b, at a planet–star separation of 23 au (Thorsett et al. 1993; Sigurdsson et al. 2003; Beer et al. 2004; Sigurdsson and Thorsett 2005), MOA-2010-BLG-477L b, at a planet–star separation of a few au (Blackman et al. 2021), and WD 1856+534 b, at a planet–star separation of just 0.02 au (Munõz and Petrovich 2020; Vanderburg et al. 2020; Alonso et al. 2021; Merlov et al. 2021; Xu et al. 2021).

These five planets orbiting white dwarfs span five orders of magnitude in star-planet separation (0.02-2500 au), which already reveals to us that their dynamical origins are diverse. MOA-2010-BLG-477L b's location at a few au might have just exceeded its progenitor star's critical engulfment distance, and the planet has since remained undisturbed. WD 0806-661 b (at 2500 au) was either gravitationally scattered outward, or represents a captured free-floating planet. Both WD J0914+1914 b (0.07 au) and WD 1856+534 b (0.02 au) must have been either dragged inwards in their progenitor's envelope—and survived the process—or were more suddenly scattered inwards, and then dynamically settled due to tidal influence with the star. This variety of possibilities has prompted theoretical investigations about planet–white dwarf interactions (Veras and Fuller 2019; Veras et al. 2019a; O'Connor and Lai 2020) as well as the possibility of forming new planets only after the star has become a white dwarf (Bear and Soker 2014, 2015; Schleicher and Dreizler 2014; Völschow et al. 2014; Hogg et al. 2018; Ledda et al. 2023).

Further, each of the 5 known planets (WD J0914+1914 b, PSR B1620-26 (AB) b, MOA2010-BLG-477L b, WD 1856+534 b, WD 0806-661 b) was discovered with a completely different technique: photometry, spectroscopy, imaging, microlensing, and pulsar timing. This variety of discovery methods, as well as the enticing prospect of understanding this planet population better, has prompted a large number of recent detection estimates and efforts (Faedi et al. 2011; Xu et al. 2015; Sandhaus et al. 2016; Shengbang et al. 2016; van Sluijs and Van Eylen 2018; Cortés and Kipping 2019; Dame et al. 2019; Danielski et al. 2019; Tamanini and Danielski 2019; Veras and Wolszczan 2019; Krzesinski et al. 2020; Brandner et al. 2021; Kang et al. 2021; Morris et al. 2021; Van Grootel et al. 2021; Walters et al. 2021; Lucas et al. 2022; Columba et al. 2023; Kubiak et al. 2023).

Two instruments of note which are expected to make major advances in the known population of planets orbiting white dwarfs are *Gaia* and the *James Webb Space Telescope* (JWST). Gaia is expected to discover multiple planets through yet another technique, astrometry (Perryman et al. 2014; Sanderson et al. 2022). JWST might also discover planets and help chemically characterize their atmospheres in ways which have not been possible before (Kozakis et al. 2018, 2020; Kaltenegger et al. 2020; Limbach et al. 2022).

Unlike WD J0914+1914 b, most of the known planets orbiting white dwarfs do not expel their atmospheres onto the white dwarf. Nevertheless, their positions and masses play a crucial dynamical role in shepherding smaller material towards the white dwarf. This material is what eventually enters the white dwarf's Roche radius, breaks up, and accretes onto the white dwarf's atmosphere, where the chemical autopsy can be performed best.

Figures 4 and 5 make this notion concrete. The table in Figure 4 uses the very common community term *pollutant* to indicate in the first column the type of material that enters the Roche radius. The second column then outlines the different classes of objects which can drive the pollutants to the Roche radius. The schematic in Figure 5 illustrates a particularly common pathway—gravitational scattering—through which objects can approach a white dwarf's Roche radius and be broken up into dust and gas.

Pollutants	Potential Drivers of These Pollutants	Presumed Rarity of These Pollutants
Dust, Pebbles, Boulders 10^{-6} m $- 10^3$ m	(1) White dwarf radiation (2) Asteroids (3) Terrestrial Planets (4) Giant Planets (5) Companion Stars	Common
Asteroids 10^3 m $- 10^6$ m	(1) Terrestrial Planets (2) Giant Planets (3) Companion Stars	Common
Comets 10^2 m $- 10^4$ m	(1) Terrestrial Planets (2) Giant Planets (3) Companion Stars (4) Stellar flybys (5) Galactic Tides	Uncommon
Moons 10^3 m $- 10^6$ m	(1) Terrestrial Planets (2) Giant Planets	Uncommon
Terrestrial Planets 10^6 m $- 10^7$ m	(1) Terrestrial Planets (2) Giant Planets (3) Companion Stars	Rare
Giant Planets 10^7 m $- 10^8$ m	(1) Giant Planets (2) Companion Stars	Rare

Figure 4. Summary of the different parts of planetary systems which can be transported to the Roche sphere of a white dwarf, eventually *polluting* it. The numbers in the first column refer to the typical radii of the corresponding objects. Listed in the second column are the different mechanisms or objects which drive this transport, sometimes in concert with one another.

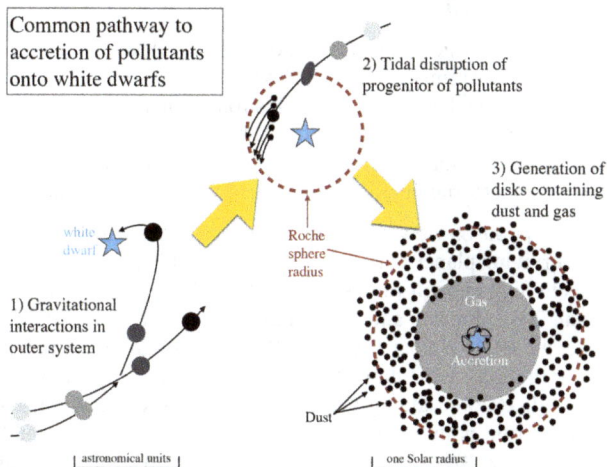

Figure 5. An assumed-to-be common method of transport of planetary material to the Roche sphere of the white dwarf, where breakup into disks of dust and gas occurs. Close to the white dwarf, dust becomes sublimated and only gas is present. This gas is accreted onto the white dwarf, polluting its atmosphere and providing us with bulk chemical constraints on the exoplanetary material.

The remaining subsections will describe each potential polluter in turn. Some are more likely than others based on theoretical considerations and our observational knowledge of the material within the Roche radius (next section). As this subsection has demonstrated, however, our current observational knowledge of material outside of the white dwarf's Roche radius is currently limited to effectively just five planets and the Helix nebula.

Delivery mechanisms: dust, pebbles and boulders as polluters

Dust, pebbles and boulders represent the only class of object which may be drawn towards the white dwarf's Roche sphere by radiation or magnetism alone (Veras et al. 2015b, 2022; Veras 2020; Zhang et al. 2021); see Figure 4. Gravitational interactions from larger objects may help, but are not necessary.

In this respect, provided that a sufficient amount of dust exists after the giant branch phases of stellar evolution, it can represent the primary polluter in some systems. These dust, pebbles and boulders may be generated from the break-up of asteroids or planets at any time, or even from smaller scale crater impacts on those bodies (Veras and Kurosawa 2020).

The timescale for radiation or magnetism to drag in dust, pebbles and boulders depends strongly on their size and the magnitude of the luminosity of the white dwarf, which relates to its cooling age. Another dependence is on the distance to the white dwarf; closer-in dust is drawn in more quickly. Hence, the breakup of minor planets at the white dwarf's Roche radius can produce fragments that reside just outside of that boundary; these are quickly enveloped (Li et al. 2021). At such close distances, these fragments may also collide with another, breaking into even smaller particles which then may be dragged in more quickly (Malamud et al. 2021; Brouwers et al. 2022, 2023a,b).

In binary systems where the binary companion to a white dwarf is emitting winds, dust may originate from this companion and reach the white dwarf's Roche radius (Perets 2011; Perets and Kenyon 2013). The amount of this dust is non-negligible when the binary separation is within about 1 au (Debes 2006; Veras et al. 2018b). For wider separations, observed pollutants must arise from a planetary system rather than from the winds of the binary star.

Delivery mechanisms: asteroids as polluters

Asteroids, or perhaps more generally, dry minor planets, represent the population of bodies which is most often cited as the primary polluters of white dwarfs (Jura 2003). This representation has become canonical over the decades due to both strong dynamical and chemical arguments. For the latter, the subsequent chapter (Xu et al. 2024, this volume) will demonstrate that the chemical compositions seen in white dwarf atmospheres are even more diverse than what we see amongst the asteroid and meteorite families in the solar system (Putirka and Xu 2021).

From a dynamical point-of-view, asteroids are effectively massless compared to planets. As a result, a single surviving planet can easily perturb asteroids gravitationally and excite their orbital eccentricities to high enough values to reach the white dwarf Roche sphere (Bonsor et al. 2011; Debes et al. 2012; Frewen and Hansen 2014; Pichierri et al. 2017; Antoniadou and Veras 2019; Veras et al. 2021, 2023a; Jin et al. 2023; McDonald and Veras 2023). Although not all orbital configurations will allow for this eccentricity excitation (Antoniadou and Veras 2016; Veras et al. 2020b), adding additional planets or companion stars increases the parameter space for asteroid pollution (Bonsor and Veras 2015; Hamers and Portegies Zwart 2016; Petrovich and Munõz 2017; Stephan et al. 2017; Mustill et al. 2018; Smallwood et al. 2018, 2021; O'Connor et al. 2022; Stock et al. 2022).

Other dynamical lines of evidence support asteroids as likely polluters. The first is that in our solar system, the pollution of the solar white dwarf will arise almost entirely from minor planets which survive break-up from the Sun's giant branch luminosity (Li et al. 2022). The second is that because a planetary system can host thousands of minor planets, the opportunities to pollute are numerous, potentially commensurate with the regularity seen in survey observations of populations of white dwarfs (Zuckerman et al. 2003, 2010; Koester et al. 2014).

Asteroids which can generate pollution do, however, require the presence of larger bodies in the system to act as perturbers. Asteroids themselves are not massive enough to gravitationally perturb one another into a white dwarf's Roche sphere, but instead require pollution drivers at least as massive as that of Earth's moon (Veras and Rosengren 2023). These larger objects need first to be fully formed and then would need to survive the giant branch phases of evolution. On the other end of the mass spectrum, there is no upper limit to the size of perturbers, meaning that no planets need to be present if a binary star can act as an efficient pollution driver (Hamers and Portegies Zwart 2016).

Delivery mechanisms: comets as polluters

Comets, or volatile-rich minor planets, have long been disfavoured as white dwarf polluters based on chemical grounds (see the subsequent chapter; Xu et al. 2024, this volume). However, repeatedly this judgement has been challenged. Observations of water-rich signatures in white dwarf atmospheres (Farihi et al. 2013; Raddi et al. 2015; Gentile Fusillo et al. 2017; Xu et al. 2017; Hoskin et al. 2020) now strongly suggest that comets occasionally pollute some white dwarfs. Although comets are still not assumed to be the primary pollutants, they provide some of the most interesting exceptions.

Unlike dry asteroids, volatile-rich comets can originate anywhere from distances of tens of au to hundreds of thousands of au away from their parent star (as well as at intermediate distances; Raymond and Armitage 2013). This wider range of locations allow for more delivery mechanisms than what is available for asteroids. Two in particular include influences from outside of the parent planetary system: stellar flybys and Galactic tides (Alcock et al. 1986; Parriott and Alcock 1998; Veras et al. 2014a); see Figure 3. The most distant comets are also strongly affected by changes to the motion of the white dwarf in the Galaxy when the white dwarf is born (Stone et al. 2015). The presence of binary stellar companions (Stephan et al. 2017) and planets (Caiazzo and Heyl 2017) can help comet pollution prospects, but planets have also been shown to hinder comet pollution depending on the system architecture (O'Connor et al. 2023b).

In the solar system, comets lose material to sublimation during each close passage of the Sun, partially altering their orbits due to conservation of linear momentum. Although this effect would also hold in white dwarf planetary systems, the orbital changes are unlikely to drive the comet into the white dwarf's Roche sphere (Veras et al. 2015a) without some other perturbing influence.

Delivery mechanisms: moons as polluters

The prospects for moons to represent potential polluters of white dwarfs primarily rely on dynamical rather than chemical arguments because chemically distinguishing asteroids from moons may be difficult. In fact, in only one case has the existence of an exo-moon been inferred based on the composition of the material accreted by a white dwarf, notably its Be content (Doyle et al. 2021).

In terms of the mass budget, in our solar system moons contain about two orders of mass more than asteroids. In terms of sheer numbers, there are dozens of moderate sized moons, which represents orders of magnitude less than the number of asteroids, but one order of magnitude more than the number of planets. The prevalence of white dwarf pollution suggests that if massive bodies are primarily responsible, then they need to be broken up in such a way that delivery can persist for a long time, and not be delivered all at once.

In order for a moon to reach the white dwarf's Roche sphere, either its parent planet has to be perturbed there (an unlikely possibility, as described in the subsequent subsections) or the moon must be stripped from its parent planet and then perturbed—as a minor planet orbiting the star in its own right—towards the white dwarf. The stripping can occur either from the moon slowly migrating beyond the planet's gravitational influence from tidal interactions, or more violently due to a gravitational scattering event. The latter scenario has been explored (Payne et al. 2016, 2017; Martinez et al. 2019; Trierweiler et al. 2022) and been found to be effective in a non-negligible number of cases. Overall, pollution from moons is less likely than from asteroids, but more likely than from planets.

Delivery mechanisms: terrestrial planets as polluters

Observations suggest that the frequency of pollution at white dwarfs (Zuckerman et al. 2003, 2010; Koester et al. 2014) is much higher than what could be explained if all the pollution was due to planets. Pollution signatures at most persist for Myr, orders of magnitude smaller than a typical ~Gyr-old white dwarf. There are simply not enough planets per system to generate all those signatures.

Nevertheless, many investigations have modelled the evolution of terrestrial planets scattering off of one another and off of giant planets as a star is transformed into a white dwarf (Debes and Sigurdsson 2002; Veras and Gänsicke 2015; Hamers and Portegies Zwart 2016; Veras et al. 2016a, 2018a; Veras 2016b; Maldonado et al. 2020a,b, 2021, 2022). They find that terrestrial planets may be scattered to the white dwarf Roche sphere at any cooling age, but often just once per system, if at all. The greater the number of initial planets, the greater the likelihood of these scattering events during the white dwarf phase.

Although these results de-emphasize the importance of terrestrial planets as polluters, they should not diminish the role of terrestrial planets as pollution drivers of smaller objects.

Delivery mechanisms: giant planets as polluters

Giant planets represent even rarer polluters. Effectively, the only two potential pollution drivers in this case are other giant planets or companion stars. As shown in the previous two chapters (Jones 2024; Mordasini and Burns 2024, both this volume), giant planets form more infrequently than terrestrial planets. Also, companion stars represent effective scatterers only if they reside sufficiently close to the white dwarf, or can change their orbit to do so.

Regardless, partly because the only detected planets orbiting white dwarfs are currently giant planets, the evolution of giant planets across all phases of stellar evolution—even excluding survival within a giant branch envelope—remains a well-investigated topic (Debes and Sigurdsson 2002; Veras et al. 2013b, 2016a, 2017b, 2018a; Mustill et al. 2014; Veras and Gänsicke 2015; Stephan et al. 2018; Munõz and Petrovich 2020; Maldonado et al. 2020a,b, 2021, 2022; O'Connor et al. 2021). Further, we know of one important case of a giant planet which actually is polluting its parent white dwarf: the evaporating atmosphere of WD J0914+1914 b is transporting detectable quantities of H, O and S into the white dwarf Roche sphere (Gänsicke et al. 2019).

STAGE 3: PROCESSING DEBRIS WITHIN THE WHITE DWARF ROCHE SPHERE

Introduction

Having reached the Roche sphere radius of a white dwarf, what then happens to planetary material? The answer is complex, and one which is greatly enhanced by abundant observations. This section will describe both the observations and theory of this narrow region extending from about 5×10^{-5} au to 5×10^{-3} au.

The mechanics of an object breaking up at a Roche sphere, and the aftermath, involves fragments and debris which are formed into structures sometimes known as disks or rings. This result is not surprising because disks and rings represent the lowest energy state that astrophysical systems which are supported by rotation settle into, and can be seen at all scales. The debris around the white dwarf can be composed of both gas and dust, and exist revolving around within the Roche sphere for potentially Myrs before being accreted onto the white dwarf's atmosphere.

Observations of this debris rely on photometry and spectroscopy, and provide us with different types of physical and chemical constraints. Figure 6 summarizes the three main types of observations in this region. Each observation type will be described in its own subsection. These observations have helped constrain theoretical models for the formation and evolution of this debris, which will also be described in two dedicated subsections.

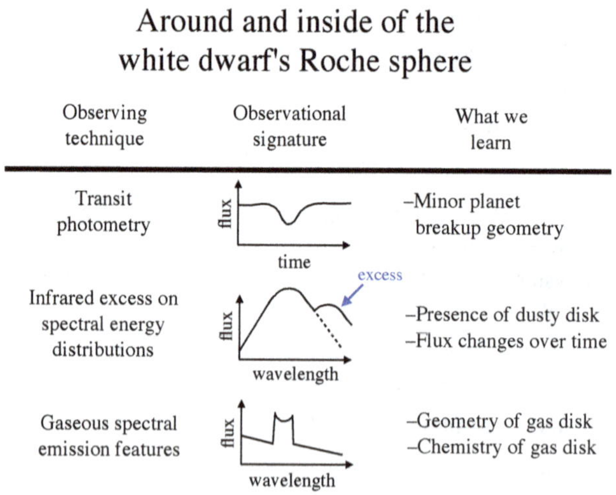

Figure 6. The three main types of observations of debris around and within a white dwarf's Roche sphere radius, located at about $1R_\odot$, or 5×10^{-3} au, from the centre of the star. These observations are independent of those of planetary material which is eventually accreted onto the star's photosphere (Xu et al. 2024, this volume), located at a distance of about 5×10^{-5} au.

Observations of debris: transits

Transit photometry refers to the dimming of the white dwarf due to material passing in front of it. Because the material orbits the white dwarf, the dimming is periodic, and any change in the amount of dimming per orbital period indicates dynamical activity. Further, a single orbiting object, such as a planet, would produce a pronounced sharp dimming feature. Alternatively, a messy collection of debris, dust and gas would produce a transit curve with lots of peaks and troughs of different extents.

When applied to Sun-like stars, transit photometry has proven to be the most successful exoplanet-hunting technique, being responsible for the discovery of the majority of the currently known exoplanets. Around white dwarfs, so far this technique has been responsible for the discovery of just one planet, WD 1856+534 b, which resides outside of the Roche sphere (Vanderburg et al. 2020). The reason for this low yield is perhaps because the technique is only sensitive to close-in planets which would be unlikely to survive the star's evolution.

However, the technique has been much more successful at charting the evolution of debris orbiting within the white dwarf's Roche sphere. This debris is commonly inferred to result from the active breakup of a minor planet, which conforms with the common assumption that white dwarfs are polluted by asteroids. Multiple debris clumps can share the same orbit around a white dwarf in a gravitationally stable fashion, even near the Roche sphere radius (Veras et al. 2016b). In few other contexts within exoplanetary science can an individual exo-minor planet be analyzed and tracked.

Over 8 white dwarf systems now have transit detections of planetary debris (Guidry et al. 2021). The first of these discoveries, for the WD 1145+017 system, was seminal (Vanderburg et al. 2015). That discovery illustrated starkly the important link between the previously theorized survival of planetary material across stellar evolution (at the time the only planets known around white dwarfs were the distant objects PSR B1620-26 (AB) b and WD 0806-661 b) and the observed signatures of metal pollution in white dwarf atmospheres. Follow-up observations and theories about the minor planet breaking up around WD 1145+017 were extensive (Rappaport et al. 2016; Zhou et al. 2016; Farihi et al. 2017b, 2018a; Gurri al. 2017; Hallakoun et al. 2017; Veras et al. 2017a; Izquierdo et al. 2018; Xu et al. 2018b, 2019a; Shestakova et al. 2019; Duvvuri et al. 2020; O'Connor and Lai 2020; Budaj et al. 2022).

Other notable white dwarfs with transiting debris signatures are ZTF J0139+5245 (Vanderbosch et al. 2020), ZTF J0328-1219 (Vanderbosch et al. 2021) and WD 1054-226 (Farihi et al. 2022). The orbital period of the transiting debris orbiting ZTF J0139+5245 is about 550 times longer than the 4.5-hour orbital period of the debris in WD 1145+017. This striking difference indicates that the debris around ZTF J0139+5245 is produced by an extended stream on a highly eccentric orbit, most of which is actually outside of the Roche sphere (Veras et al. 2020a). In the ZTF J0328-1219 system, two specific periodicities of 9.9 and 11.2 hours have been identified, representing evidence of two distinct orbiting clumps. The periodicity structure in the WD 1054-226 system is even more complex in the sense that a baseline level of flux has not yet been observed.

Overall, these transit signatures showcase an astounding variety of features and parameter ranges which remain difficult to interpret.

Observations of debris: infrared excess

An entirely different way of observing debris at and within the Roche sphere radius is with *spectral energy distributions*. These are plots of spectral flux versus wavelength for different white dwarf systems. The large bump on the plot represents the white dwarf; an accompanying smaller bump (see Fig. 6) indicates the presence of dust outside of the white dwarf. This accompanying bump is known as the *infrared excess*.

The geometric configuration of this dust cannot be identified unambiguously from the infrared excess. Instead, the mapping between the infrared excess and geometric configuration is degenerate. If one assumes that the dust resides in a flat opaque configuration, then the disk's line-of-sight inclination and its inner and outer boundary can be estimated (Jura 2003). This assumption typically yields outer boundaries which are similar to the Roche sphere radius (about $1.0\ R_\odot$), and inner boundaries which are about half of that value (about $0.5\ R_\odot$).

We now know that the assumption of the dust residing in a flat, circular opaque disk is usually poor. Jura et al. (2007a) found that a partially warped disc would better fit the infrared excess around the white dwarf GD 362, and Jura et al. (2007b) similarly found that the infrared observations of the white dwarf GD 56 cannot be reproduced with a flat disc model. Gentile Fusillo et al. (2021) was not able to use a flat disc model to reproduce the observations of WD J0846+5703. Dennihy et al. (2016) demonstrated that eccentric, rather than circular discs, can in some instances reproduce infrared excesses well. By using an independent radiative transfer-based method of analyzing the polluted white dwarf G29-38, Ballering et al. (2022) found that a vertically high, but narrow ($0.1R_\odot$), dust ring with a large opening angle is consistent with the observations. This structure has a much greater vertical extent than can be approximated by the canonical flat disk assumption.

Over 60 white dwarf systems have exhibited clear detections of infrared excess since 1987 (Zuckerman and Becklin 1987; Graham et al. 1990), including one instance of a circumbinary dusty structure (Farihi et al. 2017a). This population is now large enough to motivate reports of multiple discoveries at once and justify frequency analyses (Farihi et al. 2009, 2010; Barber et al. 2012; Bergfors et al. 2014; Rocchetto et al. 2015; Farihi 2016; Wilson et al. 2019; Chen et al. 2020; Manser et al. 2020; Rogers et al. 2020; Xu et al. 2020; Lai et al. 2021; Wang et al. 2023). The frequency analyses indicate that only a few percent of white dwarfs contain detectable infrared excess, while between 25-50% are polluted (Zuckerman et al. 2003, 2010; Koester et al. 2014). This mismatch is likely due to observational bias (Bonsor et al. 2017) because the theoretical expectation is that dust surrounds nearly every white dwarf which is polluted.

A few white dwarfs with infrared excesses are individually notable. One is G29-38, which is particularly bright and close to the Earth. As a result, it is the only dusty structure which has been chemically probed for silicate minerology (Reach et al. 2005, 2009). Another notable white dwarf is LSPM J0207+3331, because it has by far the oldest cooling age (about 3 Gyr) of any white dwarf with an infrared excess (Debes et al. 2019).

One exciting feature of white dwarf infrared excesses is the flux variability they demonstrate over yearly timescales. This variability is indicative of dynamical activity (Bear and Soker 2013). The most dramatic case so far known is that of WD 0145+234, whose infrared flux changed by one order of magnitude in under a year (Wang et al. 2019; Swan et al. 2021). Less dramatic, but still notable individual cases have been mounting (Xu and Jura 2014; Xu et al. 2018a; Farihi et al. 2018b), and now flux changes at the tens of percent level per year are assumed to be a ubiquitous feature of white dwarf infrared excesses (Swan et al. 2019b, 2020).

Observations of debris: gas emission lines

Infrared excesses indicate the presence of only dust, not gas. However, within a white dwarf Roche sphere, gas can be generated by both collisions and sublimation. Collisions may occur anywhere with a sufficient number density of fragments. Sublimation could occur only if a fragment is sufficiently close to a white dwarf; for the hottest and youngest white dwarfs, this boundary exceeds the Roche sphere, whereas for the coolest and oldest, this boundary may be as low as $0.2\ R_\odot$ (Veras et al. 2023b).

In all of these cases, a different detection technique is required than to detect dust. Gas orbiting white dwarfs is found using *emission line spectroscopy*. As a result, the chemical composition of the gas can be identified immediately, unlike for the dust. This feature helped to confirm the existence of the evaporating planet WD J0914+1914 b through the identification of H, S and O in the gas surrounding the white dwarf, as well as in the star's photosphere (Gänsicke et al. 2019).

A more easily detectable chemical element in gas surrounding white dwarfs is Ca, a very common polluter (see the subsequent chapter; Xu et al. 2024, this volume). In fact, the *Ca II triplet feature* (see Fig. 6) has become not only a way to identify gas, but also to characterize its geometry. The morphology of this observational signature was instrumental in resolving the degeneracies inherent with infrared excess for identifying the inner and outer edges of orbiting debris, and confirming that these boundaries *typically* do reside at distances of about 0.5 $R_\odot$ and 1.0 $R_\odot$ from the centre of the white dwarf (Gänsicke et al. 2006, 2007, 2008).

The known white dwarfs containing orbiting gas within the Roche sphere now number over 20 (Melis et al. 2010, 2020; Dennihy et al. 2020; Manser et al. 2020; Gentile Fusillo et al. 2021). One of these is WD 1145+017, the same white dwarf which hosts the first transit photometry detection of minor planet breakup (Xu et al. 2016; Redfield et al. 2017; Fortin-Archambault et al. 2020). Some gas signatures around white dwarfs have now been the focus of applications of spectral abundance models (Hartmann et al. 2011, 2014; Steele et al. 2021).

Like with infrared excesses, gas emission signatures can show variability (Wilson et al. 2014; Dennihy et al. 2018). In one case, the emission signatures of white dwarf SDSS J1617+1620 decreased over six years until it fell below the detectability threshold (Wilson et al. 2014). For emission signatures which maintain their strength, when a sufficient time baseline of observations of this gas is taken, then the knowledge of the structure's geometry can be derived through the use of *Doppler tomography*. Doppler tomography allows one to create a velocity map of the gas structure. This map illustrates the intensity pattern of the gas (Manser et al. 2016, 2021). One notable case of variability is with white dwarf SDSS J1228+1040 (Manser et al. 2019). This variability indicates the presence of a minor planet at a distance of 0.003 au, which is within the typical Roche sphere radius. Hence, this minor planet is thought to be composed of dense, strong material, and is likely the remnants of a planetary core instead of a rocky asteroid.

Theoretical considerations on the formation of debris

The combined observations from transit photometry, infrared excess and emission line spectroscopy paint a rich and complex picture of planetary debris evolution inside of the Roche sphere. Consolidating this information into a unified theory for formation and evolution of the debris is challenging partly because of the variety of physics which must be considered.

Five possible destruction outcomes exist for pollutants which are larger than dust (Brown, Veras and Gänsicke 2017; McDonald and Veras 2021): (1) shearing apart through overspinning just outside of the Roche sphere (Makarov and Veras 2019; Veras et al. 2020a), (2) sublimating partially or fully before reaching the Roche sphere (Steckloff et al. 2021), (3) breaking up through mutual collisions of tidal fragments before reaching the Roche sphere (Brouwers et al. 2022), (4) breaking up upon reaching the Roche sphere, or (5) more rarely, barrelling through the traditional Roche sphere and impacting onto the photosphere of the white dwarf. Some white dwarfs might only encounter one pollutant over 10 Gyr of cooling, whereas other white dwarfs might receive, e.g., many pollutants in quick succession all on the same day, and then again for thousands of times over the next Gyr (Jura 2008; Wyatt et al. 2014).

The manner of approach and entry into the Roche sphere varies for different pollutants. Because the critical engulfment distance for giant stars is on an au-scale, and only dust and pebbles may be radiatively dragged to the Roche sphere, the traditional assumption was that any pollutants larger than pebbles needed to enter the Roche sphere on a highly eccentric orbit. However, this thinking was challenged with the discovery of the minor planet breaking up around WD 1145+017 in what appears to be a near-circular orbit (Rappaport et al. 2016;

Gurri et al. 2017; Veras et al. 2017a). The first explanation posed to resolve this conundrum was by O'Connor and Lai (2020) who utilized tidal theory and ram pressure drag to demonstrate that circularization is possible before entering the Roche sphere.

Now a wider variety of entry methods into the Roche sphere is recognized. Dust and pebbles may enter on either a near-circular or highly eccentric orbit. The former is achieved through radiative drag alone (Veras et al. 2022) or with the help of freshly generated gas (Malamud et al. 2021) or a strong magnetic field (Hogg et al. 2021; Zhang et al. 2021). The latter can be prompted by gravitational interactions and lead to an extant disc capturing the pollutants (Grishin and Veras 2019).

Asteroid, comet, moon and planet polluters similarly are not required to enter the Roche sphere on eccentric orbits because tides, ram pressure drag, and Ohmic drag can first circularize the orbits of these objects (Bromley and Kenyon 2019; Veras et al. 2019a; O'Connor and Lai 2020). The possibility also exists of minor planets forming afresh from existing white dwarf discs (van Lieshout et al. 2018) on near-circular orbits just outside of the Roche sphere; these would not be far away enough to escape eventual destruction back inside of the Roche sphere. Large polluters that do enter the Roche sphere on eccentric orbits do so at an angle which is dependent on the size of the perturber (Veras et al. 2021).

Despite the diversity of possibilities, the specific case of a solar system-like asteroid entering a white dwarf Roche sphere has received significant attention. The dynamics of the breakup and the evolution of the resulting fragments has been investigated with an increasingly sophisticated set of numerical and analytical tools (Jura 2003; Debes et al. 2012; Veras et al. 2014b, 2015b, 2017a; Duvvuri et al. 2020; Malamud and Perets 2020a,b; Li et al. 2021; Brouwers et al. 2022, 2023a,b). These investigations have helped set the initial conditions for the subsequent evolution of the debris.

Theoretical considerations on the evolution of debris

After having formed into a disk-like or ring-like structure, the debris can persist for a wide variety of times (yrs to Myrs) before actually accreting onto the white dwarf (Girven et al. 2012; Veras and Heng 2020). The debris orbiting WD 1145+017 on a 4.5-hour period has exhibited different photometric transit curves each year since its discovery. Long-lasting debris dics could also allow intact embedded planetesimals to migrate within the disc when certain conditions are met (Veras et al. 2023b).

How the evolution proceeds is complicated by the dual presence of dust and gas. Their coupling admits limited analytical means to investigate their evolution. Hence, one strategy amongst researchers has been to decouple the dust and gas and focus on particular physical effects due to each. By focussing on dust, Kenyon and Bromley (2017a) investigated collisional evolution, and Farihi et al. (2017b) investigated magnetically-charged particles. By focussing on gas, Miranda and Rafikov (2018) investigated how internal pressure can drive precession of disk-like structures, Rozner et al. (2021) demonstrated how gas can mechanically erode boulder-like debris, and Trevascus et al. (2021) illustrated how the persistent injection of gas from a disrupting planet or planetesimal can maintain an eccentric disc structure.

These investigations have usefully characterized different aspects of the physics of planetary debris evolution within a white dwarf's Roche sphere. However, the most accurate estimates of the rate of accretion onto the white dwarf photosphere itself require a coupled treatment of gas and dust because all dust will eventually sublimate before reaching the photosphere. Early coupled treatments (Bochkarev and Rafikov 2011; Rafikov 2011a,b; Rafikov and Garmilla 2012; Metzger et al. 2012) emphasized the role of radiative drag on the eventual accretion rate, whereas later efforts have incorporated more collisional evolution (Kenyon and Bromley 2017b), ongoing gas generation (Malamud et al. 2021) and re-condensation (Okuya et al. 2023) in their computations.

ACKNOWLEDGEMENTS

We thank Uri Malamud, James Bryson and Judith Korth for their careful reading of the manuscript and for their expert advice, which have led to improvements. A.J.M. acknowledges funding from the Swedish Research Council (grants 2017-04945 and 202204043) and the Swedish National Space Agency (grant 120/19C), and A.B. acknowledges support from a Royal Society Dorothy Hodgkin Research Fellowship, DH150130 and a Royal Society University Research Fellowship, URF\R1\211421.

REFERENCES

Abbot DS, Webber RJ, Hadden S, Seligman D, Weare J (2021) Rare event sampling improves Mercury instability statistics. Astrophys J 923:236

Abbot DS, Hernandez DM, Hadden S, Webber RJ, Afentakis GP, Weare J (2023) Simple physics and integrators accurately reproduce Mercury instability statistics. Astrophys J 944:190

Adamów M, Niedzielski A, Villaver E, Nowak G, Wolszczan A (2012) BD+48 740—Li overabundant giant star with a planet: A case of recent engulfment? Astrophys J Lett 754:L15

Adams FC, Bloch AM (2013) Evolution of planetary orbits with stellar mass loss and tidal dissipation. Astrophys J Lett 777:L30

Adams FC, Anderson KR, Bloch AM (2013) Evolution of planetary systems with time-dependent stellar mass-loss. Mon Not R Astron Soc 432:438–454

Alcock C, Fristrom CC, Siegelman R (1986) On the number of comets around other single stars. Astrophys J 302:462–476

Alonso R, Rodríguez-Gil P, Izquierdo P, Deeg HJ, Lodieu N, Cabrera-Lavers A, Reverte-Payá D (2021) A transmission spectrum of the planet candidate WD 1856+534 b and a lower limit to its mass. Astron Astrophys 649

Andryushin AS, Popov SB (2021) Exoplanet population synthesis with account for orbit variation due to stellar evolution. Astron Rep 65:246–268

Antoniadou KI, Veras D (2016) Linking long-term planetary N-body simulations with periodic orbits: Application to white dwarf pollution. Mon Not R Astron Soc 463:4108–4120

Antoniadou KI, Veras D (2019) Driving white dwarf metal pollution through unstable eccentric periodic orbits. Astron Astrophys 629:A126

Ballering NP, Levens CI, Su KY, Cleeves LI (2022) The geometry of the G29-38 white dwarf dust disk from radiative transfer modeling. Astrophys J 939:108

Barber SD, Patterson AJ, Kilic M, Leggett SK, Dufour P, Bloom JS, Starr DL (2012) The frequency of debris disks at white dwarfs. Astrophys J 760:26

Baronett SA, Ferich N, Tamayo D, Steffen JH (2022) Stellar evolution and tidal dissipation in REBOUNDx. Mon Not R Astron Soc 510:6001–6009

Barstow MA, Barstow JK, Casewell SL, Holberg JB, Hubeny I (2014) Evidence for an external origin of heavy elements in hot DA white dwarfs. Mon Not R Astron Soc 440:1607–1625

Bear E, Soker N (2011) Evaporation of Jupiter-like planets orbiting extreme horizontal branch stars. Mon Not R Astron Soc 414:1788–1792

Bear E, Soker N (2012) A tidally destructed massive planet as the progenitor of the two light planets around the sdB star KIC 05807616. Astrophys J Lett 749:L14

Bear E, Soker N (2013) Transient outburst events from tidally disrupted asteroids near white dwarfs. New Astron 19:56–61

Bear E, Soker N (2014) First- versus second-generation planet formation in post-common envelope binary (PCEB) planetary systems. Mon Not R Astron Soc 444:1698–1704

Bear E, Soker N (2015) Planetary systems and real planetary nebulae from planet destruction near white dwarfs. Mon Not R Astron Soc 450:4233–4239

Bear E, Soker N, Harpaz A (2011) Possible implications of the planet orbiting the red horizontal branch star HIP 13044. Astrophys J 733:L44

Bear E, Merlov A, Arad Y, Soker N (2021) Rapid expansion of red giant stars during core helium flash by waves propagation to the envelope and implications to exoplanets. Mon Not R Astron Soc 507:414–420

Beer ME, King AR, Pringle JE (2004) The planet in M4: implications for planet formation in globular clusters. Mon Not R Astron Soc 355:1244–1250

Bergfors C, Farihi J, Dufour P, Rocchetto M (2014) Signs of a faint disc population at polluted white dwarfs. Mon Not R Astron Soc 444:2147–2156

Blackman JW, Beaulieu JP, Bennett DP, Danielski C, Alard C, Cole AA, Vandorou A, Ranc C, Terry SK, Bhattacharya A, Bond I (2021) A Jovian analogue orbiting a white dwarf star. Nature 598:272–275

Bochkarev KV, Rafikov RR (2011) Global modeling of radiatively driven accretion of metals from compact debris disks onto white dwarfs. Astrophys J 741:36

Bonsor A, Veras D (2015) A wide binary trigger for white dwarf pollution. Mon Not R Astron Soc 454:53–63

Bonsor A, Wyatt M (2010) Post-main-sequence evolution of A star debris discs. Mon Not R Astron Soc 409:1631–1646

Bonsor A, Xu S (2017) White dwarf planetary systems: Insights regarding the fate of planetary systems. *In:* Formation, Evolution, and Dynamics of Young Solar Systems. M Pessah, O Gressel (eds), Springer, p 229–252

Bonsor A, Mustill AJ, Wyatt MC (2011) Dynamical effects of stellar mass-loss on a Kuiper-like belt. Mon Not R Astron Soc 414:930–939

Bonsor A, Kennedy GM, Crepp JR, Johnson JA, Wyatt MC, Sibthorpe B, Su KY (2013) Spatially resolved images of dust belt(s) around the planet-hosting subgiant κ CrB. Mon Not R Astron Soc 431:3025–3035

Bonsor A, Kennedy GM, Wyatt MC, Johnson JA, Sibthorpe B (2014) Herschel observations of debris discs orbiting planet-hosting subgiants. Mon Not R Astron Soc 437:3288–3297

Bonsor A, Farihi J, Wyatt MC, van Lieshout R (2017) Infrared observations of white dwarfs and the implications for the accretion of dusty planetary material. Mon Not R Astron Soc 468:154–164

Brandner W, Zinnecker H, Kopytova T (2021) Search for giant planets around seven white dwarfs in the Hyades cluster with the Hubble Space Telescope. Mon Not R Astron Soc 500:3920–3925

Bromley BC, Kenyon SJ (2019) Ohmic heating of asteroids around magnetic stars. Astrophys J 876:17

Brouwers MG, Bonsor A, Malamud U (2022) A road-map to white dwarf pollution: tidal disruption, eccentric grind-down, and dust accretion. Mon Not R Astron Soc 509:2404–2422

Brouwers MG, Bonsor A, Malamud U (2023a) Asynchronous accretion can mimic diverse white dwarf pollutants I: Core and mantle fragments. Mon Not R Astron Soc 519:2646–2662

Brouwers MG, Bonsor A, Malamud U (2023b) Asynchronous accretion can mimic diverse white dwarf pollutants II: Water content. Mon Not R Astron Soc 519:2663–2679

Brown G, Rein H (2022) On the long-term stability of the Solar system in the presence of weak perturbations from stellar flybys. Mon Not R Astron Soc 515:5942–5950

Brown G, Rein H (2023) General relativistic precession and the long-term stability of the solar system. Mon Not R Astron Soc 521:4349–4355

Brown JC, Veras D, Gänsicke BT (2017) Deposition of steeply infalling debris around white dwarf stars. Mon Not R Astron Soc 468:1575–1593

Budaj J, Maliuk A, Hubeny I (2022) WD 1145+017: Alternative models of the atmosphere, dust clouds, and gas rings. Astron Astrophys 660:A72

Caiazzo I, Heyl JS (2017) Polluting white dwarfs with perturbed exo-comets. Mon Not R Astron Soc 469:2750–2759

Campante TL, Corsaro E, Lund MN, Mosser B, Serenelli A, Veras D, Adibekyan V, Antia HM, Ball W, Basu S, Bedding TR (2019) TESS asteroseismology of the known red-giant host stars HD 212771 and HD 203949. Astrophys J 885:31

Canup RM (2004) Dynamics of lunar formation. Annu Rev Astron Astrophys 42:441–475

Cauley PW, Farihi J, Redfield S, Bachman S, Parsons SG, Gänsicke BT (2018) Evidence for eccentric, precessing gaseous debris in the circumstellar absorption toward WD 1145+017. Astrophys J 852:L22

Chamandy L, Blackman EG, Nordhaus J, Wilson E (2021) Multiple common envelope events from successive planetary companions. Mon Not R Astron Soc 502:L110–L114

Chayer P, Fontaine G, Wesemael F (1995) Radiative levitation in hot white dwarfs: Equilibrium theory. Astrophys J Suppl Ser 99:189–221

Chen CH, Su KY L, Xu S (2020) Spitzer's debris disk legacy from main-sequence stars to white dwarfs. Nat Astron 4:328–338

Clayton GC, De Marco O, Nordhaus J, Green J, Rauch T, Werner K, Chu YH (2014) Dusty disks around central stars of planetary nebulae. Astrophys J 147:142

Columba G, Danielski C, Dorozsmai A, Toonen S, Puertas Ml (2023) Statistics of Magrathea exoplanets beyond the Main Sequence. Simulating the long-term evolution of circumbinary giant planets with TRES. Astron Astrophys 675:A156

Cortés J, Kipping D (2019) On the detectability of transiting planets orbiting white dwarfs using LSST. Mon Not R Astron Soc 488:1695–1703

Dame K, Belardi C, Kilic M, Rest A, Gianninas A, Barber S, Brown WR (2019) The DECam minute cadence survey. II. 49 variables but no planetary transits of a white dwarf. Mon Not R Astron Soc 490:1066–1075

Danielski C, Korol V, Tamanini N, Rossi EM (2019) Circumbinary exoplanets and brown dwarfs with the Laser Interferometer Space Antenna. Astron Astrophys 632:A113

Davies MB, Adams FC, Armitage P, Chambers J, Ford E, Morbidelli A, Raymond SN, Veras D (2014) The long-term dynamical evolution of planetary systems. *In:* Protostars, Planets VI, Protostars and Planets VI. H Beuther, RS Klessen, CP Dullemond, T Henning (eds.), University of Arizona Press, Tucson, p 787–808

De Marco O, Soker N (2011) The role of planets in shaping planetary nebulae. Publ Astron Soc Pac 123:402–411

Debes JH (2006) Measuring M dwarf winds with DAZ white dwarfs. Astrophys J 652:636–642

Debes JH, Sigurdsson S (2002) Are there unstable planetary systems around white dwarfs? Astrophys J 572:556–565

Debes JH, Walsh KJ, Stark C (2012) The link between planetary systems, dusty white dwarfs, and metal-polluted white dwarfs. Astrophys J 747:148

Debes JH, Thévenot M, Kuchner MJ, Burgasser AJ, Schneider AC, Meisner AM, Gagné J, Faherty JK, Rees JM, Allen M, Caselden D (2019) A 3 Gyr white dwarf with warm dust discovered via the backyard worlds: Planet 9 Citizen Science Project. Astrophys J Lett 872:L25

Dennihy E, Debes JH, Dunlap BH, Dufour P, Teske JK, Clemens JC (2016) A subtle infrared excess associated with a young white dwarf in the Edinburgh–Cape blue object survey. Astrophys J 831:31

Dennihy E, Clemens JC, Dunlap BH, Fanale SM, Fuchs JT, Hermes JJ (2018) Rapid evolution of the gaseous exoplanetary debris around the white dwarf star HE 1349–2305. Astrophys J 854:40

Dennihy E, Xu S, Lai S, Bonsor A, Clemens JC, Dufour P, Gänsicke BT, Fusillo NP, Hardy F, Hegedus RJ, Hermes JJ (2020) Five new post-main-sequence debris disks with gaseous emission. Astrophys J 905:5

Dong R, Wang Y, Lin DN, Liu XW (2010) Dusty disks around white dwarfs. I. Origin of debris disks. Astrophys J 715:1036–1049

Dosopoulou F, Kalogera V (2016a) Orbital evolution of mass-transferring eccentric binary systems. I. Phase-dependent evolution. Astrophys J 825:70

Dosopoulou F, Kalogera V (2016b) Orbital evolution of mass-transferring eccentric binary systems. II. Secular evolution. Astrophys J 825:71

Doyle AE, Desch SJ, Young ED (2021) Icy exomoons evidenced by spallogenic nuclides in polluted white dwarfs. Astrophys J Lett 907:L35

Dufour P, Bergeron P, Liebert J, Harris HC, Knapp GR, Anderson SF, Hall PB, Strauss MA, Collinge MJ, Edwards MC (2007) On the spectral evolution of cool, helium-atmosphere white dwarfs: Detailed spectroscopic and photometric analysis of DZ stars. Astrophys J 663:1291–1308

Duncan MJ, Lissauer JJ (1998) The effects of post-main-sequence solar mass loss on the stability of our planetary system. Icarus, 134:303–310

Duvvuri GM, Redfield S, Veras D (2020) Necroplanetology: Simulating the tidal disruption of differentiated planetary material orbiting WD 1145+017. Astrophys J 893:166

Faedi F, West RG, Burleigh MR, Goad MR, Hebb L (2011) Detection limits for close eclipsing and transiting substellar and planetary companions to white dwarfs in the WASP survey. Mon Not R Astron Soc 410:899–911

Farihi J (2016) Circumstellar debris and pollution at white dwarf stars. New Astron Rev 71:9–34

Farihi J, Jura M, Zuckerman B (2009) Infrared signatures of disrupted minor planets at white dwarfs. Astrophys J 694:805–819

Farihi J, Jura M, Lee JE, Zuckerman B (2010) Strengthening the case for asteroidal accretion: Evidence for subtle and diverse disks at white dwarfs. Astrophys J 714:1386–1397

Farihi J, Gänsicke BT, Koester D (2013) Evidence for water in the rocky debris of a disrupted extrasolar minor planet. Science 342:218–220

Farihi J, Parsons SG, Gänsicke BT (2017a) A circumbinary debris disk in a polluted white dwarf system. Nat Astron 1:32:1–6

Farihi J, von Hippel T, Pringle JE (2017b) Magnetospherically-trapped dust and a possible model for the unusual transits at WD 1145+017. Mon Not R Astron Soc 471:L145–L149

Farihi J, Fossati L, Wheatley PJ, Metzger BD, Mauerhan J, Bachman S, Gänsicke BT, Redfield S, Cauley PW, Kochukhov O, Achilleos N (2018a) Magnetism, X-rays and accretion rates in WD 1145+017 and other polluted white dwarf systems. Mon Not R Astron Soc 474:947–960

Farihi J, van Lieshout R, Cauley PW, Dennihy E, Su KY, Kenyon SJ, Wilson TG, Toloza O, Gänsicke BT, von Hippel T, Redfield S (2018b) Dust production and depletion in evolved planetary systems. Mon Not R Astron Soc 481:2601–2611

Farihi J, Hermes JJ, Marsh TR, Mustill AJ, Wyatt MC, Guidry JA, Wilson TG, Redfield S, Izquierdo P, Toloza O, Gänsicke BT (2022) Relentless and complex transits from a planetesimal debris disc. Mon Not R Astron Soc 511:1647–1666

Ferich N, Baronett SA, Tamayo D, Steffen JH (2022) The Yarkovsky effect in REBOUNDx. Astrophys J Suppl Ser 262:41

Fortin-Archambault M, Dufour P, Xu S (2020) Modeling of the variable circumstellar absorption features of WD 1145+017. Astrophys J 888:47

Frewen SF N, Hansen BM S (2014) Eccentric planets and stellar evolution as a cause of polluted white dwarfs. Mon Not R Astron Soc 439:2442–2458

Gänsicke BT, Marsh TR, Southworth J, Rebassa-Mansergas A (2006) A gaseous metal disk around a white dwarf. Science 314:1908–1910

Gänsicke BT, Marsh TR, Southworth J (2007) SDSSJ104341.53+085558.2: A second white dwarf with a gaseous debris disc. Mon Not R Astron Soc 380:L3539

Gänsicke BT, Koester D, Marsh TR, Rebassa-Mansergas A, Southworth J (2008) SDSS J084539. 17+ 225728.0: The first DBZ white dwarf with a metal-rich gaseous debris disc. Mon Not R Astron Soc 391:L103- L107

Gänsicke BT, Schreiber MR, Toloza O, Gentile Fusillo NP, Koester D, Manser CJ (2019) Accretion of a giant planet onto a white dwarf star. Nature 576:61–64

Gentile Fusillo NP, Gänsicke BT, Farihi J, Koester D, Schreiber MR, Pala AF (2017) Trace hydrogen in helium atmosphere white dwarfs as a possible signature of water accretion. Mon Not R Astron Soc 468:971–980

Gentile Fusillo NP, Tremblay PE, Gänsicke BT, Manser CJ, Cunningham T, Cukanovaite E, Hollands M, Marsh T, Raddi R, Jordan S, Toonen S (2019) A Gaia Data Release 2 catalogue of white dwarfs and a comparison with SDSS. Mon Not R Astron Soc 482:4570–4591

Gentile Fusillo NP, Manser CJ, Gänsicke BT, Toloza O, Koester D, Dennihy E, Brown WR, Farihi J, Hollands MA, Hoskin MJ, Izquierdo P (2021) White dwarfs with planetary remnants in the era of Gaia I: Six emission line systems. Mon Not R Astron Soc 504:2707–2726

Girven J, Brinkworth CS, Farihi J, Gänsicke BT, Hoard DW, Marsh TR, Koester D (2012) Constraints on the lifetimes of disks resulting from tidally destroyed rocky planetary bodies. Astrophys J 749:154

Goldstein J (1987) The fate of the earth in the red giant envelope of the sun. Astron Astrophys 178:283–285

Graham JR, Matthews K, Neugebauer G, Soifer BT (1990) The infrared excess of G29–38: A brown dwarf or dust? Astrophys J 357:216–223

Grishin E, Veras D (2019) Embedding planetesimals into white dwarf discs from large distances. Mon Not R Astron Soc 489:168–175

Guidry JA, Vanderbosch ZP, Hermes JJ, Barlow BN, Lopez ID, Boudreaux TM, Corcoran KA, Bell KJ, Montgomery MH, Heintz TM, Castanheira BG (2021) ISpy transits and pulsations: Empirical variability in white dwarfs using Gaia and the Zwicky Transient Facility. Astrophys J 912:125

Grunblatt SK, Huber D, Gaidos E, Hon M, Zinn JC, Stello D (2019) Giant planet occurrence within 0.2 au of low-luminosity red giant branch stars with K2. Astrophys J 158:227

Grunblatt SK, Saunders N, Chontos A, Hattori S, Veras D, Huber D, Angus R, Rice M, Breivik K, Blunt S, Giacalone S (2023) TESS giants transiting giants III: An eccentric warm Jupiter supports a period-eccentricity relation for giant planets transiting evolved stars. Astrophys J 165:44

Gurri P, Veras D, Gänsicke BT (2017) Mass and eccentricity constraints on the planetary debris orbiting the white dwarf WD 1145+017. Mon Not R Astron Soc 464:321–328

Hadjidemetriou JD (1963) Two-body problem with variable mass: A new approach. Icarus 2:440–451

Hallakoun N, Xu S, Maoz D, Marsh TR, Ivanov VD, Dhillon VS, Bours MC, Parsons SG, Kerry P, Sharma S, Su K (2017) Once in a blue moon: detection of 'bluing' during debris transits in the white dwarf WD 1145+017. Mon Not R Astron Soc 469:3113–3224

Hamers AS, Portegies Zwart SF (2016) White dwarf pollution by planets in stellar binaries. Mon Not R Astron Soc 462:L84–L87

Harrison JH D, Bonsor A, Madhusudhan N (2018) Polluted white dwarfs: constraints on the origin and geology of exoplanetary material. Mon Not R Astron Soc 479:3814–3841

Harrison JH D, Shorttle O, Bonsor A (2021) Evidence for post-nebula volatilisation in an exo-planetary body. Earth Planet Sci Lett 554:116694

Hartmann S, Nagel T, Rauch T, Werner K (2011) Non-LTE models for the gaseous metal component of circumstellar discs around white dwarfs. Astron Astrophys 530:A7

Hartmann S, Nagel T, Rauch T, Werner K (2014) Non-LTE spectral models for the gaseous debris-disk component of Ton 345. Astron Astrophys 571:A44

Hegazi A, Bear E, Soker N (2020) On the role of reduced wind mass-loss rate in enabling exoplanets to shape planetary nebulae. Mon Not R Astron Soc 496:612–619

Hoang NH, Mogavero F, Laskar J (2022) Long-term instability of the inner Solar system: numerical experiments. Mon Not R Astron Soc 514:1342–1350

Hogg MA, Wynn GA, Nixon C (2018) The galactic rate of second- and third generation disc and planet formation. Mon Not R Astron Soc 479:4486–4498

Hogg MA, Cutter R, Wynn GA (2021) The effect of a magnetic field on the dynamics of debris discs around white dwarfs. Mon Not R Astron Soc 500:2986–3001

Hon M, Huber D, Rui NZ, Fuller J, Veras D, Kuszlewicz JS, Kochukhov O, Stokholm A, Rørsted JL, Yıldız M, Orhan ZÇ (2023) A close-in Jovian planet orbiting a helium-burning red giant star. Nature 618:917–920

Hoskin MJ, Toloza O, Gänsicke BT, Raddi R, Koester D, Pala AF, Manser CJ, Farihi J, Belmonte MT, Hollands M, Gentile Fusillo N (2020) White dwarf pollution by hydrated planetary remnants: Hydrogen and metals in WDJ204713.76–125908.9. Mon Not R Astron Soc 499:171–182

Huber D, Chaplin WJ, Chontos A, Kjeldsen H, Christensen-Dalsgaard J, Bedding TR, Ball W, Brahm R, Espinoza N, Henning T, Jordán A (2019) A hot Saturn orbiting an oscillating late subgiant discovered by TESS. Astron J 157:245

Hurley JR, Pols OR, Tout CA (2000) Comprehensive analytic formulae for stellar evolution as a function of mass and metallicity. Mon Not R Astron Soc 315:543–569

Iorio L (2010) Orbital effects of Sun's mass loss and the Earth's fate. Nat Sci 2:329–337

Izquierdo P, Rodríguez-Gil P, Gänsicke BT, Mustill AJ, Toloza O, Tremblay PE, Wyatt M, Chote P, Eggl S, Farihi J, Koester D (2018) Fast spectrophotometry of WD 1145+017. Mon Not R Astron Soc 481:703–714

Jackson B, Carlberg J (2018) Accretion of planetary material onto host stars. *In:* Handbook of Exoplanets. HJ Deeg, JA Belmonte (eds) Springer, Cham, p 1895–1912

Jia S, Spruit HC (2018) Disruption of a planet spiraling into its host star. Astrophys J 864:169

Jiang C, Bedding TR, Stassun KG, Veras D, Corsaro E, Buzasi DL, Mikołajczyk P, Zhang QS, Ou JW, Campante TL, Rodrigues TS (2020) TESS asteroseismic analysis of the known exoplanet host star HD 222076. Astrophys J 896:65

Jin Z, Li D, Zhu Z-H (2023) Binary asteroid dissociation and accretion around white dwarfs. Astron Astrophys 674:A52

Jones RH (2024) Meteorites and planet formation. Rev Mineral Geochem 90:113–140

Jura M (2003) A tidally disrupted asteroid around the white dwarf G29-38. Astrophys J 584:L91–L94

Jura M (2004) Other Kuiper belts. Astrophys J 603:729–737

Jura M (2008) Pollution of single white dwarfs by accretion of many small asteroids. Astrophys J 135:1785–1792

Jura M, Xu S (2010) The survival of water within extrasolar minor planets. Astron. J 140:1129–1136

Jura M, Young ED (2014) Extrasolar cosmochemistry. Annu Rev Earth Planet Sci 42:45–67

Jura M, Farihi J, Zuckerman B, Becklin EE (2007a) Infrared emission from the dusty disk orbiting GD 362, an externally polluted white dwarf. Astron J 133:1927–1933

Jura M, Farihi J, Zuckerman B (2007b) Externally polluted white dwarfs with dust disks. Astrophys J 663:1285–1290

Jura M, Xu S, Klein B, Koester D, Zuckerman B (2012) Two extrasolar asteroids with low volatile-element mass fractions. Astrophys J 750:69

Kaltenegger L, MacDonald RJ, Kozakis T, Lewis NK, Mamajek EE, McDowell JC, Vanderburg A (2020) The white dwarf opportunity: Robust detections of molecules in earth-like exoplanet atmospheres with the James Webb space telescope. Astrophys J Lett 901:L1

Kang Y, Liu C, Shao L (2021) Prospects for detecting exoplanets around double white dwarfs with LISA and Taiji. Astrophys J 162:247

Katz JI (2018) Why is interstellar object 1I/2017 U1 ('Oumuamua) rocky, tumbling and possibly very prolate? Mon Not R Astron Soc 478:L95–L98

Kenyon SJ, Bromley BC (2017a) Numerical simulations of collisional cascades at the roche limits of white dwarf stars. Astrophys J 844:116

Kenyon SJ, Bromley BC (2017b) Numerical simulations of gaseous disks generated from collisional cascades at the Roche limits of white dwarf stars. Astrophys J 850:50

Kervella P, Homan W, Richards AM, Decin L, McDonald I, Montargès M, Ohnaka K (2016) ALMA observations of the nearby AGB star L2 Puppis. I. Mass of the central star and detection of a candidate planet. Astron Astrophys 596:A92

Koester D (2009) Accretion and diffusion in white dwarfs. New diffusion timescales and applications to GD 362 and G 29–38. Astron Astrophys 498:517–525

Koester D, Gänsicke BT, Farihi J (2014) The frequency of planetary debris around young white dwarfs. Astron Astrophys 566:A34

Kostov VB, Moore K, Tamayo D, Jayawardhana R, Rinehart SA (2016) Tatooine's future: The eccentric response of Kepler's circumbinary planets to common envelope evolution of their host stars. Astrophys J 832:183

Kozakis T, Kaltenegger L (2019) Atmospheres and UV environments of Earth-like planets throughout post-main-sequence evolution. Astrophys J 875:99

Kozakis T, Kaltenegger L, Hoard DW (2018) UV surface environments and atmospheres of Earth-like planets orbiting white dwarfs. Astrophys J 862:69

Kozakis T, Lin Z, Kaltenegger L (2020) High-resolution spectra and biosignatures of Earth-like planets transiting white dwarfs. Astrophys J 894:L6

Kratter KM, Perets HB (2012) Star hoppers: Planet instability and capture in evolving binary systems. Astrophys J 753:91

Krzesinski J, Blokesz A, Siwak M, Stachowski G (2020) The quest for planets around subdwarfs and white dwarfs from Kepler space telescope fields. I. Techniques and tests of the methods. Astron Astrophys 642:105

Kubiak S, Vanderburg A, Becker J, Gary B, Rappaport SA, Xu S, de Beurs Z (2023) TTV constraints on additional planets in the WD 1856+534 system. Mon Not R Astron Soc 521:4679–4694

Kunitomo M, Ikoma M, Sato B, Katsuta Y, Ida S (2011) Planet engulfment by ~1.5–3 $R_\odot$ red giants. Astrophys J 737:66

Lagos F, Schreiber MR, Zorotovic M, Gänsicke BT, Ronco MP, Hamers AS (2021) WD 1856 b: A close giant planet around a white dwarf that could have survived a common -envelope phase. Mon Not R Astron Soc 501:676–682

Lai S, Dennihy E, Xu S, Nitta A, Kleinman S, Leggett SK, Bonsor A, Hodgkin S, Rebassa-Mansergas A, Rogers LK (2021) Infrared excesses around bright white dwarfs from Gaia and unWISEII. Astrophys J 920:156

Lanza AF, Lebreton Y, Sallard C (2023) Residual eccentricity of an Earth-like planet orbiting a red giant Sun. Astron Astrophys 674:A176

Laskar J, Gastineau M (2009) Existence of collisional trajectories of Mercury, Mars and Venus with the Earth. Nature 459:817–819

Ledda S, Danielski C, Turrini D (2023) The quest for Magrathea planets I: Formation of second generation exoplanets around double white dwarfs. Astron Astrophys 675:A184

Li G, Naoz S, Valsecchi F, Johnson JA, Rasio FA (2014) The dynamics of the multi-planet system orbiting Kepler-56. Astrophys J 794:131

Li D, Mustill AJ, Davies MB (2021) Accretion of tidally disrupted asteroids on to white dwarfs: direct accretion versus disc processing. Mon Not R Astron Soc 508:5671–5686

Li D, Mustill AJ, Davies MB (2022) Metal pollution of the solar white dwarf by solar system small bodies. Astrophys J 924:61

Limbach MA, Vanderburg A, Stevenson KB, Blouin S, Morley C, Lustig-Yaeger J, Soares-Furtado M, Janson M (2022) A new method for finding nearby white dwarfs exoplanets and detecting biosignatures. Mon Not R Astron Soc 517:2622–2638

Lin Z, Seager S, Ranjan S, Kozakis T, Kaltenegger L (2022) H_2-dominated atmosphere as an indicator of second-generation rocky white dwarf exoplanets. Astrophys J Lett 925:L10

Lovell JB, Wyatt MC, Kalas P, Kennedy GM, Marino S, Bonsor A, Penoyre Z, Fulton BJ, Pawellek N (2022) High-resolution ALMA and HST imaging of κCrB: a broad debris disc around a post-main-sequence star with low-mass companions. Mon Not R Astron Soc 517:2546–2566

Lucas M, Bottom M, Ruane G, Ragland S (2022) An imaging search for post-main-sequence planets of Sirius B. Astron J 163:81

Luhman KL, Burgasser AJ, Bochanski JJ (2011) Discovery of a candidate for the coolest known brown dwarf. Astrophys J 730:L9

Luhn JK, Bastien FA, Wright JT, Johnson JA, Howard AW, Isaacson H (2019) Retired A stars and their companions. VIII. 15 new planetary signals around subgiants and transit parameters for California Planet search planets with subgiant hosts. Astron J, 157:149

MacLeod M, Cantiello M, Soares-Furtado M (2018) Planetary engulfment in the Hertzsprung–Russell diagram. Astrophys J Lett, 853:L1

Madappatt N, De Marco O, Villaver E (2016) The effect of tides on the population of PN from interacting binaries. Mon Not R Astron Soc 463:1040–1056

Makarov VV, Veras D (2019) Chaotic rotation and evolution of asteroids and small planets in high-eccentricity orbits around white dwarfs. Astrophys J 886:127

Malamud U, Perets HB (2016) Post-main sequence evolution of icy minor planets: implications for water retention and white dwarf pollution. Astrophys J 832:160

Malamud U, Perets HB (2017a) Post-main-sequence evolution of icy minor planets. II. Water retention and white dwarf pollution around massive progenitor stars. Astrophys J 842:67

Malamud U, Perets HB (2017b) Post-main-sequence evolution of icy minor planets. III. Water retention in dwarf planets and exomoons and implications for white dwarf pollution. Astrophys J 849:8

Malamud U, Perets HB (2020a) Tidal disruption of planetary bodies by white dwarfs–I: A hybrid SPH-analytical approach. Mon Not R Astron Soc 492:5561–5581

Malamud U, Perets HB (2020b) Tidal disruption of planetary bodies by white dwarfs–II. Debris disc structure and ejected interstellar asteroids. Mon Not R Astron Soc 493:698– 712

Malamud U, Grishin E, Brouwers M (2021) Circularization of tidal debris around white dwarfs: Implications for gas production and dust variability. Mon Not R Astron Soc 501:3806–3824

Maldonado RF, Villaver E, Mustill AJ, Chavez M, Bertone E (2020a) Dynamical evolution of two-planet systems and its connection with white dwarf atmospheric pollution. Mon Not R Astron Soc 497:4091–4106

Maldonado RF, Villaver E, Mustill AJ, Chavez M, Bertone E (2020b) Understanding the origin of white dwarf atmospheric pollution by dynamical simulations based on detected three-planet systems. Mon Not R Astron Soc 499:1854–1869

Maldonado RF, Villaver E, Mustill AJ, Chavez M, Bertone E (2021) Do instabilities in high-multiplicity systems explain the existence of close-in white dwarf planets? Mon Not R Astron Soc 501:L43–L48

Maldonado RF, Villaver E, Mustill AJ, Chávez M (2022) Disentangling the parameter space: the role of planet multiplicity in triggering dynamical instabilities on planetary systems around white dwarfs. Mon Not R Astron Soc 512:104–115

Manser CJ, Gänsicke BT, Marsh TR, Veras D, Koester D, Breedt E, Pala AF, Parsons SG, Southworth J (2016) Doppler imaging of the planetary debris disc at the white dwarf SDSSJ122859.93+104032.9. Mon Not R Astron Soc 455:4467–4478

Manser CJ, Gänsicke BT, Eggl S, Hollands M, Izquierdo P, Koester D, Landstreet JD, Lyra W, Marsh TR, Meru F, Mustill AJ (2019) A planetesimal orbiting within the debris disc around a white dwarf star. Science 364:66–69

Manser CJ, Gänsicke BT, Gentile Fusillo NP, Ashley R, Breedt E, Hollands M, Izquierdo P, Pelisoli I (2020) The frequency of gaseous debris discs around white dwarfs. Mon Not R Astron Soc 493:2127–2139

Manser CJ, Dennihy E, Gänsicke BT, Debes JH, Gentile Fusillo NP, Hermes JJ, Hollands M, Izquierdo P, Kaiser BC, Marsh TR, Reding JS (2021) Imaging the rapidly precessing planetary disc around the white dwarf HE 1349–2305 using Doppler tomography. Mon Not R Astron Soc 508:5657–5670

Marshall JP, Ertel S, Birtcil E, Villaver E, Kemper F, Boffin H, Scicluna P, Kamath D (2022) Evidence for the disruption of a planetary system during the formation of the Helix Nebula. Astron J 165:22

Martinez MA S, Stone NC, Metzger BD (2019) Orphaned exomoons: Tidal detachment and evaporation following an exoplanet–star collision. Mon Not R Astron Soc 489:5119–5135

McDonald CH, Veras D (2021) White dwarf planetary debris dependence on physical structure distributions within asteroid belts. Mon Not R Astron Soc 506:4031–4047

McDonald CH, Veras D (2023) Binary asteroid scattering around white dwarfs. Mon Not R Astron Soc 520:4009–4022

Melis C, Jura M, Albert L, Klein B, Zuckerman B (2010) Echoes of a decaying planetary system: The gaseous and dusty disks surrounding three white dwarfs. Astrophys J 722:1078–1091

Melis C, Klein B, Doyle AE, Weinberger A, Zuckerman B, Dufour P (2020) Serendipitous discovery of nine white dwarfs with gaseous debris disks. Astrophys J 905:56

Merlov A, Bear E, Soker N (2021) A red giant branch common-envelope evolution scenario for the exoplanet WD 1856 b. Astrophys J Lett 915:L34

Metzger BD, Rafikov RR, Bochkarev KV (2012) Global models of runaway accretion in white dwarf debris discs. Mon Not R Astron Soc 423:505–528

Mikkola S, Lehto HJ (2022) Overlong simulations of the solar system dynamics with two alternating step-lengths. Celestial Mech. Dyn Astron 134:20

Miranda R, Rafikov RR (2018) Fast and slow precession of gaseous debris disks around planet-accreting white dwarfs. Astrophys J 857:135

Mogavero F, Laskar J (2021) Long-term dynamics of the inner planets in the solar system. Astron Astrophys 655:A1

Mogavero F, Hoang NH, Laskar J (2023) Timescales of chaos in the inner solar system: Lyapunov spectrum and quasi-integrals of motion. Phys Rev X 13:021018

Mordasini C, Burn R (2024) Planet formation—Observational constraints, physical processes, and compositional patterns. Rev Mineral Geochem 90:55-112

Morris BM, Heng K, Brandeker A, Swan A, Lendl M (2021) ACHEOPS white dwarf transit search. Astron Astrophys 651:L12

Muñoz DJ, Petrovich C (2020) Kozai migration naturally explains the white dwarf planet WD1856 b. Astrophys J 904:L3

Mustill AJ, Villaver E (2012) Foretellings of Ragnarök: World-engulfing asymptotic giants and the inheritance of white dwarfs. Astrophys J 761:121

Mustill AJ, Veras D, Villaver E (2014) Long-term evolution of three-planet systems to the post-main sequence and beyond. Mon Not R Astron Soc 437:1404–1419

Mustill AJ, Villaver E, Veras D, Gänsicke BT, Bonsor A (2018) Unstable low-mass planetary systems as drivers of white dwarf pollution. Mon Not R Astron Soc 476:3939–3955

Nelemans G, Tauris TM (1998) Formation of undermassive single white dwarfs and the influence of planets on late stellar evolution. Astron Astrophys 335:L85–L88

Niedzielski A, Villaver E, Adamów M, Kowalik K, Wolszczan A, Maciejewski G (2021) Tracking Advanced Planetary Systems (TAPAS) with HARPS-N. VII. Elder suns with low-mass companions. Astron Astrophys 648:A58

Nordhaus J, Spiegel DS (2013) On the orbits of low-mass companions to white dwarfs and the fates of the known exoplanets. Mon Not R Astron Soc 432:500–505

O'Connor CE, Lai D (2020) High-eccentricity migration of planetesimals around polluted white dwarfs. Mon Not R Astron Soc 498:4005–4020

O'Connor CE, Liu B, Lai D (2021) Enhanced Lidov–Kozai migration and the formation of the transiting giant planet WD 1856+534 b. Mon Not R Astron Soc 501:507–514

O'Connor CE, Teyssandier J, Lai D (2022) Secular chaos in white dwarf planetary systems: origins of metal pollution and short-period planetary companions. Mon Not R Astron Soc 513:4178–4195

O'Connor CE, Bildsten L, Cantiello M, Lai D (2023a) Giant planet engulfment by evolved giant stars: light curves, asteroseismology, and survivability. Astrophys J 950:128

O'Connor CE, Lai D, Seligman DZ (2023b) On the pollution of white dwarfs by exo-Oort cloud comets. Mon Not R Astron Soc 524:6181–6197

Okuya A, Ida S, Hyodo R, Okuzumi S (2023) Modeling the evolution of silicate/volatile accretion discs around white dwarfs. Mon Not R Astron Soc 519:1657–1676

Omarov TB (1962) On differential equations for oscillating elements in the theory of variable mass movement. Izv Astrofiz Inst Acad Nauk KazSSR 14:66–71 (in Russian)

Paquette C, Pelletier C, Fontaine G, Michaud G (1986) Diffusion in white dwarfs: New results and comparative study. Astrophys J Suppl Ser 61:197–217

Parriott J, Alcock C (1998) On the number of comets around white dwarf stars: Orbit survival during the late stages of stellar evolution. Astrophys J 501:357–366

Passy J-C, Mac Low M-M, De Marco O (2012) On the survival of brown dwarfs and planets engulfed by their giant host star. Astrophys J Lett 759:L30

Payne MJ, Veras D, Holman MJ, Gänsicke BT (2016) Liberating exomoons in white dwarf planetary systems. Mon Not R Astron Soc 457:217–231

Payne MJ, Veras D, Gänsicke BT, Holman MJ (2017) The fate of exomoons in white dwarf planetary systems. Mon Not R Astron Soc 464:2557–2564

Perets HB (2011) Planets in evolved binary systems. *In:* AIP Conf Proc, Planetary Systems Beyond the Main Sequence. S Schuh, H Drechsel, U Heber (eds) 1331:56–75

Perets HB, Kenyon SJ (2013) Wind-accretion disks in wide binaries, second generation protoplanetary disks, and accretion onto white dwarfs. Astrophys J 764:169

Perryman M, Hartman J, Bakos GA, Lindegren L (2014) Astrometric exoplanet detection with Gaia. Astrophys J 797:14

Petrovich C, Muñoz DJ (2017) Planetary engulfment as a trigger for white dwarf pollution. Astrophys J 834:116

Pichierri G, Morbidelli A, Lai D (2017) Extreme secular excitation of eccentricity inside mean motion resonance. Small bodies driven into star-grazing orbits by planetary perturbations. Astron Astrophys 605:A23

Portegies Zwart S (2013) Planet-mediated precision reconstruction of the evolution of the cataclysmic variable HU Aquarii. Mon Not R Astron Soc 429:L45–L49

Putirka KD, Xu S (2021) Polluted white dwarfs reveal exotic mantle rock types on exoplanets in our solar neighborhood. Nat Commun 12:6168

Raddi R, Gänsicke BT, Koester D, Farihi J, Hermes JJ, Scaringi S, Breedt E, Girven J (2015) Likely detection of water-rich asteroid debris in a metal-polluted white dwarf. Mon Not R Astron Soc 450:2083–2093

Rafikov RR (2011a) Metal accretion onto white dwarfs caused by poynting-robertson drag on their debris disks. Astrophys J 732:L3

Rafikov RR (2011b) Runaway accretion of metals from compact discs of debris on to white dwarfs. Mon Not R Astron Soc 416:L55–L59

Rafikov RR, Garmilla JA (2012) Inner edges of compact debris disks around metal-rich white dwarfs. Astrophys J 760:123

Ramirez RM, Kaltenegger L (2016) Habitable zones of post-main sequence stars. Astrophys J 823:6

Rappaport S, Gary BL, Kaye T, Vanderburg A, Croll B, Benni P, Foote J (2016) Drifting asteroid fragments around WD 1145+017. Mon Not R Astron Soc 458:3904–3917

Raymond SN, Armitage PJ (2013) Mini-Oort clouds: compact isotropic planetesimal clouds from planet–planet scattering. Mon Not R Astron Soc 429:L99–L103

Reach WT, Kuchner MJ, von Hippel T, Burrows A, Mullally F, Kilic M, Winget DE (2005) The Dust Cloud around the White Dwarf G29-38. Astrophys J 635:L161–L164

Reach WT, Lisse C, von Hippel T, Mullally F (2009) The dust cloud around the white dwarf G 29–38. II. Spectrum from 5 to 40 μm and mid-infrared photometric variability. Astrophys J 693:697–712

Redfield S, Farihi J, Cauley PW, Parsons SG, Gänsicke BT, Duvvuri GM (2017) Spectroscopic evolution of disintegrating planetesimals: Minute to month variability in the circumstellar gas associated with WD 1145+017. Astrophys J 839:42

Reffert S, Bergmann C, Quirrenbach A, Trifonov T, Künstler A (2015) Precise radial velocities of giant stars. VII. Occurrence rate of giant extrasolar planets as a function of mass and metallicity. Astron Astrophys 574:A116

Regály Z, Fröhlich V, Vinkó J (2022) Lost in space: Companions' fatal dance around massive dying stars. Astrophys J 941:121

Rocchetto M, Farihi J, Gänsicke BT, Bergfors C (2015) The frequency and infrared brightness of circumstellar discs at white dwarfs. Mon Not R Astron Soc 449:574–587

Rodriguez DR, Zuckerman B, Melis C, Song I (2011) The ultra cool brown dwarf companion of WD 0806-661B: Age, mass, and formation mechanism. Astrophys J 732:L29

Rogers LK, Xu S, Bonsor A, Hodgkin S, Su KYL, von Hippel T, Michael J (2020) Near-infrared variability in dusty white dwarfs: tracing the accretion of planetary material. Mon Not R Astron Soc 494:2861–2874

Ronco MP, Schreiber MR, Giuppone CA, Veras D, Cuadra J, Guilera OM (2020) How Jupiters save or destroy inner Neptunes around evolved stars. Astrophys J Lett 898:L23

Röpke FK, De Marco O (2023) Simulations of common-envelope evolution in binary stellar systems: Physical models and numerical techniques. Living Rev Comput Astrophys 9:2

Rozner M, Veras D, Perets HB (2021) Rapid destruction of planetary debris around WDs through wind erosion. Mon Not R Astron Soc 502:5176–5184

Rybicki KR, Denis C (2001) On the final destiny of the Earth and the solar system. Icarus, 151:130–137

Sabach E, Soker N (2018) Accounting for planet-shaped planetary nebulae. Mon Not R Astron Soc 473:286–294

Sackmann I-J, Boothroyd AI, Kraemer KE (1993) Our Sun. III. present and future. Astrophys J Lett 418:457–468

Sanderson H, Bonsor A, Mustill A (2022) Can Gaia find planets around white dwarfs? Mon Not R Astron Soc 517:5835–5852

Sandhaus PH, Debes JH, Ely J, Hines DC, Bourque M (2016) A search for short-period rocky planets around WDs with the Cosmic Origins Spectrograph (COS). Astrophys J 823:49

Schleicher DR G, Dreizler S (2014) Planet formation from the ejecta of common envelopes. Astron Astrophys 563:A61,

Schreiber MR, Gänsicke BT, Toloza O, Hernandez M-S, Lagos F (2019) Cold giant planets evaporated by hot white dwarfs. Astrophys J 887:L4

Schröder K-P, Smith RC (2008) Distant future of the Sun and Earth revisited. Mon Not R Astron Soc 386:155–163

Sevilla J, Behmard A, Fuller J (2022) Long-term lithium abundance signatures following planetary engulfment. Mon Not R Astron Soc 516:3354–3365

Shengbang Q, Zhongtao H, Lajús EF, Wenping L, Miloslav Z, Linjia L, Voloshina I, Liang L, Jiajia H (2016) Interactions between planets and evolved stars. J Phys Conf Ser 728:042006

Shestakova LI, Demchenko BI, Serebryanskiy AV (2019) On the orbital evolution of dust grains in the sublimation region around WD1145+017. Mon Not R Astron Soc 487:3935–3945

Sigurdsson S, Richer HB, Hansen BM, Stairs IH, Thorsett SE (2003) A young white dwarf companion to pulsar B1620-26: Evidence for early planet formation. Science 301:193–196

Sigurdsson S, Thorsett SE (2005) Update on pulsar B1620-26 in M4: Observations, models, and implications. *In:* Binary Radio Pulsars. FA Rasio, IH Stairs (eds). ASP Conf Ser 328:213–223

Smallwood JL, Martin RG, Livio M, Lubow SH (2018) White dwarf pollution by asteroids from secular resonances. Mon Not R Astron Soc 480:57–67

Smallwood JL, Martin RG, Livio M, Veras D (2021) On the role of resonances in polluting white dwarfs by asteroids. Mon Not R Astron Soc 504:3375–3386

Soker N (1996) What planetary nebulae can tell us about planetary systems. Astrophys J Lett 460:L5356

Soker N (2008) Defining the termination of the asymptotic giant branch. Astrophys J Lett 674:L4952

Spiegel DS, Madhusudhan N (2012) Jupiter will become a hot Jupiter: Consequences of post-main-sequence stellar evolution on gas giant planets. Astrophys J 756:138

Staff JE, De Marco O, Wood P, Galaviz P, Passy JC (2016) Hydrodynamic simulations of the interaction between giant stars and planets. Mon Not R Astron Soc 458:832–844

Steckloff JK, Debes J, Steele A, Johnson B, Adams ER, Jacobson SA, Springmann A (2021) How sublimation delays the onset of dusty debris disk formation around white dwarf stars. Astrophys J Lett 913:L31

Steele A, Debes J, Xu S, Yeh S, Dufour P (2021) A characterization of the circumstellar gas around WD 1124–293 using Cloudy. Astrophys J 911:25

Stephan AP, Naoz S, Zuckerman B (2017) Throwing icebergs at white dwarfs. Astrophys J 844:L16

Stephan AP, Naoz S, Gaudi BS (2018) A-type stars, the destroyers of worlds: The lives and deaths of Jupiters in evolving stellar binaries. Astron J 156:128

Stephan AP, Naoz S, Gaudi BS, Salas JM (2020) Eating planets for lunch and dinner: Signatures of planet consumption by evolving stars. Astrophys J 889:45

Stephan AP, Naoz S, Gaudi BS (2021) Giant planets, tiny stars: producing short period planets around white dwarfs with the eccentric Kozai–Lidov mechanism. Astrophys J 922:4

Stern SA, Shull JM, Brandt JC (1990) Evolution and detectability of comet clouds during post-main-sequence stellar evolution. Nature 345:305–308

Stock K, Veras D, Cai MX, Spurzem R, Portegies Zwart S (2022) Birth cluster simulations of planetary systems with multiple super-Earths: Initial conditions for white dwarf pollution drivers. Mon Not R Astron Soc 512:2460–2473

Stone N, Metzger BD, Loeb A (2015) Evaporation and accretion of extrasolar comets following white dwarf kicks. Mon Not R Astron Soc 448:188–206

Su KY, Chu YH, Rieke GH, Huggins PJ, Gruendl R, Napiwotzki R, Rauch T, Latter WB, Volk K (2007) A debris disk around the central star of the Helix Nebula? Astrophys J Lett 657:L41-L45

Swan A, Farihi J, Wilson TG (2019b) Most white dwarfs with detectable dust discs show infrared variability. Mon Not R Astron Soc 484:L109-L113

Swan A, Farihi J, Wilson TG, Parsons SG (2020) The dust never settles: collisional production of gas and dust in evolved planetary systems. Mon Not R Astron Soc 496:5233–5242

Swan A, Kenyon SJ, Farihi J, Dennihy E, Gänsicke BT, Hermes JJ, Melis C, von Hippel T (2021) Collisions in a gas-rich white dwarf planetary debris disc. Mon Not R Astron Soc 506:432–440

Tamanini N, Danielski C (2019) The gravitational-wave detection of exoplanets orbiting white dwarf binaries using LISA. Nat Astron 3:858–866

Thommes EW, Duncan MJ, Levison HF (1999) The formation of Uranus and Neptune in the Jupiter–Saturn region of the Solar System. Nature 402:635–638

Thorsett SE, Arzoumanian Z, Taylor JH (1993) PSRB1620-26: A binary radio pulsar with a planetary companion? Astrophys J Lett 412:L33–L36

Trierweiler IL, Doyle AE, Melis C, Walsh KJ, Young ED (2022) Exomoons as sources of white dwarf pollution. Astrophys J 936:30

Trevascus D, Price DJ, Nealon R, Liptai D, Manser CJ, Veras D (2021) Formation of eccentric gas discs from sublimating or partially disrupted asteroids orbiting white dwarfs. Mon Not R Astron Soc 505:L2125

Tsiganis K, Gomes R, Morbidelli A, Levison HF (2005) Origin of the orbital architecture of the giant planets of the Solar System. Nature 435:459–461

Van Grootel V, Pozuelos FJ, Thuillier A, Charpinet S, Delrez L, Beck M, Fortier A, Hoyer S, Sousa SG, Barlow BN, Billot N (2021) A search for transiting planets around hot subdwarfs. I. Methods and performance tests on light curves from Kepler, K2, TESS, and CHEOPS. Astron Astrophys 650:A205

van Lieshout R, Kral Q, Charnoz S, Wyatt MC, Shannon A (2018) Exoplanet recycling in massive white-dwarf debris discs. Mon Not R Astron Soc 480:2784–2812

van Lieshout R, Rappaport SA (2018) Disintegrating rocky exoplanets. *In:* Handbook of Exoplanets. HJ Deeg, JA Belmonte (eds) Springer, Cham, p 1528–1543 van Sluijs L, Van Eylen V (2018) The occurrence of planets and other substellar bodies around white dwarfs using K2. Mon Not R Astron Soc 474:4603–4611

Vanderbosch Z, Hermes JJ, Dennihy E, Dunlap BH, Izquierdo P, Tremblay PE, Cho PB, Gänsicke BT, Toloza O, Bell KJ, Montgomery MH (2020) A white dwarf with transiting circumstellar material far outside the Roche Limit. Astrophys J 897:171

Vanderbosch ZP, Rappaport S, Guidry JA, Gary BL, Blouin S, Kaye TG, Weinberger AJ, Melis C, Klein BL, Zuckerman B, Vanderburg A (2021) Recurring planetary debris transits and circumstellar gas around white dwarf ZTFJ0328-1219. Astrophys J 917:41

Vanderburg A, Johnson JA, Rappaport S, Bieryla A, Irwin J, Lewis JA, Kipping D, Brown WR, Dufour P, Ciardi DR, Angus R (2015) A disintegrating minor planet transiting a white dwarf. Nature 526:546–549

Vanderburg A, Rappaport SA (2018) Transiting disintegrating planetary debris around WD 1145+017. *In:* Handbook of Exoplanets. HJ Deeg, JA Belmonte (eds). Springer, Cham, p 2603–2626

Vanderburg A, Rappaport SA, Xu S, Crossfield IJ, Becker JC, Gary B, Murgas F, Blouin S, Kaye TG, Palle E, Melis C (2020) A giant planet candidate transiting a white dwarf. Nature 585:363–367

Veras D (2016a) Post-main-sequence planetary system evolution. R Soc Open Sci 3:150571

Veras D (2016b) The fates of solar system analogues with one additional distant planet. Mon Not R Astron Soc 463:2958–2971

Veras D (2020) The white dwarf planet WDJ0914+1914 b: barricading potential rocky pollutants? Mon Not R Astron Soc 493:4692–4699

Veras D (2021) Planetary systems around white dwarfs. *In:* Oxford Research Encyclopedia of Planetary Science, doi:10.1093/acrefore/9780190647926.013.238

Veras D, Fuller J (2019) Tidal circularization of gaseous planets orbiting white dwarfs. Mon Not R Astron Soc 489:2941–2953

Veras D, Fuller J (2020) The dynamical history of the evaporating or disrupted ice giant planet around white dwarf WDJ0914+1914. Mon Not R Astron Soc 492:6059–6066

Veras D, Gänsicke BT (2015) Detectable close-in planets around white dwarfs through late unpacking. Mon Not R Astron Soc 447:1049–1058

Veras D, Heng K (2020) The lifetimes of planetary debris discs around white dwarfs. Mon Not R Astron Soc 496:2292–2308

Veras D, Hinkley S (2021) The post-main-sequence fate of the HR 8799 planetary system. Mon Not R Astron Soc 505:1557–1566

Veras D, Kurosawa K (2020) Generating metal-polluting debris in white dwarf planetary systems from small-impact crater ejecta. Mon Not R Astron Soc 494:442–457

Veras D, Rosengren AJ (2023) The smallest planetary drivers of white dwarf pollution. Mon Not R Astron Soc 519:6257–6266

Veras D, Scheeres DJ (2020) Post-main-sequence debris from rotation-induced YORP break-up of small bodies—II. Multiple fissions, internal strengths, and binary production. Mon Not R Astron Soc 492:2437–2445

Veras D, Tout CA (2012) The great escape—II. Exoplanet ejection from dying multiplestar systems. Mon Not R Astron Soc 422:1648–1664

Veras D, Vidotto AA (2021) Planetary magnetosphere evolution around post-main sequence stars. Mon Not R Astron Soc 506:1697–1703

Veras D, Wolszczan A (2019) Survivability of radio-loud planetary cores orbiting white dwarfs. Mon Not R Astron Soc 488:153–163

Veras D, Wyatt MC (2012) The Solar system's post-main-sequence escape boundary. Mon Not R Astron Soc 421:2969–2981

Veras D, Wyatt MC, Mustill AJ, Bonsor A, Eldridge JJ (2011) The great escape: how exoplanets and smaller bodies desert dying stars. Mon Not R Astron Soc 417:2104–2123

Veras D, Hadjidemetriou JD, Tout CA (2013a) An exoplanet's response to anisotropic stellar mass loss during birth and death. Mon Not R Astron Soc 435:2416–2430

Veras D, Mustill AJ, Bonsor A, Wyatt MC (2013b) Simulations of two-planet systems through all phases of stellar evolution: implications for the instability boundary and white dwarf pollution. Mon Not R Astron Soc 431:1686–1708

Veras D, Shannon A, Gänsicke BT (2014a) Hydrogen delivery onto white dwarfs from remnant exo-Oort cloud comets. Mon Not R Astron Soc 445:4175–4185

Veras D, Leinhardt ZM, Bonsor A, Gänsicke BT (2014b) Formation of planetary debris discs around white dwarfs—I. Tidal disruption of an extremely eccentric asteroid. Mon Not R Astron Soc 445:2244–2255

Veras D, Jacobson SA, Gänsicke BT (2014c) Post-main-sequence debris from rotation induced YORP break-up of small bodies. Mon Not R Astron Soc 445:2794–2799

Veras D, Evans NW, Wyatt MC, Tout C (2014d) The great escape—III. Placing post-main sequence evolution of planetary and binary systems in a Galactic context. Mon Not R Astron Soc 437:1127–1140

Veras D, Eggl S, Gänsicke BT (2015a) Sublimation-induced orbital perturbations of extrasolar active asteroids and comets: Application to white dwarf systems. Mon Not R Astron Soc 452:1945–1957

Veras D, Leinhardt ZM, Eggl S, Gänsicke BT (2015b) Formation of planetary debris discs around white dwarfs—II: Shrinking extremely eccentric collisionless rings. Mon Not R Astron Soc 451:3453–3459

Veras D, Eggl S, Gänsicke BT (2015c) The orbital evolution of asteroids, pebbles and planets from giant branch stellar radiation and winds. Mon Not R Astron Soc 451:2814–2834

Veras D, Mustill AJ, Gänsicke BT, Redfield S, Georgakarakos N, Bowler AB, Lloyd MJS (2016a) Full-lifetime simulations of multiple unequal-mass planets across all phases of stellar evolution. Mon Not R Astron Soc 458:3942–3967

Veras D, Marsh TR, Gänsicke BT (2016b) Dynamical mass and multiplicity constraints on co-orbital bodies around stars. Mon Not R Astron Soc 461:1413–1420

Veras D, Carter PJ, Leinhardt ZM, Gänsicke BT (2017a) Explaining the variability of WD 1145+017 with simulations of asteroid tidal disruption. Mon Not R Astron Soc 465:1008–1022

Veras D, Georgakarakos N, Dobbs-Dixon I, Gänsicke BT (2017b) Binary star influence on post-main-sequence multiplanet stability. Mon Not R Astron Soc 465:2053–2059

Veras D, Mustill AJ, Gänsicke BT (2017c) The unstable fate of the planet orbiting the A star in the HD 131399 triple stellar system. Mon Not R Astron Soc 465:1499–1504

Veras D, Georgakarakos N, Gänsicke BT, Dobbs-Dixon I (2018a) Effects of nonKozai mutual inclinations on two-planet system stability through all phases of stellar evolution. Mon Not R Astron Soc 481:2180–2188

Veras D, Xu S, Rebassa-Mansergas A (2018b) The critical binary star separation for a planetary system origin of white dwarf pollution. Mon Not R Astron Soc 473:2871–2880

Veras D, Efroimsky M, Makarov VV, Boué G, Wolthoff V, Reffert S, Quirrenbach A, Tremblay PE, Gänsicke BT (2019a) Orbital relaxation and excitation of planets tidally interacting with white dwarfs. Mon Not R Astron Soc 486:3831–3848

Veras D, Higuchi A, Ida S (2019b) Speeding past planets? Asteroids radiatively propelled by giant branch Yarkovsky effects. Mon Not R Astron Soc 485:708–724

Veras D, McDonald CH, Makarov VV (2020a) Constraining the origin of the planetary debris surrounding ZTFJ0139+5245 through rotational fission of a triaxial asteroid. Mon Not R Astron Soc 492:5291–5296

Veras D, Reichert K, Flammini Dotti F, Cai MX, Mustill AJ, Shannon A, McDonald CH, Portegies Zwart S, Kouwenhoven MB, Spurzem R (2020b) Linking the formation and fate of exo-Kuiper belts within Solar system analogues. Mon Not R Astron Soc 493:5062–5078

Veras D, Georgakarakos N, Mustill AJ, Malamud E, Cunningham T, Dobbs-Dixon I (2021) The entry geometry and velocity of planetary debris into the Roche sphere of a white dwarf. Mon Not R Astron Soc 506:1148–1164

Veras D, Birader Y, Zaman U (2022) Orbit decay of 2–100 au planetary remnants around white dwarfs with no gravitational assistance from planets. Mon Not R Astron Soc 510:3379–3388

Veras D, Georgakarakos N, Dobbs-Dixon I (2023a) High-resolution resonant portraits of a single-planet white dwarf system. Mon Not R Astron Soc 518:4537–4550

Veras D, Ida S, Grishin E, Kenyon SJ, Bromley BC (2023b) Planetesimals drifting through dusty and gaseous white dwarf debris discs: Types I, II and III-like migration. Mon Not R Astron Soc 524:1–17

Villaver E, Livio M (2007) Can planets survive stellar evolution? Astrophys J 661:1192–1201

Villaver E, Livio M (2009) The orbital evolution of gas giant planets around giant stars. Astrophys J Lett 705:L8185

Villaver E, Livio M, Mustill AJ, Siess L (2014) Hot Jupiters and cool stars. Astrophys J 794:3

Völschow M, Banerjee R, Hessman FV (2014) Second generation planet formation in NN Serpentis? Astron Astrophys 562:A19

Voyatzis G, Hadjidemetriou JD, Veras D, Varvoglis H (2013) Multiplanet destabilization and escape in post-main-sequence systems. Mon Not R Astron Soc 430:3383–3396

Walters N, Farihi J, Marsh TR, Bagnulo S, Landstreet JD, Hermes JJ, Achilleos N, Wallach A, Hart M, Manser CJ (2021) A test of the planet–star unipolar inductor for magnetic white dwarfs. Mon Not R Astron Soc 503:3743–3758

Wang TG, Jiang N, Ge J, Cutri RM, Jiang P, Sheng Z, Zhou H, Bauer J, Mainzer A, Wright EL (2019) An ongoing mid-infrared outburst in the white dwarf 0145+234: Catching in action the tidal disruption of an exoasteroid? Astrophys J 886:L5

Wang L, Zhang X, Wang J, Zhang ZX, Fang T, Gu WM, Guo J, Jiang X (2023) White dwarfs with infrared excess from LAMOST. Data Release 5. Astrophys J 944:23

Wilson DJ, Gänsicke BT, Koester D, Raddi R, Breedt E, Southworth J, Parsons SG (2014) Variable emission from a gaseous disc around a metal-polluted white dwarf. Mon Not R Astron Soc 445:1878–1884

Wilson TG, Farihi J, Gänsicke BT, Swan A (2019) The unbiased frequency of planetary signatures around single and binary white dwarfs using Spitzer and Hubble. Mon Not R Astron Soc 487:133–146

Wyatt MC, Farihi J, Pringle JE, Bonsor A (2014) Stochastic accretion of planetesimals on to white dwarfs: constraints on the mass distribution of accreted material from atmospheric pollution. Mon Not R Astron Soc 439:3371–3391

Xu S, Bonsor A (2021) Exogeology from polluted white dwarfs. Elements 17:241

Xu S, Jura M (2014) The drop during less than 300 days of a dusty white dwarf's infrared luminosity. Astrophys J 792:L39

Xu S, Ertel S, Wahhaj Z, Milli J, Scicluna P, Bertrang GM (2015) An extreme-AO search for giant planets around a white dwarf. VLT/SPHERE performance on a faint target GD 50. Astron Astrophys 579:L8

Xu S, Jura M, Dufour P, Zuckerman B (2016) Evidence for gas from a disintegrating extrasolar asteroid. Astrophys J 816:L22

Xu S, Zuckerman B, Dufour P, Young ED, Klein B, Jura M (2017) The chemical composition of an extrasolar Kuiper-belt-object. Astrophys J 836:1:L7

Xu S, Su KY, Rogers LK, Bonsor A, Olofsson J, Veras D, van Lieshout R, Dufour P, Green EM, Schlawin E, Farihi J (2018a) Infrared variability of two dusty white dwarfs. Astrophys J 866:108

Xu S, Rappaport S, van Lieshout R, Vanderburg A, Gary B, Hallakoun NA, Ivanov VD, Wyatt MC, DeVore J, Bayliss D, Bento J (2018b) A dearth of small particles in the transiting material around the white dwarf WD 1145+017. Mon Not R Astron Soc 474:4795–4809

Xu S, Hallakoun NA, Gary B, Dalba PA, Debes J, Dufour P, Fortin-Archambault M, Fukui A, Jura MA, Klein B, Kusakabe N (2019a) Shallow ultraviolet transits of WD 1145+017. Astron J157:255

Xu S, Lai S, Dennihy E (2020) Infrared excesses around bright white dwarfs from Gaia and unWISEI. Astrophys J 902:127

Xu S, Diamond-Lowe H, MacDonald RJ, Vanderburg A, Blouin S, Dufour P, Gao P, Kreidberg L, Leggett SK, Mann AW, Morley CV (2021) Gemini/GMOS transmission spectroscopy of the grazing planet candidate WD 1856+534 b. Astron J162:296

Xu S, Rogers LK, Blouin S (2024) The chemistry of extra-solar materials from white dwarf planetary systems. Rev Mineral Geochem 90:171-198

Yarza R, Razo-López NB, Murguia-Berthier A, Everson RW, Antoni A, MacLeod M, Soares-Furtado M, Lee D, Ramirez-Ruiz E (2023) Hydrodynamics and survivability during post-main-sequence planetary engulfment. Astrophys J 954:176

Zeebe RE (2015) Dynamic stability of the solar system: Statistically inconclusive results from ensemble integrations. Astrophys J 798:8

Zhang Y, Liu S-F, Lin DNC (2021) Orbital migration and circularization of tidal debris by Alfvén-wave drag: circumstellar debris and pollution around white dwarfs. Astrophys J 915:91

Zhou G, Kedziora-Chudczer L, Bailey J, Marshall JP, Bayliss DD, Stockdale C, Nelson P, Tan TG, Rodriguez JE, Tinney CG, Dragomir D (2016) Simultaneous infrared and optical observations of the transiting debris cloud around WD 1145+017. Mon Not R Astron Soc 463:4422–4432

Zink JK, Batygin K, Adams FC (2020) The great inequality and the dynamical disintegration of the outer solar system. Astron J160:232

Zotos EE, Veras D (2020) The grain size survival threshold in one-planet post-main sequence exoplanetary systems. Astron Astrophys 637:A14

Zotos EE, Veras D, Saeed T, Darriba LA (2020) Short-term stability of particles in the WDJ0914+1914 white dwarf planetary system. Mon Not R Astron Soc 497:5171–5181

Zuckerman B, Becklin EE (1987) Excess infrared radiation from a white dwarf—An orbiting brown dwarf? Nature 330:138–140

Zuckerman B, Young ED (2018) Characterizing the chemistry of planetary materials around white dwarf stars. *In:* Handbook of Exoplanets. HJ Deeg, JA Belmonte (eds). Springer, Cham, p 1546–1561

Zuckerman B, Koester D, Reid IN, Hünsch M (2003) Metal lines in DA white dwarfs. Astrophys J 596:477–495

Zuckerman B, Melis C, Klein B, Koester D, Jura M (2010) Ancient planetary systems are orbiting a large fraction of white dwarf stars. Astrophys J 722:725–736

Reviews in Mineralogy & Geochemistry
Vol. 90 pp. 171–198, 2024
Copyright © Mineralogical Society of America

6

The Chemistry of Extra-solar Materials from White Dwarf Planetary Systems

Siyi Xu (许偲艺),[1] Laura K. Rogers,[2] Simon Blouin[3]

*[1]Gemini Observatory/NSF's NOIRLab, 670 N. A'ohoku Place,
Hilo, Hawaii, 96720, USA*
*[2]Institute of Astronomy, University of Cambridge,
Madingley Road, Cambridge CB3 0HA, UK*
*[3]Department of Physics and Astronomy, University of Victoria,
Victoria, BC V8W 2Y2, Canada*

siyi.xu@noirlab.edu, lr439@ast.cam.ac.uk, sblouin@uvic.ca

INTRODUCTION

The search for life on other planets is one of the grand scientific goals of our time. The Decadal Survey on Astronomy and Astrophysics 2020 (Astro2020)[1] sets the top priority in the coming decade to be the pursuit of a new space observatory called the *Habitable World Observatory* (*HWO*), with the goal to directly detect Earth-like planets, characterize its atmosphere, and search for biosignatures. However, whether a gas is a biosignature depends on the geochemical cycles (Shahar et al. 2019). We can't fully identify a gas as a biosignature unless we know the planet's (abiotic) geochemical cycles, and that means having an inventory of the elements within that planet.

Chemical inventories of exoplanets are difficult to obtain due to the limitations of observational techniques. To directly constrain the composition of planets, accurate measurements of their masses and radii are required. The transit technique measures the flux change of a star as a planet transits the stellar disk. This flux change is proportional to $(R_P/R_*)^2$, the square of the ratio of the planet's radius R_P to the star's radius R_*; if the radius of the star is known then the radius of the planet can be inferred. The transit technique can only detect planets whose orbits are close to edge on (when the inclination of the orbit is close to 90°). For example, an Earth-like planet that transits a Sun-like star would produce a transit depth of 0.008%—such a precision has yet to be achieved from the ground due to the variable atmospheric conditions and different telescope/instrument systematics. Space telescopes are much better and the *Kepler* Mission has reached a precision of 0.002% in the best-case scenario (Gilliland et al. 2011).

Another popular technique to detect planets around stars is the radial velocity technique, which measures the wobble of a star due to the gravitational interaction with an orbiting planet. This technique provides a minimum planetary mass ($M_P \sin i$), where M_P is the mass of the planet and i is the orbital inclination. For example, the presence of Earth affects the radial velocity of the Sun, but only by 10 cm·s^{-1}. That is still beyond the reach of current Extreme Precision Radial Velocity (EPRV) Spectrographs, which can achieve a precision of $\approx$ 30 cm·s^{-1} under the best circumstances (e.g., Pepe et al. 2021; Seifahrt et al. 2022). Future instrumentation on the *Extremely Large Telescopes* (*ELTs*) will likely have even better precision. However, stellar activity will remain a major challenge in detecting radial velocity signals from an Earth-like planet (e.g., Wright 2018).

---

[1] https://nap.nationalacademies.org/catalog/26141/pathways-to-discovery-in-astronomy-and-astrophysics-for-the-2020s.

1529-6466/24/0090-0006$05.00 (print)
1943-2666/24/0090-0006$05.00 (online)   http://dx.doi.org/10.2138/rmg.2024.90.06

For planets which are detected via both the transit technique and the radial velocity technique, the planetary mass can be derived because the inclination is known ($i \approx 90°$). Using the mass and radius measurements the bulk density is inferred, $\rho_{bulk} = M_P/(4/3\pi R_P^3)$ (also see discussion in Kempton and Knutson 2024, this volume). The first planet for which this was done was HD209458. The bulk density was found to be low (0.38 g·cm^{-3}) and given the large radius (1.27 Jupiter radii), it was designated a gas giant (Charbonneau et al. 2000). Planet interior models provide theoretical mass–radius relations for different compositions and interior structures. For example, the planetary models may assume a single composition, e.g., 100% iron, or 100% hydrogen/helium in gaseous form; or multi-layered models with different compositions per layer, e.g., metallic iron core, $MgSiO_3$ mantle, water, or more exotic compositions (e.g., Seager et al. 2007; Dorn et al. 2015). The difficulty with this analysis is that the models are degenerate, with vastly different compositions and structures producing similar mass–radius curves. Therefore, to study the interior of exoplanetary bodies, further techniques are required.

White dwarf planetary systems provide a unique way to measure the bulk composition of exoplanetary material. As introduced in the previous chapter, Veras et al. (2024, this volume), extrasolar asteroids/comets/moons which have survived the evolution of their host star can end up in the atmosphere of the white dwarf. Asteroids and boulders appear to be the most common pollutants (see Veras et al. 2024, this volume) and in this chapter, we use the term "asteroids" to refer to the parent body that is polluting the atmosphere. The presence of the planetary material is detected via absorption lines of heavy elements[2]. White dwarfs with these absorption features are called 'polluted' white dwarfs. Polluted white dwarfs were expected to be rare objects because white dwarfs have high surface gravities, therefore, these heavy elements will settle out of the white dwarf's atmosphere in a short amount of time (Paquette et al. 1986), as illustrated in Figure 1. However, high-resolution spectroscopic surveys found that 25–50% of white dwarfs are polluted (Zuckerman et al. 2003, 2010; Koester et al. 2014). The mechanism responsible for making a polluted white dwarf must be common and efficient. In the early days, accretion from the interstellar medium was proposed as the source of the pollution, but this idea was rejected due to (i) the small amount of carbon pollution, which is a major element in the interstellar medium (Jura 2006), and (ii) the lack of correlation between the locations of the polluted white dwarfs and the interstellar clouds (Farihi et al. 2010).

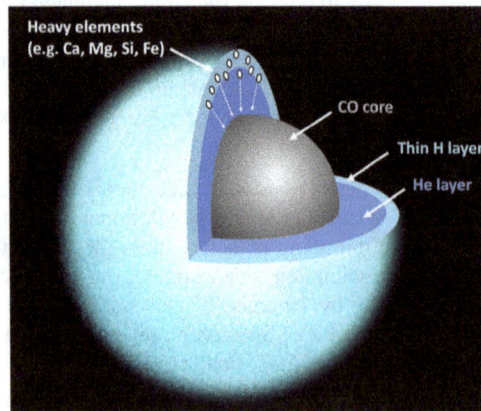

Figure 1. Cross section of a polluted white dwarf, where heavy elements (e.g., Ca, Mg, Si, Fe) arrive in the atmosphere of the white dwarf and sink down over time. Most white dwarfs have a carbon/oxygen-rich core, but only the outer hydrogen/helium layer is directly observable. This figure is for illustration purpose and the size of the different layers is not to scale.

---

[2] In this chapter, "heavy elements" refer to all elements heavier than helium.

There is strong theoretical and observational evidence that white dwarfs are accreting from planetary material. There are different mechanisms that can deliver exoplanetary material into the Roche lobe of the white dwarf, as discussed in the section *Stage 2: Delivering material to the white dwarf's Roche sphere* of the previous chapter, Veras et al. (2024, this volume). Debris disks, transits from disintegrating bodies, and intact planets have all been detected around white dwarfs (e.g., Jura et al. 2007; Vanderburg et al. 2015, 2020). Perhaps the best example supporting the asteroid tidal disruption theory is the white dwarf WD 1145+017, which has a heavily polluted atmosphere, a circumstellar disk made from dust and gas, and variable transit features from a disintegrating asteroid (Vanderburg et al. 2015; Xu et al. 2016; Rappaport et al. 2016). Since then, many more systems similar to WD 1145+017 have been detected (e.g., Vanderbosch et al. 2020). In addition, Cunningham et al. (2019) reported the first X-ray detection of the polluted white dwarf G29−38, which directly confirms that the white dwarf is actively accreting.

The previous chapter, Veras et al. (2024, this volume), presents the evolution, dynamics, and sizes of these parent bodies and describes how this material ultimately ends up in the atmosphere of these white dwarfs. This chapter will describe how the chemical autopsies of the disintegrated asteroids are conducted, and what is learnt about exoplanetary material from polluted white dwarfs.

METHODS

In this section, we describe observations of polluted white dwarfs, white dwarf modeling, and how to derive the composition of the planetary bodies from the spectrum of a polluted white dwarf.

Observation

The spectral classification of white dwarfs is based on their optical spectra. It begins with a 'D' to highlight that a white dwarf is a degenerate object, which is supported by electron degeneracy pressure. A fun fact: the more massive a white dwarf is, the smaller it is. The main spectral classifications of white dwarfs that are most relevant to this chapter are: DA, DB, DC, and DZ. Around 80% of white dwarfs are DAs, where strong and broad hydrogen absorption lines are observed in their spectra (as illustrated in Fig. 2). DB white dwarfs show strong helium absorption lines, having lost most of their hydrogen due to a late helium-shell flash (also known as "born-again" episode) during the late phases of stellar evolution (Werner and Herwig 2006) or due to interactions with a stellar companion (Reindl et al. 2014). DC (C stands for "continuum") white dwarfs display no hydrogen or helium lines in their optical spectra; this occurs when a DA white dwarf cools below 5,000 K, or when a DB white dwarf cools below 11,000 K (Saumon et al. 2022). Therefore, DC white dwarfs could have either a hydrogen or helium dominated atmosphere. The spectra of DZ white dwarfs contain atomic absorption features from heavy elements (typically Ca)[3]. Many white dwarfs have multiple classifications, and for those the letters are listed in order of which spectral features dominate. For example, DAZ and DBZ are DA and DB white dwarfs that also contain absorption features from heavy elements. DAZ, DBZ, and DZ white dwarfs are called polluted white dwarfs. Figure 2 shows example spectrum from each of these classes of polluted white dwarfs.

Spectral classification is a useful way to classify white dwarfs, but the strongest optical features are not necessarily indicative of the dominant species. All DZs fall in this category— the strongest optical features come from heavy elements, yet the white dwarf has either a

---

[3] Another common spectral type for cool white dwarfs (< 10,000 K) is DQ, which shows either carbon lines or molecular C_2 Swan bands. The carbon is believed to come from convective dredge up of material from the star's interior, rather than external accretion (Dufour et al. 2005; Bédard et al. 2022).

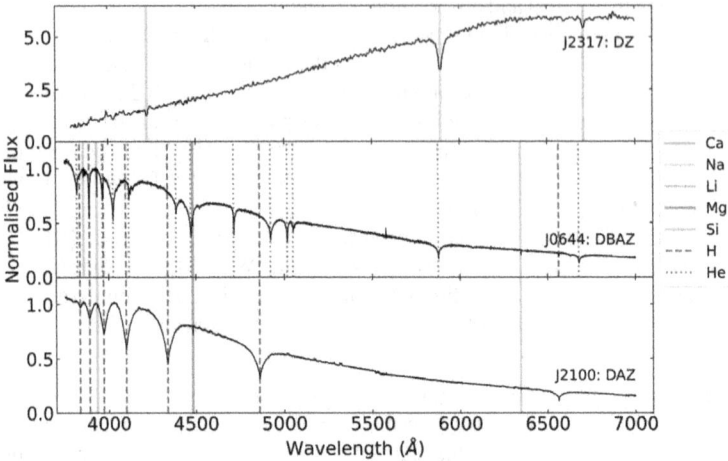

Figure 2. Normalized flux versus wavelength for three polluted white dwarfs with different spectral types: WDJ231726.74+183052.75 (T_{eff} = 4,600 K), a cool DZ white dwarf, GaiaJ0644−0352 (T_{eff} = 18,350 K), a hotter DBAZ white dwarf, and GaiaJ2100+2122 (T_{eff} = 25,570 K) a hot DAZ white dwarf. The **dashed lines** mark observed absorption lines of hydrogen, the **dotted lines** mark observed absorption lines of helium, and the **colored lines** highlight observed lines of heavy elements from Ca, Na, Li, Mg and Si. The spectrum of WDJ231726.74+183052.75 is from the OSIRIS spectrograph on the Gran Telescopio Canarias and has a spectral resolution $R \sim 1,000$ (Tremblay et al. 2020; Hollands et al. 2021), and the spectra of GaiaJ0644−0352 and GaiaJ2100+2122 are from the X-shooter instrument on the Very Large Telescope which has $R \sim 5,400$ on the blue side (< 5,600 Å) and $R \sim 8,900$ on the red side (> 5,600 Å, Rogers et al. 2024). All spectra are median filtered with a box size of five for clarity.

hydrogen or helium dominated atmosphere. Another more subtle example is the heavily polluted white dwarf GD 362, whose spectral type is DAZB, even though it has a helium dominated atmosphere. At the temperature of GD 362 (10,540 K, Xu et al. 2013), the He lines become very weak and the H lines are stronger than He, even though the H abundance is lower than He.

To obtain the abundances of planetary material polluting white dwarfs, it is crucial to obtain high quality spectra. There is a balance between wavelength range, spectral resolution, and signal-to-noise ratio (SNR). Table 1 shows the wavelengths of the strongest spectral lines for each element that have been detected in polluted white dwarfs. The detectability of an element depends on its abundance, the intrinsic strength of the line, the white dwarf parameters, and the characteristics of the instrument. The resolving power of a spectrograph, R, is defined by its ability to distinguish between two wavelengths with a difference of $\delta\lambda$, $R = \lambda/\delta\lambda$. The SNR is the ratio of the flux level to the background noise and is used to define the quality of spectra; the longer the exposure time, the larger the SNR. The SNR is roughly proportional to the square root of the exposure time. Therefore, it requires a lot of observing time to significantly improve the SNR of a spectrum. Low-resolution ($R < 5,000$) optical data is faster to obtain for a large sample of white dwarfs due to the shorter exposure times needed and the vast numbers of telescopes and instruments available; however, this will most often reveal just the strongest absorption lines (e.g., calcium H and K lines at 3933.7 Å and 3968.5 Å). In order to study the composition of planetary material, abundances from multiple elements are required. This generally needs higher resolution data ($R > 20,000$ Å) with a good SNR that spans a larger wavelength range (Table 1), which is more difficult to obtain for a large sample of white dwarfs.

Due to the strong hydrogen opacities for DA white dwarfs, absorption lines from heavy elements tend to be found in abundance in DB white dwarfs. For a given abundance, the same spectral lines tend to be stronger in DBZs than DAZs. In addition, polluted white dwarfs where more than 5 elements are discovered tend to be DBZs. Table 1 shows the wavelengths of dominant spectral lines for elements in the optical (here defined as $\lambda > 3000 \text{Å}$) versus the ultraviolet ($\lambda < 3000 \text{Å}$). For hotter white dwarfs ($T > 10,000 \text{ K}$) where there is sufficient flux from the white dwarf at ultraviolet wavelengths, spectral lines in the ultraviolet tend to be stronger and more abundant. However, ultraviolet radiation is absorbed from Earth's atmosphere, and is therefore only observable from space. The *Cosmic Origins Spectrograph* on the *Hubble Space Telescope* is the only instrument available right now with enough sensitivity and spectral resolution to study polluted white dwarfs in the ultraviolet. Additionally, the strongest lines for the volatile elements (C, N, O) are all at ultraviolet wavelengths, making ultraviolet observations crucial for questions regarding the inventory of volatile elements in exoplanetary systems.

There is a total of ≈ 1300 polluted white dwarfs with at least one element measured, according to the Montreal White Dwarf Database (Dufour et al. 2017). Most of these were first identified from the Sloan Digital Sky Survey (SDSS), which is a large multi-object spectroscopic survey that covers 3,650–10,400 Å with a resolving power of around 2,000 (e.g., Koester and Kepler 2015). Calcium is the most easily detected element in the optical, and the majority of the polluted white dwarfs only show absorption features from calcium, as shown in Table 1. About 500 polluted white dwarfs have at least two elements detected (typically Ca and Mg). GD 362 is among the most heavily polluted white dwarf and it holds the record of having 17 heavy elements detected in its atmosphere, including C, Na, Mg, Al, Si, S, Ca, Sc, Ti, V, Cr, Mn, Fe, Co, Ni, Cu, and Sr (Xu et al. 2014). It is also worth noting that Be, Li and K have only been recently detected in polluted white dwarfs (Hollands et al. 2021; Kaiser et al. 2021; Klein et al. 2021).

Modeling

Model atmospheres are used to determine elemental abundances in the atmospheres of polluted white dwarfs.

Typically, a multi-dimensional grid of models with different effective temperatures T_{eff}, surface gravities $\log g$ ($g = GM_*/R_*^2$), and compositions n_i are calculated. The predicted stellar spectra for each model in this grid are then compared to the observed white dwarf spectrum, and the parameters of the star are assumed to correspond to those of the best-fit model. This procedure of using model atmospheres to deduce elemental abundances is not unique to white dwarfs but is also similar to methods employed to derive stellar abundances in general. In this section, we briefly explain how these models are generated.

Model atmospheres rely on a number of standard approximations. First, the horizontal directions are normally ignored: the atmosphere is treated in a 1-D framework and vertically divided into $\mathcal{O}(10^2)$ layers. These layers are arranged in a way that allows one to capture all the absorption sources that contribute to the star's emerging flux. More specifically, the deepest layer is positioned well below the region where absorption lines are formed, at an optical depth $> 10^2$ (i.e., deep enough that photons are absorbed or scattered many times before escaping the atmosphere). Similarly, the uppermost layer is placed high up in the atmosphere (optical depth $< 10^{-6}$), above the line-forming region (roughly located between an optical depth of 1 and 10^{-5}). Because the atmosphere is very thin ($\sim 10^2$ m) compared to the star's radius ($\sim 10^7$ m), a plane-parallel geometry is assumed. The structure of the atmosphere is given by the hydrostatic equilibrium equation

$$\frac{dp}{dr} = -\rho(r)g \qquad (1)$$

where $p(r)$ is the pressure structure of the atmosphere, $\rho(r)$ is its density stratification, and g is the surface gravity.[4] As Equation (1) does not prescribe the temperature structure $T(r)$, the

---

[4] In practice, Equation 1 and its different profiles are parameterized as a function of the optical depth τ instead of the geometrical depth r. We use r here for simplicity.

Table 1. Heavy elements that have been detected in polluted white dwarfs, listing the number of white dwarfs with such detections and the dominant spectral lines in the optical (Op) and ultraviolet (UV) for different ionization states of that element. This list collates the most commonly observed lines for each element, with information collected from: Klein et al. (2010, 2011); Gänsicke et al. (2012); Jura et al. (2012); Hoskin et al. (2020); Hollands et al. (2021); Izquierdo et al. (2021); Kaiser et al. (2021); Johnson et al. (2022); Rogers et al. (2024). Following the convention, the wavelength of the optical and ultraviolet lines are in air and vacuum, respectively. This list is not exhaustive, but can serve as a starting point when identifying key spectral lines for each element.

Element	No.	Op/UV	Wavelength (Å)
Li	6	Op	Li I: 6707.8, 6707.9
Be	2	Op	Be II: 3130.4, 3131.1
C	41	Op	C II: 4267.3, 6578.0
		UV	C I: 1140.4, 1261.6, 1329.1, 1329.6
			C II: 1334.5, 1335.7
			C III: 1174.9, 1175.3, 1175.7, 1176.0, 1176.4
N	3	UV	N I: 1243.2
			N II: 1084.6, 1085.7
O	38	Op	O I: 7771.9, 7774.2, 7775.4, 8446.4
		UV	O I: 1152.2, 1302.2, 1304.9, 1306.0
Na	118	Op	Na I: 5890.0, 5895.9
Mg	306	Op	Mg I: 3832.3, 3838.3, 5172.7, 5183.6
		Op	Mg II: 4481.1, 4481.3, 7877.1, 7896.0
		UV	Mg II: 1239.9, 1240.9, 2795.5, 2802.7
Al	30	Op	Al I: 3944.0, 3961.5
		Op	Al II: 3586.6, 3587.1, 3587.4, 4663.1
		UV	Al II: 1191.8, 1725.0
		UV	Al III: 1379.7, 1384.1, 1854.7, 1862.8
Si	72	Op	Si I: 3906.6
		Op	Si II: 3856.0, 4128.1, 4130.9, 5056.0, 6347.1, 6371.4
		UV	Si II: 1190.4, 1193.3, 1194.5, 1197.4, 1260.4, 1264.7
		UV	Si III: 1141.6, 1161.6, 1296.7, 1298.9
		UV	Si IV: 1393.8, 1402.8
P	12	UV	P II: 1154.0, 1159.1, 1249.8
		UV	P III: 1334.8, 1344.3
S	17	UV	S I: 1316.5, 1425.0, 1425.2
		UV	S II: 1204.3, 1253.8, 1259.5
		UV	S III: 1194.0, 1194.4
K	5	Op	K I: 7664.9, 7699.0
Ca	1291	Op	Ca I: 4226.7
		Op	Ca II: 3933.7, 3968.5, 8498.0, 8542.1, 8662.1
		UV	Ca II: 1169.0, 1169.2, 1341.9, 1432.5, 1433.8
Sc	5	Op	Sc II: 3572.5, 3613.8, 3630.7
		UV	Sc II: 1418.8

Element	No.	Op/UV	Wavelength (Å)
Ti	59	Op	Ti II: 3234.5, 3236.6, 3349.0, 3349.4, 3361.2
V	5	Op	V II: 3125.3, 3267.7, 3271.1, 3276.1
		UV	V III: 1148.5, 1149.9
Cr	91	Op	Cr II: 3120.4, 3125.0, 3128.7, 3132.1, 3368.0
		UV	Cr II: 1435.0, 1435.2
		UV	Cr III: 1136.7,1146.3,1247.8,1252.6,1259.0,1261.9, 1263.6
Mn	17	Op	Mn II: 3442.0, 3460.3, 3474.0, 3474.1, 3482.9
		UV	Mn II: 1162.0,1188.5,1192.3,1192.3,1197.2,1199.4,1201.1,1234.0,1254.4
		UV	Mn III: 1174.8, 1177.6, 1179.9, 1183.3, 1183.9
Fe	286	Op	Fe I: 3570.1, 3581.2, 3719.9, 3734.9, 3749.5, 3820.4, 3859.9
		Op	Fe II: 3227.7, 5018.4, 5169.03
		UV	Fe II: 1143.2, 1144.9, 1358.9
Co	2	Op	Co II: 3754.7
Ni	30	Op	Ni I: 3524.5, 3619.4
		Op	Ni II: 3514.0
		UV	Ni II: 1317.2, 1335.2, 1370.1, 1381.3, 1411.1
Cu	1	Op	Cu I: 3247.5, 3274.0
Sr	3	Op	Sr II: 4215.5

first step in the calculation of a model atmosphere consists of guessing $T(r)$, which can be achieved using an approximate Hopf function (Mihalas 1978). Equipped with $P(r)$ and $T(r)$, an equation of state can be solved at each layer to obtain the abundances of all species (free electrons, ions, molecules). Note that the atmosphere is normally assumed to be chemically homogeneous, meaning that each layer has the same elemental composition (e.g., the ratio of Ca to H or He nuclei is constant). This elemental composition is an input parameter of the model atmosphere calculation. However, since P and T change throughout the atmosphere, the ionic and molecular populations vary (e.g., the ratio of singly ionized to neutral Ca changes). For most white dwarf atmospheres, the equation of state physics is well known (the ideal gas law and the Saha ionization equation are generally applicable)[5], but we will discuss the important exception of old, cool white dwarfs in the section *Constitutive physics uncertainties*.

Once the abundances of each species are determined, the next step consists of calculating the radiative opacity of the mixture. Many contributions must be considered: Thomson scattering from electrons, Rayleigh scattering from atoms and molecules, bound–free absorption, free–free absorption, bound–bound absorption (spectral lines), molecular bands, and collision-induced absorption. In the current context, spectral absorption lines (and to a lesser extent, molecular bands) represent the most interesting aspect of the opacity calculation, as they are the features that are directly used to infer elemental compositions. Accordingly, we will focus on this aspect of the opacity calculation in what follows. That said, we note that in general it is the other opacity sources (which do not produce salient spectral features but instead a more continuous absorption) that dominate the total opacity of the mixture and therefore control the thermodynamic structure of the atmosphere.

---

[5] Local thermodynamic equilibrium is normally assumed, which is almost always well justified.

Absorption lines are included using compilations such as the Kurucz lines list[6], the Vienna Atomic Line Database (VALD)[7], and the NIST atomic spectra database[8]. These sources provide the strengths and wavelengths of all absorption lines observed in white dwarfs, although the uncertainties on the line lists can be quite large. For each line, an absorption profile that accounts for both temperature (Doppler) and collisional broadening is calculated. We will see in the section *Constitutive physics uncertainties* that this task is still a challenge for very cool white dwarfs.

Once the opacity of each atmospheric layer is determined, the theory of radiative transfer enables the calculation of the radiative flux at each level of the vertical stratification. Since $T(r)$ was initially guessed, the energy flux[9] is generally not conserved throughout the atmosphere and radiative equilibrium is not attained. The temperature structure of the atmosphere must therefore be corrected until each layer transports a flux corresponding to $\sigma T_{\mathrm{eff}}^4$, where σ is the Stefan–Boltzmann constant (see Bergeron et al. 1991 for details on the numerical implementation of this correction procedure). The steps described above are repeated until a physical solution is reached (see Fig. 3). This final model structure is then used to generate a high-resolution synthetic spectrum that can be compared to observations after convolving it to the instrument's response function.

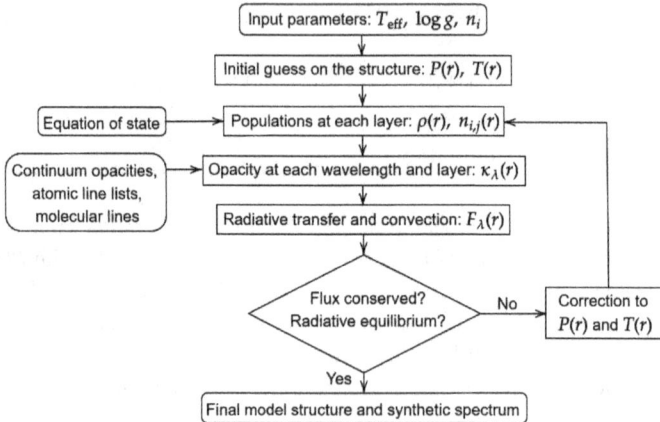

Figure 3. Schematic representation of a white dwarf model atmosphere calculation.

Once a multi-dimensional grid of model spectra has been calculated, how does one measure the elemental abundances of a given star? An iterative procedure that alternates between fitting $T_{\mathrm{eff}}/\log g$ and the individual abundances is often employed (e.g., Dufour et al. 2007; Coutu et al. 2019; Blouin 2020). T_{eff} and $\log g$ can first be estimated using photometric data (measurements of the object's intensity across specific bandpasses) and a parallax measurement (which gives the distance D separating the star from the Sun). With this approach, now widely used thanks to the precise parallaxes provided by the *Gaia* mission (Gaia Collaboration 2016), the solid angle $\pi(R_\star/D)^2$ and T_{eff} are directly adjusted to the photometric data points using a χ^2 minimization algorithm. Given the known white dwarf mass–radius relationship, this also yields the surface gravity $\log g$. Once a photometric solution is found, the individual abundances are adjusted to fit the absorption lines detected in the spectroscopic data (Fig. 4). After that, the photometric

---

[6] http://kurucz.harvard.edu/linelists.html.

[7] http://vald.astro.uu.se/.

[8] https://physics.nist.gov/PhysRefData/ASD/lines form.html.

[9] For most polluted white dwarfs (with the notable exception of those with hydrogen-dominated atmospheres and $T_{\mathrm{eff}} \gtrsim 18{,}000\,\mathrm{K}$), the energy flux in the atmosphere also includes a convective component.

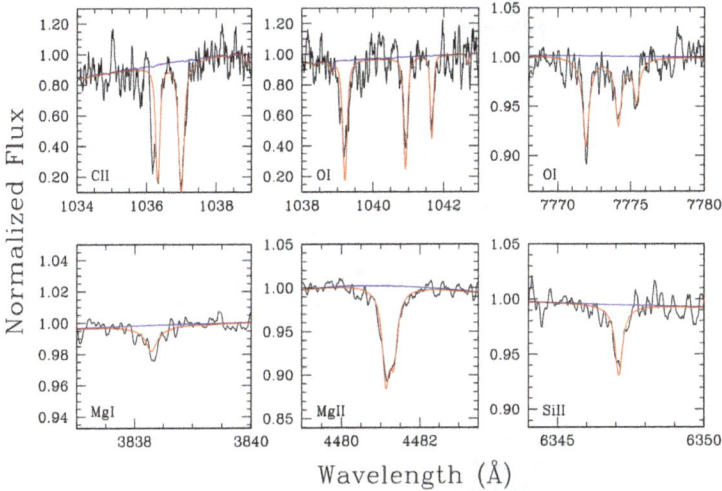

Figure 4. Illustration of the spectroscopic fitting procedure that yields the elemental abundances. The **red line** is the best-fit model to the observed polluted white dwarf spectrum (in **black**), and the **blue line** is the same model but where the abundance of the element indicated on each panel has been forced to zero. The UV data (including C II and O I) are from the FUSE satellite and the optical data are from Keck/HIRES. Figure taken from Klein et al. 2021 (© AAS, reproduced with permission).

procedure is repeated once more, as the addition of metals to the model atmosphere generally changes the overall shape of the emerging spectrum in a way that requires a different $T_{eff}/\log g$ to fit the photometry. Both steps (fitting the photometry and the spectroscopy) are repeated until internal consistency is reached.

Composition of the accreting material

Once we have the white dwarf atmospheric abundances, we can calculate the mass of the polluting material in the white dwarf's atmosphere. This mass is a lower limit because it only represents the amount of material that is currently in the white dwarf's atmosphere. For cooler DB white dwarfs, due to the transparent atmospheres, more mass is maintained in the convection zone, and this can provide more accurate estimates of the total mass accreted, albeit this is still a lower limit. The mass of heavy elements in the convection zone assuming a bulk Earth like composition is approximately 10^{20}–10^{25} g (Farihi et al. 2010; Girven et al. 2012). Harrison et al. (2021a) analyzed thirteen heavily polluted white dwarfs and estimated the parent body mass ranges from 10^{23}–7×10^{25} g, as shown in Figure 5.

In addition, we can infer the elemental composition of the planetary body. It is often assumed that the white dwarf atmosphere is dominated by one single parent body. The accretion event in the white dwarf's atmosphere is a dynamic process, with material continuously falling onto the white dwarf and settling out of the upper atmosphere that is visible to us, as illustrated in Figure 1.[10] For one parent body, the accretion stage can be divided into three distinctive phases, i.e., the build-up state, the steady state, and the declining state (Koester 2009). Figure 6 illustrates the possible accretion history for a polluted white dwarf. In the build up state, the observed atmospheric abundance is the same as the composition of the parent body. No additional correction is needed for this stage. In the steady state, the observed composition is modified by the diffusion time of each element. This typically changes the relative abundance

---

[10] In very hot white dwarfs ($T_{eff} \gtrsim 25{,}000$ K), radiative levitation can complicate this picture, as the upward force exerted by radiation on certain ions can cause them to remain at specific atmospheric depths where the radiative force counterbalances gravitational settling.

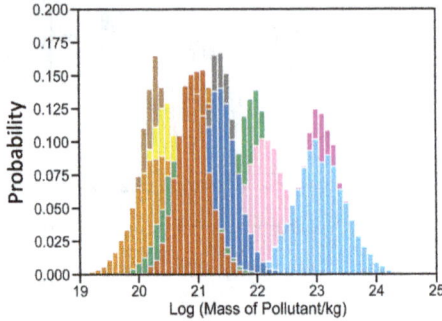

Figure 5. The posterior distribution of the total mass accreted onto thirteen polluted white dwarfs from Harrison et al. (2021a). The total mass varies from half of Vesta to the Moon.

ratios of the elements by factors of a few. While in the declining state, the observed composition can be quite different from the original composition and the abundance ratios can differ by up to several orders of magnitude. For example, if the white dwarf in Figure 6 is observed at the time 2.5×10^6 yr, one may conclude that the accreting material is O-rich and Fe-poor. In reality, this is just the result of differential settling; oxygen has the longest diffusion time and is still mostly in the atmosphere, while iron has largely diffused out of the atmosphere. Generally speaking, if a white dwarf has an infrared excess from a dust disk, we can assume that the system is either in the build up state or the steady state. If the system of interest is a hot DA white dwarf with very short settling times ($\lesssim 1$ year), we often assume that it is in a steady state. In these cases, it is straightforward to derive the composition of the parent body. It is a lot more difficult to derive the planetary composition if the system is in a declining phase. One way around this is to model all the potential compositions of the planetary body by sampling all the accretion scenarios, as has been done in, e.g., Swan et al. (2019); Doyle et al. (2020); Buchan et al. (2022). The best practice would be to try to explain the observed composition with the simplest scenario, before invoking unique and special explanations.

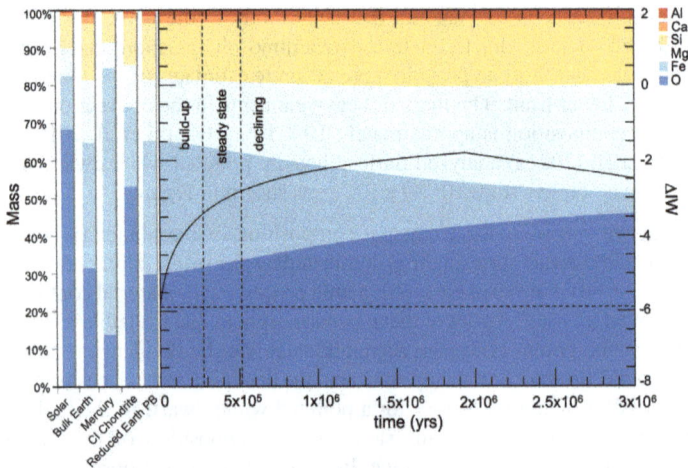

Figure 6. Possible accretion history for a polluted white dwarf assuming it is accreting from a single parent body (PB) with a reduced Earth composition. Depending on the accretion state (i.e., build-up, steady state, declining), the observed composition in the polluted white dwarfs is different. Therefore, additional corrections may be needed to infer the 'real' composition of the disintegrating parent body. Figure taken from Doyle et al. 2020 (© AAS, reproduced with permission).

It is a lot harder to derive the 'real' composition of the accreting material if a white dwarf is accreting from multiple planetary bodies at the same time. To make progress on the theoretical front, the accretion theory needs a coupled treatment of dust and gas, which is still an area of active research (Okuya et al. 2023). Strong variability appears to be common in the circumstellar dust and gas around white dwarfs (Swan et al. 2019; Dennihy et al. 2020). By analyzing a large unbiased sample of polluted white dwarfs, Wyatt et al. (2014) found that the accretion is continuous (rather than stochastic) for small planetary bodies (6.6×10^{19} g, or 35 km diameter for $3\,\text{g}\cdot\text{cm}^{-3}$), and this may be the dominant source of pollution. On the other hand, no variability in the accretion rate has been confidently detected, corroborating the picture that accretion appears to be in a steady state (Debes and Kilic 2010; Johnson et al. 2022). There could also be chemical alterations to the composition of the planetary body prior to the accretion onto the white dwarf, as discussed in the sections below and the previous chapter Veras et al. (2024, this volume).

INTERPRETATION

In this section, we describe commonly used methods to interpret the abundances of the extrasolar asteroids that have been accreted onto white dwarfs. We begin by discussing some of the key chemical, thermal, and physical processes that determine the composition of the planet, and then discuss how these processes are investigated using polluted white dwarfs.

Key processes

Planets are expected to inherit their composition from the interstellar cloud under which the star and planet formed. The composition of this planetary material may subsequently be altered by chemical, thermal, and physical processes, such as: heating during formation, post-nebula processing, differentiation, and collisional fragmentation. If no further processing occurs then the planetary material accreted would be 'primitive'; in the solar system, CI chondrites are most similar to the composition of the sun and have gone through little subsequent processing.

Heating during formation. The initial volatile budget of the planetary material is not set solely from the interstellar cloud, but additionally by its formation location with respect to the various snow lines, this leads to incomplete condensation of some elements (e.g., Pontoppidan et al. 2014). Planetary embryos can additionally accrete from a range of radial locations, called a 'feeding zone'. In the solar system, the volatile inventories of the inner terrestrial planets versus the outer gas and ice giants are broadly consistent with the positions of the snow lines.

Volatile loss. Post-nebula volatile loss occurs when a planet experiences intense heating after the protoplanetary disk has evaporated, for example, due to energetic collisions (Safronov and Zvjagina 1969), or heat released from the decay of short-lived radioisotopes (e.g., Urey 1955). This leads to a partial or full magma ocean phase in which volatiles can be degassed (e.g., Elkins-Tanton 2012). The volatile inventory from incomplete condensation in the protoplanetary disk and post-nebula volatilization are governed by different pressure, temperature, and oxidization conditions, and therefore, these volatile depletion processes imprint uniquely on the abundance pattern of the planetary body (Harrison et al. 2021c).

Core–mantle differentiation. If there is sufficient heating that large-scale melting of the parent body occurs, the segregation of the iron melt leads to the formation of a core and a mantle which form under the influence of the internal pressure and oxygen fugacity (e.g., Trønnes et al. 2019). Siderophilic elements migrate to the core, and lithophilic elements tend to the surface of the body, depending on the exact conditions under which this occurs bulk compositional differences can arise. Subsequent collisions, or other processing that leads to fragmentation can change the core to mantle ratio in a planetesimal.

Collisional fragmentation. Collisions during the formation and evolution of our solar system are fundamental to the composition of the planetary bodies. For example, the proto-Earth collided with a proto-planet approximately the size of Mars which resulted in the Earth-moon system we see today (Canup and Asphaug 2001). Collisions and fragmentation can change the relative amounts of different species, especially if the body is differentiated. Certain classes of meteorites (e.g., iron meteorites) are the collisional fragments of larger differentiated planetesimals.

Comparison with meteorites

When it became clear that polluted white dwarfs are accreting from extrasolar planetary material, the natural step was to compare the measured composition with those of rocky bodies in the solar system. The best composition database in the solar system comes from meteorites, which are fragments of minor bodies. When enough elements are detected in a polluted white dwarf, it is possible to do a direct comparison with the meteorites, look for the best match, and infer the formation scenario of the parent body. For example, GD 40 is the first polluted white dwarf for which all the major rock forming elements (i.e., Ca, Mg, Fe, Si, and O) were detected. Klein et al. (2010) found that the overall constitution of the accreting material in GD 40 is similar to bulk Earth, but the Mg/Si ratio is smaller than the value for bulk Earth and nearby stars, which may be due to accretion from a differentiated parent body, as discussed in the previous section. Using a χ^2 comparison, Xu et al. (2013) found that the best solar system analog to the material accreting onto GD 362 is mesosiderite, which is a rare type of stony-iron meteorite and is a mixture of crust and core material. G 238-44 has 10 heavy elements detected and the observed composition has no counterpart in the solar system. The best match is a mixture of iron-rich Mercury like material and an analogy of a Kuiper Belt Object (Johnson et al. 2022). Swan et al. (2019, 2023) developed methods that compare the abundances of polluted white dwarfs to solar system bodies to find the most likely compositional match, with one showing that the best match is the rare achondrite, acapulcoite.

Element ratio plots are the most widely used method to compare polluted white dwarfs with other objects. In the very early days, Jura (2006) compared the carbon-to-iron ratio between three polluted white dwarfs, the Sun, comet Halley, Earth's crust, and different meteorites. They found that carbon is depleted by more than a factor of 10 in polluted white dwarfs compared to the solar value, rejecting the interstellar accretion theory and providing a strong support for the asteroid accretion theory. Here, we revisit the commonly used element ratio figures, since the number of polluted white dwarfs has increased significantly since the last major review paper by Jura and Young (2014). The abundances of polluted white dwarfs are assembled from the literature (Koester and Wolff 2000; Zuckerman et al. 2007, 2011; Klein et al. 2011, 2021; Melis et al. 2011, 2012; Dufour et al. 2012; Gänsicke et al. 2012; Jura et al. 2012, 2015; Kawka and Vennes 2012, 2016; Farihi et al. 2013, 2016; Xu et al. 2013, 2014, 2017, 2019; Vennes and Kawka 2013; Raddi et al. 2015; Gentile Fusillo et al. 2017; Hollands et al. 2017, 2021; Melis and Dufour 2017; Blouin et al. 2018; Swan et al. 2019, 2023; Fortin-Archambault et al. 2020; Hoskin et al. 2020; Kaiser et al. 2021; González Egea et al. 2021; Izquierdo et al. 2021, 2023; Elms et al. 2022; Johnson et al. 2022; Doyle et al. 2023; Rogers et al. 2024; Vennes et al. 2024). We assume the polluted white dwarfs are all in the build up phase (see section *Composition of the accreting material*) and apply no additional correction to the observed abundances. As a comparison, we have assembled a sample of FG main-sequence stars from the Hypatia catalog (Hinkel et al. 2014). The solar values are taken from Lodders et al. (2009). We also include a large number of meteorites from Nittler et al. (2004); Alexander (2019a,b) and they are broadly separated into three categories: (1) "primitive" chondrites, which include carbonaceous chondrites, ordinary chondrites, and enstatite chondrites; (2) primitive achondrites, which include ureilites (URE), brachinites (BRA), acapulcoites and lodranites (ACA-LOD), winonaites and IAB and IICD irons (WIN-IAB); and (3) differentiated achondrites, which include angrites (ANG), aubrites

(AUB), howardite–eucrite–diogenite (HED), mesosiderite (MES), palasites (PAL), and IIE iron meteorites (see more discussions of different meteorite groups in the chapter, Jones 2024, this volume). In addition, we also included the composition of bulk earth (Allègre et al. 2001) and bulk Vesta (Toplis et al. 2013), which is assumed to be the parent body of the HED meteorite. The results are shown in Figures 7 to 10 and now we discuss some general trends among the abundance patterns.

Volatile elements (C, S). As shown in Figure 7, volatile elements are often depleted in polluted white dwarfs compared to the abundances of FG stars, similar to the values of the meteorites. The only system that has accreted solar carbon and sulfur abundances is WD 1425+540, which is also high in N and has been suggested to accrete from an analog of a Kuiper belt object (Xu et al. 2016). PG 1225−079 remains the only object that is strongly depleted in S but only moderately depleted in carbon. The observed composition does not match with any meteorite and possibly it needs a blend of two objects to explain the abundance pattern (Xu et al. 2013). Interestingly, there are two objects with an overabundance of S (i.e., PG 0843+517 and Gaia J0611−6931), and as 0843+517 also shows an overabundance of Fe (as shown in Fig. 10), it has been suggested that FeS may be a major consistent in the polluting material (Gänsicke et al. 2012).

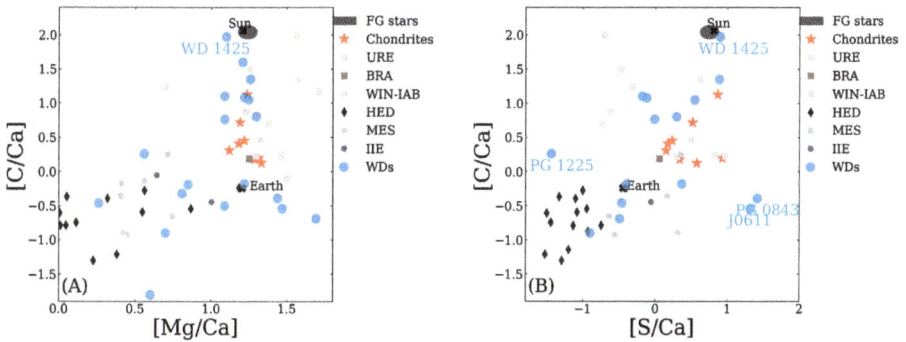

Figure 7. Logarithmic number ratios of different elements. The polluted white dwarfs are shown as **blue dots** and the average uncertainties are about ≈ 0.2 dex in the abundance ratios. The 95 percentile of the abundance ratios of the FG stars are shown as the **grey ellipse**. The chondrites are shown as **red stars** while other meteorites are shown as **squares, diamonds, and hexagons**. When the abundances are available, the solar value, bulk Earth, and bulk Vesta are also shown as **black crosses**. Most polluted white dwarfs show a depletion of volatile elements.

Fe and moderately siderophile elements (Cr, Mn, Ni). Figure 8 shows the spread of the ratios of moderately siderophile elements is pretty small, actually smaller than the observed range in meteorites. This indicates that the core formation process may be similar in both the extrasolar planetary systems and the solar system. Interestingly, WD 1145+017, a white dwarf with an actively disintegrating object in orbit, stands out as having low Mn/Fe and Cr/Fe ratios. The mass fraction of Fe is consistent with CI chondrite, and the depletion of Mn and Cr may be due to their low condensation temperatures (Fortin-Archambault et al. 2020). GD 378 stands out as having a lower Cr/Fe ratio and is also one of two white dwarfs with detection of Be (Klein et al. 2021). In fact, the Be abundance in GD 378 is about two orders of magnitudes higher than the chondritic value and it may be a result of the accretion of icy moons (Doyle et al. 2021).

Lithophile elements (Al, Ca, Ti, Na). Figure 9 shows the abundance ratios of different lithophile elements. For the Ti/Ca and Al/Ca figure, the spread in polluted white dwarfs is smaller than the range in all meteorites. Ti/Ca is much lower than in chondrites and more

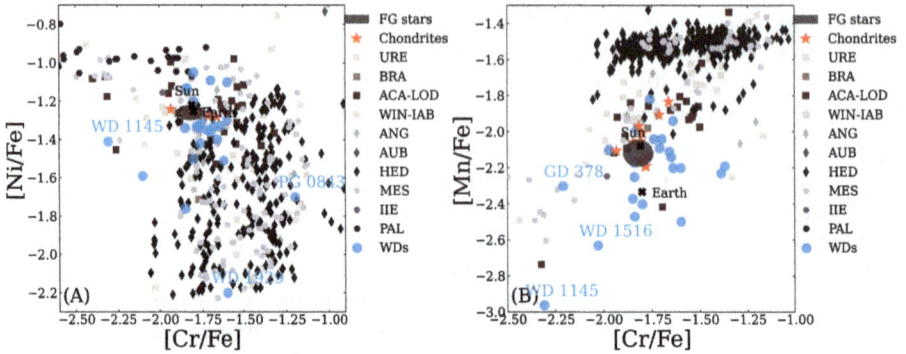

Figure 8. Similar as Figure 7 but for moderately siderophile elements. The spread observed in polluted white dwarfs is smaller than the spread of various meteorites.

comparable to other meteorites. In addition, there are a few white dwarfs with enhanced Na/Ca ratio, which could be a signature of accretion from crust material. However, this is not supported by the abundances of other elements (such as the stringent upper limit on Si, Swan et al. 2019). On the other hand, there are some very cool and old DZs with highly enhanced Li abundances and one scenario is that they have accreted crustal material (Hollands et al. 2021; Elms et al. 2022). However, it is difficult to pinpoint the exact scenario because neither Al nor Si is detected in these objects. The enhanced Li abundance relative to the other elements could be due to the other elements being more scarce when these old systems were formed (Kaiser et al. 2021). The large Na/Ca ratio observed in WD J2356–209 is interpreted as a possible accretion from a comet 67P like object (Blouin et al. 2019).

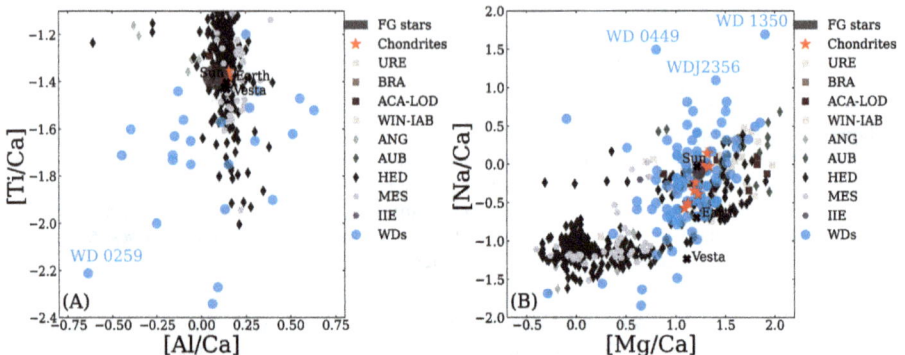

Figure 9. Similar as Figure 7 but for lithophile elements. The Ti/Ca ratios in polluted white dwarfs tend to be lower than the chondritic value and FG stars. The high Na/Ca ratios in some polluted white dwarfs cannot be simply interpreted as crust remnants; measurements of additional elements are needed to confirm this scenario.

Differentiation and collision. If a white dwarf accretes an entire asteroid that is differentiated, we may not be able to distinguish it from chondritic material. However, if some fragments of a differentiated body are accreted, it would display distinctive chemical signatures—that is what have been observed in polluted white dwarfs. As shown in Figure 10, there is a large spread in the Fe/Si and Fe/Al ratios in polluted white dwarfs, much larger than the spread in FG stars and the spread in Mg/Si and Si/Al ratios. A natural explanation for the large spread of the Fe abundance is accretion from a differentiated parent body,

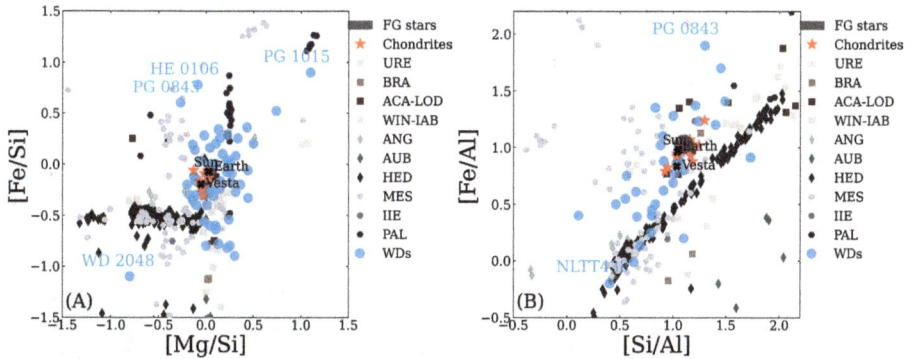

Figure 10. Similar as Figure 7 but for a different set of elements. For polluted white dwarfs, while most points cluster around the values for FG stars and chondrites, there are some objects with particularly high Fe and low Fe abundances. A natural explanation is that these white dwarfs have accreted fragments of a differentiated parent body.

as first proposed in Jura et al. (2013). NLTT 43806 has the lowest Fe/Al ratio of all polluted white dwarfs and it is a good candidate for the accretion of crust material (Zuckerman et al. 2011). Different analyses using polluted white dwarfs all found that differentiation and collision appear to be common in extrasolar planetary systems. For example, Bonsor et al. (2020) focused on the Ca/Fe measurements in 179 white dwarfs and found $66^{+4}_{-6}\%$ of the sample must have accreted remnants of differentiated bodies to explain the distribution of Ca/Fe. Doyle et al. (2020) focused on 16 white dwarfs where all the major rock forming elements (i.e., Al, Ca, Si, Mg, Fe and O) are detected and found that while most objects were formed under oxidizing conditions, about 25% were consistent with more reduced parent bodies. The prevalence of differentiation in extrasolar planetary systems can be used to provide an independent constraint on the formation timescale of planetesimals (Bonsor et al. 2023).

In summary, the element ratio plot is very useful in comparing a large number of objects and identifying order-of-magnitude differences. The main drawback is that only a few elements can be displayed at a time, and it is hard to conclude the nature of a given object without looking at all the detected elements. In addition, the figure does not show absolute numbers; so similarties may appear to exist between different groups, though it is actually different. Another issue is that meteorite measurements are typically done on a small amount of material and it is difficult to infer back the bulk composition of the parent body. For example, even though the HED meteorites occupy a large range in some element-ratio plots, they likely all come from a common parent body Vesta. However, those are still the best composition data available for comparison.

To take this one step further, Putirka and Xu (2021) attempted a mineralogy classification on the material accreted onto polluted white dwarfs. They found no evidence for accretion of continental crust in polluted white dwarfs, and found that some white dwarfs have exotic compositions with no analogs in the solar system. However, Trierweiler et al. (2023) found that the uncertainties on the abundance measurements of polluted white dwarfs may be too big to make a definitive statement. Refer to the next chapter, (Putirka 2024, this volume), for a more in-depth discussion on the exoplanet mineralogy analysis.

Modeling exoplanetary abundances and exogeology

Given that the history of a planetary body defines its composition and internal structure, considering the body from formation all the way to the accretion onto the white dwarf is crucial to truly understanding the processes affecting the abundance patterns. To move the polluted white dwarf analysis beyond directly comparing with the meteorites and solar system bodies, Harrison et al. (2018, 2021b) pioneered work that took the abundances of the planetary material and traced the history of this body back to its birth environment. Building upon this work, Buchan et al. (2022) developed the open source package PyllutedWD[11] which finds the most likely explanation for the observed composition of a planetary body that has polluted a white dwarf, incorporating all possible accretion histories in the white dwarfs' atmosphere with various processes that planetary bodies may experience. The model compares the likelihood that the abundances can be invoked by a basic primitive model, where the abundances assume stellar like material (Brewer and Fischer 2016), with a range of more complex models (listed in Fig. 11) incorporating geochemical processes such as those discussed in the section *Key processes*. These models allow for improved understanding of how the abundances in a polluted white dwarf atmosphere relate to the parent body's formation, collisional, and geological history, as well as provide a framework to statistically study a large population of polluted white dwarfs. Often the overall conclusions about any particular system are consistent with previous analysis if there are solar system analogs. The following discusses key results about these geochemical processes inferred from the modeling.

Model	Free Parameters	Schematic Diagram
(1) Initial Composition	Stellar metallicity: $[Fe/H]_{index}$ Time since accretion started: t Accretion event lifetime: t_{event} Pollution Fraction, f_{pol}	
(2) Volatile Depletion	Formation distance, $d_{formation}$	
(3) Feeding Zone	Feeding zone size, $z_{formation}$	
(4) Core-Mantle Differentiation	Fragment core fraction, f_c Core-mantle pressure, P Core-mantle oxygen fugacity, f_{O_2}	

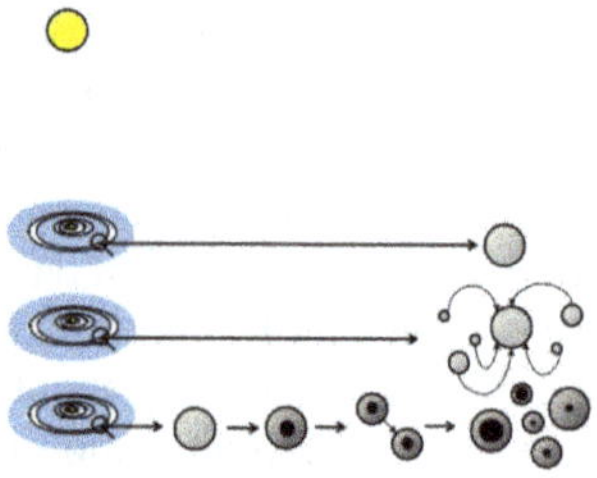

Figure 11. Table showing the models that are fitted in PyllutedWD. Model 1 is the 'primitive' model that only depends on the initial composition of the stellar nebula and the white dwarf accretion parameters; these parameters are always fitted. Models 2-4 show additional parameters that the Bayesian framework considers to explain the observed abundances. Figure adapted from Harrison et al. (2018) with updated parameters from Buchan et al. (2022).

Evidence for heating during formation. Harrison et al. (2018) found that by modeling the accreted abundances of PG 1225-079, the abundances of the refractory versus moderately volatile elements is most consistent with the accretion of a body that was extremely dry and formed in a region of the protoplanetary disk where temperatures reached above 1400 K— such an object does not exist in the solar system. It is consistent with the meteorite comparison shown in Figure 7(B). Additionally, by modeling the Hollands et al. (2017) DZ sample of 202 polluted white dwarfs, Harrison et al. (2021b) discovered that 11 white dwarfs were found to show depletion of volatiles, and the best-fitting models found that heating (described in the form of incomplete condensation of volatile species) was required to explain this abundance trend. Three of these systems required further heating such that even more moderately volatile species (e.g., Mg) did not condense.

Evidence for post-nebula volatile loss. GD 362 has the highest number of elements detected in any one polluted white dwarf. To explain the high Mn/Na ratio, Harrison et al. (2021c) showed that the body accreted must have experienced post-nebula volatilization.

---

[11] https://github.com/andrewmbuchan4/PyllutedWD Public.

(a)

(b)

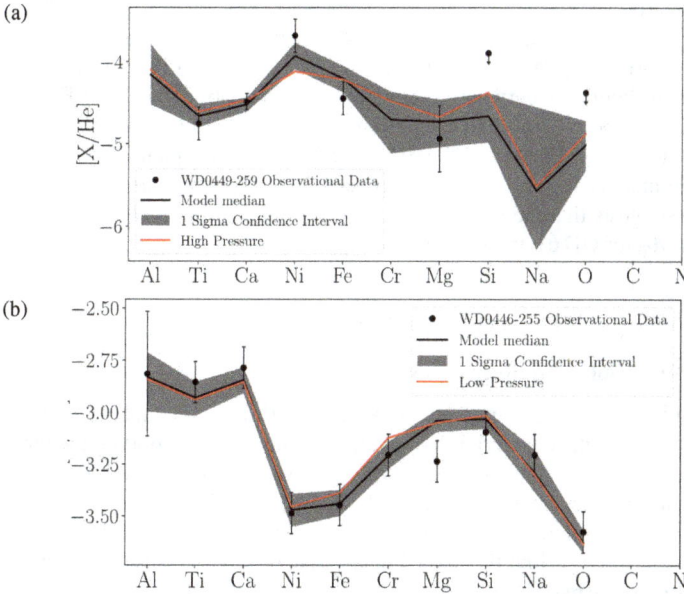

Figure 12. Abundance ratios of elements (X) relative to He for WD0449−259 (a) and WD0446−255 (b). Upper limits are shown with **arrows**. The model with the highest Bayesian evidence is plotted in **black** with 1σ errors as **grey shaded regions**. The best-fit model for WD0449−259 invokes the accretion of a core-rich fragment of a small differentiated body at low pressures; if core–mantle differentiation of the parent body occurred at high pressures instead, then the abundance pattern would follow the **red model**. The best-fit model for WD0446−255 invoked the accretion of a mantle-rich fragment of a larger parent body that underwent core–mantle differentiation at a high pressure; the **red model** shows the abundance pattern for the low pressure case. For WD0446−255, the posterior distribution on the pressure overlaps into the low-pressure region; as can be seen in the plot, it is difficult to differentiate between the two scenarios (**black line versus red line**) for this white dwarf. More data and better analysis is needed to reduce the uncertainties on the abundances such that these models can be more readily distinguished in the future. Figures adapted from Buchan et al. (2022).

This is because the volatility of Mn and Na depends on the temperature, pressure, and oxidization conditions, and so Na becomes more volatile during post-nebula volatilization processes which leads to an increased Mn/Na ratio in the planetary body.

Evidence for core–mantle differentiation. As outlined in the previous section *Comparison with meteorites*, by comparison with solar system analogues a number of white dwarfs were inferred to be accreting fragments of core–mantle differentiated bodies. By re-analysing the Hollands et al. (2017) sample, Harrison et al. (2021b) inferred that 65 out of 202 polluted white dwarfs studied have a preference for accreting a fragment of a core–mantle differentiated body, showing that differentiation and collisions appear to be commonplace in exoplanetary systems. Additionally, Buchan et al. (2022) found that from a sample of 42 white dwarfs, 14 showed evidence for the accretion of core–mantle differentiated material. The sample was selected to include those polluted white dwarfs that had a Fe detection and a detection/upper bound for one or more of: Cr, Ni, and Si, the importance of which is discussed below.

Constraining size of the parent body. Using the 14 polluted white dwarfs in the sample found to be accreting core or mantle rich fragments of larger differentiated parent bodies, Buchan et al. (2022) used the abundance pattern to constrain the sizes of these differentiated parent bodies. The amount of Ni, Cr, and Si that partition into the core or mantle is a function of the oxygen fugacity and pressure at which core–mantle differentiation occurred; therefore, studying the abundance patterns of fragments of differentiated objects that have accreted onto polluted white dwarfs can reveal the size of the parent body (pre-fragmentation). Three systems (WD 0449–259, WD 1350–162, and WD 2105–820) were found to be best explained by the accretion of a core-rich fragment, and the pressure (as constrained in Model (4) in Fig. 11) at which the parent bodies underwent core–mantle differentiation was low (Fig. 12a). This implies that the parent body was small (i.e., asteroid sized) rather than on a planet sized scale. Two polluted white dwarfs (GD 61 and WD 0446–255) were best explained by the accretion of a mantle-rich fragment at which differentiation occurred at high pressures in the parent body (Fig. 12b). This suggests that the parent bodies (pre-fragmentation) were larger, with masses 0.61 Earth mass $M_\oplus$ for GD 61 and $0.59M_\oplus$ for WD 0446–255.

LOOKING FORWARD

A large, uniform sample of polluted white dwarfs

Polluted white dwarfs have proved to be a powerful way to measure the chemical compositions of exoplanetary material. However, the sample size of polluted white dwarfs is small, and detections of heavy elements are very heterogeneous, as listed in Table 1. Using data from the *Gaia* satellite, Gentile Fusillo et al. (2021a) compiled a sample of 359,000 high-confidence white dwarfs, increasing the number of known white dwarfs tenfold. Large multi-object spectroscopic surveys like the Dark Energy Spectroscopic Instrument (DESI, Cooper et al. 2023), Large Sky Area MultiObject Fibre Spectroscopic Telescope (LAMOST, Guo et al. 2022), SDSS, and the 4-metre Multi-Object Spectroscopic Telescope (4MOST, Chiappini et al. 2019) will identify many more heavily polluted white dwarfs. Dedicated follow-up observations with a high-resolution spectrograph on the current 10-m class telescope or future 30-m class telescopes will return a much larger and more homogeneous sample of polluted white dwarfs with a large number of heavy elements. At the same time, it is important to obtain ultraviolet spectra to constrain the volatile abundances for as many heavily polluted white dwarfs as possible before the end of the *Hubble Space Telescope*.

Measuring the individual isotopic abundances of celestial bodies can yield additional precious information beyond what overall elemental abundances provide. For example, the deuterium-to-hydrogen ratio (D/H) has been measured in many objects in the solar system and is used to trace the origin of the Earth's water (Hallis 2017). Unfortunately, this is currently not feasible for polluted white dwarfs. The main challenge is that the isotopic shifts of atomic lines are very small compared to other effects at play in the white dwarf's photosphere. This includes in particular shifts resulting from interactions between the radiating atom and other charged or neutral particles in the dense atmosphere, which cause Stark (for charged particles) and van der Waals (for neutral particles) shift and broadening. Constraining the impact of these interactions on the spectral line shapes remains an area of active research (e.g., Saumon et al. 2022), meaning that potential isotopic shifts are obscured by these other spectral effects.

Improving abundance measurements: Current treatment of uncertainties

Abundance ratios published in the literature are ideally assigned an uncertainty σ that corresponds to the quadrature sum of two contributions

$$\sigma^2(Z_i/\text{He}) = \sigma_{\text{spread}}^2(Z_i/\text{He}) + \sigma_{T_{\text{eff}}}^2(Z_i/\text{He}). \tag{2}$$

The first contribution, $\sigma_{\text{spread}}^2(Z_i/\text{He})$, corresponds to the spread in the elemental abundance inferred using different spectral lines. In general, if a spectrum displays more than one spectral line for a given element, the abundance needed to reproduce each line will differ. This difference stems from uncertainties on the theoretical/experimental line strengths of the lines compiled in line lists, inaccuracies in the model structure, and/or measurement errors.

The second contribution $\sigma_{T_{eff}}^2$ (Z_i/He) has its origin in the uncertainty on the effective temperature (and $\log g$) of the star. Model atmosphere analyses generally yield uncertainties on T_{eff} of the order of ~ 5%. By varying the temperature within that range, the inferred elemental abundances will differ. Note that the T_{eff}-induced uncertainties on the abundances of different elements are generally correlated. This implies that the uncertainty on a relative abundance ratio Z_i/Z_j is often smaller than the quadrature sum of the uncertainties on Z_i/He and Z_j/He, because both Z_i/He and Z_j/He move in the same direction as T_{eff} is changed within its confidence interval. This topic is discussed in more detail in Appendix A of Klein et al. (2021).

Because Z_i/He can vary by several orders of magnitudes between different stars and elements, it is customary to report abundance measurements and their uncertainties in logarithmic terms. For simplicity, the confidence interval on the abundance is often assumed to be symmetric in this logarithmic space. This assumption can, however, complicate the subsequent analysis of the element-to-element abundance ratios (Doyle et al. 2019; Klein et al. 2021). It should be noted that this choice is somewhat arbitrary and, to date, little effort has been invested in properly characterizing the probability distribution of the measured abundances. Future work that focuses on both better understanding of the uncertainties associated with the abundances of the planetary material, and reducing the uncertainties, will enable drastic improvements in the interpretation of the abundances. Buchan et al. (2022) showed that it is possible to distinguish sizes of bodies from the fragments of core–mantle differentiated bodies using PyllutedWD; however, reduced uncertainties are required to obtain more precise sizes.

Improving abundance measurements: Constitutive physics uncertainties

Crucially, the current treatment of uncertainties outlined in the previous paragraph largely ignores systematic uncertainties resulting from gaps in our understanding of the constitutive physics of white dwarfs. The atmospheres and envelopes of white dwarfs reach physical conditions unlike those found anywhere on Earth or in most other types of stars. Therefore, modelers are often forced to rely on physics models that have not been directly validated against experimental data. A recent review describes current challenges faced on this front (Saumon et al. 2022); below we briefly discuss some of the most important ones in the context of exo-asteroid characterization.

Cool white dwarf atmospheres *($T_{eff} \approx 8,000\,K$).* Cool white dwarfs that contain little hydrogen in their outer envelopes have very transparent atmospheres. Helium has the highest ionization potential of all elements, making it very weakly ionized at low temperatures, which results in a very low opacity. As a result, the photospheres of helium-dominated atmospheres are located quite deep and reach densities of up to $1\,g\cdot cm^{-3}$. Under those liquid-like conditions, many approximations commonly used in stellar atmosphere codes (the ideal gas law, Saha ionization equilibrium, Lorentzian collisional line profiles) break down due to the incessant many-body interactions between atoms, ions, and molecules. In recent years, modern quantum chemistry simulation techniques have allowed significant progress on this problem (Kowalski et al. 2007; Kowalski 2014; Allard et al. 2016; Blouin et al. 2017, 2018). Nevertheless, current models still struggle to explain the spectra of many cool white dwarfs (Bergeron et al. 2022; Elms et al. 2022), signaling important gaps in our understanding of the equation of state and opacities under those peculiar conditions. Solving these remaining gaps has become especially pressing since the discovery of old, cool ($T_{eff} \approx 5,000\,K$) white dwarfs with Li absorption lines (such as the DZ shown in Figure 2, Hollands et al. 2021; Kaiser et al. 2021), whose interpretation depends on the availability of accurate model atmospheres in this challenging physical regime.

Diffusion timescales. A second important source of uncertainty are the diffusion timescales required to translate steady-state photospheric abundances into the composition of the accreting material (as explained in the section *Composition of the accreting material*). The diffusion coefficients of the accreted trace elements strongly depend on the ionization state of those same elements at the bottom of the envelope convection zone. The conditions encountered in this region of the star can push ionization models beyond their regime of applicability, especially in the envelopes of cool helium-atmosphere white dwarfs ($T \approx 10^6$ K and $\rho \approx 10^3$ g·cm^{-3}). The currently available diffusion timescales rely on simple heuristic ionization models that cannot be considered as reliable in this extreme regime (Paquette et al. 1986; Koester et al. 2020). New state-of-the-art simulation techniques can remedy this problem (Heinonen et al. 2020). The outputs of these improved models are not yet available, but they are eagerly awaited as they have been shown to change the inferred abundance ratios by factors of up to three. The uncertainty on the diffusion timescales affects the inferences that are made about the composition of the planetary body accreted by the white dwarf. In order for models like PyllutedWD to better determine exoplanetary composition, the uncertainties on the diffusion times need to be reduced.

Convective overshoot and thermohaline mixing. In stellar models, it is common to assume that convection zones end at a well-defined "Schwarzschild boundary" where the fluid becomes stable against convection. In reality, it is well known that such a discontinuity is unphysical (e.g., Zahn 1991; Freytag et al. 1996), and additional convective boundary mixing takes place as, for example, plumes "overshoot" past the formal stability boundary. This extra mixing is important for the interpretation of polluted white dwarfs, as it implies that the accreted material is mixed in a larger mass reservoir (by up to 2–3 orders of magnitude[12], Kupka et al. 2018; Cunningham et al. 2019), thereby implying higher accretion rates and larger accreted bodies. In addition to convective overshoot, the chemical composition gradient induced by the accretion of rocky elements can trigger thermohaline mixing (or "fingering convection", an important mixing process in the Earth's oceans). It is also thought to produce significant extra mixing in white dwarf envelopes (Bauer and Bildsten 2018; Wachlin et al. 2022). 3D hydrodynamic simulations are required to reliably determine the importance of these extra mixing processes (e.g., Tremblay et al. 2015). So far, 3D convective overshoot studies have been limited to a relatively narrow set of temperatures and compositions (Kupka et al. 2018; Cunningham et al. 2019), and no 3D simulations of thermohaline mixing in white dwarfs have been performed.

Discrepancy between ultraviolet and optical abundances. A notable discrepancy often emerges when comparing elemental Z_i/He abundances derived from ultraviolet observations to those obtained from optical measurements (Gänsicke et al. 2012; Xu et al. 2019; Rogers et al. 2024). The origin of this mismatch remains unclear. Potential culprits include inaccuracies in the model atmosphere structure, vertical composition gradients in the atmosphere, and uncertainties on the line strengths provided in the line lists used when generating synthetic spectra. Any potential solution to this problem has to contend with the fact that this discrepancy is not universal (Klein et al. 2021; Rogers et al. 2024). Fortunately, the relative Z_i/Z_j abundances appear to remain consistent between UV and optical analyses (Rogers et al. 2024), which limits the impact of this problem on planetary composition studies.

Connecting exoplanetary compositions with the solar system

Stronger collaboration between the polluted white dwarf community and the geochemistry community is urgently needed to interpret the abundance measurements. For example, Doyle et al. (2019, 2020) calculated the oxygen fugacity for a sample of polluted white dwarfs and found most extrasolar rocky bodies differentiated under oxidizing conditions compared to the solar gas. Buchan et al. (2022) proposed a new way to constrain the size of the polluting bodies from the abundances of Ni, Cr, and Si, and this size can be constrained using PyllutedWD. However, this relies on accurate and precise metal–silicate partitioning coefficients which remain uncertain for high pressures. Therefore, to improve the understanding of core–mantle differentiation and the size of the exoplanetary bodies that core–mantle fragments come from,

---

[12] The effect on relative abundances (e.g., the Ca/Mg abundance ratio) is much smaller.

further experiments on the partitioning need to be conducted. These endeavors are particularly welcomed to put planetary abundances measured from polluted white dwarfs in the context of the solar system and the more general planet formation and evolution scenarios.

Dust and gas composition

Due to the strong gravitational pull from the white dwarf, extrasolar planetesimals are tidally disrupted, making it easier to study the interior bulk compositions in comparison to indirectly inferring it. Here, we describe two other promising ways to measure the compositions of exoplanetary material using white dwarf planetary systems.

Composition of circumstellar dust. Infrared emission from dusty material within the tidal disruption radius of white dwarfs has now been observed in over one hundred systems (Lai et al. 2021). Infrared spectroscopic observations of these dust disks reveal the mineralogy of this material and provide a way to directly probe exoplanetary geology. The *Spitzer Infrared Spectrograph* (*IRS*) observed the brightest polluted white dwarfs with dust disks; all spectra show prominent silicate features around 10 μm (Jura et al. 2009). Amorphous versus crystalline silicates can be readily identified in the spectra as crystalline silicates create sharp peaks in the 10 μm feature, whereas amorphous silicates show smooth, featureless 10 μm features. The mid-infrared spectrum of G 29–38 is dominated by amorphous silicate (Reach et al. 2009), whereas the others show evidence for crystalline features related to silicate minerals. The enhanced sensitivity and resolution provided by the *James Webb Space Telescope* (*JWST*) will enable the number of white dwarf disks with mineralogy measurements to increase tenfold, as well as more in-depth studies of the mineralogy. For example, the 10 μm crystalline silicate emission feature is distinguishable dependent on whether the silicate mineralogy is olivine or pyroxene dominated, with the dominant feature shifting from ~ 9 μm for pyroxene to ~11 μm for olivine. This affects the water storage capacity of mantles and its ability to sustain plate tectonics, crucial for questions regarding habitability (e.g., Kelley et al. 2010; Lambart et al. 2016; Hinkel and Unterborn 2018; Wang et al. 2022). With a number of *JWST* proposals accepted that will study the dust mineralogy of polluted white dwarf disks, over the next few years our understanding of the mineralogy of exoplanetary material will be revolutionized (Swan et al. 2024).

Composition of circumstellar gas. A small fraction of white dwarfs with circumstellar dust also show circumstellar gas, mostly as double-peaked emission features (e.g., Gänsicke et al. 2006) but some as additional absorption features (e.g., Debes et al. 2012). Thanks to dedicated follow-up studies from the new *Gaia* white dwarfs, the number of white dwarfs with circumstellar gas detections has increased significantly (Dennihy et al. 2020; Melis et al. 2020; Gentile Fusillo et al. 2021b). WD 1145+017 has the most elements detected in circumstellar gas, including Ca, Mg, Ti, Cr, Mn, Fe, and Ni (Xu et al. 2016). There has been significant progress on the modeling front as well to understand the properties of the circumstellar gas (Gänsicke et al. 2019; Fortin-Archambault et al. 2020; Steele et al. 2021). Measuring the chemical compositions of the circumstellar gas around white dwarfs could be a powerful way to constrain the composition of the exoplanetary material and contrast with the measurements from the polluted atmospheres.

CONCLUSIONS

Spectroscopic observations of polluted white dwarfs measure the bulk compositions of extrasolar planetary material, which is not possible with any other technique. The number of heavily polluted white dwarfs has increased significantly over the past decade with many new systems having unique abundance ratios with no solar system analog. Looking to the future, improvements on white dwarf model atmospheres, a larger uniform sample of polluted white dwarfs, and a stronger connection between the white dwarf community and the geochemistry community are needed to interpret these abundance measurements. Polluted white dwarf studies are essential for assessing a planet's overall habitability from on-going and future missions such as *JWST* and *HWO*.

ACKNOWLEDGEMENT

The authors thank Rhian Jones and E. D. Young for discussions of comparing abundance measurements from polluted white dwarfs with those of the meteorites and Andy Buchan for discussions related to PyllutedWD. S. Xu is supported by the international Gemini Observatory, a program of NSF's NOIRLab, which is managed by the Association of Universities for Research in Astronomy (AURA) under a cooperative agreement with the National Science Foundation on behalf of the Gemini partnership of Argentina, Brazil, Canada, Chile, the Republic of Korea, and the United States of America. L. K Rogers acknowledges support of a Royal Society University Research Fellowship, URF\R1\211421 and an ESA Co-Sponsored Research Agreement No. 4000138341/22/NL/GLC/my = Tracing the Geology of Exoplanets. S. Blouin is a Banting Postdoctoral Fellow and a CITA National Fellow, supported by the Natural Sciences and Engineering Research Council of Canada (NSERC).

The research shown here acknowledges use of the Hypatia Catalog Database, an online compilation of stellar abundance data as described in Hinkel et al. (2014, AJ, 148, 54), which was supported by NASA's Nexus for Exoplanet System Science (NExSS) research coordination network and the Vanderbilt Initiative in Data-Intensive Astrophysics (VIDA).

REFERENCES

Alexander CMOD (2019a) Quantitative models for the elemental and isotopic fractionations in the chondrites: The non-carbonaceous chondrites. Geochim Cosmochim Acta 254:246–276, https://doi.org/10.1016/j.gca.2019.01.026

Alexander CMOD (2019b) Quantitative models for the elemental and isotopic fractionations in chondrites: The carbonaceous chondrites. Geochim Cosmochim Acta 254:277–309, https://doi.org/10.1016/j.gca.2019.02.008

Allard NF, Leininger T, Gadéa FX, Brousseau-Couture V, Dufour P (2016) Asymmetry in the triplet 3p-4s Mg lines in cool DZ white dwarfs. Astron Astrophys 588:A142, https://doi.org/10.1051/0004-6361/201527826

Allègre C, Manhès G, Lewin E (2001) Chemical composition of the Earth and the volatility control on planetary genetics. Earth Planet Sci Lett 185:49–69, https://doi.org/10.1016/S0012-821X(00)00359-9

Bauer EB, Bildsten L (2018) Increases to inferred rates of planetesimal accretion due to thermohaline mixing in metal-accreting white dwarfs. Astrophys J Lett 859:L19, https://doi.org/10.3847/2041-8213/aac492

Bédard A, Bergeron P, Brassard P (2022) On the spectral evolution of hot white dwarf stars. III. The PG 1159-DO-DB-DQ evolutionary channel revisited. Astrophys J 930:8, https://doi.org/10.3847/1538-4357/ac609d

Bergeron P, Wesemael F, Fontaine G (1991) Synthetic spectra and atmospheric properties of cool DA white dwarfs. Astrophys J 367:253, https://doi.org/10.1086/169624

Bergeron P, Kilic M, Blouin S, Bédard A, Leggett SK, Brown WR (2022) On the nature of ultracool white dwarfs: Not so cool after all. Astrophys J 934:36, https://doi.org/10.3847/1538-4357/ac76c7

Blouin S (2020) Magnesium abundances in cool metal-polluted white dwarfs. Mon Not R Astron Soc 496:1881–1890, https://doi.org/10.1093/mnras/staa1689

Blouin S, Kowalski PM, Dufour P (2017) Pressure Distortion of the H_2–He collision-induced absorption at the photosphere of cool white dwarf stars. Astrophys J 848:36, https://doi.org/10.3847/1538-4357/aa8ad6

Blouin S, Dufour P, Allard NF (2018) A new generation of cool white dwarf atmosphere models. I. Theoretical framework and applications to DZ stars. Astrophys J 863:184, https://doi.org/10.3847/1538-4357/aad4a9

Blouin S, Dufour P, Allard NF, Salim S, Rich RM, Koopmans LVE (2019) A new generation of cool white dwarf atmosphere models. III. WD J2356-209: Accretion of a planetesimal with an unusual composition. Astrophys J 872:188, https://doi.org/10.3847/1538-4357/ab0081

Bonsor A, Carter PJ, Hollands M, Gänsicke BT, Leinhardt Z, Harrison JHD (2020) Are exoplanetesimals differentiated? Mon Not R Astron Soc 492:2683-2697, https://doi.org/10.1093/mnras/stz3603

Bonsor A, Lichtenberg T, Drażkowska J, Buchan AM (2023) Rapid formation of exoplanetesimals revealed by white dwarfs. Nat Astron 7:39-48, https://doi.org/10.1038/s41550-022-01815-8

Brewer JM, Fischer DA (2016) C/O and Mg/Si ratios of stars in the solar neighborhood. Astrophys J 831:20, https://doi.org/10.3847/0004-637X/831/1/20

Buchan AM, Bonsor A, Shorttle O, Wade J, Harrison J, Noack L, Koester D (2022) Planets or asteroids? A geochemical method to constrain the masses of white dwarf pollutants. Mon Not R Astron Soc 510:3512–3530, https://doi.org/10.1093/mnras/stab3624

Canup RM, Asphaug E (2001) Origin of the Moon in a giant impact near the end of the Earth's formation. Nature 412:708–712

Charbonneau D, Brown TM, Latham DW, Mayor M (2000) Detection of planetary transits across a Sun-like star. Astrophys J Lett 529:L45–L48, https://doi.org/10.1086/312457

Chiappini C, Minchev I, Starkenburg E, Anders F, Fusillo NG, Gerhard O, Guiglion G, Khalatyan A, Kordopatis G, Lemasle B, Matijevic G (2019) 4MOST Consortium Survey 3: Milky Way disc and bulge low-resolution survey (4MIDABLE-LR). The Messenger 175:30–34, https://doi.org/10.18727/0722-6691/5122

Cooper AP, Koposov SE, Prieto CA, Manser CJ, Kizhuprakkat N, Myers AD, Dey A, Gänsicke BT, Li TS, Rockosi C, Valluri M (2023) Overview of the DESI Milky Way survey. Astrophys J 947:37, https://doi.org/10.3847/1538-4357/acb3c0

Coutu S, Dufour P, Bergeron P, Blouin S, Loranger E, Allard NF, Dunlap BH (2019) Analysis of helium-rich white dwarfs polluted by heavy elements in the Gaia Era. Astrophys J 885:74, https://doi.org/10.3847/1538-4357/ab46b9

Cunningham T, Tremblay P-E, Freytag B, Ludwig H-G, Koester D (2019) Convective overshoot and macroscopic diffusion in pure-hydrogen-atmosphere white dwarfs. Mon Not R Astron Soc 488:2503–2522, https://doi.org/10.1093/mnras/stz1759

Debes JH, Kilic M (2010) Results from a Magellan spectroscopic DAZ monitoring campaign. AIP Conf Proc 23 November 2010, 1273:488–491. https://doi.org/10.1063/1.3527870

Debes JH, Kilic M, Faedi F, Shkolnik EL, Lopez-Morales M, Weinberger AJ, Slesnick C, West RG (2012) Detection of weak circumstellar gas around the DAZ white dwarf WD 1124-293: Evidence for the accretion of multiple asteroids. Astrophys J 754:59, https://doi.org/10.1088/0004-637X/754/1/59

Dennihy E, Xu S, Lai S, Bonsor A, Clemens JC, Dufour P, Gänsicke BT, Fusillo NP, Hardy F, Hegedus RJ, Hermes JJ (2020) Five new post-main-sequence debris disks with gaseous emission. Astrophys J 905:5, https://doi.org/10.3847/1538-4357/abc339

Doyle AE, Young ED, Klein B, Zuckerman B, Schlichting HE (2019) Oxygen fugacities of extrasolar rocks: Evidence for an Earth-like geochemistry of exoplanets. Science 366:356–359, https://doi.org/10.1126/science.aax3901

Doyle AE, Klein B, Schlichting HE, Young ED (2020) Where are the extrasolar Mercuries? Astrophys J 901:10, https://doi.org/10.3847/1538-4357/abad9a

Doyle AE, Desch SJ, Young ED (2021) Icy exomoons evidenced by spallogenic nuclides in polluted white dwarfs. Astrophys J Lett 907:L35, https://doi.org/10.3847/2041-8213/abd9ba

Doyle AE, Klein BL, Dufour P, Melis C, Zuckerman B, Xu S, Weinberger AJ, Trierweiler IL, Monson NN, Jura MA, Young ED (2023) New chondritic bodies identified in eight oxygen-bearing white dwarfs. Astrophys J 950:93, https://doi.org/10.3847/1538-4357/acbd44

Dufour P, Bergeron P, Fontaine G (2005) Detailed spectroscopic and photometric analysis of DQ white dwarfs. Astrophys J 627:404–417, https://doi.org/10.1086/430373

Dufour P, Bergeron P, Liebert J, Harris HC, Knapp GR, Anderson SF, Hall PB, Strauss MA, Collinge MJ, Edwards MC (2007) On the spectral evolution of cool, helium-atmosphere white dwarfs: Detailed spectroscopic and photometric analysis of DZ stars. Astrophys J 663:1291–1308, https://doi.org/10.1086/518468

Dufour P, Kilic M, Fontaine G, Bergeron P, Melis C, Bochanski J (2012) Detailed compositional analysis of the heavily polluted DBZ white dwarf SDSS J073842.56+183509.06: A window on planet formation? Astrophys J 749:6, https://doi.org/10.1088/0004-637X/749/1/6

Dufour P, Blouin S, Coutu S, Fortin-Archambault M, Thibeault C, Bergeron P, Fontaine G (2017) The Montreal White Dwarf Database: A tool for the community, https://doi.org/10.48550/arXiv.1610.00986

Elkins-Tanton LT (2012) Magma oceans in the inner Solar System. Annu Rev Earth Planet Sci 40:113–139, https://doi.org/10.1146/annurev-earth-042711-105503

Elms AK, Tremblay P-E, Gänsicke BT, Koester D, Hollands MA, Gentile Fusillo NP, Cunningham T, Apps K (2022) Spectral analysis of ultra-cool white dwarfs polluted by planetary debris. Mon Not R Astron Soc 517:4557–4574, https://doi.org/10.1093/mnras/stac2908

Farihi J, Barstow MA, Redfield S, Dufour P, Hambly NC (2010) Rocky planetesimals as the origin of metals in DZ stars. Mon Not R Astron Soc 404:2123–2135, https://doi.org/10.1111/j.1365-2966.2010.16426.x

Farihi J, Gänsicke BT, Koester D (2013) Evidence for water in the rocky debris of a disrupted extrasolar minor planet. Science 342:218–220, https://doi.org/10.1126/science.1239447

Farihi J, Koester D, Zuckerman B, Vican L, Gänsicke BT, Smith N, Walth G, Breedt E (2016) Solar abundances of rock-forming elements, extreme oxygen and hydrogen in a young polluted white dwarf. Mon Not R Astron Soc 463:3186–3192, https://doi.org/10.1093/mnras/stw2182

Fortin-Archambault M, Dufour P, Xu S (2020) Modeling of the variable circumstellar absorption features of WD 1145+017. Astrophys J 888:47, https://doi.org/10.3847/1538-4357/ab585a

Freytag B, Ludwig H-G, Steffen M (1996) Hydrodynamical models of stellar convection. The role of overshoot in DA white dwarfs, A-type stars, and the Sun. Astron Astrophys 313:497–516

Gaia Collaboration (2016) The Gaia mission. Astron Astrophys 595:A1, https://doi.org/10.1051/0004-6361/201629272

Gänsicke BT, Marsh TR, Southworth J, Rebassa-Mansergas A (2006) A gaseous metal disk around a white dwarf. Science 314:1908, https://doi.org/10.1126/science.1135033

Gänsicke BT, Koester D, Farihi J, Girven J, Parsons SG, Breedt E (2012) The chemical diversity of exo-terrestrial planetary debris around white dwarfs. Mon Not R Astron Soc 424:333–347, https://doi.org/10.1111/j.1365-2966.2012.21201.x

Gänsicke BT, Schreiber MR, Toloza O, Fusillo NPG, Koester D, Manser CJ (2019) Accretion of a giant planet onto a white dwarf star. Nature 576:61–64, https://doi.org/10.1038/s41586-019-1789-8

Gentile Fusillo NP, Gänsicke BT, Farihi J, Koester D, Schreiber MR, Pala AF (2017) Trace hydrogen in helium atmosphere white dwarfs as a possible signature of water accretion. Mon Not R Astron Soc 468:971–980, https://doi.org/10.1093/mnras/stx468

Gentile Fusillo NP, Tremblay P-E, Cukanovaite E, Vorontseva A, Lallement R, Hollands M, Gänsicke BT, Burdge KB, McCleery J, Jordan S (2021a) A catalogue of white dwarfs in Gaia EDR3. Mon Not R Astron Soc 508:3877–3896, https://doi.org/10.1093/mnras/stab2672

Gentile Fusillo NP, Manser CJ, Gänsicke BT, Toloza O, Koester D, Dennihy E, Brown WR, Farihi J, Hollands MA, Hoskin MJ, Izquierdo P (2021b) White dwarfs with planetary remnants in the era of Gaia–I. Six emission line systems. Mon Not R Astron Soc 504:2707–2726, https://doi.org/10.1093/mnras/stab992

Gilliland RL, Chaplin WJ, Dunham EW, Argabright VS, Borucki WJ, Basri G, Bryson ST, Buzasi DL, Caldwell DA, Elsworth YP, Jenkins JM (2011) Kepler mission stellar and instrument noise properties. Astrophys J 197:6, https://doi.org/10.1088/0067-0049/197/1/6

Girven J, Brinkworth CS, Farihi J, Gänsicke BT, Hoard DW, Marsh TR, Koester D (2012) Constraints on the lifetimes of disks resulting from tidally destroyed rocky planetary bodies. Astrophys J 749:154, https://doi.org/10.1088/0004-637X/749/2/154

González Egea E, Raddi R, Koester D, Rogers LK, Marocco F, Cooper WJ, Beamín JC, Burningham B, Day-Jones A, Forbrich J, Pinfield DJ (2021) Serendipitous discovery of a dusty disc around WDJ181417.84-735459.83. Mon Not R Astron Soc 501:3916-3925, https://doi.org/10.1093/mnras/staa3836

Guo J, Zhao J, Zhang H, Zhang J, Bai Y, Walters N, Yang Y, Liu J (2022) White dwarfs identified in LAMOST Data Release 5. Mon Not R Astron Soc 509:2674-2688, https://doi.org/10.1093/mnras/stab3151

Hallis LJ (2017) D/H ratios of the inner Solar System. Philos Trans R Soc London Ser A 375:20150390, https://doi.org/10.1098/rsta.2015.0390

Harrison JHD, Bonsor A, Madhusudhan N (2018) Polluted white dwarfs: constraints on the origin and geology of exoplanetary material. Mon Not R Astron Soc 479:3814–3841, https://doi.org/10.1093/mnras/sty1700

Harrison JHD, Shorttle O, Bonsor A (2021a) Evidence for post-nebula volatilisation in an exo-planetary body. Earth Planet Sci Lett 554:116694, https://doi.org/10.1016/j.epsl.2020.116694

Harrison JHD, Bonsor A, Kama M, Buchan AM, Blouin S, Koester D (2021b) Bayesian constraints on the origin and geology of exoplanetary material using a population of externally polluted white dwarfs. Mon Not R Astron Soc 504:2853–2867, https://doi.org/10.1093/mnras/stab736

Heinonen RA, Saumon D, Daligault J, Starrett CE, Baalrud SD, Fontaine G (2020) Diffusion coefficients in the envelopes of white dwarfs. Astrophys J 896:2, https://doi.org/10.3847/1538-4357/ab91ad

Hinkel NR, Unterborn CT (2018) The star–planet connection. i. using stellar composition to observationally constrain planetary mineralogy for the 10 closest stars. Astrophys J 853:83, https://doi.org/10.3847/1538-4357/aaa5b4

Hollands MA, Koester D, Alekseev V, Herbert EL, Gänsicke BT (2017) Cool DZ white dwarfs–I. Identification and spectral analysis. Mon Not R Astron Soc 467:4970-5000, https://doi.org/10.1093/mnras/stx250

Hollands MA, Tremblay P-E, Gänsicke BT, Koester D, Gentile-Fusillo NP (2021) Alkali metals in white dwarf atmospheres as tracers of ancient planetary crusts. Nat Astron 5:451–459, https://doi.org/10.1038/s41550-020-01296-7

Hoskin MJ, Toloza O, Gänsicke BT, Raddi R, Koester D, Pala AF, Manser CJ, Farihi J, Belmonte MT, Hollands M, Gentile Fusillo N (2020) White dwarf pollution by hydrated planetary remnants: hydrogen and metals in WDJ204713.76-125908.9. Mon Not R Astron Soc 499:171–182, https://doi.org/10.1093/mnras/staa2717

Izquierdo P, Toloza O, Gänsicke BT, Rodríguez-Gil P, Farihi J, Koester D, Guo J, Redfield S (2021) GD 424—A helium-atmosphere white dwarf with a large amount of trace hydrogen in the process of digesting a rocky planetesimal. Mon Not R Astron Soc 501:4276–4288, https://doi.org/10.1093/mnras/staa3987

Izquierdo P, Gänsicke BT, Rodríguez-Gil P, Koester D, Toloza O, Gentile Fusillo NP, Pala AF, Tremblay P-E (2023) Systematic uncertainties in the characterization of helium-dominated metal-polluted white dwarf atmospheres. Mon Not R Astron Soc 520:2843-2866, https://doi.org/10.1093/mnras/stad282

Johnson TM, Klein BL, Koester D, Melis C, Zuckerman B, Jura M (2022) Unusual abundances from planetary system material polluting the white dwarf G238-44. Astrophys J 941:113, https://doi.org/10.3847/1538-4357/aca089

Jones RH (2024) Meteorites and planet formation. Rev Mineral Geochem 90:113–140

Jura M (2006) Carbon deficiency in externally polluted white dwarfs: Evidence for accretion of asteroids. Astrophys J 653:613-620, https://doi.org/10.1086/508738

Jura M, Farihi J, Zuckerman B (2007) Externally polluted white dwarfs with dust disks. Astrophys J 663:1285–1290, https://doi.org/10.1086/518767

Jura M, Farihi J, Zuckerman B (2009) Six white dwarfs with circumstellar silicates. Astron J 137:3191–3197, https://doi.org/10.1088/0004-6256/137/2/3191

Jura M, Xu S, Klein B, Koester D, Zuckerman B (2012) Two extrasolar asteroids with low volatile-element mass fractions. Astrophys J 750:69, https://doi.org/10.1088/0004-637X/750/1/69

Jura M, Xu S, Young ED (2013) ^{26}Al in the early solar system: Not so unusual after all. Astrophys J Lett 775:L41, https://doi.org/10.1088/2041-8205/775/2/L41

Jura M, Dufour P, Xu S, Zuckerman B, Klein B, Young ED, Melis C (2015) Evidence for an anhydrous carbonaceous extrasolar minor planet. Astrophys J 799:109, https://doi.org/10.1088/0004-637X/799/1/109

Kaiser BC, Clemens JC, Blouin S, Dufour P, Hegedus RJ, Reding JS, Bédard A (2021) Lithium pollution of a white dwarf records the accretion of an extrasolar planetesimal. Science 371:168–172, https://doi.org/10.1126/science.abd1714

Kawka A, Vennes S (2012) VLT/X-shooter observations and the chemical composition of cool white dwarfs. Astron Astrophys 538:A13, https://doi.org/10.1051/0004-6361/201118210

Kawka A, Vennes S (2016) Extreme abundance ratios in the polluted atmosphere of the cool white dwarf NLTT 19868. Mon Not R Astron Soc 458:325–331, https://doi.org/10.1093/mnras/stw383

Kelley KA, Plank T, Newman S, Stolper EM, Grove TL, Parman S, Hauri EH (2010) Mantle melting as a function of water content beneath the Mariana Arc. J Petrol 51:1711–1738, https://doi.org/10.1093/petrology/egq036

Kempton EM-R, Knutson HA (2024) Transiting exoplanet atmospheres in the era of JWST. Rev Mineral Geochem 90:411–464

Klein B, Jura M, Koester D, Zuckerman B, Melis C (2010) Chemical abundances in the externally polluted white dwarf GD40: Evidence of a rocky extrasolar minor planet. Astrophys J 709:950–962, https://doi.org/10.1088/0004-637X/709/2/950

Klein B, Jura M, Koester D, Zuckerman B (2011) Rocky extrasolar planetary compositions derived from externally polluted white dwarfs. Astrophys J 741:64, https://doi.org/10.1088/0004-637X/741/1/64

Klein BL, Doyle AE, Zuckerman B, Dufour P, Blouin S, Melis C, Weinberger AJ, Young ED (2021) Discovery of beryllium in white dwarfs polluted by planetesimal accretion. Astrophys J 914:61, https://doi.org/10.3847/1538-4357/abe40b

Koester D (2009) Accretion and diffusion in white dwarfs. New diffusion timescales and applications to GD362 and G29-38. Astron Astrophys 498:517–525, https://doi.org/10.1051/0004-6361/200811468

Koester D, Kepler SO (2015) DB white dwarfs in the Sloan Digital Sky Survey data release 10 and 12. Astron Astrophys 583:A86, https://doi.org/10.1051/0004-6361/201527169

Koester D, Wolff B (2000) Element abundances in cool white dwarfs. I. The DZA white dwarfs L745-46A and Ross 640. Astron Astrophys 357:587–596

Koester D, Gänsicke BT, Farihi J (2014) The frequency of planetary debris around young white dwarfs. Astron Astrophys 566:A34, https://doi.org/10.1051/0004-6361/201423691

Koester D, Kepler SO, Irwin AW (2020) New white dwarf envelope models and diffusion. Application to DQ white dwarfs. Astron Astrophys 635:A103, https://doi.org/10.1051/0004-6361/202037530

Kowalski PM (2014) Infrared absorption of dense helium and its importance in the atmospheres of cool white dwarfs. Astron Astrophys 566:L8, https://doi.org/10.1051/0004-6361/201424242

Kowalski PM, Mazevet S, Saumon D, Challacombe M (2007) Equation of state and optical properties of warm dense helium. Phys Rev B: Condens Matter 76:075112, https://doi.org/10.1103/PhysRevB.76.075112

Kupka F, Zaussinger F, Montgomery MH (2018) Mixing and overshooting in surface convection zones of DA white dwarfs: first results from ANTARES. Mon Not R Astron Soc 474:4660–4671, https://doi.org/10.1093/mnras/stx3119

Lai S, Dennihy E, Xu S, Nitta A, Kleinman S, Leggett SK, Bonsor A, Hodgkin S, Rebassa-Mansergas A, Rogers LK (2021) Infrared excesses around bright white dwarfs from Gaia and unWISE. II. Astrophys J 920:156, https://doi.org/10.3847/1538-4357/ac1354

Lambart S, Baker MB, Stolper EM (2016) The role of pyroxenite in basalt genesis: Melt-PX, a melting parameterization for mantle pyroxenites between 0.9 and 5 GPa. J Geophys Res (Solid Earth) 121:5708–5735, https://doi.org/10.1002/2015JB012762

Lodders K, Palme H, Gail H-P (2009) Abundances of the elements in the solar system. 4.4 *In:* Subvolume B: Solar System of Volume 4:Astronomy, Astrophysics, and Cosmology, Landolt-Börnstein–Group VI: Astronomy and Astrophysics, Springer Materials

Melis C, Dufour P (2017) Does a differentiated, carbonate-rich, rocky object pollute the white dwarf SDSS J104341.53+085558.2? Astrophys J 834:1, https://doi.org/10.3847/1538-4357/834/1/1

Melis C, Farihi J, Dufour P, Zuckerman B, Burgasser AJ, Bergeron P, Bochanski J, Simcoe R (2011) Accretion of a terrestrial-like minor planet by a white dwarf. Astrophys J 732:90, https://doi.org/10.1088/0004-637X/732/2/90

Melis C, Dufour P, Farihi J, Bochanski J, Burgasser AJ, Parsons SG, Gänsicke BT, Koester D, Swift BJ (2012) Gaseous material orbiting the polluted, dusty white dwarf HE 1349-2305. Astrophys J Lett 751:L4, https://doi.org/10.1088/2041-8205/751/1/L4

Melis C, Klein B, Doyle AE, Weinberger A, Zuckerman B, Dufour P (2020) Serendipitous discovery of nine white dwarfs with gaseous debris disks. Astrophys J 905:56, https://doi.org/10.3847/1538-4357/abbdfa

Mihalas D (1978) Stellar atmospheres. San Francisco: WH Freeman

Nittler LR, McCoy TJ, Clark PE, Murphy ME, Trombka JI, Jarosewich E (2004) Bulk element compositions of meteorites: A guide for interpreting remote-sensing geochemical measurements of planets and asteroids. Antarct Meteorite Res 17:231

Okuya A, Ida S, Hyodo R, Okuzumi S (2023) Modelling the evolution of silicate/volatile accretion discs around white dwarfs. Mon Not R Astron Soc 519:1657-1676, https://doi.org/10.1093/mnras/stac3522

Paquette C, Pelletier C, Fontaine G, Michaud G (1986) Diffusion coefficients for stellar plasmas. Astrophys J Suppl Ser 61:177, https://doi.org/10.1086/191111

Pepe F, Cristiani S, Rebolo R, Santos NC, Dekker H, Cabral A, Di Marcantonio P, Figueira P, Curto GL, Lovis C, Mayor M (2021) ESPRESSO at VLT. On-sky performance and first results. Astron Astrophys 645:A96, https://doi.org/10.1051/0004-6361/202038306

Pontoppidan KM, Salyk C, Bergin EA, Brittain S, Marty B, Mousis O, Öberg KI (2014) Volatiles in protoplanetary disks. Protostars Planets VI:363–385

Putirka KD (2024) Exoplanet mineralogy. Rev Mineral Geochem 90:199–258

Putirka KD, Xu S (2021) Polluted white dwarfs reveal exotic mantle rock types on exoplanets in our solar neighborhood. Nat Commun 12:6168, https://doi.org/10.1038/s41467-021-26403-8

Raddi R, Gänsicke BT, Koester D, Farihi J, Hermes JJ, Scaringi S, Breedt E, Girven J (2015) Likely detection of water-rich asteroid debris in a metal-polluted white dwarf. Mon Not R Astron Soc 450:2083–2093, https://doi.org/10.1093/mnras/stv701

Rappaport S, Gary BL, Kaye T, Vanderburg A, Croll B, Benni P, Foote J (2016) Drifting asteroid fragments around WD 1145+017. Mon Not R Astron Soc 458:3904–3917, https://doi.org/10.1093/mnras/stw612

Reach WT, Lisse C, von Hippel T, Mullally F (2009) The dust cloud around the white dwarf G 29-38. II. Spectrum from 5 to 40 μm and mid-infrared photometric variability. Astrophys J 693:697–712, https://doi.org/10.1088/0004-637X/693/1/697

Reindl N, Rauch T, Werner K, Kruk JW, Todt H (2014) On helium-dominated stellar evolution: the mysterious role of the O(He)-type stars. Astron Astrophys 566:A116, https://doi.org/10.1051/0004-6361/201423498

Rogers LK, Bonsor A, Xu S, Dufour P, Klein BL, Buchan A, Hodgkin S, Hardy F, Kissler-Patig M, Melis C, Weinberger AJ (2024) Seven white dwarfs with circumstellar gas discs I: white dwarf parameters and accreted planetary abundances. Mon Not R Astron Soc 527:6038-6054, https://doi.org/10.1093/mnras/stad3557

Safronov VS, Zvjagina EV (1969) Relative sizes of the largest bodies during the accumulation of planets. Icarus 10:109-115, https://doi.org/10.1016/0019-1035(69)90013-X

Saumon D, Blouin S, Tremblay P-E (2022) Current challenges in the physics of white dwarf stars. Phys Rep 988:1–63, https://doi.org/10.1016/j.physrep.2022.09.001

Seager S, Kuchner M, Hier-Majumder CA, Militzer B (2007) Mass–radius relationships for solid exoplanets. Astrophys J 669:1279-1297, https://doi.org/10.1086/521346

Seifahrt A, Bean JL, Kasper D, Stürmer J, Brady M, Liu R, Zechmeister M, Stefánsson GK, Montet B, White J, Tapia E (2022) MAROON-X: the first two years of EPRVs from Gemini North. Proc Vol 12184, Ground-based and Airborne Instrumentation for Astronomy IX; 121841G, https://doi.org/10.1117/12.2629428

Shahar A, Driscoll P, Weinberger A, Cody G (2019) What makes a planet habitable? Science 364:434–435, https://doi.org/10.1126/science.aaw4326

Steele A, Debes J, Xu S, Yeh S, Dufour P (2021) A characterization of the circumstellar gas of WD 1124–293 using Cloudy. Astrophys J 911:25, https://doi.org/10.3847/1538-4357/abc262

Swan A, Farihi J, Koester D, Holland M, Parsons S, Cauley PW, Redfield S, Gaensicke BT (2019) Interpretation and diversity of exoplanetary material orbiting white dwarfs. Mon Not R Astron Soc 490:202–218, https://doi.org/10.1093/mnras/stz2337

Swan A, Farihi J, Su KYL, Desch SJ (2024), The first white dwarf debris disc observed by JWST. Mon Not R Astron Sco 529:L41–L46, https://doi.org/10.1093/mnrasl/slad198

Swan A, Farihi J, Melis C, Dufour P, Desch SJ, Koester D, Guo J (2023) Planetesimals at DZ stars–I. Chondritic compositions and a massive accretion event. Mon Not R Astron Soc 526:3815–3831, https://doi.org/10.1093/mnras/stad2867

Toplis MJ, Mizzon H, Monnereau M, Forni O, McSween HY, Mittlefehldt DW, McCoy TJ, Prettyman TH, De Sanctis MC, Raymond CA, Russell CT (2013) Chondritic models of 4 Vesta: Implications for geochemical and geophysical properties. Meteorit Planet Sci 48:2300–2315, https://doi.org/10.1111/maps.12195

Tremblay P-E, Ludwig H-G, Freytag B, Fontaine G, Steffen M, Brassard P (2015) Calibration of the mixing-length theory for convective white dwarf envelopes. Astrophys J 799:142, https://doi.org/10.1088/0004-637X/799/2/142

Tremblay PE, Hollands MA, Gentile Fusillo NP, McCleery J, Izquierdo P, Gänsicke BT, Cukanovaite E, Koester D, Brown WR, Charpinet S, Cunningham T (2020) Gaia white dwarfs within 40 pc–I. Spectroscopic observations of new candidates. Mon Not R Astron Soc 497:130-145, https://doi.org/10.1093/mnras/staa1892

Trierweiler IL, Doyle AE, Young ED (2023) A chondritic solar neighborhood. Planet Sci J 4:136, https://doi.org/10.3847/PSJ/acdef3

Trønnes RG, Baron MA, Eigenmann KR, Guren MG, Heyn BH, Løken A, Mohn CE (2019) Core formation, mantle differentiation and core–mantle interaction within Earth and the terrestrial planets. Tectonophysics 760:165–198, https://doi.org/10.1016/j.tecto.2018.10.021

Urey HC (1955) The cosmic abundances of potassium, uranium, and thorium and the heat balances of the Earth, the Moon, and Mars. PNAS 41:127–144, https://doi.org/10.1073/pnas.41.3.127

Vanderbosch Z, Hermes JJ, Dennihy E, Dunlap BH, Izquierdo P, Tremblay PE, Cho PB, Gänsicke BT, Toloza O, Bell KJ, Montgomery MH (2020) A white dwarf with transiting circumstellar material far outside the Roche Limit. Astrophys J 897:171, https://doi.org/10.3847/1538-4357/ab9649

Vanderburg A, Johnson JA, Rappaport S, Bieryla A, Irwin J, Lewis JA, Kipping D, Brown WR, Dufour P, Ciardi DR, Angus R (2015) A disintegrating minor planet transiting a white dwarf. Nature 526:546–549, https://doi.org/10.1038/nature15527

Vennes S, Kawka A (2013) The polluted atmosphere of the white dwarf NLTT 25792 and the diversity of circumstellar environments. Astrophys J 779:70, https://doi.org/10.1088/0004-637X/779/1/70

Vennes S, Kawka A, Klein BL, Zuckerman B, Weinberger AJ, Melis C (2024) A cool, magnetic white dwarf accreting planetary debris. Mon Not R Astron Soc 527:3122–3138, https://doi.org/10.1093/mnras/stad3370

Veras D, Mustill AJ, Bonsor A (2024) The evolution and delivery of rocky extra-solar materials to white dwarfs. Rev Mineral Geochem 90:141–170

Wachlin FC, Vauclair G, Vauclair S, Althaus LG (2022) New simulations of accreting DA white dwarfs: Inferring accretion rates from the surface contamination. Astron Astrophys 660:A30, https://doi.org/10.1051/0004-6361/202142289

Wang HS, Quanz SP, Yong D, Liu F, Seidler F, Acuña L, Mojzsis SJ (2022) Detailed chemical compositions of planet-hosting stars–II. Exploration of the interiors of terrestrial-type exoplanets. Mon Not R Astron Soc 513:5829–5846, https://doi.org/10.1093/mnras/stac1119

Werner K, Herwig F (2006) The elemental abundances in bare planetary nebula central stars and the shell burning in AGB stars. Publ Astron Soc Pac 118:183–204, https://doi.org/10.1086/500443

Wright JT (2018) Radial velocities as an exoplanet discovery method. *In*: Handbook of Exoplanets. Deeg HJ, Belmonte JA (eds), p 4

Wyatt MC, Farihi J, Pringle JE, Bonsor A (2014) Stochastic accretion of planetesimals on to white dwarfs: constraints on the mass distribution of accreted material from atmospheric pollution. Mon Not R Astron Soc 439:3371–3391, https://doi.org/10.1093/mnras/stu183

Xu S, Jura M, Klein B, Koester D, Zuckerman B (2013) Two beyond-primitive extrasolar planetesimals. Astrophys J 766:132, https://doi.org/10.1088/0004-637X/766/2/132

Xu S, Jura M, Koester D, Klein B, Zuckerman B (2014) Elemental compositions of two extrasolar rocky planetesimals. Astrophys J 783:79, https://doi.org/10.1088/0004-637X/783/2/79

Xu S, Jura M, Dufour P, Zuckerman B (2016) Evidence for gas from a disintegrating extrasolar asteroid. Astrophys J Lett 816:L22, https://doi.org/10.3847/2041-8205/816/2/L22

Xu S, Dufour P, Klein B, Melis C, Monson NN, Zuckerman B, Young ED, Jura MA (2019) Compositions of planetary debris around dusty white dwarfs. Astron J 158:242, https://doi.org/10.3847/1538-3881/ab4cee

Zahn J-P (1991) Convective penetration in stellar interiors. Astron Astrophys 252:179–188

Zuckerman B, Koester D, Reid IN, Hünsch M (2003) Metal lines in DA white dwarfs. Astrophys J 596:477–495, https://doi.org/10.1086/377492

Zuckerman B, Koester D, Melis C, Hansen BM, Jura M (2007) The chemical composition of an extrasolar minor planet. Astrophys J 671:872–877, https://doi.org/10.1086/522223

Zuckerman B, Melis C, Klein B, Koester D, Jura M (2010) Ancient planetary systems are orbiting a large fraction of white dwarf stars. Astrophys J 722:725–736, https://doi.org/10.1088/0004-637X/722/1/725

Zuckerman B, Koester D, Dufour P, Melis C, Klein B, Jura M (2011) An aluminum/calcium-rich, iron-poor, white dwarf star: Evidence for an extrasolar planetary lithosphere? Astrophys J 739:101, https://doi.org/10.1088/0004-637X/739/2/101

Reviews in Mineralogy & Geochemistry
Vol. 90 pp. 199–258, 2024
Copyright © Mineralogical Society of America

Exoplanet Mineralogy

Keith D. Putirka

Dept. Earth & Env. Sciences
California State University
Fresno, California, 93740 U.S.A.

kputirka@csufresno.edu

INTRODUCTION

Mineral properties hold dominion over planetary behavior as their physical and thermal properties control mantle circulation, plate tectonics, melting behavior, and crustal deformation, as well as global water and CO_2 cycles. Estimating exoplanet mineralogy is thus of paramount importance for understanding planetary origins and evolution. Preliminary studies of exoplanets emphasize individual element abundances (e.g., Ca or K contents) or ratios (Mg/Si or Si/Fe), which are a valuable first step. But inter-planetary differences in, say, Si or Mg, are only relevant insofar they define a planet's mineralogy.

As will be shown, nothing definitive can yet be said about exoplanet mineralogy, largely because a number of conditions that control mineralogy are unknown, though measurements from the James Webb Space Telescope (JWST) and similar instruments might soon yield the required data for some exoplanets. Differences of opinion are perhaps, then, little more than a contest of speculation; but thoughtful speculation can be profitable, possibly accurate, and in any case essential for progress. The methods noted below will emphasize paths of speculation that minimize underlying assumptions, and several approaches to quantify uncertainties.

To calculate a planet's mineralogy, we will emphasize what geologists refer to as the "major elements" (or "major oxides") i.e., elements that occur at wt. % levels (Si, Ti, Al, Fe, Mn, Mg, Ni, Ca, Na, K, P and O). These elements, along with water, pressure (P) and temperature (T), control mineral stability. The behavior of minor or trace elements are in turn controlled by mineral assemblage. This chapter will focus on mass balance and some thermodynamic relationships that can be used to estimate mineral assemblages when an exoplanet major element composition is given. Such estimates are usually inferred from star compositions, and some caveats with regard to such approaches will be explored. Mass balance calculations are probably all that we can sensibly do to infer exoplanet mineralogy until vastly better remote-sensing methods are available. But by keeping several caution flags in mind, it should be possible in the meantime to explore possible scenarios regarding how exoplanets form and evolve. Such evolutionary scenarios inevitably reflect back on what we currently comprehend of Earth and its neighboring rocky objects, and reveal the flaws and gaps in such understanding. As we have only one planetary system to use as an analog, exoplanetary studies are tethered to what we know, and still have yet to learn about our own Solar System.

This chapter begins with some basic concepts regarding the structure and mineralogy of rocky planets, how to read and construct ternary diagrams, and why partial melting occurs when plate tectonics is operative. Partial melting is a key concept in that it governs crust and core formation, which in turn control mineralogy. These sections are for astronomers, or geologists new to the study of igneous petrology. From there, computational approaches for estimating planetary mineral assemblages will be introduced. These quantitative methods are simple, consonant with the level of information currently available on exoplanet compositions,

1529-6466/24/0090-0007$10.00 (print)
1943-2666/24/0090-0007$10.00 (online)

and while largely intended for mineralogists, should be accessible to non-specialists as well. Such methods are followed by a study of error when plotting mineral abundances in ternary diagrams, for mineralogists and petrologists who construct such diagrams. The chapter concludes with caveats, and the ways in which exoplanets might surprise us.

AN OVERVIEW OF PLANETARY STRUCTURE, MINERALOGY & PETROLOGY

Planetary structure

Rocky objects with radii greater than a few hundred km will undergo partial melting very early in their growth, due to the gravitational energy released upon accretion (Safranov 1978; Ransford 1979). In our Solar System, the starting materials that accrete to form the inner planets are thought to be similar to chondritic meteorites (so-named because they contain glass-bearing blebs called "chondrules"), although the match is imprecise (Drake and Righter 2002). The CI chondrite meteorites are of special interest as when volatile elements are ignored, they appear to be very similar to the Sun in composition (Pottasch 1964; Lodders and Fegley 2018) and may approximate the bulk materials that coalesced to form the inner planets (see Jones 2024, this volume). Whatever the parent material, accretionary heating will cause a protoplanet to become "differentiated", into a metallic core and a silicate mantle and crust (Fig. 1). This tripartite planetary structure, of an Fe–Ni alloy metal core, overlain by a silicate mantle and crust (Fig. 1), likely dominates all rocky planets in any planetary system; in our Solar system chondritic meteorites are examples of materials that have escaped such differentiation.

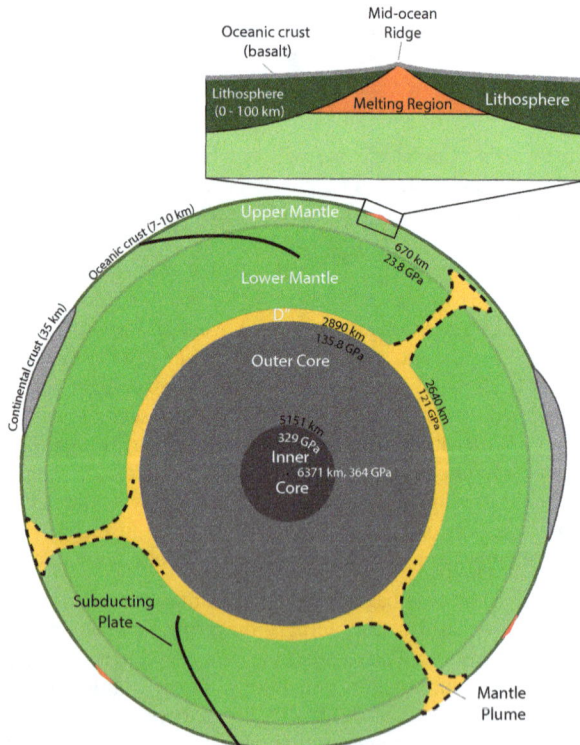

Figure 1. Cross section of Earth illustrating the core, mantle and crust (crust is not to scale), mantle plumes, and plate tectonics.

Mineralogic structure

Table 1 summarizes the minerals that are likely to be the most abundant in these differentiated planetary systems. Mineral assemblages will vary with pressure (*P*) (depth) within a planet. For example, Figure 2 shows the case for Earth, assuming constant composition, where mineral assemblages change drastically across a *P* range of 30 GPa—shifts that are significant for interpreting estimates of exoplanet densities (Dorn et al. 2015; Putirka et al. 2021). It may be challenging to infer crust compositions on exoplanets, but for any planet large enough for the silicate mantle to exceed *P* > 23 GPa, such planets are at least likely to have mineralogically distinct "upper" and "lower" mantle layers. Within Earth, the uppermost mantle (*P* < 15 GPa) is dominated by olivine and pyroxene, while above 23 GPa, bridgmanite dominates in a region called the "lower mantle" (Figs. 1–2). These two regions are separated by a "transition zone" where olivine is gradually compressed to form the polymorphs wadsleyite and ringwoodite and where Al-rich garnet is compressed to form majorite. Above 100–120 GPa, bridgmanite and ferro-periclase may transform to "post-perovskite" minerals at the base of the mantle, which within Earth might explain the so-called Dʺ layer (Mattern et al. 2005; Wicks and Duffy 2016; Kuwayama et al. 2021; Fig. 1) which has distinct seismic velocities.

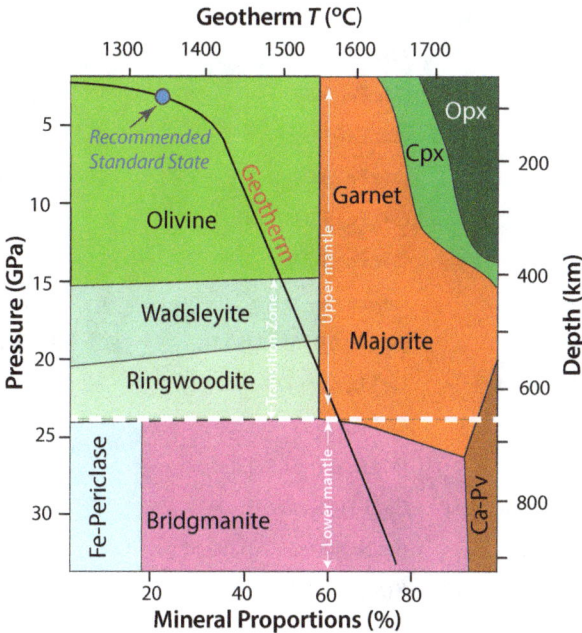

Figure 2. Cross sectional view of Earth's mantle mineralogy (Adapted from Kaminsky 2012); horizontal widths show relative mineral abundances (bottom scale); Earth's geotherm (upper horizontal scale) is also shown. Along the geotherm is a proposed standard state (2 GPa, 1350 °C) that is recommended for comparing planets due to the high abundance of and low uncertainties of experiments that bracket that state, at 1–3 GPa and 1300–1400 °C, and because diverse compositions should equilibrate rapidly at these conditions. Cpx = clinopyroxene; Opx = orthopyroxene, Ca-Pv = Ca-perovskite. See text for formulas.

Because of the challenges in performing phase equilibrium experiments at very high pressures, we have only a cursory understanding of Earth's lower mantle. And yet this region is crucial for understanding heat loss, mantle circulation and plate tectonics. For example, the core/mantle boundary (CMB) is the likely source of heat that powers mantle plumes (Putirka et al. 2007)—hot upwellings of mantle material that rise to the surface to drive large volume volcanic eruptions. Such plumes are necessarily rooted within the D" layer which might not only be mineralogically distinct, but compositionally distinct as well. The D" layer, and perhaps even larger fractions of the lowermost mantle, are a likely storehouse of subducted crust (e.g., Klein et al. 2017) that might be returned to the upper mantle via mantle plumes (Hofmann and White 1982). Mantle plumes also likely drive plate tectonics (Davies and Richards 1992), provided that the lithosphere (a planet's rocky outer shell) is sufficiently weak to allow it to break up into tectonic plates (Weller and Lenardic 2018). Absent plate tectonics, mantle plumes might provide the only manifestation of mantle convection, in what is called a "stagnant lid" mode of circulation (Ogawa et al. 1991; Solomotov and Moresi 2000; Stevenson 2003).

Table 1. A brief list of abundant minerals in Earth and other rocky objects of the Solar System.

	Mineral name	Formula	Where might it dominate?
Native elements	Kamacite, Taenite	(Fe,Ni)	core
Sulfides	Fe-sulfide (various)	FeS, Fe_2S	crust, inner core
Oxides	Ferro-periclase	$(Fe,Mg)O$	mantle
	Spinel	$MgAl_2O_4$	mantle
Silicates	Olivine	$(Mg,Fe)_2SiO_4$	mantle
	Wadsleyite	$(Mg,Fe)_2SiO_4$	transition zone (mantle)
	Ringwoodite	$(Mg,Fe)_2SiO_4$	transition zone (mantle)
	Serpentine	$(Mg,Fe)_6Si_4O_{10}(OH)_8$	crust and mantle
	Orthopyroxene	$(Mg,Fe)_2Si_2O_6$	mantle
	Majorite	$(Mg,Fe)_2Si_2O_6$	transition zone (mantle)
	Bridgmanite	$(Mg,Fe,Al)_2(Si,Al,Fe)_2O_6$	lower mantle
	Clinopyroxene	$Ca(Mg,Fe)Si_2O_6$	mantle
	Ca-perovskite	$CaSiO_3$	lower mantle
	Garnet	$(Mg,Fe,Ca)_3Al_2Si_3O1_2$	mantle
	Plagioclase feldspar	$CaAl_2Si_2O_8$–$NaAlSi_3O_8$	crust
	Hornblende	$(Ca,Na)_2(Mg,Fe,Al)_5(Al,Si)_8O_{22}(OH)_2$	crust
	Biotite	$K(Mg,Fe)_3AlSi_3O_{10}(F,OH)_2$	crust
	Alkali feldspar	$(K,Na)AlSi_3O_8$	crust
	Quartz	SiO_2	crust
Carbonates	Calcite	$CaCO_3$	crust

The driving force of planetary differentiation: partial melting

Planets larger than a few hundred km in radius will have sufficient accretional energy to partially melt. Other sources of heat include the gravitational energy released upon core formation, the latent heat of crystallization as a liquid metallic core crystallizes, and heat produced from radioactive elements. On sufficiently large planets these sources can maintain mantle circulation and partial melting—and thus the possibility of plate tectonics—for billions of years. Plate (and other types of) tectonics will be dealt in greater detail (Putirka 2024, this volume). Here we'll briefly review partial melting.

Ternary diagrams Most common rocks consist largely of just three minerals, whose abundances are used to define rock type, such as "granite", which is composed of quartz, plagioclase and alkali feldspar. Most other minerals are "accessory phases" which are used to modify the rock name; a granite that contains biotite and hornblende (and more hornblende than biotite) would be called a "biotite–hornblende granite". The three dominant minerals can be plotted in a ternary diagram where their abundances are normalized to sum to 1 or 100. And with renormalization, because only two of the abundances can vary independently, ternary diagrams can be plotted in 2-dimensions, such as Figure 3a, which shows the relative abundances of the minerals forsterite (Fo; Mg_2SiO_4; a type of olivine), diopside (Di; $CaMgSi_2O_6$; a variety of pyroxene) and anorthite (An; $CaAl_2Si_2O_8$; a variety of plagioclase feldspar). To create a ternary diagram within a 2-dimensional Cartesian coordinate system, the two following equations can be used, which translate the three mineral components, A, B and C, into x–y coordinates:

$$y = C \tag{1a}$$

$$x = \frac{B}{\cos(30°)} + C\left[\tan(30°)\right] \tag{1b}$$

Here, x is the horizontal Cartesian axis, y is the vertical Cartesian axis, C is the mineral component that plots at the top of the ternary diagram (An in Fig. 3), B is the mineral component that plots at the bottom right apex (Fo in Fig. 3), and the A component plots at the origin $(0,0)$. Figure 3a shows a ternary phase diagram that is useful for understanding mantle partial melting, and the process of "igneous differentiation".

Differentiation occurs because (a) rocks and minerals have different densities and so can be mechanically separated, and (b) because nearly all natural rocks melt incongruently, i.e., when they partially melt, the liquids do not have the same composition as the rock. To illustrate, Figure 3a shows the case of what is called "eutectic" melting, using a ternary eutectic diagram that involves three minerals that are common in the mantles of the inner planets: Fo, Di and An. The eutectic point in this diagram, at 1270 °C, represents the only point in the ternary system where the phases forsterite + diopside + anorthite + liquid coexist in equilibrium. Provided that all three minerals are present, melting will always begin at 1270 °C in this eutectic system, regardless of the relative abundances of Fo, Di and An. For example: let's say that a bulk composition falls at the green circle in Figure 3a, within the Fo + liq field, and so has a composition of approximately 40% An, 22% Di and 38% Fo. Since the rock falls into the Fo + liq field, Fo must be the last phase to be consumed during melting. And since a line drawn from the Fo apex through the bulk composition (green line) intersects the Fo + An cotectic (the dark line that separates the An + liq and the Fo + liq fields), An must be the second to the last phase to melt out of the rock. So upon heating, this rock will begin melting at 1270 °C and as heat is added to the system, each of Fo, Di and An will be consumed, in eutectic proportions, and the T remains constant at 1270 °C until one of the three phases is consumed, in this case Di; with further heating, T will now rise, and the liquid will leave the eutectic and migrate up the An + Fo cotectic, consuming both An and Fo, until An is consumed, which will happen at 1300 °C. At this point, the only mineral left is Fo and so any further heating will enrich the liquid in Fo, driving the liquid towards the Fo apex (following the green line), which means that the liquid will leave the An + Fo cotectic and head directly towards the bulk composition. Temperature will continue to rise until all of the Fo is consumed, and the last bit of Fo will be consumed and the rock will be completely melted at 1500 °C.

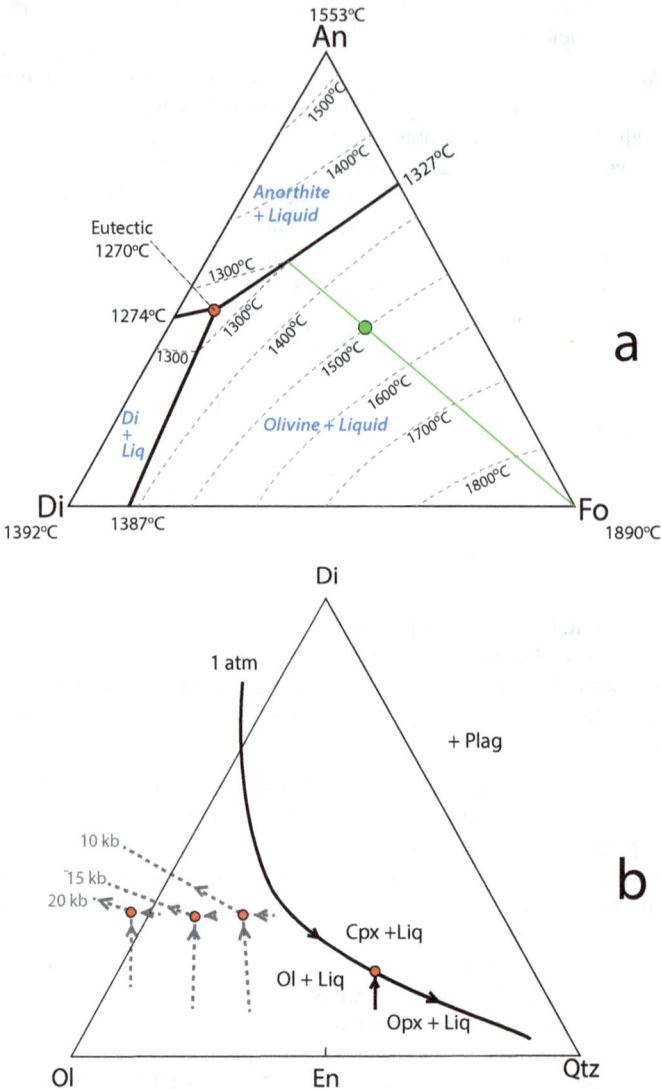

Figure 3. (a) Ternary eutectic for the case of pure diopside (Di; $CaMgSi_2O_6$), forsterite (Fo; Mg_2SiO_4) and anorthite (An; $CaAl_2Si_2O_8$), adapted from Morse (1980). Temperature contours are shown as **dashed lines**; the three **heavy dark lines** are "cotectic" curves that indicate the temperature and compositions of co-saturation of An + Ol, Di + Ol and An + Di; these intersect at the eutectic composition. For any combination of the three minerals, Di, Fo and An, melting will begin at the eutectic, at 1270 °C. **(b)** A more realistic ternary, the plagioclase-saturated ("+ plag") Di–Ol–Qtz (SiO_2) system; melting will begin at a **red dot**, the precise position of which depends upon pressure (adapted from Stolper 1980); curves show the positions of the cotectic curves at 1 atmosphere and 10, 15 and 20 kbar.

Note that in this example, the liquid only matches the bulk composition after melting is completed. The only case where the initial (and all subsequent) liquid(s) will match the bulk composition is the highly unlikely case that the bulk composition matches the eutectic (which in this diagram is at approximately 8% Fo, 43% An and 49% Di). Because rocks consist of at least 8–10 components it is even less likely than it is in Figure 3a that a planetary mantle would have a composition that precisely matches the eutectic (Walker 1986). However, if the Solar system contains an effectively countless number of rocky objects the odds that any one of them has a eutectic composition is certainly increased, which would have significant consequences for the amount of crust and the rate of heat loss for such planets.

Figure 3b is a closer approximation of partial melting within the inner planets. Melting will begin at the co-saturation point of olivine + clinopyroxene + orthopyroxene ± plagioclase at about 1100–1150 °C at 1 atm pressure; the intersection of co-saturation curves here is not a true eutectic (note that the temperature arrows do not all point towards the red dots, which represent co-saturation of Ol, Cpx, Opx and Plag): with continued heating (cooling), there is a reaction relationship as orthopyroxene (olivine) reacts with the liquid to form olivine (orthopyroxene). As P increases (i.e., as melting occurs deeper within a given planet), the initial T at which melting begins will increase, and the initial liquid compositions that are created move away from the SiO_2 apex, becoming enriched in the olivine component—i.e., they contain less SiO_2 and more MgO.

Since minerals and liquids are not equal in composition or structure, their physical properties are also unequal (e.g., viscosity, density); in the resulting slurry, there can be a rapid gravitational separation of liquid and solid phases—differentiation. As chondrite-like materials partially melt, Fe is sufficiently abundant relative to O that accretional heating yields a dense metallic liquid that sinks downward through a silicate liquid to form a core. The silicate materials that form the overlying mantle will be rich in pyroxene and olivine, and these in turn can partially melt (Fig. 3b) to yield liquids that are Si-rich and Mg-poor relative to the solid phases, and that are likely buoyant and will rise to form a crust. As can be seen from Figure 3b, the pressure at which mantle melting occurs can have a dramatic influence on the resulting crust composition.

How large must a rocky object be to acquire sufficient energy, either from accretion or radioactive elements, so as to partially melt and become differentiated? Theoretical calculations indicate that by accretional energy alone, any object that reaches a radius (r) > 1,200 km should be hot enough to exceed the "peridotite solidus" (the T at which peridotite melting begins), which is close to ca. 1,150 °C (Takahashi 1986). But in our own Solar system, Vesta (r = 263 km) and Ceres (r = 476 km) are both differentiated and Vesta appears to have a substantial metallic core. Ceres is also reported to have a distinct core, but its structure is very unclear because the reported densities indicate an ice-rich crust, and density estimates for its core range widely (1950–5150 kg/m^3; King et al. 2018), falling short of the density for an Fe–Ni alloy (7874 kg/m^3). In any case, when using bulk compositions to model exoplanet silicate mineralogy only the mantle is relevant, except for the smallest of planetary objects. Since for this chapter we are solely concerned with the solid portions of rocky planetary objects, our focus will be on mantle mineralogy.

Fortunately, while Earth is host to > 5,000 mineral species, only a small fraction is needed to describe its mantle or crust. The reason is that under the high temperatures that pertain to planetary interiors, reaction rates are rapid and minerals are able to dissolve greater amounts of trace and minor elements. The phase rule also comes into play. The rule is: $f = c - p + 2$, where c is the number of components that are needed to describe p number of phases that exist in an equilibrated system; f describes the "degrees of freedom", or the number of intensive variables (i.e., variables that do not depend upon the size of the system, such as pressure, temperature or elemental concentration) that can be varied independently without changing the total number of phases in a system. The size of c is controlled by the "major elements", since trace elements are usually too low in abundance to dictate what phases can possibly exist. Rearranging then, $p = c - f + 2$, which means that for a given f, a small value for c also means a small value for p. The end result of all these factors is that most rocks in Earth's mantle and crust are dominated by just a handful of minerals. Table 1 is by no means an exhaustive list, but provides the materials that are likely to be dominant on extrasolar planets, unless, as we note later on, the formation conditions of those planets are very different than those in our inner Solar system.

Planetary layers

Metallic core. All planets in our Solar System—even Vesta (a mere 262 km in radius)—contain a metallic core. We infer from the mineralogy of Fe meteorites that these metallic cores consist largely of Fe-Ni alloys. Two minerals are dominant in Fe meteorites, kamacite and taenite, both having the formula (Fe,Ni), with taenite containing more Ni than kamacite. These minerals are probably not relevant to planetary interiors, where in Earth's core for example, the outer core is molten and the Fe and Ni within the inner core are likely contained within a dense, hexagonal closest packed (hcp) mineral structure (Kombayashi et al. 2019). The precise stoichiometry of such inner core phases, though, is not entirely clear (Vocaldo 2010). It has long been noted (Birch 1964) that seismic velocities are too slow within Earth's core to allow it to be composed only of Fe–Ni alloys, and that "light alloying elements", such as S, Si or O likely exist at concentrations of at least 10 wt. % (Wade and Wood 2005). Such light alloying elements might exist at even greater concentrations in the cores of the Moon, Mars or Mercury (Hauck et al. 2013; Helffrich 2017; Terasaki et al. 2019). Because metal cores can represent such a large mass fraction of a planet, the amounts of these light alloying elements, especially Si, can have a significant impact on models of mantle mineralogy.

As will be emphasized later, the first and single most important choice when using bulk compositions to model bulk silicate planet (BSP) mineralogy will be how to handle Fe: how much Fe is partitioned into the core relative to the mantle? What are the total concentrations of light alloying elements in the core? What fraction of those light alloying elements is Si?

Silicate mantle. Because only a handful of minerals occur in Earth's mantle and crust, we can use a ternary diagram to classify rocks that occur there (Le Bas and Streckeisen 1991). For exoplanets, perhaps the only relevant ternary diagram that derives from the study of terrestrial samples would be the ultramafic rock diagram, shown in Figure 4a. Some additional ternary diagrams that are not likely relevant to the inner Solar System, but might be important elsewhere, are described in Putirka and Xu (2021).

The ultramafic rock diagram is used when the sum of the minerals olivine (ol) + clinopyroxene (cpx) + orthopyroxene (opx) is >90% by volume (although mineral proportions are often plotted on a molecular or wt. % basis). Minerals whose abundances are <10% can still be important, and ternary diagrams such as in Figures 3 or 4a, are perhaps best thought of as projections, as illustrated in Figures 4b-c. Figure 4b shows the case for a rock that has a small amount of garnet (gar), and so plots within the three-dimensional quadrilateral ol + cpx + opx + gar, just above the ol + cpx + opx plane. The "projection" of the sample onto the plane ol + cpx + opx is accomplished by renormalizing the sum so that ol + cpx + opx = 1 (or 100). This renormalization has the geometric effect of passing a line that emanates from the garnet apex, through the sample in 3D space, to the ol + cpx + opx plane. When ternary diagrams involve such a projection, it is customary to write (in this case) "+ gar" to indicate the phase from which the samples are being projected, and to show that all the plotted rocks contain the indicated phase. Thus, in the experiments used to produce the 1-atm (solid lines) phase boundaries of Figure 3b, every sample was found to be saturated with plagioclase, which would co-exist with either Ol, Di or Qz, or some combination. The only place where plagioclase would be absent would be at the ternary apices, which would represent case of pure Ol or Di or Qz. (Note: mineral abbreviations, such as ol, di, qz, are from Whitney and Evans (2010); most abbreviations are lower case; it customary, though hardly required, to label the apices in ternary diagrams using upper case symbols; here, both Ol and ol, will indicate olivine). Most ultramafic rocks that plot into Figure 4a are likely to be saturated with an "aluminous phase", namely either spinel, plagioclase or garnet. It is not customary to note such phases in classification diagrams, but when calculating exoplanet mineralogies, it is probably essential since such projections can involve any number of different minerals.

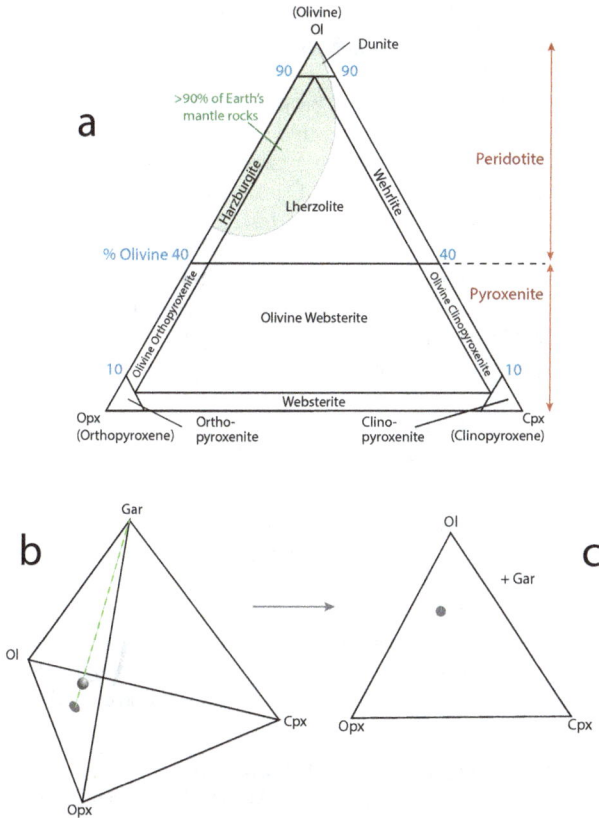

Figure 4. (a) Ultramafic rock classification diagram (Le Bas and Streckeisen 1991). **Shaded green** region shows the location of terrestrial mantle rocks, **(b)** Many ternary diagrams, such as in panel (a) or Figure 3, represent a "projection" from a 3D space of four minerals (b) into a 2D plane, that illustrates the abundances of three minerals. In (b) the rock contains a combination of olivine (Ol) + orthopyroxene (Opx) + clinopyroxene (Cpx) + garnet (Gar) and lies just above the Ol + Cpx + Opx plane. The projection, **(c)** is obtained by renormalizing so that the sum Ol + Cpx + Opx = 1 (or 100). The "+ gar" in (c) shows that the projection of the indicated sample is from the mineral garnet.

In Figure 4a, most of Earth's mantle rocks fall into the peridotite field, having >40% ol. There is some discussion that considerable amounts of pyroxenite may exist within Earth's mantle (Sobolev et al. 2005). And there is considerable geochemical evidence that pyroxenite may contribute to lavas erupted above mantle plumes (Prytulak and Elliott 2007; Putirka et al. 2018). But pyroxenite is not the dominant rock type recovered from Earth's upper mantle (McDonough and Sun 1995) and all estimates "Bulk Silicate Earth" (BSE), as well as estimates of the silicate portions of Mars, Moon, Mercury and Vesta, plot solidly within the peridotite field (Putirka et al. 2021). From here on, we will use Bulk Silicate Planet, or BSP, to refer to bulk planet compositions when their metallic cores are subtracted. For Earth-sized or larger planets the BSP will, in effect, be equivalent to the planet's mantle composition, since except for the smallest of planets, crustal mass fractions are exceedingly small.

Putirka and Rarick (2019) recommend using upper mantle *P–T* conditions as a reference or standard state, to compare planets, something close to 2 GPa and 1350°C. Adopting an upper mantle standard state has many advantages, some practical, some experimental. As to the practical advantages, the *P–T* conditions for rocky exoplanets are unknown, are hardly constant, and will vary widely: Unterborn and Panero (2019) show that core–mantle boundary conditions can range to 630 GPa and 5000 K. Most current thermodynamic models, however, barely reach pressures of about 2 GPa and experiments on a limited number of terrestrial compositions extend to only 140 GPa; and because the results of very high P experiments are controversial, even the mineralogy of Earth's lower mantle is still a bit of a mystery. On the other hand, any planet that is Moon-sized or larger will attain conditions of ca. 2 GPa and 1350 °C at some point in its cooling history. On the experimental side, experiments in the range 1–2 GPa and 1300–1400 °C are abundant, have been conducted on a very wide range of compositions, and equilibrate quickly (relative to lower T, lower P experiments), and are mostly conducted in piston-cylinder apparatus that have low calibration uncertainties (relative to diamond anvil cell experiments). Thermodynamic models, such as MELTS (Ghiroso and Sack 1995; Asimow and Ghiroso 1998), also readily apply to upper mantle conditions and can be easily updated to handle more exotic compositions as new experiments become available. Only smaller bodies, such as Vesta (whose interior does not exceed 0.4 GPa; Righter and Drake 1996), might require a lower set of P–T conditions, perhaps 1 atm and 1000 °C, but in a quest for habitable or otherwise Earth-like planets, Vesta-sized objects are of lesser interest.

Crust: oceanic and continental. Earth is the only planet to have all of the following: plate tectonics, liquid water at its surface, abundant continental (granitic) crust, and a N_2- and O_2-rich atmosphere. This combination might not be a coincidence (Campbell and Taylor 1983) and the origin and existence of continental crust is of exceeding importance for understanding planetary evolution. There are also some claims that continent-like crust has been detected outside our Solar System (Zuckerman et al. 2011; Hollands et al. 2018, 2021); these identifications are often based on the abundance of a single element (e.g., Ca, K, Li) although where more complete compositions are available (Mg, Fe and Si), such a detection is not so clear (Putirka and Xu 2021).

While Earth's mantle is dominated by olivine and pyroxene, plagioclase is dominant in the overlying crust. Accompanying plagioclase in the lower crust are olivine and pyroxenes (in the very deep crust, garnet, instead of plagioclase, is the Al-rich phase). In the upper crust, plagioclase is joined by quartz, hornblende (a variety of amphibole), alkali feldspar and biotite (a variety of mica). Micas and amphiboles contain significant amounts of water, in the form of $(OH)^-$ and so are "hydrous"; minerals that do not contain $(OH)^-$ as a part of their stoichiometry are referred to as "nominally anhydrous" (as they still can contain trace amounts of (OH^-)). During subduction (when a crust-bearing tectonic plate sinks into the mantle), amphiboles, and the mineral serpentine (a mineral that forms when water reacts with olivine), may degrade and recrystallize and so return their water back into the mantle. And when water is added to the mantle, its melting temperature is greatly lowered (Fig. 5), which means that for the same temperature, more melt, and more crust are created. Amphiboles and serpentine are thus an important part of Earth's global water cycle. But as shown by Bell et al. (1995), nominally anhydrous mantle minerals, such as garnet and pyroxene, can also contain enough $(OH)^-$ to hold an ocean's-worth or more of water in Earth's mantle (Bell et al. 1995). Guimond et al. (2023) suggest that such nominally anhydrous minerals in the transition zone (Figs. 1–2) may well control the global water cycle.

As noted, a planet's crust is created by partial melting of its mantle. For mantles that consist of peridotite or pyroxenite (e.g., any rocks that fall into the classification of Fig. 4a), the liquid (and crust) compositions will be basalt (Fig. 6a). Basalts are volcanic rocks that are not fully crystallized, and so like other types of volcanic rocks, they are classified based on their chemical composition, rather than their mineralogy (Fig. 6a). Basalts are an important crust type in that they cover 2/3 of Earth's surface and nearly the entire surfaces of Mercury, Moon, Mars, and Vesta, and most likely Venus as well. Most of Earth's basalt lies beneath the ocean basins, but Earth is unique in having large amounts of "continental crust". The average composition of the continental crust is chemically equivalent to "andesite" (Rudnick and Gao 2014), but perhaps the standout feature of Earth is the large amount of "granitic" rocks ("rhyolite" when it

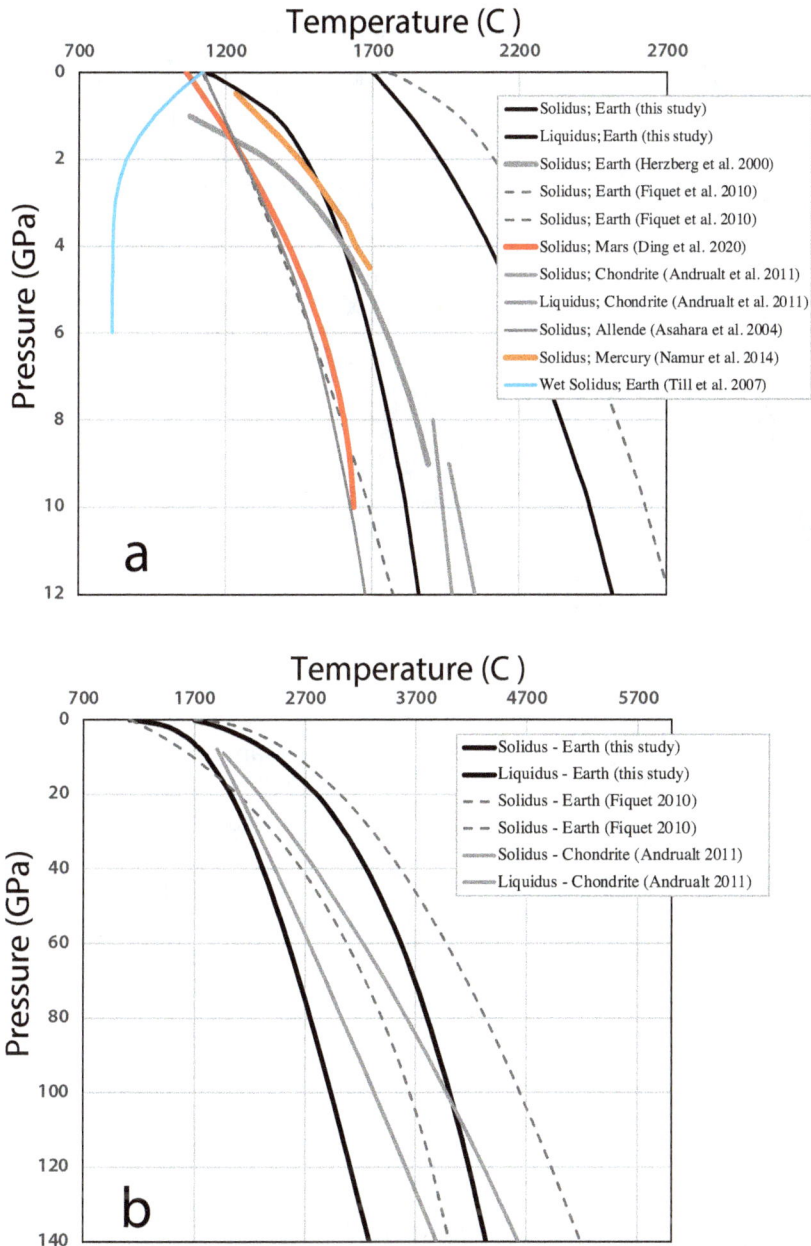

Figure 5. Solidus (*T* at which melting beings) and liquidus (*T* at which melting is completed) curves for Earth, Mars, Mercury and chondrite meteorites.

erupts from a volcano; Fig. 6a) that are common within Earth's uppermost crust; these rocks are especially low in MgO and high in SiO_2 (and high in H_2O, Na_2O and K_2O and other elements that are "incompatible" within mantle minerals; Fig. 6a–b). SiO_2, K_2O and water are sufficiently high so as to respectively crystallize large amounts of quartz, hornblende and alkali feldspar.

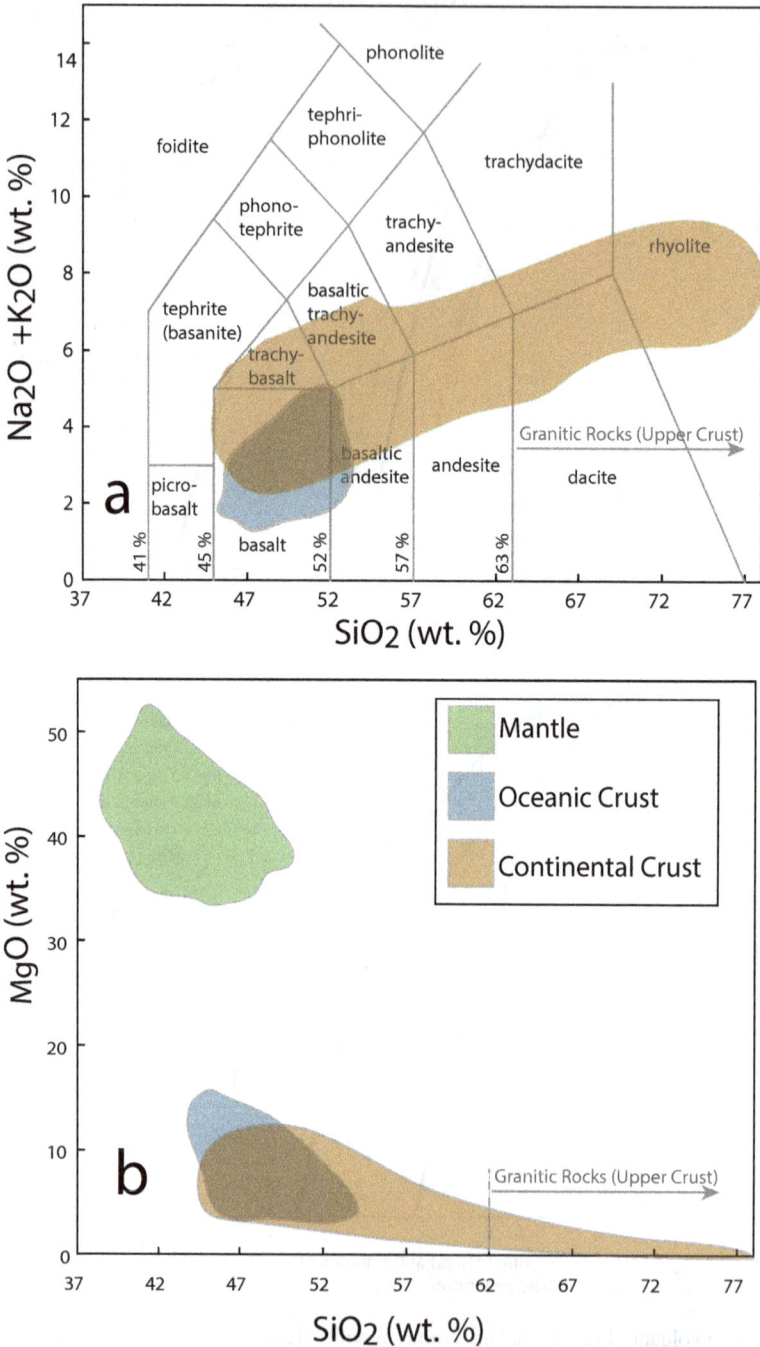

Figure 6. (a) Chemical classification diagram for volcanic rocks. "Basalts" are the dominant crust type of the inner planets and differentiated meteorites. Andesites are common above subduction zones and represent the average composition of Earth's continental crust. **(b)** Comparison of the compositions of mantle and crust rock types.

Making crust/mantle melting. Planetary crusts are produced by partial melting of the mantle. Unless melting occurs at very high pressures (e.g., Agee and Walker 1988), such partial melts are quite likely to be less dense than their associated mineral residue, and so will rise towards the surface—where they will remain, unless subduction brings them back into the mantle. The amount of crust that is produced will be proportional to the fraction of melt (F_m) that is produced during partial melting. At a given pressure, F_m is dependent upon ambient mantle T relative to the T of solidus (the T at which the mantle begins to melt) and the liquidus (the T at which the mantle is completely melted); the solidus and liquidus curves for Earth, Mercury and Mars are shown in Figure 5 (only for Earth have liquidus and solidus curves been determined at high pressures). Herzberg et al. (2000) and Hirschmann (2000) have calibrated rather precise positions for a solidus of a pyrolite mantle; pyrolite is a primitive mantle composition that is "fertile", meaning it has lots of pyroxene, especially cpx (Fig. 4), and has not had large amounts of melt extracted. Fertile mantle stands in contrast to "depleted mantle" (ol-rich and cpx-poor; Fig. 4), which have had large amounts of melt extracted. Fertile mantle can melt to a greater extent at a given T, and so yield large amounts of crust when such melts reach the surface. In general, partial melting will extract TiO_2, Al_2O_2, CaO, K_2O, Na_2O and H_2O out of the mantle, leaving a mantle residue that is depleted in these components (with slightly less SiO_2 as well) and enriched in MgO. Much experimental work has been conducted to delimit a peridotite or pyrolite solidus for the uppermost terrestrial mantle. The models of Hirschmann (2000) and Herzberg et al. (2000) are highly precise and accurate but are not calibrated to predict melting conditions in the transition zone or lower mantle. Several somewhat recent studies (Fiquet et al. 2010; Nomura et al. 2014; Kim et al. 2020) have examined partial melting to lower mantle conditions and two (Nomura et al. 2014; Kim et al. 2020) are in remarkable agreement with one another, and also consistent with the lower-P models of the solidus. Due to that agreement, a new curve is fit so as to describe the Herzberg et al. (2000) solidus to about 5 GPa, but then matches the Nomura et a. (2014) and Kim et al. (2020) solidi to 140 GPa. Following Andrault et al. (2011), a modified Simon and Gatzel equation is used, but with an added linear P-term, which not only improves the fit at high pressures, but also helps to avoid spuriously low T estimates when P approaches 1 atm:

$$T(°C)_{solidus} = 1120\left(\frac{P(GPa)}{0.25}+1\right)^{\frac{1}{8.3}} + 6.3\left[P(GPa)\right] \tag{2a}$$

This solidus is illustrated in Figure 5b, and has an error of at least ±30 °C at $P<5$ GPa and an unknown error at $P>20$ GPa, due to the very uncertain uncertainties of diamond anvil experiments. To calculate F_m we also must have an estimate of the liquidus, which was unfortunately not determined by either Nomura et al. (2014) nor Kim et al. (2020). A liquidus curve was determined by Fiquet et al. (2010), but their solidus curve falls to a considerably higher T than the solidi temperatures of either Kim et al. (2020) and Nomura et al. (2014). To calibrate a liquidus curve, we'll assume that while the Fiquet et al. temperatures are too high, that the width of their melting interval is accurate, which allows the liquidus to be constructed when the Fiquet melting interval is combined with Equation (2a). It is also assumed that the 1 atm liquidus T is 1700 °C, as in Zhang and Herzberg (1994). The resulting curve is:

$$T(°C)_{liquidus} = 1700\left(\frac{P(GPa)}{2.8}+1\right)^{\frac{1}{4.2}} \tag{2b}$$

Melting temperatures are greatly lowered in the presence of water; the following equation reproduces the solidus of Till et al. (2007, 2010), but fixes the solidus T at $P>3$ GPa to 815 °C (Grove and Till 2019):

$$T(^\circ C)_{\text{wet solidus}} = 1120 - 305\left[\text{erf}\left(0.5P(\text{GPa})\right)\right] \qquad (2c)$$

Equation (2c) (Fig. 5a) is necessarily incomplete in that the temperature of the wet solidus also increases with increasing MgO content (Grove and Till 2019; Wang et al. 2020); new experiments at very high pressures are needed to yield improved versions of Equations (2a–c), with compositional parameters being added. Variations of stellar compositions (Putirka and Rarick 2019) appear to indicate some considerable variety in exoplanet bulk compositions. Dry solidus curves have been determined for Mars (Ding et al. 2020), Mercury (Namur et al. 2014) (Fig. 5a), and Allende (a carbonaceous chondrite meteorite; Agee et al. 1995; Asahara et al. 2004). But none of the experimental studies on the dry mantle solidus extend to transition zone depths (15–23 GPa), let alone to the lower mantle (>23 GPa). Andrault et al. (2011) determine the solidus and liquidus of a chondrite meteorite (Fig. 5), but their solidus is oddly higher than that of both bulk silicate Earth (BSE), and of Allende (Fig. 5). So there are no obvious compositional corrections. The lower solidus temperatures obtained by Asahara et al. (2004) and Agee et al. (1995) for Allende are at least intuitively correct in that carbonaceous chondrites are more fertile, having a much lower Mg# ($MgO/(MgO+FeO)$) compared to BSE. But despite the challenges of DAC experiments, perhaps it would be a mistake to discount the Andrault et al. (2011) experiments entirely, since their chondrite composition is lower in CaO and Al_2O_3 compared to BSE. In any case, reproducible experiments are a critical need.

How much crust can a planet have? Besides an O_2-rich atmosphere, the identification of continental crust outside our Solar system might provide the single most compelling case for an Earth-like planet. But the potential of such an identification faces a key challenge: the mass fraction of Earth's entire crust is ca. 0.4 wt. % of its total mass, and the continental crust comprises yet a smaller fraction. Some planets have much thicker crusts relative to their planet's radius, but as a mass fraction, crusts are still a very small part of any planet (Table 2). And where the crusts are thickest (Moon, Vesta) they are basaltic, not granitic. In sum, abundant granitic crusts might only evolve on large, Earth-sized planets with plate tectonics, where water is stable on the surface. But in such cases crust mass fractions may be too small to detect. And as Venus plainly shows, Earth-sized planets will not necessarily follow Earth-like evolutionary paths.

For the sake of comparison, Table 2 shows core, mantle and crust mass fractions for the inner planets, Moon and Vesta. They are calculated using reported estimates of layer thicknesses and bulk densities of crusts, mantles and cores (Table 1, notes). These inputs yield bulk densities, (layers weighted by volume fractions) that approach very closely with those obtained from observed volumes and gravity (combining $F = GmM/r^2$ and $F = ma$, where r is the planet radius, M is the mass of the planet, a is the acceleration due to gravity at a planet's surface and m is an arbitrary mass at the surface that cancels with substitution). But this consistency check is no guarantee of accuracy as the densities and thicknesses of the various planetary components are not well known outside of Earth or Moon, where we have seismometers in place to measure seismic velocities (which can be translated readily to density). Despite undoubted errors, there is a clear trend in our Solar System to proportionally greater crust fractions as planet size decreases. A caveat is that the crust of Mercury might have been ablated by late meteorite bombardment (Helffrich et al. 2019), and perhaps Vesta has also been significantly ablated as the so-called HED meteorites appear to originate from Vesta (Mittlefehldt 2015). Earth's crust is a mere 0.4 wt. % of its total mass with the balance being a metallic, Fe–Ni alloy core that comprises 33% of the planet's mass, and remainder being Earth's mantle, which lies between the core and crust (i.e., between 35 km and 2,890 km). Earth's mantle is made of silicate minerals that thus comprise about 67% of Earth's mass (the sum of crust + oceans + atmosphere is insignificant in comparison). At 6.4×10^{23} kg, Mars is just 10% of Earth's mass (5.97×10^{24} kg) and has a crust mass fraction of 3%. Moon has 1% Earth's mass and its crust mass fraction is 9%. At 2.6×10^{20} kg, Vesta (radius = 262 km) is less than 10^{-4} Earth masses, yet it is still large enough to be differentiated into a core, mantle and crust; its crust fraction is very uncertain, but appears to be at least 20–30% of its total mass (as calculated from densities of Raymond et al. 2013). Clenet et al. (2014) argue for an even thicker crust (80 km, which implies a 40–50% mass fraction), but such thick crust seems inconsistent with inferred mantle and crust compositions (if Vesta's mantle is peridotite and its crust is basalt, then total melt fractions to produce the crust would be < 20%; at > 30% melting, the crust would be a very high Mg volcanic rock known as komatiite, rather than basalt). In any case, as we use bulk compositions to examine/model planetary mineralogy, it matters immensely whether the planets are small or large and whether and to what extent Fe has been segregated into a metallic core.

Table 2. Some planetary properties of relevance to exoplanet studies.

Rocky objects[1]	Earth	Venus	Mercury	Mars	Moon	Vesta	Phobos
Surface g (m/s²)	9.806	8.872	3.7	3.728	1.625	0.25	0.0057
Mass (kg)	5.964×10^{24}	4.868×10^{24}	3.30×10^{23}	6.42×10^{23}	7.346×10^{22}	2.581×10^{20}	1.05×10^{16}
Mass rel. to Earth	1	0.816	0.06	0.107	1.23×10^{-2}	4.3×10^{-5}	1.8×10^{-9}
Distance (AU)	1	0.7233	3871	1.5236	1	2.36	1.5326
Radius (km)	6371	6051.4	2437.6	3389.9	1737.1	262.5	11.1
Core mass %	32.5	42	72	26	1.9	16	
Mantle mass %	67	57	26	71	92	52	Undifferentiated
Crust mass %	0.5	0.8	1.9	3.0	5.4	32	

Notes: [1] Data for rocky bodies are from Lodders and Fegley (1998) and Szurgot (2015), with crust and mantle mass fractions calculated using: Mercury (Hauck et al. 2013; Sori 2018; Knibbe et al. 2021); Venus (Aitta 2012; O'Neil 2021); Earth (Lodders and Fegley 1998; Peterson and DePaolo 2007); Moon (Viswanathan et al. 2019); Mars (Knapmeyer-Endrun et al. 2021); Vesta (Raymond et al. 2013); [2] Radius calculated for equivalent volume as a sphere.

GEOCHEMISTRY OF EARTH AND THE OTHER INNER PLANETS

If we set aside volatile elements (H, He, C, O, S), which are depleted in the inner planets, the Sun and inner planets are made of what Geologists refer to as the "major elements" (or more commonly expressed as the "major oxides", since O is the dominant anion) (Table 3). These elements are, in order of decreasing abundance by weight in our Solar system: Fe, Si, Mg, Ni, Ca, Al, Na, Cr, Mn, P, K, and Ti. These are also the dominant elements in nearby stars. Just three elements, Mg + Fe + Si, account for >85% of the total in 99% of stars in the Hypatia catalog, and adding Ni, Ca and Al, we account for 96% of all cations by weight in >99% of Hypatia stars. Table 4 shows the bulk compositions of the Sun, the inner planets, Moon and Vesta, and a CI chondrite, when S and O are included, as well as when these compositions are renormalized to a S-free, O-free basis. Bulk inner planet compositions are obtained via mass balance:

$$C_i^{BP} = X^{BSP}C_i^{BSP} + X^{MC}C_i^{MC} \qquad (3)$$

Table 3a. Bulk compositions of the planets, a CI chondrite (Orgeuil) and the Sun (wt. %)[1].

	Solar	Mercury	Venus	Earth	Moon	Mars	Vesta	Orgueil
Si	7.81	7.63	16.27	15.32	20.88	16.25	19.27	15.68
Ti	0.03	0.04	0.08	0.08	0.12	0.06	0.07	0.08
Al	0.52	0.50	1.34	1.57	2.06	1.20	1.31	1.24
Cr	0.20	0.01	0.37	0.18	0.25	0.41	0.43	0.35
Fe	12.63	52.84	32.67	32.78	11.27	27.50	18.06	27.00
Mn	0.13	0.00	0.00	0.07	0.12	0.25	0.35	0.27
Mg	6.17	6.43	15.25	15.41	20.09	14.39	16.64	14.00
Ni	0.73	2.92	1.70	1.76	0.14	1.89	1.51	1.67
Ca	0.67	0.46	1.57	1.67	2.22	1.34	1.45	1.33
Na	0.32	0.28	0.14	0.18	0.04	0.33	0.04	0.74
K	0.04	0.03	0.01	0.02	0.00	0.03	0.01	0.16
P	0.06	0.00	0.01	0.01	0.00	0.05	0.10	0.00
S	3.31	15.17	0.00	0.55	0.11	3.06	0.00	
O	67.37	13.69	30.60	30.40	42.70	33.23	40.74	37.49
Total	100	100	100	100	100	100	100	100

Table 3b. Bulk planets (wt. %) on a sulfur- and oxygen-free basis.

	Solar	Mercury	Venus	Earth	Moon	Mars	Vesta	Orgueil
Si	26.66	10.72	23.44	22.19	36.52	25.51	32.53	25.09
Ti	0.10	0.05	0.11	0.12	0.20	0.10	0.12	0.12
Al	1.77	0.70	1.93	2.28	3.60	1.88	2.21	1.99
Cr	0.70	0.01	0.53	0.26	0.44	0.64	0.73	0.55
Fe	43.09	74.27	47.08	47.47	19.71	43.16	30.48	43.19
Mn	0.44	0.00	0.00	0.10	0.22	0.40	0.59	0.43
Mg	21.04	9.04	21.98	22.32	35.13	22.59	28.08	22.40
Ni	2.49	4.11	2.44	2.55	0.24	2.97	2.55	2.67
Ca	2.29	0.65	2.26	2.42	3.88	2.11	2.45	2.13
Na	1.09	0.40	0.20	0.26	0.06	0.51	0.07	1.18
K	0.13	0.04	0.01	0.02	0.01	0.04	0.01	0.26
P	0.19	0.00	0.01	0.01	0.00	0.09	0.17	0.00
Total	100	100	100	100	100	100	100	100

Note: [1] All compositions are on a H- and He-free basis. Solar bulk composition is from Magg et al. (2022); planet bulk compositions are from McDonough and Yoshizaki (2021), renormalized so as to add minor elements from the the sources shown in Table 4. Bulk compositions of Mercury and especially Venus are highly uncertain.

where the superscript BP is the "bulk planet," C_i is the concentration of i in the superscripted phase, and X is the mass fraction of either a metallic core (MC) or the bulk silicate planet (BSP). Core mass fractions are from Szurgot (2015) and to these are added the median of published estimates of the silicate portions of the planets (see note to Table 4).

In astronomy, it's typical to list elements in order of atomic number. In geology, we are less rigorous, listing the rock-forming elements roughly (often, very roughly), in order of decreasing field strength (charge divided by effective cation radius), and almost always as oxides, tacking on minor and trace elements at the end. As this chapter is designed in part to enroll geologists into the study of exoplanets, we'll follow the geological tradition, listing the oxides as: SiO_2, TiO_2, Al_2O_3, Cr_2O_3, NiO, FeO, MgO, MnO, CaO, Na_2O, K_2O, P_2O_5, and the elements in a similar order.

Later sections will introduce some approaches to mass balance to transform chemical compositions into mineral abundances, the latter of which are controlled by the major oxides. As noted, Earth's rocky mantle is dominated by the rock type peridotite (Fig. 4a); the proportion of the various minerals in this rock type are controlled by SiO_2, Al_2O_3, FeO, MgO and CaO. All other elements occur in too low a concentration (see Putirka and Rarick 2019) to greatly affect exoplanet mantle mineralogy, either because their abundances are low (e.g., Ni) or because the elements are incompatible within olivine, pyroxenes and garnet (e.g., Ti, P) and so are concentrated in the crust. Although peridotites are defined by their mineralogy, they are also distinct chemically, having high MgO (ca. 36 ± 3 wt. %) and low SiO_2 (42 ± 2 wt. %) (Fig. 6b). Basalts, which are produced by partial melting of peridotite, contain higher SiO_2 (45–52 wt. %, by definition; Le Bas et al. 1986) and much lower MgO (usually < 8 wt. %) (Fig. 6a). Basalts also contain higher Al_2O_3, CaO and Na_2O than peridotite—sufficient for basaltic liquids to precipitate plagioclase feldspar, in addition to olivine and pyroxenes.

Table 4. Bulk silicate planet compositions of the inner planets and Vesta.

	Mercury[1]	Venus[2]	Earth[3]	Moon[4]	Mars[5]	Vesta[6]
SiO_2	50.99	44.93	44.86	45.53	44.41	41.24
TiO_2	0.19	0.20	0.20	0.20	0.13	0.12
Al_2O_3	2.95	3.72	4.41	4.06	2.90	2.48
Cr_2O_3	0.04	0.79	0.38	0.44	0.76	0.63
FeOt	0.00	9.62	8.03	12.57	17.69	23.24
MnO	0.00	0.00	0.13	0.15	0.42	0.46
MgO	33.32	37.20	37.87	33.80	30.48	27.59
NiO	0.00	0.00	0.25	0.00	0.05	1.92
CaO	2.02	3.23	3.46	3.19	2.40	2.03
Na_2O	1.20	0.27	0.36	0.05	0.56	0.06
K_2O	0.12	0.02	0.03	0.01	0.04	0.01
P_2O_5	0.00	0.02	0.02	0.00	0.16	0.23
S	9.16	0.00	0.00	0.00	0.00	0.00
Total	100.00	100.00	100.00	100.00	100.00	100.00

Notes. Median compositions, from the following sources: [1] Nittler et al. (2018); [2] Shah et al. (2022), Lodders and Fegley (1998); [3] McDonough and Sun (1995), Workman and Hart (2005), Lyubetskaya and Korenaga (2007), Salters and Stracke (2004), Khan et al. (2008), Palme and O'Neil (2003); [4] Longhi (2006), Jones and Delano (1989), Kuskov et al. (2019), Khan et al. (2013), Togashi et al. (2017); [5] Yoshizaki and McDonough (2020), Khan and Connolly (2008), Wanke and Dreibus (1994), Sanloup et al. (1999), Taylor (2013). [6] Steenstra et al. (2016).

Earth's continental crust (Fig. 6b) has higher SiO_2 (on average 60.6 wt. %; Rudnick and Gao 2014) and much lower MgO (4.7 wt. %) placing it in the category of "andesite" (52–63 wt. % by definition; Le Bas et al. 1986) (Fig. 6a). So-called "granitic crust" has even higher SiO_2 (often >70 wt. %) and very low MgO (< 1 wt. %) and is chemically equivalent to "dacite" and "rhyolite" in Figure 6a. These granitic rock types appear to only be abundant on Earth. However, studies of Mercury (Vander Kaaden and McCubbin 2016; Vander Kaaden et al. 2017) and an ancient meteorite (Barrat et al. 2021) provide some evidence of andesite crust (albeit with much higher MgO) forming early and elsewhere.

Converting elements to oxides, and oxides to elements

Because O is far and away the dominant anion in the inner planets (Table 4), and because in the mantle, nearly all cations are bound to anions, it is typical that BSPs for the inner planets are reported as wt. % oxides, as are the rock compositions on which most inner planet BSP estimates are based. Table 4 shows BSP compositions for the inner planets and Vesta, taken as median values from a range of published estimates. Estimates of the Sun, in contrast, are most often reported as atomic proportions of individual elements, while star compositions are reported in the same way, but often on a log scale relative to the Sun (see Hinkel et al. 2024, this volume). Given that star compositions provide a first-order (and sometimes our only) estimate of exoplanet compositions (e.g., Hinkel and Unterborn 2018), it is clearly critical to convert back-and-forth between oxides and elemental weight %; for convenience, Table 5 provides multiplicative conversion factors for such conversions.

Table 5. Conversion factors for wt. % element-to-oxide and oxide-to-element.

	SiO_2	TiO_2	Al_2O_3	Cr_2O_3	FeOt	MnO	MgO	NiO	CaO	Na_2O	K_2O	P_2O_5
(a) Oxide weights (O)	60.08	79.88	101.96	152.00	71.85	70.94	40.30	74.69	56.0774	61.98	94.20	141.94
(b) Cation atomic wt. (C)	28.09	47.87	26.98	52.00	55.85	54.94	24.31	58.69	40.078	22.99	39.10	30.97
(c) Conv. elem to oxide (O/C)	2.14	1.67	1.89	1.46	1.29	1.29	1.66	1.27	1.40	1.35	1.20	2.29
(d) Conv. Ox to elem (C/O)	0.47	0.60	0.53	0.68	0.78	0.77	0.60	0.79	0.71	0.74	0.83	0.44

There has been speculation that some rocky material that once orbited white dwarf stars might have contained large fractions of granitic-like crust. Table 6a provides the unique compositions that comprise Earth's crust to aid such discussions. Rudnick and Gao (2014) provide the most commonly accepted values for Earth's bulk continental crust (CC in Table 6a) while Gale et al. (2013) provide the same for the Mid-Ocean Ridge Basalt (MORB) crust that covers the floor of the ocean basins. As can be seen from Table 6 and Figure 6, the continental crust is distinctly Si-rich compared to oceanic crust and would be classified as andesite, rather than basalt. But the composition of Earth's continental crust ranges much more widely than MORB (which varies quite little) and contains three compositionally distinct layers, an upper, middle and lower crust, each also characterized by Rudnick and Gao (2014) and whose upper crust composition (UC) is shown in Table 6a. For many non-geologists, Earth's crust is perhaps commonly associated with compositions similar to the SiO_2-rich upper crust, or the granitic rocks that dominate many high mountain ranges, such as the Sierra Nevada, which are even more enriched in SiO_2 (Table 6a, columns b, c). As published compositions are often presented as oxides (Table 6a, columns a–e), one can multiply these by row (d) in Table 5 to obtain elemental concentrations (Table 6a, f–j). The conversion factors of Table 5 account for the weight % of the cations of the indicated compositions; for systems where O is the only or overwhelmingly dominant anion (and in these systems it is) the O weight % can be obtained as O (wt. %) = 100 – sum of cations (wt. %). Be aware that these are minimum estimates of

Table 6a. Types of crust on Earth (CC = bulk continental crust; UCC = upper continental crust; MORB = Mid-ocean ridge basalt = oceanic crust; Plume = hot spot at Samoa).

	(a) CC	(b) UCC	(c) Granite	(d) MORB	(e) Plume		(f) CC	(g) UCC	(h) Granite	(i) MORB	(j) Plume
SiO_2	60.60	66.60	72.04	50.47	45.30	Si	28.33	31.13	33.67	23.59	21.17
TiO_2	0.70	0.64	0.42	1.68	2.50	Ti	0.42	0.38	0.25	1.01	1.50
Al_2O_3	15.90	15.40	14.32	14.70	9.20	Al	8.42	8.15	7.58	7.78	4.87
FeO_t	6.70	5.04	2.88	10.43	12.00	Fe	5.21	3.92	2.24	8.11	9.33
MnO	0.10	0.10	0.05	0.18	0.20	Mn	0.08	0.08	0.04	0.14	0.15
MgO	4.70	2.48	0.66	7.58	20.50	Mg	2.83	1.50	0.40	4.57	12.36
CaO	6.40	3.59	1.88	11.39	8.00	Ca	4.57	2.57	1.34	8.14	5.72
Na_2O	3.10	3.27	3.56	2.79	1.80	Na	2.30	2.43	2.64	2.07	1.34
K_2O	1.80	2.80	4.04	0.16	0.70	K	1.49	2.32	3.35	0.13	0.58
P_2O_5	0.10	0.15	0.18	0.18		P	0.04	0.07	0.08	0.08	0.00
						O	46.31	47.46	48.40	44.38	42.98
Total	100.10	100.07	100.03	99.57	100.20	Total	100.00	100.00	100.00	100.00	100.00

Table 6b. Convert BSE (pyrolite from McDonough and Sun 1995) from oxide to elemental weight %.

	(a) Earth's mantle [1]	(b) Earth's mantle [1] renormalized		(c) Earth's mantle [1]	(d) Earth's core[1]	(e) Earth bulk comp[1]	(f) Earth bulk comp[1]
SiO_2	45	44.90	Si	20.99	3.6	15.34	22.16
TiO_2	0.201	0.20	Ti	0.12	0	0.08	0.12
Al_2O_3	4.45	4.44	Al	2.35	0	1.59	2.29
Cr_2O_3	0.384	0.38	Cr	0.26	0	0.18	0.26
FeO_t	8.05	8.03	Fe	6.24	87.9	32.78	47.36
MnO	0.135	0.13	Mn	0.10	0	0.07	0.10
MgO	37.8	37.71	Mg	22.74	0	15.35	22.18
NiO	0.25	0.25	Ni	0.20	5.5	1.92	2.77
CaO	3.55	3.54	Ca	2.53	0	1.71	2.47
Na_2O	0.36	0.36	Na	0.27	0	0.18	0.26
K_2O	0.029	0.03	K	0.02	0	0.02	0.02
P_2O_5	0.021	0.02	P	0.01	0	0.01	0.01
Ref. [1]. Data from McDonough and Sun (1995)			O	44.17	3	30.79	
Total	100.23	100		100	100	100	

Notes. [1.] For a core composition we use the composition reported by Tronnes et al. (2019) but since they do not report S in Earth's core, we adobe the value of 1.7 wt. % S in the core from Wood et al. (2014), reducing Ni to 5 wt. % and Si to 1.8 wt. %. We assume a core mass fraction (M_c) of 0.325. And as in Table 6a, O (wt. %) is obtained by the difference between 100 and the sum of the cation wt. %, in this case, 100 − 55.83 = 44.17 wt. % O. To obtain the composition of bulk Earth we add to the BSP a core composition modified from Tronnes et al. (2019) in that we allow for 1.7 wt. % S in the core (Wood et al. 2014), as the Tronnes et al. (2019) core is curiously S-free. Assuming that mass fraction of Earth's core, M_c, is 0.325 then the composition of Bulk Earth, or BE is: BE = $M_c(C_{core})$ + (1 − $M_c)(C_{BSE})$ = 0.325(C_{core}) + 0.675(C_{BSE}), the results of which are provided in column (e) of Table 6b. Column (f) of Table 6b gives Earth's bulk composition renormalized to an O-free basis. Two notes: First, if an O-free composition is the end goal (and it is often useful to compare cations on a O-free basis) it may be tempting to renormalize BSE on an O-free basis before conducting a weighted addition of the core and mantle compositions. That procedure is fine provided that one adjusts the mass proportion of the mantle downwards to account for the loss of mass by removing O, which occurs upon such a renormalization; the result is not trivial as O is the single most abundant element by mass in Earth's mantle (Table 6, column C). Second, it is not a simple matter to use row (c) of Table 5 and either columns (e) or (f) of Table 6b to express bulk Earth as a set of oxides. For the case of Table 6a, multiplication of columns f–j by row (c) of Table 5 simply leads one back to the values in Table 6a columns (a–e). But if one multiplies the cations of column (e) in Table 6b by row (c) of Table 5, the sum will be >100%; the reason for this is because the calculation assumes that all of the elements in (e) (or f) are fully oxidized to the extent indicated by their formulas (i.e., Fe as FeO), and of course, in bulk Earth, most of the Fe occurs as Fe⁰. Finally, and as noted for the contents of Table 6a, the estimates of O wt % in Table 6b are also minimum values since within Earth's mantle at least some Fe exists as Fe_2O_3 as well as FeO.

O since Fe is known to also occur in the Fe^{3+} state, as Fe_2O_3, as well as FeO. The calculation can be made more precise by substituting estimates of total Fe as FeO (FeOt) with estimates of FeO and Fe_2O_3, adjusting the contents of Tables 5 and 6a accordingly.

Mantle rocks are Si-poor and highly Mg-enriched compared to continental crust, as can be ascertained from Figure 6, and Table 6b (column a), which provides the most-cited estimate of Earth's mantle (effectively, bulk silicate Earth or BSE, also referred to as "pyrolite") from McDonough and Sun (1995). The pyrolite, or BSE composition in Table 6b column (a), is re-normalized to 100% in column (b), and then multiplied by row (d) of Table 5 to obtain elemental weight % of cations (Table 6b, column c).

WORKING WITH STELLAR COMPOSITIONS

Two methods of estimating mantle mineralogy will be presented. The first, using matrices, has the advantage of being very easy to code into a spreadsheet or other software and very simple to edit, so as to rapidly test a range of mineralogical models. The second method is somewhat more complex, but has the advantage of yielding only positive values for mineral abundance estimates, even for rather exotic compositions. Hinkel et al. (2022) provide a thorough review of computations and notational conventions. Some notation common in the astronomical literature will be briefly reviewed, but the reader is highly recommended to read Hinkel et al. (2022) for further details as the focus here will be on converting star compositions to wt. % elements and oxides, so as to compare stars to rocks and minerals.

Star compositions are often reported as atomic abundances of a given element (A(El)) (Table 7) on a log scale, referenced to H (or sometimes to He for some white dwarfs): $A(El) = \log(Z/Y) = \log[n(Z)/n(Y)]$ where $n(Z)$ is the number of atoms of the element of interest, Z, and in the denominator, $n(Y)$ is the number of atoms of either H or He, depending upon which dominates a stellar atmosphere. Star compositions are also often reported as numbers of atoms relative to the Sun, on a log basis. Astronomers use "dex" to denote the $\log_{10}$ scale; so the amount of Fe in a star might be reported as Fe/H, where the dex value of -3.2. This means that the Fe/H ratio is $10^{-3.2} = 6.31 \times 10^{-4}$.

1. *For the case of H- or He-normalized log atomic abundances that are NOT referenced to the Solar composition*: the conversion to elemental wt. % is straightforward and is shown in Table 7: since Y is the same for all reported elements for any given star, cation fractions can be obtained by renormalization irrespective of Y, so that $10^{[Mg/Y]} + 10^{[Si/Y]} + 10^{[Ca/Y]} + 10^{[Fe/Y]} = 100\% = X_{Mg} + X_{Si} + X_{Ca} + X_{Fe}$, where X_i are the atomic % values of element i. With these atomic proportions one can multiply X_i by atomic weights and renormalize to 100 to obtain elemental weight %.

2. *For the case of H- or He-normalized log atomic abundances that ARE referenced to the Solar composition*: Stars in the immensely useful Hypatia Catalog (Hinkel et al. 2014) are further normalized to the Lodders et al. (2009) Solar composition, and reported as a difference in log concentrations relative to the Sun (as [Z/H]). It is thus essential to multiply the Hypatia compositions by the Lodders et al. (2009) Solar composition (Table 8a).

It is common in the recent literature to compare star and planet compositions using elemental ratios; while useful, comparisons on the basis of % abundances, via Tables 5–8 provide advantages. Firstly, two different objects might have the same Mg/Si ratio, for example, but only by comparing % values can one test whether they have the same Mg and Si concentrations. If the concentrations are not the same, imputed common properties or evolutionary histories may be entirely mistaken. In addition, when compositions are expressed on a % basis, mass balance calculations are facilitated, as are comparisons of exoplanetary materials to meteorites

Table 7. Sample calculation converting a stellar abundance (Magg et al. 2022) to element wt. %.

	Atomic No.	Atomic wt. (AW)	A(El) Magg (2022)	No. of atoms $n(Z) = 10A(El)$	$n(Z) \times AW$	Wt. % element
C	6	12.0107	8.56	363078054.8	4360821592	23.7722888
O	8	15.9994	8.77	588843655.4	9421145179	51.3577955
Na	11	22.98977	6.29	1949844.6	44826478.88	0.24436404
Mg	12	24.305	7.55	35481338.92	862373942.5	4.70108716
Al	13	26.981538	6.43	2691534.804	72621748.59	0.3958853
Si	14	28.0855	7.59	38904514.5	1092652742	5.95641347
P	15	30.97361	5.41	257039.5783	7961443.652	0.04340048
S	16	32.065	7.16	14454397.71	463480262.5	2.52658505
K	19	39.0983	5.14	138038.4265	5397067.809	0.02942121
Ca	20	40.078	6.37	2344228.815	93952002.46	0.51216361
Ti	22	47.867	4.94	87096.359	4169041.416	0.02272683
Cr	24	51.9961	5.74	549540.8739	28573982.23	0.15576628
Mn	25	54.938049	5.52	331131.1215	18191697.78	0.09916899
Fe	26	55.845	7.5	31622776.6	1765973959	9.62691135
Ni	28	58.6934	6.24	1737800.829	101997439.2	0.55602196
				Total	18344138580	100

and rock and mineral compositions. And of course, one can also estimate mineral abundances. Absolute concentrations thus provide paths for increasing our understanding of exoplanetary systems, and serve as an insurance policy against concluding that two things are similar when they are really quite different.

CALCULATING MINERALOGY AND ROCK TYPE: METHODS AND UNCERTAINTIES

Comparisons of elemental concentrations can be instructive (Table 3), but to truly understand any planet, whether contained within the Solar System or elsewhere, the mineralogy of the silicate fraction is essential. In this section we'll examine how to calculate a bulk silicate planet (BSP) composition using a star's bulk composition as input, and then introduce two different methods for converting a BSP into mineral abundances and rock types. Our examples will make considerable use of perhaps an otherwise obscure set of stars, known as "polluted white dwarfs". These stars are special because their atmospheres may record the remnants of rocky planetary debris (see Xu et al. 2024, this volume). Caveats accompany the use of these methods, which are detailed in a later section.

Star compositions, chondrites and planetary bulk compositions

To estimate mantle mineralogy of an exoplanet, more often than not, we will use a star's composition as a nominal planetary bulk composition, taking the refractory elements on a volatile-free basis. The foundation for such a comparison stems from within our Solar system: there is a remarkably strong 1-to-1 correlation between the concentrations of refractory

Table 8a. Solar compositions in dex, converted to element wt. %.

Reported ratio		(a) Solar phot.[1] $A(El)Sol$	(b) Solar phot. $n(Z)Sol = 10A(El)$	(c) Solar phot.[2] $AW \times 10A(El)$	(d) Solar phot. wt. %[3]
C/H	C	8.39	245470891.6	2948277237	18.79
O/H	O	8.73	537031796.4	8592186523	54.75
Na/H	Na	6.3	1995262.315	45870621.71	0.29
Mg/H	Mg	7.54	34673685.05	842743915	5.37
Al/H	Al	6.47	2951209.227	79628163.9	0.51
Si/H	Si	7.52	33113112.15	929998311.2	5.93
P/H	P	5.46	288403.1503	8932886.701	0.06
S/H	S	7.14	13803842.65	442620214.4	2.82
K/H	K	5.12	131825.6739	5154159.744	0.03
Ca/H	Ca	6.33	2137962.09	85685244.62	0.55
Ti/H	Ti	4.9	79432.82347	3802210.961	0.02
Cr/H	Cr	5.64	436515.8322	22697120.86	0.14
Mn/H	Mn	5.37	13803842.65	12878735.75	0.08
Fe/H	Fe	7.45	28183829.31	1573925948	10.03
Ni/H	Ni	6.23	1698243.652	99675693.99	0.64
			Total	15694076987	100

Note. [1] The Solar photosphere composition from Lodders et al. (2009, their Table 4). 2. Product of atomic weight (AW) and $10^{A(El)}$. 3. Column (c) renormalized to 100%.

Table 8b. Star composition from the Hypatia Catalog (Hip 32768), converted to element wt. %.

Reported ratio		(e) Hip 32768 reported dex values $A(X)32768$	(f) dex relative to Solar $10A(X)$	(g) $n(Z)32768 = 10A(X) \times n(Z)$ Sol	(h) $AW \times n(Z)32768$	(i) Element wt. %
C/H	C	−0.14	0.72	177827941	2135838051	12.63
O/H	O	−0.06	0.87	467735141.3	7483481620	44.26
Na/H	Na	0.3	2.00	3981071.706	91523922.86	0.54
Mg/H	Mg	0.21	1.62	56234132.52	1366770591	8.08
Al/H	Al	0.23	1.70	5011872.336	135228023.9	0.80
Si/H	Si	0.27	1.86	61659500.19	1731737892	10.24
P/H	P					
S/H	S	0.67	4.68	64565422.9	2070290285	12.24
K/H	K					
Ca/H	Ca	0.1	1.26	2691534.804	107871331.9	0.64
Ti/H	Ti				0	0.00
Cr/H	Cr	0.08	1.20	524807.4602	27287941.18	0.16
Mn/H	Mn					
Fe/H	Fe	0.02	1.05	29512092.27	1648102793	9.75
Ni/H	Ni	0.04	1.10	1862087.137	109292225.1	0.65
			Total		16907424677	100

elements in the Sun and a special class of meteorites, the CI carbonaceous chondrites (Lodders et al. 2009; Lodders and Fegley 2018). These meteorites, as might be expected, are C-rich; they also contain "chondrules"—glassy blebs of material that appear to be relics of the earliest stages of Solar System development (Connolly and Jones 2018). Some caveats about this foundation are worth noting. First, the meteorite database of Nittler et al (2004) contains compositions from 647 distinctly named chondrite meteorites, but for the CI chondrites, we have compositions from only 4 named samples, and most analyses of these derive from a single meteorite, Orgueil, as it is the only CI chondrite with sufficient mass to allow multiple analyses (Palme and Zipfel 2022). The Sun–CI chondrite connection thus hinges on the observation that 0.6% of all chondrites are similar to the Sun with respect to their refractory elements. On this basis, it is very commonly assumed that most of the rocky bodies of the inner Solar System have a CI chondrite bulk composition (Sanloup et al. 1999; Rubie et al. 2011; Elardo et al. 2019). However, many studies now clearly show that Earth is not chondritic with respect to its trace elements and isotopes (McDonough and Sun 1995; Drake and Righter 2002; Campbell and O'Neill 2012; Mezger et al. 2020) and may even have a non-chondritic major element content (Putirka et al. 2021). This recognition has led to the proposal that enstatite chondrites are a better fit for bulk Earth (e.g., Javoy et al. 2010; Boyet et al. 2018) since they have Earth-like ratios for many isotopes. But enstatite meteorites do not match Si isotopes (e.g., Fitoussi and Bourdon 2012) and they are not a match for various trace and major element compositions (Baedecker and Wasson 1975; Mezger et al. 2020; Putirka et al. 2021). We've known about this Earth-meteorite mismatch issue for some time, as McDonough and Sun (1995) concluded that no meteorite subclass provides a match to Earth's bulk composition. Mezger et al. (2020) provide intriguing evidence that the component that is missing from the meteorite database but that comprises Earth, may be found more heavily concentrated in the compositions of Venus and Mercury. In any case, and noting these caveats, we will proceed on the assumption that rocky planets at least roughly mimic the non-volatile compositions of the stars they orbit, which may indeed be true for the inner planets (Putirka et al. 2021, their Fig. 4).

Models for core formation (a 1st-order control on mantle mineralogy)

The amount of Fe that partitions into the core relative to the mantle has a tremendous influence on mineralogy, in large part because Fe readily partitions into silicate minerals. The recent focus on exoplanetary Mg/Si ratios is not entirely misplaced. But the ratio (Mg+Fe)/Si provides a more precise classification scheme, separating peridotites from pyroxenites, and determining when a mantle assemblage might be nominally saturated with quartz or ferro-periclase (Putirka and Rarick 2019). The mantle (Mg+Fe)/Si ratio is, of course, dependent upon the fraction of Fe that is sequestered into the core. And to obtain mantle composition from a star composition, that amount of Fe that will partition into the core will be the first decision that must be made (though in a later section we'll introduce a method to calculate Fe in the core).

There are many models of core formation but from a purely geochemical standpoint, oxygen abundance (often measured as a "fictive" oxygen fugacity, or fO_2, i.e., the partial pressure of O_2 in a vapor phase, if that vapor were in equilibrium with the core) is likely a key thermodynamic variable. And that variable will be essentially unknown. The problem is this: oxygen is so abundant in FGMK-type stars (the Sun is a G-type star) that all the planet-building major elements Si, Al, Fe, Mg, Ni and Ca are likely to be fully oxidized (to form SiO_2, Al_2O_3, FeO, MgO, NiO and CaO), with O leftover (Unterborn and Panero 2017; Putirka and Rarick 2019). However, every differentiated planetary object in the inner Solar system contains a metallic Fe-Ni alloy core that is at least roughly devoid of O, and so the inner planets are effectively O-depleted relative to the Sun. In calculating an exoplanet's mantle mineralogy we must decide, perhaps independent of star composition, the fraction of Fe that occurs in the core, as Fe^0, and what fraction occurs in the mantle, as Fe^{2+} (in mineral such as $(Mg,Fe)_2SiO_4$, and $(Mg,Fe)SiO_3$). But the actual fictive fO_2 in a planetary mantle may indeed be a function of the final calculated mineralogy: Guimond et al. (2023) show that, because pyroxenes can accommodate Fe^{3+}, but olivine cannot, mantle mineralogy can control the total amount of O that can be dissolved into the mantle, and hence control the fO_2 of the mantle, rather than our usual way of thinking, i.e., that fO_2 controls mineral proportions. N.B.: it is common in petrologic studies to describe mantle rocks as following a given "fO_2 buffer", mostly because their FeO/Fe_2O_3 ratios can mimic such trends (e.g., Ni–NiO; Rhodes and Vollinger 2005); in such discussions it is understood that no such buffers actually exist and that fO_2 values are, indeed, "fictive", in that the rocks in question have not equilibrated with a vapor phase; fO_2 is useful as it can be controlled in experiments, but in this work, all calculations deal with a defined O content, not a defined fO_2). In this work, we'll ignore Fe^{3+} (usually expressed as Fe_2O_3, which accounts for as much as 10% of all Fe in Earth's upper mantle; Rhodes and Vollinger 2005) and treat all Fe as if it were Fe^{2+} (usually expressed as FeO). To be clear that we are accounting for all Fe in the mantle (though perhaps not all O), independent of the oxidation state of Fe, we'll refer to this "total Fe" as FeOt. Since O contents are effectively unknown for exoplanets, Putirka and Rarick (2019) recommend use of $\alpha_{Fe} = Fe^{BSP}/Fe^{BP}$, where Fe^{BSP} is the cation fraction of Fe in the bulk silicate planet (crust + mantle), and Fe^{BP} is the cation fraction of Fe in the bulk planet (crust + mantle + core). This method effectively constrains the bulk mantle O content (and a bulk fictive fO_2, although fO_2 will vary widely across a planet, even within a well-mixed mantle, e.g., Stolper et al. 2020). If we assume that exoplanetary O contents are roughly Earth-like (Doyle et al. 2019), α_{Fe} is uncertain but may be close to 0.27–0.30 (see Putirka and Rarick 2019). In such a model, high-Fe planets will both have larger cores but also slightly higher FeOt in their BSP compositions. Using our Solar system as a model of how O might vary within another planetary system, Putirka and Rarick (2019) recommend exploring α_{Fe} values that range from at least 0 (Mercury-like) to 0.54 (Mars-like), but of course, our Solar System might not define the limits of α_{Fe} and exoplanetary systems might range to greater α_{Fe}.

Calculating mantle mineralogy

Normative abundances. In 1902, four geologists, Cross, Iddings, Pirsson and Washington, published a method to classify igneous rocks (Cross et al. 1902). Their method involves using rock compositions to quantify the minerals that are "capable of crystallizing from a magma" of the same composition. Their approach would come to be known as the "CIPW norm". A "norm" is a theoretical mineral assemblage, to be put into contrast with a "mode" which would refer to the actual minerals identified in a given rock. Cross et al. (1902) suggested their norms could constitute a "standard mineralogy" by which different rocks could be compared. The use of such a method is that a volcanic rock, which is incompletely crystallized, could then be compared on a mineralogical basis to a plutonic rock (an igneous rock that has fully crystallized, at some depth below the surface). The norm, or standard mineralogy, is also helpful for comparing rocks that have crystallized under very different $P–T$ conditions. This standard mineralogy is effectively a "what if" calculation: what if a volcanic rock were to crystallize completely? What if a rock that formed deep in the crust were to re-crystallize at 1 atm pressure? What minerals might form and how would those minerals differ from, say, that rock over there?

Cross and his collaborators were not fools. Cross et al. (1902) recognized that their standard mineralogy was not a substitute for mineral identification. On page 558 they write "The standard mineral composition of a rock is called its norm, and this may be *quite different from its actual mineral composition*, or mode" (emphasis altered from the original). As an example, if their method calculates that a rock is corundum (Cn; Al_2O_3) normative, that rock might not contain corundum; the $P–T$ conditions might not be conducive, or Al might be distributed differently than as calculated, or the rock contains components (e.g., water, carbon, etc.) that go unaccounted in the method. If there is error in the attempt, why make the normative estimate in the first place? Because, as Cross et al. (1902) hypothesized, the normative classification can group rocks together that follow a similar evolutionary path. Our hypothetical Cn-normative system might not contain corundum, but it's likely on a path towards corundum saturation, and knowing that path can tell us about a rock's evolutionary history (Cawthorn and Brown 1976) and its source materials (Chappell and White 1992).

In summary Cross et al. (1902) proposed a classification scheme. Their system does not substitute for examining rocks in the field or under a microscope, but it does provide insights into a system's origin and evolutionary path. Exoplanets can similarly benefit from the method. The dozens of minerals calculated in a traditional CIPW norm (Verma et al. 2003; Buckle et al. 2023), however, are largely irrelevant to our current studies of exoplanets. The abundances of Si, Mg, and Fe utterly dominate over all other cations, and our current emphasis is to estimate mantle mineral assemblages; the numbers of minerals that will control exoplanetary thermal and mechanical properties of a planet are few. The two methods described below are thus abbreviated for the specific task of estimating mineral abundances for exoplanetary mantles, and as we learn more about how S, C, O and H vary in planetary bodies, these methods will quite likely require further refinement.

Method 1: The J.B. Thompson approach (Supplement 1). Thompson (1982) proposed a method based on linear algebra that is quite different from the CIPW norm that has been classically used by geologists. But Thompson's (1982) approach has the virtue of allowing a very wide range of different mineral assemblages to be tested very quickly, which can be useful for exoplanets since we must at least hold the possibility that some may have mineral assemblages that are exotic to the inner Solar System (See Supplementary file 1 for example calculations of Method 1).

In the Thompson (1982) approach, we can calculate a normative mineral assemblage using a BSP (e.g., as in Table 4) as input. For our example, we'll consider a case where the compositions are expected to plot into Figure 4a (as peridotite or pyroxenite), so the desired outputs are the fractions of olivine (F_{Ol}), clinopyroxene (F_{Cpx}), and orthopyroxene (F_{Opx}). Some exoplanets might have sufficiently high Fe to allow wüstite to form, which we can test by calculating $F_{Wüs}$. As input, we need only the mole fractions of SiO_2, FeO, MgO and CaO, renormalized to sum to 1. For the calculation, we must also decide on specific mineral compositions (Table 9a), which for the case of Models 1 and 2 in Table 9a are obtained from experiments on peridotite compositions conducted at upper mantle *P–T* conditions (ca. 2 GPa and 1300–1400 °C; Kinzler 1997; Walter 1998).

Taking olivine as an example, both observations and experiments show that Earth's upper mantle typically has olivine that is close to 90% forsterite (Fo; Mg_2SiO_4) and 10% fayalite (Fa; Fe_2SiO_4), expressed as Fo90. The specific formula for olivine is thus $Mg_{1.8}Fe_{0.2}SiO_4$, which requires 1 unit of SiO_2, 1.8 units of MgO and 0.2 units of FeO, which are the table entries for olivine in Table 9a, Model 1. Model 1 entries for clinopyroxene and orthopyroxene similarly provide the compositions expected for a peridotite mantle, equilibrated at upper mantle conditions.

As to the calculation, in a spreadsheet, it is convenient to handle large amounts of data using row matrices, and so we can employ Model 1 using the equation:

$$\begin{bmatrix} F_{Ol} & F_{Cpx} & F_{Opx} & F_{Wüs} \end{bmatrix} = \begin{bmatrix} X^{BSP}_{SiO_2} & X^{BSP}_{FeO} & X^{BSP}_{MgO} & X^{BSP}_{CaO} \end{bmatrix} \times \begin{bmatrix} X^{Ol}_{SiO_2} & X^{Opx}_{FeO} & X^{Opx}_{MgO} & X^{Opx}_{CaO} \\ X^{Cpx}_{SiO_2} & X^{Cpx}_{FeO} & X^{Cpx}_{MgO} & X^{Cpx}_{CaO} \\ X^{Opx}_{SiO_2} & X^{Opx}_{FeO} & X^{Opx}_{MgO} & X^{Opx}_{CaO} \\ X^{Wüs}_{SiO_2} & X^{Wüs}_{FeO} & X^{Wüs}_{MgO} & X^{Wüs}_{CaO} \end{bmatrix}^{-1} \tag{4}$$

In Equation (4), the row matrix on the left-hand side is the output, which is the mole fractions of the minerals indicated in a matrix such as those in Table 9a. On the right-hand side, the row matrix contains the mole fractions of the oxides *i* for the bulk silicate planet, X^{BSP}_i,

Table 9a. Mineral compositions.

Model 1		SiO$_2$	FeO	MgO	CaO
Olivine	(Mg,Fe)$_2$SiO$_4$	1	0.2	1.8	0
Clinopyroxene	Ca(Mg,Fe)Si$_2$O$_6$	2	0.2	1.8	1
Orthopyroxene	(Mg,Fe)$_2$Si$_2$O$_6$	2	0.2	1.8	0
Wüstite	FeO	0	1	0	0
Model 2		**SiO$_2$**	**Al$_2$O$_3$**	**FmO**	**CaO**
Olivine	(Mg,Fe)$_2$SiO$_4$	1	0	2	0
Clinopyroxene	Ca(Mg,Fe)Si$_2$O$_6$	1.8	0.3	1.3	0.6
Orthopyroxene	(Mg,Fe)$_2$Si$_2$O$_6$	1.8	0.3	1.8	0.1
Garnet	(Mg,Fe,Ca)$_3$Al$_2$Si$_3$O$_{12}$	3	2	2.7	0.3
Model 3		**SiO$_2$**	**Al$_2$O$_3$**	**FmO**	**CaO**
Olivine	(Mg,Fe)$_2$SiO$_4$	1	0	2	0
Clinopyroxene	Ca(Mg,Fe)Si$_2$O$_6$	1.75	0.25	1.25	0.75
Orthopyroxene	(Mg,Fe)$_2$Si$_2$O$_6$	1.9	0.3	1.75	0.05
Garnet	(Mg,Fe,Ca)$_3$Al$_2$Si$_3$O$_{12}$	3	2	2.4	0.6
Model 4		**Si**	**Fe**	**Mg**	**Ca**
Olivine	(Mg,Fe)$_2$SiO$_4$	1.33	0.267	2.4	0
Cpx + Opx	Ca$_{0.5}$(Mg,Fe)$_{1.5}$Si$_2$O$_6$	2	0.2	1.3	0.5
Fe metal	Fe	0	1	0	0
Quartz	SiO$_2$	1	0	0	0

Table 9b. Transformation matrix.

	Olivine	Clinopyroxene	Orthopyroxene	Wüstite
SiO$_2$	−1	0	1	0
FeO	0	0	0	1
MgO	1.11	0	−0.56	−0.11
CaO	0	1	−1	0

Note. Each of the models in Table 9a, however simple, are thermodynamic models: a set of *P* and *T* conditions are assumed to apply (close to ca. 2 GPa and 1400°C), and the minerals in any given model are the phases that are presumed to have the lowest Gibbs Free Energy at such *P–T* conditions. The models in Table 9a also presume an anhydrous and largely C- and S-free bulk composition, with an Earth-like O content (e.g., fictive *f*O$_2$ close to the Ni–NiO buffer; Rhodes and Vollinger 2005). To the extent that we choose to compare exoplanets using such a standard mineralogy as Models 1 or 2 in Table 9a, these conditions act as a *de facto* standard state.

and the square matrix gives the compositions of the minerals, as in Model 1 of Table 9a. In our example in Model 1 (Table 9a) we require the mole fractions of SiO$_2$, MgO, FeO or CaO, where $X_{SiO_2} + X_{MgO} + X_{FeO} + X_{CaO} = 1$. And the square matrix on the right-hand is the inverse of the matrix of Model 1 (Table 9a), which is given in Table 9b. If we have a BSP where the mole fractions of SiO$_2$, FeO, MgO and CaO are respectively, 0.4, 0.09, 0.49 and 0.02, we have:

$$[0.144 \quad 0.020 \quad 0.108 \quad 0.036] = [0.4 \quad 0.09 \quad 0.49 \quad 0.02] \times \begin{bmatrix} -1 & 0 & 1 & 0 \\ 0 & 0 & 0 & 1 \\ 1.11 & 0 & -0.56 & -1.11 \\ 0 & 1 & -1 & 0 \end{bmatrix} \quad (5)$$

And the mineral proportions on a molecular basis, obtained from Equations (4–5), of Ol, Cpx, Opx and Wüs are respectively 0.144, 0.020, 0.108 and 0.036, which upon renormalization to 100 yields a mineral assemblage that is 46% olivine 6% clinopyroxene, 35% orthopyroxene and 12 % orthopyroxene. For a ternary diagram where Ol, Cpx and Opx lie at the apices, it is necessary to renormalize again, so that Ol + Cpx + Opx = 100 (or 1), so we have ca. 53% Ol, 7% Cpx and 40% Opx on a molecular basis. There is no reason not to use molecular proportions in ternary diagrams such as Figure 4a, and interestingly, terrestrial mantle rocks plot in effectively the same field in Figure 4a regardless of whether molecular or weight proportions are employed. However, individual compositions can plot quite differently, especially when the minerals at the apices have very different molecular weights. In any case, to obtain weight % values each mole fraction or % value would then be multiplied by the molecular formula weight of the mineral with further renormalization to 100%. Figure 4a was developed on the basis of nominal "% volume" (though really "% area", since most "volume" measurements are determined from 2D thin sections); mineral densities could be used to convert weights to volume proportions, but this additional step is not recommended.

If one prefers column matrices, then instead of Equation (4) we have:

$$
\begin{bmatrix} F_{Ol} \\ F_{Cpx} \\ F_{Opx} \\ F_{Wüs} \end{bmatrix} = \begin{bmatrix} X^{Ol}_{SiO_2} & X^{Opx}_{FeO} & X^{Opx}_{MgO} & X^{Opx}_{CaO} \\ X^{Cpx}_{SiO_2} & X^{Cpx}_{FeO} & X^{Cpx}_{MgO} & X^{Cpx}_{CaO} \\ X^{Opx}_{SiO_2} & X^{Opx}_{FeO} & X^{Opx}_{MgO} & X^{Opx}_{CaO} \\ X^{Wüs}_{SiO_2} & X^{Wüs}_{FeO} & X^{Wüs}_{MgO} & X^{Wüs}_{CaO} \end{bmatrix}^{T^{-1}} \times \begin{bmatrix} X^{BSP}_{SiO_2} \\ X^{BSP}_{FeO} \\ X^{BSP}_{MgO} \\ X^{BSP}_{CaO} \end{bmatrix} \tag{6}
$$

Equation (6) contains two critical adjustments to the right-hand side of Equation (4), namely we must take the transpose of the mineral composition matrix (Model 1, Table 9a) before calculating its inverse, and we must reverse the order of the multiplication (matrix multiplication is non-commutative).

Keep in mind that the models of Table 9a are thermodynamic models, albeit very simple ones. They assume that the minerals in question, at least where all estimates are positive (and where certain other assumptions apply; see below), are the phases that yield a minimum of the Gibbs Free Energy of a given system.

Equations (4–6) demonstrate the method, but do the very simple thermodynamic models of Table 9a have any chance of accurately predicting mineral abundances? Models 1, 2 and 3 all assume that mineral compositions are constant across all bulk compositions where the model is applied—this is highly unlikely, but apparently not fatal. Figures 7a–b and Figure 8 show the case where Model 2 is applied to predict mineral abundances for peridotites (Warren 2016) and pyroxenites (Bodinier et al. 2008), where mineral abundances are either measured to high precision or can be calculated from measured mineral and bulk compositions. The results of our tests (Fig. 7) show that Model 2 can yield estimates that are precise to ±10% and perhaps as precise as ± 3%, provided the bulk compositions used as input fall within the ultramafic rock diagram of Figure 4; it might even be useful for terrestrial peridotites. Model 2 is clearly less accurate for pyroxenites (Fig. 7b), capturing just 65% of the variation of natural samples compared to 99% for peridotites (compare R^2 values). But by adjusting the coefficients so as to obtain Model 3 (Table 9a), one can quickly devise a new Model that increases precision and accuracy for such compositions (Fig. 7c). This approach, of adjusting such matrices so as to better predict cases of measured or well-constrained mineral abundances, is highly recommended.

Note that Models 1–3 in Table 9a are only appropriate when olivine and the pyroxenes dominate, such as the rocky mantles of planetary objects; the models are not appropriate for

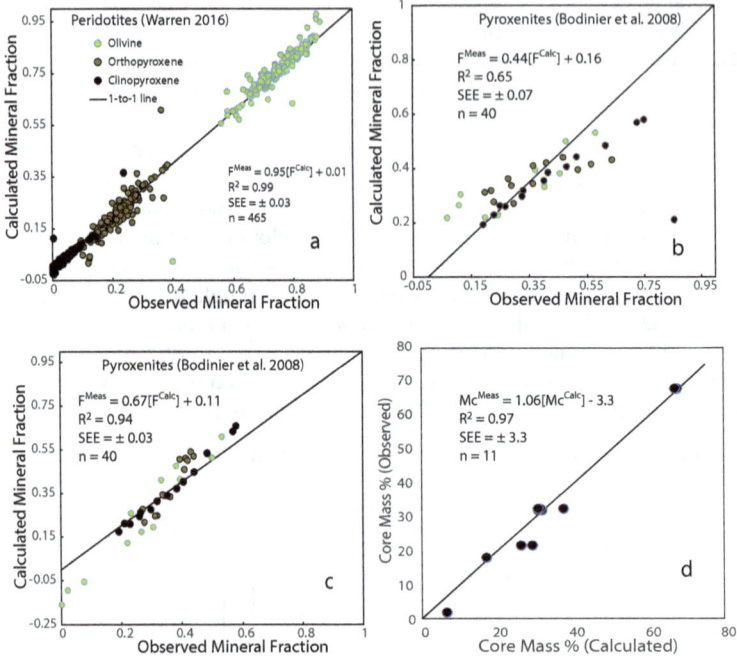

Figure 7. Test of Model 2 (Table 9a) for its ability to predict mineral proportions in natural rock samples, where mineral proportions are measured directly, or calculated by mass balance from measured mineral and whole rock compositions.

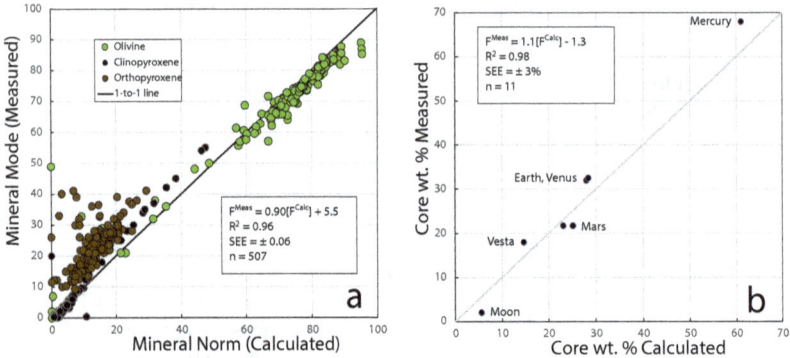

Figure 8. (a) shows a test of Model 2's ability to predict mineral abundances for the same data as shown in Figure 7. (b) shows Model 2's ability to predict core mass fractions for the inner planets, Moon and Vesta, from Fe–Si mass balance and resulting estimates of total metallic Fe.

plotting bulk planet compositions, as these are likely to have large metallic cores. Model 4 in Table 9a shows how one might address the issue: here, due to their identical stoichiometry, clinopyroxene (Cpx) and orthopyroxene (Opx) are combined to form a single pyroxene component (Pyx) and a pure-Fe metal component is added. This particular matrix describes 97% of the core mass % (M_c) values of the inner planets and Vesta (Fig. 7d), using Table 3b and bulk planet estimates from McDonough and Yoshizaki (2021) as input (measured values of M_c are from Szurgot 2015). (N.B.: the bulk compositions of Venus, and even Mercury, are highly uncertain; confidence in their compositions is low, especially for Venus).

As can be seen from these examples, the Thompson (1982) approach has advantages and disadvantages: it is ideal in quickly calculating mineral abundances for a large number of samples, and testing a wide range of possible mineral assemblages. Matrices can also be quickly edited so as to optimize performance in predicting mineral abundances in test data sets. But predicated mineral abundances can be negative, which is acceptable for classification (e.g., Thompson 1982), but not helpful if one wants to perform experiments on possible exoplanet compositions. Negative estimates show that a model does not yield a plausible mineral assemblage; this is, of course, a signal to find a new set of minerals or mineral compositions. The model also requires square matrices for transformation, which limits the range of minerals that can be explored for a given set of bulk composition and hence also limits the range of compositions and mineral assemblages for which accurate mineral abundances may be obtained.

Method 2 illustrates a new approach, analogous to the original CIPW norm, that yields only positive mineral abundances regardless of bulk composition.

Method 2: A CIPW norm analog for exoplanets (Supplement 2). The method is fashioned in the spirit of Cross et al. (1902), using a cation accounting system, but is specifically designed for exoplanet studies. Method 2 allows for (a) calculation of Fe metal (core) mass fractions, based on Fe–Si mass balance, (b) a single set of fixed, end-member mineral compositions that, without adjustment, yield positive mineral sums for any exoplanet composition, including highly exotic ones, and (c) a determination of whether a system might be saturated in ferrosilite ($Fe_2Si_2O_6$)—a mineral that is rare (because it is thermodynamically unstable) in the inner planets, but has been putatively detected in white dwarf dust disks (Reach et al. 2009). The method is as follows (See also Supplementary File 2, a spreadsheet that provides the calculations for Method 2, below):

Step 1: For input, obtain/convert a composition consisting of any of Si, Ti, Cr, Al, Fe, Mg, Ni, Ca, Na, K and P, to elemental mole % (the calculation assumes an anhydrous bulk composition; O is determined from charge balance on the cations). Any of the listed elements may be undetermined (set to 0), but useful output requires estimates for Si, Fe, and Mg at a minimum. Note: for a correct accounting, anytime a component calculates as < 0, assign a value of 0, since a negative value indicates an insufficiency of an essential ingredient for the component to form.

Step 2: Assign Cr to chromite, Chr ($FeCr_2O_4$) and eskolaite, Esk (Cr_2O_3):
(a) If either of Fe or Cr = 0, then Chr = 0
   if Fe and Cr are > 0, then
      if (Cr/Fe) > 2, Chr = Fe/(1/3), else Chr = Cr/(2/3).
(b) If Cr > Fe, then excess Cr = Cr1 = Cr − (2/3)($FeCr_2O_4$) and Cr1 is assigned to Esk, where Esk = (Cr1)/2.

Step 3: Excess Fe, Fe1 = Fe1 − (1/3)$FeCr_2O_4$.

Step 4: Assign Ti to Ilmenite, Ilm ($FeTiO_3$) and rutile, Rt (TiO_2):
(a) If Ti or Fe1 = 0, then Ilm = 0
   if Ti > 0 and Fe1 > 0, then
   if Ti < Fe1, Ilm = Ti/(1/2), else Ilm = Fe1/(1/2)
(b) Excess Ti (Ti1) is assigned to the mineral rutile, where Ti1 = Ti − (1/2)$FeTiO_3$ and Rt = Ti1

Step 5: Excess Fe, Fe2: = Fe2 − (1/2)$FeTiO_3$

Step 6: Assign Mg# (= Mg/([Mg+Fe]) to silicate minerals (0–1 possible; Earth like values are 0.80 to 0.90) and a fraction of Ni to silicates, F_{Ni} (0–1 possible; Earth-like value is ca. 0.05): The dominant silicate minerals in the mantles of the inner planets (olivine, pyroxenes, garnet) are dominantly solid solutions of Mg and Fe, with substantial amounts of Ni. For simplification, the choice of Mg# and F_{Ni} at this step will set the Mg# and F_{Ni} values for all silicates that contain such elements. The import of this step, and Steps 7–9, is to determine the amount of Fe + Ni that will be segregated into a metallic core and how much is retained by silicate minerals in the mantle.

Step 7: Calculate PFNS, the potential amount of Ni and Fe that can partition into silicate phases: if Si > (Fe2 + Mg + Ni), then PFNS = Fe2 + Ni, else PFNS = $(F_{Ni}$)(Ni) + (Mg – Mg#[Mg])/Mg#.

Step 8: Assign Fe and Ni to silicate phases (FNS) and determine any excess Fe and Ni (FeNi) relative to Si:

(a) If (Fe2 + Ni) > PFNS, then FNS = PFNS, else FNS = Fe2 + Ni.
(b) Excess Fe and Ni are assigned to FeNi as follows: if (Fe2 + Ni) > PFNS, FeNi = Fe2 + Ni – PFNS, else FeNi = 0. FeNi represents the first step in determining the amount of Fe and Ni that will partition into a metallic phase.

Step 9: Assign Fe, Ni and Mg to Silicates as Fm: Fm = FNS + Mg. (Note to astronomers: "Fm" is a shorthand for "ferro-magnesian", which indicates minerals rich in Fe and Mg, such as olivine and the pyroxenes). Here, Fm includes Ni since quantifying Ni is important to quantifying bulk planet and planetary core compositions.

Step 10: Assign P to apatite, Ap $(Ca_5(PO_4)_3(OH))$ and calculate excess Ca and P:

(a) If both P > 0 and Ca > 0, then Ap = P/(3/8), else Ap = 0.
(b) Excess Ca, Ca1 = Ca – (5/8)Ap.
(c) Excess P, if any, is assigned to phosphorous oxide in the final tally where
    P1 = P – (3/8)Ap and phosphorous oxide = P1/2.

Step 11: Assign Na, K, Al and Si to alkali feldspar, Afs $(Na,K)AlSi_3O_8)$, and nepheline, Ne $(Na,K)AlSiO_4$:

(a) Establish Si/Al ratio, SAR: If Al and Si > 0, then SAR = Si/Al, else SAR = 0
(b) Establish the amount of Si in feldspar, SiF: if SAR > 3 then SiF = 1; if SAR < 1, then
    SiF = 0. If 3 > SAR > 1, then SiF = (SAR – 1)/2
(c) Establish SiN, the amount of Si assigned to nepheline: SiN = 1 – SiF
(d) Calculate TA, total alkalis: TA = Na + K
(e) Calculate amount of Afs:
    if either TA or SiF are not > 0, then Afs = 0;
    if TA > 0 and SiF > 0, then
        if both TA < Al and 3(TA) < Si then Afs = (SiF)(TA)/(1/5), else
        if either TA is not < Al or 3(TA) is not < Si, then
        if 3(Al) < Si then Afs = (SiF)(Al)/(1/5) else Afs = (SiF)(Si)/(3/5)
(f) Calculate TA1, excess total alkalis: TA1 = TA – (1/5)Afs
(g) Calculate Al1, excess Al: Al1 = Al – (1/5)Afs
(h) Calculate Si1, excess Si: Si1 = Si – (3/5)Afs
(i) Calculate Ne, nepheline: if either of TA1 = 0 or Al1 = 0, then Ne = 0; if both TA1 > 0
    and Al1 > 0 then:
        if both TA1< Al1, and TA1 < Si1 then Ne = (SiN)(TA1)/(1/3)
            if either TA1 > Al1 or TA1 > Si1, then
            if Si1 > Al1, then Ne = (SiN)(Al1)/(1/3)
            else Ne = (SiN)(Si1)/(1/3).
(j) Calculate excess Al: Al2 = Al1 – (1/3)Ne
(k) Calculate excess Si: Si2 = Si1 – (1/3)Si5

(l) Calculate excess alkalis: $TA2 = TA1 - (1/3)Ne$

Step 12: Assign Al and Fm to garnet, PyAl (PyAl is the sum of pyrope ($Mg_3Al_2Si_3O_{12}$) and almandine ($Fe_3Al_2Si_3O_{12}$) garnet):

(a) If $3Al2 < 2Si2$ and $3Al2 < 2Fm$, then $PyAl = Al/(1/4)$
  if either $3Al2 > 2(Si2)$ or $3Al2 > 2Fm$, then
  if $Si2 < Fm$, then $PyAl = Si/(3/8)$, else
  if $Si2 > Fm$, $PyAl = Fm/(3/8)$.
(b) Calculate excess Al: $Al3 = Al2 - (1/4)PyAl$. This excess Al in the final tally is assigned to corundum (Cn; Al_2O_3), where $Cm = (Al3)/2$.
(c) Calculate excess Fm: $Fm1 = Fm - (3/8)PyAl$
(d) Calculate excess Si: $Si3 = Si2 - (3/8)PyAl$

Step 13: Assign excess Ca to Diopside + Hedenbergite, DiHd ($Ca(Mg,Fe)Si_2O_6$):

(a) If $Ca1 < (Si3/2)$ and $Ca1 < Fm1$, then $DiHd = Ca1/(1/4)$
  if either $Ca1 > (Si3/2)$ or $Ca1 > Fm1$ then
  if $Fm1 < (Si3)/2$ then $DiHd = Fm1/(1/4)$, else $DiHd = (Si3)/(1/2)$
(b) Calculate excess Ca: Ca2 is: $Ca2 = Ca1 - (1/4)Di$
(c) Calculate excess Fm: $Fm2 = Fm1 - (1/4)Di$
(d) Calculate excess Si: $Si4 = Si3 - (1/2)Di$

Step 14: Determine the potential fractions of olivine (Fol) and orthopyroxene (Fopx) (the steps that follow apportion the remaining amounts of Fm and Si between the phases olivine and orthopyroxene, based on Fm/Si ratios):

(a) If either $Fm2$ or $Si4 = 0$, then $Fol = 0$;
  if both $Fm2 > 0$ and $Si4 > 0$, then
  if $(Fm2/Si4)/2 > 1$ then $Fol = 1$, else
  if $Fm2/Si4 < 1$, then $Fol = 0$ else
  $Fol = Fm2/Si4 - 1$
(b) $Fopx = 1 - Fol$; (note: later calculations will take care of the case that either or both of $Fm2$ and $Si2 = 0$, and where $Fol = 0$ and $Fopx = 1$).
(c) Determine Si assigned to olivine (SiOl): if $Fm2 > 0$ and $Si4 > 0$ then $SiOl = (Fol)$ $(Si4)$, else $SiOl = 0$.
(d) Determine Si assigned to orthopyroxene (SiOpx): if $Fm2 > 0$ and $Si4 > 0$, then $SiOpx = (Fopx)(Si4)$, else $SiOpx = 0$.

Step 15: Determine amounts of olivine (FoFa) and orthopyroxene (EnFs)

(a) $FoFa = SiOl/(1/3)$
(b) if $Fm2 < SiOpx$, then $EnFs = Fm2/(1/2)$, else $EnFs = SiOpx/(1/2)$

(c) Calculate excess Fm: $Fm3 = Fm2 - (2/3)FoFa - (1/2)EnFs$
(d) Calculate excess Si: if $Si4 < Fm2$, then $Si5 = 0$, else
  if $Si4 > Fm2$, then $Si5 = Si4 - (1/3)FoFa - (1/2)EnFs$.

Step 16: Assign remaining Fm to ferro-periclase, FmO: If $Fm3 > 0$, $FmO = Fm3$, else $FmO = 0$

Step 17: Assign any excess FeNi and Si to ferrosilite (Fs; $FeSiO_3$):
(a) If $Si5 > FeNi$ then $Fs = FeNi/(1/2)$ else $Fs = Si3/(1/2)$
(b) $Si6 = Si5 - (1/2)Fs$
(c) $FeNi1 = FeNi - (1/2)Fs$

Step 18: Assign remaining oxides to any excess Al (corundum; Cm), Si (quartz; Qz) and alkalis (alkali oxides, AO):

(a) Corundum, Cm = Al3
(b) Quartz , Qz = Si6
(c) Alkali Oxides, AO = TA2

The above procedure yields the following estimates of mineral abundances:

Chromite	= Chr from Step 2a
Eskolaite	= Step 2b
Ilmenite	= Ilm from Step 4a
Rutile	= Rt in Step 4b
Apatite	= Ap from Step 10a
Phosphorous oxide	= P1 from Step 10c
Alkali feldspar	= Afs from Step 11e
Nepheline	= Ne from Step 11i
Pyrope + Almandine (Gar)	= PyAl from Step 12a
Diopside + Hedenbergite (Cpx)	= DiHd from Step 13a
Calcium Oxide (CaO)	= Ca2 from Step 13b
Forsterite + Fayalite (Ol)	= FoFa from Step 15a
Enstatite + Ferrosilite (Opx)	= EnFs from Step 15b
Ferropericlase	= FmO from Step 16
Ferrosilite	= Fs from Step 17
Corundum	= Cm from Step 18a
Quartz	= Qz from Step 18b
Alkali oxides	= AO from Step 18c

These minerals should sum to precisely 100%; many may be 0 but none should be negative. An example calculation is provided in Table 10, using a Solar bulk composition.

The procedure is tested on nearly 9,000 rock compositions that include peridotites (McDonough and Sun 1995), pyroxenites (GEORORC) and meteorites of all types (Nittler et al. 2004), including irons, pallasites and mesosiderites; all yield mineral sums of precisely 100%, with predicted mineral assemblages that yield little to no Fe for silicate rock samples, and significant metallic Fe for all pallasites and most mesosiderites. Sums higher or lower than 100 indicate that elements/components are either being double- or under-counted.

How well does it work? For predicting mineral modes, Method 2 is less accurate than Method 1. But as with the CIPW norm, it is designed for classification; accurate estimates of mineral abundances are possible, but only under a set of caveats and assumptions, noted in a later section. But for a small cost in accuracy we gain flexibility: in a single model, Method 2 yields positive mineral abundances for any composition, estimates of metallic core mass fractions and predicts saturation in minor phases for potentially any exotic composition. Method 2 could be refined to allow some fraction of Al and/or Na in pyroxenes, by adding an Al partitioning function, just as we added Mg–Fe–Ni partitioning in Step 6, although it's not clear that in any immediately recognizable future we will be quite ready to discuss the Ca-Tschermak's components in exoplanet clinopyroxenes. In any case, Method 2 yields a "standard mineralogy" that can describe any exoplanetary mantle, no matter how extreme or bizarre the composition, and like Method 1 (Fig. 7d), describes 98% of the variations in Mc, with a precision of ± 3 wt. %.

Table 10. Results from Method 2, using an anoxic, Solar bulk composition.

	Si	Ti	Cr	Al	Fe	Ni	Mg	Ca	Na	K	P
Solar elemental wt. %	26.78	0.10	0.70	1.78	43.29	2.44	21.15	2.30	1.10	0.12	0.23
Mole proportions	0.95	0.00	0.01	0.07	0.78	0.04	0.87	0.06	0.05	0.00	0.01
Mole %	33.60	0.08	0.47	2.32	27.31	1.47	30.66	2.02	1.68	0.11	0.26
	Step 2	Step 2	Step 3	Step 4	Step 4	Step 5	Step 6	Step 6			
	Chromite	Excess Cr	Fe1	Ilmenite	Excess Ti	Fe2	Ni #	Mg#			
	0.71	0.00	27.08	0.15	0.00	27.00	0.05	0.90			
	Step 7	Step 8	Step 8	Total Fe + Ni	Step 9						
	PNFS	FNS	FeNi		Fm						
	3.49	3.49	24.98	28.47	34.15						
	Step 10	Step 10	Step 10	Step 11	Step 11	Step 11	Step 11	Step 11			
	Apatite	Ca1	Excess P	SAR	SiF	SiN	Afs	TA1			
	0.68	1.60	0.00	14.45	1.00	0.00	8.98	0.00			
Step 11 cont.	All	Si1	Ne	TA2	Al2	Si2					
	0.53	28.21	0.00	0.00	0.53	28.21					
	Step 12	Step 12	Step 12	Step 12	Step 13	Step 13	Step 13	Step 13			
	PyAl	Al3	Fm1	Si3	DiHd	Ca2	Fm2	Si4			
	2.11	0.00	33.36	27.42	6.39	0.00	31.76	24.23			
	Step 14	Step 14	Step 14	Step 14	Step 15	Step 15	Step 15	Step 15			
	Fol	Fopx	SiOl	SiOpx	FoFa	EnFs	Fm3	Si5			
	0.31	0.69	7.53	16.69	22.60	33.39	0.00	0.00			
	Step 16	Step 17	Step 17	Step 17							
	Ferropericlase	Fs	Si6	FeNi1							
	0.00	0.00	0.00	24.98							

Final Mineral %

Chromite	Eskolaite	Ilmenite	Rutile	Apatite	P₂O₅	Alk. Fspar	Nepheline	Garnet	Clinopyroxene
0.71	0.00	0.15	0.00	0.68	0.00	8.98	0.00	2.11	6.39
CaO	**Olivine**	**Orthopyroxene**	**Ferropericlase**	**Ferrosilite**	**Fe Metal**	**Corundum**	**Quartz**	**Alkali Ox.**	**Total**
0.00	22.60	33.39	0.00	0.00	24.98	0.00	0.00	0.00	100.00

ERROR ANALYSIS IN TERNARY PROJECTIONS
(USING POLLUTED WHITE DWARFS AS EXAMPLES)

Much discussion about exoplanet compositions derives from using star compositions to predict the compositions of exoplanets (e.g., Hinkel et al. 2016). The reasoning is that the inner planets have relative Mg, Si and Fe contents that might closely approximate Solar proportions (Putirka et al. 2021; their Fig. 4). However, as is evident from Table 3, the inner planets do not precisely mimic the Sun. McDonough and Yoshizaki (2021) make a highly compelling case that Fe decreases with heliocentric distance, arguing that the shift is controlled by the Sun's magnetic field. Clearly, we need better models of how and why the inner planets vary in composition so that we might apply such understanding to other stars, to better predict exoplanet compositions. On top of this, the uncertainties reported for star compositions have their own challenges (Hinkel et al. 2016; Rogers et al. 2024). Repeat observations of given stellar objects are another highly useful task to improve estimates of error.

In the meantime, uncertain uncertainties are unlikely to halt our thinking about plausible exoplanet compositions—and neither should they. To be sure, the less we understand about error, the more our calculations are rooted in the soil of speculation rather than rising into the atmosphere of improved understanding. Such challenges lead Trierweiler et al. (2023) to conclude that errors are "hopelessly large" and that, in effect, when it comes to estimating mantle mineralogy: it's too hard; don't try. That approach also obviates all efforts that stem from mineralogy, such as studies of mantle circulation, plate tectonics, and the origin of life. Rather than give up entirely however, another option is to think more carefully about error, which provides a very precise and accurate guide to future work.

What are polluted white dwarfs? (possibilities and nominal problems)

As described by Xu et al. (2024, this volume) and by Xu and Bonsor (2021), polluted white dwarfs (PWDs) provide exceptional opportunities to examine the bulk geochemical and mineralogical characteristics of exoplanets. These stars represent a very late stage in stellar evolution, where a Sun-like star first expands to a Red Giant, obliterating planetary objects within a distance equivalent to the orbit of Mars, and then collapses to form a white dwarf, that is not much larger than Earth. Any planetary debris that survives the Red Giant phase (so the planetary analogs are, in effect, mostly outer Solar system materials) might later be absorbed into a white dwarf, yielding a "polluted" atmosphere, hence the name "polluted white dwarf". Because white dwarf atmospheres typically consist only of H or He (heavier elements sink rapidly to the stellar core), any elements that are heavier than He are likely to represent remnants of planetary debris that once orbited such stars and have very recently polluted their atmospheres. Putirka and Xu (2021) estimate mineral abundances for 23 PWDs, which represents a subset of the hundreds of white dwarfs that have been analyzed, where Si, Fe, Mg and Ca are all reported.

Putirka and Xu (2021) apply Method 1 to calculate mineral abundances for 23 PWDs and speculate that some might have mineral assemblages that are exotic to our Solar system. They note that their mineral abundance estimates are by no means definitive because, as also emphasized below, current thermodynamic models are calibrated almost solely on Earth-like compositions. Putirka and Xu (2021) thus proposed a provisional mineral classification system, to describe exoplanet mantles that fall outside the ultramafic rock triangle (Fig. 4a). Trierweiler et al. (2023) test those results and conclude that most exoplanets are like CI chondrites. How do the same data yield such differing conclusions? It might stem from a misunderstanding of certain statistical tests. Figures 9–11 will use PWDs as a case study, to examine errors more closely and to also illustrate how certain choices when plotting ternary diagrams affect plotted uncertainties.

Monte Carlo simulations

Monte Carlo simulations are especially useful when data sets are small. In geology, they are often not needed, but it's no less critical to understand how to interpret them. First let's consider the need. If geologists have a large outcrop, that outcrop can be re-sampled and re-analyzed (say, for SiO_2) many times over, perhaps using different analytical instruments. If the mean and standard deviation of repeat measurements are respectively 52.4 wt. % and ±1.2 wt. %, we would report the SiO_2 as 52.4 ± 1.2 wt. %. This means that there is a 68% (±1σ) chance that the SiO_2 content falls within the range 51.2–53.6 wt. %. We might also report a smaller "standard error", which accounts for the sample size, n. The standard error is smaller than the standard deviation, and decreases with increasing n. The standard error is useful if we think that the errors are random and only the mean value is important. For exoplanets, or PWDs, obtaining repeated observations is not as readily feasible and so a Monte Carlo simulation can be employed to "simulate" the case that many samples are available. The problem is that a mean and standard deviation are required as input, and with limited data, such inputs are uncertain, which is why Hinkel et al. (2016) report the "spread" of their dex values (e.g., for Fe/H). The hope, in any case, is that measurement errors are random and reported "spreads" are closer to a range than a standard deviation, although they might not be. Fortunately, systematic errors have not yet been documented in the astronomical literature (e.g., that Ca/H is always high relative to Fe/H, or that small stars have lower Si/H than large ones), but if such errors are ever discovered they will need to be accounted for.

With these assumptions in place, Figure 9a shows 1,000 simulations, of the PWD known as Ton 345. Figure 9a, projects PWDs into the Fe-free system: Mg_2SiO_4–$Mg_2Si_2O_6$–$CaMgSi_2O_6$, which effectively assumes that the mantle is Fe-free. This approach yields errors much larger than shown in the ternary projection of Putirka and Xu (2021), who instead use repeat analyses and independent models of certain PWD compositions (by UV and optical spectra, discussed below).

Figure 9a is a "what if" calculation: "What if we were to re-analyze Ton 345 1,000 times? Given the reported error, what range of compositions might result?". The calculations thus take the place of the analyses that we would otherwise perform if the analyses were easier to obtain.

To interpret a set of Monte Carlo simulations, we must also choose a desired level of certainty: we know that a plotted composition might plot differently if we were to re-analyze it. So what range of compositions should we consider when assessing possible mineral assemblages? The range depends upon how confident one wants to be when discussing such possibilities. The larger the level of confidence we require, the larger the range of compositions we must consider as plausibly real. In Figure 9a, the Monte Carlo results for Ton 345 are contoured at the ±1σ (68%), ±2σ (95%) and ±3σ (99.7%) confidence intervals. If we desire a 68% (±1σ) level of certainty, then we have decided to accept that Ton 345 may have a mineralogy represented by any point that falls within the 68% (red) contour. If we instead decide that only a ±3σ confidence level will do, then we must accept that Ton 345 could have a mineral assemblage that falls anywhere within the much larger 99.7% contour. Note, however, that Figure 9a says nothing about whether Ton 345 is statistically the same as any other PWD. For that test, we need a new Monte Carlo simulation for a second PWD, to test whether contours might intersect.

A critical aspect of Figure 9a is that the error inherent to Method 1 is trivial in comparison to compositional (input) error. Figure 9a shows a blue field that encompasses the ±3σ (99.7%) uncertainty region when Model 1 of Table 9 is used to predict mineral abundances when such abundances are known (Fig. 7). Trierweiler et al. (2023) imply that Method 1 (Table 9) is the source of the error but the large ellipses in Figure 9 result from the error on the input compositions. This mis-match in error is the main reason why very simple thermodynamic and mineralogic models are advocated here. Compositional precision is not yet so precise that there is any use in debating the extent to which small amounts of Al partition into Opx, or whether Ti partitions into Ol.

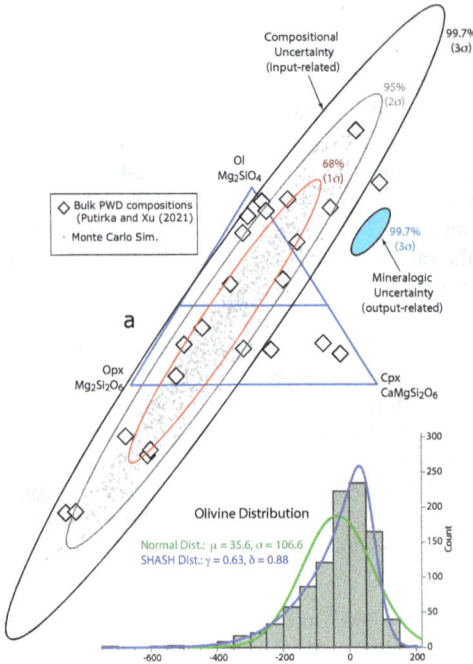

Figure 9 (a) 1,000 Monte Carlo simulations with 1σ, 2 σ and 3σ contours, of the PWD composition Ton 345, and its reported errors, along with 23 PWD compositions (Putirka and Xu 2021; see also Jura et al. 2015), projected into the Fe-free ternary system Ol + Cpx + Opx. **(b–e)** Error analysis using mean measurements uncertainties for PWDs, but projected into the ternary Ol+Pyx+Fe Metal. **(f–g)** Using the same measurement uncertainties as in Fig. 9a, panels f and g compare how errors vary depending upon the particular bulk composition that is plotted. **(h–i).** Using the same measurement uncertainties as in Fig. 9a, panels h and i show how the projected error depends upon the precise mineral formulas used to express the ternary mineral compositions.

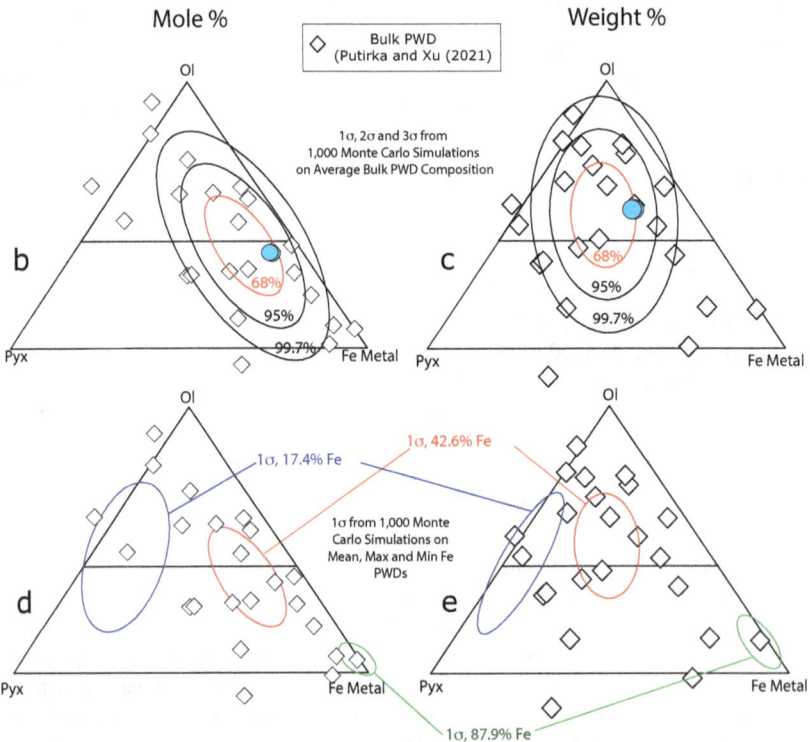

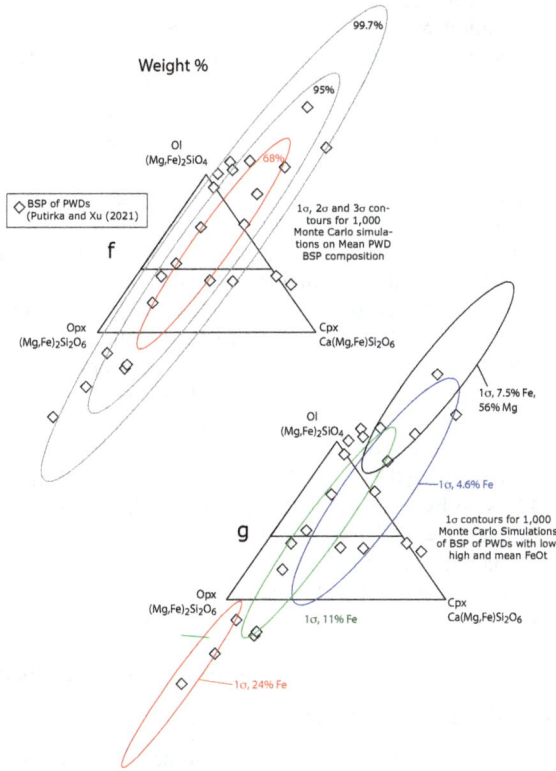

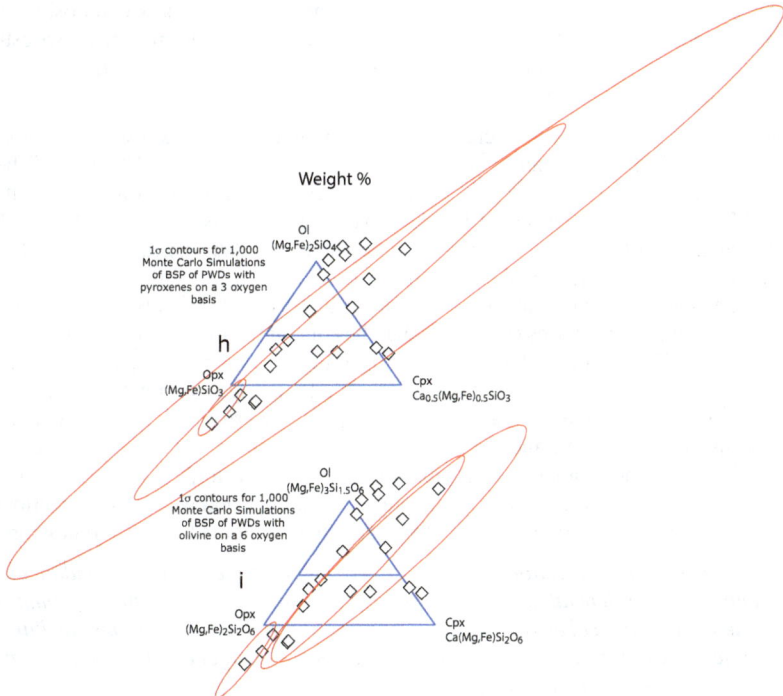

In any case, within the present context, every Monte Carlo simulation requires two inputs and a decision: a composition, an estimate of its uncertainty in the form of a standard deviation and a choice about what level of certainty is acceptable. For all subsequent Monte Carlo tests (Figs. 9b–i) the input compositions represent either a particular PWD bulk composition, or the mean of the 23 PWD compositions used by Putirka and Xu (2021), which are input as dex values (dex units are $\log_{10}$ units, so if a star has Fe/H reported as –3.2 dex, then the Fe/H ratio of the star is $10^{-3.2} = 6.31 \times 10^{-4}$). The errors are then propagated from dex inputs through to the calculation of mineral abundances. As will be shown, the shape and magnitude of the errors plotted within a ternary depend upon other items besides the input standard deviations. As the intent here is to illustrate how certain choices in constructing a ternary diagram affect the plotted errors, every Monte Carlo simulation in Figure 9 thus uses a single set of standard deviations (an average of published uncertainties across 23 PWDs), which in dex notation are: ±0.12 for Fe and Ca, ±0.14 for Mg and ±0.13 for Si (the average of the uncertainties for the 23 PWDs of Putirka and Xu 2021). Note that these errors lose their symmetry when uncertainties on the dex (log) scale are converted to a linear scale. For example, the contours of the Monte Carlo results for PWD Ton 345 (Fig. 9a) appear symmetric. But the lower panel of Fig. 9a isolates the distribution of olivine, which are skewed to higher abundances. The distribution of olivine is very well fit by a SHASH, or sinh–arcsinh, distribution (Jones and Pewsey (2009), and less so by a normal distribution (which is a special case of the SASH distribution, when $\gamma = 0$ and $\delta = 1$). Until many more PWD compositions are obtained, it's not yet clear that there is much to exploit in understanding these distributions, except to note that for any of the distributions shown in Fig. 9, high, positive olivine abundances are more likely than negative abundances, which fall under a very low probability tail (and similarly, highly positive Opx and Cpx values are under similarly under low-probability tails). Also, the greater the reported dex uncertainty, the greater the asymmetry in the probability density function.

Quirks of ternary projections, and their plotted errors

Errors within a ternary diagram are dependent on some not-so-obvious choices about how a given ternary is constructed. Figures 9b–e show the errors on PWD bulk compositions when they are projected into the ternary system using Model 4 in Table 9. This model allows estimates of metallic Fe (Figs. 9b–e) as well as silicate mineral abundances in the mantle, which it does by collapsing the two pyroxenes, Opx and Cpx, into a single pyroxene component (Pyx).

Several results are notable, especially for the very high-Fe PWD known as PG0843+517 which contains 87% Fe in its bulk composition. Errors are small for PG0843+517 because the remaining components are too low to have any great impact on error. For example, in the 1,000 Monte Carlo simulations of PG0843+517, Ca ranges between just 0–3 wt. %; the standard deviation on Ca must therefore be some value less than the range, or <±3%, even though the error on Ca is identical to that of Fe (±0.12 dex). In short, errors are weighted by their respective elemental abundances. Error on PG0843+517b is also particularly small because one of the ternary phases (Fe metal) matches one of the input components (wt. % Fe). Since Ca controls Cpx, is it possible that the error in Figure 9a is narrow perpendicular to the Cpx apex because the errors on Ca are low? This is a tempting interpretation (Trierweiler et al. 2023) but is incorrect; in every panel in Figures 9a–e, the error assigned to Ca is precisely the same as that assigned to Fe and the errors on the dex scale are not much different than for Mg or Si. Figure 9a shows a narrow distribution perpendicular to the Cpx because Ton 345, like PG0843+517, has low Ca. Incidentally, note also that mole and weight % projections have different magnitudes of uncertainty (Figs. 9b–e) and their contours have different geometries.

Tenet 1: In ternary projections, errors are weighted by the relative abundances of the elements, and thus depend upon the input composition, even when differing compositions have the same analytical errors; this also means that extreme compositions can have lower plotted ternary errors in some projections, especially when one of the input components matches a projected mineral component.

Tenet 2: Error depends upon the units used to describe the various projected phases (wt. %, mole % or volume %).

Note also that errors are elongated in the direction of the Ol–Opx tie-line (Trierweiler et al. 2023) in Figure 9a. This issue is not intrinsic to the method. To test for such error, one can apply the standard deviations of Figure 7 to a Monte Carlo simulation. Figure 7 compares the mineral abundances calculated by Model 1 to the mineral abundances of actual rocks, where bulk and mineral compositions are known and where mineral proportions can be obtained by mass balance. The blue field in Figure 9a shows that Ol/Opx and Ol/Pyx ratios can be determined with comparative accuracy relative to the spread of Ol/Opx in the PWD data set. We can safely conclude that the range in the calculated Ol/Opx ratios among PWDs is not a result of the method applied to determine the PWD mineral abundances. Figures 9b–e also show how compositional errors can be compressed along the Ol–Pyx tie-line, especially in the mole % projection (Figs. 9b–e).

Tenet 3: Errors in ternary projections reflect a competition between the plotted mineral components, and the geometry of error ellipses depend strongly on the choice of minerals being projected.

By collapsing Cpx and Opx into a single Pyx component (Figs. 9b–e), however, we've lost some critical mineralogic information. Experimental studies, for example, show that Cpx-rich systems melt at lower temperatures, and so produce greater amounts of crust at given temperature (Lambart et al. 2016). To regain such information, Figures 9f–g show projections of BSPs for the PWDs of Putirka and Xu (2021), using Model 1 of table 9. Bulk silicate compositions are also calculated using $\alpha_{Fe} = 0.27$. But in this projection, the positions of the plotted points are not affected by the choice of α_{Fe} values (Figs. 9f–g) because Wüs (FeO) is used as the apex from which the compositions are being projected. By projecting from Wüs, the positions of the plotted points are affected only by the Fe assigned to the minerals in the matrix of Model 1. Any other changes in Fe, e.g., via an assumed α_{Fe} value, will only move points along the projection line (as in Figs. 4b–c), with no impact on the positions of the points in the projected (Fe-bearing) plane, Ol–Cpx–Opx.

Tenet 4: The choice of mineral from which a ternary plot is projected affects whether a given compositional parameter affects the final plotted position.

Errors in Figure 9g are much smaller than in Figure 9a for two reasons: first, Figure 9a is an Fe-free system, whereas, by virtue of applying Model 1 in Table 9a, our mineral compositions in Figure 9g are Fe-bearing. In addition, we plot our compositions in mole %. The issue with regard to Fe is parallel to that of Figures 9b–e: by allowing Fe into the calculations, there is less room for variations in Mg and Si, especially for Fe-rich compositions. To illustrate, Figure 9g shows 1σ errors for PWDs with high Fe (PG0843+517; 24 wt. %), low Fe (NLTT 43806; 4.6 wt. %), and intermediate Fe (PWD G241-6; 7.5 wt. % Fe), the latter of which also has high Mg (56 wt. %). Also plotted is error for the average of all 23 PWD compositions (11 wt. % Fe) of Putirka and Xu (2021). Plotted errors for the high-Fe case are smaller because even at just 24 wt. % Fe in the BSP, the Fe content is still high enough to restrict variation in the remaining three elements (Si, Mg and Ca). The high-Mg case also has less error in the Ol–Opx direction, mostly by virtue of its higher Mg content. A key issue is that errors are thus dependent upon choices made regarding mineral compositions and style of projection.

One final quirk, and one that may be explored to some advantage, is that ternary errors are also affected by the choice of formulas used to describe a given phase (Figs. 9h–i). Pyroxenes are commonly expressed on the basis of 3-oxygens instead of the 6-oxygen basis used in Figures 9f–g. Figure 9h shows the results of that choice. Errors on the mean and low-Fe compositions are considerably larger than in projections that use pyroxene formulas expressed on the basis of 6 oxygens (Figs. 9f–g). But a more unconventional approach is shown in Figure 9i that reduces error, by expressing olivine compositions on the basis of 6 oxygens instead of the usual 4. This choice allows the molecular weight of olivine to more closely match that of the pyroxenes (so the formula for forsterite, usually expressed as Mg_2SiO_4, becomes $Mg_3Si_{1.5}O_6$). Error on PG0843+517 is smaller, for example, but without the concomitant amplification of error elsewhere in the ternary as occurs in Figure 9h.

Tenet 5: Error can be minimized in some projections when stoichiometric mineral formulas are written so as to equalize weight or molar proportions.

None of the plots of Figures 9a–i are intrinsically better or worse than any other, nor does any single plot provide a definitive view of error. Thus, none need be preferred—except insofar as they allow one to better describe a system, or improve tests of hypotheses.

Are any exoplanets exotic?

Understanding how to interpret the uncertainties plotted within Figure 9 is essential to considering possible exoplanet mineralogies. For example, if we require a ±3σ, or 99.7% certainty (e.g., as in Trierweiler et al. 2023), Figures 9a,f show that a wide range of mineralogies are possible: *our choice of a 3σ confidence interval requires us to consider any mineral assemblage encompassed within the 99.7% ellipse as possibly real.* One could, instead, choose a smaller level of confidence, say 1σ, but even at that level of confidence some PWDs plot outside the ternary.

The above approach treats error as if only one PWD observation is available. But more than one PWD composition has been measured. Should we average all the observed PWDs and treat them as one composition? Or are there multiple PWD compositions that we must consider? Three different tests will be presented.

First, we can compare PWDs to one another, as Figures 10a–b. Figures 10a–b show mineral abundances for 23 PWD compositions obtained using both Method 1 (Fig. 10a) and Method 2 (Fig. 10b), along with 1,000 Monte Carlo results (as in Fig. 9) using the mean PWD composition as input. If the 23 PWD compositions represent a random sampling of effectively equivalent compositions, then the distribution of the observed 23 PWDs compositions should mimic the distribution from the Monte Carlo simulations. Significantly, a large fraction of PWDs plot where the Monte Carlo simulations predict a low density of observations (Figs. 10a–b). And only four of the PWDs overlap with inner Solar System compositions, regardless of whether Method 1 or 2 is employed (Figs. 10a–b). The same result can also be seen in the Fe-free system of Figure 9a: about half of PWDs there do not fall within that ternary diagram and many plot where the Monte Carlo simulations also predict that few compositions should occur, if we are randomly sampling only one composition (note how adding Fe shifts the positions of the data compared to the Fe-free case of Fig. 9a). It is thus appears that by sampling multiple PWDs we have probably sampled different PWD compositions.

A second test is more quantitative, and can be used to calculate the probability that any PWDs fall outside the ternary diagrams of Figures 10a–b. The pertinent probability calculation involves the contours of every PWD composition (not shown in Figs. 10a–b), not just one. The simplest case is to calculate the probability that *at least one of two PWDs fall outside the ternary*, (call them A and B), where we can use the equation:

$$P(A \text{ or } B) = P(A) + P(B) - P(A \& B) \tag{7a}$$

Here, $P(A \text{ or } B)$ is the probability that either A or B fall outside the ternary, $P(A)$ is the probability that A falls outside the ternary, $P(B)$ is the probability that B falls outside the ternary, and $P(A \& B) = P(A)P(B)$, is the probability that both A and B fall outside the ternary. For the case of whether A, or B or C fall outside the ternary, we have:

$$P(A \text{ or } B \text{ or } C) = P(A) + P(B) + P(C) + P(A \& B \& C)$$
$$- P(A \& B) - P(B \& C) - P(A \& C)$$

(7b)

Equation (7b) can, of course, be expanded to cover any number of observations, using terms analogous to those in Equation (7a). If the Monte Carlo simulations for A, B and C indicate that $P(A) = 0.7$, $P(B) = 0.5$ and $P(C) = 0.3$, then the resulting probabilities that at least one PWD falls outside the ternary are: $P(A \text{ or } B) = 0.85$, and $P(A \text{ or } B \text{ or } C) = 0.9$. As should be expected, *the more PWDs that are observed to fall outside the ternary, the higher the probability that at least one PWD falls outside the ternary.*

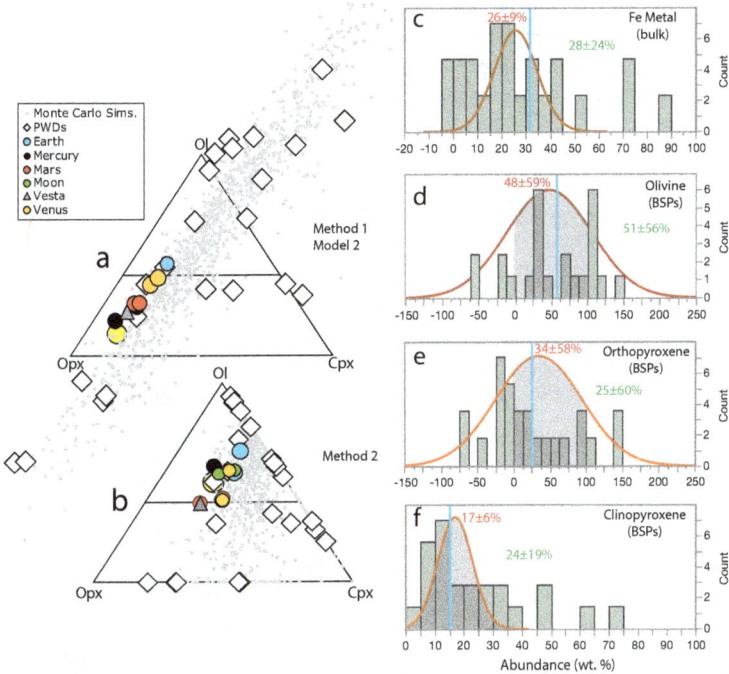

Figure 10. (a) PWD mineral abundances (from bulk compositions in Putirka and Xu (2021) are plotted using Method 1, utilizing Model 2 of Table 9a. (b) PWD mineral abundances are calculated using Method 2. Both (a) and (b) show mineral estimates for the inner planets, using compositions from Table 3b and compositions from McDonough and Yoshizaki (2021). Also shown are results from 1,000 Monte Carlo simulations using a mean PWD composition and uncertainties as input. In (b), using Method 2, most Monte Carlo simulations plot on the boundaries of the ternary diagram. (c) Comparison of the observed distribution of bulk PWD Fe metal contents (**green bars**), obtained using Model 4 of Table 9a (**green bars**) with 1,000 Monte Carlo simulations (**red curve**) assuming a Gaussian distribution, using the mean PWD composition and error as input. (d–f) mineral abundances for PWD BSPs (**green bars**) using Model 1 of Table 9a, as in Figure 10a, with Monte Carlo simulations of mean PWD (**red curve**). **Values in green** are mean and standard deviations of PWD mineral abundance estimates; **values in red** are the same for the Monte Carlo simulations. **Gray shaded areas** of (d–f) are mineral estimates that would fall within the ultramafic ternary diagram (Fig. 4a). **Blue bars** indicate Earth-like compositions.

A third test is to compare "probability density functions" (or pdf); for illustrative purposes, we'll apply a Gaussian distribution generated from a Monte Carlo simulation, using the global mean and standard deviations as parameters (Figs. 10c–f). We could also use a SHASH distribution (Fig. 9a lower panel), or other distributions, but the Central Limit Theorem states that a random sample from any distribution is likely to approximate a Gaussian distribution, so a Gaussian pdf is a good place to start, especially when the true distribution is unknown. This test examines whether the distribution of observed PWDs (Figs. 10c–f; green histogram) mimics the distribution one would expect if we have a random sample of what is, effectively, a single compositions, or a very narrow range of similar compositions (Figs. 10c–f; orange, Gaussian curve). The gray shaded areas of the Gaussian pdfs (Figs. 10d–f) show the region where mineral abundances would fall within the ultramafic ternary (Fig. 4a); blue bars (Figs. 10c–f) mark values for Earth (or nearly equivalently, CI chondrites). The mean values of the PWD compositions (green values) are not too different from Earth, but not all PWDs are Earth-like, let alone well described by the Gaussian distribution. The poorest fits are for Fe metal, when bulk PWDs are used as input (Fig. 10c) and for Clinopyroxene when BSP compositions are considered (Fig. 10f). In both cases, the observed values extend well outside even the 99.7% confidence intervals. If we applied a SHASH distribution, the observed high Cpx and high Opx cases would be predicted to be even less probable. In any case, the observed ranges for bulk Fe, Opx and Cpx might be real, since the Gaussian pdf suggests that these are very low probability events if error is random. Orthopyroxene is perhaps especially interesting as it shows an observed mode at −20 wt. % Opx (so nominally ferro-periclase normative; Putirka and Xu 2021), which is far below the predicted mean value of +34 wt. %. Olivine abundances are better predicted by a Gaussian distribution, as one mode closely approaches the predicted mean; but a second mode at > 100% Ol (also nominally ferro-periclase normative) could indicate a bi-modal distribution, or an entirely different mean that is unlike Earth.

The interpretation of Figures 10c–f is highly challenged by the small number of PWD observations; it is conceivable that new observations could well shift the observed abundances so as to better mimic a single distribution, Gaussian or otherwise. But we must interpret the data we have, not the data we wished we had. Regardless, multiple, independent observations are needed.

Comparing multiple observations of individual PWDs

Some good news is that multiple and independent analyses of PWDs are available, which indicate vastly less uncertainty than the Monte Carlo simulations based on a single observation (Fig. 9a). The independent compositional estimates of certain PWDs stem from the fact that planet-building elements, Fe, Si, Mg and Ca, can be detected from both the UV and optical portions of the electromagnetic spectrum (e.g., Klein et al. 2011; Jura et al. 2012; Xu et al. 2017, 2019). Optical and UV spectra can also yield different estimates of H and He (as H is determined either by Lyman alpha lines in the UV, or Balmer lines in the optical part of the spectrum), and estimates of H and He in turn affect estimates of the planet-building elements (see Xu et al. 2017). Optical and UV spectra thus provide highly independent tests of reproducibility. These tests are more than just differences in curves fitted to different parts of the electromagnetic spectrum. As explained by Xu et al. (2019), optical and UV results also rely on different models of white dwarf structure, as the UV and optical lines emanate from different depths in PWD atmospheres. Figure 11 shows that in some cases, though by no mean all, UV and Optical data (and some repeat observations by one or the other method) are in remarkable agreement.

It should be noted that some astronomers (e.g., Rogers et al. 2024) are greatly concerned about order-of-magnitude differences between UV and optical spectra for some elemental ratios (e.g., Si/H). But such contrasts matter less than at first appearances: while Fe/H and Si/H might differ greatly, Fe/Si ratios might still be very similar (Xu et al. 2017, 2019). The PWD PG0843+517 is illustrative: UV spectra (Gansicke et al. 2012) yield Fe/H = −4.6 dex, while optical spectra (Xu et al. 2019) yield Fe/H = −3.84 dex, and astronomers are naturally concerned about this nearly order-of-magnitude difference in Fe/H. But when estimating mineral abundances, it's not the absolute dex values of Fe/H, or any other given element, that matter so much as the relative values of Fe/H, Mg/H and Si/H. For PG 0843+517, the optical spectra not only yield higher Fe/H but also higher Mg/H and Si/H; as a result, the weight % values of Fe are not very different: 87.9 wt. % from the optical data and 73.5 wt. % from UV.

Both estimates of Fe for PG0843+517 are well above the mean Fe (42.4 ± 22.3 wt. %) of the 23 PWDs examined by Putirka and Xu (2021). Averaging the two estimates for PG 0843+517 yields 80.7 ± 10.2 wt. %. The order-of-magnitude dex-scale difference in Fe/H thus collapses to a <20% difference in Fe on a wt. % basis, and a ±10.2 wt. % uncertainty for the Fe content of PG 0843+517. On the basis of this averaging, one can reasonably conclude that PG 0843+517 is very likely enriched in Fe relative to other PWDs. The repeat observations of PG 0843-517 clearly bolster this view.

Figure 11 shows mineral abundance estimates for PG 0843+517 and six other PWDs where multiple published abundances are available. Except for PG 1015, the range of reported compositions from independent measurements is considerably less than the errors indicated in Figs. 9a. The mean and median errors shown in Figure 11 are indeed greater than the uncertainties shown in similar plots in Putirka and Xu (2021), where a similar analysis, but using only PG 1225

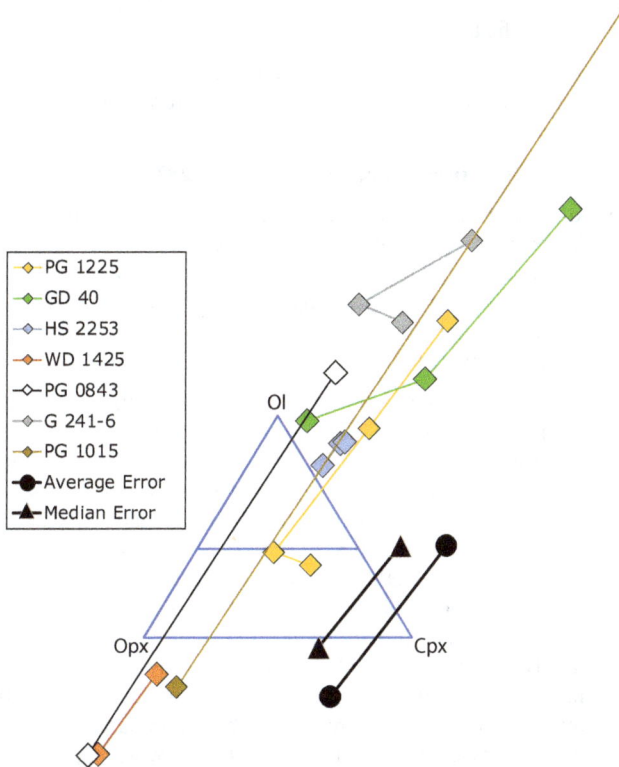

Figure 11. Comparison of multiple reported compositions for several PWDs. Average and median uncertainties for the seven PWDs are shown as **black filled symbols**. Standard deviations of the observed PWD compositions would be smaller than the mean and median uncertainties, except for PW 0843 and PG 1015. Note that with the projection from the wüstite (FeO) apex, the data in Fig. 10 are not sensitive to Fe contents; variations in Fe within the Ol–Opx–Cpx–Wüs tetrahedron occur along a line connecting a given point to the Wüs apex, as in Fig. 4d–e, if one replaces garnet with wüstite. Standard deviations of the multiple estimates for most of the PWDs shown in Fig. 10 will also yield even smaller uncertainties than indicated by the mean and median compositional ranges. Data sources are as follows: G241-6, Jura et al. (2012), Zuckerman et al. (2011); G40 Jura et al. (2012); HS 2243+8023, Klein et al. (2011); PG 0843+517 and PG 1015+161, Xu et al. (2019), Gansicke et al. (2012); PG 1225-079, Xu et al. (2013), Klein et al. (2011); WD 1425+540, Xu et al. (2017). **Note**: Gansicke et al. (2012) do not report Ca for PG 0843, so we substitute the mean of 23 PWDs where Ca = –6.94 dex.

and HS 2253, was applied. However, even in this expanded comparison, the mean and median error ranges, which include the very large errors in PG 1015, are less than even the 1σ errors from the Monte Carlo simulations (Figs. 9a–i), which implies that mean values obtained via one method (UV) can be faithfully duplicated by another (Optical). (Note that PG 0843+517 exhibits a wider mineralogical range than most, but this is not due to its <20% error on Fe, but rather the 400% difference in Mg: 4 % vs. 16%, respectively for UV and optical results).

Figure 11 also provides a test of whether any PWDs are exotic, i.e., whether they have mantles that fall outside the ultramafic ternary. PG0843 was considered as exotic by Putirka and Xu (2021), but taking the mean of the two observations, it looks to fall within the ternary. But PG1225 plotted within the ternary in Putirka and Xu (2021), but the additional estimates here indicate that it might be exotic. This example harkens back to a key point about the Monte Carlo simulations: we must be prepared to accept that some PWDs that fall within the ternary, might actually be exotic, once more analyses are obtained. In any case, four of the seven PWDs (WD 1425, GD 40, G 241-6 and HS 2253) with multiple composition estimates fall entirely outside the ultramafic ternary.

The above arguments can only be controversial to the extent that reproducibility is an unreliable measurement strategy. For the future of exoplanet research, reproducibility should, of course, be a top priority.

Do Methods 1 and 2 agree in predicting PWD mineralogy?

A final aspect of reproducibility is to test whether independent methods, such as the two introduced in this chapter, are capable of yielding similar results. And they do. Of the 23 PWDs, eight fall entirely within the ultramafic ternary using Method 1, and for these 8 compositions, the molar abundances calculated by Methods 1 and 2 correlate with R^2 values of 0.8, 0.95 and 0.93 for Ol, Cpx and Opx are respectively. Although well correlated, Method 2 yields systematically higher Cpx and lower Opx values than Method 1.

Of the 23 PWDs in Figures 9a–i, five are predicted to have Ol<0 when Method 1 (Model 1 of Table 9a) is applied. For agreement, Method 2 should then yield Ol = 0 for these cases and it does so for four of the five (80%). Putirka and Xu (2021) hypothesized that such systems could be Qz-saturated (not a necessary result of Method 1) and Method 2 also predicts Qz>0 for two of the five cases (where Ol<0 by Method 1). But Method 2 allows any excess Si (step 15) to react with excess Fe (from step 8) to form Ferrosilite (Fs; step 17), and Fs >0 for the other two cases where Ol = 0 by Method 2. If Fs were unstable for any reason (or removed from the calculation), then all 4 cases of Ol = 0 would indeed yield Qz>0 by Method 2 (and any excess Fe from step 8 would partition into a metallic Fe phase, i.e., a metal core). Another test involves the 10 PWDs that are predicted to have Opx<0 by Method 1. Method 2 should predict Opx = 0, and it does so for 80% of these cases. Putirka and Xu (2021) predicted that when Opx<0 by Method 1 such systems would be saturated with ferro-periclase, and Method 2 predicts ferro-periclase saturation for all eight cases where Opx = 0.

Methods 1 and 2 are thus likely to yield similar classifications of exoplanets, even when such planets have compositions that are exotic to the ultramafic ternary diagram (Fig. 4).

THERMODYNAMIC CONCEPTS & RELATED CAVEATS

The goal of Putirka and Xu (2021) was to show, in part, that some exoplanets may well fall outside the ultramafic ternary and that conclusion seems to be well-founded, at least thus far. But like Cross et al. (1902) we are under no illusion that our published mineral abundances represent precise estimates of exoplanetary rock types. The reasons that warrant caution are several, including the simplicity of our thermodynamic models, the potential that currently unknown phases might be stable in exotic compositions, and because even in non-exotic cases, other components, such as O, S and C, might occur in sufficient abundances to require new phases or altogether novel mineral assemblages.

Solid solution (*"All models are wrong; some are useful"* – George Box)

Minerals are defined as naturally occurring substances that have an ordered atomic arrangement and a definite chemical composition. However, "definite" does not mean "fixed". To take a contrary example, a liquid has no fixed stoichiometry, or rather no fixed ratios of certain elements, so there are no definite formulas that we use to describe them. Liquids then, are not minerals. In contrast, materials such as quartz (SiO_2), corundum (Al_2O_3) or forsterite olivine (Mg_2SiO_4) have very definite ratios of, respectively, Si/O, Al/O and Mg/Si, among others. The formulas we write are, in effect, recipes, needed to make a given mineral. But there is some flexibility in most recipes. This variability is referred to as "solid solution".

Olivine is a useful example as it is the dominant mineral in the mantles of the inner planets and its chemistry is not terribly complex. Olivine has two common end-members, the Fe-end member, fayalite (Fa; Fe_2SiO_4) and the just-mentioned Mg-end member, forsterite (Fo). For Fa, its stoichiometry indicates that Fe is in the Fe^{2+} state. Both Fe^{2+} and Mg^{2+} fill similar sites within the olivine structure, which are referred to as "M1" and "M2" (for "metal 1" and "metal 2"; Fig. 12a). The M1 and M2 sites are called "octahedral" sites because Mg and Fe are bonded to 6 oxygens that together create an 8-sided octahedron when we draw lines connecting the oxygens. Magnesium and Fe happily partition into these octahedral sites, as do other elements that have the same charge and similar ionic radii, i.e., Mn, Ni, and to a lesser extent, Ca. The smaller Si^{4+} atom occurs within a tetrahedral (or "T") site (Fig. 12a), where Si is bonded to 4 oxygens that create a 4-sided polyhedron (Fig. 12a). (Many other substitutions are possible; for example, very small amounts of P^{5+} can partition into the T site and Al^{3+} can enter both the T and M1 sites, with charge balance being taken up by vacancies or Fe^{3+} in the M1; see Shea et al. 2019 and references therein). The M1 site is a tad smaller than M2 and (so the larger Ca^{2+} will favor M2 while the smaller Ni^{2+} will favor M1), but for the case of Fe^{2+} and Mg^{2+} their radii are sufficiently similar so as to yield no energetic preference for either Fe or Mg into the M1 or M2. This ease of partitioning of Mg and Fe^{2+} allows olivine to exhibit "complete solid solution" between the Fo and Fa end-members, meaning that a natural olivine can have any composition, from pure Fo to pure Fa, or any mixture in between.

Perhaps the most glaring simplification of Method 1 and the projections of Figures 9–10 is that minerals are assigned a constant composition, regardless of the bulk composition with which they are equilibrated. So long as the magnitude of compositional error so greatly outweighs the error associated with calculating mineral abundances (Fig. 9a), the simplification will do. But if compositional error decreases, the limitations of this approach may be exposed. To give just one example, experiments by Zhao et al. (2018) indicate that the viscosity of Mars' Fe-rich mantle may be 5 times less viscous than the relatively Fe-poor terrestrial mantle—and such differences in viscosity can greatly affect whether or not we predict a planet to exhibit terrestrial-like plate tectonics (e.g., Weller and Lenardic 2018). So when applying either of Methods 1 or 2, it might be critical to use a separate model—adjusting Fe contents in the matrix of Method 1 or in the Mg# of Method 2—for each bulk composition, especially if mineralogical outputs are used as inputs into numerical models that simulate mantle circulation.

The amounts of Fe and Mg in olivine are very sensitive to temperature, as well as liquid composition, these two variables being intimately connected. And for exoplanets that equilibrate at different temperatures, *T*-dependent models could also be useful. Figure 12b shows a series of diagrams that compare Gibbs Free Energy (*G*) to composition (*X*), where *X* can range from pure Fa on the left, to pure Fo on the right. In the topmost panel, which shows the case for a high *T*, noted as T1, the liquid has a lower free energy at every possible value of *X*, and so the

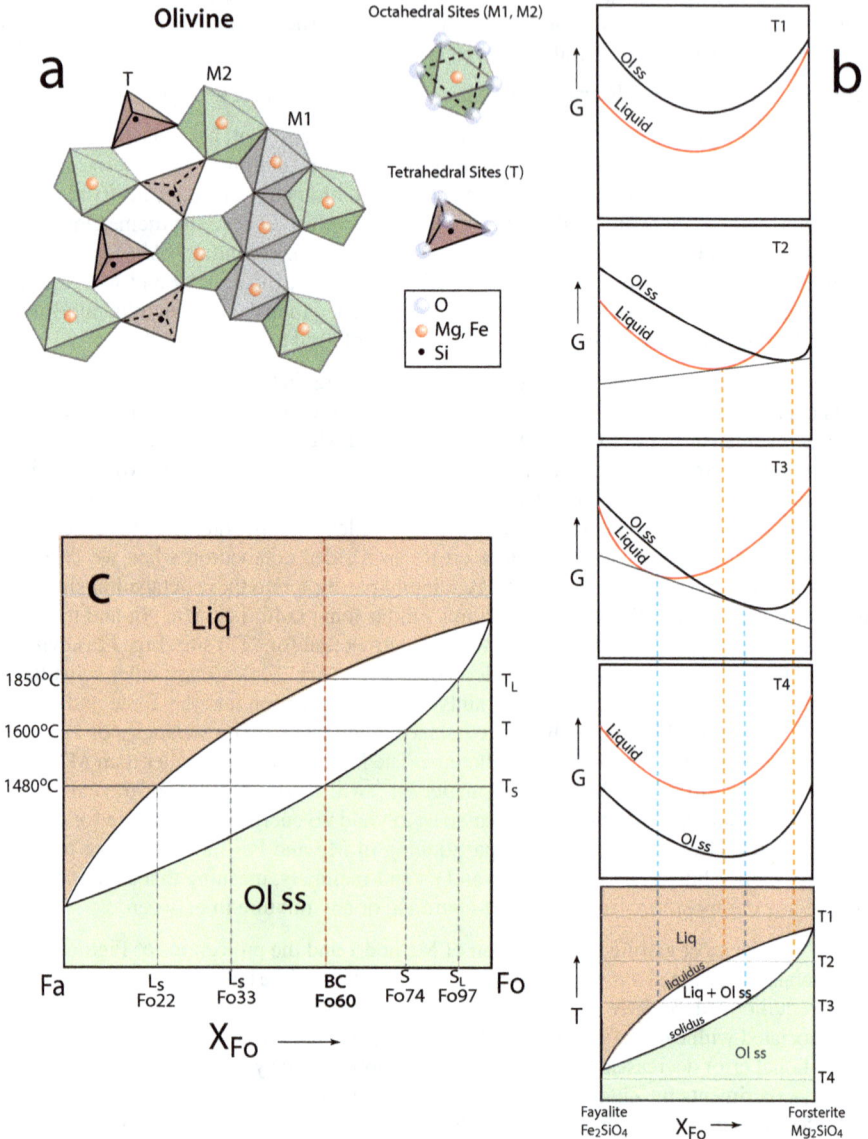

Figure 12. (a) olivine structure model showing the crystallographic sites where Fe, Mg and Si occur; (b) Gibbs Free Energy vs. Composition (*G–X* diagrams, panels T1–T4) and (c) binary temperature vs. composition phase diagram (also, *T–X* diagram, below *G–X* diagrams in b) showing the complete solid solution between Fa and Fo in olivine.

liquid is stable for all compositions. In the *T–X* phase diagram in the bottom panel, the region in red represents a range of *T–X* where only liquid is stable and olivine will not crystallize. As *T* decreases (from T2 down to T4 in Fig. 12b), the free energy curve for olivine decreases at a faster rate than for liquid, and it intersects that of the liquid at some *T* between T1 and T2. The compositions of the liquid and equilibrium olivine can be found from the common tangent of the liquid and olivine *G–X* curves (panel T2), which are projected downward onto the *T–X* diagram to show the equilibrium liquid and olivine compositions in *T–X* space. With further decreases in T (T3), the positions of the common tangent points shift towards more Fe-rich compositions. And when T is low enough (T4), olivine has a lower Gibbs Free Energy than the liquid at all compositions and only olivine is stable, as in the green area of the T–X diagram in Fig. 10b.

The G–X diagrams illustrate how the T–X, or binary phase, diagram of Figure 12b is produced, but the T–X diagram (Fig. 12c) is sufficient to describe the equilibrium state of any system. For example, the bulk composition (BC) in Figure 12c, is 60% Fo (and thus 40% Fa). Upon heating, it would begin melting at 1480°C, and the first bit of liquid to be produced would have a composition of 22% Fo (or Fo22). At 1600°C, the system is partially melted, with the liquid having a composition of Fo33, which is in equilibrium with crystals of Fo74. With further heating, the BC would be completed melted at 1850°C and the liquid would be identical in composition to BC, while the last bit of solid to dissolve (melt) into the liquid would have a composition of Fo97.

Other minerals besides olivine, but still common to Earth's mantle, exhibit solid solution. For exoplanet studies it's probably safe to ignore the minor element solid solution series and concern ourselves only with the overwhelmingly dominant Fe-Mg solid solutions that occur within olivine, the pyroxenes and garnet, but other potentially important solid solution series are shown in Table 11.

Table 11. Solid solution series.

Mineral series	Mineral names	Abbrev.	End member formulas		Exchange reaction
Olivine	Forsterite–Fayalite	Fo–Fa	Mg_2SiO_4	Fe_2SiO_4	$Mg^{2+} \Leftrightarrow Fe^{2+}$
Orthopyroxene	Enstatite–Ferrosilite	En–Fs	$Mg_2Si_2O_6$	$Fe_2Si_2O_6$	$Mg^{2+} \Leftrightarrow Fe^{2+}$
Clinopyroxene	Diopside–Hedenbergite	Di–Hd	$CaMgSi_2O_6$	$CaFeSi_2O_6$	$Mg^{2+} \Leftrightarrow Fe^{2+}$
	Diopside–Ca Tschermak	Di–CaTs	$CaMgSi_2O_6$	$CaAl_2SiO_6$	$Mg^{2+}Si^{4+} \Leftrightarrow 2Al^{3+}$
	Diopside–Jadeite	Di–Jd	$CaMgSi_2O_6$	$NaAlSi_2O_6$	$Ca^{2+}Mg^{2+} \Leftrightarrow Na^{1+}Al^{3+}$
Plagioclase feldspar	Anorthite–Albite	An–Ab	$CaAl_2Si_2O_8$	$NAlSi_3O_8$	$Ca^{2+}Al^{3+} \Leftrightarrow Na^{1+}Si^{4+}$
Alkali feldspar	Albite–Orthoclase	Ab–Or	$NaAlSi_3O_8$	$KAlSi_3O_8$	$Na^{1+} \Leftrightarrow K^{1+}$
Garnet	Pyrope–Almandine	Py–Al	$Mg_3Al_2Si_3O_{12}$	$Fe_3Al_2Si_3O_{12}$	$Mg^{2+} \Leftrightarrow Fe^{2+}$
	Pyrope–Grossular	Py–Gr	$Mg_3Al_2Si_3O_{12}$	$Ca_3Al_2Si_3O_{12}$	$Mg^{2+} \Leftrightarrow Ca^{2+}$
Oxides	Periclase–Wüstite	Per–Wüs	MgO	FeO	$Mg^{2+} \Leftrightarrow Fe^{2+}$

Although Methods 1 and 2 appear useful (Figs. 7–8, 9a), some significant fraction of the predictive error (Figs. 7–8) is assuredly due to the lack of treatment of solid solution. For example, Cpx abundances are under-predicted by both models, probably because insufficient Al is assigned to Cpx. Again, it seems vastly premature to speculate on the amounts of Al that may be partitioned into exoplanetary Cpx given the large input errors of Fig. 9a. But if the need arises, the CIPW norm-like approach of Method 2 can and should be modified to accommodate such solid solution effects. If we are ever in a position to evaluate crust and mantle compositions on exoplanets, accounting for T-sensitive partitioning may be crucial and the kind of modelling that is conducted within MELTS (Ghiorso and Sack 1995; Asimow and Ghiorso 1998; Till et al. 2012), if properly tested against known cases, should improve predictions of mineral abundances and melting behavior.

Gibbs–Duhem equation

The Gibbs–Duhem equation provides a caution flag when comparing the mineralogy of planets of different bulk compositions. The equation is a mathematical statement that all intensive parameters (pressure, temperature and chemical activities) are interconnected—a sort of geochemical Indra's Net. The equation is often expressed as a function solely of composition. For a system at equilibrium and so at constant P and T, the Gibbs–Duhem equation is:

$$\sum_i n_i d\mu_i = 0 \tag{8}$$

where n_i is the number of moles of component i in a given phase and μ_i is the chemical potential of i. The conceptual import of Equation (8) is that with any change in the number of moles of any component in a system there must be consequent changes in chemical potentials, so that the sum remains 0 for the system to approach equilibrium. Since a Gibbs–Duhem equation applies to every phase in a given equilibrium assemblage, the equation can be used to derive the phase rule, which provides the degrees of freedom (f), for a system of c components and p phases; f is the number of intensive variables (P, T, mole fraction of a given component) that can be varied independently without affecting the number of phases, p, in the system:

$$f = c - p + 2 \tag{9}$$

The Gibbs–Duhem equation thus applies to both homogenous (a single phase) and heterogenous (multi-phase) systems. Since Gibbs Free Energy can be expressed as:

$$dG = VdP - SdT + \sum_i \mu_i dn_i \tag{10}$$

several substitutions lead to the following equation that expressly illustrates the interdependencies of P, T and μ_i (see Denbigh 1981):

$$VdP - SdT = \sum_i n_i d\mu_i \tag{11}$$

and thus if an equilibrated system is perturbed in any way, by a shift in T, P or composition ($d\mu_i$) there must be a reactive perturbation in any or all these variables to maintain the equality of Equation (11).

Equations (10–11) are as a reminder that if one exoplanet has a different composition than another, then even at equivalent P–T conditions, we can be certain that there are mineralogical differences also. The recommended standard state (i.e., P–T conditions of Earth's upper mantle, or ca. 2 GPa and 1350 °C) is a recommendation to focus on the compositional—and consequently mineralogical—implications.

Small compositional shifts might be accommodated entirely by solid solution. But small differences also have the potential to precipitate a wholesale change in phase assemblage. To illustrate, Figures 13a–b show a hypothetical case where Si and Mg vary within a range that might describe Earth or the inner planets. Throughout the ranges in MgO and SiO_2 in Earth's mantle Ol and Opx are stable. Less certain (until we conduct experiments to find out) is whether other phases may be lurking in the near compositional distance. For example, at SiO_2 greater than observed in Earth's mantle (Fig. 13c) we might calculate that Qz is stable, when instead another phase (x) may have a lower Gibbs Free Energy and so would displace Qz from the equilibrium assemblage. Similarly, we might calculate that periclase is stable as MgO contents increase, but another phase (y), might replace Per (Fig. 13d). The same issue applies to any of the planet-building elements, which means that mineral assemblages could look quite different than as calculated by Methods 1 and 2.

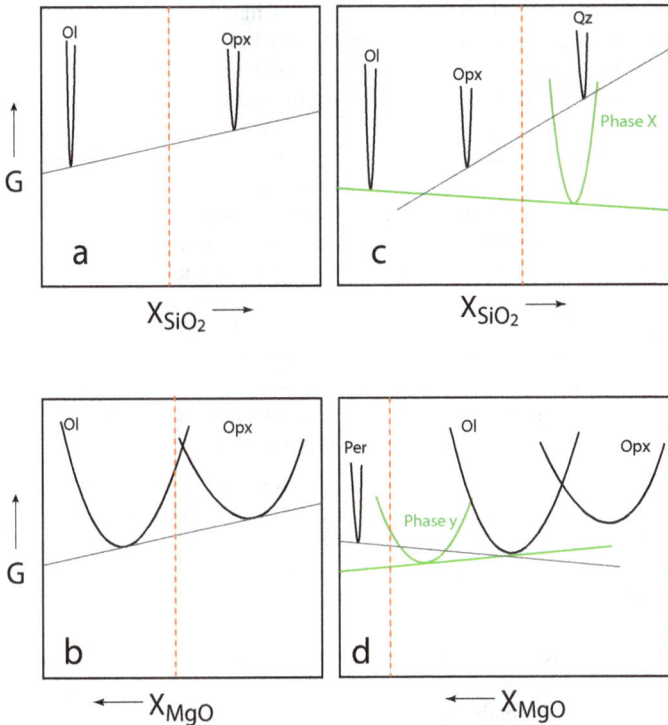

Figure 13. Hypothetical *G–X* diagrams for the cases when (**a**) SiO_2 and (**b**) MgO vary within ranges defined by Earth and the inner planets and cases that might obtain in exoplanets, where (**c**) SiO_2 and (**d**) MgO might vary more broadly, and unconventional or unknown phases with relatively low Gibbs Free Energies may be lurking.

O, C, S, H and other unknowns

Oxygen is far and away the dominant anion in the rocky materials that comprise the inner Solar System. Its abundance also provides a first-order control on the fraction of Fe that will be partitioned into a planet's metal core. But H, C, and S, are also highly abundant in stellar atmospheres and large concentrations of any of H, C, O, or S could play havoc with estimates of exoplanet mineralogy.

Oxygen (and Fe). Despite its importance, O is rarely if ever measured in rock or mineral samples. The reason is because in X-ray fluorescence (XRF) spectroscopy (the most common method of rock analysis), samples are bombarded with so-called "primary" X-rays that interact with the inner electron shells of target elements to induce "secondary" X-rays, via fluorescence. For light elements, such as O, the energies of the secondary X-rays are very low and more readily absorbed by the sample, leaving lesser amounts of energy to reach a detector. Since O is the dominant anion in nearly all rocks and minerals, though, total O contents can be calculated by charge balance, once all the major cations have been quantified. In geological contexts, O is thus largely known by calculation, the ofnly prominent exception being partial melting experiments conducted at 1-atm pressure, where an experimenter can dictate the O_2 content of the atmosphere. In experimental studies, the amount of O_2 is a system is often expressed as "oxygen fugacity" or fO_2, which is the partial pressure of O_2 in the atmosphere that is in equilibrium with a given system—usually a magma. The two quantities, fO_2 and total O content, are by no means identical, but when a system is in contact with an atmosphere of a given O content, they are certainly related to one another.

Because many systems of interest (e.g., Earth's mantle) are not in equilibrium with a vapor phase, nearly all discussions of oxygen fugacity in Geology are fictive, and hinge upon measurements of Fe. Iron occurs in two different oxidation states, Fe^{2+} and Fe^{3+}, in silicate rocks and minerals. Oxygen contents, or fO_2, are thus most commonly inferred from Fe_2O_3/FeO ratios (based on experiments where the systems are equilibrated with an atmosphere of known fO_2). Several publications provide excellent summaries of the fictive fO_2 of bulk Earth and Earth's upper mantle (e.g., Wade and Wood 2005; Corgne et al. 2008; Frost and McCammon 2008), including relevant thermodynamic theory, and Stolper et al. (2020) and Guimond et al. (2023) illustrate how mantle mineralogy and fO_2 are intimately related to one another. Doyle et al. (2019) apply this theory to interpret measurements of O contents of several white dwarfs. Interested readers are also highly encouraged to read the very accessible discussion of fugacity by its inventor, Gilbert Lewis (Lewis 1901), who devised the concept to link chemical transfer processes to Gibbs Free Energy. For the purposes of extrasolar planet studies, Fe/FeO ratios can be inferred from either Fe–Si (as in the Method 2 above) or perhaps Fe–O mass balance in some white dwarfs (Sun-like stars have an excess of O relative to rocky planets and so place no immediate constraint on rocky planet O contents; Fig. 14) .

But even here, there are significant issues in understanding Fe–O systematics in our own Solar System. Total oxygen contents appear to increase with heliocentric distance (Table 3a) while Fe decreases (Table 3). The increase in O from Mercury to Vesta is expected because O is a volatile element, i.e., its 50% condensation temperature ($T_{50\%}$ = 183 K) from the Solar nebula is well below the $T_{50\%}$ (> 1300 K) that characterizes the major elements that make up rocky planetary objects (Lodders 2003; Wood et al. 2019). Iron appears to behave independent of its $T_{50\%}$ value of 1334 K. McDonough and Yoshizaki (2021) hypothesize that the decrease in Fe with heliocentric distance might be controlled by the Sun's magnetic field during planetary accretion. The heliocentric trends for both O and Fe could be less dramatic for the early Solar system if Mercury has indeed lost a portion of its silicate mantle to late bombardment (Helffrich et a. 2019). In any case, gradients in Fe and O yield significant differences in planetary mantle compositions in our own Solar System: Mars and Vesta have considerably more FeO in their silicate mantles, for example, compared to Earth (Table 4), even though both planets have less Fe overall (Table 3). Similar patterns may apply elsewhere and so being able to predict Fe/O ratios amongst the inner planets, as well as in other planetary systems, is critical.

It should perhaps also be noted that the O contents in Table 3 are minimum estimates; those O contents, for the sake of a first-order comparison, are calculated assuming that all Fe exists as FeO. Correcting these is not a simple matter. Hawaiian lavas indicate that on a weight % basis the ratio $Fe_2O_3/(FeO + Fe_2O_3)$ is about 10% (Rhodes and Vollinger 2005). But Wood et al. (2006) posit that Earth's lower mantle might contain significant amounts of Fe^{3+}, in the form of the perovskite-structured mineral $FeAlO_3$. However, the $Fe_2O_3/(FeO + Fe_2O_3)$ ratio of the lower mantle, and even its precise bulk composition, are still unclear. Seismologists have sometimes argued that Earth's lower mantle is enriched in Fe (e.g., Anderson and Jordan 1970; van der Hilst and Karason 2000), or SiO_2 (e.g., Murakami et al. 2012), although later studies tend to contradict such hypotheses (e.g., Davies 1974; Irfune et al. 2010; Hyung et al. 2016). Nevertheless, the Wood et al (2006) hypothesis is important to exoplanet studies as it implies that final planet size can affect total O contents, even without affecting the total mass of a metallic core: high-P crystallization (high enough to stabilize perovskite-structured phases, at ca $P > 23\,GPa$) from a deep a magma ocean could have increased the oxidation state of Earth's mantle after core formation was largely completed. Their model addresses the otherwise puzzling case of why the more O-rich Mars has a mantle with even lower $Fe_2O_3/(FeO + Fe_2O_3)$ than Earth (Herd et al. 2002; Righter et al. 2008): the answer may be that Mars is too small to allow for high-P perovskite phases to crystallize to any great degree. But Mercury provides yet another confounding case: it is the most Fe-rich of the inner planets, and yet appears to have almost no FeO in its mantle (Nittler et al. 2018), despite having sufficient O to form other silicate minerals.

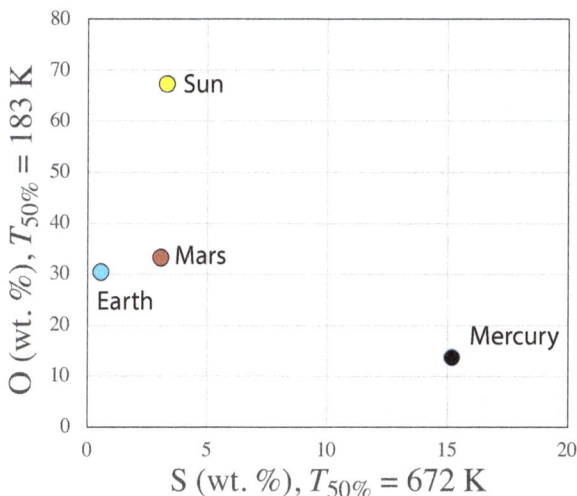

Figure 14. A comparison of O and S contents for Earth, Mercury, Mars and the Sun.

The first-order problem for estimating exoplanet mantle mineralogy is the extent to which Fe is partitioned into the core as Fe metal. This problem is closely followed by the effect of total O abundance (nominally fO_2) on the partitioning of H, C, Si, S, and other light elements into the core. One very clear prospect for future advances is to (a) better interpret spatial gradients in Fe, O and $Fe_2O_3/(FeO + Fe_2O_3)$ ratios within our own Solar System, (b) determine whether any such models could apply to other star-planet systems and (c) model such gradients, and combine these with nominal exoplanet bulk mineralogies so as to model how fO_2 might vary in exoplanet upper mantles (Stolper et al. 2020; Guimond et al. 2023).

Carbon. Carbon is heavily depleted in Earth relative to the Sun, but that pattern might not apply to all exoplanets. In a survey of 499 stars Bond et al. (2010) noted that most have C/O ratios higher than the Solar value (0.54), averaging 0.77 and ranging to > 1.0; they posit that when C/O>0.8, Si might occur largely as SiC rather than SiO_2 in circumstellar planet building materials. Suarez-Andres et al. (2018) re-surveyed the same 499 stars and concluded that none have C/O>0.8, but nonetheless show that C/O ranges well above the Solar value, approaching 0.8. Kuchner and Seager (2005) also suggest that C/O can locally approach 1.0 within a circumstellar disk, even if a bulk stellar value is much lower, allowing SiC to be a major constituent of some exoplanets. The mineralogical implications are unclear. Experiments conducted by Allen-Sutter et al. (2020) show that if water is present, SiC will, at high pressures, dissociate into SiO_2 and diamond, and so perhaps even in high C/O planets, carbide phases might be rare, although diamond might be rather abundant. No less interesting are experiments by Hakim et al. (2018) who show that the highly reducing conditions needed to stabilize carbides could be detected through the oxidation state of Fe: detection of either Fe^{2+} or Fe^{3+} (as opposed to Fe^0) at a planet's surface would, in effect, be sufficient to negate a carbide-dominated mantle.

The issue of carbide and diamond contents appears to be not quite fully settled. Most of these studies also make the mistake of using Mg/Si to estimate the ratios of olivine (Mg/Si = 2 in forsterite) to pyroxene (Mg/Si = 1 in enstatite). But as the example of Mars shows, Fe can be highly abundant in silicate mantles and so the more relevant ratio is (Mg+Fe)/Si (see Putirka and Rarick 2019). Calcium contents are also frequently not so low as to be negligible in calculating Ol/Pyx ratios. New experiments on Fe-bearing, C-rich systems, with varying O contents are likely to yield additional results on the possible mineralogical variety of exoplanetary interiors.

Sulfur and hydrogen. Relative to the Sun, Earth is depleted in all three of O ($T_{50\%}$ = 183 K), C ($T_{50\%}$ = 40 K) and S ($T_{50\%}$ = 672 K) (Table 3; Fig. 14), but the surface of Mercury is intriguingly enriched in S, containing up to 4 wt. % (Nittler et al. 2011)—despite being much closer to the Sun. Moreover, because surface rocks of Mercury have almost no Fe, it is likely that S is bound up as Mg or Ca sulfides (Nittler et al. 2011) and Lark et al. (2022) suggest that Mercury's mantle may contain 7–11 wt. % sulfides. Modelling by Boukaré et al. (2019) further indicates that an early magma ocean on Mercury could contain sulfide-rich layers that might be enriched in heat producing elements. And such mineral assemblages are likely to have mechanical and melting behaviors that are very distinct from S-poor, silicate mantles that are the basis of our current models. These findings also indicate that 50% condensation temperatures are an imperfect predictor of a planet's share of any particular planet-building element, and that S-rich and Fe-poor interiors are perhaps just one example of a range of possibilities for how planets may differ from Earth.

As to H, the inner planets of our Solar System have retained only a miniscule fraction relative to the Sun. But some early experiments (e.g., Fukai et al. 1982) indicated that at pressures as low as 3 GPa, significant amounts of H can partition into metallic Fe. Thus, Fe droplets sinking to form a metallic core could yield a metal core that contains H as a light-alloying element. Tagawa et al. (2016) showed that the outer core cold contain as much as 0.3 wt. % H, among other light-alloying elements. And experiments at very high P (Yang et al. 2022) indicate that FeH could be stable within Earth's inner core. It thus seems that a very wide range of core compositions are possible, including wide ranges in H (as well as S and O) in exoplanetary systems.

ARE WE BUILDING SAND CASTLES OR CATHEDRALS?

Two different methods for estimating mineral abundances (Methods 1 and 2) accurately predict mineral abundances for natural samples, and yield good agreement when applied to exoplanets. But there are many caveats and assumptions for the latter to be accurate. If the results are so uncertain, why bother? Because it is interesting. And we have no choice. Everything we want to know about a planet—it's ability to move heat, store water, break into mobile plates, and exsolve oceans—are dependent entirely upon the minerals that make up its interior; no modeling of any use can happen absent mineralogic estimates. And for exoplanets that do have Earth-like amounts of H, C, O and S, and whose bulk compositions fall within the ultramafic ternary, our models have an excellent chance of predicting accurate upper mantle mineral abundances and core mass fractions (Figs. 7–8, 9a). Perhaps our thermodynamic models also extrapolate better than expected, but mineralogic estimates nevertheless point us towards the kinds of experiments that need to be conducted (e.g., Zhao et al. 2018; Brugman et al. 2021). A standard mineralogy also arranges exoplanets into classes *relative to their potential to be Earth-like*. We might never know whether a given planet has plate tectonics or granitic crust—or even its true mantle mineralogy—but numerical models that are mineralogically informed can provide a list of plausible candidates of Earth-like analogues, and should inform us of how non-analogues might evolve. Mineralogical estimates also reveal the gaps in our current understanding of Earth and the inner planets.

We have also made certain progress. Now set aside is the use of "Earth-like" to refer to any exoplanet with a rocky exterior; we must appreciate that Mercury, Venus, Earth and Mars are quite different from one another. Our next focus should perhaps concern what qualifies as "exotic", and how we decide when and if an exotic planet has been identified. For example, in a survey of >4,000 star compositions from the Hypatia catalog (Hinkel et al. 2014), Putirka and Rarick (2019) showed that elements such as Al, Ti or Cr, for example, are too low in abundance relative to Mg, Fe and Si, to allow any exoplanetary mantles to consist largely of corundum or rutile or eskolaite. Rather, familiar minerals, such as olivine and the pyroxenes, are likely to dominate, perhaps augmented by sulfides or carbides. However, Putirka and Xu (2021), argue (see also Fig. 11) that some exoplanetary mantles are not contained within the ultramafic ternary diagram (Fig. 4a); that diagram likely describes most or all the inner planets (Mercury being the most plausible exception). But even a wehrlite or a websterite mantle (which fall within but at the margins of Fig. 4a) would be exotic, let alone a mantle with abundant quartz or periclase. These could exhibit very different kinds of crust, oceans and tectonic processes than any planet in our Solar System. Errors on any given polluted white dwarf (Figs. 9–11), as well as repeat analyses (Fig. 11) show that we cannot exclude more extreme and exotic compositions.

But we need a better understanding of why our own Solar system looks the way it does. When estimating exoplanet mineralogy we are, in effect, conducting "what if" calculations: what if a planet is Earth-sized, or has an Earth-like core, or an FeO/O ratio that is like Mars, or has Mercury-like S content? But will extra-solar composition gradients mimic those of our Solar System. Or are other gradients possible? And how might we predict such? Recall also that PWDs, when passing through the Red Giant phase, obliterate their inner planetary materials; PWDs thus record the equivalent of, outer, not inner Solar system materials. How well do we know the interior compositions of Jupiter or Saturn or the moons that orbit them?

The August 1990 issue of the journal *Geophysical Research Letters* provides yet another note of caution: there, authors published predictions of what the Magellan mission might find once it arrived at Venus on August 10, 1990. The resulting Magellan data would reveal a surface vastly different than expected, illustrating the challenges in extrapolating terrestrial processes to other planets—even Earth's nominal "sister planet". The MESSENGER mission to Mercury has yielded no fewer surprises, including a S-rich, and notably Fe-free surface, on the most Fe-rich of the inner planets. If Earth-centered knowledge extrapolates poorly to Venus or Mercury, how well does our Solar system extrapolate to other stellar systems? Even if our Solar system is representative, the assumption is only useful to the extent that we understand the inner planets. And we still debate whether Earth acquired its oceans from comet-like bombardments or degassing of its interior. The time range for the initiation of terrestrial plate tectonics covers 80% of Earth's history. The composition and mineralogy of Earth's lower mantle is only roughly understood, and debate still occurs as to the major light alloying elements of Earth's core. Current thoughts on exoplanets are clearly preliminary and contingent.

Exciting days, though, lie ahead. As explained in Guimond et al. (2024, this volume) the James Webb Space Telescope (JWST) may reveal to us the interior compositions of exoplanets, via measurements of their atmospheres. Fortin et al. (2022), for example, show how modeling of JWST reflectance spectra should yield very useful insights regarding both planetary surface and interior behavior. Guimond et al. (2023) are also investigating how Mg/Si ratios affect the amounts of water stored in planetary interiors. And new experiments by Brugman et al. (2021) could be used to revise thermodynamic models, to better explore exoplanet-scale ranges in composition. We also still know next to nothing of Venus, and the partial melting and rheological experiments needed to predict the behavior of Mercury and Mars have mostly yet to be performed. This all leaves much exciting work to be performed, especially if we avoid being too comfortable with existing results, models and paradigms.

ACKNOWLEDGEMENTS

Many thanks to Oliver Shorttle and Larry Nittler for their careful reading of the manuscript, their very thoughtful comments and helpful suggestions. This work was supported by NSF grant 1921182.

REFERENCES

Agee CB, Walker D (1988) Static compression and olivine flotation in ultrabasic silicate liquid. J Geophys Res 93:3437–3449

Agee CB, Li J, Shannon MC, Circone S (1995) Pressure–temperature phase diagram for the Allende meteorite. J Geophys Res 100:17725–17740

Aitta A (2012) Venus' internal structure, temperature and core composition. Icarus 218:967–974, https://doi.org/10.1016/j.icarus.2012.01.007

Allen-Sutter H, Garhart E, Leinenweber K, Prakapenka V, Greenberg E, Shim S-H (2020) Oxidation of the interiors of carbide exoplanets. Planet Sci J 1:39

Anderson DL, Jordan T (1970) The composition of the lower mantle. Phys Earth Planet Int 3:23–35

Andrault D, Bolfan-Casanova N, Nigro GL, Bouhifd MA, Garbarino G, Mezouar M (2011) Solidus and liquidus profiles of chondritic mantle: Implication for melting of the Earth across its history. Earth Planet Sci Lett 304:251–259, https://doi.org/10.1016/j.epsl.2011.02.006

Asahara Y, Kubo T, Kondo T (2004) Phase relations of a carbonaceous chondrite at lower mantle conditions. Phys Earth Planet Int 143–144:421–432

Asimow PD, Ghiorso MS (1998) Algorithmic modifications extending MELTS to calculate subsolidus phase relations. Am Mineral 83:1127–1131

Baedecker PA, Wasson JT (1975) Elemental fractionations among enstatite chondrites. Geochim Cosmochim Acta 39:735–765

Barrat J-A, Chaussidon M, Yamaguchi A, Beck P, Villeneuve J, Byrne DJ, Broadley MW, Marty B (2021) A 4,565-My-old andesite from an extinct chondritic protoplanet. PNAS 118:e2026129118

Bell DR, Ihinger PD, Rossman GR (1995) Quantitative analysis of trace OH in garnet and pyroxenes. Am Mineral 80:465–474

Birch F (1964) Density and composition of the mantle and core. J Geophys Res 69:4377–4388

Bodinier J-L, Garrido CJ, Chanefo I, Bruguier O, Gervilla F (2008) Origin of pyroxenite–peridotite veined mantle refertilization by reactions: evidence form the Ronda peridotite. J Petrol 49:999–1025

Bond JC, O'Brien DP, Lauretta DS (2010) The compositional diversity of extrasolar terrestrial planets I. In situ simulations. Astrophys J 715:1050–1070

Boukaré C-E, Parman S, Parmentier EM, Anzures BA (2019) Production and preservation of sulfide layering in Mercury's mantle. J Geophys Res Planets 124, 3354–3372, https://doi.org/10.1029/2019JE005942

Boyet M, Bouvier A, Frossard P, Hammouda T, Garcon M, Gannoun A (2018) Enstatite chondrites EL3 as building blocks for the Earth: the debate over the ^{146}Sm–^{142}Nd systematics. Earth Planet Sci Lett 488:68–78

Brugman K, Phillips MG, Till CB (2021) Experimental determination of mantle solid and melt compositions for two likely rocky exoplanet compositions. J Geophys Res Planets 126:e2020JE006371

Buckle T, Williams M, Nathwani CL, Hughes HSR (2023) WebNORM: a web application or calculating normative mineralogy. Front Earth Sci 11:1232256, https://doi.org/10.3389/feart.2023.1232256

Campbell IH, O'Neill HS (2012) Evidence against a chondritic Earth. Nature 483:553-558

Campbell IH, Taylor SR (1983) No water, no granites—no oceans, no continents. Geophys Res Lett 10:1061–1064

Cawthorn RG, Brown PA (1976) A model for the formation and crystallization of corundum-normative calc-alkaline magmas through amphibole fractionation. J Geol 84:467–476

Chappell BW, White AJR (1992) I- and S-type granites in the Lachlan fold belt. Earth Env Sci Trans Royal Soc Edinburgh 83:1–26, https://doi.org/10.1017/S0263593300007720

Clenet H, Jutzi M, Barrat, JA, Asphaug EI, Benz W, Gillet P (2014) A deep crust–mantle boundary in the asteroid 4 Vesta. Nature 511:303–306, https://doi.org/10.1038/nature13499

Connolly HC, Jones RH (2018) Chondrites: the canonical and non-canonical views. J Geophys Res Planets 121: 1885–189, https://doi.org/10.1002/2016JE005113

Corgne A, Keshav S, Wood BJ, McDonough WF, Fei, Y (2008) Metal–silicate partitioning and constraints on core composition and oxygen fugacity during Earth accretion. Geochim Cosmochim Acta 72:574–589

Cross W, Iddings JP, Pirsson LV, Washington HS (1902) A quantitative chemico-mineralogical classification and nomenclature of igneous rocks. J Geol 10:555–690

Davies GF (1974) Limits on the constitution of the lower mantle. Geophys J Astron Soc 38:479–503

Davies GF, Richards MA (1992) Mantle convection. J Geol 100:151–206

Denbigh KD (1981) The Principles of Chemical Equilibrium, 4th ed., Cambridge Univ Press, 494 p

Ding S, Dasgupta R, Tsuno K (2020) The solidus and melt productivity of nominally anhydrous Martian mantle constrained by new high pressure–temperature experiments—implications for crustal production and mantle source evolution. J Geophys Res Planets 123:e2019JE006078

Dorn C, Khan A, Heng K, Connolly JAD, Alibert Y, Benz W, Tackley P (2015) Can we constrain the interior structure of rocky exoplanets from mass and radius measurements? Astron Astrophys 577:A83, https://doi.org/10.1051/0004-6361/201424915

Doyle AE, Young ED, Klein B, Zuckerman B, Schlichting HE (2019) Oxygen fugacities of extrasolar rocks: Evidence for an Earth-like geochemistry of exoplanets. Science 366:356–359

Drake MJ, Righter K (2002) Determining the composition of the Earth. Nature 416:39–44

Elardo S, Shahar A, Mock TD, Sio CK (2019) The effect of core composition on iron isotope fractionation between planetary cores and mantles. Earth Planet Sci Lett 513:124–134

Fiquet G, Auzende AL, Siebert J, Corgne A, Bureau H, Ozawa H, Garbarino G (2010). Melting of peridotite to 140 gigapascals. Science 329:1516–1518, https://doi.org/10.1126/science.1192448

Fitoussi C, Bourdon B (2012) Silicon isotope evidence against an enstatite chondrite Earth. Science 335:1477–1480

Fortin M-A, Gazel E, Kaltenegger L, Holycros ME (2022) Volcanic exoplanet surfaces. Mon Not R Astron Soc 516:4569–4575

Frost DJ, McCammon CA (2008) The redox state of Earth's mantle. Annu Rev Earth Planet Sci 36:389–420

Fukai Y, Fuzikawa A, Watanabe K, Amano M (1982) Hydrogen in iron—its enhanced dissolution under pressure and stabilization of the γ phase. Jpn J Appl Phys 21:L318

Gale A, Dalton CA, Langmuir CH, Su Y, Schilling J-G (2013) The mean composition of ocean ridge basalts. Geochem Geophys Geosyst 14:489–518

Gansicke BT, Koester D, Farihi J, Girven J, Parsons SG, Breedt E (2012) The chemical diversity of exo-terrestrial planetary debris around white dwarfs. Mon Not R Astron Soc 424:333–347

Ghiorso MS, Sack RO (1995) Chemical mass transfer in magmatic processes. IV A revised and internally consistent thermodynamic model for the interpolation and extrapolation of liquid–solid equilibria in magmatic systems at elevated temperatures and pressures. Contra Mineral Petrol 119:197-212

Grove TL, Till CB (2019) H$_2$O-rich mantle melting near the slab–wedge interface. Contrib Mineral Petrol 174:80

Guimond CM, Shorttle S, Rudge JF (2023) Mantle mineralogy limits to rocky exoplanet water inventories. Mon Not R Astron Soc 521:2535–2552

Guimond CM, Wang H, Seidler F, Sossi P, Mahajan A, Shorttle O (2024) From stars to diverse mantles, melts, crusts, and atmospheres of rocky exoplanets. Rev Mineral Geochem 90:259–300

Hakim K, van Westrenen W, Dominik C (2018) Capturing the oxidation of silicon carbide in rocky exoplanetary interiors. Astron Astrophys 618:L6

Hauck SA III et al (2013) The curious case of Mercury's internal structure. J Geophys Res Planets 118:1204–1220

Helffrich G (2017) Mars core structure—concise review and anticipated insights from InSight. Prog Earth Planet Sci 4:24, https://doi.org/10.1186/s40645-017-0139-4

Helffrich G, Brasser R, Shahar A (2019) The chemical case for Mercury mantle stripping. Prog Earth Planet Sci 6:66, https://doi.org/10.1186/s40645-019-0312-z

Herd CD, Borg LE, Jones JH, Papike JJ (2002) Oxygen fugacity and geochemical variations in the Martian basalts: implications for Martian basalt petrogenesis and the oxidation state of the upper mantle of Mars. Geochim Cosmochim Acta 66:2025–2036

Herzberg C, Raterron P, Zhang J (2000) New experimental observations on the anhydrous solidus for peridotite KLB-1. Geochem Geophys Geosyst 1:2000GC000089

Hinkel NR, Unterborn CT (2018) The star–planet connection I: using stellar composition to observationally constrain planetary mineralogy for the ten closest stars. Astrophys J 853:83

Hinkel NR, Timmes FX, Young PA, Pagano MD Turnbull MC (2014) Stellar abundances in the solar neighborhood: the Hypatia Catalog. Astronom J 148:54

Hinkel NR, Young PA, Pagano MD, Desch SJ, Anbar AD, Adibekyan V, Blanco-Cuaresma S, Carlberg JK, Mena ED, Liu F, Nordlander T (2016) A comparison of stellar elemental abundance techniques and measurements. Astrophys J Suppl Ser 226:4

Hinkel NR, Young PA, Wheeler III CH (2022) A concise treatise on converting stellar mass fractions to abundances to molar ratios. Astron J 164:256

Hinkel NR, Youngblood A, Soares-Furtado M (2024) Host stars and how their compositions influence exoplanets. Rev Mineral Geochem 90:1–26

Hirschmann, MA (2000) Mantle solidus: Experimental constraints and the effects of peridotite composition. Geochem Geophys Geosyst 1:2000GC000070

Hofmann AW, White WM (1982) Mantle plumes form ancient oceanic crust. Earth Planet Sci Lett 57:421–436

Hollands MA, Gänsicke BT, Koester D (2018) Cool DZ white dwarfs II: compositions and evolution of old remnant planetary systems. Mon Not R Astron Soc 477:93–111

Hollands MA, Tremblay P-E, Gänsicke BT, Koester D, Gentile-Fusillo NP (2021) Alkali metals in white dwarf atmospheres as tracers of ancient planetary crusts. Nat Astron 5:451–459

Hyung E, Huang S, Petaev MI, Jacobsen SB (2016) Is the mantle chemically stratified? Insights from sound velocity modeling and isotope evolution of an early magma ocean. Earth Planet Sci Lett 440:158–168

Irfune T, Shinmei T, McCammon CA, Miyajima N, Rubie DC, Frost DJ (2010) Iron partitioning and density changes of pyrolite in Earth's lower mantle. Science 327:193–195

Javoy M, Kaminski E, Guyot F, Andrault D, Sanloup C, Moreira M, Labrosse S, Jambon A, Agrinier P, Davaille A, Jaupart C (2010) The chemical composition of the Earth: enstatite chondrite meteorites. Earth Planet Sci Lett 293:259–268

Jones RH (2024) Meteorites and planet formation. Rev Mineral Geochem 90:113–140

Jones JH, Delano JW (1989) A three-component model for the bulk composition of the Moon. Geochim Cosmochim Acta 53:513–527

Jones MC, Pewsey A (2009) Sinh–arcsinh distributions. Biometrika 96:761–780

Jura M, Xu S, Klein B, Koester D, Zuckerman B (2012) Two extrasolar asteroids with low volatile element mass fractions. Astrophys J 750:69

Jura M, Dufour P, Xu S, Zuckerman B, Klein B, Young E, Melis C (2015) Evidence for an anhydrous carbonaceous extrasolar minor planet. Astrophys J 799:109

Kaminsky F (2012) Mineralogy of the lower mantle: a review of 'super-deep' mineral inclusions in diamond. Earth Sci Rev 110:127–147

Khan A, Connolly JAD (2008) Constraining the composition and thermal state of Mars from inversion of geophysical data. J Geophys Res Planets 113:E07003

Khan A, Connolly JAD, Taylor SR (2008) Inversion of seismic and geodetic data for the major element chemistry and temperature of the Earth's mantle. J Geophys Res 113:B09308, https://doi.org/10.1029/2007JB005239

Khan A, Pommier A, Neumann GA, Mosegaard K (2013) The lunar moho and the internal structure of the Moon: A geophysical perspective. Tectonophys 609:331–352

Kim T, Ko B, Greenberg E, Prakapenka V, Shim S-H, Lee Y (2020). Low melting temperature of anhydrous mantle materials at the core–mantle boundary. Geophys Res Lett 47:e2020GL089345, https://doi.org/10.1029/2020GL089345

King SD, Castillo-Rogez JC, Toplis MJ, Bland MT, Raymond CA, Russell CT (2018) Ceres internal structure from geophysical constraints. Meteor Planet Sci 53:1999–2007

Kinzler RJ (1997) Melting of mantle peridotite at pressure approaching the spinel to garnet transition: application to mid-ocean ridge basalt petrogenesis. J Geophys Res 102:853–874

Klein B, Jura M, Koester D, Zuckerman B (2011) Rocky extrasolar planetary compositions derived from externally polluted white dwarfs. Astrophys J 741:64

Klein BZ, Jagoutz O Behn MD (2017) Archean crustal compositions promote full mantle convection. Earth Planet Sci Lett 474:516–526

Knapmeyer-Endrun B, Panning MP, Bissig F, Joshi R, Khan A, Kim D, Lekić V, Tauzin B, Tharimena S, Plasman M, Compaire N (2021) Thickness and structure of the Martian crust from InSight seismic data. Science 373:438–443

Knibbe JS, Rivoldini A, Luginbuhl SM, Namur O, Charlier B, Mezouar M, Sifre D, Berndt J, Kono Y, Neuville DR, van Westrenen W (2021) Mercury's interior structure constrained by density and P-wave velocity measurements of liquid Fe–Si–C alloys. J Geophys Planets 125:e2020JE006651, https://doi.org/10.1029/2020JE006651

Kombayashi T, Pesce G, Morard G, Anotonangeli D, Sinmyo R, Mezouar M (2019) Phase transition boundary between fcc and hcp structures in Fe–Si alloy and its implications for terrestrial planetary cores. Am Mineral 104:94–99

Kuchner MJ, Seager S (2005) Extrasolar carbon planets. arXiv:Astro-ph/0504214v2

Kuskov OL, Kronrod EV, Kronrod VA (2019) Effect of thermal state on the mantle composition and core sizes of the Moon. Geochem Int 57:605-620

Kuwayama Y, Hirose K, Cobden L, Kusakabe M, Tateno S, Oishi Y (2021) Post-perovskite phase transition in the pyrolitic lowermost mantle: Implications for ubiquitous occurrence of post-perovskite above CMB. Geophys Res Lett 49:e2021GL096219

Lambart S, Baker MB, Stolper EM (2016) The role of pyroxenite in basalt genesis: melt–PX, a melting parameterization for mantle pyroxenites between 0.9 and 5 GPa. J Geophys Res: Solid Earth 121:5708–5735, https://doi.org/ 0.1002/2015JB012762

Lark L, Parman S, Huber C, Parmentier E, Head J (2022) Sulfides in Mercury's mantle: implications for Mercury's interior as interpreted from moment of inertia. Geophys Res Lett 49:e2021GL096713

Le Bas MJ, Streckeisen AL (1991) The IUGS systematics of igneous rocks. J Geol Soc London 148:825–833

Le Bas MJ, Le Maitre RW, Streckeisen A, Zanettin B (1986) A chemical classification of volcanic rocks based on the total alkali–silica diagram. J Petrol 27:745–750

Lewis GN (1901) The law of physio-chemical change. Proc Am Acad Arts Sci 37:49–69

Lodders K (2003) Solar system abundances and condensation temperatures of the elements. Astrophys J 591:1220–1247

Lodders K, Fegley B Jr (1998) The Planetary Scientist's Companion, Oxford University Press, Oxford

Lodders K, Fegley B Jr (2018) Chemistry of the Solar System, RSC Publishing, Royal Society of Chemistry, Cambridge

Lodders K, Palme H, Gail HP (2009) Abundances of the elements in the solar system. *In:* Landolt-Börnstein, Vol. VI/4B, Trümper JE (ed.), Springer-Verlag, Berlin, p 560–630

Longhi J (2006) Petrogenesis of picritic mare magmas: constraints on the extent of lunar differentiation. Geochim Cosmochim Acta 70:5919–5934

Lyubetskaya T, Korenaga J (2007) Composition of Earth's primitive mantle and its variance 1. Methods and results. J Geophys Res 112:B03211, https://doi.org/10.1029/2005JB004223

Magg E, Bergemann M, Serenelli A, Bautista M, Plez B, Heiter U, Gerber JM, Ludwig HG, Basu S, Ferguson JW, Gallego HC (2022) Observational constraints on the origin of the elements. IV: The standard composition of the Sun. Astron Astrophys 661:A140, https://doi.org/10.1051/0004-6361/202142971

McDonough WF, Sun S-s (1995) The composition of the Earth Chem Geol 120:223–253

McDonough WF, Yoshizaki T (2021) Terrestrial planet compositions controlled by accretion disk magnetic field. Prog Earth Planet Sci 8:39, https://doi.org/10.1186/s40645-021-00429-4

Mattern E, Matas J, Ricard Y, Bass J (2005) Lower mantle composition and temperature form mineral physics and thermodynamic modeling. Geophys J Int 160:973–990

Mezger K, Schonbachler M, Bouvier A (2020) Accretion of the Earth—Missing components? Space Sci Rev 216:27, https://doi.org/10.1007/s11214-020-00649-y

Mittlefehldt DW (2015) Asteroid (4) Vesta: 1. The howardite–eucrite–diogenite (HED) clan of meteorites. Geochem 75:155–183

Morse SA (1980) Basalts and Phase Diagrams: An Introduction to the Quantitative use of Phase Diagrams in Igneous Petrology. Springer-Verlag, Berlin

Murakami M, Ohishi Y, Hirao N, Hirose K (2012) A perovskitic lower mantle inferred from high-pressure, high temperature sound velocity data. Nature 485:90–94

Namur O, Collinet M, Charlier B, Grove TL, Holtz F, McCammon C (2014) Melting processes and mantle sources of lavas on Mercury. Earth Planet Sci Lett 439:117–128

Nittler LR, McCoy TJ, Clark PE, Murphy ME, Trombka JI, Jarosewich E (2004) Bulk element compositions of meteorites: a guide for interpreting remote-sensing geochemical measurements of planets and asteroids. Antarct Met Res 17:231–251

Nittler LR, Starr RD, Weider SZ, McCoy TJ, Boynton WV, Ebel DS, Sprague AL (2011) The major-element composition of mercury's surface from messenger X-ray spectrometry. Science 333:1847–1850

Nittler LR, Chabot NL, Grove TL, Peplowski PN (2018) The chemical composition of Mercury. Chapter 2. *In:* Mercury, The View after MESSENGER, Solomon, SC, Nittler LR, Anderson B (eds), Cambridge University Press, Cambridge p 30–51

Nomura R, Hirose K, Uesugi K, Ohishi Y, Tsuchiyama A, Miyake A, Ueno Y (2014). Low core–mantle boundary temperature inferred from the solidus of pyrolite. Science 343:522–525

Ogawa M, Schubert G, Abdelfattah M (1991) Numerical simulations of three-dimensional thermal convection in a fluid with strongly temperature-dependent viscosity. J Fluid Mech 233:299–328

O'Neil C (2021) End member Venusian core scenarios: does Venus have an inner core? Geophys Res Lett 48:e2021GL095499

Palme H, O'Neill HStC (2003) Cosmochemical estimates of mantle composition. *In:* The Mantle and Core, Vol. 2 Treatise on Geochemistry, Carlson RW (ed), Elsevier–Pergamon, Oxford, p 1–38

Palme H, Zipfel J (2022) The composition of CI chondrites and their contents of chlorine and bromine: results from instrumental neutron activation analysis. Met Planet Sci 57:317–333

Peterson BT, DePaolo DJ (2007) Mass and composition of the continental crust estimated using the CRUST2.0 model. Am Geophys Union, Fall Meeting, V33A-1161

Pottasch SR (1964) A comparison of the chemical composition of the solar atmosphere with meteorites. Ann Astrophys 27:163–169

Prytulak J, Elliott T (2007) TiO_2 enrichment in ocean island basalts. Earth Planet Sci Lett 263:388–403

Putirka KD, Rarick J (2019) The composition and mineralogy of rocky exoplanets: a survey of >4000 stars from the Hypatia Catalog. Am Mineral 104:817–829

Putirka KD, Xu S (2021) Polluted white dwarfs reveal exotic mantle rock types on exoplanets in our solar neighborhood. Nat Commun 12:6168, https://doi.org/10.1038/s41467-021-26403-8

Putirka K, Perfit M, Ryerson FJ, Jackson MG (2007) Ambient and excess mantle temperatures, olivine thermometry, and active vs. passive upwelling. Chem Geol 241:177–206

Putirka K, Tao Y, Hari KR, Perfit M, Jackson M, Arevalo Jr R (2018) The mantle source of thermal plumes: minor elements in olivine and major oxides of primitive liquids (and why the olivine compositions don't matter). Am Mineral 103:1253–1270

Putirka KD, Dorn C, Hinkel NR, Unterborn CT (2021) Compositional diversity of rocky exoplanets. Elements 17:235–240

Putirka KD (2024) Some tectonic concepts relevant to the study of rocky exoplanets. Rev Mineral Geochem 90:301–322

Ransford GA (1979) A comparison of two accretional heating models. Proc Lunar Planet Sci Conf 10:1867–1879

Raymond CA, Park RS, Asmar SW, Konopliv AS, Buczkowski DL, De Sanctis MC, McSween HY, Russell CT, Jaumann R, Preusker F, and the Dawn Team (2013) The crust and mantle of Vesta's southern hemisphere. European Planet Sci Congr Abstracts. 8:EPSC2013-1002

Reach WT, Lisse C, von Hippel T, Mullally F (2009) The dust cloud around white dwarf G 29-38. II Spectrum from 5 to 40 mm and mid-infrared photometric variability. Astrophys J 693: 697–712

Rhodes JM, Vollinger MJ (2005) Ferric/ferrous ratios in 1984 Mauna Loa lavas: a contribution to understanding the oxidation state of Hawaiian magmas. Contrib Mineral Petrol 149:666–674

Righter K, Drake MJ (1996) Core formation in Earth's Moon, Mars and Vesta. Icarus 124:513–529

Righter K, Yang H, Costin CG, Downs RT (2008) Oxygen fugacity in the Martian mantle controlled by carbon: new constraints from the nakhlite MIL 03346. Met Planet Sci 43:1709–1723

Rogers LK, Bonsor A, Xu S, Dufour P, Klein BL, Buchan A, Hodgkin S, Hardy F, Kissler-Patig M, Melis C, Weinberger AJ (2024) Seven white dwarfs with circumstellar gas discs I: White dwarf parameters and accreted planetary abundances. Mon Not R Astron Soc 527:6038–6054

Rubie DC, Frost DJ, Mann U, Asahara Y, Nimmo F, Tsuno K, Kegler P, Holzheid A, Palme H (2011) Heterogeneous accretion, composition and core–mantle differentiation of the Earth. Earth Planet Sci Lett 301:31–42

Rudnick RL, Gao S (2014) Composition of the continental crust. 4.1 *In* Treatise of Geochemistry 2nd ed., Holland HD, Turekian KK (eds), Elsevier, Amsterdam, p 1–51

Safranov VS (1978) The heating of the earth during its formation. Icarus 33:3–12

Salters VJM, Stracke A (2004) Composition of the depleted mantle. Geochem Geophys Geosyst 5:Q05B07, https://doi.org/10.1029/2003GC000597

Sanloup C, Jambon A, Gillet P (1999) A simple chondritic model of Mars. Phys Earth Planet Int 112:43–54

Shah O, Helled R, Alibert Y, Mezger K (2022) Possible chemical composition and interior structure models of Venus inferred from numerical modeling. Astrophys J 926:217

Shea T, Hammer JE, Hellegrand E, Mourney AJ, Costa F, First EC, Lynn KJ, Melnik O (2019) Phosphorous and aluminum zoning in olivine: contrasting behavior of two nominally incompatible trace elements. Contrib Mineral Petrol 174:85

Sobolev AV, Hoffmann AW, Sobolev SV, Nikogosian IK (2005) An olivine-free mantle source of Hawaiian shield basalts. Nature 434:590–597

Solomotov VS, Moresi L-N (2000) Scaling of time-dependent stagnant-lid convection: application to small-scale convection on Earth and other terrestrial planets. J Geophys Res 105:21795–21817

Sori MM (2018) A thin, dense crust for Mercury. Earth Planet Sci Lett 490:92–99

Steenstra ES, Knibbe JS, Rai N, van Westrenen W (2016) Constraints on core formation in Vesta from metal–silicate partitioning of siderophile elements. Geochim Cosmochim Acta 177:48–61

Stevenson D (2003) Styles of mantle convection and their influence on planetary evolution. CR Geosci 335:99–111

Stolper E (1980) A phase diagram for mid-ocean ridge basalts: preliminary results and implications for petrogenesis. Contrib Mineral Petrol 74:13–27

Stolper E, Shorttle O, Antoshechkina PM, Asimow PD (2020) The effects of solid–solid phase equilibria on the oxygen fugacity of the upper mantle. Am Mineral 105:1445–1471

Suarez-Andres L, Israelian G, Gonzalez Hernandez JI, Adibekyan VZh, Delgado Mena E, Santos NC, Sousa SG (2018) C/O vs. Mg/Si ratios in Solar type stars: the HARPS sample. Astron Astrophys 614:A84

Szurgot M (2015) Core mass fraction and mean atomic weight of terrestrial planets, moon and protoplanet Vesta. *In:* Comparative Tectonics Geodynamics of Venus, Earth and Rocky Exoplanets, Contribution 5001, Lunar and Planetary Institute, Houston, p 35

Takahashi E (1986) Melting of a dry peridotite KLB-1 up to 14 GPa: Implications on the origin of peridotitic upper mantle. J Geophys Res 91:9367–9382

Tagawa S, Ohta K, Hirose K, Kato C, Ohishi Y (2016) Compression of Fe–Si–H alloys to core pressures. Geophys Res Lett 43:3686–3692

Taylor GJ (2013) The bulk composition of Mars. Chem der Erde 73:401–420

Terasaki H, Rivoldini A, Shimoyama Y, Nishida K, Urakawa S, Maki M, Kurokawa F, Takubo Y, Shibazaki Y, Sakamaki T, Machida A (2019) Pressure and composition effects on sound velocity and density of core-forming liquids: implication to core compositions of terrestrial planets. J Geophys Res Planets 124:2272–2293, https://doi.org/10.1029/2019JE005936

Thompson JB Jr (1982) Ration space: an algebraic and geometric approach. Chapter 2. *In:* Characterization of Metamorphism through Mineral Equilibria. Ferry JM (ed). De Gruyter, Berlin, p 33–52

Till CB, Grove TL, Withers A, Hirschmann MM, Médard E, Chatterjee N (2007) Extending the wet mantle solidus: Implications for H_2O transport and subduction zone melting processes; EOS Trans AGU Fall Meeting

Till CB, Elkins-Tanton LT, Fischer KM (2010) A mechanism for low-extent melts at the lithosphere–asthenosphere boundary. Geochem Geophys Geosyst 11:Q10015

Till CB, Grove TL, Krawczynzki MJ (2012) A melting model for variably depleted and enriched lherzolite in the plagioclase and spinel stability fields. J Geophys Res 117:B06206

Togashi S, Kita N, Tomiya A, Morshita Y (2017) Magmatic evolution of the lunar highland rocks estimated from trace elements in plagioclase: a new bulk silicate Moon model with sub-chondritic Ti/Ba, Sr/Ba and Sr/Al ratios. Geochim Cosmochim Acta 210:152–183

Trierweiler IL, Doyle AE, Young ED (2023) A chondritic solar neighborhood. arXiv:2306.03743v1 [Astro-ph.EP]

Tronnes RG, Baron MA, Eigenmann KR, Guren MG, Heyn BH, Loken A, Mohn CE (2019) Core formation, mantle differentiation and core–mantle interaction within Earth and the terrestrial planets. Tectonophys 760:165–198

Unterborn CT, Panero WR (2017) The effects of Mg/Si on the exoplanetary refractory oxygen budget. Astrophys J 845:61

Unterborn CT, Panero WR (2019) The pressure and temperature limits of likely rocky exoplanets. J Geophys Res Planets 124:1704–1716

van der Hilst RD, Karason H (2000) Compositional heterogeneity in the bottom 1000 kilometers of Earth's mantle: toward a hybrid convection model. Science 283:1885–1888

Vander Kaaden KE, McCubbin FM (2016) The origin of boninites on Mercury: an experimental study of the northern volcanic plains lavas. Geochim Cosmochim Acta 173:246–263

Vander Kaaden KE, McCubbin FM, Nittler LR, Peplowski PN, Weider, SZ, Frank, EA, McCoy TJ (2017) Geochemistry, mineralogy, and petrology of boninitic and komatiitic rocks on the Mercurian surface: Insights into the Mercurian mantle. Icarus 285:155–168

Verma SP, Torres-Alvarado IS, Velasco-Tapia F (2003) A revised CIPW norm. Schweiz Mineral Petrogr Mitt 83:197–216, https://doi.org/10.5169/seals-63145

Viswanathan V, Rambaux N, Fienga A, Laskar J, Gastineau M (2019) Observational constraint on the radius and oblateness of the lunar core–mantle boundary. Geophys Res Lett 46:7295–7303, https://doi.org/10.1029/2019GL082677

Vocaldo L (2010) New views of the Earth's inner core from computational mineral physics. *In:* New Frontiers in Integrated Solid Earth Sciences, International Year of Planet Earth, Cloetingh S, Negendank J (eds), Springer Science+Business Media, Berlin, p 397-412

Wade J, Wood BJ (2005) Core formation and the oxidation state of the Earth. Earth Planet Sci Lett 236:78-95

Walker D (1986) Melting equilibria in multicomponent systems and liquidus/solidus convergence in mantle peridotite. Contra Mineral Petrol 92:303–307

Walter MJ (1998) Melting of garnet peridotite and the origin of komatiite and depleted lithosphere. J Petrol 39:29–60

Wang J, Takahashi E, Xiong X, Chen L, Li L, Suzuki T, Walter MJ (2020) The water-saturated solidus and second critical endpoint of peridotite: implications for magma genesis within the mantle wedge. J Geophys Res Sol Earth, https://doi.org/10.1029/2020JB019452

Wanke H, Dreibus G (1994) Chemistry and accretion history of Mars. Phil Trans R Soc A 349:545–557

Warren JM (2016) Global variation in abyssal peridotite compositions. Lithos 248–251:193–219

Weller MB, Lenardic A (2018) On the evolution of terrestrial planets: Bi-stability, stochastic effects, and the non-uniqueness of tectonic states. Geosci Front 9:91–102

Whitney DL, Evans BW (2010) Abbreviations for names of rock-forming minerals. Am Mineral 95:185–187

Wicks JK, Duffy TS (2016) Crystal structures of minerals in the lower mantle. *In:* Deep Earth: Physics and Chemistry of the Lower Mantle and Core, Geophysical Monograph 217, Terasaki H, Fischer RA (eds), American Geophysical Union, John Wiley & Sons, Inc., New York, p 69–87

Wood BJ, Walter MJ, Wade J (2006) Accretion of the Earth and segregation of its core. Nature 441:825–833

Wood BJ, Kiseeva ES, Mirolo FJ (2014) Accretion and core formation: the effects of sulfur on metal–silicate partition coefficients. Geochim Cosmochim Acta 145:248–267

Wood BJ, Smythe DJ, Harrison T (2019) The condensation temperatures of the elements: a reappraisal. Am Mineral 104:844–856

Workman RK, Hart SR (2005) Major and trace element composition of the depleted MORB mantle (DMM). Earth Planet Sci Lett 231:53–72

Xu S, Bonsor A (2021) Exogeology from polluted white dwarfs. Elements 17:241–244

Xu S, Jura M, Klein B, Koester D, Zuckerman B (2013) Two beyond-primitive extrasolar planetesimals. Astrophys J 766:132

Xu S, Zuckerman B, Dufour P, Young ED, Klein B, Jura M (2017) The chemical composition of an extrasolar Kuiper-belt-object. Astrophys J Lett 836:L7

Xu S, Dufour P, Klein B, Melis C, Monson NN, Zuckerman B, Young ED, Jura M (2019) Compositions of planetary debris around dusty white dwarfs. Astronom J 158:242

Xu S, Rogers LK, Blouin S (2024) The chemistry of extra-solar materials from white dwarf planetary systems. Rev Mineral Geochem 90:171–198

Yang H, Muir JM, Zhang F (2022) Iron hydride in the Earth's inner core and its geophysical implications. Geochem Geophys Geosyst 23:e2022GC010620, https://doi.org/10.1029/2022GC010620

Yoshizaki T, McDonough WF (2020) The composition of Mars. Geochim Cosmochim Acta 273:137–162

Zhang J, Herzberg C (1994) Melting experiments on anhydrous peridotite KLB-1 from 5.0 to 22.5 GPa. J Geophys Res 99:17729–17742

Zhao Y-H, Zimmerman ME, Kohlstedt DL (2018) Effect of iron content on the creep behavior of olivine: 2 hydrous conditions. Phys Earth Planet Int 278:26–33

Zuckerman B, Koester D, Dufour P, Melis C, Klein B, Jura M (2011) An aluminum/calcium-rich, iron-poor, white dwarf star: evidence for an extrasolar planetary lithosphere. Astrophys J 739:101

Reviews in Mineralogy & Geochemistry
Vol. 90 pp. 259–300, 2024
Copyright © Mineralogical Society of America

8

From Stars to Diverse Mantles, Melts, Crusts, and Atmospheres of Rocky Exoplanets

Claire Marie Guimond[1], Haiyang Wang[2], Fabian Seidler[2], Paolo Sossi[2], Aprajit Mahajan[3], Oliver Shorttle[4,5,*]

[1]*Atmospheric, Oceanic, and Planetary Physics, University of Oxford, UK*
[2]*Institute of Geochemistry and Petrology, ETH Zurich, Clausiusstrasse 25, CH-8092, Zurich, Switzerland*
[3]*Trinity College, University of Cambridge, UK*
[4]*Department of Earth Sciences, University of Cambridge, UK*
[5]*Institute of Astronomy, University of Cambridge, UK*

[*]Corresponding author: shorttle@ast.cam.ac.uk

INTRODUCTION

The advent of radial velocity (RV) and transit photometry method has resulted in the detection of more than 5000 exoplanets since 1995. Of these detected planets, a significant fraction, ~20%,[1] may be dominantly rocky. Given the biases inherent to exoplanet detection, this implies an even larger fraction of the overall planet population as likely possessing a significant rock fraction (e.g., Fulton et al. 2017). However, owing to the difficulty in detecting Earth-sized planets, the vast majority of these rocky planets are much larger than Earth (so-called 'super-Earths'); with radii $\lesssim 1.5$ times that of Earth ($1.5\,R_E$) and with masses $2-8\,M_E$ (Earth masses; Fulton and Petigura 2018). Because the radial velocity technique yields estimates of planetary masses, and the transit method permits measurements of their radii, together, they can be used to infer bulk densities (e.g., Charbonneau et al. 1999; Valencia et al. 2007). The uncompressed densities of such super-Earths are similar to the terrestrial planets of our Solar System (Fig. 1).

To first order, similarity to solar system terrestrial planet densities may imply that the compositions and core/mantle ratios of super-Earths are also similar (e.g., Dorn et al. 2015; Dressing et al. 2015; Santos et al. 2015). However, a planet's estimated density is often degenerate with respect to its composition, and the same density may be consistent with many combinations of plausible planet-forming materials (Valencia et al. 2007; Rogers and Seager 2010): In particular, there is typically uncertainty over what the relative contribution to a planet's overall mass and radius is of dense rock/metal interior compared with low density atmosphere/envelope. Even with the knowledge that a planet has no or very little atmosphere (e.g., Kreidberg et al. 2019), degeneracies remain between the mantle to core ratio and with composition from mass–radius measurement alone (e.g., Unterborn et al. 2016; Dorn and Lichtenberg 2021). In the solar system context the polar moment of a inertia of a planet can help elucidate the mass distribution at depth; whilst for exoplanets this property of a planet may in future be inferred, it is likely to be extremely challenging to apply with useful accuracy in the case of small rocky planets (being more readily applied to fluid planets, e.g., Padovan et al. 2018; Consorzi et al. 2023). As a result of this uncertainty on planetary composition, the use of stellar elemental abundances (measured with stellar photospheric spectroscopy), have been proposed to further constrain the bulk compositions, and by extension interior structures, of rocky exoplanets on both individual and population levels (cf. Sotin et al. 2007; Grasset et al. 2009; Delgado Mena et al. 2010; Santos et al. 2015; Thiabaud et al. 2015; Dorn et al. 2017; Hinkel and Unterborn 2018; Putirka and Rarick 2019; Wang et al. 2019a, 2022a,b; Schulze et al. 2021; Spaargaren et al. 2023; Unterborn et al. 2023)

---

[1] Estimated from the proportion of confirmed planets with $R_p < 1.5\,R_E$ and periods less than 100 days, as reported by exoplanets.eu (accessed November 2023)

1529-6466/24/0090-0008$05.00 (print)
1943-2666/24/0090-0008$05.00 (online)   http://dx.doi.org/10.2138/rmg.2024.90.08

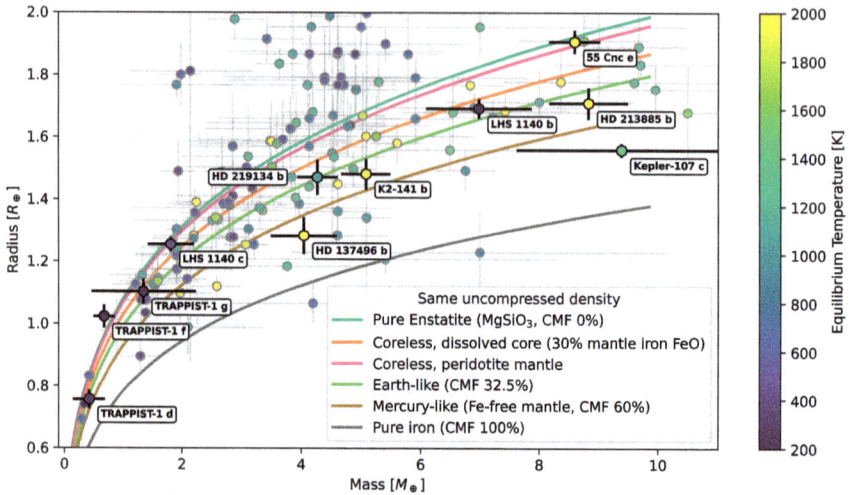

Figure 1. Mass–radius diagram for existing low-mass exoplanets, colored by their calculated equilibrium temperatures. The different curves represent mass–radius trends for various interior properties. The exoplanet data were obtained from https://exoplanet.eu/catalog/; only planets with masses and radii with uncertainty less than 70% are shown, and some interesting candidates are highlighted by name. The mantle compositions for the interior models are obtained from McDonough and Sun (1995) (coreless peridotite and Earth-like), Nittler et al. (2018) (Mercury) and Wang et al. (2022a) (coreless, dissolved core).

How the measured stellar abundances translate to those in bulk planets remains uncertain. Progress has been made by applying basic truisms from our understanding of the Solar System that are assumed to hold for exoplanetary systems; namely, that the refractory elements (of which Ca, Al and Ti are the most abundant) are present in rocky planets in roughly solar (stellar) proportions (e.g., Ringwood 1966; Wänke et al. 1974). Although the major, rock-forming elements, Mg, Si, and Fe behave in a more volatile manner than refractory Ca–Al–Ti (e.g., Lodders 2003), they are also near-solar in proportions in the Earth compared to the Sun (Palme et al. 2014; McDonough and Yoshizaki 2021, near-solar, but with important differences;). This provides the basis for estimating (refractory) rock-forming elemental compositions in rocky planets from their measured host stellar compositions for such elements. More uncertain are the abundances of volatile elements, and O in particular (Wang et al. 2019b), the depletion of which—dubbed "volatility trend"—is shown to vary across the solar system rocky bodies (relative to the Sun; Larimer and Anders 1967; Wasson and Chou 1974; Anders and Ebihara 1982; Bland et al. 2005; Moynier et al. 2011; Braukmüller et al. 2019; Wang et al. 2019b; Fegley et al. 2020; McDonough and Yoshizaki 2021; Khan et al. 2022) and is modulated by (often unknown) star-planet early interactions and planet formation history (Albarède 2009; Hin et al. 2017; Norris and Wood 2017; Sossi et al. 2019; Bitsch and Battistini 2020). Therefore, on top of assuming a stellar composition starting point, empirical constraints from solar system rocky planet compositions have been additionally proposed to constrain the bulk compositions of (rocky) exoplanets for both refractories and volatiles (Wang et al. 2019a, 2022a,b; Spaargaren et al. 2023). The simulation of physical processes during planet formation, particularly nebular condensation, have also been performed to predict exoplanetary compositions (Bond et al. 2010a,b; Moriarty et al. 2014; Dorn et al. 2019; Jorge et al. 2022; Timmermann et al. 2023).

Of all the major planet-forming elements, oxygen stands out as having first-order importance. The O budget in a rocky planet is central in determining two fundamental planetary properties, its core mass fraction and mantle Fe content: properties which, if known, would help resolving degeneracies in relating exoplanet densities to bulk compositions. Putirka and Rarick (2019) treat the planetary Fe/FeO ratio as a free parameter, an approach also adopted by Unterborn et al. (2023) and Guimond et al. (2023b), to predict the likely compositional range of rocky exoplanets (treated as a free parameter, because simply in cosmochemical terms there is almost always enough oxygen available to fully oxidise Fe and other rock forming elements, so oxygen abundances must be forced lower in planets than stellar abundances allow for cores to form at all Putirka and Rarick 2019). On this basis, these authors conclude that the mantle compositions of rocky exoplanets bear a strong resemblance to that of the Earth, with major differences being limited to Mg/Si ratios. However, uncertainty in planetary Fe/FeO ratios, together with the stochastic nature of the accretionary behaviour of volatile elements (e.g., Albarède 2009; Sossi et al. 2022) mean that the assumptions underpinning our use of the star–planet connection to model exoplanet compositions remains observationally untested: atmospheres (if present) trade off against interior mineralogy and structure (core mass fraction) to make density alone a weak constraint on planetary composition (e.g., Rogers and Seager 2010).

Consequently, it is, as yet, unknown whether the Earth-like uncompressed densities of super-Earths speak to a similarity in composition, or whether significant chemical variation exists among the population. Resolving this question is important for the field, as it has the potential to speak to a key result that has emerged from exoplanet surveys over the last decade: that there is a drop in planet occurrence rates at ~1.5 R_E (Fulton et al. 2017). This feature of the data has been linked to envelope loss from sub-Neptune-sized planets initially formed in this radius interval—a loss with several hypothesised physical drivers, e.g., photoevaporation (Owen and Wu 2013, 2017) and core-powered mass loss (Ginzburg et al. 2018; Gupta and Schlichting 2019). How similar the composition of the remaining envelope-free kernels of these sub-Neptunes are to 'born rocky' planets remains uncertain. Improved constraints on planet bulk compositions and interior structures may offer the possibility of distinguishing these two types of small ($< 1.5 R_E$) planets from one other; illuminating a dichotomous evolutionary path for small planets that we do not have an analogue for in the solar system.

Attempts to 'cut the Gordian knot' with respect to the exoplanetary compositions have used empirical trends between planets and stellar abundance data. One such trend indicates that the excess density of super-Earths scales positively with the fraction of iron in the star (Adibekyan et al. 2021), suggesting a genetic connection between the two. However, the range in densities among the terrestrial planets spans almost the entire range observed in exoplanets ($\leq 1.5 R_E$) within error (Fig. 1), such that the compositions of rocky exoplanets remain equivocal from this approach. There are however hints at planetary compositions and/or structures not represented by the solar system population, in particular those planets resolvably less dense than Earth or Mercury (Fig. 1)—although whether this is due to the contribution of atmospheres/envelopes or interior structure (such as the absence of a core) remains to be seen.

The next generation of observations are therefore needed to better constrain planetary compositions. These observations are likely to be manifest in two forms; i) detection of key gaseous species in the atmospheres of exoplanets and, ii) characterisation of surface features of (potentially) bare rocky exoplanets (Kreidberg et al. 2019; Greene et al. 2023). Because the detectability of such transmission/emission spectral features in the infrared scales proportionally with the temperature and radius of the planet, the most promising targets are those rocky super-Earths that are very close to their host star and have large, extended atmospheres and/or transparent atmospheres that facilitate observations of the surface— surfaces which may be molten due to their intense heating from their star.

This review is focused on describing the logic by which we make predictions of exoplanetary compositions and mineralogies, and how these processes could lead to compositional diversity among rocky exoplanets. We use these predictions to determine the sensitivity of present-day and future observations to detecting compositional differences between rocky exoplanets and the four terrestrial planets. First, we review data on stellar abundances and infer how changes in composition may manifest themselves in the expected bulk compositions of rocky exoplanets (see the section *Bulk chemical compositions of rocky exoplanets*). Converting this information in mass–radius relationships requires calculation of the stable mineral assemblages at a given temperature–pressure–composition (T–P–X), an exercise we describe in the section *Mantle mineralogy of rocky exoplanets*. Should the planet be hot enough to engender partial melting of the mantle, then these liquids are likely to rise to the surface and erupt to form planetary crusts; the possible compositional and mineralogical variability of which we examine in the section *Melts and crusts of rocky exoplanets*. Finally, the expected spectroscopic responses of such crusts are examined in the section *Observing exoplanet compositions and mineralogies*. This approach, taking us from star to planet, is summarised as a flow chart in Figure 2.

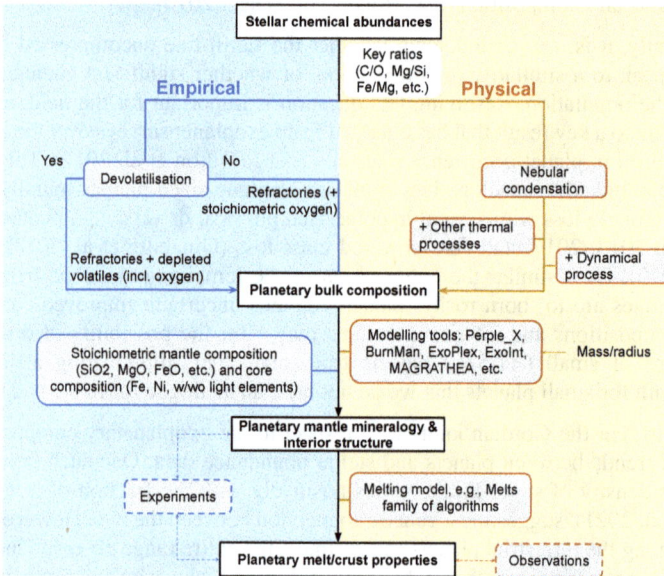

Figure 2. The outline of the review from stars to mantles, melts and crusts and from models to experiments and observations.

BULK CHEMICAL COMPOSITIONS OF ROCKY EXOPLANETS

The planet–star connection

Planets and their host stars are born from the same molecular cloud material, giving them a shared chemical origin. Gas-dust separation in the protoplanetary disk and subsequent dynamical evolution may cause the bulk chemical compositions of (particularly rocky) planets to deviate from that of their host star. However, by analogy with the composition of the Earth's mantle and that of the Sun, the *relative proportions* of refractory (rock-forming) elements between rocky exoplanets and their host stars are expected to be within ~10 % relative (Palme et al. 2014; Wang et al. 2018). This assumption has been carried over into hypothetical exoplanetary systems (e.g., Dubois et al. 2002; Sotin et al. 2007; Schulze et al. 2021). The veracity of this assumption is largely untested (with the exception of weak support coming from polluted white dwarf observations, Bonsor et al. 2021), and, in the case of non-refractory, major elements (Mg, Si and Fe), may lead to large discrepancies (>10%) between true compositions and those predicted based on the assumption that these elements are present in stellar proportions (Dauphas et al. 2015; Wang et al. 2019a; Miyazaki and Korenaga 2020). This is because the

Mg/Si ratio, for example, varies by a factor ~2 even among chondritic meteorites (Wasson and Kallemeyn 1988). Nevertheless, the advantage of adopting stellar abundances as proxies for those of potential surrounding planets is that they are measurable with stellar photospheric spectroscopy (Huang et al. 2005; Hinkel et al. 2014; Nissen et al. 2017; Delgado Mena et al. 2017; Buder et al. 2018; Liu et al. 2020; Adibekyan et al. 2021; Hourihane et al. 2023) and so provide a broad observationally-informed bound on planetary composition.

Stellar chemical compositions vary across the galaxy because stars have formed in different locations at different times and inherit the local chemical history of stellar birth, nucleosynthesis, and death (Gaidos 2000; Bond et al. 2008; Mojzsis 2022; Pignatari et al. 2023). That is, the compositions of stars are inherited from the molecular cloud from which they formed. Abundance ratios of elements with respect to H are typically reported in 'dex' (i.e., base 10 logarithmic) units: typical ranges for FGK stars in stellar surveys fall between –0.5 and +0.5 dex of the solar value; measurement errors can be as low as 0.01 dex as reported by individual groups, but are more likely to be on the order of 0.1 dex between groups given the non-standardised techniques for quantifying abundances from stellar spectra (Hinkel et al. 2016; Hinkel and Unterborn 2018). Some key elemental ratios in stars—namely, C/O, Mg/Si and Mg/Fe—have been proposed to indicate potential properties of rocky exoplanets around those stars (Bond et al. 2010a,b; Thiabaud et al. 2015; Wang et al. 2019a). Figure 3 shows the spread of these key elemental ratios for FKG stars in the solar neighbourhood in using two different stellar catalogues (GALAH, Buder et al. 2021; Hypatia, Hinkel et al. 2014). The Sun appears to be rather typical in both catalogues (aligning well with the medians of the corresponding histograms). These key ratios indicate, to first order, the propensity of mantle composition and core sizes of hypothetical rocky exoplanets around these stars (see indications in Fig. 3). The large spreads in these parameter spaces reveal the potential chemical diversity of rocky worlds around such stars.

A diversity of host star elemental abundances implies a diversity in planetary bulk compositions (e.g., Sotin et al. 2007; Bond et al. 2010b; Delgado Mena et al. 2010; Carter-

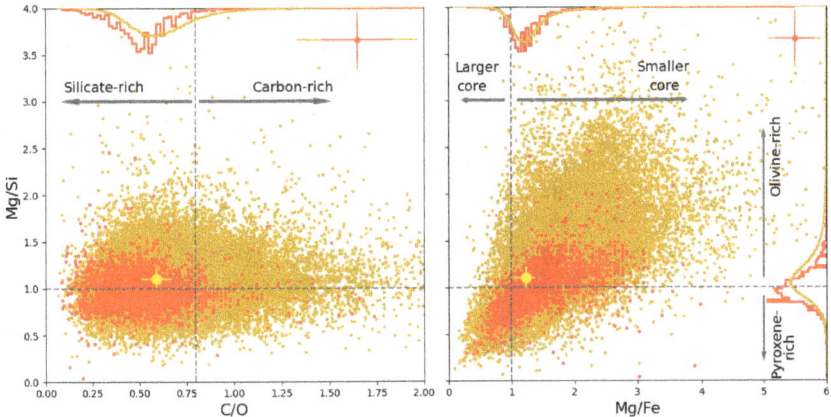

Figure 3. Key elemental ratios—C/O, Mg/Si and Mg/Fe—of stars in using both GALAH (**yellow**; Buder et al. 2021) and Hypatia (**red**; Hinkel et al. 2014) stellar catalogues. The histograms of these ratios in both databases are also shown, along with the typical error bars of these ratios. The reference solar abundances (in **bright yellow**), which have been used to normalise both databases, are from Asplund et al. (2021). The apparent fewer number of stars in the higher Mg/Si regime (> 2.0) in the **left** panel (compared to the **right** panel) is due to the fewer available measurements of carbon and oxygen abundances in the GALAH database. The **vertical and horizontal dash lines** (C/O:0.8, Mg/Fe:1.0, and Mg/Si:1.0), along with different arrows, indicate how composition may lead to mineralogical and structural propensities of hypothetical rocky exoplanets around these stars.

Bond et al. 2012a; Hazen et al. 2015; Unterborn and Panero 2017; Hinkel and Unterborn 2018; Putirka et al. 2021; Putirka and Rarick 2019; Mojzsis 2022; Wang et al. 2022a; Guimond et al. 2023a,b; Spaargaren et al. 2023). However, because the abundances of the major rock-forming elements are positively correlated in FGK stars, the resultant ratios of these elements vary by only modest amounts (factor 2–3; e.g., Hinkel et al. 2014; Young et al. 2014; Brewer et al. 2016). Consequently, should elemental abundances in a given star be a good proxy for that of the planet, then this would imply that rocky exoplanets are composed predominantly of O, Fe, Mg, Si, together with smaller amounts of the refractory elements Ca and Al—largely concordant with our experience of solar system planet composition and mineralogy (Putirka et al. 2021).

Nebular condensation

The current paradigm for the mechanisms by which the terrestrial planets formed involves a stage of nebular cooling and condensation. Collection of these nebular condensate grains into pebbles, prior to their mixing and growth into progressively more massive bodies, dubbed planetesimals ($< 10^{-3} M_E$) and embryos ($\sim 0.1 M_E$), is thought to have led to the formation of the solar system's four inner planets (e.g., Morbidelli et al. 2012; Johansen and Dorn 2022).

The composition of the solar nebular gas is closely approximated by that of the Sun (Urey 1954; Larimer and Anders 1967; Wasson and Chou 1974; Anders and Ebihara 1982; Alexander et al. 2001; Bland et al. 2005; Moynier et al. 2011; Hin et al. 2017; Norris and Wood 2017; Braukmüller et al. 2019; Sossi et al. 2019; Wang et al. 2019b; Fegley et al. 2020). Yet, this process of condensation is not isochemical, as the observed volatile abundances in solar system rocky bodies (e.g., C, O, S, P, K and Na) are depleted compared to their abundances in the Sun, whereas refractory lithophile elements are relatively enriched: in the bulk silicate Earth, this factor of refractory lithophile enrichment is 1.21 × solar, normalised to Mg (Palme et al. 2014)—normalising to the refractory lithophile element Al results in an identical baseline for refractories between the terrestrial planets and the Sun (Wang et al. 2019b; McDonough and Yoshizaki 2021; Sossi et al. 2022).

The main piece of (empirical) evidence pointing to the importance of nebular condensation in dictating the compositions of the terrestrial planets is that their abundances of volatile elements decrease as a smooth function of their 50% nebular condensation temperatures (T_c^{50}; Fig. 4). This is defined as the temperature at which 50% the mass of an element is condensed from a solar composition gas at 10^{-4} bar (e.g., Larimer and Anders 1967; Lodders 2003; Wood et al. 2019). Such a trend holds among chondritic meteorites, although they exhibit constant-abundance plateaus in the most volatile elements (with $T_c^{50} < 700$ K; Braukmüller et al. 2019). In this manner, rocky planets are, by definition, those that have lost (or never acquired) the major complement of the volatile elements present in the Sun. For example, H is depleted by a factor $\sim 10^6$ in the Earth (relative to the Sun, normalised by Al), equating to ~ 1000 µg/g (Hirschmann and Withers 2008; Marty 2012). Only O, S, C, and H are likely to be sufficiently abundant to affect mass–radius relationships in an observable manner. Though their abundances in rocky exoplanets have been yet to be measured, they must, by analogy, be depleted to similar levels to those observed in the Earth (with the possible exception of C, see below), lest the mass–radius relationships of rocky exoplanets diverge from the observed, Earth-like trend (e.g., Luque and Pallé 2022).

The difficulty in predicting the volatile budgets of rocky exoplanets lies in the complexity of the processes that lead to their accretion. Were the volatile element abundances of Earth to have resulted from nebular condensation alone, then elements that condense below a given threshold temperature, $T_{cut-off}$ should be absent from the solid material from which the Earth accreted. By contrast, the Earth should harbour its full (solar) complement of elements with $T_c^{50} > T_{cut-off}$. This is because elements typically condense over a narrow range of temperatures, passing from fully vaporised to fully condensed over a temperature range of ~ 30–40 K (e.g., Larimer 1967; Albarède 2009). That the Earth exhibits a smooth decline in volatile element abundances with

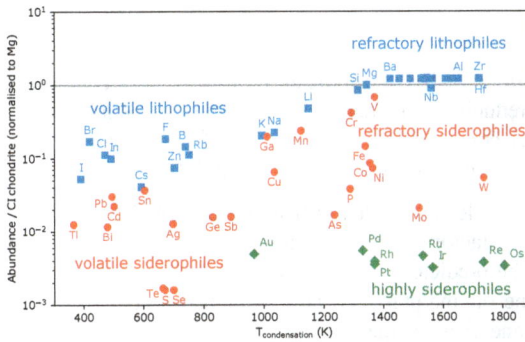

Figure 4. An example of the volatility trend of elements in rocky objects in the solar system, in this case the Earth. The Earth's composition has been normalised to CI chondrites, a model for the presumptive building blocks of rocky material in the solar system, were they not to have variably lost volatile elements. Normalisation of the Earth and CI composition to Mg accounts for the large hyper-volatile component within chondrites. Compositional data are from Palme et al. (2014) and Clay et al. (2017), half mass condensation temperatures are from Wood et al. (2019).

T_c^{50} and not a binary pattern with an inflection point at some fictive $T_{\text{cut-off}}$, indicates that these elements were brought through mixing of different materials that each experienced various $T_{\text{cut-off}}$ (Sossi et al. 2022). As a result, planetary compositions cannot be simply tied to condensation temperature, and, instead, are a complex function of their accretion histories. This problem has been circumvented by applying 'devolatilisation trends' defined for the Earth (and other telluric bodies) that quantify element abundance vs. T_c^{50} to provide limits as to the expected volatile element abundances in rocky exoplanets (Wang et al. 2019a, 2022a,b; Spaargaren et al. 2023).

It is noteworthy that there is no requirement for the condensation temperatures of a nebula around another star to conform to those for a solar composition gas. Indeed, variations in element ratios, particularly C/O (Larimer and Bartholomay 1979) may lead to drastic differences in condensation temperatures, brought about by a change in the condensing mineral assemblage. For example, carbides and sulfides (such as oldhamite, a common phase in enstatite chondrites) become stable at the expense of oxides (cf. Larimer and Bartholomay 1979; Bond et al. 2010b) for nebulae with C/O ratios higher than ~0.9. The corollary of this result is that elements that are thought of as being volatile in a Solar System context (namely C) behave as refractory elements (and vice-versa). This realisation has led to some experimental work done on reduced phases, such as SiC, which is supposed to dominate rocky exoplanets with high C/O ratios (Hakim et al. 2018, 2019; Miozzi et al. 2018; Allen-Sutter et al. 2020). As such, exotic compositions beyond silicate/pyrolitic mantles may be plausible (see the section *Beyond Earth-like mantles*).

The role of O in controlling core/mantle ratios

Among the volatile elements, oxygen is expected to be by far the largest constituent of planetary mantles, owing to its abundance in stellar nebulae and the fact that it combines readily with metals to form oxides and silicates. As a result, O plays a critical role in modulating the composition of the mantle and thus the stable mineral phases therein. This is because the finite amount of O that condenses from stellar nebulae combines sequentially with the rock-forming oxides according to their redox potentials (at 1 bar, this order is Ca > Al > Mg > Si > Fe; Chase et al. 1982), until it is exhausted. Typically, O/(Al+Ca+Mg+Si+Fe) ratios in condensing phases are such that some fraction of Fe is left over as metallic iron (Fe^0). Such a system with Fe^0 and FeO present and in equilibrium defines the oxygen fugacity (fO_2; i.e., the partial pressure of O_2 gas corrected for non-ideality) via:

$$Fe + \tfrac{1}{2}O_2 = FeO \qquad (1)$$

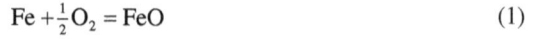

where higher proportions of FeO/Fe indicate more oxidising conditions. The reduced iron, together with other 'reducing' (or siderophile) metals (e.g., Ni and Co, and notably Si, which imparts mineralogical consequences discussed later) constitute the core of the planet, and hence determine the core mass (Wang et al. 2019a, 2022a).

Because oxygen condenses gradually from stellar nebulae as a function of temperature, there is, as yet, no satisfactory parameterisation that would permit its abundance in planet-forming materials to be predicted *a priori*. One insight is that varying the Mg/Si ratio of the nebula would alter the olivine/pyroxene ratio (which have (Mg,Fe)/O = 1/2 and (Mg,Fe)/O = 1/3, respectively) and thereby the amount of oxygen consumed during formation of the silicate portion of a planet (Unterborn and Panero 2017). However, this is only one of the routes via which oxygen may be incorporated into planets. Therefore, oxygen abundance is typically presumed to be sufficient to oxidise major silicate-building elements until Fe (i.e., Mg, Si, Ca, Al; Dorn et al. 2015; Santos et al. 2015; Brugger et al. 2017; Plotnykov and Valencia 2020; Schulze et al. 2020; Guimond et al. 2023b). This assumption is loosely supported by the observation that O/(Mg + Si + Fe + Ca + Al) is sufficiently high across the vast majority of Hypatia Catalog stars (Putirka and Rarick 2019). Meanwhile, to account for variable condensed oxygen budgets, the degree to which metals like Fe (and Ni) are oxidised is typically prescribed: as a fixed value (nil or at Earth's value; Santos et al. 2015; Schulze et al. 2020; Spaargaren et al. 2023); or, as a nominal range by introducing an additional parameter, such as mantle Mg or Fe number (Mg# = Mg/(Mg + Fe); Fe# = Fe/(Mg + Fe); Sotin et al. 2007; Brugger et al. 2017; Plotnykov and Valencia 2020, or Fe partition coefficient (Putirka and Rarick 2019; Guimond et al. 2023b; Unterborn et al. 2023), with bounds chosen to imply Fe-metal cores ubiquitously. As a result, whilst host stellar abundances may approximate *bulk* rocky planetary compositions, there is debate over how these can be used to place constraints on the interior structure and properties (i.e., densities) of rocky planets (Wang et al. 2019a; Schulze et al. 2020; Adibekyan et al. 2021; Seidler et al. 2021; Spaargaren et al. 2023; Unterborn et al. 2023).

Post-accretionary processes

Processes within planets themselves compound the challenge of linking stellar composition through to internal structure. A key process supposed to be operative early in a planet's history is Fe disproportionation, where a high pressure reaction of the form

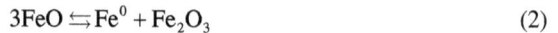

$$3FeO \leftrightarrows Fe^0 + Fe_2O_3 \qquad (2)$$

causes ferrous iron, Fe^{2+}, hosted in the silicate portion of the planet, to break down into Fe^{3+} and metallic iron (Frost et al. 2004; Armstrong et al. 2019). The metal, if it sinks to the core, will result in net oxidation of the complementary mantle with no change in the amount of O in the bulk planet. Other processes, such as the dissociation of H_2O and loss of H_2 by atmospheric escape, followed by mantle influx of the oxygen, may lead to a similar effect on the mantle oxygen budget (Catling et al. 2001; Wordsworth et al. 2018; Zahnle et al. 2019; Krissansen-Totton et al. 2021, see also Hirschmann 2023).

FeO may also be produced during core formation by the dissolution of Si into metal,

$$SiO_2^{silicate} + 2Fe^0 \leftrightarrows 2FeO^{silicate} + Si^0 \qquad (3)$$

occurring at high temperatures and pressures relevant for the base of a deep magma ocean (Javoy 1995). Reaction (3) leads to still more mantle oxidation. This reaction also proves important by increasing the Mg/Si ratio of the mantle with respect to the bulk planet, a complexity which will be returned to in the section *Mantle mineralogy of rocky exoplanets*.

These post-accretionary processes are significant for the subsequent coupled interior–atmosphere evolution of a planet. There must have been a reset of how oxidising Earth's mantle was following core formation (where it would have been highly reducing; e.g., Righter and Ghiorso 2012) to its present day more oxidising state (Frost and McCammon 2008). Iron disproportionation reactions like (2), whether driven by mineral growth or in the silicate liquid itself (Hirschmann 2022), presumably contributed to this evolution. The result is a mantle ultimately producing much more oxidising volcanic gases than it would have done immediately following core formation. Such post-accretionary shifts in mantle redox have potentially profound implications for resultant atmospheres and oxygen cycles in the planet (Ortenzi et al. 2020; Liggins et al. 2022), yet the effects of which remain hard to quantitatively predict. Improved observational constraints on the composition of Venusian magmas and volcanic gasses would provide an important new data point from the solar system on how these processes and others play out on roughly Earth-sized planets.

MANTLE MINERALOGY OF ROCKY EXOPLANETS

The major rocky reservoirs on planets are their mantles. Initially, prior to the formation of thick crusts, this mantle is a crystallised rocky layer that has the composition of the bulk planet, less those elements that partitioned into its metallic core. The stable assemblages of minerals within these mantles vary across planets according to their bulk compositions, as this section will detail. The resulting mineralogies control planetary thermochemical evolution insofar as different minerals have different material properties (e.g., density, bulk moduli and heat capacity; Spaargaren et al. 2020). Across the common mantle minerals and assemblages thereof, notable differences are suggested in, for example: viscosities (e.g., Yamazaki et al. 2009; Ammann et al. 2011; Hansen and Warren 2015; Girard et al. 2016; Thielmann et al. 2020; Cordier et al. 2023), volatile storage (see the section *Life-essential elements (CHNOPS) in rocky exoplanet mantles*; e.g., Dong et al. 2022; Wang et al. 2022b; Guimond et al. 2023b), solidii (e.g., Hirschmann 2000; Kiefer et al. 2015; Brugman et al. 2021), and ferric iron partitioning and thus upper mantle oxygen fugacity (Guimond et al. 2023a). In these ways, predicting rocky planet mineralogies represents an emerging frontier in exoplanet characterisation.

The mineralogy of a planetary mantle is set by equilibria established among its constituent phases. The stability of a given (mineral) phase assemblage for a fixed bulk composition is dictated by the minimisation of Gibbs free energy at a given $P–T$ (e.g., Connolly and Kerrick 1987). Pressure and temperature profiles through planets are predicted using equations of state and adiabatic gradients, calculated iteratively with mineralogy. The bulk composition may be estimated from host star abundances, though this inevitably introduces uncertainty: first, due to poorly-constrained partitioning processes, as detailed in the section *Bulk chemical compositions of rocky exoplanets*; and second, due to measurement error on the stellar abundances themselves. Uncertainties of 0.2 dex on two elements propagate to a fractional uncertainty on their molar ratio of 65%, for example (using the derivation in Hinkel et al. 2022). A further, underlying assumption is that this bulk oxide composition is homogeneous throughout the mantle; there is some debate about whether this is true for Earth (e.g., Anderson and Bass 1986; Ballmer et al. 2017).

Having estimates of bulk composition, mantle phase equilibria can then be modelled self-consistently with pressure-temperature profiles, most commonly using thermodynamic algorithms that minimise Gibbs free energy, a class of models including Perple_X (Connolly 2009) and HeFESTo (Stixrude and Lithgow-Bertelloni 2011). The user of these algorithms must further commit to a thermodynamic database and choice of activity composition relations for the expected minerals (e.g., Ghiorso and Sack 1995; Holland and Powell 2011; Green et al. 2016; Stixrude and Lithgow-Bertelloni 2022). Because such databases and mineral models are experimentally constrained for primarily Earth-like compositions, they necessarily require a degree of extrapolation to the exoplanetary context: this extrapolation is both in terms of composition, where, e.g., Na contents, Mg/Fe ratios, and oxygen fugacity may differ from the terrestrial mantle, and in pressure, where in super-Earth's pressures in the lower mantle may exceed 600 GPa, or five times that of Earth's lower mantle (Wagner et al. 2012; Unterborn and Panero 2019; Boujibar et al. 2020).

As a result of the above, estimates of the mineralogical diversity present in exoplanetary interiors, especially as planets become much more massive than Earth, need cautioning by the limitations of our thermodynamic models. As one example, the oft-used equation of state of crystalline Mg-postperovskite is experimentally validated to 265 GPa (Sakai et al. 2016), although higher-pressure measurements have been achieved using shock compression (Spaulding et al. 2012; Bolis et al. 2016; Fratanduono et al. 2018)—recent interior models tend to rely on *ab initio* simulations by Sakai et al. (2016) that estimate the $MgSiO_3$ equation of state to well beyond the plausible rocky exoplanet mantle pressures. Recognising these limitations and assumption, we proceed in this section to explore exoplanet mineralogy.

Possible range of mantle mineralogies for silicate planets

In the past half-decade, a gamut of theoretical studies have exploited phase equilibria models to estimate exoplanet mantle mineralogy from stellar abundances: Hinkel and Unterborn (2018) and Wang et al. (2022a) on nearby stars with higher-precision abundances; and Putirka and Rarick (2019), Guimond et al. (2023b), Spaargaren et al. (2023), Unterborn et al. (2023) and on a population level of stars from the Hypatia Catalog (Hinkel et al. 2014) and/or the GALAH survey (Buder et al. 2018). A study of mantle compositions inferred from polluted white dwarfs provides a complementary approach (Putirka and Xu 2021). These studies suggest that rocky planet mantles can, for the most part, be understood in terms of known terrestrial mineral components: i.e., familiar minerals are stabilised in these mantles, even if their relative proportions and composition differ from those in Earth's mantle.

Upper mantles. As Mg and Si are the most abundant oxide-forming elements synthesised in stars, the ferromagnesian silicates olivine (and its polymorphs; wadsleyite and ringwoodite) $(Mg,Fe)_2SiO_4$ and orthopyroxene $(Mg,Fe)SiO_3$ are the dominant minerals at pressures where they are stable, <25 GPa, depending on temperature. As pressure increases, clino- and orthopyroxene dissolve into garnet to form majorite (Akaogi and Akimoto 1977), which becomes predominant near ~15 GPa, and also contains the entirety of the Ca and Al budget of the planet at these pressures (Fig. 5). That is, no observed stellar composition appears rich enough in what would be minor oxides in Earth's mantle (e.g., CaO, Al_2O_3, TiO_2) to saturate their pure phases (Putirka and Rarick 2019). Indeed, as shown in Figure 6, 89% of Hypatia stars lead to upper mantles dominated by a mixture of olivine plus orthopyroxene, assuming for simplicity $Mg/Si_{mantle} = Mg/Si_{star}$ and 90 mol% of planets' Fe in their cores; or 85% of stars assuming instead a constant 5 wt.% Si in metal cores (e.g., Hirose et al. 2021).

Substantial diversity is nonetheless expected within these pyroxene–olivine compositions, even with relative abundances in host stars not significantly different from the sun's. Again, due to their abundance, the most important changes are predicted in the ratio of olivine to orthopyroxene, being a direct function of Mg/Si (Ringwood 1989; Bond et al. 2010b; Delgado Mena et al. 2010; Young et al. 2014; Hinkel and Unterborn 2018; Putirka and Rarick 2019; Mojzsis 2022; Wang et al. 2022a,b; Guimond et al. 2023b; Spaargaren et al. 2023). Increasing Mg/Si above unity produces more forsterite olivine and the expense of enstatite orthopyroxene—forsterite's chemical formula dictates two units MgO per unit SiO_2, versus 1:1 in enstatite (Fig. 3, right panel). The extreme ends of the stellar Mg/Si distribution see silicate mantles becoming olivine-free at Mg/Si $\lesssim 0.8$ (in the presence of some mantle FeO), or orthopyroxene-free at Mg/Si $\gtrsim 1.6$ (Fig. 6). In these cases, the oxides in excess (SiO_2 or $(Mg,Fe)O$ wustite) could form their own phases—one hypothetical avenue for "exotic" mineralogies (see the section *Beyond Earth-like mantles*).

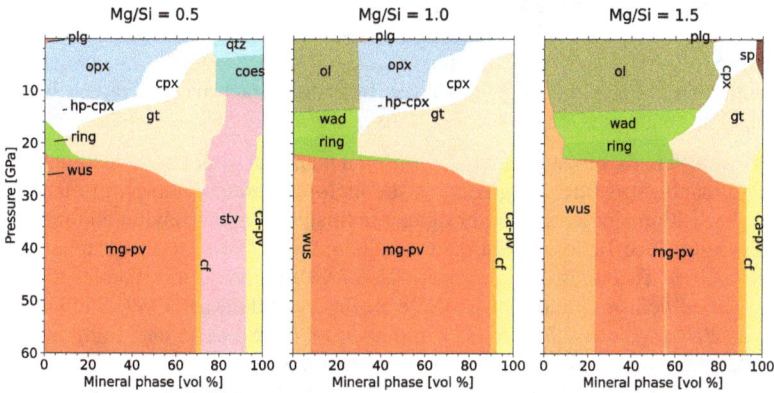

Figure 5. The effect of Mg/Si on the mineralogy of a silicate mantle. Except for Mg and Si, the compositions of other elements are kept the same as those of bulk silicate Earth (McDonough and Sun 1995). The mantle potential temperature is assumed at 1700 K. This is modelled with pyExoInt (Wang et al. 2022a) that has Perple X (Connolly 2009) integrated for mineralogy output. The abbreviations of mineral phases stand for: ol–olivine, opx–orthopyroxene, cpx–clinopyroxene, hp-cpx–high-pressure clinopyroxene, plg–plagioclase, qtz–quartz, coes–coesite, sp–spinel, wad–wadsleyite, ring–ringwoodite, gt–garnet, stv–stishovite, wus–magnesiowustite (ferropericlase), mg-pv–magnesium perovskite (bridgmanite), cf–calcium ferrite, and ca-pv–calcium perovskite.

The third-most abundant oxide in most mantles, FeO, has a secondary role in mineralogy, mostly via stabilising olivine polymorphs. Higher Fe/Si would be associated with increased ferrosilite olivine over fayalite orthopyroxene, an analogous effect to increased Mg/Si favouring forsterite. In fact, the mantle ratio (Mg+Fe)/Si is a better predictor than Mg/Si of the ratio olivine/orthopyroxene, yet it is in practice perhaps less useful because of the challenges predicting the Fe content in an exoplanet mantle (see previous section). In any case, the thermodynamic models quickly become unreliable at very FeO-rich compositions ($\gtrsim 25\,\text{wt.\%}$); such mantles' mineralogies have not been studied in detail.

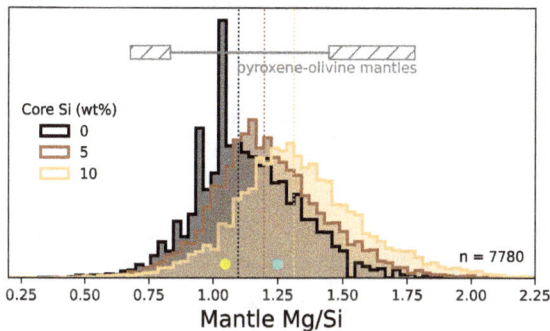

Figure 6. The distribution of mantle Mg/Si implied from stellar abundances in the Hypatia Catalog (Hinkel et al. 2014), for different assumptions about the Si content of metal cores (assumed constant for simplicity), and no devolatilisation. Calculations assume 85 mol% of the bulk planet's Fe forms a core; reasonable changes do not affect the distributions. **Vertical dotted lines** show the median of each distribution. The yellow marker indicates the solar value from Lodders et al. (2009); the **blue marker** indicates the Bulk Silicate Earth estimate. The **horizontal bar** shows where mantles are dominated by a mixture of olivine and pyroxene, with the **hatched regions** showing where only some compositions stabilise both these phases. Calculations based on the Stixrude and Lithgow-Bertelloni (2022) thermodynamic database implemented in Perple_X (version 6.9).

Lastly, higher Al_2O_3 or CaO proportions would be accommodated by increased garnet and clinopyroxene, up to a maximum at 4 GPa of about one third garnet by mass. Components beyond Mg–Si–Fe–Al–Ca are not expected to to occur in abundances high enough to notably affect mantle mineralogy—but they may affect other important material properties—solidus temperatures, for example—as the section *Melts and crusts of rocky exoplanets* will discuss.

Lower mantles. As pressures increase within a planet, several key phase transitions lead to a layered mantle structure. Ringwoodite, the high-pressure polymorph of olivine, breaks down together with majorite, to form bridgmanite (magnesium perovskite) and ferropericlase (magnesiowusite) (See Table 1 in Putirka et al. 2024, this volume for a comprehensive list of mineral formulae). This transition occurs at ~22–27 GPa in the Earth's mantle, depending on temperature, and defines the lower mantle (e.g., Ito and Takahashi 1989; Shim et al. 2001; Dong et al. 2021a, also see Fig. 5). Extrapolation of the Stixrude and Lithgow-Bertelloni (2011, 2022) thermodynamic databases suggests that the mantles of most rocky exoplanets, provided that they are comprised largely of silicates and reach sufficient pressures, will have a transition zone-like structure.

At the high pressures of lower mantles (see Fig. 5), there is typically less information available about the variation in phases and phase proportions with depth than in upper mantles, so mineralogies tend to be simplified. (Mg,Fe)SiO$_3$ perovskite and postperovskite always dominate here, with a smaller amount of (Mg,Fe)O wustite (i.e., ferropericlase) taking up any excess Mg and Fe. CaSiO$_3$ perovskite (davemaoite), Earth's third-most-common lower mantle phase, also appears as the main high-pressure host of Ca across the stellar-abundance-derived range of bulk compositions. Most Al is expected to be dissolved in perovskite (Caracas and Cohen 2005), although some models include Al in a low-abundance calcium ferrite solution as well (Stixrude and Lithgow-Bertelloni 2022). The main axis for mineralogical diversity in the lower mantle is the effect of increasing Mg/Si on increasing ferropericlase modality at the expense of perovskite (Fig. 5). Finally, perovskite will transition to postperovskite around ~120 GPa along Earth-like pressure–temperature profiles (Tsuchiya et al. 2004; Murakami et al. 2004; Oganov and Ono 2004). Because the lower mantle will dwarf the upper mantle in mass for all rocky planets of at least Earth's mass, we expect the prototypical exoplanet mantle to be mostly perovskite and post-perovskite (Fig. 7).

The existence of exoplanets more massive than Earth motivates extending silicate phase equilibria to much higher pressures (Fig. 7). Whereas Earth's core/mantle boundary (CMB) lies at 130 GPa, the highest mantle pressures reached in "rocky" planets, $R_p < 1.5\,R_E$, were calculated by Unterborn and Panero (2019) to be over five-fold greater, ~630–740 GPa, increasing with mantle FeO content (the higher value corresponds to Mars-like mantle FeO; this assumes pure solid ε-Fe cores and Mg/Si = 1). At these extreme lower mantle pressures, silicates may in fact be entirely molten, even at the 'cooler' geotherms of planets several billion years old (Stixrude 2014; Fratanduono et al. 2018; Boujibar et al. 2020). Enduring basal magma oceans are thus a ubiquitous possibility, presenting an additional challenge for interior structure modellers (and yet to be understood consequences for planetary evolution).

'Super-Earth' pressures are nonetheless accessible to shock experiments, coupled with *in-situ* synchrotron techniques, to probe states of matter at extreme conditions. One of the key mineralogical transformations occurring at high pressure is the transition from high- to low-spin iron (Badro 2014; Shim et al. 2023). Spin-transitions in Fe may result in significant changes in the elastic properties of exoplanetary mantles, particularly in oxidising, FeO-rich mantles. Further, changes in FeO and MgO crystal lattice structures have been detected in dynamic compression experiments (Coppari et al. 2013, 2021, Fig. 7). These and other possible deep-mantle phase transitions are highly relevant for mantle dynamics, via their negative, endothermic Clapeyron slopes: slowing down convection by consuming latent heat (Christensen and Yuen 1985).

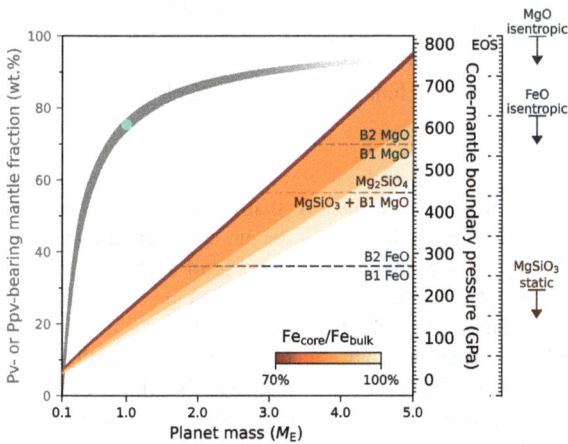

Figure 7. Left: The mass fraction of the mantle where perovskite (pv) or postperovskite (ppv) is stable (**grey curve**) increases rapidly with increasing planet mass. This mass fraction represents the lower mantle of the planet, and does not vary appreciably with Fe partitioning into the core, potential temperatures between 1600–1900 K, or diverse silicate compositions—the line width spans the 1σ distribution over these three variables. Earth's position on the left y-axis is shown with a **blue marker**. The right y-axis, for context, corresponds to the scaling of core/mantle boundary pressure with planet mass, with lines coloured by fixed partitioning of Fe into the core. Line widths span the 2σ distribution over Hypatia-derived Fe/Mg and Si/Mg, and with core Si content varying from 0 to 10% (and no other core light elements for simplicity), and at a mantle potential temperature of 1700 K. All values of Fe_{core}/Fe_{bulk} result in a similar upper limit to core/mantle boundary pressure, but a lower limit decreasing with increasing Fe in the core, due to the core/mantle boundary moving outward in radius. These core/mantle boundary pressure calculations use the ExoPlex mass–radius composition calculator (see Unterborn et al. 2023 for details), which implements a liquid Fe core, and the Stixrude and Lithgow-Bertelloni (2011) database of silicate thermodynamic properties in Perple_X, assuming as a caveat that all equations of state are valid at the relevant *T–P–X* conditions. **Dashed lines** indicate expected endmember phase transitions, not necessarily included in current interior models: in FeO and MgO as observed in ramp-compression experiments from Coppari et al. (2013, 2021), and in $MgSiO_3$ as theoretically predicted by Umemoto et al. (2017, their lower limit with Mg/Si = 2). **Right:** Experimental limits to pressure–density equations of state. $MgSiO_3$ is from static-compression experiments by Sakai et al. (2016), and MgO and FeO are again from Coppari et al. (2013, 2021). Note that shock-compression experiments, measuring irreversible compression which may not represent equilibrium states, have reached higher pressures (e.g., Root et al. 2015; Spaulding et al. 2012; Bolis et al. 2016; Sekine et al. 2016; Fratanduono et al. 2018), but are not included here.

Molecular dynamics simulations represent an alternative route to study the deep interiors of exoplanets. These simulations suggest that at extremely high pressures silicates will dissociate into their constituent oxides (MgO, FeO, SiO_2) (Umemoto et al. 2006, 2017; Tsuchiya and Tsuchiya 2011; Umemoto and Wentzcovitch 2011; Niu et al. 2015). Before complete dissociation, however, Umemoto et al. (2017) found $MgSiO_3$ postperovskite to recombine with either MgO or SiO_2 to form new phases starting at ~450–600 GPa, depending on the Mg/Si ratio. For example, the oxygen-excess phase Mg_2SiO_5 was predicted beyond 600 GPa. As reviewed in Duffy and Smith (2019), limitations of both theoretical and experimental approaches mean it will be important to reconcile results from both, if we are to constrain the mineralogy of deep exoplanet mantles.

The uncertainties on the equations of state and phase diagrams of silicate minerals at ultrahigh pressure may not be the primary source of error when interpreting mass–radius relationships of exoplanets (Unterborn and Panero 2019). Yet, understanding minerals' associated material properties at these conditions will be key to linking interior dynamics with surface observeables beyond just mass and radius. Doing so will help us answer, for example, how vigorously a deep super-Earth mantle is convecting, if it is convecting at all (e.g., Karato 2011; Stamenković et al. 2011, 2012; Tackley et al. 2013; Miyagoshi et al. 2014, 2018; Ritterbex et al. 2018; van den Berg et al. 2019; Coppari et al. 2021; Shahnas and Pysklywec 2021); and whether its likely mode of convection would help or hinder recycling at the planet surface.

Beyond Earth-like mantles

The foregoing discussion has made the implicit assumption that mantles of exoplanets are similar to those of the four terrestrial planets; that is, that they are predominantly composed of silicates. However, other mineralogies may occur, and even our solar system points to one alternate path for "rocky" planet mantles: the interior of Mercury is purported to contain an FeS layer sitting at its point of neutral buoyancy, between the core and the mantle (Malavergne et al. 2010, though see Cartier et al. 2020). A key variable in driving non-Earth-like mantles is how reducing the planet formation environment was. For disks with C/O > 0.9, Condensation sequences may diverge markedly from that of our Solar System (see the section *Bulk chemical compositions of rocky exoplanets*; Bond et al. 2010b; Moriarty et al. 2014) leading to potentially non-silicate-based rocky exoplanets. However, observations thus far suggest that such high-C/O systems are rare (~1%) among FGK dwarfs in the solar neighbourhood (Teske et al. 2013; Brewer et al. 2016).

For C/O ratios higher than ~0.9, condensation calculations predict carbides and nitrides, rather than oxides and silicates, condense from cooling protoplanetary disk gas (Larimer and Bartholomay 1979). Planets forming in such settings would be correspondingly enriched in carbon and be highly reducing. As a consequence, these planets should contain iron exclusively in its metallic form (e.g., as Fe^0 in their cores rather than as FeO in their mantles). Titanium carbide (TiC) may be an abundant minor phase in the mantles of such planets, whereas other, typically lithophile elements (Mg and Ca) may occur predominantly as sulfides; niningerite and oldhamite (Skinner and Luce 1971). Planets made from such material are therefore richer in C and S and poorer in O and Al with respect to terrestrial planets. Indeed, such mineralogies have been invoked, in part, to explain the low bulk density of some planets (e.g., Madhusudhan et al. 2012). Coupled chemical–dynamical planet formation simulations of Carter-Bond et al. (2012b) suggest that it is indeed possible to achieve nearly 50% C by mass for inner planets in the upper end of high-C/O disks. At these highly reducing conditions (fO_2 between 5 and 6.5 log units below the IW buffer, and between 2 and 10 GPa) Schmidt et al. (2014) indicate that moissanite (SiC) will be stable, and silicate phases will have Mg# between 0.993 and 0.998 (i.e., almost entirely Fe-free). More reducing still, at nine log units below IW, Si metal is stable, however, this is likely to be a lower bound, as Si will dissolve into Fe metal (Lacaze and Sundman 1991). At higher fO_2 than $\Delta IW - 5$, SiC may break down to form FeSi liquids and graphite (Hakim et al. 2018, 2019). However, a comprehensive experimental investigation of the stabilities of C- and S-rich exoplanet interiors is currently lacking.

Even among "silicate" compositions, the observed distribution of stellar Mg/Si could also imply the existence of extreme mineralogies that do not fit a pyroxenite or peridotite classification (Putirka and Rarick 2019). Indeed, the actual distribution of mantle Mg/Si across rocky exoplanets may be wider than the distribution seen in stars. Although the Earth has a superchondritic Mg/Si ratio, as inferred from peridotites of the Earth's upper mantle, yielding Mg/Si of 1.27 (by mass), there are other materials in the Solar System with markedly lower Mg/Si. These materials are represented by the enstatite chondrites, with Mg/Si = 0.64 (Wasson and Kallemeyn 1988) and therefore contain enstatite rather than forsteritic olivine as the predominant silicate phase.

Extrapolating even further might suggest exoplanetary mantles could become silica (SiO$_2$)-saturated. This condition is likely rare for two main reasons. First, Mg and Si are both major elements in the stellar nebulae, and are not readily fractionated from one another during condensation: This is because they condense into olivine and orthopyroxene, and their condensation reactions are similarly dependent on fO_2. Second, Si is more siderophile than is Mg (e.g., Ringwood and Hibberson 1991; Fischer et al. 2015) and therefore core formation will invariably result in an increase in the Mg/Si ratio of the planetary mantle relative to its bulk composition.

In summary, the major changes occurring to planetary composition, in addition to stellar compositional variability itself and where $0.3 < C/O < 0.9$, take the form of:

- Changes in oxidation state during accretion and the magma ocean stage (i.e., whilst the planet's growing metal core is still open to mass exchange with its mantle), which influence, in particular, the Fe/FeO and, under increasingly reducing conditions, Si/SiO_2 ratio of the planet, and hence ultimately the mass ratio of core to mantle.

- Fractionation of major silicate forming components Mg and Si from Fe during chemical processing in the protoplanetary disk (Dauphas et al. 2015; Miyazaki and Korenaga 2020). This may result from physical separation of their main mineral hosts olivine, orthopyroxene and Fe metal, during condensation of the nebular gas and have given rise to their variable abundances in chondrites.

- The extent of volatile element depletion, either during condensation or subsequent accretion and evolution. Owing to their high bulk densities, the class of exoplanets discussed here are not expected to contain substantial quantities ($\gg 1$ wt.%) of water (as H_2 or H_2O). Fractional vaporisation and loss of the atmospheres of planets close to their host star may also lead to changes in the chemical composition of the residual planet, because vaporisation is not chemically congruent (i.e., the vapour phase has a bulk composition distinct from that of the residue; Markova et al. 1986; Sossi et al. 2019; Wolf et al. 2022; Fegley et al. 2023). In general, volatilities of major rock-forming elements evaporating from silicate melts decrease in the order K > Na > Fe > Si > Mg > Ca > Ti > Al (Sossi et al. 2019). Therefore, the end-product of such a process would result in Ca–Al-rich silicate planets (Dorn et al. 2019). Additional models are needed to confirm whether the temperatures required for sustained loss of rocky material (with high molar mass) are likely too high, considering the large masses of super-Earths, to have resulted in significant loss (Benedikt et al. 2020; Erkaev et al. 2023).

Life-essential elements (CHNOPS) in rocky exoplanet mantles

The storage in planetary mantles of carbon, hydrogen, nitrogen, oxygen, phosphorous and sulfur is of particular interest because of their being universally required in known biology (see Krijt et al. 2022, for a recent review of the origin of these elements in planetary systems). These elements also stand out from those discussed in previous sections in purely physical terms: many of them readily form gasses or fluids at low pressures (i.e., they are 'volatile' elements) and they are some of the last elements to condense from a cooling nebular gas (if indeed they condense at the planet forming region at all). Their volatility leads to these elements partitioning into atmospheres and to their potential loss early in a planet's history (Watson et al. 1981; Kasting and Pollack 1983; Tian et al. 2009; Zahnle and Catling 2017). Partly for this reason, and partly by definition, CHNPS (but not O) may only have trace abundances in the solid mantles of rocky planets, at least for relatively-oxidised silicate planets[2]. Even at trace quantities however, the properties of the CHNOPS elements can lead to them having a profound impact on the physics and chemistry of a planet.

---

[2] Excessive amounts of accreted volatiles would form a thick liquid ocean/ice layer atop the rocky mantle, producing a planet with a measurably lower bulk density that is likely identifiable as non-rocky (e.g., Unterborn et al. 2018, see Fig. 1).

The propensity of C–H–N–O–S, especially, to be lost or only partially accreted during planet formation makes their inventories particularly hard to predict. These elements, P included now, may all also enter planetary metal cores in proportions often strongly dependent on pressure, temperature and oxygen fugacity (e.g., Rubie et al. 2011). As a result, for exoplanetary systems there is (presumably) significant inherent stochasticity in the mantle abundances of CHNOPS, even if the systemic abundance is well constrained from the stellar composition (e.g., Krijt et al. 2022; Sossi et al. 2022; Lock and Stewart 2023, see the section *Nebular condensation*).

In the absence of being able to predict accurate mantle abundances of CHNOPS, one solution for some elements (H and N in particular) is to instead consider the limits internal conditions place on their storage: specifically, how temperature and mineralogy limit the concentration of these elements that can dissolve in the mantle. Such storage limits allow a minimum diversity of elemental abundances across planets to be inferred based on properties that can be estimated. The unconstrained accretion, loss, and redistribution history of CHNOPS would then increase this diversity. We are therefore here interested in the mineral storage of CHNOPS elements in planetary mantles, and how the mineralogical variation, discussed in the section *Possible range of mantle mineralogies for silicate planets,* may in and of itself affect their presence in exoplanets.

The major mineral phases of exoplanet mantles can hold a large part of a silicate mantle's CHNOPS inventory. They do this by incorporating CHNOPS in their crystal lattices as non-stoichiometric substitutions for similarly-sized and -charged elements. Elements stored in this way have inherent solubilities in minerals, above which the mantle would 'saturate' in the element concerned and need to stabilise a new mineral (or fluid) phase specifically incorporating that element to host any more. In the case where a solubility limit leads to a fluid being stabilised then, given the (upward) mobility of fluids, such solubility constraints in principle represent upper limits to the total mantle abundance of an element. CHNOPS elements do occur as saturated phases in Earth's mantle, in primitive meteorites, and likely in exoplanets' mantles by extension: C as diamond or graphite (C^0), or in carbides (C^{4-}) or carbonates (CO_3^{2-}), for example, and S in sulfides (S^{2-}) or sulfates (S^{6+}); here the form of either element will be a strong function of the local fO_2, because of different inherent redox states in these phases. Subsequent partitioning of CHNOPS from solids to silicate melts to magmatic gases will depend further on fO_2 and pressure (see Gaillard et al. 2021, for a recent review).

The redox-dependence of CHNOPS storage is another variable to factor into predictions of their abundance and mobility in a planet. As we are far from constraining an exoplanet's mantle fO_2 profile without knowing its detailed O budget, this is an additional source of uncertainty. Nonetheless, a useful constraint can be applied: a *minimum* variability of fO_2—of several log-units at constant Fe^{2+}/Fe^{3+}—emerges as a function of silicate mineralogy variability itself (see the section *Possible range of mantle mineralogies for silicate planets*), via composition-dependent Fe activity coefficients (Guimond et al. 2023a).

By combining insights from mineralogically controlled solubility limits and fO_2's, the mantle major element compositions of exoplanets can be linked to their maximum capacity for and speciation of CHNOPS. We briefly review storage of CHNOPS in this context below.

Carbon. The behaviour of carbon is tightly coupled to redox processes (Dasgupta 2013; Stagno et al. 2019), and therefore the storage of carbon in exoplanetary mantles will be as diverse as their redox states. At low mantle carbon concentrations carbon will dissolve in major silicate minerals (e.g., olivine, orthopyroxene, clinopyroxene, and garnet), with solubilities on the order of 0.1–1ppm (Keppler et al. 2003; Rosenthal et al. 2015), yet below experimental detection limits in lower mantle minerals (Shcheka et al. 2006).

Carbon in excess leads to saturated accessory phases: carbides, graphite/diamond, and carbonates; the main carbon hosts in the terrestrial mantle, which for context may contain ~10–100ppm C (Luth et al. 1999; Dasgupta and Hirschmann 2010). The carbon phase diagram is a function of pressure, temperature, and importantly, of fO_2, as these phases' C oxidation states range from 4– to 4+. Which of these phases are actually stable in a mantle depends on where the fO_2 profile intersects the carbon phase diagram. We might generally expect fO_2 to decrease with depth because Fe^{2+}–Fe^{3+} equilibria depend on pressure (Frost and McCammon 2008). However, carbon itself may be a relevant redox buffer if it is abundant enough to overwhelm the Fe redox buffer, pointing to feedbacks between planetary C inventory and its speciation, and making predictions about exoplanet mantles complex.

Carbon cycles on exoplanets have nevertheless seen a number of recent modelling studies, picking up on the centrality of the silicate weathering stabilising climate feedback to the concept of the circumstellar habitable zone (Kasting et al. 1993). Yet no studies explicitly consider variable carbon mineral speciation in the mobility of carbon during magmatism and any related mantle outgassing. The solubility model of carbon in reduced basaltic melt from Holloway et al. (1992), assuming carbon only as graphite in the mantle magma source region, has appeared in several outgassing models (Hirschmann and Withers 2008; Tosi et al. 2017; Höning et al. 2019; Ortenzi et al. 2020; Wogan et al. 2020; Guimond et al. 2021; Baumeister et al. 2023). Here, carbon melting is controlled by redox reactions that move graphite into a dissolved carbonate component of silicate melt–melt carbon contents are therefore limited by fO_2. Meanwhile, other models (Foley and Smye 2018; Unterborn et al. 2022) employ a constant carbon partitioning coefficient (e.g., Hauri et al. 2006) between silicate minerals and melt; i.e., assuming carbon at lower concentrations where it can be accommodated as a dissolved trace element in the major silicate minerals. Lastly, in the case of more oxidised mantles with more than a few ppm C, carbonate minerals would occur. Even low carbonate modalities (~10 ppm) can suppress mantle melting temperatures significantly (leading to production of melts containing up to ~half CO_2 by mass; Dasgupta and Hirschmann 2006, 2010).

Hydrogen (and water). H is relatively soluble in silicate crystals, minerals that are otherwise nominally anhydrous; dissolved H in common mantle minerals is a major reservoir of "water" in the mantles of mostly-solid rocky planets. Upper mantle silicates tend to have a water saturation limit on the order of 10–100 ppm, generally decreasing with temperature and increasing with pressure (e.g., Kohlstedt et al. 1996; Keppler and Bolfan-Casanova 2006; Mosenfelder et al. 2006; Férot and Bolfan-Casanova 2010, 2012; Ardia et al. 2012; Tenner et al. 2012; Novella et al. 2014; Demouchy et al. 2017; Dong et al. 2021b; Andrault and Bolfan-Casanova 2022)— with the notable exception of the transition zone minerals wadsleyite and ringwoodite holding wt.% water concentrations over a narrow annulus of pressure (e.g., Kohlstedt et al. 1996; Fei and Katsura 2020; Bolfan-Casanova et al. 2023). At higher pressures, the water solubilities of nominally anhydrous perovskites, ferropericlase, and postperovskite are much more uncertain due to the technical difficulty of experiments at these conditions, but so far appear drier than olivine and orthopyroxene at the pressures investigated (Bolfan-Casanova et al. 2002, 2003; Litasov et al. 2003; Townsend et al. 2016; Chen et al. 2020; Liu et al. 2021; Ishii et al. 2022; Shim et al. 2022, cf. Murakami et al. 2002). Other factors such as fO_2 and C content may have secondary effects on the water capacities of these minerals (e.g., Bolfan-Casanova et al. 2023), though they tend to be neglected in studies of whole-mantle trends.

Integrating these water saturation limits across the expected range of silicate mineralogies and assuming a perovskite water capacity of 30 ppm (Liu et al. 2021) leads to solid mantle water capacities ranging from ~250–2000 ppm, depending mostly on mantle thickness, potential temperature, and Mg/Si ratio (Guimond et al. 2023b). These capacities could be tenfold higher if perovskite holds 1000 ppm water instead (Inoue et al. 2010; Shah et al. 2021). Regardless, because the water capacities of various silicates stable at $\lesssim 15\,GPa$ are roughly similar, many rocky exoplanets may have accordingly similar maximum water capacities in the source region of their magmas (Guimond et al. 2023b).

The stabilisation of minerals with stoichiometric water in their structure would increase the water capacity of a mantle substantially. High-pressure hydrous silicate phases carrying up to ~10 wt.% water as a stoichiometric component, such as phases A, B, D, δ, E, egg, superhydrous phase B, and δ–H solution, may also exist in Earth's and exoplanetary mantles, but generally require low temperatures; i.e., geotherms that on Earth are associated with subducting slabs and therefore plate tectonics (Ringwood and Major 1967; Liu 1987; Kanzaki 1991; Pacalo and Parise 1992; Ohtani et al. 1997; Ohtani 2021). Lastly, water appears fully miscible in silicate melts from ~4 GPa (Bureau and Keppler 1999; Mibe et al. 2007; Mookherjee et al. 2008). Melt trapped in mineral inclusions or along grain boundaries during rapid magma ocean crystallisation may retain interior water early in a planet's history (Hier-Majumder and Hirschmann 2017). If a basal magma ocean endures, it would have huge potential as a water reservoir due to virtually unlimited water solubility (Moore et al. 2023; Boley et al. 2023). However, it is unclear how much this deep reservoir would contribute to subsequent water transport—whether it would leak to the lower mantle and eventually reach the upper mantle—especially if H diffusion in postperovskite is sluggish (Peng and Deng 2023).

N–P–S. Nitrogen is soluble in mantle minerals as NH_4^+, substituting for K^+ (Hall 1999; Yokochi et al. 2009; Watenphul et al. 2010; Li et al. 2013; Johnson and Goldblatt 2015). Experiments by Li et al. (2013) found maximum NH_4^+ solubilities in enstatite orthopyroxene, diopside clinopyroxene, and pyrope garnet less than 100ppm (and much lower in forsterite olivine); solubilities markedly increase with decreasing fO_2 and increasing pressure. At sufficiently-reducing conditions, nitrides or N–Fe-alloys may occur (Daviau and Lee 2021). Hence—and because lower fO_2 also favours more N in minerals over melt and gas phases (Libourel et al. 2003; Li and Keppler 2014; Mikhail and Sverjensky 2014; Dasgupta et al. 2022)—rocky planets with reducing magma oceans and mantles might retain significant fractions of primordial nitrogen dissolved in their silicates.

Sulphur and phosphorous, like carbon, will occur in planetary mantles both in accessory phases—sulfides or sulfates for S, and phosphides or phosphates for P, depending on fO_2 and (e.g., von Gehlen 1992; O'Reilly and Griffin 2000; Truong and Lunine 2021)—and dissolved in major silicates when they are at lower concentrations. Solubilities in olivine are measured to be ~1 ppm for S (but higher in the generally less-abundant clinopyroxene) and ~1000 ppm for P (Brunet and Chazot 2001; Callegaro et al. 2020). Thus, the main (terrestrial) mantle P host is expected to be olivine (Walton et al. 2021b), also supported by findings of P associated with olivine in analogous meteorites (e.g., Davis and Olsen 1991; Walton et al. 2021a). Large differences in mantle S contents are suggested between Mercury, Earth, and Mars, possibly related to fO_2 differences (e.g., Gaillard and Scaillet 2009; Malavergne et al. 2010; Nittler et al. 2011; Franz et al. 2019). The variability of S contents between rocky exoplanet mantles may have important implications for observable features of their atmospheres, as volcanism could lead to significant transient increases in their abundance of spectrally active sulfur species (Kaltenegger and Sasselov 2010; Claringbold et al. 2023; Ostberg et al. 2023). The transport of P from planetary mantles is likely to have less direct atmospheric impacts (although the discussion of possible abiotic sources of phosphine on Venus is relevant here; Bains et al. 2022); however, it is indirectly critical for the possible presence of biosignatures, given phosphorus's universal requirement by known biology (See Walton et al. 2021b, for a recent review). Despite this lack of direct astronomical observables, phosphorus transport to the surface will readily occur via magmatism given its tendency to partition into magmas during partial melting of the mantle (Brunet and Chazot 2001). In this way, the supply of P to hypothetical biospheres is regulated by the mantle inventory and geodynamics of the planet.

MELTS AND CRUSTS OF ROCKY EXOPLANETS

The preceding discussion of CHNOPS elements in exoplanets emphasises not only their storage in mantles, but their transport out of them: it is through this connection between interior and surface/atmosphere that exoplanet mantles may impact habitability, climate, and atmospheric composition. Where this connection is strong, and surfaces/atmospheres are significantly perturbed physically and chemically by this mass exchange, it may ultimately be possible to infer interior properties and states from exoplanet observations, e.g., through volcanically produced atmospheres (Liggins et al. 2022) or surface reflectance spectra (Hu et al. 2012).

The connection between rocky exoplanet mantles and their surfaces occurs through partial melting, the process whereby changes in the pressure of the solid mineral assemblage of a mantle causes minerals to react (partially 'break down') to form a liquid. Melting may occur at various depths within mantles: melt may have potentially been left over at the base of the mantle from magma ocean solidification (Labrosse et al. 2007); melting may occur at the mantle transition zone where wet material from the upper mantle must lose its water into water-rich melts as it descends into the mineralogically dry lower mantle (Karato et al. 2020); or near the surface at low pressures where the mantle adiabat intersects the solidus (the locus of pressure-temperatures where solid mantle first begins to melt). It is this last region of melting that we will focus on here as it is the most directly connected to surface: melts from these low pressures feed almost all volcanism on Earth, whether at mid-ocean ridges, ocean islands like Hawaii and Iceland, or arc volcanoes. Adiabatic decompression melting, as this process is called, is also the mechanism of melt generation that is least contingent on the geodynamic mode of the planet. Whereas basal magma oceans may require a particular planetary composition and thermal evolution and wet melting at the transition zone may be predicated on plate tectonics, adiabatic decompression melting can take place in any convecting mantle. The youthfulness of volcanism on the small and cold Mars (Werner 2009), on the dry and hot Venus (Herrick and Hensley 2023), and plate tectonic Earth indicates the general role of adiabatic decompression melting in the dynamic homeostasis of terrestrial planets across a range of conditions and compositions. We can reasonably expect this type of melting to be a universal phenomenon on planets lacking thick volatile envelopes (i.e., rocky planets; Kite et al. 2009, 2016).

Partial melts of the mantle will move to low pressure if they are positively buoyant. This condition is met for melting in Earth's shallow upper mantle (and evidently that of Mars and Venus as well), leading to the planets' surface, 'crust', being built of the minerals these melts form on cooling. Even on Earth, where continental crust has been subsequently formed not by direct melting of the mantle, the majority of crustal surface area is ultimately the product of basalts derived from mantle melting. The mineralogy and composition of a planetary mantle is therefore closely connected to its crust, the substrate on which climatic and biological occur.

In this section we explore how the diversity of exoplanet mantles seen above will map through to their melt compositions and crusts. As discussed previously, the thermodynamic models that can calculate mineralogy are necessarily most accurate when applied to Earth-like compositions, yet here we are investigating the non-Earth-like compositions of diverse exoplanetary mantles. This uncertainty is likely magnified when calculating the composition of melts these mantles generate: the thermodynamic complexity of melts is inherently challenging even in trying to model melting of Earth-like mantle compositions. Nonetheless, qualitative insights can be gained that speak to the universality of petrological thermodynamics, even under somewhat exotic circumstances and we emphasise these below.

Mantle melting temperature

The first and most important aspect of mantle mineralogy as pertains to melt generation is the temperature the mantle melts at, its solidus temperature. The simple presence of melt, even when $\gg 1\%$, is enough to drastically change the rheology of a mantle, lowering its viscosity and thereby enhancing convection and heat transport (e.g., Kohlstedt and Zimmerman 1996). Hirschmann (2000) investigated the effect of composition on solidus temperature for a range of broadly Earth-like mantle compositions based on experiments. This made clear the importance of both the relative proportion of MgO to FeO and the alkali element (Na, K) content of mantle rocks for their solidus temperatures. As discussed previously, whilst the Mg and Fe content of a mantle can likely be predicted from the stellar composition with some accuracy, the abundance of the alkali elements will be harder to predict, as both Na and K are moderately volatile during planet formation (Lodders et al. 2009). As a result, we here consider both Na-free and Na-bearing mantles, with the inclusion of the latter qualitatively indicating the magnitude of solidus reduction that mantle alkali contents make possible. As in previous sections, the intent with the calculations presented is to provide an indication of the variability we might expect among exoplanetary mantles, rather than accurately predict their properties.

The range of solidus temperatures predicted for the mantles of rocky planets is given in Figure 8. Results are shown for Na-free (filled histograms) and Na-bearing (open histograms) mantles, for three different upper mantle pressures in each case (1, 2 and 3 GPa), and between panels the fraction of Fe in the core is varied.

The first order result of these calculations is to reproduce the effect Hirschmann (2000) identified from experiments, that the alkali content of the planetary mantle is able to cause lowering of the solidus temperature of over 100 °C. The magnitude of this effect for any given planet will be contingent on its devolatilisation history (and that of its building blocks), emphasising the likely stochasticity in this key property of exoplanet mantles. The effect is

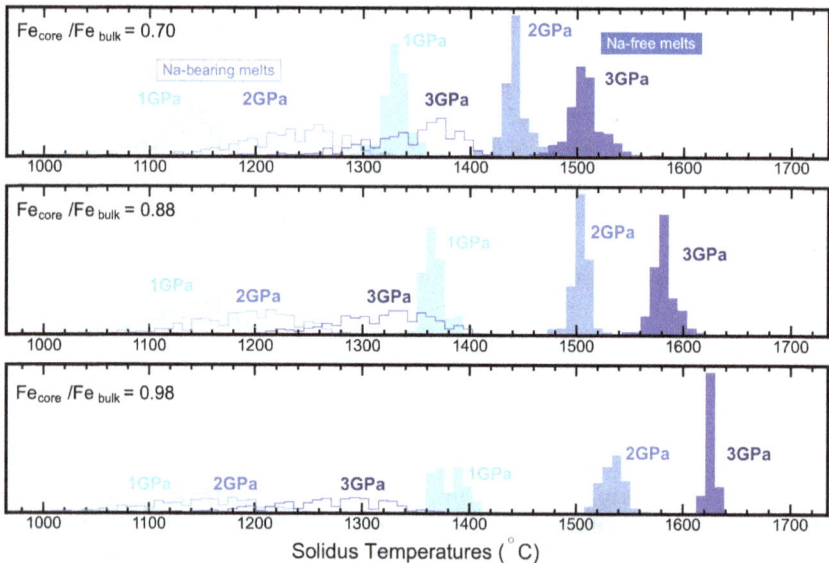

Figure 8. The solidus temperatures for sodium-bearing (**open histograms**) and sodium-absent (**filled histograms**) exoplanet mantles, calculated using pMELTS (Asimow and Ghiorso 1998; Ghiorso et al. 2002). Mantle compositions are the same as in Figure 6, drawn from the Hypatia Catalog (Hinkel et al. 2014), but now also including a set of mantles containing Na to show the effect of minor elements on solidus temperature. Mantle Na abundances are calculated assuming a constant star/planet depletion factor equal to that between the sun and Earth. The effect of mantle FeO content on solidus temperature is also shown, with panels from top to bottom recording progressively less partitioning of bulk planet Fe to the mantle (constant molar ratios Fe_{core}/Fe_{bulk} of 0.7, 0.8, and 0.98 respectively).

also broadly consistent across pressure and mantle Fe contents. Pressure variations themselves simply track the solidus to higher temperatures, as the positive volume change on reaction of producing a liquid is thermodynamically penalised at higher pressures, requiring greater temperatures to stabilise it.

If we look at single histograms in Figure 8, we see the effect of systemic (stellar) variability alone on solidus temperature. For Na-free mantles, even the quite substantial chemical variation reported in the section *Bulk chemical compositions of rocky exoplanets* translates to only modest changes in solidus temperature, of typically less than 100 °C. This is consistent with the finding of Hirschmann (2000) that variations in MgO/FeO, whilst important for solidus temperature, have much less of an effect than the alkali abundance. It is also consistent with expectations from simple model petrological systems, where melting occurs at the single eutectic temperature, irrespective of composition. This breaks down in detail for systems of real geological complexity (as modelled here, where minerals are complex solid solutions that change with bulk composition), but is the essential reason why large changes in major element composition lead to only modest changes in solidus temperature. For Na-bearing mantles however, we see that consideration of alkalis is important even for within-population variation: i.e., the variability of Na contents between planets simply inherited from the systemic composition, ignoring stochastic devolatilisation, leads to very large variation in solidus temperatures (up to 200 °C).

A final variable considered in Figure 8 is the fraction of iron that goes into the planet's core versus that which stays in the mantle as FeO. Less FeO-rich mantles lead to higher melting temperatures, by up to ~100 °C in the extreme case of almost all iron (98%) going into the core. The redox conditions of core formation therefore have a similar magnitude of effect on solidus temperature to intrinsic compositional variability at fixed core iron fraction.

Considering variation in mantle alkali content, iron content, and broader major element compositional variability together, rocky exoplanet mantle melting temperatures may vary by over 300 °C. This purely compositionally-driven change in melting behaviour of planetary mantles has important implications for geodynamics, given the stabilising role melting has in heat transport and thereby planetary thermal evolution (e.g., Nakagawa and Tackley 2012; Driscoll and Bercovici 2014).

Melt and crustal compositions

Melting is the key process whereby mantle chemistry is transported to the surface to create crusts, influence climate, and planetary habitability. Melt chemistry changes progressively during partial melting, meaning any given mantle composition can produce a wide range of melt compositions according to the depth at which melting takes place and the extent to which melting proceeds above the solidus (where melt is technically 0% of the mass of the system) to the liquidus (where the system becomes entirely molten; a temperature range of hundreds of K or more). To simply give a first indication of how planetary mantle composition might map through to melt composition, we pick a single degree of melting 20%, and show results from this alone. This degree of melting has a rationale on Earth, as for typical Earth mantle it is at this degree that the mineral clinopyroxene is lost from the melting assemblage, beyond which point melting is less productive (i.e., efficient melting stops at this point; e.g., Katz et al. 2003). Such a petrologically informed melt fraction could be identified for all exoplanetary mantles, but given the assumptions already inherent to these calculations would be beyond the scope of this review.

The mapping between mantle composition and their partial melts is shown in Figure 9. The domination of silicate melts by a few oxides, SiO_2 and MgO in particular, makes the representation of the results as oxide versus oxide in weight percent hard to evaluate. However, even viewed this way, it is clear that a systematic trend is that melts are less magnesian than their corresponding mantles and have broadly similar silica contents. Transformation of the compositions to centred log ratios (emphasising relative variation, see caption to Fig. 9) allows variation to be seen more clearly, particularly of the minor elements. Elements such as Na and Ti favor entering into the melt phase and can be seen to be systematically more abundant in melts than their corresponding mantle sources. CaO is also higher in melts than in mantles (due to its incorporation into clinopyroxene, which readily melts out), which is significant given the key role crustal Ca has in planetary carbon cycles via carbonate formation (e.g., Siever 1968).

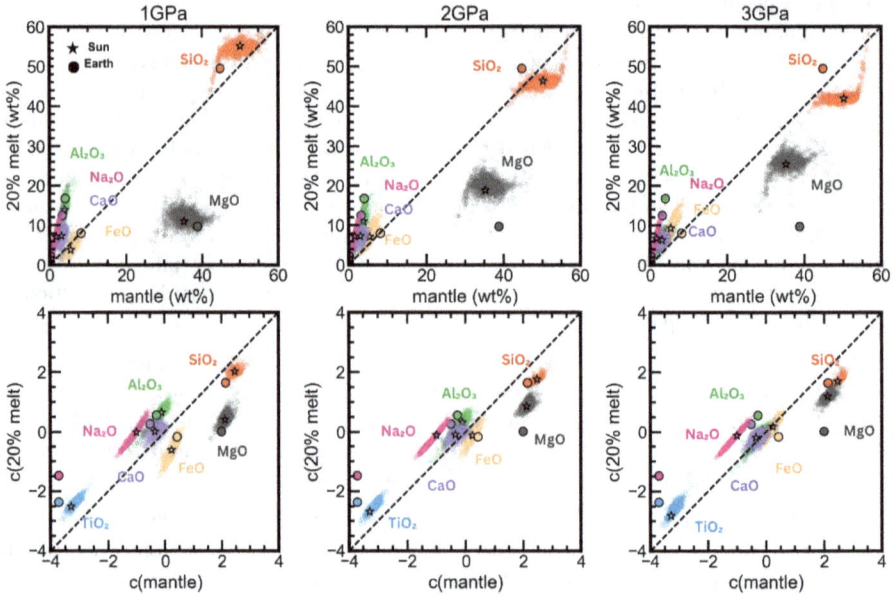

Figure 9. The diversity of melt compositions produced by 20% melting of rocky planet mantle compositions at a constant pressure, inferred from the Hypatia Catalog of stellar abundances and calculated using pMELTS as in Figure 8. Panels show melt compositions represented both as oxide weight percent (**top row**) and as centred log ratios (**bottom row;** 'c' denoting calculation of the centred log ratio from the weight percent oxide concentration). Centered log ratios preserve information on relative abundance of an element compared to the geometric mean and enable minor element variation to be seen more clearly (see Lipp et al. 2020, for a recent example of centred log ratios used and explained on geochemical data). Horizontal axes record mantle compositions; vertical axes record the compositions of their corresponding 20% melts. Equivalent calculations using the Earth's observed mantle composition (Workman and Hart 2005, **filled circle**) and a solar-composition mantle (Lodders et al. 2009, **filled star**) are shown for reference.

Another way of viewing the mapping of mantle to melt composition is by comparing the standard deviations of the respective populations: how does the variation in mantle composition (arising from stellar variability) compare to the variation that emerges once melting has taken place? Figure 10 shows this for melting calculated at three shallow upper mantle pressures. In all cases variation in SiO_2 and MgO in the melts is less than that in their sources, however for all other elements the absolute variation is greater. What this illustrates is that when it comes to creating planetary crusts, petrological thermodynamics can overprint the inherited bulk compositional characteristics of the body making prediction of the composition of exoplanetary crusts, highly non-deterministic.

Despite the high *compositional* variance, however, to a first order the mineralogies of these crusts are very similar: on cooling, these melts crystallise a similar mineral assemblage to that which they melted from in the mantle, dominantly olivine and clinopyroxene, with plagioclase feldspar incorporating most of the Al. Therefore, the compositional diversity seen

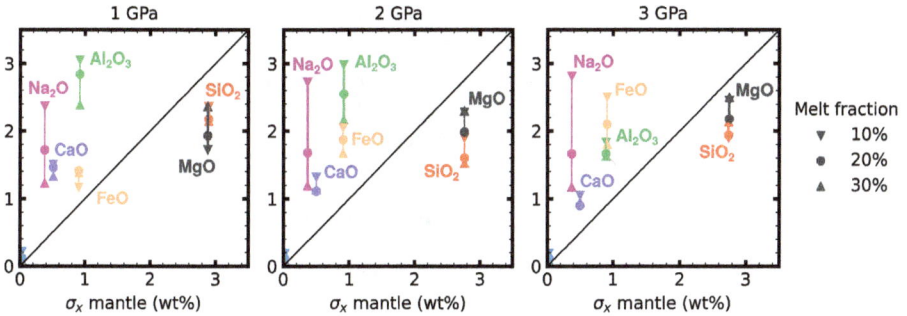

Figure 10. The variability of rock-forming oxides in mantle-derived melts compared to the variability of the same oxides in mantles, as measured by the standard deviation of their oxide weight proportions (σ_x). Data and modelling are the same as in Figures 8 and 9. Panels show calculations at three different melting pressures (1, 2, and 3 GPa; **left to right**), with symbols indicating different assumed melt fractions (10%, 20%, and 30%). **Dashed lines** indicate the 1:1 correlation between mantle σ_x and melt σ_x.

in Figures 9 and 10 does not in and of itself propagate through to predictions of exoplanets having particularly exotic silicate crustal mineralogies. As Earth has demonstrated in forming a continental crust, subsequent differentiation processes—perhaps related to tectonic mode(s)—can transform the mineralogy and composition of primary igneous crusts dramatically.

Melts, crusts and geodynamics

The crustal composition of an exoplanet will have many consequences for its evolution, although many are too contingent for useful predictions to be made. One crustal composition consequence recently subject to deterministic modelling, however, relates to the fate of liquid water on the planet surface; something with wide reaching implications for habitability and geodynamics. Water may be consumed in reactions with crustal minerals, notably the serpentinization reaction whereby ferrous iron from mafic rocks dissolves in water to produce serpentine and H_2 gas—the resulting hydrated minerals can store significant weight fractions of H_2O. Certain crustal compositions may be more readily hydrated and hence destabilise surface liquid water (Herbort et al. 2020).

Notably, increasing the FeO content of basalt stabilises these hydrated minerals (due in part to smaller volume changes during the hydration reaction), and hence on the crust's ability to sequester surface water (Wade et al. 2017; Dyck et al. 2021). On Mars, which is relatively FeO-rich but otherwise similar in composition to BSE, past serpentinization could have removed hundreds of metres of an equivalent global ocean to the crust, and eventually to the mantle, given enhanced (hydrous) amphibole stability (Wade et al. 2017). Indeed, phyllosilicates (hydrously altered crustal minerals) are observed widely on the martian surface (Carter et al. 2013; Lammer et al. 2013), and Martian D/H measurements support ~30–99% of early surface water now residing in the crust (Scheller et al. 2021). This result points to a direct effect of core formation on the surface environment of a rocky planet.

The potential fate of water on a planet emphasises that crusts act as a second silicate reservoir on a planet, and potentially a major one depending on the element concerned: particularly the 'incompatible' elements that strongly enter the melt phase, e.g., Ti and heat producing elements U, Th, and K, and elements sequestered into the crust by low temperature processes, e.g., C during formation of carbonate minerals, which helps regulate long term climate. The role of the crust in geodynamics is therefore sensitive to the amount of mantle that has been processed by melting to produced it (i.e., has the whole mantle lost its K in producing the crust, or just a small fraction) and its longevity (i.e., is the crust routinely recycled back into the mantle or is it a permanent reservoir). Both of these points indicate the strong feedbacks that will emerge on a planet between crust production/destruction and its surface and interior dynamics. This history of these processes on Earth has been hotly debated for many decades (e.g., Korenaga 2018, for a recent review). Rather than being tightly constrained in how these processes may operate, exoplanetary science ultimately has the opportunity to resolve some of these debates with new observations of rocky planet evolution.

OBSERVING EXOPLANET COMPOSITIONS AND MINERALOGIES

As discussed in the previous sections, theoretical attempts to predict the mantle mineralogy of rocky exoplanets are widespread but untested. Yet a major opportunity presented by exoplanets is empirical as well as theoretical: to extrapolate knowledge from Earth science by applying it to strange planets, but also to rewrite this knowledge using unprecedented observations. The statistical leverage of exoplanet demographics (i.e., planet mass and period distributions) has already re-hauled theories of planet formation, for example, previously only constrained by one planetary system (Mordasini 2018). Since we have seen that mass and radius are inherently limited in what they constrain about a planet's composition, even with the host star as a proxy (e.g., Sotin et al. 2007; Grasset et al. 2009; Rogers and Seager 2010; Dorn et al. 2015; Santos et al. 2015; Hinkel and Unterborn 2018; Wang et al. 2019a; Schulze et al. 2021), observations beyond mass and radius (important and hard won as they are) are needed to reveal to us the geological complexities of rocky planets.

In this section we describe some of the methods of observation that could help constrain exoplanets' mantle, melt, and crust compositions. Two of the main astrophysical signals carrying chemical information on exoplanets are: (i) the proportion of a planet-hosting star's photons, detected by a telescope, which have been absorbed by a planet's atmosphere as the planet passes in front of ("transits") that star—this technique is called transmission spectroscopy if the photon wavelengths are measured—and (ii) when the planet is instead behind the star ("in eclipse"), the proportion of detected photons now blocked by the star, which would be either emitted or reflected by the planet—this is called emission spectroscopy, and the planet's now absent contribution is identified by difference compared with when it is present just prior to entering into eclipse. A quantity often reported in these measurements is the flux ratio of photons, often measured in ppm, between the star during planetary eclipse or transit, and the star alone. In general, higher planet/star flux ratios permit better measurement precision. The James Webb Space Telescope (JWST), for example, is able to measure this ratio as a function of wavelength to unprecedented precision. Hence many of these observations have only recently become feasible given the already-proven success of JWST since its launch in 2021.

Lava worlds as a window into rocky planetary interiors

Among the potentially rocky exoplanets discovered thus far, there is a particularly interesting class that could enable observational constraints on exoplanet silicate compositions: magma ocean planets (or lava worlds). These worlds, of which no analogue exists in our present solar system, orbit their host stars on very short orbits, usually less than a few days. Such proximity to their stars implies scorching dayside temperatures hot enough to melt mantle rock at least to some depth. So far, a couple hundred such objects have been identified as candidates in transit surveys (Zilinskas et al. 2022), and a few dozen of them have mass measurements as well (see Fig. 1). Of this subset, the corresponding average densities strongly hint at silicate interiors, and consequently, silicate-dominated surfaces.

Ultra-hot lava ocean planets are typically expected around FGK-stars, as only these are bright enough to raise temperatures of close-in planets sufficiently for extensive melting. Given the relatively high intrinsic (thermal) emission of a hot lava planet, the planet-to-star photon flux ratio can reach values up to 300 ppm,[3] but are typically expected to be ~100 ppm in the mid-infrared (for a $2 R_E$ super-Earth orbiting a Sun-like star; Zilinskas et al. 2022). Additionally, their short orbital periods (≤ 2 days) imply that large numbers of orbits can be observed, resulting in improved statistics. This means that lava planets are our only opportunity to study the surfaces of rocky planets around Sun-like stars with contemporary facilities.

The intense temperatures these planets experience imply that lava oceans cover large (but not necessarily deep) surface areas of their starlit hemispheres (Léger et al. 2011). That is, planets on such close-in orbits likely have a permanently-irradiated hemisphere due to being tidally locked (as the moon is to Earth; e.g., Barnes 2017). Hence they would show strong hemispherical asymmetry in their surface conditions, with the nightside potentially being solid subject to the efficiency of heat transport through any atmosphere. Current classification of lava planets assumes a rule of thumb that surface temperatures must be in excess of ~1600 K in order to melt rock (Kite et al. 2016, see also Fig. 8). Since the main source of heat is radiation from the star, the lava ocean would be heated from the top rather than from the bottom, implying strong thermal (and potentially chemical) vertical stratification (Boukaré et al. 2022, 2023; Meier et al. 2023). In this configuration, the ocean would preferentially convect laterally, from the substellar point (where the star is directly overhead) to the terminator (where the star disappears over the horizon and the nightside begins). However, the associated heat transport is likely too inefficient to modify the surface temperature distribution by magma convection alone, meaning that, barring heat transport in an atmosphere, the hottest place on the planet's surface is still expected to be the substellar point (Kite et al. 2016).

In any case, at these hot dayside temperatures, species in the silicate melt will evaporate from the magma ocean to produce an atmosphere. The presence of this silicate vapour atmosphere is what opens up the study of lava worlds' interior compositions via emission spectroscopy (the hot atmosphere confined to the dayside implies that lava worlds have to be observed in eclipse, when the planet's starlight hemisphere is pointing towards the observer). Here, the pressures reached by rock vapour alone can be remarkable especially at the highest temperatures: whilst they are tenuous ($P \leq 10^{-5}$ bar) at $T < 2000$ K, up to Mars-like pressures are expected at $T \sim 3000$ K (Fegley and Cameron 1987; Van Buchem et al. 2022; Wolf et al. 2022). Chemically, models predict these atmospheres to be dominated by Na, K and Fe gas at low temperature, and SiO at higher temperatures (Fig. 11; Miguel et al. 2011; Van Buchem et al. 2022; Wolf et al. 2022; Zilinskas et al. 2022).

Yet because the exact gas composition depends on the chemistry of the underlying melt, the observable atmosphere is linked directly to the planets' silicate compositions. Inverting an atmospheric composition from observations should therefore allow us, in principle, to constrain a magma ocean surface composition—e.g., bulk oxide abundance (Zilinskas et al. 2022) and oxygen fugacity (Sossi et al. 2019; Wolf et al. 2022)—and by extension, potentially an underlying mantle composition as well, since the two will be linked through the process of partial melting (see the *Melts and crusts of rocky exoplanets* section).

A key feature of these atmospheres is their confinement to the partially-molten dayside. This is because a certain vapour pressure above the melt is needed to sustain the atmosphere, so without any surface melt, the atmosphere will condense and collapse. Preliminary calculations involving simple hydrodynamics—i.e., winds directed from the substellar point towards the nightside—show that atmospheres cannot extend beyond the day–night terminator, and will have the bulk of their mass contained within a few tens of degrees from the substellar point (Kite et al. 2016; Nguyen et al. 2022). Beyond this point, condensation from gas to solid/liquid will assure atmospheric collapse.

---

[3] In Zilinskas et al. (2022), larger flux ratios of up to 500 ppm are reported for TOI-1807 b, based on an overestimation of the radius. However, its radius was overestimated and has been corrected since (Hedges et al. 2022).

Inverting these observations for atmospheric composition is nevertheless complex, in part because the inversion relies on understanding the atmospheric physics involved. For example, vertical 1D radiative transfer models have hinted at pronounced thermal inversions, a result of the strong absorption by SiO in the UV. This would mean that the atmosphere exhibits emission features in the infrared. In particular, SiO would show broad emission features, allowing it to be identified (Ito et al. 2015; Zilinskas et al. 2022).

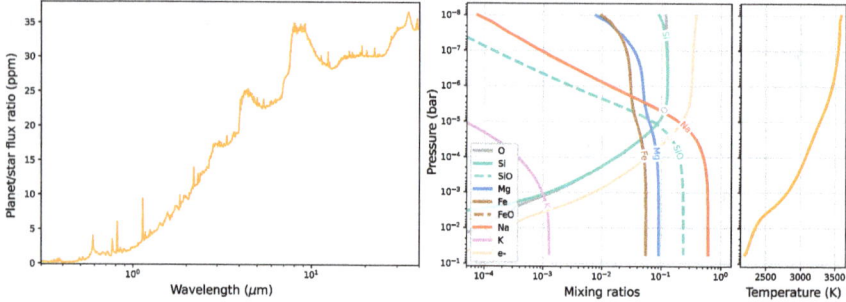

Figure 11. Properties of a silicate atmosphere vapourised from a Bulk Silicate Earth melt at 2500 K irradiation temperature from a solar-type star. **Left:** Spectrum of a lava planet of $1\,M_E$ and $1\,R_E$. **Center:** Atmospheric speciation, given in volume mixing ratio. **Right:** Temperature–pressure profile; a strong thermal inversion is pronounced in this atmosphere, leading to the emission features evident in the left plot.

Initial efforts to observe these worlds have already been made. Demory (2014), using eclipse observations with the Kepler space telescope, concluded that some hot rocky exoplanets show high albedos—consistent with cloud coverage (Mahapatra et al. 2017; Mansfield et al. 2019) or wavy magma oceans (Modirrousta-Galian et al. 2021). Using the Spitzer space telescope to observe K2-141 b, an ultra-hot rocky exoplanet orbiting a K-dwarf star, Zieba et al. (2022) found their data to be consistent with no dayside–nightside heat redistribution—ruling out a thick atmosphere of a few bars, but as expected for both a thinner ($\lesssim 0.1$ bar) rock vapour atmosphere, and no atmosphere. The resolution of Spitzer was insufficient to detect any spectral features of SiO or other mineral species.

Another ultra-hot planet studied extensively is 55 Cancri e, even though its nature as a purely rocky world might be challenged based on its low density (Dorn et al. 2017; Schulze et al. 2021). Initial considerations dismissed the presence of a primary atmosphere derived from the nebular gas, given the absence of significant H Lyman-α detection during transit observations with the Hubble Space Telescope (Ehrenreich et al. 2012), rendering a rock vapour atmosphere more likely. Despite persistent efforts, attempts to characterize this planet's atmosphere are yet to be successful. Studies by Ridden-Harper et al. (2016) and Tabernero et al. (2020) failed to detect Na (a moderately volatile element) and Ca (a refractory element). Similarly, Jindal et al. (2020) ruled out H_2O and TiO based on cross-correlation analyses with high-resolution ground-based data. A comparable outcome was reported by Keles et al. (2022), whose exhaustive high-resolution search encompassed a broad spectrum of gaseous species, yet yielded no positive detections. However, all these observations were performed in transmission spectroscopy during primary transit, which is not the ideal way to detect an atmosphere composed entirely of rock vapour.

Meanwhile, there is some evidence that 55 Cnc e harbours a substantial atmosphere, or at least a transient one (Heng 2023; Meier Valdés et al. 2023). Some studies suggest that there might be a shift in the surface's hottest point away from the substellar point, indicative of an atmosphere that redistributes heat (Demory et al. 2016; Angelo and Hu 2017; Hammond and Pierrehumbert 2017). A reanalysis of the same data could not reproduce their findings, however (Mercier et al. 2022). 55 Cnc e was subject to three observing programs in Cycle 1 of JWST.[4] By the time of release of this article, there might be further information on this enigmatic ultra-short, massive super-Earth.

Discussion above was focused on "dry" lava planets, worlds that either formed with low volatile abundances, or lost volatiles through different physical drivers (cf. Owen and Wu 2017; Ginzburg et al. 2018). If volatiles such as H_2, H_2O, CO or CO_2 are present in lava world atmospheres, they would significantly modify the spectral features of such atmospheres, and potentially wash out any signs of silicate vapour atmospheres (Piette et al. 2023). Further, the mere presence of volatiles significantly changes the chemistry of the melt and the vapour (Charnoz et al. 2023); mutual interactions modify both reservoirs (e.g., Schlichting and Young 2022). This is an area where experimental petrology of magma degassing is critical in informing the design and interpretation of observations (e.g., Bower et al. 2019): just one reason why it is crucial to continue searching for atmospheric species with JWST, and future missions, in coordination with building our experimental and theoretical understanding.

Dust tails of disintegrating planets

Small, roughly Mercury-sized lava planets may not be able to retain their silicate atmospheres and could catastrophically evaporate. This process would leave a dusty outflow similar to a cometary tail, or even a torus along the orbit, which could be detected via its effects on the shape of the planet/star flux ratio as it varies in time over an orbit (Rappaport et al. 2012; Brogi et al. 2012; Budaj 2013; Perez-Becker and Chiang 2013; Booth et al. 2023). So far, three potential candidates for catastrophically evaporating planets have been identified on this basis: K2-22 b, KIC 1255b and KOI-2700 b. Comparing both spectral and albedo-versus-orbital phase observations of the planet to synthetic observations of known composition could constrain the composition of the dust (e.g., van Lieshout et al. 2014, 2016; Booth et al. 2023; Curry et al. 2023; Estrada et al. 2023)—major mineral species in these dust tails could be detected by JWST, for example (Okuya et al. 2020). The survival time of the dust might additionally provide information on its composition (e.g., Bromley and Chiang 2023). Whilst no definite conclusions can be drawn from initial observations, Mg–Fe silicates (as opposed to pure Mg silicates or pure Al_2O_3) are a plausible match for KIC 1255 b and K2-22 b (Estrada et al. 2023).

Reflectance spectra and surface mineralogy

As emphasised in the section *Melts and crusts of rocky exoplanets*, a planet's crustal composition and mantle composition are tightly related. If a rocky planet has lost its atmosphere, then its surface composition may be accessible through reflectance and emission spectra (Hu et al. 2012). Minerals have their own intrinsic emission and absorption spectra, which may be distinguished (e.g., Sunshine et al. 1990), or even be traced for the composition of individual minerals (e.g., Lucey 1998). These methodologies have been thoroughly investigated in remote sensing on Earth and other objects of the Solar System to infer the composition of the surface (e.g., Lucey et al. 1998) or dust in Mars' thin atmosphere (e.g., Hamilton et al. 2005).

An exposed planetary surface will experience impact from micrometeorites, which will turn bulk rock into a fine porous material called regolith. Most exposed surfaces of atmosphere-less Solar System objects are covered with regolith. Due to its porosity, its reflective properties differ significantly from bulk rocks, as light is scattered multiple times between and within grains (Hapke 2012). This creates a whitening effect, and may even conjure strong back-reflection at superior conjunction (Hapke et al. 1993; Hapke 2002). Further, as the surface is subject to space weathering from energetic particles of the solar wind, its spectral features might change, alongside an overall darkening of the surface (Hapke 1973). In principle, this could allow investigation of the age of a planet's surface—with a young surface implying recent volcanism.

---

[4] JWST Cycle 1 Proposals #1952, #2084, and #2347

Nevertheless, even in regolith, the mineral grains will imprint their spectral signature on the reflected (or emitted) light. Absorption features are usually broad, but unique to the mineral and its composition. SiO_2 shows a particularly strong feature at the 9 μm wavelength band, which could be used to determine the silica content of (exo-)planetary surfaces (Hu et al. 2012; Kreidberg et al. 2019). Iron may also reveal its presence in the shift of absorption bands of olivine in the near infrared (Sunshine et al. 1990; Lucey 1998). When minerals are not in their crystalline form, for example due to the formation of glasses (either through rapid cooling of extruding melt or impact breccia), the spectral signature may still be preserved, although it may differ from mineral spectra. This effect could help determine not only the composition, but also the material state of a planetary surface (Fortin et al. 2022).

Observations of rocky planet surfaces are relatively easier to conduct for hot planets (~500–1000 K) orbiting cool stars. In these cases, the planet-to-star flux ratio will be high. The surface of LHS 3844 b, a hot (1000 K) planet orbiting an M-dwarf, has been studied with the Spitzer infrared space telescope at 4.5 μm. In this case, no evidence of a thick atmosphere was found, enabling observers to believe their flux ratio measurement contained a signal direct from the planet's surface (Whittaker et al. 2022). Even with the single wavelength band observed, a surface composed of bright feldspathic or granitic material—common on Earth's continents and on the moon—could be ruled out tentatively, favouring a dark surface with a basaltic or ultramafic lithology (the primitive product of mantle melting, see the section *Melts and crusts of rocky exoplanets*, and as seen on Mars, for example). For the hot and also likely airless rocky planet GJ 1252 b, observations again with the Spitzer telescope suggested a similarly dark surface, but were less conclusive (Crossfield et al. 2022). Further studies of planetary surfaces with JWST (which, like Spitzer, operates in infrared wavelengths) have been proposed and accepted for a few of-interest targets[5], potentially delivering detailed results by the time this chapter has been published.

SUMMARY

This review has focused on how we might predict the composition and mineralogy of exoplanet mantles and their resultant crusts. These predictions are rooted in knowledge of stellar composition, and the assumption that the composition of the star, being so much the mass of the system, is representative of the material available to build planets in the disk. As we have emphasised, deductions of composition and mineralogy become more uncertain the further we take them: the planet's composition may have been affected by devolatilisation/incomplete condensation from the nebula gas; the mantle composition is affected by an unknown core formation processes; the character of melts from that mantle are contingent on the depth and degree of melting, and so on. Whilst experiment and theory, often originating in the Earth sciences, make each step somewhat deterministic, the ensemble of branching points between a stellar composition and surface reflectance spectra, for example, make accurate prediction impossible. Nonetheless, we hope to have shown that petrological thermodynamics does introduce some certainty, even in the face of great diversity: upper and lower mantles separated by a transition zone are likely ubiquitous; and olivine and pyroxenes are likely to almost always be the major (solid) mantle silicate phases. More indirectly, the productive region of mantle melting, after incompatible minor element exhaustion, is likely to lie at a similar temperature across rocky exoplanets, and primitive crusts, derived from melting these mantles, will be broadly basaltic.

More importantly than the planetary properties we can predict, however, is that fact that our forward models of mantle mineralogy and melt chemistry are the prelude to inferring exoplanet geology from observations. Whilst detailed characterisation of rocky exoplanets is nascent,

---

[5] JWST Cycle 2 #3784 (for TOI 2445 b) and #4008 (for LHS 3844 b)

already we are probing hot planets for the compositions of any (possibly silicate) atmospheres. When such observations are secured, and the composition or mineralogy of an exoplanet is even weakly constrained, we will have the first evaluation of how the assumptions in our predictive models have fared. This sparks the exciting era of testing and revising these models in light of exoplanet observations. We look forward to new insights into the physics and chemistry of rocky planets, their evolution, and their interior processes: insights equally as valuable for developing our understanding of Earth as for searching for habitable and inhabited exoplanets.

ACKNOWLEDGEMENTS

The authors would first like to acknowledge the supreme patience, and gracious and supportive handling of the chapter by the Editorial team, and in particular an insightful review from Keith Putirka that helped improve the communication in the manuscript greatly. The research shown here acknowledges use of the Hypatia Catalog Database, an online compilation of stellar abundance data as described in Hinkel et al. (2014), which was supported by NASA's Nexus for Exoplanet System Science (NExSS) research coordination network and the Vanderbilt Initiative in Data-Intensive Astrophysics (VIDA). We also acknowledge use of the Third Data Release of the GALAH Survey. The GALAH Survey is based on data acquired through the Australian Astronomical Observatory, under programs: A/2013B/13 (The GALAH pilot survey); A/2014A/25, A/2015A/19, A2017A/18 (The GALAH survey phase 1); A2018A/18 (Open clusters with HERMES); A2019A/1 (Hierarchical star formation in Ori OB1); A2019A/15 (The GALAH survey phase 2); A/2015B/19, A/2016A/22, A/2016B/10, A/2017B/16, A/2018B/15 (The HERMES-TESS program); and A/2015A/3, A/2015B/1, A/2015B/19, A/2016A/22, A/2016B/12, A/2017A/14 (The HERMES K2-follow-up program). CMG was supported by the Science and Technology Facilities Council [grant number ST/W000903/1]. HW, FS and PS were supported by the Swiss National Science Foundation (SNSF) via an Eccellenza Professorship (203668) and the Swiss State Secretariat for Education, Research and Innovation (SERI) under contract number MB22.00033, a SERI-funded ERC Starting Grant '2ATMO' to PS. AM acknowledges support from the University of Cambridge Leverhulme Centre for Life in the Universe.

REFERENCES

Adibekyan V, Dorn C, Sousa SG, et al. (2021) A compositional link between rocky exoplanets and their host stars. Science 374:330–332, https://doi.org/10.1126/science.abg8794

Akaogi M, Akimoto S-i (1977) Pyroxene–garnet solid-solution equilibria in the systems $Mg_4Si_4O_{12}$–$Mg_3Al_2Si_3O_{12}$ and $Fe_4Si_4O_{12}$–$Fe_3Al_2Si_3O_{12}$ at high pressures and temperatures. Phys Earth Planet Sci 15:90–106

Albarède F (2009) Volatile accretion history of the terrestrial planets and dynamic implications. Nature 461:1227–1233, https://doi.org/10.1038/nature08477

Alexander CMOD, Boss AP, Carlson RW (2001) The early evolution of the inner solar system: a meteoritic perspective. Science 293:64–68, https://doi.org/10.1126/science.1052872

Allen-Sutter H, Garhart E, Leinenweber K, Prakapenka V, Greenberg E, Shim SH (2020) Oxidation of the interiors of carbide exoplanets. Planet Sci J 1:39, https://doi.org/10.3847/PSJ/abaa3e

Ammann MW, Brodholt JP, Dobson DP (2011) Ferrous iron diffusion in ferro-periclase across the spin transition. Earth Planet Sci Lett 302:393–402, https://doi.org/10.1016/j.epsl.2010.12.031

Anders E, Ebihara M (1982) Solar-System abundances of the elements. Geochim Cosmochim Acta 46:2363–2380, https://doi.org/10.1016/0016-7037(82)90208-3

Anderson DL, Bass JD (1986) Transition region of the earth's upper mantle. Nature 320:321–328, https://doi.org/10.1038/320321a0

Andrault D, Bolfan-Casanova N (2022) Mantle rain toward the Earth's surface: A model for the internal cycle of water. Phys Earth Planet Sci 322:106815, https://doi.org/10.1016/j.pepi.2021.106815

Angelo I, Hu R (2017) A case for an atmosphere on super-Earth 55 Cancri e. Astron J 154:232, https://doi.org/10.3847/1538-3881/aa9278

Ardia P, Hirschmann MM, Withers AC, Tenner TJ (2012) H_2O storage capacity of olivine at 5–8 GPa and consequences for dehydration partial melting of the upper mantle. Earth Planet Sci Lett 345–348:104–116, https://doi.org/10.1016/j.epsl.2012.05.038

Asimow PD, Ghiorso MS (1998) Algorithmic modifications extending MELTS to calculate subsolidus phase relations. Am Mineral 83:1127–1132, https://doi.org/10.2138/am-1998-9-1022

Asplund M, Amarsi AM, Grevesse N (2021) The chemical make-up of the Sun: A 2020 vision. Astron Astrophys 653:A141, https://doi.org/10.1051/0004-6361/202140445

Badro J (2014) Spin transitions in mantle minerals. Annu Rev Earth Planet Sci 42:231–248

Bains W, Shorttle O, Ranjan S, Rimmer PB, Petkowski JJ, Greaves JS, Seager S (2022) Constraints on the production of phosphine by Venusian volcanoes. Universe 8:54

Ballmer MD, Houser C, Hernlund JW, Wentzcovitch RM, Hirose K (2017) Persistence of strong silica-enriched domains in the Earth's lower mantle. Nat Geosci 10:236–240, https://doi.org/10.1038/ngeo2898

Barnes R (2017) Tidal locking of habitable exoplanets. Celest Mech Dyn Astr 129:509–536

Baumeister P, Tosi N, Brachmann C, Grenfell JL, Noack L (2023) Redox state and interior structure control on the long-term habitability of stagnant-lid planets. arXiv:2301.03466

Benedikt M, Scherf M, Lammer H, Marcq E, Odert P, Leitzinger M, Erkaev NV (2020) Escape of rock-forming volatile elements and noble gases from planetary embryos. Icarus 347:113772

Bitsch B, Battistini C (2020) Influence of sub- and super-solar metallicities on the composition of solid planetary building blocks. Astron Astrophys 633:A10, https://doi.org/10.1051/0004-6361/201936463

Bland PA, Alard O, Benedix GK, Kearsley AT, Menzies ON, Watt LE, Rogers NW (2005) Volatile fractionation in the early solar system and chondrule/matrix complementarity. Proc Natl Acad Sci 102:13755–13760, https://doi.org/10.1073/pnas.0501885102

Boley KM, Panero WR, Unterborn CT, Schulze JG, Martinez RRi, Wang J (2023) Fizzy super-Earths: Impacts of magma composition on the bulk density and structure of lava worlds. Astrophys J 954:202

Bolfan-Casanova N, Mackwell S, Keppler H, McCammon C, Rubie DC (2002) Pressure dependence of H solubility in magnesiowüstite up to 25 GPa: Implications for the storage of water in the earth's lower mantle. Geophys Res Lett 29:89-81–89-84, https://doi.org/10.1029/2001GL014457

Bolfan-Casanova N, Keppler H, Rubie DC (2003) Water partitioning at 660 km depth and evidence for very low water solubility in magnesium silicate perovskite. Geophys Res Lett 30:1905, https://doi.org/10.1029/2003GL017182

Bolfan-Casanova N, Martinek L, Manthilake G, Verdier-Paoletti M, Chauvigne P (2023) Effect of oxygen fugacity on the storage of water in wadsleyite and olivine in H and H–C fluids and implications for melting atop the transition zone. Euro J Mineral 35:549–568, https://doi.org/10.5194/ejm-35-549-2023

Bolis RM, Morard G, Vinci T, et al. (2016) Decaying shock studies of phase transitions in MgO–SiO_2 systems: Implications for the super-Earths' interiors. Geophys Res Lett 43:9475–9483, https://doi.org/10.1002/2016GL070466

Bond JC, Lauretta DS, Tinney CG, Butler RP, Marcy GW, Jones HRA, Carter BD, O'Toole SJ, Bailey J (2008) Beyond the iron peak: r- and s-process elemental abundances in stars with planets. Astrophys J 682:1234–1247, https://doi.org/10.1086/589236

Bond JC, Lauretta DS, O'Brien DP (2010a) Making the Earth: Combining dynamics and chemistry in the Solar System. Icarus 205:321–337, https://doi.org/10.1016/j.icarus.2009.07.037

Bond JC, O'Brien DP, Lauretta DS (2010b) The compositional diversity of extrasolar terrestrial planets: I. In-situ simulations. Astrophys J 715:1050–1070, https://doi.org/10.1088/0004-637X/715/2/1050

Bonsor A, Jofré P, Shorttle O, Rogers LK, Xu S, Melis C (2021) Host-star and exoplanet compositions: A pilot study using a wide binary with a polluted white dwarf. Mon Not R Astron Soc 503:1877–1883, https://doi.org/10.1093/mnras/stab370

Booth RA, Owen JE, Schulik M (2023) Dust formation in the outflows of catastrophically evaporating planets. Mon Not R Astron Soc 518:1761–1775, https://doi.org/10.1093/mnras/stac3121

Boujibar A, Driscoll P, Fei Y (2020) Super-Earth internal structures and initial thermal states. J Geophys Res: Planets 125:e2019JE006124, https://doi.org/10.1029/2019JE006124

Boukaré C-É, Cowan NB, Badro J (2022) Deep two-phase, hemispherical magma oceans on lava planets. Astrophys J 936:148, https://doi.org/10.3847/1538-4357/ac8792

Boukaré C-É, Lemasquerier D, Cowan N, Samuel H, Badro J (2023) Lava planets interior dynamics govern the long-term evolution of their magma oceans. arXiv:2308.13614

Bower DJ, Kitzmann D, Wolf AS, Sanan P, Dorn C, Oza AV (2019) Linking the evolution of terrestrial interiors and an early outgassed atmosphere to astrophysical observations. Astron Astrophys 631:A103, https://doi.org/10.1051/0004-6361/201935710

Braukmüller N, Wombacher F, Funk C, Münker C (2019) Earth's volatile element depletion pattern inherited from a carbonaceous chondrite-like source. Nat Geosci 12:564–568, https://doi.org/10.1038/s41561-019-0375-x

Brewer JM, Fischer DA, Valenti JA, Piskunov N (2016) Spectral properties of cool stars: Extended abundance analysis of 1,617 planet-search stars. Astrophys J Suppl Ser 225:32, https://doi.org/10.3847/0067-0049/225/2/32

Brogi M, Keller CU, Ovelar MdJ, Kenworthy MA, de Kok RJ, Min M, Snellen IaG (2012) Evidence for the disintegration of KIC 12557548 b. Astron Astrophys 545:L5, https://doi.org/10.1051/0004-6361/201219762

Bromley J, Chiang E (2023) Chaotic winds from a dying world: A one-dimensional map for evolving atmospheres. Mon Not R Astron Soc 521:5746–5759, https://doi.org/10.1093/mnras/stad932

Brugger B, Mousis O, Deleuil M, Deschamps F (2017) Constraints on super-Earth interiors from stellar abundances. Astrophys J 850:93, https://doi.org/10.3847/1538-4357/aa965a

Brugman K, Phillips MG, Till CB (2021) Experimental determination of mantle solidi and melt compositions for two likely rocky exoplanet compositions. J Geophys Res: Planets 126:e2020JE006731, https://doi.org/10.1029/2020JE006731

Brunet F, Chazot G (2001) Partitioning of phosphorus between olivine, clinopyroxene and silicate glass in a spinel lherzolite xenolith from Yemen Chem Geol 176:51–72, https://doi.org/10.1016/S0009-2541(00)00351-X

Budaj J (2013) Light-Curve Analysis of KIC 12557548b: An extrasolar planet with a comet-like tail. Astron Astrophys 557:A72, https://doi.org/10.1051/0004-6361/201220260

Buder S, Asplund M, Duong L, et al. (2018) The GALAH survey: Second data release. Mon Not R Astron Soc 478:4513–4552, https://doi.org/10.1093/mnras/sty1281

Buder S, Sharma S, Kos J, et al. (2021) The GALAH+ survey: Third data release. Mon Not R Astron Soc 506:150–201, https://doi.org/10.1093/mnras/stab1242

Bureau H, Keppler H (1999) Complete miscibility between silicate melts and hydrous fluids in the upper mantle: Experimental evidence and geochemical implications. Earth Planet Sci Lett 165:187–196, https://doi.org/10.1016/S0012-821X(98)00266-0

Callegaro S, Geraki K, Marzoli A, de Min A, Maneta V, Baker DR (2020) The quintet completed: The partitioning of sulfur between nominally volatile-free minerals and silicate melts. Am Mineral 105:697–707, https://doi.org/10.2138/am-2020-7188

Caracas R, Cohen RE (2005) Prediction of a new phase transition in Al_2O_3 at high pressures. Geophys Res Lett 32: L06303, https://doi.org/10.1029/2004GL022204

Carter J, Poulet F, Bibring J-P, Mangold N, Murchie S (2013) Hydrous minerals on Mars as seen by the CRISM and OMEGA imaging spectrometers: Updated global view. J Geophys Res: Planets 118:831–858, https://doi.org/10.1029/2012JE004145

Carter-Bond JC, O'Brien DP, Raymond SN (2012a) The compositional diversity of extrasolar terrestrial planets. II. migration simulations. Astrophys J 760:44, https://doi.org/10.1088/0004-637X/760/1/44

Carter-Bond JC, O'Brien DP, Delgado Mena E, Israelian G, Santos NC, González Hernández JI (2012b) Low Mg/Si planetary host stars and their Mg-depleted terrestrial planets. Astrophys J Lett 747:L2, https://doi.org/10.1088/2041-8205/747/1/L2

Cartier C, Namur O, Nittler L, Weider S, Crapster-Pregont E, Vorburger A, Frank E, Charlier B (2020) No FeS layer in Mercury? Evidence from Ti/Al measured by MESSENGER. Earth Planet Sci Lett 534:116108

Catling DC, Zahnle KJ, McKay CP (2001) Biogenic methane, hydrogen escape, and the irreversible oxidation of early Earth. Science 293:839–843, https://doi.org/10.1126/science.1061976

Charbonneau D, Brown TM, Latham DW, Mayor M (1999) Detection of planetary transits across a sun-like star. Astrophys J 529:L45

Charnoz S, Falco A, Tremblin P, Sossi P, Caracas R, Lagage P-O (2023) The effect of a small amount of hydrogen in the atmosphere of ultrahot magma-ocean planets: Atmospheric composition and escape. Astron Astrophys 674:A224, https://doi.org/10.1051/0004-6361/202245763

Chase MW, Curnutt JL, Downey JR, Mcdonald RA, Syverud AN, Valenzuela EA (1982) JANAF thermochemical tables 1982 supplement. J Phys Chem Ref Data 11:695–940, https://doi.org/10.1063/1.555666

Chen H, Leinenweber K, Prakapenka V, Prescher C, Meng Y, Bechtel H, Kunz M, Shim S-H (2020) Possible H_2O storage in the crystal structure of $CaSiO_3$ perovskite. Phys Earth Planet Sci 299:106412, https://doi.org/10.1016/j.pepi.2019.106412

Christensen UR, Yuen DA (1985) Layered convection induced by phase transitions. J Geophys Res 90:10,291–10,300, https://doi.org/10.1029/JB090iB12p10291

Claringbold A, Rimmer P, Rugheimer S, Shorttle O (2023) Prebiosignature molecules can be detected in temperate exoplanet atmospheres with JWST. arXiv:2306.02897

Clay PL, Burgess R, Busemann H, Ruzié-Hamilton L, Joachim B, Day JM, Ballentine CJ (2017) Halogens in chondritic meteorites and terrestrial accretion. Nature 551:614–618

Connolly JAD (2009) The geodynamic equation of state: What and how. Geochem Geophys Geosystems 10: Q10014, https://doi.org/10.1029/2009GC002540

Connolly JAD, Kerrick DM (1987) An algorithm and computer program for calculating composition phase diagrams. Calphad 11:1–55, https://doi.org/10.1016/0364-5916(87)90018-6

Consorzi A, Melini D, Spada G (2023) Relation between the moment of inertia and the k_2 Love number of fluid extrasolar planets. Astron Astrophys 676:A21

Coppari F, Smith RF, Eggert JH, Wang J, Rygg JR, Lazicki A, Hawreliak JA, Collins GW, Duffy TS (2013) Experimental evidence for a phase transition in magnesium oxide at exoplanet pressures. Nat Geosci 6:926–929, https://doi.org/10.1038/ngeo1948

Coppari F, Smith RF, Wang J, Millot M, Kim D, Rygg JR, Hamel S, Eggert JH, Duffy TS (2021) Implications of the iron oxide phase transition on the interiors of rocky exoplanets. Nat Geosci 14:121–126, https://doi.org/10.1038/s41561-020-00684-y

Cordier P, Gouriet K, Weidner T, Van Orman J, Castelnau O, Jackson JM, Carrez P (2023) Periclase deforms more slowly than bridgmanite under mantle conditions. Nature 613:303–307, https://doi.org/10.1038/s41586-022-05410-9

Crossfield IJ, Malik M, Hill ML, et al. (2022) GJ 1252b: A hot terrestrial super-Earth with no atmosphere. Astrophys J Lett 937:L17

Curry A, Booth R, Owen JE, Mohanty S (2023) The evolution of catastrophically evaporating rocky planets. Mon Not R Astron Soc:stae191

Dasgupta R (2013) Ingassing, storage, and outgassing of terrestrial carbon through geologic time. Rev Mineral Geochem 75:183–229, https://doi.org/10.2138/rmg.2013.75.7

Dasgupta R, Hirschmann MM (2006) Melting in the earth's deep upper mantle caused by carbon dioxide. Nature 440:659–662, https://doi.org/10.1038/nature04612

Dasgupta R, Hirschmann MM (2010) The deep carbon cycle and melting in Earth's interior. Earth Planet Sci Lett 298:1–13, https://doi.org/10.1016/j.epsl.2010.06.039

Dasgupta R, Falksen E, Pal A, Sun C (2022) The fate of nitrogen during parent body partial melting and accretion of the inner solar system bodies at reducing conditions. Geochim Cosmochim Acta 336:291–307

Dauphas N, Poitrasson F, Burkhardt C, Kobayashi H, Kurosawa K (2015) Planetary and meteoritic Mg/Si and ^{30}Si variations inherited from solar nebula chemistry. Earth Planet Sci Lett 427:236–248, https://doi.org/10.1016/j.epsl.2015.07.008

Daviau K, Lee KKM (2021) Experimental constraints on solid nitride phases in rocky mantles of reduced planets and implications for observable atmosphere compositions. J Geophys Res: Planets 126:e2020JE006687, https://doi.org/10.1029/2020JE006687

Davis AM, Olsen EJ (1991) Phosphates in pallasite meteorites as probes of mantle processes in small planetary bodies. Nature 353:637–640, https://doi.org/10.1038/353637a0

Delgado Mena E, Israelian G, González Hernández JI, Bond JC, Santos NC, Udry S, Mayor M (2010) Chemical clues on the formation of planetary systems: C/O versus Mg/Si for HARPS GTO sample. Astrophys J 725:2349–2358, https://doi.org/10.1088/0004-637X/725/2/2349

Delgado Mena E, Tsantaki M, Adibekyan VZ, Sousa SG, Santos NC, González Hernández JI, Israelian G (2017) Chemical abundances of 1111 FGK stars from the HARPS GTO planet search program. Astron Astrophys 606:A94, https://doi.org/10.1051/0004-6361/201730535

Demory B-O (2014) The albedos of Kepler's close-in super-Earths. Astrophys J Lett 789:L20, https://doi.org/10.1088/2041-8205/789/1/L20

Demory B-O, Gillon M, de Wit J, et al. (2016) A map of the large day–night temperature gradient of a super-Earth exoplanet. Nature 532:207-209, https://doi.org/10.1038/nature17169

Demouchy S, Shcheka S, Denis CMM, Thoraval C (2017) Subsolidus hydrogen partitioning between nominally anhydrous minerals in garnet-bearing peridotite. Am Mineral 102:1822–1831, https://doi.org/10.2138/am-2017-6089

Dorn C, Hinkel NR, Venturini J (2017) Bayesian analysis of interiors of HD 219134b, Kepler-10b, Kepler-93b, CoRoT-7b, 55 Cnc e, and HD 97658b using stellar abundance proxies. Astron Astrophys 597:A38, https://doi.org/10.1051/0004-6361/201628749

Dong J, Fischer RA, Stixrude LP, Lithgow-Bertelloni CR (2021a) Constraining the volume of Earth's early oceans with a temperature-dependent mantle water storage capacity model. AGU Adv 2:e2020AV000323, https://doi.org/10.1029/2020AV000323

Dong J, Fischer R, Brennan M, Daviau K, Suer T-A, Turner K, Prakapenka V, Meng Y (2021b) Determining phase transition clapeyron slopes in Mg_2SiO_4 for the mantle transition zone: A multiple logistic regression analysis on experimental data. 2021:MR45A-0071

Dong J, Fischer RA, Stixrude LP, Lithgow-Bertelloni CR, Eriksen ZT, Brennan MC (2022) Water storage capacity of the Martian mantle through time. Icarus 385:115113, https://doi.org/10.1016/j.icarus.2022.115113

Dorn C, Lichtenberg T (2021) Hidden water in magma ocean exoplanets. Astrophys J Lett 922:L4, https://doi.org/10.3847/2041-8213/ac33af

Dorn C, Khan A, Heng K, Alibert Y, Connolly JAD, Benz W, Tackley P (2015) Can we constrain interior structure of rocky exoplanets from mass and radius measurements? Astron Astrophys 577:A83, https://doi.org/10.1051/0004-6361/201424915

Dorn C, Harrison JHD, Bonsor A, Hands TO (2019) A new class of super-earths formed from high-temperature condensates: HD219134 b, 55 Cnc e, WASP-47 e. Mon Not R Astron Soc 484:712–727, https://doi.org/10.1093/mnras/sty3435

Dressing CD, Charbonneau D, Dumusque X, et al. (2015) The mass of Kepler-93b and the composition of terrestrial planets. Astrophys J 800:135, https://doi.org/10.1088/0004-637X/800/2/135

Driscoll P, Bercovici D (2014) On the thermal and magnetic histories of Earth and Venus: Influences of melting, radioactivity, and conductivity. Phys Earth Planet Sci 236:36–51

Dubois V, Mocquet A, Sotin C (2002) Effect of the chemistry of the stellar nebula on the relationship between mass and radius of silicate and metal rich exoplanets. EGS General Assembly Conf Abstr, p 4010

Duffy TS, Smith RF (2019) Ultra-high pressure dynamic compression of geological materials. Front Earth Sci 7:23

Dyck B, Wade J, Palin R (2021) The effect of core formation on surface composition and planetary habitability. Astrophys J Lett 913:L10, https://doi.org/10.3847/2041-8213/abf7ca

Ehrenreich D, Bourrier V, Bonfils X, et al. (2012) Hint of a transiting extended atmosphere on 55 Cancri b. Astron Astrophys 547:A18

Erkaev N, Scherf M, Herbort O, Lammer H, Odert P, Kubyshkina D, Leitzinger M, Woitke P, O'Neill C (2023) Modification of the radioactive heat budget of Earth-like exoplanets by the loss of primordial atmospheres. Mon Not R Astron Soc 518:3703–3721

Estrada BC, Owen JE, Jankovic MR, Wilson A, Helling C (2023) On the likely magnesium–iron silicate dusty tails of catastrophically evaporating rocky planets. Mon Not R Astron Soc:stae095

Fegley B, Cameron AGW (1987) A vaporization model for iron/silicate fractionation in the Mercury protoplanet. Earth Planet Sci Lett 82:207-222, https://doi.org/https://doi.org/10.1016/0012-821X(87)90196-8

Fegley B, Lodders K, Jacobson NS (2020) Volatile element chemistry during accretion of the earth. Chemie der Erde 80:125594, https://doi.org/10.1016/j.chemer.2019.125594

Fegley B Jr, Lodders K, Jacobson NS (2023) Chemical equilibrium calculations for bulk silicate earth material at high temperatures. Geochemistry:125961

Fei H, Katsura T (2020) High water solubility of ringwoodite at mantle transition zone temperature. Earth Planet Sci Lett 531:115987, https://doi.org/10.1016/j.epsl.2019.115987

Férot A, Bolfan-Casanova N (2010) Experimentally determined water storage capacity in the Earth's upper mantle. AGU Fall Meeting Abstr 2010:V23E-05

Férot A, Bolfan-Casanova N (2012) Water storage capacity in olivine and pyroxene to 14 GPa: Implications for the water content of the Earth's upper mantle and nature of seismic discontinuities. Earth Planet Sci Lett 349:218–230, https://doi.org/10.1016/j.epsl.2012.06.022

Fischer RA, Nakajima Y, Campbell AJ, Frost DJ, Harries D, Langenhorst F, Miyajima N, Pollok K, Rubie DC (2015) High pressure meta–silicate partitioning of Ni, Co, V, Cr, Si, and O. Geochim Cosmochim Acta 167:177–194, https://doi.org/10.1016/j.gca.2015.06.026

Foley BJ, Smye AJ (2018) Carbon cycling and habitability of Earth-size stagnant lid planets. Astrobiology 18:873–896, https://doi.org/10.1089/ast.2017.1695

Fortin M-A, Gazel E, Kaltenegger L, Holycross ME (2022) Volcanic exoplanet surfaces. Mon Not R Astron Soc 516:4569–4575, https://doi.org/10.1093/mnras/stac2198

Franz HB, King PL, Gaillard F (2019) Sulfur on Mars from the atmosphere to the core. Chapter 6 *In*: Volatiles in the Martian Crust. Filiberto J, Schwenzer SP, (eds). Elsevier, p 119–183

Fratanduono DE, Millot M, Kraus RG, Spaulding DK, Collins GW, Celliers PM, Eggert JH (2018) Thermodynamic properties of $MgSiO_3$ at super-Earth mantle conditions. Phys Rev B 97:214105, https://doi.org/10.1103/PhysRevB.97.214105

Frost DJ, McCammon CA (2008) The redox state of Earth's mantle. Annu Rev Earth Planet Sci 36:389–420, https://doi.org/10.1146/annurev.earth.36.031207.124322

Fulton BJ, Petigura EA (2018) The California-Kepler survey. VII. Precise planet radii leveraging Gaia DR2 reveal the stellar mass dependence of the planet radius gap. The Astron J 156:264

Fulton BJ, Petigura EA, Howard AW, et al. (2017) The California-Kepler survey. III. A gap in the radius distribution of small planets. The Astron J 154:109

Gaidos EJ (2000) A cosmochemical determinism in the formation of Earth-like planets. Icarus 145:637–640, https://doi.org/10.1006/icar.2000.6407

Gaillard F, Scaillet B (2009) The sulfur content of volcanic gases on Mars. Earth Planet Sci Lett 279:34–43, https://doi.org/10.1016/j.epsl.2008.12.028

Gaillard F, Bouhifd MA, Füri E, Malavergne V, Marrocchi Y, Noack L, Ortenzi G, Roskosz M, Vulpius S (2021) The diverse planetary ingassing/outgassing paths produced over billions of years of magmatic activity. Space Sci Rev 217:22, https://doi.org/10.1007/s11214-021-00802-1

Ghiorso MS, Sack RO (1995) Chemical mass transfer in magmatic processes IV. A revised and internally consistent thermodynamic model for the interpolation and extrapolation of liquid–solid equilibria in magmatic systems at elevated temperatures and pressures. Contrib Mineral Petrol 119:197–212

Ghiorso MS, Hirschmann MM, Reiners PW, Kress III VC (2002) The pMELTS: A revision of MELTS for improved calculation of phase relations and major element partitioning related to partial melting of the mantle to 3 GPa. Geochem Geophys Geosystems 3:1–35, https://doi.org/10.1029/2001GC000217

Ginzburg S, Schlichting HE, Sari Re (2018) Core-powered mass-loss and the radius distribution of small exoplanets. Mon Not R Astron Soc 476:759–765

Girard J, Amulele G, Farla R, Mohiuddin A, Karato S-i (2016) Shear deformation of bridgmanite and magnesiowüstite aggregates at lower mantle conditions. Science 351:144–147, https://doi.org/10.1126/science.aad3113

Grasset O, Schneider J, Sotin C (2009) A study of the accuracy of mass–radius relationships for silicate-rich and ice-rich planets up to 100 Earth masses. Astrophys J 693:722–733, https://doi.org/10.1088/0004-637X/693/1/722

Green E, White R, Diener J, Powell R, Holland T, Palin R (2016) Activity–composition relations for the calculation of partial melting equilibria in metabasic rocks. J Metamorph Petrol 34:845–869

Greene TP, Bell TJ, Ducrot E, Dyrek Ae, Lagage P-O, Fortney JJ (2023) Thermal emission from the Earth-sized exoplanet TRAPPIST-1 b using JWST. Nature 618:39–42, https://doi.org/10.1038/s41586-023-05951-7

Guimond CM, Noack L, Ortenzi G, Sohl F (2021) Low volcanic outgassing rates for a stagnant lid Archean Earth with graphite-saturated magmas. Phys Earth Planet Sci 320:106788, https://doi.org/10.1016/j.pepi.2021.106788

Guimond CM, Shorttle O, Rudge JF (2023a) Mantle mineralogy limits to rocky planet water inventories. Mon Not R Astron Soc 521:2535–2552, https://doi.org/10.1093/mnras/stad148

Guimond CM, Shorttle O, Jordan S, Rudge JF (2023b) A mineralogical reason why all exoplanets cannot be equally oxidizing. Mon Not R Astron Soc 525:3703–3717, https://doi.org/10.1093/mnras/stad2486

Gupta A, Schlichting HE (2019) Sculpting the valley in the radius distribution of small exoplanets as a by-product of planet formation: The core-powered mass-loss mechanism. Mon Not R Astron Soc 487:24–33, https://doi.org/10.1093/mnras/stz1230

Hakim K, van Westrenen W, Dominik C (2018a) Capturing the oxidation of silicon carbide in rocky exoplanetary interiors. Astron Astrophys 618:L6, https://doi.org/10.1051/0004-6361/201833942

Hakim K, Rivoldini A, Van Hoolst T, Cottenier S, Jaeken J, Chust T, Steinle-Neumann G (2018b) A new ab initio equation of state of hcp-Fe and its implication on the interior structure and mass–radius relations of rocky super-Earths. Icarus 313:61–78, https://doi.org/10.1016/j.icarus.2018.05.005

Hakim K, Spaargaren R, Grewal DS, Rohrbach A, Berndt J, Dominik C, van Westrenen W (2019) Mineralogy, structure, and habitability of carbon-enriched rocky exoplanets: A laboratory approach. Astrobiology 19:867–884, https://doi.org/10.1089/ast.2018.1930

Hall A (1999) Ammonium in granites and its petrogenetic significance. Earth Sci Rev 45:145–165, https://doi.org/10.1016/S0012-8252(99)00006-9

Hamilton VE, McSween Jr. HY, Hapke B (2005) Mineralogy of Martian atmospheric dust inferred from thermal infrared spectra of aerosols. J Geophys Res: Planets 110:E12006

Hammond M, Pierrehumbert R (2017) Linking the climate and thermal phase curve of 55 Cancri e. Astrophys J 849:152, https://doi.org/10.3847/1538-4357/aa9328

Hansen LN, Warren JM (2015) Quantifying the effect of pyroxene on deformation of peridotite in a natural shear zone. J Geophys Res: Solid Earth 120:2717–2738, https://doi.org/10.1002/2014JB011584

Hapke B (1973) Darkening of silicate rock powders by solar wind sputtering. The Moon 7:342–355, https://doi.org/10.1007/BF00564639

Hapke B (2002) Bidirectional reflectance spectroscopy: 5. The coherent backscatter opposition effect and anisotropic scattering. Icarus 157:523–534, https://doi.org/10.1006/icar.2002.6853

Hapke B (2012) Reflectance spectroscopy. *In*: Theory of Reflectance and Emittance Spectroscopy. Cambridge University Press, Cambridge, p 369–411

Hapke BW, Nelson RM, Smythe WD (1993) The opposition effect of the Moon: The contribution of coherent backscatter. Science 260:509–511, https://doi.org/10.1126/science.260.5107.509

Hauri EH, Gaetani GA, Green TH (2006) Partitioning of water during melting of the Earth's upper mantle at H_2O-undersaturated conditions. Earth Planet Sci Lett 248:715–734, https://doi.org/10.1016/j.epsl.2006.06.014

Hazen RM, Grew ES, Downs RT, Golden J, Hystad G (2015) Mineral ecology: chance and necessity in the mineral diversity of terrestrial planets. Can Mineral 53:295–324, https://doi.org/10.3749/canmin.1400086

Hedges C, Hughes A, Zhou G, et al. (2022) Erratum: "TOI-2076 and TOI-1807: Two young, comoving planetary systems within 50 pc identified by TESS that are ideal candidates for further follow up" (2021, AJ, 162, 54). Astron J 163:143, https://doi.org/10.3847/1538-3881/ac4477

Heng K (2023) The transient outgassed atmosphere of 55 Cancri e. Astrophys J Lett 956:L2

Herbort O, Woitke P, Helling C, Zerkle A (2020) The atmospheres of rocky exoplanets I. Outgassing of common rock and the stability of liquid water. Astron Astrophys 636:A71, https://doi.org/10.1051/0004-6361/201936614

Herrick RR, Hensley S (2023) Surface changes observed on a Venusian volcano during the Magellan mission. Science 379:1205–1208

Hier-Majumder S, Hirschmann MM (2017) The origin of volatiles in the Earth's mantle. Geochem Geophys Geosystems 18:3078–3092, https://doi.org/10.1002/2017GC006937

Hin RC, Coath CD, Carter PJ, Nimmo F, Lai Y-J, Pogge von Strandmann PAE, Willbold M, Leinhardt ZeM, Walter MJ, Elliott T (2017) Magnesium isotope evidence that accretional vapour loss shapes planetary compositions. Nature 549:511–515, https://doi.org/10.1038/nature23899

Hinkel NR, Unterborn CT (2018) The star–planet connection. I. Using stellar composition to observationally constrain planetary mineralogy for the 10 closest stars. Astrophys J 853:83, https://doi.org/10.3847/1538-4357/aaa5b4

Hinkel NR, Young PA, III, Wheeler CH (2022) A concise treatise on converting stellar mass fractions to abundances to molar ratios. Astron J 164:256, https://doi.org/10.3847/1538-3881/ac9bfa

Hinkel NR, Timmes FX, Young PA, Pagano MD, Turnbull MC (2014) Stellar abundances in the Solar neighborhood: The Hypatia Catalog. Astron J 148:54, https://doi.org/10.1088/0004-6256/148/3/54

Hinkel NR, Young PA, Pagano MD, et al. (2016) A comparison of stellar elemental abundance techniques and measurements. Astrophys J Suppl Ser 226:4, https://doi.org/10.3847/0067-0049/226/1/4

Hirose K, Wood B, Vočadlo L (2021) Light elements in the Earth's core. Nat Rev Earth Environ 2:645–658, https://doi.org/10.1038/s43017-021-00203-6

Hirschmann MM (2000) Mantle solidus: Experimental constraints and the effects of peridotite composition. Geochem Geophys Geosystems 1:1042, https://doi.org/10.1029/2000GC000070

Hirschmann MM (2022) Magma oceans, iron and chromium redox, and the origin of comparatively oxidized planetary mantles. Geochim Cosmochim Acta 328:221–241, https://doi.org/10.1016/j.gca.2022.04.005

Hirschmann MM (2023) The deep Earth oxygen cycle: Mass balance considerations on the origin and evolution of mantle and surface oxidative reservoirs. Earth Planet Sci Lett 619:118311, https://doi.org/10.1016/j.epsl.2023.118311

Hirschmann MM, Withers AC (2008) Ventilation of CO_2 from a reduced mantle and consequences for the early Martian greenhouse. Earth Planet Sci Lett 270:147–155, https://doi.org/10.1016/j.epsl.2008.03.034

Holland T, Powell R (2011) An improved and extended internally consistent thermodynamic dataset for phases of petrological interest, involving a new equation of state for solids. J Metamorph Petrol 29:333–383

Holloway JR, Pan V, Gudmundsson G (1992) High-pressure fluid-absent melting experiments in the presence of graphite; oxygen fugacity, ferric/ferrous ratio and dissolved CO_2. Eur J Mineral 4:105–114

Höning D, Tosi N, Spohn T (2019) Carbon cycling and interior evolution of water-covered plate tectonics and stagnant lid planets. Astron Astrophys 627:A48, https://doi.org/10.1051/0004-6361/201935091

Hourihane A, Franccois P, Worley CC, et al. (2023) The Gaia-ESO survey: Homogenisation of stellar parameters and elemental abundances. Astron Astrophys 676:1–33, https://doi.org/10.1051/0004-6361/202345910

Hu R, Ehlmann BL, Seager S (2012) Theoretical spectra of terrestrial exoplanet surfaces. Astrophys J 752:7

Huang C, Zhao G, Zhang HW, Chen YQ (2005) Chemical abundances of 22 extrasolar planet host stars. Mon Not R Astron Soc 363:71–78, https://doi.org/10.1111/j.1365-2966.2005.09395.x

Inoue T, Wada T, Sasaki R, Yurimoto H (2010) Water partitioning in the Earth's mantle. Phys Earth Planet Sci 183:245–251, https://doi.org/10.1016/j.pepi.2010.08.003

Ishii T, Ohtani E, Shatskiy A (2022) Aluminum and hydrogen partitioning between bridgmanite and high-pressure hydrous phases: implications for water storage in the lower mantle. Earth Planet Sci Lett 583:117441, https://doi.org/10.1016/j.epsl.2022.117441

Ito E, Takahashi E (1989) Postspinel transformations in the system Mg_2SiO_4–Fe_2SiO_4 and some geophysical implications. J Geophys Res: Solid Earth 94:10637–10646, https://doi.org/10.1029/JB094iB08p10637

Ito Y, Ikoma M, Kawahara H, Nagahara H, Kawashima Y, Nakamoto T (2015) Theoretical emission spectra of atmospheres of hot rocky super-Earths. Astrophys J 801:144, https://doi.org/10.1088/0004-637X/801/2/144

Javoy M (1995) The integral enstatite chondrite model of the Earth. Geophys Res Lett 22:2219–2222, https://doi.org/10.1029/95GL02015

Jindal A, de Mooij EJ, Jayawardhana R, Deibert EK, Brogi M, Rustamkulov Z, Fortney JJ, Hood CE, Morley CV (2020) Characterization of the atmosphere of super-Earth 55 Cancri e using high-resolution ground-based spectroscopy. Astron J 160:101

Johansen A, Dorn C (2022) Nucleation and growth of iron pebbles explains the formation of iron-rich planets akin to Mercury. Astron Astrophys 662:A19, https://doi.org/10.1051/0004-6361/202243480

Johnson B, Goldblatt C (2015) The nitrogen budget of Earth. Earth Sci Rev 148:150–173, https://doi.org/10.1016/j.earscirev.2015.05.006

Jorge DM, Kamp IEE, Waters LBFM, Woitke P, Spaargaren RJ (2022) Forming planets around stars with non-Solar elemental composition. Astron Astrophys 660:1–17, https://doi.org/10.1051/0004-6361/202142738

Kaltenegger L, Sasselov D (2010) Detecting planetary geochemical cycles on exoplanets: atmospheic signatures and the case of SO_2. Astrophys J 708:1162–1167, https://doi.org/10.1088/0004-637X/708/2/1162

Kanzaki M (1991) stability of hydrous magnesium silicates in the mantle transition zone. Phys Earth Planet Sci 66:307–312, https://doi.org/10.1016/0031-9201(91)90085-V

Karato S-i (2011) Rheological structure of the mantle of a super-Earth: Some insights from mineral physics. Icarus 212:14–23, https://doi.org/10.1016/j.icarus.2010.12.005

Karato S-i, Karki B, Park J (2020) Deep mantle melting, global water circulation and its implications for the stability of the ocean mass. Prog Earth Planet Sci 7:1–25

Kasting JF, Pollack JB (1983) Loss of water from Venus. I. Hydrodynamic escape of hydrogen. Icarus 53:479–508, https://doi.org/10.1016/0019-1035(83)90212-9

Kasting JF, Whitmire DP, Reynolds RT (1993) Habitable zones around main sequence stars. Icarus 101:108–128, https://doi.org/10.1006/icar.1993.1010

Katz RF, Spiegelman M, Langmuir CH (2003) A new parameterization of hydrous mantle melting. Geochem Geophys Geosystems 4:1073, https://doi.org/10.1029/2002GC000433

Keles E, Mallonn M, Kitzmann D, et al. (2022) The PEPSI exoplanet transit survey (PETS) I: investigating the presence of a silicate atmosphere on the super-earth 55 Cnc e. Mon Not R Astron Soc 513:1544–1556

Keppler H, Bolfan-Casanova N (2006) Thermodynamics of water solubility and partitioning. Rev Mineral Geochem 62:193–230, https://doi.org/10.2138/rmg.2006.62.9

Keppler H, Wiedenbeck M, Shcheka SS (2003) Carbon solubility in olivine and the mode of carbon storage in the Earth's mantle. Nature 424:414–416, https://doi.org/10.1038/nature01828

Khan A, Sossi PA, Liebske C, Rivoldini A, Giardini D (2022) Geophysical and cosmochemical evidence for a volatile-rich Mars. Earth Planet Sci Lett 578:117330, https://doi.org/10.1016/j.epsl.2021.117330

Kiefer WS, Filiberto J, Sandu C, Li Q (2015) The effects of mantle composition on the peridotite solidus: Implications for the magmatic history of Mars. Geochim Cosmochim Acta 162:247–258, https://doi.org/10.1016/j.gca.2015.02.010

Kite ES, Manga M, Gaidos E (2009) Geodynamics and rate of volcanism on massive Earth-like planets. Astrophys J 700:1732–1749, https://doi.org/10.1088/0004-637X/700/2/1732

Kite ES, Fegley Jr B, Schaefer L, Gaidos E (2016) Atmosphere–interior exchange on hot, rocky exoplanets. Astrophys J 828:80

Kohlstedt DL, Zimmerman ME (1996) Rheology of partially molten mantle rocks. Annu Rev Earth Planet Sci 24:41–62

Kohlstedt DL, Keppler H, Rubie DC (1996) Solubility of water in the α, β and γ phases of $(Mg,Fe)_2SiO_4$. Contrib Mineral Petrol 123:345–357, https://doi.org/10.1007/s004100050161

Korenaga J (2018) Crustal evolution and mantle dynamics through Earth history. Philos Trans R Soc A376:20170408

Kreidberg L, Koll DD, Morley C, et al. (2019) Absence of a thick atmosphere on the terrestrial exoplanet LHS 3844b. Nature 573:87–90

Krijt S, Kama M, McClure M, Teske J, Bergin EA, Shorttle O, Walsh KJ, Raymond SN (2022) Chemical habitability: supply and retention of life's essential elements during planet formation. arXiv:220310056

Krissansen-Totton J, Fortney JJ, Nimmo F, Wogan N (2021) Oxygen false positives on habitable zone planets around Sun-like stars. AGU Adv 2:e2020AV000294, https://doi.org/10.1029/2020AV000294

Labrosse S, Hernlund JW, Coltice N (2007) A crystallizing dense magma ocean at the base of the Earth's mantle. Nature 450:866–869, https://doi.org/10.1038/nature06355

Lacaze J, Sundman B (1991) An assessment of the Fe–C–Si system. Metall Trans A 22:2211–2223

Lammer H, Chassefière E, Karatekin Ö, et al. (2013) Outgassing history and escape of the Martian atmosphere and water inventory. Space Sci Rev 174:113–154, https://doi.org/10.1007/s11214-012-9943-8

Larimer JW, Anders E (1967) Chemical fractionations in meteorites-II. Abundance patterns and their interpretation. Geochim Cosmochim Acta 31:1239–1270, https://doi.org/10.1016/S0016-7037(67)80014-0

Larimer JW, Bartholomay M (1979) The role of carbon and oxygen in cosmic gases: some applications to the chemistry and mineralogy of enstatite chondrites. Geochim Cosmochim Acta 43:1455–1466

Léger A, Grasset O, Fegley B, et al. (2011) The extreme physical properties of the CoRoT-7b super-Earth. Icarus 213:1–11, https://doi.org/10.1016/j.icarus.2011.02.004

Li Y, Keppler H (2014) Nitrogen speciation in mantle and crustal fluids. Geochim Cosmochim Acta 129:13–32, https://doi.org/10.1016/j.gca.2013.12.031

Li Y, Wiedenbeck M, Shcheka S, Keppler H (2013) Nitrogen solubility in upper mantle minerals. Earth Planet Sci Lett 377–378:311–323, https://doi.org/10.1016/j.epsl.2013.07.013

Libourel G, Marty B, Humbert F (2003) Nitrogen solubility in basaltic melt. Part I. Effect of oxygen fugacity. Geochim Cosmochim Acta 67:4123–4135, https://doi.org/10.1016/S0016-7037(03)00259-X

Liggins P, Jordan S, Rimmer PB, Shorttle O (2022) Growth and evolution of secondary volcanic atmospheres: I. Identifying the geological character of hot rocky planets. J Geophys Res: Planets 127:e2021JE007123, https://doi.org/10.1029/2021JE007123

Lipp AG, Shorttle O, Syvret F, Roberts GG (2020) Major element composition of sediments in terms of weathering and provenance: implications for crustal recycling. Geochem Geophys Geosystems 21:e2019GC008758

Litasov K, Ohtani E, Langenhorst F, Yurimoto H, Kubo T, Kondo T (2003) Water solubility in Mg-perovskites and water storage capacity in the lower mantle. Earth Planet Sci Lett 211:189–203, https://doi.org/10.1016/S0012-821X(03)00200-0

Liu L-g (1987) Effects of H_2O on the phase behaviour of the forsterite–enstatite system at high pressures and temperatures and implications for the Earth. Phys Earth Planet Sci 49:142–167, https://doi.org/10.1016/0031-9201(87)90138-5

Liu F, Yong D, Asplund M, Wang HS, Spina L, Acuna L, Meléndez J, Ramírez I (2020) Detailed chemical compositions of planet-hosting stars–I. Exploration of possible planet signatures. Mon Not R Astron Soc 495:3961–3973, https://doi.org/10.1093/mnras/staa1420

Liu Z, Fei H, Chen L, McCammon C, Wang L, Liu R, Wang F, Liu B, Katsura T (2021) Bridgmanite is nearly dry at the top of the lower mantle. Earth Planet Sci Lett 570:117088, https://doi.org/10.1016/j.epsl.2021.117088

Lock SJ, Stewart ST (2023) Atmospheric loss in giant impacts depends on pre-impact surface conditions. arXiv:2309.16399

Lodders K (2003) Solar System abundances and condensation temperatures of the elements. Astrophys J 591:1220–1247, https://doi.org/10.1086/375492

Lodders K, Palme H, Gail H-P (2009) Abundances of the elements in the Solar System. Landolt Börnstein 4B:712, https://doi.org/10.1007/978-3-540-88055-4_34

Lucey PG (1998) Model near-infrared optical constants of olivine and pyroxene as a function of iron content. J Geophys Res: Planets 103:1703–1713, https://doi.org/10.1029/97JE03145

Lucey PG, Blewett DT, Hawke BR (1998) Mapping the FeO and TiO_2 content of the lunar surface with multispectral imagery. J Geophys Res: Planets 103:3679–3699, https://doi.org/10.1029/97JE03019

Luque R, Pallé E (2022) Density, not radius, separates rocky and water-rich small planets orbiting M dwarf stars. Science 377:1211–1214, https://doi.org/10.1126/science.abl7164

Luth R, Fei Y, Bertka C, Mysen B (1999) Carbon and carbonates in the mantle. Mantle petrology: Field observations and high pressure experimentation: A tribute to Francis R(Joe) Boyd. Geochem Soc Spec Publ 6:297–316

Madhusudhan N, Lee KKM, Mousis O (2012) A possible carbon-rich interior in super-Earth 55 Cancri e. Astrophys J 759:L40, https://doi.org/10.1088/2041-8205/759/2/L40

Mahapatra G, Helling C, Miguel Y (2017) Cloud formation in metal-rich atmospheres of hot super-Earths like 55 Cnc e and CoRoT7b. Mon Not R Astron Soc 472:447–464, https://doi.org/10.1093/mnras/stx1666

Malavergne V, Toplis MJ, Berthet S, Jones J (2010) Highly reducing conditions during core formation on Mercury: Implications for internal structure and the origin of a magnetic field. Icarus 206:199–209

Mansfield M, Kite ES, Hu R, Koll DDB, Malik M, Bean JL, Kempton EM-R (2019) Identifying atmospheres on rocky exoplanets through inferred high albedo. Astrophys J 886:141, https://doi.org/10.3847/1538-4357/ab4c90

Markova O, Yakovlev O, Semenov G, AN B (1986) Evaporation of natural melts in a Knudsen chamber. Geokhimiya 11:1559—1568

Marty B (2012) The origins and concentrations of water, carbon, nitrogen and noble gases on Earth. Earth Planet Sci Lett 313-314:56–66, https://doi.org/10.1016/j.epsl.2011.10.040

McDonough WF, Sun S-s (1995) The composition of the Earth. Chem Geol 120:223–253, https://doi.org/10.1016/0009-2541(94)00140-4

McDonough WF, Yoshizaki T (2021) Terrestrial planet compositions controlled by accretion disk magnetic field. Prog Earth Planet Sci 8:1–12

Meier TG, Bower DJ, Lichtenberg T, Hammond M, Tackley PJ (2023) Interior dynamics of super-Earth 55 Cancri e. Astron Astrophys 678:A29, https://doi.org/10.1051/0004-6361/202346950

Meier Valdés EA, Morris BM, Demory B-O, et al. (2023) Investigating the visible phase-curve variability of 55 cnc e. Astron Astrophys 677:A112, https://doi.org/10.1051/0004-6361/202346050

Mercier SJ, Dang L, Gass A, Cowan NB, Bell TJ (2022) Revisiting the iconic Spitzer Phase curve of 55 Cancri e: Hotter dayside, cooler nightside, and smaller phase offset. Astron J 164:204, https://doi.org/10.3847/1538-3881/ac8f22

Mibe K, Kanzaki M, Kawamoto T, Matsukage KN, Fei Y, Ono S (2007) Second critical endpoint in the peridotite–H_2O system. J Geophys Res (Solid Earth) 112:B03201, https://doi.org/10.1029/2005JB004125

Miguel Y, Kaltenegger L, Fegley B, Schaefer L (2011) Compositions of hot super-Earth atmospheres: Exploring kepler candidates. Astrophys J Lett 742:L19, https://doi.org/10.1088/2041-8205/742/2/L19

Mikhail S, Sverjensky DA (2014) Nitrogen speciation in upper mantle fluids and the origin of Earth's nitrogen-rich atmosphere. Nat Geosci 7:816–819, https://doi.org/10.1038/ngeo2271

Miozzi F, Morard G, Antonangeli D, Clark AN, Mezouar M, Dorn C, Rozel A, Fiquet G (2018) Equation of State of SiC at extreme conditions: New insight into the interior of carbon-rich exoplanets. J Geophys Res Planets 123:2295–2309, https://doi.org/10.1029/2018JE005582

Miyagoshi T, Tachinami C, Kameyama M, Ogawa M (2014) On the vigor of mantle convection in super-Earths. Astrophys J 780:L8, https://doi.org/10.1088/2041-8205/780/1/L8

Miyagoshi T, Kameyama M, Ogawa M (2018) Effects of adiabatic compression on thermal convection in super-Earths of various sizes. Earth Planets Space 70:200, https://doi.org/10.1186/s40623-018-0975-5

Miyazaki Y, Korenaga J (2020) Dynamic evolution of major element chemistry in protoplanetary disks and its implications for chondrite formation. arXiv:200413911 [astro-ph, physics:physics]

Modirrousta-Galian D, Ito Y, Micela G (2021) Exploring super-Earth surfaces: Albedo of near-airless magma ocean planets and topography. Icarus 358:114175, https://doi.org/10.1016/j.icarus.2020.114175

Mojzsis SJ (2022) Geoastronomy: Rocky planets as the Lavoisier–Lomonosov bridge from the non-living to the living world. *In*: Prebiotic Chemistry and Life's Origins. The Royal Society of Chemistry, p 21–76

Mookherjee M, Stixrude L, Karki B (2008) Hydrous silicate melt at high pressure. Nature 452:983–986, https://doi.org/10.1038/nature06918

Moore K, Cowan NB, Boukaré C-É (2023) The role of magma oceans in maintaining surface water on rocky planets orbiting M-dwarfs. Mon Not R Astron Soc 526:6235-6247

Morbidelli A, Lunine JI, O'Brien DP, Raymond SN, Walsh KJ (2012) Building terrestrial planets. Annu Rev Earth Planet Sci 40:251–275

Mordasini C (2018) Planetary Population Synthesis. arXiv:1804.01532

Moriarty J, Madhusudhan N, Fischer D (2014) Chemistry in an evolving protoplanetary disk: Effects on terrestrial planet composition. Astrophys J 787:81, https://doi.org/10.1088/0004-637X/787/1/81

Mosenfelder JL, Deligne NI, Asimow PD, Rossman GR (2006) Hydrogen incorporation in olivine from 2–12 GPa. Am Mineral 91:285–294, https://doi.org/10.2138/am.2006.1943

Moynier F, Paniello RC, Gounelle M, Albarède F, Beck P, Podosek F, Zanda B (2011) Nature of volatile depletion and genetic relationships in enstatite chondrites and aubrites inferred from Zn isotopes. Geochim Cosmochim Acta 75:297–307, https://doi.org/10.1016/j.gca.2010.09.022

Murakami M, Hirose K, Yurimoto H, Nakashima S, Takafuji N (2002) Water in Earth's lower mantle. Science 295:1885–1887, https://doi.org/10.1126/science.1065998

Murakami M, Hirose K, Kawamura K, Sata N, Ohishi Y (2004) Post-perovskite phase transition in MgSiO₃. Science 304:855–858, https://doi.org/10.1126/science.1095932

Nakagawa T, Tackley PJ (2012) Influence of magmatism on mantle cooling, surface heat flow and Urey ratio. Earth Planet Sci Lett 329:1–10

Nguyen TG, Cowan NB, Pierrehumbert RT, Lupu RE, Moores JE (2022) The impact of ultraviolet heating and cooling on the dynamics and observability of lava planet atmospheres. Mon Not R Astron Soc 513:6125-6133, https://doi.org/10.1093/mnras/stac1331

Nissen PE, Silva Aguirre V, Christensen-Dalsgaard J, Collet R, Grundahl F, Slumstrup D (2017) High-precision abundances of elements in Kepler LEGACY stars Astron Astrophys 608:A112, https://doi.org/10.1051/0004-6361/201731845

Nittler LR, Starr RD, Weider SZ, et al. (2011) The major-element composition of Mercury's surface from MESSENGER X-ray Spectrometry. Science 333:1847, https://doi.org/10.1126/science.1211567

Nittler LR, Chabot NL, Grove TL, Peplowski PN (2018) The chemical composition of Mercury. *In*: Mercury: The View after MESSENGER. Solomon SC, Nittler LR, Anderson BJE, (eds). Cambridge University Press, p 30–51

Niu H, Oganov AR, Chen X-Q, Li D (2015) Prediction of novel stable compounds in the Mg–Si–O system under exoplanet pressures. Sci Rep 5:18347, https://doi.org/10.1038/srep18347

Norris CA, Wood BJ (2017) Earth's volatile contents established by melting and vaporization. Nature 549:507–510, https://doi.org/10.1038/nature23645

Novella D, Frost DJ, Hauri EH, Bureau H, Raepsaet C, Roberge M (2014) The distribution of H₂O between silicate melt and nominally anhydrous peridotite and the onset of hydrous melting in the deep upper mantle. Earth Planet Sci Lett 400:1–13, https://doi.org/10.1016/j.epsl.2014.05.006

O'Reilly SY, Griffin WL (2000) Apatite in the mantle: Implications for metasomatic processes and high heat production in phanerozoic mantle. Lithos 53:217–232, https://doi.org/10.1016/S0024-4937(00)00026-8

Oganov AR, Ono S (2004) Theoretical and experimental evidence for a post-perovskite phase of MgSiO₃ in Earth's D'' layer. Nature 430:445–448, https://doi.org/10.1038/nature02701

Ohtani E (2021) Hydration and dehydration in Earth's Interior. Annu Rev Earth Planet Sci 49:253–278, https://doi.org/10.1146/annurev-earth-080320-062509

Ohtani E, Mizobata H, Kudoh Y, Nagase T, Arashi H, Yurimoto H, Miyagi I (1997) A new hydrous silicate, a water reservoir, in the upper part of the lower mantle. Geophys Res Lett 24:1047–1050, https://doi.org/10.1029/97GL00874

Okuya A, Okuzumi S, Ohno K, Hirano T (2020) Constraining the bulk composition of disintegrating exoplanets using combined transmission spectra from JWST and SPICA. Astrophys J 901:171, https://doi.org/10.3847/1538-4357/abb088

Ortenzi G, Noack L, Sohl F, et al. (2020) Mantle redox state drives outgassing chemistry and atmospheric composition of rocky planets. Sci Rep 10:10907, https://doi.org/10.1038/s41598-020-67751-7

Ostberg CM, Guzewich SD, Kane SR, Kohler E, Oman LD, Fauchez TJ, Kopparapu RK, Richardson J, Whelley P (2023) The prospect of detecting volcanic signatures on an ExoEarth using direct imaging. Astron J 166:199

Owen JE, Wu Y (2013) Kepler planets: a tale of evaporation. Astrophys J 775:105

Owen JE, Wu Y (2017) The evaporation valley in the Kepler planets. Astrophys J 847:29

Pacalo REG, Parise JB (1992) Crystal structure of superhydrous B, a hydrous magnesium silicate synthesized at 1400C and 20 GPa. Am Mineral 77:681–684

Padovan S, Spohn T, Baumeister P, Tosi N, Breuer D, Csizmadia S, Hellard H, Sohl F (2018) Matrix-propagator approach to compute fluid Love numbers and applicability to extrasolar planets. Astron Astrophys 620:A178

Palme H, Lodders K, Jones A (2014) Solar System Abundances of the Elements. 2.2 *In*: Treatise on Geochemistry (Second Edition). Holland HD, Turekian KK, (eds). Elsevier, Oxford, p 15–36

Peng Y, Deng J (2023) Hydrogen diffusion in the lower mantle revealed by machine learning potentials. arXiv:2311.04461

Perez-Becker D, Chiang E (2013) catastrophic evaporation of rocky planets. Mon Not R Astron Soc 433:2294–2309, https://doi.org/10.1093/mnras/stt895

Piette AAA, Gao P, Brugman K, Shahar A, Lichtenberg T, Miozzi F, Driscoll P (2023) Rocky planet or water world? Observability of low-density lava world atmospheres. arXiv:2306.10100

Pignatari M, Trueman TCL, Womack KA, et al. (2023) The chemical evolution of the Solar neighbourhood for planet-hosting stars. Mon Not R Astron Soc:stad2167, https://doi.org/10.1093/mnras/stad2167

Plotnykov M, Valencia D (2020) Chemical fingerprints of formation in rocky super-Earths' data. Mon Not R Astron Soc 499:932–947, https://doi.org/10.1093/mnras/staa2615

Putirka KD (2024) Exoplanet mineralogy. Rev Mineral Geochem 90:199–258

Putirka KD, Rarick JC (2019) The composition and mineralogy of rocky exoplanets: A survey of >4000 stars from the Hypatia Catalog. Am Mineral 104:817–829, https://doi.org/10.2138/am-2019-6787

Putirka KD, Xu S (2021) Polluted white dwarfs reveal exotic mantle rock types on exoplanets in our solar neighborhood. Nat Commun 12:6168, https://doi.org/10.1038/s41467-021-26403-8

Putirka K, Dorn C, Hinkel N, Unterborn C (2021) Compositional diversity of rocky exoplanets. arXiv:210808383 [astro-ph, physics:physics]

Rappaport S, Levine A, Chiang E, et al. (2012) Possible disintegrating short-period super-Mercury orbiting KIC 12557548. Astrophys J 752:1, https://doi.org/10.1088/0004-637X/752/1/1

Ridden-Harper A, Snellen I, Keller C, De Kok R, Di Gloria E, Hoeijmakers H, Brogi M, Fridlund M, Vermeersen B, Van Westrenen W (2016) Search for an exosphere in sodium and calcium in the transmission spectrum of exoplanet 55 Cancri e. Astron Astrophys 593:A129

Righter K, Ghiorso M (2012) Redox systematics of a magma ocean with variable pressure–temperature gradients and composition. PNAS 109:11955–11960

Ringwood AE (1966) Chemical evolution of the terrestrial planets. Geochim Cosmochim Acta 30:41–104

Ringwood AE (1989) Significance of the terrestrial Mg/Si ratio. Earth Planet Sci Lett 95:1–7, https://doi.org/10.1016/0012-821X(89)90162-3

Ringwood A, Hibberson W (1991) Solubilities of mantle oxides in molten iron at high pressures and temperatures: implications for the composition and formation of Earth's core. Earth Planet Sci Lett 102:235–251

Ringwood AE, Major A (1967) High-pressure reconnaissance investigations in the system Mg_2SiO_4–MgO–H_2O. Earth Planet Sci Lett 2:130–133, https://doi.org/10.1016/0012-821X(67)90114-8

Ritterbex S, Harada T, Tsuchiya T (2018) Vacancies in MgO at ultrahigh pressure: About mantle rheology of super-Earths. Icarus 305:350–357, https://doi.org/10.1016/j.icarus.2017.12.020

Rogers LA, Seager S (2010) A framework for quantifying the degeneracies of exoplanet interior compositions. Astrophys J 712:974-991, https://doi.org/10.1088/0004-637X/712/2/974

Root S, Shulenburger L, Lemke RW, Dolan DH, Mattsson TR, Desjarlais MP (2015) Shock response and phase transitions of MgO at planetary impact conditions. Phys Rev Lett 115:198501, https://doi.org/10.1103/PhysRevLett.115.198501

Rosenthal A, Hauri EH, Hirschmann MM (2015) Experimental determination of C, F, and H partitioning between mantle minerals and carbonated basalt, CO_2/Ba and CO_2/Nb systematics of partial melting, and the CO_2 contents of basaltic source regions. Earth Planet Sci Lett 412:77–87, https://doi.org/10.1016/j.epsl.2014.11.044

Rubie DC, Frost DJ, Mann U, Asahara Y, Nimmo F, Tsuno K, Kegler P, Holzheid A, Palme H (2011) Heterogeneous accretion, composition and core–mantle differentiation of the Earth. Earth Planet Sci Lett 301:31–42, https://doi.org/10.1016/j.epsl.2010.11.030

Sakai T, Dekura H, Hirao N (2016) Experimental and theoretical thermal equations of state of $MgSiO_3$ post-perovskite at multi-megabar pressures. Sci Rep 6:22652, https://doi.org/10.1038/srep22652

Santos NC, Adibekyan V, Mordasini C, et al. (2015) Constraining planet structure from stellar chemistry: The cases of CoRoT-7, Kepler-10, and Kepler-93. Astron Astrophys 580:L13, https://doi.org/10.1051/0004-6361/201526850

Scheller EL, Ehlmann BL, Hu R, Adams DJ, Yung YL (2021) Long-term drying of Mars by sequestration of ocean-scale volumes of water in the crust. Science 372:56–62, https://doi.org/10.1126/science.abc7717

Schlichting HE, Young ED (2022) Chemical equilibrium between cores, mantles, and atmospheres of super-Earths and sub-Neptunes and implications for their compositions, interiors, and evolution. Planet Sci J 3:127, https://doi.org/10.3847/PSJ/ac68e6

Schmidt MW, Gao C, Golubkova A, Rohrbach A, Connolly JA (2014) Natural moissanite (SiC)–a low temperature mineral formed from highly fractionated ultra-reducing COH-fluids. Prog Earth Planet Sci 1:1–14

Schulze JG, Wang J, Johnson JA, Unterborn CT, Panero WR (2020) The probability that a rocky planet's composition reflects its host star. arXiv:201108893 [astro-ph]

Schulze JG, Wang J, Johnson JA, Gaudi BS, Unterborn CT, Panero WR (2021) On the probability that a rocky planet's composition reflects its host Star. Planet Sci J 2:113, https://doi.org/10.3847/psj/abcaa8

Seidler F, Wang H, Quanz S (2021) On the importance of including devolatilized stellar abundances in determining the composition of rocky exoplanets. EGU General Assembly Confer Abstracts, p EGU21–15884

Sekine T, Ozaki N, Miyanishi K, et al. (2016) Shock compression response of forsterite above 250 GPa. Sci Adv 2:e1600157, https://doi.org/10.1126/sciadv.1600157

Shah O, Alibert Y, Helled R, Mezger K (2021) Internal water storage capacity of terrestrial planets and the effect of hydration on the M–R relation. Astron Astrophys 646:A162, https://doi.org/10.1051/0004-6361/202038839

Shahnas MH, Pysklywec RN (2021) Focused penetrative plumes: A possible consequence of the dissociation transition of post-perovskite at ~0.9 TPa in massive rocky super-Earths. Geochem Geophys Geosystems 22:e09910, https://doi.org/10.1029/2021GC009910

Shcheka SS, Wiedenbeck M, Frost DJ, Keppler H (2006) Carbon solubility in mantle minerals. Earth Planet Sci Lett 245:730–742, https://doi.org/10.1016/j.epsl.2006.03.036

Shim S-H, Duffy TS, Shen G (2001) The post-spinel transformation in Mg_2SiO_4 and its relation to the 660-km seismic discontinuity. Nature 411:571–574, https://doi.org/10.1038/35079053

Shim S-H, Chizmeshya A, Leinenweber K (2022) Water in the crystal structure of $CaSiO_3$ perovskite. Am Mineral 107:631–641, https://doi.org/10.2138/am-2022-8009

Shim S-H, Ko B, Sokaras D, et al. (2023) Ultrafast X-ray detection of low-spin iron in molten silicate under deep planetary interior conditions. Sci Adv 9:eadi6153

Siever R (1968) Sedimentological consequences of a steady-state ocean–atmosphere. Sedimentology 11:5–29

Skinner BJ, Luce FD (1971) Solid solutions of the type (Ca, Mg, Mn, Fe) S and their use as geothermometers for the enstatite chondrites. Am Mineral 56:1269–1296

Sossi PA, Klemme S, O'Neill HSC, Berndt J, Moynier Fee (2019) Evaporation of moderately volatile elements from silicate melts: experiments and theory. Geochim Cosmochim Acta 260:204–231, https://doi.org/10.1016/j.gca.2019.06.021

Sossi PA, Stotz IL, Jacobson SA, Morbidelli A, O'Neill HSC (2022) Stochastic accretion of the Earth. Nat Astron 6:951–960, https://doi.org/10.1038/s41550-022-01702-2

Sotin C, Grasset O, Mocquet A (2007) Mass–radius curve for extrasolar Earth-like planets and ocean planets. Icarus 191:337–351, https://doi.org/10.1016/j.icarus.2007.04.006

Spaargaren RJ, Ballmer MD, Bower DJ, Dorn C, Tackley PJ (2020) The influence of bulk composition on the long-term interior–atmosphere evolution of terrestrial exoplanets. Astron Astrophys 643:A44, https://doi.org/10.1051/0004-6361/202037632

Spaargaren RJ, Wang HS, Mojzsis SJ, Ballmer MD, Tackley PJ (2023) Plausible constraints on the range of bulk terrestrial exoplanet compositions in the Solar neighborhood. Astrophys J 948:53, https://doi.org/10.3847/1538-4357/acac7d

Spaulding DK, McWilliams RS, Jeanloz R, Eggert JH, Celliers PM, Hicks DG, Collins GW, Smith RF (2012) Evidence for a phase transition in silicate melt at extreme pressure and temperature conditions. Phys Rev Lett 108:065701, https://doi.org/10.1103/PhysRevLett.108.065701

Stagno V, Cerantola V, Aulbach S, Lobanov S, McCammon CA, Merlini M (2019) Carbon-bearing phases throughout earth's interior: evolution through space and time. In: Deep Carbon: Past to Present. Orcutt BN, Daniel I, Dasgupta R, (eds). Cambridge University Press, Cambridge, p 66–88

Stamenković V, Breuer D, Spohn T (2011) Thermal and transport properties of mantle rock at high pressure: applications to super-Earths. Icarus 216:572–596, https://doi.org/10.1016/j.icarus.2011.09.030

Stamenković V, Noack L, Breuer D, Spohn T (2012) The influence of pressure-dependent viscosity on the thermal evolution of super-Earths. Astrophys J 748:41, https://doi.org/10.1088/0004-637X/748/1/41

Stixrude L (2014) Melting in super-Earths. Philos Trans R Soc London Ser A 372:20130076–20130076, https://doi.org/10.1098/rsta.2013.0076

Stixrude L, Lithgow-Bertelloni C (2011) Thermodynamics of mantle minerals–II. Phase equilibria. Geophys J Int 184:1180–1213, https://doi.org/10.1111/j.1365-246X.2010.04890.x

Stixrude L, Lithgow-Bertelloni C (2022) Thermal expansivity, heat capacity and bulk modulus of the mantle. Geophys J Int 228:1119–1149, https://doi.org/10.1093/gji/ggab394

Sunshine JM, Pieters CM, Pratt SF (1990) Deconvolution of mineral absorption bands: An improved approach. J Geophys Res: Solid Earth 95:6955–6966, https://doi.org/10.1029/JB095iB05p06955

Tabernero H, Allende Prieto C, Zapatero Osorio MR, et al. (2020) HORuS transmission spectroscopy of 55 Cnc e. Mon Not R Astron Soc 498:4222–4229

Tackley PJ, Ammann M, Brodholt JP, Dobson DP, Valencia D (2013) Mantle dynamics in super-Earths: Post-perovskite rheology and self-regulation of viscosity. Icarus 225:50–61, https://doi.org/10.1016/j.icarus.2013.03.013

Tenner TJ, Hirschmann MM, Withers AC, Ardia P (2012) H_2O storage capacity of olivine and low-Ca pyroxene from 10 to 13 GPa: Consequences for dehydration melting above the transition zone. Contrib Mineral Petrol 163:297–316, https://doi.org/10.1007/s00410-011-0675-7

Teske JK, Cunha K, Schuler SC, Griffith CA, Smith VV (2013) Carbon and oxygen abundances in cool metal-rich exoplanet hosts: A case study of the C/O ratio of 55 Cancri. Astrophys J 778:132, https://doi.org/10.1088/0004-637X/778/2/132

Thiabaud A, Marboeuf U, Alibert Y, Leya I, Mezger K (2015) Elemental ratios in stars vs planets. Astron Astrophys 580:A30, https://doi.org/10.1051/0004-6361/201525963

Thielmann M, Golabek GJ, Marquardt H (2020) Ferropericlase control of lower mantle rheology: Impact of phase morphology. Geochem Geophys Geosystems 21:e2019GC008688, https://doi.org/10.1029/2019GC008688

Tian F, Kasting JF, Solomon SC (2009) Thermal escape of carbon from the early Martian atmosphere. Geophys Res Lett 36:L02205, https://doi.org/10.1029/2008GL036513

Timmermann A, Shan Y, Reiners A, Pack A (2023) Revisiting equilibrium condensation and rocky planet compositions. Introducing the ECCOplanets code. Astron Astrophys 52, https://doi.org/10.1051/0004-6361/202244850

Tosi N, Godolt M, Stracke B, Ruedas T, Grenfell JL, Höning D, Nikolaou A, Plesa A-C, Breuer D, Spohn T (2017) The habitability of a stagnant-lid Earth. Astron Astrophys 605:A71, https://doi.org/10.1051/0004-6361/201730728

Townsend JP, Tsuchiya J, Bina CR, Jacobsen SD (2016) Water partitioning between bridgmanite and postperovskite in the lowermost mantle. Earth Planet Sci Lett 454:20–27, https://doi.org/10.1016/j.epsl.2016.08.009

Truong N, Lunine JI (2021) Volcanically extruded phosphides as an abiotic source of Venusian phosphine. PNAS 118:e2021689118, https://doi.org/10.1073/pnas.2021689118

Tsuchiya T, Tsuchiya J (2011) Prediction of a hexagonal SiO_2 phase affecting stabilities of $MgSiO3$ and $CaSiO_3$ at multimegabar pressures. PNAS 108:1252–1255, https://doi.org/10.1073/pnas.1013594108

Tsuchiya T, Tsuchiya J, Umemoto K, Wentzcovitch RM (2004) Phase transition in $MgSiO_3$ perovskite in the Earth's lower mantle. Earth Planet Sci Lett 224:241–248, https://doi.org/10.1016/j.epsl.2004.05.017

Umemoto K, Wentzcovitch RM (2011) Two-stage dissociation in $MgSiO_3$ post-perovskite. Earth Planet Sci Lett 311:225–229, https://doi.org/10.1016/j.epsl.2011.09.032

Umemoto K, Wentzcovitch RM, Allen PB (2006) Dissociation of $MgSiO_3$ in the cores of gas giants and terrestrial exoplanets. Science 311:983–986, https://doi.org/10.1126/science.1120865

Umemoto K, Wentzcovitch RM, Wu S, Ji M, Wang C-Z, Ho K-M (2017) Phase transitions in $MgSiO_3$ post-perovskite in super-Earth mantles. Earth Planet Sci Lett 478:40–45, https://doi.org/10.1016/j.epsl.2017.08.032

Unterborn CT, Panero WR (2017) The effects of Mg/Si on the exoplanetary refractory oxygen budget. Astrophys J 845:61, https://doi.org/10.3847/1538-4357/aa7f79

Unterborn CT, Panero WR (2019) The pressure and temperature limits of likely rocky exoplanets. J Geophys Res: Planets 124:1704–1716, https://doi.org/10.1029/2018JE005844

Unterborn CT, Dismukes EE, Panero WR (2016) Scaling the Earth: A sensitivity analysis of terrestrial exoplanetary interior models. Astrophys J 819:32, https://doi.org/10.3847/0004-637X/819/1/32

Unterborn CT, Desch SJ, Hinkel NR, Lorenzo A (2018) Inward migration of the TRAPPIST-1 planets as inferred from their water-rich compositions. Nat Astron 2:297–302, https://doi.org/10.1038/s41550-018-0411-6

Unterborn CT, Foley BJ, Desch SJ, Young PA, Vance G, Chiffelle L, Kane SR (2022) Mantle degassing lifetimes through galactic time and the maximum age stagnant-lid rocky exoplanets can support temperate climates. Astrophys J Lett 930:L6, https://doi.org/10.3847/2041-8213/ac6596

Unterborn CT, Desch SJ, Haldemann J, Lorenzo A, Schulze JG, Hinkel NR, Panero WR (2023) The nominal ranges of rocky planet masses, radii, surface gravities, and bulk densities. Astrophys J 944:42, https://doi.org/10.3847/1538-4357/acaa3b

Urey HC (1954) On the dissipation of gas and volatilized elements from protoplanets. Astrophys J Suppl Ser 1:147, https://doi.org/10.1086/190006

Valencia D, Sasselov DD, O'Connell RJ (2007) Detailed models of super-Earths: How well can we infer bulk properties? Astrophys J 665:1413–1420, https://doi.org/10.1086/519554

Van Buchem CPA, Miguel Y, Zilinskas M, van Westrenen W (2022) LavAtmos: An open source chemical equilibrium vaporisation code for lava worlds. Meteorit Planet Sci 58:1149–1161, https://doi.org/10.1111/maps.13994

van den Berg AP, Yuen DA, Umemoto K, Jacobs MHG, Wentzcovitch RM (2019) Mass-dependent dynamics of terrestrial exoplanets using ab initio mineral properties. Icarus 317:412–426, https://doi.org/10.1016/j.icarus.2018.08.016

van Lieshout R, Min M, Dominik C (2014) Dusty tails of evaporating exoplanets–I. Constraints on the dust composition. Astron Astrophys 572:A76, https://doi.org/10.1051/0004-6361/201424876

van Lieshout R, Min M, Dominik C, Brogi M, de Graaff T, Hekker S, Kama M, Keller CU, Ridden-Harper A, van Werkhoven TIM (2016) Dusty tails of evaporating exoplanets–II. Physical modelling of the KIC 12557548b Light Curve. Astron Astrophys 596:A32, https://doi.org/10.1051/0004-6361/201629250

von Gehlen K (1992) Sulfur in the Earth's mantle—A review. *In*: Early Organic Evolution: Implications for Mineral and Energy Resources. Schidlowski M, Golubic S, Kimberley MM, McKirdy DM, Trudinger PA (eds) Springer, Berlin, Heidelberg, p 359–366

Wade J, Dyck B, Palin RM, Moore JDP, Smye AJ (2017) The divergent fates of primitive hydrospheric water on Earth and Mars. Nature 552:391–394, https://doi.org/10.1038/nature25031

Wagner FW, Tosi N, Sohl F, Rauer H, Spohn T (2012) Rocky super-Earth interiors. structure and internal dynamics of CoRoT-7b and Kepler-10b. Astron Astrophys 541:A103, https://doi.org/10.1051/0004-6361/201118441

Walton CR, Baziotis I, Černok A, Ferrière L, Asimow PD, Shorttle O, Anand M (2021a) Microtextures in the Chelyabinsk impact breccia reveal the history of phosphorus–olivine-assemblages in chondrites. Meteorit Planet Sci 56:742–766, https://doi.org/10.1111/maps.13648

Walton CR, Shorttle O, Jenner FE, Williams HM, Golden J, Morrison SM, Downs RT, Zerkle A, Hazen RM, Pasek M (2021b) Phosphorus mineral evolution and prebiotic chemistry: From minerals to microbes. Earth Sci Rev 221:103806, https://doi.org/10.1016/j.earscirev.2021.103806

Wang HS, Lineweaver CH, Ireland TR (2018) The elemental abundances (with uncertainties) of the most Earth-like planet. Icarus 299:460–474, https://doi.org/10.1016/j.icarus.2017.08.024

Wang HS, Lineweaver CH, Ireland TR (2019a) The volatility trend of protosolar and terrestrial elemental abundances. Icarus 328:287–305, https://doi.org/10.1016/j.icarus.2019.03.018

Wang HS, Liu F, Ireland TR, Brasser R, Yong D, Lineweaver CH (2019b) Enhanced constraints on the interior composition and structure of terrestrial exoplanets. Mon Not R Astron Soc 482:2222–2233, https://doi.org/10.1093/mnras/sty2749

Wang HS, Quanz SP, Yong D, Liu F, Seidler F, Acuña L, Mojzsis SJ (2022a) Detailed chemical compositions of planet-hosting stars: II. Exploration of the interiors of terrestrial-type exoplanets. Mon Not R Astron Soc, https://doi.org/10.1093/mnras/stac1119

Wang HS, Lineweaver CH, Quanz SP, Mojzsis SJ, Ireland TR, Sossi PA, Seidler F, Morel T (2022b) A model Earth-sized planet in the habitable zone of Centauri A/B. Astrophys J 927:134, https://doi.org/10.3847/1538-4357/ac4e8c

Wänke H, Baddenhausen H, Palme H, Spettel B (1974) On the chemistry of the Allende inclusions and their origin as high temperature condensates. Earth Planet Sci Lett 23:1–7

Wasson JT, Chou C-L (1974) Fractionation of moderately volatile elements in ordinary chondrites. Meteorites 9:69–84

Wasson JT, Kallemeyn GW (1988) Compositions of chondrites. Philos Trans R Soc London Ser A 325:535–544

Watenphul A, Wunder B, Wirth R, Heinrich W (2010) Ammonium-bearing clinopyroxene: A potential nitrogen reservoir in the Earth's mantle. Chem Geol 270:240–248, https://doi.org/10.1016/j.chemgeo.2009.12.003

Watson AJ, Donahue TM, Walker JCG (1981) The dynamics of a rapidly escaping atmosphere: Applications to the evolution of Earth and Venus. Icarus 48:150–166, https://doi.org/10.1016/0019-1035(81)90101-9

Werner SC (2009) The global martian volcanic evolutionary history. Icarus 201:44–68

Whittaker EA, Malik M, Ih J, Kempton EM-R, Mansfield M, Bean JL, Kite ES, Koll DDB, Cronin TW, Hu R (2022) The detectability of rocky planet surface and atmosphere composition with the JWST: The case of LHS 3844b. Astron J 164:258, https://doi.org/10.3847/1538-3881/ac9ab3

Wogan N, Krissansen-Totton J, Catling DC (2020) Abundant atmospheric methane from volcanism on terrestrial planets is unlikely and strengthens the case for methane as a biosignature. Planet Sci J 1:58, https://doi.org/10.3847/PSJ/abb99e

Wolf AS, Jäggi N, Sossi PA, Bower DJ (2022) VapoRock: Thermodynamics of vaporized silicate melts for modeling volcanic outgassing and magma ocean atmospheres. Astrophys J 947:64

Wood BJ, Smythe DJ, Harrison T (2019) The condensation temperatures of the elements: A reappraisal. Am Mineral 104:844–856

Wordsworth RD, Schaefer LK, Fischer RA (2018) Redox evolution via gravitational differentiation on low-mass planets: implications for abiotic oxygen, water loss, and habitability. Astron J 155:195, https://doi.org/10.3847/1538-3881/aab608

Workman RK, Hart SR (2005) Major and trace element composition of the depleted MORB mantle (DMM). Earth Planet Sci Lett 231:53–72

Yamazaki D, Yoshino T, Matsuzaki T, Katsura T, Yoneda A (2009) Texture of (Mg,Fe)SiO$_3$ perovskite and ferropericlase aggregate: Implications for rheology of the lower mantle. Phys Earth Planet Sci 174:138–144, https://doi.org/10.1016/j.pepi.2008.11.002

Yokochi R, Marty B, Chazot G, Burnard P (2009) Nitrogen in peridotite xenoliths: Lithophile behavior and magmatic isotope fractionation. Geochim Cosmochim Acta 73:4843–4861, https://doi.org/10.1016/j.gca.2009.05.054

Young PA, Desch SJ, Anbar AD, et al. (2014) Astrobiological stoichiometry. Astrobiology 14:603–626, https://doi.org/10.1089/ast.2014.1143

Zahnle KJ, Catling DC (2017) The cosmic shoreline: The evidence that escape determines which planets have atmospheres, and what this may mean for Proxima Centauri B. Astrophys J 843:122, https://doi.org/10.3847/1538-4357/aa7846

Zahnle KJ, Gacesa M, Catling DC (2019) Strange messenger: A new history of hydrogen on Earth, as told by Xenon. Geochim Cosmochim Acta 244:56–85, https://doi.org/10.1016/j.gca.2018.09.017

Zieba S, Zilinskas M, Kreidberg L, et al. (2022) K2 and Spitzer Phase curves of the rocky ultra-short-period planet K2-141 b hint at a tenuous rock vapour atmosphere. Astron Astrophys 664:A79, https://doi.org/10.1051/0004-6361/202142912

Zilinskas M, van Buchem CPA, Miguel Y, Louca A, Lupu R, Zieba S, van Westrenen W (2022) Observability of evaporating lava worlds. Astron Astrophys 661:A126, https://doi.org/10.1051/0004-6361/202142984

Reviews in Mineralogy & Geochemistry
Vol. 90 pp. 301–322, 2024
Copyright © Mineralogical Society of America

9

Some Tectonic Concepts Relevant to the Study of Rocky Exoplanets

Keith D. Putirka

Dept. Earth & Env. Sciences
California State University
Fresno, California, 93740 U.S.A.

INTRODUCTION

An Yin of UCLA was invited to write this chapter as he had a near unique approach to understanding tectonic processes. Unfortunately, he passed away on a field trip just a month or two before manuscripts were due. His work was informed by both field studies and theory, which were then applied to very wide-ranging systems, including the Tibetan Plateau (e.g., Murphy et al. 1997; Kapp et al. 2005), low-angle normal faults of the Basin and Range province (Yin 1989) and, in the planetary realm, the tectonics of Enceladus (Yin et al. 2016) and Mars (e.g., Yin and Wang 2023). An Yin had planned to write a chapter on planetary tectonics; the loss is massive and, in this moment at least, irreplaceable.

In this necessarily imperfect substitute we'll examine plate tectonics on Earth—its features and forces—and examine some concepts that may allow astronomers to ask useful questions regarding numeric models that putatively predict tectonic activity. But exo-planetologists should be aware that geologists are still attempting to understand: why does Earth operate as it does, and so much differently than its neighbors? Has it always operated this way and have other planets of the inner Solar System ever mimicked Earth's behavior in their past? These problems are unsolved, though some interesting speculative notions have emerged. Studies by Foley et al. et al. (2012) and Weller and Lenardic (2018), for example, attempt to distill the essential planetary properties that may influence if not dictate possible tectonic states, while Yin et al. (2016) propose a model of planetary tectonic surface features that appears remarkably precise. These studies yield some compelling expedients for analyses of planetary objects both within and outside our Solar System.

An additional underlying theme of this chapter is to address the terms "Terrestrial" or "Earth-like". In the astronomical literature, such a description often means only that a planet is covered in rocks; with such usage "Earth-like" is then synonymous with "Mars-like", or "Mercury-like" or "Moon-like"—even "asteroid-like", all of which are entirely unhelpful. Rocky bodies of the inner Solar System operate quite differently from one another—and Earth is different in the extreme. For instance, to say that Earth is unique by virtue of exhibiting plate tectonics is also to say many other things as well: Earth has large continents, and a bi-modal distribution of rock types and topography (hypsometry); volcanic and seismic activity are concentrated at plate boundaries and earthquakes can occur at very great depths (often to as great as 660 km but now recorded to 750 km, i.e., within the lower mantle; Kiser et al. 2021). Abundant surface water might be not just a signal but a qualification. To say that a planet is "Earth-like" can and perhaps should mean all these things. In isolation, these features might not guarantee that plate tectonics is operative. In combination, though, they either indicate that plate tectonics is active, or otherwise imply that very Earth-like planets can be created by other means.

1529-6466/24/0090-0009$05.00 (print)
1943-2666/24/0090-0009$05.00 (online)   http://dx.doi.org/10.2138/rmg.2024.90.09

SOME DEFINITIONS AND BASIC CONCEPTS

Plate tectonics and its basic parts

The root of "tectonics" is "tektos", meaning "to weave" or "to build", and its use can refer to earthquakes, mountain building or any deformation of a planet's surface. On Earth, tectonics takes on a special—and thus far unique—form, where the rocky surface is broken up into mobile plates and where earthquakes, volcanoes and surface deformation is concentrated at plate boundaries. Plate motions, or "plate tectonics" also control Earth's heat loss, as well as the global water and carbon cycles that affect climate and habitability (e.g., Foley and Driscoll 2016). Several other kinds of "tectonics" are also recognized, which will also be noted.

In plate tectonic theory, the mobile plates are made of *lithosphere* (literally "rocky sphere") that deform largely by brittle processes, at least at their shallowest depths (with ductile mechanisms occurring in the deeper, hotter parts). The lithospheric plates float on top of a weak layer called the "*asthenosphere*". The lithosphere can be defined in multiple ways: here, we will mostly refer to lithosphere as the "conductive lid", where heat loss is via conduction, as opposed to convection. But the lithosphere can also be defined by its seismic properties or its elastic strength, or even by the length of time that it has been physically isolated from the convecting mantle, and such definitions can lead to different estimates of its thickness. The lithosphere can be treated as a single unit based on its thermal or mechanical properties, but almost always consists of two parts in terms of composition: a crust, or top layer, that is usually richer in SiO_2 and poorer in MgO, and an underlying mantle layer that has lower SiO_2 and higher MgO, among other relative qualities. A key issue is that while the crust is compositionally distinct, the lithosphere could be similar in composition to the underlying convective mantle, though it need not be (Rychert et al. 2020). Ancient lithosphere beneath very old continents may also be quite compositionally distinct, especially with respect to its trace element and isotopic composition (Rudnick et al. 1998). The asthenosphere is a part of the convective mantle, and is weak because it is either slightly hydrated or is very close to or slightly above its solidus temperature (Rychert et al. 2020), but as will be discussed, the overlying lithospheric plates are not carried along in conveyor-belt fashion by convective currents. Convection currents are dominated by thermal upwellings called "*mantle plumes*" and downwelling currents called "*subduction zones*" (Davies and Richards 1992). Mantle plumes emanate from the core-mantle boundary, where they obtain their excess heat; they intersect the surface at random locations relative to plate boundaries. The downwelling limbs, i.e., subduction zones are considered the primary driver for mantle convection (Conrad and Lithgow-Bertelloni 2004) and are defined by a special type of plate boundary: where two plates collide, the denser one is "*subducted*" or thrust beneath the less dense plate, possibly sinking as far as the core-mantle boundary (Fig. 1). Crust and lithosphere are thus destroyed at subduction zones, but new crust is created at spreading ridges, where two plates move away from one another. Convective currents, however, have no relation to spreading ridges, except for the case when a thermal plume accidentally rises beneath a spreading ridge, as at Iceland, and the Galapagos Islands (Fig. 1). Two kinds of plate boundaries have now been described: spreading ridges and subduction zones. Examples of a spreading ridge are the Mid-Atlantic Ridge and the East Pacific Rise. A third type of plate boundary is a *transform boundary*, where two plates slide past one another; the San Andreas fault is a well-known example, but most transform faults occur in the ocean basins, such as the Romanche Fracture Zone; transform boundaries connect offset segments of spreading ridges. Subduction zones are of special interest because these are the sites where continental crust—another hallmark feature of Earth—is created. As a subducted slab sinks into the mantle it releases water that was absorbed by the rocks when it was once close to the surface. This water is released into the mantle wedge, where it lowers the melting temperature of the mantle, which then leads to the creation of a "*volcanic arc*" that occurs on top of the overriding plate (Grove et al. 2012). Earth's largest and deepest earthquakes also occur at subduction zones; the zone of earthquakes is called the Wadati–Benioff zone. The Japanese Islands are an example of an "oceanic arc"—a volcanic arc formed when one oceanic plate (usually older, colder and denser) subducts beneath another oceanic plate (usually younger, hotter and more buoyant). The Cascades are an example of a "continental arc", when an oceanic plate (denser) subducts beneath a continental plate (more buoyant, in part because it contains less dense continental crust). The Tibetan Plateau is also a site of a plate collision. The large plateau represents the collision of two continental plates, India and Eurasia; both plates are buoyant, so there is only minor subduction, and much uplift, and even some volcanic activity, as well as large earthquakes.

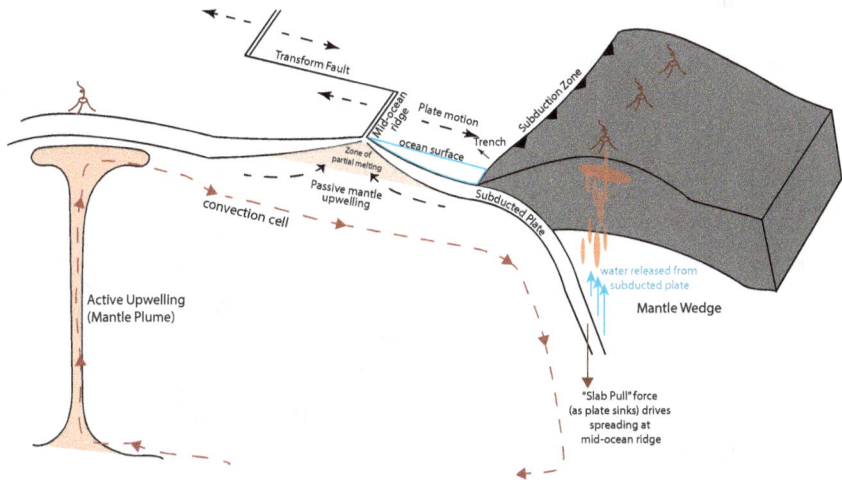

Figure 1. Illustration of plate tectonic features and mantle convection on Earth. Mid-ocean spreading ridges and the transform faults that offset such as are shown, as is a subduction zone, the dewatering of the subducted slab that drives partial melting of the mantle wedge, and the volcanic arc that is produced as a result. Also shown is a hot upwelling plume that is rooted at the core/mantle boundary (the only significant thermal boundary layer below Earth's surface) that also drives partial melting in the mantle and volcanism at the surface (at so-called "hot spots"). Note that convection is dominated by plume upwellings and down-welling tectonic plates; this whole-mantle flow has no connection to spreading ridges. Spreading ridges are a result of passive upwelling of the mantle as the plates are ripped apart by far-field forces, and are not connected to the deeper, hotter mantle. See Silver et al. (1988) and Davies and Richards (1992).

The relative motions of the plates control seismic and volcanic activity and the location of deep valleys and tall mountain ranges. New crust is created at **spreading ridges**: these are mostly oceanic, such as the East Pacific Rise, but spreading ridges can split continents (the Red Sea Rift is an example). Where two plates are pulled apart, the mantle below "passively" upwells to fill the gap (McKenzie 1967; McKenzie and Bickle 1988). We use the term "passive" to distinguish this from the active, thermally-driven upwelling that drives the formation of mantle plumes at the core/mantle boundary (Morgan 1971). The passive mantle upwelling at spreading ridges causes the mantle to partially melt, and form the volcanic rock type called basalt (Langmuir et al. 1992). That basalt layer, once formed, has a constant thickness as it moves away from the ridge, but some of the underlying mantle that was partially melted may now be less dense by virtue of having some its melt extracted (Kay and Kay 1993), and this mantle residue can become a part of the oceanic lithosphere; it is made of a rock called peridotite. The entire package of basalt crust + peridotite mantle comprises the lithosphere. The lithosphere will cool and thicken at its base (effectively increasing the mantle portion) as it moves away from a spreading ridge. What we call the "lithosphere", then, consists of both this newly formed, constant thickness crust, as well as the cooler parts of the underlying mantle that are no longer part of the deeper convective mantle. If the lithosphere thickens and cools sufficiently, and collides with less dense lithosphere, it can be subducted back into the mantle.

As to the physical properties of the tectonic plates, oceanic crust has a thickness of about 7 km and a density of 2.9 g/cm^3, and it overlies a cold, brittle mantle lithosphere that is about 3.35 g/cm^3 in density. For the continents, crustal density may be closer to 2.7 g/cm^3, and the mantle beneath it may be similar to sub-oceanic lithosphere (3.35 g/cm^3) or lower, if it is hydrated (see Kay and Kay 1993). Naturally enough, then, if oceanic and continental lithosphere collide, the denser oceanic lithosphere should sink. If two oceanic plates collide, the older (colder) of the two should subduct. The removal of partial melt from the mantle (to form crust) could also lead to a reduction in the density of the residual mantle, which could also then help isolate such residuum from the convective mantle (Kay and Kay 1993). Interestingly, though, once subduction is initiated, the downward pull of the dense, subducting slab exerts the greatest force acting on tectonic plates. This "slab pull" force is controlled by the sinking of cooler (and so denser) subducted lithosphere, facilitated by the conversion of lower-density basalt (ca. 2.8 g/cm^3) to higher-density *eclogite* (ca. 3.5–3.8 g/cm^3, made of clinopyroxene and garnet). Basalt only transforms to eclogite, though, after it has been subducted to some considerable depth (ca. 40–80 km; Stern 2002). This implies that thermal buoyancy is critical to initiate subduction. Stern (2004) posits an intervening phase of "hinged subduction", where cool lithosphere dips into the mantle.

In any case, as the lithosphere moves away from an oceanic spreading ridge, it cools, and if carried far enough from the ridge, the base of the lithosphere can be as deep as 100 km or more. Continental lithosphere could be even thicker; some estimates are as high as high as 250–400 km, for very old, cold lithosphere that has been isolated from the convecting mantle for billions of years (Pollack and Chapman 1977; Jordan 1978; Steinberger and Becker 2016), although modeling of surface stresses appears to only allow lithosphere that is mostly 100 km thick on average (Naliboff et al. 2012). The overlying continental crust is just a small portion of this, averaging 30 km in thickness, but ranging to 40–60 km in some places, such as the Tibetan Plateau. Being buoyant, this crust and it's underlying lithosphere can be resistant to subduction.

The alternatives: Delamination, stagnant lid (or plume lid or drip) and episodic lid tectonics

A *stagnant lid* "mode" is the case when planets do not exhibit a mobile lid ("mobile lid" being a synonym for plate tectonics). The lithosphere, in this case, acts as a single, globe-encircling plate. With no plate boundaries, earthquakes are few and volcanism and seismic activity are not concentrated in regional bands. Mercury, Moon and Mars are prime examples (Tosi and Padovan 2021). A current stagnant lid regime, though, does not preclude that plate tectonics operated during some earlier era.

A stagnant lid regime also does not preclude surface deformation, or the recycling of some of the lower crust or upper mantle back into a planetary interior. Bird (1979) coined the term *"delamination"* to describe the process whereby the cold mantle beneath the continents may be sufficiently dense to drip back into the mantle. Deep crustal rocks could also delaminate, if they recrystallize at high pressures to form eclogite (Fig. 2a; see also Kay and Kay 1993). Ducea and Saleeby (1996) and Saleeby et al. (2003) demonstrated the reality of such processes beneath the Sierra Nevada mountain range, which led to this new form of tectonic activity, involving gravitational instabilities, to be widely accepted. Building on work by Kay and Kay (1993) and others, Ducea and Saleeby (1996) and Saleeby et al. (2003) also show how delamination can occur as a natural consequence of crustal growth at an arc, as the roots of arc volcanoes can contain very dense mineral assemblages that are gravitationally unstable. Lee and Anderson (2015) coined the term "arclogite" to describe such dense rocks (see also Ducea et al. 2021), and De Celles et al. (2009) argue that the process is repeatable and perhaps cyclic.

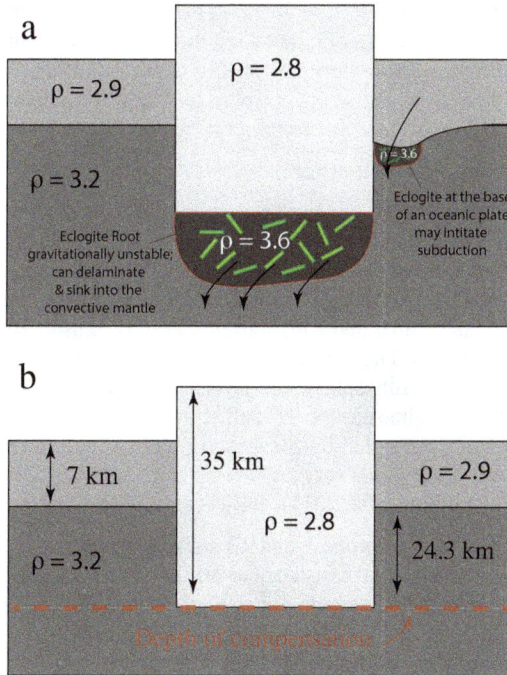

Figure 2. (a) Delamination and subduction: at the base of either the continental or oceanic crust, the rock type eclogite ($r = 3.6$ g/cm³) may form. Being more dense than the mantle below, an eclogitic root beneath a continent may be delaminated into the mantle (a form of "drip tectonics"). At the base of oceanic crust, which is already more dense than continental crust, the eclogite drip may initiate subduction, where the entire plate is pulled into the mantle (Fig. 1). **(b)** Example of isostacy. Low density continental crust ($r = 2.8$ g/cm³) will have a "root" that extends to greater depths than adjacent oceanic crust ($r = 2.9$ g/cm³) of higher density. The depth of that root is the depth at which lithostatic pressures (like hydrostatic, pressures are equal in all directions) are equal, which is called the "depth of compensation".

More importantly to the study of exoplanets are "lithospheric drips". These might provide a means to initiate subduction (e.g., Adam et al. 2021b) on an otherwise stagnant lid planet but also may be the primary means of convection when lids remain stagnant. The crusts on smaller planets might not reach sufficiently high pressures to form eclogite; Batra and Foley (2022) quantify such effects and the consequences for planetary cooling. On a stagnant lid planet, delamination, if it is mechanically possible, may provide the only means by which crustal materials can be recycled back into the mantle. This kind of return flow, and the surface deformation that ensues, has been referred to as "stagnant lid-", "plume lid-" or "drip-" tectonics (e.g., Fischer and Gerya 2016; Stern et al. 2018; McMillan and Schoenbohm 2022), but all involve the same fundamental process of delamination. Yet another variation of activity, perhaps not properly considered "tectonic", is the "heat pipe" model (Moore 2001), an idea that derives from the study of Io, a moon of Jupiter, where volcanism occurs absent large-scale mantle convection. Heat pipes might be restricted to planets with tidal heating and so would be a factor in exoplanets that are in close orbit to their stars (Jackson et al. 2008). An example would be the Trappist-1 system (Barr et al. 2018). In any case the heat pipe model will be ignored further here.

A variation on the stagnant lid regime derives from ideas about tectonism at Venus. To explain the very young surface there (< 500 Ma), Turcotte (1993) applied the term "plate tectonics" but described something different, i.e., that the Venusian crust was subject to catastrophic replacement, as the crust sinks into the mantle and the planet undergoes volcanic re-surfacing. Venus nominally exhibits a stagnant lid between these catastrophic episodes. The resurfacing involves a type of subduction, but something that is still quite different from the stable, arc-like subduction zones on Earth. Modeling of the Venusian processes (Moresi and Solomatov 1998) would lead to the term "episodic lid" (e.g., Nakagawa and Tackely 2015; Tian et al. 2023) to describe this mode of tectonics.

Why one tectonic mode and not another?

We don't know. The idea of a stagnant lid goes back at least as far as the early 1990s, as a result of experiments that simulate mantle convection (Giannandrea and Christensen 1993). In such experiments (though not necessarily in Nature), a stagnant lid regime results when colder surface layers are more than 10^3–10^4 times greater in viscosity compared to underlying, convective materials (Giannandrea and Christensen 1993; Stevenson 2003). With very large viscosity contrasts, a stagnant lid develops on top of a convecting layer; with lower viscosity contrasts, or in effect, a weaker lithosphere, the surface layers become a part of the convection process, and so a mobile lid regime ensues. This all seems simple enough, except that as shown by Solomatov (1995), and later noted by Stevenson (2003), Earth is an anomaly relative to such experimental results: Earth has a very large viscosity contrast between its lithosphere and asthenosphere (ca. 10^9; Doglioni et al. 2011), and yet still exhibits plate tectonics.

Moresi and Solomatov (1998) propose that it's not so much the viscosity of the lithosphere that matters as much as the frictional resistance to brittle failure that dictates when and where plate tectonics happens. We've long known, for example, that water can greatly reduce friction along fault surfaces (Hubbert and Rubey 1959), making such faults much weaker than their drier counterparts. The Hubbert and Rubey (1959) model would explain, at least qualitatively, why plate tectonics occurs on Earth, where water in the crust and lithosphere is abundant, but not on Venus, where the Venusian crust and lithosphere are dry. However, as noted by Bercovici et al. (2015) and emphasized by Foley (2018), the predicted lithosphere yield stresses in current numerical models (e.g., Moresi and Solomatov 1998; Tackely et al. 2000) require yield stresses that are orders of magnitude lower than observed (e.g., Zhong and Watts 2013) so as to allow plate tectonics, at least if the unidentified units employed by such models are in Pascals. It remains a challenge to bring observations and predictions of stress into agreement.

Yet another challenge is for numerical models to predict the narrow width of deformation zones. Wakabayashi (2021), for example, shows that in the Franciscan Formation of California (an archetype of an exhumed subduction zone), field and microstructural evidence indicate a zone of deformation that is a mere 300 m wide, and that this narrow deformation interval applies over a depth range of 10 to 80 km in the paleo-subduction zone. However, even very recent models appear challenged to obtain a plate boundary that is realistically narrow. Saxena et al. (2023), for example, appear to recognize that deformation zones are narrow, as they prescribe "discrete weak zones" within their numerical model, but their zones of weakness are still hundreds of km in width, orders of magnitude greater than the actual deformation widths that occur on Earth.

The effective rheological contrasts that control stagnant/mobile lid behavior also depend on gravity, and whether partial melt is present in the convecting layer, and there might also be a stochastic dependency—noted by Stevenson (2003) as "the vagaries of history" or by Lenardic and Crowley (2012) as "historical dependence". The vagaries are not entirely vague, though. Rock strength can depend upon strain history, as anyone who has tried straightening a bent metal wire can attest (when rocks are strained, or wires bent, microscopic dislocations are created to accommodate the strain, but these dislocations can later interfere with one another adding to intrinsic strength). Volcanic activity can also dehydrate a planet's interior (drier rocks are stronger) and rock strength can be affected by recrystallization and crystal size (Evans and Kohlstedt 1993). Lenardic and Crowley (2012) thus illustrate how different models, using very similar inputs, can yield different results regarding whether or not a planet exhibits plate tectonics.

In yet another approach to explaining plate tectonics, Hoink et al. (2011) suggest that the stagnant vs. mobile lid regimes are controlled not so much by the properties of the lithosphere and underlying bulk mantle, but instead depend upon the thickness of a low viscosity "channel" that directly underlies the lithosphere. If this channel is sufficiently thin then, presumably, plate tectonics, or a mobile lid regime is sustained, whereas thick channels promote a stagnant lid regime. However, this model predicts strong viscous coupling for some plates, such as at the Mid-Atlantic Ridge, despite both long-standing and recent geophysical observations to the contrary: plate motions show that slab pull accounts for 90% of the forces that drive plate tectonics, with ridge push accounting for most of the remainder, and with viscous forces being trivial (Chapple and Tullis 1977; Hager and O'Connell 1981; Lithgow-Bertelloni and Richards 1998; Conrad and Lithgow-Bertelloni 2004; Saxena et al. 2022). Conrad and Lithgow-Bertelloni (2004) do indeed describe a viscous suction force that can act on the deeper parts of subducted plates, but their viscous suction is another form of slab pull and is not an argument for viscous coupling at shallower depths. Clennett et al. (2023) propose that viscous forces might have been more important in Earth's past, but also concede that uncertainties in plate reconstructions are large enough to perhaps disallow the possibility. That lithospheric plates might be riding upon mantle convection cells is an alluring idea—but there is no evidence that the idea applies to Earth.

Finally, while mantle convection plays a key role in the modeling of plate tectonics, some questions remain about the operational aspects. Mantle plumes are the upwelling complement to subduction, and are thought to be thermally driven, with their excess heat derived from the core-mantle boundary (Davies and Richards (1992). But Frazer and Korenaga (2022) suggest that Earth's mantle plumes are much too wide (by about 400%) compared to what is predicted in numerical models. They conclude that either thermal convection models are wrong or that convection is driven more by compositional, rather than thermal contrasts. The latter result of compositionally-driven convection, though, would upend our views of how plumes nucleate and how Earth's core loses heat.

In sum, we have a mismatch between observations of plates and plumes on the one hand and numerical models that attempt to predict these on the other. And so we find ourselves at a loss to precisely explain plate tectonics and mantle convection, on a planet where observations and relevant experiments are abundant.

PLATE TECTONIC DRIVING FORCES

Several developments in the late 20th century helped to solidify our understanding of the mechanical aspects of plate tectonics. One of the most important was by McKenzie (1967) who showed that heat flow and bathymetry measurements at mid-ocean spreading ridges require no source of excess heat in the underlying mantle. Mid-ocean ridges are a product of "passive upwelling" of the mantle, where two plates are being pulled apart due to far-field forces and the mantle passively upwells to fill the gap (Fig. 1). Mid-ocean ridges (MORs) are bathymetrically high relative to surrounding ocean floor because cold, dense lithosphere is replaced by warm and buoyant mantle. That buoyant mantle provides heat sufficient to explain heat flow measurements. But MORs are not so high so as to require or allow excessively hot, buoyant mantle material beneath them (McKenzie 1967). Passive upwelling means that at a given depth (ca. 100 km) below the surface, temperatures are more or less uniform under the ocean basins, being no higher beneath a spreading ridge than laterally adjacent regions. The mantle is by no means homogeneous with respect to density and temperature (Langmuir et al. 1992; Adam et al. 2021a), but the idea of passive upwelling fits well with the cooling model for Earth's oceanic lithosphere. For the case of conductive cooling, the depth to the base of an oceanic plate, d, can be described as a square root relationship with the age of the plate, t:

$$d = \sqrt{\kappa t} \tag{1}$$

where κ is thermal diffusivity; t is determined by age dating of oceanic rocks, but can be also calculated as $t = L/v$, where L is the length of a plate, and v is plate velocity. The bathymetry of the ocean basins, at least to $t = 90$ Ma, is proportional to the square root of the age of the overlying oceanic crust, as predicted by Equation (1) (Parsons and Sclater 1977). At ages >90 Ma, the ocean floor bathymetry is shallower than predicted by Equation (1), and can be explained by small-scale convection in a narrow boundary layer beneath the oceanic lithosphere. This long-standing view has been tested repeatedly and is consistent with inferred temperature profiles of the lithosphere and observations of gravity (Morgan and Smith 1992; Crosby et al. 2006).

These observations, and modeling of such, show that tectonic plates are not driven conveyor-like on mantle convection currents, but rather are controlled by other forces. Forsyth and Uyeda (1975) presented an analysis of plate-driving forces that are still accepted today, namely ridge push, viscous drag and slab pull, with the slab pull force being the most important and viscous drag the most minor. Empirical observations (e.g., Jarrard 1986) and at least some numerical models that are specifically designed to predict modern plate motions (e.g., Conrad and Lithgow-Bertelloni 2002, 2004; Saxena et al. 2023) confirm that analysis, showing that the slab-pull force, which is generated by a subducted plate's negative buoyancy, is as great as 60%–70% of the total forces acting on a given plate. The fraction of the slab pull force increases with decreasing mantle viscosity, and the remainder is still a type of slab pull, but referred to as "slab-suction" (Conrad and Lithgow-Bertelloni 2002, 2004). The suction force is derived from mantle flow around the sinking plate, where that flow can exert some shear along the plate that exerts an additional downward force. The suction force is nominally more effective in the deeper mantle where mantle viscosity is greater. Ridge push comprises < 10% of the total force acting on plates and viscous coupling is effectively zero. These particular models are consonant with seismic observations: Ito et al. (2014) show that there is only a weak coupling of the lithosphere with underlying mantle plume beneath Hawaii, and the Jadamec (2016) analysis of mantle anisotropy shows rather complex mantle flow that is de-coupled from subducted slabs, at least to depths of 100 km.

Davies and Richards (1992) describe the issue well, noting that if mantle upwelling were the driving force at mid-ocean ridges, then all submarine ridges would look more like Iceland, with much steeper and shallower bathymetry than observed, as well as higher heat flow (McKenzie 1967) and a positive geoid anomaly (Sleep 1990; the "geoid" is the shape the Earth would have if it were covered in water, or any other fluid that lacked shear strength; above mantle plumes, as at Hawaii or Iceland, there is a positive geoid anomaly in that Earth's shape is distended upwards above the rising jet of mantle material). The idea of plates being moved about on the top of convective mantle rolls is limited to dated textbooks and some computer models; it is not a feature of the modern Earth.

Convection in the mantle, then, appears to be dominated by large plumes that upwell from the base of the mantle, receiving their excess temperatures from the core-mantle boundary (CMB), and sinking convective limbs that are defined by the subducting plates, much as illustrated in Silver et al. (1988) and Davies and Richards (1992). These plumes have no spatial correlation to the mid-ocean ridges; their surface manifestations are the so-called volcanic hot spots, such as Hawaii, or Yellowstone, where the excess heat drives partial melting and volcanic activity. Mantle plumes might, by accident, intersect a mid-ocean ridge, as appears to happen at Iceland or Galapagos, and can create thicker crust in such areas. But these are accidents that do not bear directly upon plate motion. The mantle plume theory is an essential addition to plate tectonic theory in that it explains volcanic and seismic activity that occurs far from plate boundaries, and so is otherwise anomalous in plate tectonic theory. But mantle plumes and plate tectonics are not unrelated, at least on Earth.

Buoyancy forces, rock types, and isostacy

Mantle circulation and plate tectonics are largely driven by buoyancy forces. Rocks can differ in composition, temperature or both, affecting their bulk density. Plate motions and mantle circulation can ensue when the resulting buoyancy forces are sufficient to overcome ambient rock strength. The Rayleigh number, Ra, is one measure to assess whether a planetary interior might convect. There are many different variations; this equation is from Stevenson (2003):

$$\text{Ra} = \frac{g\alpha\Delta T d^3}{\nu\kappa} \qquad (2)$$

where g is the acceleration due to gravity, α is the coefficient of thermal expansion, ΔT is the temperature drop across some depth interval d, ν is the kinematic viscosity and κ is thermal diffusivity. When Ra > 1000 the buoyancy forces in the numerator sufficiently exceed the resistive forces in the denominator and the system is likely to convect. As a planet cools, ΔT will decrease, and when convection stops (Ra < 1000), heat loss would then be dominated by conduction.

Stevenson (2003) interestingly shows, however, that smaller planets can apparently cool just as slowly as their larger counterparts—maintaining a lower T through their cooling history (Stevenson, 2003). Basalt compositions from Moon and Mars appear to confirm this view (Putirka 2016). In any case, Stevenson (2003) is largely concerned with convection as a mode of heat flow, noting that planets may instead lose heat via conduction and melt migration if very small (Moon and possibly Mercury), or via heat pipes and magma oceans if very hot (early Earth), and will lose heat by convection otherwise.

We are uncertain of the extent to which density contrasts in Earth's mantle are driven by compositional or thermal contrasts, as seismic velocities within Earth's interior can be explained by either. Adam et al. (2021a) indicate that known thermal contrasts of ca. 300 °C (as derived from thermometry of volcanic rocks at the surface) may explain most density contrasts within Earth, which would then allow for a well-mixed mantle that might be homogenous at certain spatial scales. But compositional contrasts clearly dominate near Earth's surface and appear to drive plate tectonics. There are many different rock types that drive such differences, but two, granite and basalt, are of particular importance as the former occurs only in great abundances on continents while the latter dominates the ocean basins. As noted by Campbell and Taylor (1983), only Earth contains multiple, large "batholithic masses" of granite (a recent discovery indicates a possible batholithic mass on Moon; Siegler et al. 2023).

Some definitions are needed here. Campbell and Taylor (1983) refer to "granite", *sensu lato*, i.e., any rock that is dominated by the minerals quartz (Qz; SiO_2), albite (Ab; $NaAlSi_3O_8$), orthoclase (Or; $KAlSi_3O_8$) and anorthite (An; $CaAl_2Si_2O_8$). Granitic rocks cool slowly enough that the crystals tend to be large and easily identifiable. The latter three minerals, Ab, Or and An, are "feldspars" and form two different solid solution series: the alkali feldspars (Afs), which are a mixture of Ab–Or (Na and K are close in size and have a similar charge so mix relatively easily) and the plagioclase feldspars (Plag), which are a mixture of Ab–An (Na^{1+} and Ca^{2+} are similar enough in size but differ in their charge, so they can mix together via a coupled substitution: $Na^{1+}Si^{4+} = Ca^{2+}Al^{3+}$). The larger K^+ and smaller Ca^{2+} ions are sufficiently different in size so as to preclude mixing between these two series. Other minerals are often present in granite, especially hornblende and biotite, but Qz, Afs and Plag dominate (>90%). "Granite", *sensu stricto*, represents very specific limits of the relative abundances of these minerals; when they are normalized to 100%, Qz ranges between 40–60 %, Afs is generally greater than 10%, and the remainder is Plag (Le Bas and Streckeisen 1991). In this chapter, we will consider the case of "granite" *sensu lato*, unless specified otherwise. Basalt, on the other hand, is a volcanic rock that cools quickly—sometimes so quickly that much of the rock consists of glass. But usually basalt is not entirely glass, and the minerals olivine, clinopyroxene and

An-rich feldspar are very common. Chemically, granite is rich in SiO_2 and low in MgO, while basalt has the opposite characteristics. The differences in composition and mineralogy lead to differences in density: granite and basalt have ranges in density of ca. 2.6–2.7 and 2.8–3.0 g/cm^3 respectively.

As abundant as Earth's continental crust is relative to our planetary neighbors, it still represents just 1/3 of Earth's surface. The remaining 2/3 of Earth's crust lies beneath the oceans and is made of basalt, which is created by partial melting of Earth's peridotite mantle (Langmuir et al. 1992). But while a 2/3 coverage area may seem impressive, basalt dominates to an even greater extent the surfaces of other planets, comprising effectively 100% of the crusts of Moon, Mars and Mercury, and the parent bodies of all non-chondritic meteorites. It probably also covers much of Venus as well, though the surface of Venus is still largely unknown.

In contrast to basalt, granite has a more complex origin that appears to involve water— and plate tectonics (Grove et al. 2012). Campbell and Taylor (1983) note that some of our neighboring planets might have small occurrences of granite, but none have granite in "batholithic masses" i.e., the occurrences are not so large so as to define a large mountain range let alone a continent. Again, some definitions: an individual blob of granite, perhaps a few km in diameter or smaller, is called a ***pluton***; in many places, such as the Sierra Nevada mountain range of California, many thousands of plutons of similar age are emplaced, one against the next, to form a near continuous belt of granitic rocks that dominate a mountain range of hundreds of km in extent; these collected masses of granite form a ***batholith***. The tectonic significance is that granitic batholiths appear to form beneath continental volcanic arcs (Fig. 1), and so are intimately related to subduction. The association between subduction and granitic batholiths is related to the global water cycle (see Grove et al. 2012): (a) near a mid-ocean spreading ridge, water is able to penetrate the basaltic crust and into the peridotite mantle below, creating a wide range of hydrous minerals; (b) as the hydrated lithosphere (basalt crust + underlying peridotite lithosphere) is subducted, many of the hydrous minerals break down into anhydrous minerals, releasing water to the mantle wedge (Fig. 1), which also consists of peridotite; (c) the water released to the mantle wedge decreases the melting point of peridotite and partial melting ensues; (d) the melts so produced are water-rich and and can be enriched in Si, and as these melts rise upwards, they precipitate amphibole, among other phases; (e) experiments (Sisson et al. 2005) show that the precipitation of amphibole (SiO_2-poor) yields residual liquids (magmas) that are further enriched in SiO_2, producing granite. Campbell and Taylor (1983) hypothesize that occurrences of surface water, batholiths and plate tectonics are so intimately linked that all these might have initiated on Earth together or in close temporal association. Foley and Driscoll (2016) take such connections further, showing that tectonic activity on Earth has other cascading effects, in yielding a global C-cycle and extended cooling of the core so as to respectively influence Earth's climate and perpetuate a core dynamo.

Compositional (density) contrasts also explain Earth's rather unique topography: Earth's continents sit at higher elevations than the ocean basins, giving Earth a distinctly bi-modal topographic distribution. These elevation differences can be modeled via the concept of isostacy (Fig. 2b), where the oceanic and continental crust (and its underlying rocky mantle lithosphere) floats upon a plastically deformable part of the upper mantle (asthenosphere). The asthenosphere has a lower viscosity than the lithosphere, which means that any lateral differences in pressure within the asthenosphere can be alleviated by lateral flow. The pressure at any depth within Earth is close to $P = \rho g h$, where P is pressure, g is the acceleration due to gravity and h is depth. If the mean ρ_1 for one column of rock differs from the mean ρ_2 for an adjacent column of rock, the relative heights, h_1 and h_2 of the two columns, can adjust until the pressure at the "depth of compensation" (sometimes, but not necessarily within the asthenosphere) is equalized, so that $P_1 = P_2$. Columns with lower mean ρ will rise to higher elevations compared to those with higher average ρ—provided that the effective buoyancy forces exceed ambient rock strength— and in the process yields what is known as isostatic equilibrium.

How the mantle flows and rocks break

To model plate tectonics, we study mantle circulation and lithospheric strength. The bulk properties (yield strength, viscosity) are obtained from experiments on rocks and minerals. The experimental conditions could include a fixed differential stress and strain rate. The differential stress, σ, is determined from the difference between the maximum (usually noted as σ_1) and minimum (usually noted as σ_3) principal stresses (so $\sigma = \sigma_3 - \sigma_1$) that are applied in a given experiment.

To understand mantle convection, i.e., plastic deformation within the interior of a planet, we can use equations that involve strain rate ($\dot{\varepsilon}$), which measures how rapidly a rock is deformed. For example, if a rock has an initial length of l_0 and a final length of l_f, then its total strain is $\varepsilon = (l_f - l_0)/l_0$. If that strain occurs over a time, t, then the strain rate is $\dot{\varepsilon} = \varepsilon/t$. Various "constitutive laws" describe how strain rate is related to differential stress, grain size, and other variables. For example, if deformation in a rock is controlled by diffusion (so the dominant means of plastic deformation involves the migration of "vacancies", which are unoccupied atomic sites in crystals), then a constitutive law, from Evans and Kohlstedt (1993), is:

$$\dot{\varepsilon} = A\frac{\sigma^n V}{RT}\frac{D}{d^m} \tag{3}$$

where R is the gas constant, V is molar volume, and T is temperature; σ is the differential stress as just defined, and its exponent, n, is nominally close to 1 for the case of diffusion; d is the average grain size and its exponent, m, is nominally equal to 3, if diffusion is fastest along grain boundaries, but is equal to 2 if diffusion is fastest through the grain itself; D is the diffusion coefficient, and both D and the pre-exponential coefficient A can depend upon temperature (T), pressure (P) and water content, which in some forms of the equation are explicitly shown. In Evans and Kohlstedt (1993) the values for A, n and m for diffusion creep respectively vary from 10^{-3} MPa^{-n}/s, 1.1 and 3.0 for dunite (a rock made solely of the mineral olivine) to $10^{4.9}$ MPa^{-n}/s, 1.7 and 1.7 for calcite. Earth's mantle consists of ca. 60% olivine ($[Mg,Fe]SiO_4$), and calcite ($CaCO_3$) is thought to be absent, so mantle deformation should be better described by the first set of values. Another mode of deformation is called dislocation creep, where plastic deformation is accommodated by the migration of dislocations (in effect, micro-faults at the atomic scale); a general constitutive law for dislocation (or "power law") creep, adapted from Evans and Kohlstedt (1993), can be written as:

$$\dot{\varepsilon} = A\sigma^n \exp\left(\frac{-Q}{RT}\right) \tag{4}$$

where R is the gas constant, T is temperature and Q is an activation energy; for dislocation creep, the exponent n on the σ term is generally > 1, and its value may denote a specific style of dislocation creep.

Evans and Kohlstedt (1993) tabulate experimental values of A, n and Q for rocks that are dominated by particular minerals (see Table 1). The mantle is dominated by olivine but also contains up to 20% or more each of orthopyroxene (Opx; $[Mg,Fe]_2Si_2O_6$) and clinopyroxene (Cpx; $Ca[Mg,Fe]Si_2O_6$). Some mantle rocks approach nearly 100% pyroxene, and pyroxenes have different strengths compared to olivine (Table 1). These differences in strength can depend on temperature. For example, Yamamoto et al. (2008) and Bystricky et al. (2016) both find that at low T, Opx is stronger than Ol, and Bystricky and Mackwell (2001) find that Cpx is stronger than olivine, so long as water is absent. A fascinating study by Hansen and Warren (2015) appears to at least qualitatively substantiate these experimental findings in a natural setting, where pyroxene-rich rock samples deformed at ca. 1000°C are 1.2 to 3.3 times more viscous than olivine-rich samples. However, Bystricky et al. (2016) note that the contrasts in the strengths of olivine and pyroxenes may decrease with increased T. If valid, this would imply that shifts in mantle mineralogy are less important for understanding mantle convection than for lithosphere (tectonic plate) strength.

Table 1. Power law creep constants for various minerals and rock types.

Mineral	A (MPa^{-n}/s)	n	Q (kJ/mole)	Source
Olivine	10^{-1}–$10^{5.4}$	2.1–5.1	226–540	E&K93
Clinopyroxene	$10^{10.8}$	4.7±0.2	760	B&M
Orthopyroxene	$10^{8.63}$	2.8–3.1	583–621	B2016
Quartz – dry	$10^{-11.2}$–$10^{-2.9}$	1.9–11	51–377	E&K16
Quartz – wet	$10^{-9.4}$–$10^{-1.4}$	1.4–4	134–230	E&K93
Calcite	$10^{-3.6}$–10^{8}	3.3–8.3	190–427	E&K93

Notes: E&K93 = Evans and Kohlstedt (1993), B2016 = Bystricky et al. (2016), and B&M = Bystricky and Mackwell (2001)

The very wide-ranging values reported by Evans and Kohlstedt (1993) are not just a product of experimental uncertainty: they reflect varying experimental conditions, which include contrasts in crystal growth rates and final grain size, porosity, pressure, temperature, the presence or absence of water or melt (if the experiments are conducted at high T), and the maximum value of σ_1. Recent experiments (Hansen et al. 2019), for example, show that the strength of olivine increases with decreasing grain size and that there is also significant strain hardening (as dislocations develop to a sufficient degree so as to interfere with one another, thus requiring greater stresses to induce deformation). If such potential variations are not, in and of themselves, sufficient to give one pause, imagine the havoc that ensues if any exoplanet might trap significant amounts of CO_2 so as to stabilize calcite in their otherwise olivine+pyroxene-dominated mantle. It should also be noted that deformation experiments are conducted at strain rates that are orders of magnitude faster than what occur in Earth's interior.

Consider also that olivine is not stable below depths of about 400 km, as it transitions into the minerals β-phase spinel (having the same composition as olivine but different structure), wadsleyite and ringwoodite. Olivine could be critical to understanding the brittle behavior of tectonic plates themselves, but we are only just now beginning to understand the physical properties of the remaining 2500 km of Earth's 2980 km-deep mantle. New experiments by Fei et al. (2023), for example, indicate that the lower-mantle mineral, bridgmanite, grows at a faster rate, and so produces larger crystals than other minerals that occur at mid-mantle depths (ca. 800–1,200 km). Fei et al. (2023) also find that, despite other grain-sized trends to the contrary, this particular trend to larger crystals might explain a 10^1–$10^{1.3}$ factor of viscosity increase below 800 km, close to inferred values (Rudolph et al. 2015).

Brittle deformation and the plate boundary

The deformation described by Eqns. (3–4) can be distributed over thousands of km, but in sharp contrast, deformation with the crust, and perhaps even within parts of the deep lithosphere, can occur in very narrow bands. We have already noted the findings of Wakabayashi (2021), which shows that deformation zones within subduction zones is concentrated in a zone of <300 m. This narrowness has also been documented elsewhere. The San Andreas Fault is one of the largest and best-known transform faults, and for much of its 1300 km length, fault offsets occur on sub-parallel faults that may have a total width of no more than a few hundreds of meters. Drill cores into the fault zone (the San Andreas Fault Observatory at Depth or SAFOD) show that deformation within this particular fault strand is concentrated within bands of <200 m wide, and is frequently as narrow as 2–3 m (Zoback et al. 2011). These results are remarkably similar to the very narrow deformation widths determined within fossil subduction zones. Strain hardening was noted as a possible effect in dislocation creep, but in fault zones, the narrowness of these zones may be an effect of strain weakening (e.g., Gueydan et al. 2014).

In any case, the equations that describe fault behavior are naturally different than those that describe plastic deformation during mantle convection. In the colder, brittle lithosphere, fault motion is called "stick-slip", which describes the type of discontinuous forces that are needed to overcome frictional resistance along a fault surface. Figure 3a shows a case where the maximum principal stress (σ_1) is vertical, while the intermediate (σ_2) and minimum (σ_3) principal stresses are horizontal. If the differential stress ($\sigma = \sigma_1 - \sigma_3$) is sufficiently great a failure plane can develop, as indicated by the shaded surface (Fig. 3a). Figure 3b shows the stresses in 2-dimensions, where the two principal stresses can be resolved into a normal stress, σ_n, oriented perpendicular to the failure plane, and a shear stress, τ, oriented parallel to the plane of failure. In a solid rock, with no pre-existing fractures, failure occurs when τ is greater than the shear strength of the rock (S_o), and with a correction for the increased resistance to failure as confining pressure, σ_n, increases; this last pressure-sensitive term is written as the product of the normal stress on the fault plane and an "internal coefficient of friction", μ, so the rock strength is:

$$\tau = S_o + \mu\sigma_n \tag{5}$$

When there is a pre-existing plane of weakness, then failure occurs when the shear stress is greater than the product of the normal stress and coefficient of friction along the failure plane, μ_s:

$$\tau = \mu_s\sigma_n \tag{6}$$

An analysis of experimental data (Byerlee 1978) shows that rock strength can be approximated as:

$$\tau = 0.85\,\sigma_n; \; P < 2 \text{ kbar} \tag{7a}$$

$$\tau = 0.5 + 0.6\sigma_n; \; 2 \text{ kbar} < P < 20 \text{ kbar} \tag{7b}$$

which together are known as "Byerlee's Law".

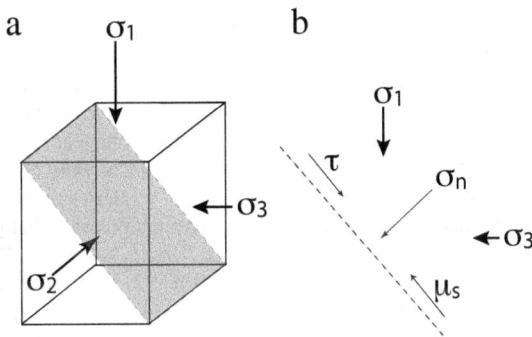

Figure 3. Illustration of brittle (Coulomb) failure. **(a)** Principal stresses σ_1, σ_2 and σ_3 act upon the shaded surface where the rock will eventually break (or perhaps already has broken). Failure is determined by the magnitude of the differential stress, $\sigma_1 - \sigma_3$ and for unbroken rock, by the intrinsic strength of the rock, often referred to as the coefficient of internal friction. **(b)** The principal stresses can be resolved into a normal stress, σ_n, and a tangential stress, τ. For a pre-existing fault surface, failure occurs when τ exceeds the frictional resistance, μ_s. That frictional resistance can be greatly reduced when water is present, or (at least typically) when the materials near the fault surface recrystallize to smaller grain size.

SO, WHEN IS A PLANET EARTH-LIKE?

From a geologic point of view, an Earth-like planet will have plate tectonics, where the tectonic plates (lithosphere) experience brittle deformation, concentrated within remarkably narrow bands. Beneath the lithosphere would be a convecting mantle that is dominated by buoyant thermal plumes and cold lithospheric plates that are subducted back into the mantle. Such a planet would have a bi-model hypsometry with high elevation continental crust and low-lying oceanic basins. The continental crust would owe much of its buoyancy to a significant fraction of low-density rocks that have andesitic and granitic compositions, with quartz and feldspar-dominated mineralogies. The ocean basins would consist of denser, basaltic rocks, with olivine and feldspar minerlaogies. The continental crust would be a product of magmatism that occurs above subduction zones (arcs) while the oceanic plates would result from volcanism at rifts in the lithosphere. Rifting would have no relation to convection within the deeper mantle but would instead be a passive response to the slab pull forces that drive subduction. Actively upwelling parts of the mantle—thermal mantle plumes—would intersect the oceanic spreading ridges only by coincidence and would drive their own form of volcanism at the surface. Earth-like plate tectonics likely requires water at a planetary surface and thus also implies a global water cycle, as water that is subducted can be returned to the surface during arc-related volcanism.

PROBLEMS OF SUBDUCTION

Subduction and the related formation of granitic crust at arcs are hallmarks of an Earth-like planet, and probably key to identifying an Earth-like planet elsewhere in the galaxy. Only Venus appears to have subduction (Schubert and Sandwell 1995; Devaille et al. 2017); it appears to be a minor feature, though Turcotte (1995) has proposed that Venus might lose heat via a catastrophic and episodic replacement of its crust. Tian et al. (2023) refer to Venus as an "episodic lid" regime. But this type of wholesale subduction, interposed by long periods of quiescence, would still yield a surface that can be near uniform in age. A varied and tectonically young surface would appear to be a terrestrial (as opposed to a Venusian) hallmark. If Earth-like subduction, with well-defined and abundant arcuate zones of volcanism and seismic activity, could be identified on any rocky planet outside of Earth, it would be a spectacular discovery and perhaps a signal of Earth-like tectonics that might not be possible via an episodic lid regime. But the question of how and why any particular long-lived subduction zone is initiated is an open question (see Stern 2004). No less crucial is the question of how and when plate tectonics is initiated.

Estimates for the initiation of terrestrial plate tectonics vary from >4 to <1 Ga. Parman et al. (2001) suggested that certain volcanic rocks from the Archean (4.0–2.5 Ga) are derived from subduction, and their study would thus place subduction initiation on Earth at least as far back as 3.5 Ga—a mere 1 Ga post accretion. Stern (2005) would later argue that the rock types and assemblages that characterize modern subduction zones do not appear in the rock record until the Neoproterozoic (1–0.54 Ga), which would make subduction a relatively recent evolutionary event. Yin et al. (2020), though, can push the rock record of modern subduction back to at least 2 Ga. These claims are not necessarily mutually exclusive: it's possible that plate tectonics was operative in the Archean, but created different rock types on a hotter Earth. Brown and Johnson (2019) suggest something along these lines, noting that that the absence of the rock types that Stern (2002) notes could be less telling than the existence of bi-modal temperature-pressure (T/P) regimes: subduction zones yield rocks with relatively cold temperatures at high pressures, whereas a "normal geotherm" yields rocks that record high temperatures at high pressures. Brown and Johnson (2019) suggest that bi-modal T/P conditions can be identified in metamorphic rocks at least as early as the Neoarchean (2.8–2.5 Ga). Putirka (2016) suggested a similar test, noting that arc-related volcanic are produced at lower temperatures, and so volcanic activity that is thermally bi-modal should indicate plate tectonics, whereas thermally unimodal volcanism would indicate a stagnant lid; Earth's volcanic rock record indicates bi-modality in the Archean.

An especially intriguing set of observations derive from study of the oldest mineral grains yet discovered on Earth. Harrison et al. (2005) examine zircon grains ($ZrSiO_4$) from Australia that have been dated at 4.0–4.4 Ga. These ancient zircons yield wide ranging isotopic ratios, and the parent and daughter isotopes are known to fractionate through crust-forming processes that occur at subduction zones. The nature and magnitude of their isotopic shifts indicate that the amounts of continental crust in the Hadean Era (4.5–4.0 Ga) could have been close to modern values, and that some residues of the crust-forming events were subducted. Other compositional aspects of these zircons indicate that they formed in a wet, low-temperature environment, not unlike modern granitic rocks that form at arcs. The data are not perfectly conclusive (Harrison 2009), but important nonetheless. A key counter-argument derives from Bernard (2018), who suggests that granitic rocks on the Archean Earth could form within a plume-lid (stagnant lid) tectonic regime, where thermal plumes yield a thickened crust that can then delaminate repeatedly to generate new granitic rocks—and a very thick continental crust, absent subduction. However, a new study by Ge et al. (2023) makes a compelling case for Archean plate tectonics as they show that Archean crustal zircons record greater amounts of H_2O and higher f_{O_2} compared to their coeval mantle counterparts; such rocks are less oxidized than modern analogs, but the crust/mantle contrasts are the same, and are easily explained if water and O are being subducted into the mantle to create the Archean crust.

It is not clear that numerical models can yet play a role in informing the issue of the start of plate tectonics on Earth. Earth was assuredly hotter in the first 1–2 Ga after accretion, and it remains unclear whether such a state of affairs has the effect of causing the viscosity contrast to increase (by decreasing mantle viscosity) or decrease (by decreasing the lithosphere viscosity via heating from the mantle). Numerical experiments by Foley (2018) indicate no dependency of plate tectonics on interior mantle temperature since convection can yield weak plate boundaries, by a form of strain weakening, at a range of temperatures. Foley (2018) interestingly finds that plate motions are more sluggish in a hotter Earth, which may be unsurprising if cold, dense slabs drive plate motion.

What does this all have to do with exoplanets? Putirka and Xu (2021) argued that the discovery of granitic rocks on an exoplanet might provide the surest evidence of Earth-like behavior. The Bedard (2006) model casts doubt on this idea. The discovery of continental crust might only tell us that we have a tectonically active planet (still interesting), but it might not tell us what type of tectonics is operable. The best tests of exoplanet tectonic styles will derive only from better understanding of the early Earth, or perhaps from the identification, via space telescopes, of water in a planetary atmosphere, that together with granite, could indicate a global water cycle.

PLATE TECTONICS ELSEWHERE IN THE SOLAR SYSTEM?

It is generally accepted that, besides Earth, plate tectonics does not operate on any of the inner planets (e.g., Devaille et al. 2017; Tosi and Padovan 2021). Katterhorn and Prockter (2014) suggest that the icy shell of Europa might have spreading ridges and subduction. All other objects besides Earth that have an observable solid surface in our Solar System would be in a stagnant lid regime (e.g., Stern et al. 2018). Smrekar et al. (2023) describe Venus as a "squishy lid"—but this is still just a subtle variation on a stagnant lid regime. The tectonic behavior of other planets, though, is inferred from surface deformation and their interpretations are non-unique. Besides being stagnant lid planets, Moon and Mercury have something else in common: their surfaces show planet-wide contractional features, which appear to record cooling and contraction of the planetary interior (e.g., Watters et al. 2016; Matsuyama et al. 2021). (This interpretation is, ironically, also the now-discarded explanation for mountain ranges on Earth). Some contractional features on Mercury might also result from mantle down-wellings (Watters et al. 2015), while on Mars, contractional deformation could also involve subduction-like activity, with lava flows piling up against one another (e.g., Yin and Wang 2023).

There has been speculation that Mars had more Earth-like tectonics very early in its history. Yin (2012) showed that subduction may have been active at a local scale to create extension in the Tharsis region. And Connerney et al. (1999) discovered magnetic lineations that appear analogous to those found at oceanic spreading ridges, while Connerney et al. (2005) appear to have discovered transform faults that might offset nominal spreading ridges. Since plate tectonics is expected to speed up heat loss, and perhaps facilitate thermal convection of the core (which nominally creates a planet's magnetic field), Nimmo and Stevenson (2000) suggested that the magnetic fields recorded in the Martian crust might also record the timing of when plate tectonics was active on this planet. But Breuer and Spohn (2003) show that the Martian crust and its magnetic field can be better explained if Mars was always in a stagnant lid regime. And new experiments by Zhou et al. (2022) indicate that the Martian crust might never have been sufficiently dense, relative to its underlying mantle, to subduct. Thermal contraction thus appears to be the cause of most tectonic features on Mars (Nahm and Schultz 2011).

Venus may be the more intriguing case study of tectonics and should provide a testing ground for numerical models that nominally predict tectonic behavior. Venus has a young surface (compared to Moon, Mars and Mercury) and, as noted earlier, the planet is possibly subject to catastrophic resurfacing (Turcotte 1995). Venus also shows distinct patterns of surface deformation that are reminiscent of subduction (DeVaille et al. 2017). Numerical models, such as those by Armann and Tackley (2012), compare a stagnant lid to an episodic lid regime and in preferring the latter, they make very specific predictions about the geology of Venus. For example, compared to the stagnant lid case, an episodic lid regime predicts the number of catastrophic re-surfacing events (between five and eight) and the length of time for each event (ca. 150 Ma), as well as a thin crust and low heat flow from the core. Tian et al. (2023) present another testable prediction, i.e., that local volcanic resurfacing intervals depend upon crustal rheology. Two questions immediately arise: (a) might any other numerical models yield similar predictions? (b) do the model predictions survive further scrutiny (i.e., geologic mapping) of the Venusian surface?

Numerical models have provided some insights; for example, they emphasize that to convert a stagnant lid into a mobile one, considerable zones of weaknesses must be introduced into the lithosphere. But we kind of knew this anyway, by observation. In any case, any model that is simultaneously capable of predicting the surface features and nominal tectonic behaviors of Earth, Venus, Europa, Moon and Mercury should stand a very good chance of providing useful insights regarding tectonic behavior on exoplanetary bodies—or even the early tectonic history of the early Earth. It is unclear that any numerical models can currently do so, but progress is possible if we aim towards such challenging goals.

PROSPECT FOR MODELING TECTONICS ON EXOPLANETS

As with mineralogy (Putirka 2024, this volume), studies of exoplanetary tectonic regimes are largely a game of "what if"; predictions of tectonic behavior in even highly simplified cases are perhaps even less clear than mineralogy estimates, and are perhaps best described as 'informed speculation'. The most profitable means to understanding tectonic regimes is mostly like that of Foley et al. (2012) and Weller and Lenardic (2018), who map the mechanical properties of planets so as to separate plate tectonic and stagnant lid regimes. Their approach has the advantage of producing testable hypotheses and consistency tests. The paper by Foley et al. (2012) is nominally about "super Earths" but is more interesting. As in earlier studies, Foley et al. (2012) argue that planets can exhibit plate tectonics when the viscosity ratio between the lithosphere and mantle, μ_l / μ_m, is sufficiently low, but in their model, Stevenson's (2003) anomaly is avoided in that they account for strain-weakening in the lithosphere. They do so by considering a critical viscosity ratio η_{lcrit}:

$$\eta_{lcrit} = \left(w \left(b Ra^{3\beta} \right)^2 \right)^{\frac{m}{p}} \qquad (8)$$

where Ra is the Rayleigh number (similar to Eqn. 1), and b, m and p are material constants, analogous to the m and n values in Eqns. (3–4) (in their preferred model, $m = 2$, $p = 3$, $b = 0.05$ and $\beta = 1/3$); w is the more critical parameter here in that it quantifies strain weakening. Plate tectonics is allowed when $\mu_l/\mu_m < \eta_{lcrit}$ (Fig. 4a). If, for whatever reason, w is sufficiently high (e.g., water infiltrates Earth's lithosphere), then so might be η_{lcrit}, which can then offset a high bulk μ_l/μ_m. Weller and Lenardic (2018) attempt a similar kind of map, using only the yield strength of the lithosphere, but they also consider the percentage of heat generated by radioactivity (Q) at any given time in a planet's history (Fig. 4b).

It is unclear to what extent predictions from the two models might differ. But as noted by Foley (2018) there nonetheless remain fundamental disagreements as to the physical models that predict how plate boundaries form. Figure 4b shows that hotter planets are more likely to exist in a stagnant lid regime. The parameter Q in Figure 4b seems, at least superficially, to preserve the idea of a viscosity ratio effect: higher Q at a constant lithosphere yield strength implies a higher ratio of μ_l/μ_m, since higher mantle temperatures would mean a lower μ_m, and higher Q expands the stagnant lid regime. Some entrepreneur need only collect mechanical estimates of the various known objects of our Solar System and compare results. But the mechanical properties of even Earth and Venus appear to be in doubt. Weller and Lenardic (2018) plot Earth and Venus at a similar (ca. 50 MPa) yield strength, though the drier Venusian lithosphere seems as if it should be placed somewhat higher (Fig. 4b). However, the authors note that the positions of the two planets are "illustrative only". Especially interesting is that Foley et al. (2012) and Weller and Lenardic (2018) both have Earth teetering at the boundary of the mobile and stagnant lid regimes, despite evidence that Earth has unreservedly exhibited a mobile lid for at least half (Yin et al. 2020), if not all its history.

Note, however, that the mobile/stagnant lid boundary in the model of Foley et al. (2012) is not sensitive to T (Fig. 4a). This contrast in results is a critical underlying issue. As Foley (2018) shows, the T-sensitivity of the boundary in Figure 4b can stem from use of a "pseudoplastic rheology" (e.g., Tackley 2000; Korenaga 2010); in such a model plate boundaries form when

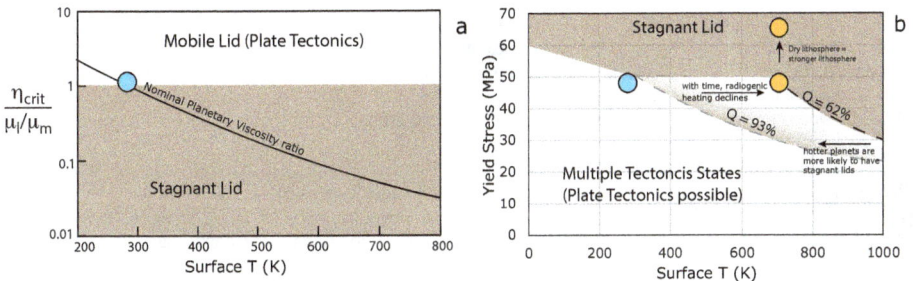

Figure 4. Thermal and mechanical criteria that allow plate tectonics. **(a)** From Foley et al. (2012), critical viscosity, η_{crit}, is normalized to the viscosity ratio, μ_l/μ_m, where μ_l is the viscosity of the lithosphere and μ_m is the viscosity of the convective mantle. When μ_l/μ_m is high (greater than a critical value as determined by η_{crit}, or $\mu_l/\mu_m > \eta_{crit}$ a stagnant lid regime ensues. The ratio is expected to vary with temperature, as indicated by the solid curve. **(b)** In the map developed by Weller and Lenardic (2018) only the lithosphere strength is used for the vertical axis, but this is compared to surface temperature and the percentage of heat that is still being produced by radioactivity (Q; as opposed to residual heat from accretion and core formation) as a planet ages.

stresses in the lithosphere exceed a value adopted for the lithosphere's yield strength; at high *T*, stresses are lower, and so may decrease below the yield strength threshold. The problem is that such models require yield strengths orders of magnitude below those observed in nature or experiment, even at low temperature conditions where stresses are higher. Bercovici and Ricard (2012) alternatively propose that plate boundaries form due to a particular style of grain diminution; their model builds on related ideas of strain localization, where a decrease in grain size yields a feedback loop that yields concentrated zones of weakness (e.g., Bercovici and Karato 2003). In the model of Foley (2018), grain diminution has the effect of increasing diffusion creep, which is highly sensitive to grain size (Eqn. 3). The result is that internal mantle temperatures do not control tectonic activity (Foley 2018). This has the salutary effect of allowing what seems likely: plate tectonic activity in the Archean (Ge et al. 2023).

A remaining challenge is to describe surface deformation from thermal states and mechanical properties. Yin et al. (2016) provide an intriguingly precise model of the Saturnian moon Enceladus. That moon has a stagnant lid that exhibits an array of strike-flip faults; the Yin et al. (2016) model, building on studies of terrestrial strike slip faults on Earth (Roy and Royden 2000), predicts the brittle and elastic thicknesses and the cohesive strength of the icy Enceladean lithosphere.

Is there some way to combine predictions of tectonic states and surface deformations? We of course have no immediate prospect of surveying exoplanet surfaces, but models that can predict many things are better than those that predict few. Or is a predictive model a chimera? Foley (2018) shows how there is no clear consensus on the physics of how plate boundaries form. Weller and Lenardic (2018) suggest that, due to feedback loops and model non-linearities, tectonic states and planetary evolutionary paths may be unpredictable. Did Earth evolve to its present state as an accident of its initial conditions, or by stringing together a series of unlikely events? Of Hazen et al.'s (2015) choice between Chance and Necessity, mineral assemblages of planetary mantles and crusts appear to fall to the latter, as the laws of thermodynamics appear determinative. But maybe deformation and tectonics is stochastic. There are many experiments in rock mechanics still to be performed and much to be explored regarding the physics of deformation before admitting defeat to the "vagaries of history".

REFERENCES

Adam C, King SG, Caddick MJ (2021a) Mantle temperature and density anomalies: the influence of thermodynamic formulation, melt, and anelasticity. Phys Earth Planet Int 319:106772

Adam C, Vidal V, Pandit B, Davaille A, Kempton PD (2021b) Lithosphere destabilization and small-scale convection constrained from geophysical data and analogical models. Geochem Geophys Geosyst 22:e2020GC009462

Armann M, Tackley PJ (2012) Simulating the thermomechanical magmatic and tectonic evolution of Venus's mantle and lithosphere: two-dimensional models. J Geophys Res 117:E12003

Barr AC, Dobos V, Kiss LL (2018) Interior structures and tidal heating in the Trappist-1 planets. Astron Astrophys 613:A37

Batra K, Foley B (2022) Scaling laws for stagnant-lid convection with buoyant crust. Geophys J Int 228:631–663

Bedard JH (2006) A catalytic delamination-driven model for coupled genesis of Archean crust and sub-continental lithospheric mantle. Geochim Cosmochim Acta 70:1188–1214

Bedard JH (2018) Stagnant lids and mntle overturns: Implications for Archean tectonics, magmagenesis, crustal growth, mantle evolution, and the start of plate tectonics. Geosci Front 9:19–49

Bercovici D, Karato S (2003) Theoretical analysis of shear localization in the lithosphere. Rev Mineral Geochem 51:387–420

Bercovici D, Ricard Y (2012) Mechanisms for the generation of plate tectonics by two phase grain-damage and pinning. Phys Earth Planet Int 202:27–55

Bercovici D, Tackley PJ, Ricard Y (2015) The generation of plate tectonics from mantle dynamics. 7.07 *In:* Treatise on Geophysics, 2nd Ed, Elsevier, p 271–317

Bird P (1979) Continental delamination and the Colorado Plateau. J Geophys Res 84:7561–7571

Brown M, Johnson T (2019) Metamorphism and the evolution of subduction on Earth. Am Mineral 104:1065–1082, https://doi.org/10.2138/am-2019-6956

Byerlee J (1978) Friction of rocks. Pure Appl Geophys 116:615–626

Breuer D, Spohn T (2003) Early plate tectonics versus single-plate tectonics on Mars: Evidence from magnetic field history and crust evolution. J Geophys Res 108:5072

Bystricky M, Mackwell S (2001) Creep in dry clinopyroxene aggregates. J Geophys Res 106:13443–13454

Bystricky M, Lawlis J, Mackwell S, Heidelbach F, Raterson P (2016) High-temperature deformation of enstatite aggregates. J Geophys Res 121:6384–6400, https://doi.org/10.1002/2016JB013011

Campbell IH, Taylor SR (1983) No water, no granites—No oceans, no continents. Geophys Res Lett 10:1061–1064

Chapple WM, Tullis TE (1977) Evaluation of the forces that drive the plates. J Geophys Res 82:1967–1984

Clennett EJ, Holt AF, Tetley MG, Becker TW, Faccenna C (2023) Assessing plate reconstruction models using plate driving force consistency tests. Sci Rep 13:10191, https://doi.org/10.1038/s41598-023-37117-w

Connerney JEP, Acuna MH, Wasilewski PJ, Ness NF, Reme H, Mazelle C, Vignes D, Lin RP, Mitchell DL, Cloutier PA (1999) Magnetic lineations in the ancient crust of Mars. Science 284:794–798

Connerney JEP, Acuna MH, Ness NF, Kleteschka G, Mitchell DL, D, Lin RP, Reme H (2005) Tectonic implications of mars crustal magmatism. Science 284:794–798

Conrad CP, Lithgow-Bertelloni C (2002) How mantle slabs drive plate tectonics. Science 109:B10407, https://doi.org/10.1029/2004JB002991

Conrad CP, Lithgow-Bertelloni C (2004) The temporal evolution of plate driving forces: importance of "slab suction" verses "slab pull" during the Cenozoic. JGeophys Res 298:207–209

Crosby AG, McKenzie D, Slater JG (2006) The relationship between depth, age and gravity in the oceans. Geophys J Int 166:553–573

Davies GF, Richards MA (1992) Mantle convection. J Geol 100:151–206

DeCelles PG, Ducea MN, Kapp P, Zandt G (2009) Cyclicity in Cordilleran orogenic systems. Nat Geosci 2, 251–257

Devaille A, Smerkar SE, Tomlinson S (2017) Experimental and observational evidence for plume-induced subduction on Venus. Nat Geosci 10:349–355

Doglioni C, Ismail-Zadeh A, Panza GRiguzzi F (2011) Lithosphere-asthenosphere viscosity contrast and decoupling. Phys Earth Planet Int 189:1–8

Ducea MN, Saleeby JB (1996) Buoyancy sources for a large, unrooted mountain range, the Sierra Nevada, California: Evidence from xenolith thermobarometry. J Geophys Res 101:8229–8244

Ducea MN, Chapman AD, Bowman E, Triantafyllou A (2021) Arclogites and their role in continental evolution: part 1: background, locations, petrography, geochemistry, chronology, and thermobarometry. Earth Sci Rev 214:103375

Evans B, Kohlstedt DL (1993) Rheology of rocks. *In:* Rocks Physics and Phase Relations: A Handbook of Physical Constants, AGU Reference Shelf Vol 3, p 148–165

Fei H, Ballmer MD, Faul U, Walte N, Weiwei C, Katsura T (2023) Variation in bridgmanite grain size accounts for the mid-mantle viscosity jump. Nature 620:794–799

Fischer R, Gerya T (2016) Early Earth plume-lid tectonics: a high-resolution 3D numerical modelling approach. J Geodyn 100:198–214

Foley BJ (2018) The dependence of planetary tectonics on mantle thermal state: applications to early Earth evolution. Phil Trans R Soc A 376:20170409

Foley BJ, Driscoll PE (2016) Whole planet coupling between climate, mantle, and core: implications for rocky planet evolution. Geochem Geophys Geosyst 17:1885–1914

Foley B, Bercovici D, Landuyt W (2012) The conditions for plate tectonics on super-Earths: Inferences from convection models with damage. Earth Planet Sci Lett 331–332:281–290

Forsyth D, Uyeda S (1975) On the relative importance of the driving forces of plate motion. Geophys J Int 43:163–200

Frazer WD, Korenaga J (2022) Dynamic topography and the nature of deep thick plumes. Earth Planet Sci Lett 578:117286

Ge F-G, Wilde SA, Zhu W-B, Wang X-L (2023) Earth's early continental crust formed from wet and oxidizing arc magmas. Nature 623:334–339

Giannandrea E, Christensen U (1993) Variable viscosity convection experiments with a stress-free upper boundary and implications for the heat transport in the Earth's mantle. Phys Earth Planet Int 78:139–152

Grove TL, Till CB, Krawczynski MJ (2012) The role of H_2O in subduction zone magmatism. Annu Rev Earth Planet Sci 40:413–439

Gueydan F, Precigout J, Montesi LGJ (2014) Strain weakening enables continental plate tectonics. Tectonophys 631:189–196

Hager BH, O'Connell RJ (1981) A simple global model of plate dynamics and mantle convection. J Geophys Res 86:4843–4867

Hansen LN, Warren JM (2015) Quantifying the effect of pyroxene on deformation of peridotite in a natural shear zone. J Geophys Res 120:2717–2738, https://doi.org/10.1002/2014JB011584

Hansen LN, Kumamoto KM, Thom CA, Wallis D, Durham WB, Goldsby DL, Breithaupt T, Meyers CD, Kohlstedt DL (2019) Low-temperature plasticity in olivine: grain size, strain hardening, and the strength of the lithosphere. J Geophys Solid Earth 124:5427–5449

Harrison TM (2009) The Hadean crust: Evidence from >4 Ga zircons. Annu Rev Earth Planet Sci 37:479–505

Harrison TM, Blichert-Toft J, Albarede F, Holden P, Mojzsis SJ (2005) Heterogeneous Hadean hafnium: Evidence of continental crust at 4.4–4.5 Ga. Science 310:1947–1950

Hazen RM, Grew, ES, Downs RT, Golden J, Hystad G (2015) Mineral ecology: Chance and necessity in the mineral diversity of terrestrial planets. Can Mineral 53:1400086

Höink T, Jellinek AM, Lenardic A (2011) Viscous coupling at the lithosphere–asthenosphere boundary. Geochem Geophys Geosyst 12:Q0AK02, https://doi.org/10.1029/2011GC003698

Hubbert MK, Rubey WW (1959) Role of fluid pressure in mechanics of overthrust faulting: I Mechanics of fluid-filled porous solids and its application to overthrust faulting. Geol Soc Am Bull 70:115–166

Ito G, Dunn, R, Li A, Wolfe CJ, Gallego A, Fu Y (2014) Seismic anisotropy and shear wave splitting associated with mantle plume-plate interaction. J Geophys Res Solid Earth 119:4923–4937, https://doi.org/10.1002/2013JB010735

Jackson B, Greenburg R, Barnes R (2008) Tidal heating of extrasolar planets. Astrophys J 681:1631, https://doi.org/10.1086/587641

Jadamec MA (2016) Insights on slab-driven mantle flow from advances in three-dimensional modelling. J Geodyn 100:51–70

Jarrard RD (1986) Relations among subduction parameters. Rev Geophys 24:217–284

Jordan TH (1978) Composition and development of the continental tectosphere. Nature 274:544–548

Kapp P, Yin A, Harrison TM, Ding L (2005) Cretaceous–Tertiary shortening, basin development, and volcanism in central Tibet. Geol Soc Am Bull 117:865–878

Katterhorn S, Prockter LM (2014) Evidence for subduction in the ice shell of Europa. Nat Geosci 7:762–767

Kay RW, Kay SM (1993) Delamination and delamination magmatism. Tectonophys 219:177–189

Kiser E, Kehoe H, Chen M, Hughes A (2021) Lower mantle seismicity following the 2015 Mw 7.9 Bonin Islands deep-focus earthquake. Geophys Res Lett 48:e2021GL09311

Korenaga J (2010) Scaling of plate tectonic convection with pseudoplastic rheology. J Geophys Res 115:B11405

Langmuir CH, Klein EM, Plank T (1992) Petrological systematics of mid-ocean ridge basalts—constraints on melt generation beneath ocean ridges. *In:* Morgan JP, Blackman DK, Sinton JM (Eds) Mantle Flow and Melt Generation at Mid-ocean Ridges. AGU Geophys Mon Ser 71:183–280

Le Bas MJ, Streckeisen AL (1991) The IUGS systematics of igneous rocks. J Geol Soc London 148:825–833

Lee C-TA, Anderson D (2015) Continental crust formation at arcs, the arclogite "delamination" cycle, and one origin for fertile melting anomalies in the mantle. Sci Bull 60:1141–1156

Lenardic A, Crowley JW (2012) On the notion of well-defied tectonics regimes for terrestrial planets in the solar system and others. Astrophys J 755:132

Lithgow-Bertelloni C, Richards MA (1998) The dynamics of Cenozoic and Mesozoic plate motions. Rev Geophys 36:27–78

Matsuyama I, Keane JT, Trinh A, Beuthe M, Watters TR (2021) Global tectonic patterns of the Moon. Icarus 358:114202

McKenzie DP (1967) Some remarks on heat flow and gravity anomalies. J Geophys Res 172:6261–6273

McKenzie D, Bickle MJ (1988) The volume and composition of melt generated by extension of the lithosphere. J Petrol 29:625–679

McMillan M, Schoenbohm LM (2022) Diverse styles of lithospheric dripping: synthesizing gravitational instability models, continental tectonics, and geologic observations. Geochem Geophys Geosyst 24:e2022GC010488

Moore WB (2001) The thermal state of Io. Icarus 154:548–550

Moresi L, Solomatov V (1998) Mantle convection with a brittle lithosphere: thoughts on the global tectonics styles of the Earth and Venus. Geophys J Int 133:669–682

Morgan WJ (1971) Convection plumes in the lower mantle. Nature 230:42–43

Morgan JP, Smith WHF (1992) Flattening of the sea-floor depth-age curve as a response to asthenospheric flow. Nature 359:524–527

Murphy MA, Yin A, Harrison TM, Durr SB, Chen Z, Reyerson FJ, Kidd WSF, Wang X, Zhou X (1997) Did the Indo-Asian collision alone create the Tibetan Plateau? Geology 25:719–722

Nahm AL, Schultz RA (2011) Magnitude of global contraction on Mars form analysis of surface faults: implications for martian thermal history. Icarus 211:389–400

Nakagawa T, Tackely PJ (2015) Influence of plate tectonics mode on the coupled thermochemical evolution of Earth's mantle and core. Geochem Geophys Geosyst 16:3400–3413

Naliboff JB, Lithgow-Bertelloni C, Ruff LJ, de Koker N (2012) The effects of lithospheric thickness and density structure on Earth's stress field. Geophys J Int 188:1–17

Nimmo F, Stevenson, D (2000) Influence of early plate tectonics on the thermal evolution and magnetic field of mars. J Geophys Res 105:11969–11979

Parsons B, Sclater JG (1977) An analysis of the variation of ocean floor bathymetry and heat flow with age. J Geophys Res 82:803–827

Parman S, Grove TL, Dann JC (2001) The production of Barberton komatiites in an Archean subduction zone. Geophys Res Lett 28:2513–2516

Pollack HN, Chapman DS (1977) On the regional variation of heat flow, geotherms, and lithospheric thickness. Tectonophysics 3–4:279–296

Putirka K (2016) Rates and styles of planetary cooling on Earth, Moon, Mars, and Vesta, using new models for oxygen fugacity, ferric–ferrous ratios, olivine–liquid Fe–Mg exchange, and mantle potential temperature. Am Mineral 101:819–840

Putirka KD (2024) Exoplanet mineralogy. Rev Mineral Geochem 90:199-258

Putirka KD, Xu S (2021) Polluted white dwarfs reveal exotic mantle rock types on exoplanets in our solar neighborhood. Nat Commun 12:6168

Roy M, Roydon LH (2000) Crustal rheology and faulting at strike-slip plate boundaries 1. An analytic model. J Geophys Res 105:5583–5597

Rudnick RL, McDonough WF, O'Connel RJ (1998) Thermal structure, thickness and composition of continental lithosphere. Chem Geol 145:395–411

Rudolph ML, Lekic V, Lithgow-Bertelloni C (2015) Viscosity jump in Earth's mid-mantle. Science 350:1349–1352

Rychert CA, Harmon N, Constable S, Wang S (2020) The nature of the lithosphere-asthenosphere boundary. J Geophys Res 125:e2018JB01646

Saleeby JB, Ducea M, Clemens-Knott D (2003) Production and loss of high-density batholitic root, southern Sierra Nevada, California. Tectonics 22:1064, https://doi.org/10.1029/2002TC001374

Saxena A, Dannberg J, Gassmoller R, Fraters M, Heister T, Styron R (2022) High-resolution mantle flow models reveal importance of plate boundary geometry and slab pull forces on generating tectonic plate motions. J Geophys Res 128:e2022JB025877

Saxena A, Dannberg J, Gassmoller R, Fraters M, Heister T, Stryon R (2023) High-resolution mantle flow models reveal importance of plate boundary geometry and slab pull forces on generating tectonic plate motions. J Geophys Res Solid Earth 128:e2022JB025877

Schubert G, Sandwell DT (1995) A global survey of possible subduction sites on Venus. Icarus 117:173–196

Siegler MA, Feng JQ, Lehman-Franco K, Andrews-Hanna JC, Economos RC, St Clair M, Million C, Head JW, Glotch TD, White MN (2023) Remote detection of a lunar granitic batholith at Compton-Belkovich. Nature 620:116–121

Silver PG, Carlson RW, Olson P (1988) Deep slabs, geochemical heterogeneity, and the large-scale structure of mantle convection: Investigation of an enduring paradox. Annu Rev Earth Planet Sci 16:477–541

Sisson TW, Ratajeski K, Hankins WB, Glazner A (2005) Voluminous granitic magmas from common basaltic sources. Contrib Mineral Petrol 148:635–661

Sleep N (1990) Hotspots and mantle plumes: some phenomenology. J Geophys Res 95, 6715- 6736

Smrekar S, Ostberg C, O'Rourke JG (2023) Earth-like lithospheric thickness and heat flow on Venus consistent with active rifting. Nat Geosci 16:13–18

Solomatov VS (1995) Scaling of temperature- and stress-dependent viscosity convection. Phys Fluids 7:266–274

Steinberger B, Becker TW (2016) A comparison of lithospheric thickness models. Tectonophysics 746:325–338

Stern RJ (2002) Subduction zones. Rev Geophys 40:1–13

Stern RJ (2004) Subduction initiation: spontaneous and induced. Earth Planet Sci Lett 226:275–292

Stern RJ (2005) Evidence from ophiolites, blueschists, and ultrahigh-pressure metamorphic terranes that the modern episode of subduction tectonics began in Neoproterozoic time. Geology 33:557–560

Stern RJ, Gerya T, Tackley PJ (2018) Stagnant lid tectonics: perspective from silicate planets, dwarf planets, large moons, and large asteroids. Geosci Front 9:103–119

Stevenson DJ (2003) Styles of mantle convection and their influence on planetary evolution. C R Geosci 335:99–111

Tackley P (2000) Self-consistent generation of tectonic plates in time-dependent, three-dimensional mantle convection simulations, 1. Pseudoplastic yielding. Geochem Geophys Geosyst 1:1026, doi:10.1029/2000GC000036

Tian J, Tackely PJ, Lourenco D (2023) The tectonics and volcanism of Venus: new modes facilitated by realistic crustal rheology and intrusive magmatism. Icarus 399:115539

Tosi N, Padovan S (2021) Mercury, Moon, Mars: Surface expressions of mantle convection and interior evolution of stagnant-lid bodies. *In:* Marquardt H, Ballmer M, Cottar S, Kontar J (Eds) Mantle Convection and Surface Expressions. AGU Geophys Monogr Ser 263:455–489

Turcotte DL (1993) An episodic hypothesis for Venusian tectonics. J Geophys Res Planet 98:17061–17068

Turcotte DL (1995) How does Venus lose heat? J Geophys Res 100:16931–16940

Wakabayashi K (2021) Subduction and exhumation slip accommodation at depths of 10–80 km inferred from field geology of exhumed rocks: evidence for temporal-spatial localization of slip. Geol Soc Am Spec Paper 552:257–296

Watters TR, Selvans MM, Banks ME, Hauck SA, Becker KJ, Robinson MS (2015) Distribution of large-scale contractional tectonic landforms on Mercury: Implications for the origin of global stresses. Geophys Res Lett 42:3755–3763

Watters TR, Daud K, Banks ME, Selvans MM, Clark R, Chapman C, Ernst CM (2016) Recent tectonics activity on Mercury revealed by small thrust fault scarps. Nat Geosci 9:743–747

Weller MB, Lenardic A (2018) On the evolution of terrestrial planets: bi-stability, stochastic effects, and the non-uniqueness of tectonic states. Geosci Front 9:91–201

Yamamoto J, Ando J-I, Kagi H, Inoue T, Yamada A, Yamazake D (2008) In situ strength measurements on natural upper mantle minerals. Phys Chem Min 35:249–257

Yin A (1989) Origin of regional, rooted low-angle normal faults – a mechanical model and its tectonic implications. Tectonics 8:469–482

Yin A (2012) An episodic slab-rollback model for the origin of the Tharsis rise on Mars: Implications for initiation of local plate subduction and final unification of a kinematically linked global plate-tectonic network on Earth. Lithosphere 4:553–593

Yin A, Wang Y-C (2023) Formation and modification of wrinkle ridges in the central Tharsis region of Mars as constrained by detailed geomorphological mapping and landsystem analysis. Earth Planet Phys 7:161–192

Yin A, Zuza AV, Papplardo RT (2016) Mechanics of evenly spaced strike-slip faults and its implications for the formation of tiger-stripe fractures on Saturn's moon Enceladus. Icarus 266:204–216

Yin A, Gunther B, Kroner A (2020) Plate-tectonic processes at ca. 2.0 Ga: Evidence from >600 km of plate convergence. Geology 48:103–107

Zhong S, Watts AB (2013) Lithospheric deformation induced by loading of the Hawaiian Islands and its implications for mantle rheology. J Geophys Res 118:6025–6048

Zhou WY, Olson PL, Shearer CK, Agee CB, Townsend JP, Hao M, Hou MQ, Zhang JS (2022) High pressure-temperature phase equilibrium studies on Martian basalts: implications for the failure of planet tectonics on Mars. Earth Planet Sci Lett 594:117751, https://doi.org/10.1016/j.epsl.2022.117751

Zoback M, Hickman S, Ellsworth W, SAFODScience Team (2011) Scientific drilling into the San Andreas Fault zone—An overview of SAFOD's first five years. Sci Drill 11:14–28, https://doi.org/10.2204/iodp.sd.11.02.2011

Reviews in Mineralogy & Geochemistry
Vol. 90 pp. 323–374, 2024
Copyright © Mineralogical Society of America

A Framework for the Origin and Deep Cycles of Volatiles in Rocky Exoplanets

Rajdeep Dasgupta[1,*], Debjeet Pathak[1], Maxime Maurice[2]

*[1]Department of Earth, Environmental and Planetary Sciences, Rice University,
MS 126, 6100 Main Street, Houston, TX 77005, USA
[2]Laboratoire de Météorologie Dynamique, IPSL, CNRS,
4 Place Jussieu, 75252 Paris, France*

*Rajdeep.Dasgupta@rice.edu

INTRODUCTION

Since the first discovery of planets in another solar system of our galaxy in the early 90s and the discovery of 51 Pegassi b, a Jovian-type planet around a Sun-like star in 1995 (Mayor and Queloz 1995), thousands of confirmed exoplanets and planetary systems have been discovered, and the total count continues to increase steadily. These discoveries have renewed our interest and hope in finding another habitable world, identifying signs of life on another planet, and finding an Earth 2.0. Although new exoplanets continue to be discovered, the focus of current astronomical research is shifting from finding and detecting new exoplanets to characterizing the known and targeted planets. Such characterizations in observational astronomy are currently feasible, if at all, only for the atmospheres of various planets. With the detection of different gas species in the atmospheres of exoplanets, the next logical question becomes what may give rise to such atmospheric compositional signals and what planetary processes can give rise to and stabilize such atmospheric compositions. Hence, in the overall spectrum of exoplanet research, the emphasis on exoplanet characterization and understanding the characterized properties will remain critical for the coming decades. Therefore, the roles of Earth and planetary scientists in exoplanet science are becoming increasingly urgent. The concepts and constraints from Earth and planetary science can help build a framework for various processes that can give rise to and stabilize a given atmospheric chemistry. Similarly, understanding the influence of different geologic and planetary processes on atmospheric chemistry may also help us use the characterized atmospheric attributes to estimate the currently unobservable characters of the exoplanets, such as their surface compositions, interior compositions, and mineralogy, the presence or absence of tectonics and such.

Based on the current observational techniques, radial velocity, transit, microlensing, and direct imaging, four broad classes of exoplanets have been determined: (1) gas giants, (2) Neptune-like, (3) Super-Earths, and (4) terrestrial planets. Among these four classes of planets, the terrestrial planets and the Super-Earths follow a mass–radius relationship that suggests that they both broadly can have compositions made up of rocks and iron (e.g., Otegi et al. 2020), i.e., similar to the inner planets in our Solar System. The Super-Earths, which can be up to ~10 times more massive than Earth (~$10 M_{\oplus}$), can, however, be more complex, with only a subset of them being truly rocky Super-Earths (e.g., Alessi et al. 2016). The presentation in this chapter will be relevant for rocky exoplanets, i.e., terrestrial exoplanets and smaller Super-Earths.

From studies on Earth, the only known inhabited planet in the universe, and other Solar System neighbors, we know that several processes can give rise to and modulate the chemistry of planetary atmospheres. These processes vary from inorganic processes at the surface to the presence or absence of life on the surface and sub-surface, as well as processes operating in

1529-6466/24/0090-0010$10.00 (print)
1943-2666/24/0090-0010$10.00 (online)

the planetary interiors. However, the timescales over which various processes can influence atmospheric chemistry and stabilize the presence of different gas species in the atmosphere vary immensely (e.g., Lee et al. 2019). Long-term, i.e., for >millions of years, stability of a given atmospheric constituent tends to rely on the bulk planetary properties and the planet's internal state. This dependence is because the gases that are present in the atmosphere also find their stability in various forms, in different parts of the planetary system, through process of planet formation (e.g., Hirschmann 2012; Füri and Marty 2015; Dasgupta and Grewal 2019; Grewal et al. 2019b, 2021b; Broadley et al. 2022; Suer et al. 2023) and subsequent planetary evolution (e.g., McGovern and Schubert 1989; Dasgupta 2013, 2022a; Korenaga et al. 2017; Fuentes et al. 2019; Lee et al. 2019). Furthermore, over planetary time scales of millions to billions of years, different processes allow exchanges of atmophile (atmosphere-loving) or volatile elements between the various planetary reservoirs, which in turn influences compositions of ocean-atmosphere systems (McGovern and Schubert 1989; Lee et al. 2013, 2019; Fuentes et al. 2019; Plank and Manning 2019). In other words, not all volatile elements are expected to be in the atmosphere for a given bulk composition of a planet or a planetary system. Given the major volatile elements, i.e., C, H, N, and S, are also essential for making life as we know it, in this presentation, we will refer to them as elements essential for life as we know it or life-essential volatile elements (LEVEs).

In this chapter, we focus on different processes that lead to the origin, long-term storage, and exchanges of LEVEs in and between different planetary reservoirs—both for systems that are young, i.e., no more than 10s of million years old, as well as those that are matured, i.e., 100s of millions to billions of years old. We use the lessons we learned from the formation and evolution of Solar Systems' rocky planets as the basis to inform our expectations for other rocky planets in our galaxy that have broad similarities to our worlds.

ORIGIN OF LEVE IN THE ROCKY PLANETS

Lessons from the Solar System—primitive building blocks

Stars and their planetary systems are known to form due to the gravitational collapse of molecular cloud cores. Our own Solar System formed this way ~4.57 billion years ago. The result was a disk of gas (99 wt.%) and dust (1 wt.%), known as the protoplanetary disk or the Solar nebular disk (Broadley et al. 2022). Given that the stars contain nearly all the mass of any solar system, the bulk compositions of planets and their building blocks largely reflect the composition of the central star. This is known to be the case for the Solar System (Fig. 1), and the concept is also being utilized for exoplanetary systems (e.g., Hinkel et al. 2024, this volume). Indeed, the first solids, which made the planetary building blocks, were primarily the result of condensation from this Solar disk, along with a small contribution of presolar matters inherited from the interstellar medium or ISM (Grossman and Larimer 1974). Primitive, undifferentiated asteroids, known to us through chondritic meteorites and recently returned samples from Hyabusa2 and comets observed in the Solar System, are composed of these early-formed solids. Therefore, an important question is when and under what condition various elements in general and life-essential volatile elements, C, N, H, and S, condensed from the solar nebular gas in the disk.

The phases condensed from the Solar gas are initially micron-sized dust particles drifting around in the protoplanetary disk. The electrostatic attraction of these dust particles results in the formation of millimeter to centimeter-sized pebbles (Morbidelli et al. 2012). When concentrated in sufficient abundance, these pebbles collapse gravitationally in a narrow annulus of rings to form bigger planetesimals (Chiang and Youdin 2010; Simon et al. 2016; Izidoro et al. 2022). The small planetesimals are not considered big enough to have undergone differentiation into a metallic core and silicate mantle due to the lack of short-lived radioactive isotopes like ^{26}Al and ^{60}Fe.

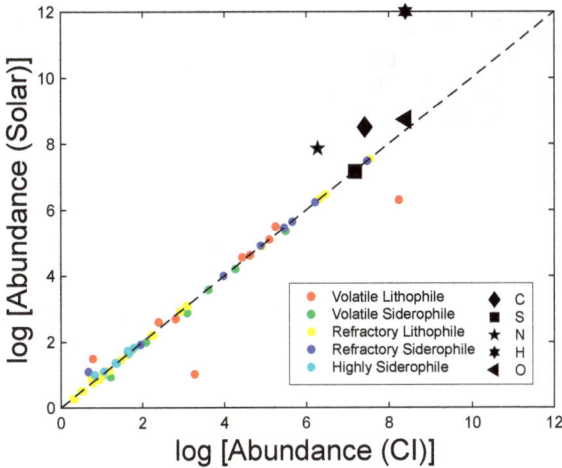

Figure 1. The Solar photosphere abundances of various elements grouped according to different geochemical classes (volatile lithophile, volatile siderophile, refractory siderophile, refractory lithophile, and highly siderophile) are plotted against the CI abundances of these classes of elements (both from Palme et al. 2014). Most of the geochemical groups of elements mimic the Solar photosphere except the highly volatile elements C, H, and N (denoted in **black symbols**). CI abundance of C, H, and N are depleted compared to their Solar abundances. The **dashed line** is a 1:1 line.

Furthermore, chondritic parent bodies are also thought to have escaped large-scale melting and differentiation, owing to the late accretion of these bodies with respect to the decay time scales of ^{26}Al. Consequently, we can sample the remnants of these undifferentiated protoplanetary bodies as various chondritic meteorites from the asteroid belt that reaches Earth. Depending on the source region of these undifferentiated objects in the protoplanetary disk, the chondrites have been classified into inner Solar System enstatite (EC), ordinary chondrites (OC), and outer Solar System carbonaceous chondrites (CC); e.g., Jones (2024, this volume). The CCs are typically more volatile-rich than the ECs. This difference in the volatile budget is likely linked to the region of formation of these parent bodies. CCs are thought to have formed at larger heliocentric distances, where volatile-rich compounds could be stable or condense. Whereas ECs are thought to have originated closer to the proto-Sun, where volatile-rich compounds were unstable and much more thermally processed. The isotopic ratios of H and N of various Solar System chondritic reservoirs have been measured and studied to understand the geochemical nature of the terrestrial planets. The isotopic signatures of the volatile elements in chondrites point unequivocally to the origin of these elements on Earth and other rocky planets in the Solar System. For example, in Figure 2, we plot the D/H vs ^{15}N/^{14}N for various geochemical reservoirs of our Solar System and compare them with various chondritic compositions. Although OCs, which sample S-type asteroids in the inner belt (2.1–2.8 AU), contain similar bulk water content as the BSE, the D/H ratio of the BSE is dissimilar to that of OCs (McNaughton et al. 1981). The BSE D/H ratio is remarkably similar to that of CCs (Robert et al. 1979) , in particular, the CI chondrites, which are representative of C-type asteroids (>2.8 AU) (Fig. 2). The ^{15}N/^{14}N ratio of the BSE also suggests an EC origin of volatiles on Earth (Pearson et al. 2006; Alexander et al. 2012; Labidi 2022). Figure 2 also shows that the Solar System's rocky planets did not get their LEVEs from the solar nebular gas capture. These volatile isotope fingerprints are essential as they suggest that the volatiles on Earth were likely sourced from materials from further reaches of the Solar System and not from the local region where the inner planets are currently situated. Interestingly, stable isotopes of several non-volatile, refractory elements provide a conflicting constraint. For example, several lithophile elements (ε^{48}Ca (Dauphas et al. 2014), ε^{50}Ti, μ^{142}Nd), moderately siderophile elements (ε^{54}Cr,

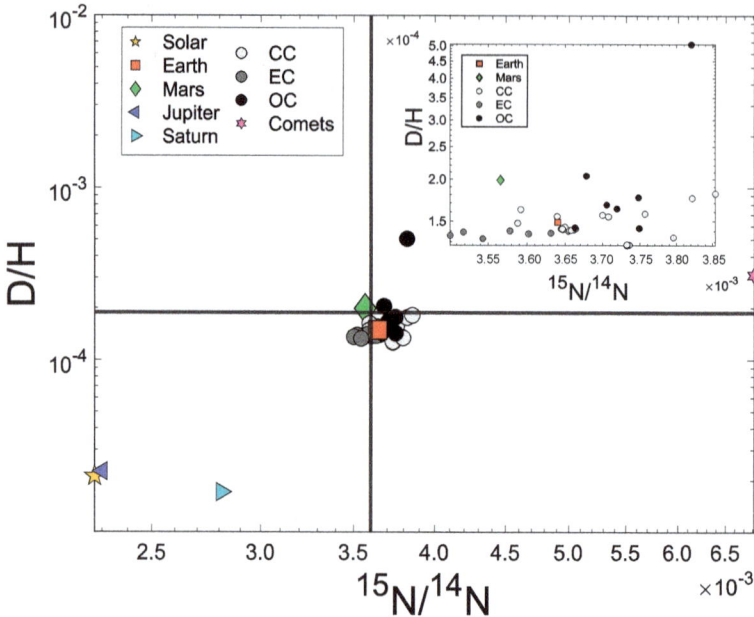

Figure 2. $^{15}N/^{14}N$ vs. D/H of various undifferentiated geochemical reservoirs (CC–Carbonaceous Chondrites, EC–Enstatite Chondrites, OC–Ordinary Chondrites) compared with those of the planets and comets of the Solar System. The relative compositions of the primitive building blocks with respect to the terrestrial planets (e.g., Earth, Mars) provide an understanding of the relative contribution of various primitive building blocks in shaping the isotopic compositions of the LEVEs of Earth. Earth–Junk and Svec (1958); Füri and Marty (2015, 2022); Mars–Mathew and Marti (2001); Hallis (2017); Jupiter, Saturn–Marboeuf et al. (2017); Chondrites–Piani et al. (2021), and references within; Comets–Balsiger et al. (1995); Eberhardt et al. (1995); Bockelée-Morvan et al. (1998); Hutsemékers et al. (2009).

$\varepsilon^{64}Ni$, $\varepsilon^{92}Mo$), and highly siderophile elements (HSEs; $\varepsilon^{100}Ru$) (Dauphas 2017; Fischer-Gödde and Kleine 2017; Kruijer et al. 2020) all point to the building blocks of Earth and Mars having the inner Solar System affinity, i.e., similar to the OCs and ECs (Warren 2011; Kruijer et al. 2020).

The isotopic fingerprints of the primitive building blocks and their comparison with the inner Solar System planetary reservoirs are important for two reasons. First, it tells us that, at least for the rocky planets in our Solar System, the primary acquisition process of LEVEs has not been nebular capture but rather through accretion of primitive solid condensates and their derivatives. Second, the geochemical similarity of the inner Solar System planets with both EC-, OC- and CC-type primitive meteorites put strong constraints on the dynamical models of Solar System formation (Burbine and O'Brien 2004; Fitoussi et al. 2016; Dauphas 2017; Brasser et al. 2018; Liebske and Khan 2019; Izidoro et al. 2022). Various models for accretion and growth of terrestrial planets, therefore, call for some modes of scattering, mixing, and implantation of outer Solar System objects to the region of rocky planet formation (Raymond and Morbidelli 2014; Izidoro et al. 2022). When placed in the context of rocky exoplanet formation, the constraints on Earth's LEVE origin are informative as they provide plausible frameworks to investigate chemical habitability of other solar systems. Indeed, the research community working on dynamics of planet formation have been working to decipher the observed architectural structure of planetary systems hosting these rocky exoplanets (Izidoro et al. 2017; Raymond et al. 2021; Hatalova et al. 2023). Finding commonalities between the early Solar System configuration and those of other young exoplanetary systems is intriguing, particularly in light of acquiring LEVEs in rocky planets.

In addition to the isotopic compositions, the concentration of major LEVEs like H, N, and C in various chondrites, both in inner Solar System (EC, OC) and outer Solar System (CC) meteorites, also suggests a spatial gradient of these volatiles in early Solar System, with outer Solar System CCs being more volatile-rich compared to the inner Solar System ECs and OCs. However, in all cases, the concentrations of C, N, H, and S even in the CCs are significantly and variably depleted in comparison to the solar abundances (Fig. 1). This observation is critical for the attempts to reconstruct rocky exoplanet compositions from their host star compositions. The comparison between Solar and CI-chondrite compositions (Fig. 1) suggests that although the non-volatile and major elements may be reasonably reconstructed for the bulk planet and some for even planetary mantles (Hinkel and Unterborn 2018; Adibekyan et al. 2021) , LEVE abundances of bulk exoplanets and exoplanetary silicate reservoirs are much more difficult to reconstruct.

The BSE abundance of elements versus 50% condensation temperature (T_{50})

Condensation temperature is a helpful concept to understand how various elements fractionate from a gaseous phase of the disk to any solid phase. It is formally defined as the temperature at which 50% of an element condenses out of a gas of Solar composition at a total pressure of 10^{-4} bar (Lodders 2003; Wood et al. 2019). Various thermodynamical models have tried to estimate these temperatures (Larimer 1967, 1973; Grossman 1972; Grossman and Larimer 1974; Boynton 1975; Wai and Wasson 1977, 1979; Sears 1978; Fegley and Lewis 1980; Saxena and Eriksson 1983; Fegley and Palme 1985; Kornacki and Fegley 1986; Palme and Fegley 1990). Studies by Lodders (2003) and Wood et al. (2019) provided more accurate and up-to-date estimates of the condensation temperatures of many elements by incorporating better thermodynamical constraints. These estimates of temperatures provide a holistic understanding of the sequence of elements that condensed out from the progressively cooling solar gas lasting a few million years after the birth of the Solar System. Moreover, it also provides an overview of a possible link between the elemental budget of the bulk Silicate Earth (BSE) and 50% condensation temperature. To summarize this, we plot the concentration of elements in bulk Silicate Earth (BSE) normalized to their values in CI chondrites (Fig. 3) against the calculated 50% condensation temperature (T_{50}) of these elements from Lodders (2003). We observe an overall trend of decreasing abundance of elements with decreasing condensation temperature. It is seen that the refractory lithophile (silicate loving) elements like Ca, Ti, Zr, Sc, and rare Earth (REE) are in chondritic concentration (Fig. 3) in silicate Earth. However, there is a strong decline in the budget of refractory lithophile elements with a decreasing 50% condensation temperature (Fig. 3). A similar declining trend is observed for refractory siderophile and volatile siderophile elements, where there is a decrease in the budget of volatile siderophile elements with a decrease in 50% condensation temperature (Fig. 3). It is observed that both refractory and volatile siderophile elements are not in chondritic proportion in the silicate mantle (Fig. 3) primarily due to their fractionation in the metallic core during core-forming processes.

Although condensation temperature provides a first-order understanding of the composition of the bulk silicate Earth compared to chondritic composition, it paints a partial picture of the major volatiles such as C and N. The low condensation temperatures for these elements are calculated assuming that the first phase to condense, containing these volatiles methane ice or methane heptahydrate ($CH_4 \cdot 7H_2O$) and ammonia ice or ammonium hydroxide ($NH_3 \cdot H_2O$) (Lodders 2003). However, several high-temperature phases contain these volatile elements. For example, calcium aluminum inclusion (CAI)-hosted osbornite, TiN is the first N-bearing phase that condenses out from the Solar gas (Meibom et al. 2007) (Table 1). Furthermore, C is also known to be incorporated into metal phases as evidenced by the presence of C in metals of chondrites (Grady et al. 1986; Mostefaoui et al. 2000) as well as in graphite and carbides (in enstatite chondrites; Grady and Wright 2003) (Table 1). On the other cooler end of the temperature spectrum in the gaseous disk, i.e., at ~100–200 K, organic matter enriched in volatiles like C, H, and S are thought

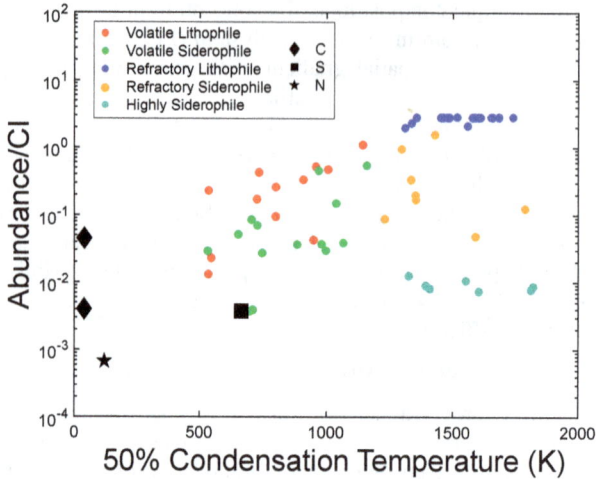

Figure 3. The CI-chondrite normalized Bulk Silicate Earth (BSE) abundances of elements (Palme et al. 2014), grouped according to their geochemical affinities (Wood et al. 2019), plotted against their 50% condensation temperatures, T_{50} (Lodders 2003). The C abundances are from Hirschmann (2018) (the lower abundance) and Marty et al. (2020) (the higher abundance). The N abundance is from Hirschmann (2018). The S abundance is from Palme et al. (2014), which is consistent with more recent mantle S estimates (e.g., Ding and Dasgupta 2017, 2018; Dasgupta and Aubaud, in press). There is an overall trend of decreasing abundance of elements with decreasing condensation temperature. The refractory lithophile elements are in chondritic proportion with the volatile lithophile having a decreasing abundance with decreasing condensation temperatures. A similar trend is observed for refractory siderophile and volatile siderophile elements. Moreover, all classes of siderophile elements (refractory, volatile, highly) are not in chondritic proportion in the BSE. This can be explained by the sequestration of the siderophile elements during core forming processes.

to have formed by reactions between various gas species in the Solar gas (Studier et al. 1965; Alexander et al. 2012). Therefore, using a low T_{50} for the major volatiles may incorrectly point to the cause of depletion of the major volatiles in the inner Solar System planets such as Earth.

An additional observation from Figure 3 is that the extent of depletion of a given LEVE in the BSE compared to the CI chondritic meteorite differs. For example, compared to the CI chondrites, the BSE nitrogen is more depleted than carbon and hydrogen (e.g., Marty 2012; Hirschmann 2016; Dasgupta and Grewal 2019; Marty et al. 2020). In other words, the C/N ratio of the BSE is superchondritic, whereas the C/H ratio is subchondritic (not shown), and the C/S ratio is chondritic to superchondritic (depending on which BSE C estimate is taken). The elemental fraction of the LEVEs from the solar to chondritic to the BSE suggests that the planet formation process from dust to pebbles to planetesimals to embryos and then to rocky planets leads to differential processing or loss of the LEVEs. For example, the superchondritic C/N ratio of the BSE has led to several models of terrestrial accretion, growth, and MO processing (Hirschmann 2016; Dasgupta and Grewal 2019; Grewal et al. 2019b; Li et al. 2023). This motivates a deeper look at the formation of rocky planets and the processing of the planetary building blocks to understand the origin of LEVEs in a fully formed planet.

THE FATE OF LEVE DURING THE ACCRETION AND DIFFERENTIATION OF ROCKY BODIES

LEVEs delivered to growing planets by various building blocks and contained in different phases (Table 1) are processed during accretion and associated planetary differentiation. The accretion process is known to lead to magma oceans (MOs), i.e., large-scale melting

Table 1. LEVE-bearing phases in the Solar System's primitive meteorites.

	C	N	S	H
CC	Organics: Soluble (e.g., amines, DNP (2,4-dinitrophenyl) derivatives, TFA (trifluoroacetic anhydride), alkyl. Insoluble (e.g., $C_{100}H_{75-79}O_{11-17}N_{3-4}S_{1-3}$, $\sim C_{100}H_{80}O_{20}N_4S_2$) Carbonates (aragonite/calcite, dolomite, breunnerite, siderite) Elemental (for CO, CV, and CK)	Organics: Soluble (e.g., pyridine carboxylic acids, diketopiperazin, hydantoins, purines) Insoluble (similar compounds like that for C)	Organics: Soluble Insoluble (aliphatic sulfur, heterocycle sulfur ($\sim C_{100}H_{70}O_{70}N_3S$) Elemental sulfur Sulfate (e.g., gypsum) Organosulfur compounds	Organics: Soluble (similar to C, N, and S) Insoluble (similar to C, N, S) Phylosilicates
OC	Organics Elemental Chondrules Presolar grains Metal	Organics Presolar grains	Sulfides (troilite) Metals	Insoluble organic matter Chondrules Mineral phases (olivine, pyroxene)
EC	Graphite Iron carbide (cohenite) Fe–Ni metal	Sinoite ($Si_2N_2O_4$) Nierite (β-Si_3N_4) Fe–Ni metal Osbornite (TiN)	Sulfides (FeS, CaS, [Mg,Mn,Fe]S, MnS), Fe–Ni metal	Insoluble organic matter Chondrule mesostases Low Ca-pyroxene
Refs.	1, 2, 3, 6, 12, 15, 17, 20, 21, 22	1–3, 10–16, 17, 22	5, 8, 9, 17, 18, 19	2, 3, 4, 6, 7, 17, 20

Notes: CC—Carbonaceous Chondrites, EC—Enstatite Chondrites, OC—Ordinary Chondrites.
Refs: 1–Simkus et al. (2019); 2–Kissel and Krueger (1987); 3–Alexander et al. (2007); 4–Piani et al. (2021); 5–Defouilloy et al. (2016); 6–Piani et al. (2016); 7–Piani et al. (2012); 8–Gao and Thiemens (1993b); 9–Gao and Thiemens (1993a); 10–Grady and Wright (2003); 11–Bannister (1941); 12–Alexander et al. (1990); 13–Lee et al. (1995); 14–Andersen et al. (1964); 15–Mostefaoui et al. (2000); 16–Martins (2018); 17–Sephton (2002); 18–Remusat (2011); 19–Naroka et al. (2023); 20–Hanon et al. (1998); 21–Grady et al. (1988); Pearson et al. (2006).

of the silicates of a rocky planet. MOs result from heating of the mantle above its solidus temperature. Various heat sources exist to control the duration and extent of the MO phase (Chao et al. 2021). Some are short-lived, such as impacts during the accretion, radioactive decay of short-lived nuclides (^{26}Al and ^{60}Fe in the case of the early solar system; Elkins-Tanton et al. 2011; Neumann et al. 2014), or the heat delivered by the core–mantle differentiation. These processes are associated with the early phases of planetary evolution, and most—if not all—terrestrial planets are thought to go through one or several episodes of MO during these early stages. This chemical differentiation, leading to layered structures similar to the rocky planets such as Earth, Mars, Moon, and Mercury, is a fundamental process that must apply to all terrestrial exoplanets and rocky Super-Earths. These MOs are short-lived and crystallize as the planet cools down after a heat perturbation, but during this stage, large planet-scale geochemical distribution is established, including those for the LEVEs.

There may also be permanent heat sources, leading to sustained MOs in rocky planets. Those include the insolation (i.e., the radiative flux received from the host star) for short-orbit exoplanets. These exoplanets have an equilibrium temperature at which a black body would be at thermal equilibrium with the received stellar flux above silicate solidus and thus host a permanent MO. Furthermore, the star's proximity required for such high flux likely results in tidal locking (particularly for planets orbiting small stars) so that the sub-stellar point is fixed. In this case, the magma ocean could be hemispherical (the night side exposed to space remains cold). However, if heat redistribution is efficient (because of the presence of an atmosphere), the MO could be global. Such planets are exposed to intense XUV stellar flux, which is likely to deplete the planet of its LEVE over a long timescale. However, a deep MO could sequester LEVE inside the planet if it is not convecting (for instance, in a cold core case; Boukaré et al. 2022).

Accretion of a large mass of nebular gas before the disk dispersal can also result in super-solidus temperature at the bottom of the primordial atmosphere. This requires essential hydrostatic heating associated with large envelope masses that only protoplanets massive enough can capture. Stökl et al. (2016) calculated that planets larger than 0.5 Earth mass could accumulate enough nebular gas to reach a surface temperature above the solidus if provided enough time before the gas disk dispersal. In such case, a Super-Earth would soon reach the runaway accretion limit, from which point it would evolve into a gas giant if supplied enough nebular gas. Timing of gas disk dispersal is then a key limiting parameter.

Permanent MOs have no like in the Solar System (although Io may have a permanent underground magma ocean sustained by tidal dissipation). As of October 2023, a few tens of confirmed exoplanets have been characterized as rocky and have a super-solidus equilibrium temperature, like CoRoT-7 b (Léger et al. 2009) or 55 Cancri e (McArthur et al. 2004) and many more could host a MO below a thick primordial atmosphere. It is noteworthy, however, that there is an observation bias favoring the detection of these planets as they orbit close to their stars and thus have a higher transit frequency and a more robust radial velocity signature. Thus, lava worlds hosting permanent MOs offer tantalizing observation opportunities, but young exoplanets hosting transient magma oceans have also been suggested to be relevant targets (Lupu et al. 2014; Bonati et al. 2019). The concepts of LEVE fractionation developed over the years for the MO stages of Solar System planets are most applicable for these exoplanetary targets.

There are two end-member MO differentiation regimes for rocky objects. (1) The internal differentiation or an internal magma ocean (IMO) scenario and (2) large-scale melting leading to an external magma ocean (EMO) scenario (Fig. 4). In the former, the early objects (e.g., asteroids or planetesimals) undergo melting chiefly owing to the decay of short-lived radionuclides, i.e., ^{26}Al and ^{60}Fe, leading to short-lived MOs (e.g., Miyamoto et al. 1981; Hevey and Sanders 2006). This may facilitate enough melting of metals and silicates such that alloy melts segregate to form cores in these bodies. However, the outer shells of planetesimals, in this case, remain unmolten and largely unmodified from the primitive state. There are suggestions that even the chondritic meteorites, which are otherwise thought to be undifferentiated objects, in some cases may derive from parent bodies that are internally differentiated, with the meteorites only sampling the primitive rubble piles at the surface regions (Elkins-Tanton 2012). It has been argued that the iron meteorite parent bodies might have undergone differentiation following an IMO (Grewal et al. 2022a). For the IMO differentiation scenario, LEVEs fractionate between two distinct geochemical reservoirs, i.e., the metallic and/or sulfide-rich core and the silicate mantle (Grewal et al. 2022a; Pathak and Dasgupta 2024). In EMO differentiation, planetesimals, protoplanets, and embryos have molten silicate surfaces. Hirschmann et al. (2021), using C–S systematics of iron meteorites, argued that the earliest-formed planetesimals in our Solar System might have experienced this differentiation scenario. Howardites-eucrites-diogenites of the HED meteorites also point to the EMO scenario for the asteroid 4 Vesta (Mandler and Elkins-Tanton 2013; Neumann et al. 2014). Otherwise, EMO is considered a widespread state of early planets during accretion and growth via impacts (e.g., Matsui and Abe 1986; Solomatov 2000; Elkins-Tanton 2012). In the EMO differentiation scenario, molten planetary objects can outgas the volatile elements, stabilizing an early atmosphere (Fig. 4), which may be lost or retained depending on several factors discussed later. In other words, for the EMO differentiation, the LEVEs fractionate

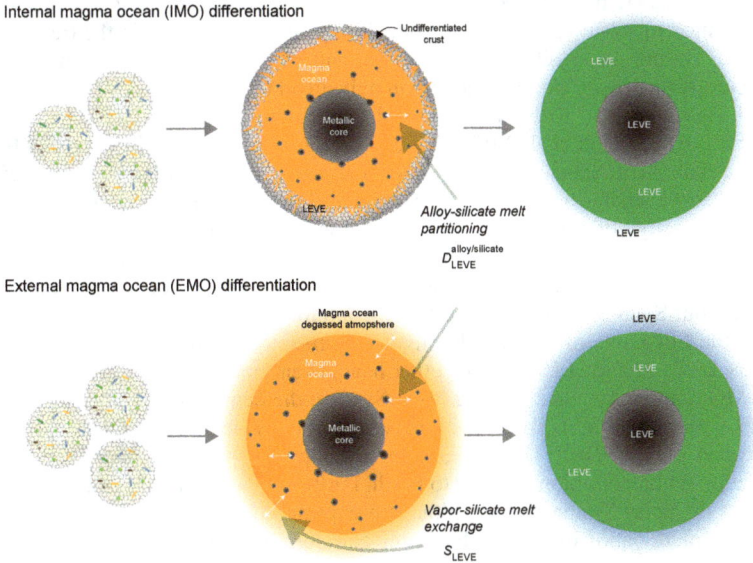

Figure 4. Two end-member regimes of rocky body accretion and differentiation. **Top panel:** Internal magma ocean differentiation (IMO). **Bottom panel:** External magma ocean differentiation (EMO). In the IMO differentiation regime, the secondary atmosphere of the planet is largely outgassed over time, via volcanism. In the EMO differentiation regime, the secondary atmosphere is first established via MO degassing and only modulated later via volcanic degassing.

between three distinct geochemical reservoirs, viz., the metallic core, silicate mantle/MO, and the overlying atmosphere. The distribution of the volatile elements between the metallic core and silicate mantle is controlled chiefly by the equilibrium partition coefficients of these elements between alloy and silicate melt, $D_{\text{LEVE}}^{\text{alloy/silicate}}$, whereas, between the MO or molten silicate mantle and the overlying atmosphere, the LEVE distribution is controlled by the solubility of these elements in silicate melt, S_{LEVE} (Fig. 4).

LEVE fractionation during alloy–silicate differentiation or core formation

The geochemical effects of core formation on the LEVE budget of rocky planets' degassable layers is understood utilizing the equilibrium partition coefficients of C, N, S, and H between Fe-rich alloy melt and silicate melt, which is given by

$$D_{\text{LEVE}}^{\text{alloy/silicate}} = \frac{\text{concentration of LEVE in alloy melt}}{\text{concentration of LEVE in silicate melt}} \tag{1}$$

If is $D_{\text{LEVE}}^{\text{alloy/silicate}} > 1$, then the LEVE of interest is expected to concentrate in the metallic core preferentially, and if it is < 1, then the LEVE is expected to concentrate in the silicate MO or mantle of the planet.

For decades, mainly from the metallurgical studies at ambient pressures, it was known that non-metals such as carbon, sulfur, and nitrogen can dissolve in molten Fe-metal and form stable intermetallic compounds (e.g., von Goldbeck 1982; Wang et al. 1991; Tsymbulov and Tsmekhman 2001). There have also been several studies exploring the presence of S, C, and H as the light elements in the Earth's outer core (Wood 1993; Li and Fei 2003; Lord et al. 2009; Badro et al. 2015; Bajgain et al. 2019, 2021). The primary motivation of these latter studies has been to explain the physical properties of the Earth's core as learned from seismological studies in general and the deficit of density of the Earth's outer core with respect to the pure molten Fe in particular. However, for a long time, the direct constraints on the equilibrium partitioning of LEVE between coexisting Fe-metallic and silicate melt systems remained primarily limited to sulfur (Li and Agee 1996, 2001). Efforts to constrain the fractionation of other LEVEs during metal–silicate equilibration and core segregation through MO through high-pressure–temperature experiments began only about ten–fifteen years ago (Rose-Weston et al. 2009; Dasgupta et al. 2013a; Roskosz et al. 2013; Boujibar et al. 2014; Chi et al. 2014; Li et al. 2015a, 2016a; Dalou et al. 2017; Suer et al. 2017; Grewal et al. 2019a,b, 2021b, 2022b; Malavergne et al. 2019; Speelmanns et al. 2019; Jackson et al. 2021; Blanchard et al. 2022; Shi et al. 2022). These studies have been conducted over a wide range of pressures, temperatures, and oxygen fugacity conditions and using various silicate and alloy compositions (Dasgupta and Grewal 2019; Suer et al. 2023). It should be noted that most of these experimental studies conducted thus far have explored intensive and extensive variable space relevant for shallow to deep MO for our Solar System's rocky planets. Therefore, these experiments are not expected to capture the full range of MO conditions for diverse rocky exoplanetary systems (Zhang and Rogers 2022; Zhang 2023). On the other hand, even though Super-Earths may have had a deeper MO than rocky planets in our Solar System, the degree of equilibrium between the impactor's core and MO is likely to decrease as proto-planets grow (Dahl and Stevenson 2010; Deguen et al. 2014). Therefore, even for Super-Earths, they may have undergone differentiation under conditions similar to Earth. Furthermore, the trends established from the experiments on Solar System compositions are expected to provide critical insight into the geochemical fractionation of the LEVEs during equilibrium core formation in a wide range of rocky exoplanets and proto exoplanets. Furthermore, there are a limited number of discovered exoplanets, such as the TRAPPIST system planets (TRAPPIST-1b, 1d, 1f; Dorn et al. 2018; Agol et al. 2021), which are comparable in size to inner Solar System planets. Hence, these planets likely have undergone MO differentiation at pressure and temperature conditions similar to Earth's core forming conditions. In the following, we present current constraints on the partitioning behavior of C, N, S, and H between Fe-metal alloy and silicate melt as a guide toward LEVE fractionation in terrestrial exoplanets and rocky Super-Earths.

Alloy–silicate partitioning of carbon

One of the most studied life-essential volatile elements, in the context of alloy–silicate equilibration and core formation is carbon (C). The partitioning of C between Fe-rich alloy and silicate melt is mainly a function of pressure, temperature, and oxygen fugacity, f_{O_2}. In Figure 5a, we plot $D_C^{alloy/silicate}$ from various experimental studies against the oxygen fugacity of the experiments. All experimental values in Figure 5a are > 1, i.e., carbon is siderophile. Figure 6c shows that $D_C^{alloy/silicate}$ decreases with an increase in f_{O_2}, especially above ~IW-2 (Armstrong et al. 2015; Li et al. 2015a, 2016a; Dalou et al. 2017). Although the $D_C^{alloy/silicate}$ parameterization in Figure 6 does not capture, the measured $D_C^{alloy/silicate}$ values in many studies also show a decrease with decreasing f_{O_2} below ~IW-2 (Fig. 5a). This behavior can be explained by changing speciation of C in silicate melt as a function of f_{O_2}. C dissolution in Fe–Mg-rich silicate melts like basalts and peridotite is featured by increasing CO_3^{2-} with increasing f_{O_2} and increasing solubility of C–H species with decreasing f_{O_2} (Ardia et al. 2013; Li et al. 2017; Dasgupta and Grewal 2019). Recent studies focusing on estimating the $D_C^{alloy/silicate}$ at pressures > 50 GPa (Fischer et al. 2020; Blanchard et al. 2022) show that carbon becomes less siderophile with increasing pressure. The effect of pressure, temperature, and oxygen fugacity are shown in modelled curves in Figure 6, based on the parameterization of $D_C^{alloy/silicate}$ from Blanchard et al. (2022), following the equation

$$\log K_D = -1(\pm 0.5) + 4842(\pm 920)/T(K) + 31(\pm 19)P(GPa)/T(K) \qquad (2)$$

where, K_D is the ratio of the alloy–silicate partition coefficient of C to that of Fe and is independent of the Fe concentration during the alloy–silicate equilibrium reaction. It is observed that there is an increase in $D_C^{alloy/silicate}$ with increasing pressure (Fig. 6a), and a decrease with temperature (Fig. 6b), and f_{O_2} (Fig. 6c).

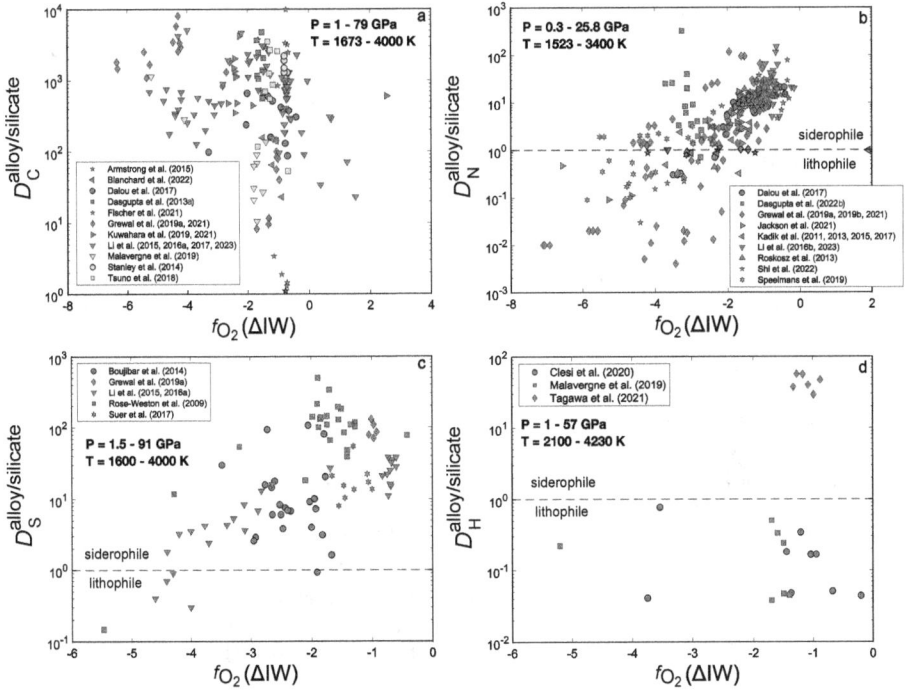

Figure 5. Fe-rich alloy melt/silicate melt partition coefficients for LEVEs, $D_{LEVE}^{alloy/silicate}$ determined using high pressure-temperature experiments in the literature: **(a)** $D_C^{alloy/silicate}$ increases with decreasing f_{O_2} till $\log f_{O_2}$ of ~IW-2, followed by a decrease below $\log f_{O_2} \sim$ IW-2. **(b)** $D_N^{alloy/silicate}$ increases with increasing f_{O_2} of the metal–silicate system. Depending on the f_{O_2} of metal–silicate equilibration, N can be either siderophile or lithophile. **(c)** $D_S^{alloy/silicate}$ also increases with increasing f_{O_2}, with S behaving as a lithophile element at $\log f_{O_2}$ < ~IW-4. **(d)** For metal–silicate partitioning of H, there is no apparent effect of f_{O_2}; however, experimental constraints on $D_H^{alloy/silicate}$ is most limited.

All data and models point to the fact that if core formation were an equilibrium process, then C would be preferentially sequestered in the metallic core of differentiated bodies. In other words, silicate portion of any rocky planet should be depleted with respect to the bulk planetary C budget available during core formation. In addition, depletion of C in the BSE with respect to chondrite can in part be explained by C partitioning into the core. If the trend of decreasing $D_C^{alloy/silicate}$ with pressure continues at higher pressures, it is possible that C would cease to be siderophile for extremely deep MO for rocky Super-Earths. However, if deep MO is more reduced, the effect of lower f_{O_2} on $D_C^{alloy/silicate}$ especially at a nominally anhydrous condition, may offset the effect of pressure and keep carbon a siderophile or core-loving element. Another way of making carbon available in the silicate portion of a rocky planet in sufficient abundance is to deliver carbon-rich materials post-dating core formation, which partly might have been the case for Earth and Mars in our own Solar System.

Alloy–silicate partitioning of nitrogen

The main control on $D_N^{alloy/silicate}$ is oxygen fugacity. In Figure 5b, the experimentally determined $D_N^{alloy/silicate}$ is plotted vs. the f_{O_2} of the metal–silicate systems (Kadik et al. 2011, 2013, 2017; Roskosz et al. 2013; Li et al. 2016b; Dalou et al. 2017; Grewal et al. 2019a,b, 2021b; Speelmanns et al. 2019; Jackson et al. 2021; Dasgupta et al. 2022b; Shi et al. 2022). It is observed that in reducing conditions, N is lithophile in nature and becomes progressively more siderophile with increasing f_{O_2} (Fig. 5b). This geochemical behavior of N is again affected by the speciation of N in silicate melts. For example, N in highly reduced, anhydrous systems is found to be usually present as N^{3-} in the silicate melt and can replace the tetrahedrally bonded O^{2-} in the silicate structure due to the similarity in size between N^{3-} and O^{2-} (Baur 1972; Grewal et al. 2020). In hydrogenated systems and reducing conditions, N is still in N^{3-} state, however, bonded to H, for example, as primary and secondary amides (Mosenfelder et al. 2019; Grewal et al. 2020). In the Fe-rich metal structure, N dissolves as interstitial N atoms (Häglund et al. 1993). However, in oxidizing conditions at $f_{O_2} > IW - 1.5$, the speciation of N in anhydrous condition is as molecular N_2 (Libourel et al. 2003; Grewal et al. 2020). Because N dissolution in silicate melts in reducing conditions is significantly greater than in oxidizing conditions, $D_N^{alloy/silicate}$ increases with increasing f_{O_2}. In addition, $D_N^{alloy/silicate}$ decreases with an increase in temperature and it increases only modestly with increasing pressure (Roskosz et al. 2013; Grewal et al. 2019a, 2021b; Jackson et al. 2021; Dasgupta et al. 2022b) whereas some studies even arguing a negative effect of pressure on $D_N^{alloy/silicate}$ (Kadik et al. 2015; Dalou et al. 2017; Grewal et al. 2019a,b). Moreover, the effect of pressure is better observed in oxidizing conditions (Grewal et al. 2019b), with almost no effect observed in reducing conditions. However, the lack of experiments at pressures >25 GPa makes it difficult to ascertain the effect of pressure on $D_N^{alloy/silicate}$ at core-forming conditions for Earth and other rocky exoplanets that are similar or larger in size. In Figure 6, the modelled curves for $D_N^{alloy/silicate}$ are shown as a function of pressure, temperature, and f_{O_2} (Fig. 5b). The model curves are based on the following parameterization from Grewal et al. (2021a), which was corrected in Grewal et al. (2022a).

$$\ln D_N^{alloy/silicate} = 7513.54 + \frac{9813.37}{T(K)} - 68.24 + 362.16 \ln(100 - X_S^{alloy})$$
$$- 39.87 \ln(100 - X_S^{alloy})^2 - 0.25 \ln(100 - X_{Si}^{alloy})^2 \qquad (3)$$
$$- 3596.39 \ln(100 - X_C^{alloy}) + 397.32 \ln(100 - X_C^{alloy})^2$$
$$+ 0.84 \, NBO/T + 1.40 \ln X_{FeO}^{alloy}$$

where $X_S^{alloy}, X_{Si}^{alloy}, X_C^{alloy}$ is wt. % S, Si, and C, respectively in the alloy, $X_{FeO}^{silicate}$ is wt.% of FeO in the silicate melt.

Alloy–silicate partitioning of sulfur

Pressure, temperature, and oxygen fugacity all play important roles in determining $D_S^{alloy/silicate}$. In Figure 5c, the experimental $D_S^{alloy/silicate}$ data (Rose-Weston et al. 2009; Boujibar et al. 2014; Li et al. 2016a; Suer et al. 2017) are plotted versus f_{O_2} of the system. It is seen that with an increase in f_{O_2}, $D_S^{alloy/silicate}$ increases. In addition, $D_S^{alloy/silicate}$ decreases with an increase in temperature. As far as the effect of pressure goes, studies like Rose-Weston et al. (2009) and Boujibar et al. (2014), considered $D_S^{alloy/silicate}$ to increase with an increase in pressure. However, a later study, which explored the effect of pressure and temperature at deep MO core forming conditions (Suer et al. 2017) of Earth (>40 GPa), concluded that $D_S^{alloy/silicate}$ initially increases with pressure in the pressure range <20 GPa; however, the increase is not as strong as the pressure increases beyond ~40 GPa and temperatures higher than >3000 K. In Figure 6, similar to $D_C^{alloy/silicate}$ and $D_N^{alloy/silicate}$, we show the modelled curves of $D_S^{alloy/silicate}$, based on the parameterization of Suer et al. (2017) given below, as a function of P, T, and f_{O_2}.

$$\log D_S^{alloy/silicate} = -3.5(\pm 0.47) + \frac{3000(\pm 1023)}{T(K)} + 33(\pm 11)\frac{P(GPa)}{T(K)}$$
$$+ 14(\pm 2)\log(1 - X_O) + \log(X_{FeO}) - \log(C_S) \qquad (4)$$

where X_O is the O content of the growing core during Earth accretion, X_{FeO} is the mole fraction of FeO in the silicate melt, C_S is the sulfide capacity of the silicate melt. We observe that $D_S^{alloy/silicate}$ has a weak positive correlation with increasing pressure (Fig. 6a), whereas shows a

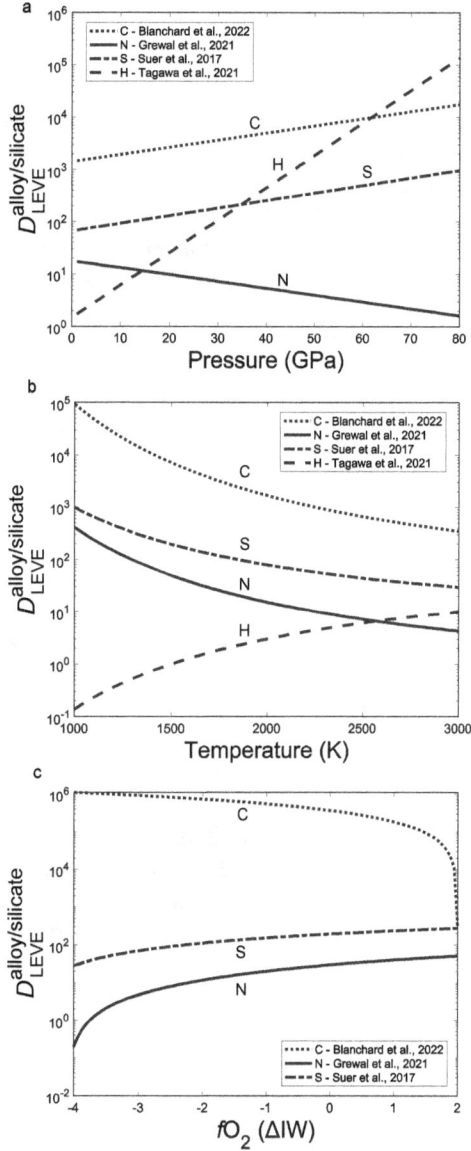

Figure 6. The modelled curves of $D_{\mathrm{LEVE}}^{\mathrm{alloy/silicate}}$, where LEVE = C, N, S, and H, are plotted against three key variables of, i.e., **(a)** pressure (at a fixed temperature of 2000 °C, and log f_{O_2} of IW − 1.5), **(b)** temperature (at a fixed pressure of 5 GPa and log f_{O_2} of IW − 1.5), and **(c)** f_{O_2} (at a fixed temperature of 2000 °C and pressure of 5 GPa). In **(a)**, we observe an increase in $D_C^{\mathrm{alloy/silicate}}$ with increasing pressure. $D_N^{\mathrm{alloy/silicate}}$ shows a small variation with pressure whereas $D_S^{\mathrm{alloy/silicate}}$ shows a very weak increase with increasing pressure at a fixed temperature. Furthermore, we observe a very strong positive correlation of $D_H^{\mathrm{metal/silicate}}$ with pressure. In **(b)**, we observe a strong decrease in $D_{\mathrm{LEVE}}^{\mathrm{alloy/silicate}}$ (LEVE = C, N, S) with increasing temperature. However, $D_H^{\mathrm{alloy/silicate}}$ increase with increasing temperature. In **(c)**, we observe a strong increase in $D_C^{\mathrm{alloy/silicate}}$ with decreasing f_{O_2}. However, $D_N^{\mathrm{alloy/silicate}}$ increase with increasing f_{O_2}. The effect of f_{O_2} on $D_H^{\mathrm{alloy/silicate}}$ is not well constrained at present. For $D_N^{\mathrm{alloy/silicate}}$, the following input parameters were used: $X_S = 5$ wt.%, $X_C = 2$ wt.%, $X_{FeO} = 8$ wt.%, $X_{Si} = 0.002$ wt.%, and NBO/T = 2.2. For $D_S^{\mathrm{alloy/silicate}}$, the following input parameters from Suer et al. (2017) were used: $C_s = -5.372$, $X_{FeO} = 0.057$, and $X_O = 0.02$.

strong anti-correlation with temperature (Fig. 6b). Finally, there is a strong positive correlation of $D_S^{\text{alloy/silicate}}$ with f_{O_2} (Figs. 5c and 6c).

Alloy–silicate partitioning of hydrogen

Among all the LEVEs, the behavior of H during core formation and metal–silicate partitioning is known with the least amount of certainty. Also, H is the least studied element in the context of metal–silicate partitioning. From the limited literature data, it appears that P and T likely have the strongest effects on $D_H^{\text{alloy/silicate}}$ although the studies diverge on their conclusions on the geochemical affinity of H. Studies such as those of Clesi et al. (2018) and Malavergne et al. (2019) argued that H is strongly lithophile in Earth's core forming condition. Whereas, Tagawa et al. (2021) claimed that H is strongly siderophile in Earth's core forming conditions. Furthermore, Tagawa et al. (2021) argued that at low pressures and temperatures (<3 GPa) H is lithophile in nature and with increasing P and T, H becomes strongly siderophile. However, there has not been a lot of work aimed at understanding the effect of f_{O_2} on $D_H^{\text{alloy/silicate}}$. In Figure 5d, we present experimental $D_H^{\text{alloy/silicate}}$ vs. f_{O_2} (Clesi et al. 2018; Malavergne et al. 2019; Tagawa et al. 2021), which shows that there is no clear trend. In Figure 6, we present modelled curves for $D_H^{\text{alloy/silicate}}$ based on the parameterization of Tagawa et al. (2021), i.e.,

$$\log K_D = 0.692(\pm 0.986) - \frac{4590(\pm 1690)}{T} + 141(\pm 43)P/T \qquad (5)$$

where K_D is the ratio of the alloy–silicate partition coefficient of H ($D_H^{\text{alloy/silicate}}$) to that of iron and is independent of the Fe concentration during the alloy–silicate equilibrium reaction, P is pressure in GPa, T is temperature in K. We observe that $D_H^{\text{alloy/silicate}}$ increases with both pressure (Fig. 6a) and temperature (Fig. 6b).

The partitioning behavior of LEVEs during metal–silicate equilibration and core formation outlined above gives a broad guidance as to what the dominant volatile element-bearing reservoirs would be if these elements were available during the core formation process of rocky exoplanets. The studies thus far attempted to obtain the ideal $D_{\text{LEVE}}^{\text{alloy/silicate}}$ value keeping in mind the expected bulk composition, average core-forming condition, and total light element budget of the Earth's core. However, many studies have shown that the presence of one of the LEVE in substantial concentration can affect the alloy–silicate partitioning behavior of another LEVE (e.g., Tsuno et al. 2018; Grewal et al. 2021b). Therefore, in volatile-rich exoplanetary systems, the behavior of a given LEVE may vary substantially from what is expected in volatile-depleted Solar System rocky planets. The prediction for enrichment or depletion of a given LEVE in the silicate portion of a planet also assumes that the rocky Super-Earth internal structures to be similar to the rocky inner Solar System planets (Hakim et al. 2018a; Boujibar et al. 2020), although the rocky exoplanets could be compositionally much more diverse. Furthermore, the expectations for enrichment or depletion based on equilibrium partitioning behavior does not consider MO dynamics. For example, depletion of C and N in the planet's silicate portion via core formation would require efficient segregation of the alloy materials. For rocky Super-Earth MOs, metal droplets maybe entrained over a long time, with turbulent flow impeding iron rain-out (Lichtenberg 2021). In addition, at Super-Earth conditions, the viscosity of the MO might get significantly higher at greater depths, which in turn would impede the segregation of the metal droplets to the core (Korenaga 2010). Furthermore, if the MO is caused by impacts, the nature of impacts, impact locations (e.g., equatorial vs polar) and strong rotation also cause significant difference in equilibration time scales between the raining metal droplets and the silicate MO (Moeller and Hansen 2013; Maas et al. 2021).

MAGMA OCEAN DEGASSING: LEVE REDISTRIBUTION AND ESTABLISHING A SECONDARY ATMOSPHERE

Synchronously with core formation or after the completion of core formation, the LEVE budget of the MO is affected by exchange with the overlying atmosphere. The viscosity of high temperature silicate melt is similar to that of water (~0.001 Pa·s; Karki et al. 2013), and the high radiative heat flux out of cooling MOs drives intense thermal convection in them, which stirs them very efficiently (Solomatov 2007). As a consequence, MOs are generally assumed to be largely homogeneous, both thermally (resulting in an adiabatic temperature profile) and chemically. Furthermore, gas–melt chemical equilibrium is easily reached at the hot surface, so LEVE contents are set to their solubility. Because of the positive effect of pressure on the solubility of gaseous species, as well as the MO homogenization, the surface solubility is set throughout the molten reservoir. One caveat, however, is that while gas ingassing should be achieved relatively quickly, bubble formation can be inefficient as pointed out by some studies (Salvador and Samuel 2023). Therefore, degassing from an MO can take a while to reach equilibrium, leading to the MO LEVE estimate based on solubility giving a minimum value.

Assuming equilibrium, the solubility is specific to each gaseous species, and in the most common form, it follows a Henry law, i.e., it is proportional to the partial pressure of the gas with which melt is at equilibrium.

For a fully molten mantle, the total budget of a given species in a system composed of the atmosphere and MO reads:

$$M_s = M_{atm} X_s^{atm} + M_{MO} X_s^{MO} \tag{6}$$

where M_s is the total mass of species s in the system, M_{atm} and M_{MO} are the mass of the atmosphere and of the MO, respectively, and X_s^{atm} and X_s^{MO} the mass fraction of s in the atmosphere and in the MO, respectively. Assuming hydrostatic equilibrium in the atmosphere yields $M_{atm} = 4\pi R_{pl}^2 P / g$ (with R_{pl} the planetary radius, P the total pressure at the surface and g the surface gravity). In the atmosphere, the mass fraction of a gas can be converted to its molar fraction, i.e. the ratio of its partial pressure to the total pressure: $X_s^{atm} = (\mu_s / \mu_{atm})(p_s / P)$, with μ_s and μ_{atm} the molecular mass of s and average molecular mass in the atmosphere, respectively, and p_s the partial pressure of s.

Chemical (e.g. redox) reactions can change the speciation of volatile elements whose masses are conserved. The endowment of a given element must hence account for all its carrying species:

$$M_e = \sum_s M_s (\mu_e / \mu_s) \lambda_s^e \tag{7}$$

where M_e is the total mass of element e in the system, μ_e the atomic mass of element e, and λ_s^e is the number of atoms of element e in one molecule of species s. The sum represents all gaseous species in the atmosphere carrying element e.

As mentioned above, X_s is generally assumed to correspond to the solubility imposed by the gas–silicate melt equilibrium taking place at the surface. It can thus be computed based on p_s (as well as other quantities like temperature, total pressure, or oxygen fugacity). Most commonly, the solubility can be expressed as

$$X_s^{MO} = \alpha_s p_s^{\beta_s} \tag{8}$$

where as α_s (also referred to as S_s in some texts and, for example, in Fig. 4) and β_s are a set of species-specific constants (the true Henrian case corresponding to $\beta_s = 1$ is often followed, except in the noticeable case of water). Hence, solubility laws generally relate the partial pressure to the molar concentrations.

Taken together, the mass conservation of element e is:

$$M_e = \sum_s \left[4\pi R_{pl}^2 g \left(\frac{\mu_s}{\mu_{atm}} \right) p_s + M_{MO} \alpha_s p_s^{\beta_s} \right] \left(\frac{\mu_e}{\mu_s} \right) \lambda_s^e \qquad (9)$$

For each species s, the first term in the square bracket corresponds to its outgassed mass and the second term to its ingassed mass.

Physical control on MO outgassing of LEVEs

Meaningful physical controls on MO LEVE outgassing can be shown by taking the partial derivatives of the above equation with respect to planetary quantities such as planetary mass, M_{pl}, and LEVE-bearing species-specific quantities such as μ_s, which is the molar mass of the gaseous species. For a simple illustrative case, we first tackle a single species-atmosphere, following a Henrian behavior ($\beta_s = 1$). The mass conservation of this species reads:

$$M_s = \left[\left(4\pi R_{pl}^2 / g \right) + \left(M_{MO} \alpha_s \right) \right] P \qquad (10)$$

The first term on the right-hand-side is the outgassed mass and the second term is the ingassed mass. They are both proportional to P. On the one hand, the radius of rocky planets scales as their mass to the power a ($0 < a < 1/3$, value for a homogeneous sphere), as a consequence of compressibility (Valencia et al. 2006; Zeng et al. 2016; Dorn and Lichtenberg 2021), and the surface gravity scales as the mass divided by the radius squared, i.e. M_{pl}^{1-2a}. Hence, at a fixed pressure, the outgassed mass scales as M_{pl}^{4a-1}. On the other hand, at a fixed pressure, the ingassed mass simply scales as M_{MO}, i.e., as M_{pl}. Thus, the out-to-ingassed mass ratio scales as $M_{pl}^{4a-2<0}$, i.e., the outgassed mass fraction decreases with increasing planetary mass (Fig. 7b).

To ensure mass conservation, the ingassed and outgassed masses add up to M_s (which scales as M_{pl} for a fixed bulk concentration). This is achieved by adjusting P. From the mass conservation equation, P is:

$$P = M_s / \left(4\pi R_{pl}^2 / g + M_{MO} \alpha_s \right) \qquad (11)$$

At a low planetary mass, P scales as $M_s / \left(4\pi R_{pl}^2 / g \right)$, i.e., M_{pl}^{2-4a}. At a high planetary mass, P scales as $M_s/(M_{MO}\alpha_s)$, i.e., goes asymptotically to X_s^{bulk} / α_s, where $X_s^{bulk} = M_s / M_{mantle}$ is the bulk mass fraction of s in the system. Notice that in the large planetary mass end-member, virtually all of s is ingassed, so X_s^{bulk} and X_s coincide asymptotically. This result can also be obtained by considering the solubility equilibrium when all of s is dissolved:

$$M_s / M_{mantle} = \alpha_s P \qquad (12)$$

Figure 7a represents the evolution of P with M_{pl}. Notice that, at $M_{pl} = 6 M_E$ (considered as an upper bound for rocky planets; Fulton et al. 2017), the outgassed mass fraction is still 0.3, so P is still far from its asymptotic value (100 bar with the values of X_s and α_s in use). For a more soluble species, however, this asymptotic value is more readily reached. In the case of water, Miyazaki and Korenaga (2022) observed that the pressure of outgassed water shows little variation with planetary mass for the Super-Earths, because it is mostly ingassed, enforcing a concentration in silicate melt close to X_s^{bulk}, and accordingly a constant equilibrium vapor pressure.

Species-specific controls are better illustrated in the case of an atmosphere with two gas species, with each species being individually conserved. To be representative of a realistic case (a reduced MO-atmosphere system), here we consider one light species similar to H_2 ($\mu_1 = 2$ amu) and one heavier species similar to CO ($\mu_2 = 28$ amu). As a first step, we can consider similar solubility laws (H_2 and CO both have low solubility), i.e. $\alpha_1 = \alpha_2 = \alpha$.

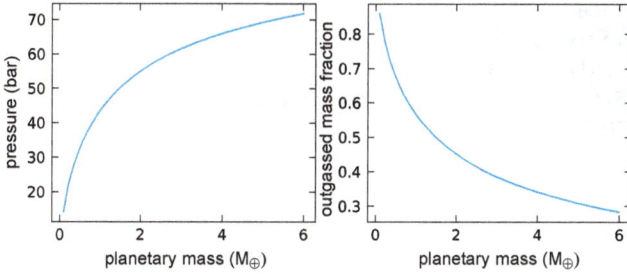

Figure 7. Effect of planetary mass on outgassing of a single-species atmosphere. The **left panel** shows the trend with pressure and the **right panel** shows the trend of outgassed mass fraction.

The mass conservation of either species reads:

$$M_{s,i} = 4\pi R_{pl}^2 g \left(\mu_i / \mu_{atm} \right) p_i + M_{MO} \alpha p_i \tag{13}$$

with $i = 1$ or 2 for the light species or the heavy one, respectively. We also assume that both species have the same bulk mass $M_{s,1} = M_{s,2} = M_s$. Figure 8 is similar to Figure 7, but shows the evolution of both species. Outgassing in the light species is much more extensive, and occurs at lower planetary mass. Indeed, due to its low molecular mass, the effect of planetary mass on outgassing previously mentioned is stronger. As a result, small planets retain more easily heavier species in their molten envelopes.

While the differential effects of other quantities like α_i and $M_{s,i}$ are linear and straightforward, they are still influenced by the effect of the difference in molecular mass between the different gaseous species present in the atmosphere, which is the main source of non-linearity (along with non-Henrian behaviors).

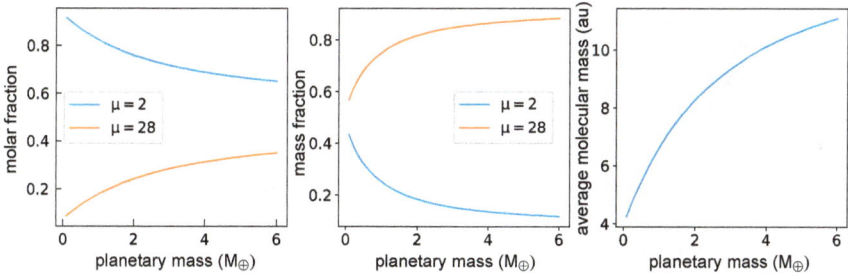

Figure 8. Effect of competitive outgassing between a light species (e.g. H_2) and a heavy one (e.g. CO) in an atmosphere composed of two gas species.

Chemical controls on magma ocean outgassing

In the purely physical picture of LEVE ingassing/outgassing introduced above the gas species are not chemically interacting with each other. However, in reality many LEVE-bearing gas species chemically interact with each other, which not only affects elemental solubilities, but also sets additional constraints on their relative abundances due to gas–gas equilibria in the atmosphere. While there is potentially a very large network of chemical reactions taking place, involving many unstable species, under the high temperatures of a MO atmosphere, chemical equilibrium is readily achieved. As a consequence, the chemistry can be simplified to a small subset of equilibrium reactions, involving a limited number of major gaseous species. These reactions are generally redox equilibria, i.e., they cause a net change in valence of various elements. Under MO conditions, H is essentially present as H_2, H_2O, SH_2, and CH_4 at very high pressures (Bower et al. 2019; Bergin et al. 2023), C as CH_4, CO, or CO_2, N as N_2 (NH_3 and HCN remaining minor species under all relevant conditions), and S as SH_2, S_2 or SO_2. The following set of reactions describes the speciation of these elements (Gaillard et al. 2022; Maurice et al. in revision):

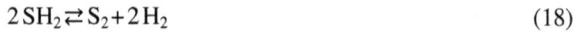

$$H_2 + 1/2\,O_2 \rightleftarrows H_2O \tag{14}$$

$$CO + 1/2\,O_2 \rightleftarrows CO_2 \tag{15}$$

$$CO + 3\,H_2 \rightleftarrows H_2O + CH_4 \tag{16}$$

$$S_2 + 2\,O_2 \rightleftarrows 2\,SO_2 \tag{17}$$

$$2\,SH_2 \rightleftarrows S_2 + 2\,H_2 \tag{18}$$

Although O_2 is only present in minute amounts in the atmosphere, it is still featured in these equilibria. The partial pressure of O_2 (or equivalently, assuming ideal gas, its fugacity f_{O_2}), is generally buffered by the presence of the MO (see below), thus setting a constraint on these chemical equilibria. Each equilibrium sets a constraint of the form:

$$\frac{\prod_p f_p^{c_p}}{\prod_r f_r^{c_r}} = K_{eq}(T) \tag{19}$$

where $f_{p/r}$ is the fugacity of species p (product) or r (reactant), and $c_{p/r}$ its stoichiometric coefficient in the equilibrium of interest, K_{eq} its equilibrium constant and T the local temperature. Assuming ideal gas behavior, the fugacities are equal to the partial pressures normalized by a reference pressure of 1 bar. Taken together, elemental mass conservations, gas–gas equilibria, and solubility laws form a system of equations that can be solved for the partial pressures of each species and their solubilities in the MO. Importantly, assuming that f_{O_2} is buffered by the presence of the MO replaces the mass conservation of O, the MO being then considered as an infinite reservoir for O.

Solubility laws

Dissolution of different gaseous species into silicate melt takes place via different mechanisms, depending on the LEVE-bearing gas molecule of interest. As mentioned above, Henry's laws relating partial pressure and solubility linearly provide an empirical description of dissolution, satisfactory for most species (H_2O being a noticeable exception). Solubility laws for many gaseous species have been investigated experimentally, mainly motivated by volcanic outgassing. However, such data are mostly limited to basalts to more evolved silicate melt compositions rather than MO-relevant ultramafic or peridotitic compositions. This is mainly due to technical difficulties of quenching glass of peridotitic composition, which is necessary to reliably quantify the content and speciation of dissolved volatiles. Below, we briefly review the known solubility laws of species relevant to MO outgassing.

H_2O dissolves in silicate melt as hydroxyl radicals (OH^-) or molecular H_2O at high pressures (>2000 bar; Stolper 1982a,b; Moore et al. 1998; Kohn 2000). The following reactions describe these two mechanisms of water dissolution in silicate melts.

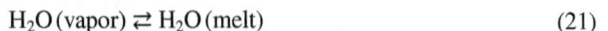

$$H_2O\,(vapor) + O^{2-}\,(melt) \rightleftarrows 2\,OH^-\,(melt) \tag{20}$$

$$H_2O\,(vapor) \rightleftarrows H_2O\,(melt) \tag{21}$$

Dissolution as hydroxyl is more important at total water content below ~2.5–5 wt.%, whereas dissolution as molecular water become increasingly more important at higher total dissolved water contents (e.g., Stolper 1982b; Moore 2008; Mysen 2014). Dissolution of water in the silicate melts as OH^- is also preferred at higher temperatures and pressures (Nowak and Behrens 1995; Mysen 2010). In the case of dissolution as OH^-, which may be more relevant for rocky exoplanets, stoichiometric dissolution yields $[OH^-]^2 \sim f_{H_2O}$. The solubility of OH^- in silicate melt is very high: 524 $ppm_w/bar^{1/2}$ for melt of peridotitic composition, relevant for MO (Stolper 1982a; Sossi et al. 2023). As a consequence, ingassing of H by H_2O dissolution is particularly efficient, especially at low f_{H_2O}. There have been many calibrations for bulk H_2O solubility in silicate melts as functions of melt compositions, pressure, temperature, and fugacity of H_2O in equilibrium fluid that can be either pure H_2O or mixed H_2O–CO_2 fluids (e.g., Moore et al. 1998; Newman and Lowenstern 2002; Papale et al. 2006; Ghiorso and Gualda 2015). One such calibration (Moore et al. 1998) that has been used in recent models of MO–atmosphere interactions (e.g., Hirschmann 2016; Dasgupta and Grewal 2019) is as follows:

$$2\ln X_{H_2O}^{melt} = a / T + \Sigma b_i X_i \left(\frac{P}{T}\right) + c\ln f_{H_2O}^{fluid} + d \qquad (22)$$

where $X_{H_2O}^{melt}$ is the mole fraction of H_2O in the melt, T in K and P in bars, $a=2565$, $c=1.171\pm0.069$, $d=-14.21\pm0.54$, $b_{Al_2O_3,FeO,Na_2O}=-1.997\pm0.706$, -0.9275 ± 0.394, and 2.736 ± 0.871. The model is calibrated for a temperature range of 700–1200 °C, pressure range of 1–3000 bar and for melts ranging from andesitic to rhyolitic in nature.

H_2 undergoes dissolution as molecular H_2. The chemical reaction that describes this form of H_2 dissolution in silicate melt is:

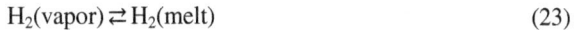

$$H_2(vapor) \rightleftarrows H_2(melt) \qquad (23)$$

Hirschmann et al. (2012) determined solubility of H_2 for a basaltic and an andesitic melt over a pressure range of 0.7–3 GPa and a temperature range of 1400–1500 °C. The calibrated solubility law for the basaltic melt is

$$\ln K = -11.403 - 0.76P \text{ (GPa)} \qquad (24)$$

where K is the equilibrium constant of Equation (23) and P is pressure in GPa. An extrapolation of the H_2 solubility measurements for basaltic and andesitic melts to peridotitic-melt like ionic porosity (controlling physical dissolution) leads to a low solubility ($\alpha_{H_2} \sim 10^{-12}$ (mol/mol)/Pa) following a Henry law. H_2O has a solubility several orders of magnitude higher, making H outgassing very sensitive to speciation between H_2O and H_2, i.e., to f_{O_2}. Sossi et al. (2023) suggested that H_2 can also dissolve as hydroxyl under mildly reduced conditions (IW–0.77–IW–1.9), leading to a non-Henrian behavor similar to water and a solubility factor $\alpha_{H_2} \sim 183 \, ppm_w/bar^{1/2}$.

CO_2 solubility in silicate melts has been extensively studied in the context of volcanic outgassing (e.g., Brooker et al. 2001a; Behrens et al. 2004; Papale et al. 2006; Iacono-Marziano et al. 2012; Ghiorso and Gualda 2015; Eguchi and Dasgupta 2018). In silicate melts CO_2 dissolves as carbonate ions (CO_3^{2-}) or molecular CO_2 (Fine and Stolper 1986; Brooker et al. 2001b; Duncan and Dasgupta 2014, 2015), with carbonate ions being the dominant species for mafic–ultramafic melts relevant for MO (Duncan et al. 2017). The following simple reactions can explain these two mechanisms of CO_2 dissolution in silicate melts:

$$CO_2 (vapor) \rightleftarrows CO_2 (melt) \qquad (25)$$

$$CO_2 (vapor) + O^{2-} (melt) \rightleftarrows CO_3^{2-} (melt) \qquad (26)$$

Among all the available CO_2 solubility laws, we present the solubility law from Eguchi and Dasgupta (2018). The model was calibrated using water-poor (≤ 1 wt% H_2O) silicate melts from graphite to CO_2-fluid-saturation over a range of pressure ($P = 0.05$–3 GPa), temperature ($T = 950$–1600 °C), and in the melt compositional range spanning from that of foidite to rhyolite. One particular advantage of this model is that it can be used for melt degassing or MO outgassing at both oxidized or vapor-saturated to reduced or vapor-undersaturated but graphite/diamond saturated conditions and allows calculation of CO_3^{2-} or CO_2^{mol} individually. The following equation describes the model of Eguchi and Dasgupta (2018):

$$\ln X_i^{\text{melt}} = -\left(\frac{\Delta H}{RT} + \frac{P\Delta V}{RT} - \frac{\Delta S}{R}\right) + \frac{\sum A_i X_i}{RT} + \frac{B\ln f_{CO_2}}{T} + \gamma_{\text{NBO}}\,\text{NBO} \tag{27}$$

where, X_i^{melt} is the mole fraction of species i (CO_3^{2-} or CO_2^{mol}) dissolved in the melt. Units for T are Kelvin, P are pascal, R are J/K·mol, X_i are mole fractions. For CO_3^{2-}, $\Delta H = -1.65 \times 10^5$ J/mol, $\Delta V = 2.38 \times 10^{-5}$ in m³/mol, ΔS in -43.64 J/K, $B = 1.47 \times 10^3$, $\gamma_{\text{NBO}} = 3.29$ and f_{CO_2} are in bars, $A_{CaO} = 1.68 \times 10^5$, $A_{Na_2O} = 1.76 \times 10^5$, $A_{K_2O} = 2.11 \times 10^5$. Whereas for CO_2^{mol}, $\Delta H = -9.02 \times 10^4$ J/mol, $\Delta V = 1.92 \times 10^{-5}$ in m³/mol, ΔS in -43.08 J/K, $B = 1.12 \times 10^3$, $\gamma_{\text{NBO}} = -7.09$. Total CO_2 solubility = CO_3^{2-} + CO_2^{mol}. For standard errors of the model parameters, refer to Table 2 of Eguchi and Dasgupta (2018).

Hirschmann (2016) proposed that C solubility can be captured by a Henry law with a coefficient that decreases with f_{O_2}, as the speciation of dissolved C evolves from carbonate to C=O and to CH and CN complexes. The generally higher solubility of H than C in silicate melts has been confirmed at a composition and temperatures more relevant for MO by Solomatova and Caracas (2021) using *ab initio* simulations. At low f_{O_2}, however, C solubility might be set by graphite (or diamond) precipitation rather than gas–melt equilibrium (Eguchi and Dasgupta 2018; Yoshioka et al. 2019). In this case, in a well-mixed MO, the concentration of dissolved C-species should correspond to their lowest value buffered by the precipitation reaction along the adiabat, and the partial pressure of C-species in the atmosphere follow the concentration by the gas–melt equilibrium.

At high to intermediate f_{O_2}, N has a low solubility, essentially via dissolution of molecular N_2 (Libourel et al. 2003; Boulliung et al. 2020; Dasgupta et al. 2022b; Gao et al. 2022). However, under reducing conditions (log $f_{O_2} \leq IW-1.5$), N readily bonds with the silicate network as N^{3-} either as a hydrogenated or non-hydrogenated species (Dalou et al. 2019; Mosenfelder et al. 2019; Grewal et al. 2020), yielding a significant increase of N solubility (Libourel et al. 2003; Boulliung et al. 2020; Bernadou et al. 2021; Dasgupta et al. 2022b). The chemical reactions that describe these two modes of N solubility in silicate melt are:

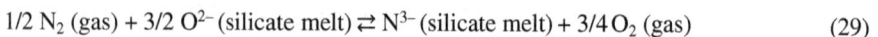

$$N_2 \text{ (gas)} \rightleftarrows N_2 \text{ (silicate melt)} \tag{28}$$

$$1/2\,N_2 \text{ (gas)} + 3/2\,O^{2-} \text{ (silicate melt)} \rightleftarrows N^{3-} \text{ (silicate melt)} + 3/4\,O_2 \text{ (gas)} \tag{29}$$

A number of work have developed parameterization of N solubility in silicate melts (Libourel et al. 2003; Bernadou et al. 2021; Dasgupta et al. 2022b; Li et al. 2023). Below we present the model from Dasgupta et al. (2022b), which is calibrated for a temperature range of 1050–2327 °C, pressure range of 1 bar to 8.2 GPa, log f_{O_2} range of IW-8.3 to IW$+8.7$, and melt composition ranging from basalts to rhyolites, include both synthetic and natural compositions, as well as hydrous and anhydrous compositions.

$$N(ppm) = P_{N_2}^{0.5} \exp\left(5908(\pm268)\frac{P_{Total}^{0.5}}{T} - 1.60(\pm0.02)\Delta IW \right)$$

$$+ P_{N_2} \exp\left(\begin{array}{l} 4.67(\pm0.80) + 7.11(\pm1.61)X_{SiO_2} \\ -13.06(\pm5.17)X_{Al_2O_3} - 120.67(\pm21.14)X_{TiO_2} \end{array} \right)$$

(30)

where the numbers in parentheses are 1σ uncertainties; T is temperature in K, P_{N_2} is the partial pressure of N_2 in GPa, P_{total} is equilibrium pressure in GPa, ΔIW is the log oxygen fugacity relative to the IW buffer, X_{SiO_2}, $X_{Al_2O_3}$, and X_{TiO_2} are the mole fractions of SiO_2, Al_2O_3, and TiO_2, respectively in the silicate melt.

Among the major LEVEs, S is the most soluble in mafic–ultramafic silicate melts, with concentrations of 10^3–10^4 ppm for sulfur fugacities less of than 0.1 MPa (Hirschmann 2016). Chemical reaction describing dissolution of S-bearing gas in silicate melt as sulfide species is:

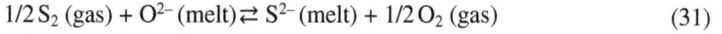

$$1/2\,S_2\,(gas) + O^{2-}\,(melt) \rightleftarrows S^{2-}\,(melt) + 1/2\,O_2\,(gas)$$

(31)

Specific solubilities are a complex function of melt and gas composition, and increase as conditions become more reducing (O'Neill and Mavrogenes 2002; Namur et al. 2016; Gaillard et al. 2022). Under moderately oxidizing conditions, where S dissolves as sulfates, it exhibits a significantly decreased solubility (Boulliung and Wood 2022).

Composition of the outgassed atmosphere

Taking into consideration the chemical aspects of vapor–silicate melt equilibria described above, Figure 9 represents the composition of an outgassed atmosphere at equilibrium with a whole-mantle MO for planets from 0.1 to 6 Earth masses ($M_\oplus$) and core–MO equilibration f_{O_2} from IW-5 to IW, containing 100 ppm of H, C, N, and S. H_2 in the atmosphere increases with decreasing planetary mass, as pointed out previously. For the smallest and most reduced planets, this alters the outgassing of other species, mainly N_2 and CO (see bottom row). The surface f_{O_2} increases with increasing planetary mass, leading to respeciation of CO into CO_2, SH_2 and S_2 into SO_2, and H_2 into H_2O. CO_2 solubility is similar to CO's so carbonated species is not directly affected by changes in f_{O_2} (it is indirectly through competitive outgassing of other f_{O_2}-sensitive species), but H_2O's solubility is orders of magnitude higher than H_2's, leading to dramatic ingassing of hydrogenated species with increasing f_{O_2}. S solubility is strongly redox sensitive and decreases with increasing f_{O_2}, at which S speciates into SO_2 in the atmosphere. This results in SO_2 being a dominant species in the atmospheres of the oxidized Super-Earths (top row, last three columns).

Considering a surface f_{O_2} of IW$+0.5$ and dissolution of H_2O and CO_2 from a H, C, N system having the BSE abundances, Sossi et al. (2020) found that the terrestrial MO was overlain by a CO-dominated atmosphere with elemental ratios H/C = 0.22 and H/N = 5.8, falling between the cases IW-2 and IW (for 1 $M_\oplus$) in Figure 9, which yield surface f_{O_2} values of IW-0.5 and IW$+1.5$, respectively. Investigating the terrestrial MO as well, but for a range of f_{O_2} extending from IW-6 to IW$+4$, Gaillard et al. (2022) distinguished between oxidized MOs, outgassing C, N and S, and reduced ones, outgassing C, H and possibly N (except near the most reduced end-member). Their model suggests a match between the MO and the atmosphere volatile budgets to the present day surficial and mantle reservoir, respectively, and hence only a modest influence of the post-MO volatile cycling.

Many studies have focused on the outgassing of either pure steam atmospheres (Zahnle et al. 1988; Hamano et al. 2013; Schaefer et al. 2016) or CO_2–H_2O ones (Elkins-Tanton 2008; Lebrun et al. 2013; Bower et al. 2019; Nikolaou et al. 2019; Miyazaki and Korenaga 2022). In the former case, only H_2O is outgassed, but its outgassing depends on the crystallization

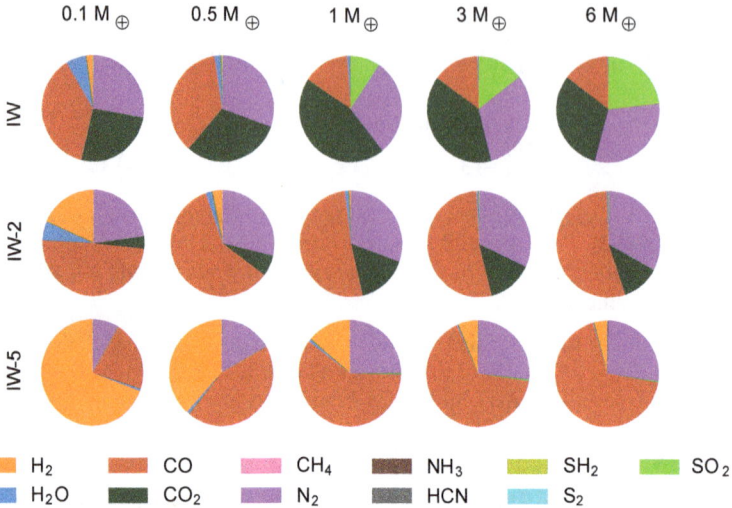

Figure 9. Compositions of C–O–H–S–N atmosphere (in molar fractions) overlying a whole mantle MO as a function of planetary mass (columns) and core–mantle equilibration f_{O_2} (rows). The figure is adapted from Maurice et al. (in press).

regime and chronology. In the latter case, the lower solubility of CO_2 compared to H_2O leads to earlier outgassing of C, while H is efficiently retained in the mantle until the late stages of crystallization. This initial state—when most of the mantle is molten—is similar to what is observed for oxidized cases in Figure 9 (where CO_2 is the dominant species in the atmosphere). If large-scale convective stirring of the MO stops before water is outgassed, then the solid mantle can retain a significant amount of H (Bower et al. 2022; Miyazaki and Korenaga 2022).

Elkins-Tanton (2012) estimated that a whole-mantle MO on the Earth containing initially 500 ppm H_2O and 100 ppm CO_2 would only become saturated in a fluid phase after 80% crystallization by volume. This would delay outgassing since bubbles do not nucleate before saturation is reached. As a consequence, the planetary-scale gas–silicate chemical equilibrium sets in late. Prior to that, the MO crystallizes without outgassing. H_2O and CO_2 are then outgassed in a catastrophic event at the end of MO crystallization. The induced greenhouse effect prolongs the crystallization of the last part of the mantle (Elkins-Tanton 2008). Other studies have assumed that the planetary-scale gas–silicate chemical equilibrium holds since the beginning of MO crystallization. This is in line with the view of a well-stirred MO, where the surface solubility provides the bottleneck for volatile concentration. Lebrun et al. (2013) and Nikolaou et al. (2019) found a delayed outgassing of water into a readily outgassed background CO_2 atmosphere, similar to Elkins-Tanton (2008), although starting at the beginning of mantle solidification. Bower et al. (2019) pointed out that the competitive outgassing of CO_2 and H_2O leads to a decrease H_2O outgassing due to the effect of CO_2 on the atmospheric molecular mass, further hampering H outgassing. Investigating the influence of planetary size, Miyazaki and Korenaga (2022) found that it had little effect on the final mass of water at the surface (water being always near the high planetary mass outgassing limit pointed in the previous section), while it did significantly affect the surficial mass of CO_2.

The most recent studies extended their MO degassing models to a wider range of compositions. Lichtenberg et al. (2021) investigated the outgassing in single-species atmospheres of H_2, H_2O, CO, CO_2, CH_4, N_2 and O_2. These authors identified three different groups of gases based on the lifetime of the MO: O_2, N_2, and CO, having a very limited greenhouse effect, yield the fastest crystallization (in ~1 kyr for a terrestrial MO), H_2O and CO_2

being strong infrared absorbers yield a longer MO lifetime (~10 kyr), and finally, H_2, because of its collision-induced absorption upon reaching high pressures, leads to a solidification in more than 1 Myr. Of note, all species except H_2O are largely or completely degassed throughout the MO crystallization, so the secondary atmosphere is established already at the beginning of mantle solidification.

Based on the magma degassing chronology for the CO_2–H_2O system calculated by Nikolau et al. (2019), Katyal et al. (2019) computed the atmospheric speciation between H_2, H_2O, CO and CO_2 for different f_{O_2} and stages of MO crystallization. They found that, depending on the value of f_{O_2} (between IW-4 and IW$+4$) as well as the C and H budget in the system, either of the four species can dominate, impacting the planetary cooling flux. Interestingly, they also found that reduced atmospheres have broader spectral features due to their larger scale height (consequence of the small molar mass of CO and H_2 compared to CO_2 and H_2O). Furthermore, reduced atmospheres are less prone to condensation and formation of clouds, which can largely hamper the spectroscopic signal.

The same C–O–H system (including CH_4) was studied in a self-consistent manner by Bower et al. (2022), accounting for concurrent outgassing and redox speciation in the atmosphere. The T-dependence of the f_{O_2} imposed by the FeO–Fe$_2$O$_3$ buffer at surface conditions leads to an increase of the f_{O_2} during MO crystallization, and accordingly, shifts from a reduced secondary atmosphere to an oxidized one as the MO crystallizes. As for other studies, C species are outgassed earlier than H ones, leading to a general evolution from CO-dominated atmospheres to a CO_2 or H_2O-dominated one, depending on the C/H ratio of the system. In many cases, the bulk of H_2O outgassing occurs after the melt fraction is reduced to 30%, i.e., after the end of the planetary-scale gas–silicate melt chemical equilibrium. The lack of a change in speciation of N_2 over the f_{O_2} span of interest (Gaillard et al. 2022) combined with the low solubility of N_2 result in a straightforward pattern for N outgassing, where N_2 is largely in the atmosphere throughout MO crystallization. Furthermore, the lack of infrared absorption feature of N_2 yields little effect on the MO lifetime.

To the authors' knowledge, only the study of Gaillard et al. (2022) has addressed the outgassing of S through MO crystallization. The high solubility of S over a large f_{O_2} range probably make it only a minor species of most secondary atmospheres. Except if f_{O_2} increases enough for SO_2 to be created and outgassed, S is likely efficiently retained in the mantle. Furthermore, if the S concentration in silicate melts exceeds the SCSS, sulfide precipitates, further isolating the planetary S budget from the atmosphere.

Retention of LEVEs in the solid mantle during MO crystallization

The outgassed atmosphere evolves as the MO cools below the liquidus and starts to crystallize. The atmospheric composition in turn also plays a crucial role in MO crystallization, as some species (like H_2O or CO_2) are powerful greenhouse gases, which can drastically reduce the cooling flux of the planet (Abe and Matsui 1985; Elkins-Tanton 2008; Salvador et al. 2017; Nikolaou et al. 2019; Lichtenberg et al. 2021). The mass of the atmosphere evolves during the solidification of the mantle because volatiles are generally incompatible in silicate minerals that is they are preferentially partitioned into melt during crystallization. As a consequence, during MO crystallization, their concentration in the remaining melts increases, affecting in turn their partial pressures and the outgassed mass fraction.

Partitioning of volatiles between the MO liquid and the crystallizing minerals is described by mineral–melt chemical equilibrium, which reads:

$$X_s^{\text{min}} = D_s^{\text{min/melt}} X_s^{\text{melt}} \tag{32}$$

where X_s^{min} and X_s^{melt} are the mass fractions of species s in the mineral and melt phase, respectively, and $D_s^{\text{min/melt}}$ the mineral–melt partition coefficient. With the appearance of a new (solid) reservoir for volatiles, the mass conservation equation is modified into:

$$M_s = \left(4\pi R_{pl}^2 / g\right)\left(\frac{\mu_s}{\mu_{atm}}\right)p_s + \left(M_{liq} + D_s^{min/melt}M_{sol}\right)\alpha_s p_s^{\beta_s}, \qquad (33)$$

where M_{liq} and M_{sol} are the masses of liquid and solid, respectively, in the MO at equilibrium with the atmosphere. In fractional crystalization, M_{sol} corresponds to the mass of the layer of crystals that just settled at the bottom of the MO, before it gets buried. In the equilibrium crystallization scenario, it corresponds to all crystals suspended in the MO, where the melt fraction exceeds the rheologically critical melt fraction (RCMF), i.e., the melt fraction below which the mantle starts to behave like a solid, preventing further equilibration. When a layer of cumulates of mass dM_{cum} becomes chemically isolated from the MO–atmosphere system, it scavenges a mass dM_s of s consisting of molecules trapped/bound in the crystals, and other dissolved in the trapped melt. $dM_s = \left(\left(1-\Phi\right)D_s^{min/melt} + \Phi\right)dM_{cum}\alpha_s p_s$, where the trapped melt fraction Φ is less than or equal to the RCMF, and is affected by the crystallization regime as well as other processes like percolation (Hier-Majumder and Hirschmann 2017).

In addition to the small amounts of volatiles entering the solid crystals (for $D_{LEVE}^{min/melt} \neq 0$), the MO cumulates can retain some trapped interstitial melts, whose concentration in incompatible species is significantly higher than that of the crystallized solids, and provide the main reservoir of volatiles in the solidified mantle. Two end-members can be distinguished in the dynamical regime of MO solidification, which ultimately control how much partial melt is retained in the solid mantle. In the fractional crystallization regime, crystals settle efficiently at the bottom of the MO, and all melt is extracted, resulting in a virtually trapped-melt-free solid mantle. In the equilibrium crystallization regime, crystals remain in suspension (due to convective mixing), and each layer of the MO solidifies when it cools down sufficiently for its melt fraction to drop below a certain threshold, at which the crystal lattice becomes interconnected and supports the strain. This threshold (noted Φ_{RCMF}, RCMF standing for rheologically critical melt fraction) is often assumed to be between a melt fraction of 30 and 40%, the remaining melt being trapped. While the incorporation of LEVE in the solidified mantle is only controlled by the values of the partition coefficients in the fractional crystallization case, the RCMF plays a very strong role in the equilibrium crystallization case. Figure 10 shows the radial profile of LEVE content in a solid mantle inherited from 4 different scenarios of MO crystallization: 1) fractional, 2) equilibrium ($\Phi_{RCMF} = 0.05$), 3) equilibrium ($\Phi_{RCMF} = 0.1$), and 4) equilibrium ($\Phi_{RCMF} = 0.3$). In the equilibrium crystallization case, the mantle budget in LEVE increases with increasing trapped melt, because those mainly accommodate LEVE. The actual contribution of the solid crystals in the cumulates of the equilibrium crystallization cases (dashed lines) is close to that of the fractional case. Parameterizing concurrent crystal settling and melt percolation, Hier-Majumder and Hirschmann (2017) showed that the latter occurs on timescale too long to significantly extract interstitial trapped melts during most of the mantle solidification. These authors showed that 77% of the bulk H_2O and 12 % of bulk CO_2 could be retained in the solid mantle at the end of the crystallization of a terrestrial MO. By controlling the efficiency of crystal settling, Bower et al. (2022) could reproduce different crystallization regimes between the equilibrium and fractional crystallization end-members. They found that the former is prone to significant volatile retention. H in particular could be retained in the mantle in excess of 90% of the initial budget, in particular at high f_{O_2}, due to the high solubility of H_2O in MO liquids. While H retention is in general more efficient than C's, graphite precipitation (not accounted for in these models) could significantly enhance the C retention in the solid mantle. Bower et al. (2022) discussed the possibility of graphite precipitating in the atmosphere for the most reduced and C-rich cases, when the CO fugacity exceeds that defined by the CCO redox buffer, but they rule out graphite/diamond precipitation in the MO, even in their most reduced cases. However, the existence of a negative pressure gradient of f_{O_2} through the MO

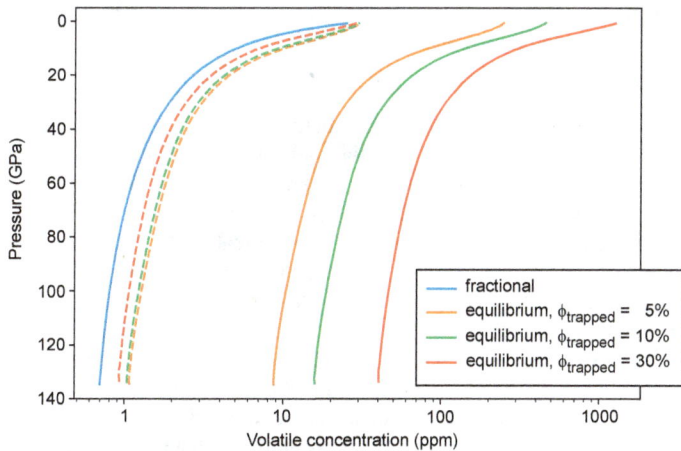

Figure 10. Profiles of volatiles retained in the MO cumulates for fractional crystyallization (**blue**) and various amount of trapped melt in equilibrium crystallization (**orange, green and red**). **Dashed lines** correspond to the concentrations effectively dissolved in solids for the corresponding cases (the rest of the budget being accounted for by trapped melts). The MO has a bulk content of 100 ppm of a LEVE species that has water-like rock–melt partitioning characteristic, i.e., $D_{LEVE}^{rock/melt}$ of 0.007 (Hirschmann et al. 2009; O'Leary et al. 2010).

(Hirschmann 2012; Armstrong et al. 2019; Deng et al. 2020) suggests that the surface f_{O_2} value they base their discussion upon could be overestimated. In any case, experimental data on C solubility at diamond saturation at higher *P–T* conditions are required for a more robust constraint on C sequestration via C-rich accessory phase in the deep MO/mantle.

As the LEVE content in the melt increases with crystallization, the shallower MO cumulates inherit a larger LEVE budget, via trapping more LEVE-rich interstitial melts. This leads to the creation of a radial gradient in the LEVE distribution of the mantle mush, with a volatile-rich upper mantle and a volatile-depleted lower mantle (Fig. 10). Because Fe is incompatible as well, a density profile similar to the volatile content one also results from the MO crystallization. A dense upper mantle overlying a less dense lower mantle is prone to overturn by Rayleigh–Taylor instability (Elkins-Tanton et al. 2005; Ballmer et al. 2017; Maurice et al. 2017). Once overturned, the layer becomes stable and can be isolated in the deep mantle. LEVE-rich, dense material could, therefore, be isolated from planetary-scale volatile cycles. Once trapped in the "non-convecting mantle", melt becomes chemically isolated from the rest of the MO. Using an equilibrium crystallization model, Miyazaki and Korenaga (2022) estimated that 99% of 400 ppm bulk H_2O is retained in the solidified mantle, while 70% of 1000 ppm bulk CO_2 is outgassed. However, these authors further considered that melt extraction of a partially molten surficial layer, on a timescale longer than MO crystallization but shorter than the onset of mantle convective mixing, would further deliver 5% of the water budget to the atmosphere. Yet, a larger fraction of H_2O retained in the mantle led these authors to estimate a significantly higher bulk H_2O inventory required for early ocean condensation.

POSSIBLE VOLATILE FLUXES IN AND OUT OF THE PLANET'S INTERIOR AND THEIR POSSIBLE VARIATION THROUGH TIME

After complete MO crystallization, interior–surface exchange of LEVEs over millions to billions to years, of any rocky planets depends on convective vigor of the planetary interiors, which in turn depends on the extent of internal and basal heating. Some rocky planets may also develop modes of tectonics that take surface and near-surface materials to the planetary interiors. These processes set up a long-term, deep cycle for LEVEs that are critical for sustaining or modulating rocky planetary habitability.

Extraction of LEVEs via mantle melting

Mantle convection aids in decompression melting and therefore, the release of volatile, incompatible elements to the extractable partial melt phase. If mantle melting takes place at shallow depths or continues to shallow depths, the generated silicate partial melts are less dense than the surrounding mantle and, therefore, moves up toward the surface. Because LEVEs often show positive pressure-dependent dissolution in silicate partial melts, decompression of melts leads to exsolution of gaseous molecules such as CO_2, H_2O, SO_2, N_2, which contribute to the secondary atmosphere of the planet. The influence of mantle outgassing on planetary atmosphere composition via the processes outlined above, however, depends on a number of properties such as thermal and redox state of the planetary interior, composition and mineralogy of the mantle, budget of LEVEs in the planetary mantles, and overlying atmospheric pressure. Here we discuss some of the key concepts and insights developed through the studies of Earth and other rocky planets in our Solar System and using principles of equilibrium thermodynamics in metal, silicate, and vapor-bearing systems.

Lithologic control on mantle melting and partial melt compositions

Extraction of LEVEs and the release to the surface via mantle melting depend on partial melt compositions, which in turn depend on the mantle compositions and mineralogy and of course the P–T–f_{O_2} conditions of mantle melting. Some LEVEs, depending on storage conditions, also directly influence mantle-equilibrated magma composition, which will be discussed in a following section. Rocky exoplanets may show a wide variety of compositions and modal mineralogy, deviating in some cases significantly from the modal mineralogy of Solar System rocky planets (e.g., Madhusudhan et al. 2012; Jontof-Hutter 2019; Putirka and Rarick 2019). Interestingly, however, the current analyses of >4000 stellar compositions with planetary systems suggest that exoplanetary mantles with exotic mineralogy should be rare (Putirka and Rarick 2019; Putirka et al. 2021, 2024, this volume). Although the exact mineralogy and composition remain uncertain partly because of uncertainty in the Fe partitioning between the metallic core and silicate mantles, most rocky exoplanetary mantles are thought to comprise olivine, orthopyroxene, and clinopyroxene ± quartz and magnesiowüstite. In other words, exoplanetary mantles may be peridotitic, pyroxenitic, or quartz- to magnesiowüstite-saturated (Putirka et al. 2021). Even for stars with high C/O ratios (Young et al. 2014) that may stabilize silicon carbide (SiC) in the mantles of such planetary systems, SiO_2 polymorph and graphite/diamond may end up being the break-down product (Hakim et al. 2018b; Allen-Sutter et al. 2020). Hence, for providing a broad guidance, here we use a range of ambient mantle peridotite and mantle heterogeneity compositions from the bulk silicate Earth and other rocky planets in the Solar System to establish a range possible melting behavior and melt compositions in terms of LEVE extraction via mantle melting. For example, the compositions and mineralogy of some of the crustal heterogeneities in the Earth's mantle, i.e., compositions ranging from basaltic eclogite/pyroxenite with or without SiO_2-polymorph to granite—introduced via subduction—could provide reasonable mantle analogues for some rocky exoplanets. Similarly, the compositional differences between the mantle of Earth, Mars, Moon, and Mercury can potentially shed light on the effect of differences, for example, in FeO* (all Fe-oxides expressed as FeO) content among others (Dreibus and Wänke 1985, 1990; McDonough and Sun 1995; Nittler et al. 2018; Yoshizaki and McDonough 2020).

Melting conditions and partial melt compositions vary significantly depending on the lithologic composition and mineralogy. Yet, in general, Fe-Mg-rich silicate lithology—similar to the Earth's mantle peridotite—exhibits highest melting temperatures (first melting temperature or the solidus and the final melting temperature or the liquidus) and quartz-bearing granitic composition exhibits the lowest melting temperatures. Basaltic compositions that yield a wide range of pyroxene and garnet-bearing lithologies with or without SiO_2 polymorphs yield intermediate melting temperatures (Fig. 11). It is worth noting that although in Figure 11, we plot a single solidus for each of the lithology of choice, the solidi locations vary strongly as a function of bulk composition for any given lithology (Putirka 2024, this volume). Limited experimental data on the solidus of plausible exoplanetary mantle composition (Brugman et al. 2021) shows that indeed it is within the range of possibilities shown in Figure 11. The solidi temperatures dictate the depth at which melting-induced release of volatile elements may commence in a convective mantle. Deeper the intersection of the mantle geotherm with a given lithology solidus, deeper is the onset of LEVE dissolution in extractable silicate melts. In other words, if a planetary or an exoplanetary mantle were convecting, more LEVEs would be extracted per unit mantle evolution time scale for a deeper mantle solidus. Some changes can, however, be expected with planetary size as well. For example, larger planets have larger g and dP/dz and, therefore, experience limited extent of mantle melting, even with the same mantle geotherm as smaller planets (Dorn et al. 2018; Miyazaki and Korenaga 2022). Other than the volatiles themselves, which is discussed later, the biggest control on solidus temperature of a given lithology are the alkali contents, Na_2O, K_2O and Mg# (molar Mg/(Mg+Fe) × 100) (Hirschmann 2000; Kogiso et al. 2004; Duncan et al. 2018; Ding et al. 2020; Collinet et al. 2021; Payré and Dasgupta 2022). Lower the Mg# and higher the bulk alkali ($Na_2O + K_2O$), lower is the solidus T at any given P, i.e., deeper is the onset of melting for any given mantle potential temperature. Therefore, even for a broadly mafic–ultramafic composition, the condition of major mantle melting and therefore LEVE extraction can vary significantly from one planetary mantle to another.

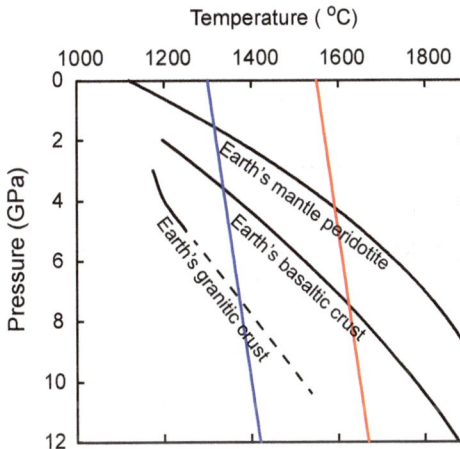

Figure 11. Solidus of various silicate lithologies in *P–T* space at volatile-free conditions—Earth's mantle peridotite (Hirschmann 2000), SiO_2-polymorph saturated mid-oceanic ridge basalt (MORB)-like eclogite (Yasuda et al. 1994), and a composition equivalent to Earth's granitic crust-derived sediment (Spandler et al. 2010). Also plotted for comparison are possible temperatures as a function of pressure for two mantle potential temperatures, 1300 (**blue**) and 1550 °C (**red**), with gradient similar to that of the Earth's interior. In a convective, upwelling mantle, different lithologies or mineral assemblages would undergo onset of melting at different depths and therefore, the efficiency of release of LEVEs via mantle melting could vary significantly depending on dominant interior composition.

MODE OF LEVE STORAGE AND EFFECTS ON MANTLE MELTING

Storage mechanisms of LEVEs in the rocky mantles

Different volatile elements are known to be stored in largely silicate mantle of rocky planets differently. Table 2, following Dasgupta et al. (2022a), shows how different LEVEs are known to be stored in the convective and near-surface, conductive silicate mantles of Earth and Solar System planets and how such storage is expected to vary as a function of oxygen fugacity, f_{O_2}, which can vary both as a function of planetary building blocks, MO processes, and depth. The information in Table 2 is chiefly given assuming that the volatile budgets and major element compositions of the planetary mantles are similar to those for rocky planets in

Table 2. LEVE-bearing phases in the rocky reservoirs of Earth and planets arranged as a function of oxygen fugacity.

f_{O_2} (ΔIW) in the shallow mantle	Example planetary mantles	C storage	H storage	S storage	N storage
>+4	Earth – subduction zones?	CO_2 fluid/dolomite/ magnesite	Ambient convective mantle: NAMs; Near surface thermal boundary layer: NAMs, hydrous silicate minerals	Anhydrite, Fe–Ni sulfide	NAMs, hydrous silicate minerals
~3.5	Earth	Ambinet convective mantle: CO_2 fluid/ dolomite/magnesite, graphite/diamond, Fe–Ni metal, Fe–Ni–S alloy melt, Fe-carbide, Si-carbide; Near-surface thermal boundary layer: dolomite	Ambient convective mantle: NAMs; Near surface thermal boundary layer: NAMs, hydrous silicate minerals	Fe–Ni sulfide	NAMs, graphite/diamond, Fe–Ni alloy, Fe–Ni–S alloy melt
−1	Mars and the Moon, polluted White Dwarf planets	graphite/diamond, Fe–Ni metal, Fe–Ni–S alloy melt, Fe-carbide, Si-carbide	Ambient convective mantle: NAMs; Near-surface thermal boundary layer: NAMs, hydrous silicate minerals	Fe–Ni sulfide	graphite/diamond, Fe–Ni metal, Fe–Ni–S alloy melt, Si-nitride
<−4	Mercury, Aubrite parent body, polluted White Dwarf planets	graphite/diamond, Fe–Ni–S alloy melt, silicon carbide, nominally C–free minerals (?)	NAMs	Ca–Mg–Fe sulfide, nominally S-free minerals (?)	NAMs, graphite/diamond, Sioxynitride (sinoite), Si-nitride

Note: Storage of C, H, S, and N in different planetary bodies in the Solar System and how such storage mechanism varies as a function of mantle f_{O_2}. The noted f_{O_2} values are of the shallow mantles and are expected to diminish with depth following iron-dominated redox equilibria as on Earth.

the inner Solar System. However, wider range of storage possibilities remain if the volatile budgets far exceed those we estimate for our rocky worlds and/or the bulk compositions and the redox state differ significantly (Table 2). The two main modes of LEVE storage in the mantle is (1) via incorporation in mantle silicate minerals and (2) as accessory phases.

The first mode of storage puts a limit on the concentration of LEVEs that may be incorporated in the solid mantle. A good example of the first mode of volatile storage is for water or hydrogen storage, where at high $P-T$, hydrogen can be incorporated in nominally anhydrous minerals, NAMs as point defects or via various substitution mechanisms (e.g., Bell and Rossman 1992; Kohlstedt et al. 1996; Bolfan-Casanova 2005; Skogby 2006). The water or hydrogen storage capacity of mantle depends on the mineralogy and mineral phase proportions along with mantle potential temperatures. The hydrogen storage capacity for the Earth's mantle in the present day and how it may have evolved with time has been heavily explored (e.g., Hirschmann 2006; Dong et al. 2021). The total water storage through NAMs and its temporal evolution has also now been explored for other rocky planets in the Solar System (e.g., Mars; Dong et al. 2022) and beyond (Guimond et al. 2023). One key finding is that the water storage capacity through storage in the NAMs increases with decreasing mantle temperature. Therefore, with secular cooling of any rocky exoplanetary mantles, more water could be sequestered in the solid reservoir than those stored during the early evolution of the planet. In other words, for a given planetary water budget, surface liquid water ocean volume is expected to be larger during early history of solid planet.

Similar to hydrogen, other elements such as carbon, sulfur, and nitrogen can also be stored, in limited quantities, in dominant mantle minerals (e.g., Keppler et al. 2003; Shcheka et al. 2006; Li et al. 2013; Yoshioka et al. 2018; Callegaro et al. 2020; Johnson et al., 2024). However, at least for conditions similar to the Earth's upper mantle, storage of these LEVEs in the silicate mantle minerals is not thought to be the dominant storage mechanisms (Dasgupta et al. 2022a). However, for more reducing mantle conditions, storage of some of these elements in the mantle silicates may be more important. This is known for N, where the solubility and partitioning of N into nominally N-free mantle silicates seem to increase at more reducing conditions (Li et al. 2013; Yoshioka et al. 2018; Dasgupta et al. 2022b). N incorporation in silicate minerals may take place via NH_4^+ substituting for cations such as K^+ or at reducing conditions, N^{3-} substituting for O^{2-}.

The second mode of storage of LEVEs in the rocky mantles is as minor to trace accessory phases. For the LEVEs that are stored chiefly this way in the planetary interiors, the bulk mantle concentrations are directly proportional to the mass fractions of the accessory phases that house them. For the Solar System rocky mantles, the elements that are known to be stored this way are carbon, sulfur, and locally hydrogen (Lorand 1991; Dasgupta and Hirschmann 2010; Ding et al. 2014; Tsuno and Dasgupta 2015; Lorand and Luguet 2016; Ding and Dasgupta 2017; Dasgupta 2018; Dasgupta and Aubaud, in press). Hydrogen-bearing accessory phases that may be stable in the silicate mantle are hydrous silicate minerals such as amphibole ($NaCa_2(Mg_4Al)$) $(Si_6Al_2)O_{22}(OH)_2$), phlogopite $KMg_3AlSi_3O_{10}(F,OH)_2$, chlorite, and a number of dense hydrous magnesium silicates (DHMS). Hydrogen storage in these phases is possible in the lithospheric mantle that is within the thermal boundary layer (Mandler and Grove 2016; Dasgupta 2018; Saha and Dasgupta 2019; Saha et al. 2021; Juriček and Keppler 2023) or within a cooler regions within the mantle. Cooler regions in the mantle are possible if cooler, near-surface rocks are recycled into the mantle, i.e., within the regions of cold downwellings (Frost 1999; Ohtani et al. 2000; Litasov and Ohtani 2003; Eguchi and Dasgupta 2022). The stability of these hydrous silicate phases increase the water storage capacity of rocky planetary mantles. Therefore, exoplanets with thick surface thermal boundary layer and those having cold downwellings into their mantle would sequester a greater mass fraction of their water inventory in the solid silicate reservoirs.

Unlike water, which is dominantly stored as a trace element in the NAMs in the silicate mantle, the main inventory of carbon and sulfur is stored chiefly in accessory phases. For sulfur, the accessory phase changes from sulfide to sulfate mineral or melt phase with increasing oxidation of the mantle. For all rocky planets of the Solar System, the solid mantle chiefly stores sulfur in monosulfide solid solution (MSS) or liquid solution (Lorand 1991; Zhang et al. 2015; Zhang and Hirschmann 2016; Brenan et al. 2019; Dasgupta et al. 2022a; Dasgupta and Aubaud, in press). Only in highly oxidized mantle domains, such as those in the subduction zone conditions on Earth, the stable S-bearing phase maybe sulfate minerals (e.g., anhydrite, $CaSO_4$) (Jégo and Dasgupta 2014; Chowdhury and Dasgupta 2019). The sulfide composition is thought to be chiefly Fe–Ni±Cu–S solution for a wide range of mantle f_{O_2} conditions. For example, in the Earth's shallow upper mantle, the sulfide is known to be of composition with total metal, M over sulfur, S ratio close to 1 (M/S ~ 1) such as $Fe_{0.69}Ni_{0.23}Cu_{0.01}S_{1.00}$. Although with increasing depth in the mantle, sulfides may become more metal-rich as iron disprportionation in silicates become important (Tsuno and Dasgupta 2015). FeS-rich sulfides are expected to be the stable S-bearing accessory phase over a wide range of planetary mantle conditions (Ding et al. 2014, 2015, 2018; Namur et al. 2016; Dasgupta et al. 2022a). For example, in the Solar System, the mantle conditions include those of the terrestrial mantle, lunar mantle, martian mantle, and Mercurian mantle with oxygen fugacity varying from IW + 3.5 to IW − 5, where IW refers to the log f_{O_2} equivalent of iron-wüstite equilibrium (Fe + 0.5 O_2 = FeO). At extremely reduced, Mercurian mantle conditions, however, the equilibrium sulfides start showing the presence of Ca and Mg (Evans et al. 2012; Namur et al. 2016; Cartier and Wood 2019). Given the estimated and expected f_{O_2} of the rocky exoplanets seem to fall within the same range as known for the Solar System planets (Doyle et al. 2019, 2020), sulfur storage in similar range of sulfide phase compositions can be expected for extrasolar mantles as well.

Carbon storage, similar to sulfur storage, also depends on mantle f_{O_2}. At f_{O_2} > ~IW + 3, C is stored in CO_2-rich fluids at shallow depths and carbonate minerals (calcite, $CaCO_3$; dolomite, $Ca(Mg,Fe)(CO_3)_2$; magnesite, $MgCO_3$) at deeper depths (Luth 1999; Dasgupta and Hirschmann 2010). For peridotitic mineralogy, i.e., with olivine, orthopyroxene, and clinopyroxene, the transition from CO_2-rich fluid to carbonate mineral stability takes place at ~2 GPa at the solidus (Falloon and Green 1989). For more pyroxene- and garnet-rich lithologies, the transition from CO_2-rich fluid stability to carbonate mineral stability is somewhat deeper, i.e., at pressures of 3–4 GPa (Luth 1993, 1995). At conditions that are more reducing, the stable carbon-bearing phase is chiefly graphite or diamond depending on depths (Frost and Wood 1997; Luth 1999; Dasgupta et al. 2022a; Dasgupta and Aubaud, in press). At deeper mantle depths, as the Fe–Ni metal starts precipitating and mantle sulfides become more metal-rich, carbon maybe dissolved in metallic melt or metal sulfide melts (Rohrbach et al. 2014; Tsuno and Dasgupta 2015; Zhang et al. 2019). In addition, Fe-carbides (e.g., $(Fe,Ni)_3C$ and $(Fe,Ni)_7C_3$) may also be stable depending on the metal/carbon ratio of the mantle subsystems (Dasgupta and Hirschmann 2010; Rohrbach and Schmidt 2011; Dasgupta 2013). The relative contributions of Fe–Ni alloy, Fe-carbides, sulfides, and diamond in hosting the deep mantle C inventory depends on the mantle carbon content, f_{O_2}, and Fe-content. Extremely reduced domain of silicate mantles also have been documented to contain silicon carbide (moissanite, SiC) as a stable carbon-bearing phase, e.g., Leung et al. (1990) (Gorshkov et al. 1997; Trumbull et al. 2009; Schmidt et al. 2014). Exoplanetary mantles from high C/O ratio stellar systems may also contain silicon carbide, SiC as a stable C-bearing phase (Allen-Sutter et al. 2020). However, widespread presence of SiC instead of silicates likely would require complete reduction of Fe^{2+} to Fe^0 in planetary mantles (Hakim et al. 2018b).

Extraction of LEVEs via mantle melting

Even if present in trace concentrations, many LEVEs impart significant control on changing melting behavior of silicate mantles. The influences range from affecting the solidus temperature of mantle rocks (e.g., Dasgupta and Hirschmann 2006; Médard and Grove 2006; Hirschmann et al. 2009; Grove et al. 2012), modifying the melting reactions and changing mantle-generated partial melt compositions (e.g., Wallace and Green 1988; Gaetani and Grove 1998;

Dasgupta and Hirschmann 2007; Dasgupta et al. 2007; Pintér et al. 2021; Lara and Dasgupta 2023). These effects remain central to ongoing differentiation of a largely solid mantle, outgassing of LEVEs, and modulating composition of secondary atmosphere.

Mantle volatiles, which can dissolve in significant concentrations in partial melts, impart major control in lowering the mantle solidus. Incorporation of LEVEs in near-solidus partial melts happens via one of two mechanisms. (1) Through partitioning of the LEVE from nominally LEVE-free mantle silicates to the partial melt (example of this is partitioning of H and N during mantle melting) and (2) via melting or dissolution of LEVE-rich accessory phase into mantle melts (example of this is incorporation of C and S in mantle melts). In the case of (1), it is possible to define and determine the partition coefficient of the LEVE in question between the mantle rock and the generated partial melt, $D_{LEVE}^{rock/melt}$ (= concentration of the LEVE in the rock/concentration of the LEVE in the partial melt). This partition coefficient influences the concentration of the LEVE in the partial melt (C_{LEVE}^{l}) along with the original mantle abundance of the LEVE (C_{LEVE}^{0}) and the degree of partial melting (F). The relevant equation that can be used to calculate the concentration of a LEVE in the mantle-derived partial melt is:

$$C_{LEVE}^{l} / C_{LEVE}^{0} = 1 / \left(F + D_{LEVE}^{rock/melt} \left(1 - F \right) \right) \tag{34}$$

Although the equation above is for a scenario, where generated partial melt remains in contact with the residual mantle mineralogy throughout the melting process (batch or equilibrium melting), the composition of melts calculated this way largely follow the composition of aggregate melts in fractional melting scenario (melt efficiently separates from the mantle residue).

In the case of (2), the above equation (Eqn. 34) will only hold if the solubility of the LEVE in question in the mantle partial melt at the saturation of the LEVE-bearing accessory phase (e.g., graphite, sulfide, carbonate, or sulfate) exceeds the value given by $C_{LEVE}^{0} / \left(F + D_{LEVE}^{rock/melt} \left(1 - F \right) \right)$. Otherwise, the LEVE concentration of the partial melt is set by the saturation limit of the LEVE in question.

Melting of nominally anhydrous mantle

The extent of depression of the mantle solidus in the first mechanism, i.e., (1) depends on the bulk concentration of the LEVE in question and $D_{LEVE}^{rock/melt}$. Higher the abundance of the LEVE in the mantle and lower the $D_{LEVE}^{rock/melt}$, greater is the abundance of the LEVE in the near-solidus partial melts (Eqn. 34). Greater the abundance of the LEVE in the partial melt, greater is its influence in increasing the entropy of the system and therefore, greater is its effect in lowering the solidus temperature. The most of well-known effect of this style of LEVE influence on mantle melting is that of 'water' that is stored in nominally anhydrous mantle lithologies. Partitioning of 'water' from minerals to near-solidus melt phase causes lowering of the mantle solidus via stabilizing a hydrous or water-bearing silicate partial melts. The water content of the near-solidus partial melt, for a given bulk H_2O content of the mantle, depends on $D_{H_2O}^{rock/melt}$, which in turn depends on mantle mineralogy. Lower the $D_{H_2O}^{rock/melt}$, greater is the influence of water in depressing the mantle solidus. Figure 12 shows the estimated solidi of Earth's dominant mantle rock peridotite, with different bulk water content. It shows that especially for hot mantles, partial melting should commence at significantly deeper depths with 100s of ppm water in the mantle rock. If a planetary mantle is pyroxene-rich, the effect of 'water' in influencing deeper mantle melting should be less as $D_{H_2O}^{pyroxene/melt} > D_{H_2O}^{olivine/melt}$ (Aubaud et al. 2004; Hauri et al. 2006; Hirschmann et al. 2009; O'Leary et al. 2010). One interesting point is that over a wide range of f_{O_2}, the effect of water in inducing mantle melting is expected to remain similar. Taken together, these suggest that hotter and wetter mantle will be processed through the solidus more efficiently, leading to more efficient release of water to the planetary surface reservoirs. Pyroxene-rich lithologies, i.e., planetary systems with lower Mg/Si ratios, should retain more and release less water via mantle melting.

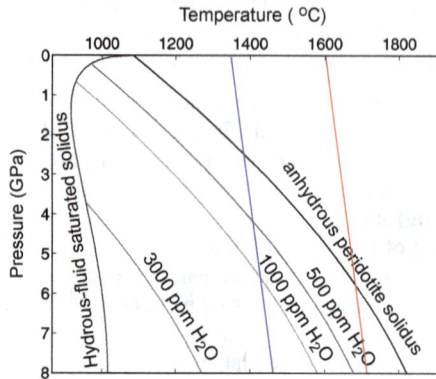

Figure 12. Estimated solidi of Earth's mantle-peridotite as a function of hundreds to thousands of ppm water being stored in nominally anhydrous mantle minerals and the hydrous fluid-saturated peridotite solidus, adapted from Katz et al. (2003). Also plotted for reference are the two possible mantle adiabats estimated for the present-day Earth's mantle, with corresponding mantle potential temperatures of 1350 and 1600 °C. Because of the slopes of the mantle solidi and how they change with pressure, hotter mantle would experience much deeper melting than the cooler mantle for a given concentration of trace water.

For many LEVEs such as C and S, and maybe even H and N under specific conditions, mantle melting and volatile extraction are controlled by stability, melting, and dissolution behavior of various accessory phases. The stable accessory phase and/or its composition may also vary depending on oxidation state of the mantle. These concepts were discussed at length in a recent synthesis by Dasgupta et al. (2022a). Some of the salient points are presented here in the context of rocky exoplanets.

Dehydration and decarbonation melting

For different range of concentrations, water and carbon dioxide maybe stored in the silicate mantles in the form of hydrous silicate and carbonate mineral phases, respectively. Water storage in hydrous silicate minerals takes place within the stability field of various hydrous silicate minerals and for bulk water content exceeding the storage capacity of NAMs. Given a wide range of hydrous silicate minerals can be stable, depending on the bulk major element composition of the mantle, the dehydration solidus location can vary widely. For example, in K-rich, hydrous domain of the Earth's mantle, the melting should commence across the dehydration solidus of phlogopite–peridotite (e.g., Sekine and Wyllie 1982; Mallik et al. 2015). Whereas for K-depleted mantle domain, dehydration melting may be controlled by the dehydration solidus of amphibole peridotite (Ulmer 2001; Mandler and Grove 2016; Dasgupta 2018; Saha and Dasgupta 2019; Fig. 13). Both amphibole and phlogopite may also coexist in mantle lithologies, depending on the bulk composition, chiefly the Na/K ratio (Saha et al. 2018; Dasgupta et al. 2022a). In other words, in addition to the major mantle minerals, accessory hydrous silicate minerals may be stable at the solidus of the cooler mantle domains.

Deep mantle carbon storage in oxidizing conditions takes place via carbonate minerals and across the decarbonation solidus, melting is dominated by the carbonate phase, which produces partial melts with CO_2 content as high as ~40 wt.%. Fig 13b compares the location of the carbonated peridotite solidus with that of amphibole peridotite (Fig. 13a) for Earth's mantle-relevant bulk composition. It can be seen in Figure 13 that the carbonated peridotite solidus and amphibole peridotite are similar in temperature at a given pressure. However, for amphibole peridotite solidus to be applicable, the mantle needs to have thousands of ppm water. Whereas the carbonated peridotite solidus is applicable even with CO_2 content down to tens of ppm. Hence, in oxidizing conditions, the onset of mantle melting via decompression

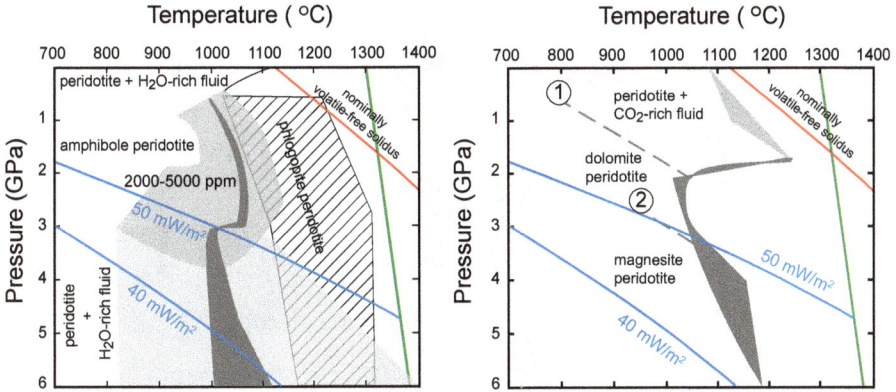

Figure 13. Solidi of fluid-absent hydrous peridotite (**a**) and carbonated peridotite (**b**) for compositions similar to the Earth's present-day ambient mantle. C–O–H volatiles in rocky planetary mantles affect the rocks' melting conditions, thereby affecting the efficiency of mobilizing LEVEs from the interiors to the planetary surfaces. The **hatched field** in (**a**) is the loci of the onset of melting for phlogopite-bearing peridotite, whereas the **grey band** marks the solidi of fluid-absent peridotite with water stored amphibole. In (**b**), the **dashed lines** marked 1 and 2 are carbonation reactions that lock in CO_2-rich fluid in mantle peridotite as mineral carbonates as pressure increases. Plotted for reference are the convective mantle geotherm in **green** for the mantle potential temperature of 1300 °C and conductive mantle geotherms in **blue** for surface heat flux of .40 and 50 mW/m². The figure panels are taken and modified from Figure 1 and 2 of Dasgupta et al. (2022a).

is governed chiefly by the presence of trace carbonate minerals for silicate mantles that are depleted in CO_2 and H_2O. This leads to efficient extraction and outgassing of CO_2 from a vast region of the Earth's mantle (Dasgupta and Hirschmann 2006; Dasgupta et al. 2013b; Fuentes et al. 2019; Hauri et al. 2019). No rocky planetary mantles—in the Solar System or in extrasolar planets—with upper mantle oxidation similar to that of the present-day Earth have been identified yet. Mantles of all rocky bodies are known to be more reduced. Hence, extraction of CO_2 from solid silicate mantles with efficiency similar to that observed for the present-day Earth may be not observed from current population of known rocky planets. Furthermore, unlike Earth's present-day mantle that are depleted in H_2O and CO_2, if rocky exoplanetary mantles contain modestly higher budget of H_2O, the effects of H_2O in inducing deeper mantle melting could be similarly or more important than the effect of CO_2.

Fate of carbon in reduced mantles

For all known planetary mantles, other than Earth's present-day shallow mantle, carbon is expected to be present in reduced form. In such cases, carbon's effect on freezing point depression of the rocky mantle solidus is muted. The release of carbon via melting in such cases is expected to be controlled by f_{O_2}-dependent oxidation of graphite or diamond (e.g., $C + 0.5 O_2 = CO$; $C + O_2 = CO_2$) and subsequent dissolution of CO_2 (as molecular CO_2 or CO_3^{2-}) or CO in silicate partial melts (Holloway et al. 1992; Hirschmann and Withers 2008; Stanley et al. 2011; Eguchi and Dasgupta 2018; Dasgupta and Grewal 2019; Dasgupta et al. 2022a). Importantly, such dissolution only takes place at conditions where decompression melting of nominally anhydrous mantle commences. In other words, for a given mantle potential temperature, carbon dissolution would take place over a much smaller mantle volume compared to what would be the case for an oxidized mantle with C stored in carbonate minerals. C dissolution into mantle partial melts taking place over a larger mantle volume is only possible if mantle potential temperature is hotter. Figure 14a, following Dasgupta et al. (2022a), shows equilibrium concentration of carbon in basaltic melts as a function of mantle

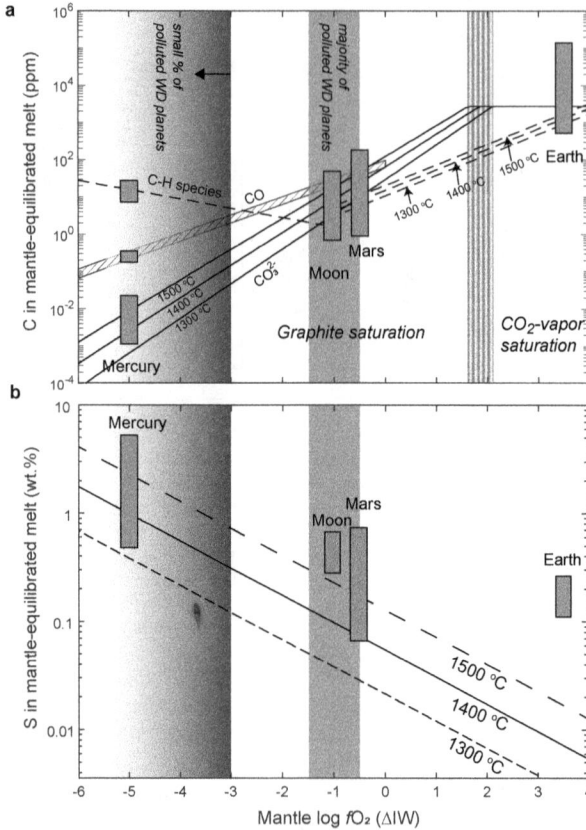

Figure 14. Modeled carbon content of mantle-derived, silicate partial melts at graphite saturation **(a)** and sulfur content of mantle-derived, silicate partial melts at sulfide saturation **(b)** as a function of mantle f_{O_2}. Plotted for comparison are the calculated C and S contents of mantle-derived basaltic melts for Solar System rocky planets, i.e, Earth, Mars, Moon, and Mercury. Also plotted for comparison, as vertical bands, are the f_{O_2} state of the majority (Doyle et al. 2019) and a small percentage of (Doyle et al. 2020) polluted White Dwarf (WD) planetary mantles. The CO content of partial melts in **(a)** is based on Yoshioka et al. (2019). The figure suggests that for all other known planetary mantles, supplying C to the surface via mantle melting is expected to be far less efficient than that for the Earth's present-day mantle. Whereas more sulfur maybe supplied to the crust for more reduced planetary mantles. The figure is adapted and modified from Figure 13 of Dasgupta et al. (2022a).

f_{O_2} and temperatures. Figure 14a shows that with reduction, mantle melting should extract less and less carbon into the graphite-saturated anhydrous silicate melts. CO_3^{2-} dissolution remains most effective down to $\log f_{O_2}$ of ~IW whereas CO dissolution is more effective at lower f_{O_2}. Accordingly, for the Solar System rocky bodies, mantle melting is expected to extract most carbon for Earth, followed by Mars and the Moon and least for Mercury. Only possibility of dissolving high C contents in basaltic melts at extremely reduced planetary mantles is via dissolution of C–H species such as CH_4 (Fig. 14a), which would require high fugacity of H_2 or relatively hydrous magma generation. Figure 14a also shows the presently available estimates of redox states of planetary mantles from around White Dwarf stars. The current estimates from the polluted White Dwarfs suggest that the extrasolar rocky planets may have similar range of f_{O_2} as observed or expected in the rocky planets of our own Solar System, after core formation. If these estimates are reliable and if there are no further mechanisms for mantle oxidation, carbon ventilation from majority of exoplanetary mantles maybe muted.

In addition to intrinsic f_{O_2} variation related to the redox states of building blocks and conditions of core formation setting the residual mantles at different redox states, rocky mantles are expected to vary in f_{O_2} as a function of depth as well. This is owing to disproportionation of iron ($Fe^{2+} = Fe^{3+} + Fe^0$) as a function of depth and greater incorporation of Fe^{3+} in deeper mantle silicate mineral phases such as garnet and bridgmanite. Therefore, as long as rocky exoplanetary mantles are Al-bearing and have broad compositional similarities with inner Solar System silicate mantles, iron disproportionation should lead to gradual reduction with depth (Frost and McCammon 2008; Stagno et al. 2013). Therefore, deeper mantles are expected to host C in the form of graphite/diamond even if shallower mantles house C in the form of C^{4+} species (Rohrbach and Schmidt 2011; Stagno et al. 2013). This leads to a proposition that a hotter mantle with a given Fe^{3+}/Fe_{total} bulk ratio undergoing decompression melting should melt at a relatively reducing conditions compared to a cooler mantle (Gaillard et al. 2015) (Fig. 15). Therefore, if exoplanetary mantles are hotter, with similar bulk compositions as terrestrial planets, C extraction would initially take place via diamond-present, deep silicate melting. Therefore, if such deep-seated melts extract from the mantle without much shallower equilibration, C extraction may be restricted. Cooler exoplanetary mantles, on the other hand, have the possibility of experiencing carbonated mantle melting. Therefore, inherently more reduced nature and hotter conditions of exoplanetary mantles both may lead to less efficient extraction of carbon into the partial melts and therefore less carbon being available for melt degassing (Dasgupta et al. 2022a). At the minimum, hotter mantles may not be able to extract carbon any more efficiently than cooler mantles, owing to negative f_{O_2} gradient with depth.

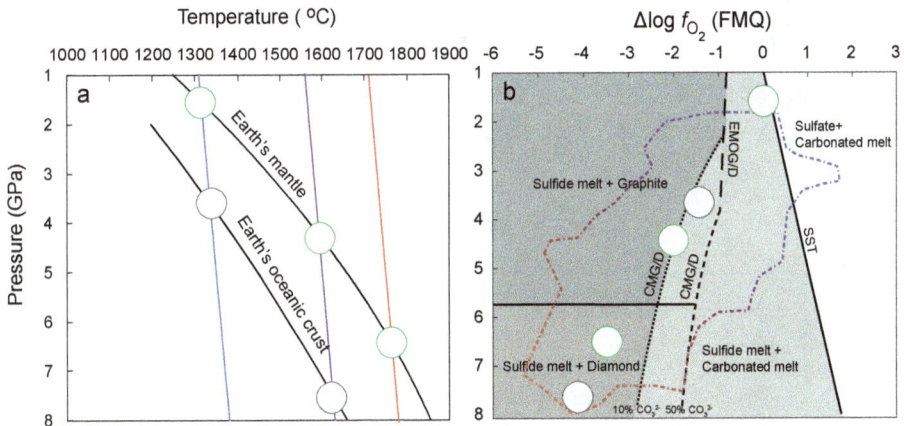

Figure 15. Pressure–temperature (**a**) and pressure–oxygen fugacity space (**b**) showing how the melting condition of a silicate planetary interior could vary as a function of mantle potential temperature. The mantle solidi for Earth's mantle peridotite and MORB-oceanic crust and the depth–f_{O_2} trend (field outlined by **dashed line**) are based on the Earth's mantle peridotite and eclogite compositions. The **circles** in both panels mark the P–T–f_{O_2} conditions of silicate melt generation at the onset of melting of peridotite (circles with **green outlines**) and eclogite (circles with **grey outlines**). In panel (**b**) the fields of C- and S-bearing accessory phases in the mantle are marked for a mantle potential temperature of 1350 °C, based on Chowdhury and Dasgupta (2020) and Dasgupta and Aubaud (in press). Onset of melting for three different mantle potential temperatures are marked: 1300, 1550, and 1700 °C. Hotter mantle should start generating melt, via decompression, at higher pressures and therefore at lower f_{O_2}, if mantle f_{O_2} are controlled by iron redox. Therefore, melting induced release of CO_2 via near-solidus melting is favored for cooler and less reduced mantle, where CO_2-rich partial melts can be generated. The figure is motivated and adapted from Gaillard et al. (2015).

DEGASSING OF LEVE FROM MANTLE MELTS

The key mechanism of degassing of volatile elements from mantle-derived partial melts is depressurization. Vapor–melt equilibria involving major LEVE-bearing gas species (e.g., H_2O, CO_2, N_2) show positive pressure-dependent solubility in silicate melts. Therefore, rising mantle-derived partial melts would exsolve dissolved LEVEs in the form of various gases upon decompression, if solubility limit is reached (Fig. 16). The extent of degassing as a function of depth depends on melt composition, initial LEVE content of the magma, and volatile species under consideration. Figure 16 shows solubility of the key volatile species as a function of pressure for a basaltic melt composition. It is observed that for any given LEVE budget of the primary basaltic melt, not all LEVE would start to degas at the same depth. For example, at a relatively oxidized condition, degassing of CO_2 would commence at a much greater depth (even for a lower abundance of CO_2 in the melt) compared to the exsolution of water vapor (Fig. 16a,c). Furthermore, some volatile species show negative pressure dependence of solubility depending on the redox state of the magma. The key examples are dissolution of sulfur as S^{2-} for sulfide-saturated silicate melts and dissolution of carbon as CO_3^{2-} for graphite- or diamond-saturated silicate melts (Fig. 16a). In other words, magmatic degassing of carbon and sulfur at reducing conditions and from dry magmas are expected to be minimal, if they remain sulfide and graphite/diamond saturated, and will rely on partitioning of these elements into the vapor phase rich in other species such as H_2O. Carbon release through magmatic degassing can, however, be higher if hydrogen fugacity is high and the chief C-bearing species is CH_4 (Ardia et al. 2013; Li et al. 2017; Dasgupta and Grewal 2019).

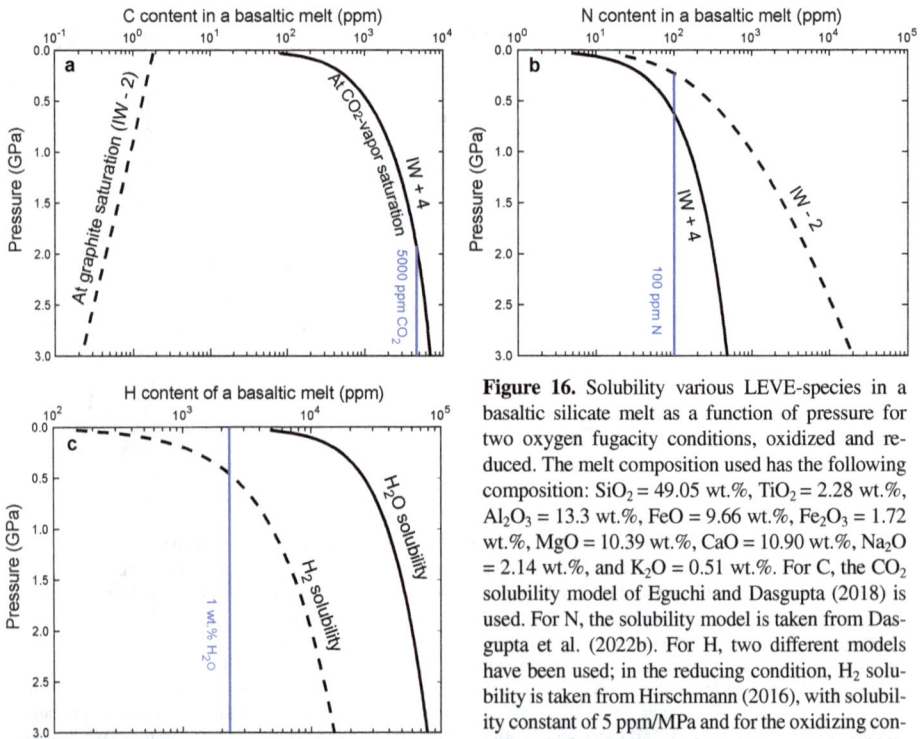

Figure 16. Solubility various LEVE-species in a basaltic silicate melt as a function of pressure for two oxygen fugacity conditions, oxidized and reduced. The melt composition used has the following composition: $SiO_2 = 49.05$ wt.%, $TiO_2 = 2.28$ wt.%, $Al_2O_3 = 13.3$ wt.%, $FeO = 9.66$ wt.%, $Fe_2O_3 = 1.72$ wt.%, $MgO = 10.39$ wt.%, $CaO = 10.90$ wt.%, $Na_2O = 2.14$ wt.%, and $K_2O = 0.51$ wt.%. For C, the CO_2 solubility model of Eguchi and Dasgupta (2018) is used. For N, the solubility model is taken from Dasgupta et al. (2022b). For H, two different models have been used; in the reducing condition, H_2 solubility is taken from Hirschmann (2016), with solubility constant of 5 ppm/MPa and for the oxidizing condition, H_2O solubility is from Moore et al. (1998). The **blue vertical lines** represent typical volatile contents for different LEVEs that may be expected in a basaltic melt generated from an Earth-like mantle composition. The blue vertical lines and their intersections with various solubility curves show the depth at which the basaltic melt of a given LEVE content would start degassing if vapor bubble nucleation is not inhibited.

The gaseous species that gets released during magmatic degassing depends on redox state of the magma, imposed by fugacity of oxygen and that of other volatiles. Several studies have explored how depth of magma emplacement can alter the magma redox state and in turn the oxidized to reduced species of various C–O–H–S gases (e.g., Gaillard et al. 2011; Sun and Lee 2022). Release of more reduced gases can affect the build-up oxygen (O_2) critical for complex life, as we know on Earth. Therefore, among many suggestions on the cause Great Oxygenation Event (GOE) on Earth (e.g., Holland 2002; Aulbach and Stagno 2016; Duncan and Dasgupta 2017; Eguchi et al. 2020; O'Neill and Aulbach 2022), one suggestion is that the released volcanic gases changed from dominantly reduced species to oxidized species. While evolution of mantle f_{O_2} from reduced to less reduced condition can indeed affect the composition of gases (Holland 2002; Aulbach and Stagno 2016; O'Neill and Aulbach 2022), a change in the thickness of the crust across the GOE, associated with the growth of continental crust could also have played a role (e.g., Gaillard et al. 2011; Sun and Lee 2022). Magmatic differentiation beneath thicker crust releases more oxidized gases compared to gases released from silicate melts at shallower depths.

SEQUESTRATION OF VOLATILES IN THE CRUST AND POSSIBLE RECYCLING INTO THE MANTLE

Magmatism involves not only partial melting of the solid mantle but also emplacement and differentiation of mantle-derived melts in the crust. Depending on the nature of crust building, i.e., via eruption of magma to the surface (extrusive magmatism) or emplacement of the magma into and beneath the crust (intrusive magmatism), differential extent of degassing of LEVE is expected. For example, extrusive volcanism or eruption to near-surface pressures leads to near-complete degassing of water and carbon dioxide, whereas emplacement of magma at elevated pressures may retain a significant portion of sulfur (as sulfides; Ding et al. 2015) and even water (as hydrous silicate minerals such as amphibole; (e.g., Davidson et al. 2007; Larocque and Canil 2010; Krawczynski et al. 2012)). Therefore, magmatic processes may retain some volatiles, in significant proportion, in near surface solid crustal reservoirs. In planets having an active ocean-atmosphere system, this budget of magmatic, crustal LEVEs adds to what may be sequestered in the crust via weathering processes (e.g., Walker et al. 1981; Alt 1995; Alt and Teagle 1999; Berner 2003; Lee et al. 2019).

If a rocky exoplanet poses an active rock cycle similar to that of Earth, LEVEs sequestered in crustal reservoirs can be re-introduced into the planetary interiors. The processes may include plate-tectonic cycles of the present-day Earth, which is known to cycle many LEVEs, including carbon, sulfur, water, and nitrogen back to Earth (e.g., Schmidt and Poli 1998; Dasgupta 2013; Jégo and Dasgupta 2014; Plank and Manning 2019). Processes such as delamination or lithospheric foundering may be more relevant for deep cycling of LEVEs in stagnant-lid planets (e.g., Elkins-Tanton et al. 2007). The fate of the recycled LEVEs depend on the pressure–temperature path of the rocks that are reintroduced into the mantle, the detailed devolatilization and/or melting phase relations of the rocks in question (e.g., Dasgupta et al. 2004; Grassi and Schmidt 2011; Tsuno and Dasgupta 2012; Dasgupta 2013; Kiseeva et al. 2013; Jégo and Dasgupta 2014; Walters et al. 2020; Eguchi and Dasgupta 2022), and the dissolution efficiency of various LEVEs in the generated fluids or melts (e.g., Duncan and Dasgupta 2015; Li et al. 2015b; Mallik et al. 2018; Chowdhury and Dasgupta 2019; Zajacz and Tsay 2019). The latter two can vary significantly as a function of bulk composition of the lithology and their LEVE contents and speciation (e.g., sulfate vs sulfide; carbonate vs graphite/diamond, NH_4^+ vs N_2). The dissolution efficiency of LEVEs such as C, N, S in released fluids and melts also depend on the solid-state storage of the LEVEs in downwelling rocks, which in turn varies with the oxygen fugacity of the downwelling rocks (e.g., Mikhail and Sverjensky 2014; Duncan and Dasgupta 2017; Walters et al. 2021; Chowdhury et al. 2022). Given the uncertainties in the composition of the rocky exoplanets and consequently their surface or crustal compositions, wide range possibilities of volatile recycling may be expected.

DEEP VOLATILE CYCLES IN OTHER ROCKY WORLDS: OUTLOOK AND NEED FOR NEWER CONSTRAINTS

Processes for rocky planet formation, differentiation, and subsequent evolution all affect the origin, evolution, and surface availability of life-essential volatile elements in rocky planets. The outline of process-based fractionation of LEVEs at different stages of planetary evolution as provided here comes largely from the available constraints guided by the history of Earth and other rocky planets in the Solar System. However, observing Earth-like exoplanets in the act of forming and in the habitable zone of their host stars remain difficult today and will continue to be difficult for many more years. The currently observable population of rocky exoplanetary systems demonstrate much larger variability in terms of parameter spaces and therefore conditions for LEVE cycling (Kite et al. 2021). For example, even for relatively small exoplanets such as those with radius $< 3.5 R_\oplus$, the existing population exhibit atmosphere–rocky (molten or solid) surface P–T conditions that are distinct from the current Solar System planets. Among these there are lava world or magma sea planets, with pressure reaching 1–30 GPa at the interface between H_2-dominated atmosphere and the molten rock surface (Kite et al. 2021). Solubilities of LEVEs in mafic–ultramafic molten silicates and partitioning of LEVEs between alloys and silicates mostly do not exist at these conditions—especially for ultramafic silicate melts. New experiments and computational constraints will be needed to shed light whether distinct melt–vapor interfaces are expected or whether complete miscibility between gas and silicate melt is more likely. Similarly, magma–vapor cycling of minor LEVEs such as C, N, and S may also be more complicated in these conditions. For example, MO–atmosphere exchange for these conditions will likely not be restricted to the volatile elements, but the atmosphere will include aerosols composed of silicates and salts (Gao et al. 2021). A subset of Super-Earths and Sub-Neptunes are also thought to be water worlds, i.e., bodies rich in water and poor in H_2 gas. The interface between liquid water and rock surface in these bodies may reach tens of GPa pressures and high hundreds to thousands K temperatures. Under such conditions, there may be complete miscibility between molten silicates and water (Kovačević et al. 2022). In addition, it is worth noting that growth of many of these planets via episodes of giant impacts could also create conditions where conditions of alloy–silicate immiscibility is inhibited, and core–mantle separation may not be complete until later stages of planetary evolution. All of these possibilities suggest that applying the framework developed for and learned from the Solar System planets to rocky exoplanets may require careful modifications.

Current activities in exoplanetary science are dominated by astronomical characterization followed by modeling efforts of various kinds and interdisciplinary effort of integrating different modeling platforms. However, understanding chemical habitability of rocky exoplanets would require deeper look at cycling of LEVEs across reservoirs and through time. Such efforts are only possible through proper assessment of available material properties and their limitations (e.g., melting condition and melt composition, physical properties of melts, solubility and saturation of LEVEs as a function of P–T–X, storage mechanisms of LEVEs at different P–T–X, partitioning of LEVEs between various phases as a function of P–T–X–f_{O_2}). The required domain knowledge is with geologists (e.g., petrologists, volcanologists, geochemists, mineral physicists) working on various processes and properties of Earth and by extension Solar System rocky planets. Constraining various modes of LEVE cycling in various exoplanetary systems would require observational astronomers and astrophysical modelers working closely with observational Earth Scientists.

ACKNOWLEDGMENTS

The authors thank Colin Jackson and Yoshinori Miyazaki for their thoughtful and thorough reviews. RD thanks the volume editors Keith Putirka, Natalie Hinkel, and Siyi Xu for the invitation to contribute to this volume and for their patient handling of the chapter. This work received support from NASA grants 80NSSC18K0828, 80NSSC22K0635, and 80NSSC18K1314 and NSF grant EAR-1763226.

REFERENCES

Abe Y, Matsui T (1985) The formation of an impact-generated H_2O atmosphere and its implications for the early thermal history of the Earth. J Geophys Res Solid Earth 90:C545–C559

Adibekyan V, Dorn C, Sousa SG, Santos NC, Bitsch B, Israelian G, Mordasini C, Barros SCC, Delgado Mena E, Demangeon ODS, Faria JP, Figueira P, Hakobyan AA, Oshagh M, Soares BMTB, Kunitomo M, Takeda Y, Jofré E, Petrucci R, Martioli E (2021) A compositional link between rocky exoplanets and their host stars. Science 374:330–332

Agol E, Dorn C, Grimm SL, Turbet M, Ducrot E, Delrez L, Gillon M, Demory B-O, Burdanov A, Barkaoui K, Benkhaldoun Z, Bolmont E, Burgasser A, Carey S, Wit Jd., Fabrycky D, Foreman-Mackey D, Haldemann J, Hernandez DM, Ingalls J, Jehin E, Langford Z, Leconte J, Lederer SM, Luger R, Malhotra R, Meadows VS, Morris BM, Pozuelos FJ, Queloz D, Raymond SN, Selsis F, Sestovic M, Triaud AHMJ, Grootel VV (2021) Refining the transit-timing and photometric analysis of TRAPPIST-1: Masses, radii, densities, dynamics, and Ephemerides. Planet Sci J 2:1

Alessi M, Pudritz RE, Cridland AJ (2016) On the formation and chemical composition of super Earths. Mon Not R Astron Soc 464:428–452

Alexander CMOD, Arden JW, Ash RD, Pillinger CT (1990) Presolar components in the ordinary chondrites. Earth Planet Sci Lett 99:220–229

Alexander CMOD, Fogel M, Yabuta H, Cody GD (2007) The origin and evolution of chondrites recorded in the elemental and isotopic compositions of their macromolecular organic matter. Geochim Cosmochim Acta 71:4380–4403

Alexander CMOD, Bowden R, Fogel ML, Howard KT, Herd CDK, Nittler LR (2012) The provenances of asteroids, and their contributions to the volatile inventories of the terrestrial planets. Science 337:721–723

Allen-Sutter H, Garhart E, Leinenweber K, Prakapenka V, Greenberg E, Shim SH (2020) Oxidation of the interiors of carbide exoplanets. Planet Sci J 1:39

Alt JC (1995) Subseafloor processes in mid-ocean ridge hydrothermal systems: Physical, chemical, biological, and geological interactions. *In:* SE Humphris RE Zierenberg LS Mullineaux RE Thomson (Eds) Geophysical Monograph 95, American Geophysical Union, Washington DC, p 85–114

Alt JC, Teagle DAH (1999) The uptake of carbon during alteration of ocean crust. Geochim Cosmochim Acta 63:1527–1535

Andersen CA, Keil K, Mason B (1964) Silicon oxynitride: A meteoritic mineral. Science 146:256–257

Ardia P, Hirschmann MM, Withers AC, Stanley BD (2013) Solubility of CH_4 in a synthetic basaltic melt, with applications to atmosphere–magma ocean–core partitioning of volatiles and to the evolution of the Martian atmosphere. Geochim Cosmochim Acta 114:52–71

Armstrong LS, Hirschmann MM, Stanley BD, Falksen EG, Jacobsen SD (2015) Speciation and solubility of reduced C–O–H–N volatiles in mafic melt: Implications for volcanism, atmospheric evolution, and deep volatile cycles in the terrestrial planets. Geochim Cosmochim Acta 171:283–302

Armstrong K, Frost DJ, McCammon CA, Rubie DC, Boffa Ballaran T (2019) Deep magma ocean formation set the oxidation state of Earth's mantle. Science 365:903–906

Aubaud C, Hauri EH, Hirschmann MM (2004) Hydrogen partition coefficients between nominally anhydrous minerals and basaltic melts. Geophys Res Lett 31:L20611, https://doi.org/10.1029/2004GL021341

Aulbach S, Stagno V (2016) Evidence for a reducing Archean ambient mantle and its effects on the carbon cycle. Geology 44:751–754

Badro J, Brodholt JP, Piet H, Siebert J, Ryerson FJ (2015) Core formation and core composition from coupled geochemical and geophysical constraints. PNAS 112:12310–12314

Bajgain S, Mookherjee M, Dasgupta R, Ghosh DB, Karki BB (2019) Nitrogen content in the Earth's outer core. Geophys Res Lett 46:89–98, https://doi.org/10.1029/2018GL080555

Bajgain SK, Mookherjee M, Dasgupta R (2021) Earth's core could be the largest terrestrial carbon reservoir. Commun Earth Environ 2:165

Ballmer MD, Lourenço DL, Hirose K, Caracas R, Nomura R (2017) Reconciling magma-ocean crystallization models with the present-day structure of the Earth's mantle. Geochem Geophys Geosyst 18:2785–2806

Balsiger H, Altwegg K, Geiss J (1995) D/H and $^{18}O/^{16}O$ ratio in the hydronium ion and in neutral water from in situ ion measurements in comet Halley. J Geophys Res Space Phys 100:5827–5834

Bannister FA (1941) Osbornite, meteoritic titanium nitride. Mineral Mag 26:36–44

Baur WH (1972) Occurrence of nitride nitrogen in silicate minerals. Nature 240:461–462

Behrens H, Ohlhorst S, Holtz F, Champenois M (2004) CO_2 solubility in dacitic melts equilibrated with H_2O–CO_2 fluids: Implications for modeling the solubility of CO_2 in silicic melts. Geochim Cosmochim Acta 68:4687–4703

Bell DR, Rossman GR (1992) Water in Earth's mantle: the role of nominally anhydrous minerals. Science 255:1391–1397

Bergin EA, Kempton EMR, Hirschmann M, Bastelberger ST, Teal DJ, Blake GA, Ciesla FJ, Li J (2023) Exoplanet volatile carbon content as a natural pathway for haze formation. Astrophys J Lett 949:L17

Bernadou F, Gaillard F, Füri E, Marrocchi Y, Slodczyk A (2021) Nitrogen solubility in basaltic silicate melt—Implications for degassing processes. Chem Geol 573:20192

Berner RA (2003) The long-term carbon cycle, fossil fuels and atmospheric composition. Nature 426:323–326

Blanchard I, Rubie DC, Jennings ES, Franchi IA, Zhao X, Petitgirard S, Miyajima N, Jacobson SA, Morbidelli A (2022) The metal–silicate partitioning of carbon during Earth's accretion and its distribution in the early solar system. Earth Planet Sci Lett 580:117374

Bockelée-Morvan D, Gautier D, Lis DC, Young K, Keene J, Phillips T, Owen T, Crovisier J, Goldsmith PF, Bergin EA, Despois D, Wootten A (1998) Deuterated water in comet C/1996 B2 (Hyakutake) and its implications for the origin of comets. Icarus 133:147–162

Bolfan-Casanova N (2005) Water in the Earth's mantle. Mineral Mag 69:229–257

Bonati I, Lichtenberg T, Bower DJ, Timpe ML, Quanz SP (2019) Direct imaging of molten protoplanets in nearby young stellar associations. Astron Astrophys 621:A125

Boujibar A, Andrault D, Bouhifd MA, Bolfan-Casanova N, Devidal J-L, Trcera N (2014) Metal–silicate partitioning of sulphur, new experimental and thermodynamic constraints on planetary accretion. Earth Planet Sci Lett 391:42–54

Boujibar A, Driscoll P, Fei Y (2020) Super-Earth internal structures and initial thermal states. J Geophys Res Planets 125:e2019JE006124

Boukaré C-É, Cowan NB, Badro J (2022) Deep two-phase, hemispherical magma oceans on lava planets. Astrophys J 936:148

Boulliung J, Wood BJ (2022) SO_2 solubility and degassing behavior in silicate melts. Geochim Cosmochim Acta 336:150–164

Boulliung J, Füri E, Dalou C, Tissandier L, Zimmermann L, Marrocchi Y (2020) Oxygen fugacity and melt composition controls on nitrogen solubility in silicate melts. Geochim Cosmochim Acta 284:120–133

Bower DJ, Kitzmann D, Wolf AS, Sanan P, Dorn C, Oza AV (2019) Linking the evolution of terrestrial interiors and an early outgassed atmosphere to astrophysical observations. Astron Astrophys 631:A103

Bower DJ, Hakim K, Sossi PA, Sanan P (2022) Retention of water in terrestrial magma oceans and carbon-rich early atmospheres. Planet Sci J 3:93

Boynton WV (1975) Fractionation in the solar nebula: condensation of yttrium and the rare earth elements. Geochim Cosmochim Acta 39:569–584

Brasser R, Dauphas N, Mojzsis SJ (2018) Jupiter's influence on the building blocks of Mars and Earth. Geophys Res Lett 45:5908–5917

Brenan JM, Mungall JE, Bennett NR (2019) Abundance of highly siderophile elements in lunar basalts controlled by iron sulfide melt. Nat Geosci 12:701–706

Broadley MW, Bekaert DV, Piani L, Füri E, Marty B (2022) Origin of life-forming volatile elements in the inner Solar System. Nature 611:245–255

Brooker RA, Kohn SC, Holloway JR, McMillan PF (2001a) Structural controls on the solubility of CO_2 in silicate melts Part I: Bulk solubility data. Chem Geol 174:225–239

Brooker RA, Kohn SC, Holloway JR, McMillan PF (2001b) Structural controls on the solubility of CO_2 in silicate melts Part II: IR characteristics of carbonate groups in silicate glasses. Chem Geol 174:241–254

Brugman K, Phillips MG, Till CB (2021) Experimental determination of mantle solidi and melt compositions for two likely rocky exoplanet compositions. J Geophys Res Planets 126:e2020JE006731

Burbine TH, O'Brien KM (2004) Determining the possible building blocks of the Earth and Mars. Meteorit Planet Sci 39:667–681

Callegaro S, Geraki K, Marzoli A, De Min A, Maneta V, Baker DR (2020) The quintet completed: The partitioning of sulfur between nominally volatile-free minerals and silicate melts. Am Mineral 105:697–707

Cartier C, Wood BJ (2019) The role of reducing conditions in building Mercury. Elements 15:39–45

Chao K-H, deGraffenried R, Lach M, Nelson W, Truax K, Gaidos E (2021) Lava worlds: From early earth to exoplanets. Geochemistry 81:125735

Chi H, Dasgupta R, Duncan M, Shimizu N (2014) Partitioning of carbon between Fe-rich alloy melt and silicate melt in a magma ocean—Implications for the abundance and origin of volatiles in Earth, Mars, and the Moon. Geochim Cosmochim Acta 139:447–471

Chiang E, Youdin AN (2010) Forming planetesimals in solar and extrasolar nebulae. Annu Rev Earth Planet Sci 38:493–522

Chowdhury P, Dasgupta R (2019) Effect of sulfate on the basaltic liquidus and Sulfur Concentration at Anhydrite Saturation (SCAS) of hydrous basalts—Implications for sulfur cycle in subduction zones. Chem Geol 522:162–174

Chowdhury P, Dasgupta R (2020) Sulfur extraction via carbonated melts from sulfide-bearing mantle lithologies—Implications for deep sulfur cycle and mantle redox. Geochim Cosmochim Acta 269:376–397

Chowdhury P, Dasgupta R, Phelps PR, Costin G, Lee C-TA (2022) Oxygen fugacity range of subducting crust inferred from fractionation of trace elements during fluid-present slab melting in the presence of anhydrite versus sulfide. Geochim Cosmochim Acta 325:214–231

Clesi V, Bouhifd MA, Bolfan-Casanova N, Manthilake G, Schiavi F, Raepsaet C, Bureau H, Khodja H, Andrault D (2018) Low hydrogen contents in the cores of terrestrial planets. Sci Adv 4:e1701876

Collinet M, Plesa A-C, Grove TL, Schwinger S, Ruedas T, Breuer D (2021) MAGMARS: A melting model for the Martian mantle and FeO-rich peridotite. J Geophys Res: Planets 126:e2021JE006985

Dahl TW, Stevenson DJ (2010) Turbulent mixing of metal and silicate during planet accretion — And interpretation of the Hf–W chronometer. Earth Planet Sci Lett 295:177–186

Dalou C, Hirschmann MM, von der Handt A, Mosenfelder J, Armstrong LS (2017) Nitrogen and carbon fractionation during core–mantle differentiation at shallow depth. Earth Planet Sci Lett 458:141–151

Dalou C, Hirschmann MM, Jacobsen SD, Le Losq C (2019) Raman spectroscopy study of C–O–H–N speciation in reduced basaltic glasses: Implications for reduced planetary mantles. Geochim Cosmochim Acta 265:32–47

Dasgupta R (2013) Ingassing, storage, and outgassing of terrestrial carbon through geologic time. Rev Mineral Geochem 75:183–229

Dasgupta R (2018) Volatile bearing partial melts beneath oceans and continents—Where, how much, and of what compositions? Am J Sci 318:141–165

Dasgupta R, Aubaud C (in press) Major volatiles in the Earth's mantle beneath mid-ocean ridges and intraplate ocean islands, *In:* C Chauvel (Ed) Earth's Interior, Treat Geochem

Dasgupta R, Grewal DS (2019) Origin and early differentiation of carbon and associated life-essential volatile elements on Earth. *In:* B Orcutt, I Daniel, R Dasgupta (Eds) Deep Carbon: Past to Present, Cambridge University Press, Cambridge, p 4–39, https://doi.org/10.1017/9781108677950

Dasgupta R, Hirschmann MM (2006) Melting in the Earth's deep upper mantle caused by carbon dioxide. Nature 440:659–662

Dasgupta R, Hirschmann MM (2007) A modified iterative sandwich method for determination of near-solidus partial melt compositions. II. Application to determination of near-solidus melt compositions of carbonated peridotite. Contrib Mineral Petrol 154:647–661

Dasgupta R, Hirschmann MM (2010) The deep carbon cycle and melting in Earth's interior. Earth Planet Sci Lett 298:1–13

Dasgupta R, Hirschmann MM, Withers AC (2004) Deep global cycling of carbon constrained by the solidus of anhydrous, carbonated eclogite under upper mantle conditions. Earth Planet Sci Lett 227:73–85

Dasgupta R, Hirschmann MM, Smith ND (2007) Partial melting experiments of peridotite + CO_2 at 3 GPa and genesis of alkalic ocean island basalts. J Petrol 48:2093–2124

Dasgupta R, Chi H, Shimizu N, Buono A, Walker D (2013a) Carbon solution and partitioning between metallic and silicate melts in a shallow magma ocean: implications for the origin and distribution of terrestrial carbon. Geochim Cosmochim Acta 102:191–212

Dasgupta R, Mallik A, Tsuno K, Withers AC, Hirth G, Hirschmann MM (2013b) Carbon-dioxide-rich silicate melt in the Earth's upper mantle. Nature 493:211–215

Dasgupta R, Chowdhury P, Eguchi J, Sun C, Saha S (2022a) Volatile-bearing partial melts in the lithospheric and sub-lithospheric mantle on Earth and other rocky planets. Rev Mineral Geochem 87:575–606

Dasgupta R, Falksen E, Pal A, Sun C (2022b) The fate of nitrogen during parent body partial melting and accretion of the inner solar system bodies at reducing conditions. Geochim Cosmochim Acta 336:291–307

Dauphas N (2017) The isotopic nature of the Earth's accreting material through time. Nature 541:521

Dauphas N, Chen JH, Zhang J, Papanastassiou DA, Davis AM, Travaglio C (2014) Calcium-48 isotopic anomalies in bulk chondrites and achondrites: Evidence for a uniform isotopic reservoir in the inner protoplanetary disk. Earth Planet Sci Lett 407:96–108

Davidson J, Turner S, Handley H, Macpherson C, Dosseto A (2007) Amphibole "sponge" in arc crust? Geology 35:787–790

Defouilloy C, Cartigny P, Assayag N, Moynier F, Barrat JA (2016) High-precision sulfur isotope composition of enstatite meteorites and implications of the formation and evolution of their parent bodies. Geochim Cosmochim Acta 172:393–409

Deguen R, Landeau M, Olson P (2014) Turbulent metal–silicate mixing, fragmentation, and equilibration in magma oceans. Earth Planet Sci Lett 391:274–287

Deng J, Du Z, Karki BB, Ghosh DB, Lee KKM (2020) A magma ocean origin to divergent redox evolutions of rocky planetary bodies and early atmospheres. Nat Commun 11:2007

Ding S, Dasgupta R (2017) The fate of sulfide during decompression melting of peridotite—Implications for sulfur inventory of the MORB-source depleted upper mantle. Earth Planet Sci Lett 459:183–195

Ding S, Dasgupta R (2018) Sulfur inventory of ocean island basalt source regions constrained by modeling the fate of sulfide during decompression melting of a heterogenous mantle. J Petrol 59:1281–1308

Ding S, Dasgupta R, Tsuno K (2014) Sulfur concentration of martian basalts at sulfide saturation at high pressures and temperatures—implications for deep sulfur cycle on Mars. Geochim Cosmochim Acta 131:227–246

Ding S, Dasgupta R, Lee C-TA, Wadhwa M (2015) New bulk sulfur measurements of Martian meteorites and modeling the fate of sulfur during melting and crystallization—Implications for sulfur transfer from Martian mantle to crust–atmosphere system. Earth Planet Sci Lett 409:157–167

Ding S, Hough T, Dasgupta R (2018) New high pressure experiments on sulfide saturation of high-FeO* basalts with variable TiO_2 contents—Implications for the sulfur inventory of the lunar interior. Geochim Cosmochim Acta 222:319–339

Ding S, Dasgupta R, Tsuno K (2020) The solidus and melt productivity of nominally anhydrous Martian mantle constrained by new high pressure-temperature experiments—Implications for crustal production and mantle source evolution. J Geophys Res Planets 123:e2019JE006078

Dong J, Fischer RA, Stixrude LP, Lithgow-Bertelloni CR (2021) Constraining the volume of Earth's early oceans with a temperature-dependent mantle water storage capacity model. AGU Adv 2:e2020AV000323

Dong J, Fischer RA, Stixrude LP, Lithgow-Bertelloni CR, Eriksen ZT, Brennan MC (2022) Water storage capacity of the martian mantle through time. Icarus 385:115113

Dorn C, Lichtenberg T (2021) Hidden water in magma ocean exoplanets. Astrophys J Lett 922:L4

Dorn C, Mosegaard K, Grimm SL, Alibert Y (2018) Interior characterization in multiplanetary systems: TRAPPIST-1. Astrophys J 865:20

Doyle AE, Young ED, Klein B, Zuckerman B, Schlichting HE (2019) Oxygen fugacities of extrasolar rocks: Evidence for an Earth-like geochemistry of exoplanets. Science 366:356–359

Doyle AE, Klein B, Schlichting HE, Young ED (2020) Where are the extrasolar Mercuries? Astrophys J 901:10

Dreibus G, Wänke H (1985) Mars, a volatile-rich planet. Meteoritics 20:367–381

Dreibus G, Wänke H (1990) Comparison of the chemistry of moon and mars. Adv Space Res 10:7–16

Duncan MS, Dasgupta R (2014) CO_2 solubility and speciation in rhyolitic sediment partial melts at 1.5–3 GPa—Implications for carbon flux in subduction zones. Geochim Cosmochim Acta 124:328–347, https://doi.org/10.1016/j.gca.2013.09.026

Duncan MS, Dasgupta R (2015) Pressure and temperature dependence of CO_2 solubility in hydrous rhyolitic melt—Implications for carbon transfer to mantle source of volcanic arcs via partial melt of subducting crustal lithologies. Contrib Mineral Petrol 169:1–19

Duncan MS, Dasgupta R (2017) Rise of Earth's atmospheric oxygen controlled by efficient subduction of organic carbon. Nat Geosci 10:387–392

Duncan MS, Dasgupta R, Tsuno K (2017) Experimental determination of CO_2 content at graphite saturation along a natural basalt–peridotite melt join: Implications for the fate of carbon in terrestrial magma oceans. Earth Planet Sci Lett 466:115–128

Duncan MS, Schmerr NC, Bertka CM, Fei Y (2018) Extending the solidus for a model iron-rich Martian mantle composition to 25 GPa. Geophys Res Lett 45:10,211–210,220

Eberhardt P, Reber M, Krankowsky D, Hodges RR (1995) The D/H and $^{18}O/^{16}O$ ratios in water from comet P/Halley. Astron Astrophys 302:301–316

Eguchi J, Dasgupta R (2018) A CO_2 solubility model for silicate melts from fluid saturation to graphite or diamond saturation. Chem Geol 487:23–38

Eguchi J, Dasgupta R (2022) Cycling of CO_2 and H_2O constrained by experimental investigation of a model ophicarbonate at deep subduction zone conditions. Earth Planet Sci Lett 600:117866

Eguchi J, Seales J, Dasgupta R (2020) Great Oxidation and Lomagundi events linked by deep cycling and enhanced degassing of carbon. Nat Geosci 13:71–76

Elkins-Tanton LT (2008) Linked magma ocean solidification and atmospheric growth for Earth and Mars. Earth Planet Sci Lett 271:181–191

Elkins-Tanton LT (2012) Magma oceans in the inner Solar System. Annu Rev Earth Planet Sci 40:113–139, https://doi.org/10.1146/annurev-earth-042711-105503

Elkins-Tanton LT, Hess PC, Parmentier EM (2005) Possible formation of ancient crust on Mars through magma ocean processes. J Geophys Res Planets 110:E12S01, https://doi.org/10.1029/2005JE002480

Elkins-Tanton LT, Smrekar SE, Hess PC, Parmentier EM (2007) Volcanism and volatile recycling on a one-plate planet: Applications to Venus. J Geophys Res 112:E04S06, https://doi.org/10.1029/2006JE002793

Elkins-Tanton LT, Weiss BP, Zuber MT (2011) Chondrites as samples of differentiated planetesimals. Earth Planet Sci Lett 305:1–10

Evans LG, Peplowski PN, Rhodes EA, Lawrence DJ, McCoy TJ, Nittler LR, Solomon SC, Sprague AL, Stockstill-Cahill KR, Starr RD, Weider SZ, Boynton WV, Hamara DK, Goldsten JO (2012) Major-element abundances on the surface of Mercury: Results from the MESSENGER gamma-ray spectrometer. J Geophys Res Planets 117: E00L07

Falloon TJ, Green DH (1989) The solidus of carbonated, fertile peridotite. Earth Planet Sci Lett 94:364–370

Fegley B, Lewis JS (1980) Volatile element chemistry in the solar nebula: Na, K, F, Cl, Br, and P. Icarus 41:439–455

Fegley B, Palme H (1985) Evidence for oxidizing conditions in the solar nebula from Mo and W depletions in refractory inclusions in carbonaceous chondrites. Earth Planet Sci Lett 72:311–326

Fine G, Stolper E (1986) Dissolved carbon dioxide in basaltic glasses: concentrations and speciation. Earth Planet Sci Lett 76:263–278

Fischer-Gödde M, Kleine T (2017) Ruthenium isotopic evidence for an inner Solar System origin of the late veneer. Nature 541:525

Fischer RA, Cottrell E, Hauri E, Lee KKM, Le Voyer M (2020) The carbon content of Earth and its core. PNAS 117:8743–8749

Fitoussi C, Bourdon B, Wang X (2016) The building blocks of Earth and Mars: A close genetic link. Earth Planet Sci Lett 434:151–160

Frost DJ (1999) The stability of dense hydrous magnesium silicates in Earth's transition zone and lower mantle. *In:* Y Fei C Bertka BO Mysen (Eds) Mantle Petrology: Field Observations and High Pressure Experimentation: A Tribute to Francis R (Joe) Boyd 6, The Geochemical Society, p 283–296

Frost DJ, McCammon CA (2008) The redox state of Earth's mantle. Annu Rev Earth Planet Sci 36:389–420

Frost DJ, Wood BJ (1997) Experimental measurements of the fugacity of CO_2 and graphite/ diamond stability from 35 to 77 kbar at 925 to 1650 °C. Geochim Cosmochim Acta 61:1565–1574

Fuentes JJ, Crowley JW, Dasgupta R, Mitrovica JX (2019) The influence of plate tectonic style on melt production and CO_2 outgassing flux at mid-ocean ridges. Earth Planet Sci Lett 511:154–163

Fulton BJ, Petigura EA, Howard AW, Isaacson H, Marcy GW, Cargile PA, Hebb L, Weiss LM, Johnson JA, Morton TD, Sinukoff E, Crossfield IJM, Hirsch LA (2017) The California-Kepler Survey. IIIA Gap in the Radius Distribution of Small Planets*. Astron J 154:109

Füri E, Marty B (2015) Nitrogen isotope variations in the Solar System. Nat Geosci 8:515–522

Gaetani GA, Grove TL (1998) The influence of water on melting of mantle peridotite. Contrib Mineral Petrol 131:323–346

Gaillard F, Scaillet B, Arndt NT (2011) Atmospheric oxygenation caused by a change in volcanic degassing pressure. Nature 478:229–232

Gaillard F, Scaillet B, Pichavant M, Iacono-Marziano G (2015) The redox geodynamics linking basalts and their mantle sources through space and time. Chem Geol 418:217–233

Gaillard F, Bernadou F, Roskosz M, Bouhifd MA, Marrocchi Y, Iacono-Marziano G, Moreira M, Scaillet B, Rogerie G (2022) Redox controls during magma ocean degassing. Earth Planet Sci Lett 577:117255

Gao X, Thiemens MH (1993a) Isotopic composition and concentration of sulfur in carbonaceous chondrites. Geochim Cosmochim Acta 57:3159–3169

Gao X, Thiemens MH (1993b) Variations of the isotopic composition of sulfur in enstatite and ordinary chondrites. Geochim Cosmochim Acta 57:3171–3176

Gao P, Wakeford HR, Moran SE, Parmentier V (2021) Aerosols in exoplanet atmospheres. J Geophys Res Planets 126:e2020JE006655

Gao Z, Yang Y-N, Yang S-Y, Li Y (2022) Experimental determination of N_2 solubility in silicate melts and implications for N_2–Ar–CO_2 fractionation in magmas. Geochim Cosmochim Acta 326:17–40

Ghiorso MS, Gualda GAR (2015) An H_2O–CO_2 mixed fluid saturation model compatible with rhyolite-MELTS. Contrib Mineral Petrol 169:53

Gorshkov AI, Bao YN, Bershov LV, Ryabchikov ID, Sivtsov AV, Lapina MI (1997) Inclusions of native metals and other minerals in diamond from Kimberlite Pipe 50, Liaoning, China. Int Geol Rev 8:794–804

Grady MM, Wright IP (2003) Elemental and isotopic abundances of carbon and nitrogen in meteorites. Space Sci Rev 106:231–248

Grady MM, Wright IP, Carr LP, Pillinger CT (1986) Compositional differences in enstatite chondrites based on carbon and nitrogen stable isotope measurements. Geochim Cosmochim Acta 50:2799–2813

Grady MM, Wright IP, Swart PK, Pillinger CT (1988) The carbon and oxygen isotopic composition of meteoritic carbonates. Geochim Cosmochim Acta 52:2855–2866

Grassi D, Schmidt MW (2011) The melting of carbonated pelites from 70 to 700 km depth. J Petrol 52:765–789

Grewal DS, Dasgupta R, Holmes AK, Costin G, Li Y, Tsuno K (2019a) The fate of nitrogen during core–mantle separation on Earth. Geochim Cosmochim Acta 251:87–115

Grewal DS, Dasgupta R, Sun C, Tsuno K, Costin G (2019b) Delivery of carbon, nitrogen and sulfur to the silicate Earth by a giant impact. Sci Adv 5:eaau3669, https://doi.org/10.1126/sciadv.aau3669

Grewal DS, Dasgupta R, Farnell A (2020) The speciation of carbon, nitrogen, and water in magma oceans and its effect on volatile partitioning between major reservoirs of the Solar System rocky bodies. Geochim Cosmochim Acta 280:281–301

Grewal DS, Dasgupta R, Aithala S (2021a) The effect of carbon concentration on its core–mantle partitioning behavior in inner Solar System rocky bodies. Earth Planet Sci Lett 571:117090

Grewal DS, Dasgupta R, Hough T, Farnell A (2021b) Rates of protoplanetary accretion and differentiation set nitrogen budget of rocky planets. Nat Geosci 14:369–376

Grewal DS, Seales JD, Dasgupta R (2022a) Internal or external magma oceans in the earliest protoplanets— Perspectives from nitrogen and carbon fractionation. Earth Planet Sci Lett 598:117847

Grewal DS, Sun T, Aithala S, Hough T, Dasgupta R, Yeung LY, Schauble EA (2022b) Limited nitrogen isotopic fractionation during core–mantle differentiation in rocky protoplanets and planets. Geochim Cosmochim Acta 338:347–364

Grossman L (1972) Condensation in the primitive solar nebula. Geochim Cosmochim Acta 36:597–619

Grossman L, Larimer JW (1974) Early chemical history of the solar system. Rev Geophys 12:71–101

Grove TL, Till CB, Krawczynski M (2012) The role of H_2O in subduction zone magmatism. Annu Rev Earth Planet Sci 40:413–439, https://doi.org/10.1146/annurev-earth-042711-105310

Guimond CM, Shorttle O, Rudge JF (2023) Mantle mineralogy limits to rocky planet water inventories. Mon Not R Astron Soc 521:2535–2552

Häglund J, Fernández Guillermet A, Grimvall G, Körling M (1993) Theory of bonding in transition-metal carbides and nitrides. Phys Rev B 48:11685–11691

Hakim K, Rivoldini A, Van Hoolst T, Cottenier S, Jaeken J, Chust T, Steinle-Neumann G (2018a) A new ab initio equation of state of hcp-Fe and its implication on the interior structure and mass–radius relations of rocky super-Earths. Icarus 313:61–78

Hakim K, van Westrenen W, Dominik C (2018b) Capturing the oxidation of silicon carbide in rocky exoplanetary interiors. Astron Astrophys 618:L6

Hallis LJ (2017) D/H ratios of the inner Solar System. Philos Trans R Soc A 375:20150390

Hamano K, Abe Y, Genda H (2013) Emergence of two types of terrestrial planet on solidification of magma ocean. Nature 497:607–610

Hanon P, Robert F, Chaussidon M (1998) High carbon concentrations in meteoritic chondrules: A record of metal–silicate differentiation. Geochim Cosmochim Acta 62:903–913

Hatalova P, Brasser R, Mamonova E, Werner SC (2023) Forming rocky exoplanets around K-dwarf stars. Astron Astrophys 676:A131

Hauri EH, Gaetani GA, Green TH (2006) Partitioning of water during melting of the Earth's upper mantle at H_2O-undersaturated conditions. Earth Planet Sci Lett 248:715–734

Hauri EH, Cottrell E, Kelley KA, Tucker JM, Shimizu K, Voyer ML, Marske J, Saal AE (2019) Carbon in the convecting mantle. *In:* BN Orcutt, I Daniel, R Dasgupta (Eds) Deep Carbon: Past to Present, Cambridge University Press, Cambridge, p 237–275

Hevey PJ, Sanders IS (2006) A model for planetesimal meltdown by ^{26}Al and its implications for meteorite parent bodies. Meteorit Planet Sci 41:95–106

Hier-Majumder S, Hirschmann MM (2017) The origin of volatiles in the Earth's mantle. Geochem Geophys Geosyst 18:3078–3092

Hinkel NR, Unterborn CT (2018) The star–planet connection. I Using stellar composition to observationally constrain planetary mineralogy for the 10 closest stars*. Astrophys J 853:83

Hinkel NR, Youngblood A, Soares-Furtado M (2024) Host stars and how their compositions influence exoplanets. Rev Mineral Geochem 90:1–26

Hirschmann MM (2000) The mantle solidus: experimental constraints and the effect of peridotite composition. Geochem Geophys Geosyst 1:2000GC000070

Hirschmann MM (2006) Water, melting, and the deep Earth H_2O cycle. Annu Rev Earth Planet Sci 34:629–653

Hirschmann MM (2012) Magma ocean influence on early atmosphere mass and composition. Earth Planet Sci Lett 341–344:48–57

Hirschmann MM (2016) Constraints on the early delivery and fractionation of Earth's major volatiles from C/H, C/N, and C/S ratios. Am Mineral 101:540–553

Hirschmann MM (2018) Comparative deep Earth volatile cycles: The case for C recycling from exosphere/mantle fractionation of major (H_2O, C, N) volatiles and from H_2O/Ce, CO_2/Ba, and CO_2/Nb exosphere ratios. Earth Planet Sci Lett 502:262–273, https://doi.org/10.1016/j.epsl.2018.08.023

Hirschmann MM, Withers AC (2008) Ventilation of CO_2 from a reduced mantle and consequences for the early Martian greenhouse. Earth Planet Sci Lett 270:147–155

Hirschmann MM, Tenner T, Aubaud C, Withers AC (2009) Dehydration melting of nominally anhydrous mantle: The primacy of partitioning. Phys Earth Planet Sci 176:54–68, https://doi.org/10.1016/j.pepi.2009.04.001

Hirschmann MM, Withers AC, Ardia P, Foley NT (2012) Solubility of molecular hydrogen in silicate melts and consequences for volatile evolution of terrestrial planets. Earth Planet Sci Lett 345:38–48

Hirschmann MM, Bergin EA, Blake GA, Ciesla FJ, Li J (2021) Early volatile depletion on planetesimals inferred from C–S systematics of iron meteorite parent bodies. PNAS 118:e(2026779118)

Holland HD (2002) Volcanic gases, black smokers, and the great oxidation event. Geochim Cosmochim Acta 66:3811–3826

Holloway JR, Pan V, Gudmundsson G (1992) High-pressure fluid-absent melting experiments in the presence of graphite; oxygen fugacity, ferric/ferrous ratio and dissolved CO_2. Eur J Mineral 4:105–114

Hutsemékers D, Manfroid J, Jehin E, Arpigny C (2009) New constraints on the delivery of cometary water and nitrogen to Earth from the $^{15}N/^{14}N$ isotopic ratio. Icarus 204:346–348

Iacono-Marziano G, Morizet Y, Le Trong E, Gaillard F (2012) New experimental data and semi-empirical parameterization of $H_2O–CO_2$ solubility in mafic melts. Geochim Cosmochim Acta 97:1–23

Izidoro A, Ogihara M, Raymond SN, Morbidelli A, Pierens A, Bitsch B, Cossou C, Hersant F (2017) Breaking the chains: hot super-Earth systems from migration and disruption of compact resonant chains. Mon Not R Astron Soc 470:1750–1770

Izidoro A, Dasgupta R, Raymond SN, Deienno R, Bitsch B, Isella A (2022) Planetesimal rings as the cause of the Solar System's planetary architecture. Nat Astron 6:357–366

Jackson CRM, Cottrell E, Du Z, Bennett NR, Fei Y (2021) High pressure redistribution of nitrogen and sulfur during planetary stratification. Geochem Perspect Lett 18:37–42

Jégo S, Dasgupta R (2014) The fate of sulfur during fluid-present melting of subducting basaltic crust at variable oxygen fugacity. J Petrol 55:1019–1050, https://doi.org/10.1093/petrology/egu016

Johnson A, Dasgupta R, Costin G, Tsuno K (2024) Electron probe microanalysis of trace sulfur in experimental basaltic glasses and silicate minerals. Am Mineral. In Press, https://doi.org/10.2138/am-2023-9140

Jones RH (2024) Meteorites and planet formation. Rev Mineral Geochem 90:113–140

Jontof-Hutter D (2019) The compositional diversity of low-mass exoplanets. Annu Rev Earth Planet Sci 47:141–171

Junk G, Svec HJ (1958) The absolute abundance of the nitrogen isotopes in the atmosphere and compressed gas from various sources. Geochim Cosmochim Acta 14:234–243

Juriček MP, Keppler H (2023) Amphibole stability, water storage in the mantle, and the nature of the lithosphere-asthenosphere boundary. Earth Planet Sci Lett 608:118082

Kadik AA, Kurovskaya NA, Ignat'ev YA, Kononkova NN, Koltashev VV, Plotnichenko VG (2011) Influence of oxygen fugacity on the solubility of nitrogen, carbon, and hydrogen in $FeO–Na_2O–SiO_2–Al_2O_3$ melts in equilibrium with metallic iron at 1.5 GPa and 1400 °C. Geochem Int 49:429–438

Kadik AA, Litvin YA, Koltashev VV, Kryukova EB, Plotnichenko VG, Tsekhonya TI, Kononkova NN (2013) Solution behavior of reduced N–H–O volatiles in $FeO–Na_2O–SiO_2–Al_2O_3$ melt equilibrated with molten Fe alloy at high pressure and temperature. Phys Earth Planet Sci 214:14–24

Kadik AA, Koltashev VV, Kryukova EB, Plotnichenko VG, Tsekhonya TI, Kononkova NN (2015) Solubility of nitrogen, carbon, and hydrogen in $FeO–Na_2O–Al_2O_3–SiO_2$ melt and liquid iron alloy: Influence of oxygen fugacity. Geochem Int 53:849–868

Kadik AA, Kurovskaya NA, Lukanin OA, Ignat'ev YA, Koltashev VV, Kryukova EB, Plotnichenko VG, Kononkova NN (2017) Formation of N–C–O–H molecules and complexes in the basalt–basaltic andesite melts at 1.5 GPa and 1400 °C in the presence of liquid iron alloys. Geochem Int 55:151–162

Karki BB, Zhang J, Stixrude L (2013) First principles viscosity and derived models for $MgO–SiO_2$ melt system at high temperature. Geophys Res Lett 40:94–99

Katyal N, Nikolaou A, Godolt M, Grenfell JL, Tosi N, Schreier F, Rauer H (2019) Evolution and spectral response of a steam atmosphere for early Earth with a coupled climate–interior model. Astrophys J 875:31

Katz RF, Spiegelman M, Langmuir CH (2003) A new parameterization of hydrous mantle melting. Geochem Geophys Geosyst 4:1073, https://doi.org/10.1029/2002GC000433

Keppler H, Wiedenbeck M, Shcheka SS (2003) Carbon solubility in olivine and the mode of carbon storage in the Earth's mantle. Nature 424:414–416

Kissel J, Krueger FR (1987) The organic component in dust from comet Halley as measured by the PUMA mass spectrometer on board Vega 1. Nature 326:755–760

Kiseeva ES, Litasov KD, Yaxley GM, Ohtani E, Kamenetsky VS (2013) Melting and phase relations of carbonated eclogite at 9–21 GPa and the petrogenesis of alkali-rich melts in the deep mantle. J Petrol 54:1555–1583

Kite E, Kreidberg L, Schaefer L, Caracas R, Hirschmann MM (2021) "Earth cousins" are new targets for planetary materials research. EOS 102, https://doi.org/10.1029/2021EO159473

Kogiso T, Hirschmann MM, Pertermann M (2004) High-pressure partial melting of mafic lithologies in the mantle. J Petrol 45: 2407–2422

Kohlstedt DL, Keppler H, Rubie DC (1996) Solubility of water in the α, β and γ phases of $(Mg,Fe)_2SiO_4$. Contrib Mineral Petrol 123:345–357

Kohn SC (2000) The dissolution mechanisms of water in silicate melts; a synthesis of recent data. Mineral Mag 64:389–408

Korenaga J (2010) On the likelihood of plate tectonics on Super-Earths: Does size matter? Astrophys J Lett 725:L43

Korenaga J, Planavsky NJ, Evans DAD (2017) Global water cycle and the coevolution of the Earth's interior and surface environment. Philos Trans R Soc A 375:20150393

Kornacki AS, Fegley B (1986) The abundance and relative volatility of refractory trace elements in Allende Ca,Al-rich inclusions: implications for chemical and physical processes in the solar nebula. Earth Planet Sci Lett 79:217–234

Kovačević T, González-Cataldo F, Stewart ST, Militzer B (2022) Miscibility of rock and ice in the interiors of water worlds. Sci Rep 12:13055

Krawczynski MJ, Grove TL, Behrens H (2012) Amphibole stability in primitive arc magmas: Effects of temperature, H_2O content, and oxygen fugacity. Contrib Mineral Petrol 164:317–339

Kruijer TS, Kleine T, Borg LE (2020) The great isotopic dichotomy of the early Solar System. Nat Astron 4:32–40

Labidi J (2022) The origin of nitrogen in Earth's mantle: Constraints from basalts $^{15}N/^{14}N$ and $N_2/^3He$ ratios. Chem Geol 597:120780

Lara M, Dasgupta R (2023) Effects of $H_2O–CO_2$ fluids, temperature, and peridotite fertility on partial melting in mantle wedges and generation of primary arc basalts. J Petrol 64:egad047

Larimer JW (1967) Chemical fractionations in meteorites–I Condensation of the elements. Geochim Cosmochim Acta 31:1215–1238

Larimer JW (1973) Chemical fractionations in meteorites–VII Cosmothermometry and cosmobarometry. Geochim Cosmochim Acta 37:1603–1623

Larocque J, Canil D (2010) The role of amphibole in the evolution of arc magmas and crust: the case from the Jurassic Bonanza arc section, Vancouver Island, Canada. Contrib Mineral Petrol 159:475–492

Lebrun T, Massol H, Chassefière E, Davaille A, Marcq E, Sarda P, Leblanc F, Brandeis G (2013) Thermal evolution of an early magma ocean in interaction with the atmosphere. J Geophys Res Planets 118:1155–1176

Lee MR, Russell SS, Arden JW, Pillinger CT (1995) Nierite (Si_3N_4), a new mineral from ordinary and enstatite chondrites. Meteoritics 30:387–398

Lee C-TA, Shen B, Slotnick B, Liao K, Dickens G, Yokoyama Y, Lenardic A, Dasgupta R, Jellinek MS LJ, Schneider T, Tice M (2013) Continent-island arc fluctuations, growth of crustal carbonates, and long-term climate change. Geosphere 9:21–36, https://doi.org/10.1130/GES00822.1

Lee C-TA, Jiang H, Dasgupta R, Torres M (2019) A framework for understanding whole-Earth carbon cycling. *In:* BN Orcutt, I Daniel, R Dasgupta (Eds) Deep Carbon: Past to Present, Cambridge University Press, Cambridge, p 313–357

Léger A, Rouan D, Schneider J, Barge P, Fridlund M, Samuel B, Ollivier M, Guenther E, Deleuil M, Deeg HJ, Auvergne M, Alonso R, Aigrain S, Alapini A, Almenara JM, Baglin A, Barbieri M, Bruntt H, Bordé P, Bouchy F, Cabrera J, Catala C, Carone L, Carpano S, Csizmadia S, Dvorak R, Erikson A, Ferraz-Mello S, Foing B, Fressin F, Gandolfi D, Gillon M, Gondoin P, Grasset O, Guillot T, Hatzes A, Hébrard G, Jorda L, Lammer H, Llebaria A, Loeillet B, Mayor M, Mazeh T, Moutou C, Pätzold M, Pont F, Queloz D, Rauer H, Renner S, Samadi R, Shporer A, Sotin C, Tingley B, Wuchterl G, Adda M, Agogu P, Appourchaux T, Ballans H, Baron P, Beaufort T, Bellenger R, Berlin R, Bernardi P, Blouin D, Baudin F, Bodin P, Boisnard L, Boit L, Bonneau F, Borzeix S, Briet R, Buey J-T, Butler B, Cailleau D, Cautain R, Chabaud P-Y, Chaintreuil S, Chiavassa F, Costes V, Cuna Parrho V, De Oliveira Fialho F, Decaudin M, Defise J-M, Djalal S, Epstein G, Exil G-E, Fauré C, Fenouillet T, Gaboriaud A, Gallic A, Gamet P, Gavalda P, Grolleau E, Gruneisen R, Gueguen L, Guis V, Guivarc'h V, Guterman P, Hallouard D, Hasiba J, Heuripeau F, Huntzinger G, Hustaix H, Imad C, Imbert C, Johlander B, Jouret M, Journoud P, Karioty F, Kerjean L, Lafaille V, Lafond L, Lam-Trong T, Landiech P, Lapeyrere V, Larqué T, Laudet P, Lautier N, Lecann H, Lefevre L, Leruyet B, Levacher P, Magnan A, Mazy E, Mertens F, Mesnager J-M, Meunier J-C, Michel J-P, Monjoin W, Naudet D, Nguyen-Kim K, Orcesi J-L, Ottacher H, Perez R, Peter G, Plasson P, Plesseria J-Y, Pontet B, Pradines A, Quentin C, Reynaud J-L, Rolland G, Rollenhagen F, Romagnan R, Russ N, Schmidt R, Schwartz N, Sebbag I, Sedes G, Smit H, Steller MB, Sunter W, Surace C, Tello M, Tiphène D, Toulouse P, Ulmer B, Vandermarcq O, Vergnault E, Vuillemin A, Zanatta P (2009) Transiting exoplanets from the CoRoT space mission. Astron Astrophys 506:287–302

Leung I, Guo W, Friedman I, Gleason J (1990) Natural occurrence of silicon carbide in a diamondiferous kimberlite from Fuxian. Nature 346:352–354

Li J, Agee CB (1996) Geochemistry of mantle–core differentiation at high pressure. Nature 381:686–689

Li J, Agee CB (2001) Element partitioning constraints on the light element composition of the Earth's core. Geophys Res Lett 28:81–84

Li J, Fei Y (2003) Experimental constraints on core composition. *In:* RW Carlson (Ed) The Mantle and Core, Treatise on Geochemistry, Elsevier, Amsterdam, p 521–546

Li Y, Wiedenbeck M, Shcheka S, Keppler H (2013) Nitrogen solubility in upper mantle minerals. Earth Planet Sci Lett 377–378:311–323

Li Y, Dasgupta R, Tsuno K (2015a) The effects of sulfur, silicon, water, and oxygen fugacity on carbon solubility and partitioning in Fe-rich alloy and silicate melt systems at 3 GPa and 1600°C: Implications for core–mantle differentiation and degassing of magma oceans and reduced planetary mantles. Earth Planet Sci Lett 415:54–66

Li Y, Huang R, Wiedenbeck M, Keppler H (2015b) Nitrogen distribution between aqueous fluids and silicate melts. Earth Planet Sci Lett 411:218–228

Li Y, Dasgupta R, Tsuno K, Monteleone B, Shimizu N (2016a) Carbon and sulfur budget of the silicate Earth explained by accretion of differentiated planetary embryos. Nature Geosci 9:781–785

Li Y, Marty B, Shcheka S, Zimmermann L, Keppler H (2016b) Nitrogen isotope fractionation during terrestrial core–mantle separation. Geochem Perspect Lett 2:138–147

Li Y, Dasgupta R, Tsuno K (2017) Carbon contents in reduced basalts at graphite saturation: Implications for the degassing of Mars, Mercury, and the Moon. J Geophys Res Planets 122:1300–1320

Li Y, Wiedenbeck M, Monteleone B, Dasgupta R, Costin G, Gao Z, Lu W (2023) Nitrogen and carbon fractionation in planetary magma oceans and origin of the superchondritic C/N ratio in the bulk silicate Earth. Earth Planet Sci Lett 605:118032

Libourel G, Marty B, Humbert F (2003) Nitrogen solubility in basaltic melt. Part I. Effect of oxygen fugacity. Geochim Cosmochim Acta 67:4123–4135

Lichtenberg T (2021) Redox hysteresis of super-Earth exoplanets from magma ocean circulation. Astrophys J Lett 914:L4

Lichtenberg T, Bower DJ, Hammond M, Boukrouche R, Sanan P, Tsai S-M, Pierrehumbert RT (2021) Vertically resolved magma ocean–protoatmosphere evolution: H_2, H_2O, CO_2, CH_4 CO, O_2, and N_2 as primary absorbers. J Geophys Res Planets 126:e2020JE006711

Liebske C, Khan A (2019) On the principal building blocks of Mars and Earth. Icarus 322:121–134

Litasov K, Ohtani E (2003) Stability of various hydrous phases in CMAS pyrolite–H_2O system up to 25 GPa. Phys Chem Mineral 30:147–156

Lodders K (2003) Solar system abundances and condensation temperatures of the elements. Astrophys J 591:1220–1247, https://doi.org/10.1086/375492

Lorand J-P (1991) Sulphide petrology and sulphur geochemistry of orogenic lherzolites: A comparative study of the Pyrenean bodies (France) and the Lanzo massif (Italy). J Petrol Spec Vol, p 77–95

Lorand J-P, Luguet A (2016) Chalcophile and siderophile elements in mantle rocks: Trace elements controlled by trace minerals. Rev Mineral Geochem 81:441–488

Lord OT, Walter MJ, Dasgupta R, Walker D, Clark SM (2009) Melting in the Fe–C system to 70 GPa. Earth Planet Sci Lett 284:157–167

Lupu RE, Zahnle K, Marley MS, Schaefer L, Fegley B, Morley C, Cahoy K, Freedman R, Fortney JJ (2014) The atmospheres of Earthlike planets after giant impact events. Astrophys J 784:27

Luth RW (1993) Diamonds, eclogites, and oxidation state of the Earth's mantle. Science 261:66–68

Luth RW (1995) Experimental determination of the reaction dolomite + 2 coesite = diopside + $2CO_2$ to 6 GPa. Contrib Mineral Petrol 122:152–158

Luth RW (1999) Carbon and carbonates in the mantle. *In:* Fei Y, Bertka CM, Mysen BO (Eds) Mantle Petrology: Field Observations and High Pressure Experimentation: A Tribute to Francis R (Joe) Boyd 6, Geochemical Society, p 297–316

Maas C, Manske L, Wünnemann K, Hansen U (2021) On the fate of impact-delivered metal in a terrestrial magma ocean. Earth Planet Sci Lett 554:116680

Madhusudhan N, Lee KKM, Mousis O (2012) A possible carbon-rich interior in Super-Earth 55 Cancrie. Astrophys J Lett 759:L40

Malavergne V, Bureau H, Raepsaet C, Gaillard F, Poncet M, Surblé S, Sifré D, Shcheka S, Fourdrin C, Deldicque D, Khodja H (2019) Experimental constraints on the fate of H and C during planetary core–mantle differentiation. Implications for the Earth. Icarus 321:473–485

Mallik A, Nelson J, Dasgupta R (2015) Partial melting of fertile peridotite fluxed by hydrous rhyolitic melt at 2–3 GPa: implications for mantle wedge hybridization by sediment melt and generation of ultrapotassic magmas in convergent margins. Contrib Mineral Petrol 169:1–24

Mallik A, Li Y, Wiedenbeck M (2018) Nitrogen evolution within the Earth's atmosphere–mantle system assessed by recycling in subduction zones. Earth Planet Sci Lett 482:556–566

Mandler BE, Grove TL (2016) Controls on the stability and composition of amphibole in the Earth's mantle. Contrib Mineral Petrol 171:1–20

Mandler BE, Elkins-Tanton LT (2013) The origin of eucrites, diogenites, and olivine diogenites: Magma ocean crystallization and shallow magma chamber processes on Vesta. Meteorit Planet Sci 48:2333–2349

Marboeuf U, Thiabaud A, Alibert Y, Benz W (2017) Isotopic ratios D/H and $^{15}N/^{14}N$ in giant planets. Mon Not R Astron Soc 475:2355–2362

Martins Z (2018) The nitrogen heterocycle content of meteorites and their significance for the origin of life. Life 8:28

Marty B (2012) The origins and concentrations of water, carbon, nitrogen and noble gases on Earth. Earth Planet Sci Lett 313–314:56–66

Marty B, Almayrac M, Barry PH, Bekaert DV, Broadley MW, Byrne DJ, Ballentine CJ, Caracausi A (2020) An evaluation of the C/N ratio of the mantle from natural CO_2-rich gas analysis: Geochemical and cosmochemical implications. Earth Planet Sci Lett 551:116574

Mathew KJ, Marti K (2001) Early evolution of Martian volatiles: Nitrogen and noble gas components in ALH84001 and Chassigny. J Geophys Res Planets 106:1401–1422

Matsui T, Abe Y (1986) Evolution of an impact-induced atmosphere and magma ocean on the accreting Earth. Nature 319:303–305

Maurice M, Dasgupta R, Hassanzadeh P (2024) Atmospheres of lava worlds. Astron Astrophys. In Press

Maurice M, Tosi N, Samuel H, Plesa A-C, Hüttig C, Breuer D (2017) Onset of solid-state mantle convection and mixing during magma ocean solidification. J Geophys Res Planets 122:577–598

Mayor M, Queloz D (1995) A Jupiter-mass companion to a solar-type star. Nature 378:355–359

McArthur BE, Endl M, Cochran WD, Benedict GF, Fischer DA, Marcy GW, Butler RP, Naef D, Mayor M, Queloz D, Udry S, Harrison TE (2004) Detection of a Neptune-mass planet in the ρ1 Cancri system using the Hobby-Eberly Telescope. Astrophys J 614:L81

McDonough WF, Sun S-s (1995) The composition of the Earth. Chem Geol 120:223–253

McGovern PJ, Schubert G (1989) Thermal evolution of the Earth: Effects of volatile exchange between atmosphere and interior. Earth Planet Sci Lett 96:27–37

McNaughton NJ, Borthwick J, Fallick AE, Pillinger CT (1981) Deuterium/hydrogen ratios in unequilibrated ordinary chondrites. Nature 294:639–641

Médard E, Grove TL (2006) Early hydrous melting and degassing of the Martian interior. J Geophys Res Planets 111:E11003

Meibom A, Krot AN, Robert F, Mostefaoui S, Russell SS, Petaev MI, Gounelle M (2007) Nitrogen and carbon isotopic composition of the Sun inferred from a high-temperature solar nebular condensate. Astrophys J 656:L33

Mikhail S, Sverjensky DA (2014) Nitrogen speciation in upper mantle fluids and the origin of Earth's nitrogen-rich atmosphere. Nat Geosci 7:816–819

Miyamoto M, Fujii N, Takeda H (1981) Ordinary chondrite parent body—An internal heating model. Lunar Planet Sci Conf Proc Sect 2:1145–1152

Miyazaki Y, Korenaga J (2022) Inefficient water degassing inhibits ocean formation on rocky planets: An insight from self-consistent mantle degassing models. Astrobiology 22:713–734

Moeller A, Hansen U (2013) Influence of rotation on the metal rain in a Hadean magma ocean. Geochem Geophys Geosyst 14:1226–1244

Moore G (2008) Interpreting H_2O and CO_2 contents in melt inclusions: constraints from solubility experiments and modeling. Rev Mineral Geochem 69:333–362

Moore G, Vennemann T, Carmichael ISE (1998) An empirical model for the solubility of H_2O in magmas to 3 kilobars. Am Mineral 83:36–42

Morbidelli A, Lunine JI, O'Brien DP, Raymond SN, Walsh KJ (2012) Building terrestrial planets. Annu Rev Earth Planet Sci 40:251–275

Mosenfelder JL, Von Der Handt A, Füri E, Dalou C, Hervig RL, Rossman GR, Hirschmann MM (2019) Nitrogen incorporation in silicates and metals: Results from SIMS, EPMA, FTIR, and laser-extraction mass spectrometry. Am Mineral 104:31–46

Mostefaoui S, Perron C, Zinner E, Sagon G (2000) Metal-associated carbon in primitive chondrites: structure, isotopic composition, and origin. Geochim Cosmochim Acta 64:1945–1964

Mysen B (2010) Structure of H_2O-saturated peralkaline aluminosilicate melt and coexisting aluminosilicate-saturated aqueous fluid determined in-situ to 800 °C and ~800 MPa. Geochim Cosmochim Acta 74:4123–4139

Mysen B (2014) Water-melt interaction in hydrous magmatic systems at high temperature and pressure. Prog Earth Planet Sci 1:4–4

Namur O, Charlier B, Holtz F, Cartier C, McCammon C (2016) Sulfur solubility in reduced mafic silicate melts: Implications for the speciation and distribution of sulfur on Mercury. Earth Planet Sci Lett 448:102–114

Naraoka H, Hashiguchi M, Okazaki R (2023) Soluble sulfur-bearing organic compounds in carbonaceous meteorites: Implications for chemical evolution in primitive asteroids. ACS Earth Space Chem 7:41–48

Newman S, Lowenstern JB (2002) VolatileCalc: A silicate melt–H_2O–CO_2 solution model written in Visual Basic for excel. Computers Geosci 28:597–604

Neumann W, Breuer D, Spohn T (2014) Differentiation of Vesta: Implications for a shallow magma ocean. Earth Planet Sci Lett 395:267–280

Nikolaou A, Katyal N, Tosi N, Godolt M, Grenfell JL, Rauer H (2019) What factors affect the duration and outgassing of the terrestrial magma ocean? Astrophys J 875:11

Nittler LR, Chabot NL, Grove TL, Peplowski PN (2018) The chemical composition of Mercury. *In:* BJ Anderson LR Nittler, SC Solomon (Eds), Mercury: The View after MESSENGER, Cambridge Planetary Science, Cambridge University Press, Cambridge, p 30–51

Nowak M, Behrens H (1995) The speciation of water in haplogranitic glasses and melts determined by in situ near-infrared spectroscopy. Geochim Cosmochim Acta 59:3445–3450

O'Leary JA, Gaetani GA, Hauri EH (2010) The effect of tetrahedral Al^{3+} on the partitioning of water between clinopyroxene and silicate melt. Earth Planet Sci Lett 297:111–120

O'Neill C, Aulbach S (2022) Destabilization of deep oxidized mantle drove the Great Oxidation Event. Sci Adv 8:eabg1626

O'Neill HSC, Mavrogenes JA (2002) The sulfide capacity and the sulfur content at sulfide saturation of silicate melts at 1400°C and 1 bar. J Petrol. 43:1049–1087

Ohtani E, Mizobata H, Yurimoto H (2000) Stability of dense hydrous magnesium silicate phases in the systems Mg_2SiO_4–H_2O and $MgSiO_3$–H_2O at pressures up to 27 GPa. Phys Chem Mineral 27:533–544

Otegi JF, Bouchy F, Helled R (2020) Revisited mass–radius relations for exoplanets below 120 $M_\oplus$. Astron Astrophys 634:A43

Palme H, Fegley B (1990) High-temperature condensation of iron-rich olivine in the solar nebula. Earth Planet Sci Lett 101:180–195

Papale P, Moretti R, Barbato D (2006) The compositional dependence of the saturation surface of H_2O+CO_2 fluids in silicate melts. Chem Geol 229:78–95

Palme H, Lodders K, Jones A (2014) Solar System abundances of the elements, 2.2 *In:* HD Holland KK Turekian, (Eds) Treatise on Geochemistry (Second Edition), Elsevier, Oxford, p 15–36

Pathak D, Dasgupta R (2024) Nitrogen inventory of iron meteorite parent bodies constrained by nitrogen partitioning between Fe-rich solid and liquid alloys. Geochim Cosmochim Acta 371:199–213, doi:10.1016/j.gca.2024.02.012

Payré V, Dasgupta R (2022) Effects of phosphorus on partial melting of the Martian mantle and compositions of the Martian crust. Geochim Cosmochim Acta 327:229–246

Pearson VK, Sephton MA, Franchi IA, Gibson JM, Gilmour I (2006) Carbon and nitrogen in carbonaceous chondrites: Elemental abundances and stable isotopic compositions. Meteorit Planet Sci 41:1899–1918

Piani L, Robert F, Beyssac O, Binet L, Bourot-Denise M, Derenne S, Le Guillou C, Marrocchi Y, Mostefaoui S, Rouzaud J-N, Thomen A (2012) Structure, composition, and location of organic matter in the enstatite chondrite Sahara 97096 (EH3). Meteorit Planet Sci 47:8–29

Piani L, Marrocchi Y, Libourel G, Tissandier L (2016) Magmatic sulfides in the porphyritic chondrules of EH enstatite chondrites. Geochim Cosmochim Acta 195:84–99

Piani L, Marrocchi Y, Vacher LG, Yurimoto H, Bizzarro M (2021) Origin of hydrogen isotopic variations in chondritic water and organics. Earth Planet Sci Lett 567:117008

Pintér Z, Foley SF, Yaxley GM, Rosenthal A, Rapp RP, Lanati AW, Rushmer T (2021) Experimental investigation of the composition of incipient melts in upper mantle peridotites in the presence of CO_2 and H_2O. Lithos 396–397:106224

Plank T, Manning CE (2019) Subducting carbon. Nature 574:343–352

Putirka KD (2024) Exoplanet mineralogy. Rev Mineral Geochem 90:199–258

Putirka KD, Rarick JC (2019) The composition and mineralogy of rocky exoplanets: A survey of >4000 stars from the Hypatia Catalog. Am Mineral 104:817–829

Putirka KD, Dorn C, Hinkel NR, Unterborn CT (2021) Compositional diversity of rocky exoplanets. Elements 17:235–240

Raymond SN, Morbidelli A (2014) The Grand Tack model: A critical review. Proc Int Astron Union 9:194–203

Raymond SN, Izidoro A, Bolmont E, Dorn C, Selsis F, Turbet M, Agol E, Barth P, Carone L, Dasgupta R, Gillon M, Grimm SL (2021) An upper limit on late accretion and water delivery in the TRAPPIST-1 exoplanet system. Nat Astron 6:80–88

Remusat L (2011) Organic and volatile elements in the solar system. EPJ Web Conf 18:05002

Robert F, Merlivat L, Javoy M (1979) Deuterium concentration in the early Solar System: hydrogen and oxygen isotope study. Nature 282:785

Rohrbach A, Schmidt MW (2011) Redox freezing and melting in the Earth's deep mantle resulting from carbon–iron redox coupling. Nature 472:209–212

Rohrbach A, Ghosh S, Schmidt MW, Wijbrans CH, Klemme S (2014) The stability of Fe–Ni carbides in the Earth's mantle: Evidence for a low Fe–Ni–C melt fraction in the deep mantle. Earth Planet Sci Lett 388:211–221

Rose-Weston L, Brenan JM, Fei Y, Secco RA, Frost DJ (2009) Effect of pressure, temperature, and oxygen fugacity on the metal–silicate partitioning of Te, Se, and S: Implications for earth differentiation. Geochim Cosmochim Acta 73:4598–4615

Roskosz M, Bouhifd MA, Jephcoat AP, Marty B, Mysen BO (2013) Nitrogen solubility in molten metal and silicate at high pressure and temperature. Geochim Cosmochim Acta 121:15–28

Saha S, Dasgupta R (2019) Phase relations of a depleted peridotite fluxed by a CO_2–H_2O fluid– Implications for the stability of partial melts versus volatile-bearing mineral phases in the cratonic mantle. J Geophys Res Solid Earth 124:10089–10106

Saha S, Dasgupta R, Tsuno K (2018) High pressure phase relations of a depleted peridotite fluxed by CO_2–H_2O bearing siliceous melts and the origin of mid-lithospheric discontinuity. Geochem Geophys Geosyst 19:595–620

Saha S, Peng Y, Dasgupta R, Mookherjee M, Fischer KM (2021) Assessing the presence of volatile-bearing mineral phases in the cratonic mantle as a possible cause of mid-lithospheric discontinuities. Earth Planet Sci Lett 553:116602

Salvador A, Samuel H (2023) Convective outgassing efficiency in planetary magma oceans: Insights from computational fluid dynamics. Icarus 390:115265

Salvador A, Massol H, Davaille A, Marcq E, Sarda P, Chassefière E (2017) The relative influence of H_2O and CO_2 on the primitive surface conditions and evolution of rocky planets. J Geophys Res Planets 122:1458–1486

Saxena SK, Eriksson G (1983) Theoretical computation of mineral assemblages in pyrolite and lherzolite. J Petrol 24:538–555

Schaefer L, Wordsworth RD, Berta-Thompson Z, Sasselov D (2016) Predictions of the atmospheric composition of GJ 1132b. Astrophys J 829:63

Schmidt MW, Poli S (1998) Experimentally based water budgets for dehydrating slabs and consequences for arc magma generation. Earth Planet Sci Lett 163:361–379

Schmidt MW, Gao C, Golubkova A, Rohrbach A, Connolly JAD (2014) Natural moissanite (SiC)—a low temperature mineral formed from highly fractionated ultra-reducing COH-fluids. Prog Earth Planet Sci 1:27

Sears DW (1978) Condensation and the composition of iron meteorites. Earth Planet Sci Lett 41:128–138

Sekine T, Wyllie PJ (1982) The system granite–peridotite–H_2O at 30 kbar, with applications to hybridization in subduction zone magmatism. Contrib Mineral Petrol 81:190–202

Sephton MA (2002) Organic compounds in carbonaceous meteorites. Nat Prod Rep 19:292–311

Shcheka SS, Wiedenbeck M, Frost DJ, Keppler H (2006) Carbon solubility in mantle minerals. Earth Planet Sci Lett 245:730–742

Shi L, Lu W, Kagoshima T, Sano Y, Gao Z, Du Z, Liu Y, Fei Y, Li Y (2022) Nitrogen isotope evidence for Earth's heterogeneous accretion of volatiles. Nat Commun 13:4769

Simkus DN, Aponte JC, Elsila JE, Hilts RW, McLain HL, Herd CDK (2019) New insights into the heterogeneity of the Tagish Lake meteorite: Soluble organic compositions of variously altered specimens. Meteorit Planet Sci 54:1283–1302

Simon JB, Armitage PJ, Li R, Youdin AN (2016) The mass and size distribution of planetesimals formed by the streaming instability. I The role of self-gravity. Astrophys J 822:55

Skogby H (2006) Water in natural mantle minerals I: Pyroxenes. Rev Mineral Geochem 62:155–167

Solomatov VS (2000) Fluid dynamics of a terrestrial magma ocean. *In:* R Canup K Righter (Eds) Origin of the Earth and Moon, University of Arizona Press, Tucson, Arizona, p 323–338

Solomatov V (2007) Magma oceans and primordial mantle differentiation. 9.04 *In:* G Schubert (Ed) Treatise on Geophysics, Elsevier, Amsterdam, p 91–119

Solomatova NV, Caracas R (2021) Genesis of a CO_2-rich and H_2O-depleted atmosphere from Earth's early global magma ocean. Sci Adv 7:eabj0406

Sossi PA, Burnham AD, Badro J, Lanzirotti A, Newville M, O'Neill HSC (2020) Redox state of Earth's magma ocean and its Venus-like early atmosphere. Sci Adv 6:eabd1387

Sossi PA, Tollan PME, Badro J, Bower DJ (2023) Solubility of water in peridotite liquids and the prevalence of steam atmospheres on rocky planets. Earth Planet Sci Lett 601:117894

Spandler C, Yaxley G, Green DH, Scott D (2010) Experimental phase and melting relations of metapelite in the upper mantle: implications for the petrogenesis of intraplate magmas. Contrib Mineral Petrol 160:569–589

Speelmanns IM, Schmidt MW, Liebske C (2019) The almost lithophile character of nitrogen during core formation. Earth Planet Sci Lett 510:186–197

Stagno V, Ojwang DO, McCammon CA, Frost DJ (2013) The oxidation state of the mantle and the extraction of carbon from Earth's interior. Nature 493:84–88

Stanley BD, Hirschmann MM, Withers AC (2011) CO_2 solubility in Martian basalts and Martian atmospheric evolution. Geochim Cosmochim Acta 75:5987–6003

Stökl A, Dorfi EA, Johnstone CP, Lammer H (2016) Dynamical accretion of primordial atmospheres around planets with masses between 0.1 and 5 $M_{\oplus}$ in the habitable zone. Astrophys J 825:86

Stolper E (1982a) The speciation of water in silicate melts. Geochim Cosmochim Acta 46:2609–2620

Stolper E (1982b) Water in silicate glasses: An infrared spectroscopic study. Contrib Mineral Petrol 81:1–17

Studier MH, Hayatsu R, Anders E (1965) Organic compounds in carbonaceous chondrites. Science 149:1455–1459

Suer T-A, Siebert J, Remusat L, Menguy N, Fiquet G (2017) A sulfur-poor terrestrial core inferred from metal–silicate partitioning experiments. Earth Planet Sci Lett 469:84–97

Suer T-A, Jackson C, Grewal DS, Dalou C, Lichtenberg T (2023) The distribution of volatile elements during rocky planet formation. Front Earth Sci 11:1159412

Sun C, Lee C-TA (2022) Redox evolution of crystallizing magmas with C–H–O–S volatiles and its implications for atmospheric oxygenation. Geochim Cosmochim Acta 338:302–321

Tagawa S, Sakamoto N, Hirose K, Yokoo S, Hernlund J, Ohishi Y, Yurimoto H (2021) Experimental evidence for hydrogen incorporation into Earth's core. Nat Commun 12:2588

Trumbull RB, Yang J-S, Robinson PT, Di Pierro S, Vennemann T, Wiedenbeck M (2009) The carbon isotope composition of natural SiC (moissanite) from the Earth's mantle: New discoveries from ophiolites. Lithos 113:612–620

Tsuno K, Dasgupta R (2012) The effect of carbonates on near-solidus melting of pelite at 3 GPa: Relative efficiency of H_2O and CO_2 subduction. Earth Planet Sci Lett 319–320:185–196

Tsuno K, Dasgupta R (2015) Fe–Ni–Cu–C–S phase relations at high pressures and temperatures—The role of sulfur in carbon storage and diamond stability at mid- to deep-upper mantle. Earth Planet Sci Lett 412:132–142

Tsuno K, Grewal DS, Dasgupta R (2018) Core–mantle fractionation of carbon in Earth and Mars: The effects of sulfur. Geochim Cosmochim Acta 238:477–495

Tsymbulov LB, Tsmekhman LS (2001) Solubility of carbon in sulfide melts of the system Fe–Ni–S. Russ J Appl Chem 74:925–929

Ulmer P (2001) Partial melting in the mantle wedge—The role of H_2O in the genesis of mantle-derived 'arc-related' magmas. Phys Earth Planet Sci 127:215–232

Valencia D, O'Connell RJ, Sasselov D (2006) Internal structure of massive terrestrial planets. Icarus 181:545–554

von Goldbeck OK (1982) Iron—Nitrogen Fe—N, IRON—Binary Phase Diagrams, Springer Berlin Heidelberg, p 67–70

Wai CM, Wasson JT (1977) Nebular condensation of moderately volatile elements and their abundances in ordinary chondrites. Earth Planet Sci Lett 36:1–13

Wai CM, Wasson JT (1979) Nebular condensation of Ga, Ge and Sb and the chemical classification of iron meteorites. Nature 282:790–793

Walker JCG, Hays PB, Kasting JF (1981) A negative feedback mechanism for the long-term stabilization of Earth's surface temperature. J Geophys Res 86:9776–9782

Wallace ME, Green DH (1988) An experimental determination of primary carbonatite magma composition. Nature 335:343–346

Walters JB, Cruz-Uribe AM, Marschall HR (2020) Sulfur loss from subducted altered oceanic crust and implications for mantle oxidation. Geochem Persp Lett 13:36–41

Wang C, Hirama J, Nagasaka T, Ban-Ya S (1991) Phase equilibria of liquid Fe–S–C ternary system. ISIJ Int 31:1292–1299

Warren PH (2011) Stable-isotopic anomalies and the accretionary assemblage of the Earth and Mars: A subordinate role for carbonaceous chondrites. Earth Planet Sci Lett 311:93–100

Wood BJ (1993) Carbon in the core. Earth Planet Sci Lett 117:593–607

Wood BJ, Smythe DJ, Harrison T (2019) The condensation temperatures of the elements: A reappraisal. Am Mineral 104:844–856

Yasuda A, Fujii T, Kurita K (1994) Melting phase relations of anhydrous mid-ocean ridge basalt from 3 to 20 GPa: implications for the behavior of subducted oceanic crust in the mantle. J Geophys Res 99:9401–9414

Yoshizaki T, McDonough WF (2020) The composition of Mars. Geochim Cosmochim Acta 273:137–162

Yoshioka T, Wiedenbeck M, Shcheka S, Keppler H (2018) Nitrogen solubility in the deep mantle and the origin of Earth's primordial nitrogen budget. Earth Planet Sci Lett 488:134–143

Yoshioka T, Nakashima D, Nakamura T, Shcheka S, Keppler H (2019) Carbon solubility in silicate melts in equilibrium with a CO–CO_2 gas phase and graphite. Geochim Cosmochim Acta 259:129–143

Young PA, Desch SJ, Anbar AD, Barnes R, Hinkel NR, Kopparapu R, Madhusudhan N, Monga N, Pagano MD, Riner MA, Scannapieco E, Shim SH, Truitt A (2014) Astrobiological stoichiometry. Astrobiology 14:603–626

Zahnle KJ, Kasting JF, Pollack JB (1988) Evolution of a steam atmosphere during earth's accretion. Icarus 74:62–97

Zajacz Z, Tsay A (2019) An accurate model to predict sulfur concentration at anhydrite saturation in silicate melts. Geochim Cosmochim Acta 261:288–304

Zeng L, Sasselov DD, Jacobsen SB (2016) Mass–radius relation for rocky planets based on PREM. Astrophys J 819:127

Zhang X (2023) JWST's eyes on an alien world. The Innovation 4:100428

Zhang J, Rogers LA (2022) Thermal evolution and magnetic history of rocky planets. Astrophys J 938:131

Zhang Z, Hirschmann MM (2016) Experimental constraints on mantle sulfide melting up to 8 GPa. Am Mineral 101:181–192

Zhang Z, Lentsch N, Hirschmann MM (2015) Carbon-saturated monosulfide melting in the shallow mantle: Solubility and effect on solidus. Contrib Mineral Petrol 170:47

Zhang Z, Qin T, Pommier A, Hirschmann MM (2019) Carbon storage in Fe–Ni—S liquids in the deep upper mantle and its relation to diamond and Fe–Ni alloy precipitation. Earth Planet Sci Lett 520:164–174

Reviews in Mineralogy & Geochemistry
Vol. 90 pp. 375–410, 2024
Copyright © Mineralogical Society of America

Exoplanet Magnetic Fields

David A. Brain

Laboratory for Atmospheric and Space Physics
University of Colorado Boulder
Boulder, CO 80303, U.S.A.

david.brain@colorado.edu

Melodie M. Kao

Department of Astronomy and Astrophysics
University of California Santa Cruz
Santa Cruz, CA 95064

melodie.kao@ucsc.edu

Joseph G. O'Rourke

School of Earth and Space Exploration
Arizona State University
Tempe, Arizona 85287, U.S.A.

jgorourk@asu.edu

INTRODUCTION

Planetary magnetic fields are important indicators of planetary processes and evolution, from a planet's outer core to its surface (if it possesses one) to its atmosphere and near-space environment. Magnetic fields are most directly measured in situ, and determining whether distant planetary objects possess magnetic fields can be challenging. At present we have no unambiguous measurements of magnetic fields on exoplanets. Nevertheless, it would be surprising if at least some exoplanets did not generate a magnetic field, like many planetary bodies in the solar system.

This chapter provides an overview of the current understanding of exoplanetary magnetic fields and their consequences. In the next section we review the current understanding of planetary dynamo generation as it applies to solar system objects and discuss the implications for exoplanetary magnetic field generation. Following this, we describe seven methods for determining the existence and strength of an exoplanetary magnetic field and discuss the near-term prospects for each method. We close by highlighting four main consequences of exoplanetary magnetic fields for a planet and its evolution.

THEORY OF EXOPLANET MAGNETIC DYNAMOS

A dynamo turns fluid motion into a magnetic field via electromagnetic induction. In (exo)planetary science, theory aims to answer three questions arising from this definition. First, what planetary bodies host large reservoirs of electrically conductive fluid? The discussions below will largely focus on dynamos in terrestrial planets, but the basic principles apply to gas giants as well. Second, what processes can provoke this fluid into motion?

In Earth, our long-lived dynamo arises from thermo-chemical convection in the deep interior (e.g., Landeau et al. 2022). Third, what is the strength and geometry of the resulting magnetic fields? Models of exoplanets often assume that the dipole moment of a dynamo-generated field is aligned with the rotational axis of the planet. On average over geologic timescales, Earth indeed has a geocentric axial dipole (e.g., Meert 2009). However, short-term variations in orientation are common ($\sim 10°$ for Earth now)—and the present-day dynamos of Neptune and Uranus are severely misaligned with their rotational axes today.

Ultimately, answering questions about planetary dynamos requires a diversity of disciplines—not only the physics of electromagnetism, but also mineral physics, geochemistry, and geodynamics. Progress arises from theory, numerical simulations, and laboratory experiments. Beyond the short introduction here, readers are also encouraged to consult more comprehensive reviews of the theory of (exo)planetary dynamos (e.g., Stevenson 2003, 2010; Christensen 2010; Schubert and Soderlund 2011; Laneuville et al. 2020).

Fundamental dynamo equations

Mathematically, dynamo theory connects the fluid velocity field (**v**) to the magnetic field (**B**). Three variables describe the thermodynamic properties of the fluid: density (ρ), pressure (P), and temperature (T). At least five equations with vectors (more for scalar quantities) are thus needed to solve for these variables. First, we can consider the induction equation, which combines Maxwell's equations with the Lorentz force law and Ohm's law (e.g., Schubert and Soderlund 2011):

$$\frac{\partial \mathbf{B}}{\partial t} = \nabla \times (\mathbf{v} \times \mathbf{B}) - \nabla \times (\eta \nabla \times \mathbf{B}) \tag{1}$$

Here, the magnetic diffusivity is $\eta = 1/(\mu_0 \sigma)$, where $\mu_0 = 4\pi \times 10^{-7}$ H/m and σ is the electrical conductivity (in units of S/m). The first and second terms on the left side, respectively, relate to the convection and diffusion of the magnetic field.

Next, we can express the conservation of mass as

$$\frac{\partial P}{\partial t} + \nabla \cdot (\rho \mathbf{v}) = 0 \tag{2}$$

meaning that mass can move around but not be spontaneously created or destroyed. Then, the conservation of momentum is

$$\rho \frac{D\mathbf{v}}{Dt} + 2\rho \Omega \times \mathbf{v} = -\nabla P - \rho \Omega \times (\Omega \times \mathbf{r}) + \frac{1}{\mu_0}(\nabla \times \mathbf{B}) \times \mathbf{B} + \rho g + \frac{1}{3}\rho \mathbf{n}\nabla(\nabla \cdot \mathbf{v}) + \rho \mathbf{n}\nabla^2 \mathbf{v} \tag{3}$$

where the rotation rate of the reference frame is Ω; the position vector is **r**; the gravitational acceleration is g; and the kinematic viscosity is v (Schubert and Soderlund 2011). From left to right, there are eight terms in Equation (3) (Schubert and Soderlund 2011): inertial acceleration, Coriolis acceleration, pressure gradient, centrifugal force, Lorentz force, buoyancy force, and two terms related to viscous diffusion under the so-called Stokes assumption (i.e., neglecting the bulk viscosity).

Fourth, we can express the conservation of energy as

$$\rho C_P \frac{DT}{Dt} - \alpha T \left(\frac{DP}{Dt} \right) = \Phi + \nabla \cdot (k\nabla T) \tag{4}$$

where time is t; the coefficient of thermal expansion is α; the specific heat at constant pressure is C_p; the total dissipation (viscous and ohmic heating) is Φ; and the thermal conductivity is k (Schubert and Soderlund 2011). The right side of this equation is the dissipation associated with the dynamo. The left side is the change in thermal energy, including the effects of adiabatic (de)compression.

Last, we need an equation of state:

$$\rho = f(P,T) \tag{5}$$

where f is whatever function is typically used for the material in question (e.g., metallic hydrogen, molten silicates, iron alloys, etc.). Solving these equations is difficult, even with modern numerical approaches. To train our intuition for planetary dynamos, we should first analyze non-dimensional parameters that arise from these fundamental equations.

Key non-dimensional parameters

Non-dimensional parameters are ratios between different forces or other quantities. They reveal the terms that are the most important in the governing equations. Numerical models that solve Equations (1–5) need five control parameters for a dynamo. Two other parameters provide criteria for the existence (or lack thereof) of dynamos in different types of planets (e.g., Stevenson 2003, 2010).

Control parameters for numerical models of dynamos. Control parameters help characterize numerical models of dynamos. For example, we can define five non-dimensional parameters (the Rayleigh, Reynolds, Ekman, Prandtl, and magnetic Prandtl numbers) that are popular among modelers (e.g., Schubert and Soderlund 2011):

$$\mathrm{Ra} = \frac{\alpha g \Delta T D^3}{\nu \kappa}; \ \ \mathrm{Re} = \frac{\nu D}{\nu}; \ \ \mathrm{E} = \frac{\nu}{\Omega D^2}; \ \ \mathrm{Pr} = \frac{\nu}{\kappa}; \ \ \mathrm{Pr_m} = \frac{\nu}{\eta} \tag{6}$$

The Rayleigh number (Ra) is the ratio of the timescales for the transport of heat via diffusion and convection. High Ra (e.g., $> 10^3$) indicates vigorous convection. Here, ΔT is a temperature difference that drives convection; D is a length scale; and κ is the thermal diffusivity of the fluid. The Reynolds number (Re) is the ratio of inertial and viscous forces in the fluid. Dynamo-generating flows tend to be turbulent with $\mathrm{Re} > \sim 10^3$–10^4, typically. In contrast, the Ekman number (E) is the ratio of viscous and Coriolis forces in the fluid. The Prandtl number (Pr) is the ratio of the diffusivity of momentum (kinematic viscosity) and of heat, while the magnetic Prandtl number ($\mathrm{Pr_m}$) is the kinematic viscosity divided by the magnetic diffusivity. While these parameters control the properties of an extant dynamo, they do not quantify the prerequisites for a dynamo to exist.

Criteria for the existence and strength of a planetary dynamo. A dynamo requires a continual source of power, such as the flux of heat from (hot) planets to (cold) space. The five non-dimensional parameters in Equation (6) control the properties of an active dynamo. To determine if a dynamo should exist at all, we need to define two other parameters (e.g., Stevenson 2003, 2010):

$$\mathrm{Re_m} = \frac{\nu D}{\eta}; \ \ \mathrm{Ro} = \frac{\nu}{\Omega D} \tag{7}$$

Here, the magnetic Reynolds number ($\mathrm{Re_m}$) is the ratio of magnetic induction and diffusion. The Rossby number (Ro) is the ratio of inertial and Coriolis forces. A dynamo requires $\mathrm{Re_m} > 10$–100 (so the field is strengthened faster than it diffuses away) and $\mathrm{Ro} \ll 1$ (so the Coriolis force is dynamically important) (e.g., Stevenson 2003, 2010). In practice, the Rossby-number criterion is satisfied for even the slowest-rotating planet in the Solar System (Venus) and thus also for tidally locked exoplanets with close-in orbits (such as hot Jupiters). Therefore, the first criterion is most important (e.g., Stevenson 2003, 2010). A dynamo is expected when a large (big D) volume of conductive fluid (small η) is undergoing convection (non-zero ν).

Figure 1 shows how Re_m yields a "regime diagram" for dynamos that explains why they are marginal in terrestrial planets but widespread in gas and ice giants. If convection stops, then the dynamo dies on the timescale of magnetic diffusion (e.g., Stevenson 2003, 2010), i.e., $\tau \sim D^2/(\pi^2\eta)$ (e.g., $\sim 10^4$ and $\sim 10^8$ years for Earth and Jupiter, respectively). Since that timescale is a geologic eyeblink, observation of a live dynamo basically means that convection is currently occurring. Overall, theory has taken us from the induction equation to the realization that studying planetary dynamos means analyzing when and where convection is likely to happen.

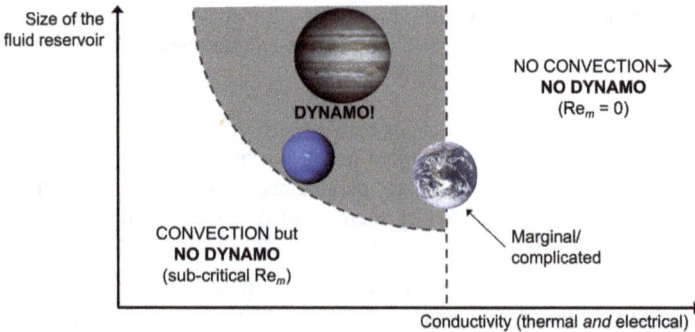

Figure 1. Prospects for a dynamo based on the magnetic Reynolds number. Only in the **grey, shaded** region is Re_m large enough for the dynamo. Below and to the left, Re_m is non-zero but too small for a dynamo. The fluid is not moving fast enough and/or the reservoir is too small. The absence of a dynamo in the rightmost region is perhaps counter-intuitive. A fluid is unlikely to host a dynamo if it is too electrically conductive (e.g., if Earth's core were made of copper or gold) because it would be very thermally conductive as well—and thus likely to transport heat via conduction without fluid motions. Therefore, gas and ice giant planets plot in more favorable regions, even though their interiors are less electrically conductive than Earth's metallic core. Terrestrial planets are marginal in their capacity to host a dynamo (e.g., Earth plots at the same location as dynamo-less Venus). Created based on Figure 1 in Stevenson (2003).

Application of theory to Earth's dynamo

We can estimate the control parameters for perhaps the best-studied planetary dynamo: the one that exists today in Earth's outer core. For liquid iron alloys at the relevant pressures and temperatures, $\nu \sim 10^{-6}\,\mathrm{m^2/s}$, $\eta \sim 1\,\mathrm{m^2/s}$, and $\kappa \sim 10^{-5}\,\mathrm{m^2/s}$ (e.g., Schubert and Soderlund 2011). Therefore, $Pr_m \sim 10^{-6}$ and $Pr \sim 0.1$ for Earth's outer core. Because $Pr \gg Pr_m$, magnetic fields vary over much longer length scales than variations in the velocity field. Earth's 24-hour day means $\Omega \sim 7.3 \times 10^{-5}\,\mathrm{s^{-1}}$, so $E \sim 10^{-15}$ for our core with $D \sim 3000$ km. Flow velocities in Earth's outer core are inferred to be $v \sim 0.5$ mm/s (e.g., Bloxham and Jackson 1991), meaning that $Re > 10^7$. As discussed below, ΔT for the core is unknown—and perhaps the relevant buoyancy is chemical, not thermal. However, estimates for Ra have ranged from $\sim 10^{22}$–10^{30} (e.g., Schubert and Soderlund 2011). Ultimately, convection in Earth's outer core is vigorous and turbulent—and the Coriolis force governs the patterns of fluid motion.

Numerical simulations would require impossibly small spatial and temporal resolution to implement realistic control parameters. For example, a cutting-edge study of Earth's dynamo used $Ra \sim 5 \times 10^6$; $E = 3 \times 10^{-4}$; $Pr = 1$; and $Pr_m = 10$ (Driscoll 2016). Those assumed control parameters were at least ~ 1–15 orders of magnitude away from the best estimates of their true values.

These numerical studies are still worthwhile! Published dynamo models are "Earth-like" in many respects, especially regarding the dipolarity of the magnetic field and its strength at the surface (e.g., Kuang and Bloxham 1997; Christensen et al. 2010). People can extrapolate scaling laws for dynamo properties, using scaling laws derived by running dynamo models over the accessible ranges of control parameters. However, we cannot use even "Earth-like" dynamo models to simulate the billion-year lives of planetary dynamos. Resource limitations mean that they can model no more than a few million years at a time (e.g., Matsui et al. 2016). As shown in Figure 2, scientists use three-dimensional models of dynamos to provide "snapshots" of their behavior (e.g., Driscoll 2016). We must turn to the simpler criteria to study how Earth's dynamo changed over geologic time as the core evolved.

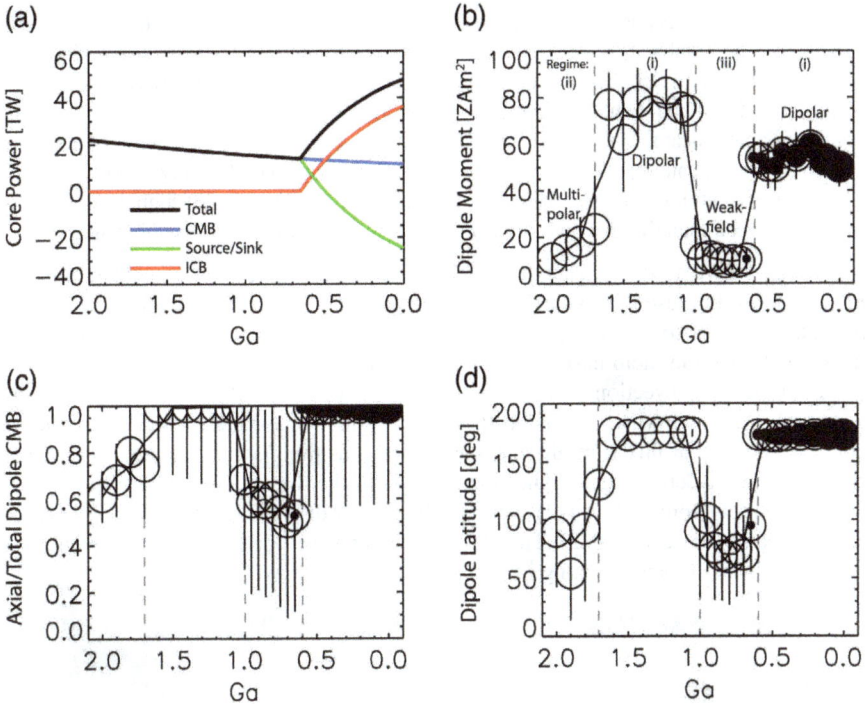

Figure 2. Combining thermal evolution models with numerical simulations of dynamos. **(a)** One-dimensional, parameterized models calculate how fast the core cools over time; the size of the inner core; and the magnitude of the dissipation (power) available for a dynamo over billions of years. Dynamo models then predict the strength **(b)**, dipolarity **(c)**, and tilt **(d)** of the magnetic field at representative times. This is Figure 2 from Driscoll (2016), reproduced with permission.

We know that Earth's dynamo exists today because we are immersed in the resulting magnetic field. Using the parameters quoted above, $Re_m \sim 1500$ and $Ro \sim 2 \times 10^{-6}$, which easily satisfies the criteria from the subsection above. The magnetic diffusion time for Earth's dynamo is $\sim 10^4$ years (e.g., Stevenson 2003, 2010), while we know from paleomagnetism that the dynamo has existed for at least ~ 3.5 billion years (e.g., Tarduno et al. 2010)—and possibly since soon after Earth accreted (cf., Borlina et al. 2020). However, the difficulty in understanding Earth's dynamo lies in explaining why fluid velocities in the liquid outer core have been non-zero.

Earth's dynamo can live if convection remains the dominant mode of heat transfer in the fluid. In planetary interiors, the first enemy of convection is thermal conduction. The mode that "wins" is the one that can transport heat most rapidly in a given situation. Thus, a dynamo will die if the cooling rate of the fluid is low enough that conduction can transport the associated heat flux. Quantitatively, convection in a low-viscosity fluid will tend to maintain an adiabatic thermal gradient, i.e., $(dT/dr)_{ad} \sim -\alpha T g / C_p$, where r is distance from the planetary center (e.g., Stevenson 2003, 2010). By Fourier's law, the associated "adiabatic heat flux" is $Q_{ad} = -k(dT/dr)_{ad}$, where k is the thermal conductivity of the fluid. In thermal evolution models, a simple way to assess the likelihood of a dynamo is to compare the actual heat flux out of the fluid region (Q) to the adiabatic heat flux. If $Q > Q_{ad}$, then thermal convection can sustain a dynamo, assuming the fluid is homogeneous and not undergoing any chemical reactions. Otherwise, convection will cease as heat moves towards the top of the fluid and creates thermal stratification. Although this criterion is easy to impose in models, we do not know if it holds for Earth's core at present day.

Earth's outer core is possibly cooling too slowly today for thermal convection alone to sustain our dynamo. The adiabatic thermal gradient in the core is roughly $(dT/dr)_{ad} \sim 0.5$ K/km (e.g., Stevenson 2003, 2010; Schubert and Soderlund 2011). The thermal conductivity of Earth's core remains debated, with published estimates ranging from ~30–200 W/(m·K) (e.g., Williams 2018). Thus, the adiabatic heat flux might be ~15–100 mW/m². In comparison, the vigor of convection in the overlying, silicate mantle determines Q for the core. Estimates span $Q \sim 30$–100 mW/m² (e.g., Lay et al. 2008). Obviously, these estimates of Q and Q_{ad} overlap. If higher estimates of k are proved correct, then thermal convection cannot power Earth's dynamo without assistance.

Fortunately, we need not panic that Earth's dynamo might die soon. The inner core is our dynamo's savior because it provides chemical buoyancy to power convection (Fig. 3). As the core cools, the inner core slowly freezes and grows, excluding "light elements" from its solid structure and injecting them into the bottom of the outer core. This compositional buoyancy is enough to drive convection (e.g., Labrosse 2015). Even if $Q < Q_{ad}$, chemical convection can move hot fluid downwards and cold fluid upwards. Scaling laws show that the total energy flux—any heat flux from thermal convection plus the gravitational energy associated with chemical convection—determines the strength of the dynamo (e.g., Christensen 2010). However, the inner core is almost certainly much younger than Earth's dynamo (e.g., Labrosse 2015; Landeau et al. 2017, 2022). Therefore, explaining the operation of the dynamo in the past is more difficult than accepting its present-day existence.

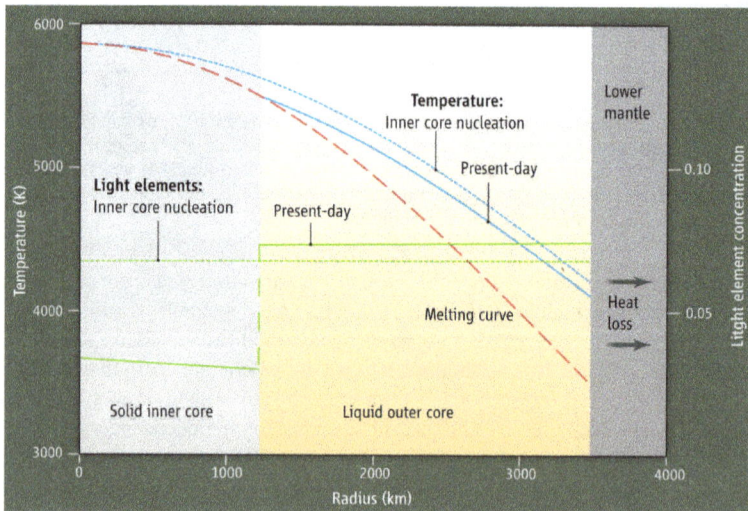

Figure 3. Earth's center is the hottest place in the planet, but it is slowly freezing because of the extreme pressure. Growth of the inner core (**left, grey region**) excludes light elements into the outer core (**green lines**). The inner core boundary is the intersection between the adiabatic temperature profile (**blue curves**) and the melting temperature (**red curve**). Earth has an inner core because the melting curve and the adiabat first cross at the center. The fluxes of heat and light elements both provide power for the dynamo. If the total heat flux is sub-adiabatic at present day, a thermally stratified layer could exist at the top of the core (not shown here). This is the Figure from Olson (2013), reproduced with permission.

Scientists coined the term "the new core paradox" to encapsulate the seeming disconnect between the inner core's relative youth and its (perhaps) singular role in maintaining the dynamo (Olson 2013). Several solutions are under active investigation (e.g., Landeau et al. 2022; Driscoll and Davies 2023). Briefly, other chemical processes (such as the precipitation of light species such as MgO, SiO_2, and/or FeO) could provide chemical buoyancy in the core prior to the nucleation of the inner core (e.g., Badro et al. 2016; O'Rourke and Stevenson 2016; Hirose et al. 2017). Additionally, the metallic core is perhaps not the only place that could have hosted a dynamo. Earth's silicate mantle might have solidified from the middle outwards after accretion (e.g., Labrosse et al. 2007; Coltice et al. 2011; Pachhai et al. 2022; Ferrick and Korenaga 2023), rather than from the bottom upwards as traditionally assumed. A "basal magma ocean" could have survived for roughly half of Earth's life at the base of the mantle (e.g., Labrosse et al. 2007). Scientists recently learned that molten silicates are electrically conductive under deep mantle pressures and temperatures (e.g., Stixrude et al. 2020). If the basal magma ocean was vigorously convecting, then it could have hosted a dynamo early in Earth's history (Ziegler and Stegman 2013; Blanc et al. 2020), when a core-hosted dynamo seems less likely. Recent studies suggested that a basal magma ocean exists today in Mars (Khan et al. 2023; Samuel et al. 2023), in the past in the Moon (Scheinberg et al. 2018; Hamid et al. 2023), and in the past (and perhaps today) in Venus (O'Rourke 2020). Basal magma oceans are probably even larger and longer-lived in massive, terrestrial exoplanets, known as super-Earths (e.g., Soubiran and Militzer 2018). There are no true paradoxes, only gaps in our scientific understanding.

Astute readers might realize that a new core paradox implies the existence of an "old core paradox." Indeed, half a century ago, scientists recognized a "paradox" because mineral physics experiments indicated that the core should not first freeze at its center, if the liquid were vigorously convecting (Kennedy and Higgins 1973). Since the existence of the inner core (from seismology) and the dynamo (from daily experience) were irrefutable, scientists were not surprised to find an error with the experiments (e.g., Olson 2013). Modern measurements of the adiabatic gradient and the melting curve of iron alloys are now self-consistent (e.g., Anzellini et al. 2013). Still, the lesson remains that our understanding of planetary dynamos can hinge on quantities that are extraordinarily difficult to determine. Depending on the relative slopes of the melting curve and the adiabat in a particular fluid, crystallization can occur at the top, middle, or bottom of a fluid reservoir (Fig. 4). Some critical constraints (such as the existence of an inner core) are likely impossible to obtain for exoplanets in the foreseeable future.

Application of theory to the Solar System

Many questions await answers about Earth's dynamo, but theory must also attempt to explain a puzzling array of observations from across the Solar System (Table 1). For example, gas and ice giants in the Solar System all have dynamos today. The dynamos in the ice giants are strikingly multipolar and non-axisymmetric (e.g., Stanley and Bloxham 2004; Soderlund and Stanley 2020), raising many questions that await answers from new missions (e.g., Cohen et al. 2022). Jupiter's dynamo field is actually better mapped than Earth's dynamo field (e.g., Stevenson 2010; Moore et al. 2018; Connerney et al. 2022), because Earth's magnetic field includes contributions from magnetized minerals in the crust. This crustal remanent magnetism is how we know that Earth had a dynamo in the past. Obviously, studying remanent magnetism requires a solid piece of a planetary body, either in situ in the crust or as a meteorite. So, we cannot prove that the giant planets had dynamos before human history, but we have no reason to believe that they did not. Based on the reasoning that underlies Figure 1, we expect near-ubiquitous dynamos in giant exoplanets, although their strength, orientation, and dipolarity are uncertain.

Empirically, dynamos in terrestrial planets are fragile things. No planetary property clearly correlates with where dynamos are found in terrestrial bodies in the Solar System—neither size nor bulk composition nor distance from the Sun. It seems almost random that Mercury and Ganymede host dynamos today while Venus and Mars do not. The Solar System also lacks

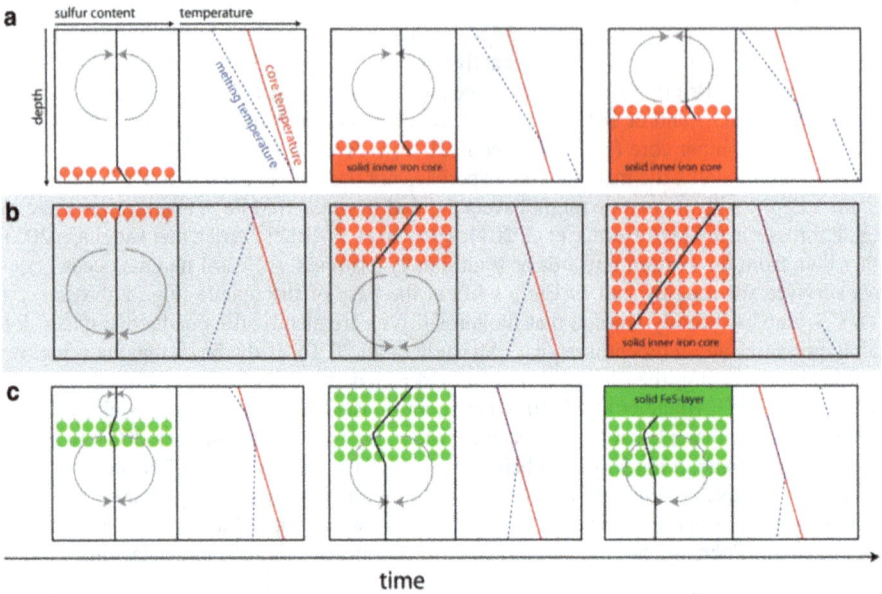

Figure 4. Metallic cores do not always solidify from the center outwards. **(a)** An Earth-like regime with an inner core is probably the most favorable for a dynamo. **(b)** Iron snow happens if the adiabat first crosses the liquidus at the top of the core. This regime might be common in smaller terrestrial worlds. **(c)** In some cores, the crystallizing solid might be less dense than the residual fluid. For example, a core with a high sulfur content, above the Fe–FeS eutectic composition, could crystallize solid FeS that rises to the top. Other regimes are possible. Overall, chemistry is critical to the chances for a dynamo in terrestrial worlds. This is a truncation of Figure 1 from Breuer et al. (2015), reproduced under CC BY 4.0.

an example of large terrestrial exoplanets (super-Earths), which are apparently common in the galaxy. Super-Earths could be relatively likely to host dynamos simply due to their size (e.g., van Summeren et al. 2013; Boujibar et al. 2020; Blaske and O'Rourke 2021). However, a higher mass leads to greater pressures in their deep interiors. High pressure could decrease the convective vigor of the solid mantle (by making it more viscous)—and thus slow the cooling rate of any deeper fluid region(s) (e.g., Tachinami et al. 2011). Overall, we should expect lots of dynamo-related diversity even among exoplanets with similar sizes, compositions, and orbits.

We can tell myriad stories about individual worlds. Frustratingly, explanations for the presence or absence of a dynamo often hinge on relatively minor differences. Convection has many enemies. For example, a canonical explanation for Venus's lack of a dynamo is that its slightly lower internal pressures, relative to Earth, fatally delayed the nucleation of the inner core (Stevenson et al. 1983). Or perhaps Venus's core formed with a stabilizing chemical gradient that would inhibit convection even if it were cooling faster than Earth's core (Jacobson et al. 2017). Radiogenic heating in the silicate mantle (from the slow decay of K, U, and Th isotopes) can also kill a dynamo by slowing the cooling rate of the deep interior (e.g., Nimmo et al. 2020). Tidal heating promotes geologic activity near the surface but can also suppress the cooling of the core, which is likely why Io does not have a dynamo today (e.g., Wienbruch and Spohn 1995). Ultimately, understanding dynamos in terrestrial planets means understanding terrestrial planets as complex systems. The number of potentially pivotal parameters is larger than the number of terrestrial planets in the Solar System, so observational constraints from exoplanets will be invaluable.

Table 1. Terrestrial bodies in the Solar System, grouped based on the(non)existence of a dynamo now and/or in the past. Searches for remanent magnetization in meteorites and in the crusts of terrestrial bodies allow us to learn if a planetary body had a dynamo in the past, even if one does not exist today. The existence of metallic cores in many icy satellites, such as Europa, is uncertain.

Observational constraints	Planetary body
Dynamo exists today	Mercury, Earth, Ganymede, Jupiter, Saturn, Uranus, Neptune
Dynamo existed in the past	Mercury, Earth, Moon, Mars, sundry meteorite parent bodies
Metallic core, but no known dynamo	Venus, Io, Europa (?)

Takeaway messages for studies of exoplanets

The theory of planetary dynamos offers perhaps only two firm predictions for future studies of exoplanets. First, dynamos are likely quite common, if not universal, in gas and ice giant planets. Measuring the dipolarity and orientation of such dynamos in exoplanets would yield new insights into our understanding of the internal structure and dynamics of these worlds—in parallel with the ongoing exploration of Jupiter, by the NASA Juno mission, and a future mission to an ice giant. Second, some terrestrial planets will host dynamos while others will not. Understanding dynamos in terrestrial planets requires as much geochemistry as electrodynamics, given the critical role of chemical convection. Exoplanets offer a large sample size that will enable statistical tests of which planetary properties most influence the likelihood of a dynamo. Dynamos are thus one of the best examples of how studying exoplanets will teach us about the Solar System, as highlighted in the newest decadal survey, "Origins, Worlds, and Life: A Decadal Strategy for Planetary Science and Astrobiology 2023–2032." Our ignorance about the dynamos of terrestrial planets in particular might be the best motivation for an ambitious observational campaign in the coming decades.

ASSESSING EXOPLANET MAGNETIC FIELDS

Methods for detecting and assessing exoplanet magnetic fields broadly fall into two categories: direct measurements and indirect inferences of magnetic properties. Table 2 summarizes these methods. In principle, most may apply to terrestrial planets except where noted. In practice, these methods largely favor gas giants for current and/or next generation instruments. The following two subsections give brief overviews on the relevant theory, advantages, limitations, and notable observational efforts. This discussion is by no means a complete review but aimed toward giving the interested reader a general sense of the current measurement landscape and a starting point for a deeper dive into the literature.

Table 2. Summary of exoplanet magnetic field measurement methods.

Method	Planet type	Information
Direct		
Exoplanet aurorae	all	local strength
He 1083 nm spectropolarimetry	transiting hot Jupiter	l.o.s. averaged strength
Radiation belt emission	all	dipole magnetic component
Indirect		
Star–planet interactions	close-in	magnetopause size
Ohmic dissipation	transiting hot Jupiter	
Magnetospheric bow shocks	transiting	magnetopause size
Atmospheric outflow transit spectroscopy	transiting close-in	strongly or weakly magnetized

Direct measurements

Exoplanet aurorae. The discovery of Jupiter's radio aurorae[1] by Burke and Franklin (1955) precipitated a decades-long search for exoplanet radio aurorae that continues today (Zarka 2007; Turner et al. 2021, 2023, and references therein). Solar System planets demonstrate several mechanisms for producing aurorae: Most familiar are aurorae on Earth, Jupiter, and Saturn driven by incident plasma flow from the solar wind. An analogous mechanism also occurs on the Galilean moons, where Jupiter's circumplanetary plasma torus supplies the incident plasma (de Kleer et al. 2023). Additionally, magnetospheric plasma departing from rigid co-rotation couples Jupiter's outer magnetosphere with its ionosphere to produce its main aurorae (Cowley and Bunce 2001; Nichols and Cowley 2003). Finally, close-in satellites like Io, Europa, Ganymede, Enceladus and perhaps Callisto can excite aurorae on their hosts (Clarke et al. 2002, 2011).

Each of these mechanisms produce multiwavelength aurorae, but it is the radio component of auroral emissions that offers a direct measurement of an object's magnetic field strength. In rarified plasmas where the plasma frequency ($\propto n_e^{1/2}$) is much less than the cyclotron frequency ($\propto B$), upward traveling and mildly relativistic electron populations with energy anisotropies can excite the electron cyclotron maser instability rather than plasma emission (Treumann 2006). This instability produces periodically repeating and extremely bright, coherent radio emissions easily distinguished from incoherent (gyro)synchrotron emissions via high brightness temperatures and exceptionally strong circular polarization (Dulk 1985; Hallinan et al. 2008).

When plasma electron number densities n_e are low, radio aurorae resulting from the electron cyclotron maser instability occur at frequencies f very near the fundamental electron cyclotron frequency (Treumann 2006):

$$f = \frac{eB}{2\pi m_e} \tag{8}$$

For magnetic field strengths B in units of Gauss, the emission frequency in megahertz is approximated as $f_{[MHz]} \approx 2.8\, B_{[Gauss]}$, and exoplanets with 10 Gauss fields such magnetic Jupiter analogs are expected to produce emissions at ~28 MHz, while magnetic Earth analogs will be a factor of ~ten weaker.

Above ~10 MHz, astrophysical radio emissions including exoplanet aurorae can reach the ground. The Square Kilometer Array (SKA) is expected to attain sufficient sensitivity to detect some exoplanet radio aurorae from strongly magnetized gas giants (Grießmeier et al. 2007). However, the Earth's ionosphere is opaque to radio emissions at frequencies below ~10 MHz. Detecting terrestrial exoplanet radio aurorae below this ionospheric cutoff will therefore require space-based arrays such as the Great Observatory at Long Wavelengths (GO-LoW; Knapp et al. 2024) and the lunar Farside Array for Radio Science Investigations of the Dark ages and Exoplanets (FARSIDE) (Burns et al. 2019).

The luminosities of auroral radio emissions depend in part on key magnetic properties of exoplanet systems. The power available to drive aurorae by the breakdown of plasma co-rotation scales with the planet's magnetic field B_p, its angular velocity Ω_p, and its radius R_p (Saur et al. 2021):

$$S \propto B_p^2\, \Omega_p^2\, R_p^2 \tag{9}$$

---

[1] The term "aurorae" typically refers to atomic and molecular emissions in a planet's upper atmosphere in response to energy deposition from current systems driven by the described mechanisms. Here, "radio aurorae" refer to radio emissions from electrons in these auroral current systems.

favoring objects with strong magnetic fields and rapid rotation, such as brown dwarfs. For numerical treatments of the magnetospheric–ionospheric coupling currents that arise, we refer the interested reader to Nichols et al. (2012) and Turnpenney et al. (2017).

Power from aurorae driven by plasma flows scale with incident magnetic flux, which increases for close-in planets when all else is equal (Zarka et al. 2018). This motivates searches for radio aurorae from hot Jupiters, for which some dynamo theories predict stronger magnetic fields (e.g., Yadav and Thorngren 2017). Stronger fields offer two advantages: they can lead to larger magnetospheric cross-sectional areas and therefore higher dissipated flux and expected auroral power, and they may produce aurorae at higher frequencies where today's low frequency arrays, such as LOFAR, are more sensitive.

However, no detections of exoplanet aurorae have been confirmed to date (Turner et al. 2021, and references therein), despite an initial promising detection of excess radio emission from the τ Boötis system at 21–30 MHz with the beam-formed LOFAR array (Turner et al. 2021, 2023). Indeed, initial modeling finds that hot Jupiters may suffer from dense, plasma-filled magnetospheres that inhibit electron cyclotron maser emission, though instrument sensitivities remain a limiting factor.

Instead, aurorae have been conclusively detected on brown dwarfs and low mass M dwarfs, collectively known as ultracool dwarfs (Fig. 5; e.g., Hallinan et al. 2007, 2008, 2015; Kao et al. 2016; Route and Wolszczan 2012; Vedantham et al. 2020a, 2023, and references therein). Such objects can serve as laboratories for understanding extrasolar auroral physics (Pineda et al. 2017; Saur et al. 2021) as well as planetary dynamo models (Kao et al. 2016, 2018).

While radio aurorae offer direct measurements of exoplanet magnetic fields, they probe only the local emitting region. As such, they provide lower bounds on a surface-averaged

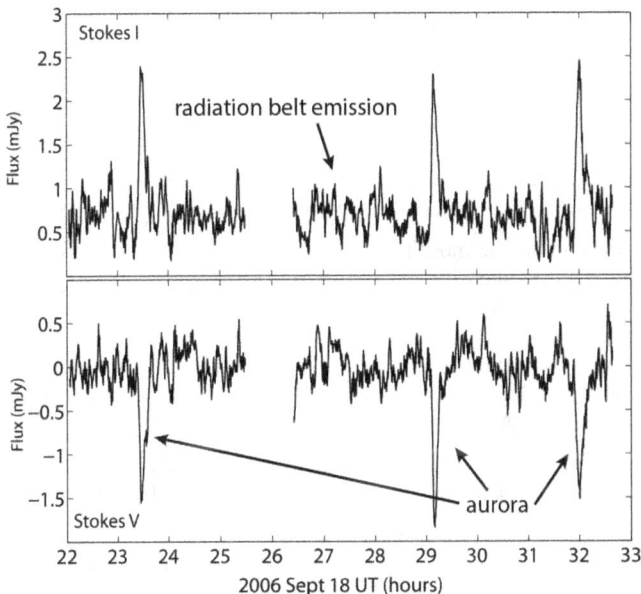

Figure 5. Stokes I and V radio timeseries at 8.44 GHz of the ultracool dwarf LSR J1835+3259 showing ~100% circularly polarized and rotationally periodic electron cyclotron maser bursts later identified as aurora by Hallinan et al. (2015). Using multi-epoch resolved imaging of this same object, Kao et al. (2023) showed that the non-auroral radio emissions are extrasolar Jovian radiation belt analogs (Radiation Belt Emissions subsection; see also Fig. 8). Credit: adapted from Figure 1 of Hallinan et al. (2008) with permission.

magnetic energy (Kao et al. 2016, 2018). Even so, these lower bound measurements can set stringent constraints on exoplanet dynamo models. Numerical dynamo simulations finding that convected energy flux sufficiently determines the magnetic field strengths of planets and brown dwarfs (Christensen et al. 2009) have been extended to models arguing that hot Jupiters may support strong magnetic fields (Yadav and Thorngren 2017, see also discussion in the *Star–planet interactions* subsection). However, direct measurements of brown dwarf magnetic field strengths via their radio aurorae demonstrate otherwise[2] (Kao et al. 2016, 2018).

Obtaining topological information from aurorae requires dynamic spectra paired with careful modeling, as has been attempted for ultracool dwarfs with inconclusive results (Yu et al. 2011; Lynch et al. 2015). Furthermore, radio aurorae can be emitted along an object's magnetic field lines, resulting in broadband emission that increases with frequency as emitting regions approach strong magnetic fields close to the dynamo surface and then cut off sharply above an object's upper atmosphere at the high frequency end of the radio spectrum (Fig. 6; Zarka 2007). While the broadband nature of auroral radio emissions allow for relatively lenient frequency search spaces, obtaining the strongest constraints on surface-averaged magnetic field strengths requires broadband observations that can identify this cutoff frequency. However, care must be taken when interpreting non-detections, which can be attributed to many factors that are unrelated to objects' magnetic field strengths (Kao and Shkolnik 2024).

A conclusive detection of radio aurorae from a free-floating exoplanet or planet-mass brown dwarf (e.g., Kao et al. 2016, 2018) requires demonstrating electron cyclotron maser emissions originating from the system that are periodic on the rotational timescale of the planet[3] or the orbital timescale of a suitable satellite (for the latter, see the *Star–planet interactions* subsection). Radio aurorae from an exoplanet that is gravitationally bound to its host star may exhibit similar behaviors, though the additional possibility of aurorae driven by stellar winds can relax phenomenological requirements: persistent electron cyclotron maser emissions that disappear during eclipse would also be sufficient. When combined with a broadband search for the auroral cutoff frequency, diagnosing exoplanet magnetic fields with radio aurorae can require significant observational investment, lending themselves well to large multi-frequency all-sky surveys.

Helium 1083 nm transmission spectropolarimetry. Recent theoretical developments of the detailed radiative transfer properties of helium absorption in escaping atmospheres (Oklopčić and Hirata 2018; Oklopčić 2019) have led to one of the most promising new methods for assessing gas giant magnetic fields: spectropolarimetric transit observations of He 1083 nm absorption in hot Jupiters (Oklopčić et al. 2020).

As helium atoms in the thermosphere and exosphere of a hot Jupiter absorbs background light from its host star, the intensity and spectrum of the incident stellar radiation can excite neutral helium to a 2^3S_1 triplet state. This excited state is metastable because transitions to the ground state are exclusively highly forbidden, allowing sufficient populations of these excited helium atoms to accumulate and facilitating detections of absorption at He 1083 nm in a handful of transiting hot Jupiters (e.g., Allart et al. 2018, 2019; Nortmann et al. 2018; Salz et al. 2018; Spake et al. 2018; Alonso-Floriano et al. 2019).

Polarization in He 1083 nm absorption features diagnoses exoplanet magnetic fields. The polarization arises from the Hanle effect, where atoms irradiated by incident linearly polarized

---

[2] These models apply only to rapidly rotating objects with dipole-dominated magnetic fields (Christensen et al. 2009). The *Theory of exoplanet magnetic dynamos* section comments on the former requirement, while the confirmation of strong dipole fields traced by bright radiation belts around aurorae-emitting ultracool dwarfs (Climent et al. 2023; Kao et al. 2023) suggest that at least some aurorae emitting brown dwarfs may meet magnetic energy partition requirements.

[3] An object's rotation period measured from cloud variability can differ from the rotation period of its deep interior as measured from its radio aurorae. This effect has been observed on a cold T6.5 spectral type brown dwarf as well as on Jupiter (Allers et al. 2020, and references therein).

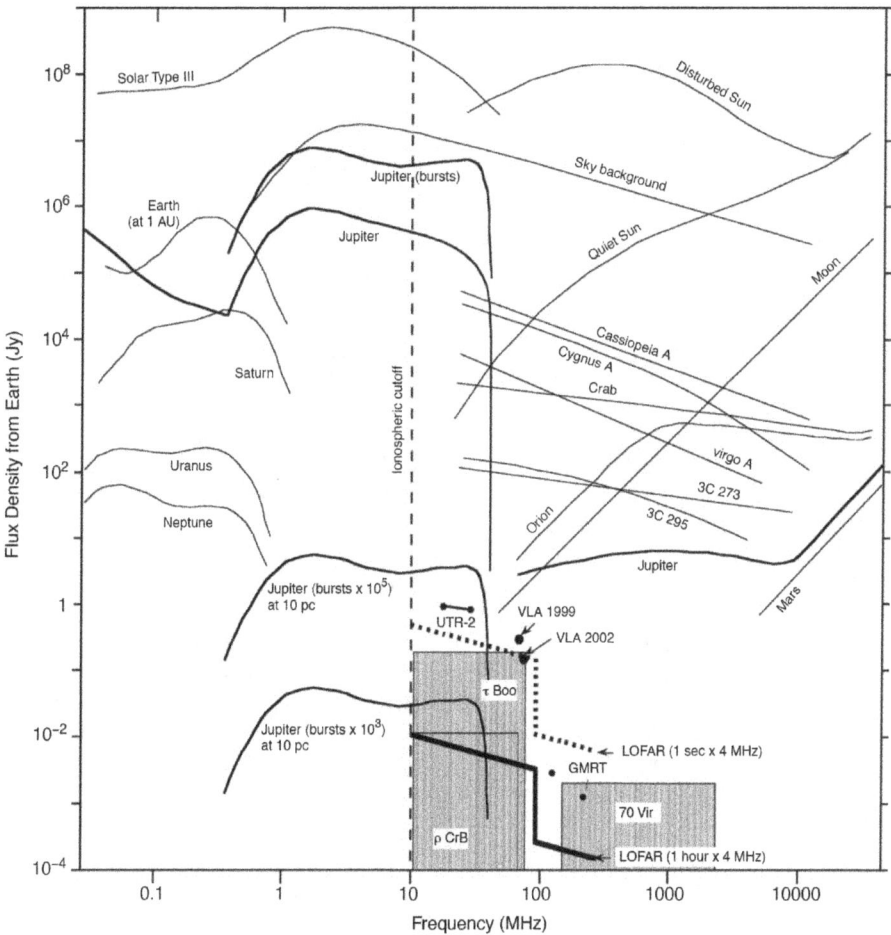

Figure 6. Radio spectra of auroral emissions from Solar System planets compared to other astrophysical radio sources and instruments. Aurorae are broadband and increase in frequency as magnetic field strengths increase near the emitting planet. Sharp cut-offs occur at the high frequency end of their spectra that correspond with emitting regions nearest the upper atmosphere of the planet. Credit: Zarka (2007).

light, such as from a stellar radiation field, preferentially align with and precess about an ambient magnetic field. Fine structure splitting of the He 1083 nm triplet then selectively absorbs specific linear polarizations of the incident light (Oklopčić et al. 2020). The resulting absorption feature exhibits circular or linear polarization (from the Zeeman and Hanle effects, respectively) depending on the orientation of the ambient magnetic field with respect to the observer's line of sight.

One advantage of applying the Hanle effect as a diagnostic of exoplanet magnetic fields is that it occurs for magnetic fields stronger than a few $\times 10^{-4}$ Gauss. In contrast, the Zeeman effect produces circular polarization but requires very strong magnetic fields $B > 3(\Delta v/10 \text{ km} \cdot \text{s}^{-1})$ kiloGauss to exceed the Doppler broadening of the line Δv (~600 Gauss for HD 209458 b).

The linear polarization signal depends on optical depth and the line-of-sight components of the ambient magnetic field strength (Fig. 7), with an expected signal between 0.1% and 0.01% depending on assumed geometries (Oklopčić et al. 2020). Accordingly, He 1083 nm spectropolarimetry favors systems with large transit depths and bright host stars.

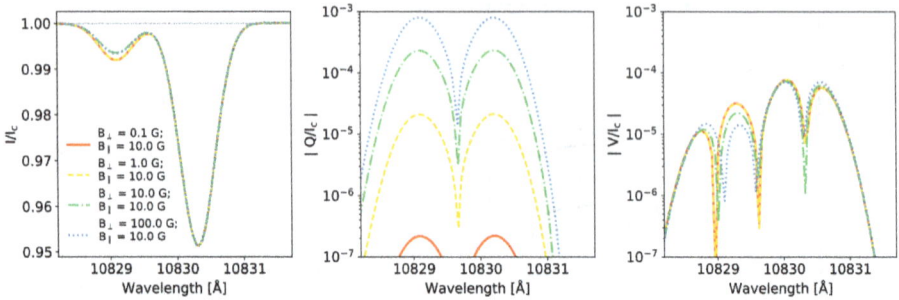

Figure 7. Line-of-sight magnetic field strengths impact the (**left**) total intensity profile of He 1083 nm absorption, (**center**) the linear polarization signal, and (**right**) circular polarization signal, where I, Q, V are the corresponding Stokes parameters. Credit: Oklopčić et al. (2020).

Radiation belt emission. The previous two methods yield local (Helium 1083 nm transmission spectropolarimetry) or line-of-sight averaged magnetic field strengths (*Exoplanet aurorae*). In contrast, the presence or lack of radiation belt emission can directly assess an object's magnetic topology.

All of the Solar System planets with large-scale magnetic fields maintain high energy plasma populations, with energies of tens of MeV, in the equatorial regions of their magnetic dipole components (Fig. 8; Mauk and Fox 2010). Electrons populating planetary radiation belts produce radio emissions that differ from aurorae: instead of coherent electron cyclotron maser emissions near the fundamental cyclotron frequency, Solar System radiation belts produce incoherent synchrotron emissions (e.g., Bolton et al. 2002), which occur at hundreds to thousands of times the local electron cyclotron frequency depending on emitting electron energies (Rybicki and Lightman 1986). Radiation belt emissions vary more gradually than auroral emissions (Kao et al. 2023), so they are particularly well-suited to snapshot searches (e.g., Kao et al. 2019; Richey-Yowell et al. 2020; Kao and Pineda 2022). Finally, their broadband emission at high cyclotron frequency harmonics allows the use of radio arrays tuned to a broad range of frequencies that can exceed the fundamental cyclotron frequencies of exoplanet magnetic fields.

In addition to Earth, Jupiter, Saturn, Uranus, and Neptune (Mauk and Fox 2010), all brown dwarfs that produce radio aurorae at gigahertz frequencies also exhibit non-auroral radio emissions (Pineda et al. 2017; Kao et al. 2019; Kao and Shkolnik 2024, and references therein). Such emissions can trace extrasolar analogs to Jovian radiation belts (Fig. 8; Climent et al. 2023; Kao et al. 2023).

Solar system and brown dwarf radiation belts extend well beyond several times the object's radius (Bolton et al. 2004; Kollmann et al. 2018; Climent et al. 2023; Kao et al. 2023). Despite these extended sizes, resolved radiation belt imaging will be limited to all but the nearest exoplanets until space-based radio arrays can offer baselines exceeding the diameter of Earth. However, confirming that the quasi-quiescent component of brown dwarf radio emissions traces their radiation belts opens a pathway to identifying extrasolar radiation belts using *unresolved* radio timeseries with future arrays. For exoplanets bound to their host stars, distinguishing between stellar and exoplanet radio emissions in unresolved radio timeseries will add an additional layer of difficulty. Free-floating planets are untroubled by this concern, as two isolated planetary mass brown dwarfs exhibiting radiation belt emissions showcase (Kao et al. 2016).

At present, topological information from detections of extrasolar radiation belts is limited: they confirm only that an object has a dipole field component but offer no further information on the partition of magnetic energies between dipole and higher-order field components. For instance, Uranus and Neptune host radiation belts (Mauk and Fox 2010) though dipoles do

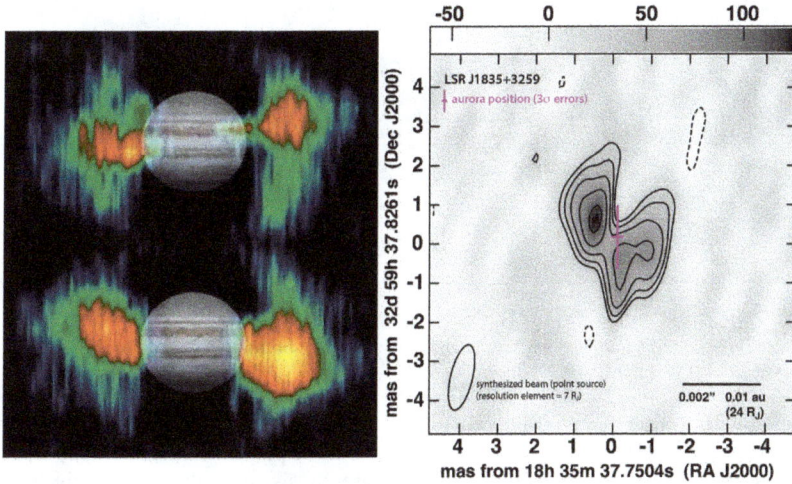

Figure 8. (**Left**) 2 GHz imaging of Jupiter's inner electron radiation belts at different rotational phases. Jupiter's dipole moment is offset from its rotation axis. Credit: NASA/JPL. (**Right**) 8.4 GHz imaging of an electron radiation belt around the aurora-emitting ultracool dwarf LSR J1835+3259. Credit: Kao et al. (2023).

not dominate their magnetic fields (Stevenson 2010). In contrast, Mars and Venus do not have global magnetic fields or radiation belts.

Finally, brown dwarf radiation belt emissions are fainter than radio aurorae by factors of at least a few (Kao et al. 2016, 2018, 2023; Pineda et al. 2017, and references therein). If exoplanet radiation belt emissions scale similarly, they will likely be too faint to be detectable with current or even next generation instruments like the next generation Very Large Array (ngVLA), Square Kilometre Array (SKA), Deep Synoptic Array 2000 (DSA 2000). If astronomers wish to access the model-independent topological information offered by radiation belts, investing in future generations of high sensitivity radio arrays with very long baseline interferometry capabilities will be necessary.

Indirect inferences

Star–planet interactions. Close-in companions can excite aurorae on their hosts, which offer model-dependent means of indirectly inferring basic magnetospheric properties of the companions. In the Solar System, companion-driven aurorae occur on Jupiter (Fig. 9; see also Mura et al. 2017) and scaling up theory grounded in Solar System objects suggests that they can also occur on host stars with close-in planets. In particular, magnetized and/or ionized companions must reside within their hosts' sub-Alfvénic magnetospheric regions, such that hostward-traveling Alfvén waves perturbed by these companions can exceed companion-ward plasma velocities to reach the host (Saur et al. 2013, and references therein).

In sub-Alfvénic interactions, Poynting flux (electromagnetic energy flux) can radiate away from the companion toward the host. The total dissipated power S in companion-driven aurorae depends on the host star's wind properties and the obstacle size presented by the companion, R_o:

$$S \propto R_o^2 B_{\mathrm{wind}} \Delta u^2 \sin^2 \theta \sqrt{\rho_{\mathrm{wind}}} \tag{10}$$

where B_{wind} is the host star's magnetic field at the planet's location, Δu is the apparent wind velocity with respect to the companion, θ is the angle between the stellar magnetic field and the apparent wind velocity, and ρ_{wind} is the wind mass density (Fig. 10; Saur et al. 2013). The obstacle size folds in the companion's magnetic field. All other factors being equal,

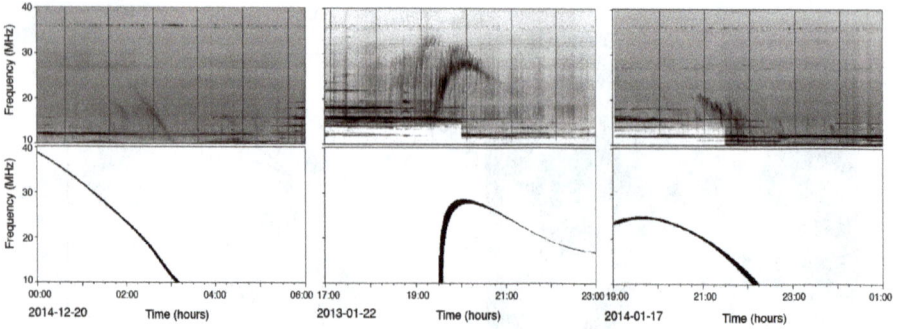

Figure 9. Io-induced radio auorae from Jupiter and modeling (**top, bottom**). Credit: adapted from Figure 5 in Louis et al. (2019) under CC BY 4.0.

a companion with a large magnetosphere will present as a larger obstacle than one possessing only an ionosphere, and it will consequently dissipate more power. This model, known as the Alfvén wing model, can produce flux values that are consistent with aurorae driven by Io, Europa, Ganymede and Callisto on Jupiter (Saur et al. 2013). However, it has not yet been validated for any exoplanet system.

Crucially, star–planet interaction radio emissions induced on the host will occur at frequencies corresponding to the host's magnetic field. This offers an incredible advantage: although radio aurorae from terrestrial planets cannot be observed from the ground (see the *Exoplanet aurorae* subsection above), magnetized terrestrial planets may induce aurorae on their more strongly magnetized hosts, which can produce radio emissions at frequencies that are accessible today.

However, diagnosing exoplanet magnetic properties with star–planet interaction emissions requires addressing several challenges. First, one must demonstrate that the observed emission is unequivocally attributable to a star–planet interaction by showing that it is periodic on the orbital timescale of the responsible planet and cannot be associated with other stellar magnetic activity. Second, one must understand how the total dissipated power partitions into emissions at different wavelengths to convert S to the auroral power observed at the wavelength(s) of interest. Third, one must have a reasonably accurate stellar wind model. Fourth, one must know the ephemerides of the companion. Even with all this information in hand, one can only infer an obstacle size. Obtaining a magnetospheric size requires comparing this obstacle size to a well-characterized planet radius (see the *Ohmic dissipation* subsection below for additional considerations), such as one derived from transit data. Finally, one will need to account for the impact of the host star's plasma pressure on the magnetopause of the planet.

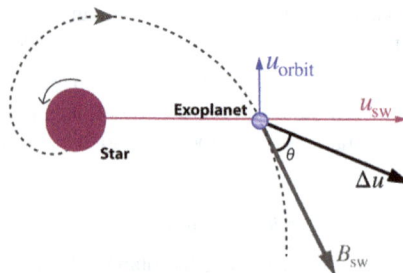

Figure 10. Schematic of quantities relevant to the Alfvén wing model of star–planet interaction by Saur et al. (2013) and summarized in Equation (10). The **dashed line** represents a Parker spiral stellar magnetic field. Credit: adapted from Figure 1 of Turnpenney et al. (2018). CC BY 3.0

As recent radio detections of candidate star–planet interaction emissions demonstrate, the first requirement necessitates significant investment of observing resources as well as a mature understanding of stellar flare behaviors. Distinguishing auroral radio emissions from plasma emissions can be difficult, and we refer the reader to Villadsen and Hallinan (2019) and Vedantham et al. (2020b) for in-depth discussions. Additionally, follow-up radio campaigns have not yielded a repeating detection of 150 MHz emission from the M dwarf GJ 1151 that initial modeling suggested a hypothetical close-in terrestrial companion could plausibly excite (Vedantham et al. 2020b). Radial velocity campaigns rule out massive close-in companions around this star (Pope et al. 2020; Blanco-Pozo et al. 2023) but cannot rule-in a terrestrial companion. Pineda and Villadsen (2023) reported repeating candidate detections of 2–4 GHz star–planet interaction emission from the known planet host YZ Ceti, but the emissions exhibit some scatter when phase-folded with the orbital period of YZ Ceti b. System geometries may explain this scatter (Pineda et al. 2017; Trigilio et al. 2023), but such effects are not yet well understood (Kavanagh and Vedantham 2023). Finally, ordinary stellar magnetic activity processes can produce radio emissions that may masquerade as star–planet interactions (Villadsen and Hallinan 2019; Yu et al. 2023) or other radio emissions interpreted as aurorae on active stars (Villadsen and Hallinan 2019; Zic et al. 2019). Here, the old adage "know thy star, know thy planet" strikes true.

Brown dwarfs again may offer a laboratory for testing star–planet interaction models. As close-in companions around brown dwarfs are discovered, we may find that some drive radio aurorae on their hosts. Companion interactions show uniquely characteristic arcs in the dynamic spectra (time- and frequency-dependent flux densities) of radio aurorae (Fig. 9). Detailed models of these radio arcs can yield information about the orbits, rotation, and magnetic fields of companions (Hess et al. 2008). Similar models are being developed for future detections of exoplanet radio emission.

This section would be incomplete without a discussion of orbitally modulated magnetically active stellar chromospheric lines. The prevailing interpretation for such emissions are star–planet magnetospheric interactions and they have been observed from hot Jupiter systems including HD 179949 b, HD 189733 b, τ Boötis, and ν Andromedae (Fig. 11; Shkolnik et al. 2003, 2005, 2008; Gurdemir et al. 2012; Cauley et al. 2018, and references therein). Several theories have been proposed to explain the power observed in these emissions, including reconnection events between stellar and planetary magnetic field lines as the latter travels through the stellar magnetosphere (Cuntz et al. 2000), reconnection events triggered by the planet on the stellar surface (Lanza 2009, 2012), the Alfvén wing model (Saur et al. 2013), and the stretching of magnetic flux tubes between the star and the planet as the latter orbits (Lanza 2013).

Of these, only the last model can account for the total power P derived from detections of star–planet chromospheric emissions from hot Jupiters:

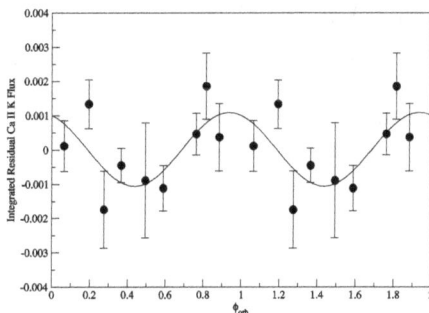

Figure 11. Orbitally modulated Ca II K emission attributed to star–planet interactions from the hot Jupiter system τ Böotis. Credit: Figure 13 in Shkolnik et al. (2008).

$$P \approx \frac{2\pi}{\mu} f_\mathrm{p} R_\mathrm{p}^2 B_\mathrm{p}^2 v_\mathrm{rel} \qquad (11)$$

where f_p is the fractional hemispheric flux tube coverage for the planet and depends on the ratio between the planet's surface polar magnetic field B_p and the stellar magnetic field at the location of the planet, R_p is the planet radius, and v_rel is the relative velocity between the planet and stellar magnetic field lines at the planet's location. Although this model can explain observations, the magnetic field strengths that Cauley et al. (2019) derive using it remain unconfirmed by an independent means. As with the Alfvén wing model, this model also relies on an accurate understanding of energy partitions; inferred exoplanet magnetic fields can differ by several orders of magnitude depending on the fraction of the total star–planet interaction power that is radiated in the observed chromospheric lines. Tuning this energy partition allow Cauley et al. (2019) to infer magnetic field strengths for HD 179949 b, HD 189733 b, τ Boötis, and ν Andromedae that agree with predictions from magnetic dynamo scaling relationships relying primarily on a planet's convected thermal energy (Christensen et al. 2009; Yadav and Thorngren 2017). However, direct measurements of brown dwarf magnetic field strengths call this tuning into question (Kao et al. 2018), underscoring the necessity of independently validating these inferred magnetic field strengths with other measurement methods and/or detailed energy partition studies.

Ohmic dissipation. Above 1000 K equilibrium temperatures, many hot Jupiters exhibit inflated radii relative to predictions from planetary evolution models (e.g., Thorngren and Fortney 2018). One proposed explanation for this "radius anomaly" is heating from interior Ohmic dissipation in electrical currents induced by ionized winds in magnetized planets (Batygin and Stevenson 2010; Batygin et al. 2011).

Ohmic dissipation cannot account for all observed radius anomaly behaviors (Wu and Lithwick 2013; Ginzburg and Sari 2016), and interior heating may also arise from tidal dissipation and shear instabilities or vertical mixing. As such, inferring inflated hot Jupiter magnetic fields from Ohmic dissipation models (e.g., Rauscher and Menou 2013; Wu and Lithwick 2013) requires carefully accounting for all such interior heating processes. We refer the interested reader to Fortney et al. (2021) for additional discussion on hot Jupiter interior heating mechanisms.

Magnetospheric bow shocks. Stellar winds interacting with an exoplanet's magnetic field can form bow shocks when they reach supersonic speeds and are observed around Solar System planets. Under certain conditions, these magnetospheric shocks may absorb excess stellar radiation ahead of a planet's orbit, resulting in both deeper transits and earlier transit ingresses, and perhaps even pre-transit absorption (Fig. 12), at wavelengths where the shock is optically thick (e.g., Vidotto et al. 2010).

Magnetospheric stand-off distances inferred from pre-transit absorption can then place lower-bound limits on exoplanet magnetic field strengths by balancing the stellar wind magnetic pressure B_w at the location of the planet against the pressure of the planet's magnetic field B_p at its magnetopause and neglecting ram and thermal pressure terms (Vidotto et al. 2010):

$$\frac{B_\mathrm{w}^2}{8\pi} + \rho_\mathrm{w}\,\Delta u_\mathrm{w}^2 + p_\mathrm{w} = \frac{B_\mathrm{p}^2}{8\pi} + p_\mathrm{p} \qquad (12)$$

Magnetospheric bow shocks are only one of several proposed interpretations (Lai et al. 2010; Vidotto et al. 2010; Llama et al. 2011, 2013) for enhanced UV and/or Balmer line transit depths and early ingress detections compared to optical transits observed on WASP-12b (Fossati et al. 2010; Haswell et al. 2012) and HD 189733b (e.g., Ben-Jaffel and Ballester 2013;

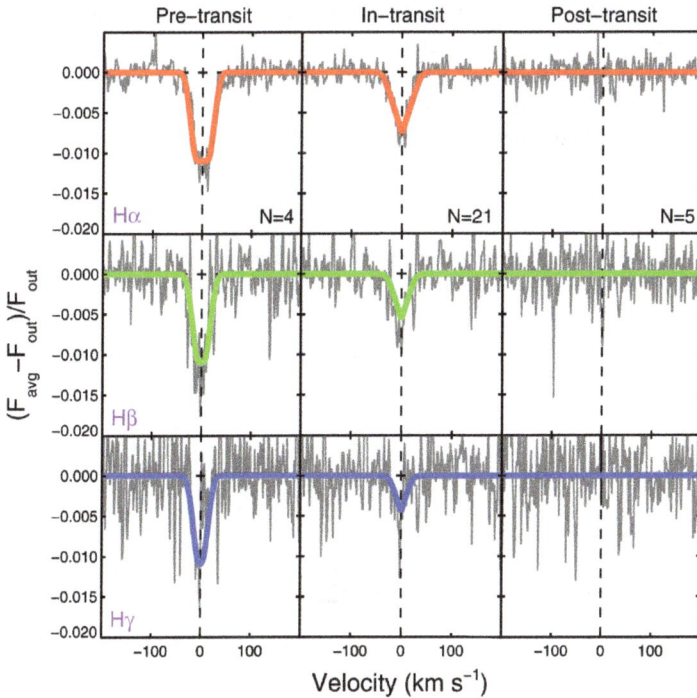

Figure 12. Pre-transit Balmer line absorption from averaged transmission spectra of the hot Jupiter HD 189733 b (**grey**: averaged data; **red/green/blue**: averaged model), initially attributed as a thin bow-shock though later modeling cast this interpretation into doubt. Credit: adapted from Figure 1 in Cauley et al. (2015).

Bourrier et al. 2013; Cauley et al. 2015, 2016). More detailed numerical hydrodynamical simulations find optical depths requiring rapid cooling of the shocked wind that is difficult to achieve (Alexander et al. 2016), casting doubt on but not ruling out the magnetospheric bow shock interpretation for WASP-12b. Other interpretations, such as overflow of the planet's Roche lobe (Lai et al. 2010; Bisikalo et al. 2013) and charge exchange between stellar winds and planetary outflows (Tremblin and Chiang 2013), can also reasonably but incompletely reproduce observed transit features (Cauley et al. 2016, and references therein).

Atmospheric outflow transit spectroscopy. Stellar radiation can drive highly ionized outflowing winds on close-in planets (See the *Atmospheric escape* subsection below). In regions of a planet's magnetosphere where magnetic pressure dominates thermal and ram pressures from its photoevaporative winds, atmospheric outflow is effectively "anchored" along the planet's magnetic field lines. In contrast to the magnetically confined strong field regime, in the unconfined weak field regime where thermal and ram pressures dominate, day-to-night winds can occur instead (e.g., Batygin et al. 2013; Adams and Owen 2015).

Three dimensional magnetohydrodynamic modeling that calculates magnetic drag from local conditions—most importantly, temperature—find that highly irradiated ultra hot Jupiters exhibit magnetically sensitive circulation patterns in their upper atmospheres. Neglecting a localized treatment of magnetic drag results in day-to-night winds that observationally manifest as blue shifted transmission spectra. In contrast, the introduction of even a weak magnetic field can shift dayside winds poleward and introduce sufficient magnetic drag to somewhat suppress blueshifts and/or introduce mid-transit redshifts for certain species (Beltz et al. 2023). Magnetic drag may also influence emission spectra and phase curves (Beltz et al. 2022a,b) and in some cases magnetic models fit phase curves better than the non-magnetic model (Coulombe et al. 2023). Similarly, two-dimensional modeling finds blue-shifted He 1083 nm transits in the unconfined wind regime that do not occur when planet winds are magnetically confined (Fig. 13). Instead, these transits are deeper than in the magnetically unconfined regime, though many parameters can influence transit depths (Schreyer et al. 2024).

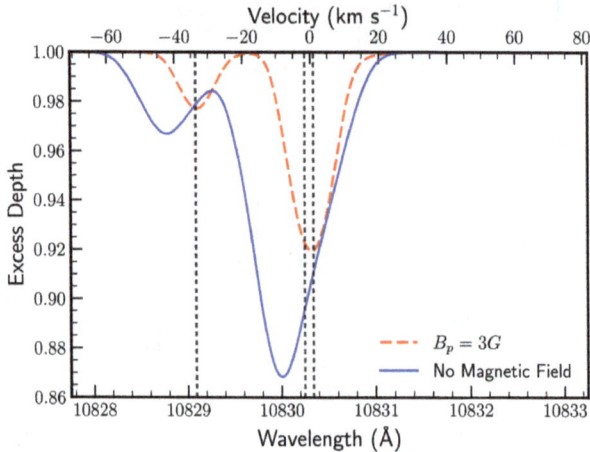

Figure 13. Modeled He 1083 nm absorption for a magnetized and unmagnetized transiting hot Jupiter. In the unmagnetized case, day-to-night winds result in a net blueshift. Credit: Figure 4 from Schreyer et al. (2024). CC BY 4.0

Comparing He 1083 nm transit spectroscopy to models can, in theory, distinguish between the confined and unconfined regimes, but diagnosing actual exoplanet magnetic field strengths requires more careful interpretation. In particular, the confined and unconfined regimes depend on the magnetic strength, topology, and orientation of the planet, as well as the stellar wind environment and ionizing flux. Additionally, the orbital eccentricity must be well known, as eccentric orbits introduce line-of-sight velocity shifts that linearly depend on the eccentricity deviation from zero. For typical eccentricity uncertainties of order ~0.1, the resulting velocity shift is comparable to the effect of magnetically sensitive outflow geometries (Schreyer et al. 2024). Finally, as is the case for other indirect methods discussed in this chapter, independent cross-validation of these modeled effects with other measurements of exoplanet magnetic fields remains to be seen.

A promising future

The reader may notice that no exoplanet magnetic field strengths are quoted in this section. This choice is deliberate, as no confirmed direct measurements of exoplanet magnetic field strengths exist at the time of writing, reflecting the aspirational nature of most of the discussed methods and the instrumentation and interpretative challenges that they must overcome. Nevertheless, tantalizing observational hints of exoplanet magnetism have inspired exciting theoretical developments and showcase a future ripe with possibility. Even as we look toward the forthcoming era of exoplanet magnetic field measurements, we urge readers to look to brown dwarfs today as laboratories that can test and refine models of aurorae, radiation belts, and star–planet interactions—and the information we may learn from them in the future when they are detected in exoplanet systems.

INFLUENCES OF EXOPLANET MAGNETIC FIELDS

Knowing whether an exoplanet possesses a magnetic field can influence our knowledge about other important properties of the planet and the processes that occur there. Perhaps the most obvious information comes about the interior of the planet (as discussed in the *Theory of exoplanet magnetic dynamos* section, including its internal heat and its potential, in the case of rocky exoplanets, for geologic activity). But other important information can be inferred about: (1) the manner in which the planet interacts with its space environment and the extent of this interaction region; (2) whether the planet is likely to retain an atmosphere; (3) particle precipitation into the planet's atmosphere and/or surface and its consequences; and (4) the evolution of the planet's rotation rate and orbit. We discuss each of these four areas in the sections below.

Interaction with space environment

The nature of a planet's magnetic field determines how it interacts with its space environment. Planetary bodies possessing internally generated magnetic fields form *intrinsic* magnetospheres, where incident flowing plasma (i.e., the stellar wind) is deflected around the object by the planet's field. In our own solar system, Mercury, Earth, Jupiter, Saturn, Neptune, and Uranus all have intrinsic magnetospheres, along with Jupiter's moon Ganymede (see Table 1). Planetary bodies possessing a sufficiently conducting layer (e.g., an ionosphere or a salty ocean) may form *induced* magnetospheres, where the incident flowing plasma induces a current in the conductors, which in turn generates a magnetic field that deflects the incident plasma. Venus, Mars, Pluto, comets, Jupiter's moon Europa, and Saturn's moon Titan all have induced magnetospheres. Other objects, such as Earth's Moon, have no significant internally generated magnetic field and no significant conducting layer, and incident plasma strikes the surface directly.

Induced magnetospheres are typically smaller than intrinsic magnetospheres relative to the size of the planet (Fig. 14). One metric for magnetospheric size is the 'standoff distance' of the solar wind, determined to first order through considering the balance between the pressure from the incident stellar wind ($\rho_{sw}v_{sw}^2$) and the magnetic pressure generated at the planet ($B_P^2/(2\mu_0)$). The standoff distances for Earth, Jupiter, and Neptune are about 10, 50, and 25 planetary radii respectively (see Kivelson and Bagenal 2007), while Venus, Mars, and Pluto have standoff distances of about 1.05, 1.3, and 3 planetary radii (see Brain et al. 2016). This pattern does not always hold; in some cases, an intrinsic planetary field can be relatively weak (e.g., Mercury, with a small intrinsic magnetosphere) or an induced planetary field can be relatively strong or be complemented by dynamic ($\rho_P v_P^2$) and thermal plasma (nkT) pressure

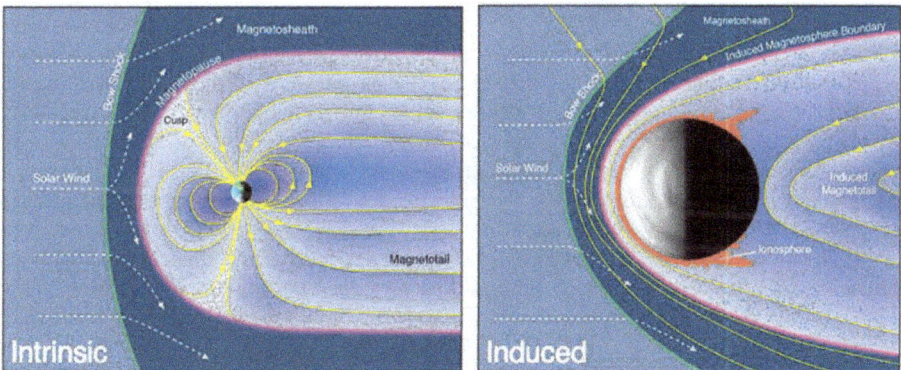

Figure 14. Schematics of characteristic induced (**top**) and intrinsic (**bottom**) planetary magnetospheres. Induced magnetospheres tend to be smaller relative to the size of the planet. Credit: Planetary Magnetic Fields and Climate Evolution" from *Comparative Climatology of Terrestrial* © 2013 The Arizona Board of Regents. Reprinted by permission of the University of Arizona Press.

from the planetary object (e.g., active comets, with large induced magnetospheres). But given the same external conditions and no significant planetary atmospheric outflow we generally expect intrinsic magnetospheres to be larger than induced magnetospheres. Therefore, exoplanets without outflows and with internally generated fields should influence a larger region of space than if they were unmagnetized.

The presence and nature of a significant internally generated magnetic field also partially determines the processes that occur near a planet and in its upper atmosphere. Intrinsic magnetospheres may trap radiation belts similar to those at Earth and Jupiter, allowing the magnetic field to be detected (see the *Radiation belt emission* subsection above). Aurorae are not unique to intrinsic magnetospheres (Schneider et al. 2015). However, aurorae in intrinsic magnetospheres are usually spatially confined to regions near the magnetic poles and are brighter (in our own solar system) than the more global aurora that occur in induced magnetospheres. Uranus and Neptune possess magnetic fields that have significant non-dipolar components and are highly titled relative to the planet's rotation axis (Ness et al. 1986; Connerney et al. 1991). Both features have consequences for the access of stellar wind particles into the magnetosphere. Close-in magnetized exoplanets may be directly connected to the stellar magnetic field, resulting in a so-called star–planet interaction (SPI) that enables direct exchange of plasma along open magnetic flux tubes from the planet (see the *Star–planet interactions* subsection above).

Atmospheric escape

A possible consequence of a planet possessing a magnetic field is the influence of the magnetic field on the escape of atmospheric particles. This topic is much debated in recent years. On one hand, it has often been thought that planetary magnetic fields act as shields for their atmospheres, preventing stellar winds from accessing the atmosphere and driving escape (e.g., Hutchins et al. 1997; Dehant et al. 2007; Lundin et al. 2007). On the other hand, atmospheric escape does not necessarily require direct access of a stellar wind to a planet's atmosphere; energy can transfer to atmospheric atoms and molecules via a variety of processes, and a stellar wind can transfer energy to an atmosphere indirectly (Moore and Horwitz 2007; Brain et al. 2013, 2016; Tarduno et al. 2014; Del Genio et al. 2020; Gronoff et al. 2020; Brain 2021; Maggiolo and Gunell 2021; Vidotto 2021). Whether (and how) a planet's magnetic field influences its escape rate likely depends upon many characteristics of the planet and its host star. Additionally, the influence on escape rates depends upon which escape process and which atmospheric species is being considered.

Hydrodynamic escape. At present most analyses of atmospheric escape from exoplanets consider hydrodynamic escape, sometimes called photoevaporation. Hydrodynamic escape occurs in the fluid limit of Jeans (thermal) escape, when a fraction of the thermal distribution of atmospheric particles has energies that exceed the energy required to escape from the planet. When this fraction becomes significant, the escaping flow can be approximated as a fluid. A recent review of hydrodynamic escape from exoplanets was provided by Owen (2019). An estimate of the hydrodynamic loss rate for planets, Φ, can be obtained if one has estimates for the planet's mass (M_p) and size (R_p) as well as the flux at EUV wavelengths ($\sim$10–120 nm) from the star (F_{EUV}), the radius where the EUV fluxes are absorbed (R_{EUV}^2), and some efficiency factor (ε) for the conversion of incident EUV energy into atmospheric particle heating (e.g., Erkaev et al. 2007):

$$\Phi = \frac{\varepsilon \pi F_{EUV} R_{EUV}^2 R_P}{GM_P} \qquad (13)$$

This equation represents the "energy-limited" hydrodynamic escape rate, where the available incident EUV energy (as opposed to supply of atmospheric particles from the lower atmosphere) controls the escaping particle flux.

Hydrodynamic escape most likely occurs when the planet's atmosphere contains light species (such as hydrogen), the gravity of the planet is weak, or the stellar EUV flux is large (as is the case for active or young stars). For example, hydrodynamic escape is thought to have stripped the primordial H/He atmospheres of the terrestrial planets early in solar system history, when stellar EUV fluxes were considerably larger (Watson et al. 1981). But the reduced EUV flux experienced by the giant planets and their deep gravitational potentials have allowed them to retain significant H/He atmospheres. Hydrodynamic escape may also be responsible for a dearth of close-in exoplanets intermediate in size between Earth and Neptune (e.g., Fulton et al. 2017). Given the above, the focus on hydrodynamic escape over other escape mechanisms reflects current transit spectroscopy, often of exoplanets orbiting active stars at close distances). To first order, hydrodynamic escape is a thermal process driven by stellar EUV fluxes, resulting in the loss of predominantly neutral atmospheric particles. Planetary magnetic fields might thus be considered to have little influence on hydrodynamic escape rates. However, recent studies have considered the idea that outflowing hydrodynamic winds from planets undergo ionization from the same intense stellar EUV fluxes that heated them initially (e.g., Trammell et al. 2011; Kislyakova et al. 2014; Khodachenko et al. 2015; Erkaev et al. 2017; Owen and Adams 2019). Once ionized, the outflowing particles can interact with a planet's magnetic field. Atmospheric particles that are ionized on closed planetary magnetic field lines may be trapped and re-impact the atmosphere, while those on open field lines are likely to escape the planet but may be reconfigured into a magnetotail shape. The orientation of the magnetic field with respect to the field carried by the stellar wind is therefore important since it influences the topology (open vs. closed nature) of fields close to the planet.

Overall, however, one expects hydrodynamic escape rates to be reduced if a planetary magnetic field is present, and the fraction of open magnetic field lines (approximated by the extent of the polar cap) provides an important metric for this attenuation (e.g., Trammell et al. 2011; Khodachenko et al. 2015). The conversion of escaping neutral hydrogen to protons should modify the shape of Lyman-alpha absorption lines in the spectrum of a transiting planet. Several studies (e.g., Kislyakova et al. 2014; Erkaev et al. 2017; Villarreal D'Angelo et al. 2018; Ben-Jaffel et al. 2022) have suggested that the shape of the line can couple with models for the stellar wind interaction with a magnetized planet to infer both the outflow rate and the strength of the magnetic field (e.g., Fig. 15), as discussed in the *Atmospheric outflow transit spectroscopy* subsection above.

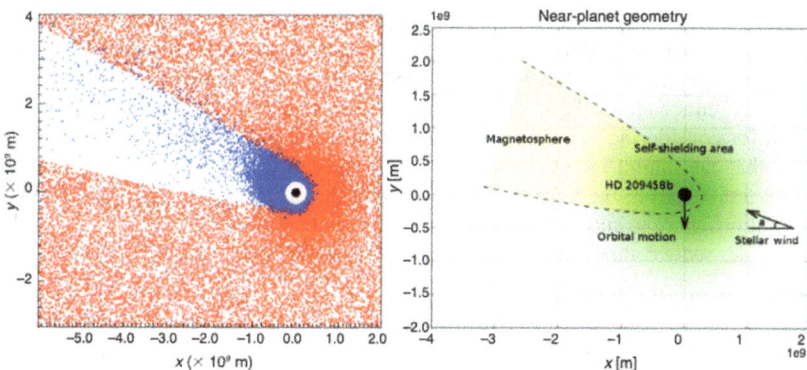

Figure 15. (**Left**) The simulated hydrogen exosphere around transiting exoplanet HD 209458b, with neutral atoms indicated in **blue** and ions in **red**. (**Right**) Cartoon showing the size and orientation of the planet's magnetosphere in the simulation. The situation is consistent with observed absorption in the red and blue wings of stellar Lyman-alpha during transit periods. From Kislyakova et al. (2014). Reprinted with permission from AAAS.

Non-thermal escape. Planets that are not experiencing hydrodynamic escape through some combination of larger size, sufficiently small stellar EUV flux, and more massive atmospheric species characteristic of a secondary atmosphere are still subject to classic Jeans escape and to non-thermal escape processes including ion escape, photochemical escape, and sputtering. Ion escape results when charged particles are accelerated (via a combination of electric fields) to the energy required for escape. Photochemical escape occurs when exothermic chemical reactions are driven in the planet's upper atmosphere, usually by solar UV fluxes and often involving the dissociative recombination of molecular ions. Sputtering occurs when particles incident upon an atmosphere (either stellar wind particles or escaping atmospheric ions) collide with target atmospheric particles, "splashing" them out of the atmosphere. A notable aspect of non-thermal escape processes is that they can be more effective at removing massive atmospheric species (e.g., oxygen, carbon, nitrogen, noble gases) than thermal escape. Thus, they are important to evaluate, especially for planets possessing secondary atmospheres.

A planetary magnetic field can influence all three of the non-thermal processes identified above, since all three involve charged particles—and these influences may be important (e.g., Airapetian et al. 2017). The influence of a planetary magnetic field on photochemical escape is likely to be small. This partly because the energy that results from photochemical reactions may not be sufficient for escape on large planets where global magnetic fields are more likely. Sputtering may be more strongly influenced by the presence of a magnetic field, which deflects the stellar wind at larger distances from the planet and reduces the flux of stellar wind particles that could cause sputtering.

The connection of planetary magnetic fields to ion escape is the most direct and is therefore the most often considered. Multiple studies have considered the influence of a magnetic field on ion escape (and sometimes other non-thermal processes). Several studies have taken semi-analytic approaches. Most do not consider exoplanets specifically but are either general or generalize from solar system planets. For example, Kulikov et al. (2007) examined the influence of an early Martian magnetic field on escape, concluding that the presence of a magnetic field likely decreased escape rates due to the larger standoff distance of the magnetic field. Blackman and Tarduno (2018) used a purely theoretical approach to consider the increased capture of energy from the stellar wind by a planet's magnetic field (compared to the case with no field), the slowing of the incident solar wind by the planetary field, and the recapture of escaping particles. They concluded that the impact of a planetary magnetic field on atmospheric escape depends upon the details of each of these three processes. Gunell et al. (2018) constructed semi-analytic models for several escape processes and applied them to Venus, Earth, and Mars (Fig. 16). They found that the total atmospheric escape rate is comparable for the three planets and inferred that the presence of a planetary magnetic field does not reduce atmospheric escape. Finally, Ramstad and Barabash (2021) provided a formalism for considering atmospheric escape from many planets that accounts for various processes that can inhibit the escape of ions: the supply of ions from the lower atmosphere, the energy transfer from the solar wind to ions, and the efficiency of transport of ions from the upper atmosphere to space. They concluded that there is no solid evidence that planetary magnetic fields reduce atmospheric ion escape. Instead, they may increase escape rates.

There is observational support for the idea that planetary magnetic fields do not necessarily reduce atmospheric escape rates. In situ spacecraft observations of ion escape from terrestrial planets show that the escape rates from Venus, Earth, and Mars agree to within a factor of a few (Barabash 2010; Strangeway et al. 2010; Ramstad and Barabash 2021), with *greater* ion escape fluxes from Earth. There is some question about what fraction of the measured ion fluxes at Earth actually escape the planet (similar to the discussion of hydrodynamic escape above), since Earth's magnetic field may redirect upward-moving atmospheric ions measured in the magnetospheric polar regions back to the planet (Seki et al. 2001). However, recent measurements of ion fluxes

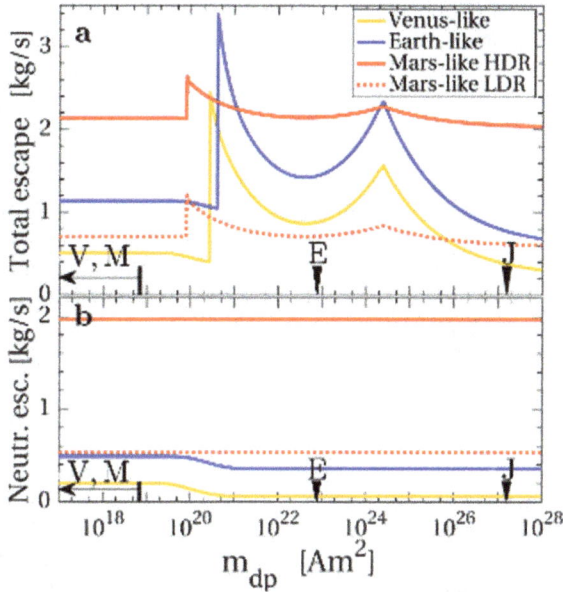

Figure 16. Semi-analytic model of atmospheric escape rates from Venus-, Earth-, and Mars-like planets as a function of planetary magnetic field strength. Calculations for Mars are performed for high and low dissociative recombination rates of molecular oxygen ions. the current dipole field strengths of Venus, Mars, Earth, and Jupiter are indicated. Credit: Gunell et al., Astron Astrophys **614** (2018) L3, reproduced with permission © ESO.

in different regions of Earth's magnetosphere suggest that a significant fraction (>50%) escape (Slapak et al. 2017). Earth also appears to be more responsive to changes in the solar wind pressure than either Venus or Mars (Ramstad and Barabash 2021).

Detailed plasma simulations of the interaction of a stellar wind with a planetary atmosphere (e.g., Fig. 17) suggest that there is not a monotonic relationship between the strength of a planet's magnetic field and its escape rate (Egan et al. 2019; Sakata et al. 2020, 2022). Instead, there is an intermediate planetary magnetic field strength for which ion escape rates are maximized. At lower planetary field strengths, a significant fraction of newly born ions is carried back into the planet's atmosphere by the incoming stellar wind. At higher field strengths, newly born ions are increasingly likely to be trapped in a large planetary magnetic field, unable to escape.

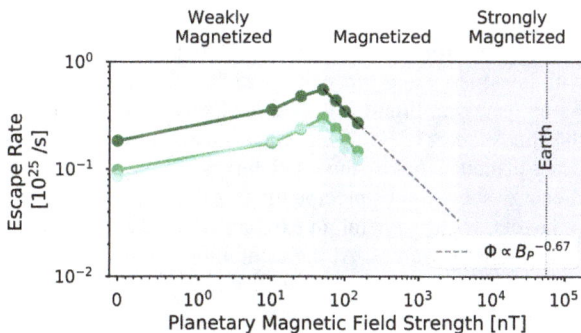

Figure 17. Ion escape rates increase with magnetic field strength before quenching for strongly magnetized planets. The dashed line indicates an extrapolated escape rate (Φ) based on the simulations. The **vertical dotted line** indicates Earth's current dipole field strength. From Egan et al. (2019) © The Authors.

A few studies have considered non-thermal ion escape and related processes from specific exoplanetary systems. Cohen et al. (2014) explored Earth-like planets orbiting M Dwarf stars, and noted that potentially habitable planets will orbit close to the star where they should transition from sub-Alfvenic to superAlfvenic regimes on every planetary orbit. Joule heating in the atmosphere of such a magnetized planet is significant (albeit lower than an unmagnetized planet), especially when the planet is an a sub-Alfvenic regime. At Earth, Joule heating is generally correlated with increased atmospheric ion escape. Dong et al. (2017) simulated an Earth-like planet orbiting orbiting Proxima Centauri-b, and found that the presence of a magnetic field generally lowers atmospheric ion escape rates. And Dong et al. (2020) simulated magnetized and unmagnetized Venus and Earth analogs orbiting TOI-700d. The magnetized Earth-like planets in their simulations had escape rates that were intermediate between the unmagnetized Earth analog and the unmagnetized Venus analog. Collectively, the studies above suggest that planetary magnetic fields influence escape rates for at least some planets, but not necessarily in obvious and consistent ways.

Particle precipitation

Several different populations of charged particles can precipitate into a planet's atmosphere or onto its surface. These include Galactic Cosmic Rays (GCRs), stellar wind particles, Stellar Energetic Particles (StEPs), and particles accelerated within the planet's magnetosphere. Each population of particles has a different range of energies, with typically different consequences for the planet. GCRs originate from a variety of sources outside of the astrosphere of the planet's host star, are highly energetic, and are commonly associated with supernovae and active galactic nuclei. The stellar wind originates in the stellar corona of the host star, contains both electrons and ions, and is less energetic than GCRs. StEPs also originate in the stellar corona as well as in the stellar wind, are more energetic than the stellar wind, and are associated with energetic events from the star such as coronal mass ejections and flares. To this point, many exoplanets have been identified around active stars, orbiting at small distances; this is likely to make the flux of StEPs large relative to the fluxes experienced at Earth (F. Fraschetti and R. Jolitz, personal correspondence 2023). Finally, particles within a planet's magnetosphere (induced or intrinsic) can be accelerated by processes such as field-aligned currents, magnetic reconnection within the magnetosphere, or interactions with orbiting moons. Some of these particles may be accelerated into the planet's atmosphere, causing heating, ionization, or aurora. Collectively, precipitating particles can have energies ranging from $\sim 10^0 \, \text{eV}$ up to $\sim 10^{20} \, \text{eV}$ or more.

Magnetic fields alter the trajectory of a charged particle, with turning radius given by

$$r = \frac{m v_\perp}{qB} \tag{14}$$

Thus, the incident energy spectrum of precipitating particles is modified by the presence of a magnetic field, with low energy and low mass particles more strongly attenuated than high energy particles. As an example, Figure 18 shows calculated energy spectra of GCR fluxes precipitating into the atmospheres of Earth-like planets with different magnetic field strengths. For sufficiently strong planetary fields, the GCR fluxes are reduced by a factor of 1000 or more. The detailed energy spectrum of precipitating GCRs, stellar wind particles, and StEPs depend upon many factors, but in general an exoplanetary magnetic field should reduce the precipitating fluxes from these sources. At the same time, locally accelerated particles may be more abundant and more energetic near magnetized planets, due to the stronger fields and currents associated with intrinsic magnetospheres.

Precipitating particles can have several consequences for planets, from their magnetosphere to their surface. Sufficiently strongly magnetized planets such as Jupiter (or even Earth) have radiation belts, comprised mostly of energetic charged particles captured from the stellar wind.

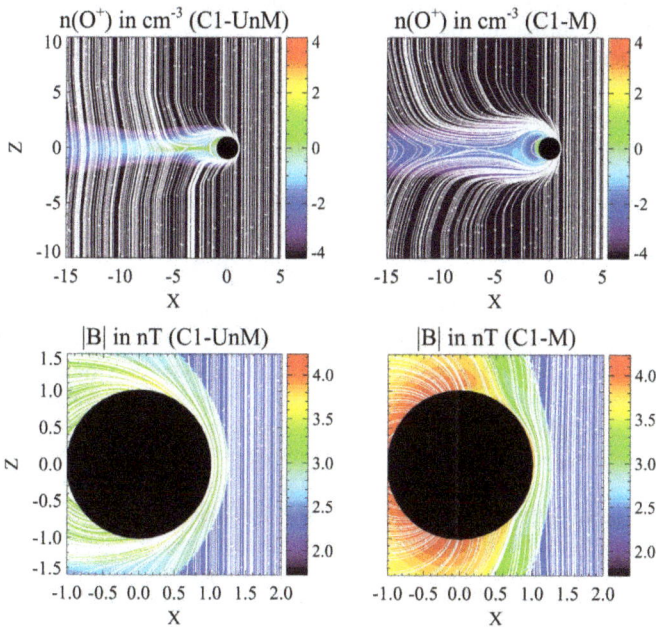

Figure 18. MHD simulation results for an unmagnetized (**left panels**) and magnetized (**right panels**) approximately Venus-like planet orbiting Proxima Centauri b. The assumed magnetic field strength for the results in the right panels is 1/3 Earth's. The star is to the left in all panels. The **top panels** show the atomic oxygen ion density, with **white** magnetic field **lines**. The **bottom panels** show close-ups of the magnetic field strength near the planet. From Dong et al. (2017) reproduced by permission of the AAS.

If a planet has a magnetic field, then it may have a radiation belt. In some cases, this may be how an exoplanet's magnetic field is measured (see the *Radiation belt emission* subsection above).

Precipitating particles can also influence the chemistry of planetary atmospheres, affecting the abundance of climatically important trace gases. An example is the production of nitrogen oxides in Earth-like atmospheres by StEP protons with energies of several 10's of MeV (Segura et al. 2010). During stellar flares, production of nitrogen oxides should increase, as it does on Earth (Rodger et al. 2008). This is associated with a corresponding decrease in oxide molecules, including ozone (Segura et al. 2010). One may expect similar effects in the atmospheres of water worlds, since flares at Earth are also associated with the production of hydrogen oxides (Rodger et al. 2008).

The radiation reaching the surface of a rocky exoplanet (and therefore its habitability) can be significantly reduced by the presence of a planetary field due to the deflection of incident precipitating particles. This is not guaranteed, however. The amount of radiation reaching the surface depends upon both the strength of the field and the thickness of the planet's atmosphere. For a sufficiently thick atmosphere (or a sufficiently weak field), the surface radiation will not be significantly influenced by a planetary magnetic field. For example, Molina-Cuberos et al. (2001) calculated that the presence of an Earth-like magnetic field at Mars would not alter the radiation dose at the surface.

An indirect influence of precipitating particles on surface radiation may occur because of the chemistry discussed above. Increase production of nitrogen oxides, causing decreases in ozone abundance, would enable greater fluxes of damaging UVB radiation (Tilley et al. 2019) to reach a planet's surface, influencing habitability.

Lastly, rocky planets without significant atmospheres will experience surface sputtering in the absence of a sufficiently strong magnetic field. Surface sputtering occurs at both the Moon and Mercury today, as well as at asteroids and many outer planet satellites. The interaction between the surface and charged particles can create an observable exosphere as at the Moon or Mercury (Lammer and Bauer 1997; Wurz et al. 2007), alter surface chemistry as at Europa (Ip et al. 1998), and change the color of the surface as at Ganymede (Ip et al. 1997). Ganymede presents an especially interesting case because it possesses a global magnetic field, and the surface sputtering occurs where charged particles are accelerated from Jupiter's magnetosphere along open magnetic field lines to the surface.

Rotation rate and orbit

An exoplanetary magnetic field can influence the motion of the planet itself. It is thought that giant planets should form with high rotation rates, near their 'breakup velocity', as they accumulate angular momentum from the material that accretes to form the planet. The rotation rates of gas giant exoplanets are likely to be reduced by the presence of a magnetic field, if the field formed while a partially ionized gas disk was still present near the planet on which the planet's magnetic field can exert torque, dissipating momentum into the disk. This magnetic braking effect was first proposed for Jupiter and Saturn, which rotate much more slowly than their breakup velocity (Takata and Stevenson 1996). The idea was expanded upon by Batygin (2018), who asserted that the internal energy forming giant planets would both contribute to planetary magnetic field generation and provide luminosity that partially ionizes the disk, allowing efficient and rapid magnetic braking to occur. Observations of planetary mass objects support these ideas (Fig. 19), with typical rotation rates well below the breakup velocity (Bryan et al. 2020; Fig. 20). Ginzburg and Chiang (2020) showed that the timescale for magnetic braking is shorter than the timescale for giant planet contraction, so that giant planets never approach their breakup velocity. In fact, giant planets may increase their rotation somewhat after the surrounding gas disk dissipates and further contraction occurs.

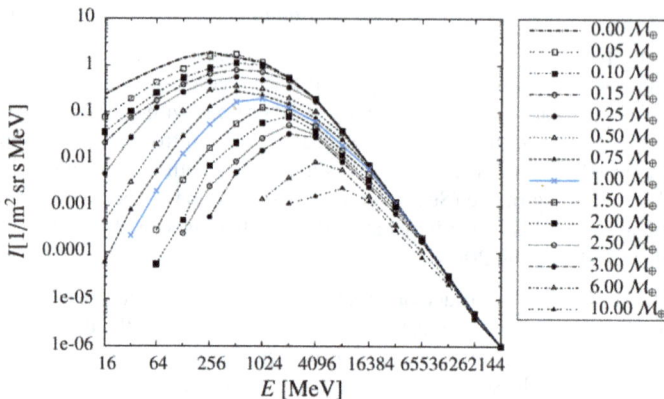

Figure 19. Flux of GCRs incident upon a planet orbiting a Sun-type star as a function of particle energy. Different curves correspond to different planetary dipole field strengths. Credit: Grießmeier et al., Astron Astrophys, **581** (2015) A44, reproduced with permission © ESO.

Magnetic torques can be exerted between a close-in planet's magnetic field and its host star or stellar wind. When the wind is sub-Alfvénic, such planets can exert torques directly on the star. The resulting exchange of angular momentum can change the rotation of the star, causing the planet to migrate inward or outward (depending upon the distance of the planet from the star). In some cases, this effect is sufficiently strong that it dominates tidal effects between the planet and star (Strugarek et al. 2017). This situation is applicable to the kind of systems discussed in the *Star–planet interactions* subsection above: close-in planets magnetically connected to their star.

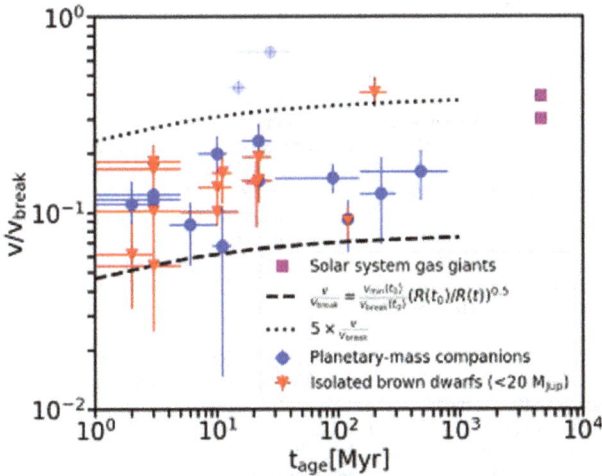

Figure 20. Rotation velocities relative to the break-up velocity as a function of age for a variety of objects, including Jupiter and Saturn, brown dwarfs, and planetary mass objects. Rotation rates are determined from spectral line broadening. From Bryan et al. (2020), reproduced by permission of the AAS.

ACKNOWLEDGMENTS

We thank J.S. Pineda, H. Beltz, E. Schreyer, and R. Murray-Clay for valuable feedback and discussions. The authors led sections as follows: D.A. Brain (Introduction and Influences sections); M.M. Kao (Assessing section); J.G O'Rourke (Theory section). D.A. Brain acknowledges support from NASA grants 80NSSC20K0594 and 80NSSC23K1358. M.M. Kao acknowledges support from the Heising-Simons Foundation through the 51 Pegasi b Fellowship grant 2021-2943.

REFERENCES

Adams FC, Owen JE (2015) Magnetically controlled mass loss from exoplanets. *In:* 18th Cambridge Workshop on Cool Stars, Stellar Systems, and the Sun. Vol 18 of Cambridge Workshop on Cool Stars, Stellar Systems, and the Sun, p 713–722

Airapetian VS, Glocer A, Khazanov GV, Loyd ROP, France K, Sojka J, Danchi WC, Liemohn MW (2017) How hospitable are space weather affected habitable zones? The role of ion escape. Astrophys J Lett 836:L3

Alexander RD, Wynn GA, Mohammed H, Nichols JD, Ercolano B (2016) Magnetospheres of hot Jupiters: Hydrodynamic models and ultraviolet absorption. Mon Not R Astron Soc 456:2766–2778

Allart R, Bourrier V, Lovis C, Ehrenreich D, Spake JJ, Wyttenbach A, Pino L, Pepe F, Sing DK, Lecavelier des Etangs A (2018) Spectrally resolved helium absorption from the extended atmosphere of a warm Neptune-mass exoplanet. Science 362:1384–1387

Allart R, Bourrier V, Lovis C, Ehrenreich D, Aceituno J, Guijarro A, Pepe F, Sing DK, Spake JJ, Wyttenbach A (2019) High-resolution confirmation of an extended helium atmosphere around WASP-107b. Astron Astrophys 623:A58

Allers KN, Vos JM, Biller BA, Williams PKG (2020) A measurement of the wind speed on a brown dwarf. Science 368:169–172

Alonso-Floriano FJ, Snellen IAG, Czesla S, Bauer FF, Salz M, Lampón M, Lara LM, Nagel E, López-Puertas M, Nortmann L, Sánchez-López A, Sanz-Forcada J, Caballero JA, Reiners A, Ribas I, Quirrenbach A, Amado PJ, Aceituno J, Anglada-Escudé G, Béjar VJS, Brinkmöller M, Hatzes AP, Henning T, Kaminski A, Kürster M, Labarga F, Montes D, Pallé E, Schmitt JHMM, Zapatero Osorio MR (2019) He I λ 10 830 Å in the transmission spectrum of HD209458 b. Astron Astrophys 629:A110

Anzellini S, Dewaele A, Mezouar M, Loubeyre P, Morard G (2013) Melting of iron at Earth's inner core boundary based on fast X-ray diffraction. Science 340:464–466

Badro J, Siebert J, Nimm F (2016) An early geodynamo driven by exsolution of mantle components from Earth's core. Nature 536:326–328

Barabash S (2010) Venus, Earth, Mars: Comparative ion escape rates. *In:* EGU General Assembly Conf Abstr, p 5308

Batygin K (2018) On the terminal rotation rates of giant planets. Astronom J 155:178

Batygin K, Stanley S, Stevenson DJ (2013) Magnetically controlled circulation on hot extrasolar planets. Astrophys J 776:53

Batygin K, Stevenson, DJ (2010) Inflating hot Jupiters with ohmic dissipation. Astrophys J Lett 714:L238–L243

Batygin K, Stevenson DJ, Bodenheimer PH (2011) Evolution of ohmically heated hot Jupiters. Astrophys J 738:1

Beltz H, Rauscher E, Kempton EMR, Malsky I, Ochs G, Arora M, Savel A (2022a) Magnetic drag and 3D effects in theoretical high-resolution emission spectra of ultrahot Jupiters: The case of WASP-76b. Astronom J 164:140

Beltz H, Rauscher E, Roman, MT, Guilliat A (2022b) Exploring the effects of active magnetic drag in a general circulation model of the ultrahot Jupiter WASP-76b. Astronom J 163:35

Beltz H, Rauscher E, Kempton EMR, Malsky I, Savel AB (2023) Magnetic effects and 3D structure in theoretical high-resolution transmission spectra of ultrahot Jupiters: The case of WASP-76b. Astronom J 165:257

Ben-Jaffel L, Ballester GE (2013) Hubble Space Telescope detection of oxygen in the atmosphere of exoplanet HD 89733b. Astron Astrophys 553:A52

Ben-Jaffel L, Ballester GE, Garćia Muñoz A, Lavvas P, Sing, DK, Sanz-Forcada J, Cohen O, Kataria T, Henry GW, Buchhave L, Mikal-Evans T, Wakeford HR, López-Morales M (2022) Signatures of strong magnetization and a metal-poor atmosphere for a Neptune-sized exoplanet. Nat Astron 6:141–153

Bisikalo D, Kaygorodov P, Ionov D, Shematovich V, Lammer H, Fossati L (2013) Three-dimensional gas dynamic simulation of the interaction between the exoplanet WASP-12b and its host star. Astrophys J 764:19

Blackman EG, Tarduno JA (2018) Mass, energy, and momentum capture from stellar winds by magnetized and unmagnetized planets: implications for atmospheric erosion and habitability. Mon Not R Astron Soc 481:5146–5155

Blanc NA, Stegman DR, Ziegler LB (2020) Thermal and magnetic evolution of a crystallizing basal magma ocean in Earth's mantle. Earth Planet Sci Lett 534:116085

Blanco-Pozo J, Perger M, Damasso M, Anglada Escudé G, Ribas I, Baroch D, Caballero JA, Cifuentes C, Jeffers SV, Lafarga M, Kaminski A, Kaur S, Nagel E, Perdelwitz V, Pérez-Torres M, Sozzetti A, Viganò D, Amado PJ, Andreuzzi G, Béjar VJS, Brown EL, Del Sordo F, Dreizler S, Galadí-Enríquez D, Hatzes AP, Kürster M, Lanza AF, Melis A, Molinari E, Montes D, Murgia M, Pallé E, Peña-Moñino L, Perrodin D, Pilia M, Poretti E, Quirrenbach A, Reiners A, Schweitzer A, Zapatero Osorio MR, Zechmeister M (2023) The CARMENES search for exoplanets around M dwarfs. A long-period planet around GJ 1151 measured with CARMENES and HARPS-N data. Astron Astrophys 671:A50

Blaske CH, O'Rourke JG (2021) Energetic requirements for dynamos in the metallic cores of super-Earth and super-Venus exoplanets. J Geophys Res Planets 126:e2020JE006739

Bloxham J, Jackson A (1991) Fluid flow near the surface of Earth's outer core. Rev Geophys 29:97–120

Bolton SJ, Janssen M, Thorne R, Levin S, Klein M, Gulkis S, Bastian T, Sault R, Elachi C, Hofstadter M, Bunker A, Dulk G, Gudim E, Hamilton G, Johnson WTK, Leblanc Y, Liepack O, McLeod R, Roller J, Roth L, West R (2002) Ultra-relativistic electrons in Jupiter's radiation belts. Nature 415:987–991

Bolton SJ, Thorne RM, Bourdarie S, de Pater I, Mauk B (2004) Jupiter's inner radiation belts. *In:* Bagenal F, Dowling TE, McKinnon WB (eds) Jupiter. The Planet Satellites and Magnetosphere, vol 1, p 671–688

Borlina CS, Weiss BP, Lima EA, Tang F, Taylor RJM, Einsle JF, Harrison RJ, Fu, RR, Bell EA, Alexander EW, Kirkpatrick HM, Wielicki MM, Harrison TM, Ramezani J, Maloof AC (2020) Reevaluating the evidence for a Hadean–Eoarchean dynamo. Sci Adv 6:eaav9634

Boujibar A, Driscoll P, Fei Y (2020) Super-Earth internal structures and initial thermal states. J Geophys Res Planets 125:e2019JE006124

Bourrier V, Lecavelier des Etangs A, Dupuy H, Ehrenreich D, Vidal-Madjar A, Hébrard G, Ballester GE, Désert JM, Ferlet R, Sing DK, Wheatley PJ (2013) Atmospheric escape from HD 189733b observed in H I Lyman-α: detailed analysis of HST/STIS September 2011 observations. Astron Astrophys 551:A63

Brain DA (2021) Induced magnetospheres: Atmospheric escape. *In:* Maggiolo R, André N, Hasegawa H, Welling DT (eds) Magnetospheres in the Solar System, vol 2, p 441

Brain DA, Leblanc F, Luhmann JG, Moore TE, Tian F (2013) Planetary magnetic fields and climate evolution. *In:* Mackwell SJ, Simon-Miller AA, Harder JW, Bullock MA (eds) Comparative Climatology of Terrestrial Planets, p 487–503

Brain DA, Bagenal F, Ma, YJ, Nilsson H, Stenberg-Wieser G (2016) Atmospheric escape from unmagnetized bodies. J Geophys Res Planets 121:2364–2385

Breuer D, Rueckriemen T, Spohn T (2015) Iron snow, crystal floats, and inner-core growth: modes of core solidification and implications for dynamos in terrestrial planets and moons. Prog Earth Planet Sci 2:39

Bryan ML, Ginzburg S, Chiang E, Morley C, Bowler BP, Xuan JW, Knutson HA (2020) As the worlds turn: Constraining spin evolution in the planetary-mass regime. Astrophys J 905:37

Burke BF, Franklin KL (1955) Observations of a variable radio source associated with the planet Jupiter. J Geophys Res 60:213–217

Burns JO, Hallinan G, Lux J, Teitelbaum L, Kocz J, MacDowall R, Bradley R, Rapetti D, Wu, W, Furlanetto S, Austin A, Romero-Wolf A, Chang T-C, Bowman J, Kasper J, Anderson M, Zhen Z, Pober J, Mirocha J (2019) NASA probe study report: Farside Array for Radio Science Investigations of the Dark ages and Exoplanets (FARSIDE). arXiv:1911.08649

Cauley PW, Redfield S, Jensen AG, Barman T, Endl M, Cochran WD (2015) Optical hydrogen absorption consistent with a thin bow shock leading the hot Jupiter HD 189733b. Astrophys J 810:13

Cauley PW, Redfield S, Jensen AG, Barman T (2016) Variation in the pre-transit Balmer line signal around the hot Jupiter HD 189733b. Astronom J 152:20

Cauley PW, Shkolnik EL, Llama J, Bourrier V, Moutou C (2018) Evidence of magnetic star–planet interactions in the HD 189733 system from orbitally phased Ca II K variations. Astronom J 156:262

Cauley PW, Shkolnik EL, Llama J, Lanza AF (2019) Magnetic field strengths of hot Jupiters from signals of star–planet interactions. Nat Astron 3:1128–1134

Christensen UR (2010) Dynamo scaling laws and applications to the planets. Space Sci Rev 152:565–590

Christensen UR, Holzwarth V, Reiners A (2009) Energy flux determines magnetic field strength of planets and stars. Nature 457:167–169

Christensen UR, Aubert J, Hulot G (2010) Conditions for Earth-like geodynamo models. Earth Planet Sci Lett 296:487–496

Clarke JT, Ajello J, Ballester G, Ben Jaffel L, Connerney J, Gérard JC, Gladstone GR, Grodent D, Pryor W, Trauger J, Waite JH (2002) Ultraviolet emissions from the magnetic footprints of Io, Ganymede and Europa on Jupiter. Nature 415:997–1000

Clarke JT, Wannawichian S, Hernandez N, Bonfond B, Gerard JC, Grodent D (2011) Detection of auroral emissions from Callisto's magnetic footprint at Jupiter. *In:* EPSC-DPS Joint Meeting 2011, p 1468

Climent JB, Guirado JC, Pérez-Torres M, Marcaide JM, Peña-Moñino L (2023) Evidence for a radiation belt around a brown dwarf. Science 381:1120–1124

Cohen O, Drake JJ, Glocer A, Garraffo C, Poppenhaeger K, Bell JM, Ridley Astronom J, Gombosi TI (2014) Magnetospheric structure and atmospheric Joule heating of habitable planets orbiting M-dwarf stars. Astrophys J 790:57

Cohen IJ, Beddingfield C, Chancia R, DiBraccio G, Hedman M, MacKenzie S, Mauk B, Sayanagi KM, Soderlund KM, Turtle E, Ahrens C, Arridge CS, Brooks SM, Bunce E, Charnoz S, Coustenis A, Dillman RA, Dutta S, Fletcher LN, Harbison R, Helled R, Holme R, Jozwiak L, Kasaba Y, Kollmann P, Luszcz-Cook S, Mandt K, Mousis O, Mura A, Murakami G, Parisi M, Rymer A, Stanley S, Stephan K, Ronald JVervack J, Wong MH, Wurz P (2022) The case for a new Frontiers-class Uranus orbiter: System science at an underexplored and unique world with a mid-scale mission. Planet Sci J 3:58

Coltice N, Moreira M, Hernlund J, Labrosse S (2011) Crystallization of a basal magma ocean recorded by helium and neon. Earth Planet Sci Lett 308:193–199

Connerney JEP, Acuna MH, Ness NF (1991) The magnetic field of Neptune. J Geophys Res 96:19023–19042

Connerney JEP, Timmins S, Oliversen RJ, Espley JR, Joergensen JL, Kotsiaros S, Joergensen PS, Merayo JMG, Herceg M, Bloxham J, Moore KM, Mura A, Moirano A, Bolton SJ, Levin SM (2022) A new model of Jupiter's magnetic field at the completion of Juno's prime mission. J Geophys Res Planets 127:e2021JE007055

Coulombe L-P, Benneke B, Challener R, Piette AAA, Wiser LS, Mansfield M, MacDonald RJ, Beltz H, Feinstein AD, Radica M, Savel AB, Dos Santos LA, Bean JL, Parmentier V, Wong I, Rauscher E, Komacek TD, Kempton EMR, Tan X, Hammond M, Lewis NT, Line MR, Lee EKH, Shivkumar H, Crossfield IJM, Nixon MC, Rackham BV, Wakeford HR, Welbanks L, Zhang X, Batalha NM, Berta-Thompson ZK, Changeat Q, Désert J-M, Espinoza N, Goyal JM, Harrington J, Knutson HA, Kreidberg L, López-Morales M, Shporer A, Sing DK, Stevenson KB, Aggarwal K, Ahrer E-M, Alam MK, Bell TJ, Blecic J, Caceres C, Carter AL, Casewell SL, Crouzet N, Cubillos PE, Decin L, Fortney JJ, Gibson NP, Heng K, Henning T, Iro N, Kendrew S, Lagage P-O, Leconte J, Lendl M, Lothringer JD, Mancini L, Mikal-Evans T, Molaverdikhani K, Nikolov NK, Ohno K, Pallé E, Piaulet C, Redfield S, Roy P-A, Tsai S-M, Venot O, Wheatley PJ (2023) A broadband thermal emission spectrum of the ultra-hot Jupiter WASP-18b. Nature 620:292–298

Cowley SWH, Bunce EJ (2001) Origin of the main auroral oval in Jupiter's coupled magnetosphere–ionosphere system. Planet Space Sci 49:1067–1088

Cuntz M, Saar SH, Musielak ZE (2000) On stellar activity enhancement due to interactions with extrasolar giant planets. Astrophys J Lett 533:L151–L154

de Kleer K, Milby Z, Schmidt C, Camarca M, Brown ME (2023) The optical aurorae of Europa, Ganymede, and Callisto. Planet Sci J 4:37

Dehant V, Lammer H, Kulikov YN, Grießmeier JM, Breuer D, Verhoeven O, Karatekin Ö, van Hoolst T, Korablev O, Lognonné P (2007) Planetary magnetic dynamo effect on atmospheric protection of early Earth and Mars. Space Sci Rev 129:279–300

Del Genio AD, Brain D, Noack L, Schaefer L (2020) The inner Solar System's habitability through time. *In:* Meadows VS, Arney GN, Schmidt BE, Des Marais DJ (eds) Planetary Astrobiology, p 419

Dong C, Lingam M, Ma Y, Cohen O (2017) Is Proxima Centauri b habitable? A study of atmospheric loss. Astrophys J Lett 837:L26

Dong C, Jin M, Lingam M (2020) Atmospheric escape from TOI-700 d: Venus versus Earth Analogs. Astrophys J Lett 896:L24

Driscoll, PE (2016) Simulating 2 Ga of geodynamo history. Geophys Res Lett 43:5680–5687

Driscoll P, Davies C (2023) The "New Core Paradox": Challenges and potential solutions. J Geophys Res Solid Earth 128:e2022JB025355

Dulk GA (1985) Radio emission from the sun and stars. Annu Rev Astron Astrophys 23:169–224

Egan H, Jarvinen R, Ma, Y, Brain D (2019) Planetary magnetic field control of ion escape from weakly magnetized planets. Mon Not R Astron Soc 488:2108– 2120

Erkaev NV, Kulikov YN, Lammer H, Selsis F, Langmayr D, Jaritz GF, Biernat HK (2007) Roche lobe effects on the atmospheric loss from "Hot Jupiters". Astron Astrophys 472:329–334

Erkaev NV, Odert P, Lammer H, Kislyakova KG, Fossati L, Mezentsev AV, Johnstone CP, Kubyshkina DI, Shaikhislamov IF, Khodachenko ML (2017) Effect of stellar wind induced magnetic fields on planetary obstacles of non-magnetized hot Jupiters. Mon Not R Astron Soc 470:4330– 4336

Ferrick AL, Korenaga J (2023) Defining Earth's elusive thermal budget in the presence of a hidden reservoir. Earth Planet Sci Lett 601:117893

Fortney JJ, Dawson RI, Komacek TD (2021) Hot Jupiters: Origins structure atmospheres. J Geophys Res Planets 126:e06629

Fossati L, Haswell CA, Froning CS, Hebb L, Holmes S, Kolb U, Helling C, Carter A, Wheatley P, Collier Cameron A, Loeillet B, Pollacco D, Street R, Stempels HC, Simpson E, Udry S, Joshi YC, West RG, Skillen I, Wilson D (2010) Metals in the exosphere of the highly irradiated planet WASP-12b. Astrophys J Lett 714:L222–L227

Fulton BJ, Petigura EA, Howard AW, Isaacson H, Marcy GW, Cargile PA, Hebb L, Weiss LM, Johnson JA, Morton TD, Sinukoff E, Crossfield IJM, Hirsch LA (2017) The California-Kepler Survey. IIIA Gap in the radius distribution of small planets. Astronom J 154:109

Ginzburg S, Chiang E (2020) Breaking the centrifugal barrier to giant planet contraction by magnetic disc braking. Mon Not R Astron Soc 491:L34–L39

Ginzburg S, Sari R (2016) Extended heat deposition in hot Jupiters: Application to ohmic heating. Astrophys J 819:116

Grießmeier JM, Zarka P, Spreeuw H (2007) Predicting low-frequency radio fluxes of known extrasolar planets. Astron Astrophys 475:359–368

Grießmeier JM, Tabataba-Vakili F, Stadelmann A, Grenfell JL, Atri D (2015) Galactic cosmic rays on extrasolar Earth-like planets. I Cosmic ray flux. Astron Astrophys 581:A44

Gronoff G, Arras P, Baraka S, Bell JM, Cessateur G, Cohen O, Curry SM, Drake JJ, Elrod M, Erwin J, Garcia-Sage K, Garraffo C, Glocer A, Heavens NG, Lovato K, Maggiolo R, Parkinson CD, Simon Wedlund C, Weimer DR, Moore WB (2020) Atmospheric escape processes and planetary atmospheric evolution. J Geophys Res Space Phys 125:e27639

Gunell H, Maggiolo R, Nilsson H, Stenberg Wieser G, Slapak R, Lindkvist J, Hamrin M, De Keyser J (2018) Why an intrinsic magnetic field does not protect a planet against atmospheric escape. Astron Astrophys 614:L3

Gurdemir L, Redfield S, Cuntz M (2012) Planet-induced emission enhancements in HD 179949: Results from McDonald observations. Publ Astron Soc Aust 29:141–149

Hallinan G, Bourke S, Lane C, Antonova A, Zavala RT, Brisken WF, Boyle RP, Vrba FJ, Doyle JG, Golden A (2007) Periodic bursts of coherent radio emission from an ultracool dwarf. Astrophys J Lett 663:L25–L28

Hallinan G, Antonova A, Doyle JG, Bourke S, Lane C, Golden A (2008) Confirmation of the electron cyclotron maser instability as the dominant source of radio emission from very low mass stars and brown dwarfs. Astrophys J 684:644–653

Hallinan G, Littlefair SP, Cotter G, Bourke S, Harding LK, Pineda JS, Butler RP, Golden A, Basri G, Doyle JG, Kao MM, Berdyugina SV, Kuznetsov A, Rupen MP, Antonova A (2015) Magnetospherically driven optical and radio aurorae at the end of the stellar main sequence. Nature 523:568–571

Hamid SS, O'Rourke JG, Soderlund KM (2023) A long-lived lunar magnetic field powered by convection in the core and a basal magma ocean. Planet Sci J 4:88

Haswell CA, Fossati L, Ayres T, France K, Froning CS, Holmes S, Kolb UC, Busuttil R, Street RA, Hebb L, Collier Cameron A, Enoch B, Burwitz V, Rodriguez J, West RG, Pollacco D, Wheatley PJ, Carter A (2012) Near-ultraviolet absorption chromospheric activity, and star–planet interactions in the WASP-12 system. Astrophys J 760:79

Hess S, Cecconi B, Zarka P (2008) Modeling of Io–Jupiter decameter arcs, emission beaming and energy source. Geophys Res Lett 35:L13107

Hirose K, Morard G, Sinmyo R, Umemoto K, Hernlund J, Helffrich G, Labrosse S (2017) Crystallization of silicon dioxide and compositional evolution of the Earth's core. Nature 543:99–102

Hutchins KS, Jakosky BM, Luhmann JG (1997) Impact of a paleomagnetic field on sputtering loss of Martian atmospheric argon and neon. J Geophys Res 102:9183–9190

Ip WH, Williams DJ, McEntire RW, Mauk B (1997) Energetic ion sputtering effects at Ganymede. Geophys Res Lett 24:2631–2634

Ip WH, Williams DJ, McEntire RW, Mauk BH (1998) Ion sputtering and surface erosion at Europa. Geophys Res Lett 25:829–832

Jacobson SA, Rubie DC, Hernlund J, Morbidelli A, Nakajima M (2017) Formation, stratification, and mixing of the cores of Earth and Venus. Earth Planet Sci Lett 474:375–386

Kao MM, Pineda JS (2022) Radio emission from binary ultracool dwarf systems. Astrophys J 932:21

Kao MM, Hallinan G, Pineda JS, Escala I, Burgasser A, Bourke S, Stevenson D (2016) Auroral radio emission from late L and T dwarfs: A New constraint on dynamo theory in the substellar regime. Astrophys J 818:24

Kao MM, Hallinan G, Pineda JS, Stevenson D, Burgasser A (2018) The strongest magnetic fields on the coolest brown dwarfs. Astrophys JS 237:25

Kao MM, Hallinan G, Pineda JS (2019) Constraints on magnetospheric radio emission from Y dwarfs. Mon Not R Astron Soc 487:1994–2004

Kao MM, Mioduszewski Astronom J, Villadsen J, Shkolnik EL (2023) Resolved imaging confirms a radiation belt around an ultracool dwarf. Nature 619:272–275

Kao MM, Shkolnik EL (2024) The occurrence rate of quiescent radio emission for ultracool dwarfs using a generalized semi-analytical Bayesian framework. Mon Not R Astron Soc 527:6835–6866

Kavanagh RD, Vedantham HK (2023) Hunting for exoplanets via magnetic star–planet interactions: geometrical considerations for radio emission. Mon Not R Astron Soc 524:6267–6284

Kennedy GC, Higgins GH (1973) The core paradox. J Geophys Res (18961977), 78:900–904

Khan A, Huang D, Durán C, Sossi PA, Giardini D, Murakami M (2023) Evidence for a liquid silicate layer atop the Martian core. Nature 622:718–723

Khodachenko ML, Shaikhislamov IF, Lammer H, Prokopov PA (2015) Atmosphere expansion and mass loss of close-orbit giant exoplanets heated by stellar XUVII effects of planetary magnetic field; structuring of inner magnetosphere. Astrophys J 813:50

Kislyakova KG, Holmström M, Lammer H, Odert P, Khodachenko ML (2014) Magnetic moment and plasma environment of HD 209458b as determined from Lyα observations. Science 346:981–984

Kivelson MG, Bagenal F (2007) Planetary magnetospheres. *In:* McFadden LAA, Weissman PR, Johnson TV(eds) Encyclopedia of the Solar System, p 519–540

Knapp M, Paritsky L, Kononov E, Kao M (2024) NASA NIAC Report: Great Observatory at Long Wavelengths (GO-LoW). arXiv:2404.08432

Kollmann P, Roussos E, Paranicas C, Woodfield EE, Mauk BH, Clark G, Smith DC, Vandegriff J (2018) Electron acceleration to MeV energies at Jupiter and Saturn. J Geophys Res Space Phys 123:9110–9129

Kuang W, Bloxham J (1997) An Earth-like numerical dynamo model. Nature 389:371–374

Kulikov YN, Lammer H, Lichtenegger HIM, Penz T, Breuer D, Spohn T, Lundin R, Biernat HK (2007) A comparative study of the influence of the active young Sun on the early atmospheres of Earth Venus, and Mars. Space Sci Rev 129:207–243

Labrosse S (2015) Thermal evolution of the core with a high thermal conductivity. Phys Earth Planet Sci 247:36–55

Labrosse S, Hernlund JW, Coltice N (2007) A crystallizing dense magma ocean at the base of the Earth's mantle. Nature 450:866–869

Lai D, Helling C, van den Heuvel EPJ (2010) Mass transfer transiting stream, and magnetopause in close-in exoplanetary systems with applications to WASP-12. Astrophys J 721:923–928

Lammer H, Bauer SJ (1997) Mercury's exosphere: origin of surface sputtering and implications. Planet Space Sci 45:73–79

Landeau M, Aubert J, Olson P (2017) The signature of inner-core nucleation on the geodynamo. Earth Planet Sci Lett 465:193– 204

Landeau M, Fournier A, Nataf H-C, Cébron D, Schaeffer N (2022) Sustaining Earth's magnetic dynamo. Nat Rev Earth Environ 3:255–269

Laneuville M, Dong C, O'Rourke JG, Schneider AC (2020) Magnetic fields on rocky planets. *In:* Planetary Diversity: Rocky planet processes and their observational signatures. EJ Tasker, C Unterborn, M Laneuville, Y Fujii, SJ Desch, HE Hartnett (eds) IOP Publishing

Lanza AF (2009) Stellar coronal magnetic fields and star–planet interaction. Astron Astrophys 505:339–350

Lanza AF (2012) Star–planet magnetic interaction and activity in late-type stars with close-in planets. Astron Astrophys 544:A23

Lanza AF (2013) Star–planet magnetic interaction and evaporation of planetary atmospheres. Astron Astrophys 557:A31

Lay T, Hernlund J, Buffett BA (2008) Core–mantle boundary heat flow. Nat Geosci 1:25–32

Llama J, Wood K, Jardine M, Vidotto AA, Helling C, Fossati L, Haswell CA (2011) The shocking transit of WASP-12b: modelling the observed early ingress in the near-ultraviolet. Mon Not R Astron Soc 416:L41–L44

Llama J, Vidotto AA, Jardine M, Wood K, Fares R, Gombosi TI (2013) Exoplanet transit variability: bow shocks and winds around HD 189733b. Mon Not R Astron Soc 436:2179–2187

Louis CK, Hess SLG, Cecconi B, Zarka P, Lamy L, Aicardi S, Loh A (2019) ExPRES: An Exoplanetary and Planetary Radio Emissions Simulator. Astron Astrophys 627:A30

Lundin R, Lammer H, Ribas I (2007) Planetary magnetic fields and solar forcing: Implications for atmospheric evolution. Space Sci Rev 129:245–27

Lynch C, Mutel RL, Güdel M (2015) Wideband dynamic radio spectra of two ultra-cool dwarfs. Astrophys J 802:106

Maggiolo R, Gunell H (2021) Does a magnetosphere protect the ionosphere? *In:* Maggiolo R, André N, Hasegawa H, Welling DT (eds) Magnetospheres in the Solar System, vol 2, p 729

Matsui H, Heien E, Aubert J, Aurnou JM, Avery M, Brown B, Buffett BA, Busse F, Christensen UR, Davies CJ, Featherstone N, Gastine T, Glatzmaier GA, Gubbins D, Guermond J-L, Hayashi Y-Y, Hollerbach R, Hwang LJ, Jackson A, Jones CA, Jiang W, Kellogg LH, Kuang W, Landeau M, Marti P, Olson P, Ribeiro A, Sasaki Y, Schaeffer N, Simitev RD, Sheyko A, Silva L, Stanley S, Takahashi F, Takehiro S-i, Wicht J, Willis AP (2016) Performance benchmarks for a next generation numerical dynamo model. Geochem Geophys Geosyst 17:1586–1607

Mauk BH, Fox NJ (2010) Electron radiation belts of the solar system. J Geophys Res Space Phys 115:A12220

Meert JG (2009) In GAD we trust. Nat Geosci 2:673–674

Molina-Cuberos GJ, Stumptner W, Lammer H, Kömle NI, O'Brien K (2001) Cosmic ray and UV radiation models on the ancient Martian surface. Icarus 154:216–222

Moore TE, Horwitz JL (2007) Stellar ablation of planetary atmospheres. Rev Geophys 45:RG3002

Moore KM, Yadav RK, Kulowski L, Cao H, Bloxham J, Connerney JEP, Kotsiaros S, Jørgensen JL, Merayo JMG, Stevenson DJ, Bolton SJ, Levin SM (2018) A complex dynamo inferred from the hemispheric dichotomy of Jupiter's magnetic field. Nature 561:76–78

Mura A, Adriani A, Altieri F, Connerney JEP, Bolton SJ, Moriconi ML, Gérard JC, Kurth WS, Dinelli BM, Fabiano F, Tosi F, Atreya SK, Bagenal F, Gladstone GR, Hansen C, Levin SM, Mauk BH, McComas DJ, Sindoni G, Filacchione G, Migliorini A, Grassi D, Piccioni G, Noschese R, Cicchetti A, Turrini D, Stefani S, Amoroso M, Olivieri A (2017) Infrared observations of Jovian aurora from Juno's first orbits: Main oval and satellite footprints. Geophys Res Lett 44:5308–5316

Ness NF, Acuna MH, Behannon KW, Burlaga LF, Connerney JEP, Lepping RP, Neubauer FM (1986) Magnetic fields at Uranus. Science 233:85–88

Nichols JD, Cowley SWH (2003) Magnetosphere–ionosphere coupling currents in Jupiter's middle magnetosphere: dependence on the effective ionospheric Pedersen conductivity and iogenic plasma mass outflow rate. Ann Geophys 21:1419–1441

Nichols JD, Burleigh MR, Casewell SL, Cowley SWH, Wynn GA, Clarke JT, West AA (2012) Origin of electron cyclotron maser induced radio emissions at ultracool dwarfs: Magnetosphere–ionosphere coupling currents. Astrophys J 760:59

Nimmo F, Primack J, Faber SM, Ramirez-Ruiz E, Safarzadeh M (2020) Radiogenic heating and its influence on rocky planet dynamos and habitability. Astrophys J Lett 903:L37

Nortmann L, Pallé E, Salz M, Sanz-Forcada J, Nagel E, Alonso-Floriano FJ, Czesla S, Yan F, Chen G, Snellen IAG, Zechmeister M, Schmitt JHMM, López-Puertas M, Casasayas-Barris N, Bauer FF, Amado PJ, Caballero JA, Dreizler S, Henning T, Lampón M, Montes D, Molaverdikhani K, Quirrenbach A, Reiners A, Ribas I, Sánchez-López A, Schneider PC, Zapatero Osorio MR (2018) Ground-based detection of an extended helium atmosphere in the Saturn-mass exoplanet WASP-69b. Science 362:1388–1391

Oklopčić A (2019) Helium absorption at 1083 nm from extended exoplanet atmospheres: Dependence on stellar radiation. Astrophys J 881:133

Oklopčić A, Hirata CM (2018) A new window into escaping exoplanet atmospheres: 10830 Å line of helium. Astrophys J Lett 855:L11

Oklopčić A, Silva M, Montero-Camacho P, Hirata CM (2020) Detecting magnetic fields in exoplanets with spectropolarimetry of the helium line at 1083 nm. Astrophys J 890:88

Olson P (2013) The new core paradox. Science 342:431–432

O'Rourke JG (2020) Venus: A thick basal magma ocean may exist today. Geophys Res Lett 47:e2019GL086126

O'Rourke JG, Stevenson DJ (2016) Powering Earth's dynamo with magnesium precipitation from the core. Nature 529:387–389

Owen JE (2019) Atmospheric escape and the evolution of close-in exoplanets. Annu Rev Earth Planet Sci 47:67–90

Owen JE, Adams FC (2019) Effects of magnetic fields on the location of the evaporation valley for low-mass exoplanets. Mon Not R Astron Soc 490:15–20

Pachhai S, Li, M, Thorne MS, Dettmer J, Tkalčić H (2022) Internal structure of ultralow-velocity zones consistent with origin from a basal magma ocean. Nat Geosci 15:79–84

Pineda JS, Villadsen J (2023) Coherent radio bursts from known M-dwarf planet-host YZ Ceti. Nat Astron 7:569–578

Pineda JS, Hallinan G, Kao MM (2017) A panchromatic view of brown dwarf aurorae. Astrophys J 846:75

Pope BJS, Bedell M, Callingham JR, Vedantham HK, Snellen IAG, Price-Whelan AM, Shimwell TW (2020) No massive companion to the coherent radio-emitting M dwarf GJ 1151. Astrophys J Lett, 890:L1

Ramstad R, Barabash S (2021) Do intrinsic magnetic fields protect planetary atmospheres from stellar winds? Space Sci Rev 217:36

Rauscher E, Menou K (2013) Three-dimensional atmospheric circulation models of HD 189733b and HD 209458b with consistent magnetic drag and ohmic dissipation. Astrophys J 764:103

Richey-Yowell T, Kao MM, Pineda JS, Shkolnik EL, Hallinan G (2020) On the correlation between L dwarf optical and infrared variability and radio aurorae. Astrophys J 903:74

Rodger CJ, Verronen PT, Clilverd MA, Seppälä, A, Turunen E (2008) Atmospheric impact of the Carrington event solar protons. J Geophys Res Atmos 113(D23):D23302

Route M, Wolszczan A (2012) The Arecibo detection of the coolest radio-flaring brown dwarf. Astrophys J Lett 747:L22

Rybicki GB, Lightman AP (1986) Radiative Processes in Astrophysics. Wiley-VCH Verlag GmbH & Co. KGaA

Sakata R, Seki K, Sakai S, Terada N, Shinagawa H, Tanaka T (2020) Effects of an intrinsic magnetic field on ion loss from ancient Mars based on multispecies MHD simulations. J Geophys Res Space Phys 125:e26945

Sakata R, Seki K, Sakai S, Terada N, Shinagawa H, Tanaka T (2022) Multispecies MHD study of ion escape at ancient Mars: Effects of an intrinsic magnetic field and solar XUV radiation. J Geophys Res Space Phys 127:e30427

Salz M, Czesla S, Schneider PC, Nagel E, Schmitt JHMM, Nortmann L, Alonso-Floriano FJ, López-Puertas M, Lampón M, Bauer FF, Snellen IAG, Pallé E, Caballero JA, Yan F, Chen G, Sanz-Forcada J, Amado PJ, Quirrenbach A, Ribas I, Reiners A, Béjar VJS, Casasayas-Barris N, Cortés-Contreras M, Dreizler S, Guenther EW, Henning T, Jeffers SV, Kaminski A, Kürster M, Lafarga M, Lara LM, Molaverdikhani K, Montes D, Morales JC, Sánchez-López A, Seifert W, Zapatero Osorio MR, Zechmeister M (2018) Detection of He I λ10830 Å absorption on HD 189733 b with CARMENES high-resolution transmission spectroscopy. Astron Astrophys 620:A97

Samuel H, Drilleau M, Rivoldini A, Xu, Z, Huang Q, Garcia RF, Lekíc V, Irving JCE, Badro J, Lognonné PH, Connolly JAD, Kawamura T, Gudkova T, Banerdt WB (2023) Geophysical evidence for an enriched molten silicate layer above Mars's core. Nature 622:712–717

Saur J, Grambusch T, Duling S, Neubauer FM, Simon S (2013) Magnetic energy fluxes in sub-Alfvénic planet star and moon planet interactions. Astron Astrophys 552:A119

Saur J, Willmes C, Fischer C, Wennmacher A, Roth L, Youngblood A, Strobel DF, Reiners A (2021) Brown dwarfs as ideal candidates for detecting UV aurora outside the Solar System: Hubble Space Telescope observations of 2MASS J1237+6526. Astron Astrophys 655:A7

Scheinberg AL, Soderlund KM, Elkins-Tanton LT (2018) A basal magma ocean dynamo to explain the early lunar magnetic field. Earth Planet Sci Lett 492:144–151

Schneider NM, Deighan JI, Jain SK, Stiepen A, Stewart AIF, Larson D, Mitchell DL, Mazelle C, Lee CO, Lillis RJ, Evans JS, Brain D, Stevens MH, McClintock WE, Chaffin MS, Crismani M, Holsclaw GM, Lefevre F, Lo DY, Clarke JT, Montmessin F, Jakosky BM (2015) Discovery of diffuse aurora on Mars. Science 350:0313

Schreyer E, Owen JE, Spake JJ, Bahroloom Z, Di Giampasquale S (2024) Using helium 10 830 Å transits to constrain planetary magnetic fields. Mon Not R Astron Soc 527:5117–5130

Schubert G, Soderlund KM (2011) Planetary magnetic fields: Observations and models. Phys Earth Planet Sci 187:92–108

Segura A, Walkowicz LM, Meadows V, Kasting J, Hawley S (2010) The effect of a strong stellar flare on the atmospheric chemistry of an Earth-like planet orbiting an M dwarf. Astrobiology 10:751–771

Seki K, Elphic RC, Hirahara M, Terasawa T, Mukai T (2001) On atmospheric loss of oxygen ions from Earth through magnetospheric processes. Science 291:1939–1941

Shkolnik E, Walker GAH, Bohlender DA (2003) Evidence for planet-induced chromospheric activity on HD 179949. Astrophys J 597:1092–1096

Shkolnik E, Walker GAH, Bohlender DA, Gu, PG, Kürster M (2005) Hot Jupiters and hot spots: The short- and long-term chromospheric activity on stars with giant planets. Astrophys J 622:1075–1090

Shkolnik E, Bohlender DA, Walker GAH, Collier Cameron A (2008) The on/off nature of star–planet interactions. Astrophys J 676:628–638

Slapak R, Schillings A, Nilsson H, Yamauchi M, Westerberg L-G, Dandouras I (2017) Atmospheric loss from the dayside open polar region and its dependence on geomagnetic activity: implications for atmospheric escape on evolutionary timescales. Ann Geophys 35:721–731

Soderlund KM, Stanley S (2020) The underexplored frontier of ice giant dynamos. Philos Trans R Soc A: 378:20190479

Soubiran F, Militzer B (2018) Electrical conductivity and magnetic dynamos in magma oceans of super-Earths. Nat Commun 9:3883

Spake JJ, Sing DK, Evans TM, Oklopčić, , A, Bourrier V, Kreidberg L, Rackham BV, Irwin J, Ehrenreich D, Wyttenbach A, Wakeford HR, Zhou Y, Chubb KL, Nikolov N, Goyal JM, Henry GW, Williamson MH, Blumenthal S, Anderson DR, Hellier C, Charbonneau D, Udry S, Madhusudhan N (2018) Helium in the eroding atmosphere of an exoplanet. Nature 557:68–70

Stanley S, Bloxham J (2004) Convective-region geometry as the cause of Uranus' and Neptune's unusual magnetic fields. Nature 428:151–153

Stevenson DJ (2003) Planetary magnetic fields. Earth Planet Sci Lett 208:1–11

Stevenson DJ (2010) Planetary magnetic fields: Achievements and prospects. Space Sci Rev 152:651–664

Stevenson DJ, Spohn T, Schubert G (1983) Magnetism and thermal evolution of the terrestrial planets. Icarus 54:466–489

Stixrude L, Scipioni R, Desjarlais MP (2020) A silicate dynamo in the early Earth. Nat Commun 11:935

Strangeway RJ, Russell CT, Luhmann JG, Moore TE, Foster JC, Barabash SV, Nilsson H (2010) Does a planetary-scale magnetic field enhance or inhibit ionospheric plasma outflows? *In:* AGU Fall Meeting Abstr, p SM33B–1893

Strugarek A, Bolmont E, Mathis S, Brun AS, Réville V, Gallet F, Charbonnel C (2017) The fate of close-in planets: Tidal or magnetic migration? Astrophys J Lett 847:L16

Tachinami C, Senshu H, Ida S (2011) Thermal evolution and lifetime of intrinsic magnetic fields of super-Earths in habitable zones. Astrophys J 726:70

Takata T, Stevenson DJ (1996) Despin mechanism for protogiant planets and ionization state of protogiant planetary disks. Icarus 123:404–421

Tarduno JA, Blackman EG, Mamajek EE (2014) Detecting the oldest geodynamo and attendant shielding from the solar wind: Implications for habitability. Phys Earth Planet Sci 233:68–87

Tarduno JA, Cottrell RD, Watkeys MK, Hofmann A, Doubrovine PV, Mamajek EE, Liu D, Sibeck DG, Neukirch LP, Usui Y (2010) Geodynamo solar wind, and magnetopause 3.4 to 3.45 billion years ago. Science 327:1238–1240

Thorngren DP, Fortney JJ (2018) Bayesian analysis of hot-jupiter radius anomalies: Evidence for ohmic dissipation? Astronom J 155:214

Tilley MA, Segura A, Meadows V, Hawley S, Davenport J (2019) Modeling repeated M dwarf flaring at an Earth-like planet in the habitable zone: Atmospheric effects for an unmagnetized planet. Astrobiology 19:64–86

Trammell GB, Arras P, Li Z-Y (2011) Hot Jupiter magnetospheres. Astrophys J 728:15

Tremblin P, Chiang E (2013) Colliding planetary and stellar winds: charge exchange and transit spectroscopy in neutral hydrogen. Mon Not R Astron Soc 428:2565–2576

Treumann RA (2006) The electron-cyclotron maser for astrophysical application. Astron Astrophys Rev 13:229–315

Trigilio C, Biswas A, Leto P, Umana G, Busa I, Cavallaro F, Das B, Chandra P, Perez-Torres M, Wade GA, Bordiu C, Buemi CS, Bufano F, Ingallinera A, Loru S, Riggi S (2023) Star–planet interaction at radio wavelengths in YZ Ceti: Inferring planetary magnetic field. arXiv:2305.00809

Turner JD, Zarka P, Grießmeier J-M, Lazio J, Cecconi B, Emilio Enriquez J, Girard JN, Jayawardhana R, Lamy L, Nichols JD, de Pater I (2021) The search for radio emission from the exoplanetary systems 55 Cancri, υ Andromedae, τ Boötis using LOFAR beam-formed observations. Astron Astrophys 645:A59

Turner JD, Zarka P, Griessmeier J-M, Mauduit E, Lamy L, Kimura T, Cecconi B, Girard JN, Koopmans LVE (2023) Follow-up radio observations of the τ Boötis exoplanetary system: Preliminary results from NenuFARarXiv. arXiv:2310.05363

Turnpenney S, Nichols JD, Wynn GA, Casewell SL (2017) Auroral radio emission from ultracool dwarfs: A Jovian model. Mon Not R Astron Soc 470:4274– 4284

Turnpenney S, Nichols JD, Wynn GA, Burleigh MR (2018) Exoplanet-induced radio emission from M Dwarfs. Astrophys J 854:72

van Summeren J, Gaidos E, Conrad CP (2013) Magnetodynamo lifetimes for rocky Earth-mass exoplanets with contrasting mantle convection regimes. J Geophys Res Planets 118:938–951

Vedantham HK, Callingham JR, Shimwell TW, Dupuy T, Best WMJ, Liu MC, Zhang Z, De, K, Lamy L, Zarka P, Röttgering HJA, Shulevski A (2020a) Direct radio discovery of a cold brown dwarf. Astrophys J Lett 903:L33

Vedantham HK, Callingham JR, Shimwell TW, Tasse C, Pope BJS, Bedell M, Snellen I, Best P, Hardcastle MJ, Haverkorn M, Mechev A, O'Sullivan SP, Röttgering HJA, White GJ (2020b) Coherent radio emission from a quiescent red dwarf indicative of star–planet interaction. Nat Astron 4:577–583

Vedantham HK, Dupuy TJ, Evans EL, Sanghi A, Callingham JR, Shimwell TW, Best WMJ, Liu MC, Zarka P (2023) Polarised radio pulsations from a new T-dwarf binary. Astron Astrophys 675:L6

Vidotto AA (2021) The evolution of the solar wind. Living Rev Sol Phys 18:3

Vidotto AA, Jardine M, Helling C (2010) Early UV Ingress in WASP-12b: Measuring planetary magnetic fields. Astrophys J Lett 722:L168–L172

Villadsen J, Hallinan G (2019) Ultra-wideband detection of 22 coherent radio bursts on M dwarfs. Astrophys J 871:214

Villarreal D'Angelo C, Esquivel A, Schneiter M, Sgró MA (2018) Magnetized winds and their influence in the escaping upper atmosphere of HD 209458b. Mon Not R Astron Soc 479:3115–3125

Watson AJ, Donahue TM, Walker JCG (1981) The dynamics of a rapidly escaping atmosphere: Applications to the evolution of Earth and Venus. Icarus 48:150–166

Wienbruch U, Spohn T (1995) A self sustained magnetic field on Io? Planet Space Sci 43:1045–1057

Williams Q (2018) The thermal conductivity of Earth's core: A key geophysical parameter's constraints and uncertainties. Annu Rev Earth Planet Sci 46:47–66

Wu Y, Lithwick Y (2013) Ohmic heating suspends not reverses, the cooling contraction of hot Jupiters. Astrophys J 763:13

Wurz P, Rohner U, Whitby JA, Kolb C, Lammer H, Dobnikar P, Martín-Fernández JA (2007) The lunar exosphere: The sputtering contribution. Icarus 191:486–496

Yadav RK, Thorngren DP (2017) Estimating the magnetic field strength in hot Jupiters. Astrophys J Lett 849:L12

Yu S, Chen B, Sharma R, Bastian TS, Mondal S, Gary DE, Luo Y, Battaglia M (2023) Detection of long-lasting aurora-like radio emission above a sunspot. Nat Astron 8:50–59

Yu S, Hallinan G, Doyle JG, MacKinnon AL, Antonova A, Kuznetsov A, Golden A, Zhang ZH (2011) Modelling the radio pulses of an ultracool dwarf. Astron Astrophys 525:A39

Zarka P (2007) Plasma interactions of exoplanets with their parent star and associated radio emissions. Planet Space Sci 55:598–617

Zarka P, Marques MS, Louis C, Ryabov VB, Lamy L, Echer E, Cecconi B (2018) Jupiter radio emission induced by Ganymede and consequences for the radio detection of exoplanets. Astron Astrophys 618:A84

Zic A, Stewart A, Lenc E, Murphy T, Lynch C, Kaplan DL, Hotan A, Anderson C, Bunton JD, Chippendale A, Mader S, Phillips C (2019) ASKAP detection of periodic and elliptically polarized radio pulses from UV Ceti. Mon Not R Astron Soc 488:559–571

Ziegler LB, Stegman DR (2013) Implications of a long-lived basal magma ocean in generating Earth's ancient magnetic field. Geochem Geophys Geosyst 14:4735–4742

Reviews in Mineralogy & Geochemistry
Vol. 90 pp. 411–464, 2024
Copyright © Mineralogical Society of America

12

Transiting Exoplanet Atmospheres in the Era of JWST

Eliza M.-R. Kempton[1] and Heather A. Knutson[2]

*[1]Department of Astronomy, University of Maryland,
College Park, MD 20742, USA*

ekempton@umd.edu

*[2]Division of Geological and Planetary Sciences,
California Institute of Technology, Pasadena, CA 91125, USA*

INTRODUCTION

The field of exoplanet atmospheric characterization has recently made considerable advances with the advent of high-resolution spectroscopy from large ground-based telescopes and the commissioning of the James Webb Space Telescope (JWST). We have entered an era in which atmospheric compositions, aerosol properties, thermal structures, mass loss, and three-dimensional effects can be reliably constrained. While the challenges of remote sensing techniques imply that individual exoplanet atmospheres will likely never be characterized to the degree of detail that is possible for solar system bodies, exoplanets present an exciting opportunity to characterize a diverse array of worlds with properties that are not represented in our solar system. This review article summarizes the current state of exoplanet atmospheric studies for transiting planets. We focus on how observational results inform our understanding of exoplanet properties and ultimately address broad questions about planetary formation, evolution, and diversity. This review is meant to provide an overview of the exoplanet atmospheres field for planetary- and geo-scientists without astronomy backgrounds, and exoplanet specialists, alike. We give special attention to the first year of JWST data and recent results in high-resolution spectroscopy that have not been summarized by previous review articles.

A historical perspective

As soon as the first exoplanets were discovered in the 1990s, the quest to characterize these unique objects in more detail began in earnest. Images from science fiction movies come to mind when we picture alien planets, but these discoveries provided us with an exciting new opportunity to actually *measure* the atmospheric properties of extrasolar worlds. The first transiting exoplanet was discovered in 2000 (Charbonneau et al. 2000). These are planets with orbits that track directly in front of their host stars as viewed by an Earthbound observer, producing a small dip in the amount of light received, known as a transit. We focus on transiting planets in this review article because they present special opportunities for measuring atmospheric properties. The clever techniques for transiting exoplanet atmospheric characterization that have been developed by the astronomical community (described in detail below under *Observational techniques for atmospheric characterization*) are all premised on using the known orbital geometry of the system to extract the planetary signal from the combined light of the planet and host star. These techniques have been applied with great success over the last two decades to measure a host of atmospheric properties.

1529-6466/24/0090-0012$10.00 (print)
1943-2666/24/0090-0012$10.00 (online)

http://dx.doi.org/10.2138/rmg.2024.90.12

The first detection of an exoplanet atmosphere occurred in 2002 (Charbonneau et al. 2002). The planet, HD 209458 b, was the only known transiting planet at the time (although that was not the case for long), and it belongs to a broader class of exoplanets referred to as 'hot Jupiters'. Such planets are aptly named for their large sizes and small orbital separations—HD 209458 b orbits its Sun-like host star every 3.5 days at an orbital distance of 0.05 AU, and it has a radius of 1.35 Jupiter radii (R_J). By measuring a small amount of excess absorption during transit at the wavelength of the sodium resonance doublet (589.3 nm) with the STIS instrument on the Hubble Space Telescope (HST), Charbonneau et al. (2002) inferred the presence of gaseous sodium in the planet's atmosphere. Although subsequent studies of HD 209458 b's sodium absorption signal with ground-based high-resolution spectrographs have revealed that this measurement may be biased by deformations in the stellar line shape due to the planetary transit (Casasayas-Barris et al. 2020, 2021), this first atmospheric measurement unquestionably marked the birth of a new field of exoplanet atmospheric characterization studies.

Not long after, came the first measurements of exoplanetary thermal emission via *secondary eclipse* (Charbonneau et al. 2005; Deming et al. 2005), which occurs when a planet passes *behind* its host star. Then, in 2007, the first phase curve observations of thermal emission versus orbital phase were obtained for the hot Jupiter HD 189733 b (Knutson et al. 2007). The thermal emission measurements were all made with NASA's Spitzer Space Telescope, which became a workhorse for infrared (IR) characterization of exoplanet atmospheres before it was decommissioned in late 2020. Other important firsts include the measurement of escaping gas from an exoplanet atmosphere (Vidal-Madjar et al. 2003), and the first robust detections of molecules and (more tentatively) high-altitude winds using a novel cross-correlation spectroscopy technique with high-resolution spectrographs on ground-based telescopes (Snellen et al. 2010). More details on all of these observational techniques can be found under *Observational techniques for atmospheric characterization*, below. All of the aforementioned observations were of hot Jupiter targets. The first atmospheric spectrum of an object smaller than Neptune was obtained in 2010 for the planet GJ 1214 b (Bean et al. 2010), ultimately indicating the presence of a thick layer of clouds or haze (Kreidberg et al. 2014a). In 2018 the first thermal emission measurement was made for a rocky, terrestrial exoplanet, LHS 3844 b, disappointingly indicating the lack of any atmosphere at all (Kreidberg et al. 2019).

Today, exoplanet atmospheric characterization has become its own *bona fide* sub-field of astronomy. Detections of several dozen atomic and molecular species along with clouds and hazes have been claimed in the literature for over 100 individual exoplanets[1] (see e.g., Burrows 2014; Madhusudhan et al. 2016; Deming and Seager 2017; Madhusudhan 2019). We note that some of these detections have been made at high statistical significance, whereas others are more tentative or ambiguous, so we encourage the casual reader of the exoplanet atmospheres literature to do so with a critical eye. All of these atmospheric characterization studies have been helped along by the discovery of thousands of transiting exoplanets[2] with ground-based (e.g., HAT, WASP, MEarth, Speculoos) and space-based (e.g., CoRoT, Kepler, TESS) surveys. On the population level, tantalizing hints of planetary diversity have been uncovered, and well-founded attempts are being made to tie statistical trends in atmospheric properties to underlying theories of planet formation and evolution (e.g., Sing et al. 2016; Tsiaras et al. 2018; Welbanks et al. 2019; Goyal et al. 2021; Mansfield et al. 2021; Changeat et al. 2022; Brande et al. 2023; Deming et al. 2023; Gandhi et al. 2023).

In late 2021, the James Webb Space Telescope (JWST) launched successfully, and scientific operations began in the summer of 2022. The telescope's large aperture and IR observing capabilities have opened the door to studies of smaller and colder planets than had

---

[1] At the time of publication, these two websites provide useful lists of published exoplanet atmospheric characterization results: http://research.iac.es/proyecto/exoatmospheres/index.php and https://exoplanetarchive. ipac.caltech. edu/cgi-bin/atmospheres/nph-firefly?atmospheres.
[2] A database that maintains a list of all known exoplanets: https://exoplanetarchive.ipac.caltech.edu/.

previously been possible (e.g., Greene et al. 2023; Kempton et al. 2023; Madhusudhan et al. 2023; Zieba et al. 2023). The high signal-to-noise (S/N) spectra delivered by JWST for larger and hotter planets additionally enable analyses of processes that had remained hidden in earlier datasets such as inhomogeneous cloud formation (Feinstein et al. 2023) and photochemistry (Tsai et al. 2023). This new era of exoplanet characterization with JWST is accompanied by a windfall of ground-based exoplanet data using recently commissioned high-resolution spectrographs (e.g., CARMENES, ESPRESSO, CRIRES+, IGRINS, GIANO, MAROONX, etc.) that are providing detailed compositional measurements for hot and ultra-hot giant planets (e.g., Birkby 2018; Giacobbe et al. 2021; Gandhi et al. 2023; Pelletier et al. 2023). The first JWST exoplanet observations have already been transformative, as have studies that have detected a slew of atomic, ionic, and molecular species in hot Jupiter atmospheres from the ground. These recent results will be summarized in the following sections along with the pre-existing context from two decades of exoplanet atmospheric characterization studies.

Exoplanet demographics

We currently know of more than 10,000 extrasolar planets and planet candidates[3] (Fig. 1), most of which orbit stars with masses ranging from 0.5–1.5× the mass of the Sun (for astronomers, this corresponds to F through early M spectral types). If we exclude planets detected using the microlensing technique (a small fraction of this total), nearly all of these planets orbit stars in our local neighborhood[4] of the Milky Way galaxy. This means that when we discuss the properties of extrasolar planets in subsequent sections, we are implicitly focusing on planets orbiting relatively nearby and (unless specified otherwise) Sun-like stars. In this section, we provide a brief overview of this exoplanet population for the non-expert reader. We begin by briefly summarizing the two detection techniques most commonly used to find exoplanets and their corresponding sensitivities to different kinds of planets. For readers interested in learning more about complementary microlensing and direct imaging techniques, we recommend reviews by Gaudi (2022) and Currie et al. (2023). For a more comprehensive overview of exoplanet demographics, we recommend the review by Gaudi et al. (2021).

Detection techniques

The first planet orbiting a Sun-like star was detected using the radial velocity technique (Mayor and Queloz 1995). This technique relies on the fact that a star and planet will orbit around their mutual center of mass. This causes the star's spectrum to be Doppler shifted as it moves towards and then away from the observer. The semi-amplitude of this Doppler shift is largest for massive planets with short orbital periods (e.g., Fischer et al. 2014); smaller planets on more distant orbits have smaller radial velocity semi-amplitudes and are correspondingly harder to detect[5]. By measuring a planet's radial velocity semi-amplitude, we can place constraints on its mass (technically $M_p \sin(i)$, where M_p is the planet mass and i is the orbital inclination), orbital period, and orbital eccentricity. We can then convert this orbital period to an orbital semi-major axis using Kepler's third law.

---

[3] For the latest numbers see Footnote 2 above. Most unconfirmed candidates were detected using transit surveys and it is likely that this sample contains some false positives, which are typically multiple star systems where one stellar component eclipses another. Transiting planet candidates can be validated statistically using the transit light curve shapes and other complementary information, such as adaptive optics imaging to resolve nearby stars (e.g., Morton et al. 2016; Giacalone et al. 2021), or they can be confirmed directly by radial velocity measurements of the planet masses.

[4] The distance from Earth to the center of the Milky Way is approximately 8.2 kpc (Bland-Hawthorn and Gerhard 2016), while most known exoplanets are located within a few hundred pc of the Earth's location (see https://exoplanetarchive. ipac.caltech.edu/).

[5] The orbital motion of the Earth around the Sun produces a sinusoidal radial velocity signal with a semi-amplitude of 8.95 cm·s⁻¹. We can use Equation (1) in Fischer et al. (2014) to calculate that a Jovian planet orbiting a Sun-like star with an orbital period of a few days would have a radial velocity semi-amplitude that is a factor of ~10^3 larger.

Over the past decade the radial velocity technique has been overtaken by the transit technique, which is responsible for identifying most of the exoplanets known today. This technique focuses on planetary systems where the planet passes in front of its host star as seen from the Earth. During a transit, the planet will block part of the star's light. The amount of light blocked tells us the radius of the planet relative to that of the star, and the intervals between transits tell us the planet's orbital period. If we assume that the planet orbits are randomly oriented, the probability of seeing a transit P is given by $P = R_*/a$, where R_* is the stellar radius and a is the planet's orbital semi-major axis (Winn 2010). This means that transit surveys are biased towards close-in planets; this bias is even stronger than that of radial velocity surveys. Transit surveys also detect large planets more easily than small planets, as they block more of the star's light.

It is very challenging to detect Earth analogs orbiting Sun-like stars in current radial velocity and transit surveys. Fortunately, the size of both the transit and radial velocity signals increase with decreasing stellar mass. As a result, it is significantly easier to detect small planets orbiting small stars ('M dwarfs'). Small stars are also significantly less luminous than the Sun, and the orbital periods corresponding to Earth-like insolations are much closer in. This means that most small (approximately 1–2 $R_\oplus$) transiting planets that are amenable to atmospheric characterization orbit low-mass stars. This has important implications for our understanding of the population-level properties of small rocky exoplanets.

Planet types and order-of-magnitude occurrence rates

As noted above in *A historical perspective*, the close-in gas giant exoplanets known as 'hot Jupiters' were the first type of exoplanet detected in orbit around nearby Sun-like stars. These planets are relatively rare, with an order-of-magnitude occurrence rate of approximately 1% for Sun-like stars (e.g., Petigura et al. 2018; Dattilo et al. 2023). Gas giant planets at intermediate orbital distances (orbital periods of ~10–100 days) are often referred to as 'warm Jupiters', and have a moderately enhanced occurrence rate relative to hot Jupiters (e.g., Fernandes et al. 2019; Fulton et al. 2021). Gas giant planets at larger separations (orbital periods greater than several hundred days) are typically referred to as 'cold Jupiters'. The most precise estimates of the occurrence rates of cold Jupiters currently come from radial velocity surveys, as there are very few transiting gas giant planets at these separations (e.g., Foreman-Mackey et al. 2016). These surveys indicate that the occurrence rate of gas giant planets rises dramatically as we move farther away from the star (e.g., Fernandes et al. 2019; Fulton et al. 2021), with 14 ± 2% of Sun-like stars hosting a gas giant planet between 2–8 AU (Fulton et al. 2021). Current radial velocity surveys of bright nearby stars have baselines as long as ~30 years; this means that our knowledge of the occurrence rates of gas giant planets in these data sets is limited to planets with orbital semi-major axes comparable to or less than that of Saturn in our own solar system.

Exoplanets smaller than Neptune (typically defined as $< 4 R_\oplus$ or $\lesssim 10 M_\oplus$) are often found on close-in orbits around Sun-like stars. Such planets have an overall much higher occurrence than the gas giant planets: ~50% for orbital periods of less than 100 days (just outside the orbit of Mercury in the solar system; e.g., Fulton and Petigura 2018; Hsu et al. 2019). This population is observed to have a bimodal radius distribution, with peaks at 1.3 and 2.4 $R_\oplus$ (e.g., Fulton et al. 2017; Fulton and Petigura 2018; Van Eylen et al. 2018; Hardegree-Ullman et al. 2020; Petigura et al. 2022). The smaller planets (radii between 1.0–1.7 $R_\oplus$) have bulk densities consistent with Earth-like compositions (e.g., Lozovsky et al. 2018; Dai et al. 2019), and are therefore termed as 'super-Earths'. The larger planets (radii between 1.7–3.5 $R_\oplus$) have lower bulk densities, consistent with the presence of modest (a few percent of the total planet mass) hydrogen- and helium-rich gas envelopes (e.g., Lozovsky et al. 2018; Lee 2019; Neil et al. 2022). These planets are therefore termed as 'sub-Neptunes', although some may also have water-rich envelopes (see discussion below). The location of the bimodal radius 'gap' moves towards smaller radii at larger orbital separations (Fulton et al. 2017; Fulton and Petigura 2018; Van Eylen et al. 2018; Hardegree-Ullman et al. 2020; Petigura et al. 2022). This suggests that the gap was carved out by either photoevaporative (e.g., Owen and Wu 2017) or core-powered (e.g., Ginzburg et al. 2018; Gupta and Schlichting 2019) mass loss[6], although Lee et al. (2022) proposed that the division between the two populations might instead be largely primordial.

The order-of-magnitude occurrence rates stated above apply to planets orbiting Sun-like stars (meaning F/G/K main-sequence stars, for astronomers). These values change with decreasing stellar mass; gas giant planets are a factor of 2–3 less common around low-mass stars (e.g., Montet et al. 2014; Bryant et al. 2023), while small planets on close-in orbits appear to be a factor of a few more common (e.g., Dressing and Charbonneau 2015; Mulders et al. 2015; Hardegree-Ullman et al. 2019; Hsu et al. 2020).

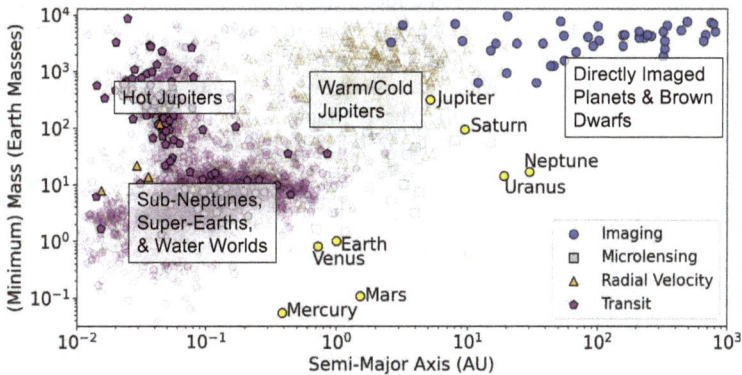

Figure 1. Distribution of confirmed exoplanets in mass–period space. Planets with spectroscopic measurements that constrain their atmospheric properties are shown as **dark points**, those without are shown as **light points**. This review article focuses on atmospheric characterization of *transiting* exoplanets, i.e., the **dark purple pentagon** symbols. Solar system planets are shown as **yellow circles** for context. Both the radial velocity and transit techniques are most sensitive to detecting massive planets on close-in orbits, while the direct imaging technique is most sensitive to young, self-luminous planets on relatively wide orbits. Figure adapted from Currie et al. (2023).

Composition constraints from bulk density measurements

For planets with measured masses and radii, we can obtain a constraint on their bulk densities and corresponding bulk compositions. The bulk densities of gas giant exoplanets are relatively low, indicating that they possess thick, hydrogen-dominated gas envelopes (e.g., Thorngren et al. 2016). These bulk densities can be used to place an upper limit on the abundance of hydrogen and helium relative to heavier elements in the planet's atmosphere (often referred to as the planet's 'atmospheric metallicity' by astronomers; Thorngren et al. 2019). Exoplanets smaller than Neptune exhibit widely varying bulk densities, which reflect their varying bulk compositions. Rocky super-Earths have relatively high bulk densities, while sub-Neptunes with puffy hydrogen-rich atmospheres have much lower bulk densities (e.g., Lozovsky et al. 2018; Neil et al. 2022).

Small planets with intermediate densities are more ambiguous, as their masses and radii can be equally well fit with either water-rich or hydrogen-rich envelopes (e.g., Mousis et al. 2020; Turbet et al. 2020; Aguichine et al. 2021, see Fig. 2). For lower-mass stars, which

---

[6] In photoevaporative mass loss models the atmospheric outflow is driven by heating from high-energy (extreme ultraviolet and X-ray) stellar irradiation. In core-powered mass loss models, the heat source driving the outflow is cooling of the planetary core. However, the predicted mass loss rates in core-powered mass loss models still depend on the total irradiation received by the planet, which determines the temperature of the atmosphere and the corresponding sound speed.

are less luminous, the water ice line is located much closer to the star. This means that even relatively close-in planets forming around low-mass stars may still be able to accrete significant quantities of ice-rich solids (e.g., Kimura and Ikoma 2022). Although some of these 'water worlds' may subsequently accrete hydrogenrich envelopes, such envelopes are more difficult to retain when they orbit low-mass stars. These stars are more magnetically active than their solar counterparts, which means that they emit more high energy photons, and also have more frequent flares and coronal mass ejections, all of which can drive atmospheric outflows (e.g., Atri and Mogan 2021; Harbach et al. 2021). It is therefore thought that planets with water-dominated envelopes may be more common around low-mass stars. Luque and Pallé (2022) plotted all of the currently known planets orbiting low-mass stars with precisely measured masses and radii in mass–radius space and identified a sub-population of low-density planets whose densities appear to be well-matched by water-rich compositions (for alternative hydrogen-rich models, see Rogers et al. 2023). Upcoming observations of candidate water worlds using JWST will soon provide the first direct constraints on their atmospheric water content (see *Atmospheric composition*). These atmospheric characterization studies should provide a much clearer picture of the relative frequency of water worlds around low-mass stars.

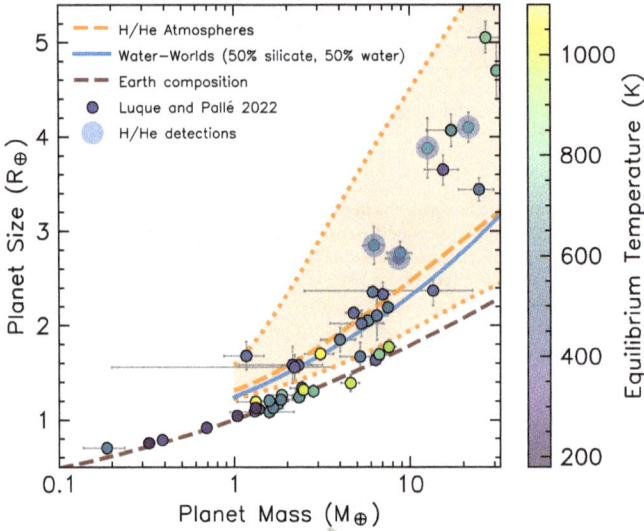

Figure 2. Measured masses and radii for small planets orbiting small (M dwarf) stars from Luque and Pallé (2022). This sample only includes the subset of planets with small fractional uncertainties in mass and radius. Theoretical mass–radius relations for planets with an Earth-like bulk composition (**dashed brown line**), half water and half Earth-like (**solid blue line**), and Earth-like with a few percent hydrogen-rich atmosphere (**dashed orange line**) are overplotted for comparison. The **light orange shading** denotes the range of possible hydrogen-rich atmospheres that can be retained by these planets under a range of possible starting assumptions. Figure adapted from Rogers et al. (2023).

Observational techniques for atmospheric characterization

There are multiple complementary techniques that can be used to detect and characterize the atmospheric compositions of transiting extrasolar planets, as detailed below. All of these techniques leverage knowledge of the transiting planet's orbit to disentangle the combined (unresolved) light from the planet and its much brighter host star (Fig. 3). This is distinct from the approach used to characterize directly imaged planets and brown dwarfs, whose thermal emission can be spatially resolved from that of their host stars (for a recent review, see Currie et al. 2023).

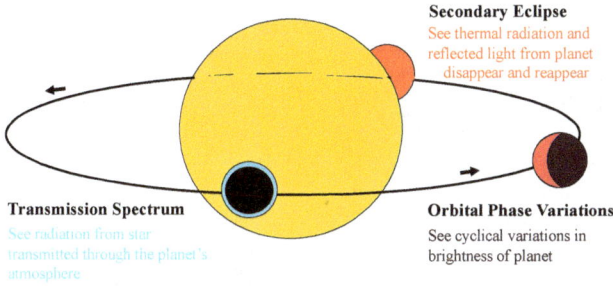

Figure 3. Schematic diagram illustrating three techniques that can be used to characterize the atmospheric properties of transiting exoplanets. Adapted from a figure originally created by Sara Seager (private commun.).

When a transiting planet passes in front of its host star, it will block more of the star's light and therefore appear larger in wavelengths at which the planet's atmosphere is strongly absorbing.

Conversely, the planet will appear smaller and block less of the star's light in wavelengths at which its atmosphere is relatively transparent. This wavelength-dependent transit depth is called a 'transmission spectrum', and is the most widely used method for characterizing the atmospheric compositions of transiting extrasolar planets. We can calculate the relative size of this wavelength-dependent change in transit depth δD_{tr} using the following expression:

$$\delta D_{tr} \simeq \frac{2R_P \delta R_P}{R_*^2} \tag{1}$$

where R_P is the planet radius, R_* is the stellar radius, and δR_P is the wavelength-dependent change in the planet radius. The approximation holds so long as $\delta R_P \ll R_P$, which is true even for hot giants with very low densities. This change can be approximated as a multiple of the atmospheric scale height H:

$$\delta R_P \simeq sH = s\left(\frac{kT_{eq}}{\mu g}\right) \tag{2}$$

where k is the Boltzmann constant, T_{eq} is the predicted atmospheric equilibrium temperature, μ is the atmospheric mean molecular weight, and g is the planet's surface gravity. The scaling factor s typically ranges between 1–5, with lower values more representative of weak absorption and/or atmospheres with significant aerosol opacity, and higher values more representative of strong absorption in a cloud-free atmosphere (e.g., Seager and Sasselov 2000; Miller-Ricci et al. 2009; Benneke and Seager 2012, 2013). For atmospheres with very high aerosol opacity, this signature may be completely obscured (see *Aerosols*). If we assume that the star and planet both radiate as blackbodies, we can calculate the planet's predicted equilibrium temperature as:

$$T_{tr} = T_*\sqrt{\frac{R_*}{a}}\left[\frac{1}{4}(1 - A_B)\right]^{1/4} \tag{3}$$

where T_* is the effective temperature of the host star, a is the planet's semi-major axis, and A_B is its Bond albedo (defined as the fraction of incident radiation that is reflected back to space; Seager 2010). This equation assumes that the planet's atmosphere efficiently redistributes heat from the day side to the night side. In the limit of no heat redistribution (i.e., instantaneous radiative equilibrium at each longitude and latitude point), the effective hemisphere-integrated dayside equilibrium temperature can be calculated by replacing the factor of with a factor of (Hansen 2008), and intermediate values are possible between these two extremes[7]. As demonstrated by these expressions, the overall strength of absorption during the transit can vary by more than an order of magnitude when comparing hydrogen-dominated (low mean molecular weight) atmospheres to those with higher mean molecular weights (e.g., water, carbon dioxide, methane). The presence of high-altitude aerosols from photochemical hazes or condensate clouds can also attenuate the amplitude of gas absorption features by scattering the stellar photons as they pass through the atmosphere. The geometry of transmission spectroscopy means that we are primarily sensitive to the properties of the planet's atmosphere near the day–night terminator. This is particularly relevant when determining cloud properties, which can vary significantly with longitude (see *Three-dimensional atmospheric structure*). Even for clear atmospheres without significant cloud opacity, the relatively long path length of starlight passing through the planet's atmosphere means that this technique is primarily sensitive to atmospheric pressures between 0.001–0.1 bars for typical hot Jupiter atmospheres observed at near-infrared wavelengths (e.g., Fortney 2005; Sing et al. 2016).

If we wait approximately half an orbit we can also observe the planet passing behind its host star (the 'secondary eclipse'). By measuring the relative decrease in light during this eclipse, we can determine the amount of light reflected (at optical wavelengths) or emitted (at IR wavelengths) by the planet. If we assume that the star and the planet both radiate as blackbodies and take the long-wavelength (Rayleigh–Jeans) limit, we can write a simple expression for the depth of the secondary eclipse D_{sec}:

$$D_{\text{sec}} \simeq \left(\frac{R_P}{R_*} \right)^2 \frac{T_{\text{eq}}}{T_*} \tag{4}$$

The planet's emission spectrum also contains information about its atmospheric composition, as well as the average temperature as a function of pressure in its dayside atmosphere. For cloud-free hot Jupiter atmospheres observed at near-infrared wavelengths, the shorter path length of light emitted from the deeper layers of the atmosphere means that we can also potentially probe somewhat higher (a factor of a few) pressures as compared to transmission spectroscopy (e.g., Fortney 2005; Showman et al. 2009).

Close-in exoplanets are expected to be tidally locked, and as a result can exhibit large daynight temperature gradients. This means that the temperature, chemistry, and cloud properties on the daysides of these planets can differ from those measured at the terminator via transmission spectroscopy. We can obtain a global view of these atmospheres by measuring changes in the planet's brightness as a function of orbital phase (the planet's 'phase curve'). By measuring the planet's phase curve at IR wavelengths where its spectrum is dominated by thermal emission, we can map its emission spectrum as a function of longitude (Cowan and Agol 2008, 2011; Rauscher et al. 2018; Morris et al. 2022). These phase curves provide invaluable information about the atmospheric circulation patterns of tidally locked exoplanets; see *Three-dimensional atmospheric structure* for more details. We can also spatially resolve the dayside atmosphere using a second technique called 'eclipse mapping' (Williams et al. 2006; Majeau et al. 2012; de Wit et al. 2012). This technique utilizes the measured changes in brightness during secondary eclipse ingress and egress (defined as the periods when the planet is only partially occulted by the star; for definitions of these terms see Winn 2010) to map the planet's dayside brightness as a function of longitude and latitude. This technique is complementary to phase curve observations, which can characterize the planet's night side but can only measure changes in the planet's atmospheric properties as a function of longitude.

---

[7] Intermediate regimes are typically parameterized by replacing this fraction with an unknown redistribution parameter *f*. Along with the Bond albedo, this redistribution parameter can be directly constrained by infrared phase curve observations (e.g., Schwartz and Cowan 2015; Schwartz et al. 2017); see discussion below.

To date, most published observations of exoplanet atmospheres have been obtained at low spectral resolution using space telescopes (Spitzer, HST, and/or JWST). Because all of these techniques rely on measurements of very small changes in the star's brightness over multi-hour timescales, it is often difficult to achieve the required stability and precision using ground-based observatories. This is because the properties of Earth's atmosphere also vary on similar timescales. However, recent advances in instrumentation on ground-based telescopes have opened up new venues for atmospheric characterization at higher spectral resolution ($R > 20,000$, where $R = \delta\lambda/\lambda$). At these resolutions, spectral features from the star, planet, and Earth's atmosphere can all be readily differentiated from one another. Crucially, the planet's spectral features are Doppler shifted by its orbital motion, while those of the star and Earth's atmosphere remain approximately constant in wavelength over several hour timescales. This means that we can use this wavelength-dependent shift to uniquely identify the absorption features from the planet's transmission or emission spectrum. Notably, this technique is not limited to transiting exoplanets and can also be used to detect spectral features in the emission spectra of non-transiting planets. For more details see the review by Birkby (2018).

Common model frameworks for interpreting exoplanet spectra

When fitting transmission and emission spectra, we must necessarily make a range of simplifying assumptions in order to build simple parametric models that can be used in atmospheric retrieval frameworks. Although we are only sensitive to the atmospheric properties in a narrow range of pressures (typically $0.001-1$ bars for hot Jupiter atmospheres observed at infrared wavelengths at low to moderate spectral resolution), most retrievals typically assume that the inferred elemental abundances are representative of the bulk atmosphere (i.e., there is no net gradient in elemental abundances over the range of pressures, latitudes, or longitudes probed). Similarly, fits to transmission spectra often make the simplifying assumption that the atmosphere is isothermal, while fits to emission spectra typically utilize a simple parametric vertical temperature profile with up to six free parameters (e.g., Madhusudhan and Seager 2009; Line et al. 2013). Many models assume that the atmospheric chemistry is in local thermal equilibrium, or retrieve for the abundances of individual molecules assuming a single fixed abundance for each molecule as a function of pressure. For an overview of the exoplanet retrieval codes commonly in use and the corresponding assumptions made by each, see MacDonald and Batalha (2023). It is worth noting that the high quality of recent JWST observations of hot Jupiters has forced modelers to revisit many of these assumptions, some of which have proven to be too simple for the sensitivity of these new data sets. For more background on exoplanet atmosphere modeling, the reader is encouraged to refer to the following review articles: Marley and Robinson (2015), Madhusudhan (2019), and Fortney et al. (2021).

High-level scientific questions

Our ability to characterize transiting exoplanet atmospheres is fundamentally limited by our great distance from these systems and the fact that the planet is viewed as an unresolved object, blended with the light from its host star. Despite immense improvements in remote sensing capabilities, it is safe to say that we will never in any of our lifetimes characterize an individual exoplanet atmosphere to the degree that we have for planets within our solar system. This is a crucial piece of context for the non-astronomer to understand when formulating a realistic vision for the types of questions that exoplanet studies can address. Yet exoplanets also present an immense opportunity—that of studying a myriad of planetary systems at a *population* level. Exoplanets also provide access to types of planetary environments that do not exist in our solar system (e.g., hot Jupiters, sub-Neptunes, super-Earths, and perhaps water-worlds). A simple summary is that exoplanets allow for coarse measurements for many objects, whereas solar system studies provide detailed data on the outcome of a single instance of planet formation. Leveraging both types of information together equips us with a more complete view of planetary systems and the processes that give rise to them.

Given this context, the types of questions that exoplanet atmosphere studies aim to address are typically those that relate to bulk properties or large-scale atmospheric structure, or those that tie a collection of rough measurements to our understanding of the exoplanet population (or a subpopulation, thereof). Below, we provide an illustrative list of major open scientific questions that can be targeted through exoplanet atmospheric studies. These questions span the planet size and temperature range represented by transiting exoplanetary systems. Meaningful movement toward answering any of these questions would represent a major advance for (exo)planetary science.

- Did close-in gas giant planets form *in situ* or migrate in from farther out in the disk?
- What are the large scale atmospheric dynamics for hot Jupiters, and how do they differ with respect to solar system giant planets and young, hot, directly-imaged planets on wide orbits?
- What are the aerosols in exoplanet atmospheres made of and how do they form?
- How much hydrogen and helium gas can small planets accrete, and which planets keep (or lose) their primordial hydrogen-rich atmospheres?
- Do water worlds exist, and if so, how common are they?
- How do interactions with magma oceans shape the observed atmospheric compositions of subNeptunes and terrestrial exoplanets?
- What kinds of outgassed, high mean molecular weight atmospheres do terrestrial planets have, and what does that mean for their potential habitability?
- Which terrestrial exoplanets lose their outgassed atmospheres? What determines their total atmospheric masses?

ATMOSPHERIC COMPOSITION

Composition as a signpost of formation and atmospheric chemistry

The atmosphere is the outermost layer of a planet and the *only* component of an exoplanet that can readily have its composition directly measured using remote sensing techniques[8]. We therefore rely on observations of an exoplanet's atmosphere as a window into its history and the processes that shape its present-day state. For example, atmospheric observations can be used to inform our understanding of unseen features and processes such as surface-atmosphere interactions or interior structure. High H_2S or SO_2 concentrations in a terrestrial habitable zone planet could be indicative of surface volcanism (Kaltenegger and Sasselov 2010); atmospheric O_2 and O_3 could signify the possible presence of surface life, especially when accompanied by disequilibrium biosignature pairs such as CH_4 (Lovelock 1965; Domagal-Goldman et al. 2014); and a water world might be distinguished from a sub-Neptune with a dry rocky interior via an elevated abundance of water in its atmosphere (e.g., Rogers and Seager 2010).

As with solar system planets, the present-day state of an exoplanet atmosphere is the outcome of its entire history of planet formation and evolution. By measuring an exoplanet's atmospheric composition, one can attempt to decode the processes that gave rise to that planet in the first place. On a single-planet basis, such an analysis is nearly impossible due to vast degeneracies in the range of histories that can all produce similar outcomes, in addition to fundamental uncertainties in the planet formation process and the evolution of protoplanetary disks. But on a population level, we can hope to link trends in atmospheric properties to simple theories for how planets form and evolve, and anchor those theories with measurements, analogously to how our understanding of the history of our solar system stems from observations of the many bodies orbiting the Sun. Several examples for how trends in exoplanet atmospheric observations might be tied back to planet formation and evolution theories are listed, below.

---

[8] Spectroscopic characterization of rocky exoplanet surfaces might also be possible with JWST under ideal conditions (Hu et al. 2012; Whittaker et al. 2022).

Giant planet mass–metallicity relation. Solar system giant planets exhibit a tight anticorrelation between their mass and atmospheric metallicity[9] (Fig. 4, left panel). A similar relation is predicted to be a general outcome of planet formation via core accretion, although there may also be considerable intrinsic scatter in the trend due to the stochastic nature of planet formation (Fortney et al. 2013; Venturini et al. 2016).

Carbon-to-oxygen ratios. The composition of a planet depends on its formation location relative to various snow lines in the protoplanetary disk. The abundant volatiles oxygen and carbon are expected to be especially critical to forming planets due to their roles in delivering icy materials. Measuring the C/O ratio in exoplanetary atmospheres is therefore useful for linking present-day envelope composition to the planet's birth location and the relative import of accreting solids vs. gas during envelope formation (Öberg et al. 2011; Madhusudhan et al. 2014a). It has been difficult to measure C/O in solar system giant planets because they are all cold enough for oxygen to be sequestered out of the observable atmosphere via condensation processes (e.g., Helled and Lunine 2014). Transiting exoplanets, which are typically highly irradiated, provide an excellent opportunity to directly measure atmospheric C/O without relying on model extrapolations (Madhusudhan 2012).

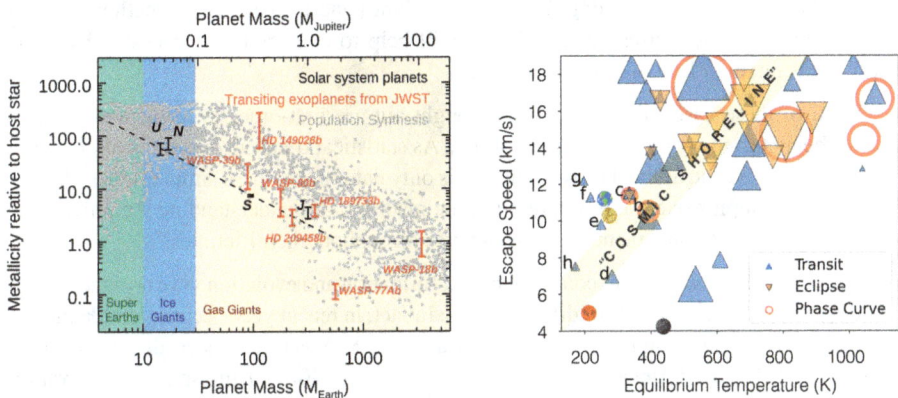

Figure 4. Examples of statistical comparative planetology approaches (Bean et al. 2017) to constrain planet formation and evolution processes via ensemble observations of exoplanet atmospheres. **Left:** Atmospheric metallicity vs. mass for solar system planets (**black symbols**) and for exoplanets that have detections of carbon- and/or oxygen-bearing species using JWST (**red symbols**). Overlaid are the predictions from population synthesis models from Fortney et al. (2013) showing a rise and then a plateau in metallicity as planetary mass decreases (**gray dots**). The solar system giant planets are observed to follow a very tight mass–metallicity correlation (**dashed line**), with the caveat that oxygen is undetected in the atmospheres of these planets, as it is sequestered in condensates below the photosphere. Figure adapted from Mansfield et al. (2018). **Right:** The "cosmic shoreline" (Zahnle and Catling 2017) is denoted (**yellow diagonal band**), which is an observed delineation in escape speed and insolation between solar system bodies that do and do not possess gaseous atmospheres (toward the upper left and toward the lower right of the plot, respectively). Transiting exoplanets that will be observed in Cycles 1 and 2 of JWST are over-plotted in this same parameter space. Symbol size denotes the expected S/N of a single transit or eclipse observation using the methods of Kempton et al. (2018). The letters b–h denote the planets in the TRAPPIST-1 system, which are all slated for JWST observations, and the terrestrial solar system planets are shown for reference. By identifying which terrestrial exoplanets possess atmospheres and whether they are bounded by the same "shoreline" as for the solar system, astronomers can constrain the processes by which planets lose or retain their atmospheres. Figure courtesy of Jegug Ih.

---

[9] Metallicity here and throughout this review article is defined as $(N_X/N_H)_{planet}/(N_X/N_H)_{Sun}$, where N_X/N_H is the ratio of the number of some metal species (X) relative to hydrogen. Species X is selected differently across the literature, depending on what is most readily observable, leading to an inherent inconsistency in how metallicity is measured in different studies.

Other elemental abundance ratios. As with C/O, measuring elemental abundance ratios of various volatile and/or refractory species (e.g., Si/O, Si/C, Fe/O, etc.) provides a tracer for formation location and conditions (Piso et al. 2016; Lothringer et al. 2021; Chachan et al. 2023; Crossfield 2023). Relative abundance measurements can also be used to constrain physical and chemical processes such as condensation or transport (e.g., Gibson et al. 2022; Pelletier et al. 2023).

Atmospheric composition straddling the sub-Neptune to super-Earth radius 'gap'. A strong dip in exoplanetary occurrence for planets with radii $\approx 1.6\,R_\oplus$ has been explained as being the dividing line between two populations of low-mass exoplanets: rocky super-Earths and gas-rich sub-Neptunes (Fulton et al. 2017, and see discussion in *Exoplanet demographics*). Theories of photoevaporative (Owen and Wu 2017) and core-powered (Ginzburg et al. 2018; Gupta and Schlichting 2019) mass loss both posit that the sub-Neptunes are planets that have succeeded in retaining their primordial nebular gas atmospheres, while super-Earths have lost their hydrogen entirely and have secondary high mean molecular weight atmospheres.

The presence or absence of atmospheres on terrestrial exoplanets. In the solar system, a "cosmic shoreline" in escape velocity and insolation separates bodies with atmospheres from those without (Zahnle and Catling 2017, Fig. 4, right panel). Identifying whether a similar dividing line exists for terrestrial exoplanets will help to constrain the processes by which (exo)planets retain or lose their atmospheres.

In all of these cases, the trends being sought out are first-order to begin with, and the theories being tested are often highly simplified. As statistical trends in exoplanet atmospheres data are uncovered and as the data warrant it, it is only natural that these simpler ideas will give way to more complex ones, and progress will be made toward understanding the universality of the processes that shape planetary atmospheres throughout their lifetimes.

An even more direct way to constrain planet formation and evolution via exoplanet studies would be to observe exoplanets of different ages. In fact, in recent years a considerable number of exoplanets orbiting young stars (i.e., with ages ≲100 Myr) have been discovered (e.g., David et al. 2016, 2019; Benatti et al. 2019; Plavchan et al. 2020). Atmospheric observations of younger planets could reveal atmospheric escape or degassing processes while they are still ongoing (Zhang et al. 2022b) and might even show us what true primordial atmospheres look like. Unfortunately, the practical challenges to characterizing atmospheres of young planets are considerable. Young stars tend to be quite active. The resulting stellar variability hinders our ability to detect the minute atmospheric signatures of exoplanets orbiting these stars (Cauley et al. 2018; Hirano et al. 2020; Pallé et al. 2020a; Rackham et al. 2023). We are also fundamentally limited by the number of nearby young stars that are bright enough to present sufficient SNR for atmospheric characterization studies.

Finally, measurements of atmospheric composition provide a direct indication of the chemical processes unfolding in a planet's atmosphere. For example, the measured abundances of molecules, atoms, and ions can be cross-checked against the predictions of thermochemical equilibrium for a given elemental mixture (e.g., Burrows 2014; Lodders and Fegley 2002; Schaefer and Fegley 2010). Departures from equilibrium are then attributed to disequilibrium processes such as vertical or horizontal mixing, or photochemistry (Tsai et al. 2023). Furthermore, the detection of any aerosol species (see *Aerosols*) can be related back to the chemical and physical conditions that gave rise to them in the first place. We therefore turn to spectroscopic measurements of exoplanet atmospheric composition as a powerful tool for probing the physics, chemistry, and history of exoplanetary environments.

Water, water everywhere

The first molecule to be reliably detected in a large number of exoplanet atmospheres was H_2O. Water has many vibration–rotation absorption bands across the near-to-mid IR, and oxygen and hydrogen are cosmically abundant, making this an ideal molecule to search for. Furthermore, water is stable in gas phase from ~370–2200 K—at lower temperatures it condenses into clouds, droplets, or ice; and at higher temperatures it thermally dissociates. Fortunately, most transiting exoplanets have temperatures within the range in which gas-phase H_2O is the expectation.

The search for H_2O in transiting exoplanet atmospheres from space was enabled by the installation of the Wide Field Camera 3 (WFC3) instrument on board HST during its 2009 servicing mission. WFC3 carries a grism centered on the strong 1.4 µm water absorption band with sufficient spectral resolution to resolve the shape of the band, and a novel spatial scanning procedure was developed to spread the exoplanetary spectrum across many detector pixels so as to minimize concerns about detector systematics (McCullough and MacKenty 2012). The first detection of the 1.4 µm water feature in a giant planet atmosphere with the WFC3 spatial scanning mode was made by Deming et al. (2013), and many more soon followed (e.g., Wakeford et al. 2013; Kreidberg et al. 2014b; Sing et al. 2016; Fu et al. 2017; Tsiaras et al. 2018; Changeat et al. 2022, etc; Fig. 5).

The ease with which the 1.4 µm water feature became detectable with HST turned this absorption band into a powerful diagnostic for the chemistry of exoplanet atmospheres. Compared to the baseline expectation of solar composition, a weaker than expected water feature in transmission can be attributed to a low water abundance, a high mean molecular weight atmosphere, or an obscuring cloud deck that mutes the underlying spectral features (Miller-Ricci et al. 2009; Benneke and Seager 2013). In thermal emission, the strength of the water feature depends on the H_2O abundance and on the temperature gradient in the planet's atmosphere. Subtle differences in the shape of a spectrum resulting from each of these various scenarios can potentially be disentangled with sufficiently high S/N and broad enough wavelength coverage (Benneke and Seager 2012). By combining WFC3 measurements and longer wavelength observations with Spitzer's IRAC instrument (Fig. 5), one can furthermore obtain constraints on a planet's metallicity and C/O ratio, by assuming that the atmosphere resides in a state of thermochemical equilibrium (e.g., Wakeford et al. 2018; Zhang et al. 2020). However, without simultaneous detection of the major carbon- and oxygen-bearing species in an atmosphere, these two properties remain degenerate with one another. To quantify water abundances and their associated uncertainties, novel retrieval techniques have ultimately been brought to bear on WFC3 transmission and emission spectra (e.g., Kreidberg et al. 2014b; Madhusudhan et al. 2014b; Line et al. 2016).

Ultra-hot Jupiters, which have equilibrium temperatures in excess of ~2000 K, have presented a particularly interesting case for interpreting H_2O detections. Weak or absent H_2O features in dayside thermal emission spectra were noted for multiple ultra-hot Jupiters, and several hypotheses were posed to explain these observations (Evans et al. 2017; Sheppard et al. 2017; Kreidberg et al. 2018). Initially, retrievals were run that indicated either very low metallicities or very high C/O ratios (Sheppard et al. 2017; Pinhas et al. 2019; Gandhi et al. 2020b). The former reduces the abundances of all 'metal'-bearing species, and the latter reduces the H_2O abundance by tying up nearly all atmospheric oxygen in the CO molecule. Either of these abundance patterns would be surprising though, especially since slightly cooler hot Jupiters are not observed to have similarly weak H_2O features (Mansfield et al. 2021). A more natural explanation was posed by Arcangeli et al. (2018) and Parmentier et al. (2018) who pointed out that thermal dissociation of H_2O at temperatures in excess of ~2200 K coupled with the onset of continuum opacity from the hydrogen anion (H^-) around the same

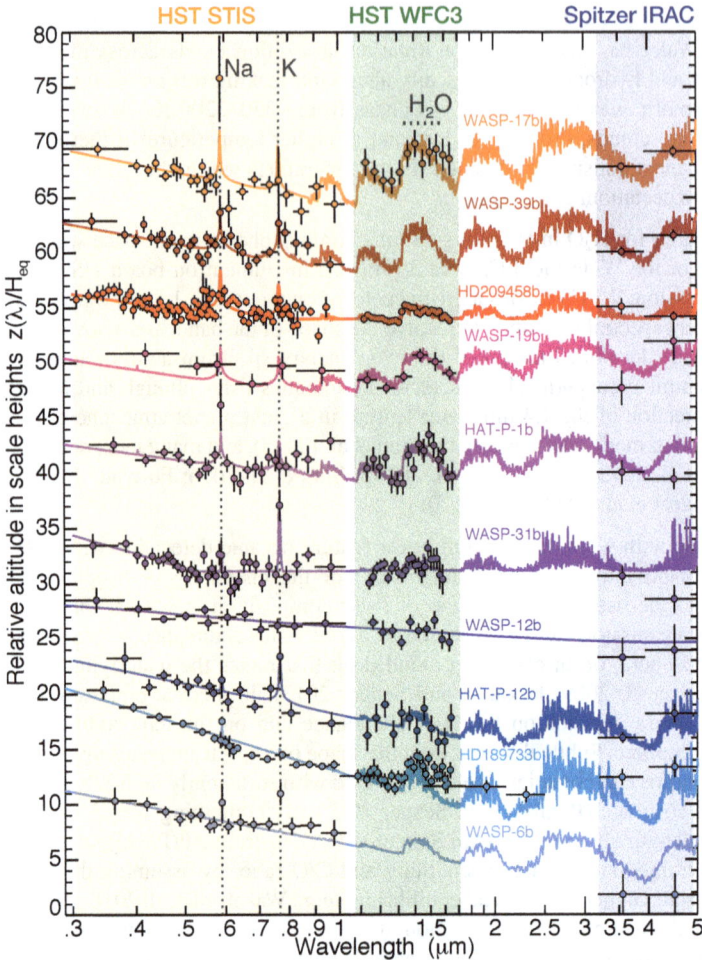

Figure 5. Transmission spectra for various hot Jupiters plotted in units of the planet's atmospheric scale height. Data are shown from the HST STIS and WFC3 instruments, and Spitzer IRAC channel 1 and 2 (3.6 and 4.5 μm, respectively), as indicated. The WFC3 data cover a strong water band at 1.4 μm, and STIS covers absorption lines from Na and K. The two Spitzer IRAC photometric channels are sensitive to CH_4, CO, and CO_2, although the lack of spectroscopic information over this wavelength range makes it difficult to fully constrain the atmospheric carbon chemistry. Muted spectral features and strongly sloping optical and near-IR spectra for the planets plotted toward the bottom of the figure are attributed to aerosol obscuration. Figure adapted from Sing et al. (2016).

temperature provided a high-quality fit to available data without resorting to elemental abundance patterns that differed dramatically from the planets' host stars. Still, the precision and wavelength coverage from HST and Spitzer alone were not sufficient to unambiguously resolve the question of why the ultra-hot planets have muted water features.

On the other end of the planetary 'spectrum', sub-Neptunes and super-Earths have also been observed to have muted or absent water features in transmission. In this case, the interpretation is different because these smaller planets can have high mean molecular weight atmospheres, and they also tend to be colder planets, which makes them potentially amenable to aerosol formation. (Aerosols are not generally considered to be a major atmospheric constituent for ultra-hot planets because we do not know of any cloud or haze species that can form and persist at such high temperatures.) The spectra of rocky super-Earths will be discussed in more detail in *The challenge of terrestrial planets*. A key science question that astronomers aimed to address with the initial atmospheric observations of sub-Neptunes was to break degeneracies between 'mini-Neptune' and water world scenarios by measuring the atmospheric composition and ascertaining whether it was hydrogen- or water-dominated (Miller-Ricci et al. 2009; Miller-Ricci and Fortney 2010; Rogers and Seager 2010). Unfortunately, degeneracies between aerosols and high mean molecular weight made such distinctions extremely challenging with available instruments prior to the launch of JWST (e.g., Bean et al. 2010; Berta et al. 2012; Knutson et al. 2014; Guo et al. 2020; Mikal-Evans et al. 2021, 2023a). Perhaps the most famous among sub-Neptunes is the planet GJ 1214b, which was observed to have a staggeringly flat transmission spectra with 12 stacked transits with the HST+WFC3 instrument (Kreidberg et al. 2014a). The spectrum is so featureless that the only viable interpretation is a very thick and high-altitude layer of clouds or haze (see *Aerosols*), which obscures any direct indications of the atmosphere below.

One challenge to interpreting any claimed detections of water in low-mass exoplanets is that many such planets that are amenable to atmospheric characterization necessarily orbit low-mass Mdwarf stars. Small host stars are required to produce large transit depths and therefore sufficiently high S/N transmission spectra. But low-mass stars also have water in their *own* spectra due to their correspondingly low temperatures. What's worse, the water is not expected to be uniformly distributed throughout the stellar atmosphere and instead to preferentially lie in cooler star-spot regions. The result is that H_2O features can be spuriously imprinted on transmission spectra for planets orbiting M-dwarfs that do not originate in the planetary atmosphere but actually in the star itself (e.g., Deming and Sheppard 2017; Rackham et al. 2018; Zhang et al. 2018; Lim et al. 2023). Techniques for mitigating stellar contamination in the transmission spectra of these systems is an ongoing area of research.

Detection of H_2O from the ground has also been enabled via high resolution spectroscopy. The first such measurement was made by Birkby et al. (2013) using the high-resolution CRIRES spectrograph on the Very Large Telescope (VLT) to capture water absorption lines in the dayside emission spectrum of the hot Jupiter HD 189733b. As the observing techniques have matured and more near-IR high resolution spectrographs have come online, the rate of ground-based water detections has accelerated (see Fig. 6). One particular advantage of high-resolution water detections over space-based measurements with HST+WFC3 is that the former are often simultaneously sensitive to oxygen- and carbon-bearing molecules, enabling direct constraints on the atmospheric C/O ratio (Line et al. 2021; Pelletier et al. 2021; Brogi et al. 2023). Such measurements have indicated C/O values for various hot Jupiters ranging from near-solar (the solar value is 0.55) to super-solar values near 1. With significant scatter in the results to-date, the implications for hot Jupiter formation are murky, but the picture should solidify in the coming years with many more direct C/O measurements enabled by JWST. Ground-based measurements of H_2O have been attempted for sub-Neptunes and super-Earths as well, typically at lower spectral resolution, but to-date all have resulted in nondetections (e.g., Bean et al. 2010, 2011; Cáceres et al. 2014; Diamond-Lowe et al. 2018, 2020a,b).

Refractory species in hot Jupiter atmospheres

Hot and ultra-hot Jupiters have high enough temperatures that most refractory species are rendered in the gas phase, and some can even be ionized via thermal or non-thermal processes. This is advantageous for exoplanet studies because it means that many elements that would otherwise be sequestered deep within a colder giant planet like Jupiter or Saturn are accessible to direct detection. Measured abundance patterns can then be compared to theories of planet formation or used to identify various chemical processes, as discussed above in *High-level scientific questions*. Another goal of refractory species detections in hot Jupiter

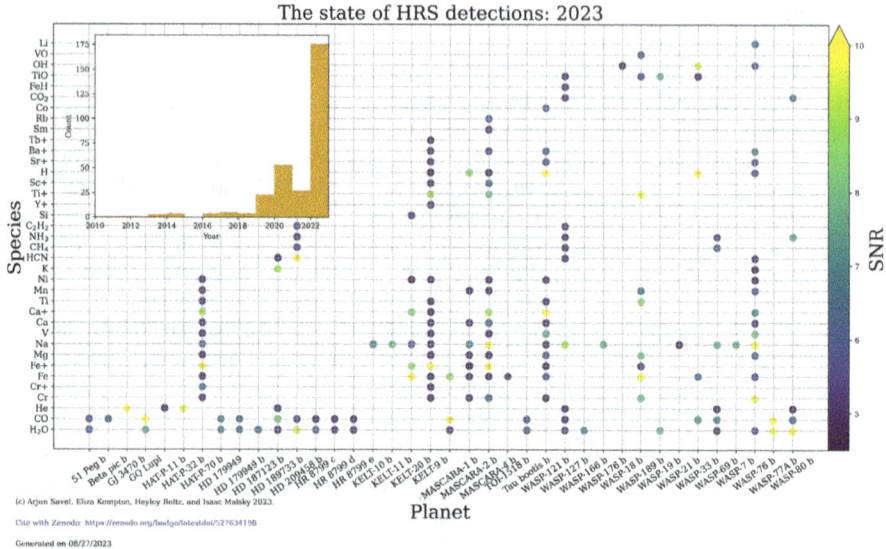

Figure 6. Current state of detections of ions, atoms, and molecules with high-resolution spectroscopy, as of summer 2023. The significance of each claimed detection is indicated by the **symbol color**. The embedded histogram shows the number of high-resolution spectroscopy atmospheric characterization papers published per year, revealing a steep acceleration. The uptick corresponds to multiple new instruments coming online as well as the maturation of the observing technique. Figure courtesy of Arjun Savel.

atmospheres has been to identify the optical and UV absorbers that drive thermal inversions in these planets (see *Dayside temperature structure* for more details). TiO and VO were initially proposed as likely species to drive stratospheric inversions in hot Jupiters (Fortney et al. 2008). More recently, Lothringer et al. (2018) pointed out that a whole host of atomic metals, metal hydrides, and oxides should be in the gas phase in ultra-hot planets and would serve as even stronger optical and UV opacity sources.

Motivated by goals of measuring refractory abundances and identifying key optical absorbers, a large and increasing number of chemical species have been detected in hot Jupiter atmospheres in recent years. This has mostly been made possible by ground-based high-resolution spectrographs that observe at optical wavelengths, as well as the STIS instrument and WFC3/UVIS instrument mode aboard HST. Space-based observations with STIS and WFC3/UVIS jointly provide broad near-UV to optical wavelength coverage (~200–1000 nm) but only at relatively low spectral resolution, which presents a challenge for uniquely identify chemical species. For example, with low-resolution optical spectra, it has often been difficult to disentangle the causes of slopes in transmission spectra, which can be attributed to some combination of aerosol scattering (see *Aerosols*), stellar activity, or optical absorbers (e.g., Pont et al. 2008; McCullough et al. 2014; Evans et al. 2018). At near-UV wavelengths certain ultrahot planets have been observed to have sharply increased transit depths, consistent with the presence of SiO, SH, Mg, and/or Fe, which would serve to drive thermal inversions or act as condensate cloud precursors (Evans et al. 2018; Fu et al. 2021; Lothringer et al. 2022). Individual strong lines due to atomic (e.g., Na, K; Sing et al. 2016) and ionic (e.g., Fe^+ and Mg^+; Sing et al. 2019) species have been easier to uniquely identify, albeit the line profiles are typically not fully resolved by HST, resulting in degenerate interpretations of abundances vs. broadening mechanisms. Sodium and potassium in particular have been identified in a large number of hot Jupiter spectra with STIS (see Fig. 5). In some planets just one of these two species is detected, whereas others produce clear detections of both. Identifying abundance patterns in Na and K vs. fundamental parameters such as equilibrium temperature has so far been elusive.

The optical opacity 'bumps' that have been observed with HST can be fully resolved via high-resolution spectroscopy in order to uniquely identify the species present. To date, well over a dozen elements and 37 individual molecular, atomic, and ionic species have been identified in hot Jupiter atmospheres with high-resolution techniques, spanning a broad portion of the periodic table (e.g., Wyttenbach et al. 2015; Hoeijmakers et al. 2018; Ehrenreich et al. 2020; Tabernero et al. 2021; Kesseli et al. 2022; Langeveld et al. 2022; Flagg et al. 2023; Pelletier et al. 2023, Fig. 6). Of these species, iron and sodium have so far proven the most readily detectable in a large number of hot and ultrahot atmospheres due to their especially strong and unique optical opacity patterns.

Much of the focus of high-resolution studies initially was on the *detection* of individual species. Papers reporting detection significances have recently been giving way to those that quantify relative and/or absolute abundances via high-resolution retrieval techniques (e.g., Gibson et al. 2020; Pelletier et al. 2021; Kasper et al. 2023; Maguire et al. 2023; Gandhi et al. 2023). Such studies have revealed a range of solar and non-solar abundance patterns. For instance, in a retrieval study of six high S/N ultra-hot Jupiters, Gandhi et al. (2023) found iron abundances to be well-matched to the planets' host stars. However, other refractories such as Mg, Ni, and Cr presented more variable abundance patterns; and several species such as Na, Ti, and Ca were found to be uniformly under-abundant relative to stellar, implying some sort of depletion process such as condensation or ionization. In a detailed study of the ultra-hot Jupiter WASP-76b, which measured abundances of 14 individual refractory species, Pelletier et al. (2023) similarly found abundances broadly consistent with solar (and stellar), with some notable exceptions. Elements with high condensation temperatures were found to be depleted, potentially implying condensation cold-trapping on the planet's night side, whereas Ni was over-abundant, perhaps indicating that WASP-76b accreted a differentiated planetary core during its late stages of formation. Studies such as these highlight the power of systematic investigations of gas-phase refractory elements in hot Jupiters to reveal the physics, chemistry, and history of these planets' atmospheres.

The JWST landscape

The first JWST spectrum of a transiting exoplanet was released on July 12, 2022 as part of a handful of 'early release observations' (EROs) meant to demonstrate to the public the power of the newly commissioned space telescope[10] (Pontoppidan et al. 2022). The ~1300 K hot Jupiter WASP-96b was targeted with the NIRISS instrument (Fig. 7). The resulting spectrum spanning 0.6–2.8 μm had exactly the intended effect. It revealed a full rainbow of water features along with evidence for obscuring aerosols, and beyond that it gave the astronomical community a small taste of what was to come from exoplanet studies in the JWST era.

Just a month and a half later, the first peer-reviewed scientific exoplanet result from JWST revealed the striking first-time discovery of CO_2 in an exoplanet atmosphere (JWST Transiting Exoplanet Community Early Release Science Team et al. 2023, Fig. 7). Carbon dioxide, which had previously been out of reach for spectroscopic studies due to the wavelength coverage of available instruments, was detected at a staggering significance of 26σ. Chemically, the CO_2 molecule is especially interesting because it serves as a metallicity indicator in hot hydrogen-rich atmospheres (Fortney et al. 2010). The strong CO_2 absorption feature identified in the hot Jupiter WASP-39b indicates that the planet has ~10× enhanced metallicity relative to its host star. The planet's high metallicity and low mass (relative to Jupiter), intriguingly place it right along the solar system giant planet mass–metallicity relation (Constantinou et al. 2023; Fig. 4).

Further studies of WASP-39b by the The JWST Transiting Exoplanet Community (JTEC) Early Release Science (ERS) program have since produced a full panchromatic transmission

---

[10] https://archive.stsci.edu/hlsp/jwst-ero

spectrum of the planet from 0.6–5.2 µm (JWST Transiting Exoplanet Community Early Release Science Team et al. 2023; Ahrer et al. 2023; Alderson et al. 2023; Feinstein et al. 2023; Rustamkulov et al. 2023; Fig. 7). Spectral features from H_2O, SO_2 (Tsai et al. 2023), and CO (Esparza-Borges et al. 2023; Grant et al. 2023) have been identified, in addition to the aforementioned CO_2, as well as signatures of patchy aerosol coverage. The discovery of SO_2 at 4.05 µm is especially intriguing because this molecule was not predicted in observable amounts by any chemical equilibrium models. Instead, it is believed to be the byproduct of *photochemical* alteration of the atmosphere (Polman et al. 2023; Tsai et al. 2023). The strength

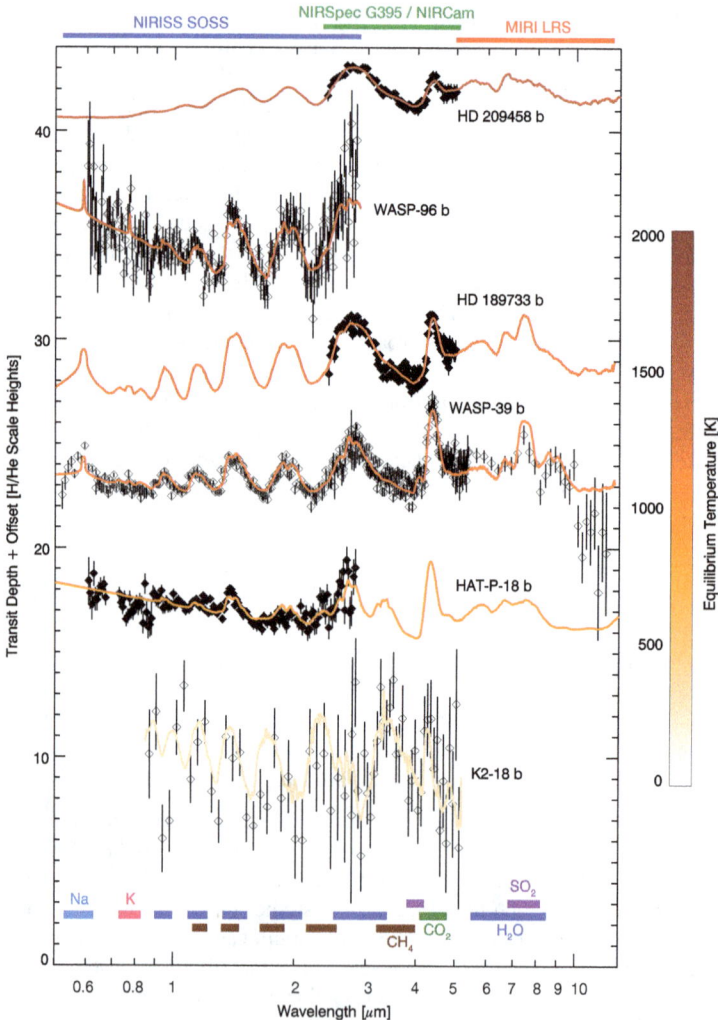

Figure 7. A selection of transmission spectra analyzed to-date from JWST. Wavelength coverage of the various instrument modes used for transmission spectroscopy are indicated above. Atomic and molecular opacity sources that have been identified in the planets shown are indicated below. Several other planets that have been observed with JWST but do not readily reveal any identifiable absorbers are not shown. Comparing against Figure 5, one can see the benefits of the expanded wavelength coverage and improved precision of JWST relative to HST and Spitzer for exoplanet atmospheric characterization. Figure data from Fu et al. 2022, Madhusudhan et al. 2023, Radica et al. 2023, Xue et al. 2023, Carter et al. (submitted), and Fu et al. (submitted).

of the observed feature is well-matched by hot Jupiter photochemistry models, and it is now anticipated that SO_2 might appear in many JWST hot Jupiter observations. The discovery of photochemically derived species opens the door to a whole host of disequilibrium chemistry studies, which will be an exciting new arena for JWST.

Another possible hint of disequilibrium chemistry comes from attempts to detect methane, which is expected to be abundant in hydrogen-rich atmospheres below ~1000 K. Exoplanet atmosphere observations prior to the launch of JWST already hinted at a 'missing methane' problem, with cooler planets not showing obvious signs of methane absorption in HST or Spitzer data (e.g., Stevenson et al. 2010; Kreidberg et al. 2018; Benneke et al. 2019a), although some ground-based detections had been reported (Guilluy et al. 2019, 2022; Giacobbe et al. 2021; Carleo et al. 2022). Methane should be readily observable with JWST, as it has multiple strong absorption bands over the 1–8 μm wavelength range. However, the molecule is notably absent from the transmission spectrum of the ~850 K planet HAT-P-18b with JWST's NIRISS instrument (Fu et al. 2022). Recently, methane was finally detected by JWST in yet colder planets: the ~825 K 'warm' Jupiter WASP-80b (Bell et al. 2023a) and the ~360 K sub-Neptune K2-18b (Madhusudhan et al. 2023). In the latter case, the JWST measurement resolves previous ambiguity from HST+WFC3 observations as to which gas had been detected, H_2O or CH_4 (Benneke et al. 2019b; Tsiaras et al. 2019; Bézard et al. 2022). The accompanying detection of CO_2 and non-detection of water vapor in K2-18b, also perhaps indicates the presence of a liquid water ocean below the planet's thick atmosphere (Madhusudhan et al. 2023). Further JWST observations will map out the parameter space over which methane exists in hydrogen-rich planetary atmospheres and will hopefully hint at the underlying mechanisms behind the missing methane problem such as hot planetary interiors coupled with efficient vertical mixing (Fortney et al. 2020), horizontal quenching (Cooper and Showman 2006; Zamyatina et al. 2023), or photochemistry (Line et al. 2011; Miller-Ricci Kempton et al. 2012b).

It is still early days for JWST, and the observatory has just begun to reveal its prowess in characterizing exoplanet atmospheric composition. Along with metallicities, the reliable measurement of C/O ratios in exoplanet atmospheres has been highly anticipated, enabled by the broad wavelength coverage of the JWST instruments. For example, the 0.6–12 μm wavelength range covered by the JWST exoplanet instrument suite spans spectral features from H_2O, CH_4, CO_2, and CO, which allows for direct measurement of the atmospheric C/O ratio under the assumption of thermochemical equilibrium (e.g., Batalha and Line 2017). The first constraints on metallicities and C/O ratios reveal the diverse outcomes of planet formation processes (Figs. 4 and 8). Derived metallicities in hot Jupiters range from sub-solar (August et al. 2023) to highly super-solar (Bean et al. 2023). Whereas WASP-39b is found to lie directly along the solar system mass–metallicity relation, several other planets do not, implying either that this is not a universal correlation for giant planets, or that there is considerable scatter in the underlying trend. Measurements of C/O ratios for hot Jupiters have have also recovered a range of values from sub-solar (August et al. 2023; Coulombe et al. 2023; JWST Transiting Exoplanet Community Early Release Science Team et al. 2023), to solar (Radica et al. 2023), to super-solar (Bean et al. 2023). Published hot Jupiter studies with JWST are still in the small number statistics regime. However, planned observations of several dozen such planets with JWST in its first two years of operation[11] should go a long way toward revealing trends in metallicities and C/O ratios and establishing whether abundance patterns align with specific theories of giant planet formation.

Another arena in which JWST has already made its mark is to resolve previous ambiguity over the atmospheric composition of ultrahot Jupiters (see *Water, water everywhere*). Whereas with HST alone it had been challenging to determine the cause of weakened H_2O features

---

[11] A list of all approved transiting exoplanet observations with JWST is maintained here: https://www.stsci.edu/~nnikolov/TrExoLiSTS/JWST/trexolists.html

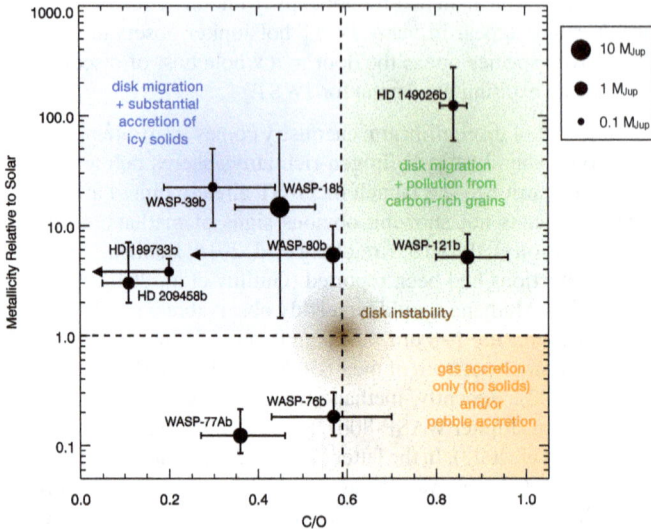

Figure 8. Metallicity vs. C/O ratio for all planets with measurements of both carbon- and oxygen-bearing species from JWST or high-resolution ground-based spectrographs as of the writing of this article. Formation scenarios consistent with different combinations of metallicity and C/O are a summary of work described in Öberg et al. (2011) , Booth et al. (2017), Madhusudhan et al. (2017) and Reggiani et al. (2022). Figure data from August et al. (2023), Bean et al. (2023), Bell et al. (2023a), Brogi et al. (2023), Xue et al. (2023), Fu et al. (submitted.), Mansfield et al. (submitted.), Welbanks et al. (in prep.), and Pelletier et al. (in prep.).

in ultrahot Jupiter emission spectra, the wavelength coverage and precision of JWST data for the planet WASP-18b has allowed for a robust measurement of the planet's underlying atmospheric composition (Coulombe et al. 2023). The NIRISS secondary eclipse spectrum of WASP-18b clearly shows evidence for weakened, but still significant, H_2O features in emission. The detailed shape of the spectrum is best-fit by models with near-solar metallicity, sub-solar C/O, H^- continuum opacity, and water depleted in the observable atmosphere via thermal dissociation. This composition is in line with 'vanilla' predictions of an unaltered nebular gas atmosphere in thermochemical equilibrium and rules out more exotic scenarios.

Attempts to characterize even smaller exoplanets with JWST are also just beginning in earnest. Terrestrial planets will be discussed in more detail below, but the first JWST investigations of subNeptunes are also taking shape. Following on years of ambiguous characterization of sub-Neptunes that have produced degenerate interpretations of atmospheric composition and aerosols (see *Water, water everywhere* and *Aerosols*), the first phase curve observation of a sub-Neptune (the planet GJ 1214 b) has revealed clear evidence that the planet has a high mean molecular weight atmosphere (Kempton et al. 2023). The planet's dayside and nightside thermal emission spectra additionally show spectroscopic signs of H_2O and perhaps CH_4 (the two are partially degenerate with one another over the mid-IR wavelength range observed). The derived composition of the planet is consistent with GJ 1214 b either being a water world or 'gas dwarf', i.e., an initially hydrogen-rich planet that has experienced considerable loss of lighter elements throughout its lifetime. Approximately 20 additional sub-Neptunes are scheduled for transmission spectrum observations with JWST during Cycles 1 and 2[12], opening the door to further compositional characterization of this intriguing class of planets, so long as spectral features are not entirely obscured by aerosols.

---

[12] JWST (and HST) observations are scheduled in annual cycles. Cycle 1 can therefore be thought of as the first year of JWST operations, Cycle 2 as the second year, etc.

The challenge of terrestrial planets

Terrestrial planets are especially challenging targets for atmospheric characterization due to their small sizes and also the expectation that their atmospheres will typically have high mean molecular weight, which reduces the size of spectral features observed in transmission (see *Observational techniques for atmospheric characterization* and also a review article by Wordsworth and Kreidberg (2022) for more background on terrestrial exoplanets). The first attempts to measure the atmospheric composition of rocky exoplanets typically resulted in non-detections of spectral features, which in turn could rule out cloud-free hydrogen dominated atmospheres but left open a wide range of plausible atmospheric compositions and cloud properties (e.g., de Wit et al. 2016, 2018; Diamond-Lowe et al. 2018, 2020b, 2023; Mugnai et al. 2021; Libby-Roberts et al. 2022). To this day, there have not yet been any robust detections of atmospheric species in terrestrial exoplanets. The small number of works that have claimed the detection of atmospheric gases for rocky planets via transmission spectroscopy have been called into question or have not been reproduced (e.g., Southworth et al. 2017; Swain et al. 2021).

An alternative approach to characterizing rocky planet atmospheres was first demonstrated by Kreidberg et al. (2019) and Crossfield et al. (2022). The former measured the phase curve of the terrestrial exoplanet LHS 3844 b, and the latter observed the secondary eclipse of GJ 1252 b, both with the Spitzer Space Telescope. The goal of both observations was to determine whether the planet in question has a thick atmosphere or is an airless barren rock. The technique is discussed in more detail under *The dayside temperatures of sub-Neptunes and super-Earths*, but briefly, the premise is that, for tidally-locked exoplanets (as is expected to be the case for these and most other terrestrial planet atmospheric characterization targets orbiting M-dwarfs), an atmosphere serves to transport heat away from the planet's hot dayside to its colder nightside (Seager and Deming 2009; Koll 2022). In both cases, the planets' high observed dayside temperatures were found to be consistent with the lack of a substantial atmosphere, although Earth-thickness 1-bar atmospheres could not be ruled out. The inferred limits on atmospheric thickness are also composition-dependent due to the differing abilities of various gases to absorb light and transport heat, governed by their wavelength-dependent opacities (Whittaker et al. 2022; Ih et al. 2023; Lincowski et al. 2023). With Spitzer, such dayside thermal emission measurements were only possible for the few hottest and most favorable targets, but JWST has much expanded capabilities in this arena (Koll et al. 2019).

The large aperture and IR observing capabilities of JWST have long promised to extend the parameter space of observable exoplanet atmospheres to terrestrial planets (e.g., Deming et al. 2009; Beichman et al. 2014; Batalha et al. 2015). Nearly 30 such planets (i.e., rocky super-Earths to subEarths) are already approved for observation in Cycle 1 and 2 of JWST operations. The list of planned observations includes all 7 of the planets in the TRAPPIST-1 system, as well as numerous terrestrial planets orbiting earlier (larger, warmer, and more massive) M stars, and a handful of ultrashort-period (USP) rocky planets with periods shorter than 1 day orbiting G, K, and M stars. The TRAPPIST-1 system is of particular interest for habitability studies aiming to identify biosignature gases because the late (i.e., small and cool) M-dwarf host star brings the habitable zone to very short orbital periods and produces large transit depths and thus atmospheric signal sizes (e.g., Barstow and Irwin 2016; Krissansen-Totton et al. 2018; Lustig-Yaeger et al. 2019). TRAPPIST-1 e, f, and g are all potentially habitable environments, and are largely seen as the best prospects for characterizing potentially habitable worlds within the next decade (Gillon et al. 2017).

The first year of JWST data has so far been marked by more non-detections of terrestrial atmospheres, but now with the vastly improved capabilities of the new facility, such measurements are more meaningfully constraining. Transmission spectra to-date are consistent with flat lines (LustigYaeger et al. 2023), or with the possibility that spectral features are caused by the host star and not the planet (Lim et al. 2023; Moran et al. 2023). Thermal emission studies of terrestrial planets (so far limited to TRAPPIST-1 b and c) have been more revealing. As with previous Spitzer thermal emission studies, the goal with these observations has been to measure the planets' dayside temperatures and infer the presence or lack of an atmosphere. For both planets, the measured dayside temperatures are again consistent with no atmosphere being present (Greene et al. 2023; Zieba et al. 2023). The mid-IR capabilities of JWST has allowed for these measurements to be made at much longer wavelengths than Spitzer could access. By observing at 15 microns in the center of an expected strong CO_2 absorption band, the JWST measurements can rule out very thin atmospheres (for TRAPPIST-1 b down to even Mars thickness), under the assumption that CO_2 would be a dominant gas in any moderately-irradiated terrestrial environment (Ih et al. 2023; Lincowski et al. 2023).

Still, a key promise of JWST is to deliver spectra of smaller and cooler exoplanets than what was previously possible with HST. In light of the flat-line spectra and dayside thermal emission results, a question that has supplanted the characterization of rocky exoplanets has been to identify whether such planets possess atmospheres at all. Figure 4 (right panel) shows that Cycle 1 and 2 JWST targets cover the parameter space of planets that would and would not be expected to host atmospheres, based on solar system considerations. If some of the less-irradiated and/or higher surface gravity terrestrial exoplanets are found to have atmospheres, multiple modeling studies have shown JWST's capabilities to spectroscopically characterize such environments under optimal conditions of minimal cloud obscuration, large scale heights, and stacking multiple transits to improve S/N (e.g., Barstow and Irwin 2016; Morley et al. 2017; Batalha et al. 2018; Krissansen-Totton et al. 2018; Fauchez et al. 2019; Lustig-Yaeger et al. 2019; Pidhorodetska et al. 2020; Suissa et al. 2020). For the subset of rocky planets without atmospheres, mid-IR emission spectroscopy measurements with JWST offer an exciting opportunity to characterize their surface compositions for the first time (e.g., Hu et al. 2012; Whittaker et al. 2022; Ih et al. 2023).

AEROSOLS

Terminology and background

Aerosols in this work are defined to be any kind of particle suspended in a gaseous atmosphere, regardless of their composition or formation pathway[13]. Aerosols can be broken up into sub-categories: clouds are defined as solid particles or liquid droplets formed by *condensation* processes, hazes are involatile particles produced by chemical (and often photochemical) processes, and dust is made up of solid particles suspended in an atmosphere that originated elsewhere (e.g., particles kicked up from the surface or those that originated from a meteorite breaking up as it entered a planet's atmosphere). These are all process-based definitions. If the formation mechanism for the particles in question is unknown, we revert to the blanket term 'aerosol'. We warn the reader that some published papers in the exoplanetary literature employ the term 'haze' when referring to small particles ($\lesssim 1$ μm) and 'clouds' when referring to larger particles, but we prefer the process-based definitions for the physical insight they bring. All solar system planets and moons with significant atmospheres have some sort of aerosol layer. For example, Earth has water clouds, surface dust, and technology-derived haze (i.e., smog). Venus has sulfuric acid clouds and haze. Titan has clouds and haze formed from organic compounds. It stands to reason that exoplanets too should have aerosol layers and that these will be a fundamental component of their atmospheres, governing energy balance, thermal structures, and observed spectra; as is the case for the solar system planets. As we will see, there is indeed plentiful observational evidence for exoplanet aerosols. For a more detailed review of aerosols in exoplanet atmospheres, we refer the reader to a recent article by Gao et al. (2021).

The observational signatures of clouds and hazes in exoplanet transmission spectra are primarily muted (or absent) spectral features or strong blue-ward spectral slopes at

---

[13] This definition and those that follow for clouds, haze, and dust are all attributed to an online article written for the Planetary Society by Sarah Hörst: https://www.planetary.org/articles/0324-clouds-and-haze-and-dust-oh-my.

optical wavelengths[14]. These come about due to the propensity of the aerosol particles to scatter or absorb starlight. Rayleigh-like scattering slopes arise for small particles, whereas flatter spectra result when particle sizes are larger or the clouds are very thick. The aerosol species themselves also have their own spectral signatures, but these are typically weak features with wavelength-dependent shapes that depend on particle size distribution (e.g., Wakeford and Sing 2015). This results in degeneracies among spectra associated with distinct aerosol species, making the aerosol composition difficult to uniquely constrain spectroscopically. In thermal emission, the signatures of aeorols tend to be even more subtle. Aerosol layers can impact the thermal structure of an atmosphere (for example, causing thermal inversions in the case of very absorptive clouds or hazes, thus altering the shape of the planet's emission spectrum; Morley et al. 2015; Arney et al. 2016; Lavvas and Arfaux 2021), and optically thick aerosols can mute spectral features; but these effects are not uniquely attributable to clouds and therefore can be challenging to interpret. The result is that one can often tell from an observation that aerosols are present, and inferences can be made about the vertical distribution of the clouds or haze, but concluding anything robustly about the aerosol composition on a planet-by-planet basis is exceedingly difficult. Forward models are useful to motivate which types of aerosols are consistent with a specific observation, providing probabilistic arguments on the aerosol composition. This approach can be especially powerful at the population level.

Oftentimes in the exoplanet literature, aerosols are treated as a 'nuisance' parameter, due to their impact of hindering the detection the underlying gaseous atmosphere. Modeling the complex microphysics and chemical processes that lead to aerosol formation is a challenging task, so parameterized studies that reduce the aerosols to as few defining properties as possible are common (e.g., Ackerman and Marley 2001; Benneke and Seager 2012). Yet it is only by understanding the aeorols themselves, including their composition, formation, and optical properties, that we can gain a holistic picture of the planetary atmosphere in question. Studies of exoplanet aerosols additionally provide us with unique laboratories for probing cloud and haze formation in conditions that are not accessible within the solar system.

Hot Jupiter aerosols

For hydrogen-rich atmospheres hotter than $\sim$1000 K, the types of clouds that are able to form due to condensation processes are those that are more commonly thought of as refractory species. For example, based on chemical equilibrium calculations one would expect clouds of Fe, Ni, Al_2O_3,

Mg_2SiO_4, TiO_2, and MnS for a solar-composition gas mixture (e.g., Burrows and Sharp 1999; Mbarek and Kempton 2016; Woitke et al. 2018; Kitzmann et al. 2023). Because such clouds are expected to only form at very high temperatures, and they incorporate trace species, it can be tempting to ignore the impacts of aerosols on hot Jupiter studies. Yet it was shown early on that transmission spectroscopy geometry, specifically the oblique geometric path taken by stellar photons through the exoplanetary atmosphere on their way to the observer, can result in cloud optical depths considerably in excess of unity, even for trace species (Fortney 2005). Obscuration by clouds was one explanation immediately put forth for the weaker than expected sodium absorption signal seen in the very first exoplanet transmission observation (Charbonneau et al. 2002). These early studies indicated that cloud modeling would need to be an integral component of interpreting hot Jupiter atmospheric observations.

The presence of aerosol layers has been inferred from multiple hot Jupiter transmission spectroscopy studies, starting with the benchmark planet HD 189733 b (Pont et al. 2008). For that planet, a strong spectral slope over optical wavelengths accompanied by non-detections of sodium and potassium, which should have been present under clear atmosphere conditions, constituted strong evidence for cloud or haze obscuration. However, later work showed that the optical slope could equivalently be the signature of unocculted starspots on the surface of the planet's active host star, leading to an ambiguity in how to interpret the observational result.

---

[14] Additionally, since many cloud species contain oxygen, the rainout process can alter the C/O of the gas-phase atmosphere, which can impact abundance interpretations if not properly accounted for (e.g., Helling et al. 2019).

Since then, many other hot Jupiters have revealed optical spectral slopes and/or muted spectral features over IR wavelengths, indicating that aerosol coverage is a likely culprit across the population (e.g., Sing et al. 2016; Wakeford et al. 2017, and see Fig. 5). Other observational indications of clouds in hot Jupiters come from optical and IR phase curve observations, which will be discussed further under *Complications from clouds, chemical gradients, magnetic fields*.

On a population level, the aerosol composition and formation mechanism can become more apparent. By comparing the strength of transmission spectral features to a suite of aerosol microphysics forward models, Gao et al. (2020) argued that the muted spectral features for transiting gas giant planets are primarily caused by silicate clouds for planets hotter than 950 K and hydrocarbon hazes for cooler planets (Fig. 9). Their argument hinges on the relatively high abundances of Si, Mg, and C, their three main aerosol-forming species, and that other species of comparable abundance (e.g., Fe and Na) don't readily form clouds due to high nucleation energy barriers inhibiting particle formation.

JWST is expected to improve our understanding of hot Jupiter aerosols by providing higher-precision spectra and broader wavelength coverage, allowing for degeneracies to be broken between aerosol coverage and high mean molecular weight or non-solar abundance patterns (Batalha and Line 2017). Already this enhanced precision has led to the finding of *partial* cloud coverage of the terminator of WASP-39b due to subtle departures in the shape of its transmission spectrum from a fully-clouded planet (Feinstein et al. 2023). The access to longer wavelengths with JWST also presents the opportunity to directly measure spectral features from aerosols (e.g., Wakeford and Sing 2015), such as the silicate features that have been observed in mid-IR spectra of brown dwarfs and directly-imaged giant planets (e.g., Burningham et al. 2021; Miles et al. 2023). As for high-resolution spectroscopy studies, the signatures of aerosols are difficult to distinguish because high resolution data processing techniques typically remove the spectral continuum, which is where most of the aerosol information is contained (e.g., Snellen et al. 2010; de Kok et al. 2013). An advantage of high-resolution studies though is that the sharply peaked cores of spectral lines tend to extend above cloud decks, resulting in an ability to measure gas-phase composition even for aerosol enshrouded planets (Kempton et al. 2014; Gandhi et al. 2020a; Hood et al. 2020).

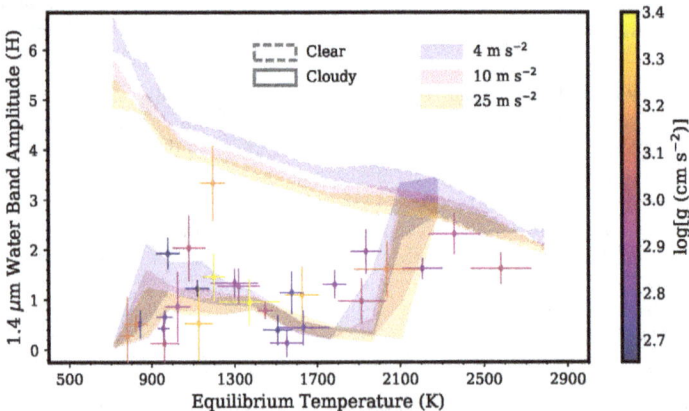

Figure 9. Clear and cloudy atmosphere model tracks compared with transmission spectroscopy measurements of the 1.4 µm water feature amplitude as a function of planetary equilibrium temperature for transiting gas giant planets. Hot Jupiter transmission spectra for planets colder than ~2100 K generally have weaker absorption features than what is predicted for clear solar-composition gas mixtures. The cloud microphysics models shown here provide a good overall fit to the observed trend. In addition to cloud condensation, the models include the formation of hydrocarbon hazes, which increasingly dominate at equilibrium temperatures below 950 K. The clouds are primarily formed from Mg_2SiO_4, with minor contributions from Al_2O_3 and TiO_2. Some variation in the degree of aerosol coverage is expected based on surface gravity, metallicity, and C/O ratio, which is likely driving the intrinsic scatter in the observed data points. Figure from Gao et al. (2020).

Another particularly promising avenue for further constraining aerosol composition in hot Jupiter atmospheres is by using 3-D diagnostics to determine *where* on the planet (i.e., as a function of longitude and/or latitude) the aerosols are located. The aerosol spatial distribution and the physical conditions derived at those locations (e.g., temperature, UV irradiation, wind speeds) can then be directly linked to a proposed aerosol formation mechanism and composition. Such analyses can be accomplished on high S/N spectra and phase curves from JWST or high-resolution spectra from ground-based telescopes (e.g., Kempton et al. 2017; Ehrenreich et al. 2020; Espinoza and Jones 2021; Parmentier et al. 2021; Roman et al. 2021). We discuss 3-D diagnostics for aerosols further in *Complications from clouds, chemical gradients, and magnetic fields.*

Aerosols in sub-Neptunes and super-Earths

Because they tend to orbit smaller stars and thus have cooler temperatures, transiting planets smaller than Neptune are especially likely to host aerosol layers. This was heavily implied by the first investigations of sub-Neptune exoplanets, which returned featureless transmission spectra (e.g., Bean et al. 2010; Berta et al. 2012; Knutson et al. 2014). To date, most flat and muted sub-Neptune and super-Earth transmission spectra are consistent with either aerosols or high mean molecular weight atmospheres, leading to degenerate interpretation. However, in the case of the planet GJ 1214 b, the data were obtained at high enough precision by stacking multiple transits with HST that the degeneracy could be broken, and a thick aerosol layer remains as the only viable explanation (Kreidberg et al. 2014a). JWST should similarly provide the precision to break the aerosol vs. mean molecular weight degeneracy in a *single* transit for many sub-Neptunes, allowing for improved aerosol characterization for smaller planets.

Even without an unambiguous detection of aerosols on a planet-by-planet basis, the ubiquity of flattened transmission spectra and indications that the flatness correlates with planetary equilibrium temperature (Crossfield and Kreidberg 2017; Libby-Roberts et al. 2020; Gao et al. 2021) point to aerosol coverage being a defining characteristic of sub-Neptune exoplanets. Hints that the aerosol coverage may clear at high temperatures ($\gtrsim 900$ K) and lower temperatures ($\lesssim 400$ K) provide hints as to the dominant particle composition and formation pathway (Fig. 10).

As for the aerosol formation mechanism, it is hypothesized that hydrocarbon-based hazes readily form in hydrogen-rich sub-Neptune exoplanets below a temperature of ~850 K (e.g., Morley et al. 2015; Kawashima and Ikoma 2019). Under such conditions, methane is expected to be plentiful in chemical equilibrium. Methane is readily photolyzed by the UV radiation from the host star, producing a rich collection of higher-order hydrocarbons that can continue to polymerize and ultimately form large involatile haze particles (e.g., Miller-Ricci Kempton et al. 2012b; Morley et al. 2013; Kawashima and Ikoma 2018; Lavvas et al. 2019). This is analogous to how we believe hydrocarbon 'tholin' haze forms in Titan's atmosphere. The propensity for hazes to form under sub-Neptune conditions is supported by lab work, in which an ensemble of gases is irradiated by a UV or plasma energy source, and the resulting solid particles are collected and analyzed (He et al. 2018; Hörst et al. 2018). Interestingly, lab studies are also able to form hazes in gas mixtures without methane (He et al. 2020), indicating that haze formation in exoplanet atmospheres may come about via diverse chemical pathways that have yet to be characterized.

Candidates for condensate clouds in sub-Neptunes include sulfides (ZnS, Na_2S), sulfates (K_2SO_4), salts (KCl), and graphite (Miller-Ricci Kempton et al. 2012b; Morley et al. 2013; Mbarek and Kempton 2016). These are the expected equilibrium condensates over the temperature range of most sub-Neptunes studied to date, although some of these species may not form clouds due to their high nucleation-limited energy barriers (Gao et al. 2020). Arguments for haze being dominant over clouds in sub-Neptune atmospheres also hinge on how thick and high up the aerosol layers must be in order to fully flatten transmission spectra, especially in the case of the well-studied planet GJ 1214 b. It is difficult to build models in which low-abundance species such as ZnS or KCl are able to provide sufficient opacity to match existing observational data (Morley et al. 2013, 2015).

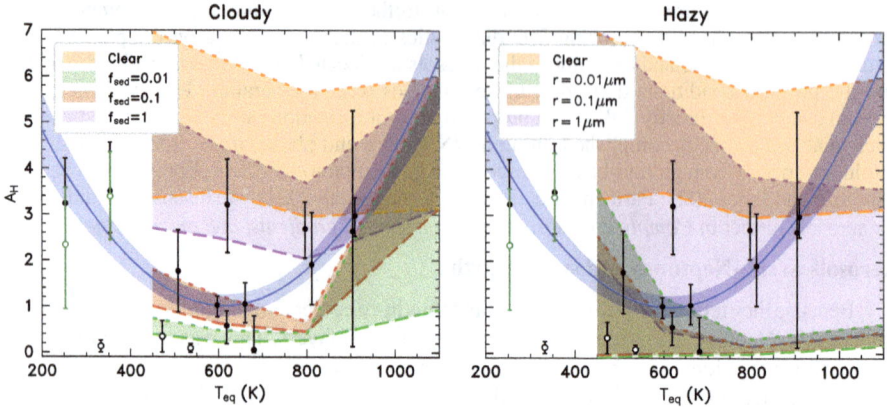

Figure 10. Strength of the 1.4 μm water feature in units of scale heights for transmission spectra of planets 2–6 $R_\oplus$ in size. The **blue parabola** is a best-fit second order polynomial trend, implying a minimum in the strength of transmission spectral features around an equilibrium temperature of 600 K, perhaps indicating the conditions for maximal aerosol coverage. **Left: Colored lines** indicate clear and cloudy models from Morley et al. (2015). The **dotted and dashed contours** are for 100× and 300× solar metallicity, respectively. **Right: Colored lines** indicate clear and hazy models from Morley et al. (2015). The hazy models include soot hazes only. The **dotted and dashed contours** are for 1% haze precursor conversion into soots while the dashed contours are 10%. Figure from Brande et al. (2024).

Recent JWST phase curve observations of GJ 1214 b have thrown another surprise into our evolving understanding of sub-Neptune exoplanets. The very high measured Bond albedo of the planet based on its global thermal emission ($A_B \approx 0.5$) implies that the planet's aerosol layer is highly reflective (Kempton et al. 2023). This is in tension with our understanding of hydrocarbon hazes (e.g., soots and tholins), which are primarily believed to be dark and absorptive (Khare et al. 1984; Morley et al. 2013, 2015). Additional lab and theoretical work is urgently needed to understand how such reflective and abundant aerosols are formed in sub-Neptune environments. Some possibilities are a more reflective type of hydrocarbon or darker particles coated in high-albedo condensates (e.g., Lavvas et al. 2019). Upcoming JWST observations should shed light on whether sub-Neptune aerosols share a universal set of properties and whether transitions from clear to aerosol-enshrouded conditions occur at expected levels of insolation.

ATMOSPHERIC MASS LOSS

Transiting exoplanets are particularly vulnerable to atmospheric mass loss as a result of their close-in orbits. The smallest and most highly irradiated exoplanets may lose their entire atmosphere (see *Exoplanet demographics*), while the population of close-in gas giant planets appears to be minimally altered by atmospheric mass loss (e.g., Vissapragada et al. 2022; Lampón et al. 2023). For giant planets, atmospheric outflows are driven by high energy irradiation from the host star, which causes the uppermost layers of the planet's atmosphere to expand until they become unbound; this process is often referred to as 'photoevaporation' (for a comprehensive review of theoretical work on this topic, see Owen 2019). For smaller planets, core-powered mass loss (Ginzburg et al. 2018) might also play an important role. For a more detailed discussion of current constraints on these processes from the measured radius-period distribution of sub-Neptune-sized exoplanets, see Rogers et al. (2021) and Owen et al. (2023).

We can directly observe the atmospheric outflows of transiting planets by measuring the depth of the transit in strong atomic absorption lines. The large planet–star radius ratios of transiting gas giant exoplanets make them particularly favorable targets for these observations.

The first atmospheric outflows from close-in gas giant planets were detected by measuring the strength of the absorption in the Lyman α line of hydrogen (Vidal-Madjar et al. 2003; Lecavelier Des Etangs et al. 2010). This line is located in the UV and can therefore only be accessed by space telescopes like HST. Because the core of this line is masked by absorption from the interstellar medium, these observations are only sensitive to absorption in the line wings and are limited to relatively nearby (distances of ~100 pc or less) stars. This absorption corresponds to the higher velocity components of the outflow, which are located farther from the planet (e.g., Owen et al. 2023). To date, there are seven planets whose outflows have been measured in this line; see Figure 11 for a visualization of where these planets are located in mass–period space.

Recent theoretical (Oklopčić and Hirata 2018) and observational work (Nortmann et al. 2018; Spake et al. 2018) revealed that atmospheric outflows could also be detected using metastable helium absorption at 1083 nm. Unlike Lyman α, this line can be readily observed using high resolution spectrographs on ground-based telescopes. Because we can measure absorption in the line core, this line provides a complementary tool to probe the lower velocity components of the outflow, which are located closer to the planet. To date, atmospheric outflows have been measured for 20 planets using this line (see Fig. 11).

Outflows have also been detected in the optical Hα, Hβ, and Hγ lines (e.g., Jensen et al. 2012; Yan and Henning 2018; Casasayas-Barris et al. 2019; Wyttenbach et al. 2020), as well as UV lines of other atomic species (e.g., Vidal-Madjar et al. 2004; Sing et al. 2019; Dos Santos et al. 2023). Some of the refractory atomic species detected in optical high spectral resolution data sets (see *Refractory species in hot Jupiter atmospheres*) likely also probe unbound regions of the atmosphere, but more detailed models are needed in order to interpret these absorption signals (Linssen and Oklopčić 2023). By combining the information from multiple lines together, we can obtain a more detailed picture of the overall structure and thermodynamics of the outflow (Lampón et al. 2021; Yan et al. 2022a; Huang et al. 2023; Linssen and Oklopčić 2023).

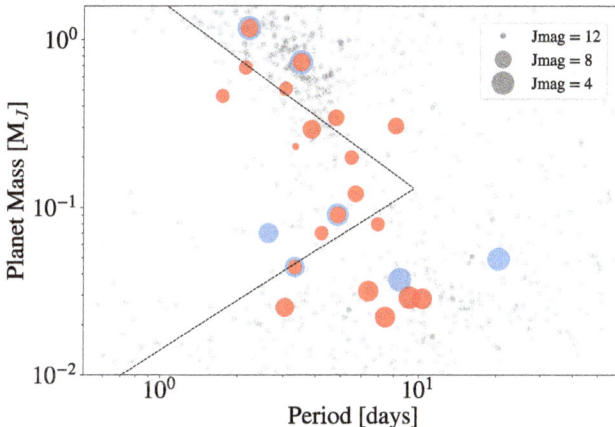

Figure 11. Orbital period versus mass distribution for the current sample of planets with measured mass loss rates ($> 3\sigma$ significance; for a complete list see review by Dos Santos 2022) using either Lyman α (**blue filled circles**; we also include a detection of AU Mic b by Rockcliffe et al. 2023), metastable helium (**red filled circles**, we also include detections for HAT-P-67 b, TOI-1268 b, TOI-1420 b, and TOI-2134 b from Gully-Santiago et al. 2023; Pérez González et al. 2023; Zhang et al. 2023a), or both (**blue circles with red fill**). The size of the points is scaled according to the host star's brightness in *J* band (infrared), which depends on the star's distance and mass; brighter stars (smaller *J* magnitudes) are generally located closer to the Earth and/or have larger masses. The full sample of confirmed planets is shown as **grey circles** for comparison. There is a deficit of sub-Saturn-sized planets on close-in orbits; this region is called the 'Neptune desert', and its approximate boundaries as defined in Mazeh et al. (2016) are shown as **black dashed lines**. Figure courtesy of M. Saidel.

The magnitude of the atmospheric absorption signal during transit can be converted into a mass loss rate by modeling the outflow as a spherically symmetric isothermal Parker wind (Oklopčić and Hirata 2018; Lampón et al. 2020; Dos Santos et al. 2022; Linssen et al. 2022). If the outflow is not spherical but instead is sculpted into a comet-like tail by the stellar wind, we would expect to see an extended absorption signal after the end of the transit egress (e.g., Ehrenreich et al. 2015; Lavie et al. 2017; Kirk et al. 2020; Spake et al. 2021). If there is outflowing material orbiting just ahead of the planet, we may also see absorption prior to the planet's ingress, or even absorption extending many hours before and/or after the planet transit (Gully-Santiago et al. 2023; Zhang et al. 2023c, see Fig. 12). The time-dependent absorption signal, as well as its spectroscopically resolved velocity structure, therefore provide us with important information about the three-dimensional structure of the atmospheric outflow (e.g., Wang and Dai 2021a,b; MacLeod and Oklopčić 2022). In addition to stellar winds, these outflow geometries may also be shaped by the planetary and stellar magnetic field geometries (Owen and Adams 2014; Fossati et al. 2023; Schreyer et al. 2023).

It is more challenging to detect atmospheric outflows from sub-Neptune-sized planets. There are currently only three sub-Neptune-sized planets orbiting mature (>1 Gyr) stars with published detections (Bourrier et al. 2018; Ninan et al. 2020; Pallé et al. 2020b; Orell-Miquel et al. 2022; Zhang et al. 2023a), one of which is disputed (GJ 1214 b; see discussion in Spake et al. 2022). Fortunately, these outflows are more easily observable if we focus on the subset of small planets orbiting young stars. Young stars are more active and have enhanced high energy fluxes (e.g., Johnstone et al. 2021; King and Wheatley 2021), while young planets have radii that are still inflated by leftover heat from their formation. As a result, young sub-Neptunes are expected to have enhanced mass loss rates as compared to their more evolved counterparts.

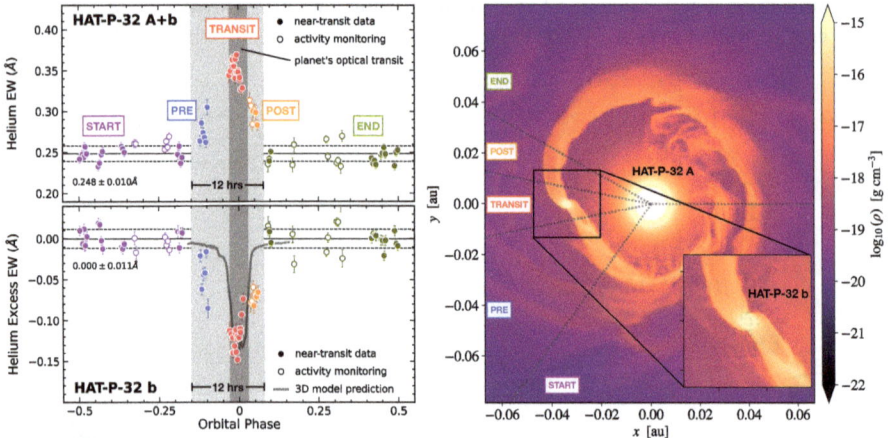

Figure 12. Measurement of hot Jupiter HAT-P-32 b's 1083 nm metastable helium absorption signal as a function of orbital phase from Zhang et al. (2023c). The unusually extended nature of this planet's outflow is distinct from that of most other hot Jupiters with published helium detections, which tend to have more narrowly confined outflows. **Left, upper panel:** Helium line equivalent width (EW) for HAT-P-32 b as a function of orbital phase. **Solid circles** indicate data taken in conjunction with a transit event, while **open circles** indicate data taken as part of a stellar monitoring program. The phased data are divided into five sections marked by gray shading and colored accordingly. The period where the planet is transiting the star is shown with **dark gray shading**. Results from a 3D hydrodynamic model are overplotted as a **solid gray line**. **Left, lower panel:** Equivalent width values after subtracting the average stellar spectrum. **Right:** Slice through the orbital plane of a 3D hydrodynamic simulation of a system with properties similar to that of HAT-P-32. The outflowing gas expands into long tails that lead and trail the planet's orbit, resulting in strong helium absorption before and after the transit. Approximate viewing angles for each colored time segment are shown with **colored labels**. The logarithmic gas density distribution is indicated using the **color bar** on the right. Figure from Zhang et al. (2023c).

Observations of young transiting sub-Neptunes have revealed the presence of atmospheric outflows in both Lyman α (Zhang et al. 2022c) and metastable helium (Zhang et al. 2022a, 2023b; Orell-Miquel et al. 2023). These observations can be used to test the predictions of atmospheric mass loss models seeking to explain the origin of the bimodal radius distribution of small close-in planets (see *Exoplanet demographics*).

DAYSIDE TEMPERATURE STRUCTURE

The physics of thermal inversions

Measuring the thermal structure of exoplanet atmospheres provides critical insight into how energy is transported and deposited in planetary envelopes. In the solar system, for example, we know that Earth has a stratospheric thermal inversion due to UV/optical absorption by the O_3 molecule. Venus has a lower equilibrium temperature than the Earth, despite receiving nearly twice as much energy from the Sun as the Earth does, due to its high Bond albedo. Mercury has a scalding hot dayside and a frigid nightside because it lacks a thick atmosphere to transport heat. All of these types of processes can be assessed by measuring the dayside temperatures and vertical temperature gradients in exoplanet atmospheres.

Thermal inversions in particular have been an interesting phenomenon accessed via dayside thermal emission spectra. Planetary atmospheres that are strongly absorbing at the wavelengths at which their host stars put out most of their energy will experience heating at the location where the stellar energy is deposited. To ensure global energy balance, this heating comes at a cost of cooling regions deeper in the atmosphere, thus creating a thermal inversion in which temperature *increases* with altitude, peaking around the region where the starlight is absorbed (i.e., the $\tau \sim 1$ surface, where τ is the optical depth). Spectroscopically, thermal inversions are identified by observing spectral lines in emission, as opposed to absorption lines, which are seen when temperature decreases outwardly. The shape of spectral features relative to the surrounding continuum is therefore used to assess the temperature gradient in the observable portion of an exoplanet atmosphere via thermal emission measurements. By detecting a thermal inversion and simultaneously measuring atmospheric composition, astronomers can also attempt to infer which absorber(s) are responsible for the upper-atmosphere heating. As discussed in *Refractory species in hot Jupiter atmospheres*, TiO, VO, and a variety of refractory species have been proposed as optical and UV absorbers that can generate thermal inversions in hot Jupiters (e.g., Fortney et al. 2008; Lothringer et al. 2018). Other opacity sources such as hazes, clouds, or even water vapor for planets orbiting M-dwarfs have been proposed to similarly drive thermal inversions in cooler planets (e.g., Morley et al. 2015; Arney et al. 2016; Malik et al. 2019; Lavvas and Arfaux 2021; Roman et al. 2021). Because thermal inversions are generated by absorption of incident stellar energy, they are primarily expected to be a dayside phenomenon, although efficient horizontal heat exchange can cause them to persist away from the sub-stellar point and even around to a planet's nightside (e.g., Komacek et al. 2022).

The hot-to-ultrahot Jupiter transition

Forward models of hot Jupiter emission spectra have long predicted a transition in dayside thermal structure from planets with inversions to those without, as a function of decreasing planetary temperature. The thermal inversions would be driven by gas-phase optical and UV absorbers that condense out of the atmosphere at lower temperatures, thus rendering the atmosphere more transparent to stellar irradiation (and therefore producing un-inverted temperature profiles) at lower equilibrium temperatures (e.g., Hubeny et al. 2003). Fortney et al. (2008) initially proposed that TiO and VO should be the key drivers of thermal inversions, resulting in a transition to inverted temperature profiles around a planetary equilibrium temperature of 1500 K. Evidence of thermal inversions from secondary eclipse spectra probing planets around this cutoff temperature with Spitzer observations was initially mixed (e.g., Richardson et al. 2007; Charbonneau et al. 2008; Knutson et al. 2008, 2009; Todorov et al. 2010, 2012, 2013; Deming et al. 2011; Baskin et al. 2013; Diamond-Lowe et al. 2014). Ultimately, improved spectroscopic investigations with the HST+WFC3 instrument and high-resolution ground-based spectrographs clearly demonstrated un-inverted temperature profiles in various hot Jupiters around and above the predicted 1500 K cutoff temperature, via spectral features appearing in absorption (Birkby et al. 2013; Kreidberg et al. 2014b; Schwarz et al. 2015; Line et al. 2016, 2021). This was accompanied by failures to definitively detect gas-phase TiO and VO in transmission spectra of some of the same planets, implying removal via nightside condensation cold-trapping or some other disequilibrium chemistry process, or perhaps a more mundane explanation such as inaccurate line lists (e.g., Désert et al. 2008; Hoeijmakers et al. 2015).

Even hotter planets were ultimately required to produce definitive evidence for thermal inversions. 'Ultra-hot' Jupiters, as discussed in *Water, water everywhere* are those that are so hot that water dissociates in their atmospheres, and various refractory elements (not just Ti and V) are predicted to be in the gas phase (Lothringer et al. 2018; Parmentier et al. 2018). In these planets, temperature inversions are predicted to be helped along by gas-phase metals and oxides such as Fe, Mg, SiO, etc. Formally the cutoff between hot and ultrahot Jupiters occurs around $T_{eq} = 2200$ K. The first ultra-hot Jupiter to produce a clear detection of a dayside thermal inversion was WASP-121b (Evans et al. 2017). The 1.4 μm water feature in this planet's secondary eclipse spectrum appears in emission, although the feature is quite subtle. Other ultrahot Jupiters, as mentioned in *Water, water everywhere*, produced nearly featureless secondary eclipse spectra across the WFC3 bandpass, leading to ambiguous interpretation as to whether water was simply absent from these atmospheres or the dayside temperature profiles were isothermal, thus masking any spectral features (Sheppard et al. 2017; Kreidberg et al. 2018; Mansfield et al. 2018, 2021).

The picture of a transition to ultra-hot planets with thermal inversions becomes clearer with population-level studies. When looking at WFC3 thermal emission spectra vs. the planets' measured dayside temperatures, Mansfield et al. (2021) identify a clear trend from un-inverted temperature profiles at lower dayside temperatures, to inverted profiles at dayside temperatures above ~2500 K (Fig. 13). This is in line with predictions from forward models, although such models still predict the transition to thermal inversions to occur at somewhat lower temperatures. Interestingly, both the models and the data reveal a shift back to featureless spectra with WFC3 at even higher dayside temperatures ($T_{day} \gtrsim 3000$ K), corresponding to full removal of atmospheric H_2O via thermal dissociation. Another population-level prediction is that ultra-hot planets orbiting earlier-type (i.e., hotter) host stars should produce even larger thermal inversions because the peak of the stellar spectral energy distribution (SED) aligns particularly well with the expected UV/optical opacity sources in the planets' atmospheres. This prediction has played out in secondary eclipse observations of the ultra-hot Jupiter KELT-20b, which orbits a hot A-type host star. For this planet, the 1.4 μm H_2O feature appears strongly in emission, much more so than for comparably irradiated planets orbiting later-type G stars (Fu et al. 2022).

Recent JWST and high-resolution emission spectroscopy studies have solidified our understanding of the hot-to-ultra-hot Jupiter transition by providing increased precision and wavelength coverage. For example, JWST has the power to resolve the subtle shape of spectral features that were previously hidden in the noise of HST observations. In the case of the the ultra-hot Jupiter WASP-18b, the planet's emission spectrum, which was nearly featureless in HST observations (Sheppard et al. 2017; Arcangeli et al. 2018) is now revealed by JWST to contain very subtle water features in emission, thus confirming the presence of a thermal inversion (Coulombe et al. 2023). In contrast, the cooler 'normal' hot Jupiter HD 149026 b shows spectral features in absorption, including clear detections of H_2O and a strong CO_2 feature implying high metallicity (Bean et al. 2023). With high-resolution observations from

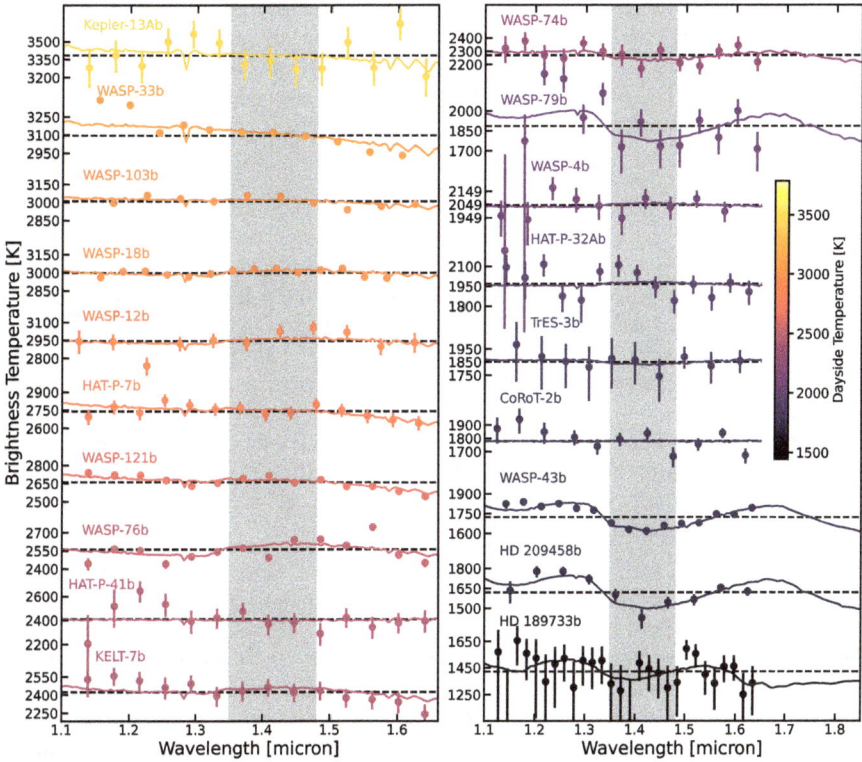

Figure 13. Brightness temperature vs. wavelength for hot Jupiter secondary eclipse spectra observed by HST with the WFC3 instrument. Brightness temperature is a measure of the approximate temperature of the photosphere at the wavelength being observed. The 1.4 μm water band is indicated by the **gray shaded** region. For less irradiated planets (toward the bottom right of the plot), the 1.4 μm water band appears in absorption, indicating atmospheres with non-inverted temperature profiles. For hotter planets, the absorption features disappear, and in some cases (e.g., WASP-76b, WASP-121b, WASP-12b), the 1.4 μm water band subtly inverts into emission, indicating a possible thermal inversion. When compared against forward models of hot Jupiter emission spectra, these observations align well with predictions that thermal inversions occur for planets hotter than ~2000 K, and water dissociation reduces the abundance of H_2O in the dayside atmosphere. Figure adapted from Mansfield et al. (2021).

the ground, the emission spectra of various ultrahot Jupiters also provide clear indications of emission lines, demonstrating inverted thermal structures (e.g., Kasper et al. 2021; Yan et al. 2022b, 2023; Brogi et al. 2023; van Sluijs et al. 2023). In at least one case, the detection of a thermal inversion is (finally) accompanied by a high-confidence detection of TiO in the transmission spectrum of the same planet (Yan et al. 2020; Prinoth et al. 2022).

In summary, with newer and better data and population-level studies, astronomers are now finding that the dayside thermal structures of hot and ultra-hot Jupiters appear to align with the basic predictions of forward models, albeit with a transition to inverted temperature profiles occurring at somewhat higher equilibrium temperatures than what is predicted for solar composition atmospheres. The forward models assume thermochemical equilibrium and 1-D radiative-convective energy balance, with gas-phase metals and metal oxides serving as strong optical and UV absorbers at high temperatures. Additional new frontiers that will be opened with JWST in the near term include studies of the thermal structures of even colder giant planets. At lower temperatures, dayside clouds or even haze might play a primary role in mediating the deposition of stellar energy in the planets' atmospheres. Hints of such effects already exist in the WFC3 emission spectra of the coolest hot Jupiters investigated to-date (Crouzet et al. 2014; Mansfield et al. 2021). JWST will also enable more detailed studies of the *3-D* structure of giant planet daysides, which will be discussed in more detail in *Three-dimensional atmospheric structure*.

Hot Jupiter albedos

Efforts to measure the reflected light from hot Jupiters at optical wavelengths began soon after the discovery of such planets, in order to constrain their albedos. These studies initially resulted in non-detections and upper limits, some of which were quite constraining (e.g., Charbonneau et al. 1999; Rowe et al. 2008; Winn et al. 2008). It was quickly realized that the implied low albedos were in line with the predictions from radiative transfer models for such planets. In the absence of dayside clouds, strong optical absorption lines such as those from Na, K, TiO, etc. absorb out much of the incident stellar radiation, while the only source of reflected light is Rayleigh scattering from the gaseous atmosphere (Seager et al. 2000; Burrows et al. 2008). For cooler planets, in which reflective dayside clouds are expected, geometric albedos should be higher (e.g., Cahoy et al. 2010; Adams et al. 2022), but the general trend of shallower secondary eclipses with lower levels of insolation typically makes it more challenging to detect such signals.

Space telescopes such as CoRoT, Kepler, and TESS were ultimately able to detect the optical secondary eclipses of a number of hot and ultrahot Jupiters, although the broad photometric bandpasses of these facilities has meant that it is typically not possible to fully disentangle the relative contributions of thermal emission vs. scattered light, resulting in model-dependent albedo inferences (e.g., Alonso et al. 2009; Christiansen et al. 2010; Demory et al. 2011). A compilation of optical secondary eclipse measurements for 21 planets with CoRoT, Kepler, and TESS reveals that most such detections have been made at less than 3σ confidence, with inferred geometric albedos ranging between 0 and $\sim$0.3 (Wong et al. 2020). One notable exception is the planet Kepler-7b, which has an inferred albedo of $\sim$0.25–0.35, measured at high confidence (Demory et al. 2011; Wong et al. 2020).

This is consistent with the planet's (relatively) low dayside temperature of $\sim$1000 K and the expectation that such conditions are conducive to the formation of reflective clouds. More recently, the European CHEOPS satellite has demonstrated its ability to produce well-constrained measurements of hot Jupiter geometric albedos (Brandeker et al. 2022; Krenn et al. 2023). The inferred values for the hot Jupiters HD 209458 b and HD 189733 b from CHEOPS light curves are 0.096 ± 0.016 and 0.076 ± 0.016. These albedos are far lower than for any solar system planets but in line with models of hot Jupiters having cloud-free dayside atmospheres. In summary, hot Jupiters are dark, but cooler giant planets may be more reflective.

The dayside temperatures of sub-Neptunes and super-Earths

Detecting the thermal emission from smaller and typically cooler sub-Neptunes and super-Earths is a much more technically challenging endeavor than for hot Jupiters. Because of this, such studies have mostly been limited to simply detecting a secondary eclipse and measuring an associated brightness temperature, as opposed to full spectroscopic characterization. Once measured, the dayside temperature of the planet can then be used to obtain a combined constraint on both daynight heat redistribution and albedo. All tidally-locked planets have a maximum dayside temperature that can be achieved if the planet's only energy source is the irradiation from its host star:

$$T_{\max} = T_* \sqrt{\frac{R_*}{d}} \left(\frac{2}{3}\right)^{1/4} \tag{5}$$

This is simply Equation (3) taken in the limit of no day–night heat redistribution (instantaneous reradiation) and zero albedo. Lower measured dayside temperatures are indicative of either a reflective planet or considerable day–night heat transport (or some combination thereof; Koll et al. 2019; Mansfield et al. 2019; Koll 2022).

To date there have only been successful thermal emission detections for two sub-Neptunes: the planets TOI-824b (Roy et al. 2022) and GJ 1214b (Kempton et al. 2023). The former is a hot dense sub-Neptune, whereas the latter is a cooler planet that was already known to have a thick aerosol layer from transmission spectroscopy measurements (see *Aerosols in sub-Neptunes and super-Earths*). The dayside temperature of TOI-824b is consistent with its T_{max}, whereas GJ 1214b is significantly colder. For sub-Neptunes, a maximally hot dayside, implying poor day–night heat redistribution, requires a high mean molecular weight atmosphere. This result comes from 3-D general circulation models, which demonstrate that heat transport efficiency decreases as a function of increasing mean molecular weight (e.g., Kataria et al. 2014; Charnay et al. 2015; Zhang and Showman 2017). Hydrogen-rich, solar-composition subNeptune atmospheres are predicted to transport heat very efficiently, resulting in cooler daysides and nearly homogeneous global temperatures. Conversely, GJ 1214b's dayside temperature is colder than even its zero-albedo temperature in the limit of fully efficient day–night heat transport, meaning the planet must have a non-zero albedo. This interpretation is confirmed by a full-orbit phase curve with JWST that is best fit by a high mean molecular weight atmosphere coupled with the presence of highly reflective aerosols (see *Aerosols in sub-Neptunes and super-Earths* and *Three-dimensional atmospheric structure*).

GJ 1214b is also the only planet smaller than Neptune to have spectral features identified in its dayside thermal emission spectrum. Subtle departures from a blackbody shape imply the presence of gaseous water in this planet's atmosphere and a non-inverted temperature profile (Kempton et al. 2023). Interestingly, for planets orbiting M-dwarf host stars, water vapor can actually serve as a source of thermal inversions (Malik et al. 2019; Selsis et al. 2023). This is because its strong near-IR opacity efficiently absorbs stellar light, which in this case peaks at red to near-IR wavelengths. The predicted thermal inversions are fairly weak and high up in the planets' atmospheres though, making their observable consequences negligible for low-resolution spectroscopy with JWST.

Rocky planet thermal emission measurements with Spitzer and more recently JWST have focused on measuring dayside temperatures (as well as phase curves in certain cases) to constrain the presence or absence of an atmosphere. Rocky planets without atmospheres have no mechanism by which to transport heat to their nightsides (Seager and Deming 2009; Koll et al. 2019; Koll 2022). Furthermore, many kinds of rocks that are known to form planetary surfaces in the solar system are very dark[15] (Hu et al. 2012; Mansfield et al. 2019). It therefore can be concluded that a terrestrial planet with a maximally hot dayside temperature is unlikely to have an atmosphere, whereas colder dayside temperatures imply the presence of an atmosphere. Several terrestrial planets to-date have been subjected to this 'secondary eclipse test' to measure their dayside temperatures, with the conclusion in the majority of cases being to rule out the presence of thick atmospheres to varying degrees of confidence (Kreidberg et al. 2019; Crossfield et al. 2022; Whittaker et al. 2022; Greene et al. 2023; Ih et al. 2023, see *The challenge of terrestrial planets*). The coldest terrestrial planet yet observed in secondary eclipse is TRAPPIST-1c. For that planet, its dayside temperature is only ~2σ consistent with its T_{max} value (Zieba et al. 2023). In this case, the presence of a thick atmosphere is not definitively ruled out, implying that perhaps less irradiated planets are more likely to retain their atmospheres, even if they orbit active M-dwarf stars. Further measurements of rocky planet secondary

---

[15] The assumption of dark planetary surfaces breaks down in the habitable zone and at very small orbital separations, where reflective surfaces are possible (Mansfield et al. 2019).

eclipses with JWST will continue to map out the parameter space of which planets do and do not possess atmospheres, with many such observations already planned for Cycle 2.

THREE-DIMENSIONAL ATMOSPHERIC STRUCTURE

Close-in exoplanets are expected to be tidally locked, with permanent day and night sides. As a result, they can exhibit relatively large day–night temperature gradients, along with corresponding gradients in their atmospheric chemistries and cloud properties. Importantly, tidally locked planets will have relatively slow rotation periods (on the order of days) compared to the gas giant planets in the solar system. This means that the typical length scales for their atmospheric circulation patterns will be much larger (~hemisphere-scale) than those of planets like Earth, Jupiter, or Saturn. For a review of the fundamental principles and relevant dynamical regimes for atmospheric circulation on close-in gas giant planets, we recommend Showman et al. (2010, 2020).

Fundamentals of day–night heat transport on hot Jupiters

There is a considerable body of observational constraints on the atmospheric circulation patterns of hot Jupiters. During its sixteen years of operation, the Spitzer Space telescope measured broadband infrared secondary eclipse depths for more than a hundred close-in gas giant exoplanets (e.g., Baxter et al. 2020; Wallack et al. 2021; Deming et al. 2023). It also measured broadband infrared phase curves for several dozen gas giant exoplanets (e.g., Bell et al. 2021; May et al. 2022). There are only a few planets with spectroscopic phase curves measured with HST (Stevenson et al. 2014; Kreidberg et al. 2018; Arcangeli et al. 2019, 2021; Mikal-Evans et al. 2022) and (more recently) JWST (Bell et al. 2023b; Kempton et al. 2023; Mikal-Evans et al. 2023b). Lastly, there are currently two published secondary eclipse maps of the dayside atmospheres of these planets, one from Spitzer (de Wit et al. 2012; Majeau et al. 2012) and one from JWST (Coulombe et al. 2023).

There are several big-picture takeaways that have emerged from the current body of observations. First, both models (Perez-Becker and Showman 2013; Komacek and Showman 2016; Komacek et al. 2017) and observations (e.g., Bell et al. 2021; Wallack et al. 2021; May et al. 2022; Deming et al. 2023) agree that the most highly irradiated gas giant exoplanets have a lower day–night heat redistribution efficiency (defined as the fraction of energy incident on the dayside that is transported to the night side by atmospheric winds) than their more moderately irradiated counterparts. This means that the most highly irradiated gas giant exoplanets have relatively large day–night temperature contrasts, while their cooler counterparts tend to have more uniform temperature distributions (see Fig. 14).

These same data also indicate that most close-in gas giant exoplanets have a super-rotating (eastward) equatorial band of wind that transports energy from the day side to the night side, in good agreement with predictions from general circulation models (see Fig. 15 and review by Showman et al. 2020). This is readily apparent in infrared Spitzer phase curve observations (Bell et al. 2021; May et al. 2022), which show that the hottest region on the day side is shifted eastward of the substellar point for most hot Jupiters (this corresponds to a phase curve that peaks just before the secondary eclipse). There are several notable exceptions to this trend, which we discuss in more detail later in this section. We can also see the effects of atmospheric circulation in high resolution emission and transmission spectroscopy, where we can directly measure the Doppler shift induced by the planet's atmospheric winds. This can manifest as either a net shift in the lines for a single coherent flow direction, or an overall broadening of the lines for observations that integrate over multiple flow directions (e.g., Miller-Ricci Kempton and Rauscher 2012a; Showman et al. 2013; Beltz et al. 2021, 2022). Doppler shifts due to atmospheric winds have been seen in high resolution transmission spectroscopy, which probes

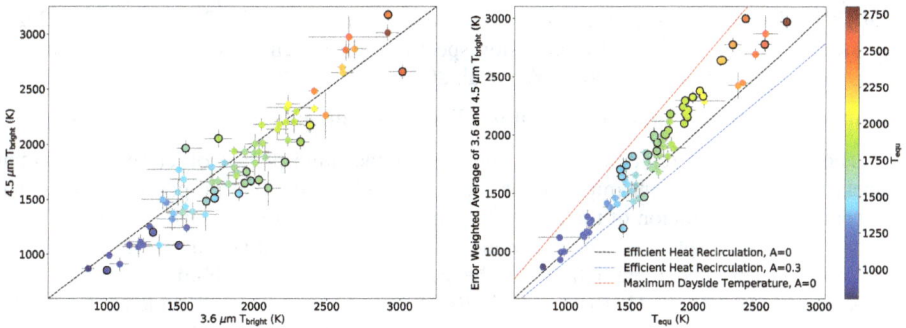

Figure 14. Dayside brightness temperatures for the sample of hot Jupiters with measured eclipse depths in the 3.6 and 4.5 μm bands with Spitzer as a function of their predicted equilibrium temperatures. Planets with relatively inefficient day–night recirculation will lie closer to the **red dashed line** (maximum dayside temperature assuming zero recirculation and zero albedo), while planets with relatively efficient day–night circulation will lie closer to the **black dashed line** (complete day–night recirculation of energy, zero albedo). The subset of hot Jupiters whose dayside albedos are enhanced by reflective silicate clouds (equilibrium temperatures near 1500 K) can also lie below the **black line**, as indicated by the **blue dashed line**. Planets with **black circles** have spectral slopes that are inconsistent with that of a blackbody across the 3.6 to 4.5 μm band, indicating the presence of strong molecular features. Figure from Wallack et al. (2021).

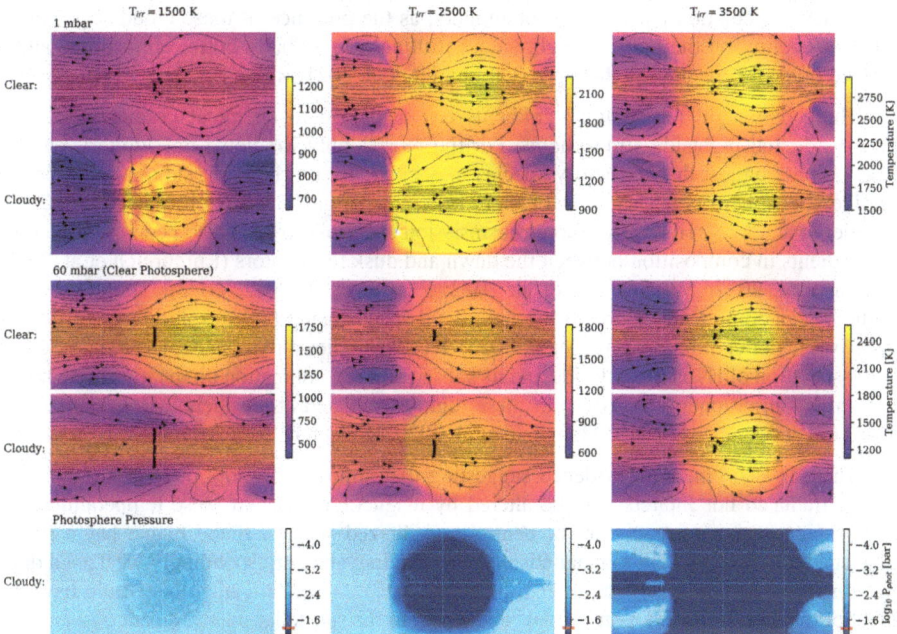

Figure 15. Properties of hot Jupiters with three different incident flux levels from general circulation models with and without radiatively active clouds included. The **top two rows** show the temperature distribution at the top of the atmosphere (1 mbar), and the **middle two rows** show the temperature distribution slightly deeper down, at the approximate level of the infrared photosphere. The **bottom panel** shows how the presence of clouds alters the level of the photosphere by increasing the atmospheric opacity; more cloudy regions have lower photospheric pressures, meaning that we do not see as deep into these cloudy regions. The approximate photospheric pressure for the clear atmosphere (60 mbar) is indicated by a **red line** on the color bar. Figure from Roman et al. (2021).

the day–night terminator region (e.g., Snellen et al. 2010; Louden and Wheatley 2015; Brogi et al. 2016; Flowers et al. 2019; Seidel et al. 2021; Gandhi et al. 2022; Kesseli et al. 2022; Pai Asnodkar et al. 2022), and in emission spectroscopy, which integrates over the dayside atmosphere (e.g., Lesjak et al. 2023; Yan et al. 2023).

Complications from clouds, chemical gradients, and magnetic fields

The non-uniform temperature distributions in the atmospheres of close-in gas giant planets also have important implications for their condensate cloud properties. Clouds that can condense in one region of the atmosphere may not be able to condense in other regions; this can lead to hemisphere-sized cloudy and clear regions in the atmospheres of these planets (Fig. 15, and for model predictions of the irradiation-dependent cloud distributions, see Parmentier et al. 2018, 2021; Roman et al. 2021). We can see empirical evidence for patchy clouds in the optical phase curves of close-in planets, which exhibit localized regions of high albedo due to the presence of reflective silicate clouds (Demory et al. 2013). When viewed in transmission, the properties of clouds and/or hazes are also expected to differ between the dawn and dusk terminators. This effect will cause the shape of the transit ingress (when the planet is entering the disk of the star) to differ from that of the transit egress (when the planet is exiting the disk of the star), and should be detectable with JWST (Kempton et al. 2017; Powell et al. 2019; Espinoza and Jones 2021; Steinrueck et al. 2021; Carone et al. 2023). Close-in gas giant planets also appear to have surprisingly uniform nightside temperatures, and it has been suggested that this may be due to the presence of nightside clouds (Keating et al. 2019; Gao and Powell 2021). If confirmed by JWST, this would have important consequences for atmospheric circulation patterns on hot Jupiters, as the presence of these clouds can inhibit radiative cooling on the planet's night side, resulting in a globally hotter atmosphere and a reduced offset for the dayside hot spot (Parmentier et al. 2021; Roman et al. 2021).

These day–night temperature gradients can also lead to chemical gradients between the dayside and nightside atmospheres. For the most highly irradiated hot Jupiters, current observations suggest that refractory species may condense on the night side, even when the dayside atmosphere is hot enough for them to remain in gas phase (Lothringer et al. 2022; Pelletier et al. 2023). High resolution transmsission spectroscopy has also been used to argue for gradients in composition between the dawn and dusk terminators (Ehrenreich et al. 2020; Gandhi et al. 2022; Mikal-Evans et al. 2022; Prinoth et al. 2022, 2023). These chemical gradients can complicate efforts to measure wind speeds using high resolution spectroscopy (e.g., Wardenier et al. 2023; Savel et al. 2023). Recent studies have also explored the role that the dissociation of H_2 on the day side and its subsequent recombination on the night side might play in day–night energy transport in these highly irradiated atmospheres (Bell and Cowan 2018; Tan and Komacek 2019; Mansfield et al. 2020; Roth et al. 2021; Changeat 2022).

There is also emerging evidence suggesting that atmospheric flow patterns on the most highly irradiated hot Jupiters may be altered by magnetic effects. At these temperatures, the atmosphere consists of a mixture of neutral and ionized species. If the planet has a strong magnetic field, this can lead to magnetically induced drag and correspondingly weakened day–night energy transport (Perna et al. 2010; Menou 2012). Observationally, this would have the effect of moving the hot spot on the day side closer to the substellar point. Recent observations of the ultra-hot Jupiter WASP-18b with JWST indicate that its relatively small dayside hot spot offset is best matched by circulation models with enhanced drag due to MHD effects (Coulombe et al. 2023). Other ultra-hot Jupiters also appear to have similarly small hot spot offsets in their Spitzer phase curves (Bell et al. 2021; May et al. 2022). As the magnetic field increasingly dominates the atmospheric flow patterns, it may cause the location of the dayside hot spot to vary from orbit to orbit, perhaps even shifting it westward of the substellar point (i.e., opposite of the predicted wind direction for a neutral atmosphere; Rogers 2017; Hindle et al. 2021a,b). This may explain the westward and/or time-varying hot spot offsets of several hot Jupiters (e.g., Dang et al. 2018; Bell et al. 2019). Several planets also appear to have time-varying optical phase curves (Armstrong et al. 2016; Jackson et al. 2019a,b), although in some cases these variations may be the result of stellar and/or instrumental variability (Lally and Vanderburg 2022).

Circulation patterns of sub-Neptune-sized planets

There are currently only a few sub-Neptune-sized planets with phase curve observations. As discussed under *The dayside temperatures of sub-Neptunes and super-Earths*, for rocky planets these phase curves can be used to infer the presence or absence of a thick atmosphere based on the observed day–night temperature gradient (Seager and Deming 2009). For planets with thick atmospheres, the shape of the phase curve can also be used to constrain the planet's atmospheric composition. A recent JWST observation of the mid-IR phase curve of the sub-Neptune GJ 1214 b by Kempton et al. (2023) indicates that this planet likely possesses a high mean molecular weight atmosphere with highly reflective clouds or hazes. Spitzer phase curves of hot rocky super-Earths K2-141b and LHS 3844 b indicate that these planets have large day–night temperature gradients, suggesting that their atmospheres must be relatively tenuous, if they possess one at all (Kreidberg et al. 2019; Zieba et al. 2022). Although the Spitzer IR phase curve of the super-Earth 55 Cnc e initially appeared to require the presence of a thick atmosphere (Demory et al. 2016b), a subsequent re-analysis of these data resulted in a larger day–night temperature gradient more in line with those observed for other hot rocky super-Earths (Mercier et al. 2022). Puzzlingly, this planet also appears to have a time-varying infrared flux from its dayside (Demory et al. 2016a), along with a variable optical phase curve (Meier Valdés et al. 2023). This may indicate the presence of a tenuous, time-varying outgassed high mean molecular weight atmosphere (Heng 2023). JWST will soon observe phase curves for multiple additional rocky exoplanets and sub-Neptunes, expanding our understanding of their atmospheric properties.

CONCLUSIONS

With the launch of JWST and the advent of high-resolution spectrographs on large ground-based telescopes, the exoplanet atmospheres field has entered a new era. As detailed in the previous sections, we are now reliably measuring chemical abundances and abundance ratios, global temperature fields, wind speeds, and atmospheric escape rates. At the time of writing this article we are just over one year into the JWST mission, and we are already seeing results that are fundamentally shifting our understanding of exoplanet atmospheres. Some early takeaways include the diversity of chemical inventories in giant planet atmospheres and an apparent lack of atmospheres on at least some rocky planets orbiting M-dwarfs. At the same time, a multitude of results from ground-based high-resolution spectroscopy are revealing the richness of (ultra)hot Jupiter chemistry.

With these new observational results come new scientific questions. It is already clear that at this improved level of measurement precision, the 3-D nature of exoplanet atmospheres will need to be carefully taken into account to not bias any scientific conclusions. This challenge is accompanied by new opportunities to directly infer properties of 3-D circulation and weather in exoplanet atmospheres. Measurements of individual exoplanets' spectra are also telling a story about how those planets formed and evolved, but backing out the correct narrative is a truly challenging endeavor, which can be helped along somewhat by population-level investigations. The characterization of smaller exoplanets is one of the key promises of the JWST mission, but new questions have arisen about what subset of such planets even host atmospheres at all and how to disentangle the signatures of stellar activity from atmospheric absorption. Investigations of sub-Neptunes aimed at distinguishing those with primordial atmospheres from a potential population of water worlds must still contend with the confounding influence of aerosols on spectroscopic observations. Along the way, the properties of the aerosols themselves are presenting their own surprises.

The pace of new observational results from JWST and ground-based high-resolution studies is only accelerating. We are in a regime in which our state of knowledge of exoplanet atmospheres in each successive year expands substantially. As such, this review article serves as a snapshot in time of the state of exoplanet observations following the first year of JWST science. We anticipate that some of the open questions presented in this article will be resolved in the near term, whereas others will take future generations of telescopes and scientists to fully answer. What we can surely say is that exoplanet atmospheres have yet to reveal all of their surprises to us.

ACKNOWLEDGEMENTS

E.M.R.K. and H.A.K. would like to acknowledge Jacob Bean, Yoni Brande, Thayne Currie, Peter Gao, Jegug Ih, Megan Mansfield, James Rogers, Michael Roman, Morgan Saidel, Arjun Savel, Nicole Wallack, and Zhoujian Zhang who graciously contributed figures to this review. H.A.K. is also grateful to the Woods Hole Geophysical Fluid Dynamics Program, which provided a thoughtful and interactive venue for developing ideas incorporated into several sections of this review. E.M.R.K. would like to thank Jacob Bean and Tad Komacek for insightful discussions while developing this manuscript.

REFERENCES

Ackerman AS, Marley MS (2001) Precipitating condensation clouds in substellar atmospheres. Astrophys J 556:872–884, https://doi.org/10.1086/321540

Adams DJ, Kataria T, Batalha NE, Gao P, Knutson HA (2022) Spatially resolved modeling of optical albedos for a sample of six hot Jupiters. Astrophys J 926:157, https://doi.org/10.3847/1538-4357/ac3d32

Aguichine A, Mousis O, Deleuil M, Marcq E (2021) Mass–radius relationships for irradiated ocean planets. Astrophys J 914:84, https://doi.org/10.3847/1538-4357/abfa99

Ahrer E-M, Stevenson KB, Mansfield M, et al. (2023) Early release science of the exoplanet WASP-39b with JWST NIRCam. Nature 614:653–658, https://doi.org/10.1038/s41586-022-05590-4

Alderson L, Wakeford HR, Alam MK, et al. (2023) Early release science of the exoplanet WASP-39b with JWST NIRSpec G395H. Nature 614:664–669, https://doi.org/10.1038/s41586-022-05591-3

Alonso R, Guillot T, Mazeh T, Aigrain S, Alapini A, Barge P, Hatzes A, Pont F (2009) The secondary eclipse of the transiting exoplanet CoRoT-2b. Astron Astrophys 501:L23–L26, https://doi.org/10.1051/0004-6361/200912505

Arcangeli J, Désert J-M, Line MR, Bean JL, Parmentier V, Stevenson KB, Kreidberg L, Fortney JJ, Mansfield M, Showman AP (2018) H⁻ opacity and water dissociation in the dayside atmosphere of the very hot gas giant WASP-18b. Astrophys J Lett 855:L30, https://doi.org/10.3847/2041-8213/aab272

Arcangeli J, Désert J-M, Parmentier V, Stevenson KB, Bean JL, Line MR, Kreidberg L, Fortney JJ, Showman AP (2019) Climate of an ultra hot Jupiter. Spectroscopic phase curve of WASP-18b with HST/WFC3. Astron Astrophys 625:A136, https://doi.org/10.1051/0004-6361/201834891

Arcangeli J, Désert J-M, Parmentier V, Tsai S-M, Stevenson KB (2021) A new approach to spectroscopic phase curves. The emission spectrum of WASP-12b observed in quadrature with HST/WFC3. Astron Astrophys 646:A94, https://doi.org/10.1051/0004-6361/202038865

Armstrong DJ, de Mooij E, Barstow J, Osborn HP, Blake J, Saniee NF (2016) Variability in the atmosphere of the hot giant planet HAT-P-7 b. Nat Astron 1:0004, https://doi.org/10.1038/s41550-016-0004

Arney G, Domagal-Goldman SD, Meadows VS, Wolf ET, Schwieterman E, Charnay B, Claire M, Hébrard E, Trainer MG (2016) The pale orange dot: The spectrum and habitability of hazy archean Earth. Astrobiology 16:873–899, https://doi.org/10.1089/ast.2015.1422

Atri D, Mogan SRC (2021) Stellar flares versus luminosity: XUV-induced atmospheric escape and planetary habitability. Mon Not R Astron Soc 500:L1–L5, https://doi.org/10.1093/mnrasl/slaa166

August PC, Bean JL, Zhang M, Lunine J, Xue Q, Line M, Smith P (2023) Confirmation of sub-solar metallicity for WASP-77Ab from JWST thermal emission spectroscopy. Astrophys J Lett 953:L24, https://doi.org/ 10.3847/2041-8213/ace828

Barstow JK, Irwin PGJ (2016) Habitable worlds with JWST: transit spectroscopy of the TRAPPIST-1 system? Mon Not R Astron Soc 461:L92–L96, https://doi.org/10.1093/mnrasl/slw109

Baskin NJ, Knutson HA, Burrows A, et al. (2013) Secondary eclipse photometry of the exoplanet WASP-5b with warm Spitzer. Astrophys J 773:124, https://doi.org/10.1088/0004-637X/773/2/124

Batalha NE, Line MR (2017) Information content analysis for selection of optimal JWST observing modes for transiting exoplanet atmospheres. Astron J 153:151, https://doi.org/10.3847/1538-3881/aa5faa

Batalha N, Kalirai J, Lunine J, Clampin M, Lindler D (2015) Transiting exoplanet simulations with the James Webb Space Telescope. arXiv:1507.02655, https://doi.org/10.48550/arXiv.1507.02655

Batalha NE, Lewis NK, Line MR, Valenti J, Stevenson K (2018) Strategies for constraining the atmospheres of temperate terrestrial planets with JWST. Astrophys J 856:L34, https://doi.org/10.3847/2041-8213/aab896

Baxter C, Désert J-M, Parmentier V, Line M, Fortney J, Arcangeli J, Bean JL, Todorov KO, Mansfield M (2020) A transition between the hot and the ultra-hot Jupiter atmospheres. Astron Astrophys 639:A36, https://doi.org/10.1051/0004-6361/201937394

Bean JL, Miller-Ricci Kempton E, Homeier D (2010) A ground-based transmission spectrum of the super-Earth exoplanet GJ 1214b. Nature 468:669–672, https://doi.org/10.1038/nature09596

Bean JL, Désert J-M, Kabath P, Stalder B, Seager S, Miller-Ricci Kempton E, Berta ZK, Homeier D, Walsh S, Seifahrt A (2011) The optical and near-infrared transmission spectrum of the super-Earth GJ 1214b: Further evidence for a metal-rich atmosphere. Astrophys J 743:92, https://doi.org/10.1088/0004-637X/743/1/92

Bean JL, Abbot DS, Kempton EM-R (2017) A statistical comparative planetology approach to the hunt for habitable exoplanets and life beyond the Solar System. Astrophys J Lett 841:L24, https://doi.org/10.3847/2041-8213/aa738a

Bean JL, Xue Q, August PC, et al. (2023) High atmospheric metal enrichment for a Saturn-mass planet. Nature 618:43–46, https://doi.org/10.1038/s41586-023-05984-y

Beichman C, Benneke B, Knutson H, et al. (2014) Observations of transiting exoplanets with the James Webb Space Telescope (JWST) Publ Astron Soc Pac 126:1134, https://doi.org/10.1086/679566

Bell TJ, Cowan NB (2018) Increased heat transport in ultra-hot Jupiter atmospheres through H_2 dissociation and recombination. Astrophys J Lett 857:L20, https://doi.org/10.3847/2041-8213/aabcc8

Bell TJ, Zhang M, Cubillos PE, et al. (2019) Mass loss from the exoplanet WASP-12b inferred from Spitzer phase curves. Mon Not R Astron Soc 489:1995–2013, https://doi.org/10.1093/mnras/stz2018

Bell TJ, Dang L, Cowan NB, et al. (2021) A comprehensive reanalysis of Spitzer's 4.5 micron phase curves, and the phase variations of the ultra-hot Jupiters MASCARA-1b and KELT-16b. Mon Not R Astron Soc 504:3316–3337, https://doi.org/10.1093/mnras/stab1027

Bell TJ, Kreidberg L, Kendrew S, Bean J, Crouzet N, Ducrot E, Dyrek A, Gao P, Lagage P-O, Moses JI (2023a) A first look at the JWST MIRI/LRS phase curve of WASP-43b. arXiv:2301.06350, https://doi.org/10.48550/arXiv.2301.06350

Bell TJ, Welbanks L, Schlawin E, et al. (2023b) Methane throughout the atmosphere of the warm exoplanet WASP-80b. Nature. 623:709–712, https://doi.org/10.1038/s41586-023-06687-0

Beltz H, Rauscher E, Brogi M, Kempton EM-R (2021) A significant increase in detection of high-resolution emission spectra using a three-dimensional atmospheric model of a hot Jupiter. Astron J 161:1, https://doi.org/10.3847/1538-3881/abb67b

Beltz H, Rauscher E, Kempton EM-R, Malsky I, Ochs G, Arora M, Savel A (2022) Magnetic drag and 3D effects in theoretical high-resolution emission spectra of ultrahot Jupiters: the case of WASP-76b. Astron J 164:140, https://doi.org/10.3847/1538-3881/ac897b

Benatti S, Nardiello D, Malavolta L, et al. (2019) A possibly inflated planet around the bright young star DS Tucanae A. Astron Astrophys 630:A81, https://doi.org/10.1051/0004-6361/201935598

Benneke B, Seager S (2012) Atmospheric retrieval for super-Earths: Uniquely constraining the atmospheric composition with transmission spectroscopy. Astrophys J 753:100, https://doi.org/10.1088/0004-637X/753/2/100

Benneke B, Seager S (2013) How to distinguish between cloudy mini-neptunes and water/volatile-dominated super-Earths. Astrophys J 778:153, https://doi.org/10.1088/0004-637X/778/2/153

Benneke B, Wong I, Piaulet C, et al. (2019a) Water vapor and clouds on the habitable-zone sub-Neptune exoplanet K2-18b. Astrophys J Lett 887:L14, https://doi.org/10.3847/2041-8213/ab59dc

Benneke B, Knutson HA, Lothringer J, et al. (2019b) A sub-Neptune exoplanet with a low-metallicity methane-depleted atmosphere and Mie-scattering clouds. Nat Astron 3:813–821, https://doi.org/10.1038/s41550-019-0800-5

Berta ZK, Charbonneau D, Désert J-M, Miller-Ricci Kempton E, McCullough PR, Burke CJ, Fortney JJ, Irwin J, Nutzman P, Homeier D (2012) The flat transmission spectrum of the super-Earth GJ1214b from Wide Field Camera 3 on the Hubble Space Telescope. Astrophys J 747:35, https://doi.org/10.1088/0004-637X/747/1/35

Bézard B, Charnay B, Blain D (2022) Methane as a dominant absorber in the habitable-zone sub-Neptune K2-18 b. Nat Astron 6:537–540, https://doi.org/10.1038/s41550-022-01678-z

Birkby JL (2018) Spectroscopic direct detection of exoplanets. In: Handbook of Exoplanets. Deeg HJ, Belmonte JA, (eds) Springer Charm, p 16

Birkby JL, de Kok RJ, Brogi M, de Mooij EJW, Schwarz H, Albrecht S, Snellen IAG (2013) Detection of water absorption in the day side atmosphere of HD 189733 b using ground-based high-resolution spectroscopy at 3.2 micron. Mon Not R Astron Soc 436:L35–L39, https://doi.org/10.1093/mnrasl/slt107

Bland-Hawthorn J, Gerhard O (2016) The galaxy in context: Structural, kinematic, and integrated properties. Annu Rev Astron Astrophys 54:529–596, https://doi.org/10.1146/annurev-astro-081915-023441

Booth RA, Clarke CJ, Madhusudhan N, Ilee JD (2017) Chemical enrichment of giant planets and discs due to pebble drift. Mon Not R Astron Soc 469:3994–4011, https://doi.org/10.1093/mnras/stx1103

Bourrier V, Lecavelier des Etangs A, Ehrenreich D, et al. (2018) Hubble PanCET: an extended upper atmosphere of neutral hydrogen around the warm Neptune GJ 3470b. Astron Astrophys 620:A147, https://doi.org/10.1051/0004-6361/201833675

Brande J, Crossfield IJM, Kreidberg L, et al. (2023) Clouds and clarity: Revisiting atmospheric feature trends in Neptune-size exoplanets. Astrophys J Lett 961:L23–L30, https://doi.org/10.3847/2041-8213/ad1b5c

Brandeker A, Heng K, Lendl M, et al. (2022) CHEOPS geometric albedo of the hot Jupiter HD 209458 b. Astron Astrophys 659:L4, https://doi.org/10.1051/0004-6361/202243082

Brogi M, de Kok RJ, Albrecht S, Snellen IAG, Birkby JL, Schwarz H (2016) Rotation and winds of exoplanet HD 189733b measured with high-dispersion transmission spectroscopy. Astrophys J 817:106, https://doi.org/10.3847/0004-637X/817/2/106

Brogi M, Emeka-Okafor V, Line MR, et al. (2023) The roasting marshmallows program with IGRINS on Gemini South I: Composition and climate of the ultrahot Jupiter WASP-18 b. Astron J 165:91, https://doi.org/10.3847/1538-3881/acaf5c

Bryant EM, Bayliss D, Van Eylen V (2023) The occurrence rate of giant planets orbiting low-mass stars with TESS. Mon Not R Astron Soc 521:3663–3681, https://doi.org/10.1093/mnras/stad626

Burningham B, Faherty JK, Gonzales EC, Marley MS, Visscher C, Lupu R, Gaarn J, Fabienne Bieger M, Freedman R, Saumon D (2021) Cloud busting: enstatite and quartz clouds in the atmosphere of 2M2224-0158. Mon Not R Astron Soc 506:1944–1961, https://doi.org/10.1093/mnras/stab1361

Burrows AS (2014) Highlights in the study of exoplanet atmospheres. Nature 513:345–352, https://doi.org/10.1038/nature13782

Burrows A, Sharp CM (1999) Chemical equilibrium abundances in brown dwarf and extrasolar giant planet atmospheres. Astrophys J 512:843–863, https://doi.org/10.1086/306811

Burrows A, Budaj J, Hubeny I (2008) Theoretical spectra and light curves of close-in extrasolar giant planets and comparison with data. Astrophys J 678:1436–1457, https://doi.org/10.1086/533518

Cáceres C, Kabath P, Hoyer S, Ivanov VD, Rojo P, Girard JH, Miller-Ricci Kempton E, Fortney JJ, Minniti D (2014) Ground-based transit observations of the super-Earth GJ 1214 b. Astron Astrophys 565:A7, https://doi.org/10.1051/0004-6361/201321087

Cahoy KL, Marley MS, Fortney JJ (2010) Exoplanet albedo spectra and colors as a function of planet phase, separation, and metallicity. Astrophys J 724:189–214, https://doi.org/10.1088/0004-637X/724/1/189

Carleo I, Giacobbe P, Guilluy G, et al. (2022) The GAPS programme at TNG XXXIX. Multiple molecular species in the atmosphere of the warm giant planet WASP-80 b unveiled at high resolution with GIANO-B. Astron J 164:101, https://doi.org/10.3847/1538-3881/ac80bf

Carone L, Lewis DA, Samra D, Schneider AD, Helling C (2023) WASP-39b: exo-Saturn with patchy cloud composition, moderate metallicity, and underdepleted S/O. arXiv:2301.08492, https://doi.org/10.48550/arXiv.2301.08492

Casasayas-Barris N, Pallé E, Yan F, et al. (2019) Atmospheric characterization of the ultra-hot Jupiter MASCARA-2b/KELT-20b. Detection of CaII, FeII, NaI, and the Balmer series of H (H-alpha, H-beta, and H-gamma) with high-dispersion transit spectroscopy. Astron Astrophys 628:A9, https://doi.org/10.1051/0004-6361/201935623

Casasayas-Barris N, Pallé E, Yan F, et al. (2020) Is there Na I in the atmosphere of HD 209458b? Effect of the centre-to-limb variation and Rossiter–McLaughlin effect in transmission. Astron Astrophys 635:A206, https://doi.org/10.1051/0004-6361/201937221

Casasayas-Barris N, Pallé E, Stangret M, et al. (2021) The atmosphere of HD 290458b seen with ESPRESSO: No detectable planetary absorptions at high resolution. Astron Astrophys 647:A26, https://doi.org/10.1051/0004-6361/202039539

Cauley PW, Kuckein C, Redfield S, Shkolnik EL, Denker C, Llama J, Verma M (2018) The effects of stellar activity on optical high-resolution exoplanet transmission spectra. Astron J 156:189, https://doi.org/10.3847/1538-3881/aaddf9

Chachan Y, Knutson HA, Lothringer J, Blake GA (2023) Breaking degeneracies in formation histories by measuring refractory content in gas giants. Astrophys J 943:112, https://doi.org/10.3847/1538-4357/aca614

Changeat Q (2022) On spectroscopic phase-curve retrievals: H_2 dissociation and thermal inversion in the atmosphere of the ultrahot Jupiter WASP-103. Astron J 163:106, https://doi.org/10.3847/1538-3881/ac4475

Changeat Q, Edwards B, Al-Refaie AF, et al. (2022) Five key exoplanet questions answered via the analysis of 25 hot-Jupiter atmospheres in eclipse. Astrophys J 260:3, https://doi.org/10.3847/1538-4365/ac5cc2

Charbonneau D, Noyes RW, Korzennik SG, Nisenson P, Jha S, Vogt SS, Kibrick RI (1999) An upper limit on the reflected light from the planet orbiting the star Tau Bootis. Astrophys J Lett 522:L145–L148, https://doi.org/10.1086/312234

Charbonneau D, Brown TM, Latham DW, Mayor M (2000) Detection of planetary transits across a Sun-like star. Astrophys J Lett 529:L45–L48, https://doi.org/10.1086/312457

Charbonneau D, Brown TM, Noyes RW, Gilliland RL (2002) Detection of an extrasolar planet atmosphere. Astrophys J 568:377–384, https://doi.org/10.1086/338770

Charbonneau D, Allen LE, Megeath ST, et al. (2005) Detection of thermal emission from an extrasolar planet. Astrophys J 626:523–529, https://doi.org/10.1086/429991

Charbonneau D, Knutson HA, Barman T, Allen LE, Mayor M, Megeath ST, Queloz D, Udry S (2008) The broadband infrared emission spectrum of the exoplanet HD 189733b. Astrophys J 686:1341–1348, https://doi.org/10.1086/591635

Charnay B, Meadows V, Misra A, Leconte J, Arney G (2015) 3D Modeling of GJ1214b's Atmosphere: Formation of inhomogeneous high clouds and observational implications. Astrophys J Lett 813:L1, https://doi.org/10.1088/2041-8205/813/1/L1

Christiansen JL, Ballard S, Charbonneau D, Madhusudhan N, Seager S, Holman MJ, Wellnitz DD, Deming D, A'Hearn MF, EPOXI Team (2010) Studying the atmosphere of the exoplanet HAT-P-7b via secondary eclipse measurements with EPOXI, Spitzer, and Kepler. Astrophys J 710:97–104, https://doi.org/10.1088/0004-637X/710/1/97

Constantinou S, Madhusudhan N, Gandhi S (2023) Early insights for atmospheric retrievals of exoplanets using JWST transit spectroscopy. Astrophys J Lett 943:L10, https://doi.org/10.3847/2041-8213/acaead

Cooper CS, Showman AP (2006) Dynamics and disequilibrium carbon chemistry in hot Jupiter atmospheres, with application to HD 209458b. Astrophys J 649:1048–1063, https://doi.org/10.1086/506312

Coulombe L-P, Benneke B, Challener R, et al. (2023) A broadband thermal emission spectrum of the ultra-hot Jupiter WASP-18b. Nature 620:292–298, https://doi.org/10.1038/s41586-023-06230-1

Cowan NB, Agol E (2008) Inverting phase functions to map exoplanets. Astrophys J Lett 678:L129, https://doi.org/10.1086/588553

Cowan NB, Agol E (2011) A model for thermal phase variations of circular and eccentric exoplanets. Astrophys J 726:82, https://doi.org/10.1088/0004-637X/726/2/82

Crossfield IJM (2023) Volatile-to-sulfur ratios can recover a gas giant's accretion history. Astrophys J Lett 952:L18, https://doi.org/10.3847/2041-8213/ace35f

Crossfield IJM, Kreidberg L (2017) Trends in atmospheric properties of Neptune-size exoplanets. Astron J 154:261, https://doi.org/10.3847/1538-3881/aa9279

Crossfield IJM, Malik M, Hill ML, et al. (2022) GJ1252b: A hot terrestrial super-Earth with no atmosphere. Astrophys J Lett 937:L17, https://doi.org/10.3847/2041-8213/ac886b

Crouzet N, McCullough PR, Deming D, Madhusudhan N (2014) Water vapor in the spectrum of the extrasolar planet HD 189733b. II. The eclipse. Astrophys J 795:166, https://doi.org/10.1088/0004-637X/795/2/166

Currie T, Biller B, Lagrange A, Marois C, Guyon O, Nielsen EL, Bonnefoy M, De Rosa RJ (2023) Direct imaging and spectroscopy of extrasolar planets. *In*: Protostars and Planets VII. Inutsuka S, Aikawa Y, Muto T, Tomida K, Tamura M (eds), https://ui.adsabs.harvard.edu/abs/2023ASPC..534..799C/abstract

Dai F, Masuda K, Winn JN, Zeng L (2019) Homogeneous analysis of hot Earths: Masses, sizes, and compositions. Astrophys J 883:79, https://doi.org/10.3847/1538-4357/ab3a3b

Dang L, Cowan NB, Schwartz JC, et al. (2018) Detection of a westward hotspot offset in the atmosphere of hot gas giant CoRoT-2b. Nat Astron 2:220–227, https://doi.org/10.1038/s41550-017-0351-6

Dattilo A, Batalha NM, Bryson S (2023) A unified treatment of kepler occurrence to trace planet evolution I: methodology. Astron J 166:122, https://doi.org/10.3847/1538-3881/acebc8

David TJ, Hillenbrand LA, Petigura EA, et al. (2016) A Neptune-sized transiting planet closely orbiting a 5–10-million-year-old star. Nature 534:658–661, https://doi.org/10.1038/nature18293

David TJ, Cody AM, Hedges CL, et al. (2019) A warm Jupiter-sized planet transiting the pre-main-sequence star V1298 Tau. Astron J 158:79, https://doi.org/10.3847/1538-3881/ab290f

de Kok RJ, Brogi M, Snellen IAG, Birkby J, Albrecht S, de Mooij EJW (2013) Detection of carbon monoxide in the high-resolution day-side spectrum of the exoplanet HD 189733b. Astron Astrophys 554:A82, https://doi.org/10.1051/0004-6361/201321381

de Wit J, Gillon M, Demory B-O, Seager S (2012) Towards consistent mapping of distant worlds: secondary-eclipse scanning of the exoplanet HD 189733b. Astron Astrophys 548:A128, https://doi.org/10.1051/0004-6361/201219060

de Wit J, Wakeford HR, Gillon M, et al. (2016) A combined transmission spectrum of the Earth-sized exoplanets TRAPPIST-1 b and c. Nature 537:69–72, https://doi.org/10.1038/nature18641

de Wit J, Wakeford HR, Lewis NK, et al. (2018) Atmospheric reconnaissance of the habitable-zone Earth-sized planets orbiting TRAPPIST-1. Nat Astron 2:214–219, https://doi.org/10.1038/s41550-017-0374-z

Deming LD, Seager S (2017) Illusion and reality in the atmospheres of exoplanets. J Geophys Res (Planets) 122:53–75, https://doi.org/10.1002/2016JE005155

Deming D, Sheppard K (2017) Spectral resolution-linked bias in transit spectroscopy of extrasolar planets. Astrophys J Lett 841:L3, https://doi.org/10.3847/2041-8213/aa706c

Deming D, Seager S, Richardson LJ, Harrington J (2005) Infrared radiation from an extrasolar planet. Nature 434:740–743, https://doi.org/10.1038/nature03507

Deming D, Seager S, Winn J, et al. (2009) Discovery and characterization of transiting super Earths using an all-sky transit survey and follow-up by the James Webb Space Telescope. Publ Astron Soc Pac 121:952, https://doi.org/10.1086/605913

Deming D, Knutson H, Agol E, et al. (2011) Warm Spitzer photometry of the transiting exoplanets CoRoT-1 and CoRoT-2 at secondary eclipse. Astrophys J 726:95, https://doi.org/10.1088/0004-637X/726/2/95

Deming D, Wilkins A, McCullough P, et al. (2013) Infrared transmission spectroscopy of the exoplanets HD 209458b and XO-1b using the Wide Field Camera-3 on the Hubble Space Telescope. Astrophys J 774:95, https://doi.org/10.1088/0004-637X/774/2/95

Deming D, Line MR, Knutson HA, Crossfield IJM, Kempton EM-R, Komacek TD, Wallack NL, Fu G (2023) Emergent spectral fluxes of hot Jupiters: An abrupt rise in dayside brightness temperature under strong irradiation. Astron J 165:104, https://doi.org/10.3847/1538-3881/acb210

Demory B-O, Seager S, Madhusudhan N, et al. (2011) The high albedo of the hot Jupiter Kepler-7 b. Astrophys J Lett 735:L12, https://doi.org/10.1088/2041-8205/735/1/L12

Demory B-O, de Wit J, Lewis N, et al. (2013) Inference of inhomogeneous clouds in an exoplanet atmosphere. Astrophys J Lett 776:L25, https://doi.org/10.1088/2041-8205/776/2/L25

Demory B-O, Gillon M, Madhusudhan N, Queloz D (2016a) Variability in the super-Earth 55 Cnc e. Mon Not R Astron Soc 455:2018–2027, https://doi.org/10.1093/mnras/stv2239

Demory B-O, Gillon M, de Wit J, et al. (2016b) A map of the large day–night temperature gradient of a super-Earth exoplanet. Nature 532:207–209, https://doi.org/10.1038/nature17169

Désert J-M, Vidal-Madjar A, Lecavelier Des Etangs A, Sing D, Ehrenreich D, Hébrard G, Ferlet R (2008) TiO and VO broad band absorption features in the optical spectrum of the atmosphere of the hot-Jupiter HD 209458b. Astron Astrophys 492:585–592, https://doi.org/10.1051/0004-6361:200810355

Diamond-Lowe H, Stevenson KB, Bean JL, Line MR, Fortney JJ (2014) New analysis indicates no thermal inversion in the atmosphere of HD 209458b. Astrophys J 796:66, https://doi.org/10.1088/0004-637X/796/1/66

Diamond-Lowe H, Berta-Thompson Z, Charbonneau D, Kempton EM-R (2018) Ground-based optical transmission spectroscopy of the small, rocky exoplanet GJ 1132b. Astron J 156:42, https://doi.org/10.3847/1538-3881/aac6dd

Diamond-Lowe H, Charbonneau D, Malik M, Kempton EM-R, Beletsky Y (2020a) Optical transmission spectroscopy of the terrestrial exoplanet LHS 3844b from 13 ground-based transit observations. Astron J 160:188, https://doi.org/10.3847/1538-3881/abaf4f

Diamond-Lowe H, Berta-Thompson Z, Charbonneau D, Dittmann J, Kempton EM-R (2020b) Simultaneous optical transmission spectroscopy of a terrestrial, habitable-zone exoplanet with two ground-based multiobject spectrographs. Astron J 160:27, https://doi.org/10.3847/1538-3881/ab935f

Diamond-Lowe H, Mendonça JM, Charbonneau D, Buchhave LA (2023) Ground-based optical transmission spectroscopy of the nearby terrestrial exoplanet LTT 1445Ab. Astron J 165:169, https://doi.org/10.3847/1538-3881/acbf39

Domagal-Goldman SD, Segura A, Claire MW, Robinson TD, Meadows VS (2014) Abiotic ozone and oxygen in atmospheres similar to prebiotic Earth. Astrophys J 792:90, https://doi.org/10.1088/0004-637X/792/2/90

Dos Santos LA (2023) Observations of planetary winds and outflows. Proc Int Astron Union 370:56–71, https://doi.org/10.1017/S1743921322004239

Dos Santos LA, Vidotto AA, Vissapragada S, Alam MK, Allart R, Bourrier V, Kirk J, Seidel JV, Ehrenreich D (2022) p-winds: An open-source Python code to model planetary outflows and upper atmospheres. Astron Astrophys 659:A62, https://doi.org/10.1051/0004-6361/202142038

Dos Santos LA, García Muñoz A, Sing DK, et al. (2023) Hydrodynamic atmospheric escape in HD 189733 b: signatures of carbon and hydrogen measured with the Hubble Space Telescope. Astron J 166:89, https://doi.org/10.3847/1538-3881/ace445

Dressing CD, Charbonneau D (2015) The occurrence of potentially habitable planets orbiting M dwarfs estimated from the full Kepler dataset and an empirical measurement of the detection sensitivity. Astrophys J 807:45, https://doi.org/10.1088/0004-637X/807/1/45

Ehrenreich D, Bourrier V, Wheatley PJ, et al. (2015) A giant comet-like cloud of hydrogen escaping the warm Neptune-mass exoplanet GJ 436b. Nature 522:459–461, https://doi.org/10.1038/nature14501

Ehrenreich D, Lovis C, Allart R, et al. (2020) Nightside condensation of iron in an ultrahot giant exoplanet. Nature 580:597–601, https://doi.org/10.1038/s41586-020-2107-1

Esparza-Borges E, López-Morales M, Adams Redai JI, et al. (2023) Detection of carbon monoxide in the atmosphere of WASP-39b applying standard cross-correlation techniques to JWST NIRSpec G395H Data. Astrophys J Lett 955:L19, https://doi.org/10.3847/2041-8213/acf27b

Espinoza N, Jones K (2021) Constraining mornings and evenings on distant worlds: A new semianalytical approach and prospects with transmission spectroscopy. Astron J 162:165, https://doi.org/10.3847/1538-3881/ac134d

Evans TM, Sing DK, Kataria T, et al. (2017) An ultrahot gas-giant exoplanet with a stratosphere. Nature 548:58–61, https://doi.org/10.1038/nature23266

Evans TM, Sing DK, Goyal JM, et al. (2018) An optical transmission spectrum for the ultra-hot Jupiter WASP-121b measured with the Hubble Space Telescope. Astron J 156:283, https://doi.org/10.3847/1538-3881/aaebff

Fauchez TJ, Turbet M, Villanueva GL, et al. (2019) Impact of clouds and hazes on the simulated JWST transmission spectra of habitable zone planets in the TRAPPIST-1 System. Astrophys J 887:194, https://doi.org/10.3847/1538-4357/ab5862

Feinstein AD, Radica M, Welbanks L, et al. (2023) Early release science of the exoplanet WASP-39b with JWST NIRISS. Nature 614:670–675, https://doi.org/10.1038/s41586-022-05674-1

Fernandes RB, Mulders GD, Pascucci I, Mordasini C, Emsenhuber A (2019) Hints for a turnover at the snow line in the giant planet occurrence rate. Astrophys J 874:81, https://doi.org/10.3847/1538-4357/ab0300

Fischer DA, Howard AW, Laughlin GP, Macintosh B, Mahadevan S, Sahlmann J, Yee JC (2014) Exoplanet detection techniques. *In*: Protostars and Planets VI. Beuther H, Klessen RS, Dullemond CP, Henning T (eds), https://ui.adsabs.harvard.edu/abs/2014prpl.conf..715F/abstract

Flagg L, Turner JD, Deibert E, Ridden-Harper A, de Mooij E, MacDonald RJ, Jayawardhana R, Gibson N, Langeveld A, Sing D (2023) ExoGemS detection of a metal hydride in an exoplanet atmosphere at high spectral resolution. Astrophys J Lett 953:L19, https://doi.org/10.3847/2041-8213/ace529

Flowers E, Brogi M, Rauscher E, Kempton EM-R, Chiavassa A (2019) The high-resolution transmission spectrum of HD 189733b interpreted with atmospheric doppler shifts from three-dimensional general circulation models. Astron J 157:209, https://doi.org/10.3847/1538-3881/ab164c

Foreman-Mackey D, Morton TD, Hogg DW, Agol E, Schölkopf B (2016) The population of long-period transiting exoplanets. Astron J 152:206, https://doi.org/10.3847/0004-6256/152/6/206

Fortney JJ (2005) The effect of condensates on the characterization of transiting planet atmospheres with transmission spectroscopy. Mon Not R Astron Soc 364:649–653, https://doi.org/10.1111/j.1365-2966.2005.09587.x

Fortney JJ, Lodders K, Marley MS, Freedman RS (2008) A unified theory for the atmospheres of the hot and very hot Jupiters: Two classes of irradiated atmospheres. Astrophys J 678:1419–1435, https://doi.org/10.1086/528370

Fortney JJ, Shabram M, Showman AP, Lian Y, Freedman RS, Marley MS, Lewis NK (2010) Transmission spectra of three-dimensional hot Jupiter model atmospheres. Astrophys J 709:1396–1406, https://doi.org/10.1088/0004-637X/709/2/1396

Fortney JJ, Mordasini C, Nettelmann N, Kempton EM-R, Greene TP, Zahnle K (2013) A framework for characterizing the atmospheres of low-mass low-density transiting planets. Astrophys J 775:80, https://doi.org/10.1088/0004-637X/775/1/80

Fortney JJ, Visscher C, Marley MS, Hood CE, Line MR, Thorngren DP, Freedman RS, Lupu R (2020) Beyond equilibrium temperature: How the atmosphere/interior connection affects the onset of methane, ammonia, and clouds in warm transiting giant planets. Astron J 160:288, https://doi.org/10.3847/1538-3881/abc5bd

Fortney JJ, Dawson RI, Komacek TD (2021) Hot Jupiters: Origins, structure, atmospheres. J Geophys Res E: Planets 126:e06629, https://doi.org/10.1029/2020JE006629

Fossati L, Pillitteri I, Shaikhislamov IF, Bonfanti A, Borsa F, Carleo I, Guilluy G, Rumenskikh MS (2023) Possible origin of the non-detection of metastable He I in the upper atmosphere of the hot Jupiter WASP-80b. Astron Astrophys 673:A37, https://doi.org/10.1051/0004-6361/202245667

Fu G, Deming D, Knutson H, Madhusudhan N, Mandell A, Fraine J (2017) Statistical analysis of Hubble/WFC3 transit spectroscopy of extrasolar planets. Astrophys J Lett 847:L22, https://doi.org/10.3847/2041-8213/aa8e40

Fu G, Deming D, Lothringer J, et al. (2021) The Hubble PanCET program: Transit and eclipse spectroscopy of the strongly irradiated giant exoplanet WASP-76b. Astron J 162:108, https://doi.org/10.3847/1538-3881/ac1200

Fu G, Espinoza N, Sing DK, et al. (2022) Water and an escaping helium tail detected in the hazy and methane-depleted atmosphere of HAT-P-18b from JWST NIRISS/SOSS. Astrophys J Lett 940:L35, https://doi.org/10.3847/2041-8213/ac9977

Fulton BJ, Petigura EA (2018) The California–Kepler survey. VII. Precise planet radii leveraging Gaia DR2 reveal the stellar mass dependence of the planet radius gap. Astron J 156:264, https://doi.org/10.3847/1538-3881/aae828

Fulton BJ, Petigura EA, Howard AW, et al. (2017) The California–Kepler survey. III. A gap in the radius distribution of small planets. Astron J 154:109, https://doi.org/10.3847/1538-3881/aa80eb

Fulton BJ, Rosenthal LJ, Hirsch LA, et al. (2021) California Legacy survey. II. Occurrence of giant planets beyond the ice line. Astrophys. J.s 255:14, https://doi.org/10.3847/1538-4365/abfcc1

Gandhi S, Brogi M, Webb RK (2020a) Seeing above the clouds with high-resolution spectroscopy Mon Not R Astron Soc 498:194, https://doi.org/10.1093/mnras/staa2424

Gandhi S, Madhusudhan N, Mandell A (2020b) H⁻ and dissociation in ultra-hot Jupiters: A retrieval case study of WASP-18b. Astron J 159:232, https://doi.org/10.3847/1538-3881/ab845e

Gandhi S, Kesseli A, Snellen I, Brogi M, Wardenier JP, Parmentier V, Welbanks L, Savel AB (2022) Spatially resolving the terminator: variation of Fe, temperature, and winds in WASP-76 b across planetary limbs and orbital phase. Mon Not R Astron Soc 515:749–766, https://doi.org/10.1093/mnras/stac1744

Gandhi S, Kesseli A, Zhang Y, et al. (2023) Retrieval survey of metals in six ultrahot Jupiters: Trends in chemistry, rain-out, ionization, and atmospheric dynamics. Astron J 165:242, https://doi.org/10.3847/1538-3881/accd65

Gao P, Powell D (2021) A universal cloud composition on the nightsides of hot Jupiters. Astrophys J Lett 918:L7, https://doi.org/10.3847/2041-8213/ac139f

Gao P, Thorngren DP, Lee EKH, Fortney JJ, Morley CV, Wakeford HR, Powell DK, Stevenson KB, Zhang X (2020) Aerosol composition of hot giant exoplanets dominated by silicates and hydrocarbon hazes. Nat Astron 4:951–956, https://doi.org/10.1038/s41550-020-1114-3

Gao P, Wakeford HR, Moran SE, Parmentier V (2021) Aerosols in exoplanet atmospheres. J Geophys Res (Planets) 126:e06655, https://doi.org/10.1029/2020JE006655

Gaudi BS (2022) The demographics of wide-separation planets. *In*: Demographics of Exoplanetary Systems, Lecture Notes of the 3rd Advanced School on Exoplanetary Science. 466:237–291, https://doi.org/10.1007/978-3-030-88124-5 4

Gaudi BS, Meyer M, Christiansen J (2021) The demographics of exoplanets. *In*: ExoFrontiers; Big Questions in Exoplanetary Science, Madhusudhan N (ed) 2.1–2.21, https://doi.org/10.1088/2514-3433/abfa8fch2

Giacalone S, Dressing CD, Jensen ELN, et al. (2021) Vetting of 384 TESS objects of interest with TRICERATOPS and statistical validation of 12 planet candidates. Astron J 161:24, https://doi.org/10.3847/1538-3881/abc6af

Giacobbe P, Brogi M, Gandhi S, et al. (2021) Five carbon- and nitrogen-bearing species in a hot giant planet's atmosphere. Nature 592:205–208, https://doi.org/10.1038/s41586-021-03381-x

Gibson NP, Merritt S, Nugroho SK, et al. (2020) Detection of Fe I in the atmosphere of the ultra-hot Jupiter WASP-121b, and a new likelihood-based approach for Doppler-resolved spectroscopy. Mon Not R Astron Soc 493:2215–2228, https://doi.org/10.1093/mnras/staa228

Gibson NP, Nugroho SK, Lothringer J, Maguire C, Sing DK (2022) Relative abundance constraints from high-resolution optical transmission spectroscopy of WASP-121b, and a fast model-filtering technique for accelerating retrievals. Mon Not R Astron Soc 512:4618–4638, https://doi.org/10.1093/mnras/stac091

Gillon M, Triaud AHMJ, Demory B-O, et al. (2017) Seven temperate terrestrial planets around the nearby ultracool dwarf star TRAPPIST-1. Nature 542:456–460, https://doi.org/10.1038/nature21360

Ginzburg S, Schlichting HE, Sari Re (2018) Core-powered mass-loss and the radius distribution of small exoplanets. Mon Not R Astron Soc 476:759–765, https://doi.org/10.1093/mnras/sty290

Goyal JM, Lewis NK, Wakeford, HR, MacDonald RJ, Mayne NJ (2021) Why is it so hot in here? Exploring population trends in Spitzer thermal emission observations of hot Jupiters using planet-specific, self-consistent atmospheric models. Astrophys J 923:242, https://doi.org/10.3847/1538-4357/ac27b2

Grant D, Lothringer JD, Wakeford HR, et al. (2023) Detection of carbon monoxide's 4.6 micron fundamental band structure in WASP-39b's atmosphere with JWST NIRSpec G395H. Astrophys J Lett 949:L15, https://doi.org/10.3847/2041-8213/acd544

Greene TP, Bell TJ, Ducrot E, Dyrek A, Lagage P-O, Fortney JJ (2023) Thermal emission from the Earth-sized exoplanet TRAPPIST-1 b using JWST. Nature 618:39–42, https://doi.org/10.1038/s41586-023-05951-7

Guilluy G, Sozzetti A, Brogi M, et al. (2019) Exoplanet atmospheres with GIANO. II. Detection of molecular absorption in the dayside spectrum of HD 102195b. Astron Astrophys 625:A107, https://doi.org/10.1051/0004-6361/201834615

Guilluy G, Giacobbe P, Carleo I, et al. (2022) The GAPS programme at TNG. XXXVIII. Five molecules in the atmosphere of the warm giant planet WASP-69b detected at high spectral resolution. Astron Astrophys 665:A104, https://doi.org/10.1051/0004-6361/202243854

Gully-Santiago M, Morley CV, Luna J, et al. (2023) A large and variable leading tail of helium in a hot Saturn undergoing runaway inflation. arXiv:2307.08959, https://doi.org/10.48550/arXiv.2307.08959

Guo X, Crossfield IJM, Dragomir D, et al. (2020) Updated parameters and a new transmission spectrum of HD 97658b. Astron J 159:239, https://doi.org/10.3847/1538-3881/ab8815

Gupta A, Schlichting HE (2019) Sculpting the valley in the radius distribution of small exoplanets as a by-product of planet formation: the core-powered mass-loss mechanism. Mon Not R Astron Soc 487:24–33, https://doi.org/10.1093/mnras/stz1230

Hansen BMS (2008) On the absorption and redistribution of energy in irradiated planets. Astrophys J 179:484–508, https://doi.org/10.1086/591964

Harbach LM, Moschou SP, Garraffo C, Drake JJ, Alvarado-Gómez JD, Cohen O, Fraschetti F (2021) Stellar winds drive strong variations in exoplanet evaporative outflow patterns and transit absorption signatures. Astrophys J 913:130, https://doi.org/10.3847/1538-4357/abf63a

Hardegree-Ullman KK, Cushing MC, Muirhead PS, Christiansen JL (2019) Kepler planet occurrence rates for mid-type M dwarfs as a function of spectral type. Astron J 158:75, https://doi.org/10.3847/1538-3881/ab21d2

Hardegree-Ullman KK, Zink JK, Christiansen JL, Dressing CD, Ciardi DR, Schlieder JE (2020) Scaling K2. I. Revised parameters for 222,088 K2 stars and a K2 planet radius valley at 1.9. Astrophys J 247:28, https://doi.org/10.3847/1538-4365/ab7230

He C, Hörst SM, Lewis NK, et al. (2018) Photochemical haze formation in the atmospheres of super-Earths and mini-Neptunes. Astron J 156:38, https://doi.org/10.3847/1538-3881/aac883

He C, Hörst SM, Lewis NK, et al. (2020) Haze formation in warm H_2-rich exoplanet atmospheres. Planet Sci J 1:51, https://doi.org/10.3847/PSJ/abb1a4

Helled R, Lunine J (2014) Measuring Jupiter's water abundance by Juno: the link between interior and formation models. Mon Not R Astron Soc 441:2273–2279, https://doi.org/10.1093/mnras/stu516

Helling C, Gourbin P, Woitke P, Parmentier V (2019) Sparkling nights and very hot days on WASP-18b: The formation of clouds and the emergence of an ionosphere. Astron Astrophys 626:A133, https://doi.org/10.1051/0004-6361/201834085

Heng K (2023) The transient outgassed atmosphere of 55 Cancri e. Astrophys J Lett 956:L20, https://doi.org/10.3847/2041-8213/acfe05

Hindle AW, Bushby PJ, Rogers TM (2021a) Observational consequences of shallow-water magnetohydrodynamics on hot Jupiters. Astrophys J Lett 916:L8, https://doi.org/10.3847/2041-8213/ac0fec

Hindle AW, Bushby PJ, Rogers TM (2021b) The magnetic mechanism for hotspot reversals in hot Jupiter atmospheres. Astrophys J 922:176, https://doi.org/10.3847/1538-4357/ac0e2e

Hirano T, Krishnamurthy V, Gaidos E, et al. (2020) Limits on the spin-orbit angle and atmospheric escape for the 22 Myr old planet AU Mic b. Astrophys J Lett 899:L13, https://doi.org/10.3847/2041-8213/aba6eb

Hoeijmakers HJ, de Kok RJ, Snellen IAG, Brogi M, Birkby JL, Schwarz H (2015) A search for TiO in the optical high-resolution transmission spectrum of HD 209458b: Hindrance due to inaccuracies in the line database. Astron Astrophys 575:A20, https://doi.org/10.1051/0004-6361/201424794

Hoeijmakers HJ, Ehrenreich D, Heng K, et al. (2018) Atomic iron and titanium in the atmosphere of the exoplanet KELT-9b. Nature 560:453–455, https://doi.org/10.1038/s41586-018-0401-y

Hood CE, Fortney JJ, Line MR, Martin EC, Morley CV, Birkby JL, Rustamkulov Z, Lupu RE, Freedman RS (2020) Prospects for characterizing the haziest sub-Neptune exoplanets with high-resolution spectroscopy. Astron J 160:198, https://doi.org/10.3847/1538-3881/abb46b

Hörst SM, He C, Lewis NK, Kempton EM-R, Marley MS, Morley CV, Moses JI, Valenti JA, Vuitton V (2018) Haze production rates in super-Earth and mini-Neptune atmosphere experiments. Nat Astron 2:303–306, https://doi.org/10.1038/s41550-018-0397-0

Hsu DC, Ford EB, Ragozzine D, Ashby K (2019) Occurrence rates of planets orbiting FGK stars: Combining Kepler DR25, Gaia DR2, and Bayesian inference. Astron J 158:109, https://doi.org/10.3847/1538-3881/ab31ab

Hsu DC, Ford EB, Terrien R (2020) Occurrence rates of planets orbiting M stars: Applying ABC to Kepler DR25, Gaia DR2, and 2MASS data. Mon Not R Astron Soc 498:2249–2262, https://doi.org/10.1093/mnras/staa2391

Hu R, Ehlmann BL, Seager S (2012) Theoretical spectra of terrestrial exoplanet surfaces. Astrophys J 752:7, https://doi.org/10.1088/0004-637X/752/1/7

Huang C, Koskinen T, Lavvas P, Fossati L (2023) A hydrodynamic study of the escape of metal species and excited hydrogen from the atmosphere of the hot Jupiter WASP-121b. Astrophys J 951:123, https://doi.org/10.3847/1538-4357/accd5e

Hubeny I, Burrows A, Sudarsky D (2003) A possible bifurcation in atmospheres of strongly irradiated stars and planets. Astrophys J 594:1011–1018, https://doi.org/10.1086/377080

Ih J, Kempton EM-R, Whittaker EA, Lessard M (2023) Constraining the thickness of TRAPPIST-1 b's atmosphere from its JWST secondary eclipse observation at 15 microns. Astrophys J Lett 952:L4, https://doi.org/10.3847/2041-8213/ace03b

Jackson B, Adams E, Sandidge W, Kreyche S, Briggs J (2019) Variability in the atmosphere of the hot Jupiter Kepler-76b. Astron J 157:239, https://doi.org/10.3847/1538-3881/ab1b30

Jensen AG, Redfield S, Endl M, Cochran WD, Koesterke L, Barman T (2012) A detection of H-alpha in an exoplanetary atmosphere. Astrophys J 751:86, https://doi.org/10.1088/0004-637X/751/2/86

Johnstone CP, Bartel M, Güdel M (2021) The active lives of stars: A complete description of the rotation and XUV evolution of F, G, K, and M dwarfs. Astron Astrophys 649:A96, https://doi.org/10.1051/0004-6361/202038407

JWST Transiting Exoplanet Community Early Release Science Team, Ahrer E-M, Alderson L, et al. (2023) Identification of carbon dioxide in an exoplanet atmosphere. Nature 614:649–652, https://doi.org/10.1038/s41586-022-05269-w

Kaltenegger L, Sasselov D (2010) Detecting planetary geochemical cycles on exoplanets: Atmospheric signatures and the case of SO_2. Astrophys J 708:1162–1167, https://doi.org/10.1088/0004-637X/708/2/1162

Kasper D, Bean JL, Line MR, Seifahrt A, Stürmer J, Pino L, Désert J-M, Brogi M (2021) Confirmation of iron emission lines and nondetection of TiO on the dayside of KELT-9b with MAROON-X. Astrophys J Lett 921:L18, https://doi.org/10.3847/2041-8213/ac30e1

Kasper D, Bean JL, Line MR, et al. (2023) Unifying high- and low-resolution observations to constrain the dayside atmosphere of KELT-20b/MASCARA-2b. Astron J 165:7, https://doi.org/10.3847/1538-3881/ac9f40

Kataria T, Showman AP, Fortney JJ, Marley MS, Freedman RS (2014) The atmospheric circulation of the super Earth GJ 1214b: Dependence on composition and metallicity. Astrophys J 785:92, https://doi.org/10.1088/0004-637X/785/2/92

Kawashima Y, Ikoma M (2018) Theoretical transmission spectra of exoplanet atmospheres with hydrocarbon haze: Effect of creation, growth, and settling of haze particles. I. Model description and first results. Astrophys J 853:7, https://doi.org/10.3847/1538-4357/aaa0c5

Kawashima Y, Ikoma M (2019) Theoretical transmission spectra of exoplanet atmospheres with hydrocarbon haze: Effect of creation, growth, and settling of haze particles. II. Dependence on UV irradiation intensity, metallicity, C/O ratio, eddy diffusion coefficient, and temperature. Astrophys J 877:109, https://doi.org/10.3847/1538-4357/ab1b1d

Keating D, Cowan NB, Dang L (2019) Uniformly hot nightside temperatures on short-period gas giants. Nat Astron 3:1092–1098, https://doi.org/10.1038/s41550-019-0859-z

Kempton EM-R, Perna R, Heng K (2014) High resolution transmission spectroscopy as a diagnostic for Jovian exoplanet atmospheres: Constraints from theoretical models. Astrophys J 795:24, https://doi.org/10.1088/0004-637X/795/1/24

Kempton EM-R, Bean JL, Parmentier V (2017) An observational diagnostic for distinguishing between clouds and haze in hot exoplanet atmospheres. Astrophys J Lett 845:L20, https://doi.org/10.3847/2041-8213/aa84ac

Kempton EM-R, Bean JL, Louie DR, et al. (2018) A framework for prioritizing the TESS planetary candidates most amenable to atmospheric characterization. Publ Astron Soc Pac 130:114401, https://doi.org/10.1088/1538-3873/aadf6f

Kempton EM-R, Zhang M, Bean JL, et al. (2023) A reflective, metal-rich atmosphere for GJ 1214b from its JWST phase curve. Nature 620:67–71, https://doi.org/10.1038/s41586-023-06159-5

Kesseli AY, Snellen IAG, Casasayas-Barris N, Mollière P, Sánchez-López A (2022) An atomic spectral survey of WASP-76b: Resolving chemical gradients and asymmetries. Astron J 163:107, https://doi.org/10.3847/1538-3881/ac4336

Khare BN, Sagan C, Arakawa ET, Suits F, Callcott TA, Williams MW (1984) Optical constants of organic tholins produced in a simulated Titanian atmosphere: From soft X-ray to microwave frequencies. Icarus 60:127–137, https://doi.org/10.1016/0019-1035(84)90142-8

Kimura T, Ikoma M (2022) Predicted diversity in water content of terrestrial exoplanets orbiting M dwarfs. Nat Astron 6:1296–1307, https://doi.org/10.1038/s41550-022-01781-1

King GW, Wheatley PJ (2021) EUV irradiation of exoplanet atmospheres occurs on Gyr time-scales. Mon Not R Astron Soc 501:L28–L32, https://doi.org/10.1093/mnrasl/slaa186

Kirk J, Alam MK, López-Morales M, Zeng L (2020) Confirmation of WASP-107b's Extended helium atmosphere with Keck II/NIRSPEC. Astron J 159:115, https://doi.org/10.3847/1538-3881/ab6e66

Kitzmann D, Stock JW, Patzer ABC (2023) FastChem Cond: Equilibrium chemistry with condensation and rainout for cool planetary and stellar environments. Mon Not R Astron Soc 527:7263–7283, https://doi.org/10.1093/mnras/stad3515

Knutson HA, Charbonneau D, Allen LE, Fortney JJ, Agol E, Cowan NB, Showman AP, Cooper CS, Megeath ST (2007) A map of the day–night contrast of the extrasolar planet HD 189733b. Nature 447:183–186, https://doi.org/10.1038/nature05782

Knutson HA, Charbonneau D, Allen LE, Burrows A, Megeath ST (2008) The 3.6–8.0 micron broadband emission spectrum of HD 209458b: Evidence for an atmospheric temperature inversion. Astrophys J 673:526–531, https://doi.org/10.1086/523894

Knutson HA, Charbonneau D, Burrows A, O'Donovan FT, Mandushev G (2009) Detection of a temperature inversion in the broadband infrared emission spectrum of TrES-4. Astrophys J 691:866–874, https://doi.org/10.1088/0004-637X/691/1/866

Knutson HA, Dragomir D, Kreidberg L, Kempton EM-R, McCullough PR, Fortney JJ, Bean JL, Gillon M, Homeier D, Howard AW (2014) Hubble Space Telescope near-IR transmission spectroscopy of the super-Earth HD 97658b. Astrophys J 794:155, https://doi.org/10.1088/0004-637X/794/2/155

Koll DDB (2022) A scaling for atmospheric heat redistribution on tidally locked rocky planets. Astrophys J 924:134, https://doi.org/10.3847/1538-4357/ac3b48

Koll DDB, Malik M, Mansfield M, Kempton EM-R, Kite E, Abbot D, Bean JL (2019) Identifying candidate atmospheres on rocky M dwarf planets via eclipse photometry. Astrophys J 886:140, https://doi.org/10.3847/1538-4357/ab4c91

Komacek TD, Showman AP (2016) Atmospheric circulation of hot Jupiters: Dayside–nightside temperature differences. Astrophys J 821:16, https://doi.org/10.3847/0004-637X/821/1/16

Komacek TD, Showman AP, Tan X (2017) Atmospheric circulation of hot Jupiters: Dayside–nightside temperature differences. II. Comparison with observations. Astrophys J 835:198, https://doi.org/10.3847/1538-4357/835/2/198

Komacek TD, Tan X, Gao P, Lee EKH (2022) Patchy nightside clouds on ultra-hot Jupiters: General circulation model simulations with radiatively active cloud tracers. Astrophys J 934:79, https://doi.org/10.3847/1538-4357/ac7723

Kreidberg L, Bean JL, Désert J-M, Benneke B, Deming D, Stevenson KB, Seager S, Berta-Thompson Z, Seifahrt A, Homeier D (2014a) Clouds in the atmosphere of the super-Earth exoplanet GJ 1214b. Nature 505:69–72, https://doi.org/10.1038/nature12888

Kreidberg L, Bean JL, Désert J-M, et al. (2014b) A precise water abundance measurement for the hot Jupiter WASP-43b. Astrophys J Lett 793:L27, https://doi.org/10.1088/2041-8205/793/2/L27

Kreidberg L, Line MR, Parmentier V, et al. (2018) Global climate and atmospheric composition of the ultra-hot Jupiter WASP-103b from HST and Spitzer phase curve observations. Astron J 156:17, https://doi.org/10.3847/1538-3881/aac3df

Kreidberg L, Koll DDB, Morley C, et al. (2019) Absence of a thick atmosphere on the terrestrial exoplanet LHS 3844b. Nature 573:87, https://doi.org/10.1038/s41586-019-1497-4

Krenn AF, Lendl M, Patel JA, et al. (2023) The geometric albedo of the hot Jupiter HD 189733b measured with CHEOPS. Astron Astrophys 672:A24, https://doi.org/10.1051/0004-6361/202245016

Krissansen-Totton J, Garland R, Irwin P, Catling DC (2018) Detectability of biosignatures in anoxic atmospheres with the James Webb Space Telescope: A TRAPPIST-1e Case Study. Astron J 156:114, https://doi.org/10.3847/1538-3881/aad564

Lally M, Vanderburg A (2022) Reassessing the evidence for time variability in the atmosphere of the exoplanet HAT-P-7b. Astron J 163:181, https://doi.org/10.3847/1538-3881/ac53a8

Lampón M, López-Puertas M, Lara LM, et al. (2020) Modelling the He I triplet absorption at 10,830 Anstroms in the atmosphere of HD 209458 b. Astron Astrophys 636:A13, https://doi.org/10.1051/0004-6361/201937175

Lampón M, López-Puertas M, Czesla S, et al. (2021) Evidence of energy-, recombination-, and photon-limited escape regimes in giant planet H/He atmospheres. Astron Astrophys 648:L7, https://doi.org/10.1051/0004-6361/202140423

Lampón M, López-Puertas M, Sanz-Forcada J, et al. (2023) Characterisation of the upper atmospheres of HAT-P-32 b, WASP-69 b, GJ 1214 b, and WASP-76 b through their He I triplet absorption. Astron Astrophys 673:A140, https://doi.org/10.1051/0004-6361/202245649

Langeveld AB, Madhusudhan N, Cabot SHC (2022) A survey of sodium absorption in 10 giant exoplanets with high-resolution transmission spectroscopy. Mon Not R Astron Soc 514:5192–5213, https://doi.org/10.1093/mnras/stac1539

Lavie B, Ehrenreich D, Bourrier V, et al. (2017) The long egress of GJ436b's giant exosphere. Astron Astrophys 605:L7, https://doi.org/10.1051/0004-6361/201731340

Lavvas P, Arfaux A (2021) Impact of photochemical hazes and gases on exoplanet atmospheric thermal structure. Mon Not R Astron Soc 502:5643–5657, https://doi.org/10.1093/mnras/stab456

Lavvas P, Koskinen T, Steinrueck ME, García Muñoz A, Showman AP (2019) Photochemical hazes in sub-Neptunian atmospheres with a focus on GJ1214b. Astrophys J 878:118, https://doi.org/10.3847/1538-4357/ab204e

Lecavelier Des Etangs A, Ehrenreich D, Vidal-Madjar A, Ballester GE, Désert J-M, Ferlet R, Hébrard G, Sing DK, Tchakoumegni K-O, Udry S (2010) Evaporation of the planet HD189733b observed in H I Lyman-Alpha. Astron Astrophys 514:A72, https://doi.org/10.1051/0004-6361/200913347

Lee EJ (2019) The boundary between gas-rich and gas-poor planets. Astrophys J 878:36, https://doi.org/10.3847/1538-4357/ab1b40

Lee EJ, Karalis A, Thorngren DP (2022) Creating the radius gap without mass loss. Astrophys J 941:186, https://doi.org/10.3847/1538-4357/ac9c66

Lesjak F, Nortmann L, Yan F, et al. (2023) Retrieval of the dayside atmosphere of WASP-43b with CRIRES+. Astron Astrophys 678:A23, https://doi.org/10.1051/0004-6361/202347151

Libby-Roberts JE, Berta-Thompson ZK, Désert J-M, et al. (2020) The featureless transmission spectra of two super-puff planets. Astron J 159:57, https://doi.org/10.3847/1538-3881/ab5d36

Libby-Roberts JE, Berta-Thompson ZK, Diamond-Lowe H, et al. (2022) The featureless HST/WFC3 transmission spectrum of the rocky exoplanet GJ1132b: No evidence for a cloud-free primordial atmosphere and constraints on starspot contamination. Astron J 164:59, https://doi.org/10.3847/1538-3881/ac75de

Lim O, Benneke B, Doyon R, et al. (2023) Atmospheric reconnaissance of TRAPPIST-1b with JWST/NIRISS: Evidence for strong stellar contamination in the transmission spectra. Astrophys J Lett 955:L22, https://doi.org/10.3847/2041-8213/acf7c4

Lincowski AP, Meadows VS, Zieba S, et al. (2023) Potential atmospheric compositions of TRAPPIST-1 c constrained by JWST/MIRI Observations at 15 Microns. Astrophys J Lett 955:L7, https://doi.org/10.3847/2041-8213/acee02

Line MR, Vasisht G, Chen P, Angerhausen D, Yung YL (2011) Thermochemical and photochemical kinetics in cooler hydrogen-dominated extrasolar planets: A methane-poor GJ436b? Astrophys J 738:32, https://doi.org/10.1088/0004-637X/738/1/32

Line MR, Wolf AS, Zhang X, et al. (2013) A systematic retrieval analysis of secondary eclipse spectra. I. A comparison of atmospheric retrieval techniques. Astrophys J 775:137, https://doi.org/10.1088/0004-637X/775/2/137

Line MR, Stevenson KB, Bean J, Désert J-M, Fortney JJ, Kreidberg L, Madhusudhan N, Showman AP, Diamond-Lowe H (2016) No thermal inversion and a solar water abundance for the hot Jupiter HD209458b from HST/WFC3 spectroscopy. Astron J 152:203, https://doi.org/10.3847/0004-6256/152/6/203

Line MR, Brogi M, Bean JL, et al. (2021) A solar C/O and sub-solar metallicity in a hot Jupiter atmosphere. Nature 598:580–584, https://doi.org/10.1038/s41586-021-03912-6

Linssen DC, Oklopčić A (2023) Expanding the inventory of spectral lines used to trace atmospheric escape in exoplanets. Astron Astrophys 675:A193, https://doi.org/10.1051/0004-6361/202346583

Linssen DC, Oklopčić A, MacLeod M (2022) Constraining planetary mass-loss rates by simulating Parker wind profiles with Cloudy. Astron Astrophys 667:A54, https://doi.org/10.1051/0004-6361/202243830

Lodders K, Fegley B (2002) Atmospheric chemistry in giant planets, brown dwarfs, and low-mass dwarf stars. I. Carbon, nitrogen, and oxygen. Icarus 155:393–424, https://doi.org/10.1006/icar.2001.6740

Lothringer JD, Barman T, Koskinen T (2018) Extremely irradiated hot Jupiters: Non-oxide inversions, H⁻ opacity, and thermal dissociation of molecules. Astrophys J 866:27, https://doi.org/10.3847/1538-4357/aadd9e

Lothringer JD, Rustamkulov Z, Sing DK, Gibson NP, Wilson J, Schlaufman KC (2021) A new window into planet formation and migration: Refractory-to-volatile elemental ratios in ultra-hot Jupiters. Astrophys J 914:12, https://doi.org/10.3847/1538-4357/abf8a9

Lothringer JD, Sing DK, Rustamkulov Z, Wakeford HR, Stevenson KB, Nikolov N, Lavvas P, Spake JJ, Winch AT (2022) UV absorption by silicate cloud precursors in ultra-hot Jupiter WASP-178b. Nature 604:49–52, https://doi.org/10.1038/s41586-022-04453-2

Louden T, Wheatley PJ (2015) Spatially resolved eastward winds and rotation of HD189733b. Astrophys J Lett 814:L24, https://doi.org/10.1088/2041-8205/814/2/L24

Lovelock JE (1965) A physical basis for life detection experiments. Nature 207:568–570, https://doi.org/10.1038/207568a0

Lozovsky M, Helled R, Dorn C, Venturini J (2018) Threshold radii of volatile-rich planets. Astrophys J 866:49, https://doi.org/10.3847/1538-4357/aadd09

Luque R, Pallé E (2022) Density, not radius, separates rocky and water-rich small planets orbiting M dwarf stars. Science 377:1211–1214, https://doi.org/10.1126/science.abl7164

Lustig-Yaeger J, Meadows VS, Lincowski AP (2019) The detectability and characterization of the TRAPPIST-1 exoplanet atmospheres with JWST. Astron J 158:27, https://doi.org/10.3847/1538-3881/ab21e0

Lustig-Yaeger J, Fu G, May EM, et al. (2023) A JWST transmission spectrum of the nearby Earth-sized exoplanet LHS 475 b. Nat Astron 7:1317–1328, https://doi.org/10.1038/s41550-023-02064-z

Macdonald RJ, Batalha NE (2023) A catalog of exoplanet atmospheric retrieval codes. Res Notes Am Astron Soc 7:54, https://doi.org/10.3847/2515-5172/acc46a

MacLeod M, Oklopčić A (2022) Stellar wind confinement of evaporating exoplanet atmospheres and its signatures in 1083 nm observations. Astrophys J 926:226, https://doi.org/10.3847/1538-4357/ac46ce

Madhusudhan N (2012) C/O ratio as a dimension for characterizing exoplanetary atmospheres. Astrophys J 758:36, https://doi.org/10.1088/0004-637X/758/1/36

Madhusudhan N (2019) Exoplanetary atmospheres: Key insights, challenges, and prospects. Annu Rev Astron Astrophys 57:617–663, https://doi.org/10.1146/annurev-astro-081817-051846

Madhusudhan N, Seager S (2009) A temperature and abundance retrieval method for exoplanet atmospheres. Astrophys J 707:24, https://doi.org/10.1088/0004-637X/707/1/24

Madhusudhan N, Amin MA, Kennedy GM (2014a) Toward chemical constraints on hot Jupiter migration. Astrophys J Lett 794:L12, https://doi.org/10.1088/2041-8205/794/1/L12

Madhusudhan N, Crouzet N, McCullough PR, Deming D, Hedges C (2014b) H_2O abundances in the atmospheres of three hot Jupiters. Astrophys J Lett 791:L9, https://doi.org/10.1088/2041-8205/791/1/L9

Madhusudhan N, Agúndez M, Moses JI, Hu Y (2016) Exoplanetary atmospheres—Chemistry, formation conditions, and habitability. Space Sci Rev 205:285–348, https://doi.org/10.1007/s11214-016-0254-3

Madhusudhan N, Bitsch B, Johansen A, Eriksson L (2017) Atmospheric signatures of giant exoplanet formation by pebble accretion. Mon Not R Astron Soc 469:4102–4115, https://doi.org/10.1093/mnras/stx1139

Madhusudhan N, Sarkar S, Constantinou S, Holmberg Mr, Piette AAA, Moses JI (2023) Carbon-bearing molecules in a possible Hycean atmosphere. Astrophys J Lett 956:L13, https://doi.org/10.3847/2041-8213/acf577

Maguire C, Gibson NP, Nugroho SK, Ramkumar S, Fortune M, Merritt SR, de Mooij E (2023) High-resolution atmospheric retrievals of WASP-121b transmission spectroscopy with ESPRESSO: Consistent relative abundance constraints across multiple epochs and instruments. Mon Not R Astron Soc 519:1030–1048, https://doi.org/10.1093/mnras/stac3388

Majeau C, Agol E, Cowan NB (2012) A two-dimensional infrared map of the extrasolar planet HD189733b. Astrophys J Lett 747:L20, https://doi.org/10.1088/2041-8205/747/2/L20

Malik M, Kempton EM-R, Koll DDB, Mansfield M, Bean JL, Kite E (2019) Analyzing atmospheric temperature profiles and spectra of M dwarf rocky planets. Astrophys J 886:142, https://doi.org/10.3847/1538-4357/ab4a05

Mansfield M, Bean JL, Line MR, Parmentier V, Kreidberg L, Désert J-M, Fortney JJ, Stevenson KB, Arcangeli J, Dragomir D (2018) An HST/WFC3 thermal emission spectrum of the hot Jupiter HAT-P-7b. Astron J 156:10, https://doi.org/10.3847/1538-3881/aac497

Mansfield M, Kite ES, Hu R, Koll DDB, Malik M, Bean JL, Kempton EM-R (2019) Identifying atmospheres on rocky exoplanets through inferred high albedo. Astrophys J 886:141, https://doi.org/10.3847/1538-4357/ab4c90

Mansfield M, Bean JL, Stevenson KB, et al. (2020) Evidence for H_2 dissociation and recombination heat transport in the atmosphere of KELT-9b. Astrophys J Lett 888:L15, https://doi.org/10.3847/2041-8213/ab5b09

Mansfield M, Line MR, Bean JL, et al. (2021) A unique hot Jupiter spectral sequence with evidence for compositional diversity. Nat Astron 5:1224–1232, https://doi.org/10.1038/s41550-021-01455-4

Marley MS, Robinson TD (2015) On the cool side: Modeling the atmospheres of brown dwarfs and giant planets. Annu Rev Astron Astrophys 53:279, https://doi.org/10.1146/annurev-astro-082214-122522

May EM, Stevenson KB, Bean JL, et al. (2022) A new analysis of eight Spitzer phase curves and hot Jupiter population trends: Qatar-1b, Qatar-2b, WASP-52b, WASP-34b, and WASP-140b. Astron J 163:256, https://doi.org/10.3847/1538-3881/ac6261

Mayor M, Queloz D (1995) A Jupiter-mass companion to a solar-type star. Nature 378:355–359, https://doi.org/10.1038/378355a0

Mazeh T, Holczer T, Faigler S (2016) Dearth of short-period Neptunian exoplanets: A desert in period–mass and period–radius planes. Astron Astrophys 589:A75, https://doi.org/10.1051/0004-6361/201528065

Mbarek R, Kempton EM-R (2016) Clouds in super-Earth atmospheres: Chemical equilibrium calculations. Astrophys J 827:121, https://doi.org/10.3847/0004-637X/827/2/121

McCullough P, MacKenty J (2012) Considerations for using spatial scans with WFC3. Instrument Science Report WFC3 2012-08. Space Telescope Science Institute, https://www.stsci.edu/files/live/sites/www/files/home/hst/instrumentation/wfc3/documentation/instrument-science-reports-isrs/_documents/2012/WFC3-2012-08.pdf

McCullough PR, Crouzet N, Deming D, Madhusudhan N (2014) Water vapor in the spectrum of the extrasolar planet HD189733b. I. The transit. Astrophys J 791:55, https://doi.org/10.1088/0004-637X/791/1/55

Meier Valdés EA, Morris BM, Demory B-O, et al. (2023) Investigating the visible phase-curve variability of 55 Cnc e. Astron Astrophys 677:A112, https://doi.org/10.1051/0004-6361/202346050

Menou K (2012) Magnetic scaling laws for the atmospheres of hot giant exoplanets. Astrophys J 745:138, https://doi.org/10.1088/0004-637X/745/2/138

Mercier SJ, Dang L, Gass A, Cowan NB, Bell TJ (2022) Revisiting the iconic Spitzer phase curve of 55 Cancri e: Hotter dayside, cooler nightside, and smaller phase offset. Astron J 164:204, https://doi.org/10.3847/1538-3881/ac8f22

Mikal-Evans T, Crossfield IJM, Benneke B, et al. (2021) Transmission spectroscopy for the warm sub-Neptune HD3167c: Evidence for molecular absorption and a possible high-metallicity atmosphere. Astron J 161:18, https://doi.org/10.3847/1538-3881/abc874

Mikal-Evans T, Sing DK, Barstow JK, et al. (2022) Diurnal variations in the stratosphere of the ultrahot giant exoplanet WASP-121b. Nat Astron 6:471–479, https://doi.org/10.1038/s41550-021-01592-w

Mikal-Evans T, Madhusudhan N, Dittmann J, Günther MN, Welbanks L, Van Eylen V, Crossfield IJM, Daylan T, Kreidberg L (2023a) Hubble Space Telescope transmission spectroscopy for the temperate sub-Neptune TOI-270 d: A possible hydrogen-rich atmosphere containing water vapor. Astron J 165:84, https://doi.org/10.3847/1538-3881/aca90b

Mikal-Evans T, Sing DK, Dong J, et al. (2023b) A JWST NIRSpec phase curve for WASP-121b: Dayside emission strongest eastward of the substellar point and nightside conditions conducive to cloud formation. Astrophys J Lett 943:L17, https://doi.org/10.3847/2041-8213/acb049

Miles BE, Biller BA, Patapis P, et al. (2023) The JWST early-release science program for direct observations of exoplanetary systems II: A 1 to 20 micron spectrum of the planetary-mass companion VHS 1256-1257b. Astrophys J Lett 946:L6, https://doi.org/10.3847/2041-8213/acb04a

Miller-Ricci E, Fortney JJ (2010) The nature of the atmosphere of the transiting super-Earth GJ 1214b. Astrophys J Lett 716:L74–L79, https://doi.org/10.1088/2041-8205/716/1/L74

Miller-Ricci E, Seager S, Sasselov D (2009) The atmospheric signatures of super-Earths: How to distinguish between hydrogen-rich and hydrogen-poor atmospheres. Astrophys J 690:1056–1067, https://doi.org/10.1088/0004-637X/690/2/1056

Miller-Ricci Kempton E, Rauscher E (2012a) Constraining high-speed winds in exoplanet atmospheres through observations of anomalous doppler shifts during transit. Astrophys J 751:117, https://doi.org/10.1088/0004-637X/751/2/117

Miller-Ricci Kempton E, Zahnle K, Fortney JJ (2012b) The atmospheric chemistry of GJ 1214b: Photochemistry and clouds. Astrophys J 745:3, https://doi.org/10.1088/0004-637X/745/1/3

Montet BT, Crepp JR, Johnson JA, Howard AW, Marcy GW (2014) The TRENDS high-contrast imaging survey. IV. The occurrence rate of giant planets around M dwarfs. Astrophys J 781:28, https://doi.org/10.1088/0004-637X/781/1/28

Moran SE, Stevenson KB, Sing DK, et al. (2023) High tide or riptide on the cosmic shoreline? A water-rich atmosphere or stellar contamination for the warm super-Earth GJ 486b from JWST observations. Astrophys J Lett 948:L11, https://doi.org/10.3847/2041-8213/accb9c

Morley CV, Fortney JJ, Kempton EM-R, Marley MS, Visscher C, Zahnle K (2013) Quantitatively assessing the role of clouds in the transmission spectrum of GJ 1214b. Astrophys J 775:33, https://doi.org/10.1088/0004-637X/775/1/33

Morley CV, Fortney JJ, Marley MS, Zahnle K, Line M, Kempton E, Lewis N, Cahoy K (2015) Thermal emission and reflected light spectra of super Earths with flat transmission spectra. Astrophys J 815:110, https://doi.org/10.1088/0004-637X/815/2/110

Morley CV, Kreidberg L, Rustamkulov Z, Robinson T, Fortney JJ (2017) Observing the atmospheres of known temperate Earth-sized planets with JWST. Astrophys J 850:121, https://doi.org/10.3847/1538-4357/aa927b

Morris BM, Heng K, Jones K, Piaulet C, Demory B-O, Kitzmann D, Jens Hoeijmakers H (2022) Physically-motivated basis functions for temperature maps of exoplanets. Astron Astrophys 660:A123, https://doi.org/10.1051/0004-6361/202142135

Morton TD, Bryson ST, Coughlin JL, et al. (2016) False positive probabilities for all Kepler objects of interest: 1284 Newly validated planets and 428 likely false positives. Astrophys J 822:86, https://doi.org/10.3847/0004-637X/822/2/86

Mousis O, Deleuil M, Aguichine A, Marcq E, Naar J, Aguirre LA, Brugger B, Gonçalves T (2020) Irradiated ocean planets bridge super-Earth and sub-Neptune populations. Astrophys J Lett 896:L22, https://doi.org/10.3847/2041-8213/ab9530

Mugnai LV, Modirrousta-Galian D, Edwards B, et al. (2021) ARES. V. No evidence for molecular absorption in the HST WFC3 spectrum of GJ 1132 b. Astron J 161:284, https://doi.org/10.3847/1538-3881/abf3c3

Mulders GD, Pascucci I, Apai D (2015) An increase in the mass of planetary systems around lower-mass stars. Astrophys J 814:130, https://doi.org/10.1088/0004-637X/814/2/130

Neil AR, Liston J, Rogers LA (2022) Evaluating the evidence for water world populations using mixture models. Astrophys J 933:63, https://doi.org/10.3847/1538-4357/ac609b

Ninan JP, Stefansson G, Mahadevan S, et al. (2020) Evidence for He I 10,830 angstrom absorption during the transit of a warm Neptune around the M-dwarf GJ 3470 with the habitable-zone Planet Finder. Astrophys J 894:97, https://doi.org/10.3847/1538-4357/ab8559

Nortmann L, Pallé E, Salz M, et al. (2018) Ground-based detection of an extended helium atmosphere in the Saturn-mass exoplanet WASP-69b. Science 362:1388–1391, https://doi.org/10.1126/science.aat5348

Öberg KI, Murray-Clay R, Bergin EA (2011) The effects of snowlines on C/O in planetary atmospheres. Astrophys J Lett 743:L16, https://doi.org/10.1088/2041-8205/743/1/L16

Oklopčić A, Hirata CM (2018) A new window into escaping exoplanet atmospheres: 10,830 Angstrom line of helium. Astrophys J Lett 855:L11, https://doi.org/10.3847/2041-8213/aaada9

Orell-Miquel J, Murgas F, Pallé E, et al. (2022) A tentative detection of He I in the atmosphere of GJ 1214 b. Astron Astrophys 659:A55, https://doi.org/10.1051/0004-6361/202142455

Orell-Miquel J, Lampón M, López-Puertas M, et al. (2023) Confirmation of an He I evaporating atmosphere around the 650-Myr-old sub-Neptune HD235088 b (TOI-1430 b) with CARMENES. Astron Astrophys 677:A56, https://doi.org/10.1051/0004-6361/202346445

Owen JE (2019) Atmospheric escape and the evolution of close-in exoplanets. Annu Rev Earth Planet Sci 47:67—90, https://doi.org/10.1146/annurev-earth-053018-060246

Owen JE, Adams FC (2014) Magnetically controlled mass-loss from extrasolar planets in close orbits. Mon Not R Astron Soc 444:3761–3779, https://doi.org/10.1093/mnras/stu1684

Owen JE, Wu Y (2017) The evaporation valley in the Kepler planets. Astrophys J 847:29, https://doi.org/10.3847/1538-4357/aa890a

Owen JE, Murray-Clay RA, Schreyer E, Schlichting HE, Ardila D, Gupta A, Loyd ROP, Shkolnik EL, Sing DK, Swain MR (2023) The fundamentals of Lyman-alpha exoplanet transits. Mon Not R Astron Soc 518:4357–4371, https://doi.org/10.1093/mnras/stac3414

Pai Asnodkar A, Wang J, Eastman JD, Cauley PW, Gaudi BS, Ilyin I, Strassmeier K (2022) Variable and supersonic winds in the atmosphere of an ultrahot giant planet. Astron J 163:155, https://doi.org/10.3847/1538-3881/ac51d2

Pallé E, Oshagh M, Casasayas-Barris N, et al. (2020a) Transmission spectroscopy and Rossiter–McLaughlin measurements of the young Neptune orbiting AU Mic. Astron Astrophys 643:A25, https://doi.org/10.1051/0004-6361/202038583

Pallé E, Nortmann L, Casasayas-Barris N, et al. (2020b) A He I upper atmosphere around the warm Neptune GJ 3470 b. Astron Astrophys 638:A61, https://doi.org/10.1051/0004-6361/202037719

Parmentier V, Line MR, Bean JL, et al. (2018) From thermal dissociation to condensation in the atmospheres of ultra hot Jupiters: WASP-121b in context. Astron Astrophys 617:A110, https://doi.org/10.1051/0004-6361/201833059

Parmentier V, Showman AP, Fortney JJ (2021) The cloudy shape of hot Jupiter thermal phase curves. Mon Not R Astron Soc 501:78–108, https://doi.org/10.1093/mnras/staa3418

Pelletier S, Benneke B, Darveau-Bernier A, et al. (2021) Where is the water? Jupiter-like C/H Ratio but strong H_2O depletion found on Tau Boötis b Using SPIRou. Astron J 162:73, https://doi.org/10.3847/1538-3881/ac0428

Pelletier S, Benneke B, Ali-Dib M, et al. (2023) Vanadium oxide and a sharp onset of cold-trapping on a giant exoplanet. Nature 619:491–494, https://doi.org/10.1038/s41586-023-06134-0

Pérez González J, Greklek-McKeon M, Vissapragada S, Saidel M, Knutson HA, Linssen D, Oklopčić A (2023) Detection of an atmospheric outflow from the young hot Saturn TOI-1268b. arXiv:2307.09515, https://doi.org/10.48550/arXiv.2307.09515

Perez-Becker D, Showman AP (2013) Atmospheric heat redistribution on hot Jupiters. Astrophys J 776:134, https://doi.org/10.1088/0004-637X/776/2/134

Perna R, Menou K, Rauscher E (2010) Magnetic drag on hot Jupiter atmospheric winds. Astrophys J 719:1421–1426, https://doi.org/10.1088/0004-637X/719/2/1421

Petigura EA, Marcy GW, Winn JN, Weiss LM, Fulton BJ, Howard AW, Sinukoff E, Isaacson H, Morton TD, Johnson JA (2018) The California–Kepler survey. IV. Metal-rich stars host a greater diversity of planets. Astron J 155:89, https://doi.org/10.3847/1538-3881/aaa54c

Petigura EA, Rogers JG, Isaacson H, et al. (2022) The California–Kepler survey. X. The radius gap as a function of stellar mass, metallicity, and age. Astron J 163:179, https://doi.org/10.3847/1538-3881/ac51e3

Pidhorodetska D, Fauchez TJ, Villanueva GL, Domagal-Goldman SD, Kopparapu RK (2020) Detectability of molecular signatures on TRAPPIST-1e through transmission spectroscopy simulated for future space-based observatories. Astrophys J Lett 898:L33, https://doi.org/10.3847/2041-8213/aba4a1

Pinhas A, Madhusudhan N, Gandhi S, MacDonald R (2019) H_2O abundances and cloud properties in ten hot giant exoplanets. Mon Not R Astron Soc 482:1485–1498, https://doi.org/10.1093/mnras/sty2544

Piso A-MA, Pegues J, Öberg KI (2016) The role of ice compositions for snowlines and the C/N/O ratios in active disks. Astrophys J 833:203, https://doi.org/10.3847/1538-4357/833/2/203

Plavchan P, Barclay T, Gagné J, et al. (2020) A planet within the debris disk around the pre-main-sequence star AU Microscopii. Nature 582:497–500, https://doi.org/10.1038/s41586-020-2400-z

Polman J, Waters LBFM, Min M, Miguel Y, Khorshid N (2023) H_2S and SO_2 detectability in hot Jupiters. Sulphur species as indicators of metallicity and C/O ratio. Astron Astrophys 670:A161, https://doi.org/10.1051/0004-6361/202244647

Pont F, Knutson H, Gilliland RL, Moutou C, Charbonneau D (2008) Detection of atmospheric haze on an extrasolar planet: the 0.55–1.05 micron transmission spectrum of HD 189733b with the HubbleSpaceTelescope. Mon Not R Astron Soc 385:109–118, https://doi.org/10.1111/j.1365-2966.2008.12852.x

Pontoppidan KM, Barrientes J, Blome C, et al. (2022) The JWST early release observations. Astrophys J Lett 936:L14, https://doi.org/10.3847/2041-8213/ac8a4e

Powell D, Louden T, Kreidberg L, Zhang X, Gao P, Parmentier V (2019) Transit signatures of inhomogeneous clouds on hot Jupiters: Insights from microphysical cloud modeling. Astrophys J 887:170, https://doi.org/10.3847/1538-4357/ab55d9

Prinoth B, Hoeijmakers HJ, Kitzmann D, et al. (2022) Titanium oxide and chemical inhomogeneity in the atmosphere of the exoplanet WASP-189 b. Nat Astron 6:449–457, https://doi.org/10.1038/s41550-021-01581-z

Prinoth B, Hoeijmakers HJ, Pelletier S, et al. (2023) Time-resolved transmission spectroscopy of the ultra-hot Jupiter WASP-189 b. Astron Astrophys 678:A182, https://doi.org/10.1051/0004-6361/202347262

Rackham BV, Apai D, Giampapa MS (2018) The transit light source effect: False spectral features and incorrect densities for M-dwarf transiting planets. Astrophys J 853:122, https://doi.org/10.3847/1538-4357/aaa08c

Rackham BV, Espinoza N, Berdyugina SV, et al. (2023) The effect of stellar contamination on low-resolution transmission spectroscopy: needs identified by NASA's exoplanet exploration program study analysis Group 21. RAS Techniques and Instruments 2:148–206, https://doi.org/10.1093/rasti/rzad009

Radica M, Welbanks L, Espinoza N, et al. (2023) Awesome SOSS: transmission spectroscopy of WASP-96b with NIRISS/SOSS. Mon Not R Astron Soc 524:835–856, https://doi.org/10.1093/mnras/stad1762

Rauscher E, Suri V, Cowan NB (2018) A more informative map: Inverting thermal orbital phase and eclipse light curves of exoplanets. Astron J 156:235, https://doi.org/10.3847/1538-3881/aae57f

Reggiani H, Schlaufman KC, Healy BF, Lothringer JD, Sing DK (2022) Evidence that the hot Jupiter WASP-77 A b formed beyond its parent protoplanetary disk's H_2O ice line. Astron J 163:159, https://doi.org/10.3847/1538-3881/ac4d9f

Richardson LJ, Deming D, Horning K, Seager S, Harrington J (2007) A spectrum of an extrasolar planet. Nature 445:892–895, https://doi.org/10.1038/nature05636

Rockcliffe KE, Newton ER, Youngblood A, Duvvuri GM, Plavchan P, Gao P, Mann AW, Lowrance PJ (2023) The variable detection of atmospheric escape around the young, hot Neptune AU Mic b. Astron J 166:77, https://doi.org/10.3847/1538-3881/ace536

Rogers TM (2017) Constraints on the magnetic field strength of HAT-P-7 b and other hot giant exoplanets. Nat Astron 1:0131, https://doi.org/10.1038/s41550-017-013

Rogers LA, Seager S (2010) Three possible origins for the gas layer on GJ 1214b. Astrophys J 716:1208-1216, https://doi.org/10.1088/0004-637X/716/2/1208

Rogers JG, Gupta A, Owen JE, Schlichting HE (2021) Photoevaporation versus core-powered mass-loss: model comparison with the 3D radius gap. Mon Not R Astron Soc 508:5886–5902, https://doi.org/10.1093/mnras/stab2897

Rogers JG, Schlichting HE, Owen JE (2023) Conclusive evidence for a population of water worlds around M dwarfs remains elusive. Astrophys J Lett 947:L19, https://doi.org/10.3847/2041-8213/acc86f

Roman MT, Kempton EM-R, Rauscher E, Harada CK, Bean JL, Stevenson KB (2021) Clouds in three-dimensional models of hot Jupiters over a wide range of temperatures. I. Thermal structures and broadband phase-curve predictions. Astrophys J 908:101, https://doi.org/10.3847/1538-4357/abd549

Roth A, Drummond B, Hébrard E, Tremblin P, Goyal J, Mayne N (2021) Pseudo-2D modelling of heat redistribution through H_2 thermal dissociation/recombination: consequences for ultra-hot Jupiters. Mon Not R Astron Soc 505:4515–4530, https://doi.org/10.1093/mnras/stab1256

Rowe JF, Matthews JM, Seager S, et al. (2008) The very low albedo of an extrasolar planet: MOST space-based photometry of HD 209458. Astrophys J 689:1345–1353, https://doi.org/10.1086/591835

Roy P-A, Benneke B, Piaulet C, et al. (2022) Is the hot, dense sub-Neptune TOI-824 b an exposed Neptune mantle? Spitzer detection of the hot dayside and reanalysis of the interior composition. Astrophys J 941:89, https://doi.org/10.3847/1538-4357/ac9f18

Rustamkulov Z, Sing DK, Mukherjee S, et al. (2023) Early Release Science of the exoplanet WASP-39b with JWST NIRSpec PRISM. Nature 614:659–663, https://doi.org/10.1038/s41586-022-05677-y

Savel AB, Kempton EM-R, Rauscher E, Komacek TD, Bean JL, Malik M, Malsky I (2023) Diagnosing limb asymmetries in hot and ultrahot Jupiters with high-resolution transmission spectroscopy. Astrophys J 944:99, https://doi.org/10.3847/1538-4357/acb141

Schaefer L, Fegley B (2010) Chemistry of atmospheres formed during accretion of the Earth and other terrestrial planets. Icarus 208:438–448, https://doi.org/10.1016/j.icarus.2010.01.026

Schreyer E, Owen JE, Spake JJ, Bahroloom Z, Di Giampasquale S (2024) Using helium 10,830 Angstrom transits to constrain planetary magnetic fields. Mon Not R Astron Soc 527:5117–5130, https://doi.org/10.1093/mnras/stad3528

Schwartz JC, Cowan NB (2015) Balancing the energy budget of short-period gas giant planets: Evidence for reflective clouds and optical absorbers. Mon Not R Astron Soc 449:4192, https://doi.org/10.1093/mnras/stv470

Schwarz H, Brogi M, de Kok R, Birkby J, Snellen I (2015) Evidence against a strong thermal inversion in HD 209458b from high-dispersion spectroscopy. Astron Astrophys 576:A111, https://doi.org/10.1051/0004-6361/201425170

Schwartz JC, Kashner Z, Jovmir D. Cowan NB (2017) Phase offsets and the energy budgets of hot Jupiters. Astrophys J 850:154, https://doi.org/10.3847/1538-4357/aa9567

Seager S (2010) Exoplanet Atmospheres: Physical Processes. Princeton University Press

Seager S, Deming D (2009) On the method to infer an atmosphere on a tidally locked super Earth exoplanet and upper limits to GJ 876d. Astrophys J 703:1884–1889, https://doi.org/10.1088/0004-637X/703/2/1884

Seager S, Sasselov DD (2000) Theoretical transmission spectra during extrasolar giant planet transits. Astrophys J 537:916–921, https://doi.org/10.1086/309088

Seager S, Whitney BA, Sasselov DD (2000) Photometric light curves and polarization of close-in extrasolar giant planets. Astrophys J 540:504–520, https://doi.org/10.1086/309292

Seidel JV, Ehrenreich D, Allart R, et al. (2021) Into the storm: diving into the winds of the ultra-hot Jupiter WASP-76 b with HARPS and ESPRESSO. Astron Astrophys 653:A73, https://doi.org/10.1051/0004-6361/202140569

Selsis F, Leconte J, Turbet M, Chaverot G, Bolmont É (2023) A cool runaway greenhouse without surface magma ocean. Nature 620:287–291, https://doi.org/10.1038/s41586-023-06258-3

Sheppard KB, Mandell AM, Tamburo P, Gandhi S, Pinhas A, Madhusudhan N, Deming D (2017) Evidence for a dayside thermal inversion and high metallicity for the hot Jupiter WASP-18b. Astrophys J Lett 850:L32, https://doi.org/10.3847/2041-8213/aa9ae9

Showman AP, Fortney JJ, Lian Y, et al. (2009) Atmospheric circulation of hot Jupiters: Coupled radiative-dynamical general circulation model simulations of HD 189733b and HD 209458b. Astrophys J 699:564, https://doi.org/10.1088/0004-637X/699/1/564

Showman AP, Cho JY-K, Menou K (2010) Atmospheric circulation of exoplanets. *In:* Exoplanets. Seager S (ed) Space Science Series. University of Arizona Press, p 471–516

Showman AP, Fortney JJ, Lewis NK, Shabram M (2013) Doppler signatures of the atmospheric circulation on hot Jupiters. Astrophys J 762:24, https://doi.org/10.1088/0004-637X/762/1/24

Showman AP, Tan X, Parmentier V (2020) Atmospheric dynamics of hot giant planets and brown dwarfs. Space Sci Rev 216:139, https://doi.org/10.1007/s11214-020-00758-8

Sing DK, Fortney JJ, Nikolov N, et al. (2016) A continuum from clear to cloudy hot-Jupiter exoplanets without primordial water depletion. Nature 529:59–62, https://doi.org/10.1038/nature16068

Sing DK, Lavvas P, Ballester GE, et al. (2019) The Hubble Space Telescope PanCET program: Exospheric Mg II and Fe II in the near-ultraviolet transmission spectrum of WASP-121b using jitter decorrelation. Astron J 158:91, https://doi.org/10.3847/1538-3881/ab2986

Snellen IAG, de Kok RJ, de Mooij EJW, Albrecht S (2010) The orbital motion, absolute mass and high-altitude winds of exoplanet HD209458b. Nature 465:1049–1051, https://doi.org/10.1038/nature09111

Southworth J, Mancini L, Madhusudhan N, Mollière P, Ciceri S, Henning T (2017) Detection of the atmosphere of the 1.6 exoplanet GJ 1132 b. Astron J 153:191, https://doi.org/10.3847/1538-3881/aa6477

Spake JJ, Sing DK, Evans TM, et al. (2018) Helium in the eroding atmosphere of an exoplanet. Nature 557:68–70, https://doi.org/10.1038/s41586-018-0067-5

Spake JJ, Oklopčić A, Hillenbrand LA (2021) The posttransit tail of WASP-107b observed at 10,830 angstroms. Astron J 162:284, https://doi.org/10.3847/1538-3881/ac178a

Spake JJ, Oklopčić A, Hillenbrand LA, Knutson HA, Kasper D, Dai F, Orell-Miquel J, Vissapragada S, Zhang M, Bean JL (2022) Non-detection of He I in the atmosphere of GJ 1214b with Keck/NIRSPEC, at a time of minimal telluric contamination. Astrophys J Lett 939:L11, https://doi.org/10.3847/2041-8213/ac88c9

Steinrueck ME, Showman AP, Lavvas P, Koskinen T, Tan X, Zhang X (2021) 3D simulations of photochemical hazes in the atmosphere of hot Jupiter HD 189733b. Mon Not R Astron Soc 504:2783–2799, https://doi.org/10.1093/mnras/stab1053

Stevenson KB, Harrington J, Nymeyer S, Madhusudhan N, Seager S, Bowman WC, Hardy RA, Deming D, Rauscher E, Lust NB (2010) Possible thermochemical disequilibrium in the atmosphere of the exoplanet GJ 436b. Nature 464:1161–1164, https://doi.org/10.1038/nature09013

Stevenson KB, Désert J-M, Line MR, et al. (2014) Thermal structure of an exoplanet atmosphere from phase-resolved emission spectroscopy. Science 346:838–841, https://doi.org/10.1126/science.1256758

Suissa G, Mandell AM, Wolf ET, Villanueva GL, Fauchez T, Kopparapu Rk (2020) Dim prospects for transmission spectra of ocean earths around M stars. Astrophys J 891:58, https://doi.org/10.3847/1538-4357/ab72f9

Swain MR, Estrela R, Roudier GM, Sotin C, Rimmer PB, Valio A, West R, Pearson K, Huber-Feely N, Zellem RT (2021) Detection of an atmosphere on a rocky exoplanet. Astron J 161:213, https://doi.org/10.3847/1538-3881/abe879

Tabernero HM, Zapatero Osorio MR, Allart R, et al. (2021) ESPRESSO high-resolution transmission spectroscopy of WASP-76 b. Astron Astrophys 646:A158, https://doi.org/10.1051/0004-6361/202039511

Tan X, Komacek TD (2019) The atmospheric circulation of ultra-hot Jupiters. Astrophys J 886:26, https://doi.org/10.3847/1538-4357/ab4a76

Thorngren DP, Fortney JJ, Murray-Clay RA, Lopez ED (2016) The mass–metallicity relation for giant planets. Astrophys J 831:64, https://doi.org/10.3847/0004-637X/831/1/64

Thorngren DP, Marley MS, Fortney JJ (2019) An empirical mass–radius relation for cool giant planets. Res Notes Am Astron Soc 3:128, https://doi.org/10.3847/2515-5172/ab4353

Todorov K, Deming D, Harrington J, Stevenson KB, Bowman WC, Nymeyer S, Fortney JJ, Bakos GA (2010) Spitzer IRAC secondary eclipse photometry of the transiting extrasolar planet HAT-P-1b. Astrophys J 708:498–504, https://doi.org/10.1088/0004-637X/708/1/498

Todorov KO, Deming D, Knutson HA, et al. (2012) Warm Spitzer observations of three hot exoplanets: XO-4b, HAT-P-6b, and HAT-P-8b. Astrophys J 746:111, https://doi.org/10.1088/0004-637X/746/1/111

Todorov KO, Deming D, Knutson HA, et al. (2013) Warm Spitzer photometry of three hot Jupiters: HAT-P-3b, HAT-P-4b and HAT-P-12b. Astrophys J 770:102, https://doi.org/10.1088/0004-637X/770/2/102

Tsai S-M, Lee EKH, Powell D, et al. (2023) Photochemically produced SO_2 in the atmosphere of WASP-39b. Nature 617:483–487, https://doi.org/10.1038/s41586-023-05902-2

Tsiaras A, Waldmann IP, Zingales T, et al. (2018) A population study of gaseous exoplanets. Astron J 155:156, https://doi.org/10.3847/1538-3881/aaaf75

Tsiaras A, Waldmann IP, Tinetti G, Tennyson J, Yurchenko SN (2019) Water vapour in the atmosphere of the habitable-zone eight-Earth-mass planet K2-18 b. Nat Astron 3:1086–1091, https://doi.org/10.1038/s41550-019-0878-9

Turbet M, Bolmont E, Ehrenreich D, Gratier P, Leconte J, Selsis F, Hara N, Lovis C (2020) Revised mass–radius relationships for water-rich rocky planets more irradiated than the runaway greenhouse limit. Astron Astrophys 638:A41, https://doi.org/10.1051/0004-6361/201937151

Van Eylen V, Agentoft C, Lundkvist MS, Kjeldsen H, Owen JE, Fulton BJ, Petigura E, Snellen I (2018) An asteroseismic view of the radius valley: stripped cores, not born rocky. Mon Not R Astron Soc 479:4786–4795, https://doi.org/10.1093/mnras/sty1783

van Sluijs L, Birkby JL, Lothringer J, Lee EKH, Crossfield IJM, Parmentier V, Brogi M, Kulesa C, McCarthy D, Charbonneau D (2023) Carbon monoxide emission lines reveal an inverted atmosphere in the ultra hot Jupiter WASP-33 b consistent with an eastward hot spot. Mon Not R Astron Soc 522:2145–2170, https://doi.org/10.1093/mnras/stad1103

Venturini J, Alibert Y, Benz W (2016) Planet formation with envelope enrichment: new insights on planetary diversity. Astron Astrophys 596:A90, https://doi.org/10.1051/0004-6361/201628828

Vidal-Madjar A, Lecavelier des Etangs A, Désert J-M, Ballester GE, Ferlet R, Hébrard G, Mayor M (2003) An extended upper atmosphere around the extrasolar planet HD209458b. Nature 422:143–146, https://doi.org/10.1038/nature01448

Vidal-Madjar A, Désert J-M, Lecavelier des Etangs A, Hébrard G, Ballester GE, Ehrenreich D, Ferlet R, McConnell JC, Mayor M, Parkinson CD (2004) Detection of oxygen and carbon in the hydrodynamically escaping atmosphere of the extrasolar planet HD 209458b. Astrophys J Lett 604:L69–L72, https://doi.org/10.1086/383347

Vissapragada S, Knutson HA, Greklek-McKeon M, et al. (2022) The upper edge of the Neptune desert is stable against photoevaporation. Astron J 164:234, https://doi.org/10.3847/1538-3881/ac92f2

Wakeford HR, Sing DK (2015) Transmission spectral properties of clouds for hot Jupiter exoplanets. Astron Astrophys 573:A122, https://doi.org/10.1051/0004-6361/201424207

Wakeford HR, Sing DK, Deming D, et al. (2013) HST hot Jupiter transmission spectral survey: detection of water in HAT-P-1b from WFC3 near-IR spatial scan observations. Mon Not R Astron Soc 435:3481–3493, https://doi.org/10.1093/mnras/stt1536

Wakeford HR, Stevenson KB, Lewis NK, et al. (2017) HST PanCET program: A cloudy atmosphere for the promising JWST target WASP-101b. Astrophys J Lett 835:L12, https://doi.org/10.3847/2041-8213/835/1/L12

Wakeford HR, Sing DK, Deming D, et al. (2018) The complete transmission spectrum of WASP-39b with a precise water constraint. Astron J 155:29, https://doi.org/10.3847/1538-3881/aa9e4e

Wallack NL, Knutson HA, Deming D (2021) Trends in Spitzer secondary eclipses. Astron J 162:36, https://doi.org/10.3847/1538-3881/abdbb2

Wang L, Dai F (2021a) Metastable helium absorptions with 3D hydrodynamics and self-consistent photochemistry. II. WASP-107b, stellar wind, radiation pressure, and shear instability. Astrophys J 914:99, https://doi.org/10.3847/1538-4357/abf1ed

Wang L, Dai F (2021b) Metastable helium absorptions with 3D Hydrodynamics and self-consistent photochemistry. I. WASP-69b, dimensionality, X-ray and UV flux level, spectral types, and flares. Astrophys J 914:98, https://doi.org/10.3847/1538-4357/abf1ee

Wardenier JP, Parmentier V, Line MR, Lee EKH (2023) Modelling the effect of 3D temperature and chemistry on the cross-correlation signal of transiting ultra-hot Jupiters: A study of five chemical species on WASP-76b. Mon Not R Astron Soc 525:4942–4961, https://doi.org/10.1093/mnras/stad2586

Welbanks L, Madhusudhan N, Allard NF, Hubeny I, Spiegelman F, Leininger T (2019) Mass–metallicity trends in transiting exoplanets from atmospheric abundances of H_2O, Na, and K. Astrophys J Lett 887:L20, https://doi.org/10.3847/2041-8213/ab5a89

Whittaker EA, Malik M, Ih J, Kempton EM-R, Mansfield M, Bean JL, Kite ES, Koll DDB, Cronin TW, Hu R (2022) The detectability of rocky planet surface and atmosphere composition with the JWST: The Case of LHS 3844b. Astron J 164:258, https://doi.org/10.3847/1538-3881/ac9ab3

Williams PKG, Charbonneau D, Cooper CS, Showman AP, Fortney JJ (2006) Resolving the surfaces of extrasolar planets with secondary eclipse light curves. Astrophys J 649:1020–1027, https://doi.org/10.1086/506468

Winn JN (2010) Exoplanet transits and occultations. *In*: Exoplanets. Seager S (ed) Space Science Series. University of Arizona Press, p 55–77

Winn JN, Holman MJ, Shporer A, Fernández J, Mazeh T, Latham DW, Charbonneau D, Everett ME (2008) The transit light curve project. VIII. Six occultations of the exoplanet TrES-3. Astron J 136:267–271, https://doi.org/10.1088/0004-6256/136/1/267

Woitke P, Helling C, Hunter GH, Millard JD, Turner GE, Worters M, Blecic J, Stock JW (2018) Equilibrium chemistry down to 100 K. Impact of silicates and phyllosilicates on the carbon to oxygen ratio. Astron Astrophys 614:A1, https://doi.org/10.1051/0004-6361/201732193

Wong I, Shporer A, Daylan T, et al. (2020) Systematic phase curve study of known transiting systems from year one of the TESS mission. Astron J 160:155, https://doi.org/10.3847/1538-3881/ababad

Wordsworth R, Kreidberg L (2022) Atmospheres of rocky exoplanets. Annu Rev Astron Astrophys 60:159-201, https://doi.org/10.1146/annurev-astro-052920-125632

Wyttenbach A, Ehrenreich D, Lovis C, Udry S, Pepe F (2015) Spectrally resolved detection of sodium in the atmosphere of HD 189733b with the HARPS spectrograph. Astron Astrophys 577:A62, https://doi.org/10.1051/0004-6361/201525729

Wyttenbach A, Mollière P, Ehrenreich D, et al. (2020) Mass-loss rate and local thermodynamic state of the KELT-9 b thermosphere from the hydrogen Balmer series. Astron Astrophys 638:A87, https://doi.org/10.1051/0004-6361/201937316

Xue Q, Bean JL, Zhang M, Welbanks L, Lunine J, August P (2023) JWST transmission spectroscopy of HD 209458b: A super-solar metallicity, a very low C/O, and no evidence of CH_4, HCN, or C_2H_2. arXiv:2310.03245, https://doi.org/10.48550/arXiv.2310.03245

Yan F, Henning T (2018) An extended hydrogen envelope of the extremely hot giant exoplanet KELT-9b. Nat Astron 2:714–718, https://doi.org/10.1038/s41550-018-0503-3

Yan F, Pallé E, Reiners A, Molaverdikhani K, Casasayas-Barris N, Nortmann L, Chen G, Mollière P, Stangret M (2020) A temperature inversion with atomic iron in the ultra-hot dayside atmosphere of WASP-189b. Astron Astrophys 640:L5, https://doi.org/10.1051/0004-6361/202038294

Yan D, Seon K-i, Guo J, Chen G, Li L (2022a) Modeling the H-alpha and He 10830 transmission spectrum of WASP-52b. Astrophys J 936:177, https://doi.org/10.3847/1538-4357/ac8793

Yan F, Reiners A, Pallé E, et al. (2022b) Detection of iron emission lines and a temperature inversion on the dayside of the ultra-hot Jupiter KELT-20b. Astron Astrophys 659:A7, https://doi.org/10.1051/0004-6361/202142395

Yan F, Nortmann L, Reiners A, et al. (2023) CRIRES$^+$ detection of CO emissions lines and temperature inversions on the dayside of WASP-18b and WASP-76b. Astron Astrophys 672:A107, https://doi.org/10.1051/0004-6361/202245371

Zahnle KJ, Catling DC (2017) The cosmic shoreline: The evidence that escape determines which planets have atmospheres, and what this may mean for Proxima Centauri B. Astrophys J 843:122, https://doi.org/10.3847/1538-4357/aa7846

Zamyatina M, Hébrard E, Drummond B, Mayne NJ, Manners J, Christie DA, Tremblin P, Sing DK, Kohary K (2023) Observability of signatures of transport-induced chemistry in clear atmospheres of hot gas giant exoplanets. Mon Not R Astron Soc 519:3129–3153, https://doi.org/10.1093/mnras/stac3432

Zhang Z, Zhou Y, Rackham BV, Apai D (2018) The near-infrared transmission spectra of TRAPPIST-1 planets b, c, d, e, f, and g and stellar contamination in multi-epoch transit spectra. Astron J 156:178, https://doi.org/10.3847/1538-3881/aade4f

Zhang M, Chachan Y, Kempton EM-R, Knutson HA, Chang WH (2020) PLATON II: New capabilities and a comprehensive retrieval on HD 189733b transit and eclipse data. Astrophys J 899:27, https://doi.org/10.3847/1538-4357/aba1e6

Zhang M, Knutson HA, Wang L, Dai F, Barragán O (2022a) Escaping helium from TOI 560.01, a young mini-Neptune. Astron J 163:67, https://doi.org/10.3847/1538-3881/ac3fa7

Zhang M, Cauley PW, Knutson HA, France K, Kreidberg L, Oklopčić A, Redfield S, Shkolnik EL (2022b) More evidence for variable helium absorption from HD 189733b. Astron J 164:237, https://doi.org/10.3847/1538-3881/ac9675

Zhang M, Knutson HA, Wang L, et al. (2022c) Detection of ongoing mass loss from HD 63433c, a young mini-Neptune. Astron J 163:68, https://doi.org/10.3847/1538-3881/ac3f3b

Zhang M, Dai F, Bean JL, Knutson HA, Rescigno F (2023a) Outflowing helium from a mature mini-Neptune. Astrophys J Lett 953:L25, https://doi.org/10.3847/2041-8213/aced51 Zhang X, Showman AP (2017) Effects of bulk composition on the atmospheric dynamics on close-in exoplanets. Astrophys J 836:73, https://doi.org/10.3847/1538-4357/836/1/73

Zhang M, Knutson HA, Dai F, Wang L, Ricker GR, Schwarz RP, Mann C, Collins K (2023b) Detection of atmospheric escape from four young mini-Neptunes. Astron J 165:62, https://doi.org/10.3847/1538-3881/aca75b

Zhang Z, Morley CV, Gully-Santiago M, et al. (2023c) Giant tidal tails of helium escaping the hot Jupiter HAT-P-32 b. Sci Adv 9:eadf8736, https://doi.org/10.1126/sciadv.adf8736

Zieba S, Zilinskas M, Kreidberg L, et al. (2022) K2 and Spitzer phase curves of the rocky ultra-short-period planet K2-141 b hint at a tenuous rock vapor atmosphere. Astron Astrophys 664:A79, https://doi.org/10.1051/0004-6361/202142912

Zieba S, Kreidberg L, Ducrot E, et al. (2023) No thick carbon dioxide atmosphere on the rocky exoplanet TRAPPIST-1c. Nature 620:746–749, https://doi.org/10.1038/s41586-023-06232-z

Reviews in Mineralogy & Geochemistry
Vol. 90 pp. 465–514, 2024
Copyright © Mineralogical Society of America

13

An Overview of Exoplanet Biosignatures

Edward W. Schwieterman and Michaela Leung

Department of Earth and Planetary Sciences
University of California, Riverside
Riverside, CA, 92521
U.S.A.

eschwiet@ucr.edu, mleun019@ucr.edu

INTRODUCTION: THE SEARCH FOR LIFE BEYOND THE SOLAR SYSTEM

Habitable exoplanets and the context for biosignatures

In the last three decades, knowledge of planetary systems other than our own has increased rapidly with the discovery of over 5,000 exoplanets (Christiansen 2022). These worlds vary greatly in mass, composition, and insolation, encompassing planets ranging in size from smaller than Mercury (Barclay et al. 2013) to more than twice the radius of Jupiter (Crouzet et al. 2017). Most recently, exoplanetary science has entered a new era of planetary characterization with the launch of the James Webb Space Telescope (JWST), which has already unveiled the composition and chemical processes of exoplanet atmospheres in unprecedented detail (e.g., Ahrer et al. 2023; Alderson et al. 2023; Tsai et al. 2023). The study of exoplanet atmospheres is advancing at an unprecedented rate, but current reviews can provide a foundational basis for understanding their formation, dynamics, climates, chemistries, and observables (Helling 2019; Jontof-Hutter 2019; Pierrehumbert and Hammond 2019; Shields 2019; Madhusudhan 2019; Wordsworth and Kreidberg 2022; Kempton and Knutson 2024, this volume).

A subset of discovered exoplanets orbit within the so-called "habitable zone" (HZ), most often defined as the range of distances from a star where a geologically active rocky planet with an N_2–CO_2–H_2O atmosphere can maintain temperatures suitable for surface liquid water (Kasting et al. 1993; Kopparapu et al. 2013, 2014; Kane et al. 2016). The HZ can also be defined more broadly and may encompass super-Earths with H_2-dominated atmospheres and deep global oceans, called Hycean (hydrogen ocean) worlds (Madhusudhan et al. 2021, 2023). While liquid water is a key requirement for life as we know it, habitability as a concept encompasses other potential requirements, including the availability of energy, nutrients, and other hospitable physiochemical conditions beyond clement temperatures (Cockell et al. 2016; Hoehler et al. 2020).

The first step in searching for evidence of life is characterizing the habitability of the star-planet system. The HZ can predict the planetary targets most likely to host surface oceans, allowing for the robust exchange of gases between a potential biosphere and the atmosphere (Kasting et al. 2014) and potentially those targets for which access to information about the planetary surface is obtainable remotely. The HZ as a conceptual tool is intended to guide the design of telescopes and surveys to find habitable worlds with remotely characterizable atmospheres and surfaces and does not (and was never intended to) circumscribe all—or possibly even most—environments in which life could arise or is currently present. For example, icy moons, also termed ocean worlds, such as analogs to our solar system moons Europa and Enceladus, may be common habitable environments in the universe (Hendrix et al. 2019). Still, their potentially habitable subsurface environments are inaccessible for remote characterization.

1529-6466/24/0090-0013$05.00 (print)
1943-2666/24/0090-0013$05.00 (online)

http://dx.doi.org/10.2138/rmg.2024.90.13

The search for life beyond the solar system is a central goal of the NASA Astrobiology Program (Hays et al. 2015) and one of the driving motivations behind the National Academy of Sciences's recommendations for a proposed infrared (IR) / optical / ultraviolet (UV) Surveyor to succeed JWST and the Hubble Space Telescope (National Academies of Sciences, Engineering, and Medicine 2021). Such evidence for life, termed "biosignatures," could be determined from the spectral characterization of these distant worlds.

What is an exoplanet biosignature?

A biosignature is generally defined as "substances, structures, patterns, or processes, or ensembles of these features" that indicate the current or former presence of life and can be separated from abiotic sources (Des Marais and Walter 1999; Des Marais et al. 2008; Hays et al. 2015). In the context of exoplanets, a biosignature is a remotely observable indication of living processes influencing a planet's atmosphere or surface. A biosignature can be a gaseous molecule (or suite of molecules), a surface feature (or suite of features), or time-dependent modulations of gases or surface features that can be linked to life (Fig. 1). We narrow this definition to exclude signs of technological life, called technosignatures (Tarter 2001, 2006; Wright et al. 2022), which are beyond the scope of this chapter. (However, see Haqq-Misra et al. (2022) for an overview of potential planetary technosignatures).

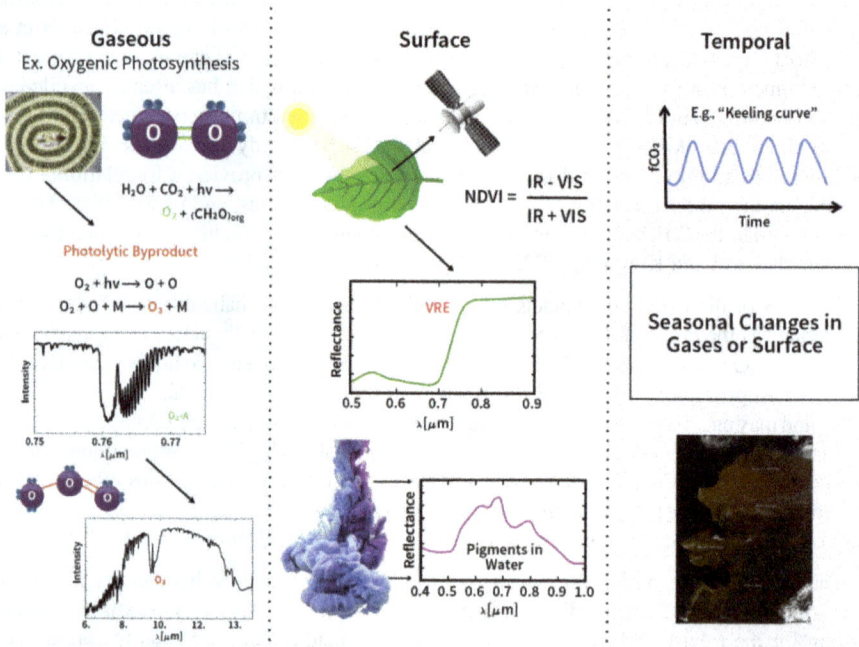

Figure 1. Summary of gaseous (**left**), surface (**center**), and temporal (**right**) biosignatures. Gaseous biosignatures include spectrally active, volatile metabolic products, such as O_2 produced by oxygenic photosynthesis, and potential photochemical byproducts, such as ozone (O_3) from O_2 photochemistry. Surface biosignatures include the vegetation red edge (VRE), which results from the sharp contrast between chlorophyll absorption at visible wavelengths and scattering at infrared wavelengths in photosynthetic organisms. Temporal biosignatures include time-dependent modulation of gases linked to life—such as seasonal changes in CO_2 consumed by photosynthesis and released by the decay of organic materials—or variations in albedo from the growth and decay of vegetation. This figure is reproduced from Schwieterman (2021) under Creative Commons Attribution License CC-BY 4.0. Sub image credits: NASA and the Encyclopedia of Life (EOL).

Any purported exoplanet biosignature will require further vetting beyond the first observation to confirm a biological origin. So, it is best to designate all exoplanet biosignatures as "potential biosignatures" until such vetting has brought reasonable certainty to that designation (Schwieterman et al. 2018a; Meadows et al. 2022). We note in passing that some authors have expressed skepticism that any remote signature could robustly indicate (non-technological-)life (e.g., Smith and Mathis 2023) or take issue with the term "biosignature" as typically applied to potential signs of life on exoplanets and seek to redefine it (e.g., Gillen et al. 2023). While these perspectives are important and should be discussed (Malaterre et al. 2023), we will not litigate them here, and our remit is to provide an overview of exoplanet biosignatures as proposed in the literature.

A compelling biosignature has the following attributes: detectability, survivability, and specificity (Meadows 2017; Meadows et al. 2018b). Detectability refers to the intrinsic absorption, scattering, or emission properties of the molecule or feature. To be observed, a biosignature must influence the information embedded within light that is reflected or emitted from or transmitted through the target body. For example, the O_2 molecule has a strong absorption feature at 0.76 μm that is apparent in the reflected light spectrum of Earth (Fig. 2). Survivability refers to the biosignature molecule or substance's ability to endure rapid depletion or destruction by planetary or astrophysical factors such as the host star's ultraviolet (UV) radiation. A useful biosignature must have the capacity to accumulate to levels that can meaningfully impact the planet's spectrum, an outcome controlled by factors such as chemical kinetics and dissociation cross-sections in the context of atmospheric biosignatures. However, robustness to environmental depletion can cut both ways if small abiotic sources of potential biosignature molecules can accumulate over long timescales. Finally, specificity refers to the separability of biotic and abiotic sources, which will always depend on the available context. Earth's CH_4, for example, is in a strong kinetic and thermodynamic disequilibrium with our planet's temperate O_2-rich atmosphere (Sagan et al. 1993). In contrast, methane is an abundant gas in Jupiter, Saturn, Uranus, and Neptune, as it is an expected equilibrium product in massive, primordial H_2-rich atmospheres with high-temperature interiors (Lodders and Fegley 2002; Moses et al. 2013).

Here, we provide an overview of proposed exoplanet biosignatures, including their biological origins, observable features, and potentially confounding abiotic sources. Additional information is available within the cited sources herein and past comprehensive reviews (Seager et al. 2012; Seager and Bains 2015; Grenfell 2017; Kaltenegger 2017; Catling et al. 2018; Fujii et al. 2018; Schwieterman et al. 2018a; Walker et al. 2018). We aim to emphasize material published since these reviews while providing a foundational understanding of each named biosignature.

ATMOSPHERIC (GASEOUS) BIOSIGNATURES

Atmospheric, or gaseous, biosignatures are volatile molecules that are either direct products of life or secondary products from the environmental processing of biogenic compounds. Biogenic gas emissions can include those directly related to the primary energy-yielding metabolism (such as O_2 from oxygenic photosynthesis) or incidental products from other cellular processes. Biogenic substances such as DNA and RNA make poor remotely detectable biosignatures due to their high molecular weight, low volatility, significant fragility, and broad and ambiguous spectral features. Thus, the search for atmospheric biosignatures is limited to small, volatile molecules with a reasonable chance of accumulating to detectable concentrations. Because all small molecules have abiotic sources—though some are much more limited than others—the interpretation of any of these molecules as a sign of life highly depends on the context in which they are found (Krissansen-Totton et al. 2022). It is crucial to emphasize that planetary atmospheres and surfaces can consume molecules at high rates, so the mere existence of molecules in a non-planetary context, such as the interstellar medium (e.g., McGuire 2018), does not (necessarily) negate their potential utility for fingerprinting life in the proper context.

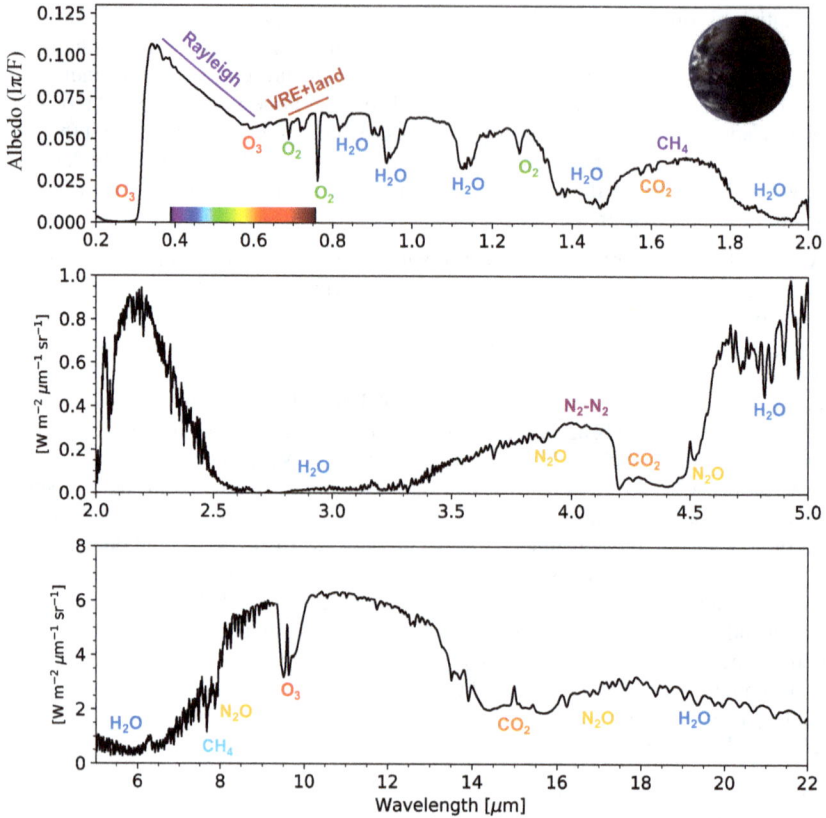

Figure 2. Simulated spectrum of Earth from 0.2–22 μm at quadrature-phase (half illumination) showing various spectral features that include biosignature gases O_2, O_3, and CH_4; the vegetation red edge (VRE) surface feature; and habitability marker gases H_2O, CO_2, and N_2. Rayleigh scattering is indicative of atmospheric pressure. **Top:** Apparent spectral albedo in reflected light from 0.2–2 μm (ultraviolet/visible/near-infrared) at quadrature phase. **Middle:** Near-infrared (2–5 μm) spectral radiance (units: $W \cdot m^{-2} \cdot \mu m^{-1} \cdot sr^{-1}$), including reflected and emitted light components. **Bottom:** Thermal infrared (5-22 μm) spectral radiance. The synthetic spectrum was calculated using the Virtual Planetary Laboratory 3D spectral Earth model (Robinson et al. 2011; Schwieterman et al. 2015b). This figure is reproduced with minor modifications from Schwieterman et al. (2018a) under Creative Commons Attribution License CC-BY 4.0. Modifications include additional gas feature labels, an inset of a simulated Earth at half illumination, and an **inset color bar** indicating the visible wavelength range.

Gaseous biosignatures may be observed via transmitted, reflected, or emitted light from a planetary atmosphere (Figs. 2, 3). Therefore, biosignature gases must interact with light via dissociation, electronic, or vibrotational transitions to be spectrally observable. Figure 4 shows the intrinsic near-to-far infrared (IR) opacities for several proposed biosignature molecules with data sourced primarily from the 2020 HITRAN database and original data sources (Sharpe et al. 2004; Gordon et al. 2022). Tables 1 and 2 summarize several key gaseous biosignatures, including their modern concentration on Earth, the environments and organisms that produce them, the wavelengths of their major absorption features, and their known abiotic sources. Below, we describe each gas in more detail, including its sources, sinks, and potentially observable features.

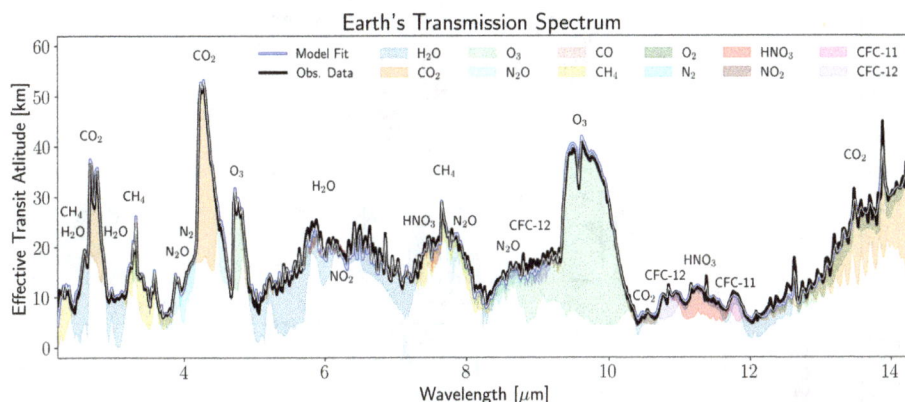

Figure 3. Earth's infrared (2–15 μm) cloud-free transmission spectrum from Macdonald and Cowan (2019; **black line**) and fit using the SMARTER retrieval model (**blue line**; Lustig-Yaeger et al. 2023). This spectrum shows gaseous habitability markers CO_2 (2.7, 4.3, 15 μm), H_2O (6 μm), and N_2 (4.2 μm); biosignatures CH_4 (3.3, 7.7 μm), O_3 (4.7, 9.7 μm) and N_2O (4, 7.7, 8.5 μm), and industrial pollutants CFC-11 (CCl_3F; 11.8 μm) and CFC-12 (CCl_2F_2; 10.8 μm), and NO_2 (6.2 μm). This figure is taken from Lustig-Yaeger et al. (2023) under Creative Commons Attribution License CC-BY 4.0.

Oxygen and ozone (O_2 and O_3)

Oxygen (O_2). The most referenced gaseous biosignature is molecular oxygen (O_2), a product of oxygenic photosynthesis on Earth. Photosynthesis is a means by which life can convert light energy, inorganic carbon in the form of carbon dioxide (CO_2), and water into organic biomass. The net equation for oxygenic photosynthesis can be given as:

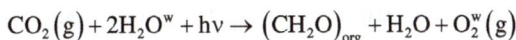

$$CO_2(g) + 2H_2O^w + h\nu \rightarrow (CH_2O)_{org} + H_2O + O_2^w(g)$$

where hν is the energy of the photon(s) involved in the reaction (where h is Planck's constant and ν is the photon's frequency) and $(CH_2O)_{org}$ represents organic matter (biomass). Here we have used a w superscript to indicate that the O_2 resulting from the net reaction has been sourced from H_2O rather than CO_2. We can generally describe oxygenic photosynthesis as a reduction-oxidation reaction where electrons are transferred from H_2O, the reductant, through complex biochemical reactions, to carbon dioxide, thus reducing the carbon and evolving a waste oxidant (O_2). Notably, the simplified net reaction presented here belies significant metabolic complexity beyond the scope of this chapter (see, e.g., Hohmann-Marriott and Blankenship 2012 for a detailed treatment). Oxygenic photosynthesis is an immensely productive metabolism because it leverages common molecules H_2O and CO_2 and bountiful photons from the Sun. Other forms of photosynthesis exist but are limited by reductants less common on terrestrial planets, such as H_2, H_2S, and ferrous iron (Fe^{2+}). (On an H_2-rich planet, CO_2 may instead be limiting.) Therefore, oxygenic photosynthesis may be the most productive metabolism on any planet orbiting a star (Kiang et al. 2007a,b).

Oxygenic photosynthesis relies on Photosynthetically Active Radiation (PAR) photons in the ~0.4–0.72 μm range, while anoxygenic photosynthesis can use longer wavelength, lower energy photons into the near-infrared (NIR; Hohmann-Marriott and Blankenship 2012). The long-wavelength limit for oxygenic photosynthesis depends on many factors, including the primary photosynthetic chlorophyll pigments, but practical limits such as photon flux may mean oxygenic photosynthesis could be light-limited around cool M dwarf stars (Lehmer et al. 2018).

Importantly, the accumulation of oxygen in Earth's atmosphere involves complex long-term biogeochemical cycles and requires the burial of organic carbon to prevent the back

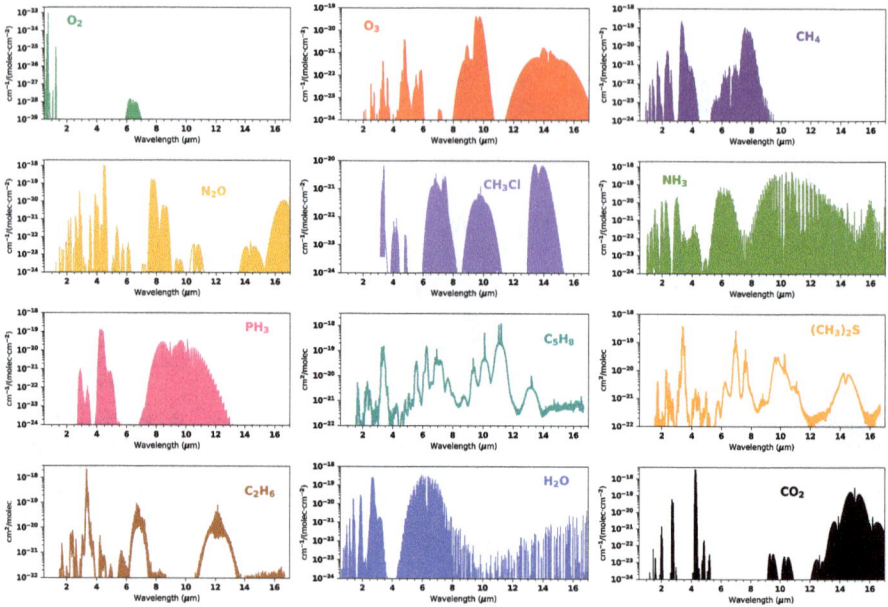

Figure 4. Absorption features for potential biosignature molecules and other gases important for terrestrial planet characterization. Shown are line intensities for the most abundant isotopologue from the HITRAN 2020 database (Gordon et al. 2022) for O_2, O_3, CH_4, N_2O, CH_3Cl, NH_3, PH_3, H_2O, and CO_2 in units of $cm^{-1}/(molec \cdot cm^{-2})$ and absorption cross-sections from the PNNL database (Sharpe et al. 2004) for C_5H_8, $(CH_3)_2S$ (DMS), and C_2H_6 in units of $cm^2/molecule$. Note that opacities are often incomplete for $\lambda < 2$ µm either because these data are not in HITRAN but are available elsewhere (e.g., O_3) or because they haven't been measured (e.g., CH_3Cl).

reaction of the equation above for net oxygenation to occur (Catling 2014). Without resupply from photosynthesis (and organic carbon burial), the oxygen in Earth's atmosphere would be depleted by oxidative weathering and reactions with reduced volcanic gases over geologically short timescales (Lécuyer and Ricard 1999). See Lyons et al. (2014, 2021), Meadows et al. (2018), and Stüeken et al. (2024) for more detailed histories of the oxidation and oxygenation of Earth's atmosphere and accompanying implications for the remote detectability of oxygenic photosynthesis over geologic time.

It has long been recognized that the near-infrared and optical spectral bands of oxygen at 0.76 um (Fraunhoffer-A) and 0.69 µm (Fraunhoffer-B) could provide remote indications of Earth's atmospheric oxygen and, by extension, its photosynthetic biosphere (Owen 1980; Sagan et al. 1993). Oxygen is an interesting molecule as it is a symmetric, homonuclear molecule that nonetheless possesses significant electronic transitions in the optical and near-infrared spectrum, in contrast to N_2, which possesses few potentially detectable bands and only through collisionally induced absorption and dimer features in the infrared (Lafferty et al. 1996; Schwieterman et al. 2015b). Oxygen also absorbs meaningfully at 0.63 µm (O_2-γ) and 1.27 µm. In addition, there are O_2-O_2 collisionally induced absorption features (sometimes called O_4 dimer features, or $(O_2)_2$ features) at UV/Vis/NIR wavelengths at 0.345, 0.36, 0.38, 0.445, 0.475, 0.53, 0.57, 0.63, 1.06, and 1.27 µm (Greenblatt et al. 1990; Hermans et al. 1999; Maté et al. 1999; Karman et al. 2019) and in the mid-infrared (MIR) at 6.4 µm (Timofeev and Tonkov 1979; Maté et al. 2000; Fauchez et al. 2020).

Detecting O_2. The observational manifestation of oxygen (O_2) biosignatures will depend on observing mode, wavelength coverage, and spectral resolving power. (This is generally true for all spectrally active gas species.) The 0.76 μm band is a commonly discussed target in reflected light due to its intrinsic strength, co-location with the high spectral fluxes from main sequence stars (and thus high planet flux in reflected light), and proximity to other absorption features of interest, such as water vapor. The Science Technology Definition Teams (STDT) for future reflected light direct-imaging mission proposals have used this band as a fiducial target in a terrestrial exoplanet atmosphere (The LUVOIR Team 2019; Gaudi et al. 2020).

Unfortunately, the O_2-A band's features are too narrow and weak to be detected by JWST via transit transmission observations at modern Earth-like abundances (Pidhorodetska et al. 2020; Tremblay et al. 2020). However, other molecules may be characterized in terrestrial atmospheres (Meadows et al. 2023). In some scenarios, abiotic O_2 may accumulate to substantially higher levels than found on modern Earth, which may be more detectable in transmission or reflected light spectra (see below). High-resolution (R~100,000) ground-based observations by Extremely Large Telescopes (ELTs) may also search for O_2 in exoplanetary atmospheres (Snellen et al. 2013). However, such searches present many challenges and may require decades or more observing time for a positive result (Currie et al. 2023; Hardegree-Ullman et al. 2023).

The most viable approach for detecting Earth-like levels of O_2 in exoplanetary atmospheres is via reflected light direct imaging with a space-based telescope, such as the Infrared/Optical/Ultraviolet (IR/O/UV) Surveyor recommended by the 2020 Astronomy & Astrophysics Decadal Survey (National Academies of Sciences, Engineering, and Medicine 2021). NASA's current designation for this mission concept is the Habitable Worlds Observatory (HWO; https://habitableworldsobservatory.org/home), whose specific architecture and capabilities have yet to be determined.

Ozone (O_3). Ozone is a photochemical by-product of atmospheric oxygen. Most of Earth's O_3 is generated in the stratosphere (15–50 km altitude) due to the photolysis of O_2. The dominant photochemical reactions that produce and consume O_3 in Earth's atmosphere are often called the "Chapman reactions" (Chapman 1930):

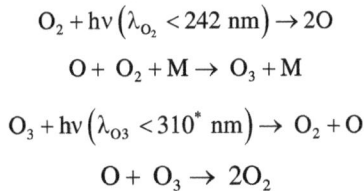

$$O_2 + h\nu \left(\lambda_{O_2} < 242 \text{ nm}\right) \rightarrow 2O$$

$$O + O_2 + M \rightarrow O_3 + M$$

$$O_3 + h\nu \left(\lambda_{O3} < 310^* \text{ nm}\right) \rightarrow O_2 + O$$

$$O + O_3 \rightarrow 2O_2$$

Ozone concentrations peak in Earth's stratosphere near 10 parts-per-million (ppm) at 30 km altitude, though both the peak concentration and the altitude at which that peak occurs vary spatially and temporally (Camp et al. 2003). The absorption of shortwave (UV/Visible) radiation by O_3 heats the upper atmosphere and creates a temperature inversion in Earth's stratosphere.

The concentration of O_3 is strongly sensitive to the intensity and spectral energy distribution of the incident UV flux, and so predicted O_3 concentrations can be markedly different assuming non-Sun-like host stars and/or different flux-equivalent distances from those stars even for planets with identical atmospheric O_2 levels (Segura et al. 2003; Grenfell et al. 2007, 2014), and these predictions are very sensitive to the input stellar spectra (Cooke et al. 2023). In addition to UV radiation, stellar energic particles (StEP) from flare events can deplete O_3 columns, depending on the intensity and frequency of these events (Tabataba-Vakili et al. 2016; Tilley et al. 2019). There is a growing recognition that StEPs are an important area of study to fully understand the atmospheric chemistry of exoplanets (Airapetian et al. 2020; Chen et al. 2020; Garcia-Sage et al. 2023).

Tropospheric (0–15 km) O_3 exists at substantially lower concentrations (~10–20 parts-per-billion, ppb) than in the stratosphere but is very important for tropospheric chemistry. The formation of O_3 is temperature-dependent (Burkholder et al. 2019; Pidhorodetska et al. 2021), and radical species, including OH, Cl, and Br, can catalytically destroy ozone (Solomon 1999), both factors that may further complicate predicting O_3 abundances from O_2 levels alone (Kozakis et al. 2022).

Ozone possesses significant absorption features in the UV/Vis/NIR and mid-infrared wavelength regimes (Figs. 2–4). The strongest mid-infrared feature is the 9.65 μm, which has long been proposed as an indirect indicator of O_2 in the mid-infrared (e.g., Leger et al. 1993), where O_2 bands are non-existent or too weak to detect. The UV O_3 Hartley band is centered near 0.25 μm and extends from 0.2–0.31 μm; this band's absorption shields Earth's surface from deleterious UV radiation (ozone and other molecules such as O_2 and CO_2 absorb at shorter wavelengths). The weaker, structured Huggins UV band spans the range 0.31–0.36 μm. Ozone also absorbs in the visible Chappuis band between 0.4 and 0.7 μm with a peak near 0.6 μm. The weaker near-IR Wulf band extends from the long wavelength end of the Chappuis band to beyond 1 μm. Because O_3 absorption is continuous (though varying in opacity) throughout the UV/Vis/NIR, the literature varies concerning the wavelength cutoffs between each named region. O_3 has weaker features—that nonetheless vary greatly in strength—throughout the NIR and MIR, including at 2.05, 2.15, 2.5, 2.7, 3.3, 3.6, 4.7, 5.5, 5.8, 7.8, 8.9, and 14.0 μm (Gordon et al. 2022).

Detecting O_3. Detecting ozone in the transmission spectra of temperate rocky exoplanets may be challenging. Existing studies find that over 100 transits with JWST's NIRSpec or MIRI instruments would be required to detect O_3 at 3σ significance on TRAPPIST-1e if it were an Earth-like planet (Lustig-Yaeger et al. 2019; Wunderlich et al. 2019, 2020; Gialluca et al. 2021), which is at or beyond the upper range of the transits that could be viewed in JWST's nominal five-year lifetime. These results depend on how the O_3 formation is modeled, which is critically dependent on the input stellar spectra, and studies of O_3 detectability that assume Earth's O_3 profile have found fewer transits would be required to detect O_3 on TRAPPIST-1e. For example, Lustig-Yaeger et al. (2023) find that the 4.7 μm O_3 feature can be constrained twice as precisely with NIRSpec than the 9.65 μm O_3 band can be constrained with MIRI. However, these results assume the actual Earth's spectrum rather than one based on self-consistent photochemistry with the host star. Notably, the detectability of O_3 (and other gases) in transmission spectroscopy is related to their intrinsic molecular opacities and the backlighting effect from the host star. Because the star is brighter in the NIR, features with lower intrinsic opacities can, therefore, be more detectable in practice. It is important to note that O_3 detectability in transmission spectra also depends on scale height, with higher temperatures allowing for greater transit signatures but, on the other hand, suppressing the chemical formation of O_3 (Pidhorodetska et al. 2021; Harman et al. 2022).

O_3 could be detected in reflected light by a future space-based UV/Vis/IR telescope such as HWO via its UV Hartley absorption at ~0.3 μm and broad visible absorption resulting from its Chappuis band (The LUVOIR Team 2019; Gaudi et al. 2020; Damiano et al. 2023). Additionally, the 9.65 μm O_3 band is one of the primary target molecules for missions designed to directly image planets in the MIR, such as ESA's Large Interferometer For Exoplanets (LIFE) mission concept (Alei et al. 2022; Konrad et al. 2022; Quanz et al. 2022). Because the formation of O_3 is non-linear with O_2 abundance, it could be a sensitive indicator of O_2 levels too low to detect directly (Reinhard et al. 2017, 2019; Olson et al. 2018; Schwieterman et al. 2018b), such as may have been the case during parts of Earth's Proterozoic Eon (0.5–2.5 Ga). However, caution is warranted given the dependence of O_3 formation and destruction on numerous other factors (Kozakis et al. 2022; Cooke et al. 2023), and the dependent of O_3 detectability in the IR on the temperature structure of the atmosphere.

False positives for oxygen and ozone biosignatures

Oxygenic photosynthesis produces essentially all the free O_2 in Earth's atmosphere today. Decades ago, there was a general view in the astrobiology community that detectable free O_2 was unlikely in the atmosphere of a habitable planet with liquid water oceans, supporting its status as a target molecule for remotely searching for life elsewhere (e.g., Des Marais et al. 2002). Over the past decade, numerous studies have challenged this view by proposing potential mechanisms for the abiotic accumulation of atmospheric O_2 on exoplanets (e.g., reviews in Meadows 2017; Meadows et al. 2018). We summarize these mechanisms and the potential spectral features that could disambiguate potential "false positives" from true positive biosignatures. Because O_2 is the most well-studied potential remote biosignature, our treatment of its potential false positives accompanying spectral signatures is more comprehensive than we present for other potential biosignature gases.

Figure 5 provides a graphical summary of proposed O_2 false positives and the atmospheric context that could identify them. It is important to note that, as of this writing, the characterization of the atmospheric composition of terrestrial exoplanets is just beginning. Therefore, we have limited information about the prevalence of abundant abiotic O_2 on exoplanets, just as we are limited in understanding the prevalence of alien biospheres.

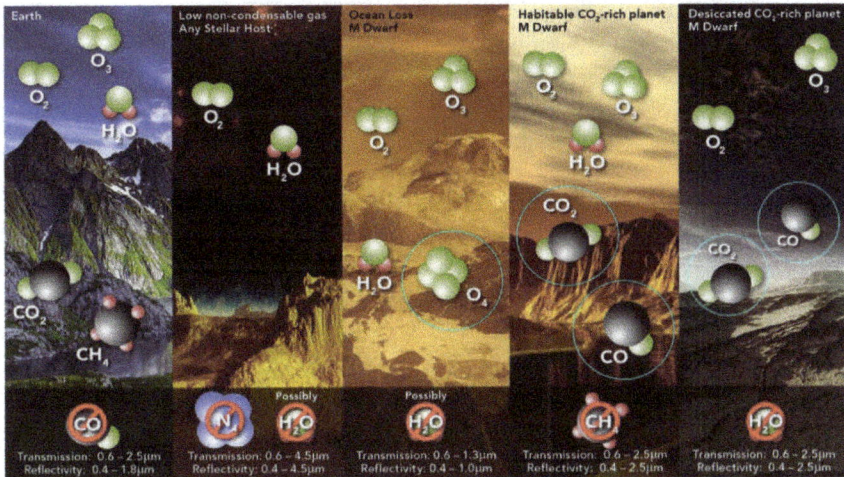

Figure 5. Graphical illustration of potential scenarios leading to potentially detectable abiotic O_2 and/or O_3. Circle molecules identify spectrally active species such as O_4 (O_2–O_2 collisionally induced absorption) that could be used to fingerprint the origin(s) of O_2 and/or O_3, while the red strike-through indicates species that should be absent. Modern Earth is distinguished by the combination of spectrally active O_2, O_3, H_2O, CO_2, and CH_4 together with low levels of CO. See section *False positives for oxygen and ozone biosignatures* in this text for additional details. This figure is reproduced from Meadows et al. (2018b) under Creative Commons Attribution License CC-BY. Graphic artist: R. Hasler.

Photochemical O_2 and O_3 in CO_2-rich atmospheres. The photolysis of O-bearing molecules such as CO_2 can instantiate a series of photochemical reactions that generate O_2 and O_3 molecules. Indeed, the dry atmosphere of Mars has low but detectable (from Earth and interplanetary space) levels of O_2 and O_3 from oxygen liberated by CO_2 photolysis (Noxon et al. 1976; Fast et al. 2006). The direct recombination of CO and O into CO_2 ($CO + O + M \rightarrow CO_2 + M$) is spin forbidden, so it occurs slowly. The photolysis products from abundant water vapor ($H_2O + h\nu \rightarrow H + OH$) can effectively catalyze this recombination, however:

$$CO + OH \rightarrow CO_2 + H$$

$$O_2 + H + M \rightarrow HO_2 + M$$

$$HO_2 + O \rightarrow O_2 + OH$$

$$\text{Net: } CO + O \rightarrow CO_2$$

The Martian atmosphere's water vapor prevents the decomposition of its CO_2 into CO and O_2 (McElroy and Donahue 1972; Krasnopolsky 2011). On Venus, the CO and O recombination is catalyzed by volcanic HCl (DeMore and Yung 1982). A very dry terrestrial exoplanet with a CO_2-rich atmosphere, extremely low H_2O content, and the lack of other chemical catalysts could hypothetically decompose into a CO and O_2-rich atmosphere, depending on other factors (Gao et al. 2015). Some authors have found that the stellar spectrum of M dwarf host stars can drive the production of abiotic O_2 (and O_3) even on planets with surface oceans and abundant atmospheric water vapor (Domagal-Goldman et al. 2014; Tian et al. 2014; Harman et al. 2015). M dwarf host stars have more intense FUV radiation, which drives CO_2 photolysis, but less of the NUV radiation that is critical for photolyzing water and initiating the catalytic cycles that recombine CO and O in otherwise anoxic atmospheres. However, these results are sensitive to model assumptions, including the input disassociation cross-sections (Ranjan et al. 2020) and the lightning flash rate (Harman et al. 2018). NO_x species (e.g., NO, NO_2) produced by lightning can drive the catalytic recombination of CO and O (Harman et al. 2018). The time-dependent UV radiation and particle fluxes from flares are also important, though currently understudied, considerations when modeling potential false positives for O_2 and O_3 (France et al. 2013, 2016; Chen et al. 2020).

Ranjan et al. (2023) have found that some past results predicting robust abiotic O_2 accumulation from CO_2 decomposition on terrestrial planets orbiting M dwarf host stars were erroneous and caused by instituting a model "top" at too low of an altitude (too high of a pressure), which effectively confined CO_2 photolysis to one numerical layer. They argue that abiotic O_2 accumulation via CO_2 photolysis will not generate O_2 levels above 1%, and perhaps much lower, assuming abundant tropospheric water vapor (i.e., habitable conditions). However, these O_2 levels can generate spectrally apparent O_3, which may be confounding when using O_3 as a tracer for biotic O_2. This false positive scenario could be identified via spectrally active CO (with features at 1.6, 2.35, and 4.7 µm), which accumulates before abiotic O_2 begins to rise (Schwieterman et al. 2016). Very dry atmospheres that facilitate abiotic O_2 accumulation via lack of HO_x (e.g., OH, HO_2) catalysts could be identified via the lack of water vapor absorption (Gao et al. 2015). Future theoretical work—and, ultimately, observations—will further elucidate the extent of abiotic O_2 and O_3 that can be maintained in terrestrial atmospheres from the photolysis of other O-bearing species.

Hydrogen loss and oxygen retention during early runaway greenhouses. The evolution of the atmospheres of terrestrial planets can be significantly influenced by the early, pre-main-sequence (PMS) phase of stellar evolution. During the PMS, protostars have yet to begin fusing hydrogen in their cores, and their luminosity is instead powered by gravitational contraction (Kippenhahn et al. 1990). As the radii of the stars shrink, their luminosity can decrease by one to two orders of magnitude before entering the main sequence, the core hydrogen-burning phase. The PMS phase for M dwarf stars ($0.08\,M_\odot \leq M \leq 0.6\,M_\odot$) can last hundreds of millions of years, compared to ~50 Myr for the Sun (Baraffe et al. 1998; Reid and Hawley 2005). Consequently, planets in the habitable zones of such stars today likely experienced extended runaway greenhouse states, which would have enriched their upper atmospheres with water vapor.

In addition to their extended PMS phases, due to their fully convective interiors and consequent magnetic activity, M dwarfs emit much more X-ray and extreme ultraviolet (XUV) radiation than Sun-like stars. This XUV radiation can contribute to substantial atmospheric loss, potentially ablating away entire atmospheres (Lammer et al. 2003, 2008; Johnstone et al. 2019) and—coupled with photodissociation of atmospheric water vapor—leading to the equivalent loss of many oceans of water (Ramirez and Kaltenegger 2014; Luger and Barnes 2015; Tian 2015). Because hydrogen atoms are less massive than oxygen atoms, hydrogen can more easily escape to space, though hydrodynamic flows can also carry away oxygen. Each Earth Ocean is equivalent to ~240 bars of O_2, so, depending on initial conditions, hundreds to thousands of bars of O_2 can be left behind to oxidize the planet's interior, surface, and atmosphere (Luger and Barnes 2015; Tian 2015).

Potential planetary sinks for this atmospheric oxygen (other than hydrodynamic loss) include dissolution in a magma ocean coupled with oxidation of reduced iron (Zahnle et al. 1988; Chassefière et al. 2012; Hamano et al. 2013), non-thermal ion escape (Persson et al. 2020), and reaction with reducing volcanic gases over time. Venus went through a runaway greenhouse at some point in its history, which could have removed more than an Earth ocean of water (Kasting et al. 1984; Kasting 1988; Chassefière 1996; Gillmann et al. 2009). If so, Venus would have at one time accumulated abundant abiotic atmospheric O_2, which was then consumed through some combination of the sinks enumerated above (Chassefière et al. 2012; Kane et al. 2019). Given the likely broad distribution in initial conditions among terrestrial planets, it was predicted that some Venus-like post-runaway atmospheres may have maintained elevated O_2 levels today (Chassefière et al. 2012).

Luger and Barnes (2015) demonstrated that planets currently within the habitable zones of M dwarf stars are particularly susceptible to abiotic O_2 accumulation during their star's pre-main sequence phase when the stars' luminosities and XUV fluxes are substantially enhanced relative to later times. Depending on their initial water endowments, these planets could become completely desiccated or retain substantial H_2O, complicating habitability and biosignature assessments. Many additional authors have predicted the possibility of high levels (tens, hundreds, or thousands of bars) of abiotic O_2 in the atmospheres of terrestrial planets orbiting M dwarf stars both within and outside the habitable zone (Schaefer et al. 2016; Wordsworth et al. 2018; Krissansen-Totton and Fortney 2022). Near-future characterization of "Venus zone" exoplanets (Kane et al. 2014) may elucidate the prevalence of post-runaway O_2-rich atmospheres in the solar neighborhood, which will have important implications for future biosignature surveys (Ostberg et al. 2023).

Abiotic O_2-rich post-runaway atmospheres could be identified by O_2 abundances that are too large to be biologically produced. On Earth, O_2 levels are self-regulated over time by negative feedbacks, including combustion of terrestrial vegetation (i.e., fires) to levels $\lesssim 0.35$ bar (Lenton 2013). High levels of O_2 could be fingerprinted by O_2–O_2 collisionally induced absorption (CIA), sometimes called O_4, which is strongly dependent on density (Misra et al. 2014; Schwieterman et al. 2016). O_2–O_2 CIA has prominent absorption features at 0.345, 0.36, 0.38, 0.445, 0.475, 0.53, 0.57, 0.63, 1.06, 1.27, and 6.4 μm (Hermans et al. 1999; Maté et al. 2000; Richard et al. 2012; Karman et al. 2019). At optical and near-infrared wavelengths, these O_2–O_2 bands would generate large absorption features in reflected light (Fig. 6; Meadows et al. 2018). In transit, the 1.06 and 1.27 μm bands would be the best target for identifying O_2–O_2 CIA (Lustig-Yaeger et al. 2019). The 6.4 μm band could also fingerprint high-O_2 atmospheres in transit, assuming minimal interference from water vapor absorption (Fauchez et al. 2020).

Hydrogen loss and oxygen retention in atmospheres with limited non-condensable gases. Abiotic oxygen could accumulate in a terrestrial planet's atmosphere without going through a runaway greenhouse. On Earth, almost all water vapor is confined to the troposphere via the "cold trap" at the tropopause, which is the altitude of the temperature minimum where

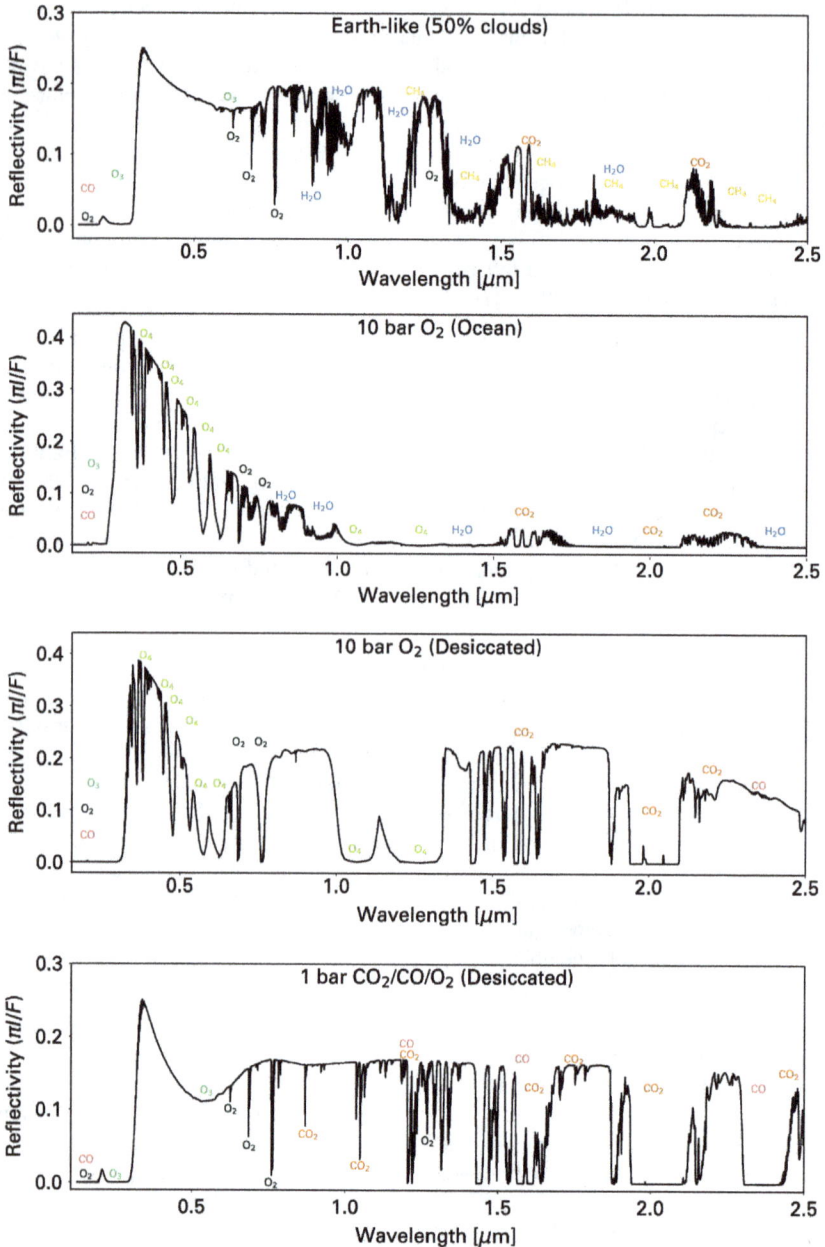

Figure 6. Spectra of hypothetical atmospheres for Proxima Centauri b in reflected light, illustrating spectral features of oxygen consistent with an inhabited planet (topmost panel) or various false positive mechanisms (all other panels). **First panel:** photochemically self-consistent Earth, with enhanced CH_4 spectral features consistent with longer photochemical lifetimes predicted for planets orbiting M dwarf host stars. **Second panel:** 10 bar abiotic O_2-dominated atmosphere with a remaining surface ocean. **Third panel:** 10 bar abiotic O_2-dominated atmosphere after complete desiccation and loss of surface ocean. **Final panel:** desiccated, CO_2-rich atmosphere photochemically decomposed into $CO_2/CO/O_2$ due to lack of catalysts requiring H. Compare with conceptual illustrations in Figure 5. This figure is sourced from Meadows et al. (2018a,b) under Creative Commons Attribution License CC-BY.

the troposphere transitions into the stratosphere. The tropopause altitude depends on latitude and ranges from 9 km at the poles to 17 km at the equator. Water condenses at the tropopause's temperature and pressure conditions, resulting in a relatively dry stratosphere. However, the efficacy of the cold trap relies on the presence of non-condensing gases like N_2 and argon (Ar). Wordsworth and Pierrehumbert (2014) show that reducing the non-condensing gas abundances below ~0.2 bar lifts the cold trap and leads to the accumulation of water vapor in the stratosphere, where UV photons can more readily photolyze H_2O. The hydrogen liberated from this H_2O photolysis can diffuse through the upper atmosphere and escape to space, resulting in the net oxidation of the planetary surface and/or atmosphere. Once O_2 accumulates such that the total pressure of non-condensing gases is $\gtrsim 0.2$ bar, the cold trap is (re-)established, limiting further H loss and O_2 build-up. (The rate at which O_2 builds up in the atmosphere depends on additional factors, including crustal oxidation and the outgassing of volcanic gases.) This amount of O_2 is practically identical to Earth's total O_2 abundance of 0.21 bar.

This abiotic O_2 scenario could be positively identified or ruled out by constraining the abundance of non-condensing gases. N_2 is difficult to detect because it is a homonuclear molecule with no transitional dipole moment; however, N_2 could be directly identified via its N_2–N_2 CIA band centered at 4.3 μm (Lafferty et al. 1996; Schwieterman et al. 2015b), which produces potentially detectable signatures in Earth's transmission spectrum (Lustig-Yaeger et al. 2023) and could plausibly be detected in the transmission spectra of exoplanets (Kaltenegger et al. 2020). Directly detecting N_2 would be difficult in reflected light because of the absence of significant absorption features. However, N_2 could be inferred indirectly in reflected light by characterizing the total atmospheric pressure via Rayleigh scattering and/or pressure broadening and ruling out other absorbing gases through a process of elimination, given sufficient spectral coverage (Hall et al. 2023).

Perpetual greenhouses, water worlds, and desert worlds. Krissansen-Totton et al. (2021) proposed three other scenarios for the abiotic accumulation of O_2 on terrestrial worlds orbiting sun-like stars, all of which relate to the initial volatile inventories of the planets in question. They find that if the initial CO_2:H_2O inventory exceeds one by mass, the planet's atmosphere becomes trapped in a perpetual greenhouse state that does not abate over several Gys for worlds at 1 AU, a scenario also proposed by prior authors (Marcq et al. 2017; Salvador et al. 2017). Because the atmospheric water is supercritical, silicate weathering reactions requiring liquid water that would otherwise draw down CO_2 cannot operate effectively. As in the runaway greenhouse scenarios described above, atmospheric water vapor can be photolyzed, followed by hydrogen escape and atmospheric oxidation. The lack of liquid water further arrests oxidative weathering that would consume this abiotic O_2 in the atmosphere, potentially allowing it to accumulate to high concentrations (i.e., $p_{O_2} > 0.02$ bar). However, this scenario requires a low crustal oxidation efficiency, as atmospheric O_2 could directly oxidize fresh crust (Krissansen-Totton et al. 2021). Strong CO_2 bands and lack of a habitable surface could identify this scenario. Because the structure of absorption bands is temperature-dependent, their shape could also indicate temperatures too high for habitability (Young et al. 2024).

A water world scenario, assuming initial H_2O inventories of over 10 Earth oceans, may also lead to the accumulation of abiotic O_2 (Krissansen-Totton et al. 2021). This is caused by the pressure overburden of a massive ocean arresting the creation of new crust, which strongly limits surface oxygen sinks (Noack et al. 2016; Kite and Ford 2018). In this case, O_2 is net produced slowly due to hydrogen escape but accumulates over geologic time due to strongly reduced sinks. This false positive scenario could be indicated by the absence of continents, which could be revealed via rotational mapping techniques (Cowan et al. 2009; Kawahara and Fujii 2011; Fujii et al. 2017; Lustig-Yaeger et al. 2018).

Finally, the last scenario presented by Krissansen-Totton et al. (2021) is a desert world scenario, predicated on very low initial volatile inventories, which we briefly summarize as follows. In this case, the low levels of volatiles generate a weak early greenhouse, allowing the underlying magma ocean to cool in less than ~10^5 years. Consequently, abiotically generated atmospheric O_2 sourced from XUV-driven hydrogen escape cannot react with the reductants in the magma ocean. This O_2 could persist over Gyrs because the low volatile inventory implies a small source of volcanic reductants, but this scenario also requires inefficient crustal oxidation. This desert world scenario could be identified by the lack of ocean glint (Gaidos and Williams 2004; Williams and Gaidos 2008; Zugger et al. 2010; Robinson et al. 2014; Lustig-Yaeger et al. 2018; Ryan and Robinson 2022).

Importantly, for each scenario listed above, the extent of O_2 accumulation depends on many factors that can differ drastically for exoplanets, and many combinations of these variables do not lead to predicted abiotic O_2 accumulation. Moreover, each scenario is accompanied by predicted diagnostic features that could either identify or rule out the scenario in question. Finally, in every case, the co-existence of O_2 with complementary reduced biosignature gases, such as CH_4 and/or N_2O, would be a distinguishing factor that would exclude an abiotic origin for O_2.

Methane (CH_4)

Methanogenesis is an ancient metabolism that arose early in the history of life on Earth and likely contributed to an early methane-rich atmosphere in the Archean Eon (Ueno et al. 2006; Catling and Zahnle 2020; Stüeken et al. 2024). Methanogenesis involves the chemotrophic generation of energy via anaerobic respiration that results in the production of methane gas. Most commonly, CO_2 is used as a terminal electron acceptor, or acetic acid (CH_3COOH) is disproportionated into CH_4 and CO_2. These pathways can be simplified as:

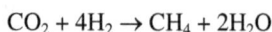

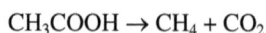

$$CO_2 + 4H_2 \rightarrow CH_4 + 2H_2O$$

$$CH_3COOH \rightarrow CH_4 + CO_2$$

where H_2 is hydrogen gas. Methanogenesis also requires catalysis by specialized enzymes in complex biochemical processes not captured by these simplified net equations (Ferry 1999).

Methanogenesis occurs in various anaerobic environments today, including in oxygen-poor soils, in anoxic lakes, at hydrothermal vents, and in the gastrointestinal tracts of animals (Martin et al. 2008; Thauer et al. 2008). Importantly, CH_4 can be oxidized aerobically and anaerobically by other organisms, such as *Methylococcus* spp. (Hanson and Hanson 1996) and sulfate reducers (Cui et al. 2015), respectively, before it is released into the atmosphere. Therefore, the total biogenic production of CH_4 on Earth is greater than its net atmospheric release. This net flux on Earth is approximately ~30–40 Tmol/year (Dlugokencky et al. 2011), inclusive of biotic, industrial, and abiotic sources. This results in a modern CH_4 concentration of ~1.9 ppm (Lan et al. 2023), though pre-industrial concentrations were likely closer to 500 ppb (Solomon et al. 2007). Sources of abiotic methane on Earth include water–rock reactions such as serpentinization, which could account for as much as 10% of the non-anthropogenic methane production (Etiope and Sherwood-Lollar 2013).

CH_4 is a biosignature in Earth's atmosphere because it has strong thermodynamic and kinetic disequilibrium with atmospheric O_2 (Lovelock 1965; Hitchcock and Lovelock 1967; Sagan et al. 1993). The atmospheric lifetime of methane on Earth is about 10 years. Its major photochemical sink is the hydroxyl radical (OH), which in Earth's O_2-rich atmosphere is primarily sourced from the photolysis of tropospheric ozone ($O_3 + h\nu \rightarrow O(^1D) + O_2$ followed by $O(^1D) + H_2O \rightarrow 2OH$; Jacob 1999). The atmospheric lifetime of CH_4 is a strong function of the incident stellar spectrum, especially in the context of O_2-rich atmospheres, and the same CH_4 flux could result in CH_4 concentrations ~1,000 times greater for Earth-like terrestrial planets orbiting M dwarf stars (Segura et al. 2005; Rugheimer et al. 2015; Wunderlich et al. 2019). This is because M dwarfs emit substantially less energy at NUV wavelengths that photolyze ozone compared to sunlike stars, corresponding to diminished production of OH radicals to react with and remove CH_4 (Segura et al. 2005).

Methane has strong absorption features in the NIR at 1.65, 2.3, and 3.3 µm (with weaker bands at 0.9, 1, 1.15, and 1.35 µm) and in the MIR most strongly in a wide band centered at ~7.7 µm and extending from ~7.2–8.4 µm. These bands often overlap strongly with H_2O, so moderate to high spectral resolution is required to distinguish these species. We can search for CH_4 in terrestrial planets' reflection, transmission, or emission spectra, with NIR bands more favorable for reflected or transmitted light observations and MIR bands most favorable for emitted light observations. Importantly, Earth's modern concentration of methane may be undetectable in reflected light, because of its low concentration and correspondingly weak absorption bands; however, this was not the case for the Archean or perhaps the Proterozoic Eons, with higher methane concentrations (Reinhard et al. 2017; The LUVOIR Team 2019; Gaudi et al. 2020). M and K dwarf host stars, with their corresponding stellar spectra and consequent implications for photochemistry, may promote detectable concentrations of CH_4 in modern Earth-like atmospheres in reflected light (e.g., Segura et al. 2005; Arney 2019).

Of course, plentiful abiotic methane also exists in the atmospheres of the solar system's gas planets, where it is made by equilibrium processes in high-temperature layers of H_2-rich atmospheres and transported to higher altitudes with lower temperatures (Lodders and Fegley 2002; Madhusudhan et al. 2016). On a temperate planet, such equilibrium reactions cannot happen in the atmosphere on a geologically relevant timescale, so an active surface source is necessary to maintain atmospheric methane. In an anoxic terrestrial atmosphere like the Archean Earth, CH_4 is destroyed by OH radicals sourced from water vapor photolysis or by direct photolysis of CH_4 by FUV radiation (Kasting et al. 1983). CH_4 is also prevalent in the atmosphere of Saturn's moon Titan, and while the origin of this CH_4 is debated (e.g., Glein 2015), its photochemical stability is aided by the cold temperatures (~90 K), which preclude abundant water vapor and downstream photochemical sinks like OH.

Krissansen-Totton et al. (2018b) suggest that the disequilibrium between CH_4 and CO_2 would constitute a biosignature if CH_4 mixing ratios exceed 10^{-3} to 10^{-2} and if both gases are observed without abundant CO. This is because CO_2 and CH_4 are two vastly different oxidation states of carbon. Any abiotic source (e.g., a mantle with oxygen fugacity orders of magnitude lower than modern Earth) that produces both should also generate abundant CO, which has an intermediate oxidation state. Indeed, the CO_2–CH_4 disequilibrium biosignature may have been observable in Earth's atmosphere for longer than observable CH_4–O_2 disequilibrium (Krissansen-Totton et al. 2018b). Coupled with the kinetic disequilibrium generated from the photochemical destruction of CH_4 in temperate atmospheres and the need for continuous sources, these factors motivate the potential of CH_4 to be an exoplanet biosignature even within atmospheres lacking observable O_2 (Thompson et al. 2022). It is important to note that in some scenarios, volcanic production could result in both CO_2 and CH_4 production, though CO would also be produced (Schaefer and Fegley 2010; Liggins et al. 2022, 2023). In addition, photochemical processes can produce CO_2 even in H_2-rich atmospheres that contain both CH_4 and H_2O. In this scenario, the carbon in CH_4 is photo-oxidized using the O liberated from H_2O photolysis (Hu 2021); however, yet again, abundant CO would be predicted in this scenario as an intermediate product. It is also often assumed that CO would be drawn down by life via acetogenesis, as CO affords both a reductant and carbon source (Ragsdale 2004; Wang et al. 2016; Catling et al. 2018). Therefore, the observable absence (or low abundance) of CO underpins the potential utility of the CO_2–CH_4 biosignature pair in anoxic atmospheres. This factor, too, is complicated by the observation that life produces CO through various direct and indirect processes and does not need to consume CO at the maximum biotic rate (Schwieterman et al. 2019; Zhan et al. 2022).

High levels of methane can lead to the production of organic hazes, which may have existed during portions of the Archean (Trainer et al. 2006; Zerkle et al. 2012; Arney et al. 2016) and may have been important for generating prebiotic molecules for Hadean Earth (Pearce et al. 2024). These hazes would have had both strong climatic and spectral impacts on Earth's atmosphere and could therefore be common on temperate exoplanets in a similar biogeochemical regime to Earth's Archean (Arney et al. 2017) or Hadean eons. For hydrocarbon haze to accumulate, the ratio of CH_4 to CO_2 must exceed ~0.1 (the exact ratio depends on the inherent photochemistry of the planet-star system), so it has been suggested that hydrocarbon haze could be an indirect exoplanet biosignature (Arney et al. 2018) given that it may be implausible for abiotic CH_4 production rates alone to generate haze in atmospheres with Archean-like levels of CO_2. The presence of haze at an unexpectedly low CH_4/CO_2 ratio could imply the presence of other organic molecules fueling haze formation (e.g. organic sulfur gases), even if these gases are not themselves detectable in the spectrum, which could strengthen a biological interpretation (Arney et al. 2018). However, organic haze is predicted for a wide variety of planetary scenarios, including those with no biological inputs (He et al. 2020; Moran et al. 2020; Gao et al. 2021). Nonetheless, photochemical models continue to predict that the distribution of carbon species (CO_2, CO, CH_4) will differ between abiotic and scenarios with a robust biosphere (Akahori et al. 2023; Watanabe and Ozaki 2024), though there may be substantial overlap, particularly for temperate anoxic planets orbiting M dwarf hosts. Figure 7 shows planetary spectra of hypothetical Archean-like planets orbiting various

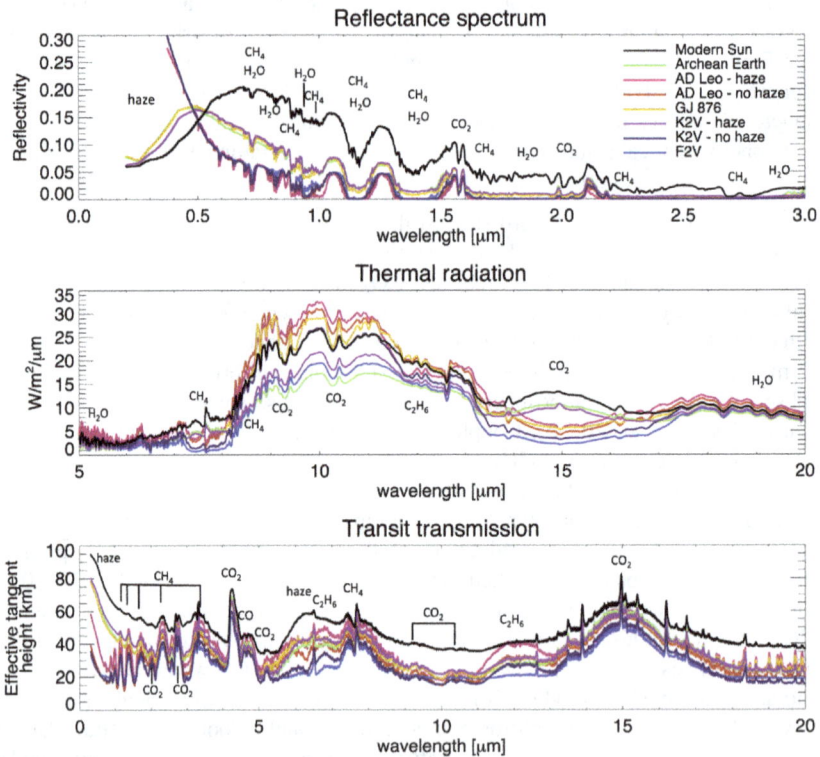

Figure 7. Synthetic spectra of anoxic Archean-like exoplanets orbiting other stars modeled with self-consistent photochemistry and climate. The top panel contains reflectance spectra, the middle panel contains thermal emission spectra, and the bottom panel contains transmission spectra. The K2V-haze case has a CH_4/CO_2 ratio 0.3, the AD Leo-haze case has a CH_4/CO_2 ratio of 0.9, and all other cases have CH_4/CO_2 ratios of 0.2. Used by permission of AAS, from Arney et al. (2017), The Astrophysical Journal, Vol. 836, Fig. 6, p. 49.

stars, displaying the simultaneous presence of CH_4, CO_2, organic haze, and ethane (C_2H_6) that results from the photochemical processing of CH_4 and organosulfur gases.

Though the interpretation of CH_4 biosignatures is complex—particularly in the absence of O_2 or O_3—it is nonetheless a key target in the search for life beyond Earth. Indeed, the CO_2–CH_4 biosignature couple may be the most observable potential biosignature in the TRAPPIST-1 planetary system (Krissansen-Totton et al. 2018a; Mikal-Evans 2021; Meadows et al. 2023).

Nitrous oxide (N_2O)

N_2O is overwhelmingly produced by life on Earth due to microbial nitrogen metabolism in terrestrial and aquatic environments. Through the processes of nitrogen fixation, nitrification, and denitrification, atmospheric N_2 gas is converted into bioavailable forms of nitrogen such as ammonium (NH_4^+), oxidized to form nitrite (NO_2^-) and nitrate (NO_3^-), and then reduced back to N_2 gas, with N_2O as an intermediate product (Thamdrup 2012; Tian et al. 2015). N_2O is also produced through direct ammonia oxidation by some microorganisms (Prosser and Nicol 2012).

The modern concentration of atmospheric N_2O is ~330 ppb, with a preindustrial concentration of ~270 ppb (Myhre et al. 2013). The primary atmospheric sink for N_2O is photolysis, which converts N_2O into N_2 and O_2, though reaction with atomic oxygen radicals also contributes to the photochemical destruction of N_2O. The photochemical survival of N_2O is dependent on the incident stellar spectrum and oxygen abundance, as overlying O_2 (and accompanying O_3) can provide a shielding effect (Grenfell et al. 2014; Rugheimer and Kaltenegger 2018). The photochemical lifetime of N_2O in Earth's atmosphere today is about 120 years (Prather et al. 2015). The net production of N_2O by terrestrial and marine sourced on Earth today is ~0.4 Tmol/year (Tian et al. 2020). Figures 8 and 9 illustrate how N_2O concentrations depend on molecular flux, O_2 concentration, and the incident stellar spectrum and provide a general example of how these factors can influence biosignature gas abundances.

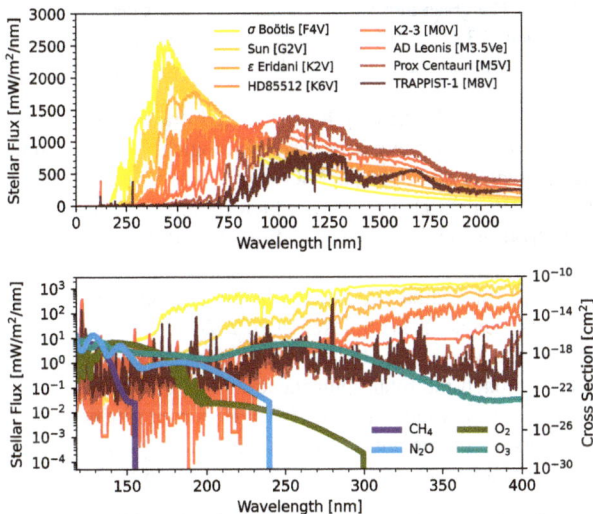

Figure 8. Top panel: Spectral flux received at an Earth-like insolation from various stellar host types in the UV–Vis–NIR. Bottom panel: the identical stellar spectra focused on the far-UV to near-UV with the dissociation cross-section of O_2, O_3, N_2O, and CH_4 overplotted (right y-axis). The different energy distributions from each star will impact the photolysis rate and, thus, the steady-state concentration given a specific molecular production rate (see Fig. 9). This figure was taken from Schwieterman et al. (2022) under Creative Commons Attribution License CC-BY 4.0.

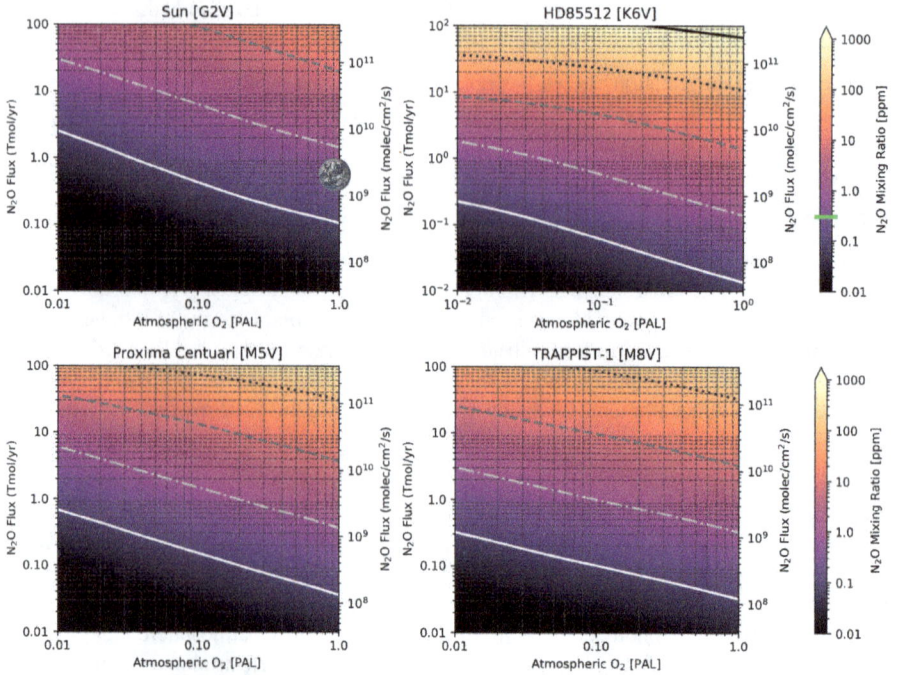

Figure 9. N_2O concentrations in parts-per-million (ppm) as a function of N_2O flux (*y*-axes) and atmospheric O_2 (*x*-axes) predicted by a photochemical model for an Earth-like planet orbiting within the habitable zone of each named star. The stars listed here correspond to those whose spectra are shown in Figure 8. The inset Earth image in the **top left panel** indicates Earth's N_2O molecular flux (0.4 Tmol/yr; 1.5×10^9 molecules/cm^2/s). The **green horizontal line** indicates Earth's current N_2O atmospheric mixing ratio (~330 parts-per-billion; 3.3×10^{-7}). Data subpanels were adapted from Schwieterman et al. (2022) under Creative Commons Attribution License CC-BY 4.0.

The N_2O concentration on Earth may have changed drastically over geologic time. The biogenic flux of N_2O to the atmosphere depends on biological and environmental factors. The final step of the nitrogen cycle that converts N_2O into N_2 gas is mediated by an enzyme called nitrous oxide reductase (Pauleta et al. 2013), which requires a copper catalyst. It has been suggested that H_2S-rich oceans in the Proterozoic Eon (2.5 to 0.5 Ga) would have been deficient in copper, reducing the efficiency of the last step in the denitrification cycle that converts N_2O to N_2, which could have had substantial climatic implications given N_2O's status as a greenhouse gas (Buick 2007; Roberson et al. 2011; Stüeken et al. 2024). In the absence of metal co-factors or the evolution of the nitrous oxide reductase enzyme (or an analog), the end member N_2O emission flux would approach the total denitrification flux of the biosphere, which is on the order of ~20 Tmol N_2 per year (Canfield et al. 2010). However, nitrification and denitrification depend on various factors, including nutrient availability and oxygen concentrations so that this value could be smaller or larger on an exoplanet (Schwieterman et al. 2022).

N_2O is advantageous as a potential remote biosignature due to its overwhelming biological origin on Earth and distinct spectral features in the near- and mid-IR (Sagan et al. 1993; Kaltenegger and Selsis 2007; Rauer et al. 2011). N_2O can be seen weakly in Earth's near-infrared reflected light spectrum (Fig. 2; Sagan et al. 1993), and more strongly in Earth's emitted mid-infrared spectrum (Fig. 2; Mettler et al. 2023), and in Earth's transmitted light spectrum (Fig. 3; Lustig-Yaeger et al. 2023). N_2O has significant absorption bands in the near-

IR at 2.25, 2.9, 4.0, and 4.5 μm, with weaker bands at 1.5, 1.6, 1.7, 1.8, 2.6, and 3.7 μm (yet weaker bands are present at shorter wavelengths). In the mid-IR, N_2O's strongest bands reside at 7.8, 8.5, and 17 μm, and would be compelling targets for space-based direct imaging of terrestrial worlds (Fig. 10; Angerhausen et al. 2024). Importantly, many of these absorption bands overlap with CH_4; for example, the 7.8 μm N_2O band overlaps with the 7.7 μm CH_4 band, so the actual detectability of the gas will depend on a confluence of factors, including the concentrations of overlapping absorbing species.

Figure 10 shows the resulting planetary spectra of Earth-like planets with varied N_2O production rates and stellar hosts. These scenarios correspond to the ones shown in Figures 8 and 9, and together, this sequence of three figures traces the vertically integrated connection between the intrinsic properties of the N_2O molecule (i.e., dissociation cross-sections), incident stellar spectra, biosignature molecular fluxes, and resulting spectral observables. These spectra also illustrate that N_2O may be a more detectable biosignature on planets orbiting cooler stars than the Sun. A hypothetical inhabited TRAPPIST-1e with an N_2O flux constituting a significant fraction of Earth's planetary denitrification flux (see Stüeken et al. 2024) would produce N_2O transit signatures equivalent to other major gas absorbers (e.g., CO_2, CH_4), and could be detectable by JWST (Schwieterman et al. 2022).

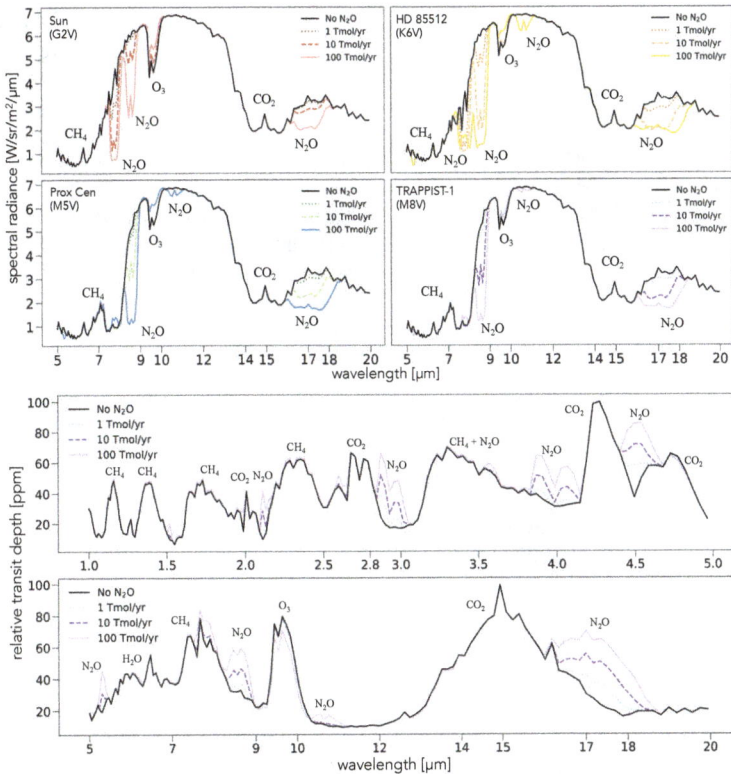

Figure 10. Spectra of Earth-like planets showing N_2O features as a function of N_2O production rate. The top set of four panels shows the IR emission spectra of Earth-like planets with 21% O_2, 50% cloud cover, and net N_2O fluxes of 1, 10, and 100 Tmol/yr. The bottom two panels show a hypothetical transmission spectrum of TRAPPIST-1e under the same assumptions. Major absorbing species are indicated. Each scenario's abundance is determined from the photochemical simulations shown in Figure 9. These figures were adapted from Schwieterman et al. (2022) under Creative Commons Attribution License CC-BY 4.0.

The abiotic sources of N_2O are limited on Earth, with a very small amount produced by lightning (Schumann and Huntrieser 2007) and chemodenitrification (Samarkin et al. 2010; Jones et al. 2015) processes. Chemodenitrification refers to the abiotic reduction of oxidized aqueous nitrogen species (e.g., NO_3^-) by reductants such as ferrous iron (Fe^{2+}), which can produce N_2O gas abiotically (Stanton et al. 2018). However, N_2O production via chemodenitrification requires the simultaneous presence of reductants and oxidants in ocean layers. This is implausible with an abiotic O_2 atmosphere, which requires a completely oxidized ocean for long-term stability. Lightning can produce small amounts of NO_x species that can be chemodenitrified without biotic O_2. However, while chemodenitrification could have maintained N_2O concentrations in the ~10 ppb range in the Archean (Buessecker et al. 2022), remotely detectable concentrations likely require direct or indirect biological production (Schwieterman et al. 2022).

Photochemical simulations show that an active star can produce abundant StEPs that could split the N_2 molecule and initiate a series of photochemical reactions that could produce abiotic N_2O (Airapetian et al. 2016, 2020; Chen et al. 2020). However, such a scenario could be identified by the characterization of the star and the additional spectrally active gaseous species that would be produced, such as NO_2, HNO_3, and HCN. The simultaneous presence of N_2O and O_2 would be a stronger biosignature than either alone, akin to the O_2 and CH_4 biosignature couple.

Schwieterman et al. (2022) present a more comprehensive summary of the potential for N_2O biosignature detection on exoplanets, including the plausible range of biological fluxes, photochemical accumulation, and resulting spectral signatures in transmission and emission spectra. This work also includes a more extensive discussion of possible abiotic sources for N_2O and the possible distinguishing spectral features of those scenarios.

Alternative biosignature gases

Above, we have reviewed the potential atmospheric biosignatures that exist at spectrally significant abundances ($\gtrsim$ 100 ppb) in the modern atmosphere and likely generated remotely detectable signatures for a substantial fraction of Earth's geologic history ($\gtrsim$ 500 Myr; also see Table 1). Life produces more volatile compounds that do not accumulate to greater concentrations in Earth's atmosphere due to lower net biological production and/or shorter photochemical lifetimes. However, the net production of biogenic gases can be highly sensitive to the chemical nature of the local environment and evolutionary contingencies (e.g., Garcia et al. 2020; Parsons et al. 2021), and the photochemical lifetime is a strong function of the host star (e.g., Segura et al. 2005; France et al. 2016). Given the likely diverse compositions, evolutionary histories, and astrophysical settings of rocky exoplanets, biosignature gases at trace concentrations on Earth could have significant abundances elsewhere. Moreover, alternative biochemistries (i.e., "life as we don't know it") may further expand the range of potential biosignatures (Davila and McKay 2014; Seager et al. 2016; Chan et al. 2019; Petkowski et al. 2020). Below, we describe proposed alternative biosignature gases (Table 2) and provide the most recent references for comprehensive evaluations of their biological production, sources of abiotic false positives, and potential for atmospheric accumulation and detectability on exoplanets.

Ammonia (NH_3). Nitrogen (N_2) is fixed to bioavailable forms like NH_3 by microorganisms called diazotrophs in an endothermic process (Gruber 2008). (Because nitrogen is triple bonded, this process is particularly energy-demanding.) These bioavailable forms of nitrogen are precursors to all N-bearing biological molecules, including amino acids, proteins, and nucleic acids. Diazotrophy is the source of most fixed nitrogen on Earth, with small abiotic sources from lightning (Navarro-González et al. 2001; Schumann and Huntrieser 2007). NH_3 can also result from the abiotic or biologically mediated degradation of organic matter, though most of this fixed nitrogen originated via biologically mediated N_2 fixation. NH_3 is volatile and enters the atmosphere in gaseous form. However, the atmospheric lifetime and concentration of NH_3 are low due to its high solubility and short photochemical lifetime due to photolysis and reactions with radical species (Kasting 1982). The photochemical lifetime would be longer in an H_2-rich atmosphere (Seager et al. 2013a; Huang et al. 2022a; Ranjan et al. 2022).

Table 1. Abundant and/or prominent Earth atmospheric biosignature molecules.

Biosignature	Production environment	Conc. on Earth	False positives and secondary outcomes	Main spectral features (μm) (strongest bolded) via HITRAN and NIST	Citation (see references therein)
Oxygen (O_2)	Photic terrestrial and marine environments, produced by cyanobacteria, algae, and plants	20.95%	Low noncondensable inventories (Wordsworth and Pierrehumbert 2014); Ocean loss (Luger and Barnes 2015; Tian 2015); CO_2 photolysis (Domagal-Goldman et al. 2014; Gao et al. 2015; Harman et al. 2015); High CO_2/H_2O inventories or extremely low initial H_2O inventories (Krissansen-Totton et al. 2021); Atmospheric exchange/Exogenous delivery (Felton et al. 2022)	O_2 at 0.63, 0.69, **0.76**, 1.27; CIA/O_4 at 0.445, 0.475, 0.53, 0.57, 0.63, 1.06, **1.27**, **6.4**	Meadows 2017; Meadows et al. 2018b
Ozone (O_3)	Photochemical product of O_2, H_2O, SO_2, CO_2	0.07 ppm	Abiotic O_2 or CO_2 can produce O_3 features (Domagal-Goldman et al. 2014; Gao et al. 2015; Harman et al. 2015)	0.25 (Hartley), 0.4–0.7 (Chappuis), 2.6, 4.7, **9.65**, 14.6	Meadows et al. 2018b; Kozakis et al. 2022
Methane (CH_4)	Wetlands, rice paddies, livestock emissions, biomass burning & decomposition	1.90 ppm	Serpentinization; Fischer–Troph reactions; High-temperature mantle reactions; Equilibrium reactions in high-T regions of gas giants	1.65, 2.4, 3.3, **7.7** (broad)	Etiope and Sherwood-Lollar 2013; Thompson et al. 2022
Nitrous oxide (N_2O)	Marine settings, some soils	0.33 ppm	Production by StEPs; Chemodenitrification trace production by lightning and NO_x reactions	1.5, 1.6, 1.7, 1.8, 2.3, 2.6, 2.9, 3.7, **4.0**, **4.5**, **7.8**, **8.5**, 17	Schwieterman et al. 2022

Table 2. Alternative atmospheric biosignatures.

Biosignature	Production environment	Conc. on Earth	False positives and secondary outcomes	Main Spectral Features (μm) (strongest bolded) via HITRAN and NIST	Citation (see references therein)
Organic sulfur gases e.g. DMS ($(CH_3)_2S$) DMDS ($(CH_3)_2S_2$)	Marine/lacustrine settings	~100 ppt	Secondary ethane signature may be produced abiotically, if sulfur gases do not reach detectable levels	2.3, 3.4, ~7 (DMS, DMDS), ~9 (DMDS), ~10 (DMS, CH3SH), 14 (DMS, CH3SH),18 (DMDS)	Pilcher 2003; Domagal-Goldman et al. 2011
Halomethanes: Methyl chloride (CH_3Cl) Methyl bromide (CH_3Br)	Marine algae, terrestrial and marine algae; salt marches and wetlands	~500 ppt (CH_3Cl) ~9 ppt (CH_3Br)	High T perchlorate pyrolysis (limited in planetary context), Exogenous delivery, Limited hydrothermal/high T processes	3.3, 7.0, **9.9**, **13.7** (Cl) 3.5, 6.9, **10.2**, **16.4** (Br)	Segura et al. 2005; Leung et al. 2022
Organic haze	Photochemical product of CH_4, CH_3X species	Not present at high altitude on modern Earth due to oxidizing conditions	Abiotic methane can lead to haze production	NIR spectral slope, ~6	Arney et al. 2018
Phosphine (PH_3)	Strictly anoxic environments including paddy fields, lakes/rivers, wetlands/marshes	Ppq–ppb levels, spatially and temporally variable	Phosphite & phosphate degradation, lightning, volcanism, exogenous delivery	2.9, 3.4, 4.4, 5.0, 8.9, **9.5** (broad)	Sousa-Silva et al. 2020
Isoprene (C_5H_8)	Deciduous trees and land plants	~1–5 ppb localized and time-varying	No known (energetically unfavorable)	3.4, 5.6, 6.3, 9.4, **10.2**, **11.4**	Zhan et al. 2021
Methanol (CH_3OH)	Plants via demethylation of pectin	10 ppb (surface)	Abiotic photochemistry, comets & primitive material	2.9, 3.3, 4.6, 7.5, **9.7**	Huang et al. 2022b
Carbonyls, e.g. formaldehyde (CH_2O)	80% of all biological compounds contain carbonyls	~0.4 ppb (CDC)	Detectable signature is CO, which can be produced abiotically (see Schwieterman et al., 2019; Wogan & Catling 2020)	(CO) 1.18, 1.57, 2.35, **4.6**	Zhan et al. 2022
Ammonia (NH_3)	Anerobic nitrogen fixing organisms, ammonification and nitrate reduction	Spatially and temporally variable on Earth	Volcanism, photochemistry, high temperature synthesis, equilibrium thermochemistry	1.2, 1.3, 1.5, 2.0, 2.3, 3, 3.9, 6.1, **10.5** (broad)	Seager et al. 2013a; Huang et al. 2022a

Seager et al. (2013) propose a hypothetical metabolism for NH_3 production in H_2-rich atmospheres ($N_2 + 3H_2 \rightarrow 2NH_3$), which is energy-yielding but kinetically inhibited at habitable temperatures and unknown to exist on Earth. A larger production rate and longer photochemical lifetime could result in detectable amounts of NH_3 in super-Earth atmospheres (Huang et al. 2022a). At sufficiently high fluxes, NH_3 could enter a "photochemical runaway" in analogy to O_2 on Earth (Ranjan et al. 2022). NH_3 absorbs strongly at 2.0, 2.3, 3.0, 5.5–6.5, and 9–13 μm. It would be more strongly detectable in H_2 atmospheres in transit transmission spectroscopy due to the extended scale height of H_2-rich atmospheres and, consequently, larger spectral signatures (Phillips et al. 2021). Abiotic sources of NH_3 include equilibrium reactions in giant planet atmospheres (Madhusudhan et al. 2016) and outgassing of highly reduced planetary interiors (Schaefer and Fegley 2010; Liggins et al. 2022). Strong kinetic disequilibrium on a temperate planet with an H_2O-rich troposphere would indicate biotic NH_3, in analogy to the logic applied for interpreting CH_4 biosignatures (Thompson et al. 2022). Huang et al. (2022) provide a recent comprehensive treatment of the context for NH_3 as an exoplanet biosignature.

Phosphine (PH₃). Phosphorous is an essential and often limiting nutrient for life (Syverson et al. 2021). The production of phosphine (PH_3) by life is associated with anaerobic environments, as PH_3 interferes with aerobic metabolism (Nath et al. 2011), and high concentrations of O_2 efficiently oxidize PH_3. The exact mechanism for the biological production of PH_3 is currently unknown (and may involve indirect production via organic decay); however, it is confidently associated with life as it is produced in some microbial cultures (Jenkins et al. 2000; Liu et al. 2008) and has dynamic diurnal variability in terrestrial and marine environments, suggesting links to overall metabolic activity (Roels et al. 2005; Zhu et al. 2007). Bains et al. (2019) show that in some circumstances, phosphine could be produced by coupling bacterial phosphate reduction and phosphite disproportionation. In a planetary atmosphere, PH_3 is destroyed by reactions with radical species (e.g., $PH_3 + OH \rightarrow PH_2 + H_2O$) and photolysis (Cao et al. 2000; Glindemann et al. 2005). At sufficient fluxes, PH_3 can potentially build up to detectable levels on anoxic (H_2-rich or CO_2-rich) planets (Sousa-Silva et al. 2020; Angerhausen et al. 2023). PH_3 has strong infrared absorption bands at 2.9, 4.4, and 9.5 μm (the latter includes broad absorption 8–11 μm; Sousa-Silva et al. 2015; Gordon et al. 2022). Notably, PH_3 is produced abiotically in high-temperature layers of H_2-rich planetary atmospheres (Visscher et al. 2006) and could be introduced at limited rates via volcanism or micrometeorite collisions (Omran et al. 2021; Truong and Lunine 2021). As is the case for CH_4 and NH_3, PH_3's interpretation as a biosignature relies on its strong kinetic disequilibrium in temperate atmospheres and the consequently high surface fluxes required to maintain a detectable concentration (Sousa-Silva et al. 2020). It has been claimed that phosphine has been detected in the Venusian atmosphere (Greaves et al. 2020, 2022); though others dispute the detection (Akins et al. 2021; Lincowski et al. 2021; Villanueva et al. 2021; Cordiner et al. 2022). This tentative PH_3 detection has spurred an extensive community effort to understand the potential abiotic and biological sources of PH_3 (e.g., Bains et al. 2021). Sousa-Silva et al. (2020) present an extensive investigation of the potential for PH_3 to serve as an exoplanet biosignature.

Isoprene (C₅H₈). Isoprene (C_5H_8) is a significant biotic volatile organic compound on Earth, produced at a rate comparable to methane (Müller et al. 2008; Zhan et al. 2021). Deciduous trees and other organisms, including bacteria and animals, generate C_5H_8 (Fuentes et al. 1996; King et al. 2010). In the Earth's atmosphere, isoprene undergoes rapid destruction primarily through reactions with OH radicals and subsequent reactions with O_2, forming various reactive products and contributing significantly to aerosol formation (Palmer 2003; Medeiros et al. 2018). Consequently, the photochemical lifetime of C_5H_8 in Earth's modern atmosphere is less than three hours (Zhan et al. 2021), which explains its low concentration (spatially and temporally dynamic, but ranging from ppt to the ppb levels) despite its large production rate. This photochemical lifetime would be longer in an H_2-rich atmosphere, allowing for greater accumulation; however, Zhan et al. (2021) find it would require a C_5H_8 flux ~100 times greater than Earth to be detectable. The strongest spectral features of C_5H_8 are at 3.4, 5.6, 6.3, 9.4, 10.2, and 11.4 μm with significant overlap between adjacent bands (Sharpe et al. 2004; Gordon et al. 2022). While it would be difficult for C_5H_8 to accumulate to uniquely identifiable concentrations (particularly given its absorption features overlap with other C-bearing species), it has no known abiotic false positives in a planetary context, which motivates adding this gas to our list of potential exoplanet biosignatures. Zhan et al. (2021) provide a detailed assessment of the biosignature potential of C_5H_8.

Biogenic sulfur gases. Organosulfur gases like dimethyl sulfide (DMS; $(CH_3)_2S$), dimethyl disulfide (DMDS; $(CH_3)_2S_2$), and methanethiol (CH_3SH) are produced by various organisms, including marine algae and photosynthetic bacteria (Visscher et al. 1991, 2003; Hu et al. 2007). Many of these organosulfur gases are indirect products of metabolism via the degradation of sulfur-containing biomolecules such as dimethyl sulfoniopropionate (DMSP; Stefels et al. 2007). However, it has recently been claimed that diverse species, including bacteria, haloarchaea, and algae, can methylate inorganic H_2S into DMS as a detoxification strategy (Li et al. 2023), demonstrating these gases are sometimes direct, rather than incidental, biogenic products. Other sulfur gases such as OCS, H_2S, and CS_2 are produced by life but have robust abiotic sources, such as volcanism (Arney et al. 2014). The concentrations of these sulfur species are spatially and temporally variable (Zhang et al. 2020). Above surface ocean DMS concentrations can range up to over 100 ppt and are maintained by a global flux of about ~0.4 Tmol/year (Hulswar et al. 2022). DMS, DMDS, and CH_3SH are efficiently destroyed by radical species, including OH, in Earth's O_2-rich atmosphere (Fung et al. 2022). The photochemical lifetimes of these species are consequently short, with a lifetime of ~1 day for DMS (Fung et al. 2022).

Pilcher (2003) suggested organosulfur gases could serve as remote biosignatures on anoxic worlds similar to the Archean Earth because photochemical lifetimes would be longer, and a reducing biosphere may be more conducive to producing these gases. Domagal-Goldman et al. (2011) investigated this scenario with a photochemical model. They find it difficult to accumulate spectrally relevant concentrations of observable DMS, DMDS, or CH_3SH except for the case of a low-activity M dwarf host star. They find that C_2H_6 could be a secondary biosignature of a sulfur biosphere, as it is a photochemical product from the CH_3 radicals sourced from these gases. Importantly, Domagal-Goldman et al. (2011) investigated anoxic, CO_2-rich scenarios with no haze; alternative atmospheric scenarios may be more amenable to the accumulation of sulfur gases. Biogenic sulfur gases may be a compelling biosignature target on H_2-rich Hycean worlds and more easily detectable than on planets with high molecular weight atmospheres (Madhusudhan et al. 2021, 2023). Meadows et al. (2023) find that of all the biogenic sulfur gases, CH_3SH is the most detectable in the TRAPPIST-1 planetary system, and show that a sulfagenic biosphere could produce a CH_3SH signature detectable in 50 transits with JWST's NIRSpec instrument.

All three gases have strong but overlapping features throughout the IR (Sharpe et al. 2004). DMS has significant absorption features at 2.3, 3.4, 6–7, 10, and 14 μm. DMDS and CH_3SH have features that are only somewhat offset from these values (see Table 2). Methylated organosulfur gases are not made in equilibrium reactions; however, they can be generated by electric discharge in atmospheric mixtures of CH_4 and H_2S (Raulin and Toupance 1975). Further work is required to fully elucidate and quantify the potential abiotic false positives for methylated organosulfur gases.

Halomethanes (CH_3Cl, CH_3Br, etc.). Halomethanes, including CH_3Cl and CH_3Br, are biologically produced by many organisms, including marine and terrestrial micro- and macroalgae, some plants, bacteria, and fungi (e.g., Saini et al. 1995; Manley et al. 2006; Paul and Pohnert 2011; see Table 1 in Leung et al. 2022). Halomethanes include methyl halides (CH_3X, where X is F, Cl, Br, or I) and polyhalomethanes with the generalized formula

$CH_{4-a,b,c,d}F_aCl_bBr_cI_d$; e.g., CH_2I_2, CH_2Br_2, $CHCl_3$, CBr_4). Notably, halomethanes containing F are not known to be produced biologically (Leung et al. 2022). The most abundant halomethane in Earth's atmosphere is CH_3Cl, with a concentration of ~500 ppt (Seinfeld and Pandis 2016). CH_3Cl and other halomethanes are primarily destroyed by reactions with OH radicals and photolysis (Yang et al. 2005). Segura et al. (2005) first suggested CH_3Cl may be a compelling biosignature on Earth-like planets orbiting M dwarfs due to the reduced NUV radiation that would result in sharply reduced photochemical sinks and, thus, higher steady-state abundances (i.e., ~1 ppm vs. ~500 ppt for the same flux). Subsequently, additional authors have evaluated the potential photochemical accumulation and detectability of CH_3Cl on exoplanets (e.g., Rugheimer et al. 2013; Wunderlich et al. 2021). However, CH_3Cl is just one of many biogenic halomethane species, most of which have not been robustly examined as potential exoplanet biosignatures.

Leung et al. (2022) provide an overview of the potential for halomethanes as a general biosignature class and include a quantitative assessment of CH_3Cl and CH_3Br biosignatures over a wide range of molecular fluxes and stellar host star spectral types. They find that introducing multiple halomethane species results in co-additive photochemical and spectral effects since these species compete for the same photochemical sinks and absorb at similar wavelengths. CH_3Cl has its strongest features at 3.3, 7.0, 9.9, and 13.7 μm, while CH_3Br has its strongest features at 3.5, 6.9, 10.2, and 16.4 μm. JWST is unlikely to detect methylated halogens because their strongest absorption features are in the MIR where signal-to-noise ratios are poor (Meadows et al. 2023). However, halomethanes may be detectable in MIR-emitted light spectra by the LIFE mission (Angerhausen et al. 2024), assuming source fluxes that are somewhat higher than Earth's global average but within the range observed in local environments. Potential abiotic sources of halomethanes include exogenous delivery of cometary material, trace levels of volcanic emission, and high-temperature perchlorate pyrolysis (Frische et al. 2006; Keppler et al. 2015; Fayolle et al. 2017), but it is unlikely these sources would approach the biogenic fluxes observed on Earth (Leung et al. 2022).

Methylation as a general biosignature. Methylation is a fundamental biochemical process performed by organisms across all domains of life on Earth and often results in the formation of volatile and spectrally active products. These methylated gaseous products tend to have limited abiotic sources. Additional potential biosignature gases include—but are not limited to, methylated chalcogens, metals, and metalloids such as $(CH_3)_2Se$ (DMSe), $(CH_3)_2Se_2$ (DMDSe), $(CH_3)_2Te$ (DMTe), $(CH_3)_2Hg$, $(CH_3)_2As$, $(CH_3)_3Sb$, and $(CH_3)_3Bi$ (Fatoki 1997; Basnayake et al. 2001; Bentley and Chasteen 2002; Thayer 2002; Meyer et al. 2008; Ellwood et al. 2016; Yang et al. 2016). These species absorb strongly in the IR (e.g., Gutowsky 1949); however, kinetic rate and opacity data are incomplete, and/or the most recent measurements are highly outdated. Another challenge for these gases as biosignatures is their low global production rates on Earth. Additional work on measuring fundamental spectral and chemical inputs could be highly impactful in exploring this group of potential biosignatures.

Additional gases. The list above is not exhaustive. A vast array of small molecules can be produced by life, including many heretofore unexamined in the voluminous and expanding corpus of exoplanet biosignature literature. Seager et al. (2016) propose systematically examining these small molecules for their biosignature potential. Such an effort would require an in-depth examination of their biological and abiotic sources, photochemical longevity in diverse atmospheres, and spectral features (i.e., detectability, survivability, and specificity). This would, in turn, involve studies of laboratory kinetics (reaction rates, dissociation cross-sections), fundamental spectroscopic measurements in the Vis/NIR/MIR, and end-to-end planetary simulations using atmospheric photochemistry and spectral models in addition to instrumental models for future observatories. Moreover, abiotic production due to planetary and/or astrophysical processes would have to be examined rigorously. Due to the vastness of the undertaking, our guidebook for finding life elsewhere remains incomplete, though tremendous progress has been made to enhance our knowledge of possibilities in recent years.

Frameworks for assessing potential atmospheric biosignatures

As discussed above, detecting any single gaseous species is likely insufficient evidence for life without adequate planetary context (Krissansen-Totton et al. 2022; Meadows et al. 2022). This context may—but not necessarily—be provided by the detection of multiple biosignature features. We also lack a complete understanding of false positives and secondary processes which may influence biosignature detection and interpretation. Therefore, many have suggested possible "agnostic" approaches to biosignatures, i.e., searches for general patterns that may indicate living processes without overreliance on our understanding of Earth life (Johnson et al. 2018; Walker et al. 2018). However, any agnostic approach to exoplanet biosignatures must grapple with the limitations of remote data. Below, we briefly summarize the logical basis for each proposed framework and refer the curious reader to studies that describe them more comprehensively.

Thermodynamic disequilibrium and biosignature pairs. Earlier sections discussed thermodynamic and kinetic disequilibria as applied to specific biosignature gases. Here we generalize these concepts as examples of agnostic biosignatures. The thermodynamic disequilibrium between atmospheric O_2 and CH_4 has long been considered a marker of biology's influence on Earth's atmosphere because, in equilibrium, these constituents react together to form CO_2 and H_2O (Lovelock 1965, 1975; Hitchcock and Lovelock 1967; Sagan et al. 1993). N_2O and other trace biotic species (e.g., CH_3Cl) would also effectively disappear from Earth's atmosphere if reacted to equilibrium. A weakness of thermodynamic disequilibrium biosignatures is that temperate terrestrial atmospheres do not foster equilibrium reactions without catalysts. For example, the reaction $2O_2 + CH_4 \rightarrow CO_2 + 2H_2O$ that underpins the $O_2 + CH_4$ biosignature pair does not occur directly under these conditions but through intermediate photochemical steps ultimately initiated by photolysis. On the other hand, the abundances of atmospheric species used to quantify thermodynamic disequilibrium are potentially observable and quantifiable, while the kinetic processes that affect them are sometimes less likely to be so.

Total atmospheric disequilibrium can be quantified by comparing the difference in Gibbs free energy between the observed and equilibrium states, equivalent to the untapped chemical energy in the planet's atmosphere, often given in joules per mole of atmosphere (Krissansen-Totton et al. 2016). The largest disequilibrium in the Earth system, and the solar system in general, is the co-existence of abundant N_2, O_2, and liquid water, much larger than Earth's O_2–CH_4 atmospheric disequilibrium (Krissansen-Totton et al. 2016). This N_2–O_2–$H_2O_{(aq)}$ disequilibrium has likely existed since the oxidation of Earth's atmosphere and was preceded by biogenic disequilibrium between CO_2 and CH_4 (Krissansen-Totton et al. 2018b). Wogan and Catling (2020) distinguish "edible" and "non-edible" disequilibria. For example, the photochemically produced disequilibrium between CO and O_2 in the atmosphere of Mars leaves easily exploitable free energy ("edible") on the table. This concept has been used to set an upper limit on the chemosynthetic biosphere that could exist in the Martian subsurface (Sholes et al. 2019). In contrast, Earth's CH_4–O_2 and N_2–O_2–$H_2O_{(aq)}$ disequilibria are largely kinetically inhibited (i.e., not edible), while maintained by oxygenic photosynthesis.

So long as the abundances of major gas species (or surface features in the case of an ocean) can be retrieved via spectral observations, planetary atmospheric disequilibria can be quantified. Krissansen-Totton et al. (2018a) demonstrated that CH_4–CO_2 disequilibria could be retrieved on a version of TRAPPIST-1e similar to the Archean Earth. Similarly, Young et al. (2024) showed that chemical disequilibrium biosignatures can be inferred from Proterozoic Earth-like planets with sufficient O_2 levels. Table 3 provides a summary of potential biosignature disequilibrium pairs.

Table 3. Disequilibrium biosignature combinations.

Pairing	Indication of	Observational strategy (μm) (e.g.)	Citation
$O_2/O_3 + CH_4$ (oxic atmospheres)	"redox disequilibrium", since oxidation to CO_2 and H_2O does not occur, sources of O_2 and CH_4 must be present	Combined detections of O_2: 0.76, O_3: 9.65, CH_4: 1.65, 3.3, 7.7	Lovelock 1965; Hitchcock and Lovelock 1967; Sagan et al. 1993
$CO_2 + CH_4$ (anoxic/ Archean-like atmospheres)	non-volcanic source of CH_4 if observed without significant CO	Combined detections of CH_4: 1.65, 3.3, 7.7, CO_2: 1.6, 2.0, 4.3, 15	Krissansen-Totton et al. 2018a,b
$N_2 + O_2 + H_2O$	largest thermodynamic equilibrium on the modern Earth	Combined detections of oceans (via glint or polarimetry), N_2 (via N_2–N_2 CIA centered at 4.3 μm or scattering), O_2: 0.76	Krissansen-Totton et al. 2016
$N_2O + O_2/O_3$ and/or CH_4	"redox disequilibrium," since oxidation states differ between N_2O, O_2, and CH_4	Combined IR detection of N_2O: 7.8, 8.5, O_3: 9.65, and CH_4: 7.7	Kaltenegger 2017; Schwieterman et al. 2022

Kinetic disequilibrium. As noted above, in temperate (i.e., "habitable") planetary atmospheres, gaseous constituents are not determined by equilibrium processes but by chemical kinetics (Catling and Kasting 2017). CH_4 is destroyed in our O_2-rich atmosphere by radical species such as OH, which are downstream photolysis products dependent on UV photons from the Sun (Jacob 1999). Other gases, such as N_2O, are primarily destroyed directly by photolysis or can be consumed via wet or dry deposition. These sinks' relative contributions strongly depend on the planetary context, such as the background bulk atmospheric composition and the spectral characteristics of the planet's host star (Figs. 8–9). Given an understanding of these environments and a retrieved gas abundance, one can use a photochemical model to infer the production rate (molecular flux) needed to sustain that abundance. Sufficiently high production rates can fingerprint biological activity if no known abiotic source can plausibly sustain that flux. A biochemical model can further estimate the biomass needed to generate this production rate (Seager et al. 2013b). This exercise can also be inverted to preemptively assess the plausibility of potential biosignatures given the biomass needed to produce a detectable amount of that biosignature (Seager et al. 2013b). A downside of kinetic biosignatures is that they can require environmental information that may be observationally inaccessible, such as surface or subsurface chemical sinks. However, observable or otherwise inferable sinks can be used to determine a lower limit on gas production rates, which may be sufficient to infer the biogenicity of a putative biosignature gas.

Probabilistic approaches. As detailed in the previous sections, the detection and interpretation of biosignatures could be a daunting task as many biosignature gases suffer from false positives and potentially false negatives. The statistical assessment of biosignatures is an approach designed to delineate uncertainties in their detection and interpretation and assign confidence levels to the putative detection of life. Catling et al. (2018) put forward a framework for assessing exoplanet biosignatures based on Bayes' theorem and included several confidence levels for biosignature detection. Bayesian approaches to biosignatures evaluate new information conditioned on both contextual information (e.g., astrophysical and planetary properties) and prior probabilities (i.e., the likelihood of life originating in a certain environment). Walker et al. (2018) further extend this concept to interrogate how Bayesian formalism could guide searches for life on exoplanets and inform our understanding of the likelihood of the origin of life. Additional alternative probabilistic approaches include Signal Detection Theory (Pohorille and Sokolowska 2020). The difficulty in probabilistic biosignature assessment is that it relies in part on information that is hard to quantify on absolute terms, such as the probability of abiogenesis (Walker et al. 2018). Nevertheless, statistical assessments demonstrate that a single robust biosignature detection would imply biospheres are common in the universe, providing compelling motivation despite the challenge (Balbi and Grimaldi 2020). Probabilistic methods continue to be developed and refined to advance the search for life elsewhere (Bixel and Apai 2021; Madau 2023; Kipping and Wright 2024). A review of probabilistic biosignatures is available in Walker et al. (2020).

Network-based biosignatures. Atmospheric chemistry can be depicted graphically with networks, where the nodes symbolize chemical species, and the connections represent reactions (Solé and Munteanu 2004; Fisher et al. 2022). The atmospheric chemical reaction networks of planets in the solar system have distinct structures, with Earth's network standing out as the least random among planetary networks by various metrics (Holme et al. 2011; Wong et al. 2023). Indeed, Earth's atmospheric reaction network shows similarities to the topologies found in biological metabolic networks (Solé and Munteanu 2004; Fisher et al. 2022). This suggests network topologies could serve as agnostic exoplanet biosignatures (Fisher et al. 2022; Wong et al. 2023). Network-based biosignatures are an emerging area that may lead to unique insights facilitating the identification of planetary biospheres; however, additional theoretical work is required to quantify the observability of these metrics.

SURFACE AND TEMPORAL BIOSIGNATURES

Surface and temporal biosignatures present additional avenues for the remote detection of life (Table 4). A surface biosignature is created when a component or product of living organisms, such as a pigment, introduces a detectable pattern like a spectral or polarization signature onto a planetary surface. These signatures are then recorded in reflected, transmitted, or scattered light. To accurately interpret surface reflectance signatures, one must consider not just the interaction of light with living materials (such as transmission and scattering in cellular or community structures) but also how the host star's radiation interacts with the planet's atmosphere, reaches its surface, and is then reflected or scattered to an observer. Often complementary to surface signatures, a temporal biosignature refers to changes over time in observables traceable back to a living process. This could manifest as a seasonal variation in the intensity or location of a surface biosignature or as fluctuations in a detectable gas like CO_2, resulting from shifts in the global balance of photosynthesis and respiration due to seasonal changes in environmental conditions (Hall et al. 1975; Meadows 2006, 2008), i.e., the seasonal component of the CO_2 "Keeling curve" (Keeling et al. 1996).

Surface biosignature detection requires observations of reflected light, which remain out of reach for JWST but may be obtained by HWO or a yet more advanced future observatory (e.g., Apai et al. 2019). An additional complication for current observational strategies is that temporal biosignatures require multi-epoch observations to capture changes over time, while transmission spectra of exoplanets are always obtained at the same orbital phase. Given these challenges, we offer a briefer overview of surface and temporal biosignatures than their gaseous counterparts. These types of biosignatures will likely only come into play as additional characterization of an already promising biosphere candidate, provided sufficient technological capabilities from a reflected light direct-imaging mission such as HWO (The LUVOIR Team 2019; Gaudi et al. 2020; Meadows et al. 2022). Schwieterman (2018) provides a more extensive review solely focused on potential surface and temporal biosignatures.

Table 4. Surface and temporal biosignatures.

Biosignature	Origin (e.g.)	False positives/Unknowns	Observational strategies	Citations
Photosynthetic pigments	Chlorophyll and bacteriochlorophyll pigments	Non biological pigment mimics such as minerals	Visible & NIR reflectance spectra via direct imaging	Kiang et al. 2007a, b
Non-photosynthetic pigments	Microorganism reflectance from pigments with diverse functions	Non biological pigment mimics such as minerals	Visible & NIR reflectance spectra via direct imaging	Schwieterman et al. 2015a
Retinal pigments	*Haloarchaea* bacteriorhodopsin	Nonbiological pigment mimics such as minerals	Visible & NIR reflectance spectra via direct imaging	DasSarma and Schwieterman 2021
Vegetation red edge	Plant leaves; phototroph cell structure	Surface materials with sharp spectral breaks	Visible & NIR reflectance spectra via direct imaging	Seager et al. 2005
Chirality	Nearly all biological molecules on the Earth have L-amino acids and D-sugars	Unclear how universal this characteristic is, enantiomeric excess also seen in carbonaceous chondrites. Weak signal	Can be revealed via spectropolarimetry although signal is weak	Sterzik et al. 2012; Patty et al. 2018; Sparks et al. 2021; Gleiser 2022
Florescence & bioluminescence	Reprocessing or direct production of light from biological organisms	Requires very high signal to noise and probably low cloud cover; non biological processes can also fluoresce; bioluminescence produces very weak signatures	Reflectance spectra via direct imaging, peaks in visual wavelengths	Papageorgiou 2007; Haddock et al. 2010; O'Malley-James and Kaltenegger 2018
Seasonality	Temporal changes in gas levels due to productivity changes as the planet orbits	Requires significant time investment and retrieval accuracy	Time series retrievals of abundance of gases linked to biological outputs or inputs, such as O_3 or CO_2 over time	Olson et al. 2018; Mettler et al. 2023

Surface signatures of photosynthesis

Most of Earth's biomass is produced via photosynthesis, so it stands to reason that photosynthetic biomass may afford the most extensive surface signature of life on an inhabited planet. We offered a brief review of oxygenic photosynthesis in an earlier section and here consider the photosynthetic pigments and resulting surface biosignatures. Light absorption is facilitated by pigments called chlorophylls in oxygenic phototrophs (e.g., plants, algae, cyanobacteria) Chlorophyll *a* (Chl *a*) and *b* (Chl *b*) are common in vegetation, and absorb light in the red and blue wavelengths, with peaks at 435 and 660–670 nm for Chl *a*, and 460 and 650 nm for Chl *b* (see Table 1 in Schwieterman et al. 2018a for absorption maxima of all photosynthetic pigments). The green color of chlorophylls arises from weaker absorption in the green spectrum and reflection of this color to the observer.

However, photosynthesis can also occur without producing oxygen in a variety of environmental settings. This process is more evolutionarily ancient than oxygenic photosynthesis and is called anoxygenic photosynthesis (Olson 2006). It can be represented by the following general formula:

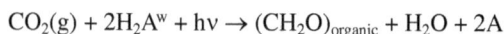

$$CO_2(g) + 2H_2A^w + h\nu \rightarrow (CH_2O)_{organic} + H_2O + 2A$$

where CO_2 is carbon dioxide gas, H_2A is an electron source (such as H_2S or H_2, $h\nu$ is light energy, $(CH_2O)_{organic}$ represents biomass, and $2A$ is an oxidized waste product.

Photosynthesis is the conversion of light energy into chemical energy and involves a complex series of oxidation/reduction reactions. Light is absorbed by photosynthetic pigments in both oxygenic and anoxygenic phototrophs, and electrons in special types of pigments called reaction centers are excited and donated to receptor molecules. Those electrons are ultimately captured and stored in energy-rich compounds such as ATP and NADPH, which are in turn used to reduce CO_2 to CH_2O. The electron donated by the pigment is replaced by the external reductant H_2A or in the case of oxygenic photosynthesis, H_2O (Blankenship 2002).

Anoxygenic phototrophs use special types of pigments called bacteriochlorophyll (Bchls), that absorb light primarily in the near-infrared (see Table 1 in Schwieterman et al. 2018a). For example, Bchl *a* in purple bacteria has two characteristic absorption maxima at 790–810 nm and 830–920 nm. The pigment that absorbs furthest into the NIR is Bchl *b* at 1015–1040 nm, which near the theoretical limit of ~1100 nm for photon absorption in electronic transitions (Kiang et al. 2014). Phototrophs also employ accessory pigments, like carotenoids, to absorb light at shorter wavelengths and that energy is transferred to the reaction center, allowing effective harvesting of light across the visible spectrum (Allakhverdiev et al. 2016). The absorbance spectra of chlorophyll and bacteriochlorophyll pigments are detailed in Figure 11.

The vegetation red edge (VRE) and other photosynthetic "edge" signatures

The VRE is a possible exoplanet biosignature (Seager et al. 2005) detectable in the reflectance spectrum of a planet, and is identified by a sharp increase in reflectance at the boundary between visible and near-infrared wavelengths (~700 nm), most strongly seen in green vascular plants today (though not limited to them). This effect arises from the contrast in chlorophyll absorption in the red (~650–700 nm) and scattering properties of cellular structures in the near-infrared (~750–1100 nm) (Gates et al. 1965; Knipling 1970). The VRE is evident in various oxygenic phototrophs, including plants, algae, and cyanobacteria, with wavelengths clustered between 690–730 nm, and is stronger than the chlorophyll "green bump" near 550 nm (Kiang et al. 2007b). Figure 12 illustrates the VRE along with common surface materials, such as snow and soil, other biogenic materials, including a bacterial mat. Figure 12 also shows slope change features from materials such as cinnabar and sulfur, which may constitute false positives for biogenic "edge" signatures (Seager et al. 2005).

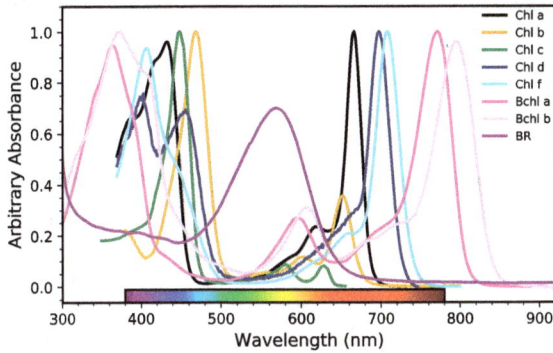

Figure 11. Chlorophyll (Chl), bacteriochlorophyll (Bchl), and bacteriorhodopsin (BR) absorbance spectra. The figure was taken from DasSarma and Schwieterman (2021) under Creative Commons Attribution License CC-BY 4.0.

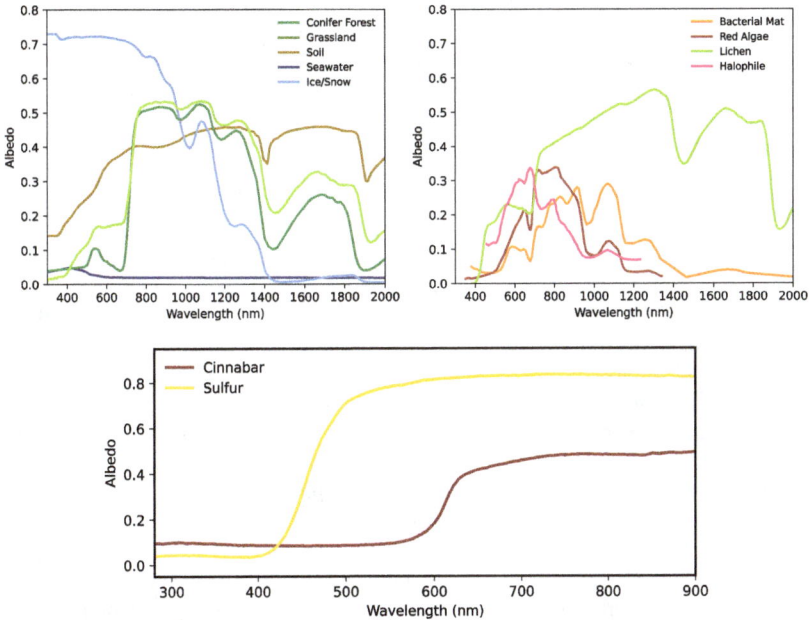

Figure 12. Spectral albedos of various living and non-living materials. Sulfur and cinnabar may be "false positives" for edge-like biosignatures (bottom panel) but lack the hydration bands present in the living material. The halophile spectrum is from Dalton et al. (2009), the conifer forest spectrum is sourced from the ASTER spectral library (Baldridge et al. 2009), and all other spectral albedos are sourced from the USGS spectral library (Clark et al. 2007).

The VRE's spectral break is commonly used in the Earth sciences to assess vegetation health and coverage of the planet through the normalized difference vegetation index (NDVI; Tucker 1979; Tucker et al. 2005). The NDVI is calculated as:

$$NDVI = \frac{NIR - VIS}{NIR + VIS}$$

where NIR is near-infrared albedo and VIS is visible band albedo. NDVI values range from –1 to 1, with higher values indicating vegetation. However, using a broad metric like NDVI for exoplanet biosignature analysis has limitations, as it struggles to distinguish between the red-edge and mineral slopes. Minerals such as cinnabar and elemental sulfur have steep slopes from visible to red wavelengths that can mimic vegetation signals (Fig. 12). Studies suggest that while NDVI scores the lunar surface similarly to Earth's vegetation, the VRE could aid in the initial identification of Earth-like worlds when combined with other data (Livengood et al. 2011).

The strength of the VRE likely varied significantly throughout Earth's history. Land plants responsible for the modern VRE have only existed for about 400 million years (Kenrick and Crane 1997). Algae may have supported VRE signatures for up to 1.0–1.6 Ga (Brocks et al. 2023). Before that, photosynthetic life was bacterial, dating back to at least 3.5 Ga (Stüeken et al. 2024). The reflectance spectra of these early surface microbial communities would have differed from vegetation and would be likely dominated by microbial mats or stromatolites in marine intertidal and continental settings. Sanromá et al. (2013) determined that oxygenic cyanobacteria in microbial mats, if covering 50% of early Earth's land, could be distinguishable from vegetation due to dimmer reflectance at wavelengths beyond ~1.3 μm. Furthermore, Sanromá et al. (2014) observed that mats of anoxygenic phototrophs, like purple bacteria, would show a red-shifted VRE effect due to the longer wavelength absorption of bacteriochlorophylls, with NIR edges beyond around 1 μm. However, significant coverage would be necessary for detection. Tinetti et al. (2006) found that NIR analogs to the VRE may be more easily detected in reflected light through the clouds and atmosphere. Future missions such as HWO could search for photosynthetic edge-like features analogous to the VRE on exoplanets in reflected light, which could be detected given sufficient surface coverage (Gomez Barrientos et al. 2023).

Rhodopsin-based phototrophy

In the search for extraterrestrial life, pigments other than those used in photosynthesis, such as bacteriorhodopsin produced by haloarchaea, are gaining attention as potential biosignatures (DasSarma 2006; DasSarma et al. 2020). Unlike chlorophyll, bacteriorhodopsin is not linked to the production of organic matter. It is a light-driven proton pump that generates cellular energy, absorbing light in the green wavelength range of the visible spectrum (~570 nm), complementary to chlorophylls and bacteriochlorophylls (Fig. 11). DasSarma and Schwieterman (2021) speculated that this spectral complementarity may reveal the timing of rhodopsin's evolutionary origins with respect to chlorophyll pigments and offer a possible alternative surface biosignature to search for on exoplanets. Sephus et al. (2022) further explored this concept, analyzing the evolutionary history of microbial rhodopsins. Initially functioning as light-driven proton pumps and primarily absorbing green light, these pigments likely evolved in response to Earth's early photic environments. This indicates that inhabited planets could show diverse spectral signatures based on their dominant pigments, which could be centered on rhodopsin-based phototrophy (DasSarma et al. 2020). Figure 13 shows hypothetical planetary spectra, contrasting those dominated by VRE-producing forests with a surface dominated by salt ponds hosting pigmented halophilic archaea with rhodopsin and carotenoid pigments. This example is for illustrative purposes, as the surface coverage fraction of these biogenic materials is unlikely to be close to 100%.

Predicting photosynthetic surface signatures around other stars

A star's spectral energy distribution (SED) varies with its temperature, influencing the wavelength of pigment absorption in photosynthesis. Kiang et al. (2007a) suggest that the evolution of pigment absorption might adapt to the most efficient wavelength based on the star's temperature, resulting in predictable 'edge' wavelengths. This shift is linked more strongly to surface photon irradiance than energy irradiance, considering photosynthesis requires specific photon amounts. For instance, planets orbiting cooler stars (like M-dwarfs) might

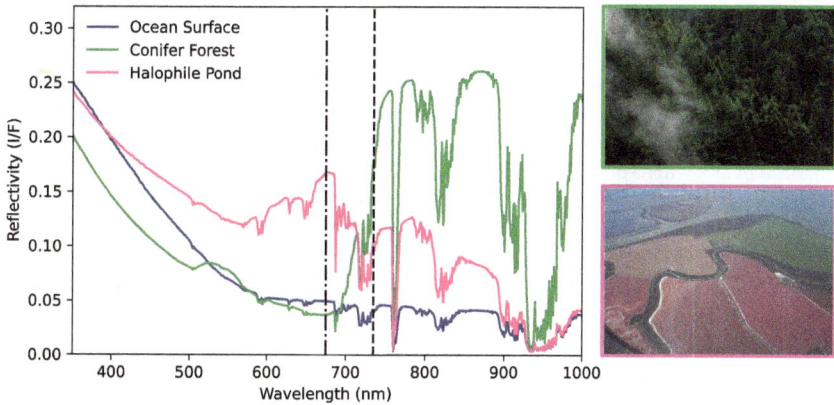

Figure 13. Differentiating between the surface biosignatures of vegetation and halophiles. **Left:** This panel illustrates simulated reflection spectra of planets similar to Earth, with surfaces predominantly covered by oceans, forests, or pigmented halophiles. A **dot–dashed line** at 675 nm indicates a distinct peak in the reflectance of halophile ponds. The Vegetation Red Edge (VRE) characteristic of forests is observer 700 and 750 nm (**dashed line**). The forest surface's albedo was calculated using data from the ASTER spectral library (Baldridge et al. 2009), while the halophile surface data was sourced from the spectral analysis of San Francisco Bay's salt ponds (Dalton et al. 2009). **Right:** The images display a conifer forest (**top**, from Andrew Coehlo under an Unsplash Licence) and salt ponds dominated by halophiles in San Francisco Bay (**bottom**, from Grombo, CC BY-SA 3.0 Wikipedia Commons).

exhibit biospheres with NIR-shifted red edges, while those around hotter stars (like F-dwarfs) could have blue-shifted edges. Several workers have developed theoretical models to predict photosynthetic absorption based on the SED of the host star (Covone et al. 2021; Lehmer et al. 2021; Lingam et al. 2021; Marcos-Arenal et al. 2022; Battistuzzi et al. 2023; Illner et al. 2023). However, the adaptation of 'edge' wavelengths might not directly correspond to the host star's spectrum or the wavelengths of light at the surface of the planet. This is because phototrophs, at least on Earth, are usually not limited by photon availability but rather by resources like water, electron donors, nitrogen, iron, or phosphorus (Bristow et al. 2017). Additionally, continental and marine environments offer a wide variety of light environments for which phototrophs are adapted (e.g., Kurashov et al. 2019), which are only partly influenced by the star. As such, the relationship between star spectrum and photosynthetic pigment evolution is not necessarily straightforward or fully understood.

Non-photosynthetic and other surface biosignatures

Pigments can play a variety of crucial roles in organisms beyond light capture for photosynthesis, which can sometimes be completely decoupled from the light environment (Schwieterman et al. 2015a). These functions include protection against ultraviolet radiation, vital for species in environments with high UV exposure and contributions to temperature regulation by reflecting sunlight or aiding in heat absorption. Additionally, pigments are involved in visual signaling, playing a role in processes like camouflage, mating, and warning signals. Moreover, some pigments have antioxidative properties, protecting cells from damage caused by oxidative stress. Many pigments display edge features analogous to the edges seen in photosynthetic organisms and efforts have been undertaken to categorize the wide diversity of these features (Hegde et al. 2015; Coelho et al. 2022). A large and robust chemosynthetic biosphere could potentially be identified via non-photosynthetic pigments. However, just as is the case with photosynthetic signatures, the surface coverage would have to be sufficient to allow for remote detectability in planetary disk-averaged spectra.

Polarization and chirality

Polarization measurements, both linear and circular, have been proposed as surface biosignatures within and outside the solar system (Sparks et al. 2009b, 2012; Berdyugina et al. 2016; Patty et al. 2017; Gleiser 2022). Polarization helps discern Earth-like atmospheres and is influenced by pigments like chlorophyll (Sterzik et al. 2012, 2014, 2019). Studies show linear polarization peaks where pigment absorption is highest, contrasting with abiotic materials like sand or rock (Berdyugina et al. 2016). Circular polarization signatures can arise from chiral biological molecules. Chirality (mirror image biomolecules that can't be superimposed), and specifically homochirality (the exclusive use of one "handedness" of a biomolecule over the other), is considered a universal agnostic biosignature and is a highly specific indicator of living material (Sparks et al. 2009a, 2021; Patty et al. 2018; Gleiser 2022; though see Avnir 2021 for a qualification of this view). However, circular polarization features related to chiral centers in photosynthetic pigments and interactions between pigments are weaker than linear polarization features and challenging to link definitively to surface microbial communities or vegetation in remote observations. Additional work is underway to comprehensively examine polarization signatures on Earth and their possible manifestation elsewhere in the universe (Patty et al. 2019, 2022; Gordon et al. 2023).

Fluorescence and bioluminescence

Vegetation and microorganisms can emit light through fluorescence and bioluminescence (Papageorgiou 2007; Haddock et al. 2010). Chlorophyll autofluorescence, occurring when the pigments are excited by UV light and the absorbed photons are re-emitted at lower energy levels, is observable via Earth-orbiting satellites. The main spectral range for chlorophyll autofluorescence is in the red 640–800 nm, with characteristic peaks at 685 and 740 nm, though the overall signals are small and require high-resolution spectroscopy to observe (Joiner et al. 2011; Sun et al. 2017). Some authors have proposed biological autofluorescence could be a temporal surface biosignature, with fluorescence from organisms responding to stellar flare activity (O'Malley-James and Kaltenegger 2018, 2019). Komatsu et al. (2023) quantitatively examine the potential of detecting fluorescence biosignatures on exoplanets in reflected light. They find bacteriochlorophyll autofluorescence emission at 1000–1100 nm is plausibly detectable (given sufficient surface coverage of fluorescing vegetation) on planets orbiting ultracool dwarfs with an HWO-like telescope due to the coincidence of fluorescence wavelengths with atomic and molecular absorption in the stellar spectrum. Potential false positives include minerals such as fluorite and calcite, which can also be remotely detected on Earth from space (Köhler et al. 2021), though their fluorescence profiles may vary from biological fluorescence.

Complementary to florescence, bioluminescence is the active production (not reprocessing) of light by organisms. Known life accomplishes this through the oxidation of a class of molecules called luciferins and are found in diverse forms of life encompassing bacteria, single cell eukaryotes, and animals (Haddock et al. 2010). Bioluminescence of some marine microbes can be observed by Earth-observing satellites and patches of luminescent plankton can encompass more than 10,000 km^2 (Miller et al. 2005), leading to its suggestion as a potential future exoplanet biosignature (Seager et al. 2012). However, the overall luminosity of these luminous patches is low, and no study has thus far quantitatively examined its detectability on exoplanets.

Seasonal biosignatures

Earth's atmosphere displays seasonal variations in gases like CO_2, O_2, and CH_4, which are influenced by biological processes. CO_2 levels fluctuate due to plant growth and decay, decreasing in spring and summer and increasing in fall and winter (Hall et al. 1975; Keeling et al. 1996). O_2 variation is closely related to CO_2 changes, influenced by photosynthesis and organic matter decay (i.e., $CO_2 + H_2O \rightarrow (CH_2O)_{org} + O_2$). While O_2's absolute variability is greater than CO_2's (Keeling and Shertz 1992), partly due to CO_2's higher solubility in ocean water, its proportional change relative to background concentrations is much smaller. CH_4 shows a more complex seasonal pattern, with lowest levels in northern summer, a smaller dip in winter, and peaks in late fall and early spring (Rasmussen and Khalil 1981). CH_4's primary non-human source is methanogenic microbes in wetlands, but its seasonal changes are largely governed by interactions with OH, mainly derived from tropospheric water and thus varying with surface temperature (Khalil and Rasmussen 1983).

The seasonal oscillations in Earth's atmospheric gases vary by hemisphere and latitude; CO_2 changes are more pronounced in the northern hemisphere due to greater land coverage (Keeling et al. 1996). These patterns suggest that detecting seasonal cycles on exoplanets will depend on factors like viewing angle and planet tilt. Planets with greater obliquity or eccentricity than Earth might have more detectable seasonal signatures, though one must consider viewing geometry and mixing hemispheres in a disk average. Even a maximum CO_2 variation of 1–3% would be difficult to detect, given weak or saturated CO_2 bands and competing effects from clouds and changes to atmospheric temperature structure (Mettler et al. 2023).

Olson et al. (2018) provide a detailed review of the general phenomenology underlying atmospheric seasonality as a potential exoplanet biosignature and the opportunities and challenges associated with observing it. They describe the possible impacts of obliquity, eccentricity, and viewing geometry on the observed signal. These authors proposed that O_2 seasonality may have been magnified on Proterozoic Earth, due to the smaller background O_2 concentrations and therefore larger potential for proportional changes (cf. Wogan et al. 2022). This O_2 would not be directly observable but could be inferred by seasonality in the UV signatures of O_3, which can be observed at lower O_2 abundances than directly (cf. Kozakis et al. 2022).

Mettler et al. (2023) examine the seasonality of atmospheric gases N_2O, CH_4, O_3, and CO_2 in the IR using Earth-observing data. Figure 14 shows the equivalent width (EW) variation in IR O_3, CO_2, CH_4 and N_2O bands at 9.6, 4.3, 7.7, and 4.5 μm, respectively, from a "north pole on" viewing geometry (Mettler et al. 2023). The spectral variations shown are small and impacted by cloud formation, changes in atmospheric temperature structure, and other seasonal changes in atmospheric and surface conditions. Nonetheless, these data demonstrate that variations in the spectral signatures of gases reveal seasonal processes, in principle. Ultimately, the observed seasonality in these gases depends on these complex factors and viewing geometry, in addition to their intrinsic changes in abundance.

Seasonal biosignatures need not be limited to gases, as the surface pigmentation of vegetation also changes seasonally, though such a biosignature would have compounding challenges. Nevertheless, seasonal biosignatures may confirm candidate biospheres once identified, perhaps in the distant future.

CONCLUSIONS

Earth's oxygenic photosynthetic biosphere has left an indelible mark on the remote spectrum of our planet. If other planets in our local universe are inhabited, their biospheres could similarly alter the chemical composition of their planetary atmospheres and their resulting remotely detectable spectral signatures. Here, we have provided an overview of potential exoplanet biosignatures, including their sources and sinks, observable features, and possible false positives. Confirming that potential biosignatures fingerprint real biospheres requires planetary context, and no single signature is likely sufficient on its own for one to claim the presence of life. We have given particular emphasis to atmospheric (gaseous) biosignatures, as it is becoming feasible to search for them in the near to intermediate term. Potential gaseous biosignatures include those that dominate on Earth today (O_2, O_3, CH_4, and N_2O) and those

Atmospheric Seasonality - O3, CO2, CH4, N2O

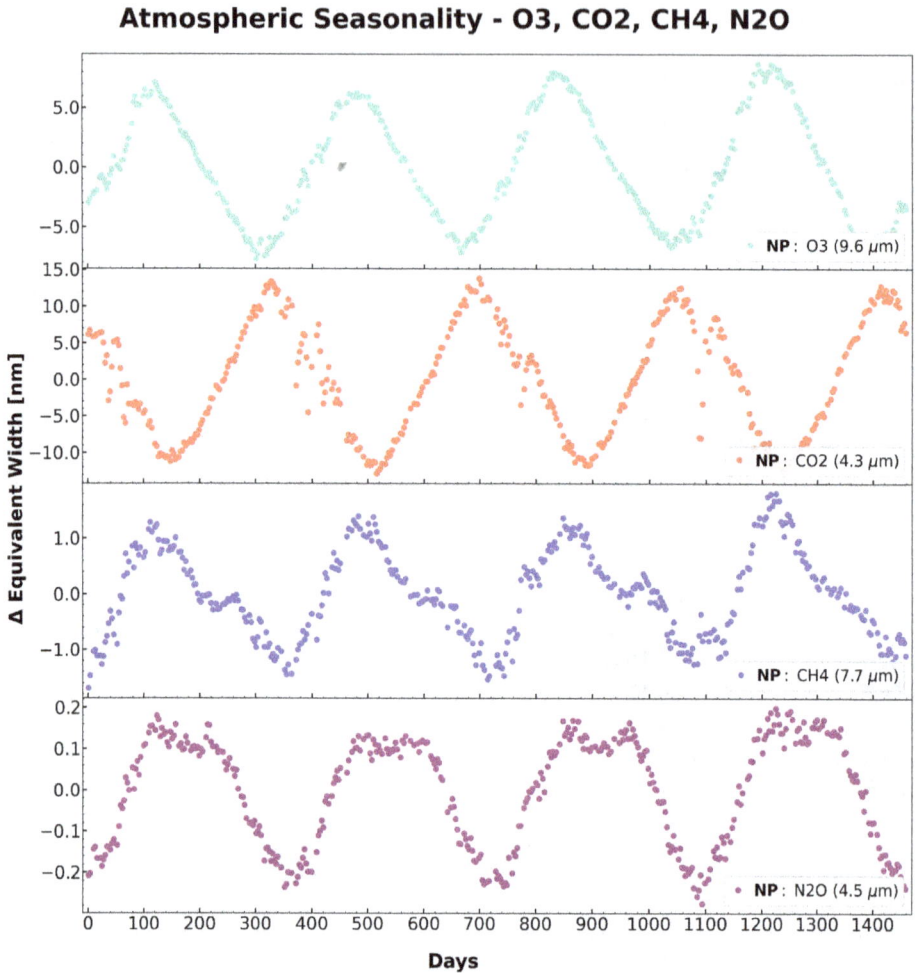

Figure 14. Variations in the Equivalent Width (EW; in nm) of IR spectral bands of O_3, CO_2, CH_4, and N_2O from reconstructed disk-averaged thermal IR data from the Atmospheric Infrared Sounder (AIRS) onboard the Earth-observing Aqua satellite. The data span four years and are based on a "North Pole" view. Additional details are available in Mettler et al. (2023). This figure was sourced from Mettler et al. (2023) under Creative Commons Attribution License CC-BY 4.0.

known to be produced by life but with small concentrations in Earth's modern atmosphere (e.g., $(CH_3)_2S$, CH_3Cl, PH_3, C_5H_8), though these could accumulate to higher levels in other planetary environments. The changing oxygenation of Earth's atmosphere over geologic time likely influenced the concentrations of biogenic gases, especially CH_4 and N_2O. Thermodynamic and kinetic disequilibria can be useful metrics for evaluating potential atmospheric biosignatures. Additionally, surface biosignatures, such as the vegetation red edge, and temporal biosignatures, such as atmospheric seasonality, may offer independent lines of evidence supporting the identification of putative biospheres, though they are less easily characterized. Bayesian frameworks for biosignatures may potentially provide statistical robustness to this search and the development of further frameworks for characterizing biosignatures continues. The search for life beyond the solar system is a compelling and monumental undertaking, and we are privileged to live in a time when that search is becoming possible.

ACKNOWLEDGMENTS

ES and ML gratefully acknowledge support from the NASA Exobiology Program under grant no. 80NSSC20K1437, the NASA Exoplanet Research Program under grant no. 80NSSC22K0235, the NASA Interdisciplinary Consortium for Astrobiology Research (ICAR) under grant nos. 80NSSC21K0594, 80NSSC23K1399, and 80NSSC23K1398, and The Kavli Foundation. We thank Giada Arney, Ravi Kopparapu, and Niki Parenteau for their helpful comments and suggestions that improved this chapter.

REFERENCES

Ahrer EM, Alderson L, Batalha NM, et al. (2023) Identification of carbon dioxide in an exoplanet atmosphere. Nature 614:649–652, https://doi.org/10.1038/s41586-022-05269-w

Airapetian VS, Glocer A, Gronoff G, et al. (2016) Prebiotic chemistry and atmospheric warming of early Earth by an active young Sun. Nat Geosci 9:452–455, https://doi.org/10.1038/ngeo2719

Airapetian VS, Barnes R, Cohen O, et al. (2020) Impact of space weather on climate and habitability of terrestrial-type exoplanets. Int J Astrobiol 19:136–194, https://doi.org/10.1017/S1473550419000132

Akahori A, Watanabe Y, Tajika E (2023) Controls of atmospheric methane on early Earth and inhabited Earth-like terrestrial exoplanets. arxiv preprint 231106882

Akins AB, Lincowski AP, Meadows VS, Steffes PG (2021) Complications in the ALMA detection of phosphine at Venus. Astrophys J 907:L27, https://doi.org/10.3847/2041-8213/abd56a

Alderson L, Wakeford HR, Alam MK, et al. (2023) Early release science of the exoplanet WASP-39b with JWST NIRSpec G395H. Nature 614:664–669, https://doi.org/10.1038/s41586-022-05591-3

Alei E, Konrad BS, Angerhausen D, et al. (2022) Large Interferometer For Exoplanets (LIFE) V. Diagnostic potential of a mid-infrared space-interferometer for studying Earth analogs. Astron Astrophys 665:A106, https://doi.org/10.1051/0004-6361/202243760

Allakhverdiev SI, Kreslavski VD, Zharmukhamedov SK, et al. (2016) Chlorophylls *d* and *f* and their role in primary photosynthetic processes of cyanobacteria. Biochem 81:201–212, https://doi.org/10.1134/S0006297916030020

Angerhausen D, Ottiger M, Dannert F, et al. (2023) Large Interferometer For Exoplanets: VIII. Where is the phosphine? Observing exoplanetary PH$_3$ with a space-based mid-infrared nulling interferometer. Astrobiology 23:183–194, https://doi.org/10.1089/ast.2022.0010

Angerhausen D, Pidhorodetska D, Leung M, et al. (2024) Large Interferometer For Exoplanets (LIFE): XII. The detectability of capstone biosignatures in the mid-infrared - Sniffing exoplanetary laughing gas and methylated halogens. Astron J in press: https://doi.org/10.48550/arXiv.2401.08492

Apai D, Milster TD, Kim DW, et al. (2019) A thousand Earths: A very large aperture, ultralight space telescope array for atmospheric biosignature surveys. Astron J 158:83, https://doi.org/10.3847/1538-3881/ab2631

Arney GN (2019) The K dwarf advantage for biosignatures on directly imaged exoplanets. Astrophys J Lett 873:L7, https://doi.org/10.3847/2041-8213/ab0651

Arney G, Meadows V, Crisp D, et al. (2014) Spatially-resolved measurements of H$_2$O, HCl, CO, OCS, SO$_2$, cloud opacity, and acid concentration in the Venus near-infrared spectral windows. J Geophys Res E: Planets 119:1860–1891, https://doi.org/10.1002/2014JE004662

Arney G, Domagal-Goldman SD, Meadows VS, et al. (2016) The pale orange dot: The spectrum and habitability of hazy Archean Earth. Astrobiology 16:873–899, https://doi.org/10.1089/ast.2015.1422

Arney GN, Meadows VS, Domagal-Goldman SD, et al. (2017) Pale orange dots: The impact of organic haze on the habitability and detectability of Earthlike exoplanets. Astrophys J 836:49, https://doi.org/10.3847/1538-4357/836/1/49

Arney G, Domagal-Goldman SD, Meadows VS (2018) Organic haze as a biosignature in anoxic Earth-like atmospheres. Astrobiology 18:311–329, https://doi.org/10.1089/ast.2017.1666

Avnir D (2021) Critical review of chirality indicators of extraterrestrial life. New Astron Rev 92:101596, https://doi.org/10.1016/j.newar.2020.101596

Bains W, Petkowski JJ, Sousa-Silva C, Seager S (2019) New environmental model for thermodynamic ecology of biological phosphine production. Sci Total Environ 658:521–536, https://doi.org/10.1016/j.scitotenv.2018.12.086

Bains W, Petkowski JJ, Seager S, et al. (2021) Phosphine on Venus cannot be explained by conventional processes. Astrobiology 21:1277–1304, https://doi.org/10.1089/ast.2020.2352

Balbi A, Grimaldi C (2020) Quantifying the information impact of future searches for exoplanetary biosignatures. PNAS 117:21031–21036, https://doi.org/10.1073/pnas.2007560117

Baldridge AM, Hook SJ, Grove CI, Rivera G (2009) The ASTER spectral library version 2.0. Remote Sens Environ 113:711–715, https://doi.org/10.1016/j.rse.2008.11.007

Baraffe I, Chabrier G, Allard F, Hauschildt P (1998) Evolutionary models for solar metallicity low-mass stars: mass–magnitude relationships and color–magnitude diagrams. 1–15

Barclay T, Rowe JF, Lissauer JJ, et al. (2013) A sub-Mercury-sized exoplanet. Nature 494:452–454, https://doi.org/10.1038/nature11914

Basnayake RST, Bius JH, Akpolat OM, Chasteen TG (2001) Production of dimethyl telluride and elemental tellurium by bacteria amended with tellurite or tellurate. Appl Organomet Chem 15:499–510, https://doi.org/10.1002/aoc.186

Battistuzzi M, Cocola L, Claudi R, et al. (2023) Oxygenic photosynthetic responses of cyanobacteria exposed under an M-dwarf starlight simulator: Implications for exoplanet's habitability. Front Plant Sci 14:1–13, https://doi.org/10.3389/fpls.2023.1070359

Bentley R, Chasteen TG (2002) Microbial methylation of metalloids: Arsenic, antimony, and bismuth. Microbiol Mol Biol Rev 66:250–271, https://doi.org/10.1128/MMBR.66.2.250-271.2002

Berdyugina S V., Kuhn JR, Harrington DM, et al. (2016) Remote sensing of life: polarimetric signatures of photosynthetic pigments as sensitive biomarkers. Int J Astrobiol 15:45–56, https://doi.org/10.1017/S1473550415000129

Bixel A, Apai D (2021) Bioverse: A simulation framework to assess the statistical power of future biosignature surveys. Astron J 161:228, https://doi.org/10.3847/1538-3881/abe042

Bristow LA, Mohr W, Ahmerkamp S, Kuypers MMM (2017) Nutrients that limit growth in the ocean. Curr Biol 27:R474–R478, https://doi.org/10.1016/j.cub.2017.03.030

Brocks JJ, Nettersheim BJ, Adam P, et al. (2023) Lost world of complex life and the late rise of the eukaryotic crown. Nature 618:767–773, https://doi.org/10.1038/s41586-023-06170-w

Buessecker S, Imanaka H, Ely T, et al. (2022) Mineral-catalysed formation of marine NO and N_2O on the anoxic early Earth. Nat Geosci 15:1056–1063, https://doi.org/10.1038/s41561-022-01089-9

Buick R (2007) Did the Proterozoic Canfield Ocean cause a laughing gas greenhouse? Geobiology 5:97–100, https://doi.org/10.1111/j.1472-4669.2007.00110.x

Burkholder JB, Sander S., Abbatt J, et al. (2019) Chemical Kinetics and Photochemical Data for Use in Atmospheric Studies, Evaluation No. 19. JPL Publication No. 19-5. Pasadena

Camp CD, Roulston MS, Yung YL (2003) Temporal and spatial patterns of the interannual variability of total ozone in the tropics. J Geophys Res Atmos 108:1–17, https://doi.org/10.1029/2001JD001504

Canfield DE, Glazer AN, Falkowski PG (2010) The evolution and future of Earth's nitrogen cycle. Science 330:192–196, https://doi.org/10.1126/science.1186120

Cao H, Liu J, Zhuang Y, Dietmar G (2000) Emission sources of atmospheric phosphine and simulation of phosphine formation. Sci China Ser B Chem 43:162–168, https://doi.org/10.1007/BF03027306

Catling DC (2014) The great oxidation event transition. *In:* Turekian KK, Holland HD (eds) Treatise on Geochemistry, 2nd edn. Elsevier, Oxford, UK, p 177–195

Catling DC, Kasting JF (2017) Essentials of chemistry of planetary atmospheres. *In:* Atmospheric Evolution on Inhabited and Lifeless Worlds. Cambridge University Press, p 73–84

Catling DC, Zahnle KJ (2020) The Archean atmosphere. Sci Adv 6:eaax1420, https://doi.org/10.1126/sciadv.aax1420

Catling DC, Krissansen-Totton J, Kiang NY, et al. (2018) Exoplanet biosignatures: A framework for their assessment. Astrobiology 18:709–738, https://doi.org/10.1089/ast.2017.1737

Chan MA, Hinman NW, Potter-McIntyre SL, et al. (2019) Deciphering biosignatures in planetary contexts. Astrobiology 19:1075–1102, https://doi.org/10.1089/ast.2018.1903

Chapman S (1930) A theory of atmospheric ozone. R Meteorol Soc Mem 3:103–125

Chassefière E (1996) Hydrodynamic escape of hydrogen from a hot water-rich atmosphere: The case of Venus. J Geophys Res E: Planets 101:26039–26056, https://doi.org/10.1029/96JE01951

Chassefière E, Wieler R, Marty B, Leblanc F (2012) The evolution of Venus: Present state of knowledge and future exploration. Planet Space Sci 63–64:15–23, https://doi.org/10.1016/j.pss.2011.04.007

Chen H, Zhan Z, Youngblood A, et al. (2020) Persistence of flare-driven atmospheric chemistry on rocky habitable zone worlds. Nat Astron 5:298–310, https://doi.org/10.1038/s41550-020-01264-1

Christiansen JL (2022) Five thousand exoplanets at the NASA Exoplanet Archive. Nat Astron , https://doi.org/10.1038/s41550-022-01661-8

Clark RN, Swayze GA, Wise R, et al. (2007) USGS digital spectral library splib06a: *In:* U.S. Geological Survey, Digital Data Series 231, http://speclab.cr.usgs.gov/spectral.lib06. http://speclab.cr.usgs.gov/spectral.lib06/ds231/

Cockell CS, Bush T, Bryce C, et al. (2016) Habitability: A review. Astrobiology 16:89–117, https://doi.org/10.1089/ast.2015.1295

Coelho LF, Madden J, Kaltenegger L, et al. (2022) Color catalogue of life in ice: Surface biosignatures on icy worlds. Astrobiology 22:313–321, https://doi.org/10.1089/ast.2021.0008

Cooke GJ, Marsh DR, Walsh C, Youngblood A (2023) Degenerate interpretations of O_3 spectral features in exoplanet atmosphere observations due to stellar UV uncertainties: A 3D case study with TRAPPIST-1 e. Astrophys J 959:45, https://doi.org/10.3847/1538-4357/ad0381

Cordiner MA, Villanueva GL, Wiesemeyer H, et al. (2022) Phosphine in the Venusian atmosphere: A strict upper limit from SOFIA GREAT Observations. Geophys Res Lett 49:1–9, https://doi.org/10.1029/2022GL101055

Covone G, Ienco RM, Cacciapuoti L, Inno L (2021) Efficiency of the oxygenic photosynthesis on Earth-like planets in the habitable zone. Mon Not R Astron Soc 505:3329–3335, https://doi.org/10.1093/mnras/stab1357

Cowan NB, Agol E, Meadows VS, et al. (2009) Alien maps of an ocean-bearing world. Astrophys J 700:915–923, https://doi.org/10.1088/0004-637X/700/2/915

Crouzet N, McCullough PR, Long D, et al. (2017) Discovery of XO-6b: A hot Jupiter transiting a fast rotating F5 star on an oblique orbit. Astron J 153:94, https://doi.org/10.3847/1538-3881/153/3/94

Cui M, Ma A, Qi H, et al. (2015) Anaerobic oxidation of methane: an "active" microbial process. Microbiol Open 4:1–11, https://doi.org/10.1002/mbo3.232

Currie MH, Meadows VS, Rasmussen KC (2023) There's more to life than O_2: Simulating the detectability of a range of molecules for ground-based, high-resolution spectroscopy of transiting terrestrial exoplanets. Planet Sci J 4:83, https://doi.org/10.3847/PSJ/accf86

Dalton JB, Palmer-Moloney LJ, Rogoff D, et al. (2009) Remote monitoring of hypersaline environments in San Francisco Bay, CA, USA. Int J Remote Sens 30:2933–2949, https://doi.org/10.1080/01431160802558642

Damiano M, Hu R, Mennesson B (2023) Reflected spectroscopy of small exoplanets. III. Probing the UV band to measure biosignature gases. Astron J 166:157, https://doi.org/10.3847/1538-3881/acefd3

DasSarma S (2006) Extreme halophiles are models for astrobiology. Microbe-American Soc Microbiol 1:120–126

DasSarma S, Schwieterman EW (2021) Early evolution of purple retinal pigments on Earth and implications for exoplanet biosignatures. Int J Astrobiol 20:241–250, https://doi.org/10.1017/S1473550418000423

DasSarma S, DasSarma P, Laye VJ, Schwieterman EW (2020) Extremophilic models for astrobiology: haloarchaeal survival strategies and pigments for remote sensing. Extremophiles 24:31–41, https://doi.org/10.1007/s00792-019-01126-3

Davila AF, McKay CP (2014) Chance and necessity in biochemistry: Implications for the search for extraterrestrial biomarkers in earth-like environments. Astrobiology 14:534–540, https://doi.org/10.1089/ast.2014.1150

DeMore WB, Yung YL (1982) Catalytic processes in the atmospheres of Earth and Venus. Science 217:1209–1213, https://doi.org/10.1126/science.217.4566.1209

Des Marais DJ, Walter M. R. (1999) Astrobiology: Exploring the origins, evolution, and distribution of life in the universe. Annu Rev Ecol Evol Syst 30:397–420

Des Marais DJ, Harwit MO, Jucks KW, et al. (2002) Remote sensing of planetary properties and biosignatures on extrasolar terrestrial planets. Astrobiology 2:153–181, https://doi.org/10.1089/15311070260192246

Des Marais DJ, Nuth J, Allamandola LJ, et al. (2008) The NASA astrobiology roadmap. Astrobiology 8:715–30, https://doi.org/10.1089/ast.2008.0819

Dlugokencky EJ, Nisbet EG, Fisher R, Lowry D (2011) Global atmospheric methane: budget, changes and dangers. Philos Trans R Soc A Math Phys Eng Sci 369:2058–2072, https://doi.org/10.1098/rsta.2010.0341

Domagal-Goldman SD, Meadows VS, Claire MW, Kasting JF (2011) Using biogenic sulfur gases as remotely detectable biosignatures on anoxic planets. Astrobiology 11:419–441, https://doi.org/10.1089/ast.2010.0509

Domagal-Goldman SD, Segura A, Claire MW, et al. (2014) Abiotic ozone and oxygen in atmospheres similar to prebiotic Earth. Astrophys J 792:43, https://doi.org/10.1088/0004-637X/792/2/90

Ellwood MJ, Schneider L, Potts J, et al. (2016) Volatile selenium fluxes from selenium-contaminated sediments in an Australian coastal lake. Environ Chem 13:68–75, https://doi.org/10.1071/EN14228

Etiope G, Sherwood-Lollar BS (2013) Abiotic methane on Earth. Rev Geophys 51:276–299, https://doi.org/10.1002/rog.20011

Fast K, Kostiuk T, Espenak F, et al. (2006) Ozone abundance on Mars from infrared heterodyne spectra I. Acquisition, retrieval, and anticorrelation with water vapor. Icarus 181:419–431, https://doi.org/10.1016/j.icarus.2005.12.001

Fatoki OS (1997) Biomethylation in the natural environment: A review. S Afr J Sci 93:366–370

Fauchez TJ, Villanueva GL, Schwieterman EW, et al. (2020) Sensitive probing of exoplanetary oxygen via mid-infrared collisional absorption. Nat Astron 4:372–376, https://doi.org/10.1038/s41550-019-0977-7

Fayolle EC, Öberg KI, Jørgensen JK, et al. (2017) Protostellar and cometary detections of organohalogens. Nat Astron 1:703–708, https://doi.org/10.1038/s41550-017-0237-7

Felton RC, Bastelberger ST, Mandt KE, et al. (2022) The role of atmospheric exchange in false-positive biosignature detection. J Geophys Res E: Planets 127:1–14, https://doi.org/10.1029/2021JE006853

Ferry J (1999) Enzymology of one-carbon metabolism in methanogenic pathways. FEMS Microbiol Rev 23:13–38, https://doi.org/10.1016/S0168-6445(98)00029-1

Fisher T, Kim H, Millsaps C, et al. (2022) Inferring exoplanet disequilibria with multivariate information in atmospheric reaction networks. Astron J 164:53, https://doi.org/10.3847/1538-3881/ac6594

France K, Froning CS, Linsky JL, et al. (2013) The ultraviolet radiation environment around M dwarf exoplanet host stars. Astrophys J 763:149, https://doi.org/10.1088/0004-637X/763/2/149

France K, Loyd ROP, Youngblood A, et al. (2016) the Muscles treasury survey. I. Motivation and overview. Astrophys J 820:89, https://doi.org/10.3847/0004-637X/820/2/89

Frische M, Garofalo K, Hansteen TH, Borchers R (2006) Fluxes and origin of halogenated organic trace gases from Momotombo volcano (Nicaragua). Geochem Geophys Geosystems 7: Q05020, https://doi.org/10.1029/2005gc001162

Fuentes JD, Wang D, Neumann HH, et al. (1996) Ambient biogenic hydrocarbons and isoprene emissions from a mixed deciduous forest. J Atmos Chem 25:67–95, https://doi.org/10.1007/BF00053286

Fujii Y, Lustig-Yaeger J, Cowan NB (2017) Rotational spectral unmixing of exoplanets: Degeneracies between surface colors and geography. Astron J 154:189, https://doi.org/10.3847/1538-3881/aa89f1

Fujii Y, Angerhausen D, Deitrick R, et al. (2018) Exoplanet biosignatures: Observational prospects. Astrobiology 18:739–778, https://doi.org/10.1089/ast.2017.1733

Fung KM, Heald CL, Kroll JH, et al. (2022) Exploring dimethyl sulfide (DMS) oxidation and implications for global aerosol radiative forcing. Atmos Chem Phys 22:1549–1573, https://doi.org/10.5194/acp-22-1549-2022

Gaidos E, Williams DM (2004) Seasonality on terrestrial extrasolar planets: inferring obliquity and surface conditions from infrared light curves. New Astron 10:67–77, https://doi.org/10.1016/j.newast.2004.04.009

Gao P, Hu R, Robinson TD, et al. (2015) Stability of CO_2 atmospheres on dessicated M dwarf exoplanets. Astrophys J 806:249, https://doi.org/10.1088/0004-637X/806/2/249

Gao P, Wakeford HR, Moran SE, Parmentier V (2021) Aerosols in exoplanet atmospheres. J Geophys Res E: Planets 126:1–46, https://doi.org/10.1029/2020JE006655

Garcia AK, McShea H, Kolaczkowski B, Kaçar B (2020) Reconstructing the evolutionary history of nitrogenases: Evidence for ancestral molybdenum-cofactor utilization. Geobiology 18:394–411, https://doi.org/10.1111/gbi.12381

Garcia-Sage K, Farrish AO, Airapetian VS, et al. (2023) Star-exoplanet interactions: A growing interdisciplinary field in heliophysics. Front Astron Sp Sci 10:1–9, https://doi.org/10.3389/fspas.2023.1064076

Gates DM, Keegan HJ, Schleter JC, Weidner VR (1965) Spectral properties of plants. Appl Opt 4:11, https://doi.org/10.1364/AO.4.000011

Gaudi BS, Seager S, Mennesson B, et al. (2020) The Habitable Exoplanet Observatory (HabEx) mission concept study final report. arXiv preprint 200106683, https://doi.org/10.48550/arXiv.2001.06683

Gialluca MT, Robinson TD, Rugheimer S, Wunderlich F (2021) Characterizing atmospheres of transiting Earth-like exoplanets orbiting M dwarfs with James Webb Space Telescope. Publ Astron Soc Pacific 133:054401, https://doi.org/10.1088/1538-3873/abf367

Gillen C, Jeancolas C, McMahon S, Vickers P (2023) The call for a new definition of biosignature. Astrobiology 23:1228–1237, https://doi.org/10.1089/ast.2023.0010

Gillmann C, Chassefière E, Lognonné P (2009) A consistent picture of early hydrodynamic escape of Venus atmosphere explaining present Ne and Ar isotopic ratios and low oxygen atmospheric content. Earth Planet Sci Lett 286:503–513, https://doi.org/10.1016/j.epsl.2009.07.016

Glein CR (2015) Noble gases, nitrogen, and methane from the deep interior to the atmosphere of Titan. Icarus 250:570–586, https://doi.org/10.1016/j.icarus.2015.01.001

Gleiser M (2022) Biological homochirality and the search for extraterrestrial biosignatures. Orig Life Evol Biosph 52:93–104, https://doi.org/10.1007/s11084-022-09623-w

Glindemann D, Edwards M, Liu J, Kuschk P (2005) Phosphine in soils, sludges, biogases and atmospheric implications—A review. Ecol Eng 24:457–463, https://doi.org/10.1016/j.ecoleng.2005.01.002

Gomez Barrientos J, MacDonald RJ, Lewis NK, Kaltenegger L (2023) In search of the edge: A Bayesian exploration of the detectability of red edges in exoplanet reflection spectra. Astrophys J 946:96, https://doi.org/10.3847/1538-4357/acaf59

Gordon IE, Rothman LS, Hargreaves RJ, et al. (2022) The HITRAN2020 molecular spectroscopic database. J Quant Spectrosc Radiat Transfer 277:107949, https://doi.org/10.1016/j.jqsrt.2021.107949

Gordon KE, Karalidi T, Bott KM, et al. (2023) Polarized signatures of a habitable world: comparing models of an exoplanet Earth with visible and near-infrared Earthshine spectra. Astrophys J 945:166, https://doi.org/10.3847/1538-4357/aca7fe

Greaves JS, Richards AMS, Bains W, et al. (2020) Phosphine gas in the cloud decks of Venus. Nat Astron 5:655–664, https://doi.org/10.1038/s41550-020-1174-4

Greaves JS, Rimmer PB, Richards AMS, et al. (2022) Low levels of sulphur dioxide contamination of Venusian phosphine spectra. Mon Not R Astron Soc 514:2994–3001, https://doi.org/10.1093/mnras/stac1438

Greenblatt GD, Orlando JJ, Burkholder JB, Ravishankara a. R (1990) Absorption measurements of oxygen between 330 and 1140 nm. J Geophys Res 95:18577, https://doi.org/10.1029/JD095iD11p18577

Grenfell JL (2017) A review of exoplanetary biosignatures. Phys Rep 713:1–17, https://doi.org/10.1016/j.physrep.2017.08.003

Grenfell JL, Stracke B, von Paris P, et al. (2007) The response of atmospheric chemistry on earthlike planets around F, G and K Stars to small variations in orbital distance. Planet Space Sci 55:661–671, https://doi.org/10.1016/j.pss.2006.09.002

Grenfell JL, Gebauer S, v. Paris P, et al. (2014) Sensitivity of biosignatures on Earth-like planets orbiting in the habitable zone of cool M-dwarf Stars to varying stellar UV radiation and surface biomass emissions. Planet Space Sci 98:66–76, https://doi.org/10.1016/j.pss.2013.10.006

Gruber N (2008) The marine nitrogen cycle. *In*: Nitrogen in the Marine Environment. Elsevier, pp 1–50

Gutowsky HS (1949) The infra-red and Raman Spectra of dimethyl mercury and dimethyl zinc. J Chem Phys 17:128–138, https://doi.org/10.1063/1.1747205

Haddock SHD, Moline MA, Case JF (2010) Bioluminescence in the sea. Annu Rev Mar Sci 2:443–493, https://doi.org/10.1146/annurev-marine-120308-081028

Hall CAS, Ekdahl CA, Wartenberg DE (1975) A fifteen-year record of biotic metabolism in the Northern Hemisphere. Nature 255:136–138, https://doi.org/10.1038/255136a0

Hall S, Krissansen-Totton J, Robinson T, et al. (2023) Constraining background N_2 inventories on directly imaged terrestrial exoplanets to rule out O_2 false positives. Astron J 166:254, https://doi.org/10.3847/1538-3881/ad03e9

Hamano K, Abe Y, Genda H (2013) Emergence of two types of terrestrial planet on solidification of magma ocean. Nature 497:607–610, https://doi.org/10.1038/nature12163

Hanson RS, Hanson TE (1996) Methanotrophic bacteria. Microbiol Rev 60:439–471, https://doi.org/10.1128/mmbr.60.2.439-471.1996

Haqq-Misra J, Schwieterman EW, Socas-Navarro H, et al. (2022) Searching for technosignatures in exoplanetary systems with current and future missions. Acta Astronaut 198:194–207, https://doi.org/10.1016/j.actaastro.2022.05.040

Hardegree-Ullman KK, Apai D, Bergsten GJ, et al. (2023) Bioverse: A comprehensive assessment of the capabilities of extremely large telescopes to probe Earth-like O_2 Levels in nearby transiting habitable-zone exoplanets. Astron J 165:267, https://doi.org/10.3847/1538-3881/acd1ec

Harman CE, Schwieterman EW, Schottelkotte JC, Kasting JF (2015) Abiotic O_2 levels on planets around F, G, K, and M stars: Possible false positives for life? Astrophys J 812:137, https://doi.org/10.1088/0004-637X/812/2/137

Harman CE, Felton R, Hu R, et al. (2018) Abiotic O_2 levels on planets around F, G, K, and M stars: Effects of lightning-produced catalysts in eliminating oxygen false positives. Astrophys J 866:56, https://doi.org/10.3847/1538-4357/aadd9b

Harman CE, Kopparapu RK, Stefánsson G, et al. (2022) A snowball in hell: The potential steam atmosphere of TOI-1266c. Planet Sci J 3:45, https://doi.org/10.3847/PSJ/ac38ac

Hays LE, Archenbach L, Bailey J, et al. (2015) 2015 NASA Astrobiology Strategy. NASA Astrobiology Program, Pasadena, California

He C, Hörst SM, Lewis NK, et al. (2020) Haze formation in warm H_2-rich exoplanet atmospheres. Planet Sci J 1:51, https://doi.org/10.3847/PSJ/abb1a4

Hegde S, Paulino-Lima IG, Kent R, et al. (2015) Surface biosignatures of exo-Earths: Remote detection of extraterrestrial life. PNAS 112:3886–3891, https://doi.org/10.1073/pnas.1421237112

Helling C (2019) Exoplanet clouds. Annu Rev Earth Planet Sci 47:583–606, https://doi.org/10.1146/annurev-earth-053018-060401

Hendrix AR, Hurford TA, Barge LM, et al. (2019) The NASA roadmap to ocean worlds. Astrobiology 19:1–27, https://doi.org/10.1089/ast.2018.1955

Hermans C, Vandaele AC, Carleer M, et al. (1999) Absorption cross-sections of atmospheric constituents: NO_2, O_2, and H_2O. Environ Sci Pollut Res 6:151–158, https://doi.org/10.1007/BF02987620

Hitchcock DR, Lovelock JE (1967) Life detection by atmospheric analysis. Icarus 7:149–159, https://doi.org/10.1016/0019-1035(67)90059-0

Hoehler TM, Bains W, Davila A, et al. (2020) Life's requirements, habitability, and biological potential. *In:* Meadows VS, Arney GN, Shcmidt BE, Des Marais DJ (eds) Planetary Astrobiology. University of Arizona Press, p 37–69

Hohmann-Marriott MF, Blankenship RE (2012) Photosynthesis. Springer Netherlands, Dordrecht

Holme P, Huss M, Lee SH (2011) Atmospheric reaction systems as null-models to identify structural traces of evolution in metabolism. PLoS One 6:e19759, https://doi.org/10.1371/journal.pone.0019759

Hu R (2021) Photochemistry and spectral characterization of temperate and gas-rich exoplanets. Astrophys J 921:27, https://doi.org/10.3847/1538-4357/ac1789

Hu H, Mylon SE, Benoit G (2007) Volatile organic sulfur compounds in a stratified lake. Chemosphere 67:911–919, https://doi.org/10.1016/j.chemosphere.2006.11.012

Huang J, Seager S, Petkowski JJ, et al. (2022a) Methanol—A poor biosignature gas in exoplanet atmospheres. Astrophys J 933:6, https://doi.org/10.3847/1538-4357/ac6f60

Huang J, Seager S, Petkowski JJ, et al. (2022b) Assessment of ammonia as a biosignature gas in exoplanet atmospheres. Astrobiology 22:171–191, https://doi.org/10.1089/ast.2020.2358

Hulswar S, Simó R, Galí M, et al. (2022) Third revision of the global surface seawater dimethyl sulfide climatology (DMS-Rev3). Earth Syst Sci Data 14:2963–2987, https://doi.org/10.5194/essd-14-2963-2022

Illner D, Lingam M, Peverati R (2023) On the importance and challenges of modelling extraterrestrial photopigments via density-functional theory. Mol Phys, https://doi.org/10.1080/00268976.2023.2261563

Jacob DJ (1999) Introduction to atmospheric chemistry. Princeton University Press

Jenkins R., Morris T-A, Craig P., et al. (2000) Phosphine generation by mixed- and monoseptic-cultures of anaerobic bacteria. Sci Total Environ 250:73–81, https://doi.org/10.1016/S0048-9697(00)00368-5

Johnson SS, Anslyn EV, Graham HV, et al. (2018) Fingerprinting non-terran biosignatures. Astrobiology 18:ast.2017.1712, https://doi.org/10.1089/ast.2017.1712

Johnstone CP, Khodachenko ML, Lüftinger T, et al. (2019) Extreme hydrodynamic losses of Earth-like atmospheres in the habitable zones of very active stars. Astron Astrophys 624:L10, https://doi.org/10.1051/0004-6361/201935279

Joiner J, Yoshida Y, Vasilkov AP, et al. (2011) First observations of global and seasonal terrestrial chlorophyll fluorescence from space. Biogeosciences 8:637–651, https://doi.org/10.5194/bg-8-637-2011

Jones LC, Peters B, Lezama Pacheco JS, et al. (2015) Stable isotopes and iron oxide mineral products as markers of chemodenitrification. Environ Sci Technol 49:3444–3452, https://doi.org/10.1021/es504862x

Jontof-Hutter D (2019) The compositional diversity of low-mass exoplanets. Annu Rev Earth Planet Sci 47:141–171, https://doi.org/10.1146/annurev-earth-053018-060352

Kaltenegger L (2017) How to characterize habitable worlds and signs of life. Annu Rev Astron Astrophys 55:433–485, https://doi.org/10.1146/annurev-astro-082214-122238

Kaltenegger L, Selsis F (2007) Biomarkers set in context. *In:* Extrasolar Planets: Formation, Detection and Dynamics. p 79–98 (ArXiv: 0710.0881)

Kaltenegger L, Lin Z, Madden J (2020) High-resolution transmission spectra of earth through geological time. Astrophys J 892:L17, https://doi.org/10.3847/2041-8213/ab789f

Kane SR, Kopparapu RK, Domagal-Goldman SD (2014) On the frequency of potential Venus analogs from Kepler data. Astrophys J Lett 794:, https://doi.org/10.1088/2041-8205/794/1/L5

Kane SR, Hill ML, Kasting JF, et al. (2016) A catalog of Kepler habitable zone exoplanet Candidates. Astrophys J 830:1, https://doi.org/10.3847/0004-637X/830/1/1

Kane SR, Arney G, Crisp D, et al. (2019) Venus as a laboratory for exoplanetary science. J Geophys Res E: Planets 124:2015–2028, https://doi.org/10.1029/2019JE005939

Karman T, Gordon IE, van der Avoird A, et al. (2019) Update of the HITRAN collision-induced absorption section. Icarus 328:160–175, https://doi.org/10.1016/j.icarus.2019.02.034

Kasting JF (1982) Stability of ammonia in the primitive terrestrial atmosphere. J Geophys Res 87:3091–3098, https://doi.org/10.1029/JC087iC04p03091

Kasting JF (1988) Runaway and moist greenhouse atmospheres and the evolution of Earth and Venus. Icarus 74:472–94

Kasting JF, Zahnle KJ, Walker JCG (1983) Photochemistry of methane in the Earth's early atmosphere. *In*: Developments in Precambrian Geology, p 13–40

Kasting JF, Pollack JB, Ackerman TP (1984) Response of Earth's atmosphere to increases in solar flux and implications for loss of water from Venus. Icarus 57:335–55

Kasting JF, Whitmire DP, Reynolds RT (1993) Habitable zones around main sequence stars. Icarus 101:108–128, https://doi.org/10.1006/icar.1993.1010

Kasting JF, Kopparapu R, Ramirez RM, Harman CE (2014) Remote life-detection criteria, habitable zone boundaries, and the frequency of Earth-like planets around M and late K stars. PNAS 111:12641–12646, https://doi.org/10.1073/pnas.1309107110

Kawahara H, Fujii Y (2011) Mapping clouds and terrain of earth-like planets from photometric variability: demonstration with planets in face-on orbits. Astrophys J 739:L62, https://doi.org/10.1088/2041-8205/739/2/L62

Keeling RF, Shertz SR (1992) Seasonal and interannual variations in atmospheric oxygen and implications for the global carbon cycle. Nature 358:723–727, https://doi.org/10.1038/358723a0

Keeling CD, Chin JFS, Whorf TP (1996) Increased activity of northern vegetation inferred from atmospheric CO_2 measurements. Nature 382:146–149, https://doi.org/10.1038/382146a0

Kempton EM-R, Knutson HA (2024) Transiting exoplanet atmospheres in the era of JWST. Rev Mineral Geochem 90:411–464

Kenrick P, Crane PR (1997) The origin and early evolution of plants on land. Nature 389:33–39, https://doi.org/10.1038/37918

Keppler F, Harper DB, Greule M, et al. (2015) Chloromethane release from carbonaceous meteorite affords new insight into Mars lander findings. Sci Rep 4:7010, https://doi.org/10.1038/srep07010

Khalil MAK, Rasmussen RA (1983) Sources, sinks, and seasonal cycles of atmospheric methane. J Geophys Res: Oceans 88:5131–5144, https://doi.org/10.1029/JC088iC09p05131

Kiang NY (2014) Looking for life elsewhere: Photosynthesis and astrobiology. Biochem (Lond) 36:24–30

Kiang NY, Segura A, Tinetti G, et al. (2007a) Spectral signatures of photosynthesis. II. Coevolution with other stars and the atmosphere on extrasolar worlds. Astrobiology 7:252–74, https://doi.org/10.1089/ast.2006.0108

Kiang NY, Siefert J, Govindjee, Blankenship RE (2007b) Spectral signatures of photosynthesis. I. Review of Earth organisms. Astrobiology 7:222–51, https://doi.org/10.1089/ast.2006.0105

King J, Koc CH, Unterkofler K, et al. (2010) Physiological modeling of isoprene dynamics in exhaled breath. J Theor Biol 267:626–637, https://doi.org/10.1016/j.jtbi.2010.09.028

Kippenhahn R, Weigert A, Weiss A (1990) Stellar Structure and Evolution. Springer

Kipping D, Wright J (2024) Deconstructing alien hunting. Astron J 167:24, https://doi.org/10.3847/1538-3881/ad0cbe

Kite ES, Ford EB (2018) Habitability of exoplanet waterworlds. Astrophys J 864:75, https://doi.org/10.3847/1538-4357/aad6e0

Knipling EB (1970) Physical and physiological basis for the reflectance of visible and near-infrared radiation from vegetation. Remote Sens Environ 1:155–159, https://doi.org/10.1016/S0034-4257(70)80021-9

Köhler P, Fischer WW, Rossman GR, et al. (2021) Mineral luminescence observed from space. Geophys Res Lett 48:1–10, https://doi.org/10.1029/2021GL095227

Komatsu Y, Hori Y, Kuzuhara M, et al. (2023) Photosynthetic Fluorescence from Earthlike Planets around Sunlike and Cool Stars. Astrophys J 942:57, https://doi.org/10.3847/1538-4357/aca3a5

Konrad BS, Alei E, Quanz SP, et al. (2022) Large Interferometer For Exoplanets (LIFE)-III. Spectral resolution, wavelength range, and sensitivity requirements based on atmospheric retrieval analyses of an exo-Earth. Astron Astrophys 664:A23, https://doi.org/10.1051/0004-6361/202141964

Kopparapu RK, Ramirez R, Kasting JF, et al. (2013) Habitable zones around main-sequence stars: New estimates. Astrophys J 765:131, https://doi.org/10.1088/0004-637X/765/2/131

Kopparapu RK, Ramirez RM, SchottelKotte J, et al. (2014) Habitable zones around main-sequence stars: Dependence on planetary mass. Astrophys J 787:L29, https://doi.org/10.1088/2041-8205/787/2/L29

Kozakis T, Mendonça JM, Buchhave LA (2022) Is ozone a reliable proxy for molecular oxygen? Astron Astrophys 665:A156, https://doi.org/10.1051/0004-6361/202244164

Krasnopolsky VA (2011) Atmospheric chemistry on Venus, Earth, and Mars: Main features and comparison. Planet Space Sci 59:952–964, https://doi.org/10.1016/j.pss.2010.02.011

Krissansen-Totton J, Fortney JJ (2022) Predictions for observable atmospheres of Trappist-1 planets from a fully coupled atmosphere–interior evolution model. Astrophys J 933:115, https://doi.org/10.3847/1538-4357/ac69cb

Krissansen-Totton J, Bergsman DS, Catling DC (2016) On detecting biospheres from chemical thermodynamic disequilibrium in planetary atmospheres. Astrobiology 16:39–67, https://doi.org/10.1089/ast.2015.1327

Krissansen-Totton J, Garland R, Irwin P, Catling DC (2018a) Detectability of biosignatures in anoxic atmospheres with the James Webb Space Telescope: A TRAPPIST-1e Case Study. Astron J 156:114, https://doi.org/10.3847/1538-3881/aad564

Krissansen-Totton J, Olson S, Catling DC (2018b) Disequilibrium biosignatures over Earth history and implications for detecting exoplanet life. Sci Adv 4:eaao5747, https://doi.org/10.1126/sciadv.aao5747

Krissansen-Totton J, Fortney JJ, Nimmo F, Wogan N (2021) Oxygen false positives on habitable zone planets around Sun-like stars. AGU Adv 2: e2020AV000294, https://doi.org/10.1029/2020AV000294

Krissansen-Totton J, Thompson M, Galloway ML, Fortney JJ (2022) Understanding planetary context to enable life detection on exoplanets and test the Copernican principle. Nat Astron 6:189–198, https://doi.org/10.1038/s41550-021-01579-7

Kurashov V, Ho M-Y, Shen G, et al. (2019) Energy transfer from chlorophyll f to the trapping center in naturally occurring and engineered Photosystem I complexes. Photosynth Res 141:151–163, https://doi.org/10.1007/s11120-019-00616-x

Lafferty WJ, Solodov AM, Weber A, et al. (1996) Infrared collision-induced absorption by N(2) near 4.3 μm for atmospheric applications: measurements and empirical modeling. Appl Opt 35:5911–5917, https://doi.org/10.1364/AO.35.005911

Lammer H, Selsis F, Ribas I, et al. (2003) Atmospheric loss of exoplanets resulting from stellar X-ray and extreme-ultraviolet heating. Astrophys J 598:L121–L124, https://doi.org/10.1086/380815

Lammer H, Kasting JF, Chassefière E, et al. (2008) Atmospheric escape and evolution of terrestrial planets and satellites. Space Sci Rev 139:399–436, https://doi.org/10.1007/s11214-008-9413-5

Lan X, Thoning KW, Dlugokencky EJ (2023) Trends in globally-averaged CH_4, N_2O, and SF_6 determined from NOAA Global Monitoring Laboratory measurements, https://gml.noaa.gov/ccgg/trends_sf6/

Lécuyer C, Ricard Y (1999) Long-term fluxes and budget of ferric iron: implication for the redox states of the Earth's mantle and atmosphere. Earth Planet Sci Lett 165:197–211, https://doi.org/10.1016/S0012-821X(98)00267-2

Leger A, Pirre M, Marceau FJ (1993) Search for primitive life on a distant planet: relevance of O_2 and O_3 detections. Astron Astrophys 277:309

Lehmer OR, Catling DC, Parenteau MN, Hoehler TM (2018) The productivity of oxygenic photosynthesis around cool, M dwarf stars. Astrophys J 859:171, https://doi.org/10.3847/1538-4357/aac104

Lehmer OR, Catling DC, Parenteau MN, et al. (2021) The peak absorbance wavelength of photosynthetic pigments around other stars from spectral optimization. Front Astron Sp Sci 8:1–10, https://doi.org/10.3389/fspas.2021.689441

Lenton TM (2013) Fire feedbacks on atmospheric oxygen. *In:* Fire phenomena and the Earth system. Wiley, p 289–308

Leung M, Schwieterman EW, Parenteau MN, Fauchez TJ (2022) Alternative methylated biosignatures. I. Methyl bromide, a capstone biosignature. Astrophys J 938:6, https://doi.org/10.3847/1538-4357/ac8799

Li C-Y, Cao H-Y, Wang Q, et al. (2023) Aerobic methylation of hydrogen sulfide to dimethylsulfide in diverse microorganisms and environments. ISME J 17:1184–1193, https://doi.org/10.1038/s41396-023-01430-z

Liggins P, Jordan S, Rimmer PB, Shorttle O (2022) Growth and evolution of secondary volcanic atmospheres: I. Identifying the geological character of hot rocky planets. J Geophys Res E: Planets 127:e2021JE007123, https://doi.org/10.1029/2021JE007123

Liggins P, Jordan S, Rimmer PB, Shorttle O (2023) Growth and evolution of secondary volcanic atmospheres: 2. The importance of kinetics. J Geophys Res E: Planets 128:e2022JE007528, https://doi.org/10.1029/2022JE007528

Lincowski AP, Meadows VS, Crisp D, et al. (2021) Claimed detection of PH_3 in the clouds of Venus is consistent with mesospheric SO_2. Astrophys J Lett 908:L44, https://doi.org/10.3847/2041-8213/abde47

Lingam M, Balbi A, Mahajan SM (2021) Excitation properties of photopigments and their possible dependence on the host star. Astrophys J Lett 921:L41, https://doi.org/10.3847/2041-8213/ac3478

Liu Z, Jia S, Wang B, et al. (2008) Preliminary investigation on the role of microorganisms in the production of phosphine. J Environ Sci 20:885–890, https://doi.org/10.1016/S1001-0742(08)62142-7

Livengood TA, Deming LD, A'Hearn MF, et al. (2011) Properties of an Earth-like planet orbiting a Sun-like star: Earth observed by the EPOXI mission. Astrobiology 11:907–930, https://doi.org/10.1089/ast.2011.0614

Lodders K, Fegley B (2002) Atmospheric chemistry in giant planets, brown dwarfs, and low-mass dwarf stars. I. Carbon, nitrogen, and oxygen. Icarus 155:393–424, https://doi.org/10.1006/icar.2001.6740

Lovelock JE (1965) A physical basis for life detection experiments. Nature 207:568–570, https://doi.org/10.1038/207568a0

Lovelock JE (1975) Thermodynamics and the recognition of alien biospheres. Proc R Soc London Ser B Biol Sci 189:167–181, https://doi.org/10.1098/rspb.1975.0051

Luger R, Barnes R (2015) Extreme water loss and abiotic O_2 buildup on planets throughout the habitable zones of M dwarfs. Astrobiology 15:119–143, https://doi.org/10.1089/ast.2014.1231

Lustig-Yaeger J, Meadows VS, Tovar Mendoza G, et al. (2018) Detecting ocean glint on exoplanets using multiphase mapping. Astron J 156:301, https://doi.org/10.3847/1538-3881/aaed3a

Lustig-Yaeger J, Meadows VS, Lincowski AP (2019) The detectability and characterization of the TRAPPIST-1 exoplanet atmospheres with JWST. Astron J 158:27, https://doi.org/10.3847/1538-3881/ab21e0

Lustig-Yaeger J, Meadows VS, Crisp D, et al. (2023) Earth as a transiting exoplanet: A validation of transmission spectroscopy and atmospheric retrieval methodologies for terrestrial exoplanets. Planet Sci J 4:170, https://doi.org/10.3847/PSJ/acf3e5

Lyons TW, Reinhard CT, Planavsky NJ (2014) The rise of oxygen in Earth's early ocean and atmosphere. Nature 506:307–15, https://doi.org/10.1038/nature13068

Lyons TW, Diamond CW, Planavsky NJ, et al. (2021) Oxygenation, life, and the planetary system during Earth's middle history: An overview. Astrobiology 21:906–923, https://doi.org/10.1089/ast.2020.2418

Macdonald EJR, Cowan NB (2019) An empirical infrared transit spectrum of Earth: opacity windows and biosignatures. Mon Not R Astron Soc 489:196–204, https://doi.org/10.1093/mnras/stz2047

Madau P (2023) Beyond the Drake Equation: A time-dependent inventory of habitable planets and life-bearing worlds in the solar neighborhood. Astrophys J 957:66, https://doi.org/10.3847/1538-4357/acfe0e

Madhusudhan N, Agúndez M, Moses JI, Hu Y (2016) Exoplanetary atmospheres—Chemistry, formation conditions, and habitability. Space Sci Rev 205:285–348, https://doi.org/10.1007/s11214-016-0254-3

Madhusudhan N (2019) Exoplanetary atmospheres: Key insights, challenges, and prospects. Annu Rev Astron Astrophys 57:617–663, https://doi.org/10.1146/annurev-astro-081817-051846

Madhusudhan N, Piette AAA, Constantinou S (2021) Habitability and biosignatures of Hycean Worlds. Astrophys J 918:1, https://doi.org/10.3847/1538-4357/abfd9c

Madhusudhan N, Moses JI, Rigby F, Barrier E (2023) Chemical conditions on Hycean worlds. Faraday Discuss 245:80–111, https://doi.org/10.1039/D3FD00075C

Malaterre C, ten Kate IL, Baqué M, et al. (2023) Is there such a thing as a biosignature? Astrobiology 23:1213–1227, https://doi.org/10.1089/ast.2023.0042

Manley SL, Wang NY, Waiser ML, Cicerone RJ (2006) Coastal salt marshes as global methyl halide sources from determinations of intrinsic production by marsh plants. Global Biogeochem Cycles 20:1–13, https://doi.org/10.1029/2005GB002578

Marcos-Arenal P, Cerdán L, Burillo-Villalobos M, et al. (2022) ExoPhot: The photon absorption rate as a new metric for quantifying the exoplanetary photosynthetic activity fitness. Universe 8:1–17, https://doi.org/10.3390/universe8120624

Marcq E, Salvador A, Massol H, Davaille A (2017) Thermal radiation of magma ocean planets using a 1-D radiative-convective model of H_2O–CO_2 atmospheres. J Geophys Res E: Planets 122:1539–1553, https://doi.org/10.1002/2016JE005224

Martin W, Baross J, Kelley D, Russell MJ (2008) Hydrothermal vents and the origin of life. Nat Rev Microbiol 6:805–814, https://doi.org/10.1038/nrmicro1991

Maté B, Lugez C, Fraser GT, Lafferty WJ (1999) Absolute intensities for the O_2 1.27 continuum absorption. J Geophys Res 104:585–590

Maté B, Lugez CL, Solodov a M, et al. (2000) Investigation of the collision-induced absorption by O_2 near 6.4 μm in pure O2 and O_2/N_2 mixtures. J Geophys Res 105:22,225-22,230, https://doi.org/10.1029/2000JD900295

McElroy MB, Donahue TM (1972) Stability of the Martian atmosphere. Science 177:986–988, https://doi.org/10.1126/science.177.4053.986

McGuire BA (2018) 2018 Census of interstellar, circumstellar, extragalactic, protoplanetary disk, and exoplanetary molecules. Astrophys J Suppl Ser 239:17, https://doi.org/10.3847/1538-4365/aae5d2

Meadows VS (2006) Modelling the diversity of extrasolar terrestrial planets. Proc Int Astron Union 1:25, https://doi.org/10.1017/S1743921306009033

Meadows VS (2008) Planetary environmental signatures for habitability and life. *In:* Exoplanets. Springer Berlin Heidelberg, Berlin, Heidelberg, p 259–284

Meadows VS (2017) Reflections on O_2 as a biosignature in exoplanetary atmospheres. Astrobiology 17:1022–1052, https://doi.org/10.1089/ast.2016.1578

Meadows VS, Arney GN, Schwieterman EW, et al. (2018a) The habitability of Proxima Centauri b: Environmental states and observational discriminants. Astrobiology 18:133–189, https://doi.org/10.1089/ast.2016.1589

Meadows VS, Reinhard CT, Arney GN, et al. (2018b) Exoplanet biosignatures: Understanding oxygen as a biosignature in the context of its environment. Astrobiology 18:630–662, https://doi.org/10.1089/ast.2017.1727

Meadows V, Graham H, Abrahamsson V, et al. (2022) Community report from the biosignatures standards of evidence workshop. arXiv:2210.14293, https://doi.org/10.48550/arXiv.2210.14293

Meadows VS, Lincowski AP, Lustig-Yaeger J (2023) The feasibility of detecting biosignatures in the TRAPPIST-1 planetary system with JWST. Planet Sci J 4:192, https://doi.org/10.3847/PSJ/acf488

Medeiros DJ, Blitz MA, James L, et al. (2018) Kinetics of the reaction of OH with isoprene over a wide range of temperature and pressure including direct observation of equilibrium with the OH adducts. J Phys Chem A 122:7239–7255, https://doi.org/10.1021/acs.jpca.8b04829

Mettler J-N, Quanz SP, Helled R, et al. (2023) Earth as an exoplanet. II. Earth's time-variable thermal emission and its atmospheric seasonality of bioindicators. Astrophys J 946:82, https://doi.org/10.3847/1538-4357/acbe3c

Meyer J, Michalke K, Kouril T, Hensel R (2008) Volatilisation of metals and metalloids: An inherent feature of methanoarchaea? Syst Appl Microbiol 31:81–87, https://doi.org/10.1016/j.syapm.2008.02.001

Mikal-Evans T (2021) Detecting the proposed CH_4–CO_2 biosignature pair with the James Webb Space Telescope: TRAPPIST-1e and the effect of cloud/haze. Mon Not R Astron Soc 510:980–991, https://doi.org/10.1093/mnras/stab3383

Miller SD, Haddock SHD, Elvidge CD, Lee TF (2005) Detection of a bioluminescent milky sea from space. PNAS 102:14181–14184, https://doi.org/10.1073/pnas.0507253102

Misra A, Meadows V, Claire M, Crisp D (2014) Using dimers to measure biosignatures and atmospheric pressure for terrestrial exoplanets. Astrobiology 14:67–86, https://doi.org/10.1089/ast.2013.0990

Moran SE, Hörst SM, Vuitton V, et al. (2020) Chemistry of temperate super-Earth and mini-Neptune atmospheric hazes from laboratory experiments. Planet Sci J 1:17, https://doi.org/10.3847/PSJ/ab8eae

Moses JI, Line MR, Visscher C, et al. (2013) Compositional diversity in the atmospheres of hot Neptunes, with application to GJ 436b. Astrophys J 777:34, https://doi.org/10.1088/0004-637X/777/1/34

Müller J-F, Stavrakou T, Wallens S, et al. (2008) Global isoprene emissions estimated using MEGAN, ECMWF analyses and a detailed canopy environment model. Atmos Chem Phys 8:1329–1341, https://doi.org/10.5194/acp-8-1329-2008

Myhre G, Shindell D, Bréon F-M, et al. (2013) Anthropogenic and natural radiative forcing. *In:* Intergovernmental Panel on Climate Change (ed) Climate Change 2013—The Physical Science Basis. Cambridge University Press, Cambridge, p 659–740

Nath NS, Bhattacharya I, Tuck AG, et al. (2011) Mechanisms of phosphine toxicity. J Toxicol 2011:1–9, https://doi.org/10.1155/2011/494168

National Academies of Sciences, Engineering and M (2021) Pathways to Discovery in Astronomy and Astrophysics for the 2020s. National Academies Press, Washington, D.C.

Navarro-González R, McKay CP, Mvondo DN (2001) A possible nitrogen crisis for Archaean life due to reduced nitrogen fixation by lightning. Nature 412:61–64, https://doi.org/10.1038/35083537

Noack L, Höning D, Rivoldini A, et al. (2016) Water-rich planets: How habitable is a water layer deeper than on Earth? Icarus 277:215–236, https://doi.org/10.1016/j.icarus.2016.05.009

Noxon JF, Traub WA, Carleton NP, Connes P (1976) Detection of O_2 dayglow emission from Mars and the Martian ozone abundance. Astrophys J 207:1025, https://doi.org/10.1086/154572

O'Malley-James JT, Kaltenegger L (2018) Biofluorescent worlds: Global biological fluorescence as a biosignature. Mon Not R Astron Soc 481:2487–2496, https://doi.org/10.1093/MNRAS/STY2411

O'Malley-James JT, Kaltenegger L (2019) Biofluorescent Worlds—II. Biological fluorescence induced by stellar UV flares, a new temporal biosignature. Mon Not R Astron Soc 488:4530–4545, https://doi.org/10.1093/mnras/stz1842

Olson JM (2006) Photosynthesis in the Archean era. Photosynth Res 88:109–17, https://doi.org/10.1007/s11120-006-9040-5

Olson SL, Schwieterman EW, Reinhard CT, et al. (2018) Atmospheric seasonality as an exoplanet biosignature. Astrophys J Lett 858:L14, https://doi.org/10.3847/2041-8213/aac171

Omran A, Oze C, Jackson B, et al. (2021) Phosphine generation pathways on rocky planets. Astrobiology 21:1264–1276, https://doi.org/10.1089/ast.2021.0034

Ostberg C, Kane SR, Li Z, et al. (2023) The demographics of terrestrial planets in the Venus Zone. Astron J 165:168, https://doi.org/10.3847/1538-3881/acbfaf

Owen T (1980) The search for early forms of life in other planetary systems: Future possibilities afforded by spectroscopic techniques. *In:* Papagianni MD (ed) Strategies for the Search for Life in the Universe. Springer Netherlands, Norwell, Mass., p 177–185

Palmer PI (2003) Mapping isoprene emissions over North America using formaldehyde column observations from space. J Geophys Res 108:4180, https://doi.org/10.1029/2002JD002153

Papageorgiou GC et al. (2007) Chlorophyll a Fluorescence: A Signature of Photosynthesis, Vol. 19. Springer Science & Business Media

Parsons C, Stüeken EE, Rosen CJ, et al. (2021) Radiation of nitrogen-metabolizing enzymes across the tree of life tracks environmental transitions in Earth history. Geobiology 19:18–34, https://doi.org/10.1111/gbi.12419

Patty CHL, Visser LJJ, Ariese F, et al. (2017) Circular spectropolarimetric sensing of chiral photosystems in decaying leaves. J Quant Spectrosc Radiat Transfer 189:303–311, https://doi.org/10.1016/j.jqsrt.2016.12.023

Patty CHL, Kate IL ten, Sparks WB, Snik F (2018) Remote sensing of homochirality: A proxy for the detection of extraterrestrial life. *In:* Chiral Analysis. Elsevier, p 29–69

Patty CHL, Ten Kate IL, Buma WJ, et al. (2019) Circular spectropolarimetric sensing of vegetation in the field: Possibilities for the remote detection of extraterrestrial life. Astrobiology 19:1221–1229, https://doi.org/10.1089/ast.2019.2050

Patty CHL, Pommerol A, Kühn JG, et al. (2022) Directional aspects of vegetation linear and circular polarization biosignatures. Astrobiology 22:1034–1046, https://doi.org/10.1089/ast.2021.0156

Paul C, Pohnert G (2011) Production and role of volatile halogenated compounds from marine algae. Natural Product Rep 28:186–195, https://doi.org/10.1039/c0np00043d

Pauleta SR, Dell'Acqua S, Moura I (2013) Nitrous oxide reductase. Coord Chem Rev 257:332–349, https://doi.org/10.1016/j.ccr.2012.05.026

Pearce BKD, Hörst SM, Sebree JA, He C (2024) Organic hazes as a source of life's building blocks to warm little ponds on the Hadean Earth. Planet Sci J 5:23, https://doi.org/10.3847/PSJ/ad17bd

Persson M, Futaana Y, Ramstad R, et al. (2020) The Venusian atmospheric oxygen ion escape: Extrapolation to the early Solar System. J Geophys Res E: Planets 125:1–12, https://doi.org/10.1029/2019JE006336

Petkowski JJ, Bains W, Seager S (2020) On the potential of silicon as a building block for life. Life 10:84, https://doi.org/10.3390/life10060084

Phillips CL, Wang J, Kendrew S, et al. (2021) Detecting biosignatures in the atmospheres of gas dwarf planets with the James Webb Space Telescope. Astrophys J 923:144, https://doi.org/10.3847/1538-4357/ac29be

Pidhorodetska D, Fauchez TJ, Villanueva GL, et al. (2020) Detectability of molecular signatures on TRAPPIST-1e through transmission spectroscopy simulated for future space-based observatories. Astrophys J 898:L33, https://doi.org/10.3847/2041-8213/aba4a1

Pidhorodetska D, Moran SE, Schwieterman EW, et al. (2021) L 98-59: A benchmark system of small planets for future atmospheric characterization. Astron J 162:169, https://doi.org/10.3847/1538-3881/ac1171

Pierrehumbert RT, Hammond M (2019) Atmospheric circulation of tide-locked exoplanets. Annu Rev Fluid Mech 51:275–303, https://doi.org/10.1146/annurev-fluid-010518-040516

Pilcher CB (2003) Biosignatures of early Earths. Astrobiology 3:471–486, https://doi.org/10.1089/153110703322610582

Pohorille A, Sokolowska J (2020) Evaluating biosignatures for life detection. Astrobiology 20:1236–1250, https://doi.org/10.1089/ast.2019.2151

Prather MJ, Hsu J, DeLuca NM, et al. (2015) Measuring and modeling the lifetime of nitrous oxide including its variability. J Geophys Res Atmos 120:5693–5705, https://doi.org/10.1002/2015JD023267

Prosser JI, Nicol GW (2012) Archaeal and bacterial ammonia-oxidisers in soil: the quest for niche specialisation and differentiation. Trends Microbiol 20:523–531, https://doi.org/10.1016/j.tim.2012.08.001

Quanz SP, Ottiger M, Fontanet E, et al. (2022) Large Interferometer For Exoplanets (LIFE) I. Improved exoplanet detection yield estimates for a large mid-infrared space-interferometer mission. Astron Astrophys 664:A21, https://doi.org/10.1051/0004-6361/202140366

Ragsdale SW (2004) Life with carbon monoxide. Crit Rev Biochem Mol Biol 39:165–195, https://doi.org/10.1080/10409230490496577

Ramirez RM, Kaltenegger L (2014) The habitable zones of pre-main-sequence stars. Astrophys J 797:L25, https://doi.org/10.1088/2041-8205/797/2/L25

Ranjan S, Schwieterman EW, Harman C, et al. (2020) Photochemistry of anoxic abiotic habitable planet atmospheres: Impact of new H_2O cross sections. Astrophys J 896:148, https://doi.org/10.3847/1538-4357/ab9363

Ranjan S, Seager S, Zhan Z, et al. (2022) Photochemical runaway in exoplanet atmospheres: Implications for biosignatures. Astrophys J 930:131, https://doi.org/10.3847/1538-4357/ac5749

Ranjan S, Schwieterman EW, Leung M, et al. (2023) The importance of the upper atmosphere to CO/O_2 runaway on habitable planets orbiting low-mass stars. Astrophys J Lett 958:L15, https://doi.org/10.3847/2041-8213/ad037c

Rasmussen RA, Khalil MAK (1981) Atmospheric methane (CH_4): Trends and seasonal cycles. J Geophys Res 86:9826, https://doi.org/10.1029/JC086iC10p09826

Rauer H, Gebauer S, Paris P V., et al. (2011) Potential biosignatures in super-Earth atmospheres. Astron Astrophys 529:A8, https://doi.org/10.1051/0004-6361/201014368

Raulin F, Toupance G (1975) Formation of prebiochemical compounds in models of the primitive Earth's atmosphere. Orig Life 6:91–97, https://doi.org/10.1007/BF01372393

Reid IN, Hawley SL (2005) New Light on Dark Stars. Springer Berlin Heidelberg, Berlin, Heidelberg

Reinhard CT, Olson SL, Schwieterman EW, Lyons TW (2017) False negatives for remote life detection on ocean-bearing planets: Lessons from the early Earth. Astrobiology 17:287–297, https://doi.org/10.1089/ast.2016.1598

Reinhard CT, Schwieterman EW, Olson SL, et al. (2019) The remote detectability of Earth's biosphere through time and the importance of UV capability for characterizing habitable exoplanets. Bull Am Astron Soc 51:526

Richard C, Gordon IE, Rothman LS, et al. (2012) New section of the HITRAN database: Collision-induced absorption (CIA). J Quant Spectrosc Radiat Transfer 113:1276–1285, https://doi.org/10.1016/j.jqsrt.2011.11.004

Roberson AL, Roadt J, Halevy I, Kasting JF (2011) Greenhouse warming by nitrous oxide and methane in the Proterozoic Eon. Geobiology 9:313–20, https://doi.org/10.1111/j.1472-4669.2011.00286.x

Robinson TD, Meadows VS, Crisp D, et al. (2011) Earth as an extrasolar planet: Earth model validation using EPOXI Earth observations. Astrobiology 11:393–408, https://doi.org/10.1089/ast.2011.0642

Robinson TD, Ennico K, Meadows VS, et al. (2014) Detection of ocean glint and ozone absorption using LCROSS Earth observations. Astrophys J 787:171, https://doi.org/10.1088/0004-637X/787/2/171

Roels J, Huyghe G, Verstraete W (2005) Microbially mediated phosphine emission. Sci Total Environ 338:253–265, https://doi.org/10.1016/j.scitotenv.2004.07.016

Rugheimer S, Kaltenegger L, Zsom A, et al. (2013) Spectral fingerprints of Earth-like planets around FGK stars. Astrobiology 13:251–69, https://doi.org/10.1089/ast.2012.0888

Rugheimer S, Kaltenegger L (2018) Spectra of Earth-like planets through geological evolution around FGKM stars. Astrophys J 854:19, https://doi.org/10.3847/1538-4357/aaa47a

Rugheimer S, Kaltenegger L, Segura A, et al. (2015) Effect of UV radiation on the spectral fingerprints of earth-like planets orbiting M stars. Astrophys J 809:57, https://doi.org/10.1088/0004-637X/809/1/57

Ryan DJ, Robinson TD (2022) Detecting oceans on exoplanets with phase-dependent spectral principal component analysis. Planet Sci J 3:33, https://doi.org/10.3847/PSJ/ac4af3

Sagan C, Thompson WR, Carlson R, et al. (1993) A search for life on Earth from the Galileo spacecraft. Nature 365:715–21, https://doi.org/10.1038/365715a0

Saini HS, Attieh JM, Hanson AD (1995) Biosynthesis of halomethanes and methanethiol by higher-plants via a novel methyltransferase reaction. Plant Cell Environ 18:1027–1033, https://doi.org/10.1111/j.1365-3040.1995.tb00613.x

Salvador A, Massol H, Davaille A, et al. (2017) The relative influence of H_2O and CO_2 on the primitive surface conditions and evolution of rocky planets. J Geophys Res E: Planets 122:1458–1486, https://doi.org/10.1002/2017JE005286

Samarkin VA, Madigan MT, Bowles MW, et al. (2010) Abiotic nitrous oxide emission from the hypersaline Don Juan Pond in Antarctica. Nat Geosci 3:341–344, https://doi.org/10.1038/ngeo847

Sanromá E, Pallé E, García Munõz A (2013) On the effects of the evolution of microbial mats and land plants on the Earth as a planet. Photometric and spectroscopic light curves of paleo-Earths. Astrophys J 766:133, https://doi.org/10.1088/0004-637X/766/2/133

Sanromá E, Pallé E, Parenteau MN, et al. (2014) Characterizing the purple Earth: Modeling the globally integrated spectral variability of the Archean Earth. Astrophys J 780:52, https://doi.org/10.1088/0004-637X/780/1/52

Schaefer L, Fegley B (2010) Chemistry of atmospheres formed during accretion of the Earth and other terrestrial planets. Icarus 208:438–448, https://doi.org/10.1016/j.icarus.2010.01.026

Schaefer L, Wordsworth RD, Berta-Thompson Z, Sasselov D (2016) Predictions of the atmospheric composition of GJ 1132b. Astrophys J 829:63, https://doi.org/10.3847/0004-637X/829/2/63

Schumann U, Huntrieser H (2007) The global lightning-induced nitrogen oxides source. Atmos Chem Phys Discuss 7:2623–2818, https://doi.org/10.5194/acpd-7-2623-2007

Schwieterman EW (2018) Surface and temporal biosignatures. *In:*Deeg H, Belmont J (eds) Handbook of Exoplanets. Springer International Publishing, Cham, p 1–29

Schwieterman E (2021) Developing a Guidebook to Search for Life Beyond Earth. Scientia, https://doi.org/10.33548/SCIENTIA702

Schwieterman EW, Cockell CS, Meadows VS (2015a) Nonphotosynthetic pigments as potential biosignatures. Astrobiology 15:341–361, https://doi.org/10.1089/ast.2014.1178

Schwieterman EW, Robinson TD, Meadows VS, et al. (2015b) Detecting and constraining N_2 abundances in planetary atmospheres using collisional pairs. Astrophys J 810:57, https://doi.org/10.1088/0004-637X/810/1/57

Schwieterman EW, Meadows VS, Domagal-Goldman SD, et al. (2016) Identifying planetary biosignature impostors: Spectral features of CO and O_4 resulting from abiotic O_2/O_3 production. Astrophys J 819:L13, https://doi.org/10.3847/2041-8205/819/1/L13

Schwieterman EW, Kiang NY, Parenteau MN, et al. (2018a) Exoplanet biosignatures: A review of remotely detectable signs of life. Astrobiology 18:663–708, https://doi.org/10.1089/ast.2017.1729

Schwieterman EW, Reinhard C, Olson S, Lyons T (2018b) The importance of UV capabilities for identifying inhabited exoplanets with next generation space telescopes. arXiv preprint 180102744, https://doi.org/10.48550/arXiv.1801.02744

Schwieterman EW, Reinhard CT, Olson SL, et al. (2019) Rethinking CO antibiosignatures in the search for life beyond the Solar System. Astrophys J 874:9, https://doi.org/10.3847/1538-4357/ab05e1

Schwieterman EW, Olson SL, Pidhorodetska D, et al. (2022) Evaluating the plausible range of N₂O biosignatures on exo-Earths: An integrated biogeochemical, photochemical, and spectral modeling approach. Astrophys J 937:109, https://doi.org/10.3847/1538-4357/ac8cfb

Seager S, Bains W (2015) The search for signs of life on exoplanets at the interface of chemistry and planetary science. Sci Adv 1:e1500047, https://doi.org/10.1126/sciadv.1500047

Seager S, Turner EL, Schafer J, Ford EB (2005) Vegetation's red edge: A possible spectroscopic biosignature of extraterrestrial plants. Astrobiology 5:372–390, https://doi.org/10.1089/ast.2005.5.372

Seager S, Schrenk M, Bains W (2012) An astrophysical view of Earth-based metabolic biosignature gases. Astrobiology 12:61–82, https://doi.org/10.1089/ast.2010.0489

Seager S, Bains W, Hu R (2013a) Biosignature gases in H₂-dominated atmospheres on rocky exoplanets. Astrophys J 777:95, https://doi.org/10.1088/0004-637X/777/2/95

Seager S, Bains W, Hu R (2013b) A biomass-based model to estimate the plausibility of exoplanet biosignature gases. Astrophys J 775:104, https://doi.org/10.1088/0004-637X/775/2/104

Seager S, Bains W, Petkowski JJ (2016) Toward a list of molecules as potential biosignature gases for the search for life on exoplanets and applications to terrestrial biochemistry. Astrobiology 16:465–485, https://doi.org/10.1089/ast.2015.1404

Segura A, Krelove K, Kasting JF, et al. (2003) Ozone concentrations and ultraviolet fluxes on Earth-like planets around other stars. Astrobiology 3:689–708, https://doi.org/10.1089/153110703322736024

Segura AA, Kasting JF, Meadows V, et al. (2005) Biosignatures from Earth-like planets around M dwarfs. Astrobiology 5:706–725, https://doi.org/10.1089/ast.2005.5.706

Seinfeld JH, Pandis SN (2016) Atmospheric Chemistry and Physics: From Air Pollution to Climate Change. John Wiley & Sons

Sephus CD, Fer E, Garcia AK, et al. (2022) Earliest photic zone niches probed by ancestral microbial rhodopsins. Mol Biol Evol 39:1–16, https://doi.org/10.1093/molbev/msac100

Sharpe SW, Johnson TJ, Sams RL, et al. (2004) Gas-phase databases for quantitative infrared spectroscopy. Appl Spectrosc 58:1452–1461, https://doi.org/10.1366/0003702042641281

Shields AL (2019) The climates of other worlds: A review of the emerging field of exoplanet climatology. Astrophys J Suppl Ser 243:30, https://doi.org/10.3847/1538-4365/ab2fe7

Sholes SF, Krissansen-Totton J, Catling DC (2019) A maximum subsurface biomass on mars from untapped free energy: CO and H₂ as potential antibiosignatures. Astrobiology 19:655–668, https://doi.org/10.1089/ast.2018.1835

Smith HB, Mathis C (2023) Life detection in a universe of false positives. BioEssays 45:1–15, https://doi.org/10.1002/bies.202300050

Snellen IAG, de Kok RJ, le Poole R, et al. (2013) Finding extraterrestrial life using ground-based high-dispersion spectroscopy. Astrophys J 764:182, https://doi.org/10.1088/0004-637X/764/2/182

Solé R V, Munteanu A (2004) The large-scale organization of chemical reaction networks in astrophysics. Europhys Lett 68:170–176, https://doi.org/10.1209/epl/i2004-10241-3

Solomon S (1999) Stratospheric ozone depletion: A review of concepts and history. Rev Geophys 37:275–316, https://doi.org/10.1029/1999RG900008

Solomon S, Qin D, Manning M, Al E (2007) Contribution of Working Group I to the Fourth Assessment Report of the Intergovernmental Panel on Climate Change. Cambridge University Press, Cambridge, United Kingdom, https://archive.ipcc.ch/publications_and_data/ar4/wg1/en/contents.html

Sousa-Silva C, Al-Refaie AF, Tennyson J, Yurchenko SN (2015) ExoMol line lists–VII. The rotation–vibration spectrum of phosphine up to 1500 K. Mon Not R Astron Soc 446:2337–2347, https://doi.org/10.1093/mnras/stu2246

Sousa-Silva C, Seager S, Ranjan S, et al. (2020) Phosphine as a biosignature gas in exoplanet atmospheres. Astrobiology 20:235–268, https://doi.org/10.1089/ast.2018.1954

Sparks WB, Hough J, Germer TA, et al. (2009a) Detection of circular polarization in light scattered from photosynthetic microbes. PNAS 106:7816–7821, https://doi.org/10.1073/pnas.0810215106

Sparks WB, Hough JH, Kolokolova L, et al. (2009b) Circular polarization in scattered light as a possible biomarker. J Quant Spectrosc Radiat Transfer 110:1771–1779, https://doi.org/10.1016/j.jqsrt.2009.02.028

Sparks W, Hough JH, Germer TA, et al. (2012) Remote sensing of chiral signatures on Mars. Planet Space Sci 72:111–115, https://doi.org/10.1016/j.pss.2012.08.010

Sparks WB, Parenteau MN, Blankenship RE, et al. (2021) Spectropolarimetry of primitive phototrophs as global surface biosignatures. Astrobiology 21:219–234, https://doi.org/10.1089/ast.2020.2272

Stanton CL, Reinhard CT, Kasting JF, et al. (2018) Nitrous oxide from chemodenitrification: A possible missing link in the Proterozoic greenhouse and the evolution of aerobic respiration. Geobiology 1–13, https://doi.org/10.1111/gbi.12311

Stefels J, Steinke M, Turner S, et al. (2007) Environmental constraints on the production and removal of the climatically active gas dimethylsulphide (DMS) and implications for ecosystem modelling. Biogeochemistry 83:245–275, https://doi.org/10.1007/978-1-4020-6214-8_18

Sterzik MF, Bagnulo S, Palle E (2012) Biosignatures as revealed by spectropolarimetry of Earthshine. Nature 483:64–66, https://doi.org/10.1038/nature10778

Sterzik MF, Bagnulo S, Emde C (2014) Bio-signatures of planet Earth from spectropolarimetry. Proc Int Astron Union 10:305–312, https://doi.org/10.1017/S1743921315004962

Sterzik MF, Bagnulo S, Stam DM, et al. (2019) Spectral and temporal variability of Earth observed in polarization. Astron Astrophys 622:A41, https://doi.org/10.1051/0004-6361/201834213

Stüeken EE, Olson SL, Moore E, Foley BJ (2024) The early Earth as an analogue for exoplanetary biogeochemistry. Rev Mineral Geochem 90:515–558

Sun Y, Frankenberg C, Wood JD, et al. (2017) OCO-2 advances photosynthesis observation from space via solar-induced chlorophyll fluorescence. Science 358:eaam5747, https://doi.org/10.1126/science.aam5747

Syverson DD, Reinhard CT, Isson TT, et al. (2021) Nutrient supply to planetary biospheres from anoxic weathering of mafic oceanic crust. Geophys Res Lett 48:1–8, https://doi.org/10.1029/2021GL094442

Tabataba-Vakili F, Grenfell JL, Grießmeier J-M, Rauer H (2016) Atmospheric effects of stellar cosmic rays on Earth-like exoplanets orbiting M-dwarfs. Astron Astrophys 585:A96, https://doi.org/10.1051/0004-6361/201425602

Tarter J (2001) The Search for Extraterrestrial Intelligence (SETI). Annu Rev Astron Astrophys 39:511–548, https://doi.org/10.1146/annurev.astro.39.1.511

Tarter JC (2006) The evolution of life in the Universe: are we alone? Proc Int Astron Union 2:14–29, https://doi.org/10.1017/S1743921307009829

Thamdrup B (2012) New pathways and processes in the global nitrogen cycle. Annu Rev Ecol Evol Syst 43:407–428, https://doi.org/10.1146/annurev-ecolsys-102710-145048

Thauer RK, Kaster A-K, Seedorf H, et al. (2008) Methanogenic archaea: ecologically relevant differences in energy conservation. Nat Rev Microbiol 6:579–591, https://doi.org/10.1038/nrmicro1931

Thayer JS (2002) Review: Biological methylation of less-studied elements. Appl Organomet Chem 16:677–691, https://doi.org/10.1002/aoc.375

The LUVOIR Team (2019) The LUVOIR mission concept study final report. arXiv preprint 191206219

Thompson MA, Krissansen-Totton J, Wogan N, et al. (2022) The case and context for atmospheric methane as an exoplanet biosignature. PNAS 119:1–15, https://doi.org/10.1073/pnas.2117933119

Tian F (2015) History of water loss and atmospheric O_2 buildup on rocky exoplanets near M dwarfs. Earth Planet Sci Lett 432:126–132, https://doi.org/10.1016/j.epsl.2015.09.051

Tian F, France K, Linsky JL, et al. (2014) High stellar FUV/NUV ratio and oxygen contents in the atmospheres of potentially habitable planets. Earth Planet Sci Lett 385:22–27, https://doi.org/10.1016/j.epsl.2013.10.024

Tian H, Chen G, Lu C, et al. (2015) Global methane and nitrous oxide emissions from terrestrial ecosystems due to multiple environmental changes. Ecosyst Heal Sustain 1:1–20, https://doi.org/10.1890/EHS14-0015.1

Tian H, Xu R, Canadell JG, et al. (2020) A comprehensive quantification of global nitrous oxide sources and sinks. Nature 586:248–256, https://doi.org/10.1038/s41586-020-2780-0

Tilley MA, Segura A, Meadows V, et al. (2019) Modeling repeated M dwarf flaring at an Earth-like planet in the habitable zone: Atmospheric effects for an unmagnetized planet. Astrobiology 19:64–86, https://doi.org/10.1089/ast.2017.1794

Timofeev IUM, Tonkov MV (1979) Effect of the induced oxygen absorption band on the transformation of radiation in the 6-micron region in the earth's atmosphere. Academy of Sciences, USSR, Izvestiya, Atmospheric and Oceanic Physics 14:437–441

Tinetti G, Rashby S, Yung YL (2006) Detectability of red-edge-shifted vegetation on terrestrial planets orbiting M stars. Astrophys J 644:L129–L132, https://doi.org/10.1086/505746

Trainer MG, Pavlov AA, DeWitt HL, et al. (2006) Organic haze on Titan and the early Earth. PNAS 103:18035–18042, https://doi.org/10.1073/pnas.0608561103

Tremblay L, Line MR, Stevenson K, et al. (2020) The detectability and constraints of biosignature gases in the near- and mid-infrared from transit transmission spectroscopy. Astron J 159:117, https://doi.org/10.3847/1538-3881/ab64dd

Truong N, Lunine JI (2021) Volcanically extruded phosphides as an abiotic source of Venusian phosphine. PNAS 118:1–5, https://doi.org/10.1073/pnas.2021689118

Tsai SM, Lee EKH, Powell D, et al. (2023) Photochemically produced SO_2 in the atmosphere of WASP-39b. Nature 617:483–487, https://doi.org/10.1038/s41586-023-05902-2

Tucker CJ (1979) Red and photographic infrared linear combinations for monitoring vegetation. Remote Sens Environ 8:127–150, https://doi.org/10.1016/0034-4257(79)90013-0

Tucker CJ, Pinzon JE, Brown ME, et al. (2005) An extended AVHRR 8-km NDVI dataset compatible with MODIS and SPOT vegetation NDVI data. Int J Remote Sens 26:4485–4498, https://doi.org/10.1080/01431160500168686

Ueno Y, Yamada K, Yoshida N, et al. (2006) Evidence from fluid inclusions for microbial methanogenesis in the early Archaean era. Nature 440:516–519, https://doi.org/10.1038/nature04584

Villanueva G, Cordiner M, Irwin P, et al. (2021) No evidence of phosphine in the atmosphere of Venus from independent analyses. Nat Astron 5:631–635, https://doi.org/ 10.1038/s41550-021-01422-z

Visscher PT, Quist P, Vangemerden H (1991) Methylated sulfur-compounds in microbial mats—In situ concentrations and metabolism by a colorless sulfur bacterium. Appl Environ Microbiol 57:1758–1763

Visscher PT, Baumgartner LK, Buckley DH, et al. (2003) Dimethyl sulphide and methanethiol formation in microbial mats: potential pathways for biogenic signatures. Environ Microbiol 5:296–308

Visscher C, Lodders K, Fegley, Jr. B (2006) Atmospheric chemistry in giant planets, brown dwarfs, and low-mass dwarf stars. II. Sulfur and phosphorus. Astrophys J 648:1181–1195, https://doi.org/10.1086/506245

Walker SI, Bains W, Cronin L, et al. (2018) Exoplanet biosignatures: Future directions. Astrobiology 18:779–824, https://doi.org/10.1089/ast.2017.1738

Walker SI, Cronin L, Drew A, et al. (2020) Probabilistic biosignature frameworks. *In:* Meadows VS, Arney GN, Schmidt BE, Des Marais DJ (eds) Planetary Astrobiology. University of Arizona Press, p 477–504

Wang Y, Tian F, Li T, Hu Y (2016) On the detection of carbon monoxide as an anti-biosignature in exoplanetary atmospheres. Icarus 266:15–23, https://doi.org/10.1016/j.icarus.2015.11.010

Watanabe Y, Ozaki K (2024) Relative abundances of CO_2, CO, and CH_4 in atmospheres of Earth-like lifeless planets. Astrophys J 961:1, https://doi.org/10.3847/1538-4357/ad10a2

Williams DM, Gaidos E (2008) Detecting the glint of starlight on the oceans of distant planets. Icarus 195:927–937, https://doi.org/10.1016/j.icarus.2008.01.002

Wogan NF, Catling DC (2020) When is chemical disequilibrium in earth-like planetary atmospheres a biosignature versus an anti-biosignature? Disequilibria from dead to living worlds. Astrophys J 892:127, https://doi.org/10.3847/1538-4357/ab7b81

Wogan NF, Catling DC, Zahnle KJ, Claire MW (2022) Rapid timescale for an oxic transition during the Great Oxidation Event and the instability of low atmospheric O_2. PNAS 119:2–9, https://doi.org/10.1073/pnas.2205618119

Wong ML, Prabhu A, Williams J, et al. (2023) Toward network-based planetary biosignatures: Atmospheric chemistry as unipartite, unweighted, undirected networks. J Geophys Res E: Planets 128:e2022JE007658, https://doi.org/10.1029/2022je007658

Wordsworth R, Kreidberg L (2022) Atmospheres of rocky exoplanets. Annu Rev Astron Astrophys 60:159–201, https://doi.org/10.1146/annurev-astro-052920-125632

Wordsworth R, Pierrehumbert R (2014) Abiotic oxygen-dominated atmospheres on terrestrial habitable zone planets. Astrophys J 785:L20, https://doi.org/10.1088/2041-8205/785/2/L20

Wordsworth RD, Schaefer LK, Fischer RA (2018) Redox evolution via gravitational differentiation on low-mass planets: Implications for abiotic oxygen, water loss, and habitability. Astron J 155:195, https://doi.org/10.3847/1538-3881/aab608

Wright JT, Haqq-Misra J, Frank A, et al. (2022) The case for technosignatures: Why they may be abundant, long-lived, highly detectable, and unambiguous. Astrophys J Lett 927:L30, https://doi.org/10.3847/2041-8213/ac5824

Wunderlich F, Godolt M, Grenfell JL, et al. (2019) Detectability of atmospheric features of Earth-like planets in the habitable zone around M dwarfs. Astron Astrophys 624:A49, https://doi.org/10.1051/0004-6361/201834504

Wunderlich F, Scheucher M, Godolt M, et al. (2020) Distinguishing between wet and dry atmospheres of TRAPPIST-1 e and f. Astrophys J 901:126, https://doi.org/10.3847/1538-4357/aba59c

Wunderlich F, Scheucher M, Grenfell JL, et al. (2021) Detectability of biosignatures on LHS 1140 b. Astron Astrophys 647:A48, https://doi.org/10.1051/0004-6361/202039663

Yang X, Cox RA, Warwick NJ, et al. (2005) Tropospheric bromine chemistry and its impacts on ozone: A model study. J Geophys Res Atmos 110:1–18, https://doi.org/10.1029/2005JD006244

Yang Z, Fang W, Lu X, et al. (2016) Warming increases methylmercury production in an Arctic soil. Environ Pollut 214:504–509, https://doi.org/10.1016/j.envpol.2016.04.069

Young A V, Robinson TD, Krissansen-Totton J, et al. (2024) Inferring chemical disequilibrium biosignatures for Proterozoic Earth-like exoplanets. Nat Astron 8:101–110, https://doi.org/10.1038/s41550-023-02145-z

Zahnle KJ, Kasting JF, Pollack JB (1988) Evolution of a steam atmosphere during earth's accretion. Icarus 74:62–97, https://doi.org/10.1016/0019-1035(88)90031-0

Zerkle AL, Claire MW, Domagal-Goldman SD, et al. (2012) A bistable organic-rich atmosphere on the Neoarchaean Earth. Nat Geosci 5:359–363, https://doi.org/10.1038/ngeo1425

Zhan Z, Seager S, Petkowski JJ, et al. (2021) Assessment of isoprene as a possible biosignature gas in exoplanets with anoxic atmospheres. Astrobiology 21:765–792, https://doi.org/10.1089/ast.2019.2146

Zhan Z, Huang J, Seager S, et al. (2022) Organic carbonyls are poor biosignature gases in exoplanet atmospheres but may generate significant CO. Astrophys J 930:133, https://doi.org/10.3847/1538-4357/ac64a8

Zhang M, Park K-T, Yan J, et al. (2020) Atmospheric dimethyl sulfide and its significant influence on the sea-to-air flux calculation over the Southern Ocean. Prog Oceanogr 186:102392, https://doi.org/10.1016/j.pocean.2020.102392

Zhu R, Glindemann D, Kong D, et al. (2007) Phosphine in the marine atmosphere along a hemispheric course from China to Antarctica. Atmos Environ 41:1567–1573, https://doi.org/10.1016/j.atmosenv.2006.10.035

Zugger ME, Kasting JF, Williams DM, et al. (2010) Light scattering from exoplanet oceans and atmospheres. Astrophys J 723:1168–1179, https://doi.org/10.1088/0004-637X/723/2/1168

Reviews in Mineralogy & Geochemistry
Vol. 90 pp. 515–558, 2024
Copyright © Mineralogical Society of America

14

The Early Earth as an Analogue for Exoplanetary Biogeochemistry

Eva E. Stüeken[1,*], Stephanie L. Olson[2], Eli Moore[3], Bradford J. Foley[4]

[1]*School of Earth & Environmental Sciences, University of St Andrews,*
St Andrews, Fife, KY16 9TS, United Kingdom
[2]*Department of Earth, Atmospheric, and Planetary Science, Purdue University,*
West Lafayette, IN 47906, United States
[3]*U.S. Geological Survey, Geology, Energy & Minerals Science Center,*
Reston, VA 20192, United States
[4]*Department of Geosciences, The Pennsylvania State University,*
University Park, PA 16802, United States

* ees4@st-andrews.ac.uk

INTRODUCTION: WHY THE EARLY EARTH?

Planet Earth has evolved over the past 4.5 billion years from an entirely anoxic planet with possibly a different tectonic regime to the oxygenated world with horizontal plate tectonics that we know today. For most of this time, Earth has been inhabited by a purely microbial biosphere albeit with seemingly increasing complexity over time. A rich record of this geobiological evolution over most of Earth's history thus provides insights into the remote detectability of microbial life under a variety of planetary conditions. Here we leverage Earth's geobiological record with the aim of (a) illustrating the current state of knowledge and key knowledge gaps about the early Earth as a reference point in exoplanet science research; (b) compiling biotic and abiotic mechanisms that controlled the evolution of the atmosphere over time; and (c) reviewing current constraints on the detectability of Earth's early biosphere with state-of-the-art telescope technology. We highlight that life may have originated on a planet with a different (stagnant lid) tectonic regime and strong hydrothermal activity, and under these conditions, biogenic CH_4 gas was perhaps the most detectable atmospheric biosignature. Oxygenic photosynthesis, which is responsible for essentially all O_2 gas in the modern atmosphere, appears to have emerged concurrently with the establishment of modern plate tectonics and the emergence of continental crust, but O_2 accumulation to modern levels only occurred late in Earth's history, perhaps tied to the rise of land plants. Nutrient limitation in anoxic oceans, promoted by hydrothermal Fe fluxes, may have limited biological productivity and O_2 production. N_2O is an alternative biosignature that was perhaps significant on the redox-stratified Proterozoic Earth. We conclude that the detectability of atmospheric biosignatures on Earth was not only dependent on biological evolution but also strongly controlled by the evolving tectonic context.

Earth is currently the only known planet with a biosphere and therefore an important reference point in our search for life on other planets. The ability of Earth's modern biosphere to generate biosignatures that can be detected remotely with a space telescope was famously documented by the Galileo spacecraft in the early 1990s (Sagan et al. 1993), and this concept continues to underpin the field of Astrobiology to this day. However, it is also widely recognized that the modern biosphere is highly advanced and not necessarily a good analogue for life on other worlds. Earth's atmosphere today contains 21% O_2 gas that is generated by cyanobacteria and plants, where the latter contain chloroplasts derived from cyanobacterial ancestors. The origin of cyanobacteria may go back to a singular event in biological evolution

1529-6466/24/0090-0014$05.00 (print)
1943-2666/24/0090-0014$05.00 (online)   http://dx.doi.org/10.2138/rmg.2024.90.14

(reviewed by Sánchez-Baracaldo and Cardona 2020). Essentially all complex life today depends on O_2 gas as an electron acceptor in metabolic energy transformation (see below), but it remains unknown if (a) an O_2-producing metabolism such as oxygenic photosynthesis would evolve independently on another planet, and if (b) O_2-producing organisms on other planets (if they ever arose) would modify surface environments to a similar extent as they have on Earth, where they enabled the rise of oxygen-breathing macrofauna (Margulis and Lovelock 1974; Margulis et al. 1976). For these reasons it is conceivable, and perhaps likely, that extra-terrestrial life could be entirely microbial and largely anaerobic (Cavicchioli 2002). The search for life on other planets needs to account for this possibility and consider the effects of anoxia on nutrient availability, biological productivity, and biosignature production.

Astrobiologists over the past two decades have therefore begun to use the early Earth, prior to the rise of complex aerobic organisms, as an analogue for inhabited exoplanets (e.g., Des Marais et al. 2002; Benner et al. 2004; Kaltenegger 2017; Catling et al. 2018; Schwieterman et al. 2018; Van Kranendonk et al. 2021). Using geological and geochemical techniques, it is possible to "see through" limited availability and alteration of the ancient rock record to reconstruct environmental conditions and ecosystems in the distant past (further discussed below). For example, phylogeny and knowledge of microbial metabolisms under anoxic conditions based on modern analogues and cell cultures (see below) can provide insights into the types of biosignature gases that were likely dominating Earth's surface environments billions of years ago. These biosignatures need to be evaluated against a backdrop of tectonic and volcanic processes at that time (see below). Importantly, the modern atmosphere is strongly shaped by biological metabolisms, which have significantly sped up the production and recycling of many gaseous species (e.g., Lenton 1998). This strong biological control progressively co-evolved with abiotic planetary processes over the past four billion years. In the following, we will first describe the fundamental principles that guide this line of research. We will show how these geological and biological tools have been applied to build a knowledge base of early Earth and its biosphere. Emphasis will be placed on biological and geological processes that generate remotely detectable gases or that inhibit their production. Finally, we will leverage these records to present quantitative estimates of gas fluxes through time that can place helpful constraints in our search for extra-terrestrial microbial biospheres.

TOOLS AND PRINCIPLES OF PALEO-BIOGEOCHEMISTRY

Deciphering biosignatures from the rock record

Reconstructing the history of life and environmental conditions on the early Earth relies on a combination of geological mapping, geochemical analyses, laboratory experiments, and computational models (e.g., Oró et al. 1990). These four methodological pillars are linked to the quality of the rock record, which determines the amount and reliability of information that is preserved over billions of years and ultimately constrains boundary conditions for experiments, models, and our understanding of the evolution of life on Earth. It is therefore crucial to appraise the limitations that the rock record imposes and the tools that are available to read this biogeochemical archive.

There are broadly three categories of limitations that can hinder or prevent accurate assessments of ecosystems and habitats in deep time: diminishing preservation of rocks with increasing age; preservation bias towards a small subset of paleoenvironmental settings; and alteration of geochemical tracers by secondary processes such as metamorphism or infiltration of younger fluids. Approximately 70% of the modern Earth's surface is covered by oceanic crust that is less than 200 million years old and constantly recycled through plate tectonics (Veizer and Mackenzie 2014 and references therein). Hence with the exception of a few rare ophiolites (i.e., slivers of oceanic crust that have been obducted onto continents), we have essentially no record of the abyssal deep ocean that is older than the Mesozoic (Peters and Husson 2017). Older rocks and sediments are only preserved on continents, where they are subject to erosion, weathering and/or amalgamation during tectonic collisions. Due to these large-scale processes, rocks of Precambrian age are exceedingly rare (reviewed by Van Kranendonk et al. 2018). None of them exceed 4 Ga; only a few grains of zircon ($ZrSiO_4$) have survived from the Hadean eon (Harrison 2009), leaving more than 500 million years of Earth's history nearly unsampled. Furthermore, since the onset of modern-style plate tectonics in the late Archean (see below), marine sediments deposited on continental crust (i.e., on the inner or outer shelf) are preserved preferentially, because they are least likely to undergo erosion and subduction (Peters and Husson 2017). Early Archean greenstone successions formed under a different tectonic regime may be an exception to this rule. Apart from this exception, many inferences that we can draw about the early biosphere may thus be biased towards processes operating in less than a few hundred meters of water depth.

Another caveat in studying the ancient rock record are the effects of diagenesis, thermal maturation, and metamorphism, which arise from the increase in pressure and temperature in rocks and sediments that occur after initial microbial recycling and lithification as a consequence of continuous sedimentary burial, tectonism, and magmatic heating. Organic matter deposited in sediments first undergoes diagenesis (low-temperature degradation within unmetamorphosed sediments) and ultimatelymetamorphic heating when it progressively loses functional groups, i.e., it becomes depleted in elements other than carbon (O, H, N, P, etc.). Diagenesis quickly degrades DNA and proteins over the course of hours to months, with a few exceptions of retention for several years, and hence these molecules are not preserved over long geological time scales in sedimentary strata (e.g., Nielsen et al. 2007; Moore et al. 2012, 2014; Parducci et al. 2017; Capo et al. 2022). Following diagenesis, most remaining biological macromolecules are destroyed by metamorphic processes. Lipids are the most recalcitrant components of dead cells and can in some instances be preserved over million- to billion-year timescales as "molecular fossils" (Love and Zumberge 2021). However, a large fraction of hydrocarbon molecules are typically released when buried sediments reach the "oil window" (70–150 °C) and beyond (Horsfield and Rullkotter 1994). Residual organic matter is converted into kerogen (solid organic matter that is insoluble in non-polar organic solvents) and ultimately, with further increase in pressure, into highly inert graphite (Hayes et al. 1983). Much of the original information content of biomolecules, such as genome sequences or protein content, is therefore missing from Precambrian sedimentary rocks, limiting the geochemical toolbox to techniques that are applicable to kerogenous and graphitic material or morphological features (Gaines et al. 2009). During progressive metamorphism, hydrous minerals such as clays, gypsum or opaline silica tend to lose water and slowly convert into water-poor or anhydrous phases (Stepanov 2021). The expelled fluids may facilitate ion-exchange reactions between solids. Export of such fluids can lead to alteration of bracketing strata, a process called metasomatism (e.g., White et al. 2014).

Morphological evidence in the form of stromatolites, MISS (microbially-induced sedimentary structures) and microfossils are among the most persistent records of life that can survive metamorphism and date back to through most of the sedimentary rock record (e.g., Walter et al. 1980; Allwood et al. 2006; Noffke et al. 2013; Sugitani et al. 2015); however, their information content with regards to specific metabolisms remains limited (further discussed below). Therefore, stable isotopic analysis of major and minor elements contained in ancient biomass and sedimentary minerals is one of the primary tools used in biogeochemical studies that can sometimes shed light on metabolic evolution despite some of the issues noted above. Many of the key elements that life uses and that are cycled through the ocean-atmosphere system have multiple stable isotopes with the same number of protons (and hence identical chemical properties) but different numbers of neutrons. The neutron number changes the mass and thus influences chemical bond strengths and reaction rates in biogeochemical processes. For example, carbon exists as ^{13}C and ^{12}C (ignoring the cosmogenic, short-lived ^{14}C isotope), and the $^{13}C/^{12}C$ ratio of biological organic matter is lower than that of carbonate rocks (Fig. 1), because organisms that perform autotrophic CO_2 fixation preferentially uptake the lighter carbon isotope. The fundamental reason for this isotopic fractionation is a slightly faster reaction rate for $^{12}CO_2$ compared to $^{13}CO_2$ as the gas interacts with the enzymatic machinery of the cell (O'Leary 1981). Although this particular isotopic signature of biomass may be altered by metamorphism, as organic matter converts to kerogen and graphite, it is not necessarily lost from the record even up to high metamorphic grade (Hayes et al. 1983; Schidlowski 2001). Complications, however, may arise if the original CO_2 is derived from a non-atmospheric source and/or if abiotic processes reduce CO_2 to organic matter with similar isotopic effects (see below).

Isotopic data are typically expressed in delta notation, such as $\delta^{13}C$ [‰] = [$(^{13}C/^{12}C)_{sample}$/$(^{13}C/^{12}C)_{VPDB}$ – 1] × 1000. Here, VPDB stands for Vienna Pee Dee Belemnite, an internationally agreed-upon reference standard that is used for normalizing data from different laboratories to a common denominator. When interpreting isotopic data, an important consideration is isotopic mass balance. As biomass accumulates carbon with a relatively low $\delta^{13}C$ value, due to preferential uptake of ^{12}C, the residual CO_2 in the atmosphere becomes depleted in ^{12}C or enriched in ^{13}C. This signature is inherited by carbonate rocks that form from dissolution of CO_2 gas in water (Fig. 1). The more biomass is produced, the stronger the ^{12}C-depletion in carbonate. Time-resolved analyses of $\delta^{13}C$ in carbonates through Earth's history therefore hold information about the evolution of biological activity (e.g., Kump and Arthur 1999; Krissansen-Totton et al. 2015; Planavsky et al. 2022). Similarly, stable isotopes of sulfur (preserved in sulfide minerals and in sulfate minerals) can inform us about the evolution of the sulfur cycle through time (Canfield and Farquhar 2009), which has important implications for other nutrients whose solubility is sensitive to sulfur chemistry (see below).

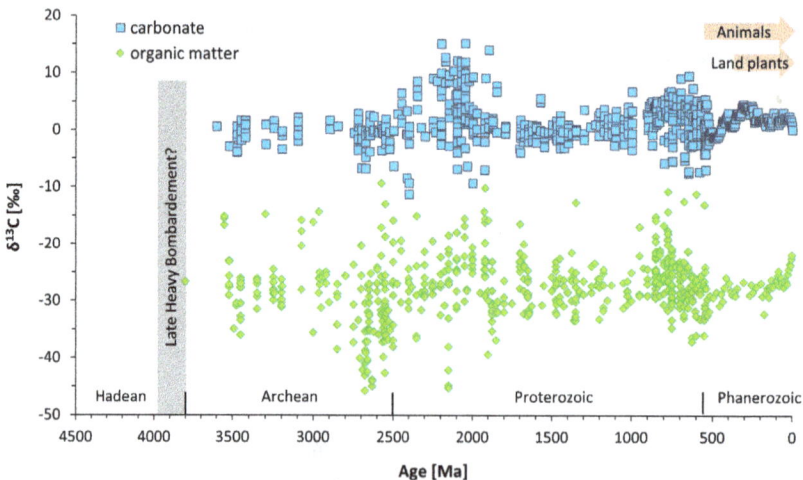

Figure 1. Inorganic (carbonate) and organic carbon isotopes versus age in millions of years, starting near the formation of Earth. Database taken from Krissansen-Totton et al. (2015), who presented each geological formation by its average composition. The oldest data point at 3.8 Ga is from the Isua Supracrustal Belt (see below) and also marks one of the oldest occurrences of meta-sedimentary rocks on Earth. No rocks are preserved from the Hadean Eon due to tectonic recycling and destruction by impacts.

In addition to stable isotopes of major and minor elements, the abundances of elements and their speciation in sedimentary rocks can also provide insights into ocean chemistry and nutrient availability. For example, the speciation of iron (i.e., the relative proportions of Fe bound to carbonates, oxides, sulfides, and silicates) has been developed as a means of reconstructing the redox state of ancient water columns locally, above the sediment–water interface of the sampling site (Poulton and Canfield 2005; Raiswell et al. 2019). A more global proxy for the redox state of the deep ocean and marine sediments is the abundance molybdenum in sediments from anoxic settings (Algeo and Lyons 2006; Scott and Lyons 2012; Little et al. 2015). The fundamental premise of this proxy is again based on mass balance. Molybdenum is well-mixed throughout the ocean today, but it is rapidly drawn down into sediments under anoxic conditions, where it reacts with reduced sulfur and becomes particle reactive (Helz et al. 2011). Hence the more expansive anoxia is across the world's oceans, the smaller the Mo reservoir in global average seawater and the lower the concentration of Mo at any given site (Anbar and Knoll 2002; Scott et al. 2008). It thus becomes possible to make predictions for the entire past ocean, even if only a few outcrops of rocks are preserved from a given time window (Reinhard et al. 2013). Other redox-sensitive elements such as U, As, V, or Cr may be used in similar ways (Tribovillard et al. 2006; Reinhard et al. 2013; Fru et al. 2019). Ultimately, the records of isotopic data, elemental speciation and abundances can be compiled and fed into computational models to derive quantitative estimates of biogeochemical cycles through deep time (see last section). In the following, we will first summarize the evidence preserved in the rock record and then converting those observations into quantitative estimates of biosignature gases.

Fundamentals of biochemistry

All life on Earth gains energy for growth and metabolism through exothermic electron transfer processes, and this fundamental principle has probably been true since life's origins (Moser et al. 1992). It is therefore conceivable that electrochemistry also forms the basis of extra-terrestrial biospheres. In order to perform electron transfer reactions, organisms must obtain electron donor and electron acceptor substrates from the environment (Moser et al. 1995). The combination of substrates fundamentally differs between chemotrophic and phototrophic organisms. Chemotrophs acquire both electron donors and acceptors from aqueous phases, gases, or minerals. For example, autotrophic methanogens (i.e., chemotrophs) use hydrogen (H_2) as the electron donor and CO_2 as the electron acceptor to gain energy and produce the waste products methane (CH_4) and H_2O (Fig. 2). The net reaction, which incorporates many intermediate biochemical processes, may be written as follows:

$$CO_2 + 4\,H_2 \rightarrow CH_4 + 2\,H_2O + E$$

The H_2 that is used in methanogenesis can come from geological sources, such as hydrothermal vents or from other organisms that produce H_2 during anaerobic degradation of organic matter (Kral et al. 1998). The energy gained from this reaction is then used to build complex organic molecules. In contrast, phototrophs use the light energy from the Sun to fix carbon and carry out other metabolic processes (Hohmann-Marriott and Blankenship 2011). The absorption of photons by light-harvesting pigments is optimized to the solar spectrum, meaning that it could differ if a similar light-dependent metabolism were to arise around other stars (Kiang et al. 2007). Oxygenic photosynthesis uses light energy to split water molecules and obtain electrons, which are then used to reduce the carbon in CO_2 from the atmosphere and synthesize simple sugars, such as glucose. Here, the net reaction is:

$$6\,CO_2 + 6\,H_2O + \text{Light Energy} \rightarrow C_6H_{12}O_6 + 6\,O_2$$

The simple sugar provides energy and fuels the synthesis of structural polymers such as cellulose, as well as all necessary cellular activities.

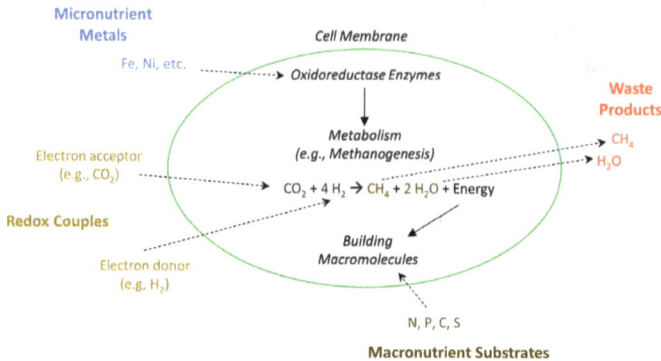

Figure 2. Schematic of a cell performing methanogenesis with energy and nutrients going in and wastes being excreted.

In aerobic respiration, a particular form of chemotrophy performed by aerobic microbes such as *E. coli* and large animals such as *H. sapiens*, organic matter is the electron donor that is consumed and oxygen (O_2) serves as the electron acceptor that is acquired in different ways by different types of organisms (inhalation, absorption, diffusion, etc.). Electrons are transferred from the organic matter (represented by the simple sugar glucose: $C_6H_{12}O_6$) to O_2 resulting in the release of energy (E) and the production of carbon dioxide (CO_2) and water (H_2O):

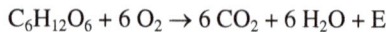

$$C_6H_{12}O_6 + 6\,O_2 \rightarrow 6\,CO_2 + 6\,H_2O + E$$

This is essentially the same process of burning wood in a fire, where the energy released from a fire is the same type of chemical energy released by the aerobic respiration of organic matter in unicellular or multicellular organisms. The CO_2 and H_2O released are the waste products of aerobic respiration. Aerobic respiration generates relatively large amounts of energy and is therefore used by all known macrofauna on Earth (Sperling et al. 2013). In anaerobic respiration, the oxidation of organic matter is coupled to electron acceptors other than O_2, such as ferric iron (Fe^{3+}), nitrate (NO_3^-), or sulfate (SO_4^{2-}), which is particularly relevant for anoxic worlds. However, anaerobic respiration yields generally less metabolic energy than aerobic respiration (Schoepp-Cothenet et al. 2012), which may limit the complexity of such organisms.

Besides thermodynamically favorable pairs of electron donors and acceptors, biological activity is also dependent on the availability of other nutrients that are required to build complex cell structures. There are six essential macronutrient elements (known as the CHNOPS elements) that make up 98% of living matter on Earth: carbon (C), hydrogen (H), nitrogen (N), oxygen (O), phosphorus (P), and sulfur (S) (Cockell et al. 2016). Other micronutrient elements are also essential, in trace amounts (e.g., Ni, Mo, Mn, Cu, and Fe). Across the tree of life, oxidoreductase enzymes are the nanomachines that perform electron transfer for obtaining energy and cellular metabolism (Falkowski 2015), and they commonly contain transition metals in their active sites to catalyze electron transfer (Williams 1981).

Transition metal cofactors used in oxidoreductases are required for various metabolic pathways, such as nickel (Ni) for methanogenesis, manganese (Mn) for oxygenic photosynthesis, or copper (Cu) for aerobic respiration. Iron (Fe) is the most widely used metal in biology and is essential for the metabolisms described above, as well as for many others (e.g., sulfate reduction, anoxygenic photosynthesis, nitrogen fixation, etc.). Indeed, large areas of the modern ocean have limited biological growth because of low Fe availability (Behrenfeld et al. 1996; Behrenfeld and Kolber 1999). When essential macronutrient CHNOPS elements or micronutrient metal cofactors are not available in the environment, they limit biological production and the excretion of waste product gases that can be crucial substrates for other organisms (Jelen et al. 2016). The ability of transition metals to catalyze electron-transfer reactions relates to fundamental chemical properties of these elements. Therefore, if life on other planets evolved that also operates on the basis of electrochemistry, where redox reactions are used to gain metabolic energy (Fig. 2), then it is conceivable that transition metals would fulfil a similar catalytic function.

While O_2 is essential for aerobic life and thus all macrofauna as we know it, it is simply a waste product of oxygenic photosynthesis. All metabolisms excrete waste products, many of which are gaseous and can leak into the atmosphere: CO_2 is a waste product of respiration; hydrogen sulfide (H_2S) is a waste product of sulfate reduction; CH_4 is a waste product of methanogenesis. When these biological waste product gases accumulate in the atmosphere, they can be identified as planetary biosignatures using remote sensing approaches. For example, the presence of abundant oxygen in Earth's atmosphere is extremely rare in the known Universe surveyed to date (Waltham 2014). Detecting oxygen or other biological waste product gases in the atmospheres of other planets could be a potential sign of life and perhaps a sign that the planet's biosphere has taken control of the residence time of many atmospheric species to a similar extent as it has on Earth. Understanding past and present planetary constraints on the biological production of these waste products (above non-biological background emissions) may therefore be crucial for detecting alien biospheres.

Fundamentals of geological constraints on Earth's atmosphere

As outlined above, biological processes can have a significant, and potentially observable, effect on a rocky planet's atmosphere. On the modern Earth, the atmospheric composition is very strongly controlled by life (Lenton 1998). However, any potential atmospheric biosignature must be disentangled from a backdrop of abiotic (geological and astrophysical) processes that also contribute to planetary atmospheres and would be dominating on lifeless worlds and on planets with a very small biosphere. Planet formation models (Ikoma and Genda 2006; Lee et al. 2014) and observations of exoplanet populations (Fulton et al. 2017) suggest that nascent planetary embryos accrete primary atmospheres from proto-planetary nebula gas, predominantly hydrogen and helium. While it is not conclusively known if Earth ever possessed such a primary atmosphere, there is evidence from noble gases of ingassing of nebula gases during formation (Williams and Mukhopadhyay 2019; Lammer et al. 2020). Moreover, Young et al. (2023) argue that the light element abundance in Earth's core, origin of Earth's water, and the oxidation state of the mantle can all be explained by the interaction of the proto-Earth with an H_2–He atmosphere. However, each of these aspects of the Earth are also well explained without primordial atmosphere ingassing (see reviews by Frost and McCammon 2008; Li and Fei 2014; Peslier et al. 2017), so should not be taken as direct evidence of an H_2–He atmosphere for the proto-Earth. The purported primary atmosphere must have been lost soon after planet formation. Proposed mechanisms for primary atmosphere loss include photoevaporation from the host star (Owen and Wu 2013), atmospheric blow off driven by the interior heat of the planet (core powered mass loss, Ginzburg et al. 2018), or impact stripping (Genda and Abe 2003; Schlichting et al. 2015).

Earth's so-called secondary atmosphere (the precursor to our current atmosphere) then formed from outgassing of volatiles stored in the interior (Schaefer and Fegley 2017; Gaillard et al. 2021). These would primarily have included different species of carbon (CO_2, CO, or CH_4), hydrogen (H_2, H_2O), nitrogen (N_2, NH_3), and sulphur (S_2, SO_2, H_2S) gases, with redox state depending on the oxidation state of Earth's mantle. Earth's mantle is highly oxidized today, and thus CO_2, N_2, H_2O and SO_2 dominate over CO, CH_4, NH_3, H_2 and H_2S (Schaefer and Fegley 2017; Gaillard et al. 2021). Geochemical evidence indicates that Earth's mantle has been oxidized since the Archean (Frost and McCammon 2008), and possibly even earlier (Trail et al. 2011). How Earth developed this oxidation state is not fully understood, but leading theories point to disproportionation of iron oxide at high pressure during the magma ocean phase, where a significant fraction of the mantle was molten due to heat from accretion, short-lived radioactive isotopes, or giant impacts (Wade and Wood 2005). Iron metal formed by disproportionation then rained down to join the core, leaving more oxidized material behind in the mantle (Wade and Wood 2005). The disproportionation mechanism applies to planets Earth-size or larger, as these planets reach high enough pressures in their lower mantles for disproportionation to occur. If this mechanism is correct, then we might expect super-Earth exoplanets to also have oxidized mantles and outgas relatively oxidized volatile species.

Significant degassing of N_2, CO_2, H_2O and SO_2 likely occurred during the early magma ocean state, in the wake of the moon forming impact, giving rise to Earth's secondary atmosphere (Sleep et al. 2014). The composition and size of the atmosphere were at that time, prior to the emergence of life, controlled by thermodynamic equilibrium between the atmosphere and the magma ocean (Elkins-Tanton 2008; Gaillard et al. 2022; Wolf et al. 2023). After the magma ocean solidified and Earth progressed towards a habitable state where H_2O was condensed as liquid water on the surface, volcanism became the main abiotic source of volatiles to the atmosphere. Volcanism has persisted on Earth throughout its history, which may have included tectonic regimes different than modern day plate tectonics, including a stagnant-lid regime similar to that which occurs on Mars (Korenaga 2013; O'Neill and Debaille 2014; Stern et al. 2018). The long-term equilibrium state of the atmosphere is controlled by volcanic and ultimately biological sources of atmospheric gases and sinks through processes such as atmospheric escape (applies to light species like H), biological uptake, or interaction with crustal rocks as occurs during weathering or hydrothermal alteration (e.g., Catling 2014).

Interaction between the atmosphere and the crust generates new volatile-rich minerals such as carbonates, hydrated silicates, ammonium-bearing clays, and sulfide or sulfate deposits. The biosphere likely enhanced some of these processes once it emerged, for example by promoting carbonate precipitation or sulfide deposition (see also Hazen et al. 2008) and by storing volatiles in the form of organic molecules. During plate tectonics, as on the modern Earth, these volatile-rich phases can be returned to the mantle by subduction or they are re-melted to drive further degassing (Bekaert et al. 2021). Tectonics thus recycles volatiles between the surface and interior, thereby providing a continual source of volatiles to the mantle to allow degassing to persist as long as volcanism is active. Life indirectly contributes to this flux.

Volcanism on the modern Earth occurs primarily at mid-ocean ridges, at hot spots, and at subduction zones, and these are likely the dominant volcanic settings on any rocky planet in a plate-tectonic regime. Volcanism will also occur on planets in a stagnant-lid regime, given sufficient mantle heat sources, as discussed below. Volcanism at mid-ocean ridges occurs when upwelling mantle, rising adiabatically, crosses the solidus and melts. This source of volcanism persists as long as the mantle temperature is higher than the mantle solidus at surface pressure. Hot spots are thought to be typically caused by upwelling mantle plumes, which are hotter than the surrounding mantle and drive volcanism at plate interiors where plumes impinge upon the base of the lithosphere. Volcanism at subduction zones occurs when down-going plates sink into the mantle and heat up to the point where the sinking crust either melts itself, or releases water that has been incorporated in the subducting plate and associated sediments by hydrothermal circulation to the overlying mantle, thereby lowering the solidus and inducing melting. Tectonics and volcanism on Earth are thus important factors in regulating habitability and atmospheric evolution over many hundreds of millions of years, i.e., on timescales longer than those over which biological metabolism operates and internally recycles volatiles.

THE GEOLOGICAL RECORD OF LIFE AND ENVIRONMENTS

Life on a fully anoxic planet (ca. 4.5–3 Ga)

Early tectonic constraints. Tectonics on the very early Earth (4.5–3.0 Ga, Condie 2007) is poorly constrained due to a sparse rock record that is difficult to uniquely interpret in terms of tectonic processes. The Jack Hills zircons, the only solid remnants from prior to ~4.0 Ga, indicate that liquid water oceans formed during the Hadean based on the oxygen isotope composition of the zircons (Mojzsis et al. 2001; Valley et al. 2002). The Jack Hills zircons were derived from felsic (silica-rich) crust, now lost to weathering. However, other regions of felsic crust are preserved from the very early Earth, including the oldest intact whole rocks on Earth, the ~4.0 Ga Acasta gneiss complex, and represent the first building blocks of the continents (Bowring et al. 1989; Reimink et al. 2019). These preserved regions of continental crust only demonstrate the minimum amount that must have existed at this time; the exact amount of continental crust present through the Hadean and Archean can only be estimated from geochemical models and is debated (e.g., see review by Harrison 2009, in particular their Fig. 1). Some authors argue for rapid growth of continents during the early Archean or even Hadean (Korenaga 2018), while others argue for a more gradual growth of continental crust during the Archean, reaching ~75% the present day volume by the end of the Archean (Harrison 2009). The early crust that formed at this time may have been largely submerged, as predicted from isostatic models (Flament et al. 2008; Lee et al. 2018). That continents on the very early Earth were mostly submerged is supported by shale-forming basins not appearing in the geologic record until the end of the Archean (Altermann and Nelson 1998; Knoll and Beukes 2009; Liebmann et al. 2022), and analysis of zircon age distributions arguing for continental emergence above sea level at ~3.0 Ga (Reimink et al. 2021). Hence the earliest life forms likely emerged and evolved on an ocean-dominated world, where the only land masses were perhaps volcanic islands above hotspots, akin to Hawai'i on the modern Earth (Van Kranendonk et al. 2015; Bada and Korenaga 2018). This carries implications for exoplanet science, because it means that an independent origin of life may not necessarily require the presence of large land masses on the scale what we experience today.

Formation of the continental crust as seen on the early Earth is thought to require melting of hydrated mafic crust (like ocean crust) at pressures of ~1–3 GPa (~30–100 km depth) (e.g., Moyen and Martin 2012). Therefore, some form of crustal recycling must have been active. However, it is not clear if this was crustal burial by volcanism, as can occur in a stagnant-lid regime (Bédard 2006; Willbold et al. 2009; Reimink et al. 2014), or by subduction (Komiya et al. 1999; Nutman et al. 2002; Moyen and Van Hunen 2012; Martin et al. 2014; Turner et al. 2014). The dominant crust generation process may have switched from a stagnant-lid to a mobile-lid, with subduction, during the Archean (Dhuime et al. 2012; Bauer et al. 2020), or even varied spatially across the Earth (Van Kranendonk 2010). As stagnant-lid tectonics is a possibility for the early Earth, we consider here how volcanism and volatile cycling would operate in such a regime. If the early Earth was in a stagnant-lid regime, the lithosphere would have been significantly thicker than in a plate-tectonic regime, forcing melting to occur at a greater depth. The mantle must be hotter in order to melt at a greater depth, as the solidus temperature increases with pressure, meaning that volcanism is impeded on stagnant-lid planets (Solomatov and Moresi 1996; Reese et al. 1999). However, stagnant-lid planets also lose heat less efficiently than plate-tectonic planets, meaning they tend to have hotter mantle temperatures. As a result, volcanism is expected to still be active on such planets given sufficient heat sources (Foley et al. 2020; Unterborn et al. 2022), and as evidenced by recent volcanism on Mars (Hartmann et al. 1999; Hauber et al. 2011), meaning volatile outgassing occurs as well. Venus also shows evidence for recent volcanism (Smrekar et al. 2010; Shalygin et al. 2015), and while it does not possess plate tectonics like the modern Earth, it is probably not in a true stagnant-lid state either (Davaille et al. 2017; Borrelli et al. 2021; Smrekar et al. 2023). Ultimately, even if the very early Earth lacked plate tectonics, there would still have been significant abiotic fluxes of CO_2, N_2, H_2O and SO_2. Substantial mantle melting and volcanism on the very early Earth is also supported by the geologic record. For example, the presence of komatiites is thought to represent very high degrees of mantle melting (Viljoen and Viljoen 1969; Brooks and Hart 1974), which is also supported by models of melt production through time based on global databases (Keller and Schoene 2012, 2018).

In theory, a planet that operates in a stagnant-lid regime and hence lacks subduction, as may have been the case for the early Earth, also loses the primary mechanism for recycling of volatiles from the surface back to the interior. This means that outgassing may be a "one-way street" in a stagnant-lid regime, where volatiles are outgassed to the surface initially and until these volatiles become depleted in the mantle. However, one possible mechanism for recycling surface volatiles to the interior in a stagnant-lid regime is burial of surface rocks by progressive eruption, which, if subaqueous, would take the form of pillow lavas. These eruptive lavas will be extensively altered and weathered by interaction with surface water, hydrating and carbonating them (Foley and Smye 2018; Marien et al. 2023). If surface rocks are buried deep enough, they can sink into the mantle, returning any volatiles acquired by weathering or hydrothermal alteration (Elkins-Tanton et al. 2007). During this burial process, though, rocks will experience increasing temperatures and pressures, and metamorphic reactions may devolatilize rock before it can founder back into the mantle (Foley and Smye 2018). Hence volatiles may have only been able to recycle between the surface and crust, rather than the mantle, if the early Earth was in a stagnant-lid regime, though outgassing of volatiles to the atmosphere would be active regardless.

Another feature of the hot, volcanically active, early Earth with significant volumes of (ultra-)mafic lavas would have been the likely presence of intensive hydrothermal activity on the seafloor, as suggested by trace element data (Viehmann et al. 2015). Hydrothermal vents are considered important sites for the origin of life, because they provide catalytic mineral surfaces and reduction potential through the process called serpentinization (Baross and Hoffman 1985; Martin et al. 2008; Russell et al. 2010). At these sites, H_2O reacts with Fe^{2+} present in ferrous minerals such as olivine and pyroxene, forming H_2 gas. In the presence of magnetite or iron sulfides, H_2 can reduce CO_2 to CH_4 and longer-chain hydrocarbons (McCollom and Seewald 2007; Proskurowski et al. 2008), and N_2 can be reduced to NH_3 (Brandes et al. 1998). Hence volcanism may have acted as an important engine of prebiotic networks. The generally reducing conditions in these settings create a redox gradient relative to the more oxidized gases released from the mantle, which may have been beneficial for driving prebiotic reactions (Martin et al. 2008). The mantle oxidation state implies that oxidized gases would have been released by volcanoes, meaning reducing gases required a different source. Serpentinizing hydrothermal vents is one possibility; another possibility is outgassing of impactors (Zahnle et al. 2020). Planetary building blocks in the inner solar system have a lower oxidation state than the modern Earth's mantle, so degassing of these materials upon impact can produce strongly reducing atmospheres that could have persisted for ~1–10 Myrs (Zahnle et al. 2020).

Traces of early life in the Archean (up to ca. 3 Ga). This framework of a volcanically active planet with limited continental exposure provides context for the oldest vestiges of life on Earth. The oldest putative signs of life in the rock record, though controversial, date back 4.1 Ga and are preserved in the form of graphite inclusions in a single zircon crystal (Bell et al. 2015). Zircons crystallize from magma in the subsurface, and so the interpretation of the carbon in the graphite inclusion is that it was originally deposited as biomass within marine sediments, which were subsequently buried to several kilometres depth where the biomass was converted to graphite as a consequence of heating and compression. During further burial and heating, the sediment package ultimately underwent melting and zircons formed that trapped some of this graphite as inclusions. The entire melt ultimately solidified into a magmatic rock, which was later exhumed and eroded, releasing the zircon grain back into the environment where it was found.

The key argument for the biogenicity of the graphite inclusions within this zircon is its carbon isotope composition (δ^{13}C) of $-27‰$, which is indistinguishable from average modern biomass (Schidlowski 2001). This carbon isotope value could possibly reflect an autotrophic lifestyle with biological CO_2 fixation, but also a heterotrophic metabolism, where living organisms were feeding on abiotically produced organic matter, cannot be ruled out from these data alone. If it was autotrophic, it would be the oldest evidence of life impacting the composition of Earth's atmosphere, because autotrophic CO_2 fixation would create an additional flux of CO_2 from the atmosphere into the lithosphere as organic matter underwent burial.

The conclusion remains controversial due to the small size and singular occurrence of Hadean zircon-hosted graphite (Nemchin et al. 2008). Also the carbon isotope value could theoretically be mimicked by abiotic processes (McCollom and Seewald 2006). However, if correct, the presence of life in the Hadean at 4.1 Ga or earlier would have important implications for exoplanet science, because it would show that life was able to emerge on Earth during the period of relatively heavy meteorite bombardment (Bottke and Norman 2017). Although it has been called into question if Earth experienced an episode of increased impact rates ("Late Heavy Bombardment") at around 4.0–3.8 Ga (Boehnke and Harrison 2016), impact rates were still likely higher in the early history of the solar system (Bottke and Norman 2017; Lowe and Byerly 2018), and the same would likely be true for other solar systems with rocky planets. If life was able to originate on Earth despite frequent impacts at that time, it would thus bode well for the habitability of terrestrial planets elsewhere. As noted above, meteorite bombardment may create a linkage between the origins of life and the reducing power imported by impact events (Zahnle et al. 2020). The development of life at this time may also imply that the biosphere had enough refugia available to survive impact events, for example in the subsurface (Abramov and Mojzsis 2009), and/or it could mean that life went extinct and re-emerged multiple times, meaning that biogenesis as such is perhaps not rare.

The next oldest and less controversial evidence of life on Earth comes from graphite-rich laminae within highly metamorphosed sedimentary rocks in several localities in the Isua supracrustal belt in Greenland, dated to approximately 3.7–3.8 Ga (Mojzsis et al. 1996; Rosing 1999; Ohtomo et al. 2014; Hassenkam et al. 2017) (Fig. 3). Although the δ^{13}C values are slightly less negative than the graphite inclusions in zircon, possibly due to metamorphic overprinting, the data are consistent with biological CO_2 fixation. Of similar age are putative microfossils preserved in jasper in the Nuvvuagittuq supracrustal belt in Quebec, Canada (Dodd et al. 2017; Papineau et al. 2022) and occurrences of stromatolites in carbonates at Isua in Greenland (Nutman et al. 2016), although the biogenicity of both of these latter occurrences has been called into question (Allwood et al. 2018; McMahon 2019; Lan et al. 2022).

The Eoarchean graphite from Isua and surrounding minerals are moderately enriched in nitrogen (Pinti et al. 2001; Papineau et al. 2005; Stüeken 2016; Hassenkam et al. 2017). These N-enrichments are further support for a biogenic origin of this graphite, because all biomass on Earth contains nitrogen in the form of amine groups as a key constituent of many essential biomolecules, and this nitrogen gets released as ammonium into sedimentary pore waters after deposition on the seafloor and is subsequently trapped in clays (Müller 1977; Schroeder and McLain 1998). Hence clays can carry nitrogen originally derived from biogenic organic matter. Biomass burial therefore constitutes a major flux of nitrogen from surface environments into the rock record and affects the pressure of N_2 gas in the atmosphere over geological timescales (Som et al. 2012, 2016; Johnson and Goldblatt 2018). Today, the conversion of N_2 to bioavailable ammonium is largely carried out by microorganisms; merely a few percent of N_2 fixation is abiotic, driven by lightning reactions in the atmosphere and by hydrothermal processes (Brandes et al. 1998; Navarro-González et al. 1998). During the origin of life, these abiotic sources of fixed N would have been dominant, and they may have involved gaseous products such as HCN, NO and N_2O that are associated with life on the modern Earth and

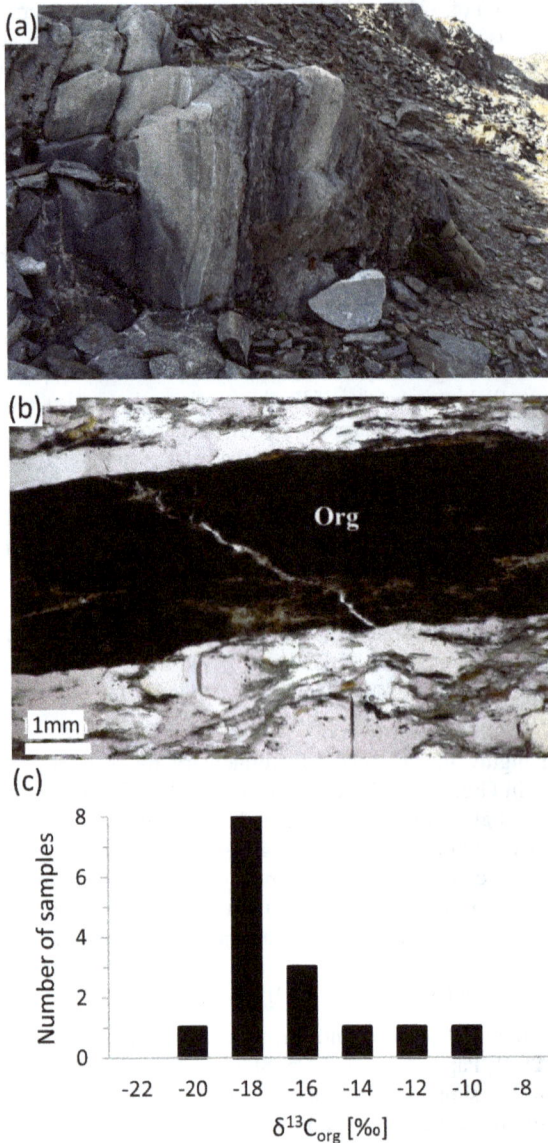

Figure 3. Organic carbon isotopes in graphitic schists as a potential biosignature. Shown here as an example is the Garbenschiefer Formation from the Isua Supracrustal Belt in Western Greenland. **(a)** Outcrop photo of the meta-turbidite (reproduced from Stüeken et al. 2021a, under Creative Commons Attribution License CC-BY). Backpack on the right side of the outcrop is approximately 50 cm tall. **(b)** Photomicrograph showing graphite lamina in quartz–mica schist (image credit: Jane Macdonald, St Andrews, 2023). **(c)** Organic carbon isotope data from these rocks, replotted from Rosing (1999) and Stüeken et al. (2021a).

that may leave signs (and potential false positives) in atmospheric spectra (Barth et al. in review.). However, at some point during early evolution, life appears to have taken control over its fixed N supply by inventing nitrogenase enzymes that convert N_2 into bioavailable NH_4^+.

It is conceivable that biological N_2 fixation (termed diazotrophy) is as old as the Last Universal Common Ancestor (LUCA) of all life on Earth and was thus potentially active at 3.8 Ga (Weiss et al. 2016; Garcia et al. 2020; Parsons et al. 2020) or even 4.1 Ga (Bell et al. 2015); however, the geochemical data from this time period are too heavily metamorphosed to test this proposition. The first geochemical evidence for diazotrophs is from rocks of low metamorphic grade at 3.4–3.2 Ga (Stüeken et al. 2015; Homann et al. 2018; Pellerin et al. 2023) and at 3 Ga and younger (Koehler et al. 2019; Ossa Ossa et al. 2019).

Although multiple lines of evidence hint at the presence of life at 3.7–3.8 Ga, the high metamorphic grade of these rocks leaves residual doubt on those results. The most widely accepted evidence of life on Earth therefore dates back to approximately 3.5 Ga (Walter et al. 1980; Buick 2007a; Baumgartner et al. 2019; Lepot 2020; Westall et al. 2023) and is based on a range of proxies, including carbon isotopes ($\delta^{13}C$), sulfur isotopes ($\delta^{34}S$) and stromatolites (Fig. 4a). The sulfur isotope data represent the oldest evidence of biological sulfate reduction, using aqueous SO_4^{2-} that was generated by photochemical conversion of volcanogenic SO_2 gas (Ueno et al. 2008; Baumgartner et al. 2020a). These data are thus evidence of volcanic and photochemical processes providing electrochemical energy for the biosphere. Chemotrophic organisms would have been able to harness this energy to generate biomass and proliferate. Perhaps most important for exoplanet science is isotopic evidence in support of biological methane production (methanogenesis) in rocks at 3.5 Ga (Ueno et al. 2006). This study found that methane was preserved in the form of gas inclusions within quartz crystals that are thought to have precipitated on the seafloor, and that it carried an isotopic fingerprint of biological formation. The stromatolites may reflect the presence of phototrophic organisms (see below). The early Archean biosphere at 3.5 Ga thus appears to already have been relatively diverse (Van Kranendonk et al. 2021). In only slightly younger rocks from 3.5–3.4 Ga, carbonaceous microfossils are preserved (Fig. 4b), providing the first physical evidence of what life looked like at that time (Westall et al. 2001, 2006; Wacey et al. 2010; Sugitani et al. 2015; Flannery et al. 2018). Furthermore, individual microfossils from ca. 3.4 Ga carry isotopic signatures indicative of methanogenesis (Flannery et al. 2018; Schopf et al. 2018). Like microbial sulfate reduction to sulfide, methanogenesis is a chemotrophic metabolism, where either CO_2 is reduced to CH_4 (chemoautotrophy) or CH_4 is liberated during the degradation of organic matter (chemoheterotrophy). In any case, life was capable of generating a greenhouse gas and potential atmospheric biosignature by at least 3.4 Ga.

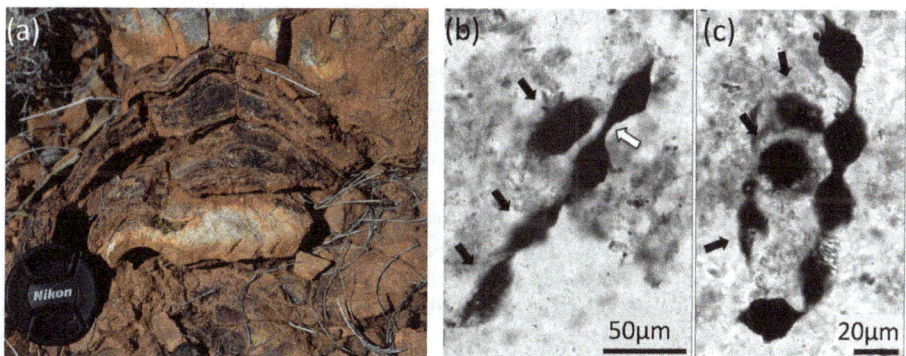

Figure 4. (a) Stromatolite from the Dresser Formation at 3.49 Ga (photo taken by Eva E. Stüeken). **(b)** and **(c)** Photomicrograph of carbonaceous microfossils from the Strelley Pool Formation at 3.4 Ga, reproduced from Sugitani et al. (2015) with editorial permission. **White arrows** point at a flange-connected pair of cells; **black arrows** point at isolated lenses.

Modelling suggests that the concentration of methane in the early Archean atmosphere was likely orders of magnitude above modern values (Catling et al. 2001) (see also below). The reasons for this are increased biological sources paired with a long methane lifetime due to very low levels of O_2, which allowed methane to accumulate in the atmosphere. Evidence for low atmospheric O_2 levels comes from the preservation of mass-independent sulfur isotope fractionation (S-MIF) in the Archean sedimentary rock record (Farquhar et al. 2000; Johnston 2011) (Fig. 5). This signature has been demonstrated to be produced by UV-photolysis of volcanogenic SO_2 gas, which requires the absence of a UV-shielding ozone layer. The products of the reaction are SO_4^{2-} (dissolved in rainwater) and elemental S. The preservation of the latter, in particular, requires low levels of O_2 in rainwater and seawater, such that oxidation of S^0 and homogenization of all sulfur as SO_4^{2-} are prevented. Computational models of this process indicate that the formation and preservation of S-MIF places an upper bound on pO_2 of 10^{-5} times present atmospheric levels (PAL) (Pavlov and Kasting 2002; Kurzweil et al. 2013).

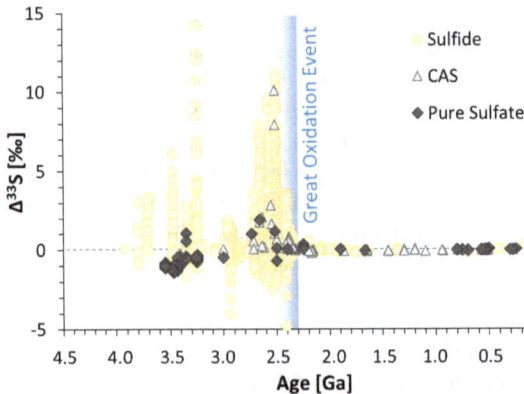

Figure 5. The sedimentary record of mass-independent sulfur isotope fractionation. Database taken from Claire et al. (2014). The disappearance of the $\Delta^{33}S$ signature at around 2.4–2.3 Ga marks the oxygenation of the atmosphere, which led to the formation of an ozone shield that blocked UV radiation and stopped SO_2 photolysis (Farquhar et al. 2000). This event is known as the Great Oxidation Event (GOE). Anoxic planets where O_2 production never occurs are more likely to resemble the early Earth prior to the GOE.

Ocean chemistry and nutrient limitation on the Archean Earth. As summarized above, from 3.8 to 3.4 billion years ago, geochemical evidence exists for the presence of various types of chemotrophy and phototrophy, which would have had specific metal cofactor requirements (Moore et al. 2017, and references therein). The rate at which biosignature gases were released into the atmosphere would therefore have been limited by the supply and solubility of nutrients and redox couples in the environment at that time. Sedimentary records of iron speciation indicate that the early ocean was anoxic and ferruginous (Fe^{2+}-rich), and this state likely persisted for most of Earth's history (i.e., until long after the evolution of oxygenic photosynthesis in the early-/late Archean, see below) (Planavsky et al. 2011; Poulton and Canfield 2011). Under widespread ferruginous marine conditions that prevailed throughout much of Earth's history, photoferrotrophy would have likely been an important form of biological primary production (Canfield et al. 2006; Kendall et al. 2012). Indeed, abundant photoferrotrophs represent a major component of the phototrophic community in ancient ferruginous ocean analogue environments (Crowe et al. 2008; Walter et al. 2014; Llirós et al. 2015). One of the main processes that has been proposed for the origin of banded iron formations (BIFs, chemical precipitates of alternating SiO_2 and iron minerals from seawater) prior to the GOE is direct Fe-oxidation by anoxygenic phototrophs (Konhauser et al. 2002; Kappler and Newman 2004).

A global redox model has indicated that the combination of Fe^{2+}-based and H_2-based anoxygenic photoautotrophy in the environment results in amplification of Earth's methane cycle and major influence on climate (Ozaki et al. 2018).

Anaerobic microbial metabolic pathways in the early ocean would have faced different availabilities of essential macro- and micronutrients compared to today (Anbar and Knoll 2002; Saito et al. 2003; Zerkle et al. 2005; Anbar 2008; Robbins et al. 2016). The micronutrient metals Fe, Mn, Ni and cobalt (Co) were present in higher concentrations, while Mo, Cu, and zinc (Zn) were present in lower concentrations (Fig. 6). Iron, Mn, Co and Ni (Rudnick and Gao 2014; White and Klein 2014; Ptáček et al. 2020) are relatively more abundant in mafic crustal rocks, and their solubilities are increased in the absence of oxygen (Fe, Mn) or not strongly affected by redox conditions (Co, Ni) (Brookins 1988). This explains their elevated marine concentrations in the early Archean when the crust was overall more mafic and leached by strong hydrothermal activity (Isley and Abbott 1999; Viehmann et al. 2015). In contrast, Mo, Cu and Zn were likely scavenged from the water column by small amounts of sulfide produced through biotic or abiotic pathways. Only a significant supply of organic ligands may have overcome extreme limitation of Cu and Zn, in particular (Robbins et al. 2016; Stüeken 2020). These trends likely impacted the biogeochemical cycles of other elements and therefore total biological activity and biosignature production in the early Archean (see also Jelen et al. 2016).

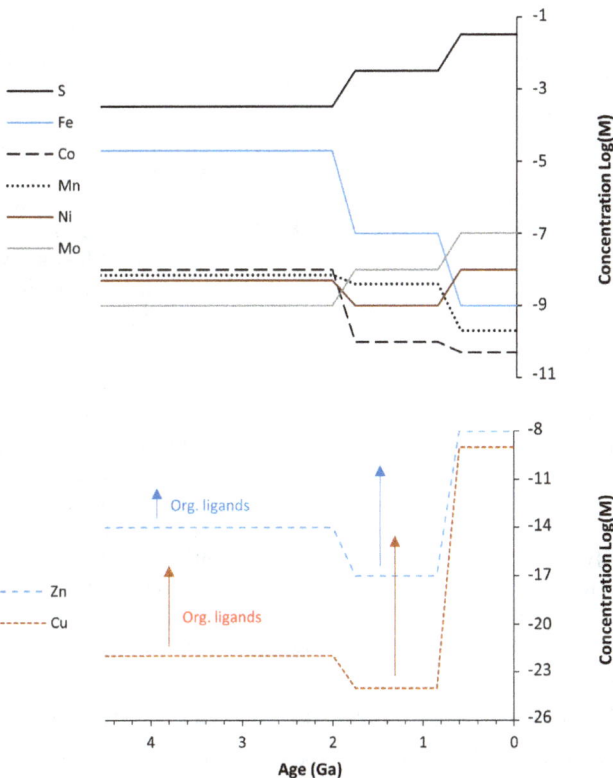

Figure 6. Modelled concentrations of trace elements through time that are important protein cofactors. The figure is primarily adapted from Saito et al. (2003) with additional incorporated results from other studies (Anbar and Knoll 2002; Zerkle et al. 2005; Anbar 2008; Robbins et al. 2016). Note that Cu (and possibly Zn) solubility could have been elevated above the indicated concentrations in the presence of organic ligands, as indicated by arrows in the lower panel (Stüeken 2020).

Regarding the major nutrient phosphorus (P), the available data are currently inconclusive. Compilations of P abundances in marine shales through time suggest lower marine P availability throughout the Precambrian until ca. 0.7 Ga (Reinhard et al. 2017b; Planavsky et al. 2023). Furthermore, P recycling from biomass would likely have been suppressed under anoxic conditions (Kipp 2022). If so, then P may have been a limiting nutrient over a large geographic region or in individual basins. In contrast, recent experiments have shown that phosphate is moderately soluble in the presence of dissolved Fe^{2+}, and therefore phosphate may have been sufficiently bioavailable to support emergent cellular systems and early microbial life (Brady et al. 2022). This interpretation is supported by elevated P levels in marine shelf-ramp carbonates of Neoarchean age, implying sufficient P availability in shallow marine habitats at that time (Ingalls et al. 2022). Similarly, thermodynamic and kinetic modelling indicates that Saturn's moon Enceladus has habitable phosphorus concentrations close to modern Earth seawater (Hao et al. 2022). Hence it is possible that P is not always the critical constraint for the production and detectability of biosignatures, although it is the limiting ingredient for biospheric productivity in most environments today.

Regarding nitrogen, the second major nutrient, redox conditions in the primitive ocean likely directly impacted nitrogen fixation and the nitrogen cycle. In particular, reduced nitrogen is a crucial component of amino acids, proteins, DNA, and RNA. Today, nitrogen makes up nearly 80% of the atmosphere in the form of dinitrogen (N_2) gas, but the two nitrogen atoms in dinitrogen gas are connected by a very strong triple bond, making dinitrogen unusable to most organisms. In the absence of significant abiotic sources, bioavailable nitrogen may have been limiting to the earliest microbial communities (Canfield et al. 2010). One may thus speculate that the early- to mid-Archean invention of the nitrogenase enzyme that can break the dinitrogen triple bond and convert N_2 into ammonium (NH_4^+) (see above) possibly represented a major turning point in the expansion of the biosphere. This hypothesis remains to be tested. Nitrogenase requires several Fe atoms in its structure (Georgiadis et al. 1992), and the most common and most efficient form of nitrogenase includes Mo as an essential metal cofactor along with Fe (Joerger et al. 1988; Garcia et al. 2022). Although Mo can be replaced by vanadium (V) or Fe, phylogenetic data suggest that the Mo-variety of nitrogenase is the most ancient (Parsons et al. 2020). In today's oxic ocean, Fe is the limiting element in the nitrogen cycle (Canfield et al. 2010). However, in the Archean, nitrogen fixation may have been limited by Mo scarcity (Anbar and Knoll 2002; Johnson et al. 2021).

Overall, the anoxic, ferruginous redox state of the early Archean ocean may have impacted total biological productivity by inhibiting biomass remineralization as a source of recycled nutrients (Kipp and Stüeken 2017) and possibly limiting the supply of P and N to primary producers. However, relatively speaking, the biological production of CH_4 was probably enhanced by heightened supplies of Ni and Co from mafic rocks. It is conceivable that these conditions are a common feature on other worlds on which biological O_2 production has not (yet) evolved or O_2 has not been able to accumulate. Mafic (Fe-rich) crustal rocks are prevalent at the surface of planets, and therefore, potential oceans on those planets may also be enriched in dissolved iron. The ferruginous early Archean ocean marks a crucial data point in assessing habitability and biosignature production. Quantitative constraints derived from the trends described above are presented below.

Microbial life in a redox-stratified world with modern plate tectonics (ca. 3–0.5 Ga)

The onset of modern plate tectonics. The early Archean Earth was a different planet from today with few emergent land masses, an anoxic atmosphere and ocean, strong volcanic activity, marine nutrient inventories buffered by mafic crust, and possibly a different tectonic style than what we have today. However, from ca. 3 Ga onwards, multiple lines of evidence indicative of plate tectonics with subduction zones, as well as cool and buoyant continental crust, appear in the geologic record (Smithies et al. 2005, 2007; Condie and Kröner 2008;

Shirey and Richardson 2011; Van Kranendonk 2011; Dhuime et al. 2012; Korenaga 2013; Brown and Johnson 2018; Bauer et al. 2020). On the modern Earth, most marine life resides near continental margins, i.e., in the vicinity of riverine nutrient sources and within the photic zone. Hence the establishment of large continental shelves in the Neoarchean would have provided an important habitat space for the biosphere.

The mantle has progressively cooled from the Archean onwards (Herzberg et al. 2010). This cooling contributed to continental emergence via isostasy (Flament et al. 2008)—changes in crustal compositional from more mafic to more felsic were important as well (e.g., Dhuime et al. 2015; Tang et al. 2016)—and may have played a role in the initiation of plate tectonics (O'Neill et al. 2016), though some geodynamical models find that mantle cooling is not a significant factor in promoting plate tectonics (Korenaga 2017; Foley 2018). In any case, the establishment of subduction and mantle cooling promoted volatile recycling into the mantle. Water, in particular, can be more readily regassed into the mantle as mantle temperature declines (Rüpke et al. 2004; Magni et al. 2014; Dong et al. 2021). Xenon isotopes indicate that Earth switched to a state of net water ingassing at the end of the Archean (Parai and Mukhopadhyay 2018).Volcanism, and therefore volatile outgassing, is affected by both mantle cooling and water ingassing. A cooling mantle lowers volcanism rates due to less extensive melting and less vigorous convection, which causes a lower flux of mantle material into the melting regions at mid-ocean ridges or subduction zones. However, water recycling can counteract this affect, as increasing mantle water content promotes melting by lowering the solidus and decreases mantle viscosity, hence increasing convective vigor (Crowley et al. 2011; Seales et al. 2022). Various models have been constructed of volatile in- and outgassing taking these competing effects into account, finding a range of behaviour from steadily decreasing outgassing rates to outgassing rates that stay relatively constant through this period of Earth's history (McGovern and Schubert 1989; Tajika and Matsui 1992; Fuentes et al. 2019). More efficient water ingassing as the mantle cools implies that the ocean volume has been declining over this period of Earth's history. However, this does not mean that the sea level relative to continents, or freeboard, changed significantly in response. The volume of ocean basins itself is not constant over time due to changes in both continental area and seafloor depth as the mantle evolves (Schubert and Reymer 1985). Korenaga et al. (2017) showed that a relatively constant freeboard could be maintained even with a net loss of water from the surface to the mantle. All told, the likely operation of plate tectonics and subduction from the Mesoarchean onwards indicates the presence of significant volcanic outgassing, hydrothermal activity on the seafloor, and deep volatile cycling into the mantle, though the former two would likely have been active even without plate tectonics, as discussed above.

Oxidation to oxygenation: the onset and consequences of oxygenic photosynthesis. Approximately concurrent with the establishment of modern-style plate tectonics, surface environments on Earth began to show the first signs of oxidation (Lyons et al. 2014). By 2.4–2.3 Ga, S-MIF, the key indicator of an anoxic atmosphere, disappeared from the sedimentary rock record (Bekker et al. 2004; Warke et al. 2020; Uveges et al. 2023). Around the same time, paleosols (fossilized soil horizons) started to retain iron in the form of Fe^{3+} minerals (Rye and Holland 1998), detrital pyrite and uraninite (reduced Fe and U minerals) disappeared from river sediments (Johnson et al. 2014), and iron oxide-stained red beds appeared in tidal flat deposits (Eriksson and Cheney 1992). This transition is known as the Great Oxidation Event (GOE, Holland 2002), when atmospheric O_2 levels permanently increased to $>10^{-5}$ PAL (Pavlov and Kasting 2002). While some photochemical processes are able to generate O_2 gas (Haqq-Misra et al. 2011), by far the major source of free O_2 on Earth is oxygenic photosynthesis (reviewed by Catling 2014). Hence the GOE is an expression of biological activity, likely facilitated by a gradual shift in tectonic regime (Van Kranendonk et al. 2012) and possibly progressive biomass burial that removed reducing potential from Earth's surface (Krissansen-Totton et al. 2015).

The metabolism was invented in the cyanobacterial phylum of bacteria and marked a turning point in the history of Earth's surface evolution.

The GOE at ca. 2.4–2.3 Ga is the latest possible time when oxygenic photosynthesis could have originated, given the convergence of numerous proxies listed above that indicate the presence of free O_2 in terrestrial and shallow-marine environments at that time. While some data and models indeed suggest that the invention of O_2-genesis occurred only shortly before the GOE (Johnson et al. 2013; Ward et al. 2016; Shih et al. 2017), other studies indicate that several tens to hundreds of millions of years passed between this biological innovation and the accumulation of O_2 in the atmosphere to appreciable levels (see reviews and models by Kurzweil et al. 2013; Lyons et al. 2014; Laakso and Schrag 2017). This time delay is thought to have been required to saturate and/or reduce the flux of various O_2 sinks (Zahnle et al. 2013). In other words, reductants in the form of crustal minerals, dissolved ferrous iron in the surface ocean and metamorphic gases first needed to be oxidized before the atmosphere could be oxygenated.

The oldest widely accepted evidence of some form of phototrophy dates back to 3.5 Ga and is preserved in the form of stromatolites and microbially-induced sedimentary structures that are reminiscent of modern photosynthetic microbial mats (Walter et al. 1980; Noffke et al. 2013; Baumgartner et al. 2020b). A larger stromatolite reef occurs in only slightly younger rocks at ca. 3.4 Ga (Allwood et al. 2006). On the modern Earth, stromatolites form when light-sensitive microbial colonies grow upwards towards the Sun. By trapping and binding suspended sediment particles and inducing precipitation of carbonate (typically by CO_2 uptake, which raises the pH around autotrophic cells), these microbial structures slowly lithify, often with characteristic laminae (Golubic et al. 2000). Today, most stromatolites are dominated by the primary producers cyanobacteria (e.g., Neilan et al. 2002). However, this was not necessarily the case at 3.5 Ga. It is conceivable that photosynthetic organisms at that time were anoxygenic, i.e., oxidizing sulfide, H_2, or ferrous iron instead of H_2O (Baumgartner et al. 2020b), and evidence of free O_2 has not been documented from these rocks. Some phylogenetic reconstructions suggest that anoxygenic photosynthesis pre-dates oxygenic photosynthesis (Raymond and Blankenship 2008), though this aspect is debated (Cardona et al. 2015). In any case, the presence of stromatolites in the Early Archean is perhaps evidence that life learned relatively early how to harvest photochemical energy from the Sun (see also Cardona et al. 2017, 2019). Early life on Earth may therefore not solely have been dependent on a supply of electrochemical energy from abiotic sources such as volcanism, UV-photolysis or lightning. Instead, a flux of reductants and stellar photons were perhaps sufficient for phototrophic cells to generate electrochemical energy internally.

Liquid water is the most abundant reductant on the planet, and hence the appearance of oxygenic photosynthesis - where H_2O is oxidized to O_2 - is thought to have spurred primary productivity by orders of magnitude (Kharecha et al. 2005; Lyons et al. 2014). The first geochemical evidence for O_2 in the surface ocean dates back to 3 Ga, where molybdenum isotopes display a covariance with Mn concentrations in sedimentary iron formations that is most parsimoniously explained by the former presence of Mn-oxides in those rocks (Planavsky et al. 2014). Hence O_2 production may have started at or shortly before 3 Ga, which is also consistent with some phylogenetic reconstructions (Sánchez-Baracaldo and Cardona 2020; Boden et al. 2021; Fig. 7) (though see alternative interpretatiosn by Shih et al. 2017). However, the near persistence of S-MIF throughout the Archean (with a possible gap around 3 Ga) indicates that O_2 levels in the global atmosphere remained below 10^{-5} PAL, meaning that O_2 would have been a trace gas and likely undetectable by remote techniques for perhaps the entire Archean (though the Mesoarchean interval with muted MIF around 3 Ga remains to be explained, possibly by the presence of O_2 or by an atmospheric haze at that time (Domagal-Goldman et al. 2008)). As noted earlier, the main reason for lack of O_2 accumulation despite ongoing production was the high abundance of reductants that first needed to be saturated (oxidized).

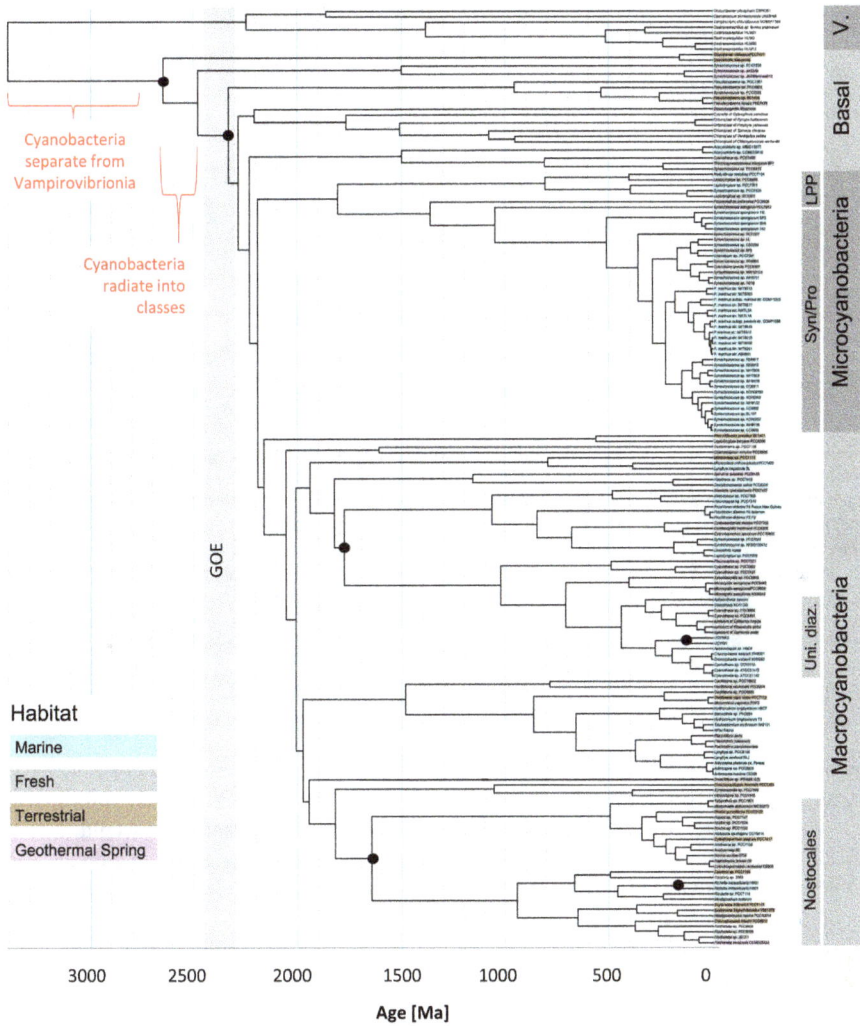

Figure 7. Phylogeny of cyanobacteria, adapted from Boden et al. (2021) under Creative Commons Attribution License CC-BY 4.0. According to this reconstruction, the origin of cyanobacteria occurred in the Archean, and diversification took place just prior to the GOE.

This is an important lesson for observations of habitable exoplanets, because the atmospheric signal of O_2-producers may be undetectable if it is swamped by the flux of biotic and abiotic reductants from the surface of the planet.

Indirect geochemical evidence of oxic surface waters in the Archean ocean becomes more widespread between 2.7–2.5 Ga (Buick 1992; Wille et al. 2007; Kendall et al. 2010; Flannery and Walter 2012; Riding et al. 2014; Koehler et al. 2018; Wilmeth et al. 2022). Some of this evidence includes nitrogen isotope data (Fig. 8) indicative of a redox-active nitrogen cycle where nitrate was able to accumulate at least transiently (Garvin et al. 2009; Godfrey and Falkowski 2009; Koehler et al. 2018). The reduction of nitrate at the interface between oxic surface waters and anoxic deep waters (where it commonly occurs today in anoxic setting, see review by Pajares and Ramos 2019) may have led to the production of N_2O gas (see below).

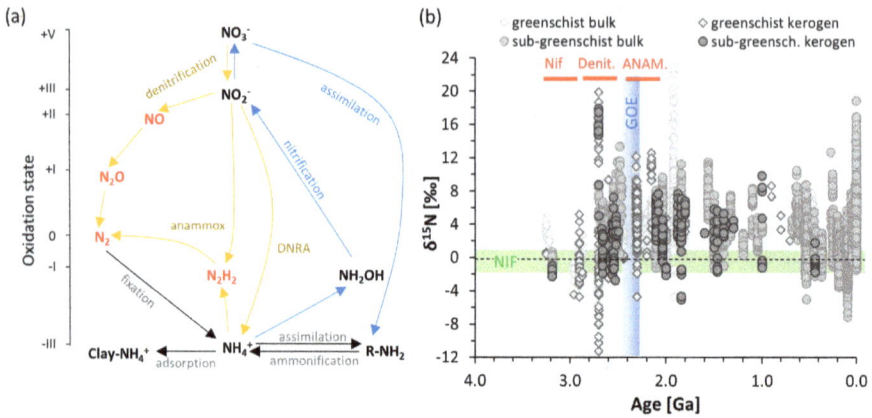

Figure 8. (a) Schematic of the biological nitrogen cycle. **Black** = entirely anoxic pathways; **orange** = pathways occurring under anoxic or suboxic conditions; **blue** = pathways that typically indicate the presence of oxic conditions. Species **highlighted in red** may occur as gases in the atmosphere. **(b)** The sedimentary $\delta^{15}N$ record through time for rocks up to greenschist metamorphic facies. Large nitrogen isotope fractionations are associated with redox reactions involving nitrification and denitrification, and therefore the increasing spread in $\delta^{15}N$ seen from ~2.8 Ga onwards may reflect the onset of these metabolisms. **Red bars** = phylogenetic constraints on the first appearance of N_2 fixation (Nif), denitrification (Denit.) and anammox (ANAM.) (Parsons et al. 2020; Liao et al. 2022). Nitrogen isotope database taken from Stüeken et al. (2016) with recent updates from the literature (Zerkle et al. 2017; Kipp et al. 2018; Koehler et al. 2018, 2019; Luo et al. 2018; Mettam et al. 2019; Ossa Ossa et al. 2019; Yang et al. 2019; Gilleaudeau et al. 2020; Hodgskiss et al. 2020; Wang et al. 2020; Stüeken et al. 2021b, 2022; Stüeken and Prave 2022). GOE = Great Oxidation Event.

Phylogenetic data suggest that biological nitrate reduction to nitrite emerged in the Neoarchean (Parsons et al. 2020), but nitrite reduction only emerged in the late Proterozoic. It is likely that high levels of ferrous iron abiotically reduced nitrite, where NO and N_2O would have been major products (Stanton et al. 2018; Buessecker et al. 2022). Hence increasing fluxes of N_2O and possibly NO from the ocean into the atmosphere could have created a novel biosignature from the Neoarchean onwards.

Although the GOE marks the time when many surface environments became oxidized, atmospheric O_2 levels may have remained as low as 0.1–1% throughout the Proterozoic (Lyons et al. 2014), possibly due to a continuously high flux of reduced iron from hydrothermal sources in the ocean that declined only slowly with mantle cooling (Planavsky et al. 2018a; Wang et al. 2022), though slightly higher levels up to 10% are achieved in some models (Laakso and Schrag 2017; Crockford et al. 2018; Fakhraee et al. 2019). In any case, the deep ocean appears to have remained largely anoxic and buffered by ferrous iron (Reinhard et al. 2013; Sperling et al. 2015). Oxygenation of the deep ocean may have occurred only periodically (Yang et al. 2017; Planavsky et al. 2018b; Stüeken et al. 2021b). Under these conditions, it is also possible that productivity and hence O_2 production were nutrient-limited (Crockford et al. 2018; Laakso and Schrag 2019).

Modern-like O_2 levels were probably only reached in the Neoproterozoic or early Phanerozoic (0.8–0.5 Ga) (Och and Shields-Zhou 2012), and possibly as late as the Devonian rise of land plants at 0.35 Ga (Lenton et al. 2016; Krause et al. 2018). This delay probably also held back the expansion of complex life, which is largely dependent on O_2 as an electrochemical energy source (Reinhard et al. 2016). Understanding biogeochemical cycles and feedback loops on this redox-stratified Proterozoic Earth may therefore be important for assessing the likelihood of complex organisms arising on other worlds.

Ocean redox stratification. Crustal differentiation and changing tectonic regimes in the Neoarchean led to a decline in the availability of (ultra-)mafic crust, which is thought to have resulted in reduced Ni availability in seawater (Konhauser et al. 2009, 2015). Nickel is a metal cofactor in methanogenesis (Diekert et al. 1981), perhaps one of the oldest metabolisms on Earth (Woese 1987), and so the decline in Ni availability may have caused a methanogen famine (Konhauser et al. 2009, 2015). In other words, Ni-based methanogenesis may have been elevated in the earlier Archean when Ni supplies were higher. Similar to Ni, Co is an important cofactor in deeply rooted metabolic pathways (hydrogenotrophic acetogenesis and methanogenesis, see Gottschalk and Thauer 2001; Ragsdale and Pierce 2008), whose availability decreased with the rise of O_2 (Moore et al. 2018). Manganese is an essential element in the oxygen evolving complex (OEC) of all oxygenic photosynthesizers (Ghanotakis and Yocum 1990; Roelofs et al. 1996). The ability of the OEC and photosystem II to split water and take the protons and electrons from the H_2O molecule to reduce the carbon in CO_2 is a crucial biological innovation, because it allows oxygenic photosynthesizers to use highly abundant water as an electron source. Manganese was more available in the Archean when geochemical evidence indicates that oxygenic photosynthesis evolved (Crowe et al. 2013; Planavsky et al. 2014; Cardona et al. 2019) (see also below). The rise of oxygen has led to decreasing Mn availability through time (Robbins et al. 2023).

Increasing O_2 levels towards the end of the Archean and across the GOE likely impacted nutrient availability in two main ways (see also Saito et al. 2003; Hao et al. 2017) (Fig. 9). First, elements or species that are more soluble in the presence of O_2 would have become more bioavailable in oxic surface waters at this time. Key examples probably included nitrate (see above) as well as Mo in the form of molybdate (Scott et al. 2008; Reinhard et al. 2013). Evidence for oxic surface waters during the Proterozoic is preserved in the form of elevated iodate concentrations in shallow marine carbonates (Hardisty et al. 2017). Second, increasing O_2 in the atmosphere would have stimulated the sulfur cycle through the onset of oxidative weathering on land with indirect effects on the solubility of sulfur-sensitive metals. In the Archean, isotopic data suggest that sulfate levels were <200 µM (less than one hundredth of present levels, Habicht et al. 2002) and possibly below 2.5 µM (Crowe et al. 2014), which resulted in supressed sulfate reduction and a carbon cycle dominated by methanogenesis. The major source of sulfur to the ocean was volcanic outgassing. Following the GOE, sulfate levels probably increased to 10 mM as indicated by the first occurrence of sulfate evaporites (Blättler et al. 2018). Lower levels down to 100 µM have been proposed for the later

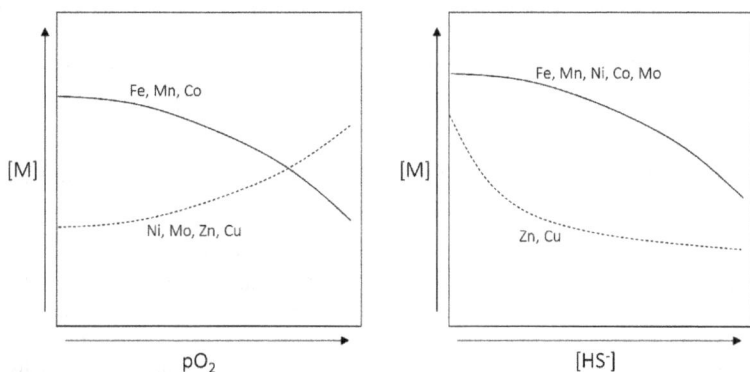

Figure 9. Trace metal concentration responses to increasing pO_2 and [HS$^-$]. The stronger impact of sulfide complexation on the solubility of Zn and Cu compared to other metals is based on differences in solubility constants (Saito et al. 2003).

Proterozoic (Fakhraee et al. 2019). In any case, the elevated flux of sulfate to the deep ocean, where conditions continued to be anoxic, resulted in the production of hydrogen sulfide (H_2S) through biological sulfate reduction. Sulfide rapidly reacts with ferrous iron, forming pyrite in sediments. Where sulfide dominates over iron in the water column, conditions are referred to as euxinic. While it was once thought that such euxinic conditions extended over the entire Proterozoic deep ocean (Canfield 1998), more recent studies on iron speciation have shown that euxina was limited to productive continental margins, akin to modern oxygen minimum zones (Planavsky et al. 2011; Poulton and Canfield 2011; Lyons et al. 2014). Quantitative models suggest that <10 % of the deep ocean was overlain by euxinic waters (Reinhard et al. 2013). Nevertheless, the residence time of transition metals in the ocean was significantly impacted (Fig. 6). In particular Cu and Zn would have been scavenged more rapidly from the water column (Saito et al. 2003), unless stabilized by organic ligands (Stüeken 2020), and also the availability of Mo was overall reduced (Reinhard et al. 2013).

Copper is a crucial metal cofactor for aerobic respiration (Klinman 1996) and denitrification (Delwiche and Bryan 1976), but unlike the marine concentrations of many metal cofactors described above, the availability of Cu probably decreased after the Paleoproterozoic GOE and only increased with ocean oxygenation in the Neoproterozoic or Phanerozoic thus supporting aerobic metabolisms (Saito et al. 2003; Ciscato et al. 2019). Similarly, the availability of Zn first decreased and only increased much later (Scott et al. 2012), which allowed for greater utilization by eukaryotes, including DNA binding zinc-finger proteins (Dupont et al. 2006, 2010). In other words, both Cu and Zn were scarce and possibly limiting for certain metabolisms throughout the Archean and most of the Proterozoic.

Persistent anoxia in the deep ocean was probably promoted by a combination of continued influx of Fe^{2+} from hydrothermal activity and a generally warm climate, which reduced oxygen solubility (Knauth 2005). Some studies propose that anoxic conditions led to sedimentary phosphate release, and the enhanced nutrient supply fuelled high biological productivity and oxygen demand resulting in further oxygen depletion and sulfide build-up via sulfate reduction (Meyer and Kump 2008). This may be consistent with the appearance of phosphorites in parts of the Proterozoic (Papineau 2010). However, as noted above, the behavior of P in the Precambrian ocean is contested (Derry 2015; Reinhard et al. 2017b; Laakso and Schrag 2019; Brady et al. 2022; Ingalls et al. 2022; Kipp 2022), and it is also possible that anoxia was maintained because low nutrient availability limited biological O_2 production (Crockford et al. 2018; Laakso and Schrag 2019; Ozaki et al. 2019; Hodgskiss et al. 2020). Remaining uncertainties about the causes of marine redox stratification during the Proterozoic make it difficult to extrapolate this scenario to other inhabited planets. However, given the longevity of this state in Earth's history, it appears plausible that similar conditions could arise elsewhere, provided that O_2 production emerges. Hence understanding the production of biosignatures on a Proterozoic-like planet is crucial in our search for life elsewhere.

QUANTITATIVE CONSTRAINTS ON ATMOSPHERIC BIOSIGNATURES THROUGH TIME

Armed with this framework of Earth's geobiological evolution over the past four billion years, we can now place constraints on the types of biosignatures that microbial biospheres are able to generate and that may be detectable on exoplanets with similar environmental conditions. We have detected 5000+ exoplanets, which is just a tiny fraction of the planets in our galaxy (Christiansen 2022). The nearest world to us, Proxima Centauri b, is 4.2 light years away (Anglada-Escudé et al. 2016). It is just one of a couple dozen of known planets that are roughly Earth-sized and in the Habitable Zone of their stars where surface liquid water and life may be possible (Hill et al. 2023). Whether any of these worlds host life is among the most fundamental questions of science. However, all of these worlds are sufficiently distant that we will not be able to search for signs life in the same ways as we have in Earth's earliest geological record, on present-day Mars, and the icy moons of our Solar System.

We must instead remotely recognize the presence of alien biospheres and characterize their biogeochemical cycles in planetary spectra obtained with large ground- and space-based telescopes (e.g., Fujii et al. 2018; Schwieterman et al. 2018). These telescopes can probe atmospheric composition by detecting absorption features associated with specific gases. We might also be able to recognize global-scale surface features, including light interaction with photosynthetic pigments (Seager et al. 2005) and 'glint' arising from specular reflection of light by a liquid ocean (Robinson et al. 2010).

Unfortunately, key details regarding marine environments will be inaccessible with telescopes (Olson et al. 2020). For example, neither the oxygen landscape of the ocean nor availability of bioessential nutrients can be directly observed. Moreover, biogenic gases produced by marine microbes may be consumed by other microbial metabolisms before entering the atmosphere, and biogenic gases do not necessarily accumulate in the atmosphere or imprint their presence in planetary spectra (Reinhard et al. 2017a). The detectability of biosignatures is also strongly influenced by instrument design and viewing geometries. The wealth of information contained within Earth's geochemical and fossil records thus does not always directly inform the signatures of Earth life that would be detectable to alien observers of our planet. It is nonetheless an important starting point. Unravelling the details of Earth's complex biogeochemical history and its relationship with remotely observable spectral signals is an important consideration for instrument design and our own search for life in the Universe (Reinhard et al. 2019).

Oxygen (O_2)

Oxygen is by far the most commonly discussed remote biosignature (Meadows et al. 2018; Schwieterman et al. 2018). Earth's oxygen-rich atmosphere that we depend on is itself the consequence of life, specifically oxygenic photosynthesis, which probably dates back to >3 Ga (see above). Hence detecting O_2 in an exoplanet atmosphere could suggest the presence of life beyond Earth. However, although O_2 is a glaring signal of life on our planet, the presence of O_2 does not uniquely indicate life. This is because there are several non-biological ways to produce O_2, potentially leading to abiotic signals that mimic life (biosignature "false positives"). Ways of abiotic oxygen production include H_2O photolysis and subsequent H escape to space, driving atmospheric oxidation. This scenario is particularly likely to play out on M dwarf planets that experience a hostile UV environment during the pre-main sequence evolution of their star, possibly leading to planetary desiccation and very high levels of abiotic oxygen (e.g., Luger and Barnes 2015). Photolysis of CO_2 can also lead to abiotic O_2 accumulation on M-dwarf planets with H_2O-depleted, CO_2-dominated atmospheres (Gao et al. 2015). A scenario for abiotic O_2 on planets orbiting Sun-like stars is the lack an effective "cold trap." H_2O condenses as rising air expands and cools, effecting trapping H_2O near Earth's surface. However, planets with low inventories of non-condensing gases such as N_2 allow H_2O to enter the upper atmosphere, enhancing H_2O photolysis, subsequent H escape, and planetary oxidation (Wordsworth and Pierrehumbert 2014). High obliquity (tilt of the rotational axis with respect to the orbital plane) may also lead to weakening of the cold trap (Kang 2019). Fortunately, these scenarios can be recognized as non-biological with appropriate context. For example, CO_2 photolysis produces CO in addition to O_2; high surface pressure can help rule out scenarios involving a weak cold trap; and many bars of O_2 or a lack of H_2O may be incompatible with life (Meadows et al. 2018). Oxygen must therefore be considered in its broader stellar, orbital, and planetary context, considering both the presence and absence of other chemical species.

Earth's history highlights the possibility of "false negatives" for life as well (Reinhard et al. 2017a), a challenge that may be more difficult to overcome. As discussed above, oxygen was much lower than today for the vast majority of Earth's history, even long after the onset of biological oxygen production during photosynthesis due to continued influx of reductants from the crust and hydrothermal sources. Oxygen was limited to "oxygen oases" in the Archean surface ocean for up to 500 million years before the GOE and atmospheric pO_2 did not approach near modern levels for another 2 billion years after that (Olson et al. 2018). Prior to the GOE, oxygenic photosynthesis would have been undetectable to a remote observer based on known Earth-based technology. For the next 2 billion years of the Proterozoic, pO_2 (0.1–10% modern levels, e.g., Laakso and Schrag 2017; Wang et al. 2022) was elevated relative to the Archean but still lower than today, as the ocean continued to experience a high influx of Fe(II) from hydrothermal sources and a lack of nutrients potentially limited oxygenic photosynthesis (see above). Detection of O_2 at the high end of estimated levels may be feasible with a future direct imaging mission but will nonetheless depend strongly on instrument design and favorable viewing geometries (Young et al. 2024). Oxygen levels at the lower end of estimates will likely be inaccessible remotely (Robinson and Reinhard 2018). We thus must be mindful that Earth-like biospheres on exoplanets may lack detectable O_2.

Fortunately, even low levels of oxygen may lead to photochemical production of O_3. Ozone absorbs strongly in the UV and mid-IR, and it is much more readily detectable than O_2. Ozone may therefore be our most reliable indicator of Proterozoic-like biospheres (Reinhard et al. 2017a, 2019), even though it is not a direct biological product. Ultimately, understanding the relationships between life, marine habitats, and atmospheric chemistry is essential to interpreting detections (or non-detections) of oxygen in exoplanet atmospheres. The absence of detectable atmospheric oxygen does not preclude aerobic ecosystems (Olson et al. 2013), but it may preclude animal-grade complexity (Catling et al. 2005). On the other hand, detection of O_2 or its photochemical product O_3 does not necessarily imply a persistently well-oxygenated ocean capable of supporting animal-grade complexity (Reinhard and Planavsky 2022). Nevertheless, O_2 remains an important target for next-generation telescopes in the search for life despite these challenges.

Methane (CH_4)

Earth's CH_4 is overwhelmingly biological, both today and in the past. Moreover, methanogenesis is an ancient metabolism on Earth, dating back to 3.5 Ga (see above). As noted above, it is formed from CO_2 and H_2, which are generated by volcanic and hydrothermal processes that are likely common in the Universe. Methane is thus widely considered a potential biosignature. However, several abiotic sources of CH_4 are also known. Volcanic outgassing, serpentinization and impacts can all produce CH_4, albeit at low rates compared to life on Earth (Thompson et al. 2022). CH_4, not unlike O_2, thus requires thorough consideration of its context. For example, the co-existence of CO_2 (C in its most oxidized form) and CH_4 (C in its most reduced form) represents chemical disequilibrium, which is most readily explained by life in the absence of their redox intermediate, CO (Krissansen-Totton et al. 2016; Thompson et al. 2022). Abiotic CH_4 is common in the solar system, and equilibrium chemistry on some exoplanets may produce co-existing CH_4 and CO_2—but all these scenarios involve production of CO alongside CH_4. Co-existence of CH_4 and CO_2, without CO, thus implies sustained surface fluxes of both endmember redox species and is a robust signature of life (Krissansen-Totton et al. 2018a).

It is widely believed that lower oxygen in the ancient ocean and atmosphere led to higher levels of atmospheric methane (e.g., Kasting 2005). The reasons are three-fold. First, lower levels of O_2 would have directly limited aerobic respiration and indirectly limited anaerobic respiration through the scarcity of alternate electron acceptors such as NO_3^- or SO_4^{2-} that are most readily produced in the presence of O_2. This limitation increases the ecological importance of less energetic metabolisms, such as methanogenesis. Second, the scarcity of electron acceptors further limits biological consumption of CH_4, increasing the fraction of biogenic CH_4 that actually reaches the atmosphere. Finally, an atmosphere that was less oxidizing could be more favorable to CH_4 accumulation. However, quantifying exactly how atmospheric CH_4 has varied through time nonetheless remains challenging.

Whereas oxygen is reactive and leaves its fingerprints all over the geochemical record (reviewed above), CH_4 leaves a more subtle signal. As noted earlier, the earliest geochemical hints of biogeochemical methane cycling included isotopically depleted organics and CH_4 gas inclusions in Archean sediments at 3.5–3.4 Ga (Ueno et al. 2006; Flannery et al. 2018). This signature becomes more conspicuous in the Neoarchean around 2.7 Ga (e.g., Eigenbrode and Freeman 2006). Conventional interpretation of these Neoarchean data implies biological assimilation of methane, which is typically isotopically depleted relative to other C substrates. This signature reflects CH_4 consumption (i.e., methanotrophy) from a methanogenic source, and although it could be compatible with high concentrations of CH_4 it only requires CH_4 production, with or without accumulation. An alternative interpretation of isotopically light organics in the Archean is that these organics were produced photochemically and subsequently deposited from the atmosphere, implying a sufficiently CH_4-rich atmosphere to trigger hydrocarbon polymerization in an organic haze (Pavlov et al. 2001). This interpretation may also find support in nuanced S isotope patterns that may arise from fluctuations in UV attenuation by an organic haze (Zerkle et al. 2012; Izon et al. 2017). Haze forms for CH_4:CO_2 ratios > 0.1 (e.g., Zerkle et al. 2012). Reasonable estimates of Archean pCO_2 imply CH_4 levels on the order of ~1000 ppm (Izon et al. 2017), which would be detectable with the James Webb Space Telescope (JWST) and next-generation telescopes (Krissansen-Totton et al. 2018b). The development of such a haze may affect our ability to characterize some planetary features, but such a haze may itself constitute a biosignature (Arney et al. 2018). It is worth noting that abiotic hazes are common in the outer solar system and may be common among close-in exoplanets with radii larger than Earth but smaller than Neptune ("sub-neptunes", Bergin et al. 2023; Gao et al. 2023). The likelihood of abiotic hazes highlights the importance of interpreting hazes in their planetary context, not unlike other biosignatures.

Atmospheric CH_4 is thought to have plummeted in association with oxygenation during the GOE, but it remains debated whether such a decline was a cause (decreased O_2 sink, Konhauser et al. 2009) or consequence (less methanogeneis, more methanotrophy, Pavlov et al. 2003). In either scenario, loss of greenhouse warming by CH_4 appears to have contributed to global glaciation in the Paleoproterozoic (Kirschvink et al. 2000; Kopp et al. 2005). The establishment of an O_3 layer following the GOE may have increased the chemical lifetime of CH_4 (Goldblatt et al. 2006), leading to at least partial recovery of pCH_4 if similar biogenic fluxes were sustained (Claire et al. 2006). In weakly oxygenated atmospheres, pCH_4 actually increases with pO_2 due to stronger UV shielding by O_3 (Fig. 10).

Higher CH_4 fluxes than today would be anticipated from the redox-stratified Proterozoic ocean (see above). However, increased SO_4^{2-} availability arising from oxidative weathering of the continents may have severely curtailed CH_4 production and preservation. Anaerobic respiration coupling organic carbon oxidation to SO_4^{2-} reduction is a more energetic metabolism than methanogenesis, limiting CH_4 production in the presence of SO_4^{2-}. At the same time, anaerobic methanotrophy that couples SO_4^{2-} reduction to CH_4 oxidation is an efficient sink for biogenic CH_4. As a result, Proterozoic pCH_4 may have only been ~10× higher than at present-day (Olson et al. 2016; Reinhard et al. 2020) (Fig. 11), much lower than early estimates of 100x present day levels (Pavlov et al. 2003; Kasting 2005).

Ironically, the decline of CH_4 biosignatures in this case is the consequence of life. One microbe's waste is another's food, and a perfectly efficient biosphere would not produce any net O_2 or CH_4 for remote observers to detect. Accumulation of these gases in the atmospheres requires inefficiency such as organic carbon burial or slow biological destruction of metabolic

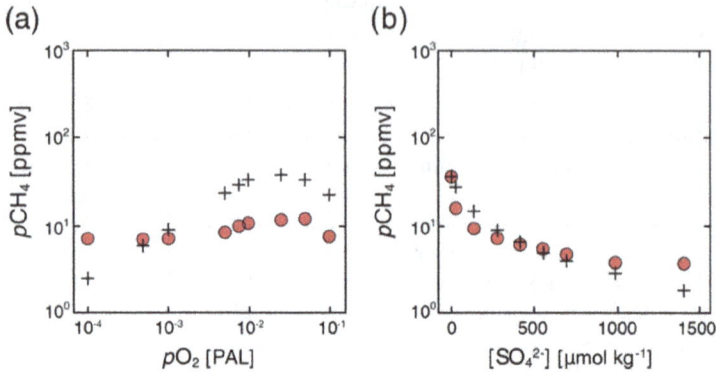

Figure 10. Sensitivity of atmospheric methane to atmospheric oxygen (**a**) and oceanic sulfate (**b**). **Black crosses** denote simulations from Olson et al. 2016 and **red circles** are from Reinhard et al. (2020). pCH$_4$ responds non-linearly to pO$_2$ due to UV attenuation by O$_3$ (Claire et al. 2006; Goldblatt et al. 2006), and pCH$_4$ drops off rapidly with increasing SO$_4^{2-}$ (Olson et al. 2016). This figure modified from Reinhard et al. (2020) under Creative Commons Attribution License CC-BY 4.0.

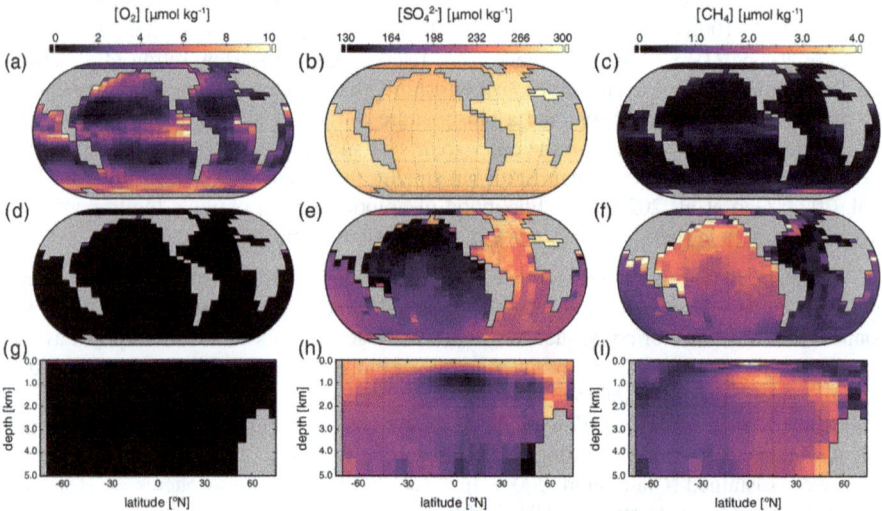

Figure 11. Simulated oxygen (**left; a,d,g**), sulfate (**middle; b,e,h**), and methane (**right; c,f,i**) concentrations in a Proterozoic-like ocean with 10–3 PAL O2. Maps show each species in surface seawater (**top; a–c**) and benthic seawater (**middle; d–f**). Also shown are zonal average vertical profiles (**bottom; g–i**). Oxygen is limited to oxygen oases in surface seawater, leading to extensive SO$_4^{2-}$ reduction and local SO$_4^{2-}$ scarcity at depth, especially underlying productive regions of the surface ocean and in the oldest bottom waters. Where SO$_4^{2-}$ concentrations are low, methanogenesis leads to CH$_4$ accumulation, but CH$_4$ must evade oxidation by both SO$_4^{2-}$ in the ocean interior and O$_2$ in surface waters before it may enter the atmosphere. This figure is reproduced from Reinhard et al. (2020) under Creative Commons Attribution License CC-BY 4.0.

waste products in the ocean relative to the rate at which they escape to the atmosphere. The strong sensitivity of pCH$_4$ to marine SO$_4^{2-}$ is a challenge for remotely characterizing biospheres because the SO$_4^{2-}$ ion does not exchange with the atmosphere like gases such as O$_2$ and CH$_4$, and SO$_4^{2-}$ does not come to equilibrium with an atmospheric phase. It is thus not possible to infer ocean SO$_4^{2-}$ levels and their consequences for marine CH$_4$ production based on atmospheric chemistry.

The downward revision of Proterozoic pCH_4 somewhat reduces the appeal of CH_4 as a biosignature for Proterozoic-like biospheres around Sun-like stars (Reinhard et al. 2017a; Olson et al. 2018), but higher CH_4 levels would be expected on planets with similarly productive biospheres orbiting cooler M dwarfs because these worlds receiver fewer UV photons (Segura et al. 2005). Moreover, as a greenhouse gas, CH_4 has a number of spectral features in the IR and is among the most accessible biosignatures in the short-term with JWST (Krissansen-Totton et al. 2018b), which is unlikely to detect even modern levels of O_2 (Fauchez et al. 2020).

Nitrous oxide (N_2O)

Nitrous oxide (N_2O) is another popular remote biosignature candidate (e.g., Rauer et al. 2011; Schwieterman et al. 2018). N_2O is produced as a waste product of incomplete denitrification in redox-stratified environments. Denitrification involves the reduction of NO_3^- to N_2 via several intermediate species, including N_2O. Some N_2O may escape full reduction to N_2 ('partial' or 'incomplete' denitrification) and enter the atmosphere. As a greenhouse gas, N_2O has several spectral features in the IR that could fingerprint life at sufficiently high levels with JWST (Schwieterman et al. 2022). Denitrification is limited to O_2-depleted environments today. Likewise, denitrification would have been limited on a broadly anoxic Archean Earth because nitrate would have been scarce outside of oxygen oases (Godfrey and Falkowski 2009). However, the redox-stratified ocean of the Proterozoic would provide optimal conditions for N_2O production (see above) (Fig. 12). There would have been enough O_2 in Proterozoic surface waters to stabilize NO_3^-, but the widely anoxic deep ocean would have allowed extensive NO_3^- reduction (Fennel et al. 2005). As noted above, geochemical evidence for the presence of nitrate in surface waters goes back to approximately 2.7 Ga and becomes more established after the GOE (Godfrey and Falkowski 2009) (Fig. 8).

Unlike CH_4, N_2O production would not be strongly curtailed by SO_4^{2-} accumulation in the ocean because denitrification is more energetically favorable than SO_4^{2-} reduction. Sulfide arising from SO_4^{2-} reduction may even enhance global N_2O production if local sulfidic conditions limited the availability of metal co-factors involved in the reduction of N_2O to N_2 (Buick 2007b; Roberson et al. 2011).

A recent study showed Proterozoic-like planets orbiting Sun-like stars may accumulate 5–50 ppm N_2O (compared to 330 ppb today) if a significant fraction of the total denitrification flux from the ocean is released as N_2O (Schwieterman et al. 2022). Around M or K dwarf stars, similar early Earth-like fluxes may lead to even higher pN_2O—perhaps a 1000+ ppm—owing to muted photochemical destruction, further increasing its appeal as a potential biosignature for

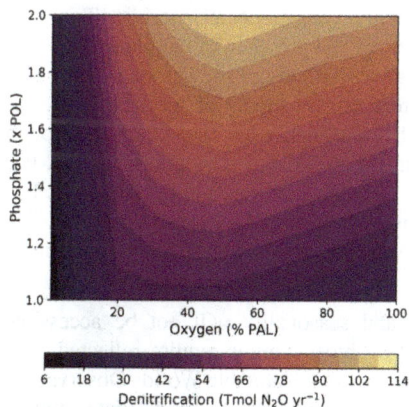

Figure 12. Denitrification rates (i.e., maximum N_2O production fluxes; contours) as a function of atmospheric oxygen (x-axis) and oceanic phosphate (y-axis). Dentrification rates are optimized for intermediate pO_2. Denitrification is further sensitive to nutrient availability, which dictates the flux of particulate organics available to denitrifiers. Figure reproduced from Schwieterman et al. (2022) under Creative Commons Attribution License CC-BY 4.0.

weakly oxygenated worlds (Segura et al. 2005; Schwieterman et al. 2022). N_2O may be less vulnerable to non-biological false positives than CH_4 and O_2. As mentioned above, reduction of nitrite by Fe(II) can produce N_2O without direct involvement of life (Stanton et al. 2018; Buessecker et al. 2022). However, the oxidized N species required for this process are an indirect consequence of oxygenic photosynthesis. A scenario where an abiotically oxygenated atmosphere leads to oxidized N that is then reduced by Fe is unlikely because known pathways to abiotic O_2 are likely incompatible with an Fe(II)-rich ocean (Schwieterman et al. 2022). N_2O arising from this abiotic pathway is therefore a potential biosignature for exoplanet atmospheres, much like O_3 is a potential biosignature despite its abiotic production in the atmosphere.

Other signatures of life

There are a number of trace biogenic gases on Earth that are not detectable remotely, today or in the distant past, largely due to rapid destruction in the atmosphere. Examples include dimethyl sulfide (DMS), which is produced by marine photosynthesizers (Domagal-Goldman et al. 2011); isoprene, which is primarily produced by deciduous trees but is common across the tree of life (Zhan et al. 2021); and methyl chloride or bromide (CH_3Cl and CH_3Br, respectively), which are widely produced by bacteria, algae, and fungi as a detoxification strategy in a variety of marine and terrestrial environments (Leung et al. 2022). On other worlds, especially those orbiting M-dwarf host stars with lower UV fluxes, these species may accumulate to higher, remotely detectable levels (e.g., Segura et al. 2005). Investigating the detectability of trace biogenic gases and identifying novel biosignature gases remains an active area of research (Seager et al. 2016).

Absorption features arising from biogenic gases are not the only signal of life that we may be able to remotely detect. We may also be able to directly identify biological interaction with light. Most notably, absorption of visible photons and enhanced in the near-IR by the photosynthetic pigment chlorophyll is detectable in satellite observations of Earth from space and in observations of earthshine (Seager et al. 2005). This signal is largely the consequence of land vegetation today, but similar signals may arise from lichen or photosynthetic marine microbes on planets more similar to early Earth (O'Malley-James and Kaltenegger 2019). The manifestation of these "surface biosignatures" may vary on shorter timescales as well. Photosynthesis, whether by land plants or microbes, varies seasonally on Earth as light, temperature, and nutrient conditions evolve.

This seasonal variability also affects the production/consumption of gases that interact with life (Olson et al. 2018). Atmospheric O_2, CH_4, and N_2O all oscillate seasonally as photosynthesis waxes and wanes. Carbon dioxide is not itself a biosignature, but it is fixed into biomass by photosynthesis and released to the environment during respiration. As a result, CO2 levels vary seasonally, and these oscillations are biogenic in origin. Temporal variability in atmospheric composition could thus reveal the presence of exoplanet life. The detectability of this biosignature depends on both the magnitude of the seasonal change as well as the average abundance of the gas, which affects the sensitivity of spectral features to changes in concentration. The detectability of Earth's seasonality has thus changed through time as the biosphere and atmospheric composition has co-evolved (Olson et al. 2018). For example, O_2 seasonality is negligible compared to a background of 21% by volume today, but O_2 seasonality could have been more dramatic in the past when O_2 levels were lower. The expression of seasonality may also differ among Earth-like worlds with differing obliquities or eccentricities (Jernigan et al. 2023).

Unlike CH_4 and N_2O, surface biosignatures and seasonality will not be accessible in transmission spectra with JWST. Access to surface biosignatures require reflected light spectroscopy with a direct imaging telescope, such as NASA's Habitable Worlds Observatory. As mentioned above, surface biosignatures like the red edge of vegetation are detectable in the visible and near-IR. Seasonality is also most readily characterized with direct imaging, but may be accessible in both reflected visible light and emitted light in the mid-IR with a telescope such as the Large Interferometer for Exoplanets (LIFE) mission concept (Quanz et al. 2022; Mettler et al. 2023). Consideration of how these phenomena have varied through Earth history and how their detectability may differ under a diversity of stellar, orbital, and planetary scenarios remains an exciting opportunity for future work.

CONCLUSIONS

We conclude by summarizing the most important lessons that the early Earth can teach us for our quest of life on other planets:

(1) Earth's atmosphere was not accreted from the solar nebular in its current form; it is the product of volcanic outgassing, ingassing, loss of light volatiles to space and biological metabolisms. The redox state of outgassing volatiles has likely been oxic throughout Earth's history (i.e., dominated by N_2, H_2O, CO_2 and SO_2 as opposed to NH_3, H_2, CO, CH_4 and H_2S), which carries implications for the utility of certain gases as biosignatures. Ingassing, i.e., recycling of volatiles back into the mantle, is not dependent on the establishment of modern-style plate tectonics. Biological gas production is dependent on metabolic innovations but also on the supply of nutrients, which ties biological activity to abiotic supplies of substrates. General trends in atmospheric evolution are summarized in Figure 13.

(2) The record of life is fragmented by loss and alteration of the rock record, but remaining evidence dates back to 3.5 Ga and possibly 3.7 Ga. Life may thus have originated on a water world where continents were largely submerged and land exposure was perhaps limited to volcanic islands. The tectonic regime may have resembled a stagnant lid, akin to Venus or Mars very early in Earth history, but subduction commenced by at least 3.7 Ga and modern-style plate tectonics was in operation by ca. 3 Ga. Abundant volcanism and hydrothermal activity likely generated large fluxes of CO_2 and H_2, which are substrates for biological methanogenesis. Biogenic CH_4 may have been abundant enough to be remotely detectable at that time. The predominance of (ultra-)mafic volcanic rocks led to elevated concentrations of Fe, Ni and Co in seawater, which was fully anoxic in the early Archean.

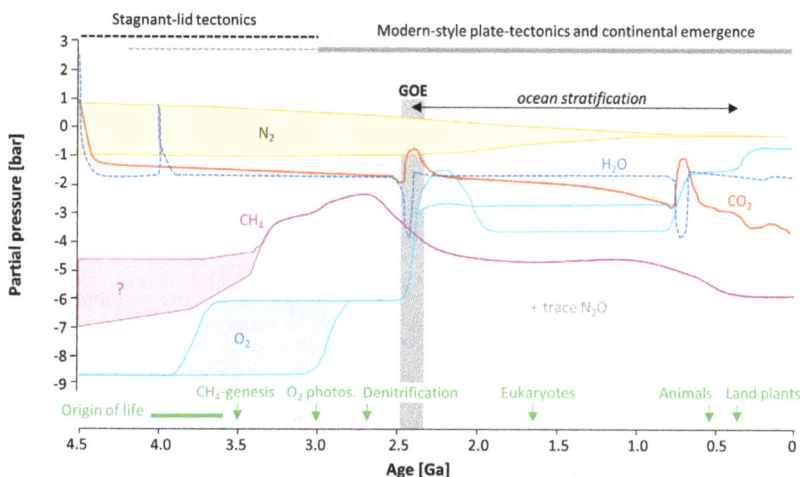

Figure 13. Trends in the abundances of major atmospheric gases over Earth's history (modified after Goldblatt 2017), along with key biological innovations (annotations in **green** at the bottom) and tectonic regimes (**black and gray bars** at the top). See text for references.

(3) Oxygenic photosynthesis likely evolved in the early to mid-Archean; however, due to continued input of reductants and perhaps nutrient limitation, biogenic O_2 was not able to accumulate in the atmosphere until 2.4 Ga (the Great Oxidation Event, GOE) and therefore not detectable remotely. Around the time when oxygenic photosynthesis emerged, the rock record shows evidence for widespread modern-style plate tectonics and continental emergence. It is conceivable that the provision of large continental shelves created new habitat space for biological innovation.

(4) From the GOE onwards, the Earth looked tectonically similar to today; however, the ocean was redox stratified with an oxygenated surface layer above anoxic bottom waters. Atmospheric pO_2 was likely intermediate between Archean and modern values. Anoxia was buffered by a continued supply of ferrous iron from hydrothermal sources. Seawater sulfate levels increased, which may have reduced fluxes of CH_4 into the atmosphere and rendered Earth's biosphere overall less detectable during its 'middle age'. A potentially important biosignature at this time may have been N_2O, whose production would have increased with increasing nitrate levels in surface waters and high potential for biological denitrification to N_2O along the redox-cline.

It is important to note that much of these inferences rest on assumptions about total atmospheric pressure, global average surface temperature, and the major ion chemistry of seawater over time. These parameters are very difficult to constrain from existing geochemical proxies. The early Earth can therefore provide only limited insights into biosignatures on planets with widely different boundary conditions. However, if nothing else, the Earth teaches us that the establishment, maintenance and detection of life on any planet is contingent upon a favorable interplay between geological and biological processes.

ACKNOWLEDGMENTS

We thank Eddie Schwieterman and Martin van Kranendonk for helpful comments that improved the manuscript. We also thank Clint Scott for constructive feedback on an earlier version. The editors of this issue are thanked for the invitation to contribute. EES was financially supported by a NERC Frontiers grant (NE/V010824/1) and a Leverhulme Trust grant (RPG–2022–313) during this project. SLO acknowledges support from NASA Exobiology, NASA Habitable Worlds, NASA ICAR, and the Heising-Simons Foundation. EM was supported financially by the Energy Resources Program of the U.S. Geological Survey. Any use of trade, firm, or product names is for descriptive purposes only and does not imply endorsement by the U.S. Government.

REFERENCES

Abramov O, Mojzsis SJ (2009) Microbial habitability of the Hadean Earth during the late heavy bombardment. Nature 459:419–422

Algeo TJ, Lyons TW (2006) Mo–total organic carbon covariation in modern anoxic marine environments: Implications for analysis of paleoredox and paleohydrographic conditions. Paleoceanography 21:PA1016, https://doi.org/10.1029/2004PA001112

Allwood AC, Walter MR, Kamber BS, Marshall CP, Burch IW (2006) Stromatolite reef from the Early Archaean era of Australia. Nature 441:714–718

Allwood AC, Rosing MT, Flannery DT, Hurowitz JA, Heirwegh CM (2018) Reassessing evidence of life in 3,700-million-year-old rocks of Greenland. Nature 563:241–244

Altermann W, Nelson DR (1998) Sedimentation rates, basin analysis and regional correlations of three Neoarchaean and Palaeoproterozoic sub-basins of the Kaapvaal craton as inferred from precise U Pb zircon ages from volcaniclastic sediments. Sediment Geol 120:225–256

Anbar A (2008) Elements and evolution. Science 322:1481–1483

Anbar AD, Knoll AH (2002) Proterozoic ocean chemistry and evolution: a bioinorganic bridge? Science 297:1137–1142

Anglada-Escudé G, Amado PJ, Barnes J, Berdiñas ZM, Butler RP, Coleman GA, de La Cueva I, Dreizler S, Endl M, Giesers B, Jeffers SV (2016) A terrestrial planet candidate in a temperate orbit around Proxima Centauri. Nature 536:437–440

Arney G, Domagal-Goldman SD, Meadows VS (2018) Organic haze as a biosignature in anoxic Earth-like atmospheres. Astrobiology 18:311–329

Bada JL, Korenaga J (2018) Exposed areas above sea level on Earth> 3.5 Gyr ago: Implications for prebiotic and primitive biotic chemistry. Life 8:55

Baross J, Hoffman SE (1985) Submarine hydrothermal vents and associated gradient environments as sites for the origin and evolution of life. Origins of Life and Evolution of Biospheres 15:327–345

Barth P, Stüeken EE, Helling C, Schwieterman EW, Telling J (in review) The effect of lightning on the atmospheric chemistry of exoplanets and potential biosignatures. Astronom Astrophys, https://doi.org/10.48550/arXiv.2305.05697

Bauer AM, Reimink JR, Chacko T, Foley BJ, Shirey SB, Pearson DG (2020) Hafnium isotopes in zircons document the gradual onset of mobile-lid tectonics. Geochem Perspect Lett 14:1–6

Baumgartner R, van Kranendonk MJ, Wacey D, Fiorentini M, Saunders M, Caruso S, Pagès A, Homann M, Guagliardo P (2019) Nano-porous pyrite and organic matter in 3.5 billion-year-old stromatolites record primordial life. Geology 47:1039–1043

Baumgartner R, Caruso S, Fiorentini ML, van Kranendonk MJ, Martin L, Jeon H, Wacey D, Pagès A (2020a) Sulfidization of 3.48 billion-year-old stromatolites of the Dresser Formation, Pilbara Craton: Constraints from in-situ sulfur isotope analysis of pyrite. Chem Geol 538:119488

Baumgartner RJ, Van Kranendonk MJ, Pagès A, Fiorentini ML, Wacey D, Ryan C (2020b) Accumulation of transition metals and metalloids in sulfidized stromatolites of the 3.48 billion–year–old Dresser Formation, Pilbara Craton. Precambrian Res 337:105534

Bédard JH (2006) A catalytic delamination-driven model for coupled genesis of Archaean crust and sub-continental lithospheric mantle. Geochim Cosmochim Acta 70:1188–1214

Behrenfeld MJ, Kolber ZS (1999) Widespread iron limitation of phytoplankton in the South Pacific Ocean. Science 283:840–843

Behrenfeld MJ, Bale AJ, Kolber ZS, Aiken J, Falkowski PG (1996) Confirmation of iron limitation of phytoplankton photosynthesis in the equatorial Pacific Ocean. Nature 383:508–511

Bekaert DV, Turner SJ, Broadley MW, Barnes JD, Halldórsson SA, Labidi J, Wade J, Walowski KJ, Barry PH (2021) Subduction-driven volatile recycling: A global mass balance. Annu Rev Earth Planet Sci 49:37–70

Bekker A, Holland HD, Wang P-L, Rumble III D, Stein HJ, Hannah JL, Coetzee LL, Beukes NJ (2004) Dating the rise of atmospheric oxygen. Nature 427:117–120

Bell EA, Boehnke P, Harrison TM, Mao WL (2015) Potentially biogenic carbon preserved in a 4.1 billion-year-old zircon. PNAS 112:14518–14521

Benner SA, Ricardo A, Carrigan MA (2004) Is there a common chemical model for life in the universe? Curr Opinion Chem Biol 8:672–689

Bergin EA, Kempton EM-R, Hirschmann M, Bastelberger ST, Teal DJ, Blake GA, Ciesla FJ, Li J (2023) Exoplanet volatile carbon content as a natural pathway for haze formation. Astrophys J Letters 949:L17

Blättler CL, Claire MW, Prave AR, Kirsimäe K, Higgins JA, Medvedev PV, Romashkin AE, Rychanchik DV, Zerkle AL, Paiste K, Kreitsmann T (2018) Two-billion-year-old evaporites capture Earth's great oxidation. Science 360:320–323

Boden JS, Konhauser KO, Robbins LJ, Sánchez-Baracaldo P (2021) Timing the evolution of antioxidant enzymes in cyanobacteria. Nat Commun 12:4742, https://doi.org/10.1038/s41467-021-24396-y

Boehnke P, Harrison TM (2016) Illusory late heavy bombardments. PNAS 113:10802–10806

Borrelli ME, O'Rourke JG, Smrekar SE, Ostberg CM (2021) A global survey of lithospheric flexure at steep-sided domical volcanoes on Venus reveals intermediate elastic thicknesses. J Geophys Res: Planets 126:e2020JE006756

Bottke WF, Norman MD (2017) The late heavy bombardment. Annu Rev Earth Planet Sci 45:619–647

Bowring SA, Williams IS, Compston W (1989) 3.96 Ga gneisses from the slave province, Northwest Territories, Canada. Geology 17:971–975

Brady MP, Tostevin R, Tosca NJ (2022) Marine phosphate availability and the chemical origins of life on Earth. Nat Commun 13:5162, https://doi.org/10.1038/s41467-022-32815-x

Brandes JA, Boctor NZ, Cody GD, Cooper BA, Hazen RM, Yoder Jr HS (1998) Abiotic nitrogen reduction on the early Earth. Nature 395:365–367

Brookins DG (1988) Eh–pH Diagrams for Geochemistry. Springer-Verlag, New York

Brooks C, Hart SR (1974) On the significance of komatiite. Geology 2:107–110

Brown M, Johnson T (2018) Secular change in metamorphism and the onset of global plate tectonics. Am Mineral 103:181–196

Buessecker S, Imanaka H, Ely T, Hu R, Romaniello SJ, Cadillo-Quiroz H (2022) Mineral-catalysed formation of marine NO and N2O on the anoxic early Earth. Nat Geosci 15:1056–1063

Buick R (1992) The antiquity of oxygenic photosynthesis: evidence from stromatolites in sulphate-deficient Archaean lakes. Science 255:74–77

Buick R (2007a) The earliest records of life on Earth. *In*: Planets and Life: The Emerging Science of Astrobiology. Sullivan WTI, Baross J (eds). Cambridge University Press, p 237–264

Buick R (2007b) Did the Proterozoic 'Canfield Ocean' cause a laughing gas greenhouse? Geobiology 5:97–100

Canfield DE (1998) A new model for Proterozoic ocean chemistry. Nature 396:450–453

Canfield DE, Farquhar J (2009) Animal evolution, bioturbation, and the sulfate concentration of the oceans. PNAS 106:8123–8127

Canfield DE, Rosing MT, Bjerrum C (2006) Early anaerobic metabolisms. Philos Trans R Soc B 361:1819–1836

Canfield DE, Glazer AN, Falkowski PG (2010) The evolution and future of Earth's nitrogen cycle. Science 330:192–196

Capo E, Monchamp ME, Coolen MJ, Domaizon I, Armbrecht L, Bertilsson S (2022) Environmental paleomicrobiology: using DNA preserved in aquatic sediments to its full potential. Environ Microbiol 24:2201–2209

Cardona T, Murray JW, Rutherford AW (2015) Origin and evolution of water oxidation before the last common ancestor of the cyanobacteria. Mol Biol Evol 32:1310–1328

Cardona T, Sanchez-Baracaldo P, Rutherford AW, Larkum A (2017) Molecular evidence for the early evolution of photosynthetic water oxidation. bioRxiv, https://doi.org/10.1101/109447

Cardona T, Sánchez-Baracaldo P, Rutherford AW, Larkum AW (2019) Early Archean origin of photosystem II. Geobiology 17:127–150

Catling D (2014) The Great Oxidation Event transition. Treatise Geochem 6:177–195

Catling DC, Zahnle KJ, McKay CP (2001) Biogenic methane, hydrogen escape, and the irreversible oxidation of early Earth. Science 293:839–843

Catling D, Glein CR, Zahnle KJ, McKay CP (2005) Why O$_2$ is required by complex life on habitable planets and the concept of planetary 'oxygenation time'. Astrobiology 5:415–438

Catling DC, Krissansen-Totton J, Kiang NY, Crisp D, Robinson TD, DasSarma S, Rushby AJ, Del Genio A, Bains W, Domagal-Goldman S (2018) Exoplanet biosignatures: A framework for their assessment. Astrobiology 18:709–738

Cavicchioli R (2002) Extremophiles and the search for extraterrestrial life. Astrobiology 2:281–292

Christiansen JL (2022) Five thousand exoplanets at the NASA Exoplanet Archive. Nat Astron 6:516–519

Ciscato ER, Bontognali TRR, Poulton SW, Vance D (2019) Copper and its isotopes in organic-rich sediments: From the modern Peru margin to Archean shales. Geosciences 9:325, https://doi.org/10.3390/geosciences9080325

Claire MW, Catling DC, Zahnle KJ (2006) Biogeochemical modelling of the rise in atmospheric oxygen. Geobiology 4:239–269

Claire MW, Kasting JF, Domagal-Goldman SD, Stüeken EE, Buick R, Meadows VS (2014) Modeling the signature of sulfur mass-independent fractionation produced in the Archean atmosphere. Geochim Cosmochim Acta 141:365–380

Cockell CS, Bush T, Bryce C, Direito S, Fox-Powell M, Harrison JP, Lammer H (2016) Habitability: A review. Astrobiology 16:89–117

Condie K (2007) The distribution of Paleoarchean crust. Developments in Precambrian Geology 15:9–18

Condie KC, Kröner A (2008) When did plate tectonics begin? Evidence from the geologic record. *In*: When did plate tectonics begin on planet Earth. Vol 440. Geol Soc Am Spec Pap, p 281–294

Crockford PW, Hayles JA, Bao H, Planavsky NJ, Bekker A, Fralick PW, Halverson GP, Bui TH, Peng Y, Wing BA (2018) Triple oxygen isotope evidence for limited mid-Proterozoic primary productivity. Nature 559:613–616

Crowe SA, Jones C, Katsev S, Magen C, O'Neill AH, Sturm A, Canfield DE, Haffner GD, Mucci A, Sundby B, Fowle DA (2008) Photoferrotrophs thrive in an Archean Ocean analogue. PNAS 105:15938–15943

Crowe SA, Døssing LN, Beukes NJ, Bau M, Kruger SJ, Frei R, Canfield DE (2013) Atmospheric oxygenation three billion years ago. Nature 501:535–538

Crowe SA, Paris G, Katsev S, Jones C, Kim ST, Zerkle AL, Nomosatryo S, Fowle DA, Adkins JF, Sessions AL, Farquhar J (2014) Sulfate was a trace constituent of Archean seawater. Science 346:735–739

Crowley JW, Gérault M, O'Connell RJ (2011) On the relative influence of heat and water transport on planetary dynamics. Earth Planet Sci Lett 310:380–388

Davaille A, Smrekar SE, Tomlinson S (2017) Experimental and observational evidence for plume-induced subduction on Venus. Nat Geosci 10:349–355

Delwiche CC, Bryan BA (1976) Denitrification. Annu Rev Microbiol 30:241–262

Derry LA (2015) Causes and consequences of mid-Proterozoic anoxia. Geophys Res Lett 42:8538–8546

Des Marais DJ, Harwit MO, Jucks KW, Kasting JF, Lin DNCC, Lunine JI, Schneider J, Seager S, Traub WA, Woolf NJ (2002) Remote sensing of planetary properties and biosignatures on extrasolar terrestrial planets. Astrobiology 2:153–181

Dhuime B, Hawkesworth CJ, Cawood PA, Storey CD (2012) A change in the geodynamics of continental growth 3 billion years ago. Science 335:1334–1336

Dhuime B, Wuestefeld A, Hawkesworth CJ (2015) Emergence of modern continental crust about 3 billion years ago. Nat Geosci 8:552–555

Diekert G, Konheiser U, Piechulla K, Thauer RK (1981) Nickel requirement and Factor F430 content of methanogenic bacteria. J Bacteriol 148:459–464

Dodd MS, Papineau D, Grenne T, Slack JF, Rittner M, Pirajno F, O'Neil J, Little CT (2017) Evidence for early life in Earth's oldest hydrothermal vent precipitates. Nature 543:60–64

Domagal-Goldman SD, Kasting JF, Johnston DT, Farquhar J (2008) Organic haze, glaciations and multiple sulfur isotopes in the Mid-Archean Era. Earth Planet Sci Lett 269:29–40

Domagal-Goldman SD, Meadows VS, Claire MW, Kasting JF (2011) Using biogenic sulfur gases as remotely detectable biosignatures on anoxic planets. Astrobiology 11:419–441

Dong J, Fischer RA, Stixrude LP, Lithgow-Bertelloni CR (2021) Constraining the volume of Earth's early oceans with a temperature-dependent mantle water storage capacity model. AGU Adv 2:e2020AV000323

Dupont CL, Yang S, Palenik B, Bourne PE (2006) Modern proteomes contain putative imprints of ancient shifts in trace metal geochemistry. PNAS 103:17822–17827

Dupont CL, Butcher A, Valas RE, Bourne PE, Caetano-Anollés G (2010) History of biological metal utilization inferred through phylogenomic analysis of protein structures. PNAS 107:10567–10572

Eigenbrode JL, Freeman KH (2006) Late Archean rise of aerobic microbial ecosystems. PNAS 103:15759–15764

Elkins-Tanton LT (2008) Linked magma ocean solidification and atmospheric growth for Earth and Mars. Earth Planet Sci Lett 271:181–191

Elkins-Tanton LT, Smrekar SE, Hess PC, Parmentier EM (2007) Volcanism and volatile recycling on a one-plate planet: Applications to Venus. J Geophys Res: Planets 112:E04S06

Eriksson PG, Cheney ES (1992) Evidence for the transition to an oxygen-rich atmosphere during the evolution of red beds in the lower Proterozoic sequences of southern Africa. Precambrian Res 54:257–269

Fakhraee M, Hancisse O, Canfield DE, Crowe SA, Katsev S (2019) Proterozoic seawater sulfate scarcity and the evolution of ocean–atmosphere chemistry. Nat Geosci 12:375–380

Falkowski PG (2015) Life's Engines. Princeton University Press, Princeton, USA

Farquhar J, Bao H, Thiemens MH (2000) Atmospheric influence of Earth's earliest sulfur cycle. Science 289:756–758

Fauchez TJ, Villanueva GL, Schwieterman EW, Turbet M, Arney G, Pidhorodetska D, Kopparapu RK, Mandell A, Domagal-Goldman SD (2020) Sensitive probing of exoplanetary oxygen via mid-infrared collisional absorption. Nat Astron 4:372–376

Fennel K, Follows M, Falkowski PG (2005) The co-evolution of the nitrogen, carbon and oxygen cycles in the Proterozoic ocean. Am J Sci 305:526–545

Flament N, Coltice N, Rey PF (2008) A case for late-Archaean continental emergence from thermal evolution models and hypsometry. Earth Planet Sci Lett 275:326–336

Flannery DT, Walter MR (2012) Archean tufted microbial mats and the Great Oxidation Event: new insights into an ancient problem. Austral J Earth Sci 59:1–11

Flannery DT, Allwood AC, Summons RE, Williford KH, Abbey W, Matys ED, Ferralis N (2018) Spatially-resolved isotopic study of carbon trapped in ~3.43 Ga Strelley Pool Formation stromatolites. Geochim Cosmochim Acta 223:21–35

Foley BJ (2018) The dependence of planetary tectonics on mantle thermal state: applications to early Earth evolution. Philos Trans R Soc A 376:20170409

Foley BJ, Smye AJ (2018) Carbon cycling and habitability of Earth-size stagnant lid planets. Astrobiology in press:arXiv preprint arXiv:1712.03614

Foley B, Houser C, Noack L, Tosi N (2020) The heat budget of rocky planets. *In*: Planetary Diversity: Rocky planet processes and their observational signatures. Vol 4. EJ Tasker, C Unterborn, M Laneuville, Y Fujii, SJ Desch, HE Hartnett (eds) IOP Publishing, p4-1–4-70

Frost DJ, McCammon CA (2008) The redox state of Earth's mantle. Annu Rev Earth Planet Sci 36:389–420

Fru EC, Somogyi A, El Albani A, Medjoubi K, Aubineau J, Robbins LJ, Lalonde SV, Konhauser KO (2019) The rise of oxygen-driven arsenic cycling at ca. 2.48 Ga. Geobiology 47:243–246

Fuentes JJ, Crowley JW, Dasgupta R, Mitrovica JX (2019) The influence of plate tectonic style on melt production and CO_2 outgassing flux at mid-ocean ridges. Earth Planet Sci Lett 511:154–163

Fujii Y, Angerhausen D, Deitrick R, Domagal-Goldman S, Grenfell JL, Hori Y, Kane SR, Pallé E, Rauer H, Siegler N, Stapelfeldt K (2018) Exoplanet biosignatures: Observational prospects. Astrobiology 18:739–778

Fulton BJ, Petigura EA, Howard AW, Isaacson H, Marcy GW, Cargile PA, Hebb L, Weiss LM, Johnson JA, Morton TD, Sinukoff E (2017) The California-Kepler survey. III. A gap in the radius distribution of small planets. Astron J 154:147, https://doi.org/10.3847/1538-3881/abd806

Gaillard F, Bouhifd MA, Füri E, Malavergne V, Marrocchi Y, Noack L, Ortenzi G, Roskosz M, Vulpius S (2021) The diverse planetary ingassing/outgassing paths produced over billions of years of magmatic activity. Space Sci Rev 217:1–54

Gaillard F, Bernadou F, Roskosz M, Bouhifd MA, Marrocchi Y, Iacono-Marziano G, Moreira M, Scaillet B, Rogerie G (2022) Redox controls during magma ocean degassing. Earth Planet Sci Lett 577:117255

Gaines SM, Eglinton G, Rullkotter J (2009) Echoes of Life: What Fossil Molecules Reveal about Earth History. Oxford University Press

Gao P, Hu R, Robinson TD, Li C, Yung YL (2015) Stability of CO_2 atmospheres on dessicated M dwarf exoplanets. Astrophys J 806:249

Gao P, Piette, A.A., Steinrueck ME, Nixon MC, Zhang M, Kempton EM, Bean JL, Rauscher E, Parmentier V, Batalha NE, Savel AB (2023) The hazy and metal-rich atmosphere of GJ 1214 b constrained by near and mid-infrared transmission spectroscopy. Astrophys J 5:951–996

Garcia AK, McShea H, Kolaczkowski B, Kacar B (2020) Reconstructing the evolutionary history of nitrogenases: evidence for ancestral molybdenum-cofactor utilization. Geobiology 18:394–411

Garcia AK, Kolaczkowski B, Kaçar B (2022) Reconstruction of nitrogenase predecessors suggests origin from maturase-like proteins. Genome Biol Evol 14:evac031, https://doi.org/10.1093/gbe/evac031

Garvin J, Buick R, Anbar AD, Arnold GL, Kaufman AJ (2009) Isotopic evidence for an aerobic nitrogen cycle in the latest Archean. Science 323:1045–1048

Genda H, Abe Y (2003) Survival of a proto-atmosphere through the stage of giant impacts: the mechanical aspects. Icarus 164:149–162

Georgiadis MM, Komiya H, Chakrabarti P, Woo D, Kornuc JJ, Rees DC (1992) Crystallographic structure of the nitrogenase iron protein from *Azotobacter vinelandii*. Science 257:1653–1659

Ghanotakis DF, Yocum CF (1990) Photosystem II and the oxygen-evolving complex. Annu Rev Plant Physiol Plant Mol Biol 41:255–276

Gilleaudeau GJ, Sahoo SK, Ostrander CM, Owens JD, Poulton SW, Lyons TW, Anbar A (2020) Molybdenum isotope and trace metal signals in an iron-rich Mesoproterozoic ocean: A snapshot from the Vindhyan Basin, India. Precambrian Res 343:105718, https://doi.org/10.1016/j.precamres.2020.105718

Ginzburg S, Schlichting HE, Sari RE (2018) Core-powered mass-loss and the radius distribution of small exoplanets. Mon Not R Astron Soc 476:759–765

Godfrey LV, Falkowski PG (2009) The cycling and redox state of nitrogen in the Archaean ocean. Nat Geosci 2:725–729

Goldblatt C (2017) Atmospheric evolution. *In*: Encyclopedia of Geochemistry. White WM (ed) Springer, p 62–96

Goldblatt C, Lenton TM, Watson AJ (2006) Bistability of atmospheric oxygen and the Great Oxidation. Nature 443:683–686

Golubic S, Seong-Joo L, Browne KM (2000) Cyanobacteria: architects of sedimentary structures. *In*: Microbial Sediments. Riding R, Awramik SM (eds). Springer, Berlin, p 57–67

Gottschalk G, Thauer RK (2001) The Na^+-translocating methyltransferase complex from methanogenic archaea. Biochim Biophys Acta (BBA), Bioenerg 1505:28–36

Habicht KS, Gade M, Thamdrup B, Berg P, Canfield DE (2002) Calibration of sulfate levels in the Archean Ocean. Science 298:2372–2374

Hao J, Sverjensky DA, Hazen RM (2017) Mobility of nutrients and trace metals during weathering in the late Archean. Earth Planet Sci Lett 471:148–159

Hao J, Glein CR, Huang F, Yee N, Catling DC, Postberg F, Hillier JK, Hazen RM (2022) Abundant phosphorus expected for possible life in Enceladus's ocean. PNAS 119:e2201388119, https://doi.org/10.1073/pnas.2201388119

Haqq-Misra J, Kasting JF, Lee S (2011) Availability of O_2 and H_2O_2 on pre-photosynthetic Earth. Astrobiology 11:293–302

Hardisty DS, Lu Z, Bekker A, Diamond CW, Gill BC, Jiang G, Kah LC, Knoll AH, Loyd SJ, Osburn MR, Planavsky NJ (2017) Perspectives on Proterozoic surface ocean redox from iodine contents in ancient and recent carbonate. Earth Planet Sci Lett 463:159–170

Harrison TM (2009) The Hadean crust: Evidence from >4 Ga zircons. Annu Rev Earth Planet Sci 37:479–505, https://doi.org/10.1146/annurev.earth.031208.100151

Hartmann WK, Malin M, McEwen A, Carr M, Soderblom L, Thomas P, Danielson E, James P, Veverka J (1999) Evidence for recent volcanism on Mars from crater counts. Nature 397:586–589

Hassenkam T, Andersson MP, Dalby KN, Mackenzie DMA, Rosing MT (2017) Elements of Eoarchean life trapped in mineral inclusions. Nature 548:78–81

Hauber E, Brož P, Jagert F, Jodłowski P, Platz T (2011) Very recent and wide-spread basaltic volcanism on Mars. Geophys Res Lett 38:L10201

Hayes JM, Kaplan IR, Wedeking KW (1983) Precambrian organic geochmistry, preservation of the record. *In*: Earth's Earliest Biosphere—Its origin and evolution. Schopf JW (ed) Princeton University Press, Princeton, NJ, p 93–134

Hazen RM, Papineau D, Bleeker W, Downs RT, Ferry JM, McCoy TJ, Sverjensky DA, Yang H (2008) Mineral evolution. Am Mineral 93:1693–1720

Helz GR, Bura-Nakic E, Mikac N, Ciglenecki I (2011) New model for molybdenum behavior in euxinic waters. Chem Geol 284:323–332

Herzberg C, Condie K, Korenaga J (2010) Thermal history of the Earth and its petrological expression. Earth Planet Sci Lett 292:79–88

Hill ML, Bott K, Dalba PA, Fetherolf T, Kane SR, Kopparapu R, Li Z, Ostberg C (2023) A catalog of habitable zone exoplanets. Astron J 165:34, https://doi.org/10.3847/1538-3881/aca1c0

Hodgskiss MS, Sansjofre P, Kunzmann M, Sperling EA, Cole DB, Crockford PW, Gibson TM, Halverson GP (2020) A high-TOC shale in a low productivity world: The late Mesoproterozoic Arctic Bay Formation, Nunavut. Earth Planet Sci Lett 544:116384, https://doi.org/10.1016/j.epsl.2020.116384

Hohmann-Marriott MF, Blankenship RE (2011) Evolution of photosynthesis. Annu Rev Plant Biol 62:515–548

Holland HD (2002) Volcanic gases, black smokers, and the Great Oxidation Event. Geochim Cosmochim Acta 66:3811–3826

Homann M, Sansjofre P, Van Zuilen M, Heubeck C, Gong J, Killingsworth B, Foster IS, Airo A, Van Kranendonk MJ, Ader M, Lalonde SV (2018) Microbial life and biogeochemical cycling on land 3,220 million years ago. Nat Geosci 11:665–671

Horsfield B, Rullkotter J (1994) Diagenesis, catagenesis, and metagenesis of organic matter: Chapter 10: Part III. Processes. *In*: The Petroleum System: From Source to Trap. Magoon LB, Dow WG (eds). American Association of Petroleum Engineers, Washington, USA, p 189–199

Ikoma M, Genda H (2006) Constraints on the mass of a habitable planet with water of nebular origin. Astrophys J 648:696

Ingalls M, Grotzinger JP, Present T, Rasmussen B, Fischer WW (2022) Carbonate-associated phosphate (CAP) indicates elevated phosphate availability in Neoarchean shallow marine environments. Geophys Res Lett 49: e2022GL098100, https://doi.org/10.1029/2022GL098100

Isley AE, Abbott DH (1999) Plume-related mafic volcanism and the deposition of banded iron formation. J Geophys Res: Solid Earth 104:15461–15477

Izon G, Zerkle AL, Williford KH, Farquhar J, Poulton SW, Claire MW (2017) Biological regulation of atmospheric chemistry en route to planetary oxygenation. PNAS 114:2571–2579

Jelen BI, Giovannelli D, Falkowski PG (2016) The role of microbial electron transfer in the coevolution of the biosphere and geosphere. Annu Rev Microbiol 70:45–62

Jernigan J, Laflèche É, Burke A, Olson S (2023) Superhabitability of high-obliquity and high-eccentricity planets. Astrophys J 944:205, https://doi.org/10.3847/1538-4357/acb81c

Joerger RD, Bishop PE, Evans HJ (1988) Bacterial alternative nitrogen fixation systems. CRC Crit Rev Microbiol 16:1–14

Johnson JE, Webb SM, Thomas K, Ono S, Kirschvink JL, Fischer WW (2013) Manganese-oxidizing photosynthesis before the rise of cyanobacteria. PNAS 110:11238–11243

Johnson JE, Gerpheide A, Lamb MP, Fischer WW (2014) O_2 constraints from Paleoproterozoic detrital pyrite and uraninite. Geol Soc Am Bull 126:813–830, https://doi.org/10.1130/B30949.1

Johnson BW, Goldblatt C (2018) EarthN: a new Earth system nitrogen model. Geochem Geophys Geosystems 19:2516–2542

Johnson AC, Ostrander CM, Romaniello SJ, Reinhard CT, Greaney AT, Lyons TW, Anbar AD (2021) Reconciling evidence of oxidative weathering and atmospheric anoxia on Archean Earth. Sci Adv 7:eabj0108, https://doi.org/10.1126/sciadv.abj0108

Johnston DT (2011) Multiple sulfur isotopes and the evolution of Earth's surface sulfur cycle. Earth Sci Rev 106:161–183

Kaltenegger L (2017) How to characterize habitable worlds and signs of life. Annu Rev Astron Astrophys 55:433–485

Kang W (2019) Wetter stratospheres on high-obliquity planets. Astrophys J Lett 877:L6

Kappler A, Newman DK (2004) Formation of Fe (III)-minerals by Fe (II)-oxidizing photoautotrophic bacteria. Geochim Cosmochim Acta 68:1217–1226

Kasting JF (2005) Methane and climate during the Precambrian era. Precambrian Res 137:119–129

Keller CB, Schoene B (2012) Statistical geochemistry reveals disruption in secular lithospheric evolution about 2.5 Gyr ago. Nature 485:490–493

Keller B, Schoene B (2018) Plate tectonics and continental basaltic geochemistry throughout Earth history. Earth Planet Sci Lett 481:290–304

Kendall B, Anbar AD, Kappler A, Konhauser KO (2012) The global iron cycle. *In*: Fundamentals of Geobiology. Knoll AH, Canfield DE, Konhauser KO (eds). Blackwell Publishing Ltd,

Kendall B, Reinhard CT, Lyons TW, Kaufman AJ, Poulton SW, Anbar A (2010) Pervasive oxygenation along late Archaean ocean margins. Nat Geosci 3:647–652

Kharecha P, Kasting J, Siefert J (2005) A coupled atmosphere–ecosystem model of the early Archean Earth. Geobiol 3:53–76

Kiang NY, Segura A, Tinetti G, Blankenship RE, Cohen M, Siefert J, Crisp D, Meadows VS (2007) Spectral signatures of photosynthesis. II. Coevolution with other stars and the atmosphere on extrasolar worlds. Astrobiology 7:252–274

Kipp MA (2022) A Double-Edged Sword: The role of sulfate in anoxic marine phosphorus cycling through Earth history. Geophys Res Lett 49:e2022GL099817, https://doi.org/10.1029/2022GL099817

Kipp MA, Stüeken EE (2017) Biomass recycling and Earth's early phosphorus cycle. Sci Adv 3:eaao4795, https://doi.org/10.1126/sciadv.aao4795

Kipp MA, Stüeken EE, Yun M, Bekker A, Buick R (2018) Pervasive aerobic nitrogen cycling in the surface ocean across the Paleoproterozoic era. Earth Planet Sci Lett 500:117–126

Kirschvink JL, Gaidos EJ, Bertani LE, Beukes NJ, Gutzmer J, Maepa LN, Steinberger RE (2000) Paleoproterozoic snowball Earth: Extreme climatic and geochemical global change and its biological consequences. PNAS 97:1400–1405

Klinman JP (1996) Mechanisms whereby mononuclear copper proteins functionalize organic substrates. Chem Rev 96:2541–2561

Knauth LP (2005) Temperature and salinity history of the Precambrian ocean: implications for the course of microbial evolution. Palaeogeogr Palaeoclimatol,Palaeoecol 219:53–69

Knoll AH, Beukes NJ (2009) Introduction: Initial investigations of a Neoarchean shelf margin–basin transition (Transvaal Supergroup, South Africa). Precambrian Res 169:1–14

Koehler MC, Buick R, Kipp MA, Stüeken EE, Zaloumis J (2018) Transient surface oxygenation recorded in the ~2.66 Ga Jeerinah Formation, Australia. PNAS 115:7711–7716

Koehler MC, Buick R, Barley ME (2019) Nitrogen isotope evidence for anoxic deep marine environments from the Mesoarchean Mosquito Creek Formation, Australia. Precambrian Res 320:281–290

Komiya T, Maruyama S, Masuda T, Nohda S, Hayashi M, Okamoto K (1999) Plate tectonics at 3.8–3.7 Ga: Field evidence from the Isua accretionary complex, southern West Greenland. J Geol 107:515–554

Konhauser KO, Hamade T, Raiswell R, Morris RC, Ferris FG, Southam G, Canfield DE (2002) Could bacteria have formed the Precambrian banded iron formations? Geology 30:1079–1082

Konhauser KO, Pecoits E, Lalonde SV, Papineau D, Nisbet EG, Barley ME, Arndt NT, Zahnle KJ, Kamber BS (2009) Oceanic nickel depletion and a methanogen famine before the Great Oxidation Event. Nature 458:750–753

Konhauser KO, Robbins LJ, Pecoits E, Peacock C, Kappler A, Lalonde SV (2015) The Archean nickel famine revisited. Astrobiology 15:804–815

Kopp RE, Kirschvink JL, Hilburn IA, Nash CZ (2005) The Paleoproterozoic snowball Earth: a climate disaster triggered by the evolution of oxygenic photosynthesis. PNAS 102:11131–11136

Korenaga J (2013) Initiation and evolution of plate tectonics on Earth: theories and observations. Annu Rev Earth Planet Sci 41:117–151

Korenaga J (2017) Pitfalls in modeling mantle convection with internal heat production. J Geophys Res: Solid Earth 122:4064–4085

Korenaga J (2018) Crustal evolution and mantle dynamics through Earth history. Philos Trans R Soc A 376:20170408

Korenaga J, Planavsky N, Evans DA (2017) Global water cycle and the coevolution of the Earth's interior and surface environment. Philos Trans R Soc 375:20150393

Kral TA, Brink KM, Miller SL, McKay CP (1998) Hydrogen consumption by methanogens on the early Earth. Origins of Life and Evolution of the Biosphere 28:311–319

Krause AJ, Mills BJ, Zhang S, Planavsky NJ, Lenton TM, Poulton SW (2018) Stepwise oxygenation of the Paleozoic atmosphere. Nat Commun 9:4081, https://doi.org/10.1038/s41467-018-06383-y

Krissansen-Totton J, Buick R, Catling DC (2015) A statistical analysis of the carbon isotope record from the Archean to Phanerozoic and implications for the rise of oxygen. Am J Sci 315:275–316

Krissansen-Totton J, Bergsman DS, Catling DC (2016) On detecting biospheres from chemical thermodynamic disequilibrium in planetary atmospheres. Astrobiology 16:39–67

Krissansen-Totton J, Olson S, Catling DC (2018a) Disequilibrium biosignatures over Earth history and implications for detecting exoplanet life. Sci Adv 4:eaao5747, https://doi.org/10.1126/sciadv.aao5747

Krissansen-Totton J, Garland R, Irwin P, Catling DC (2018b) Detectability of biosignatures in anoxic atmospheres with the James Webb space telescope: A TRAPPIST-1e Case Study. Astron J 156:114, https://doi.org/10.3847/1538-3881/aad564

Kump LR, Arthur MA (1999) Interpreting carbon-isotope excursions: carbonates and organic matter. Chem Geol 161:181–198

Kurzweil F, Claire MW, Thomazo C, Peters M, Hannington M, Strauss H (2013) Atmospheric sulfur rearrangement 2.7 billion years ago: Evidence for oxygenic photosynthesis. Earth Planet Sci Lett 366:17–26

Laakso TA, Schrag DP (2017) A theory of atmospheric oxygen. Geobiology 15:366–384

Laakso TA, Schrag DP (2019) A small marine biosphere in the Proterozoic. Geobiology 17:161–171

Lammer H, Leitzinger M, Scherf M, Odert P, Burger C, Kubyshkina D, Fossati L, Pilat-Lohinger E, Ragossnig F, Dorfi EA (2020) Constraining the early evolution of Venus and Earth through atmospheric Ar, Ne isotope and bulk K/U ratios. Icarus 339:113551

Lan Z, Kamo SL, Roberts NM, Sano Y, Li XH (2022) A Neoarchean (ca. 2500 Ma) age for jaspilite–carbonate BIF hosting purported micro-fossils from the Eoarchean (≥ 3750 Ma) Nuvvuagittuq supracrustal belt (Québec, Canada). Precambrian Res 377:106728, https://doi.org/10.1016/j.precamres.2022.106728

Lee C-TA, Caves J, Jiang H, Cao W, Lenardic A, McKenzie R, Shorttle O, Yin Q-Z, Dyer B (2018) Deep mantle roots and continental emergence: Implications for whole-Earth elemental cycling, long-term climate, and the Cambrian explosion. Int Geol Rev 60:431–448

Lee EJ, Chiang E, Ormel CW (2014) Make super-Earths, not Jupiters: Accreting nebular gas onto solid cores at 0.1 AU and beyond. Astrophys J 797:95

Lenton TM (1998) Gaia and natural selection. Nature 394:439–447

Lenton TM, Dahl TW, Daines SJ, Mills BJ, Ozaki K, Saltzman MR, Porada P (2016) Earliest land plants created modern levels of atmospheric oxygen. PNAS 113:9704–9709

Lepot K (2020) Signatures of early microbial life from the Archean (4 to 2.5 Ga) eon. Earth Sci Rev 209:103296, https://doi.org/10.1016/j.earscirev.2020.103296

Leung M, Schwieterman EW, Parenteau MN, Fauchez TJ (2022) Alternative methylated biosignatures. I. Methyl bromide, a capstone biosignature. Astrophys J 938:6

Li J, Fei Y (2014) Experimental constraints on core composition. Treatise Geochem 3:527–557

Liao T, Wang S, Stüeken EE, Luo H (2022) Phylogenomic evidence for the origin of obligately anaerobic *Anammox* bacteria around the Great Oxidation Event. Mol Biol Evol 39:msac170, https://doi.org/10.1093/molbev/msac170

Liebmann J, Spencer CJ, Kirkland CL, Ernst RE (2022) Large igneous provinces track fluctuations in subaerial exposure of continents across the Archean–Proterozoic transition. Terra Nova 33:465–474, https://doi.org/10.1111/ter.12531

Little SH, Vance D, Lyons TW, McManus J (2015) Controls on trace metal authigenic enrichment in reducing sediments: Insights from modern oxygen-deficient settings. Am J Sci 315:77–119

Llirós M, García–Armisen T, Darchambeau F, Morana C, Triadó–Margarit X, Inceoğlu Ö, Borrego CM, Bouillon S, Servais P, Borges AV, Descy JP (2015) Pelagic photoferrotrophy and iron cycling in a modern ferruginous basin. Sci Rep 5:13803, https://doi.org/10.1038/srep13803

Love GD, Zumberge JA (2021) Emerging patterns in proterozoic lipid biomarker records. Cambridge University Press, Cambridge, UK

Lowe DR, Byerly GR (2018) The terrestrial record of late heavy bombardment. New Astron Rev 81:39–61

Luger R, Barnes R (2015) Extreme water loss and abiotic O_2 buildup on planets throughout the habitable zones of M dwarfs. Astrobiology 15:119–143

Luo G, Junium CK, Izon G, Ono S, Beukes NJ, Algeo TJ, Cui Y, Xie S, Summons RE (2018) Nitrogen fixation sustained productivity in the wake of the Palaeoproterozoic Great Oxygenation Event. Nat Commun 9:1–9

Lyons TW, Reinhard CT, Planavsky NJ (2014) The rise of oxygen in Earth's early ocean and atmosphere. Nature 506:307–315

Magni V, Bouilhol P, Van Hunen J (2014) Deep water recycling through time. Geochem Geophys Geosystems 15:4203–4216

Margulis L, Lovelock JE (1974) Biological modulation of the Earth's atmosphere. Icarus 21:471–489

Margulis L, Walker JCG, Rambler M (1976) Reassessment of roles of oxygen and ultraviolet light in Precambrian evolution. Nature 264:620–624

Marien CS, Jäger O, Tusch J, Viehmann S, Surma J, Van Kranendonk MJ, Münker C (2023) Interstitial carbonates in pillowed metabasaltic rocks from the Pilbara Craton, Western Australia: A vestige of Archean seawater chemistry and seawater-rock interactions. Precambrian Res 394:107109

Martin W, Baross J, Kelley D, Russell MJ (2008) Hydrothermal vents and the origin of life. Nat Rev Microbiol 6:805–814

Martin H, Moyen JF, Guitreau M, Blichert-Toft J, Le Pennec JL (2014) Why Archaean TTG cannot be generated by MORB melting in subduction zones. Lithos 198:1–13

McCollom TM, Seewald JS (2006) Carbon isotopic composition of organic compounds produced by abiotic synthesis under hyhdrothermal conditions. Earth Planet Sci Lett 243:74–84

McCollom TM, Seewald JS (2007) Abiotic synthesis of organic compounds in deep-sea hydrothermal environments. Chem Rev 107:382–401

McGovern PJ, Schubert G (1989) Thermal evolution of the Earth: effects of volatile exchange between atmosphere and interior. Earth Planet Sci Lett 96:27–37

McMahon S (2019) Earth's earliest and deepest purported fossils may be iron-mineralized chemical gardens. Proc R Soc B 286:20192410, https://doi.org/10.1098/rspb.2019.2410

Meadows VS, Reinhard CT, Arney GN, Parenteau MN, Schwieterman EW, Domagal-Goldman SD, Lincowski AP, Stapelfeldt KR, Rauer H, DasSarma S, Hegde S (2018) Exoplanet biosignatures: Understanding oxygen as a biosignature in the context of its environment. Astrobiology 18:630–662

Mettam C, Zerkle AL, Claire MW, Prave AR, Poulton SW, Junium CK (2019) Anaerobic nitrogen cycling on a Neoarchaean ocean margin. Earth Planet Sci Lett 527:115800

Mettler JN, Quanz SP, Helled R, Olson SL, Schwieterman EW (2023) Earth as an exoplanet. II. Earth's time-variable thermal emission and its atmospheric seasonality of bioindicators. Astrophys J 946:82

Meyer KM, Kump LR (2008) Oceanic euxinia in Earth history: Causes and consequences. Annu Rev Earth Planet Sci 36:251–288

Mojzsis SJ, Arrhenius G, McKeegan KD, Harrison TM, Nutman AP, Friend CR (1996) Evidence for life on Earth before 3,800 million years ago. Nature 384:55–59

Mojzsis SJ, Harrison M, Pidgeon RT (2001) Oxygen-isotope evidence from ancient zircons for liquid water at the Earth's surface 4,300 Myr ago. Nature 409:178–181

Moore EK, Nunn BL, Goodlett DR, Harvey HR (2012) Identifying and tracking proteins through the marine water column: Insights into the inputs and preservation mechanisms of protein in sediments. Geochim Cosmochim Acta 83:324–359

Moore EK, Harvey HR, Faux JF, Goodlett DR, Nunn BL (2014) Electrophoretic extraction and proteomic characterization of proteins buried in marine sediments. Chromatography 1:176–193

Moore EK, Jelen BI, Giovannelli D, Raanan H, Falkowski PG (2017) Metal availability and the expanding network of microbial metabolisms in the Archaean eon. Nat Geosci 10:629–636

Moore EK, Hao J, Prabhu A, Zhong H, Jelen BI, Meyer M, Hazen RM, Falkowski PG (2018) Geological and chemical factors that impacted the biological utilization of cobalt in the Archean eon. J Geophys Res: Biogeosci 123:743–759

Moser CC, Page CC, Farid R, Dutton PL (1995) Biological electron transfer. J Bioenerg Biomembranes 27:263–274

Moser CC, Keske JM, Warncke K, Farid RS, Dutton PL (1992) Nature of biological electron transfer. Nature 355:796–802

Moyen JF, Martin H (2012) Forty years of TTG research. Lithos 148:312–336

Moyen JF, Van Hunen J (2012) Short-term episodicity of Archaean plate tectonics. Geology 40:451–454

Müller PJ (1977) CN ratios in Pacific deep-sea sediments: Effect of inorganic ammonium and organic nitrogen compounds sorbed by clays. Geochim Cosmochim Acta 41:765–776

Navarro-González R, Molina MJ, Molina LT (1998) Nitrogen fixation by volcanic lightning in the early Earth. Geophys Res Lett 25:3123–3126

Neilan BA, Burns BP, Relman DA, Lowe DR (2002) Molecular identification of cyanobacteria associated with stromatolites from distinct geographical locations. Astrobiology 2:271–280

Nemchin AA, Whitehouse MJ, Menneken M, Geisler T, Pidgeon RT, Wilde SA (2008) A light carbon reservoir recorded in zircon-hosted diamond from the Jack Hills. Nature 454:92–95

Nielsen KM, Johnsen PJ, Bensasson D, Daffonchio D (2007) Release and persistence of extracellular DNA in the environment. Environ Biosafety Res 6:37–53

Noffke N, Christian D, Wacey D, Hazen RM (2013) Microbially induced sedimentary structures recording an ancient ecosystem in the ca. 3.48 billion-year-old Dresser Formation, Pilbara, Western Australia. Astrobiology 13:1103–1124

Nutman AP, Friend CR, Bennett VC (2002) Evidence for 3650–3600 Ma assembly of the northern end of the Itsaq Gneiss Complex, Greenland: implication for early Archaean tectonics. Tectonics 21:1–5

Nutman AP, Bennett VC, Friend CR, van Kranendonk MJ, Chivas AR (2016) Rapid emergence of life shown by discovery of 3,700-million-year-old microbial structures. Nature 537:535–538

O'Leary MH (1981) Carbon isotope fractionation in plants. Phytochemistry 20:553–567

O'Malley-James JT, Kaltenegger L (2019) Expanding the timeline for Earth's photosynthetic red edge biosignature. Astrophys J 879:L20, https://doi.org/10.3847/2041-8213/ab2769

O'Neill C, Debaille V (2014) The evolution of Hadean–Eoarchaean geodynamics. Earth Planet Sci Lett 406:49–58

O'Neill C, Lenardic A, Weller M, Moresi L, Quenette S, Zhang S (2016) A window for plate tectonics in terrestrial planet evolution? Phys Earth Planet Sci 255:80–92

Och LM, Shields-Zhou GA (2012) The Neoproterozoic oxygenation event: environmental perturbations and biogeochemical cycling. Earth Sci Rev 110:26–57

Ohtomo Y, Kakegawa T, Ishida A, Nagase T, Rosing MT (2014) Evidence for biogenic graphite in early Archaean Isua metasedimentary rocks. Nat Geosci 7:25–28

Olson SL, Kump LR, Kasting JF (2013) Quantifying the areal extent and dissolved oxygen concentrations of Archean oxygen oases. Chem Geol 362:35–43

Olson SL, Reinhard CT, Lyons TW (2016) Limited role for methane in the mid-Proterozoic greenhouse. PNAS 113:11447–11452

Olson SL, Schwieterman EW, Reinhard CT, Lyons TW (2018) Earth: Atmospheric evolution of a habitable planet. *In*: Handbook of Exoplanets. HJ Deeg, JA Belmonte (eds). Springer International Publishing, p 2817–2853

Olson SL, Jansen M, Abbot DS (2020) Oceanographic considerations for exoplanet life detection. Astrophys J 895:19, https://doi.org/10.3847/1538-4357/ab88c9

Oró J, Miller SL, Lazcano A (1990) The origin and early evolution of life on Earth. Annu Rev Earth Planet Sci 18:317–356

Ossa Ossa F, Hofmann A, Spangenberg JE, Poulton SW, Stüeken EE, Schoenberg R, Eickmann B, Wille M, Butler M, Bekker A (2019) Limited oxygen production in the Mesoarchean ocean. PNAS 116:6647–6652

Owen JE, Wu Y (2013) Kepler planets: a tale of evaporation. Astrophys J 775:105

Ozaki K, Tajika E, Hong PK, Nakagawa Y, Reinhard CT (2018) Effects of primitive photosynthesis on Earth's early climate system. Nat Geosci 11:55–59

Ozaki K, Reinhard CT, Tajika E (2019) A sluggish mid-Proterozoic biosphere and its effect on Earth's redox balance. Geobiology 17:3–11

Pajares S, Ramos R (2019) Processes and microorganisms involved in the marine nitrogen cycle: knowledge and gaps. Front Marine Sci 6:739, https://doi.org/10.3389/fmars.2019.00739

Papineau D (2010) Global biogeochemical changes at both ends of the Proterozoic: Insights from phosphorites. Astrobiology 10:165–181

Papineau D, Mojzsis SJ, Karhu JA, Marty B (2005) Nitrogen isotopic composition of ammoniated phyllosilicates: case studies from Precambrian metamorphosed sedimentary rocks. Chem Geol 216:37–58

Papineau D, She Z, Dodd MS, Iacoviello F, Slack JF, Hauri E, Shearing P, Little CT (2022) Metabolically diverse primordial microbial communities in Earth's oldest seafloor-hydrothermal jasper. Sci Adv 8:eabm2296, https://doi.org/10.1126/sciadv.abm2296

Parai R, Mukhopadhyay S (2018) Xenon isotopic constraints on the history of volatile recycling into the mantle. Nature 560:223–227

Parducci L, Bennett KD, Ficetola GF, Alsos IG, Suyama Y, Wood JR, Pedersen MW (2017) Ancient plant DNA in lake sediments. New Phytolog 214:924–942

Parsons C, Stüeken EE, Rosen C, Mateos K, Anderson R (2020) Radiation of nitrogen-metabolizing enzymes across the tree of life tracks environmental transitions in Earth history. Geobiology 19:18–34, https://doi.org/10.1111/gbi.12419

Pavlov AA, Kasting JF (2002) Mass-independent fractionation of sulfur isotopes in Archean sediments: strong evidence for an anoxic Archean atmosphere. Astrobiology 2:27–41

Pavlov AA, Brown LL, Kasting JF (2001) UV shielding of NH_3 and O_2 by organic hazes in the Archean atmosphere. J Geophys Res 106:23267–23287

Pavlov AA, Hurtgen MT, Kasting JF, Arthur MA (2003) Methane-rich Proterozoic atmosphere. Geology 31:87–90

Pellerin A, Thomazo C, Ader M, Marin-Carbonne J, Alleon J, Vennin E, Hofmann A (2023) Iron-mediated anaerobic ammonium oxidation recorded in the early Archean ferruginous ocean. Geogiology 21:277–289

Peslier AH, Schönbächler M, Busemann H, Karato SI (2017) Water in the Earth's interior: distribution and origin. Space Sci Rev 212:743–810

Peters SE, Husson JM (2017) Sediment cycling on continental and oceanic crust. Geology 45:323–326

Pinti DL, Hashizume K, Matsuda JI (2001) Nitrogen and argon signatures in 3.8 to 2.8 Ga metasediments: Clues on the chemical state of the Archean ocean and the deep biosphere. Geochim Cosmochim Acta 65:2301–2315

Planavsky NJ, McGoldrick P, Scott CT, Li C, Reinhard CT, Kelly AE, Chu X, Bekker A, Love GD, Lyons TW (2011) Widespread iron-rich conditions in the mid-Proterozoic ocean. Nature 477:448–451

Planavsky NJ, Asael D, Hofmann A, Reinhard CT, Lalonde SV, Knudsen A, Wang X, Ossa Ossa F, Pecoits E, Smith AJ, Beukes NJ (2014) Evidence for oxygenic photosynthesis half a billion years before the Great Oxidation Event. Nat Geosci 7:283–286

Planavsky NJ, Cole DB, Isson TT, Reinhard CT, Crockford PW, Sheldon ND, Lyons TW (2018a) A case for low atmospheric oxygen levels during Earth's middle history. Emerging Top Life Sci 2:149–159

Planavsky NJ, Slack JF, Cannon WF, O'Connell B, Isson TT, Asael D, Jackson JC, Hardisty DS, Lyons TW, Bekker A (2018b) Evidence for episodic oxygenation in a weakly redox-buffered deep mid-Proterozoic ocean. Chem Geol 483:581–594

Planavsky NJ, Fakhraee M, Bolton EW, Reinhard CT, Isson TT, Zhang S, Mills BJ (2022) On carbon burial and net primary production through Earth's history. Am J Sci 322:413–460

Planavsky NJ, Asael D, Rooney AD, Robbins LJ, Gill BC, Dehler CM, Cole DB, Porter SM, Love GD, Konhauser KO, Reinhard CT (2023) A sedimentary record of the evolution of the global marine phosphorus cycle. Geobiology 21:168–174

Poulton SW, Canfield DE (2005) Development of a sequential extraction procedure for iron: implications for iron partitioning in continentally derived particulates. Chem Geol 214:209–221

Poulton SW, Canfield DE (2011) Ferruginous conditions: A dominant feature of the ocean through Earth's history. Elements 7:107–112

Proskurowski G, Lilley MD, Seewald JS, Früh-Green GL, Olson EJ, Lupton JE, Sylva SP, Kelley DS (2008) Abiogenic hydrocarbon production at Lost City hydrothermal field. Science 319:604–607

Ptáček MP, Dauphas N, Greber ND (2020) Chemical evolution of the continental crust from a data-driven inversion of terrigenous sediment compositions. Earth Planet Sci Lett 539:116090, https://doi.org/10.1016/j.epsl.2020.116090

Quanz SP, Ottiger M, Fontanet E, Kammerer J, Menti F, Dannert F, Gheorghe A, Absil O, Airapetian VS, Alei E, Allart R (2022) Large Interferometer For Exoplanets (LIFE)-I. Improved exoplanet detection yield estimates for a large mid-infrared space-interferometer mission. Astron Astrophys 664:A21

Ragsdale SW, Pierce E (2008) Acetogenesis and the Wood–Ljungdahl pathway of CO_2 fixation. Biochimica Biophysica Acta (BBA), Proteins Proteomics 1784:1873–1898

Raiswell R, Hardisty DS, Lyons TW, Canfield DE, Owens J, Planavsky N, Poulton SW, Reinhard CT (2019) The iron paleoredox proxies: A guide to the pitfalls, problems and proper practice. Am J Sci 318:491–526

Rauer H, Gebauer SV, Paris PV, Cabrera J, Godolt M, Grenfell JL, Belu A, Selsis F, Hedelt P, Schreier F (2011) Potential biosignatures in super-Earth atmospheres-I. Spectral appearance of super-Earths around M dwarfs. Astron Astrophys 529:A8, https://doi.org/10.1051/0004-6361/201014368

Raymond J, Blankenship RE (2008) The origin of the oxygen-evolving complex. Coord Chem Rev 252:377–383

Reese CC, Solomatov VS, Moresi LN (1999) Non-Newtonian stagnant lid convection and magmatic resurfacing on Venus. Icarus 139:67–80

Reimink JR, Chacko T, Stern RA, Heaman LM (2014) Earth's earliest evolved crust generated in an Iceland-like setting. Nat Geosci 7:529–533

Reimink JR, Bauer AM, Chacko T (2019) The Acasta gneiss complex. *In*: Earth's Oldest Rocks. Elsevier, p 329–348

Reimink JR, Davies JH, Ielpi A (2021) Global zircon analysis records a gradual rise of continental crust throughout the Neoarchean. Earth Planet Sci Lett 554:116654

Reinhard CT, Planavsky NJ (2022) The history of ocean oxygenation. Annu Rev Mar Sci 14:331–353

Reinhard CT, Planavsky NJ, Robbins LJ, Partin CA, Gill BC, Lalonde SV, Bekker A, Konhauser KO, Lyons TW (2013) Proterozoic ocean redox and biogeochemical stasis. PNAS 110:5357–5362

Reinhard CT, Planavsky N, Olson SL, Lyons TW, Erwin DH (2016) Earth's oxygen cycle and the evolution of animal life. PNAS 113:8933–8938, https://doi.org/10.1073/pnas.1521544113

Reinhard CT, Olson SL, Schwieterman EW, Lyons TW (2017a) False negatives for remote life detection on ocean-bearing planets: Lessons from the early Earth. Astrobiology 17:287–297

Reinhard CT, Planavsky NJ, Gill BC, Ozaki K, Robbins LJ, Lyons TW, Fischer WW, Wang C, Cole DB, Konhauser KO (2017b) Evolution of the global phosphorus cycle. Nature 541:386–389

Reinhard CT, Schwieterman EW, Olson SL, Planavsky NJ, Arney GN, Ozaki K, Som S, Robinson TD, Domagal-Goldman SD, Lisman D, Mennesson B (2019) The remote detectability of Earth's biosphere through time and the importance of UV capability for characterizing habitable exoplanets. arXiv:preprint arXiv:1903.05611

Reinhard CT, Olson SL, Kirtland Turner S, Pälike C, Kanzaki Y, Ridgwell A (2020) Oceanic and atmospheric methane cycling in the cGENIE Earth system model–release v0. 9.14. Geosci Model Dev 13:5687–5706

Riding R, Fralick P, Liang L (2014) Identification of an Archean marine oxygen oasis. Precambrian Res 251:232–237

Robbins LJ, Lalonde SV, Planavsky NJ, Partin CA, Reinhard CT, Kendall B, Scott C, Hardisty DS, Gill BC, Alessi DS, Dupont CL (2016) Trace elements at the intersection of marine biological and geochemical evolution. Earth Sci Rev 163:323–348

Robbins LJ, Fakhraee M, Smith AJ, Bishop BA, Swanner ED, Peacock C, Wang CL, Planavksy NJ, Reinhard CT, Crowe SA, Lyons TW (2023) Manganese oxides, Earth surface oxygenation, and the rise of oxygenic photosynthesis. Earth Sci Rev 239:104368, https://doi.org/10.1016/j.earscirev.2023.10436

Roberson AL, Roadt J, Halevy I, Kasting JF (2011) Greenhouse warming by nitrous oxide and methane in the Proterozoic Eon. Geobiology 9:313–320

Robinson TD, Reinhard CT (2018) Earth as an exoplanet. *In*: Planetary Astrobiology. Meadows V, Arney G, Schmidt B, Des Marais DJ (eds), p 379–418

Robinson TD, Meadows VS, Crisp D (2010) Detecting oceans on extrasolar planets using the glint effect. Astrophys J 721:L67-L71

Roelofs TA, Liang W, Latimer MJ, Cinco RM, Rompel A, Andrews JC, Sauer K, Yachandra VK, Klein MP (1996) Oxidation states of the manganese cluster during the flash-induced S-state cycle of the photosynthetic oxygen-evolving complex. PNAS 93:3335–3340

Rosing MT (1999) ^{13}C-depleted carbon microparticles in >3700-Ma sea-floor sedimentary rocks from West Greenland. Science 283:674–676

Rudnick RL, Gao S (2014) Composition of the continental crust. Treatise Geochem 4:1–51

Rüpke LH, Morgan JP, Hort M, Connolly JA (2004) Serpentine and the subduction zone water cycle. Earth Planet Sci Lett 223:17–34

Russell MJ, Hall AJ, Martin W (2010) Serpentinization as a source of energy at the origin of life. Geobiology 8:355–371

Rye R, Holland HD (1998) Paleosols and the evolution of atmospheric oxygen: a critical review. Am J Sci 298:621–672

Sagan C, Thompson WR, Carlson R, Gurnett D, Hord C (1993) A search for life on Earth from the Galileo spacecraft. Nature 365:715–721

Saito MA, Sigman DM, Morel FMM (2003) The bioinorganic chemistry of the ancient ocean: the co-evolution of cyanobacterial metal requirements and biogeochemical cycles at the Archean–Proterozoic boundary? Inorganica Chimica Acta 356:308–318

Sánchez-Baracaldo P, Cardona T (2020) On the origin of oxygenic photosynthesis and Cyanobacteria. New Phytolog 225:1440–1446

Schaefer L, Fegley B (2017) Redox states of initial atmospheres outgassed on rocky planets and planetesimals. Astrophys J 843:120

Schidlowski M (2001) Carbon isotopes as biogeochemical recorders of life over 3.8 Ga of Earth history: Evolution of a concept. Precambrian Res 106:117–134

Schlichting HE, Sari RE, Yalinewich A (2015) Atmospheric mass loss during planet formation: the importance of planetesimal impacts. Icarus 247:81–94

Schoepp-Cothenet B, Van Lis R, Atteia A, Baymann F, Capowiez L, Ducluzeau A-L, Duval S, Brink FT, Russell MJ, Nitschke W (2012) On the universal core of bioenergetics. Biochim Biophys Acta, Bioenerg 1827:79–93

Schopf JW, Kitajima K, Spicuzza MJ, Kudryavtsev AB, Valley JW (2018) SIMS analyses of the oldest known assemblage of microfossils document their taxon-correlated carbon isotope compositions. PNAS 115:53–58

Schroeder PA, McLain AA (1998) Illite-smectites and the influence of burial diagenesis on the geochemical cycling of nitrogen. Clay Minerals 33:539–546

Schubert G, Reymer APS (1985) Continental volume and freeboard through geological time. Nature 316:336–339

Schwieterman EW, Kiang NY, Parenteau MN, Harman CE, DasSarma S, Fisher TM, Arney GN, Hartnett HE, Reinhard CT, Olson SL, Meadows VS (2018) Exoplanet biosignatures: A review of remotely detectable signs of life. Astrobiology 18:663–708

Schwieterman EW, Olson SL, Pidhorodetska D, Reinhard CT, Ganti A, Fauchez TJ, Bastelberger ST, Crouse JS, Ridgwell A, Lyons TW (2022) Evaluating the plausible range of N_2O biosignatures on exo-Earths: An integrated biogeochemical, photochemical, and spectral modeling approach. Astrophys J 937:109, https://doi.org/10.3847/1538-4357/ac8cfb

Scott C, Lyons TW (2012) Contrasting molybdenum cycling and isotopic properties in euxinic versus non-euxinic sediments and sedimentary rocks: refining the paleoproxies. Chem Geol 324:19–27

Scott C, Lyons TW, Bekker A, Shen Y, Poulton SW, Chu X, Anbar AD (2008) Tracing the stepwise oxygenation of the Proterozoic ocean. Nature 452:456–459

Scott C, Planavsky NJ, Dupont CL, Kendall B, Gill BC, Robbins LJ, Husband KF, Arnold GL, Wing BA, Poulton SW, Bekker A (2012) Bioavailability of zinc in marine systems through time. Nat Geosci 6:125–128

Seager S, Turner EL, Schafer J, Ford EB (2005) Vegetation's red edge: a possible spectroscopic biosignature of extraterrestrial plants. Astrobiology 5:372–390

Seager S, Bains W, Petkowski JJ (2016) Toward a list of molecules as potential biosignature gases for the search for life on exoplanets and applications to terrestrial biochemistry. Astrobiology 16:465–485

Seales J, Lenardic A, Richards M (2022) Buffering of mantle conditions through water cycling and thermal feedbacks maintains magmatism over geologic time. Commun Earth Environ 3:293

Segura A, Kasting JF, Meadows V, Cohen M, Scalo J, Crisp D, Butler RA, Tinetti G (2005) Biosignatures from Earth-like planets around M dwarfs. Astrobiology 5:706–725

Shalygin EV, Markiewicz WJ, Basilevsky AT, Titov DV, Ignatiev NI, Head JW (2015) Active volcanism on Venus in the Ganiki Chasma rift zone. Geophys Res Lett 42:4762–4769

Shih PM, Hemp J, Ward LM, Matzke NJ, Fischer WW (2017) Crown group oxyphotobacteria postdate the rise of oxygen. Geobiology 15:19–29

Shirey SB, Richardson SH (2011) Start of the Wilson cycle at 3 Ga shown by diamonds from subcontinental mantle. Science 333:434–436

Sleep NH, Zahnle KJ, Lupu RE (2014) Terrestrial aftermath of the Moon-forming impact. Philos Trans R Soc A 372:20130172

Smithies RH, Champion DC, Van Kranendonk MJ, Howard HM, Hickman AH (2005) Modern-style subduction processes in the Mesoarchaean: geochemical evidence from the 3.12 Ga Whundo intraoceanic arc. Earth Planet Sci Lett 231:221–237

Smithies RH, Van Kranendonk MJ, Champion DC (2007) The Mesoarchaean emergence of modern style subduction. Gondwana Res 11:50–68

Smrekar SE, Stofan ER, Mueller N, Treiman A, Elkins-Tanton L, Helbert J, Piccioni G, Drossart P (2010) Recent hotspot volcanism on Venus from VIRTIS emissivity data. Science 328:605–608

Smrekar SE, Ostberg C, O'Rourke JG (2023) Earth-like lithospheric thickness and heat flow on Venus consistent with active rifting. Nat Geosci 16:13–18

Solomatov VS, Moresi LN (1996) Stagnant lid convection on Venus. J Geophys Res: Planets 101:4737–4753

Som SM, Catling DC, Harnmeijer JP, Polivka PM, Buick R (2012) Air density 2.7 billion years ago limited to less than twice modern levels by fossil raindrop imprints. Nature 484:359–362

Som SM, Buick R, Hagadorn JW, Blake TS, Perreault JM, Harnmeijer JP, Catling D (2016) Earth's air pressure 2.7 billion years ago constrained to less than half of modern levels. Nat Geosci 9:448–451

Sperling EA, Frieder CA, Raman AV, Girguis PR, Levin LA, Knoll AH (2013) Oxygen, ecology, and the Cambrian radiation of animals. PNAS 110:13446–13451

Sperling EA, Wolock CJ, Morgan AS, Gill BC, Kunzmann M, Halverson GP, Macdonald FA, Knoll AH, Johnston DT (2015) Statistical analysis of iron geochemical data suggests limited late Proterozoic oxygenation. Nature 523:451–454

Stanton CL, Reinhard CT, Kasting JF, Ostrom NE, Haslun JA, Lyons TW, Glass JB (2018) Nitrous oxide from chemodenitrification: A possible missing link in the Proterozoic greenhouse and the evolution of aerobic respiration. Geobiology 16:597–609

Stepanov AS (2021) A review of the geochemical changes occurring during metamorphic devolatilization of metasedimentary rocks. Chem Geol 568:120080, https://doi.org/10.1016/j.chemgeo.2021.120080

Stern RJ, Gerya T, Tackley PJ (2018) Stagnant lid tectonics: Perspectives from silicate planets, dwarf planets, large moons, and large asteroids. Geosci Front 9:103–119

Stüeken EE (2016) Nitrogen in ancient mud: a biosignature? Astrobiology 16:730–735

Stüeken EE (2020) Hydrothermal vents and organic ligands sustained the Precambrian copper budget. Geochem Perspect Lett 16:12–16

Stüeken EE, Prave AR (2022) Diagenetic nutrient supplies to the Proterozoic biosphere archived in divergent nitrogen isotopic ratios between kerogen and silicate minerals. Geobiology 20:623–633

Stüeken EE, Buick R, Guy BM, Koehler MC (2015) Isotopic evidence for biological nitrogen fixation by Mo-nitrogenase at 3.2 Gyr. Nature 520:666–669

Stüeken EE, Kipp MA, Koehler MC, Buick R (2016) The evolution of Earth's biogeochemical nitrogen cycle. Earth Sci Rev 160:220–239

Stüeken EE, Boocock T, Szilas K, Mikhail S, Gardiner NJ (2021a) Reconstructing nitrogen sources to Earth's earliest biosphere at 3.7 Ga. Front Earth Sci 9:675726, https://doi.org/10.3389/feart.2021.675726

Stüeken EE, Kuznetsov AB, Vasilyeva IM, Krupenin MT, Bekker A (2021b) Transient deep-water oxygenation recorded by rare Mesoproterozoic phosphorites, South Urals. Precambrian Res 360:106242

Stüeken EE, Viehmann S, Hohl SV (2022) Contrasting nutrient availability between marine and brackish waters in the late Mesoproterozoic: Evidence from the Paranoá Group, Brazil. Geobiology 20:159–174

Sugitani K, Mimura K, Takeuchi M, Lepot K, Ito S, Javaux EJ (2015) Early evolution of large micro-organisms with cytological complexity revealed by microanalyses of 3.4 Ga organic-walled microfossils. Geobiology 13:507–521

Tajika E, Matsui T (1992) Evolution of terrestrial proto-CO_2 atmosphere coupled with thermal history of the Earth. Earth Planet Sci Lett 113:251–266

Tang D, Shi X, Wang X, Jiang G (2016) Extremely low oxygen concentration in mid-Proterozoic shallow seawaters. Precambrian Research 276:145–157

Thompson MA, Krissansen-Totton J, Wogan N, Telus M, Fortney JJ (2022) The case and context for atmospheric methane as an exoplanet biosignature. PNAS 119:e2117933119

Trail D, Watson EB, Tailby ND (2011) The oxidation state of Hadean magmas and implications for early Earth's atmosphere. Nature 480:79–82

Tribovillard N, Algeo TJ, Lyons J, Riboulleau A (2006) Trace metals as paleoredox and paleoproductivity proxies: an update. Chem Geol 232:12–32

Turner S, Rushmer T, Reagan M, Moyen JF (2014) Heading down early on? Start of subduction on Earth. Geology 42:139–142

Ueno Y, Ono S, Rumble III D, Maruyama S (2008) Quadruple sulfur isotope analysis of ca. 3.5 Ga Dresser Formation: New evidence for microbial sulfate reduction in the early Archean. Geochim Cosmochim Acta 72:5675–5691

Ueno Y, Yamada K, Yoshida N, Maruyama S, Isozaki Y (2006) Evidence from fluid inclusions for microbial methanogenesis in the early Archaean era. Nature 440:516–519

Unterborn CT, Foley BJ, Desch SJ, Young PA, Vance G, Chiffelle L, Kane SR (2022) Mantle degassing lifetimes through galactic time and the maximum age stagnant-lid rocky exoplanets can support temperate climates. Astrophys J Letters 930:L6

Uveges BT, Izon G, Ono S, Beukes NJ, Summons RE (2023) Reconciling discrepant minor sulfur isotope records of the Great Oxidation Event. Nat Commun 14:279, https://doi.org/10.1038/s41467-023-35820-w

Valley JW, Peck WH, King EM, Wilde SA (2002) A cool early Earth. Geology 30:351–354

Van Kranendonk MJ (2010) Two types of Archean continental crust: Plume and plate tectonics on early Earth. Am J Sci 310:1187–1209

Van Kranendonk MJ (2011) Onset of plate tectonics. Science 333:413–414

Van Kranendonk MJ, Altermann W, Beard BL, Hoffman PF, Johnson CJ, Kasting JF, Melezhik VA, Nutman AP, Papineau D, Pirajno F (2012) A chronostratigraphic division of the Precambrian: possibilities and challenges. *In*: The Geologic Time Scale. Gradstein FM, Ogg JG, Schmitz MD, Ogg GJ (eds). Elsevier, Boston, USA, p 299–392

Van Kranendonk MJ, Smithies RH, Griffin WL, Huston DL, Hickman AH, Champion DC, Anhaeusser CR, Pirajno F (2015) Making it thick: a volcanic plateau origin of Palaeoarchean continental lithosphere of the Pilbara and Kaapvaal cratons. Geol Soc London, Spec Publ 389:83–111

Van Kranendonk MJ, Bennett V, Hoffmann E (2018) Earth's Oldest Rocks. Elsevier, Amsterdam

Van Kranendonk MJ, Djokic T, Baumgartner R, Bontognali T, Sugitani K, Kiyokawa S, Walter MR (2021) Life analogue sites for Mars from early Earth: Diverse habitats from the Pilbara Craton and Mount Bruce Supergroup, Western Australia. *In*: Mars Geological Enigmas: From the Late Noachian Epoch to the Present Day. Soare RJ, Conway SJ, Oehler DZ, Williams J-P (eds). Elsevier, USA, p 357–403

Veizer J, Mackenzie FT (2014) The evolution of sedimentary rocks. Treatise Geochem 9:399–435

Viehmann S, Bau M, Hoffmann JE, Münker C (2015) Geochemistry of the Krivoy Rog Banded Iron Formation, Ukraine, and the impact of peak episodes of increased global magmatic activity on the trace element composition of Precambrian seawater. Precambrian Res 270:165–180

Viljoen MJ, Viljoen RP (1969) The geology and geochemistry of the lower ultramafic unit of the Onverwacht Group and a proposed new class of igneous rock. Spec Publ Geol S S Afr 2:55–85

Wacey D, McLoughlin N, Whitehouse MJ, Kilburn MR (2010) Two coexisting sulfur metabolisms in a ca. 3400 Ma sandstone. Geology 38:1115–1118

Wade J, Wood BJ (2005) Core formation and the oxidation state of the Earth. Earth Planet Sci Lett 236:78–95

Walter MR, Buick R, Dunlop JSR (1980) Stromatolites 3,400–3,500 Myr old from the North Pole area, Western Australia. Nature 284:443–445

Walter XA, Picazo A, Miracle MR, Vicente E, Camacho A, Aragno M, Zopfi J (2014) Phototrophic Fe (II)-oxidation in the chemocline of a ferruginous meromictic lake. Front Microbiol 5:713, https://doi.org/10.3389/fmicb.2014.00713

Waltham D (2014) Lucky Planet: Why Earth is Exceptional and What That Means for Life in the Universe. Basic Books, New York

Wang Z, Wang X, Shi X, Tang D, Stüeken EE, Song H (2020) Coupled nitrate and phosphate availability facilitated the expansion of eukaryotic life at circa 1.56 Ga. J Geophys Res: Biogeosci 125: e2019JG005487, https://doi.org/10.1029/2019JG005487

Wang C, Lechte MA, Reinhard CT, Asael D, Cole DB, Halverson GP, Porter SM, Galili N, Halevy I, Rainbird RH, Lyons TW (2022) Strong evidence for a weakly oxygenated ocean–atmosphere system during the Proterozoic. PNAS 119:e2116101119

Ward LM, Kirschvink JL, Fischer WW (2016) Timescales of oxygenation following the evolution of oxygenic photosynthesis. Origins of Life and Evolution of Biospheres 46:51–65

Warke MR, Di Rocco T, Zerkle AL, Lepland A, Prave AR, Martin AP, Ueno Y, Condon DJ, Claire MW (2020) The Great Oxidation Event preceded a Paleoproterozoic "Snowball Earth". PNAS 117:13314–13320

Weiss MC, Sousa FL, Mrnjavac N, Neukirchen S, Roettger M, Nelson-Sathi S, Martin WF (2016) The physiology and habitat of the last universal common ancestor. Nat Microbiol 1:16116, https://doi.org/10.1038/nmicrobiol.2016.116

Westall F, de Wit MJ, Dann J, van der Gaast S, de Ronde CE, Gerneke D (2001) Early Archean fossil bacteria and biofilms in hydrothermally-influenced sediments from the Barberton greenstone belt, South Africa. Precambrian Res 106:93–116

Westall F, de Vries ST, Nijman W, Rouchon V, Orberger B, Pearson V, Watson J, Verchovsky A, Wright I, Rouzaud JN, Marchesini D (2006) The 3.466 Ga "kitty's gap chert," an early archean microbial ecosystem. *In*: Processes on the Early Earth. Reimold WU, Gibson DG (eds). Geol Soc Am, Boulder, CO, p 105–131

Westall F, Brack A, Fairén AG, Schulte MD (2023) Setting the geological scene for the origin of life and continuing open questions about its emergence. Front Astron Space Sci 9:1095701

White AJR, Legras M, Smith RE, Nadoll P (2014) Deformation-driven, regional-scale metasomatism in the Hamersley Basin, Western Australia. J Metamorph Petrol 32:417–433

White WM, Klein EM (2014) Composition of the oceanic crust. Treatise Geochem 4:457–496

Willbold M, Hegner E, Stracke A, Rocholl A (2009) Continental geochemical signatures in dacites from Iceland and implications for models of early Archaean crust formation. Earth Planet Sci Lett 279:44–52

Wille M, Kramers JD, Naegler TF, Beukes NJ, Schroeder S, Meisel T, Lacassie JP, Voegelin AR (2007) Evidence for a gradual rise of oxygen between 2.6 and 2.5 Ga from Mo isotopes and Re-PGE signatures in shales. Geochim Cosmochim Acta 71:2417–2435

Williams RJP (1981) The Bakerian Lecture, 1981 Natural selection of the chemical elements. Proc R Soc London B 213:361–397

Williams CD, Mukhopadhyay S (2019) Capture of nebular gases during Earth's accretion is preserved in deep-mantle neon. Nature 565:78–81

Wilmeth DT, Lalonde SV, Berelson WM, Petryshyn V, Celestian AJ, Beukes NJ, Awramik SM, Spear JR, Mahseredjian T, Corsetti FA (2022) Evidence for benthic oxygen production in Neoarchean lacustrine stromatolites. Geology 50:907–911

Woese CR (1987) Bacterial evolution. Microbiol Rev 51:221–271

Wolf AS, Jäggi N, Sossi PA, Bower DJ (2023) VapoRock: Thermodynamics of vaporized silicate melts for modeling volcanic outgassing and magma ocean atmospheres. Astrophys J 947:64

Wordsworth R, Pierrehumbert R (2014) Abiotic oxygen-dominated atmospheres on terrestrial habitable zone planets. Astrophys J Letters 785:L20, https://doi.org/10.1088/2041-8205/785/2/L20

Yang S, Kendall B, Lu X, Zhang F, Zheng W (2017) Uranium isotope compositions of mid-Proterozoic black shales: Evidence for an episode of increased ocean oxygenation at 1.36 Ga and evaluation of the effect of post-depositional hydrothermal fluid flow. Precambrian Res 298:187–210

Yang J, Junium CK, Grassineau NV, Nisbet EG, Izon G, Mettam C, Martin A, Zerkle AL (2019) Ammonium availability in the Late Archaean nitrogen cycle. Nat Geosci 12:553–557

Young ED, Shahar A, Schlichting HE (2023) Earth shaped by primordial H_2 atmospheres. Nature 616:306–311

Young AV, Robinson TD, Krissansen-Totton J, Schwieterman EW, Wogan NF, Way MJ, Sohl LE, Arney GN, Reinhard CT, Line MR, Catling DC (2024) Inferring chemical disequilibrium biosignatures for Proterozoic Earth-like exoplanets. Nat Astron 8, 101–110 (2024). https://doi.org/10.1038/s41550-023-02145-z

Zahnle KJ, Catling DC, Claire MW (2013) The rise of oxygen and the hydrogen hourglass. Chem Geol 362:26–34

Zahnle KJ, Lupu R, Catling DC, Wogan N (2020) Creation and evolution of impact-generated reduced atmospheres of early Earth. Planet Sci J 1:11, https://doi.org/10.3847/PSJ/ab7e2c

Zerkle A, House CH, Brantley SL (2005) Biogeochemical signatures through time as inferred from whole microbial genomes. Am J Sci 305:467–502

Zerkle AL, Claire MW, Domagal-Goldman SD, Farquhar J, Poulton SW (2012) A bistable organic-rich atmosphere on the Neoarchaean Earth. Nat Geosci 5:359–363

Zerkle AL, Poulton SW, Newton RJ, Mettam C, Claire MW, Bekker A, Junium CK (2017) Onset of the aerobic nitrogen cycle during the Great Oxidation Event. Nature 542:465–467

Zhan Z, Seager S, Petkowski JJ, Sousa-Silva C, Ranjan S, Huang J, Bains W (2021) Assessment of isoprene as a possible biosignature gas in exoplanets with anoxic atmospheres. Astrobiology 21:765–792

Reviews in Mineralogy & Geochemistry
Vol. 90 pp. 559–594, 2024
Copyright © Mineralogical Society of America

15

Exoplanet Geology: What Can We Learn from Current and Future Observations?

Bradford J. Foley

Department of Geosciences, Center for Exoplanets and Habitable Worlds
Pennsylvania State University
University Park, PA 16802, U.S.A.

bjf5382@psu.edu

OVERVIEW

Nearly 30 years after the discovery of the first exoplanet around a main sequence star, thousands of planets have now been confirmed. These discoveries have completely revolutionized our understanding of planetary systems, revealing types of planets that do not exist in our solar system but are common in extrasolar systems, and a wide range of system architectures. Our solar system is clearly not the default for planetary systems. The community is now moving beyond basic characterization of exoplanets (mass, radius, and orbits) towards a deeper characterization of their atmospheres and even surfaces. With improved observational capabilities there is potential to now probe the geology of rocky exoplanets; this raises the possibility of an analogous revolution in our understanding of rocky planet evolution. However, characterizing the geology or geological processes occurring on rocky exoplanets is a major challenge, even with next generation telescopes. This chapter reviews what we may be able to accomplish with these efforts in the near-term and long-term. In the near-term, the James Webb Space Telescope (JWST) is revealing which rocky planets lose versus retain their atmospheres. This chapter discusses the implications of such discoveries, including how even planets with no or minimal atmospheres can still provide constraints on surface geology and long-term geological evolution. Longer-term possibilities are then reviewed, including whether the hypothesis of climate stabilization by the carbonate–silicate cycle can be tested by next generation telescopes. New modeling strategies sweeping through ranges of possibly evolutionary scenarios will be needed to use the current and future observations to constrain rocky exoplanet geology and evolution.

INTRODUCTION

Exoplanets have revolutionized our understanding of the formation of planetary systems, continuing a long progression of scientific thought from ancient ideas of geocentrism towards the modern understanding of Earth's place in the galaxy and universe (e.g., see Dick 1993, for a historical review). In the early twentieth century it was unclear whether planetary systems were common or rare, formed only by fluke events like stellar close encounters as Jeans (1919) argued. By the mid-twentieth century, the sheer number of stars and prominence of the nebula collapse theory for planet formation popularized the view that planetary systems should be relatively common (e.g., Dick 1993). Before the discovery of the first exoplanets, though, we lacked any broader context for what planetary systems could look like, so our solar system was often assumed to be the default (e.g., Pfalzner et al. 2015; Dawson and Johnson 2018). However, with > 5000 confirmed exoplanets now discovered, it has become clear that our solar system is not the default for planetary systems. The first exoplanets discovered around sun-like stars were a type not seen in our solar system and one that strongly challenged models of

1529-6466/24/0090-0015$05.00 (print)
1943-2666/24/0090-0015$05.00 (online)

http://dx.doi.org/10.2138/rmg.2024.90.15

planet formation: "hot Jupiters," gas giant planets orbiting very close to their host stars (Mayor and Queloz 1995). The radial velocity method, used to detect the first exoplanets and still one of the most fruitful methods used today, is biased towards finding large planets that orbit close to their host star, so close-in gas giant planets, if they existed, were going to be the first planets discovered. Work on exoplanet population statistics has now shown that hot Jupiters are not common, with frequencies of ~1% (e.g., Wright et al. 2012; Dawson and Johnson 2018; Zhu and Dong 2021). However, that they exist at all demonstrates that gas giant planets are not confined to the outer regions of planetary systems, as in our own.

In addition to hot Jupiters, there are other types of planets commonly found in exoplanet systems that are absent in our solar system. "Super-Earths," predominantly rocky planets that are larger in mass and radius than the Earth, and "mini-Neptunes" (or "sub-Neptunes"), planets larger than Earth but smaller than the ice giants in our solar system, and with a bulk density much lower than expected for a rocky planet, are common (e.g., Batalha 2014; Kane et al. 2021; Zhu and Dong 2021). The mini-Neptunes may be ice and volatile rich or possess thick H_2/He_2 atmospheres (e.g., Bean et al. 2021). Exoplanet demographics show that the super-Earths and mini-Neptunes make up separate and distinct planet populations, separated by a radius "gap" at around 1.5–2.0 Earth radii (Fulton et al. 2017). That is, planets with a radius of $\approx$ 1.5–2.0 Earth radii are rare; planets are either smaller and rocky or larger and volatile rich and/or possess thick atmospheres comprised of nebular gas (e.g., see Bean et al. 2021, for a recent review). Exoplanets also show a wide range of system architectures compared to that seen in our solar system, further highlighting that ours is not the default structure for planetary systems (Kane et al. 2021; Zhu and Dong 2021).

Now armed with a broad statistical sample of the diversity of planets and planetary systems, the theory of planet formation is undergoing significant revision and new modeling approaches are being enabled (e.g., Drazkowska et al. 2023). As planet formation is an inherently stochastic process, studies of the formation of our solar system haven typically taken a statistical approach, running a large suite of models each with slightly different initial conditions, and looking at the statistical distribution of planets formed (e.g., Chambers 2001). Much has been learned from this approach on the proto-planetary disk characteristics and processes that can result in systems like ours. Naturally, though, we can be stuck with "just so" stories when trying to explain only our solar system. In fact, given the stochastic nature of planet formation, at least some aspects of our solar system inevitably are the result of specific events that occurred during formation, and therefore really are explained by seemingly ad hoc hypotheses. The small size of Mars may be an example of this. The current leading model is the "Grand Tack," where inward migration of Jupiter scattered solids away from the region where Mars would later form, leaving less mass available and hence a small planet (Walsh et al. 2011). Whether this (if it indeed is the correct explanation for Mars's size) is a general process that would happen broadly in planetary systems, or just a result of the random quirks that went into forming our own system is not currently known. However, exoplanets provide a whole population of planets and systems for models to try to match. With this population it will be easier to work out which aspects of planetary systems are a result of general physics and processes occurring during formation, and which are due to random chance. Population synthesis studies are embarking on this line of research now and are already reshaping our understanding of planet formation (e.g., Mordasini et al. 2009).

The work of discovering exoplanets and quantifying the demographics of planetary systems will of course continue for years to come and is vital to answering the question of whether our solar system architecture is rare or common. However, with the launch of the James Webb Space Telescope (JWST), and plans for future missions, such as the direct imaging Habitable Worlds Observatory (HWO) or the proposed mid-infrared Large Interferometer For Exoplanets (LIFE) (Quanz et al. 2022), we are now entering an era of better characterization of exoplanets. With these new observational capabilities comes new opportunities to constrain not just basic planet properties like size and bulk density, but to probe their atmospheres or even surfaces. These new observations may then be used to infer something about the geological characteristics, such as the surface lithology, or geological processes, such as tectonics or volcanism, of rocky exoplanets.

As a result, exoplanets have the potential, at least seemingly, to spur a revolution in our understanding of planetary geological evolution, the same way they have revolutionized our understanding of planetary system formation and architecture. Like the field of planet formation before the discovery of exoplanets, geoscientists and planetary scientists only have a limited sample of planets and moons to study in detail: there are only four rocky planets (the topic of this volume) in our solar system. Each of these planets has been studied well enough to constrain to first order surface and interior composition, surface features, and tectonic processes operating today, though of course many details remain topics of intense study and debate. The four rocky planets of our solar system show significant diversity: a range of sizes from Mercury to Earth, different core sizes, tectonic states, atmosphere sizes and compositions, and magnetic fields or lack thereof for Venus and Mars. Earth is also the only planet where plate tectonics is known to operate. Mars and Mercury are likely in a "stagnant lid" mode of tectonics, where convection operates in the mantle but is unable to "break" the lithosphere into discrete plates that can move with respect to each other (e.g., Bercovici et al. 2015; Breuer and Moore 2015). Venus does not possess a global network of mobile plates like the modern Earth does, as evidenced by the lack of hallmark topographic features like ridges at zones of plate divergence or subduction zone trenches at regions of convergence. Venus may not be in a stagnant-lid state, however, as it does show evidence for localized subduction (Sandwell and Schubert 1992; Davaille et al. 2017) and regions of thin lithosphere and high heat flow that are not consistent with stagnant-lid convection (Borrelli et al. 2021; Smrekar et al. 2023). Venus may therefore operate in a tectonic regime intermediary to plate tectonics and stagnant-lid tectonics.

Geoscientists and planetary scientists have sought to explain the differences between the current states and evolutionary histories of the four rocky planets for decades. In particular, how Earth evolved into a habitable planet with liquid water, a temperate climate, plate tectonics, and a magnetic field, while the other rocky planets in our solar system did not, has been a long-standing question. Further motivating this question is the potential importance of both plate tectonics (e.g., Kasting and Catling 2003) and the magnetic field (e.g., Cockell et al. 2016) for Earth's habitability. Much progress has been made on elucidating the basic mechanics behind key processes and characteristics of the solar system rocky planets, like the physics behind plate tectonics and mantle convection, generating a magnetic field, or atmospheric evolution and retention.

However, as with solar system formation, explanations of Earth's fundamentally different evolution in comparison to its neighboring planets can ultimately lead to similar "just so" stories, due to the lack of a broader sample of planets with which to compare the Earth. With only one planet as an example of a habitable planet with plate tectonics and a magnetic field, and only a handful of counter examples, it is hard to generalize and test theories for why Earth ended up the way it did. Multiple hypotheses can explain the scant available data, and it is hard to disentangle cause and effect for some of Earth's unique features. For example, is plate tectonics caused by the presence of liquid water oceans, through the rheological weakening effects of water (Tozer 1985; Mian and Tozer 1990; Lenardic and Kaula 1994; Moresi and Solomatov 1998; Regenauer-Lieb et al. 2001; Richards et al. 2001; Korenaga 2007, 2010), or are liquid water oceans themselves caused by plate tectonics, and the climate stabilization it helps to provide through the carbonate–silicate cycle (Landuyt and Bercovici 2009)? A larger sample of planets, if they can be well characterized, could help answer these questions.

However, characterizing geological processes or properties of exoplanets is a significant challenge. Therefore, the ability of exoplanet studies to revolutionize our understanding of rocky planet diversity and evolution is far less certain than for planetary system formation. Finding exoplanets and constraining their sizes and orbits was enough to completely change our view of how planetary systems formed. For geological processes, far more detailed information about exoplanets is needed. This chapter therefore reviews the prospects for constraining geological properties and processes on exoplanets, and how these constraints might affect broader understanding of rocky planet evolution. The chapter focuses on a few key areas and discusses current or future observations that might help constrain them. First the chapter examines how compositional diversity of exoplanets might be constrained using current or near future observations. Then it discusses how current and future observations of rocky exoplanet atmospheres can inform their geological processes. In particular, this section looks at current discoveries of planets that appear to lack atmospheres and reviews what the lack of an atmosphere implies about the release of volatiles from the interior ("outgassing") on such planets. The following section then looks at what we might learn from future direct imaging missions, focusing on how scientists might test whether the carbonate–silicate cycle operates on rocky exoplanets to regulate their climates, and what implications this has for rocky planet interiors. Finally, the chapter concludes with summary thoughts and discussion.

TESTING MODELS OF THE COMPOSITIONAL DIVERSITY OF ROCKY EXOPLANETS

Rocky exoplanets can come in a wide variety of compositions, which will influence factors like the size of the core relative to the silicate mantle, the mineralogical makeup of the mantle and crust, as well as important material properties that control the dynamics of the mantle and core, how they evolve over time, and whether processes like plate tectonics or core dynamos can operate. However, directly constraining the composition of rocky exoplanets is clearly difficult, as we cannot directly sample and analyze their surface and mantle rocks. Several approaches have been developed to provide more indirect constraints on the plausible range of exoplanet compositions, and hence to supply geodynamicists with a framework to work within exploring how planet composition influences geophysical processes (e.g., Spaargaren et al. 2020). One approach is to construct interior structure models of planets with different compositions, and then use the measured mass and radius of individual planets to assess which compositional model they best match (Fortney et al. 2007; Seager et al. 2007; Valencia et al. 2007a,b; Zeng and Sasselov 2013). However, rocky planet composition cannot be uniquely inferred from mass and radius information alone (Dorn et al. 2015), so additional constraints are necessary. One popular avenue is to use star compositions to infer planet compositions (Dorn et al. 2015; Unterborn et al. 2016). As planets and the stars they orbit originally form from the same proto-stellar nebula, then planets should roughly match the composition of their host stars for refractory (rock-forming) elements (Thiabaud et al. 2015; Lodders 2020; Jorge et al. 2022). Measuring the composition of refractory elements in stars therefore provides an indication of the range of plausible rocky planet compositions. A number of studies have used compilations of stellar compositions, from databases such as the Hypatia catalog (Hinkel et al. 2014), to estimate this range of planet compositions, and construct models of resulting interior structures, mantle mineralogies, and heat budgets (Unterborn et al. 2015, 2022; Hinkel and Unterborn 2018; Putirka and Rarick 2019; Spaargaren et al. 2023). This approach of using stellar composition to estimate planet composition is covered in Hinkel et al. (2024) and Putirka (2024), both this volume).

Using stellar compositions to infer planet composition of course relies on the assumption that planets will approximately match the composition of their host stars, which ultimately needs to be tested observationally. Determinations of resulting interior structure and mantle mineralogy also rely on models which inevitably have limitations and may not always capture reality. As such it is critical to find additional ways to constrain the composition of rocky exoplanets more directly. This section focuses on other observations that provide such a constraint. These observations can only be attained in certain situations, meaning they cannot be used to constrain the full range of compositions across the observed exoplanet population, the way mineralogical models based on stellar composition can. However, even if just a few planets can have their crust, mantle, or even whole planet bulk composition estimated from direct observations, this will serve as a valuable test of models based on stellar compositions.

Direct measures of rocky planet composition

Four potential avenues for more directly constraining the composition of rocky exoplanets are: 1) pollution of white dwarf stars; 2) measuring the chemical composition of dust grains derived from "disintegrating planets,"; 3) measuring the chemistry of silicate vapor atmospheres for planets with magma oceans at their sub-stellar points ("lava worlds") or the chemistry of the lava itself; and 4) estimating surface compositions from thermal emission of airless rocky bodies. White dwarf pollution is a powerful tool that provides a direct measure of the elemental abundances of rocky exoplanetary material falling onto ("polluting") a white dwarf star. This material is likely the remnants of rocky planets that once orbited the star. White dwarf pollution has already been used to constrain the redox state of exoplanetary silicate materials, finding states similar to Earth and Mars (Doyle et al. 2019); this result has important implications for exoplanet mineralogy, interior structure, and atmospheres. Veras et al. (2024, this volume) and Xu et al. (2024, this volume) thoroughly discuss white dwarf pollution and what constraints observations so far place on exoplanet geology, so this section will focus on the latter three methods.

Disintegrating planets

Ultra-short period (USP) planets are planets that orbit their host star with extremely short orbital periods and hence extremely small orbital separations (Sanchis-Ojeda et al. 2014). Some rocky USP planets orbit so close to their host star, and therefore receive such high levels of stellar flux, that the rocky surface is evaporating and being lost to space (e.g., Rappaport et al. 2012; Sanchis-Ojeda et al. 2015; van Lieshout and Rappaport 2018). The result is a rocky planet with dust clouds that can either lead or trail the planet, or both, causing a distinct shape to the transit light curve that can also vary significantly over time (e.g., Rappaport et al. 2012; van Werkhoven et al. 2014; Sanchis-Ojeda et al. 2015). For vaporized silicate to escape to space, the planet itself must be small, on the order of Mercury-sized to the size of Earth's moon (Perez-Becker and Chiang 2013), and these are the sizes inferred for the 3 disintegrating planets found so far: Kepler-1520b, KOI-2700b, and K2-22b (Rappaport et al. 2012, 2014; Sanchis-Ojeda et al. 2015).

Disintegrating planets provide an excellent opportunity to constrain their composition, because the widely scattered dust grains around the planet are well suited for study with "transmission spectroscopy" (Bodman et al. 2018). In transmission spectroscopy, light from the star passes through a planet's atmosphere, or in this case dust clouds, during transit. Comparing the signal during transit and outside of transit allows the contribution from the planet to be isolated. In the case of disintegrating planets, the spectroscopic signal from the dust grains can be used to infer their composition (Okuya et al. 2020), with K2-22b probably the planet best suited for this analysis (Bodman et al. 2018). Interpreting the observed dust composition to constrain the composition of the disintegrating planet will not be without its challenges. It is not immediately clear if the crust, mantle, or core of the planet is disintegrating, or if dust will reflect some mixture of all three potential compositional layers. While the composition of the dust itself once measured can help shed light on this, as more silica-rich minerals would be expected from the crust, and significant iron abundance would indicate disintegration from the core, there will be some degree of degeneracy between the bulk planet composition and the presumed layer or layers that dust is derived from.

Disintegrating planets are rare and may only spend ~10–100 Myrs during the phase of active disintegration (Perez-Becker and Chiang 2013). Such planets will therefore not form a large dataset of planet compositions with which to test compositional models like those presented in Putirka (2024, this volume). Furthermore, the origin of disintegrating planets is not well understood. Whether they were born as rocky planets in the inner proto-planetary disk (either in-situ or at least relatively close to where they are currently found) or formed further out as mini-Neptunes or even gas giant planets that have migrated to their current ultra-short period orbits and lost their gas envelopes is not clear (e.g., Jackson et al. 2013; Lopez 2017; Winn et al. 2018). Observations of the dust being lost from these planets can potentially shed light on this formation history as well, which is important for interpreting what they mean for rocky exoplanet compositions more generally. If disintegrating planets began their lives as mini-Neptunes or gas giants, then their composition may not be as reflective of rocky exoplanets more broadly. However, they nonetheless can help complement observations from e.g., white dwarf pollution and, when viewed in the proper context of their potential formation mechanisms, help test models of exoplanet compositional diversity.

Lava worlds

A less extreme type of USP planet than disintegrating planets are those that receive enough stellar radiation flux to have dayside temperatures exceeding the melting temperature of the crust or underlying mantle, leading to large lava ponds or magma oceans on their surfaces. However, stellar flux is not so high, or the planet mass is large enough that silicate vapor is not being lost to space at substantial rates, as in the case of disintegrating planets. These planets are "lava worlds" (see review by Chao et al. 2021), and have been recognized dating back to the discovery of some of the first rocky exoplanets (Léger et al. 2009). Most lava worlds are expected to be tidally locked, meaning they would have a permanent day side (presumably molten) and a permanent night side (presumably solid). Some of the most well-known lava planets are CoRoT-7b (Léger et al. 2011), 55 Cancri e (Demory et al. 2011; Winn et al. 2011), Kepler-10b (Batalha et al. 2011), and Kepler-78b (Howard et al. 2013; Sanchis-Ojeda et al. 2013).

Theoretical models predict that lava worlds could have atmospheres composed of vaporized silicates, in equilibrium with the lava ponds or oceans at the surface of the dayside of these planets (Schaefer and Fegley 2009; Miguel et al. 2011; Schaefer et al. 2012; Ito et al. 2015). These silicate vapor atmospheres are expected to be dominated by Na, O_2, Si, SiO, or K (Schaefer and Fegley 2009; Ito et al. 2015), depending on the composition of the surface experiencing melting and vaporization (Miguel et al. 2011; Schaefer et al. 2012), though volatiles like Na and K may be lost leaving SiO or SiO_2 as dominant observable constituents (Nguyen et al. 2020; Ito and Ikoma 2021). If the composition of these silicate vapor atmospheres can be measured, then they provide a direct window into the composition of the interior (Zilinskas et al. 2022; Piette et al. 2023). As a result, studying the atmospheres and geodynamics of lava worlds has become a major focus of activity, with papers considering details of atmospheric structure and transport (Hammond and Pierrehumbert 2017; Nguyen et al. 2022), dynamics of the magma ocean (Kite et al. 2016), and convection in the whole mantle (Meier et al. 2023).

55 Cancri e is the most well studied lava world. Demory et al. (2016) mapped thermal emission from the planet by measuring the host star's light curve as the planet passed around behind the star. As a transiting planet orbits towards its "secondary eclipse," where it passes directly behind the host star, more and more of the dayside of the planet comes into view from our reference frame. As a result, the observed light includes both the star's light and light emitted by the planet. However, when the planet reaches secondary eclipse, the planet's contribution drops away, leaving only the star's light. Comparison of the star plus planet emission throughout an orbit to that of just the star during secondary eclipse allows for the contribution from the planet to be isolated. Demory et al. (2016) found a large day/night temperature contrast, indicating limited redistribution of heat from day to night side. However, the point of peak emission was offset from the substellar point, indicating flow of either atmosphere or lava redistributing heat on the day side. The offset in peak emission has been interpreted as being indicative of an atmosphere (Angelo and Hu 2017; Hammond and Pierrehumbert 2017), though stronger confirmation is still lacking. In fact, reanalysis of the data by Mercier et al. (2022) finds a negligible peak emission point offset, indicating minimal planetary heat redistribution; clearly additional observation and

analysis is needed. Using transmission spectroscopy, an abundant low mean molecular weight atmosphere can be ruled out, but these observations can't distinguish between clouds, a high mean molecular weight atmosphere like a silicate vapor atmosphere, or no atmosphere at all (Deibert et al. 2021). The lack of any evidence for escaping Helium also indicates that 55 Cancri e lacks a low mean molecular weight atmosphere today, either because it has been lost already or never accreted one to begin with (Zhang et al. 2021).

Observations with JWST will hopefully confirm the nature of 55 Cancri e and other lava worlds atmospheres one way or another. As of yet lava world atmospheres are still awaiting detailed characterization, and thus their ability to constrain rocky planet composition is still mostly theoretical. Zieba et al. (2022) measured thermal emission of a different lava world, K2-141 b, and found emission consistent with a thin silicate vapor atmosphere. This might be our first tantalizing glimpse at the characterization of lava world atmospheres, that hopefully will be refined in the coming years. Another possibility for constraining lava world compositions is using the emission or reflection of light from their magma or crystal mush surfaces (Fortin et al. 2022), similar to what can be done for solid surfaces as discussed in the next section. This work is also in its infancy but could provide another promising avenue for rocky planet compositional characterization.

Emission from solid surface planets

Moving back to larger orbital distances are planets cool enough to have solid surfaces, but that still receive such high radiation fluxes from their host stars that they lack atmospheres. Thermal emission is the primary observation used to look for these "airless" or "bare rock" planets. Planets lacking atmospheres should lack heat redistribution from dayside to nightside, resulting in high temperatures on the dayside and strong day-night temperature contrasts. Measuring the full thermal phase curve of a planet can show this temperature contrast and hence whether there is significant heat redistribution, though an even simpler method is to measure dayside thermal emission, constrain the temperature, and compare with models that account for heat redistribution by different assumed atmospheres and an airless case (Koll et al. 2019). A handful of planets that are good candidates to be bare rocks have been identified so far based on thermal emission observations, as outlined in more detail below in *Observations of rocky planet atmospheres to date*.

Bare rock planets present a rare opportunity for characterizing surface geology, since the observed thermal emission comes from the surface rocks. Different types of crust (e.g., mafic crust like the basalt making up Earth's seafloor, or felsic crust like our continents, predominantly formed of granitic-type rock) produce different emission as a function of wavelength. Observing thermal emission from a bare rock planet in different wavelengths can therefore allow the surface rock composition to be constrained (Hu et al. 2012).

Thermal emission observations have been made for planets thought to be bare rocks: LHS 3844b (Kreidberg et al. 2019; Whittaker et al. 2022), GJ 1252b (Crossfield et al. 2022), Trappist-1b (Greene et al. 2023; Ih et al. 2023), and Trappist-1c (Zieba et al. 2023). It should be noted, though, that the evidence supporting the inferred lack of atmosphere for these planets is of varying robustness, e.g., the observations of Trappist-1c are permissive of some hypothetical thin atmospheres (Lincowski et al. 2023). More observations and better models will be needed to confirm which planets are truly airless. Results obtained so far assuming these planets are indeed bare rocks have found surface compositions broadly consistent with mafic or ultramafic crusts, similar to the basalt ocean crust on Earth today. Another possible crust composition is a highly oxidized, hematite rich crust which would be produced by photolysis of water, escape of H to space and oxidation of the surface by remaining O (Hu et al. 2012). Current observations are not able to distinguish between the above compositions; more detailed measurements at multiple wavelengths will be needed to more uniquely constrain surface compositions.

However, for the most part felsic crusts have been found to not match the data. Mafic or ultramafic crusts would be expected to form via partial melting of a mantle with a similar composition to Earth's, so the crust compositions inferred so far appear consistent with an Earth-like composition (see Guimond et al. 2024; Purika 2024, both this volume). Observations of a felsic crust would be exciting, as felsic crust on Earth predominantly forms due to hydrous melting (e.g., Campbell and Taylor 1983). Felsic crust on an exoplanet could indicate that water was present at least at some point during the planet's history, even if water has since been lost. Liquid water is not expected on planets receiving the high stellar fluxes that the candidate bare rocks do, but even unlikely possibilities should be considered and tested given the lack of definitive constraints. Felsic crust could potentially also form from mantle melting and distillation on planets with non-Earth-like (probably more silica-rich) bulk compositions (Putirka and Rarick 2019; Brugman et al. 2021). Combining stellar composition constraints and thermal emission where possible could therefore act as a valuable test of exoplanet composition models. In particular targeting planets where stellar compositions would imply a mantle that would produce something other than mafic or ultramafic crust could help confirm predictions from compositional models.

CONSTRAINTS ON EXOPLANET GEOLOGY FROM THE PRESENCE OR ABSENCE OF AN ATMOSPHERE

One of the top priorities for exoplanet science, both currently and in the coming decades, is to characterize exoplanet atmospheres to constrain their prospects for habitability, look for potential signs of life, learn about planet formation and evolution, and even potentially constrain geological processes (e.g., National Academies of Sciences, Engineering, and Medicine 2021). Attempts to characterize rocky planet atmospheres are still in their infancy. The Hubble and Spitzer space telescopes have been used to look at a handful of rocky planets so far (see e.g., Wordsworth and Kreidberg 2022, for a review), and now with JWST we are entering a new era of rocky exoplanet atmospheric observing capabilities. Observations of Trappist-1 and other systems will hopefully provide our first close look at the atmospheres of habitable zone rocky planets.

Given the wealth of exoplanet atmosphere data that will be collected in the coming years, the science of exoplanet geology must be poised to utilize these observations to learn as much about rocky exoplanet interiors as possible to move forward. The challenge facing researchers interested in exoplanet interiors, however, is that atmospheres are only an indirect result of rocky planet interior processes, and thus only indirectly constrain these processes. Tectonic and geological processes on solar system planets are largely studied by detailed mapping of surface topography, composition, and geophysical quantities like gravity; such detailed surface characterization is not feasible in the foreseeable future for exoplanets. It will therefore be difficult to tightly or uniquely constrain geological processes based on atmospheric observations. However, given the wide diversity possible for exoplanets and the dearth of information about their geology, any constraint that can at least rule out some scenarios is still highly valuable.

In that spirit this section will focus on what can potentially be constrained about rocky exoplanet geological processes from the most basic, and robust, aspect of atmospheric characterization conducted so far, which is simply determining which planets do or do not have atmospheres at all. While the ultimate goal of rocky exoplanet atmospheric characterization is detailed mapping of the abundances of different atmospheric gasses, as outlined in more detail in *Observations of rocky planet atmospheres to date* below, this has proved challenging so far. However, a handful of rocky exoplanets planets have been inferred to various levels of confidence to be airless. These observations motivate the question of what factors control atmosphere formation and retention on rocky planets, a question which is likely to drive much of the work in the field in the near-term. A key aspect of atmosphere formation and retention is exoplanet geology, as volcanism and release of volatiles from the interior (called outgassing) is a major source of atmospheric gases for rocky planets. This section will therefore focus on the constraints provided on volatile outgassing by knowledge of whether a planet does or doesn't have an atmosphere.

Airless exoplanets also provide some information about the composition of rocks at the surface, integrated over the surface area from which emission on the planet is observed. This surface composition constraint provides a window into the overall planet composition (see *Emission from solid surface planets*) and also informs whether volcanism has taken place at least at some point in time during the planet's history as discussed more in this section. Future telescopes may also be able to constrain brightness variations across a planet's surface as it rotates through direct imaging, which can be used to estimate lateral compositional differences or the distinction between e.g., oceans and continents (e.g., Ford et al. 2001; Cowan et al. 2009; Zugger et al. 2010). Thus, surface mapping of exoplanets may be possible in the future. However, these very coarse constraints are the limit of exoplanet surface characterization for the coming decades, and yet are still far below the level of detail geologists and planetary scientists are used to when studying the geological processes on solar system planets. Hence the clear need to use atmosphere observations as at least a complementary tool to help constrain exoplanet geology.

This section will first briefly review the methods for atmosphere characterization, then observations that have been made so far, with a focus on the discovery of a number of presumably airless planets. It then reviews the key sources and sinks of atmospheric gases, volcanic outgassing, and atmospheric escape. The section closes by discussing how the observation of a lack of atmosphere on a planet can be combined with models of escape to constrain outgassing rates and evolutionary history.

Methods for atmosphere characterization on rocky exoplanets

Exoplanet atmospheres can be constrained by transmission spectroscopy and thermal emission, or in future missions by direct imaging (e.g., Wordsworth and Kreidberg 2022). Transmission spectroscopy and emission were briefly introduced in *Testing models of the compositional diversity of rocky exoplanets* but will be more fully explained here. Transmission spectroscopy refers to measuring the starlight as it passes through a transiting planet's atmosphere, and comparing this to the starlight observed when the planet is not transiting (e.g., Seager and Sasselov 2000; Brown 2001; Ehrenreich et al. 2006; Kaltenegger and Traub 2009). From comparing these two cases, the signal from the planet's atmosphere can be isolated. In particular, measuring the flux in different wavelengths allows potential absorption features for key atmospheric gases to be detected, and hence characterization of the species making up the planet's atmosphere. Atmospheres can also be characterized by the thermal emission the plant gives off. Here the combined radiation emitted by both the planet and star during the planet's orbit is measured (e.g., Deming et al. 2005). For a transiting planet, the night side of the planet will be facing observers on Earth when it passes in front of the star, while the day side of the planet will be facing observers as it orbits behind the host star before the planet disappears behind the star during "secondary eclipse." Tracking how the combined star plus planet emission changes over the course of a planet's orbit allows the difference between day and night side temperatures to be constrained, which informs whether heat is being redistributed by an atmosphere or not (e.g., Knutson et al. 2007). Looking at emission in different wavelength bands further allows astronomers to compare observations with different atmosphere, or for airless planets, surface, composition models (e.g., Sudarsky et al. 2003; Hu et al. 2012; Stevenson et al. 2014; Kreidberg et al. 2019; Piette et al. 2022). Finally, next generation space telescopes will use direct imaging for atmosphere characterization, where light from the planet is directly observed. These telescopes will use choronographs to block out the star's light, allowing light from the planet to be observed and analyzed (National Academies of Sciences, Engineering, and Medicine 2021).

Detailed characterization of rocky exoplanet atmospheres is challenging. Obtaining sufficient signal from the planet requires combining observations from multiple transits for either transmission spectroscopy or mapping thermal emission. Both techniques are also more successful for smaller, cooler stars as the planet will have a larger impact on the observed starlight in such cases. These practical considerations favor planets in short period orbits around M-dwarf stars as the most readily observable. Planets in such systems are more likely to transit from our reference frame, the short period orbits allow for multiple transits to be observed with limited observing time, and M-dwarf stars are abundant and frequently host rocky planets (e.g., Wordsworth and Kreidberg 2022). Moreover, the cool temperature of M-dwarfs means that the habitable zone is much closer to the star than for a sun-like star, so even habitable zone planets have short period orbits (e.g., Kasting et al. 1993b; Kopparapu et al. 2013).

However, stellar contamination appears to be a major problem for transit spectroscopy for planets around M-dwarfs (e.g., Lim et al. 2023). Moreover, M-dwarf stars also differ from sun-like stars in ways that may be detrimental to habitability. M-dwarf stars take up to ~ 1 Gyr to reach the main sequence, and during this time they are highly luminous and emit high energy radiation. This high luminosity and extreme radiation can potentially desiccate the planets that will eventually end up in the habitable zone of these stars after they reach the main sequence (Ramirez et al. 2014; Luger and Barnes 2015). Planetary systems around M dwarfs after they reach the main sequence may therefore be left with only desiccated planets, including those in the habitable zone, and planets too cold to support liquid surface water beyond the habitable zone outer edge. Nevertheless, the favorable observing conditions for planets around M-dwarfs mean they have been the primary focus of attempts to characterize rocky exoplanet atmospheres thus far, including some of the first observing programs with JWST.

Observations of rocky planet atmospheres to date

A handful of rocky planets (~ 10) have been observed with transmission spectroscopy (Wordsworth and Kreidberg 2022), most of them in short period orbits around M-dwarf stars. These planets have typically shown "featureless" spectra, lacking any strong signal from candidate gases like H_2O or CO_2 (e.g., de Wit et al. 2016, 2018; Diamond-Lowe et al. 2018, 2020; Libby-Roberts et al. 2022; Lustig-Yaeger et al. 2023); featureless spectra have also been seen for some mini-Neptunes, such as GJ 1214 b, as well (e.g., Bean et al. 2010). Although seemingly disappointing, the observed featureless spectra allow cloud-free H_2/He_2 atmospheres to be ruled out, meaning these rocky planets likely have either no atmosphere or higher mean molecular weight atmospheres. An atmosphere dominated by high molecular mass gases, like H_2O or CO_2, which are the dominant products of volcanic outgassing on the modern Earth (e.g., Gaillard et al. 2021), is gravitationally bound closer to the planet's surface than a low mean molecular mass atmosphere (e.g., one dominated by H_2 or He_2). With less atmospheric cross section for starlight to pass through, high mean molecular mass atmospheres produce smaller signals than, e.g., H_2 or He_2 dominated atmospheres. The cloud-free caveat is also important, however, as another possibility for the featureless spectra is clouds masking any signal from atmospheric gases. The possibility of clouds then provides a note of caution for the prospects of observing higher mean molecular weight atmospheres with next generation telescopes; even with highly sensitive observing capabilities, clouds might mask our ability to determine the makeup of such atmospheres or search for biosignature gases (e.g., Komacek et al. 2020).

A few rocky planets, LHS 3844 b (Kreidberg et al. 2019), GJ 1252 b (Crossfield et al. 2022), and now with JWST Trappist-1 b (Greene et al. 2023) and Trappist-1 c (Zieba et al. 2023), have been observed in thermal emission. These planets all show strong dayside to nightside temperature contrasts consistent with either very thin atmospheres or airless rocky bodies. While the finding of presumably atmosphereless planets may also seem disappointing, determining which planets do and do not have atmospheres is a fundamental question for planetary evolution. Zahnle and Catling (2017) dub the boundary between airless planets and planets that retain atmospheres as the "cosmic shoreline." Zahnle and Catling (2017) empirically fit this boundary to solar system bodies with simple power-law relationships. The relationships are formulated a few different ways, which may be applicable for different stellar environments. In one case the cosmic shoreline is defined as a relationship between the incident stellar flux (I) and a planet's escape velocity (v_{esc}), which is determined by its surface gravity, where $I \propto v_{esc}^4$ marks the critical incident flux above which atmospheres are lost. For systems where high energy stellar radiation consisting of X-ray and extreme ultraviolet, together referred to as "XUV," radiation is dominant, then instead the integrated XUV flux (I_{XUV}) may be more important, leading to the scaling $I_{XUV} \propto v_{esc}^4$ for the cosmic shoreline. The presence of absence of an atmosphere is thus assumed to be determined to first-order solely by atmospheric escape, where high escape velocity and low stellar flux means a planet can gravitationally bind its atmosphere and resist loss to space, while low escape velocity or high stellar flux means atmosphere can be easily lost.

Current work characterizing which planets do and do not have atmospheres will help test the cosmic shoreline proposed by Zahnle and Catling (2017). The candidate airless planets discussed so far experience high incident flux as they orbit very close to their host stars, much too close to lie within the liquid water habitable zone. They also likely experienced high XUV fluxes as is expected for low mass stars (Johnstone et al. 2021). In terms of incident stellar flux these four planets straddle the proposed cosmic shoreline, with GJ 1252 b and LHS 3844 b on the airless side, and Trappist-1 b and c on the side where atmospheres could potentially be retained (Fig. 1). It is therefore at least somewhat surprising that Trappist-1 b and c apparently lack atmospheres today. Trappist-1 c in particular receives a comparable incident flux as modern Venus, yet instead of possessing a thick CO_2 dominated atmosphere, it is consistent with a bare rock or a thin CO_2 atmosphere. Specifically, the upper limit for a CO_2 poor atmosphere (10 ppm CO_2) is 10 bar, and for a pure CO_2 atmosphere the upper limit is 0.1 bar (Zieba et al. 2023).

However, in terms of integrated XUV flux all four planets sit in the airless regime, indicating the potentially important role of high energy radiation for atmosphere loss, as discussed further in *Thermal vs non-thermal escape and implications for the geological processes of airless bodies*. The different predictions from the two different formulations for

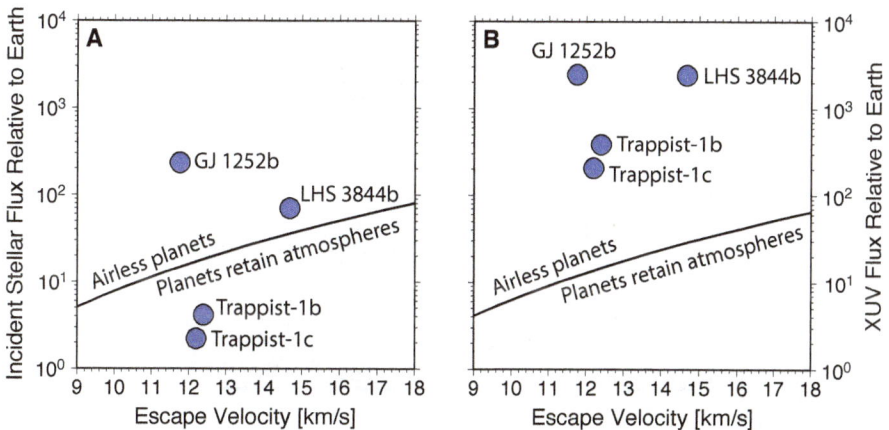

Figure 1. The cosmic shoreline proposed by Zahnle and Catling (2017) in terms of present-day stellar flux (**A**), $I \propto v_{esc}^4$, and integrated XUV flux (**B**), $I_{XUV} \propto v_{esc}^4$, with regions where planets should be airless and should retain their atmospheres labelled. The four planets interpreted to be airless based on thermal emission are plotted. Depending on the chosen formulation for the cosmic shoreline, the two Trappist planets may fall into either the regime where they would be predicted to still retain atmospheres or to be airless.

the cosmic shoreline also highlight how system specific factors can be important for whether planets end up retaining or losing atmospheres. In fact, even planet specific factors could be important, especially for planets falling near the cosmic shoreline, where secondary effects can tip the balance between losing or retaining an atmosphere. Such secondary effects could include additional escape processes not considered in the Zahnle and Catling (2017) model, or geological processes and characteristics of planets, like their abundance of volatiles that can form atmospheric gases or their long-term rates of volcanism and outgassing. Ultimately then a full accounting of different atmospheric sources and sinks is needed to study how airless planets come to lose their atmospheres, which can be accomplished using detailed models of planetary evolution (e.g., Kane et al. 2020; Crossfield et al. 2022; Krissansen-Totton and Fortney 2022; Krissansen-Totton 2023; Teixeira et al. 2023). Such an approach further allows constraints to be placed on airless planet's atmospheric sources and sinks, with at least some of these constraints tying back to exoplanet geology. To discuss how this can be accomplished it is necessary to first briefly review the major sources and sinks of atmospheric gases.

Atmosphere sources and sinks

The evolution of an atmosphere's size and composition is a result of the sources and sinks of gases. Therefore, knowledge of an exoplanet's atmosphere provides at least some constraint on these sources and sinks. Sources of atmospheric gasses include capture of nebula gas during planet formation, which would largely consist of H_2 and He_2, outgassing from the interior, or outgassing from impactors (Fig. 2). Sinks include atmospheric loss to space by various escape processes and incorporation of atmospheric gases into surface rocks, as occurs via weathering on Earth (see e.g., Lammer et al. 2018, for a review). This section focuses on the rocky exoplanets currently inferred to lack atmospheres. These planets are all very hot, receiving stellar radiation flux higher than the runaway greenhouse limit; they thus are expected to lack liquid water oceans, meaning that weathering will not be a significant sink for atmospheric gases. These planets also will have already lost any primordial atmosphere acquired from nebula gas, and as we are looking at planets well after the formation stage, impact outgassing will be a minimal source as well. For airless rocky planets the dominant source, if any sources exist, will therefore be interior outgassing and the dominant sink will be escape to space.

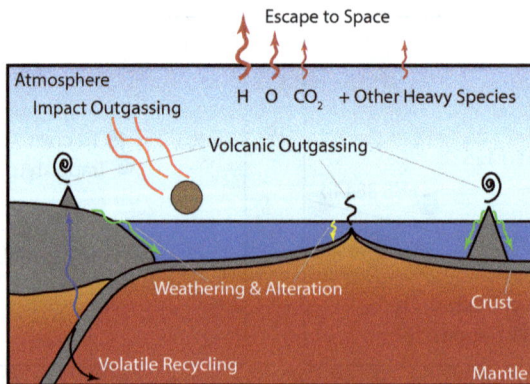

Figure 2. Schematic diagram of atmosphere sources and sinks. Adapted from Foley (2015), Foley and Driscoll (2016).

On a solid surface planet, outgassing from the interior predominantly occurs by volcanism, releasing gases derived from volatiles previously stored in minerals in the crust and mantle. The discussion in this section will focus on volcanism on solid surface planets like occurs on the rocky planets in our solar system. Solid surface planets are those whose dayside temperatures fall below the melting temperature of their constituent surface rocks. Planets hotter than this will have large magma ponds or oceans on their surfaces and can possess atmospheres produced by melt–vapor equilibrium (see *Lava worlds* for a discussion of these planets). Atmospheres of lava world planets are certainly interesting, as they can help probe interior composition as discussed above, but they fall under a very different regime than outgassing produced by volcanism on a solid surface planet.

Considering volcanic outgassing as the dominant atmosphere source and loss to space as the dominant sink, then the volcanic outgassing rate could in principle be constrained for an airless body if the rate at which atmosphere could escape can be estimated or modeled. Constraints on outgassing rate in turn then constrain both volatile abundance in the planet's interior and the integrated rate of surface volcanism (see *Controls on interior outgassing*). Outgassing rate constraints will be upper bounds, and can be derived because, on an airless body the outgassing rate must be less than the atmospheric loss rate, otherwise gases would build up on the planet and an atmosphere would be present. However, both atmospheric escape and volatile outgassing are complex topics. Escape involves multiple different loss mechanisms that can dominate under different conditions (e.g., Tian 2015; Catling and Kasting 2017). The structure of the planet's atmosphere and mixing ratios of constituent gases is important in many cases, meaning that models must be able to track these properties to first order to derive reasonably accurate atmosphere loss rates. Outgassing is a result of volcanism which also occurs by a variety of different mechanisms, depends on the chemical and thermal state of a planet's interior, its style of tectonics, and conditions at which melt erupts at the surface, among other factors. In order to illustrate how knowledge of escape can be used to constrain outgassing on airless planets, it is therefore necessary to first review the basics of both outgassing and escape.

Controls on interior outgassing

Crustal and mantle rocks contain volatiles stored either in specific volatile-bearing minerals or as trace elements that take the form of defects in a mineral crystal lattice (see e.g., Hirschmann 2006, for a review of water storage in mantle minerals). When a rock melts, these volatiles are released and become incorporated into the resulting magma. As magma rises to the surface and erupts the volatiles in the magma can form into gas bubbles that eventually degas upon eruption, with the exact species and efficiency of this degassing depending on magma composition and eruption temperature and pressure conditions.

Melting is therefore an essential part of outgassing to form an atmosphere. Regions of a rocky planet can melt when the constituent rocks exceed their "solidus," or the temperature where a rock first begins to melt (see e.g., Foley et al. 2020, for an overview of this process). As rocks are mixtures of different minerals, each of which has different properties, a rock does not completely melt at one temperature. Instead, the solidus marks the temperature below which all components of the rock are solid, the "liquidus" marks the temperature above which the rock is entirely molten, and in between these temperatures the rock is partially molten, with some percentage melt and the remainder staying solid. For an Earth-like mantle composition the solidus is $\approx 1100\,°C$ at surface pressure and liquidus is $\approx 1800\,°C$ at surface pressure under dry conditions (Takahashi 1986; Katz et al. 2003). When the temperature first exceeds the solidus, minerals with the lowest melting points melt, followed by those with higher melting points as temperature increases, and finally reaching complete melting when the liquidus temperature is reached. On the modern Earth, regions experiencing melting rarely reach the liquidus, and instead only partially melt with melt fractions ranging between a few to a few 10s % (e.g., at mid-ocean ridges the mantle experiences about 10–25 % melting Spiegelman and McKenzie 1987; Kushiro 2001).

Common volatile elements that contribute to outgassing to the atmosphere on Earth are H, C, O, N, and S (e.g., Gaillard et al. 2021), and these will presumably be the primary volatile elements on rocky exoplanets as well barring planets with radically different compositions than Earth's (e.g., Lichtenberg et al. 2023). Outgassing on the modern Earth is specifically dominated by CO_2 and H_2O (Holland 1984; Gaillard et al. 2021). Volatile elements are typically "incompatible" during melting, which means that they preferentially go into magma during melting rather than remaining behind in the residual solid. As a result, the area of mantle or crust experiencing partial melting will be devolatilized, with volatiles then becoming concentrated in the melt (see e.g., Hirth and Kohlstedt 1996; Hirschmann 2006; Korenaga 2006, for discussion of this behavior for water). This incompatible nature means that magmatism is an efficient process for removing volatiles from the interior to the surface, with some caveats discussed below.

The specific mixture of gases these volatiles make and then release to the atmosphere depends on many factors, including mantle chemistry and surface pressure (e.g., Gaillard and Scaillet 2014; Foley et al. 2020; Ortenzi et al. 2020; Lichtenberg et al. 2023). If the mantle is more oxidized, meaning more oxygen is available to form gases, then oxygen-bearing gases like H_2O and CO_2 dominate, as is the case on the modern Earth. Under more reducing conditions gases like H_2, CO, or CH_4 dominate instead (e.g., Holland 1984; Kasting et al. 1993a; Gaillard and Scaillet 2014; Ortenzi et al. 2020). The oxidation state of the mantle also influences the partitioning behavior of mantle volatiles, with more oxidizing conditions leading to an enhanced ability of volatiles to preferentially partition into the melt during magmatism (Guimond et al. 2021).

The flux of a gas brought to the surface by volcanism can be quantified as the product of the melt eruption rate, $\dot{M}$, given here with units of $m^3 \cdot s^{-1}$, and the concentration of volatile species i in the melt, C_i, with units of either $mol \cdot m^{-3}$ or $kg \cdot m^{-3}$. A certain fraction, f_{outgas}, of the gas in the magma will then outgas, with the remainder staying soluble in the melt. The outgassing flux of gas species i, F_i, is therefore (e.g., McGovern and Schubert 1989; Grott et al. 2011; Foley and Smye 2018)

$$F = f_{outgas} \dot{M} C_i \tag{1}$$

Petrological models constrained by melting experiments can determine the speciation of different gases and the concentration of volatiles in the melt resulting from melting a rock of a given composition (Gaillard and Scaillet 2014). To determine the melt eruption rate, though, we must consider the tectonic setting where melting can occur.

On Earth most volcanism is caused by partial melting of the mantle, which occurs in a few different settings (Fig. 3). One significant setting is at mid-ocean ridges, where mantle upwells to fill in the space left by the divergent spreading of surface plates on either side of the ridge. In this setting, upwelling mantle melts by "decompression melting," where a mantle parcel rising along an adiabat can begin to melt at shallow pressure where the solidus is lower. Most materials have lower melting temperatures at low pressure than at higher pressure, meaning a rising parcel can melt simply by moving to lower pressures rather than having its temperature rise. Decompression melting only occurs if the mantle is hot enough for the adiabat to cross the solidus before reaching the base of the planet's lithosphere, where temperatures are colder. At mid-ocean ridges, the lithosphere is very thin, allowing rising mantle to reach low pressures while still following the mantle interior adiabat, therefore facilitating melting. A different but related setting is melting associated with hot, active upwelling plumes. Plumes have an elevated temperature compared to the average mantle interior, and therefore can melt more easily and extensively during ascent. Finally, melting on Earth also occurs at subduction zones, where surface tectonic plates sink back into the interior. In these settings water incorporated into hydrated minerals in the crust and mantle of the sinking plate is released by metamorphic reactions at increasing temperature and pressure. The release of this water decreases the mantle and crustal solidus, leading to melting even at relatively low temperatures that would have been below the solidus without water present. In all these settings the rate of melt production can be calculated by estimating the flux of mantle into the region where melting occurs (where the solidus is exceeded), $\dot{m}$, and the fraction of melt in this melting region. The melt fraction, ϕ, can be simply calculated based on the temperature within the melting region, T as

$$\phi = \frac{T - T_{sol}}{T_{liq} - T_{sol}} \tag{2}$$

where T_{sol} and T_{liq} are the solidus and liquidus temperatures, respectively (Grott et al. 2011; Morschhauser et al. 2011; Tosi et al. 2017). The melt eruption rate, $\dot{M}$, is then

$$\dot{M} = f_{erupt} \phi \dot{m} \tag{3}$$

where f_{erupt} is the fraction of melt produced that erupts at the surface.

The tectonic settings where melting occurs on Earth discussed here are in many ways tied to plate tectonics. Mid-ocean ridges, where significant volcanism occurs by decompression melting are unique to plate tectonics, as are subduction zones where melting occurs by fluid release. Upwelling mantle plumes can occur regardless of a planet's tectonic setting but are facilitated by plate tectonics. Plate tectonics leads to efficient mantle interior cooling, therefore increasing the temperature contrast at the core–mantle boundary which drives plume formation (Jellinek et al. 2002). However, plate tectonics may not be common for rocky planets if our solar system is any indication. Stagnant-lid tectonics, as present on Mars and Mercury today (Breuer and Moore 2015), represents another end-member style of tectonics where the surface is rigid and immobile, with mantle convection taking place beneath this thick "stagnant-lid" (e.g., Ogawa et al. 1991; Davaille and Jaupart 1993; Moresi and Solomatov 1995; Solomatov 1995; Stern et al. 2018). With stagnant-lid tectonics, there are no mid-ocean ridges where plates spread apart or subduction zones where plates come together. This difference in tectonics

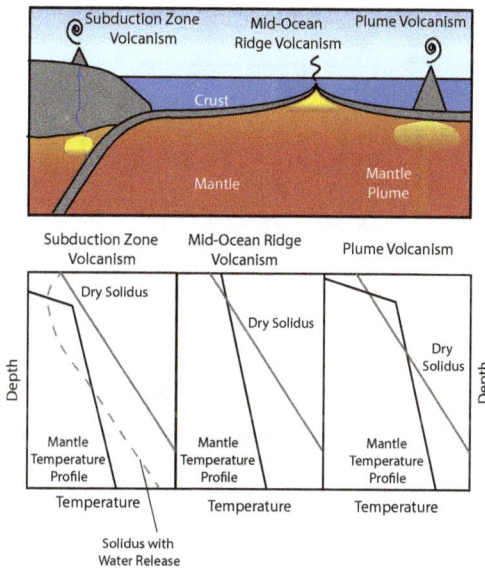

Figure 3. Schematic diagram of melting environments on the modern Earth. **Bottom panels** show temperature profiles with depth through the mantle in the different settings, along with the melting curve, or solidus. In subduction zones melting occurs because release of water lowers the solidus, at mid-ocean ridges hot mantle reaches the near surface and melts by decompression, and at mantle plumes elevated temperatures cause mantle to melt.

probably limits the rate and longevity of volcanism and outgassing on stagnant-lid planets but does not eliminate it for otherwise Earth-like planets (Kite et al. 2009; Grott et al. 2011; Tosi et al. 2017; Foley et al. 2020; Unterborn et al. 2022). Very large planets > 3–4 Earth masses ($M_\oplus$) may have volcanism entirely suppressed due to effects of pressure discussed below, however (Noack et al. 2017; Dorn et al. 2018).

Volcanism on stagnant-lid planets can occur via processes that are analogous to those that operate on the plate-tectonic Earth, but with key differences due to the change in tectonic mode. Melting can occur in mantle upwelling zones when material crosses the solidus before reaching the base of the lithosphere, as on the plate-tectonic Earth. Upwelling can be both active, as in a mantle plume, or passive, where ambient mantle rises to compensate for mantle flowing down into the interior at focused, active downwellings (this is analogous to the style of upwelling at mid-ocean ridges on Earth) (e.g., Reese et al. 1999; Hauck and Phillips 2002; Fraeman and Korenaga 2010; Morschhauser et al. 2011; Tosi et al. 2017; Foley and Smye 2018). However, on a stagnant-lid planet the lithosphere overlying the actively convecting mantle interior is far thicker than on a plate tectonic planet, so upwelling mantle must be at much hotter temperature in order to cross the solidus before reaching the base of the lithosphere (Fig. 4). This effect acts to suppress melting and volcanism but is compensated at least somewhat by stagnant-lid convection leading to higher mantle temperatures due to less efficient heat loss when compared to a plate-tectonic planet, all else equal (see Foley et al. 2020, for a review and summary of these effects). Another mechanism for volcanism on a stagnant-lid planet is foundering of the lower crust or lithosphere. This can occur by gravitational instability of regions of the lid that are denser than their surroundings due to differences in their chemical composition or phase. Foundering drives volcanism by upwelling of mantle into the space left by sinking crust or lithosphere, causing decompression melting, and potentially melting of the sinking crust itself. If the foundering crust contains volatiles, potentially due to interaction with a surface hydrosphere or atmosphere before crust was buried to mid- to lower-lithosphere depths by lava flows, then volatiles can potentially be transported back to the interior of stagnant-lid planets, or drive mantle melting by volatile release, analogous to subduction zones on the modern Earth (Elkins-Tanton et al. 2007).

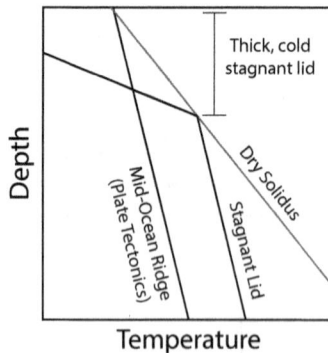

Figure 4. Schematic diagram showing the coldest temperature profile that can still intersect the solidus for a stagnant-lid planet and for a plate tectonic planet at a mid-ocean ridge setting. The thick, cold stagnant lid requires hotter temperatures in the mantle for melting to occur.

Volatiles stored in crustal rocks can also be released directly through metamorphism, without melting the host rock, in some cases. During continental collision crust is brought to high temperatures and pressures, which are sufficient to cause metamorphic breakdown of volatile bearing minerals. These volatiles can then be incorporated into fluids and released to the atmosphere (Tajika and Matsui 1992; Sleep and Zahnle 2001). On a stagnant-lid planet, burial of crust under repeated lava flows can similarly take volatile bearing minerals to the temperature and pressure conditions where they break down and release their volatiles (Foley and Smye 2018; Höning et al. 2019). If released volatiles can percolate to the surface through cracks and pore space, then this metamorphic breakdown contributes an additional source of atmospheric gases.

Volcanism and subsequent outgassing of volatiles to form an atmosphere should be a common process on rocky planets, regardless of tectonic regime (see also Dasgupta et al. 2024; Guimond et al. 2024, both this volume, for a more detailed discussion). However, there are important caveats. On large (> 3–$4\ M_\oplus$) stagnant-lid planets melt production and eruption may be inhibited due to the effects of increased pressure in the interior. With increasing planet mass, the pressure gradient in the mantle increases. As a result, the temperature required to cause melting of the mantle beneath a lithosphere of a given thickness is higher on a larger planet than a smaller one, which tends to suppress melting. Moreover, even if melt forms, above pressures of ~ 10 GPa silicate melts can be denser than solids, meaning melt would sink rather than rise; volcanism would not occur in this case. Both effects have been argued to suppress volcanism on large stagnant-lid exoplanets (Noack et al. 2017; Dorn et al. 2018). However, the result is sensitive to assumptions about the viscosity structure of the mantle, which controls the convective vigor that in turn sets the thickness of the stagnant lid. Larger planets retain heat better than smaller planets, leading to higher temperatures which also tend to promote thinner lithospheres, both effects acting to promote volcanism. As a result, volcanism may still be active even on large rocky exoplanets in a stagnant-lid regime, at least with sufficient heat production by radioactive decay or tidal heating (Unterborn et al. 2022). Miyazaki and Korenaga (2022) argue that partitioning of volatiles like H_2O and CO_2 during magma ocean solidification will also influence later outgassing, with planets that retain water in their interiors likely to fall into stagnant-lid states and experience limited outgassing of water. These planets in their models would still develop CO_2 atmospheres, however.

Another important caveat is that volcanic outgassing is suppressed at higher surface pressures, as volatiles are generally more soluble in melt with increasing pressure; this means volatiles erupted at high surface pressure will remain dissolved in lava rather than being released as gases. One result of this solubility effect is that planets with thick atmospheres or water oceans may experience limited outgassing (Krissansen-Totton et al. 2021). Suppression of outgassing beneath a thick atmosphere or ocean is important for considering the potential habitability and evolution of such exoplanets, but not relevant to the discussion in this section focused on airless bodies, and what they tell us about rocky planet interiors.

Thermal vs non-thermal escape and implications for the geological processes of airless bodies

Escape can remove a substantial mass of atmospheric gases from a planet, depending on atmosphere composition, planet size, and stellar radiation, among other factors. The gap between rocky super-Earths and sub-Neptunes with thick H_2 and He_2 dominated atmospheres may be explained by atmospheric loss on the rocky planets, meaning the loss of up to a few percent of the planet's mass by escape (e.g., Owen and Wu 2013; Ginzburg et al. 2018). Atmospheric escape can occur by two primary groups of mechanisms, thermal escape and non-thermal escape (Catling and Kasting 2017). Thermal escape encompasses mechanisms by which stellar radiation heats an atmosphere providing enough energy for molecules to escape, and non-thermal escape involves chemical reactions or ionic interactions providing the necessary energy for molecules to escape. Non-thermal escape is often associated with stripping of atmospheric molecules by interactions with stellar winds, and models of stellar wind-atmosphere interactions can provide first order estimates of non-thermal escape rates for exoplanets (e.g., Dong et al. 2017, 2018; Garraffo et al. 2017). Impacts can also cause significant atmospheric stripping (Schlichting et al. 2015). Impact fluxes are high during and just after planet formation, so impact erosion can be important for helping remove any primordial atmosphere or atmosphere formed by very early interior outgassing but will be a minor loss process for the majority of planets' lifetimes.

Both thermal escape and non-thermal escape via stellar wind stripping can be significant sources for atmospheric loss, especially for planets around M-dwarf stars like the handful of presumably airless planets considered here. M-dwarf stars emit high energy XUV radiation at an elevated rate for longer than higher mass stars (e.g., Johnstone et al. 2021). This high XUV flux can drive rapid thermal escape of hydrogen, especially when the planet is in a runaway greenhouse state where water vapor is abundant in the upper atmosphere (Luger and Barnes 2015). The airless rocky exoplanets all receive stellar radiation flux above the runaway greenhouse limit. Thus, if water vapor was present in the atmosphere or being outgassed from the interior, it would remain in vapor form and could well mix into the stratosphere for rapid, XUV driven escape. This rapid thermal escape can also potentially drag away heavier gases like O_2 or CO_2, leading to significant loss of all major atmospheric gases (Zahnle and Kasting 1986, 2023; Odert et al. 2018; Krissansen-Totton and Fortney 2022; Krissansen-Totton 2023). Higher levels of stellar activity also enhance stellar wind stripping, so planets around M-dwarf stars will also experience elevated rates of non-thermal escape (e.g., Dong et al. 2018).

We can explore how these different escape mechanisms would constrain present-day outgassing rates on LHS 3844 b and Trappist-1 b and c by recognizing that present-day outgassing must be less than atmosphere loss, otherwise an atmosphere would be accumulating on these planets. In other words, estimating present day atmospheric escape rates then also gives the maximum allowable present day outgassing rate (Fig. 5). For illustrative purposes only non-thermal escape due to stellar wind stripping is considered first. Three-dimensional magneto-hydro-dynamic models have been used to infer mass loss rates via stellar wind stripping of ~ 170 kg·s^{-1} and ~ 60 kg·s^{-1} for Trappist-1 b and c, respectively (Dong et al. 2018). Kreidberg et al. (2019) estimated the present-day stellar stripping rate for LHS 3844 b at 30–300 kg·s^{-1} based on simulations of stellar winds for Proxima Cenatauri (same stellar class as LHS 3844) from Dong et al. (2017). Garraffo et al. (2017) argue for higher stripping rates by about an order of magnitude for the Trappist-1 system. All told estimates range from order of 10–1000 kg·s^{-1} mass loss rates.

However, compared to outgassing on the modern Earth, these atmospheric stripping rates from stellar winds are small (Fig. 5A). Outgassing on Earth is dominated by CO_2 and H_2O. The rate of CO_2 outgassing is estimated at $\sim 10^{12}$–10^{13} mol·yr^{-1} or $\sim 10^3$–10^4 kg·s^{-1} (Marty and Tolstikhin 1998), while outgassing of H_2O is estimated at ~ 1–4.5×10^{13} mol·yr^{-1}, or $\approx 5 \times 10^3$–2.5×10^4 kg·s^{-1} (e.g., Parai and Mukhopadhyay 2012; Peslier et al. 2017). Summing these estimates outgassing releases atmospheric gases at a rate of $\approx 6 \times 10^3$–3.15×10^4 kg·s^{-1} on the modern Earth, ranging from a factor of a few to up to 3 orders of magnitude larger than the atmospheric stripping rates, even the highest estimate for Trappist-1 b. Even removing outgassing at subduction zones, which primarily reflects release of CO_2 and H_2O incorporated onto subducting plates by weathering, likely to be absent on the planets considered here as they are too hot for liquid water oceans, the result remains the same. CO_2 degassing decreases by about 1/3, and H_2O degassing is decreased to ~ 1–1.5×10^{13} mol·yr^{-1}. These reduced outgassing flux ranges still lead to total rates of $\sim 5.6 \times 10^3$–1.5×10^4 kg·s^{-1}, much larger than the range of stellar wind stripping rates. Therefore, with only stellar wind stripping as an atmosphere loss process outgassing would have to be much lower than the present day Earth's on an airless body, as otherwise an atmosphere would accumulate.

Thermal escape rates can be much higher, leading to a much larger range of allowable present day outgassing rates (Fig. 5B). Estimating thermal escape requires knowledge of the mixing ratios of different atmospheric species as well atmospheric structure, and therefore requires a model coupling outgassing to atmosphere evolution and escape; a simple general estimate cannot be provided. However, based on measured or modeled XUV fluxes, an upper

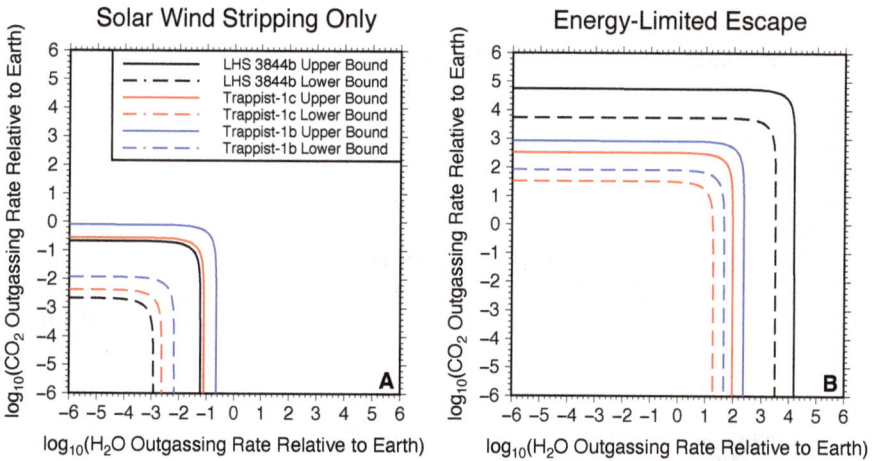

Figure 5. Contour lines representing the combined maximum allowable present day outgassing rates of CO_2 and H_2O, relative to the Earth's present day outgassing rates, for LHS 3844 b, Trappist-1 b, and Trappist-1 c. Estimates based on mass loss rates from solar wind stripping only (**A**), and from energy-limited escape (**B**) are shown. Upper bounds on the maximum allowable outgassing rates are calculated using the upper limits on mass loss rates and lower limits on Earth's present-day CO_2 and H_2O outgassing rates, while lower bounds use the lower limits on mass loss rates and upper limits on Earth's outgassing rates.

bound can be estimated by assuming energy-limited escape, where all the XUV energy absorbed by the atmosphere contributes to causing escape (e.g., equating the absorbed XUV flux with the energy carried by escaping molecules, Catling and Kasting 2017). Diamond-Lowe et al. (2021) estimated the modern-day energy-limited escape flux for LHS 3844 b as $\sim 7.7 \times 10^7 \, kg \cdot s^{-1}$ and Crossfield et al. (2022) estimated a similar value, $\sim 2 \times 10^7 \, kg \cdot s^{-1}$, for the airless exoplanet GJ 1252 b. Wheatley et al. (2017) estimated energy-limited escape rates of $1.18 \times 10^6 \, kg \cdot s^{-1}$ for Trappist-1 b, and $4.7 \times 10^5 \, kg \cdot s^{-1}$ for Trappist-1 c. Clearly under this energy-limited escape regime atmosphere loss is significantly higher than when considering only stellar wind stripping. Energy-limited escape strictly only applies to H, but rapid H escape can also drag away heavier species like CO_2 (Krissansen-Totton 2023), so it is reasonable for at least a first order estimate to consider this loss rate applying to the whole atmosphere. Outgassing rates up to 4–5 orders of magnitude higher than modern Earth's would be needed to still buildup an atmosphere for LHS 3844 b in the face of such high atmosphere loss rates. For the Trappist planets outgassing rates up to 2–3 orders of magnitude larger than modern Earth's would still keep the planets airless (Fig. 5B). It should be noted that these energy-limited atmosphere loss rates are upper bounds, and real rates could be much lower if H is not abundant in the upper atmosphere or significant a fraction of the XUV energy is not available to drive thermal escape.

The simple calculations presented here show that for an airless rocky planet, the lower the present-day rate of atmospheric loss (if loss rates can be well estimated), the tighter the constraint on present-day outgassing. If energy-limited escape is a first-order accurate estimate of loss rates on the planets considered, then even outgassing much more rapid than occurs on the modern Earth would still result in airless planets. Thus, a wide range of outgassing rates would be compatible with the lack of atmosphere, and the constraint is weaker. However, further context for cases where atmosphere mass loss rates are high can be provided by considering how large outgassing rates could conceivably get on rocky planets. Rocky planets generally cool over time as they lose heat from formation and radioactive heat sources decay (e.g., Davies 2007; Breuer and Moore 2015; Foley et al. 2020). As a result, early in a planet's history volcanism, and hence outgassing, rates are higher due to more extensive melt production (i.e., both larger melt fractions, ϕ, and fluxes of mantle material into the melting region, $\dot{m}$, see Equation (3). Outgassing rates of both H_2O and CO_2 can easily exceed 2–3 orders of magnitude larger than present day rates during a planet's early evolution when heat sources were larger (e.g., McGovern and Schubert 1989; Tajika and Matsui 1992; Sleep and Zahnle 2001; Crowley et al. 2011; Driscoll and Bercovici 2013; Foley and Smye 2018). Therefore, even the high mass loss rates for energy-limited escape provide a meaningful constraint for present day outgassing on Trappist-1 b and c, as some geologically plausible outgassing rates can be excluded. In fact, both Trappist-1 b and c likely experience significant tidal heating, estimated to provide a heat flux of ≈ 1 W·m^{-2} and ≈ 0.6 W·m^{-2} for b and c, respectively (Dobos et al. 2019). These heat fluxes are huge compared to the Earth; Earth's heat flux through oceanic plates is ≈ 0.1 W·m^{-2} (Jaupart et al. 2015). Tidal heating is therefore sufficient to drive extensive volcanism, and if interior volatile abundances were Earth-like, extensive outgassing, potentially large enough to outpace atmospheric escape. The lack of an atmosphere on Trappist-1 b and c may therefore require that the planets are volatile poor today. Indeed Krissansen-Totton and Fortney (2022) argues that even higher rates of escape during the early evolution of the Trappist-1 system can thoroughly deplete planets b and c of volatiles, such that even with active volcanism today outgassing rates would be muted. For the energy-limited escape rates estimated for LHS 3844 b, even a planet significantly more enriched in volatiles than the Earth would still have its volatiles stripped away once outgassed, as found by Crossfield et al. (2022) for GJ 1252 b, a planet with similar energy-limited escape rates.

Although thermal escape is the more efficient, and therefore probably dominant escape process on the airless planets discussed here, it is still illustrative to consider the implications for these planets if stellar wind stripping, with its relatively lower mass loss rates, were the only major escape process. Future observations of additional planets may reveal airless bodies where total escape rates, even including thermal escape, are comparable to the stellar wind stripping rates discussed here. Such cases provide the tightest constraints on present day outgassing rates, and thus implications for planetary volcanism rates or volatile budgets. As an example, take the lower bounds on the maximum allowable present day outgassing rates based solely on solar wind stripping for LHS 3844 b as the total atmospheric loss rate. In this case, outgassing rates would need to be 2–3 orders of magnitude lower than the present-day Earth's for LHS 3844 b to lack an atmosphere. This can be achieved either through much lower rates of interior melting and subsequent volcanism, much lower interior volatile abundances, or a combination of the two. For LHS 3844 b either option would be plausible. As LHS 3844 b is the only known planet in its system, planet-planet interactions are not available to pump up the orbital eccentricity and drive high rates of tidal heating like in the Trappist system. Given the age of the system (7.8 ± 1.6 Gyr), it is plausible that LHS 3844 b has cooled to the point where volcanism has ceased (Kane et al. 2020; Unterborn et al. 2022), meaning no present day outgassing even with volatile stores in the interior. A volatile poor interior, either due to loss of volatiles early in the planet's history or a volatile poor formation, is also always a possibility for any planet where present-day outgassing rates can be inferred to be low. In fact, Kane et al. (2020) argued that with only stellar wind stripping as an atmospheric loss process for LHS 3844 b, the planet would need to have formed volatile poor, as outgassing earlier in the planet's history would build up an atmosphere too large to be stripped away by the present day with an Earth-like volatile budget. For Trappist-1 b and c, significant tidal heating means volcanism is likely. As a result, if stellar wind stripping were the only loss process, then Trappist-1 b and c's interiors would need to be volatile depleted to keep outgassing rates low enough to be airless (Teixeira et al. 2023).

Implications for outgassing history and future directions

The discussion in the previous section focused on present day outgassing and atmospheric loss rates. However, an airless planet is not just a snapshot of the present-day state, but also a function of the evolutionary history of the planet up to that point. For example, a planet that is airless today could be a result of having lost its atmosphere and planetary volatile store during an earlier period of rapid escape, as may be the case for Trappist-1 b and c. The best avenue for constraining the geology and geologic history of planets currently observed to be airless is through models coupling interior outgassing, atmosphere evolution, and escape using broad sweeps through the relevant parameter space, as in e.g., Krissansen-Totton and Fortney (2022); Krissansen-Totton (2023). These papers argue that Trappist-1 b and c experience rapid XUV driven thermal escape during their early evolution that left them volatile depleted and airless. Trappist-1 b and c receive stellar radiation fluxes higher than the runaway greenhouse limit, leading to high mixing ratios of H in the upper atmosphere and hence rapid escape. Meanwhile, Trappist-1 e and f could still host atmospheres today because they formed further out and receive radiation less than the runaway greenhouse limit. As a result, water can condense and H mixing ratios in the upper atmosphere are low, limiting escape. It is important that early escape not only remove atmosphere but fully deplete the interior volatile stores on Trappist-1 b and c as well for them to still be airless today; without such complete volatile depletion later volcanism could re-form atmospheres (Teixeira et al. 2023). However, with complete volatile depletion then even with significant later volcanism there will be little to outgas, limiting later atmosphere development. Observations from JWST will hopefully soon determine whether Trappist-1 e or f have atmospheres. If they are instead airless, though, then much tighter constraints on these planets' geologies would be placed, as they would likely need to be either volatile poor or have muted volcanism in order to not still have observable atmospheres (Krissansen-Totton 2023).

The scenario of hydrodynamic loss of H dragging away heavier species like CO_2 to leave behind a volatile depleted and airless planet may be complicated by additional factors, though, like the planet's composition and interactions between the atmosphere and planetary interior. These complicating factors, however, allow tighter constraints to be placed on planetary evolution for airless planets, as scenarios where even high XUV fluxes still leave an atmosphere behind, or volatiles in the interior that can outgas to form an atmosphere, after the star settles on the main sequence can then be ruled out. If rocky planets are in a magma ocean phase when experiencing high stellar fluxes that lead to a runaway greenhouse climate, then the abundance of gases in the atmosphere will be determined by equilibrium with the rapidly convecting magma. Sossi et al. (2020) and Bower et al. (2022) argue that the high solubility of H_2O in magma means that magma ocean atmospheres are dominated by CO or CO_2, depending on oxidation state, rather than steam. Steam atmospheres only form at the end of magma ocean solidification and only if magma–atmosphere equilibrium can be maintained until the end of solidification. If a planet in a runaway greenhouse state has a relatively dry, CO or CO_2 dominated atmosphere, then H escape could be slower as H will have to diffuse through the dominant species to reach the upper atmosphere. This in turn can slow loss of the heavier species dragged away by H as well, leading to more scenarios where planets can retain atmospheres. More work will be needed to determine just where such scenarios occur and what can then be constrained about planets that are inferred to be airless. Planetary volatile budget and oxidation state will likely be important as well, and therefore some ranges of these parameters may be able to be ruled out for airless planets. As explained in *Controls on interior outgassing* above, mantle oxidation state controls whether outgassing is dominated by reduced species like H_2, CO, or CH_4, or oxidized species like H_2O and CO_2. Moreover, bulk interior volatile inventories will further dictate whether e.g., H-bearing, C-bearing, or even more exotic gases like S-bearing species dominate released gases. As high H mixing ratios in the upper atmosphere are needed for rapid escape, some planets whose compositions lead to atmospheres dominated by other gases may be able to retain atmospheres more readily. Therefore, even in environments that favor high escape rates, some combinations of planetary volatile inventories and oxidation states might still leave atmospheres behind, and such parameter combinations can thus be ruled out for airless bodies.

Overall, airless exoplanets actually present an excellent opportunity to learn about geologic characteristics and geologic history of these planets. Building a large sample of airless rocky planets will not only help define the cosmic shoreline, but also potentially help test models of interior outgassing discussed in *Controls on interior outgassing*. For example, Noack et al. (2017); Dorn et al. (2018) propose that high mass rocky exoplanets in a stagnant-lid regime will experience limited outgassing. Determining whether planets operate in stagnant-lid or plate tectonic regimes will be difficult, and potentially implausible, as there are no known signatures of tectonic regime that can be realistically observed with current and upcoming missions. However, if a large database of airless exoplanets is built up, then seeing a preference for high mass planets to be airless would lend credence to the models of Noack et al. (2017) and Dorn et al. (2018). Moreover, a large sample of airless exoplanets could also test models of the longevity of volcanic activity (e.g., Unterborn et al. 2022). The age distribution of airless rocky planets, combined with models or constraints on escape rates, could help estimate when rocky planet outgassing wanes. The seemingly disappointing discovery of airless rocky exoplanets thus actually provides an excellent opportunity to study exoplanet geology.

FUTURE DIRECTIONS:
TESTING THE CARBONATE–SILICATE CYCLE FEEDBACK

This chapter so far has focused on current observations or those feasible in the near-term with current resources like JWST. However, the ability to characterize rocky exoplanets will improve greatly in the medium to long-term through new direct imaging missions like the proposed HWO or mid-IR LIFE mission. These new capabilities will open up a wide range of new questions that can be asked and models for exoplanet dynamics and evolution that can be tested. One of the most important hypotheses that can potentially be tested is the carbonate–silicate cycle weathering feedback, thought to be essential for sustaining the long-term habitability of Earth (e.g., Walker et al. 1981; Kasting and Catling 2003; Abbot et al. 2012; Foley and Driscoll 2016).

The carbonate–silicate cycle is the long-term cycling of carbon between Earth's interior, atmosphere, ocean, and crust, which acts to stabilize climate against changes in solar luminosity or internal perturbations like varying CO_2 outgassing rates (Berner 2004; Foley and Driscoll 2016). CO_2 is released to the atmosphere by outgassing, primarily from volcanism. This CO_2 is then removed by silicate weathering, the breaking down of silicate rocks on the surface by acidic rainwater to form ions that flow through groundwater to rivers and eventually the oceans. In the oceans carbonate minerals and SiO_2 precipitate, locking CO_2 away in seafloor rocks. Eventually this geologically stored CO_2 reaches a subduction zone, where a fraction of the CO_2 is liberated from the sinking plate by metamorphic reactions and returns to the atmosphere, while the remainder is subducted into the mantle interior, closing the cycle (Fig. 6). The rate of silicate weathering, which removes CO_2 from the atmosphere, depends on temperature, soil pH, and the rate at which water flows through the subsurface, leading to a negative climate feedback (Kump et al. 2000; Brantley and Olsen 2014): at higher temperatures and atmospheric CO_2 levels, weathering rates increase acting to lower CO_2 and cool the climate, while at low temperatures weathering rates decrease allowing outgassing to build CO_2 up in the atmosphere and warm the climate.

The negative feedback mechanism inherent in the carbonate–silicate cycle acts to keep the rate of atmospheric CO_2 removal in balance with the rate of CO_2 release to the atmosphere by outgassing (Berner and Caldeira 1997), so the system can be well understood by looking at climate states that result from a steady state between weathering and outgassing (Walker et al. 1981; Berner et al. 1983; Sleep and Zahnle 2001). A simple form of the rate at which weathering on land removes CO_2 from the atmosphere (F_w) can be given by (e.g., Foley and Driscoll 2016)

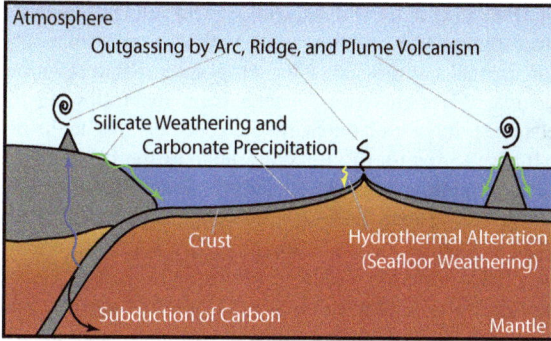

Figure 6. Schematic diagram of the carbonate–silicate cycle as it operates on the modern-day, plate tectonic Earth. Adapted from Foley (2015), Foley and Driscoll (2016).

$$F_w = F_w^* \exp\left[\frac{E}{R_g}\left(\frac{1}{T^*} - \frac{1}{T_s}\right)\right]\left(\frac{P}{P^*}\right)^\beta \tag{4}$$

where $F_w \approx 6 \times 10^{12}$ mol·yr^{-1} is the present-day silicate weathering flux, E is an activation energy for silicate weathering on land, R_g is the universal gas constant, T_s is the surface temperature, $T^* \approx 288$ K is the preindustrial surface temperature of Earth, P is the partial pressure of atmospheric CO_2, $P^* \approx 30$ Pa is the preindustrial atmospheric CO_2 on Earth, and β is a constant. Some models explicitly include a term for runoff, which is related to the precipitation rate. However, precipitation scales with climate, as warmer climates tend towards higher precipitation rates, so runoff can essentially be grouped into the temperature and atmospheric CO_2 dependent terms (e.g., Berner 2004). Key parameters like E and β are inferred from lab experiments and models (e.g., Walker et al. 1981; Brantley and Olsen 2014; Krissansen-Totton and Catling 2017) but are still uncertain. In fact, the formulation itself for weathering flux is uncertain, as Equation (4) does not take into account other potentially important factors like uplift and erosion (e.g., Kump and Arthur 1997; Maher and Chamberlain 2014), limits to weathering imposed by the formation of thick soils (West et al. 2005; West 2012; Foley 2015; Kump 2018), weathering of ocean crust on the seafloor (Krissansen-Totton and Catling 2017; Krissansen-Totton et al. 2018; Coogan and Gillis 2013; Coogan and Dosso 2015), or effects of continental versus ocean coverage (Abbot et al. 2012; Hayworth and Foley 2020).

Nevertheless, Equation (4) still helps illustrate how the carbonate–silicate cycle operates to regulate climate, and the trends in atmospheric CO_2 that would be expected for planets with this cycle in operation that could be looked for with next generation space telescopes. Specifically, planets within the habitable zone should show an increase in atmospheric CO_2 moving from those near the inner edge of the habitable zone towards those near the outer edge (Bean et al. 2017; Turbet 2020; Lehmer et al. 2020; Lustig-Yaeger et al. 2022). At the simplest level, this is due to the need for higher concentrations of atmospheric CO_2 to keep the climate warm when received stellar radiation is low, as found near the outer edge of the habitable zone, and less atmospheric CO_2 at high incident stellar flux as found near the inner edge. Bean et al. (2017) and Turbet (2020) assumed a constant temperature for planets through the habitable zone in order to calculate how atmospheric CO_2 would vary as a function of orbital distance or incident stellar flux. Clearly planets further from their host star would need higher atmospheric CO_2 levels to maintain the same temperature as planets closer in, all else equal.

Although it does not drastically change the overall trend of atmospheric CO_2 versus orbital distance or incident stellar flux, Lehmer et al. (2020) pointed out that planets within the habitable zone with an active carbonate–silicate cycle will not sit at the same surface temperature, all other factors equal, but instead show decreasing temperatures with increasing orbital distance. Atmospheric CO_2 still increases with orbital distance, but with a different functional form than if one assumes constant surface temperatures prevail throughout the habitable zone. The reason for this can be seen from the following thought experiment: Consider two planets, one lying within the habitable zone but close to the inner edge (planet b), and a second lying further out, closer to the outer edge of the habitable zone (planet c). The two planets have the same geologic features including the same rate of CO_2 outgassing. As the carbonate–silicate cycle acts to keep the CO_2 weathering flux in balance with CO_2 outgassing, then both planets will have the same weathering flux, F_w^0. Planet c receives less incident stellar flux than b, meaning that c would be colder if atmospheric CO_2 levels were the same. This cooler surface temperature would result in a lower a weathering flux, which the carbonate–silicate cycle would correct for by allowing more CO_2 to build up in the atmosphere, until the weathering flux for planets b and c are again equal at F_w^0. Equating the two planets' weathering fluxes and for the sake of generality assuming that both their atmospheric CO_2 levels and surface temperatures are different gives:

$$\frac{1}{T_c} - \frac{1}{T_b} = \frac{R_g}{E} \ln\left(\frac{P_c}{P_b}\right)^\beta \tag{5}$$

where subscripts b and c denote planets b and c, respectively. With $P_c > P_b$, as it should be based on orbital distance and the carbonate–silicate cycle feedback, then the right-hand side of Equation (5) is positive. For the left-hand side to also be positive, $T_c < T_b$. Equation (5) also shows that if $\beta = 0$, that is there is no direct influence of atmospheric CO_2 on weathering, then $T_c = T_b$. Or in other words, with $\beta = 0$ the only way for two planets at two different orbital distances to have the same weathering flux, F_w^0, is for them to have the same temperature; CO_2 would adjust to whatever value was needed to accomplish this in such a case.

Atmospheric CO_2 and surface temperature as a function of incident stellar flux is given in Figure 7, using Equation (4) and the simple climate parameterization from Abbot et al. (2012). The calculation assumes $E = 30$ kJ·mol^{-1} and $\beta = 0.33$ (Krissansen-Totton and Catling 2017), and shows the expected increase in atmospheric CO_2 with decreasing incident stellar flux through the habitable zone. However, as Lehmer et al. (2020) demonstrates, a sample of real exoplanets would be likely to have significant variations in their geological features, which will lead to significant variability in quantities like CO_2 outgassing rate, activation energy for weathering, β, or other factors. These differences in geological characteristics from planet to planet will cause a significant amount of scatter about the trend of increasing atmospheric CO_2 with decreasing flux, meaning that a larger number of planets would need to be observed to detect the trend.

A simplified version of the analysis from Lehmer et al. (2020) is presented here (Fig. 8). 1000 calculations, each representing a hypothetical planet, are run, sampled from uniform distributions of the parameters E from 15–30 kJ·mol^{-1}, β from 0.24–0.44 (both as estimated from Krissansen-Totton and Catling 2017), and the CO_2 outgassing rate relative to the Earth in logspace ranging from 0.3 to 3 times the nominal present day Earth's value of $\approx 6 \times 10^{12}$ mol·yr^{-1}. The outgassing rate variations in particular are small compared to range of outgassing rates possible over a planet's lifetime (see *Thermal vs non-thermal escape and implications for the geological processes of airless bodies*). However, even with this underestimate of the true variability, modeled planets scatter by orders of magnitude in atmospheric CO_2 around the expected trend of increasing CO_2 with decreasing incident stellar flux (Fig. 8A). From the complete set of 1000 modeled planets, random subsamples of planets, ranging from 5–100, are drawn and fit to a linear relationship between $\log_{10}(P)$ and incident stellar flux. The slope in this space should be negative reflecting the decrease in atmospheric CO_2 with increasing flux. For samples of planets less than ~10, the slope found from randomly sampling planets shows a huge

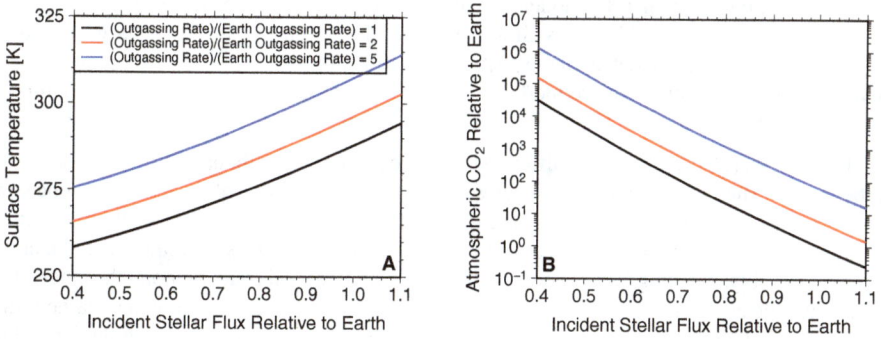

Figure 7. Surface temperature **(A)** and atmospheric CO_2 **(B)** as a function of relative stellar flux received resulting from a steady state between silicate weathering and CO_2 outgassing. Models presented assume $E = 30$ kJ·mol^{-1}, $\beta = 0.33$, and CO_2 outgassing rates relative to the modern-day Earth value as indicated by the legend.

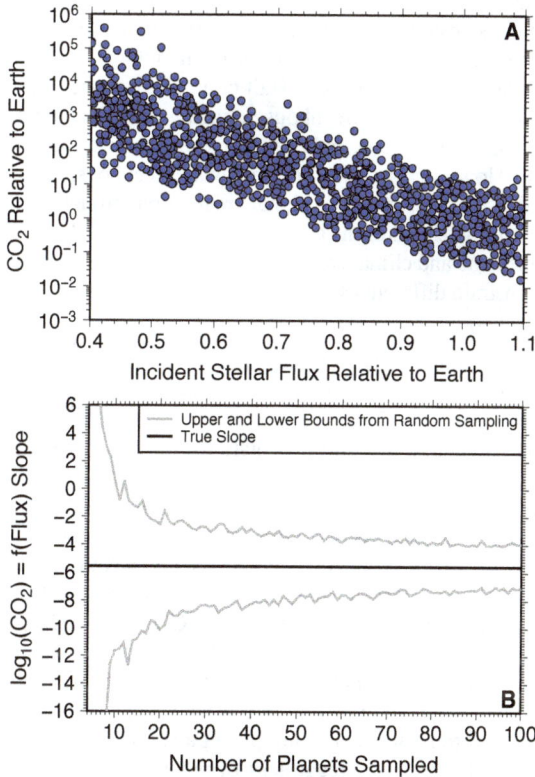

Figure 8. Atmospheric CO_2 as a function of incident stellar flux relative to what the modern-day Earth receives for 1000 modeled planets where key parameters for the carbonate–silicate cycle are randomly varied as described in the text **(A)**. A given number of planets, ranging from 5 to 100, are then randomly drawn from the full set of 1000 models and a linear fit to their $\log_{10}(P)$ as a function of incident stellar flux is performed. The resulting range of slopes from this linear fitting of a random sample of planets is plotted as a function of the number of planets sampled, along with the "true" slope obtained from fitting the entire sample **(B)**.

variability, from very strongly negative slopes, much lower than the "true" slope determined from a fit to the entire set of 1000 modeled planets, to even obtaining positive slopes, the reverse of the expected trend (Fig. 8B). This simple exercise shows that > 10s of planets will need to be observed to confidently detect the expected trend of decreasing CO_2 with increasing flux, as shown with a more comprehensive model in Lehmer et al. (2020). The real variability is almost certainly larger than shown in Figure 8, as only narrow ranges of the key parameters were used.

Turbet et al. (2019) proposed a different but related test of the carbonate–silicate cycle based on the onset of runaway greenhouse climates inward of the inner edge of the habitable zone (see also Schlecker et al. 2024). Planets inward of the habitable zone inner edge would experience runaway greenhouse climates with steam atmospheres. Such a climate state inflates the planetary radius measurably, meaning a shift from inflated to not inflated radii could pinpoint the inner edge of the habitable zone. At the same time, the carbonate–silicate cycle would predict a large jump in atmospheric CO_2 across the habitable zone inner edge, from low levels inside the habitable zone but near the inner edge, to Venus-like thick CO_2 atmospheres inward of the habitable zone inner edge (Turbet 2020). The approach of looking for major climate and atmosphere transitions across the habitable zone inner edge is probably more feasible than detecting an atmospheric CO_2 versus stellar flux trend. Lustig-Yaeger et al. (2022) showed that changes in transmission spectra with atmospheric CO_2 are small, requiring up to 100 transits per planet to resolve with JWST. Detecting a full population level trend of CO_2 versus flux is therefore not feasible, though null hypotheses instead could be tested. Moreover, atmospheric CO_2 partial pressure may be difficult even with HWO, as recent studies testing retrieval methodologies find Earth-like CO_2 levels difficult to constrain (Damiano and Hu 2022; Hall et al. 2023; Robinson and Salvador 2023). Looking for the transition to runaway greenhouse climates at the habitable zone inner edge is feasible with basic transit observations, rather than more intensive transmission spectroscopy (Schlecker et al. 2024). However, just observing runaway greenhouse climates at the habitable zone inner edge would also not test whether climate regulation throughout the habitable zone is a widespread process. Ultimately all of the approaches discussed here should be followed to test the idea of the habitable zone and climate regulation via the carbonate–silicate cycle, as they are complementary and constrain different aspects of the problem.

If we are able to develop a large dataset of characterized rocky planet atmospheres across the habitable zone in the coming decades, additional information about rocky planet geology could potentially be gleaned. One of these pieces of information is active volcanism and volatile cycling as a function of planet age. The carbonate–silicate cycle relies on planets having active volcanism and volatile cycling, either through plate tectonics (e.g., Sleep and Zahnle 2001) or stagnant-lid tectonics (e.g., Foley and Smye 2018; Valencia et al. 2018). However, once volcanism on a planet ceases due to mantle cooling, so does an active carbonate–silicate cycle. It is not clear how atmospheric CO_2 would evolve on a planet after the cessation of volcanism; Foley (2019) assumed the climate would cool as erosion and weathering would continue until topography is smoothed out and surface rocks are fully weathered. If this is true, then planets without active volcanism may have low atmospheric CO_2 levels. One could test this idea by looking at whether the expected trend of atmospheric CO_2 with stellar flux breaks down moving from a sample of younger planets to a sample of older planets, with the older planets showing lower atmospheric CO_2 and potentially a weaker relationship with flux. In fact, any breakdown of the expected trend for increasing planet age could be a sign of the cessation of the carbonate–silicate cycle feedback, regardless of whether older planets show lower than expected CO_2, higher than expected, or just increased random variability in either direction, given the uncertainty of how atmospheric CO_2 will respond to the end of volcanism.

In addition to planets geologically "dying" as they age, future work will likely identify additional factors that can cause planets to lose their carbonate–silicate cycle feedbacks. In a similar way to the discussion above, these factors can also be tested by looking to see if planets with these factors preferentially show larger deviations from the expected CO_2-flux trend. As such, a large dataset of rocky planets atmospheres covering as wide a range of environments and geological characteristics as possible will enable significant advances in our understanding of rocky planet evolution. The search for habitable exoplanets and signs of life therefore dovetails nicely with the goal of learning about rocky exoplanet geological evolution. However, even planets that are not likely to be habitable could still be valuable in helping constrain planetary evolution, so such planets should not be overlooked when prioritizing future observations.

SUMMARY

The number of rocky exoplanets discovered to date dwarfs the number of rocky bodies in our solar system, thereby massively enhancing our sample size of rocky planet geological outcomes. However, in comparison to our study of solar system bodies, the information we can get from rocky exoplanets is severely limited, and typically only indirectly constraints the geological features or processes of these planets. It is therefore questionable how much we will actually be able to learn about rocky planet geological evolution from exoplanets. This chapter focused on a few aspects of rocky exoplanet geology that can potentially be constrained with current or near future observational capabilities. However, some of the most promising techniques involve looking at planets in exotic, compared to those in solar system, states or situations, or carefully extracting constraints from observations that are seemingly only loosely related to geological processes. Planetary scientists looking to learn about the geological evolution of exoplanets will therefore need to shift perspective to the types of planets that are most readily observed, even if they are drastically different than Earth or other solar system planets.

One example is the most promising ways to directly constrain the composition of rocky exoplanets involve planets that have either already been destroyed, in the case of white dwarf pollution (see Veras et al. 2024; Xu et al. 2024, both this volume), or are being destroyed, in the case of disintegrating planets. Lava planets with magma oceans at the sub-stellar points or hot, but solid, airless planets are also promising avenues for direct characterization of the planet's crust or interior composition. Determining the range of exoplanet compositions is a vital goal in assessing all aspects of their geological evolution. Currently models of rocky exoplanet compositions are typically based on the abundances of refractory elements in their host stars, assuming that these compositions will be linked. While this is a reasonable approach, it is important to find ways directly measure rocky exoplanet compositions, to help benchmark these models and inform the diversity of rocky exoplanets.

Another example focused on how planets that lack atmospheres entirely can actually inform volcanic outgassing history, and potentially interior volatile inventory, of rocky planets. Airless rocky planets may seem like a disappointing discovery, in particular in the framework of astrobiology and searching for biosignatures. However, they are scientifically important for helping map the "cosmic shoreline" where planets are able to retain atmospheres, which is itself astrobiologically important. Moreover, as discussed in this chapter, an airless rocky planet constrains volatile outgassing rates, both at the present day and potentially over the planet's history, if atmospheric escape rates can be constrained. An airless planet must have escape rates that exceed outgassing rates at the present day, otherwise an atmosphere would be accumulating. Moreover, during the planet's history escape must have been able to remove whatever atmosphere may have formed during the planet's evolution. As a result, there is a clear need for collaboration between studies of atmospheric escape and stripping by stellar winds and studies of rocky planet interior evolution. Coupling these two processes in models of planetary evolution is essential for mapping which geological characteristics (like planet composition, starting volatile abundances, or interior oxidation state) can be ruled out by the finding of a planet lacking any atmosphere.

While current atmospheric characterization for rocky exoplanets is mostly only able to determine whether an atmosphere is present or not, future missions should allow more detailed characterization of atmospheric compositions. Such abilities will open up a wide range of new questions that can be answered about rocky planet evolution. This chapter focused on the possibility of testing whether the carbonate–silicate cycle climate feedback operates on rocky exoplanets in the same way it is thought to stabilize Earth's climate. Such a test will require measuring atmospheric CO_2 concentrations for 10s of rocky exoplanets, given the likely intrinsic variability on planets for weathering and outgassing. Moreover, if such a large dataset of rocky planet atmosphere characterizations can be built, additional geological features can potentially be inferred. For example, the typical lifetime of volcanism can potentially be constrained if it is found that beyond a certain range of ages the relationship between atmospheric CO_2 and incident stellar flux expected for an active carbonate–silicate cycle breaks down. The constraints discussed in this chapter are basic. However, given how little is confidently known about rocky exoplanet geology, any additional constraints that can be developed still represent a significant advance. Whether we can go beyond the very basic constraints towards assessing things like the tectonic mode of a rocky exoplanet or directly observing active volcanism remains to be seen. Detecting volcanism, though, appears promising. Sulfate aerosols have been proposed as a potential marker of active volcanism (Kaltenegger et al. 2010). Explosive volcanism on Earth launches sulfate aerosols into the stratosphere, where they can exist for months to years before raining out. Stratospheric sulfate aerosols are readily detectable in transmission spectroscopy with instruments like JWST, and seeing a time-varying signature would plausibly indicate active volcanic eruptions on a planet (Misra et al. 2015). Ostberg et al. (2023) proposed an analogous method for detecting volcanism with direct imaging. The applicability of the sulfate aerosol volcanism signature, though, depends on the broader context of the planet, like the mantle and atmosphere composition, while detectability further depends on atmosphere composition and structure. In the end coupled models of atmospheric chemistry and outgassing would likely be needed to determine how well any proposed signature can be used as a marker of active volcanism. Nevertheless, any potential signature of exovolcanism is exciting and should be pursued, as it would be highly relevant for assessing exoplanet geology and habitability.

Inferring tectonic mode is likely to be even more difficult. There is no known unique signature of either plate tectonics or stagnant-lid tectonics that can be reasonably observed by proposed future telescopes. That the theory of plate tectonics was not developed until the 1960s despite hundreds of years of directly studying Earth's geology is a sobering reminder of the difficulty of determining tectonic mode. However, the more we can learn about rocky planets in any context, the more potential avenues for placing any type of constraint on tectonics. While there may be no one "smoking gun" signature, an accumulation of different lines of evidence all pointing towards one tectonic mode may eventually allow for strong inferences of exoplanet tectonics. The same logic holds for any other key aspect of exoplanet geology we might wish to learn about. The future will clearly require interdisciplinary, collaborative work to combine the different observations we are able to make into constraints on exoplanet geology.

ACKNOWLEDGEMENTS

I thank Dimitri Veras and Josh Krissansen-Totton for thorough reviews that greatly improved the final paper. The Center for Exoplanets and Habitable Worlds is supported by the Pennsylvania State University and the Eberly College of Science.

REFERENCES

Abbot DS, Cowan NB, Ciesla FJ (2012) Indication of insensitivity of planetary weathering behavior and habitable zone to surface land fraction. Astrophys J 756:178

Angelo I, Hu R (2017) A Case for an atmosphere on super-earth 55 Cancri e. Astronom J 154:232

Batalha NM (2014) Exploring exoplanet populations with NASA's Kepler Mission. Proc Natl Acad Sci 111:12647–12654

Batalha NM Borucki WJ, Bryson ST, Buchhave LA, Caldwell DA, Christensen-Dalsgaard J, Ciardi D, Dunham EW, Fressin F, Gautier I, Thomas N, Gilliland RL, Haas MR, Howell SB, Jenkins JM, Kjeldsen H, Koch DG, Latham DW, Lissauer JJ, Marcy GW, Rowe JF, Sasselov DD, Seager S, Steffen JH, Torres G, Basri GS, Brown TM, Charbonneau D, Christiansen J, Clarke B, Cochran WD, Dupree A, Fabrycky DC, Fischer D, Ford EB, Fortney J, Girouard FR, Holman MJ, Johnson J, Isaacson H, Klaus TC, Machalek P, Moorehead AV, Morehead RC, Ragozzine D, Tenenbaum P, Twicken J, Quinn S, VanCleve J, Walkowicz LM, Welsh WF, Devore E, Gould A (2011) Kepler's first rocky planet: Kepler-10b. Astrophys J 729:27

Bean JL, Kempton EMR, Homeier D (2010) A ground-based transmission spectrum of the super-Earth exoplanet GJ 1214b. Nature 468:669–672

Bean JL, Abbot DS, Kempton EMR (2017) A statistical comparative planetology approach to the hunt for habitable exoplanets and life beyond the Solar System. Astrophys J Lett 841:L24

Bean JL, Raymond SN, Owen, JE (2021) The nature and origins of sub-Neptune size planets. J Geophys Res Planets 126:e06639

Bercovici D, Tackley P, Ricard Y (2015) The generation of plate tectonics from mantle dynamics. *In:* Treatise on Geophysics. Vol 7. Bercovici D, Schubert G (eds) Elsevier, New York p 271–318

Berner R (2004) The Phanerozoic carbon cycle: CO_2 and O_2, Oxford University Press, New York

Berner RA, Caldeira K (1997) The need for mass balance and feedback in the geochemical carbon cycle. Geology 25:955

Berner RA, Lasaga AC, Garrels RM (1983) The carbonate–silicate geochemical cycle and its effect on atmospheric carbon dioxide over the past 100 million years. Am J Sci 283:641–683

Bodman EHL, Wright JT, Desch SJ, Lisse CM (2018) Inferring the composition of disintegrating planet interiors from dust tails with future James Webb Space Telescope observations. Astron J 156:173

Borrelli ME, O'Rourke SE, Ostberg CM (2021) A global survey of lithospheric flexure at steep-sided domical volcanoes on Venus reveals intermediate elastic thicknesses. J Geophys Res Planets 126:e06756

Bower DJ, Hakim K, Sossi PA, Sanan P (2022) Retention of water in terrestrial magma oceans and carbon-rich early atmospheres. Planet Sci J 3:93

Brantley S, Olsen A (2014) Reaction kinetics of primary rock-forming minerals under ambient condition. 7.3 *In:* Treatise on Geochemistry (Second Edition) Vol 7. Drever JI, Turekian KK, Holland HD (eds) Elsevier, New York p 69–113

Breuer D, Moore W (2015) 10.08–Dynamics, thermal history of the terrestrial planets, the Moon, and Io. *In:* Treatise on Geophysics (Second Edition) Vol 10. Schubert G (ed) Elsevier, New York p 255–305

Brown T M (2001) Transmission spectra as diagnostics of extrasolar giant planet atmospheres. Astrophys J 553:1006–1026

Brugman K, Phillips MG, Till CB (2021) Experimental determination of mantle solidi and melt compositions for two likely rocky exoplanet compositions. J Geophys Res Planets 126:e06731

Campbell IH, Taylor SR (1983) No water, no granites—No oceans, no continents. Geophys Res Lett 10:1061–1064

Catling DC, Kasting JF (2017) Atmospheric Evolution on Inhabited And Lifeless Worlds, Cambridge University Press, Cambridge

Chambers JE (2001) Making more terrestrial planets. Icarus 152:205–224

Chao K-H, de Graffenried R, Lach M, Nelson W, Truax K, Gaidos E (2021) Lava worlds: From early earth to exoplanets. Chemie Erde / Geochem 81:125735

Cockell CS, Bush T, Bryce C, Direito S, Fox-Powell M, Harrison JP, Lammer H, Landenmark H, Martin-Torres J, Nicholson N, Noack L, O'Malley-James J, Payler SJ, Rushby A, Samuels T, Schwendner P, Wadsworth J, Zorzano MP (2016) Habitability: A review. Astrobiology 16:89–117

Coogan LA, Dosso SE (2015) Alteration of ocean crust provides a strong temperature dependent feedback on the geological carbon cycle and is a primary driver of the Sr-isotopic composition of seawater. Earth Planet Sci Lett 415:38–46

Coogan LA, Gillis KM (2013) Evidence that low-temperature oceanic hydrothermal systems play an important role in the silicate–carbonate weathering cycle and long-term climate regulation. Geochem Geophys Geosyst 14:1771–1786

Cowan NB, Agol E, Meadows VS, Robinson T, Livengood TA, Deming D, Lisse CM, A'Hearn MF, Wellnitz DD, Seager S, Charbonneau D, EPOXI Team (2009) Alien maps of an ocean-bearing world. Astrophys J 700:915–923

Crossfield IJ, Malik M, Hill ML, Kane SR, Foley B, Polanski AS, Coria D, Brande J, Zhang Y, Wienke K, Kreidberg L (2022) GJ 1252b: a hot terrestrial super-earth with no atmosphere. Astrophys J Lett 937:L17

Crowley JW, Gérault M, O'Connell RJ (2011) On the relative influence of heat and water transport on planetary dynamics. Earth Planet Sci Lett 310:380–388

Damiano M, Hu R (2022) Reflected spectroscopy of small exoplanets II: Characterization of terrestrial exoplanets. Astronom J 163:299

Dasgupta R, Pathak D, Maurice M (2024) A framework for the origin and deep cycles of volatiles in rocky exoplanets. Rev Mineral Geochem 90:323–374

Davaille A, Jaupart C (1993) Transient high-Rayleigh-number thermal convection with large viscosity variations. J Fluid Mech 253:141–166

Davaille A, Smrekar SE, Tomlinson S (2017) Experimental and observational evidence for plume-induced subduction on Venus. Nat Geosci 10:349–355

Davies GF (2007) Thermal evolution of the mantle. *In:* Treatise on Geophysics. Vol 9. Stevenson D, Schubert G (eds) Elsevier, New York, p 197–216

Dawson RI, Johnson JA (2018) Origins of hot Jupiters. Annu Rev Astron Astrophys 56:175–221

de Wit J, Wakeford HR, Gillon M, Lewis NK, Valenti JA, Demory B-O, Burgasser AJ, Burdanov A, Delrez L, Jehin E, Lederer SM, Queloz D, Triaud AHMJ, Van Grootel V (2016) A combined transmission spectrum of the Earth-sized exoplanets Trappist-1 b and c. Nature 537:69–72

de Wit J, Wakeford HR, Lewis NK, Delrez L, Gillon M, Selsis F, Leconte J, Demory BO, Bolmont E, Bourrier V, Burgasser AJ, Grimm S, Jehin E, Lederer SM, Owen JE, Stamenković V, Triaud AHMJ (2018) Atmospheric reconnaissance of the habitable-zone Earth-sized planets orbiting Trappist-1. Nat Astron 2:214–219

Deibert EK, de Mooij EJW, Jayawardhana R, Ridden-Harper A, Sivanandam S, Karjalainen R, Karjalainen M (2021) A near-infrared chemical inventory of the atmosphere of 55 Cancri e. Astronom J 161:209

Deming D, Seager S, Richardson LJ, Harrington J (2005) Infrared radiation from an extrasolar planet. Nature 434:740–743

Demory BO, Gillon M, Deming D, Valencia D, Seager S, Benneke B, Lovis C, Cubillos P, Harrington J, Stevenson KB, Mayor M, Pepe F, Queloz D, Ségransan D, Udry S (2011) Detection of a transit of the super-Earth 55 Cancri e with warm Spitzer. Astron Astrophys 533:A114

Demory BO, Gillon M, de Wit J, Madhusudhan N, Bolmont E, Heng K, Kataria T, Lewis N, Hu R, Krick J, Stamenković V, Benneke B, Kane S, Queloz D (2016) A map of the large day-night temperature gradient of a super-Earth exoplanet. Nature 532:207–209

Diamond-Lowe H, Berta-Thompson Z, Charbonneau D, Kempton EMR (2018) Ground-based optical transmission spectroscopy of the small, rocky exoplanet GJ 1132b. Astronom J 156: 42

Diamond-Lowe H, Charbonneau D, Malik M, Kempton EMR, Beletsky Y (2020) Optical transmission spectroscopy of the terrestrial exoplanet LHS 3844b from 13 ground-based transit observations. Astronom J 160:188

Diamond-Lowe H, Youngblood A, Charbonneau D, King G, Teal DJ, Bastelberger S, Corrales L, Kempton, EMR (2021) The high-energy spectrum of the nearby planet-hosting inactive mid-M dwarf LHS 3844. Astronom J 162:10

Dick SJ (1993) The search for extraterrestrial intelligence and the NASA High Resolution Microwave Survey / HRMA—Historical perspectives. Space Sci Rev 64:93–139

Dobos V, Barr AC, Kiss LL (2019) Tidal heating and the habitability of the Trappist-1 exoplanets. Astron Astrophys 624:A2

Dong C, Lingam M, Ma Y, Cohen O (2017) Is Proxima Centauri b habitable? A study of atmospheric loss. Astrophys J Lett 837:L26

Dong C, Jin M, Lingam M, Airapetian VS, Ma Y, van der Holst B (2018) Atmospheric escape from the Trappist-1 planets and implications for habitability. Proc Natl Acad Sci 115:260–265

Dorn C, Khan A, Heng K, Connolly JAD, Alibert Y, Benz W, Tackley P (2015) Can we constrain the interior structure of rocky exoplanets from mass and radius measurements? Astron Astrophys 577:A83

Dorn C, Noack L, Rozel AB (2018) Outgassing on stagnant-lid super-Earths. Astron Astrophys 614:A18

Doyle AE, Young ED, Klein B, Zuckerman B, Schlichting HE (2019) Oxygen fugacities of extrasolar rocks: Evidence for an Earth-like geochemistry of exoplanets. Science 366:356–359

Drazkowska J, Bitsch B, Lambrechts M, Mulders GD, Harsono D, Vazan A, Liu B, Ormel CW, Kretke K, Morbidelli A (2023) Planet formation theory in the era of ALMA, Kepler: from pebbles to exoplanets. *In:* Protostars and Planets VII Vol 534. Inutsuka S, Aikawa Y, Muto T, Tomida K, Tamura M (eds) Astron Soc Pac Conf Ser, p 717

Driscoll P, Bercovici D (2013) Divergent evolution of Earth and Venus: Influence of degassing, tectonics, and magnetic fields. Icarus 226:1447–1464

Ehrenreich D, Tinetti G, Lecavelier Des Etangs A, Vidal-Madjar A, Selsis F (2006) The transmission spectrum of Earth-size transiting planets. Astron Astrophys 448:379–393

Elkins-Tanton LT, Smrekar SE, Hess PC, Parmentier EM (2007) Volcanism and volatile recycling on a one-plate planet: Applications to Venus. J Geophys Res Planets 112:E04S06

Foley BJ (2015) The role of plate tectonic–climate coupling and exposed land area in the development of habitable climates on rocky planets. Astrophys J 812:36

Foley, BJ (2019) Habitability of earth-like stagnant lid planets: Climate evolution and recovery from snowball states. Astrophys J 875:72

Foley BJ, Driscoll PE (2016) Whole planet coupling between climate, mantle, and core: Implications for rocky planet evolution. Geochem Geophys Geosyst 17:1885–1914

Foley BJ, Smye AJ (2018) Carbon cycling and habitability of Earth-size stagnant lid planets. Astrobiology 18:873–896

Foley BJ, Houser C, Noack L, Tosi N (2020) Chapter 4: The heat budget of rocky planets. *In:* Planetary diversity: Rocky planet processes, their observational signatures. Tasker EJ, Unterborn C, Laneuville M, Fuji Y, Desch SJ, Hartnett HE (eds) IOP Publishing, Bristol, p 4-1–4-60

Ford EB, Seager S, Turner EL (2001) Characterization of extrasolar terrestrial planets from diurnal photometric variability. Nature 412:885–887

Fortin M-A, Gazel E, Kaltenegger L, Holycross ME (2022) Volcanic exoplanet surfaces. Mon Not R Astron Soc 516:4569–4575

Fortney JJ, Marley MS, Barnes JW (2007) Planetary radii across five orders of magnitude in mass and stellar insolation: Application to transits. Astrophys J 659:1661–1672

Fraeman AA, Korenaga J (2010) The influence of mantle melting on the evolution of Mars. Icarus 210:43–57

Fulton BJ, Petigura EA, Howard AW, Isaacson H, Marcy GW, Cargile PA, Hebb L, Weiss LM, Johnson JA, Morton TD, Sinukoff E, Crossfield IJM, Hirsch LA (2017) The California–Kepler Survey. III A gap in the radius distribution of small planets. Astronom J 154:109

Gaillard F, Scaillet B (2014) A theoretical framework for volcanic degassing chemistry in a comparative planetology perspective and implications for planetary atmospheres. Earth Planet Sci Lett 403:307–316

Gaillard F, Bouhifd MA, Füri E, Malavergne V, Marrocchi Y, Noack L, Ortenzi G, Roskosz M, Vulpius S (2021) The diverse planetary ingassing/outgassing paths produced over billions of years of magmatic activity. Space Sci Rev 217:22

Garraffo C, Drake JJ, Cohen O, Alvarado-Gómez JD, Moschou SP (2017) The threatening magnetic and plasma environment of the Trappist-1 planets. Astrophys J Lett 843:L33

Ginzburg S, Schlichting HE, Sari R (2018) Core-powered mass-loss and the radius distribution of small exoplanets. Mon Not R Astron Soc 476:759–765

Greene TP, Bell TJ, Ducrot E, Dyrek A, Lagage P-O, Fortney JJ (2023) Thermal emission from the Earth-sized exoplanet Trappist-1 b using JWST Nature 618:39–42

Grott M, Morschhauser A, Breuer D, Hauber E (2011) Volcanic outgassing of CO_2 and H_2O on Mars. Earth Planet Sci Lett 308:391–400

Guimond CM, Noack L, Ortenzi G, Sohl F (2021) Low volcanic outgassing rates for a stagnant lid Archean earth with graphite-saturated magmas. Phys Earth Planet Int 320:106788

Guimond CM, Wang H, Seidler F, Sossi P, Mahajan A, Shorttle O (2024) From stars to diverse mantles, melts, crusts, and atmospheres of rocky exoplanets. Rev Mineral Geochem 90:259–300

Hall S, Krissansen-Totton J, Robinson T, Salvador A, Fortney JJ (2023) Constraining background N_2 inventories on directly imaged terrestrial exoplanets to rule out O_2 false positives. Astronom J 166:254

Hammond M, Pierrehumbert RT (2017) Linking the climate and thermal phase curve of 55 Cancrie. Astrophys J 849:152

Hauck SA, Phillips RJ (2002) Thermal and crustal evolution of Mars. J Geophys Res Planets 107: 6–1

Hayworth BPC, Foley BJ (2020) Waterworlds may have better climate buffering capacities than their continental counterparts. Astrophys J Lett 902:L10

Hinkel NR, Unterborn CT (2018) The star–planet connection. I Using stellar composition to observationally constrain planetary mineralogy for the 10 closest stars. Astrophys J 853:83

Hinkel NR, Timmes FX, Young PA, Pagano MD, Turnbull MC (2014) Stellar abundances in the solar neighborhood: The Hypatia Catalog. Astronom J 148:54

Hinkel NR, Youngblood A, Soares-Furtado M (2024) Host stars and how their compositions influence exoplanets. Rev Mineral Geochem 90:1–26

Hirschmann MM (2006) Water, melting, and the deep earth H_2O cycle. Annu Rev Earth Planet Sci 34:629–53

Hirth G, Kohlstedt D (1996) Water in the oceanic upper mantle: implications for rheology, melt extraction and the evolution of the lithosphere. Earth Planet Sci Lett 144:93–108

Holland HD (1984) The chemical evolution of the atmosphere and oceans. Princeton University Press, Princeton

Höning D, Tosi N, Spohn T (2019) Carbon cycling and interior evolution of water covered plate tectonics and stagnant-lid planets. Astron Astrophys 627:A48

Howard AW, Sanchis-Ojeda R, Marcy GW, Johnson JA, Winn JN, Isaacson H, Fischer DA, Fulton BJ, Sinukoff E, Fortney JJ (2013) A rocky composition for an Earth-sized exoplanet. Nature 503:381–384

Hu R, Ehlmann BL, Seager S (2012) Theoretical spectra of terrestrial exoplanet surfaces. Astrophys J 752:7

Ih J, Kempton EMR, Whittaker EA, Lessard M (2023) Constraining the thickness of Trappist-1 b's atmosphere from its JWST secondary eclipse observation at 15 μm. Astrophys J Lett 952:L4

Ito Y, Ikoma M (2021) Hydrodynamic escape of mineral atmosphere from hot rocky exoplanet. I Model description. Mon Not R Astron Soc 502:750–771

Ito Y, Ikoma M, Kawahara H, Nagahara H, Kawashima Y, Nakamoto T (2015) Theoretical emission spectra of atmospheres of hot rocky super-Earths. Astrophys J 801:144

Jackson B, Stark CC, Adams ER, Chambers J, Deming D (2013) A Survey for very short-period planets in the Kepler data. Astrophys J 779:165

Jaupart C, Labrosse S, Lucazeau F, Mareschal J-C (2015) Temperatures, heat, energy in the mantle of the Earth. 7.06 *In:* Treatise on Geophysics (Second Edition) Vol 7. Bercovici D, Schubert G (eds) Elsevier, New York, p 223–270

Jeans J (1919) Problems of Cosmogony and Stellar Dynamics. Cambridge University Press, Cambridge

Jellinek AM, Lenardic A, Manga M (2002) The influence of interior mantle temperature on the structure of plumes: Heads for Venus, tails for the Earth. Geophys Res Lett 29:1532

Johnstone CP, Bartel M, Güdel M (2021) The active lives of stars: A complete description of the rotation and XUV evolution of F, G, K, and M dwarfs. Astron Astrophys 649:A96

Jorge DM, Kamp IEE, Waters LBFM, Woitke P, Spaargaren RJ (2022) Forming planets around stars with non-solar elemental composition. Astron Astrophys 660:A85

Kaltenegger L, Traub WA (2009) Transits of Earth-like planets. Astrophys J 698:519–527

Kaltenegger L, Henning WG, Sasselov DD (2010) Detecting volcanism on extrasolar planets. Astronom J 140:1370–1380

Kane SR, Roettenbacher RM, Unterborn CT, Foley BJ, Hill ML (2020) A Volatile-poor formation of LHS 3844b based on its lack of significant atmosphere. Planet Sci J 1:36

Kane SR, Arney GN, Byrne PK, Dalba PA, Desch SJ, Horner J, Izenberg NR, Mandt KE, Meadows VS, Quick LC (2021) The fundamental connections between the Solar System and exoplanetary science. J Geophys Res Planets 126:e06643

Kasting JF, Catling D (2003) Evolution of a habitable planet. Annu Rev Astron Astrophys 41:429–463

Kasting JF, Eggler DH, Raeburn SP (1993a) Mantle redox evolution and the oxidation state of the Archean atmosphere. J Geol 101:245–257

Kasting JF, Whitmire DP, Reynolds RT (1993b) Habitable zones around main sequence stars. Icarus 101:108–128

Katz RF, Spiegelman M, Langmuir CH (2003) A new parameterization of hydrous mantle melting. Geochem Geophys Geosyst 4:1073

Kite ES, Manga M, Gaidos E (2009) Geodynamics and rate of volcanism on massive Earth-like planets. Astrophys J 700:1732–1749

Kite ES, Fegley B, Schaefer L, Gaidos E (2016) Atmosphere–interior exchange on hot, rocky exoplanets. Astrophys J 828:80

Knutson HA, Charbonneau D, Allen LE, Fortney JJ, Agol E, Cowan NB, Showman AP, Cooper CS, Megeath ST (2007) A map of the day-night contrast of the extrasolar planet HD 189733b. Nature 447:183–186

Koll DDB, Malik M, Mansfield M, Kempton EMR, Kite E, Abbot D, Bean JL (2019) Identifying candidate atmospheres on rocky M Dwarf planets via eclipse photometry. Astrophys J 886:140

Komacek TD, Fauchez TJ, Wolf ET, Abbot DS (2020) Clouds will likely prevent the detection of water vapor in JWST transmission spectra of terrestrial exoplanets. Astrophys J Lett 888: L20

Kopparapu RK, Ramirez R, Kasting JF, Eymet V, Robinson TD, Mahadevan S, Terrien RC, Domagal-Goldman S, Meadows V, Deshpande R (2013) Habitable zones around main-sequence stars: New estimates. Astrophys J 765:131

Korenaga J (2006) Archean geodynamics, the thermal evolution of Earth. *In:* Archean Geodynamics and Environments. Vol 164. Benn K, Mareschal J-C, Condie K (eds) AGU Geophys Monogr Ser, p 7–32

Korenaga J (2007) Thermal cracking and the deep hydration of oceanic lithosphere: A key to the generation of plate tectonics? J Geophys Res 112:B05408, https://doi.org/10.1029/2006JB004502

Korenaga J (2010) On the likelihood of plate tectonics on super-earths: Does size matter? Astrophys J 725:L43

Kreidberg L, Koll DD, Morley C, Hu R, Schaefer L, Deming D, Stevenson KB, Dittmann J, Vanderburg A, Berardo D, Guo X, Stassun K, Crossfield I, Charbonneau D, Latham DW, Loeb A, Ricker G, Seager S, Vanderspek R (2019) Absence of a thick atmosphere on the terrestrial exoplanet LHS 3844b. Nature 573:87–90

Krissansen-Totton J (2023) Implications of atmospheric nondetections for Trappist-1 inner planets on atmospheric retention prospects for outer planets. Astrophys J Lett 951:L39

Krissansen-Totton J, Catling DC (2017) Constraining climate sensitivity and continental versus seafloor weathering using an inverse geological carbon cycle model. Nat Commun 8:15423

Krissansen-Totton J, Fortney JJ (2022) Predictions for observable atmospheres of Trappist-1 planets from a fully coupled atmosphere–interior evolution model. Astrophys J 933:115

Krissansen-Totton J, Arney GN, Catling DC (2018) Constraining the climate and ocean pH of the early Earth with a geological carbon cycle model. Proc Nat Acad Sci 115:4105–4110

Krissansen-Totton J, Galloway ML, Wogan N, Dhaliwal JK, Fortney JJ (2021) Waterworlds probably do not experience magmatic outgassing. Astrophys J 913:107

Kump LR (2018) Prolonged Late Permian–Early Triassic hyperthermal: failure of climate regulation? Phil Trans R Soc A 376:20170078

Kump LR, Arthur MA (1997) Global chemical erosion during the cenozoic: Weatherability balances the budgets. *In:* Tectonic Uplift, Climate Change. Ruddiman WF (ed), Springer, p 399–426

Kump LR, Brantley SL, Arthur MA (2000) Chemical weathering, atmospheric CO_2, and climate. Annu Rev Earth Planet Sci 28:611–667

Kushiro I (2001) Partial melting experiments on peridotite and origin of mid-ocean ridge basalt. Annu Rev Earth Planet Sci 29:71–107

Lammer H, Zerkle A L, Gebauer S, Tosi N, Noack L, Scherf M, Pilat-Lohinger E, Güdel M, Grenfell J L, Godolt M, Nikolaou A (2018) Origin and evolution of the atmospheres of early Venus, Earth and Mars. Astron Astrophys Rev 26:2

Landuyt W, Bercovici D (2009) Variations in planetary convection via the effect of climate on damage. Earth Planet Sci Lett 277:29–37

Léger A, Rouan D, Schneider J, Barge P, Fridlund M, Samuel B, Ollivier M, Guenther E, Deleuil M, Deeg HJ, Auvergne M, Alonso R, Aigrain S, Alapini A, Almenara JM, Baglin A, Barbieri M, Bruntt H, Bordé P, Bouchy F, Cabrera J, Catala C, Carone L, Carpano S, Csizmadia S, Dvorak R, Erikson A, Ferraz-Mello S, Foing B, Fressin F, Gandolfi D, Gillon M, Gondoin P, Grasset O, Guillot T, Hatzes A, Hébrard G, Jorda L, Lammer H, Llebaria A, Loeillet B, Mayor M, Mazeh T, Moutou C, Pätzold M, Pont F, Queloz D, Rauer H, Renner S, Samadi R, Shporer A, Sotin C, Tingley B, Wuchterl G, Adda M, Agogu P, Appourchaux T, Ballans H, Baron P, Beaufort T, Bellenger R, Berlin R, Bernardi P, Blouin D, Baudin F, Bodin P, Boisnard L, Boit L, Bonneau F, Borzeix S, Briet R, Buey JT, Butler B, Cailleau D, Cautain R, Chabaud PY, Chaintreuil S, Chiavassa F, Costes V, Cuna Parrho V, de Oliveira Fialho F, Decaudin M, Defise JM, Djalal S, Epstein G, Exil GE, Fauré C, Fenouillet T, Gaboriaud A, Gallic A, Gamet P, Gavalda P, Grolleau E, Gruneisen R, Gueguen L, Guis V, Guivarc'h V, Guterman P, Hallouard D et al. (2009) Transiting exoplanets from the CoRoT space mission. VIII CoRoT-7b: the first super-Earth with measured radius. Astron Astrophys 506:287–302

Léger A, Grasset O, Fegley B, Codron F, Albarede AF, Barge P, Barnes R, Cance P, Carpy S, Catalano F, Cavarroc C, Demangeon O, Ferraz-Mello S, Gabor P, Grießmeier JM, Leibacher J, Libourel G, Maurin AS, Raymond SN, Rouan D, Samuel B, Schaefer L, Schneider J, Schuller PA, Selsis F, Sotin C (2011) The extreme physical properties of the CoRoT-7b super-Earth. Icarus 213:1–11

Lehmer OR, Catling DC, Krissansen-Totton J (2020) Carbonate–silicate cycle predictions of Earth-like planetary climates and testing the habitable zone concept. Nat Commun 11:6153

Lenardic A, Kaula W (1994) Self-lubricated mantle convection: Two-dimensional models. Geophys Res Lett 21:1707–1710

Libby-Roberts JE, Berta-Thompson ZK, Diamond-Lowe H, Gully-Santiago MA, Irwin JM, Kempton EMR, Rackham BV, Charbonneau D, Désert J-M, Dittmann JA, Hofmann R, Morley CV, Newton ER (2022) The featureless HST/WFC3 transmission spectrum of the rocky exoplanet GJ 1132b: No evidence for a cloud-free primordial atmosphere and constraints on starspot contamination. Astronom J 164:59

Lichtenberg T, Schaefer LK, Nakajima M, Fischer RA (2023) Geophysical evolution during rocky planet formation. *In:* Astron Soc Pac Conf Ser Vol 534. Inutsuka S, Aikawa Y, Muto T, Tomida K, Tamura M (eds), p 907

Lim O, Benneke B, Doyon R, MacDonald RJ, Piaulet C, Artigau É, Coulombé L-P, Radica M, L'Heureux A, Albert L, Rackham BV, de Wit J, Salhi S, Roy P-A, Flagg L, Fournier-Tondreau M, Taylor J, Cook NJ, Lafrenière D, Cowan NB, Kaltenegger L, Rowe JF, Espinoza N, Dang L, Darveau-Bernier A (2023) Atmospheric reconnaissance of trappist-1 b with JWST/NIRISS: Evidence for strong stellar contamination in the transmission spectra. Astrophys J Lett 955:L22

Lincowski AP, Meadows VS, Zieba S, Kreidberg L, Morley C, Gillon M, Selsis F, Agol E, Bolmont E, Ducrot E, Hu R, Koll DDB, Lyu X, Mandell A, Suissa G, Tamburo P (2023) Potential atmospheric compositions of Trappist-1 c constrained by JWST/MIRI observations at 15 μm. Astrophys J Lett 955:L7

Lodders K (2020) Solar elemental abundances. *In:* Oxford Research Encyclopedia of Planetary Science. Oxford University Press, https://oxfordre.com/planetaryscience/view/10.1093/acrefore/9780190647926.001.0001/acrefore-9780190647926-e-145

Lopez ED (2017) Born dry in the photoevaporation desert: Kepler's ultra-short-period planets formed water-poor. Mon Not R Astron Soc 472:245–253

Luger R, Barnes R (2015) Extreme water loss and abiotic O_2 buildup on planets throughout the habitable zones of M dwarfs. Astrobiology 15:119–143

Lustig-Yaeger J, Sotzen KS, Stevenson KB, Luger R, May EM, Mayorga LC, Mandt K, Izenberg NR (2022) Hierarchical Bayesian atmospheric retrieval modeling for population studies of exoplanet atmospheres: A case study on the habitable zone. Astronom J 163:140

Lustig-Yaeger J, Fu G, May EM, Ceballos KNO, Moran SE, Peacock S, Stevenson KB, Kirk J, López-Morales M, MacDonald RJ, Mayorga LC, Sing DK, Sotzen KS, Valenti JA, Redai JIA, Alam MK, Batalha NE, Bennett KA, Gonzalez-Quiles J, Kruse E, Lothringer JD, Rustamkulov Z, Wakeford HR (2023) A JWST transmission spectrum of the nearby Earth-sized exoplanet LHS 475 b. Nat Astron 7:1317–1328

Maher K, Chamberlain C (2014) Hydrologic regulation of chemical weathering and the geologic carbon cycle. Science 343:1502–1504

Marty B, Tolstikhin IN (1998) CO_2 fluxes from mid-ocean ridges, arcs and plumes. Chem Geol 145:233–248

Mayor M, Queloz D (1995) A Jupiter-mass companion to a solar-type star. Nature 378:355–359

McGovern P, Schubert G (1989) Thermal evolution of the earth: effects of volatile exchange between atmosphere and interior. Earth Planet Sci Lett 96, 27–37

Meier TG, Bower DJ, Lichtenberg T, Hammond, M, Tackley PJ (2023) Interior dynamics of super-Earth 55 Cancri e. Astron Astrophys 678:A29

Mercier SJ, Dang L, Gass A, Cowan NB, Bell TJ (2022) Revisiting the iconic Spitzer Phase curve of 55 Cancri e: hotter dayside, cooler nightside, and smaller phase offset. Astronom J 164: 204

Mian ZU, Tozer DC (1990) No water, no plate tectonics: convective heat transfer and the planetary surfaces of Venus and Earth. Terra Nova 2:455–459

Miguel Y, Kaltenegger L, Fegley B, Schaefer L (2011) Compositions of hot superearth atmospheres: Exploring Kepler candidates. Astrophys J Lett 742:L19

Misra A, Krissansen-Totton J, Koehler MC, Sholes S (2015) Transient sulfate aerosols as a signature of exoplanet volcanism. Astrobiology 15:462–477

Miyazaki Y, Korenaga J (2022) Inefficient water degassing inhibits ocean formation on rocky planets: An insight from self-consistent mantle degassing models. Astrobiology 22:713–734

Mordasini C, Alibert Y, Benz W (2009) Extrasolar planet population synthesis. I Method, formation tracks, and mass-distance distribution. Astron Astrophys 501:1139–1160

Moresi LN, Solomatov VS (1995) Numerical investigation of 2D convection with extremely large viscosity variations. Phys Fluids 7:2154–2162

Moresi L, Solomatov V (1998) Mantle convection with a brittle lithosphere: Thoughts on the global tectonic style of the Earth and Venus. Geophys J Int 133:669–682

Morschhauser A, Grott M, Breuer D (2011) Crustal recycling, mantle dehydration, and the thermal evolution of Mars. Icarus 212:541–558

National Academies of Sciences, Engineering, and Medicine (2021) Pathways to Discovery in Astronomy and Astrophysics for the 2020s. Washington, DC: The National Academies Press, https://doi.org/10.17226/26141.

Nguyen TG, Cowan NB, Banerjee A, Moores JE (2020) Modelling the atmosphere of lava planet K2–141b: implications for low- and high-resolution spectroscopy. Mon Not R Astron Soc 499:4605–4612

Nguyen TG, Cowan NB, Pierrehumbert RT, Lupu RE, Moores JE (2022) The impact of ultraviolet heating and cooling on the dynamics and observability of lava planet atmospheres. Mon Not R Astron Soc 513:6125–6133

Noack L, Rivoldini A, Van Hoolst T (2017) Volcanism and outgassing of stagnant-lid planets: Implications for the habitable zone. Phys Earth Planet Int 269:40–57

Odert P, Lammer H, Erkaev NV, Nikolaou A, Lichtenegger HIM, Johnstone CP, Kislyakova KG, Leitzinger M, Tosi N (2018) Escape and fractionation of volatiles and noble gases from Mars-sized planetary embryos and growing protoplanets. Icarus 307:327–346

Ogawa M, Schubert G, Zebib A (1991) Numerical simulations of three-dimensional thermal convection in a fluid with strongly temperature-dependent viscosity. J Fluid Mech 233:299–328

Okuya A, Okuzumi S, Ohno K, Hirano T (2020) Constraining the bulk composition of disintegrating exoplanets using combined transmission spectra from JWST and SPICA Astrophys J 901:171

Ortenzi G, Noack L, Sohl F, Guimond CM, Grenfell JL, Dorn C, Schmidt JM, Vulpius S, Katyal N, Kitzmann D, Rauer H (2020) Mantle redox state drives outgassing chemistry and atmospheric composition of rocky planets. Sci Rep 10:10907

Ostberg CM, Guzewich SD, Kane SR, Kohler E, Oman LD, Fauchez TJ, Kopparapu RK, Richardson J, Whelley P (2023) The prospect of detecting volcanic signatures on an exoearth using direct imaging. Astron J 166:199

Owen JE, Wu Y (2013) Kepler Planets: A tale of evaporation. Astrophys J 775:105

Parai R, Mukhopadhyay S (2012) How large is the subducted water flux? new constraints on mantle regassing rates. Earth Planet Sci Lett 317:396–406

Perez-Becker D, Chiang E (2013) Catastrophic evaporation of rocky planets. Mon Not R Astron Soc 433:2294–2309

Peslier AH, Schönbächler M, Busemann H, Karato S-I (2017) Water in the Earth's interior: Distribution and origin. Space Sci Rev 212:743–810

Pfalzner S, Davies MB, Gounelle M, Johansen A, Münker C, Lacerda P, Portegies Zwart S, Testi L, Trieloff M, Veras D (2015) The formation of the solar system. Phys Scr 90:068001

Piette AAA, Madhusudhan N, Mandell AM (2022) HyDRo: atmospheric retrieval of rocky exoplanets in thermal emission. Mon Notices Royal Astron Soc 511:2565–2584

Piette AAA, Gao P, Brugman K, Shahar A, Lichtenberg T, Miozzi F, Driscoll P (2023) Rocky planet or water world? Observability of low-density lava world atmospheres. Astrophys J 954:29

Putirka KD (2024) Exoplanet mineralogy. Rev Mineral Geochem 90:199–258

Putirka KD, Rarick JC (2019) The composition and mineralogy of rocky exoplanets: A survey of > 4000 stars from the Hypatia Catalog. Am Mineral 104:817–829

Quanz SP, Ottiger M, Fontanet E, Kammerer J, Menti F, Dannert F, Gheorghe A, Absil O, Airapetian VS, Alei E, Allart R, Angerhausen D, Blumenthal S, Buchhave LA, Cabrera J, Carrión-González Ó, Chauvin G, Danchi WC, Dandumont C, Defrére D, Dorn C, Ehrenreich D, Ertel S, Fridlund M, García Muñoz A, Gascón C, Girard JH, Glauser A, Grenfell JL, Guidi G, Hagelberg J, Helled R, Ireland MJ, Janson M, Kopparapu RK, Korth J, Kozakis T, Kraus S, Léger A, Leedjärv L, Lichtenberg T, Lillo-Box J, Linz H, Liseau R, Loicq J, Mahendra V, Malbet F, Mathew J, Mennesson B, Meyer MR, Mishra L, Molaverdikhani K, Noack L, Oza AV, Pallé E, Parviainen H, Quirrenbach A, Rauer H, Ribas I, Rice M, Romagnolo A, Rugheimer S, Schwieterman EW, Serabyn E, Sharma S, Stassun KG, Szulágyi J, Wang HS, Wunderlich F, Wyatt MC, LIFE Collaboration (2022) Large Interferometer For Exoplanets (LIFE) I Improved exoplanet detection yield estimates for a large mid-infrared space-interferometer mission. Astron Astrophys 664:A21

Ramirez RM, Kopparapu R, Zugger ME, Robinson TD, Freedman R, Kasting JF (2014) Warming early Mars with CO_2 and H_2. Nat Geosci 7:59–63

Rappaport S, Levine A, Chiang E, El Mellah I, Jenkins J, Kalomeni B, Kite ES, Kotson M, Nelson L, Rousseau-Nepton L, Tran K (2012) Possible disintegrating short-period super-Mercury orbiting KIC 12557548. Astrophys J 752:1

Rappaport S, Barclay T, DeVore J, Rowe J, Sanchis-Ojeda R, Still M (2014) KOI-2700b—A planet candidate with dusty effluents on a 22 hr orbit. Astrophys J 784:40

Reese CC, Solomatov VS, Moresi LN (1999) Non-newtonian stagnant lid convection and magmatic resurfacing on Venus. Icarus 139:67–80

Regenauer-Lieb K, Yuen D, Branlund J (2001) The initiation of subduction: Criticality by addition of water? Science 294:578–580

Richards M, Yang W-S, Baumgardner J, Bunge H-P (2001) Role of a low-viscosity zone in stabilizing plate tectonics: Implications for comparative terrestrial planetology. Geochem Geophys Geosyst 2:2000GC000115

Robinson TD, Salvador A (2023) Exploring and validating exoplanet atmospheric retrievals with Solar System analog observations. Planet Sci J 4:10

Sanchis-Ojeda R, Rappaport S, Winn JN, Levine A, Kotson MC, Latham DW, Buchhave LA (2013) Transits and occultations of an Earth-sized planet in an 8.5 hr Orbit. Astrophys J 774:54

Sanchis-Ojeda R, Rappaport S, Winn JN, Kotson MC, Levine A, El Mellah I (2014) A Study of the shortest-period planets found with Kepler. Astrophys J 787:47

Sanchis-Ojeda R, Rappaport S, Pallè E, Delrez L, DeVore J, Gandolfi D, Fukui A, Ribas I, Stassun KG, Albrecht S, Dai F, Gaidos E, Gillon M, Hirano T, Holman M, Howard AW, Isaacson H, Jehin E, Kuzuhara M, Mann AW, Marcy GW, Miles-Páez PA, Montañés-Rodríguez P, Murgas F, Narita N, Nowak G, Onitsuka M, Paegert M, Van Eylen V, Winn JN, Yu L (2015) The K2-ESPRINT project I: Discovery of the disintegrating rocky planet K2–22b with a cometary head and leading tail. Astrophys J 812:112

Sandwell DT, Schubert G (1992) Flexural ridges, trenches, and outer rises around coronae on Venus. J Geophys Res 97:16069

Schaefer L, Fegley B (2009) Chemistry of silicate atmospheres of evaporating super Earths. Astrophys J Lett 703:L113–L117

Schaefer L, Lodders K, Fegley B (2012) Vaporization of the Earth: Application to exoplanet atmospheres. Astrophys J 755:41

Schlecker M, Apai D, Lichtenberg T, Bergsten G , Salvador A, Hardegree-Ullman KK (2024). Bioverse: The habitable zone inner edge discontinuity as an imprint of runaway greenhouse climates on exoplanet demographics. Planet Sci J 5:3

Schlichting HE, Sari R, Yalinewich A (2015) Atmospheric mass loss during planet formation: The importance of planetesimal impacts. Icarus 247:81–94

Seager S, Sasselov DD (2000) Theoretical transmission spectra during extrasolar giant planet transits. Astrophys J 537:916–921

Seager S, Kuchner M, Hier-Majumder CA, Militzer B (2007) Mass–radius relationships for solid exoplanets. Astrophys J 669:1279–1297

Sleep NH, Zahnle K (2001) Carbon dioxide cycling and implications for climate on ancient Earth. J Geophys Res 106:1373–1400

Smrekar SE, Ostberg C, O'Rourke JG (2023) Earth-like lithospheric thickness and heat flow on Venus consistent with active rifting. Nat Geosci 16:13–18

Solomatov V (1995) Scaling of temperature- and stress-dependent viscosity convection. Phys Fluids 7:266–274

Sossi PA, Burnham AD, Badro J, Lanzirotti A, Newville M, O'Neill HSC (2020) Redox state of Earth's magma ocean and its Venus-like early atmosphere. Sci Adv 6:eabd1387

Spaargaren RJ, Ballmer MD, Bower DJ, Dorn C, Tackley PJ (2020) The influence of bulk composition on the long-term interior-atmosphere evolution of terrestrial exoplanets. Astron Astrophys 643:A44

Spaargaren RJ, Wang HS, Mojzsis SJ, Ballmer MD, Tackley PJ (2023) Plausible constraints on the range of bulk terrestrial exoplanet compositions in the solar neighborhood. Astrophys J 948:53

Spiegelman M, McKenzie D (1987) Simple 2-D models for melt extraction at mid-ocean ridges and island arcs. Earth Planet Sci Lett 83:137–152

Stern RJ, Gerya T, Tackley PJ (2018) Stagnant lid tectonics: Perspectives from silicate planets, dwarf planets, large moons, and large asteroids. Geosci Front 9:103–119

Stevenson KB, Désert J-M, Line MR, Bean JL, Fortney JJ, Showman AP, Kataria T, Kreidberg L, McCullough PR, Henry GW, Charbonneau D, Burrows A, Seager S, Madhusudhan N, Williamson MH, Homeier D (2014) Thermal structure of an exoplanet atmosphere from phase-resolved emission spectroscopy. Science 346:838–841

Sudarsky D, Burrows A, Hubeny I (2003) Theoretical spectra and atmospheres of extrasolar giant planets. Astrophys J 588:1121–1148

Tajika E, Matsui T (1992) Evolution of terrestrial proto-CO_2 atmosphere coupled with thermal history of the Earth. Earth Planet Sci Lett 113:251–266

Takahashi E (1986) Melting of a dry peridotite KLB-1 up to 14 GPa—Implications on the origin of peridotitic upper mantle. J Geophys Res 91:9367–9382

Teixeira KE, Morley CV, Foley BJ, Unterborn CT (2023) The carbon-deficient evolution of Trappist-1c. Astrophys J 960:44

Thiabaud A, Marboeuf U, Alibert Y, Leya I, Mezger K (2015) Elemental ratios in stars vs planets. Astron Astrophys 580:A30

Tian F (2015) Atmospheric escape from solar system terrestrial planets and exoplanets. Annu Rev Earth Planet Sci 43:459–476

Tosi N, Godolt M, Stracke B, Ruedas T, Grenfell JL, Höning D, Nikolaou A, Plesa A-C, Breuer D, Spohn T (2017) The habitability of a stagnant-lid Earth. Astron Astrophys 605:A71

Tozer D (1985) Heat transfer and planetary evolution. Geophys Surv 7:213–246

Turbet M (2020) Two examples of how to use observations of terrestrial planets orbiting in temperate orbits around low mass stars to test key concepts of planetary habitability. arXiv:2005.06512

Turbet M, Ehrenreich D, Lovis C, Bolmont E, Fauchez T (2019) The runaway greenhouse radius inflation effect. An observational diagnostic to probe water on Earth-sized planets and test the habitable zone concept. Astron Astrophys 628:A12

Unterborn CT, Johnson JA, Panero WR (2015) Thorium abundances in solar twins and analogs: Implications for the habitability of extrasolar planetary systems. Astrophys J 806:139

Unterborn CT, Dismukes EE, Panero WR (2016) Scaling the Earth: A sensitivity analysis of terrestrial exoplanetary interior models. Astrophys J 819:32

Unterborn CT, Foley BJ, Desch SJ, Young PA, Vance G, Chiffelle L, Kane SR (2022) Mantle degassing lifetimes through galactic time and the maximum age stagnant-lid rocky exoplanets can support temperate climates. Astrophys J Lett 930:L6

Valencia D, Sasselov DD, O'Connell RJ (2007a) Detailed models of super-Earths: How well can we infer bulk properties? Astrophys J 665:1413–1420

Valencia D, Sasselov DD, O'Connell RJ (2007b) Radius and structure models of the first super-Earth planet. Astrophys J 656:545–551

Valencia D, Tan VYY, Zajac Z (2018) Habitability from tidally induced tectonics. Astrophys J 857:106

van Lieshout R, Rappaport SA (2018) Disintegrating rocky exoplanets. *In:* Handbook of Exoplanets. Deeg HJ, Belmonte JA (eds), Springer Charm, p 15

van Werkhoven TIM, Brogi M, Snellen IAG, Keller CU (2014) Analysis and interpretation of 15 quarters of Kepler data of the disintegrating planet KIC 12557548 b. Astron Astrophys 561:A3

Veras D, Mustill AJ, Bonsor A (2024) The evolution and delivery of rocky extra-solar materials to white dwarfs. Rev Mineral Geochem 90:141–170

Walker J, Hayes P, Kasting J (1981) A negative feedback mechanism for the long-term stabilization of Earth's surface temperature. J Geophys Res 86:9776–9782

Walsh KJ, Morbidelli A, Raymond SN, O'Brien DP, Mandell AM (2011) A low mass for Mars from Jupiter's early gas-driven migration. Nature 475:206–209

West AJ (2012) Thickness of the chemical weathering zone and implications for erosional and climatic drivers of weathering and for carbon-cycle feedbacks. Geology 40:811–814

West AJ, Galy A, Bickle M (2005) Tectonic and climatic controls on silicate weathering. Earth Planet Sci Lett 235:211–228

Wheatley PJ, Louden T, Bourrier V, Ehrenreich D, Gillon M (2017) Strong XUV irradiation of the Earth-sized exoplanets orbiting the ultracool dwarf Trappist-1. Mon Not R Astron Soc 465:L74–L78

Whittaker EA, Malik M, Ih J, Kempton EMR, Mansfield M, Bean JL, Kite ES, Koll DDB, Cronin TW, Hu R (2022) The detectability of rocky planet surface and atmosphere composition with the JWST: The case of LHS 3844b. Astronom J 164:258

Winn JN, Matthews JM, Dawson RI, Fabrycky D, Holman MJ, Kallinger T, Kuschnig R, Sasselov D, Dragomir D, Guenther DB, Moffat AFJ, Rowe JF, Rucinski S, Weiss WW (2011) A super-Earth transiting a naked-eye star. Astrophys J Lett 737:L18

Winn JN, Sanchis-Ojeda R, Rappaport S (2018) Kepler-78 and the ultra-short period planets. New Astron Rev 83:37–48

Wordsworth R, Kreidberg L (2022) Atmospheres of rocky exoplanets. Annu Rev Astron Astrophys 60:159–201

Wright JT, Marcy GW, Howard AW, Johnson JA, Morton TD, Fischer DA (2012) The frequency of hot Jupiters orbiting nearby Solar-type Stars. Astrophys J 753:160

Xu S, Rogers LK, Blouin S (2024) The chemistry of extra-solar materials from white dwarf planetary systems. Rev Mineral Geochem 90:171–198

Zahnle KJ, Kasting JF (1986) Mass fractionation during transonic escape and implications for loss of water from Mars and Venus. Icarus 68:462–480

Zahnle KJ, Catling DC (2017) The cosmic shoreline: The evidence that escape determines which planets have atmospheres, and what this may mean for Proxima Centauri B. Astrophys J 843:122

Zahnle KJ, Kasting JF (2023) Elemental and isotopic fractionation as fossils of water escape from Venus. Geochim Cosmochim Acta 361:228–244

Zeng L, Sasselov D (2013) A detailed model grid for solid planets from 0.1 through 100 earth masses. Publ Astron Soc Pac 125:227

Zhang M, Knutson HA, Wang L, Dai F, Oklopcic A, Hu R (2021) No escaping helium from 55 Cnc e. Astronom J 161:181

Zhu W, Dong S (2021) Exoplanet statistics and theoretical implications. Annu Rev Astron Astrophys 59:291–336

Zieba S, Zilinskas M, Kreidberg L, Nguyen TG, Miguel Y, Cowan NB, Pierrehumbert R, Carone L, Dang L, Hammond M, Louden T, Lupu R, Malavolta L, Stevenson KB (2022) K2 and Spitzer phase curves of the rocky ultra-short-period planet K2–141 b hint at a tenuous rock vapor atmosphere. Astron Astrophys 664:A79

Zieba S, Kreidberg L, Ducrot E, Gillon M, Morley C, Schaefer L, Tamburo P, Koll DD, Lyu X, Acuña L, Agol E, Iyer AR, Hu R, Lincowski AP, Meadows VS, Selsis F, Bolmont E, Mandell AM, Suissa G (2023) No thick carbon dioxide atmosphere on the rocky exoplanet Trappist-1 c. Nature 620:746–749

Zilinskas M, van Buchem CPA, Miguel Y, Louca A, Lupu R, Zieba S, van Westrenen W (2022) Observability of evaporating lava worlds. Astron Astrophys 661:A126

Zugger ME, Kasting JF, Williams DM, Kane TJ, Philbrick CR (2010) Light scattering from exoplanet oceans and atmospheres. Astrophys J 723:1168–1179

www.ingramcontent.com/pod-product-compliance
Lightning Source LLC
Chambersburg PA
CBHW060418220326
41598CB00021BA/2213